ANALYTICAL INSTRUMENTATION HANDBOOK

ANALYTICAL INSTRUMENTATION HANDBOOK

Edited by
GALEN WOOD EWING
New Mexico Highlands University
Las Vegas, New Mexico

MARCEL DEKKER, INC. New York and Basel

Library of Congress Cataloging-in-Publication Data

Analytical instrumentation handbook / edited by Galen Wood Ewing.
p. cm.
Includes bibliographical references.
ISBN 0-8247-8184-8 (alk. paper)
1. Instrumental analysis. 2. Chemistry, Analytic--Instruments.
I. Ewing, Galen Wood.
QD79.I5A49 1990
543' .07--dc20 90-3091
CIP

This book is printed on acid-free paper.

MARCEL DEKKER, INC.
270 Madison Avenue, New York, New York 10016

Current printing (last digit):
10 9 8 7 6 5 4 3 2 1

PRINTED IN THE UNITED STATES OF AMERICA

Preface

This *Analytical Instrumentation Handbook* is intended primarily to be a guide for workers in the field of analytical chemistry who are called on to decide what approach to pursue in solving specific problems.

No one person can be expected to be expert in more than a few of the techniques that are available in a modern analytical laboratory. Hence there is a tendency to adopt a particular method simply because it is more familiar to the operator (or perhaps less expensive) than other alternative methods, some of which might be better adapted to the problem at hand. Rather than choose a method arbitrarily, the chemist is urged to consult this handbook for an overview of other possible techniques. Once a technique has been selected, the handbook will present pertinent information about options within each domain, including the pros and cons of different kinds of instrumentation.

The handbook will also be of value as a library reference for all persons interested in this basic area of chemical science, from college and university students through the technician and research levels, to industrial management.

The reader will find considerable variation in format from chapter to chapter. I feel that this is not only inevitable, but desirable, as the experts writing the various chapters are the best judges of the needs in particular fields. Some writers have listed sources of equipment by brand names, while others have not felt this to be necessary. Some have relied heavily on block diagrams of instruments rather than detailed descriptions; this approach is especially appropriate in a field where most of the instruments in use have been assembled by the user. No attempt has been made to list all manufacturers; the numerous buyers' guides should be consulted if this information is needed.

I acknowledge with thanks the painstaking efforts of the chapter authors. Their cooperation is greatly appreciated.

Galen Wood Ewing

Contents

Part V: MISCELLANEOUS METHODS

Contributors

H. J. BOWLEY BP Research Centre, Sunbury-on-Thames, Middlesex, England

HARRY G. BRITTAIN Squibb Institute for Medical Research, New Brunswick, New Jersey

DAVID BROWN Baxter Health Care Corporation, Round Lake, Illinois

THOMAS L. CHESTER The Procter & Gamble Company, Cincinnati, Ohio

JOHN COATES* Spectra-Tech, Inc., Stamford, Connecticut

DAVID DOLLIMORE Department of Chemistry, University of Toledo, Toledo, Ohio

GARETH R. EATON Department of Chemistry, University of Denver, Denver, Colorado

SANDRA S. EATON Department of Chemistry, University of Colorado—Denver, Denver, Colorado

GALEN WOOD EWING† Department of Chemistry, New Mexico Highlands University, Las Vegas, New Mexico

EDWARD M. EYRING Department of Chemistry, University of Utah, Salt Lake City, Utah

SHERRY O. FARWELL Department of Chemistry, University of Idaho, Moscow, Idaho

D. L. GERRARD BP Research Centre, Sunbury-on-Thames, Middlesex, England

WILLARD A. HARELAND Sandia National Laboratories, Albuquerque, New Mexico

DENNIS JENKE Baxter Health Care Corporation, Round Lake, Illinois

---

*_Current affiliation_: Nicolet Instrument Corporation, Madison, Wisconsin.
†Retired.

STANISLAW J. KOMOROWSKI Department of Calorimetry, Institute of Physical Chemistry of The Polish Academy of Sciences, Warsaw, Poland

WALTER E. KUPPER Mettler Instrument Corp., Hightstown, New Jersey

NELIA F. LEITE* Department of Chemistry, University of Utah, Salt Lake City, Utah

TRUMAN S. LIGHT† The Foxboro Company, Foxboro, Massachusetts

T. S. MA† Department of Chemistry, Brooklyn College of The City University of New York, Brooklyn, New York

TSUTOMU MASUJIMA Institute of Pharmaceutical Sciences, Hiroshima University, Hiroshima, Japan

LINDA B. McGOWN Department of Chemistry, Duke University, Durham, North Carolina

KASEM NITHIPATIKOM‡ Department of Chemistry, Duke University, Durham, North Carolina

ANDREW F. PALUS EG&G Princeton Applied Research, Princeton, New Jersey

M. L. PARSONS Pacific Scientific Co., Duarte, California

CHARLES J. PATTON§ Alpkem Corporation, Clackamas, Oregon

MATTHEW PETERSHEIM Department of Chemistry, Seton Hall University, South Orange, New Jersey

LARRY K. SVEUM Department of Chemistry, New Mexico Highlands University, Las Vegas, New Mexico

MAARTEN van SWAAY Department of Computing and Information Sciences, Kansas State University, Manhattan, Kansas

NOEL H. TURNER Naval Research Laboratory, Washington, DC

ADRIAN P. WADE Department of Chemistry, University of British Columbia, Vancouver, British Columbia, Canada

JOSEPH WANG Department of Chemistry, New Mexico State University, Las Cruces, New Mexico

---

**Current affiliation*: Instituto de Pesquisa Espaciais, São Paolo, Brazil.
†Retired.
‡*Current affiliation*: Texas Tech University, Lubbock, Texas.
§*Current affiliation*: U.S. Geological Survey, Arvada, Colorado.

RAYMOND J. WEIGAND Baxter Health Care Corporation, Round Lake, Illinois

FREDERICK L. YARGER Department of Physics, New Mexico Highlands University, Las Vegas, New Mexico

ANALYTICAL INSTRUMENTATION HANDBOOK

Introduction

GALEN WOOD EWING* / Department of Chemistry, New Mexico Highlands University, Las Vegas, New Mexico

One of the principal thrusts of this handbook is to assist persons new to a field in making an intelligent choice of instruments to purchase. It is always important to follow a well-thought-out plan in attempting to reach a decision about what analytical tool to adopt. To assist in this task, we present herewith a checklist of questions, the answers to which should go far to clarify one's thinking along these lines.†

- What is the physical state of the samples to be analyzed (i.e., gas, liquid, solid, paste, etc.)?
- Are measurements required for one, several, or many constituents?
- Is the desired information qualitative, semiquantitative, or fully quantitative?
- What degree of measurement precision is required?
- What degree of accuracy is required?
- Is chemical speciation required?
- What are the expected ranges of analyte concentration?
- What is the probable sample matrix, and which components of the matrix are possible interferents?
- What is the required selectivity of response?
- How much sample will be available for each determination?
- Must the treatment of the sample be nondestructive?
- What sample-handling and preparation steps are necessary?
- What is the expected total number of samples and the desired sample throughput per day?
- How quickly will the results be needed, and how much time can be spent on method development?
- Are high-purity reference standards available at the appropriate concentration levels?
- Will the measurements be performed in a laboratory, in a plant, or at a field site?
- What is the level of academic training of those who will perform the measurement?
- What safety practices are required?
- Is automation required or desirable?

---

*Retired.
†These questions have been formulated by Professor Sherry O. Farwell.

- Are there governmental or quasi-governmental regulations or methods that must be followed (e.g., ASTM, USP, or other)?
- What is the desired form of output data and storage?
- What economic restraints must be observed with respect to equipment and salaries?

These questions are not to be considered exclusive; other equally valid ones could be raised. Usually some degree of compromise is necessary: high speed of operation may not be compatible with the required precision, for example. The analyst must assign priorities to the various factors according to the overall objectives of the particular problem to be addressed.

No claim is made that the coverages of the individual chapters are all-inclusive. The intent has been the description of the major types of instruments, rather than all possible modifications. Since the emphasis is on *instrumentation*, sufficient theory has been included only to explain the functions of the various instrumental features; references to in-depth treatments have been given as appropriate.

1

The Use of Computers in the Laboratory

MAARTEN van SWAAY/ Department of Computing and Information Sciences, Kansas State University, Manhattan, Kansas

I. INTRODUCTION

The term computer calls up a largely monolithic view of some clever box that performs magic on demand. In reality the introduction of a computer into the laboratory has effects that reach in many different directions.

The digital nature of the machine introduces the need to convert the largely analog world of the laboratory into digital form, which raises concerns about resolution, conversion speed, and sampling. Not only must the signals from various transducers be converted into digital form; the ubiquitous time variable is also converted into discrete time increments. Consequently, the record of an experiment will usually consist of long strings of digitized data collected on some time schedule. Even for short experiments the length of such strings can become quite large because measurements may have to be scheduled on very short time intervals to capture rapidly varying signals. Often it will be necessary to transfer data to a mass storage device as they are collected. The computer must then handle the disk transfer without missing any of the incoming experimental data.

Scheduling of sampling introduces the need for accurate time-keeping and for reproducible timing and synchronization. Sampling also implies a deliberate discarding of information, namely, the shape of the sampled signal between sample points. We must take care that the discarded information is not part of what we wish to record. Furthermore, sampling can introduce artifacts in the reconstructed signal, known as aliases. Because it is very difficult to discriminate against alias signals we must guard against their introduction.

The digital nature and the flexibility of the computer enable it to handle enormously complex relationships and calculations with apparently simple commands. Many laboratory systems now come with a rich assortment of tools such as Fourier transform packages, smoothing and filtering operators, etc. It is tempting to use such tools, but casual application can result in spectacular abuse of the data. A computer can be the ultimate poker player!

Finally, the high speed of even low-cost computers enables the machines to manage many activities seemingly simultaneously. What may appear to the human user as simultaneous activities will usually still be the result of sequential actions taken by a single CPU (central processing unit). To ensure that time-critical work is performed on time, a resource manager must allocate CPU time in such a way that even under worst-case conditions no time-critical demands will be overlooked. We will find that the resource manager forms a crucial but largely invisible part of the operating system.

The experimenter who wants to make efficient use of the power of a computer must stand ready to invest considerable time in understanding the way in which many apparently unrelated domains can and will interact. To bring some order into the matter we will subdivide this chapter into a number of topics:

- Operating system
  - Resource management
  - File management
  - User interface
- User terminal
  - Text and graphic interaction
- Data transfer and communication
  - Data transfer hardware
    - RS232,RS422/423/449,IEEE488/GPIB,Ethernet
  - Networks
    - Protocol models and standards
    - Shared printers and mass storage devices
    - Compatibility between data representations
- Programming considerations
  - Languages and tools
- Signal processing and signal/noise discrimination
  - Averaging and filtering
  - Correlation
  - Transforms
- Signal conversion
  - Analog-digital converters
  - Track/hold amplifiers
  - Sampling and sampling artifacts
  - Digital-analog converters
  - Shaft encoders
- Literature
  - References and sources for further reading

Operating systems—For a chapter on computers it is only fitting that the discussion start with the part that makes a computer functional and usable: its operating system. The central part of the operating system is its resource manager; that part largely defines what activities the system can support. Another important part of the operating system is the file manager that allocates and organizes space for all stored data. The most visible part is the user interface: the software that defines the way in which the system interacts with its user(s).

User terminal—The user terminal is the hardware through which the user interacts with the system. It must handle both text and graphics input and output, and its behavior and appearance must be comfortable to the user.

Data transfer and communication—Invariably, data of interest have to be moved between parts of a system and/or between systems. Communication issues range from the simple need to move data between a laboratory instrument and its host computer to the sharing of data between systems located in several laboratories and offices. Communication will obviously require some physical connection between systems, but we will find that communication also requires several layers of intricate software.

Programming considerations—Any command typed into a computer terminal can be seen as either a program or a part of a program. Thus one cannot be a computer user without also being a programmer. The extent of the programming activity depends on the level of detail at which a user wishes to control the system.

Signal processing and signal/noise discrimination—Experimental data are rarely obtained in a form that is suitable for a final report. Operations on those data range from simple reformatting and scaling through elaborate transformations. Important mathematical operations are the discrimination between signal and noise, correction for nonlinearity of input sensors, and the transformation of data into a form matching the interest of the investigator.

Signal conversion—The vast bulk of laboratory data is available only in analog form. Before analog data can be accepted by a computer for further processing they must be converted into digital form. Similarly, digital data produced by a computer must often be converted into analog signals before they can be used to control an experiment.

Literature—The rising popularity of computers in the laboratory is due primarily to major cost reductions. Much of the underlying theory and implementation strategy was developed more than ten years ago. Items of current interest can be found in the commercial literature (sales brochures, application notes, equipment catalogs), rather than in the scientific journals. For that reason no literature references are placed in the text. Instead, a summary for further reading is placed at the end of the chapter.

II. OPERATING SYSTEM

For many people the term *operating system* is synonymous with what they see of the system, that is, with the user interface. That user interface defines such things as the repertory of commands the system is willing to accept from the user and the way unacceptable user input is reported.

In reality the user interface is a relatively small part of the operating system, and for stable use patterns it is quite possible to replace or hide the existing user interface by installation of another software layer that is tuned to the requirements of the user(s). Other equally important parts are the file manager, the resource allocator, the command interpreter, and the device handlers. These less visible parts have a much larger impact on the capabilities of the system, and it is generally impractical to modify or replace them. Finally, it is reasonable to include a group of utilities with the operating system, even though those utilities should be seen as conveniences, rather than as indispensable parts of the system. Among such utilities one may count an editor, a file directory reporter, a file transfer handler, a linker, a librarian, etc.

Two operating systems may present similar or even identical user interfaces even though the underlying structures are entirely different and often not even comparable. Especially in the laboratory environment we will find that the underlying structure may be far more important than the user interface. On the other hand, one should note that a poorly designed user interface can make even a good operating system difficult to use. A short description of the structure and purpose of an operating system is in order, so that its services can be put in proper perspective.

It is useful to see an operating system primarily as a resource manager. Resources in this context include the CPU, available memory, mass storage devices, available terminals, printers and communication lines, and any input and output ports that may be used to connect the system to various pieces of laboratory equipment.

Systems connected to laboratory instruments must often remain aware of the demands of an experiment over long periods of time, even though those demands require only a small portion of the available CPU time. Under those conditions it is unattractive to assign the CPU full-time to the demands of the experiment: the CPU will then spend most of its time waiting for the experiment to request its attention. Under such circumstances it becomes particularly important that the machine resources are managed in such a way that the CPU can be allocated to other productive tasks between demands imposed by the laboratory equipment. Examples of such tasks are program development, text processing, and data work-up and plotting.

In many cases we will find that the demands of the experiment must be met on a tight time schedule, but that the other activities are far more tolerant to occasional delays. Thus the operating system must treat the needs of the experiment as high-priority requests that can preempt the CPU from any other tasks. Such a hierarchical scheduling is often called a foreground-background schedule: the high-priority task is the foreground, and one or more tasks that are not time-critical can proceed in the background. Please note that this view may appear strange at first sight: to the user the foreground task may be largely invisible, while the background task, such as a text editing program, may be very visible to the user.

Another common scheduling policy allocates CPU time slices to each of several or many users in a round-robin fashion. As long as those time slices are allocated frequently enough to each user, all users can be given the feeling that they are sole master of the machine. Such a scheduling strategy is known as a time-shared system. It is very popular on large machines, but it is poorly suited for a laboratory environment, primarily because laboratory experiments rarely have the patience we may expect from human beings.

Figure 1 shows a schematic view of the structure of an operating system. At its heart is the resource allocator or kernel. The kernel has the task of allocating the CPU to the various demands placed on it. These demands include not only the tasks described earlier, but also such activities as moving text to a printer, handling data streams arriving or departing over a network or phone line, etc. Each data stream between memory and an external device (printer, disk, terminal, phone line) is managed

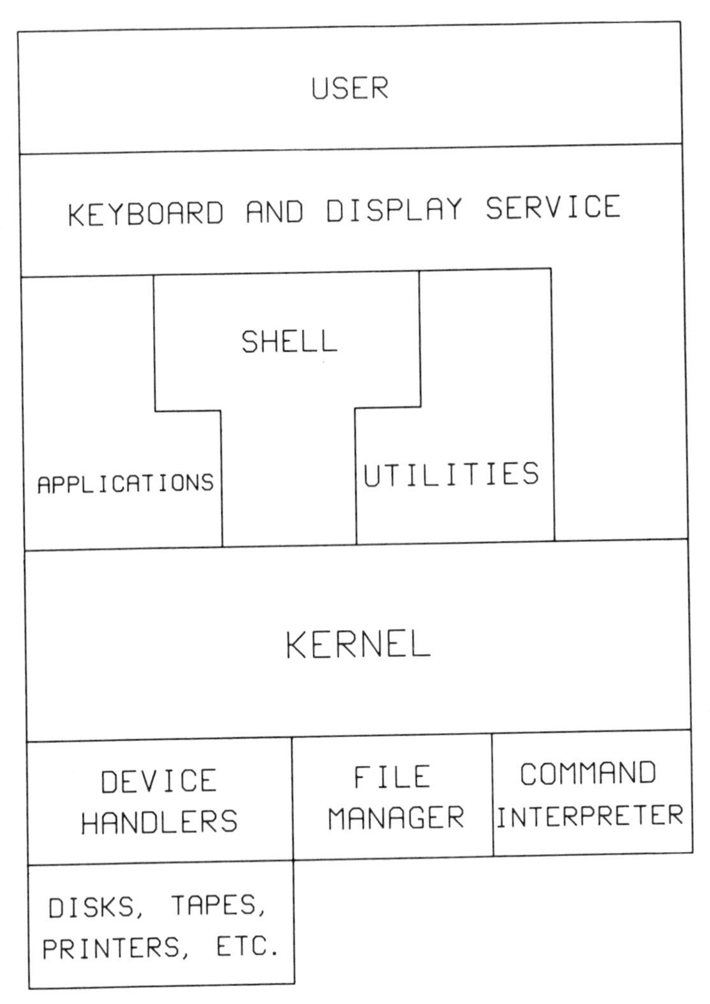

Figure 1. The layered structure of an operating system. The parts of the system are drawn in such a way that information exchange between parts of the system may cross boundaries only in the vertical direction.

by a handler for that device. The handler in turn may place demands on the CPU whenever its device issues a request for service. Because a program, such as an editor, cannot be aware of the activities of other programs in the system, no programs can be allowed to interact directly with any of the external devices. Instead, each program must place requests for input and output operations to the kernel. The kernel in turn maintains a list of requests for each device and passes those requests to the corresponding handler in an appropriate sequence.

Space in main memory is generally allocated by the kernel. In contrast, space on mass storage devices is normally handled by a relatively autonomous utility: the file management system. The file manager maintains an inventory of available space on the mass storage devices connected to the system, and also maintains a catalog of the various blocks of information stored on those devices. Although it is possible that some of the management data maintained by the file manager are kept in main memory, most of that information will normally be kept on the disks. To shield programs from the need to be aware of the nature and organization of the mass storage devices, all references to data on those devices are made by name, rather than by location. The file manager receives requests in symbolic form from the kernel and translates these requests into physical addresses by referring to its catalog. The physical addresses in turn are passed onto the disk handler that manages the actual transfer of data to or from the disk.

All nontrivial programs deal with files and/or other data streams. Normally it will be convenient if the source and destination of those data streams can be defined at run time. That implies that the specification for the data streams must be supplied by the user in symbolic form as part of a command. That command in turn must be interpreted, and relevant parts must be extracted from it. The need to interpret command lines will be common to all programs, and it will be convenient if a software component handling that work is built into the operating system. We may call that component a command line interpreter. Please note that the command line interpreter may receive its input not only from the keyboard, but also from running programs.

Lastly, there will be a software component that accepts input from the keyboard and recognizes a set of user commands such as RUN, DATE, LISTFILES, PRINT, etc. This is the layer that defines the appearance of the system to the user: the user interface. If user commands contain file specifications, those specifications can be validated and interpreted by the command line interpreter.

From the overall structure it becomes obvious that the part of the operating system that is visible to the user is far removed from the heart of the system. It is quite possible to install a software package on top of one operating system to make it appear to be an entirely different operating system. That may put the user at ease, but it has little effect on the capabilities of the operating system: those are defined largely by the structure of the kernel.

A kernel that cannot allocate CPU time to at least several tasks according to their instantaneous demands will be a poor choice for a laboratory computer system. Not only will such a system be incapable of handling simultaneous experiment control and editing or other activities; the system may not even be able to write experimental data to disk without conflict with the timing constraints of the experiment.

Recently the term "windowing" has become a popular buzzword: it refers to software packages that allow a user to activate several programs so that they apparently run simultaneously. Each of the programs is visible to the user via a marked-off portion on a display screen, much as one would work on several notepads scattered on a desk. The appearance is misleading: only one of the visible programs is active at any time. The user switches between programs either with keyboard commands or with the help of a pointing device. The window presentation is undoubtedly an effective strategy to give the system the familiar feel and appearance of a desk, but again the impressive window display is far removed from the heart of the operating system.

Even though the user interface is located on the outside of a multi-layered software structure, it goes without saying that the user interface performs an important function. The user interface handles all communication between the user and the system. If that communication interferes with the thought patterns of the user, the system will not be comfortable to use. Ease of use depends less on choice of words or graphical tools than on consistency of system behavior. Standard user requests should require standard user input, independent of the task the system may be performing. If some commands require that the ENTER key be struck to transmit the command to the system, then all commands on the system should be implemented that way. If the system supports on-line help, then a single HELP key should be assigned for help requests under all conditions, and the help request should yield only the appropriate help messages in a uniform format. In practice it may be difficult to enforce such simple rules, because various utilities on the system may have been designed by a multitude of independent software teams.

Invariably the average user will occasionally issue commands that are unacceptable to the system. The system should respond with a message that leads the user to the correct command. All too often the messages returned by the system consist of cryptic diagnostics for the system designer, rather than helpful hints for the user. Unless the user happens to be the system designer, such cryptic messages will be frightening and disruptive, instead of helpful. For commercial software packages it may not be possible to control the nature of the responses built into the package, but designers of home-grown software packages will do well to choose the messages they build into the software on the basis of user productivity.

For the laboratory scientist it is unfortunate that the large market that has driven the cost of small computers down is almost entirely unaware of the concept of "real time." As long as the system can keep up with the fingers of a single user it satisfies most of the market. Consequently one of the most popular operating systems, MS-DOS, provides no support for any real-time operations. It is possible to attach code to the existing operating system, but the task is far from trivial. Other operating systems are available for machines based on the 8088 and 8086 processors, but they are not nearly as popular, and few, if any, applications are available under other operating systems. Manufacturers such as Hewlett-Packard and Digital Equipment Corporation have marketed lab-oriented systems for many years. The market for those systems is much smaller than the market for personal computers. Consequently, the cost of lab-oriented systems from these manufacturers is much higher than the cost of a personal computer of comparable processing power.

A number of manufacturers offer "laboratory front ends" that can be attached to various popular microcomputers. The front end may be a fully independent processor with its own operating system, data memory, and laboratory devices. But the operating system is dedicated to the lab environment, and interacts with the user only through a program running on the host system. It is debatable which of the two systems should be called the front end: the lab system may well cost considerably more than the system to which it is attached. The lab system is not designed to support any tasks other than managing the lab experiment so that the full power of the processor is available for that task. Furthermore, the operating system of the front end need not support the rich variety of services expected from a general-purpose system.

III. FILES AND DATA STORAGE

Any nontrivial use of a computer in the laboratory will require the creation and maintenance of data files. The data obtained from an experiment must be archived, and they must remain available in machine-accessible form for subsequent review and processing. Furthermore, some experiments will produce data in a quantity beyond what can be held in the memory of the machine. Until recently it was not practical to provide large amounts of memory on a system: memory was relatively expensive and few processors could support more than 64 KB (kilobyte, 1 K = 1024) of address space. Long data streams therefore had to be transferred to disk during the experiment. The current generation of processors supports a memory space of at least 1 MB (megabyte, 1 M = 1,048,576), and several processors support an address space as large as 1 GB (gigabyte, 1 G = 1024M). Memory costs have also fallen rapidly, so that data acquisition systems with 4-16 MB of memory are no longer out of reach.

Because access to disk data is much slower than access to memory data, a disk is usually organized to handle data in blocks varying in size from 64 bytes to 4 KB. The disk surface is subdivided into concentric "tracks," and each track is further subdivided into several sectors. The sector is the smallest unit of data that can be moved to or from the disk.

The number of bytes in one sector may range from 64 bytes for an 8-inch single-density floppy disk to 4 KB for some large high-performance disks. If the size of a sector is not convenient for an operating system (e.g., if the operating system must deal with several disks with different characteristics) the operating system may handle data in blocks that correspond to several disk sectors.

The file manager is responsible for the allocation of disk space for the files created by various programs. The file manager also maintains a directory of existing files, and uses that directory to map the name of a file to its placement on the disk.

If an experiment requires that data be transferred to a disk during the experiment, several issues must be addressed. The system must be able to manage both the experimental data stream and the disk transfer operation without losing any of the incoming data. Because the disk transfer almost invariably operates on blocks of data, while most experiments tend to produce data in smaller units (bytes or words), staging buffers must be maintained between the data stream coming from the experiment and the data stream going to the disk. If a disk transfer cannot be completed in the time interval between data collections on the experiment, at least two transfer buffers will be required, so that at any time at least one buffer can be allocated to each data stream. If both data transfers require CPU activity, the allocation of CPU time to the two demands must be scheduled in such a way that both data streams maintain their timing obligations. In practice, that means that the experimental data must be collected on a clock schedule and that the disk transfers must be able to keep up with the experiment.

On most modern machines the task of moving data between disk and main memory is delegated to a "direct-memory-access" (DMA) controller, so that the CPU remains available for other tasks. The disk transfer will still slow down other activities, however, because very few activities can proceed without access to memory. Access to memory implies traffic on the data bus; that traffic will have to compete with the traffic handled by the DMA controller.

Disk transfers are subject to two timing constraints. Access to a disk requires that the desired disk sector be located underneath the disk read/write head. That may take as long as one whole revolution of the disk, although on average the "rotational latency" will be only a half revolution. For a standard floppy disk ($5\frac{1}{4}$ inch) the worst-case rotational latency is 160 msec. That implies that a disk transfer buffer can remain unavailable at least that long while experimental data continue to arrive.

If the data transfers can be organized so that multiple data blocks can be transferred to disk in a single operation the situation may improve. Successive blocks can be placed on the disk in such a way that the time gap between block transfers will always be much smaller than the time for one revolution. It is not practical to place successive data blocks on adjacent disk sectors: the disk service code will usually require some time to prepare for the next transfer. In practice one often finds a placement strategy in which sequential blocks of data are placed two sectors apart on the disk: on one revolution all the even-numbered sectors are used, and on the next revolution all the odd-numbered sectors are used. With that strategy the initial access to the disk may still suffer a latency of one revolution, but subsequent accesses will have much smaller latency. Unfortunately, the strategy requires that data are laid down in a contiguous area on the disk. That is no problem on an empty disk, but if several files are created and discarded on a disk the available space on the disk will inevitably become a collection of scattered small areas that cannot be used for a large contiguous file. The data on the disk must then be moved around: the disk must be "squeezed." That is not a difficult operation, but the disk will be inaccessible during the squeeze operation. Furthermore, the squeeze operation must be invoked by the user and thus requires additional attention on the part of the user.

The need to squeeze the disk can be avoided if files can be placed in randomly located areas on the disk. Of course, that implies that the file manager must maintain a record of all occupied and free areas on the disk. The operating system for the IBM PC follows this strategy: a map of the available disk areas is maintained on each disk, and areas are allocated to a file as the need arises. Files can then grow to arbitrary length as long as disk space is available, and the disk never needs to be squeezed. The convenience comes at a price, however. Sequential access to a file will require more head movement and longer latency between sectors. In addition, random access to various parts of a file will require considerably more effort and time than random access to a contiguous file, because the position of the next block of the file cannot be predicted from the position of its predecessor.

In its simplest form the blocks of a randomly placed file are connected by means of pointers in each block that contain the disk address of the next block. Such a file can only be accessed sequentially. More commonly, the locations of all blocks of the file are collected into an index table for that file. The index table does allow random access to the file, but it requires additional storage space.

Disk storage has become extremely reliable, even on removable disks. Many users take this reliability for granted and spend little time worrying about the consequences of the loss of data stored on a disk. Needless to say, the practice is dangerous and invites disaster, usually at a most inopportune moment. Backing up large amounts of disk data is a tedious chore, and selective backup of modified data requires careful bookkeeping to insure that no modified data are overlooked. As the cost of large disks has come down, the need for backup support has grown: most modern operating systems maintain a marker with each entry in the disk directory. All programs modifying disk data are expected to set the marker for those data. Then a backup program can scan the disk directory for marked files, copy each marked file to a backup device, and finally remove the mark.

On a system with two or more units of the same type of disk it is tempting to create duplicate copies of each disk for backup. The strategy leaves something to be desired: if the disk drives fail both the original and the backup copies become unreadable. If possible, data should be backed up onto independent storage devices, so that even when one device fails the data will remain accessible via the other device.

Because large disks have become quite affordable, it has become attractive to store reference data on disk, rather than on paper. Even with no more than the global search capability of a standard editor one can often find text patterns faster from a disk than from printed text. Designing the structure of the stored data for efficient retrieval is a large topic by itself and falls outside the scope of the present discussion.

IV. THE USER TERMINAL

To interact with a system the user needs a device that will accept and return both text and graphic information. Information produced by the system is normally placed on a video display. The video display has the advantage that selected portions of the image can be modified cleanly and rapidly, but it has the disadvantage that it can display only a limited portion of, for example, a large text file. A keyboard accepts text input from the user. Graphic input can be supported by a pointing device. For both types of input it is essential that the user can see the result of his/her actions. A few users may be able to enter text without visual feedback, but graphic interaction is impossible without some indication of the current position of the pointing device.

The conventional terminal display has room for 24 lines of 80 characters. Almost invariably the display is "painted" sequentially in the order in which the text is read. The data stream for a screen full of text can then be limited to little more than one byte per displayed character, or less than 20,000 bits. In contrast, a graphics display may consist of as many as 1000×1000 individual dots, or 1,000,000 bits. Not only is the amount of information contained in a graphic display far larger than what is displayed on a text display; it will often be necessary to paint the display in a random sequence. In that case the data stream sent to the screen must not only contain the value of each dot, but also the position of that dot. We conclude then that the transmission of a single graphic image will usually require a much longer or much wider data stream than the transmission of a single page of text.

A text terminal can be connected to its host by an inexpensive two-wire cable over which characters are transmitted serially. At the commonly used speed of 960 characters/sec an entire screen of text can be transmitted in about 2 sec over a cable of several hundred feet. Such a data path will be inadequate for any but the simplest graphic displays: the transmission of the data for a simple stripchart curve may take several seconds.

A graphics display must rely on memory in which the displayed image is stored. It is possible to build that memory into the display screen itself (storage oscilloscopes and long-persistence scopes), but most modern display devices are based on what is known as a bit-mapped display. The display image is stored in random-access memory that is accessible to display-scanning hardware and to a digital data bus. For a very simple monochrome display each dot (pixel) on the screen will be associated with a single bit of display memory. More elaborate displays may control each pixel with as many as three groups of 4-8 bits to control the intensity of three primary display colors. Beyond that, many display devices contain several "display planes," so that the presented display can be switched instantaneously between several images. The link between the display tube and the bitmap scanning hardware is based on standard video technology; low-cost coaxial cable can carry the analog video signal over distances of 1000 ft or more. The bitmap scanning hardware must be located close to the associated bitplane memory, however.

As noted before, it is impractical to control a graphics display via a data path of low bandwidth. If the display memory must be controlled from a remote host, that host must be capable of filling the display memory. Here again, a high bandwidth will be required unless the data stream can be compressed by elimination of "uninteresting" portions of the data stream. As an example, a circle segment can be defined entirely by origin, radius, and start-stop angles. A graphics display device should possess the intelligence to reconstruct the circle segment from that information. A graphics instruction set will normally include commands to draw straight-line and circle segments, and commands to fill bounded display areas with a color and/or a pattern. A standard graphics instruction set known as GKS (graphic kernel set) is now widely adopted, even though it is primarily aimed at the creation of display images, rather than at the support of interactive graphics. With a graphics instruction set it becomes practical to control a display device via the relatively low-speed serial link capable of handling 960 characters/sec. A data link supporting higher speeds, such as the IEEE488 link, can maintain graphics displays on less intelligent display devices, but the transmission distance will be limited to about 50 ft. Finally, the display memory can be made part of the main memory of the host processor. In that case the display speed will be limited only by the speed at which the host can compute and store the image.

In summary then, we find that the performance of a graphics display is limited either by the rate at which data can be transmitted to it from a remote host, or by the rate at which display data can be computed by a local host. Systems intended for high-quality interactive graphics are known as graphics workstations; they are based on a high-performance processor sharing memory with a display. Graphics workstations are normally designed as single-user systems, primarily because it is not practical to include more than one graphics display memory in the system. In contrast, several graphics terminals may be connected to a single host computer, but the rate at which graphic images can be updated will generally be limited by the transmission bandwidth of the link to the host computer, as well as by the computation load placed on the host by other users.

User interaction with a graphics display may take one of two forms. An input device can be used to supply a pair of coordinates to the system. The supplied coordinates are then used by the system to place a small icon on the screen at a corresponding position. Input devices supporting this strategy are the joy stick, track ball, "mouse," and drafting tablet. With all these devices the link between the input signal and the display position is maintained by the display processor. Alternatively, the screen itself may be made touch-sensitive, so that the user can point directly at the screen. In this case no feedback information needs to be displayed, but the sensing equipment will be more elaborate. Trackballs, joysticks, and mice can easily support a resolution of several hundred increments along both display axes, but it is difficult to build a touch-sensitive screen with a resolution beyond 100 increments on each axis. A touch-sensitive screen can impose a much smaller load on the display processor however.

All of the popular microcomputers have a bit-mapped display; the IBM PC and its look-alikes contain four display planes in the standard configuration. An extended graphics card can be installed to support color graphics and higher resolution than the 640 × 200 pixel resolution of the standard configuration.

Many of the machines support the use of either a joystick or a mouse; the Apple Macintosh has become famous for its extensive use of a mouse as a control device. Much has been written about the virtue of a mouse as the interface of choice. One must be careful about the value of some testimonials, however. It is human nature to extol the virtues of whatever one owns. Without doubt a pointing device is the device of choice for interaction with graphic information. On the other hand, the use of a pointing device requires that each item to be selected must be made visible before it can be pointed to. There are many examples of good user interfaces based on only a keyboard, and there are also many examples of bad user interfaces based primarily on a pointing device.

There is good reason to expect that recent advances in the area of graphics hardware will lead to major improvements and/or cost reductions of graphic displays in the near future. With a conventional processor graphic operations such as moving part of a display are quite time-consuming, because the processor is not aware of the regularity of graphics display data. It will then require several instructions to move one byte of display data. In contrast, a dedicated graphics processor can be designed with an instruction set that is optimized for the handling of display data, so that entire blocks of display data can be moved with only a few instructions. One may encounter the acronym BITBLT (bit block transfer) for such processors. A graphics processor can easily be an order of magnitude faster than a general-purpose processor for the manipulation of display data. The arrival on the market of high-performance graphics processors and the rapidly falling prices of large memory arrays form the basis for the expected jump in graphics capability.

V. DATA TRANSFER AND COMMUNICATION

Invariably there comes a time when data acquired on one machine have to be transported to another machine, be it for further processing, for archiving, or for some other reason. The compact machine that can be moved close to a laboratory experiment will rarely have either the storage devices required for archiving or the processing power required for extensive data manipulation. Even for the experiment itself data may have to be moved between measuring instruments (digital voltmeter, frequency meter, etc.) and a computer system that captures and stores the data.

In the simplest form, data can be transported between machines on a storage medium that can be read and written by both machines, such as a floppy disk. The strategy does have its limitations: most systems tend to include storage devices that match the performance of the system. Thus a small lab computer can be served adequately by floppy disk storage, but it is unlikely that the same floppy disk drive is part of a large mainframe system. Furthermore, the transport of data will require substantial amounts of user effort, and any dialog between the two machines will be hampered by a huge latency as the information is hand-carried from one machine to another.

Direct communication between machines is now supported by a variety of mechanisms ranging from simple user-initiated transactions to fully distributed systems. Many buzzwords are currently fashionable, such as RS232, RS422, RS423, RS449, IEEE488, IEEE802.x, LAN, OSI, ISO, TCP/IP, CSMA/CD, X.25, modem, GPIB,HPIB,Ethernet,DNA,SNA,KERMIT, etc. Before we can discuss various strategies we will need to lay down some fundamental concepts.

We will define a *node* as any machine or instrument to which one or more links to other nodes are connected. A *link* will be any type of hardware that can transport data between nodes.

If communication has to be established between only a small number of nodes, and if most of those nodes need to communicate with only a small subset of the nodes in the network, then point-to-point links can be installed as needed. But if information must be shared routinely and widely among several nodes, the point-to-point connections soon become impractical. With the addition of a central node we create a star network: data can be transmitted from all nodes to the central node, from where they can be retransmitted to any other node. The routing task of the central node may well consume a substantial fraction of its time.

For point-to-point connections each link will be owned by the transmitter connected to that link. The link will always be available to the transmitter, and any data placed on the link will arrive at a single receiver. Thus there will be no need for address labels on messages, and there will be no contention for access to the link. A star network is still based on point-to-point links so there will be no access contention on the links. But the messages will now have to carry a destination address that will be interpreted by the central node, and messages sent out by the central node will have to carry an address identifying the origin of the message.

When traffic density increases on a star network, the central node will eventually become a choke point. Instead of constructing the network with a shared central node, one can build a network based on a shared data path. That distributes the routing and addressing work over all the nodes, but it also places higher demands on the network intelligence that must be built into each node. Specifically, each node must now be able to recognize when it may claim access to the shared data path, and each node must be able to identify which messages are addressed to it.

A. ISO and OSI

The management of communication between nodes can be divided into a hierarchy of several layers of abstraction. By a confusing coincidence, the International Standards Organization (ISO) has defined a hierarchy known as the open systems interconnect (OSI). This model consists of seven layers:

Application layer: services that provide access to the communication facilities (remote file access, remote log-in, network management)
Presentation layer: interpretation of message content in a form that is meaningful to the user
Session layer: system-dependent details at source and destination nodes
Transport layer: end-to-end control of message and data format
Network layer: addressing and routing of messages
Data link layer: integrity and sequencing of data
Physical layer: physical interconnections and signal levels

On each node one will find a layered set of software and hardware services corresponding to the levels defined in the OSI model. Information then flows downward through the layers of one node, across the interconnecting data path, and back up through the layers of another node.

B. TCP/IP

TCP/IP (Transmission Control Protocol/Internet Protocol) is a standard that covers part of the Internet model defined by the Defense Department. The model is intentionally very similar to the OSI model. TCP/IP covers the lowest four layers of the OSI model. The TCP/IP standard is strictly enforced by the Defense Department, and must therefore be vendor-independent. The entire TCP/IP package is usually implemented as an intelligent board that can be inserted as a peripheral on the backplane of its host system. (The set of signal lines on which a CPU interacts with its memory and input/output devices is often implemented as a flat structure with connectors accepting memory, CPU and other circuit cards. That flat structure may be called the backplane. If implemented as a circuit board the structure may be called a motherboard. The signal lines located on the backplane are collectively called the system bus.) One may expect that TCP/IP will eventually be built into an emerging ISO standard, or that the emerging ISO standard will replace TCP/IP. In the meantime, TCP/IP is available, fully defined and a close fit to the OSI model.

C. IEEE802

In parallel with the TCP/IP standard, the IEEE (institute of Electrical and Electronic Engineers) is developing a set of standards known as IEEE802.x. These standards will cover the same OSI layers as the TCP/IP standard. The intent of both TCP/IP and IEEE802 is the definition of a standard that will allow interconnection of machines from any combination of vendors. The difference lies in the fact that TCP/IP is imposed by the Defense Department, whereas adoption of IEEE802 must rely on voluntary acceptance by the industry.

One might describe the hierarchy of communication layers in terms of a set of wrappers or envelopes that contain the actual data. On transmission, each layer accepts a message from the layer above, encloses the message in an envelope, writes instructions on the outside of the envelope, and then hands the resulting package to the layer below. Eventually the package is transported across a data link to another node. Now the message travels upward; at each layer the outer envelope is removed and the revealed content is inspected, formatted as needed, and handed upward. Eventually the actual data are unwrapped and presented to the user or to the user program.

The OSI model is not a standard, but it allows for the definition of standards at each level of the hierarchy. Its major virtue lies in the fact that the various levels of the model can be made largely opaque, so that the implementation of a given level can be modified without significant effect on other levels.

Well-established standards exist only for the lower levels of the OSI model. Three standards for the physical layer have become well-known: RS232 (a recommended standard proposed by the Electrical Industries Association, EIA) defines hardware and signal levels for point-to-point links, GPIB (general-purpose interface bus) defines hardware and signal levels for a short-range (16M) bus structure that can link various pieces of measuring equipment, and Ethernet defines hardware and signal levels for a bus structure that can interconnect computers over distances of up to 2800M. Each of these standards has its own virtues and drawbacks; we will discuss them briefly in the following sections.

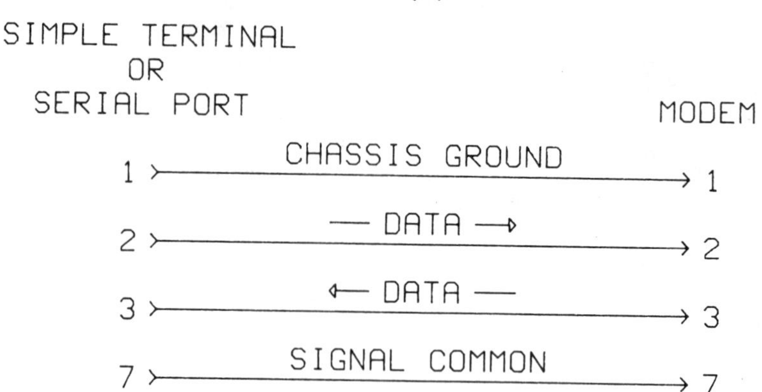

Figure 2. Typical RS232 connections. ≻— represents a socket, → represents a pin. Figure 2E shows both pin numbers and the associated RS232 signal names. Signal names beginning with "A" serve as chassis and signal common lines. Signal names beginning with "B" carry data. Signal names beginning with "C" handle equipment housekeeping. (A) Connection between a simple terminal or serial port and a modem. (B) Connection between a simple terminal and a simple serial port. Both DTE devices should have pin connectors. The null modem has socket connectors on both sides. In practice the standard may be violated on one side to eliminate the null modem. (C) Fully interlocked connection between a terminal and a modem. (D) Connection between a terminal with hardware handshake signals and a simple modem or serial port. Note that the serial port would violate the pin/socket rule. (E) Connection between a serial port with full modem control and a fully controllable modem.

D. RS232

RS232 is the oldest of the standards; it is still widely used to connect terminals to host computers, either directly by wire or via the public telephone network. In the latter case the RS232 data signals must be transformed into audio signals by means of a modem (modulator-demodulator) at each end. Ironically, RS232 is probably best known for the 25-pin connector that users associate with it, even though the shape of the connector is emphatically excluded from the standard.

RS232 describes specifications for DTE (data terminal equipment) devices that produce and/or consume data, and for DCE (data communication equipment) devices that transport data. A computer and a terminal are examples of DTE devices; a modem is an example of a DCE device. Because the link between two DTE devices may include DCE devices, a DTE device must be able to test the readiness of the link. The RS232 standard recognizes three types of signals: the A group consists of chassis and signal ground, the B group consists of data signals, and the C group consists of control signals. The readiness of the DCE is reported by the CC (data

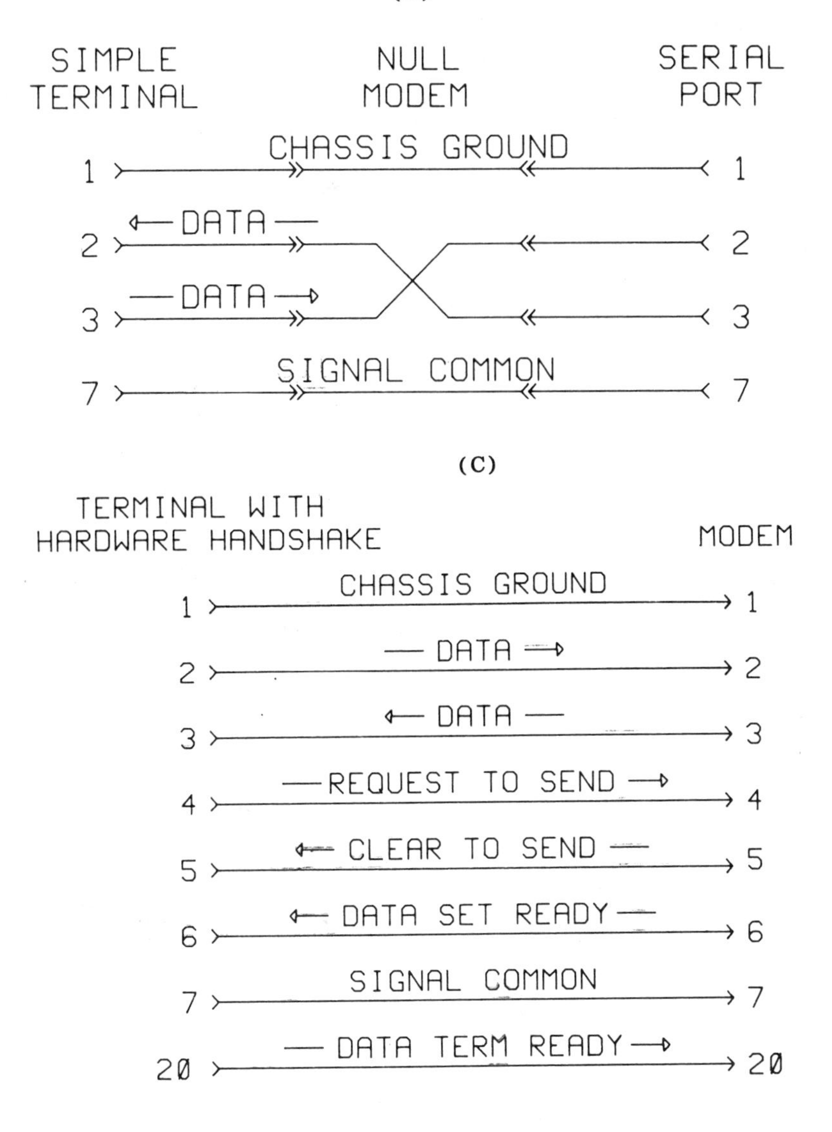

set ready) signal; the presence of an incoming data carrier signal is reported by the CF (carrier detect) signal. The presence of a functioning DTE device at the far end can only be tested by its response to a wake-up message transmitted to it. RS232 provides for a full-duplex link: data can flow in both directions simultaneously on a single RS232 link. Furthermore, RS232 includes provisions for a secondary data channel and a tertiary data channel. Consequently, a full RS232 implementation may involve more than 20 signals. In practice, an RS232 link may require as few as three wires. Figure 2 shows some of the commonly used RS232 implementations, with descriptive names for the signals involved.

(D)

TERMINAL WITH HARDWARE HANDSHAKE

MODEM OR SERIAL PORT

Terminal pin	Signal	Modem pin
1	CHASSIS GROUND	1
2	— DATA →	2
3	← DATA —	3
4	— REQUEST TO SEND →	
5	← CLEAR TO SEND —	
6	← DATA SET READY —	
7	SIGNAL COMMON	7
8	← CARRIER DETECT —	
20	— DATA TERM READY →	

(E)

COMPUTER PORT

MODEM

	Computer pin	Signal	Modem pin
AA	1	CHASSIS GROUND	1
BA	2	— DATA →	2
BB	3	← DATA —	3
CA	4	— REQUEST TO SEND →	4
CB	5	← CLEAR TO SEND —	5
CC	6	← DATA SET READY —	6
AB	7	SIGNAL COMMON	7
CF	8	← CARRIER DETECT —	8
CD	20	— TERMINAL READY →	20
CE	22	← RING INDICATOR —	22

For the three-wire implementation (transmit, receive, common) one must assume that the entire link, including any DCE devices, will be unconditionally available. Synchronization of the DTE devices must be based on the transmission of a set of reserved code patterns. These control codes are not identified in the RS232 standard. In the ASCII character code* we find control codes DC1 and DC3 (11 hex and 13 hex), which are widely used as "enable transmit" and "disable transmit" commands. Occasionally one may find a synchronization protocol based on the ASCII ACK and NACK (06 hex and 15 hex) control codes.

E. RS422/423/449

RS422, RS423, and RS449 form a set of new standards that can be seen as a superset of RS232. The standard represents a substantial improvement over RS232, but it is not widely used. A possible reason is the growing popularity of local area networks based on Ethernet, which has reduced the demand for point-to-point links.

F. IEEE488

The IEEE488 standard had its origin in the HPIB (Hewlett-Packard interface bus). Hewlett-Packard developed the bus for use on its line of instruments and published its specifications. Subsequently the specifications were incorporated into an industry-wide standard. Hewlett-Packard still uses the term GPIB (general purpose interface bus) in its literature; GPIB may be considered a synonym for IEEE488.

The IEEE488 bus consists of a 24-wire cable carrying 16 signals. A group of eight signals carries data and device addresses. Three lines carry bus synchronization signals, and five lines carry bus housekeeping signals. The bus can interconnect up to 15 devices and can extend over a length of up to 20M. Data rates on the IEEE488 bus tend to be limited by the capability of the devices attached to it, but may be as high as 1 MB/sec. The length specification on the bus limits its use to the interconnection of various instruments for a single experiment, but it serves that purpose very well.

G. Ethernet

Ethernet is probably the most widely used buzzword of the day. The specification was developed jointly between Digital Equipment Corporation, Xerox, and Intel to provide an industry-wide interconnection medium. Ethernet is a bus structure in which all signals are carried sequentially on a single wire. The original specification calls for a rather heavy coaxial cable, to allow the installation of taps for new nodes even when the cable is in use. Lately some alternatives, such as "thin-wire" and optical cable, have appeared on the market. On these alternative cables power must be removed for the installation of new taps, but even on the original cable most users prefer to bring the network down for any cable modification.

---

*American Standard Code for Information Interchange.

The Ethernet standard extends only over hardware and signal levels; it does not include specifications for the resolution of access contention. The Ethernet standard does include specifications for the assignment of network addresses to devices connected to a network. An Ethernet network can operate under control of a variety of network protocols; it is even possible to support two different network protocols simultaneously on a single Ethernet cable.

H. CSMA/CD

A popular protocol based on Ethernet is defined in the X.25 standard, which covers the lowest three layers of the OSI model. The X.25 standard includes Ethernet for the bottom layer, and superimposes two layers of network protocol on top of the physical layer. Access contention is handled by the CSMA/CD protocol (carrier sense multiple access/collision detection). Each node on the net has the obligation to listen for the absence of traffic before the node can start a transmission. That still leaves the possibility that two or more nodes start a transmission simultaneously. Such a collision must be recognized by the transmitting nodes: each node must listen to its own transmission to detect contamination by another transmission. If any node detects interference with its transmission it transmits a burst of noise and then releases the bus for some prescribed time that is uniquely defined for each node on the network. The burst of noise serves to inform the other transmitter that its transmission was probably contaminated as well.

I. TOKEN-RING

The collision-handling strategy does not give a transmitter an absolute guarantee that it will acquire the bus, but statistical arguments indicate that "starving a transmitter" is extremely unlikely. In practice the traffic on an Ethernet cable is usually limited by the nodes to a small fraction of the traffic that could be supported by the cable. An absolute guarantee can be provided by a "token-ring" protocol: a single "right-to-transmit" token is passed sequentially to all nodes in cyclic order. A node wishing to transmit a message waits until it receives a token, which grants ownership of the bus to that node. The node then places the message on the bus, and finally sends the token to the next node on the bus. A node without waiting messages simply passes the token to the next node.

The CSMA/CD lends itself well to an office environment, where network traffic tends to come in randomly spaced bursts with considerable tolerance for small network delays. The token-ring strategy is better suited to manufacturing automation, where many nodes must be guaranteed access to the network on a regular schedule.

A network based on Ethernet requires a considerable amount of intelligence near the hardware level (address creation and recognition, collision handling, error detection, etc.). An Ethernet controller often contains its own processor and memory, even though the controller may be inserted as a peripheral on the backplane of its host. A simple Ethernet controller may cost as much as a small personal computer.

It should be clear that communication between machines almost invariably involves the use of matching software packages on both sides. That software is quite elaborate; in spite of generally high license fees, the purchase of proved commercial packages will be far cheaper than any attempt at home design.

J. File Servers and Print Servers

The nodes on a network may all support similar activities, or one or more nodes may be dedicated to specific tasks. Thus one may find a node that manages all printing needs of the entire network (print server), or one may find a node that manages the storage of all programs and data sets (file server). When data sets are shared by an entire network under control of a file server some provision must be made to ensure single ownership of data during modification. That provision usually is implemented by means of data locks and a lock manager that is part of the file server. Finally, one may find a node that serves as a connection to another network. Such a node is known as a gateway if the two networks are different, or a bridge if the two networks are similar.

For an occasional dial-up session between a small computer and a remote host one may find an exception to the rule of matched software. A utility on the small machine makes that machine appear as a simple terminal to the remote host. In effect the software on the small machine is adapted to the existing environment on the remote host. Such utilities usually include provisions to feed a local text file into an editor of the remote host, and to capture text streams from the remote host into a local file. With such terminal emulators the burden of validation of the transmitted or received text usually falls on the user. The popularity of such emulators attests to the quality of data transmission over the public phone network.

A terminal emulator is poorly suited to the transfer of nontext files: those cannot easily be validated by a human user, and they may contain data that are interpreted as special characters by the emulator. A simple solution is based on the conversion of binary data into printable hexadecimal form before transmission, and back conversion of the transmitted file. The conversion will double the transmission time because each 4-bit group in the binary stream will be converted into an 8-bit character. Furthermore, it will still be impossible for a human reader to validate the data.

K. KERMIT

KERMIT is the name of a very popular package that has been available in the public domain for some years; versions exist for most of the popular small machines and for all the major minicomputers and mainframes. KERMIT operates in two modes: as a terminal emulator and as a file transfer protocol. Normally KERMIT is started on a small machine as a terminal emulator, so that a session on a remote host may be initiated. Then a command is issued to the remote host to start its copy of KERMIT for the reception (or transmission) of a specified file or group of files. Next a command is given to the local machine to transmit (or receive) a local file or group of files. Both copies of KERMIT then make some allowance for local housekeeping, after which they cooperate in the transfer of the data. The data stream is cut into packages by the transmitter, and a checksum is appended to each package. The receiver inspects the incoming package for validity and either acknowledges receipt or requests a retransmission. The strategy is quite reliable: the probability for undetected errors is less than one in 65,000. A complete set of KERMIT packages for all supported machines and operating systems can be obtained from Columbia University for a handling fee. Alternatively, any KERMIT implementation may be freely copied if the copyright headers are included with the copy.

Whenever nontext data are moved between machines one should be aware of possible subtle difficulties. The representation of printable text has been well standardized. Even though two major forms of character encoding are in use (ASCII for most of the world, EBCDIC for large IBM systems), both codes are based on one byte per character. Systems based on EBCDIC encoding almost invariably support conversion to ASCII for all data that travel to and from remote terminals, so that the EBCDIC encoding is largely invisible to the users at such terminals. As a result text can be reliably exchanged between dissimilar machines, and the characters will remain in proper sequence.

In contrast, the encoding of larger data objects, such as integers, real numbers, and complex numbers, varies from machine to machine. If such data objects are transmitted as byte streams, the sequencing of bytes at the transmitting machine may not match the sequencing of bytes at the receiving machine, and the number of bytes per data object may differ between machines as well. To handle this problem the communication package must include a layer that enforces a uniform data format at the transport layer. Network packages should provide that service, but terminal emulators and KERMIT packages do not.

VI. SOFTWARE, LANGUAGES, AND TOOLS

In spite of some exaggerated sales literature, even the simplest use of a computer involves programming. The computer takes no initiatives: it waits for commands from its user. Every command entered into a computer must be recognized as a program or as a part of a program. Thus the question is not whether a given computer system requires programming capabilities of its users, but what programming capabilities will be required of the users. In particular, the amount of programming effort required is largely controlled by the amount of program detail a user must provide to make the system perform a specified task. At one extreme a system dedicated to a well-defined and stable task may require no more than a single RUN command. At the other extreme a system may require extensive programming in assembler language to control an interface at the bit level. As usual, simplicity has its price. A system can be controlled with simple commands only if the designer of the system has predefined the behavior of the system with only a limited amount of user-controlled flexibility. Conversely, a flexible system will require a correspondingly rich command language to specify the many options it supports.

The demands imposed by the modern laboratory make the term "standard lab interface" almost a contradiction in terms. Certainly any usable lab interface must include such items as a set of analog input channels, a set of analog output channels, a programmable clock, and a reasonable number of digital input and output lines. But beyond that common list we must make choices about such things as analog resolution, conversion speed, single-ended or differential inputs, isolation amplifiers, low-level, high-level, or programmable gain input channels, etc. To a certain extent it is possible to overspecify the interface so that it will meet all but the most stringent demands, but that will almost certainly price the equipment out of the market. In practice we can therefore expect that few if any of the commercially available components will be exactly tailored to any specific application. Consequently, the owner of a laboratory system must stand ready to adapt some hardware and/or some software to the needs of the application. It is generally not practical to build significant improvements into existing hardware. Therefore the hardware will have to be carefully specified to meet the expected applications. Fortunately, that is not an insurmountable task for a person with a reasonable amount of technical experience.

The situation is quite different when it comes to software adaptation. In effect the software defines an abstract instrument with a flexibility that dwarfs the flexibility of even the most elaborate interfacing hardware. That makes the temptation to fine-tune the software of the system to one's desires almost irresistible.

Software activities can be broadly classified into two groups: those that deal with the experiment interface and those that deal with operations on the stored data. Operations on stored data tend to be well-defined mathematical manipulations such as averaging, integration/differentiation, Fourier transform, etc. Standard mathematical packages are readily available; they only require that the data on which they operate are accessible in some conventional format. In contrast, the software associated with the experiment interface is invariably specific for that interface and can rarely be transported from one system to another.

In principle an experimenter can attempt to build the software for an experiment interface from the ground up. That is a time-consuming task that should only be attempted by persons with considerable experience in engineering, in low-level software, and in event-scheduling software. Engineering experience is required for proper interpretation of the hardware specifications, low-level software is required for the code that interacts with the hardware interface at bit level, and event scheduling will be required to insure that all the demands placed on the processor can be met under all conditions.

A properly designed software package can be subdivided into several layers, with the innermost layer responsible for direct control of the system hardware, and with subsequent layers dealing in progressively higher levels of abstraction. Thus the innermost layer will be aware of such things as port addresses, interface status bits, data formats of analog-digital converters, clock control, etc. The next layer may deal with time in standard engineering units, with analog inputs encoded as integers in units of mV or PSI, etc. The various software layers should be organized in a hierarchy so that no layer can exert control over any higher-up layer. If a software layer is properly designed and constructed it will be opaque, so that no details of the underlying layers are visible from layers higher up. At the outermost layer the user deals only with the "shell" of the operating system, and may even perform an entire experiment with a single keyboard command. At the next level below the user would deal with one of several standard programming languages such as Pascal, BASIC, FORTRAN, etc., extended to support a set of calls to yet lower levels. The preferred method to provide that support is in the form of a library of callable procedures, although companies such as Hewlett-Packard offer well-designed extensions of BASIC that support direct manipulation of laboratory interfaces by means of additional keywords built into the language. A recent development in this area is the announcement by Lotus of a software package under the name Measure, which supports direct insertion of live laboratory data into cells of a Lotus-1-2-3 spreadsheet. The spreadsheet software in turn offers excellent data-handling support and graphic representation.

At even lower levels one would deal with such concerns as the conversion of data between several internal representations, such as signed and unsigned integers, long and short integers, interface-specific formats, etc. High-level languages are not designed to support such activities. The "C" language is a notable exception in this respect: it has the structural power of Pascal, but also provides access to machine-level details such as memory addresses and individual data bits. Finally, assembler language remains the tool of choice for much of the code that is responsible for the control of interface hardware.

It will be convenient if various portions of a program can be written in languages that serve the purpose of each portion. Thus code for a mathematical operation may be written in FORTRAN, record-keeping functions may be written in Pascal or BASIC, and machine control may be written in assembler language. All those language processors or "translators" should then be designed to produce output in a common format, usually called an "object" file. All object files must then be collected into a single file representing the executable image of the program. That task is performed by a "linker." The linker reads all object files specified to it, scans them for cross references, and resolves their addresses. If any unresolved references remain, the linker attempts to match those to procedures and data items stored in a library, and extracts portions from the library as needed. Finally, the linker writes a file containing the executable image of the program to disk.

A good laboratory system should come with several layers of carefully constructed and well-documented software. That will give the owner the capability to perform relatively standard experiments with a minimum of programming at a level of abstraction that is close to the context of the experiment. As the demands of the experiment grow, the owner can then invade lower layers of software with the help of the documentation to further adapt the system to those demands. Run-time variables and options can be inserted by means of user menus or by means of a setup dialog. Programmers should be careful to ensure that the level of detail managed by such a user interface matches the level of abstraction at which that interface is placed. A request for a device address may be appropriate at the inner levels, but it will frighten a user accustomed to think only in terms of choosing between 1-ml and 10-ml samples.

As was stated earlier in this chapter, the operating systems available on personal computers generally will not support real-time operations. The same observation holds for utilities and application programs. Translators (compilers) are available for a variety of languages, but without help from the operating system none of those languages can be expected to support real-time operations. For operations on stored data a rich choice of software is available, however. In addition to the LOTUS-1-2-3 spreadsheet and similar products, one may find packages specifically aimed at the handling of laboratory data, such as the DAD-iSP package from Keithley Instruments. Most general-purpose software will carry a much lower price tag than software aimed at laboratory applications, because the laboratory market is considerably smaller than the general market.

VII. SIGNAL PROCESSING AND SIGNAL/NOISE DISCRIMINATION

Mathematical operations on computer-acquired experimental data fall into three broad classes: real-time "on-the-fly" processing into control signals that are returned to the experiment, operations on stored data to enhance the signal-to-noise ratio, and operations that transform measured data into a more meaningful form.

The discussion below is aimed primarily at the user of mathematical tools. It does not maintain the rigor that would be needed to confirm the validity of the operations or conclusions presented here. Neither does this discussion contain the implementation details that would be of interest to a programmer.

Examples of the first class can be found primarily in applications for process control. Laboratory experiments still tend to be considerably less elaborate than production streams in terms of process control, although one may find functions such as computer controlled wavelength scanning in spectrophotometry. Computer control on the basis of real-time measurements is comparatively rare in the laboratory, however. This section will therefore deal primarily with the remaining two classes: improvements in signal-to-noise ratio, and operations that transform measured data into desired information.

Improvements in signal-to-noise ratio must be based on recognizable differences between the nature of the signal and the nature of the noise. In the majority of cases those differences can be expressed in terms of time or in terms of frequency.

Discrimination on the basis of time is the basis for all stimulus/response experiments: any response returned by the experiment is expected to show a clear correlation with the preceding stimulus. Examples of such experiments are chromatography (time-spaced peaks resulting from a preceding sample application) and wavelength-scanned spectrophotometry. When such experiments are repeated, the signal components of the returned signal can be expected to be consistent from run to run, but noise contributions in the returned signal will vary randomly between runs. Upon coaddition of the results from multiple runs we will expect the magnitude of the signal component to increase linearly with the number of runs, whereas the noise contribution will increase with the square root of the number of runs. The coaddition is made possible by the capability of the computer to store the result of an experiment in machine-accessible form.

Discrimination on the basis of frequency is possible with both analog and digital data systems. An analog system can include bandpass, lowpass, or highpass filters in the signal chain. Because many experiments produce signals in a frequency band starting at very low frequency the prevalent analog filter is a lowpass filter. Such a filter can also be regarded as a short-term history keeper: the output from the filter is determined not only by the current input to the filter, but also by the recent history of the input signal. A digital system can simulate such a lowpass filter, but the digital system can also do something that an analog system can never achieve. The digital system can store the experimental data, and it can subsequently "look into the future" on those stored data. This allows the digital system to produce the effect of a lowpass filter without the attendant time skew that is unavoidable with analog systems. It is important to note here that the filter operations to be described cannot replace the anti-alias filters described in the section on sampling. In the absence of an anti-alias filter a set of sampled data may contain spurious components that cannot be removed by any amount of subsequent digital filtering. Operations on stored data can only extract or remove signal components that have been properly sampled according to the Nyquist sampling theorem.

Three major approaches have become popular for digital lowpass filters. By far the simplest filter is known as the moving-window average. The value of each data point is replaced by the average of that point and some reasonable number of its neighbors on both sides. The required arithmetic operations are simple and fast, and they can be encoded in a few statements.

A much better result can be obtained by a curve-fitting strategy. A subset of the data string centered around the data point of interest is treated as a polynomial function. Then the center value of that function is calculated by a least-squares fit of the coefficients of the polynomial and subsequent evaluation at its midpoint. For uniformly spaced data points the entire operation can be reduced to the calculation of a weighted average with fixed weighting factors. The values of the weighting factors depend only on the number of data points included in the calculation. The calculation requires one multiplication per included data point, but currently popular processors can perform those multiplications almost as fast as additions.

Subject to time constraints, either filtering strategy described above could be performed on the incoming data stream. Each operates only on data collected over a small time interval spanning recent history. In contrast, the next strategy requires access to the entire data set. It can therefore be applied only to stored data after the experiment has been completed.

The most powerful tool to control the frequency content of data recorded as a function of time is based on the Fourier transform. The indiscriminate use of the Fourier transform carries substantial risks, however. We will discuss those after a brief review of the nature of the Fourier transform.

The Fourier transform is a well-established mathematical transformation that maps a periodic function of some variable X into an equivalent function of the variable 1/X. The theory and background of the transform may be found in many books; it will not be discussed here. In the laboratory we often measure variables as a function of time; the Fourier transform of such a variable will then be a variable expressed as a function of reciprocal time, that is, as a function of frequency. A second Fourier transform will again yield a variable as a function of time. In other words, the Fourier transform is symmetric.

Let us now apply the Fourier transform to a function f(t) to produce the corresponding function $F(\omega)$, where t has the dimension of time, and ω has the dimension of frequency. If f(t) includes signals over a frequency band 0–S and noise covering a frequency band from S upward, then the region below S in $F(\omega)$ will represent the measurement of interest, and the region above S will represent the noise contribution. We can then discard the region above S in $F(\omega)$. A subsequent inverse Fourier transform of the resulting truncated $F(\omega)$ function should yield the signal of interest without the contaminating noise.

The abrupt boundaries of signal and noise frequencies are unlikely in practice. We should therefore do something less drastic than truncating the high-frequency part of the frequency function. Without supplying a detailed proof, we will state that we can achieve any bandpass function by multiplying the frequency function with the gain/frequency function representing the desired bandpass and applying the inverse Fourier transform. Thus multiplication of the frequency function with a function $X = 1/\omega$ will be equivalent to the effect of a first-order lowpass filter, or an integrator. Similarly, multiplication of the Fourier transform function with a function $X = \omega$ and back transformation will be equivalent to the action of a first-order highpass filter, or a differentiator. These operations are illustrated in Figure 3.

The Fourier transform has become a popular and powerful tool by virtue of the fact that under certain conditions the transform can be performed very rapidly by an algorithm originally published by Cooley and Tukey. If sampled values of a function are spaced in uniform increments of its independent variable, the Fourier transform of that function can be calculated with a relatively small number of multiplications and additions. The efficiency of the algorithm is particularly impressive if the data set contains a number of entries that is an integral power of two. The algorithm has become known as the fast Fourier transform (FFT). It is available as a built-in function for many lab-oriented software packages.

As stated before, the casual use of the Fourier transform for data manipulation carries substantial risk. The risk arises from the fact that the Fourier transform requires that the input function be periodic. Few, if any, sampled signals in the laboratory are periodic. We acquire data on a single run of an experiment. Submission of that data set for FFT carries the implicit assumption that the data set represents one cycle of a periodic function. In other words, the FFT operation assumes that the front of the data set mates with the back. In reality there will often be a big difference between the values of the first and last data points. Even if those values are similar, the slope and/or higher derivatives at the front and tail end of the data set may have widely differing values. The

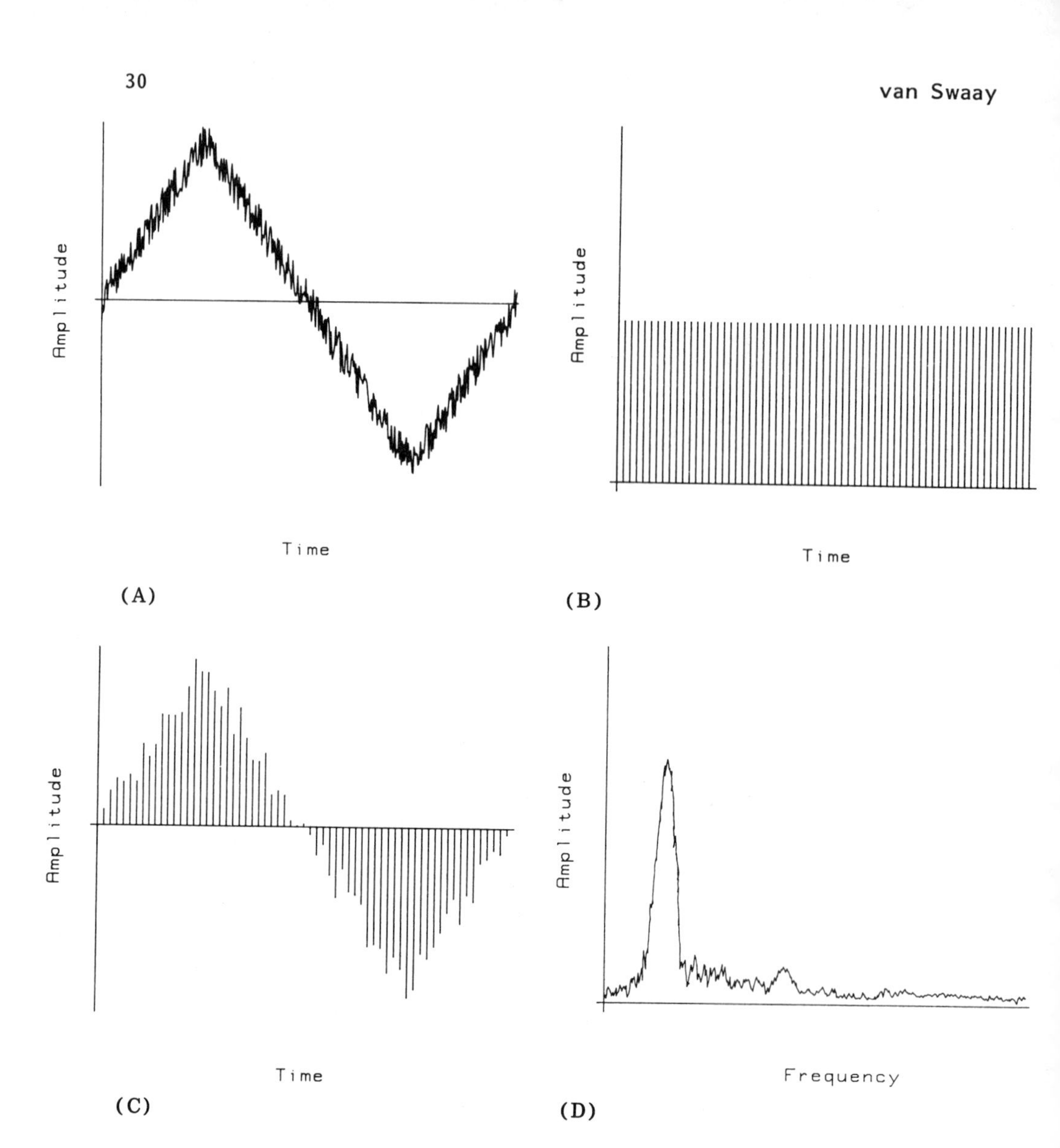

Figure 3. Sampling and Fourier-transform operations. (A) Noisy input signal. (B) Sampling waveform. (C) Resulting sampled data. (D) Fourier transform of (C). The complete spectrum consists of a set of harmonics of the sampling frequency, each of which is broadened by sidebands representing the spectrum of the input signal. For clarity, only a single right sideband is shown. The sampling frequency itself is absent because the input contains no DC component. (E, G, K) Weighting functions, plotted in log-log format to emphasize the relationship to corresponding analog filter functions. (E, F) Lowpass filter and resulting time function. The breakpoint should be chosen to pass significant components of the input signal and to reject higher frequencies. (G, H) Integrator and resulting time function. (K, L) Differentiator with high-frequency noise rejection and resulting time function. Again the breakpoint should be chosen to

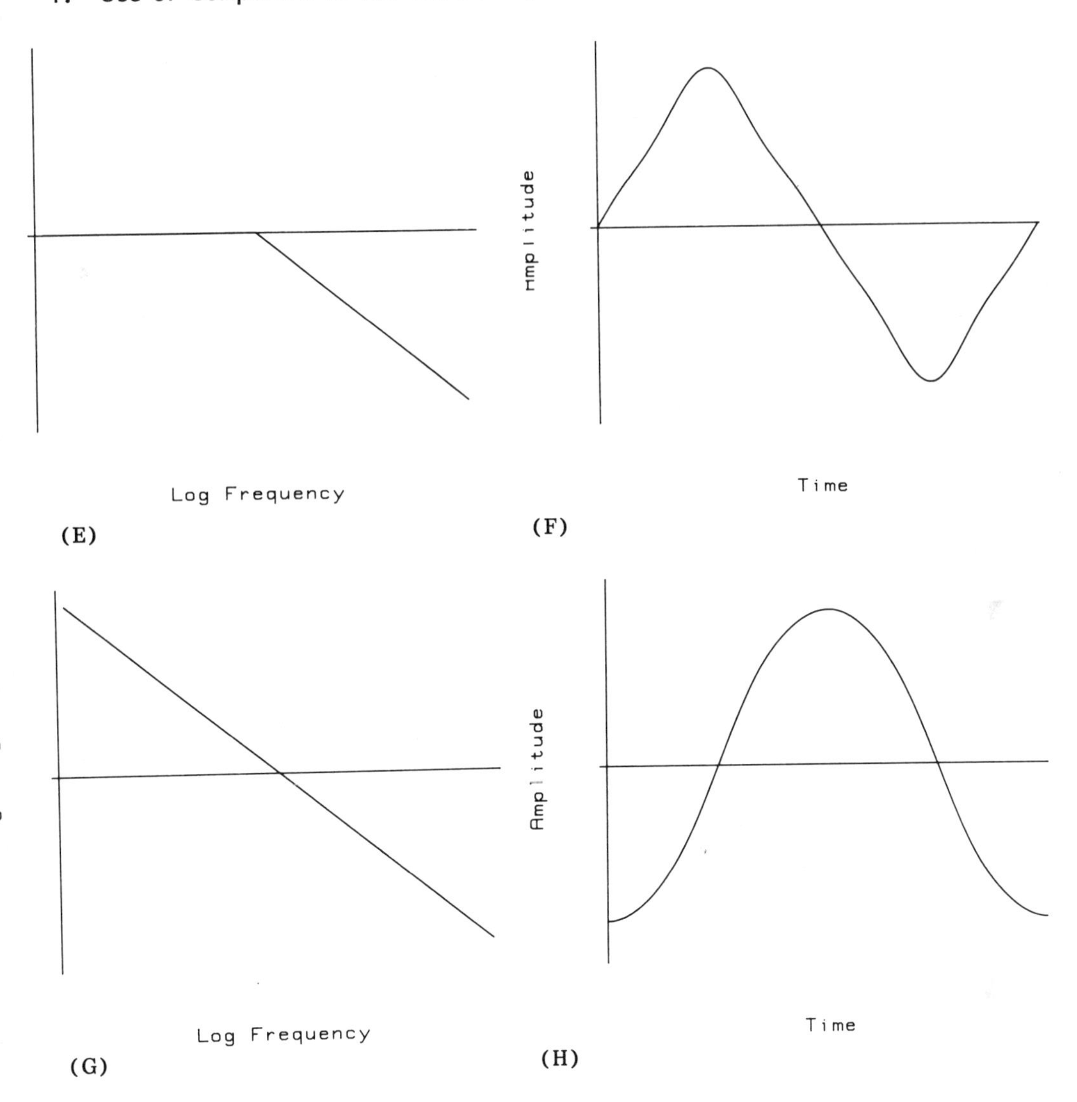

---

discriminate between signal and noise frequencies. Maximum discrimination between signal and noise energy is achieved if the FT of a clean input signal is used as the filter function, but the application of such a "matched filter" tends to introduce considerable distortion in the reconstructed waveform.

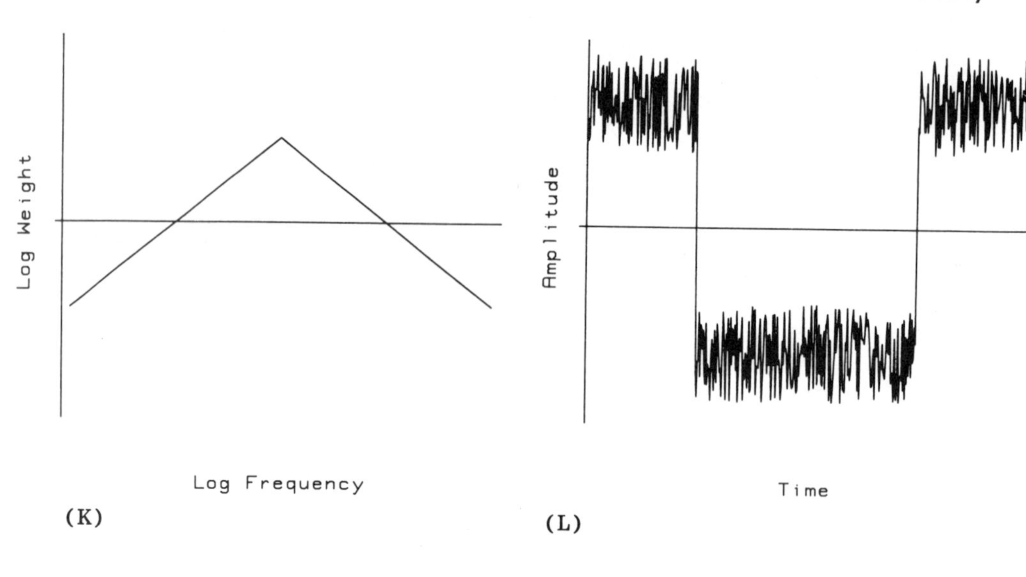

frequency transform will then have high-frequency components that represent the large transient between the first and last data points of the input data. It should be obvious that these high-frequency components are unwanted artifacts.

Several approaches are available to reduce or avoid the introduction of artifacts. Detrending is a term used to describe the addition of a linear function to the input data, chosen to yield equal values for the first and last data point. Detrending eliminates the zero-order transient, but has little effect on higher-order transients (abrupt changes in slope or higher derivatives). More effective approaches must all rely on some knowledge of the expected shape of the input function. The underlying goal will be the reduction of transients between the first and last data points and between as many derivatives as possible at those points. Figure 4 illustrates the detrending operation, as well as inversion through the origin. The inversion will remove the slope transient at both ends of a detrended function.

The Fourier transform can be used to enhance or reduce the contributions of various frequency regions in a signal. At its logical extreme the Fourier transform can be used to extract the contribution of a single frequency component of a signal. The same effect can be achieved with much less effort, however. We need not calculate the contribution of all frequency components if we are interested in only a single component. This leads us to correlation techniques.

Let us consider an incoming data stream that is expected to contain some specified frequency component synchronized to a known point on the time axis. We can create a data set that represents the amplitude of a reference function representing the frequency component of interest. The reference function must not contain a DC component: its average must be zero. Multiplication of corresponding points of the incoming stream and the reference set and summation of the products will yield a nonzero value only if the incoming data stream does indeed contain the frequency component of interest. Furthermore, the actual value of the summed products will represent the amplitude of the sought frequency in the incoming signal. The strategy can be extended to signals of arbitrary waveform and is known by the name signal correlation.

A special form of correlation forms the basis for a digital bandpass filter. The filter consists of a shift register through which successive samples of the digitized input signal are propagated. Taps at selected positions on the shift register serve to extract corresponding data elements from recent history. The output from each tap is multiplied by an appropriate coefficient, and the outputs of all multipliers are summed to produce the output of the filter. Current technology makes it possible to place all components of such a filter onto a single integrated circuit, with signal-handling capabilities ranging up to the megahertz region.

Some experiments yield signals that require a Fourier transform to convert the data into usable form. Examples of such experiments are pulsed nuclear magnetic resonance (NMR) and spectrophotometers based on interferometry. If the sample in an NMR experiment is subjected to a short burst of radiofrequency (RF) radiation at the appropriate frequency, the shortness of the burst makes the radiation appear to be broadband to the sample. Consequently, all sample resonances are excited. Subsequent decay of the excited sample is accompanied by radiation at the resonant frequencies of the sample. This radiation can be collected as a function of amplitude versus time. The desired amplitude versus frequency function is then obtained by Fourier transform.

Interferometric spectrophotometry is based on a mapping of optical frequencies into audio frequencies by means of a moving-mirror interferometer. The resulting audio signal is collected as a function of amplitude versus time. Again a Fourier transform serves to convert the measured function of amplitude versus time into the desired function of amplitude versus frequency.

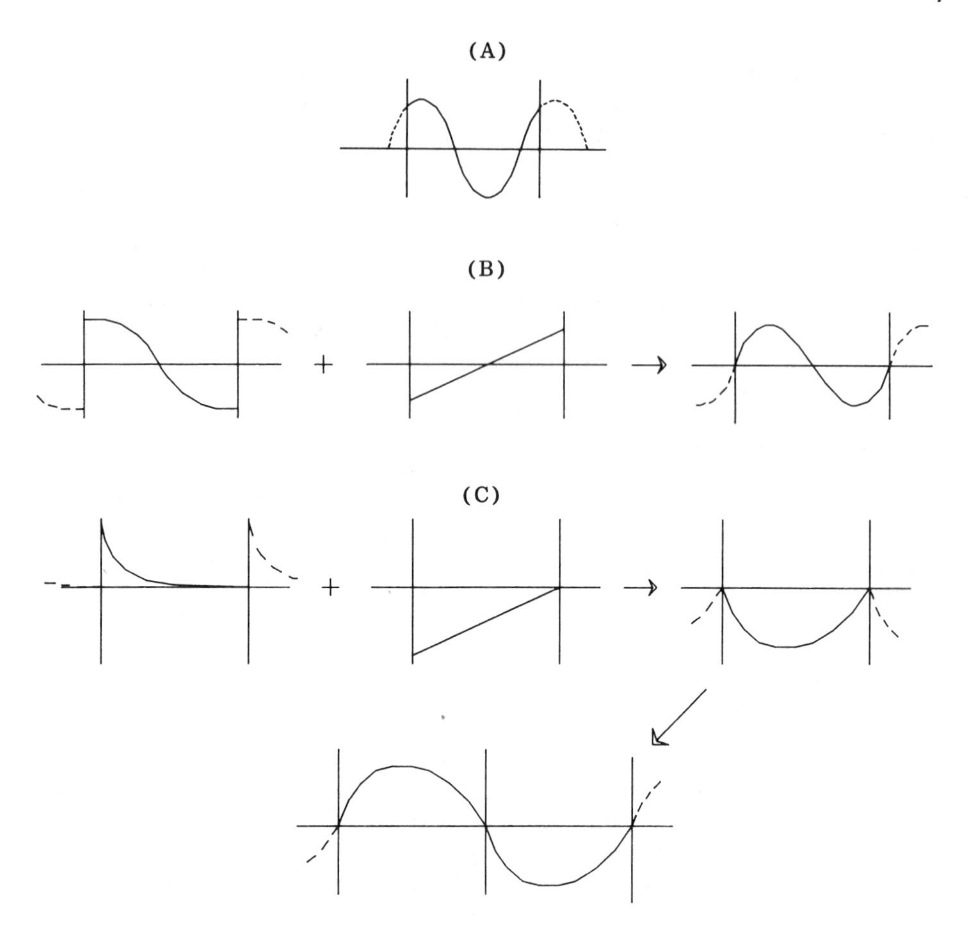

Figure 4. Strategies to reduce the effect of end-to-end transients before Fourier transformation. (A) Input signal without end-to-end transient. (B) Simple detrending. (C) Detrending combined with curve inversion around the origin.

The advantage in both cases lies in the fact that no signal discrimination is necessary. All signal energy available from the experiment is collected during the entire experiment. In contrast, the scanning spectrophotometer would discard all energy except the energy in the currently active resolution element of the spectrum. The improvement in signal-to-noise ratio obtained can be traded for either sensitivity or speed. The extensive processing of experimental data would make the approach impractical without the help of a digital computer to store and manipulate the data.

Although the Fourier transform is widely used to manipulate and transform data, it is by no means the only operation of interest. A transformation based on Chebyshev polynomials has been proposed for the smoothing of signal functions with large end-to-end transients. The Hadamard transform has been used in infrared (IR) spectroscopy. Like the Fourier transform instrument, an instrument based on the Hadamard transform can collect energy from many resolution elements simultaneously for subsequent decomposition into the individual contributions.

On a much more modest scale of complexity, we may refer to the conversion of the output of "imperfect" sensors to linear form, such as the mapping of thermocouple potentials to temperature. Such linearizing operations can be based on two major strategies. If the number of possible digital values is reasonably small (<1K) each digitized signal value can be used as an index into a table holding the corresponding temperature value. Alternatively, the temperature-potential relationship can be expressed as a polynomial, so that temperatures can be calculated by evaluation of the polynomial for the measured potential.

Table lookup takes very little time, but the speed is obtained at the cost of a large amount of memory required to hold the lookup table. Conversely, the evaluation of a polynomial may be time-consuming, but no additional data space is needed. A mixed strategy based on a coarse lookup table and polynomial (or linear) interpolation between lookup points reduces the demand for space without requiring long calculations.

As a final example of capabilities provided by collection and storage of experimental data in digital form, we may refer to "multidimensional" experiments. The term is used here to refer to experiments in which the data of interest take the form not of X-Y or X-time relationships, but of X-Y-time and more complex relationships. Examples would be the collection of an array of spectra with a stopped-flow spectrophotometer, or the collection of an array of flame emission spectra after injection of a pulse of sample. Within the constraints of achievable data throughput a computer can organize a single stream of sequential data into a two-dimensional array or even into a higher-dimensioned array. It then becomes a simple matter to extract data from that stored array in any of several cross sections.

VIII. SIGNAL CONVERSION

The vast majority of signals available from laboratory equipment are presented in analog form. Acceptance of such signals by a computer will require conversion into digital form. The conversion not only reduces the continuous range of input values into a discrete set; it also reduces the continuous record of signals into a discrete set of samples spread over time. Both the quantization of signal value and the discrete sampling over time result in a loss of information. Thus we must make sure that the lost information is not part of what we wish to observe.

In this section we will deal only with the analog signals presented to the system. In many cases those signals are generated at considerable distance from the system, and provision must be made to transport the analog signals with a minimum of degradation. The transport of analog signals and the precautions necessary to minimize the injection of noise are subjects that could easily fill another chapter. Because that topic is not unique to the use of computers, it falls outside the scope of this chapter.

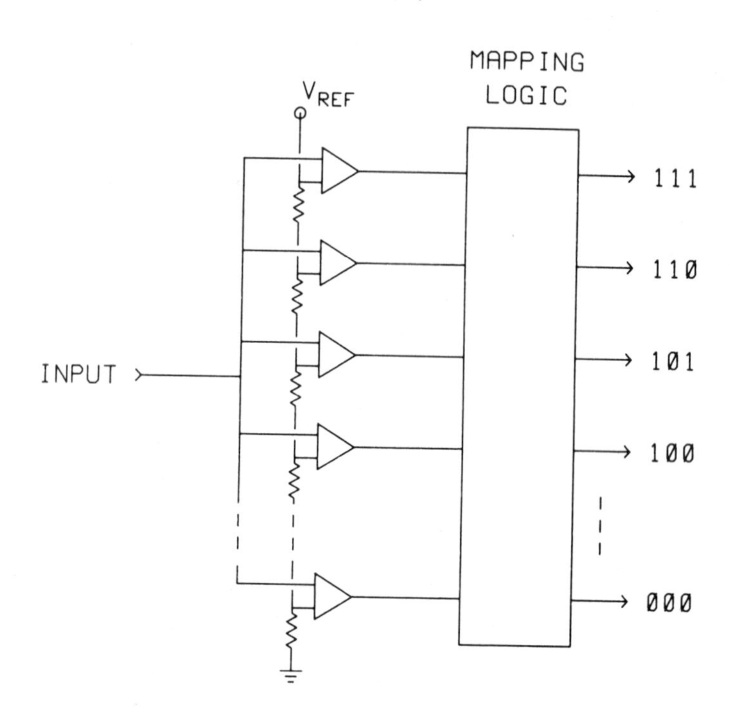

Figure 5. Popular analog-to-digital converters. (A) Flash converter; one comparator is required between each pair of resolution elements. If overrange and underrange detection is required, two additional comparators will be needed at the ends of the allowed signal range. (B) Successive-approximation converter. This is the most popular choice for medium-speed data acquisition because it represents a good compromise between speed and cost. (C) Dual-slope integrating converter. This device is intrinsically slow, but also supplies superior noise rejection.

Although the analog signals produced by various sensors can have a continuous range of values, the signal invariably contains a noise component. For a given value of a signal, the fluctuations of that signal caused by noise do not carry information. Consequently, we can convert the signal into digital form without loss of information if the resolution (step size) of the digital signal is smaller than the amplitude of the noise component of the analog signal. Conversely, we cannot hope to "create" information with a converter than can resolve signals smaller than the noise level riding on the input signal.

Analog-to-digital converters (ADCs) can be divided into three major categories: parallel or "flash" converters, counting converters, and integrating converters. Figure 5 illustrates some of these. The parallel or "flash" converter compares the analog signal with a set of reference values in parallel and extracts the closest match with digital logic.

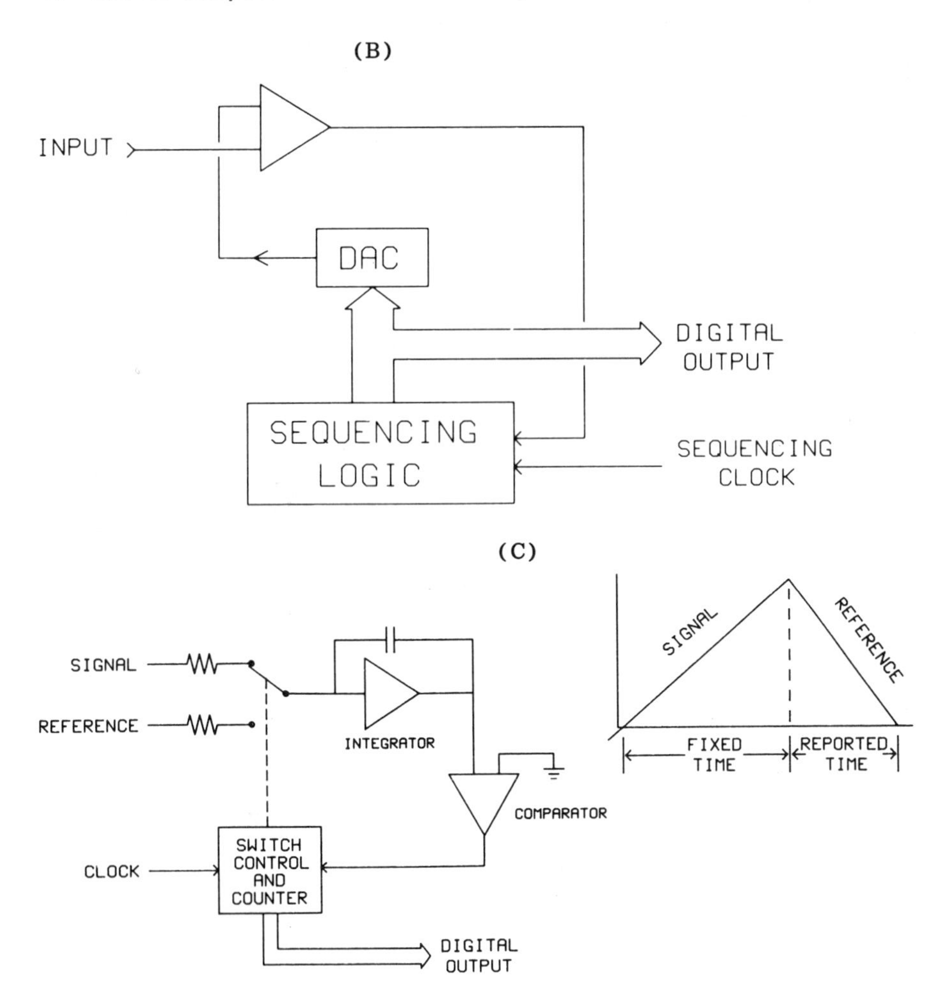

Counting converters adjust an analog reference signal in steps until the reference signal matches the input signal. The number of steps will then be a digital representation of the value of the input signal. The simplest form of a counting converter generates a staircase reference signal. The successive approximation (S-A) converter is far more efficient: it is based on a variable step size, with the first step equal to half the range of the input signal, and with subsequent steps one half the size of their predecessor.

Integrating converters integrate the input signal over a fixed time interval by collecting charge on a capacitor. Then the capacitor is discharged at a fixed rate controlled by a reference signal. Finally the discharge time is reported in digital form.

Flash converters are very fast: conversion speeds of the order of 1 nsec are possible. The flash converter requires one comparator for each possible value of the digitized signal; resolution is limited to 6-8 bits by space and cost constraints. Flash converters are widely used in systems that can digitize an entire video image as it is transmitted.

S-A converters contain only a single comparator. The input signal is compared against a reference signal that is constructed bit by bit; at each stage the comparator controls whether the corresponding bit in the digital representation should be turned on or off. Modern S-A converters can perform a 14-bit conversion in as little as 1 μsec, although resolution of 12 bits and conversion times of the order of 10 μsec are more common. The S-A converter represents an attractive compromise for use in data acquisition systems. The conversion time is fixed and reasonably short, and adequate resolution can be achieved at reasonable cost.

Integrating converters are intrinsically low-speed devices; they typically require 10-100 msec per conversion. Because the input signal is integrated over time, such converters reject high-frequency noise. Resolution of 16 bits is not uncommon, and resolution up to 20 bits can be achieved under favorable conditions. Digital multimeters are almost invariably based on an integrating converter. If the noise riding on the input signal contains a dominant component at a known frequency (e.g., power-line noise) the integration time of the input signal should be set at an integral number of cycles of the dominant noise source.

The successive-approximation converter requires that the input signal remain unchanged for the duration of the conversion. A 10-bit converter accepting an input signal with a range of 10 V achieves a resolution of 10 mV. That means that the input signal presented to the converter must not change more than 10 mV during the conversion. If the conversion time is 10 μsec, the rate of change of the input signal cannot be more than 1 mV/μsec or 1000 V/sec. This corresponds to a maximum frequency of about 16 Hz for a full-scale sine wave.

Much better performance can be achieved if the incoming analog signal is sampled and stored in analog form, so that the converter can operate on the stored signal. The analog sampling circuit is known as a sample/hold (S/H) or track/hold amplifier (Figure 6). The hold requirement of the S/H circuit is invariably based on a capacitor; that capacitor must be charged to the value of the input signal and must then retain that charge during subsequent conversion of its voltage by the successive-approximation converter. The requirements for fast charging to the input voltage during signal acquisition and slow decay (droop) during conversion are intrinsically incompatible, so that the size of the capacitor reflects a compromise between the two requirements.

The size of the capacitor is set to the minimum value that will hold the stored voltage within one resolution element during conversion. Acquisition of a new sample from the input signal will then require some finite time because the capacitor will have to be charged via some finite impedance. In practice the capacitor is charged from a low-impedance source, such as a buffer amplifier, and the ADC senses the capacitor voltage via a high impedance circuit. Acquisition time is typically somewhat smaller than the conversion time of the ADC to which the S/H circuit is connected, even though the voltage change during the acquisition phase can be several orders larger than the voltage drop during the conversion phase.

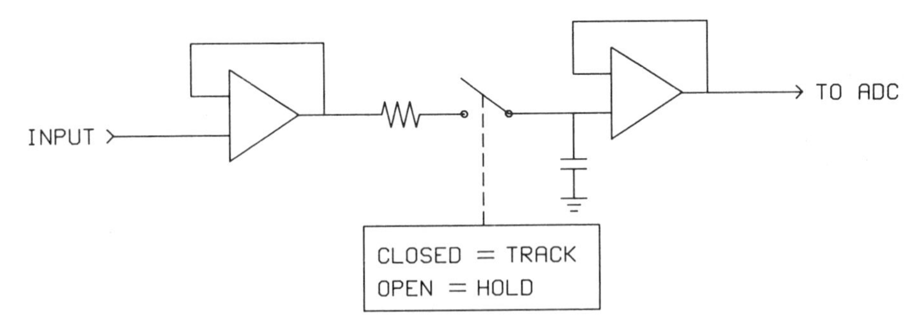

Figure 6. Sample-hold circuit. For sequential converters, such as the successive approximation converter, it is essential that the signal to be converted is held steady during the conversion. The sample-hold circuit serves as a short-term analog memory.

In many cases a lab experiment presents not one, but several analog signals. Each of those signals can be handled by its own converter, but more commonly a single converter handles all analog signals by means of analog switching circuitry that sequentially routes each of the signals to the ADC. For low sampling rates (of the order of 1 sec/sample) the resulting time skew between data channels is insignificant, but at high sampling rates this skew may become almost as large as the sampling interval. For critical applications the skew can be eliminated by the insertion of separate S/H circuits for each analog input channel. That strategy places higher demands on the quality of the S/H circuits: they must hold a stored analog value long enough to permit two (or more) analog-to-digital conversions. A further improvement in performance is possible if two or more ADCs are provided. That allows simultaneous conversion of two or more analog signals. In many cases the ADC is the slowest element in the signal path, so that the addition of a second ADC not only will eliminate sampling skew between input channels, but also will allow a higher throughput of analog signal samples.

The conversion of a continuous signal into digital form not only represents a loss of information due to quantization; it also represents a loss of information due to sampling. We wish to reconstruct the original waveform from the discrete set of sampled data. That will be possible only if the segments of the waveform between sample points contain no information.

In Figure 7 we illustrate the effect of sampling on a continuous signal. The top curve represents the frequency distribution of the sampled signal, and the center trace represents the frequency distribution of the sampling gate. The sampling gate may also be described as a comb filter that passes only DC, the sampling frequency, and all its harmonics. If the sampling gate is treated as a gain block with a gain of zero at all times except at sampling times spaced τ sec apart the frequency spectrum of the sampling gate will have the shape of a comb with tines at frequencies n/τ in which $n = 0, 1, 2, 3$, etc. The product of the two frequency spectra will have

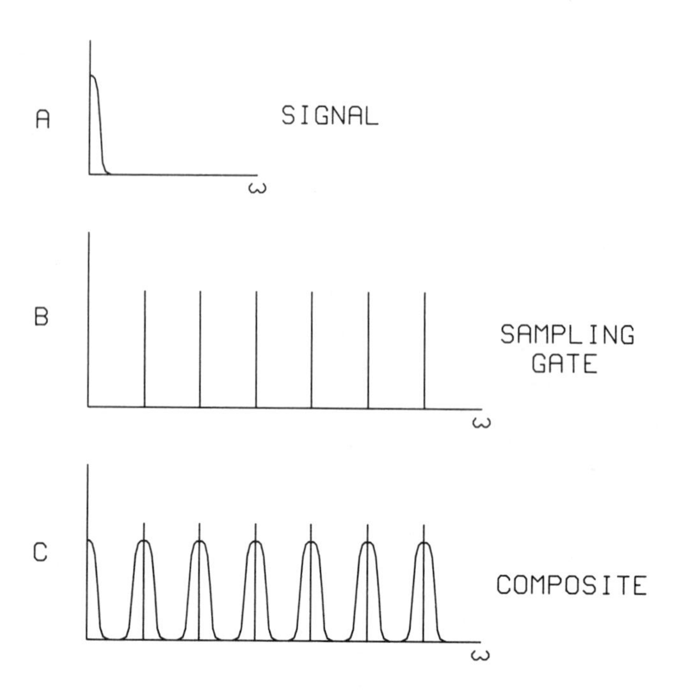

Figure 7. Frequency spectrum of a low-frequency continuous signal sampled by a sampling gate with a fundamental frequency well above the maximum frequency present in the continuous signal. The low-frequency signal reappears in the form of sidebands on each tine of the comb spectrum of the sampling gate.

the form shown as the bottom trace, with sidebands corresponding to the signal spectrum on both sides of each of the tines of the sampling spectrum.

The original signal waveform can be reconstructed by the Fourier transform of the sidebands of one of the tines of the sampling spectrum. Thus we conclude that there must be no overlap between the sideband of one tine of the sampling spectrum and the signal area of the sideband of an adjacent tine of the sampling spectrum. In Figure 8 the input to the sampling circuit has a frequency spectrum ranging from 0 to F(max), but only the region from 0 to F(sig) is of interest for the experiment. The sampling frequency must be chosen so that the tail of one sideband does not overlap with the region of interest of another sideband. In other words, if F(sample) denotes the sampling frequency, then

$$F(\text{sample}) \geq F(\text{max}) + F(\text{sig})$$

If the input signal spectrum is abruptly cut off at F(max), the expression reduces to the relation known as the Nyquist sampling theorem:

$$F(\text{sample}) > 2 \times F(\text{sig})$$

Of course it is unrealistic to expect an abrupt cutoff of the signal spectrum at F(max), but for a given sampling rate a sharp cutoff filter will reduce

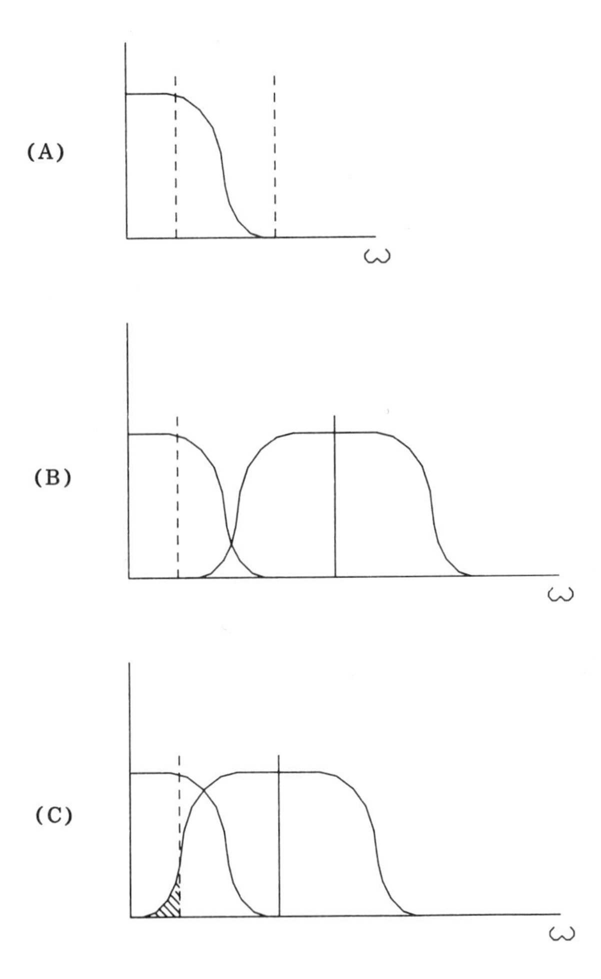

Figure 8. Alias signals produced by undersampling of a low-frequency continuous signal. (A) Frequency spectrum of the input signal. (B) Sidebands of a properly sampled signal do not overlap in the signal area. (C) Sidebands of an undersampled signal overlap in the signal area. The overlap contribution gives rise to alias components in the reconstructed signal.

the amount of sideband overlap. Alternatively, a sharp cutoff filter will allow a lower sampling rate for a given level of fidelity of the reconstructed signal.

Distortions in the reconstructed signal caused by sideband overlap are known as alias signals. These artifacts appear in the reconstructed waveform as low-frequency components for which no corresponding components in the original waveform exist. Violation of the Nyquist constraint is especially invidious because the resulting artifacts often cannot easily be recognized as such. It is essential therefore that the signals presented to an ADC be conditioned so that they contain no significant energy in the frequency band that produces alias signals. In practice, all signals presented to a sampling circuit must pass through a lowpass filter. The frequency response of the filter should be such that attenuation of all relevant signal frequencies is negligible, and so that no significant noise energy remains at a frequency F(sample) - F(signal). For a given maximum signal frequency, a sharper filter rolloff will allow a lower sampling rate. The lower sampling rate in turn will lower the throughput requirements of the ADC and subsequent signal handling. Thus the additional cost of a sharp lowpass filter may often be recouped through relaxed specifications on other parts of the system.

For most applications the significance of a signal will be controlled at least partly by the time at which the signal is acquired. Thus a set of data produced by an ADC must be associated with a set of time stamps. In practice the time stamps often take the form of addresses at which the successive ADC samples are stored in computer memory. That implies that the samples must be taken on a well-defined and precise time schedule. For low sampling rates of the order of seconds per sample it may be possible to schedule sampling under control of the time-of-day clock of the computer. Although such a strategy introduces some time jitter, the resulting time uncertainty will be insignificant compared to the sampling interval. In contrast, some high-speed experiments may require sampling intervals of 1 msec or less. In such situations a time skew of even 1 μsec will be significant. Few computers can boast instruction times as short as 1 μsec. We conclude therefore that it is impractical to trigger signal sampling by means of software instructions: the resulting time skew and jitter will be unacceptable in most cases.

Sampling of input signals must be controlled by dedicated hardware known as a real-time clock. The clock ticks must be routed directly to the sampling circuits and ADCs. Because sampling rates may vary widely between experiments, the real-time clock must be programmable. Similarly, various sampling devices must be programmable so that their connection to the clock can be enabled and disabled by software. Finally, some mechanism must be provided by which the clock and the sampling devices can be synchronized with the start of the experiment. Here again it is usually not practical to link the experiment with the sampling devices via software: the execution of the software instructions would introduce unacceptable time skews. Thus the real-time clock must have provision for external start-stop control. In addition, it will be convenient if the output ticks from the clock are accessible externally.

It is possible to construct an ADC analog switch, S/H circuits, and a real-time clock with only a handful of devices that can be mounted on a circuit board of very modest size. Boards of this type are available from many manufacturers for installation in popular microcomputers. The result is a compact package at very reasonable cost. Unfortunately, few, if any, microcomputers are built to provide a suitable environment for analog signal handling. The digital signals traveling inside a computer are far less susceptible to noise than analog signals. Thus one may be tempted to purchase an "analog front end" built around an ADC specified at 12, 14, or even 16 bits resolution, only to find that on installation on the backplane of a microcomputer the information content of several low-order bits is completely destroyed by noise injected from the ditigal environment. It is impractical, if not impossible, to combine high-resolution analog circuitry with fast digital systems in a single low-cost enclosure. ADC systems with resolution beyond 10 bits should preferably be mounted in a separate enclosure with their own power supplies. The digital signals to and from the ADC chassis can be routed to a computer system by means of a digital link that may be based on a dedicated cable or on one of several standard interconnection strategies (RS232, IEEE488).

For demanding, high-speed applications an analog front end can be combined with a dedicated processor that controls the storage of ADC output in local memory. Such a dedicated system can be designed with proper layout and shielding to hold the pollution of the analog signals to acceptable levels. If the dedicated system contains enough memory to hold the data from an entire experiment, those data can subsequently be transferred to another system without real-time or noise constraints.

Some experiments require the application of controllable analog signals, such as voltammetric techniques. The conversion of a digital voltage into analog form is simpler than the conversion from analog into digital form. In the digital domain, numeric values are almost invariably encoded as weighted bit strings. Accordingly, each of the bits can be made to control the presence or absence of a correspondingly weighted current in a current-summing network. For high-resolution converters it becomes difficult to combine resistors with widely varying values without compromising the temperature stability and aging characteristics of the network. Better performance can be achieved with a weighted attenuator, more commonly known as an R/2R ladder network. All resistors in this network have values of either R or 2R, so that good thermal stability can be achieved (Figure 9).

When the digital input to a DAC is changed, some time skew between the actions of the individual bit switches is unavoidable. This time skew can give rise to large output transients, especially at "major transitions" where one or more of the high-order bits change state. The transients can be damped with a lowpass filter, but the addition of a sample/hold stage at the output is far more effective. The S/H stage is switched to the HOLD state for the duration of the transient. This will drastically reduce the transient energy without significantly reducing the speed of the DAC.

Mechanical position, such as shaft rotation, can be encoded into digital form via an intermediate conversion into a voltage by means of a potentiometer. In general the conversion of any signal from one domain into another domain will cause some degradation of the signal. Shaft position can be encoded directly into digital form by means of a shaft encoder that consists of a digital mask. It is tempting to think of the mask as a set of concentric rings, with each of the rings containing opaque and transparent regions corresponding to the state of an associated bit in the binary string.

The use of binary code for the shaft encoder is appealing for its simplicity, but it offers no defense against the major source of error, namely, misalignment of the individual bit sensors. With a binary shaft

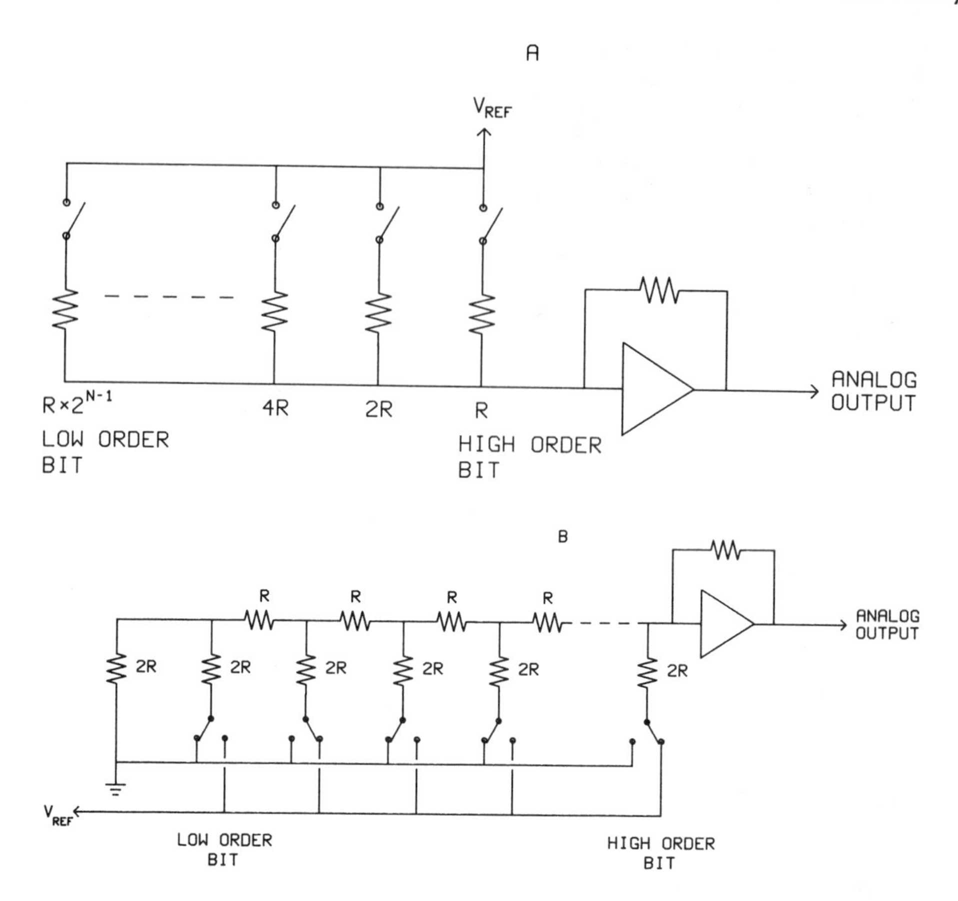

Figure 9. Resistor networks for digital-to-analog converters. (A) Weighted-resistor network. For converters beyond 8 bits it becomes difficult to maintain matched behavior for resistors of widely divergent values. (B) R-2R network. Although this circuit is more complex, it does not require a wide range of resistor values. Note that this network requires two-position switches, in contrast to the simpler on-off switches of (A).

encoder a small skew between bit sensors can give rise to a transient bit pattern that corresponds to a shaft position as much as 180 degrees removed from the actual position.

The misalignment of bit sensors can be minimized with proper mounting, but it cannot be eliminated. We should then design an encoding pattern that will minimize the consequences of sensor misalignment. That goal is met if the encoding pattern is designed so that any pair of adjacent codes differ only in the value of a single bit. Unfortunately, such a "unit-step" code cannot be a weighted code, so that it will be an impractical code for the arithmetic operations we wish to perform on it. Thus we impose one more constraint: the unit-step code must allow easy conversion into

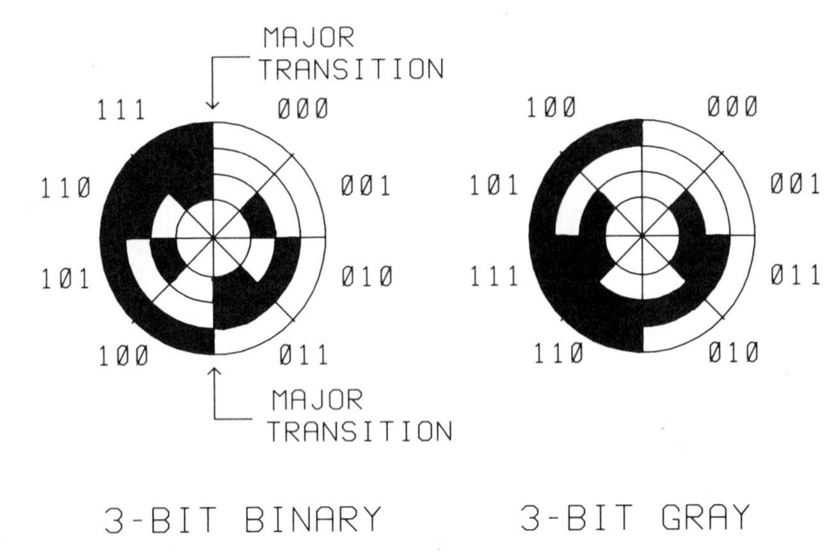

Figure 10. Binary and Gray shaft encoders. Unavoidable slight misalignment of the bit sensors can give rise to 180° position errors on the major transitions of the binary encoder. In contrast, the Gray encoder can produce at most a position error of one resolution element.

weighted code. The result is the Gray code shown in Figure 10; conversions between Gray code and binary code are logically defined in Figure 11.

In practice, the shaft encoder is often far removed from the digital part of the system, and mechanical devices are invariably slow in terms of the time scale of the digital world. Accordingly, it may be economical to transmit the bits from a shaft encoder serially via a two-wire cable, rather than broadside via a multiwire cable. Fortunately, the conversions between binary and Gray code can be performed with a minimum of hardware beyond the hardware required for the parallel-serial interconversion.

To control mechanical position with a digital system we can attach an encoder and move the mechanical device until its encoder informs the digital system that the desired position has been reached. Such feedback arrangements are quite flexible, but usually it will be necessary to add a brake mechanism that will prevent the mechanical devices from moving between control operations. Stepper motors combine the driving force with the braking function: the armature of the stepper motor is held in one of a set of preferred positions, and can be moved to the next preferred position by means of a step command that rearranges the currents applied to its set of field windings. The armature may be either a permanent magnet or an electrically excited armature driven by a fixed current. Effectively, the stepper motor is an intrinsically digital mechanical device.

In the absence of applied power the stepper motor loses its capability to maintain a fixed position. The stepper motor must therefore be initialized on power-up. The initialization may take the form of driving the stepper to a limit position and either sensing that limit or guaranteeing that it will be reached by allowing sufficient travel time.

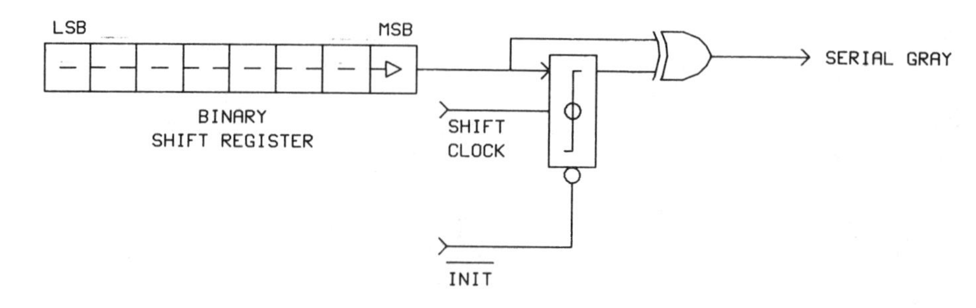

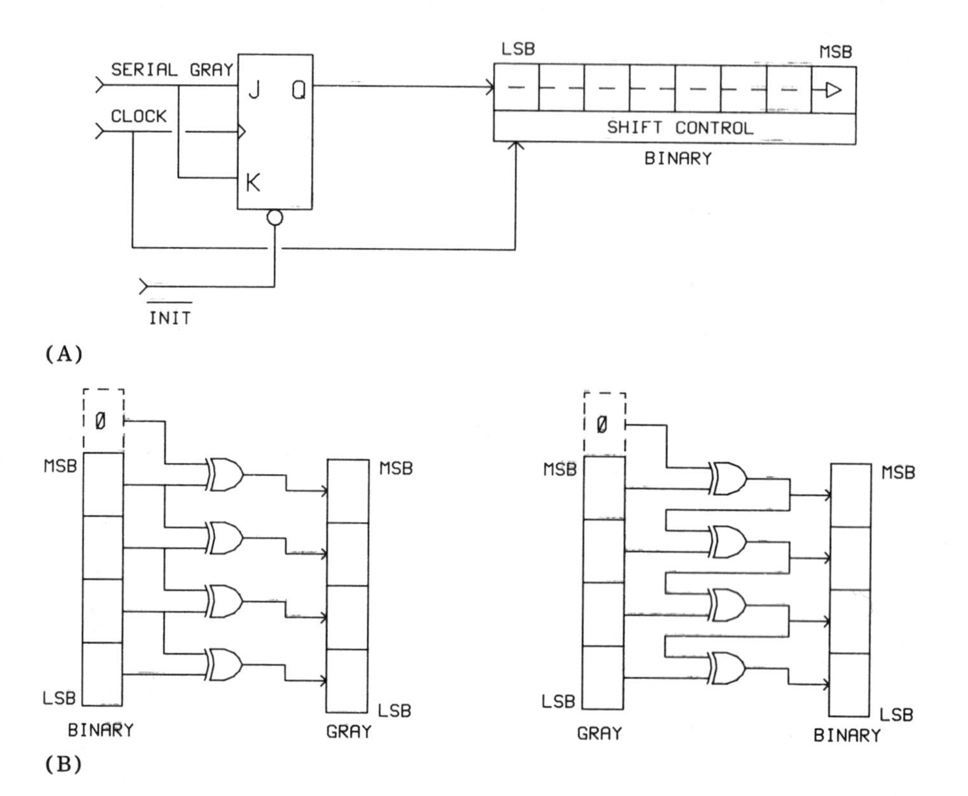

Figure 11. (A) Serial and (B) parallel conversion between Gray code and binary code. Because mechanical devices are often placed at some distance from an associated computer, the signals to and from such devices are usually transmitted serially. Then the conversion between code formats can be done serially at little additional cost.

IX. SOURCES AND REFERENCES FOR FURTHER READING

As was noted in the introduction, the use of computers in the laboratory is not new. Several books published many years ago still stand up as good reference material. Similarly, some key papers from the 1970 decade are still current. The list below does not claim to be exhaustive, but it will provide the reader with entry points for further study.

In addition to the informational literature cited, one may obtain much useful information from promotional literature available from several vendors. Among these we may list Analog Devices Inc., Burr-Brown Corporation, Digital Equipment Corporation, Hewlett-Packard Company, Keithley Instruments Inc., and Metrabyte Corporation. Inclusion in this list does not imply endorsement of a product, nor does the absence of a name imply any prejudice.

Aspnes, D. E. Digital data smoothing utilizing Chebyshev polynomials, Anal. Chem. *47*, 1181 (1975). (Smoothing based on Chebyshev polynomials is preferred over smoothing based on Fourier transform for functions with large differences between front and tail values and/or their derivatives.)

Bogner, R. E. and Constantinides, A. G., eds, *Introduction to Digital Filtering*, Wiley, New York, 1975.

Brenner, A. OSI Model Update, LAN Magazine *June*, 48 (1987). (Overview of current status of such things as ISO standard for OSI model, IEEE802, TCP/IP, XNS, NETBios, etc.)

Cane, D. High-performance microprocessors and real-time operating systems revolutionize data acquisition, Computer Design *April*, 62 (1987). (A survey of current commercial developments in the field of real-time hardware and operating systems.)

Cooley, J. W. and Tukey, J. W. An algorithm for the machine calculation of complex Fourier series, Math Comput. *19*, 297 (1965). (The key paper on the fast Fourier transform.)

Deitel, H. *An Introduction to Operating Systems*, Addison-Wesley, Reading, MA, 1984. (Deals primarily with the structure of larger multiuser systems.)

Digital Equipment Corp., Introduction to Local Area Networks, Digital Equipment Corp., Maynard, MA, 1982.

Hayes, J. W., Glover, D. E., Smith, D. E., and Overton, M. W. Some observations on digital smoothing of electroanalytical data based on the Fourier transform, Anal. Chem. *45*, 277 (1973). (Illustration of techniques to reduce the effect of front-tail discontinuities on filtering strategies based on the Fourier transform.)

KERMIT, Kermit Distribution, Columbia University Center for Computing Activities, 7th Floor, Watson Laboratory, 612 West 115th Street, New York, NY 10025. (File transfer software in the public domain for many popular systems.)

King, R. A. *The MS-DOS Handbook*, Sybex Inc., Berkeley, CA, 1985. (Contains much useful information on the internals of one of the most popular operating systems.)

LAN, The Local Area Network Magazine (monthly), 12 West 21st Street, New York, NY 10010. (A useful source of current information on networking and related areas.)

Lucas, M. S. P. The Impact of the Personal Computer on Instrumentation and Measurements, Instrumentation/Measurement Technology Conference, Boulder, Colorado, March 25-27, 1986.

Marshal, A. G. and Comisarow, M. B. Fourier and Hadamard transform methods in spectroscopy, Anal. Chem. *47*, 491A (1975). (A discussion on two strategies to collect the energy from all resolution elements of a measurement simultaneously.)

Morrison, R. *Grounding and Shielding Techniques in Instrumentation*, Wiley, New York, 1967.

Norton, P. *Programmer's Guide to the IBM PC*, Microsoft Press, Redmont, WA, 1985. (Contains much useful information on the internals of one of the most popular operating systems.)

Peatman, J. B. *Microcomputer-Based Design*, McGraw-Hill, New York, 1977.

Personal Engineering and Instrumentation News (monthly), Personal Engineering and Communications Inc., Newton Center, MA. (Deals with such things as interfaces and data acquisition/control software for personal machines.)

Rabiner, L. R. and Rader, C. M., eds., *Digital Signal Processing*, IEEE Press, Los Alamitos, CA, 1972. (A collection of papers on filter design and fast Fourier transform.)

Ramirez, R. W. *Fourier Transform Fundamentals and Concepts*, Prentice Hall, Englewood Cliffs, NJ, 1985. (Common-sense discussion of the Fourier transform, its relation to DFT and FFT, and applications in waveform analysis.)

Sargent III, M. and Shoemaker, R. L. *The IBM PC from the Inside Out*, Addison-Wesley, Reading, MA 1986 (rev. ed.). (Useful information on the hardware and system code of the IBM PC and its look-alikes.)

Savitsky, A. and Golay, M. J. E. Smoothing and differentiation of data by simplified least squares procedures, Anal. Chem. *36*, 1627 (1964).

Sheingold, D. H. *Analog-Digital Conversion Handbook*, 3rd ed., Prentice-Hall, Englewood Cliffs, NJ, 1986.

Sheingold, D. H. *Non-Linear Circuits Handbook*, 2nd ed., Analog Devices, Inc., Norwood, MA, 1976.

Taylor, J. L. *Computer-Based Data Acquisition Systems*, Instrument Society of America, 1986. (Very good survey covering such topics as error propagation analysis and error budgeting, sampling considerations, aliasing and filtering, signal transmission.)

Taylor, J. L. How to Choose Sampling Rate, Application Note 504, Neff Instrument Corporation, 700 South Myrtle Avenue, Monrovia, CA 91016.

I
THE MEASUREMENT OF MASS

2

Laboratory Balances

WALTER E. KUPPER / Mettler Instrument Corp., Hightstown, New Jersey

I. INTRODUCTION

Balances are precision weighing instruments for small loads, used primarily in professional and technical applications. The word "balance" (derived from Latin "bilanx": having two pans [1]) is sometimes used synonymously with "scale" or "scales" (old English, meaning dishes or plates). Although modern balances and scales are no longer equipped with a pair of pans, the old terms continue to be used both professionally and in everyday speech. In professional usage, a distinction is made in that "balance" is the preferred term for the more precise weighing instruments used in the laboratory, while "scales" are used in commerce, industry, transportation, health care, and household. In addition, the term "weighing machine," for any kind of balance or scale, is gaining acceptance through its use in government regulatory publications. Finally, the term "mass meter" is sometimes found in science textbooks.

Weighing is generally understood to mean the measurement of masses. However, the physical mass, which manifests itself through inertia and gravity, is seldom the direct focus of a weighing. Rather, the mass determined by weighing serves as measure for quantities of substance in commercial transactions, in the formulation of mixtures and compounds, and in chemical analyses. The numeric result indicated by a balance is conveniently taken as a direct measure of quantity. In order to obtain the true mass, the weighing result would have to be corrected for the influence of air buoyancy, but this is rarely done in chemical work.

State-of-the-art laboratory balances are fully electronic instruments that automatically equilibrate themselves within seconds and that indicate the weighing results digitally. Balances covered in this chapter range from ultramicrobalances (capacity 3 g, in 0.1-μg increments) to large precision balances (over 10 kg, in 0.1-g increments).

Most balances are still used in the traditional manner as stand-alone instruments. This is practical where one balance serves diverse and nonrepetitive weighing tasks. It becomes inefficient, however, for repetitive routine weighing procedures. Modern balances are equipped with highly developed data interfaces, making it possible to send commands to the balance as well as receive data from it. For automated processing of weight data and/or handling of the weighing samples, balances may thus be integrated into laboratory computer and robot systems.

II. MASS AND WEIGHT

We have to accept the fact that the terms mass and weight are both legitimately used to designate a quantity of matter as determined by weighing. Likewise, the instrument itself may be called either a scale or a balance. Notwithstanding the accepted usage of these terms in everyday speech, the following scientific terminology is strictly adhered to in any technical context.

Mass is an invariant measure of the quantity of matter in an object. The basic unit of mass is the kilogram, which is embodied in a standard (called kilogram prototype, a cylindrical body of platinum/iridium) at the International Bureau of Weights and Measures in Paris. In the laboratory, masses are more practically expressed in grams or milligrams, and balances generally indicate in grams.

Apparent mass is the mass of an object minus the mass of the air that the object displaces. A mass on a balance is subject to a slight lifting force due to the buoyancy of air. In accurate determinations of absolute mass, it is thus necessary to calculate and apply a correction to the weighing result. However, the difference between apparent mass and absolute (true) mass—of the order of 0.1% for most substances—is insignificant in chemical work in all but a few special applications. The chemist generally uses the term "mass" alone to mean the result of a straight (uncorrected) weighing and "apparent mass" only in contexts where the distinction is important (see Section IX).

Weight, within the context of balances and weighing, has three legitimate meanings:

1. In everyday speech, and particularly in commercial transactions, "weight" is the preferred term for a quantity of substance as determined by weighing. It is the amount in pounds, ounces, kilograms, or grams that the scale or balance indicates.
2. In physics, weight is defined as the force exerted on a body by the gravitational field of the earth and is measured in units of force (newton or dyne). The physical weight of a body varies with geographic latitude, altitude above sea level, density of the earth at the location, and to a very minute degree (about one ten-millionth of its actual value) with the time of day and the lunar and solar cycles (tides). The weight force acting on a 1-kg mass is, for example,

   9.80943 N in Paris, France
   9.80257 N in New York, NY (ground level)
   9.80144 N in New York, NY (top of World Trade Center)
   9.79524 N in Atlanta, GA
   9.79965 N in San Francisco, CA

   Electronic balances measure mass by sensing the weight force that presses an object down on the balance pan. With weight differences as large as 0.14% (as between Paris and Atlanta), it is obvious that electronic laboratory balances must be fine-calibrated on location.
3. A weight is the embodiment of a calibrated, round amount of mass in a metal weight piece. Weights are used in sets for weighing with mechanical equal-arm balances, and also to test and maintain the calibration

of self-indicating electronic and mechanical balances. While the common material for weights used to be brass (density 8.4 g/cm^3), the universal standard is now stainless steel with a density of 8.0 g/cm^3. The term "weight," with this meaning, appears to be acceptable to scientists, but "mass standard" is preferable.

For additional terms and definitions, see ref. [2].

III. CLASSIFICATION OF BALANCES

Balances are differentiated according to metrological criteria, design, and weighing principles.

In the metrological classification of weighing devices, two parameters are of primary importance, namely, the maximum load (abbreviated Max) and the graduation step or division (d) of the scale dial or digital indicator. The number of scale divisions, n = Max/d, sometimes called resolution, is useful in comparisons of weighing device performance. For example, a typical analytical balance has 2 million scale divisions (Max = 200 g, d = 0.0001 g), while a typical supermarket scale has only 3000 divisions (Max = 30 lb, d = 0.01 lb).

The diagram in Figure 1 shows how weighing devices are categorized in four accuracy classes, as recommended by the International Organization of Legal Metrology (OIML) [3] and recently adopted by the U.S. National Bureau of Standards [4]. There is a specific set of error tolerances and other criteria for each class. This structure forms the basis of uniform weights and measures regulations in most countries including the United States. However, the extent to which weighing equipment is regulated varies between countries. Laboratory balances are not subject to government Weights and Measures certification in the United States.

The category of laboratory balances, which coincides roughly with OIML classes I and II (special accuracy and high accuracy, respectively), has its own conventional nomenclature:

Name	Class	Division (d)	Max (typical)
Ultramicrobalance	I	0.0000001 g	3 g
Microbalance	I	0.000001 g	3 g
Semimicrobalance	I	0.00001 g	30 g
Macroanalytical balance	I	0.0001 g	200 g
Precision balance	II	0.001-1 g	100 g-60 kg

Although the distinction between the classes I and II is based on metrological specifications, there is also an essential design difference. All of the class I balances (analytical balances) have enclosed weighing compartments with sliding glass doors. For balances that fall into class II (precision balances) the standard design configuration has a "top-loading," stabilized weighing pan that is not enclosed in a weighing compartment.

From a technical point of view, the most important characteristic of a weighing instrument is the physical principle by which it measures weight.

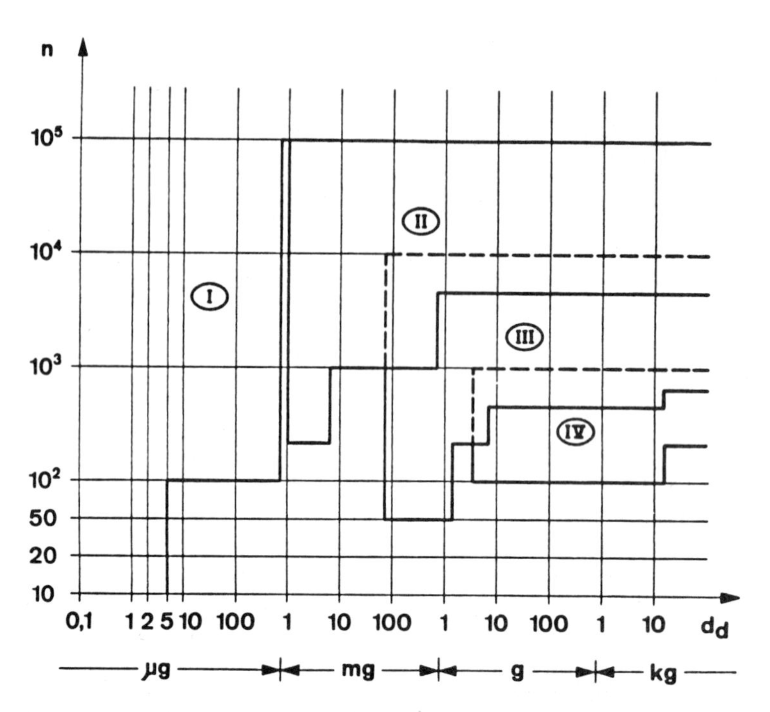

Figure 1. The OIML classification of weighing devices, based on the number n and the size d of the scale divisions. Roman numerals are used as class marks. I, Special Accuracy (e.g., analytical balances). II, High Accuracy (e.g., precision balances). III, Medium Accuracy (e.g., retail store scales). IV, Ordinary Accuracy (e.g., bathroom scales).

In this, the main dividing line is drawn between mechanical and electromechanical (commonly called "electronic") weighing devices.

The primary mechanical weighing principles are equal-arm beam balance with weight set; single-pan substitution balance; sliding poise balance; pendulous poise (angular deflection) balance; and helix spring and spiral spring balances. Many mechanical balances employ a combination of principles, such as a weight set to weigh to the nearest gram in combination with a microscopic device to determine fractions of a gram from the angular deflection of the balance beam. In addition, the same balance may incorporate a spring torque or tension device to cancel the weight of a container or other preload (tare) on the balance.

The following electromechanical (electronic) weighing principles cover close to 100% of all contemporary electronic weighing equipment: strain-gauge load cells, electromagnetic force compensation, and oscillating string sensor. The few major brands of analytical and precision balances are all based on the magnetic force compensation principle, which alone has the capability to measure small loads down to milligrams and micrograms.

IV. MECHANICAL BALANCES

In its classical form, a balance consists of a symmetric lever called a balance beam, two pans suspended from its ends, and a pivotal axis (fulcrum) at its center (Figure 2). The object whose weight is to be determined is placed on one pan, whereupon the balance is brought into equilibrium by placing the required amount of weights on the opposite pan. Thus, the weight of an object is defined as the amount represented by the calibrated standard masses that will exactly counterbalance the object on a classical equal-arm balance. Although not self-evident with modern balances and scales, the measurement of weight continues to be based on this original understanding of the term.

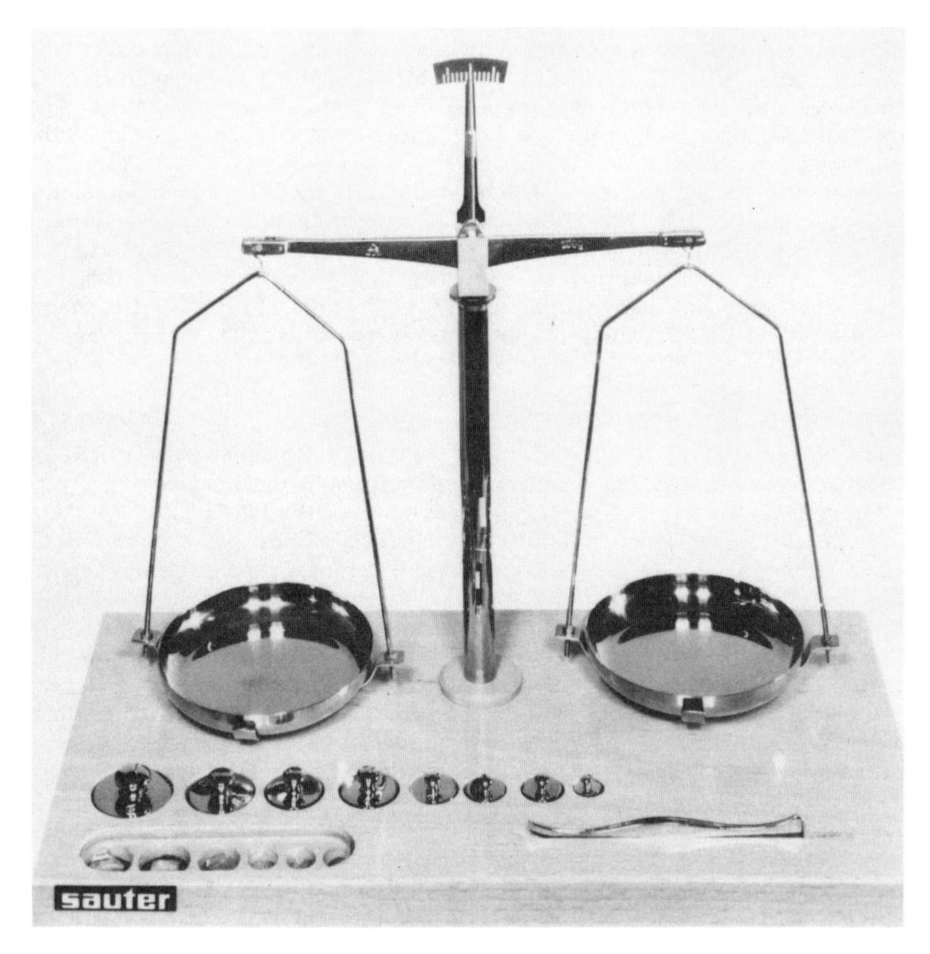

Figure 2. Traditional mechanical laboratory balance.

In the design of an equal-arm balance [5], it is critical that the pan suspension pivots be equidistant from and in a straight line with the center fulcrum. A rigid, truss-shaped construction of the beam minimizes the amount of bending when the pans are loaded. The center of gravity of the beam is located a fraction of an inch below the center fulcrum, which gives the balance the properties of a physical pendulum. With a slight difference in pan loads, the balance will come to rest at an inclined position, the angle of inclination being proportionate to the load differential (Figure 3). By reading the pointer position on a graduated angular scale, it is possible to determine differential amounts of weight, in between the discrete step values of the weight set.

In so-called "trip-balances," the beam has its center of gravity above the fulcrum. This type of balance "trips" to the side of the heavier load but will not seek nor maintain a horizontal equilibrium position with equal loads.

Notable variations and refinements of the mechanical balance include knife-edge pivots of hardened steel, agate or synthetic sapphire; air damping or eddy-current (magnetic) damping of beam oscillations; sliding weight poises or riders; built-in weight sets operated by dial knobs; microscope or microprojector reading of the angle of beam inclination; arrestment devices to disengage and protect pivots; and pan brakes to stop the swing of the balance pans.

The substitution principle was first described by the French physicist Jean Charles de Borda (1733-1799) and introduced in industrially manufactured balances by Erhard Mettler in 1946. It represented the final and conclusive step in the evolution of the mechanical balance. Substitution balances have only one pan hanger assembly, incorporating both the load pan and a set of weights on a holding rack (Figure 4). The hanger assembly

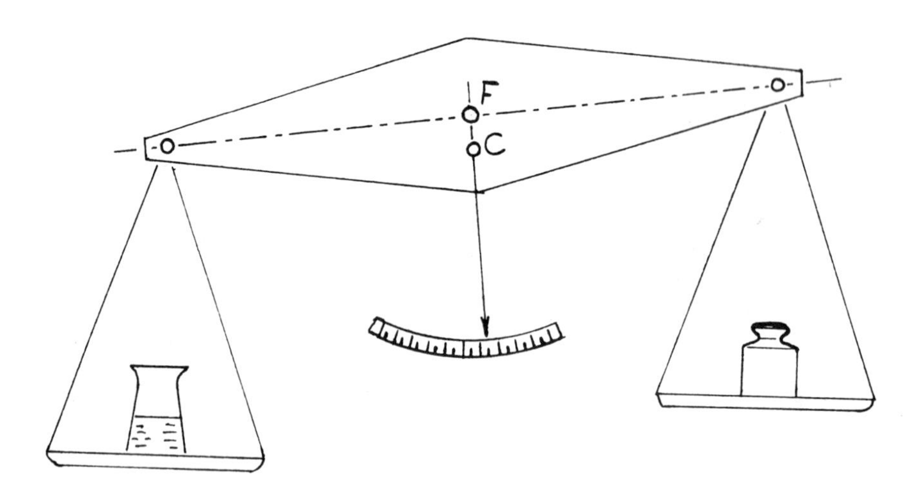

Figure 3. Determination of small weight differentials from the angle of the balance beam.

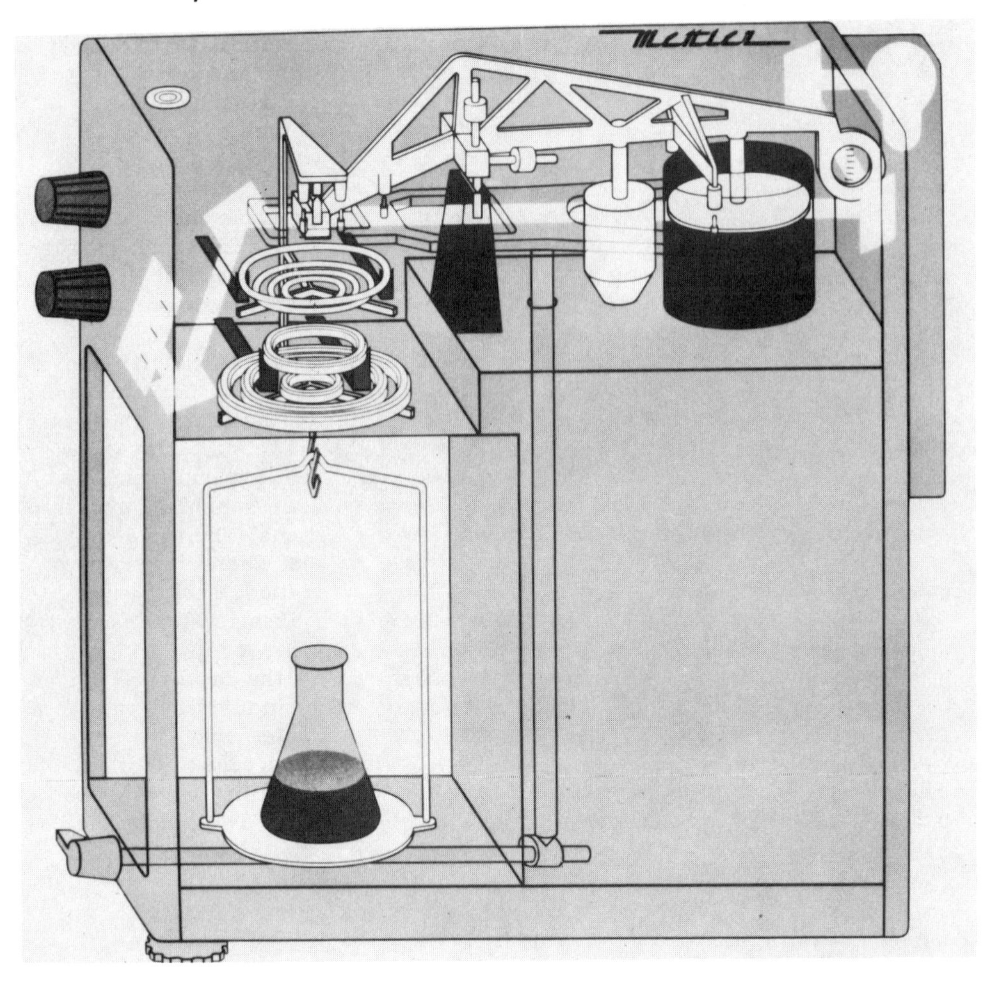

Figure 4. Cut-away side view of a substitution analytical balance.

is balanced by a counterpoise that is rigidly connected to the other end of the beam.

The weight of an object is determined by lifting weights off the holding rack until the balance returns to an equilibrium position within its angular, differential weighing range. Small increments of weight in between the discrete dial weight steps are read from the projected screen image of a graduated optical reticle, which is rigidly connected to the balance beam. With only two pivot axes, this design disposes of the problem of keeping pivots in a straight line with and equidistant from the fulcrum. Furthermore, the elastic deformation of the beam stays unchanged as the loads on the beam remain essentially constant.

Figure 5. Semimicro substitution balance, Mettler model H54AR, capacity 160 g, readable to 0.01 mg.

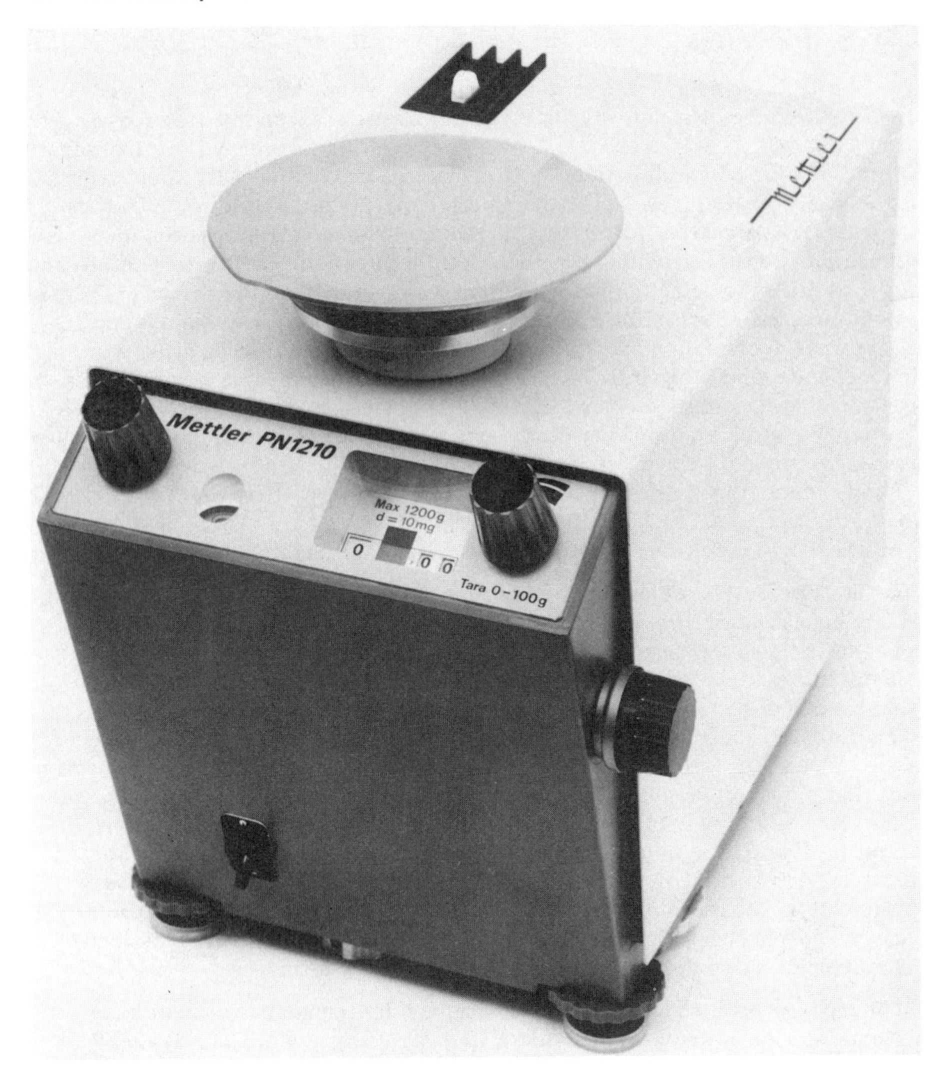

Figure 6. Mechanical precision balance, Mettler model PN 1210, capacity 1200 g, readable to 0.01 g.

Substitution balances were built as analytical balances with a hanging pan (Figure 5) and as top-loading precision balances with a stabilized pan (Figure 6). The technical development and new production of substitution balances ended several years ago, as leading balance manufacturers are now offering electronic balances of superior accuracy and operating convenience. However, substitution balances remain in widespread use in laboratories until the existing equipment is replaced by new models.

V. ELECTROMECHANICAL BALANCES

The evolution of electronic (more accurately termed "electromechanical") balances started in the late 1960s and has already extended itself over several generations of electronic technology. Among a number of technical possibilities, one operating principle, electromagnetic force compensation, emerged early as the new standard in high precision weighing. First described by K. Angstrom in 1895, the principle of electromagnetic force compensation became feasible for technical application thanks to the advancements in solid-state electronic components over the last two decades. Besides improved accuracy, reliability, and speed of operation, the main benefits from the new technology are human-engineered design for optimized interaction between operator and instrument, and numerous operating conveniences such as pushbutton zero setting, automatic calibration, built-in computing capabilities for frequently used work procedures, and data output to printers and computers.

Every electromechanical scale or balance shows a three-part structure of *load receiver*, *load transducer*, and *signal processor*. Each part serves one of the basic functions found in any weighing machine (Figure 7):

1. The weight of the object to be weighed occurs as a pressure p, which is randomly distributed over the surface of the weighing pan. The pan and the lever mechanism that supports and stabilizes it are together called the load-receiving part of the balance. Its main function is to translate the pressure p into a measurable single force F.
2. The force-measuring sensor, called load cell or transducer, produces an electrical output signal proportional to the force F. This signal may be, for example, a voltage, a current, or the frequency of an oscillation.
3. The load-cell signal is processed through analog and digital electronic circuitry, and the weighing result is displayed as a digital number, expressed in units of weight. Computations such as averaging, tare subtraction, conversion from one weight unit to another, etc. are performed on the weighing result, before the amount is displayed or transferred to peripheral equipment via a data interface.

Laboratory balances are built as compact instruments with all three functional elements in one enclosure. Modular balances with indicators separate from the load cell are used only for special applications such as vacuum balances and thermoanalytical balances.

The suspended weighing pan (Figure 8a) represents the simplest form of a load receiver. The pressure p of the load on the platform results in a single vertical force F in the pan hanger pivot. Instead of the knife edges shown, we also find torsion band pivots, especially in electronic microbalances. All other things being equal, the hanging pan design provides the highest accuracy and finest resolution of small weight differences. This traditional design continues to be used in microbalances. In order to achieve the extreme sensitivity required of these instruments, the inconvenience of a pendulous weighing pan has to be accepted.

Except for microbalances, the flexure-pivoted parallelogram linkage (Figure 8b) is the prevalent mechanism in all laboratory balances, from the semimicrobalance (typically 30 × 0.00001 g) to the precision balance range (up to 30 kg × 0.1 g). The support column of the weighing pan

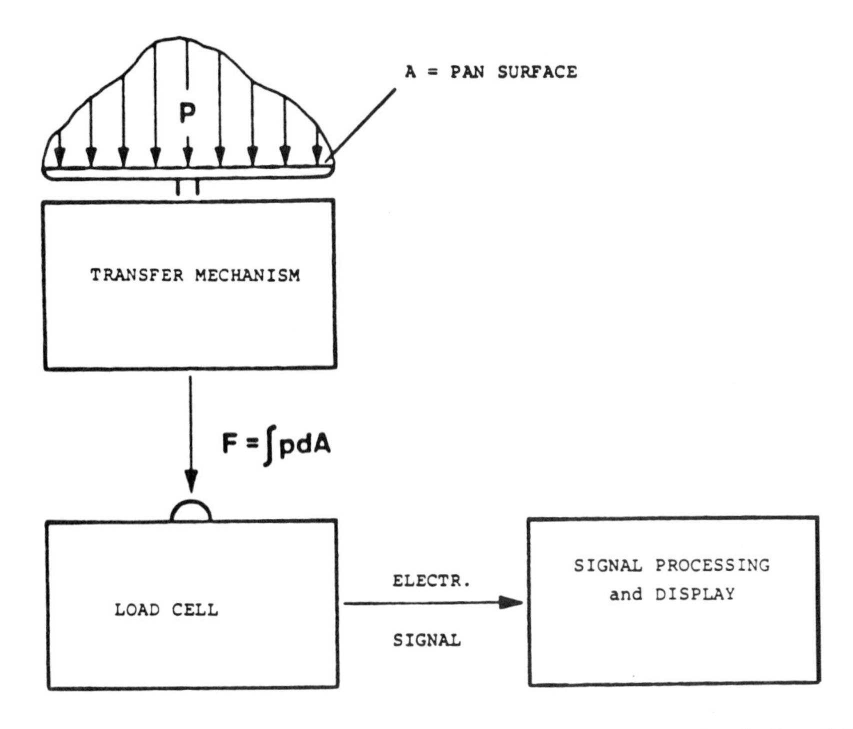

Figure 7. The three basic functions of an electromechanical weighing machine.

is guided in a vertical parallel motion by horizontal members (called checklinks) with elastic flexure pivots. The weight force is transmitted to the load cell either directly or through a reduction lever.

A basic requirement of this system is the parallelism of the checklinks, which in analytical balances needs to be accurate to a fraction of a micrometer. Nonparallel checklinks cause load shift errors, that is, variations in the weighing result when the load is moved from one place on the weighing pan to another. Load shift errors can be caused by sharp blows to the weighing pan or by severe overloads, whereby the flexure pivots may be overstressed. To prevent this kind of damage, especially the most sensitive electronic balances incorporate design features for overload and shock protection. The required degree of parallelism of the checklinks cannot be achieved through dimensional manufacturing tolerances alone. It is accomplished by means of adjustment screws, which are generally located at the point where either the top or bottom checklinks meet the structural frame of the mechanism.

The load cells in laboratory balances are based on the principle of electromagnetic force compensation [5]. This principle is capable of the finest resolution of small weights (e.g., 30 g to 0.00001 g—1 part in 3 million!) but has a practical limit of at most a few pounds of direct force measuring capacity. Where electromagnetic force compensation is used in

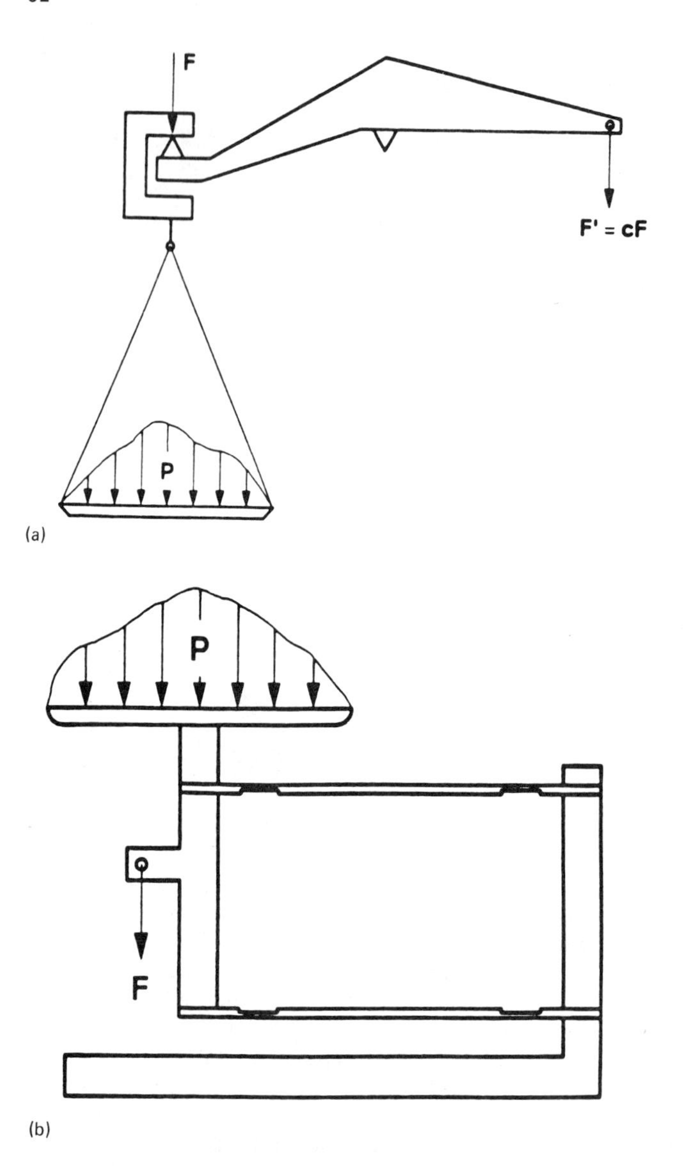

Figure 8. (a) Suspended balance pan, converts random load into a single force. (b) Parallelogram linkage stabilizes top-loading balance pan.

higher-capacity scales, up to 6000 × 0.1 kg, the weight force is brought to the load cell through multiple reduction levers.

The principle of operation is schematically illustrated in Figure 9a. The weighing load is counterbalanced (compensated) by the electromagnetic force that acts on the sensor coil as a result of the interaction between the coil current and the surrounding magnetic field. The relationship between force and current is strictly proportional according to the equation

$$F - I\ell B$$

where I represents current, ℓ stands for the total length of coil wire, and B for the magnetic flux density in the air gap.

The magnetic force compensation involves a closed-loop servo circuit. The following processes take place continuously while the balance is operating:

1. When a load is placed on the pan, the movable part of the system responds to the weight change, moving downwards by a fraction of a millimeter.
2. A photoelectric position sensor detects the downward motion of the load supporting assembly and sends an electrical signal to the servo amplifier.
3. The amplifier adjusts the amount of current flowing through the coil so as to restore the exact null position of the system.

The weight of the load on the pan is determined by a measurement of the coil current. This is possible, for example, by measuring the voltage $V = RI$ across a precision resistor wired in series with the coil.

The type of control loop employed in electromagnetic compensation sensors is in most cases a PID servo. This means that the servo response, that is, the coil current and with it the equilibrium-restoring force, consists of three components, which vary, respectively, in direct proportion, with the time integral and with the time derivative of the position sensor signal. Thanks to the integral component, the system will return to the exact null position after any transient excursion. Thus, the weighing load is counterbalanced entirely by the electromagnetic compensation force. There is no remaining deformation and, therefore, no elastic force from the flexure pivots after the balance has reached equilibrium. Also, the relative position of the coil within the magnet assembly stays exactly the same regardless of the amount of weighing load. The null-restoring property of the PID servo is essential to the high systematic accuracy of the magnetic compensation principle.

As already stated, the current flowing through the sensor coil is a direct current whose magnitude varies in proportion to the load on the balance pan. The principle of *pulse width modulation* (Figure 9b) represents a variation of the same basic idea. Square-wave direct-current pulses of constant amplitude I_0, constant frequency f, and variable time length t_p flow through the coil. I_0 is typically of the order of a few volts and f in the range of a few kilohertz. As the pulse length t_p is controlled by the servo, the effective strength of the chopped direct current

$$I_{eff} = I_0 f t_p$$

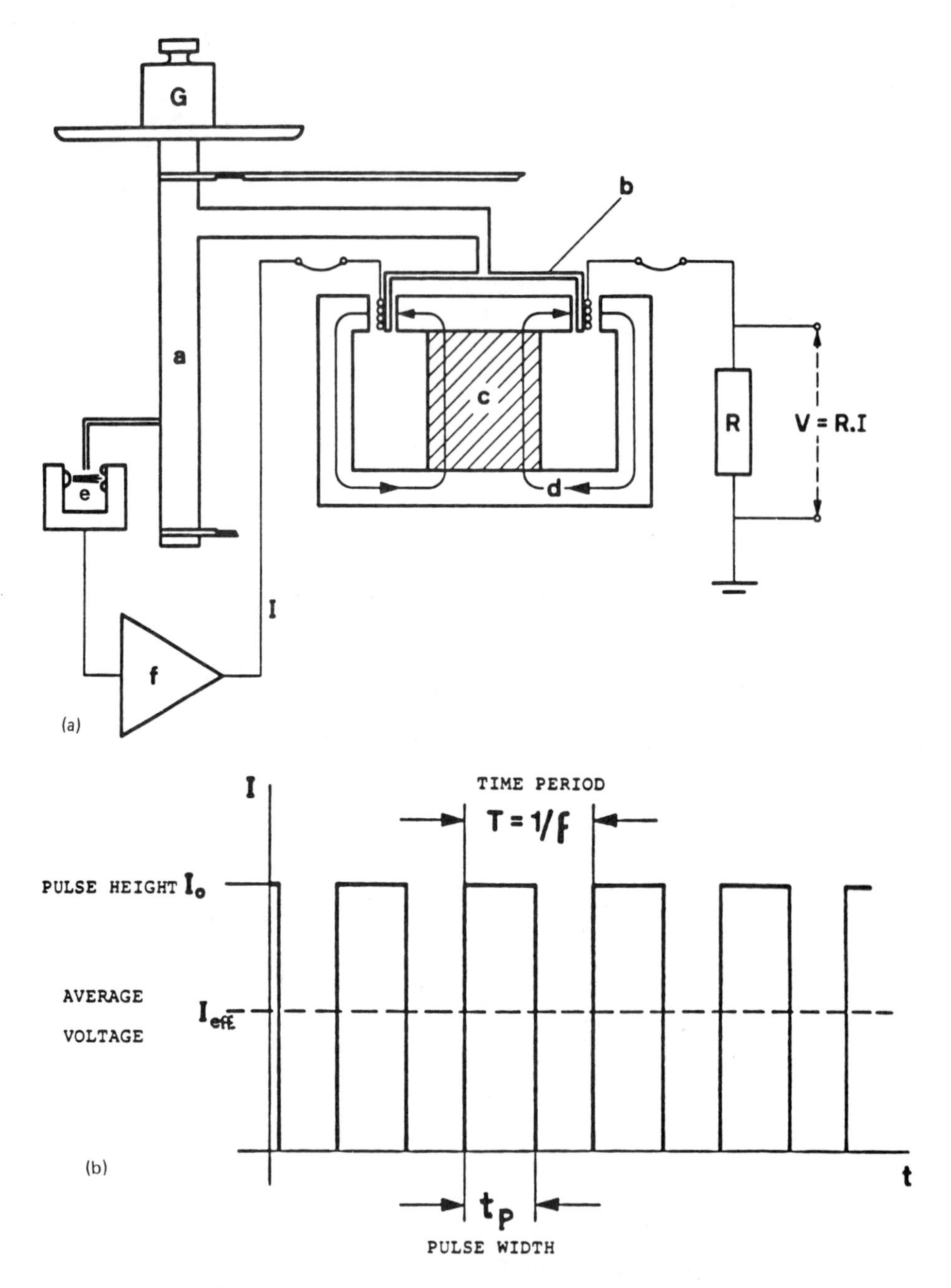

Figure 9. (a) Principle of electromagnetic force compensation: Moveable part of balance mechanism (a) with coil (b) shown in cross section. Cylindrical magnet assembly (c) shown in cross section, Alnico magnet core shaded. Magnetic field flux lines (d), photoelectric position sensor (e), servo amplifier (f). I, Coil current; R, precision resistor; V, output voltage, proportionate to weight G on pan. (b) Principle of pulse width modulation of compensating current.

is again proportionate to the weighing load. Since I_0 and f are constant parameters, the pulse length t_p is now a direct measure of the weight on the pan. In practice, the combined length of several hundred pulses is timed (with a timer clock frequency of several megahertz) in order to achieve the extremely high resolution that is characteristic of the magnetic force compensation principle.

Following is an overview of the different types of balances that are based on this principle.

Micro- and Ultramicro balances (d = 1 μg and d = 0.1 μg, respectively) are most noticeably distinguished by the extensive protection features against thermal disturbances (Figure 10). The enclosure of the Mettler model UM3 pictured has double walls and doors, and there is an additional glass shield in front. The electronic circuitry is apart from the main weighing module in separate enclosures at the front and rear of the balance. The beam fulcrum and hanger pivot are torsion bands. The electromagnetic compensation extends over a range of 150 mg in the microbalance model M3, and 15 mg in the ultramicrobalance model UM3. For larger weighing loads and tares up to the capacity of 3 g, there is a built-in set of substitution weights, operated from dial knobs.

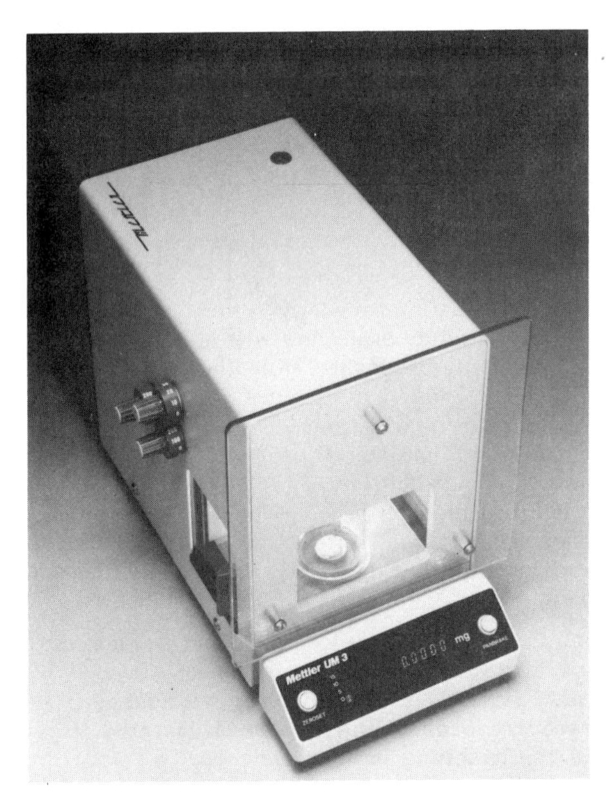

Figure 10. Ultramicrobalance Mettler UM3. Electronic weighing range 15 mg, readable to 0.1 μg; total load capacity 3000 mg, including tare.

Figure 11. Semimicrobalance Mettler AE163, shown in a typical system combination with IBM personal computer. Two electronic weighing ranges of 160 × 0.0001 and 30 × 0.00001 g.

Except for special mass comparators for metrology laboratories, micro- and ultramicrobalances are the only production models still made with a hanging pan and with built-in substitution weights. All other electronic balances have stabilized pans that are guided by flexure pivoted parallelogram mechanisms, and the magnetic force compensation is used over the entire weighing range without the additional substitution weights.

The semimicro/macroanalytical balance Mettler AE 163 (Figure 11) is a dual-range model with a semimicro range of 30 × 0.00001 g. This balance is representative in construction, appearance, and features for the entire category of macro- and semimicrobalances as they are currently being offered by several manufacturers:

Stabilized pan, accessible through three sliding doors, on both sides and on top of the weighing compartment.

Sensor, electronic circuitry, and parallelogram mechanism are housed in the rear compartment. Readout and operating key(s) are located in a slanted panel in front of the weighing compartment

The built-in calibration weight is a standard feature in all new analytical balances. This allows an instant recalibration of the balance whereby the calibration constant is calculated by the microprocessor and stored in nonvolatile memory until overwritten by the next calibration.

Top-loading precision balances with division sizes of 0.001, 0.01, and 0.1 g represent the most important category of laboratory balances, based on the number of instruments in use. The Mettler model PM200 (Figure 12) is typical in its overall design configuration for most products in this category:

Low profile, with open, unobstructed pan. The more sensitive models (d = 0.001 and 0.01 g) may be equipped with a removable draft shield if necessary.
Operating key(s) and indicator window in front.
Pan stabilized by flexure pivoted parallelogram mechanism. Magnetic compensation acts through a force reduction lever.

The basic design elements of electronic analytical and precision balances—sensor, pan guiding mechanism, overall configuration—are the result of many cycles of successive refinements and have reached a high level of perfection. Since there seems little left to improve in the physical makeup of laboratory balances, the focus for further development has shifted toward providing functional enhancements through the capabilities of microprocessors and nonvolatile memory components:

Digital filtering to achieve stable weighing results even in the presence of vibrations and air drafts.
Automatic calibration cycle. The operator is prompted by the balance display to place a calibration weight on the pan, then remove it at a second prompt. The calibration constant remains stored even while the balance is disconnected from line power.
Display in alternate or additional units besides grams. This is important in the precious metals and jewelry trades, where troy ounce, pennyweight, and carat are used.
Piece count and percentage calculation.
Auxiliary analog dial display for over/under indication.

In the balance shown in Figure 12, these functions are activated and controlled through the single key at the front of the instrument. Commands are entered into the balance through a system whereby the key is either held depressed or struck briefly. Additional and more complex functions such as weight statistics are available through separate keyboard modules (as illustrated), which are described in Section VI.

Industrial precision scales are OIML Class II weighing instruments with higher capacity ranges from about 50 kg up to several tons. As is typical for all higher-capacity scales, the indicator is separate from the weighing platform (Figure 13). The weighing force is transmitted to the sensor by means of a platform lever mechanism. Regardless of the actual scale capacity, the electromagnetically generated compensation force in the sensor coil is within a range of a few hundred grams, as the weight force on the platform is reduced up to several thousand times by multiple levers.

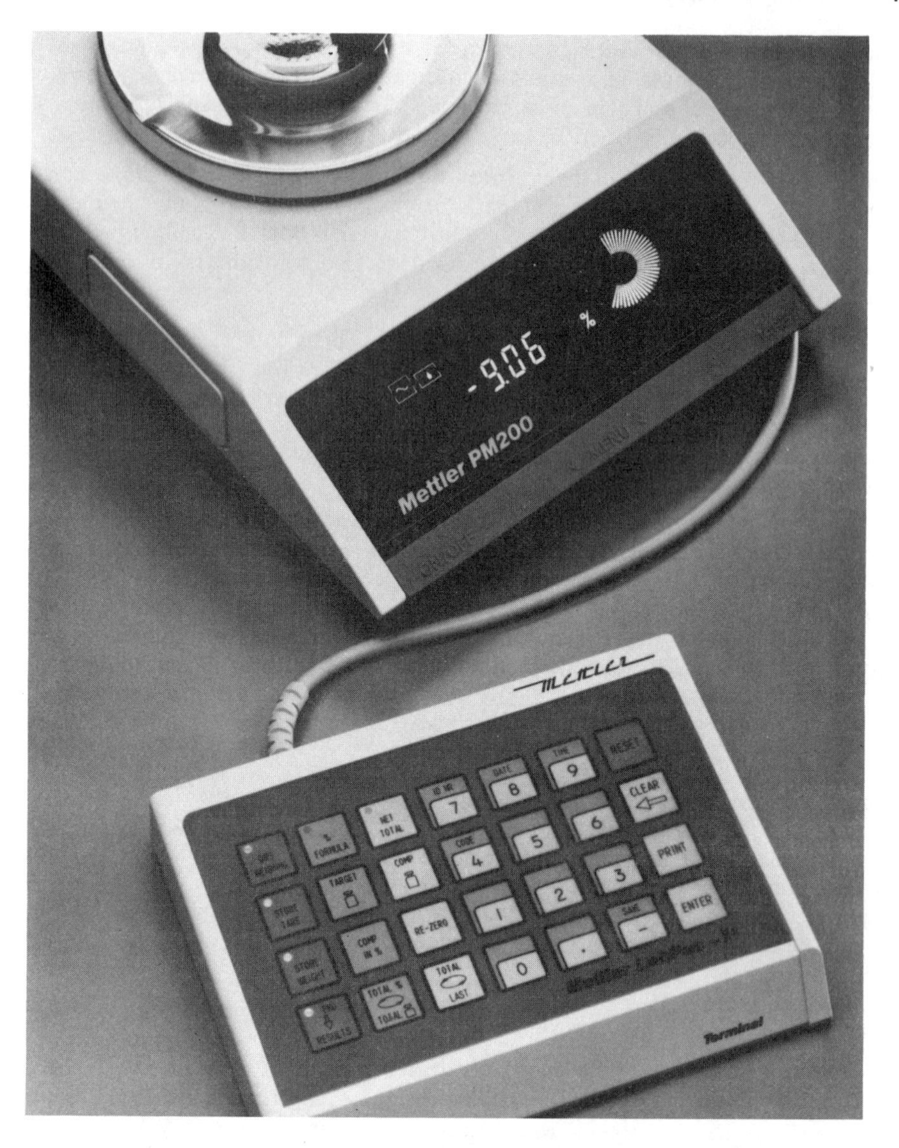

Figure 12. Precision balance Mettler PM200 working as a small system with a LabPac keyboard and a printer. Weighing range 200 × 0.001 g.

Figure 13. Industrial precision scale Sauter E1210 with a range of 120 kg × 1 g.

VI. WEIGHING SYSTEMS

Practically all electronic laboratory balances have data interfaces, either as a standard feature or an add-on option. Common standards are RS232 and 20-mA current loop (CL). This reflects a growing trend of using balances in direct connection with other equipment such as:

1. Balance printers to record weighing results instantly and reliably without transcription errors.
2. Application devices, that is, simple calculator terminals that are designed to perform some of the most common computations: weight/percent conversion, mean value and standard deviation from serial weighings; weight/piece count conversion; net total of component weights in the formulation of mixtures; division by a density value for weight to volume conversion.
3. Personal computers for flexible, user-programmed solutions to virtually any weighing procedure.
4. Larger centralized laboratory computers for structured, controlled processing of measurement data from a multitude of instruments.
5. Laboratory robots for automated mechanical handling of samples, performing weighings and other procedures without an operator.

These five types of systems differ substantially in the applications they serve and in the ways in which they are implemented.

System types 1 and 2 (printers and application devices) offer ready-made solutions from standard modules. Balances and peripherals are compatible within the same product line. They require no programming nor custom cabling. Figure 12 shows a typical example. A Mettler precision balance PM200 together with a LabPac-M keyboard performs formulation of compounds by weight percentage and other functions. In addition, a printer may be connected to the balance to document the weighing process on calculator tape.

Systems of the third type, balances connected to personal computers, are implemented in several different ways:

Epson HX-20 briefcase computers, complete with software packages and interface cables, are available from several balance manufacturers and dealers. This type of turnkey system has become very successful, although the selection of ready-made software is limited to a few of the most important applications.

Those who are familiar with personal computers and with programming in BASIC should find little difficulty in connecting a balance to a computer and writing a software program for a given application. Major balance manufacturers provide appropriate cables, interfaces, and instructions to assure the compatibility of balances with at least the most important microcomputer models (Figure 11). However, in planning a weighing system, the balance user is strongly advised to inquire beforehand whether the intended systems configuration has been tested and is recommended by the balance manufacturer.

As an alternative to writing an applications program in BASIC, there are software packages for data acquisition and processing in the laboratory with the flexibility to be adapted to most applications. For example,

C L A S S   D I S T R I B U T I O N   O F   W E I G H T S

File Name: DEMO
02/06/87 13:50
Nominal N: 0.730 gram
Class Width: 0.005 gram
Sample Size: 40

Nr	Net.Wt. W	Wt.Dev. W - N	Class, Relative to Nominal -5	-4	-3	-2	-1	+1	+2	+3	+4	+5
1	0.702	-0.028	+	.	.	.	.	.	.	.	.	.
2	0.705	-0.025	+	.	.	.	.	.	.	.	.	.
3	0.711	-0.019	.	+	.	.	.	.	.	.	.	.
4	0.713	-0.017	.	+	.	.	.	.	.	.	.	.
5	0.714	-0.016	.	+	.	.	.	.	.	.	.	.
6	0.717	-0.013	.	.	+	.	.	.	.	.	.	.
7	0.718	-0.012	.	.	+	.	.	.	.	.	.	.
8	0.719	-0.011	.	.	+	.	.	.	.	.	.	.
9	0.721	-0.009	.	.	.	+	.	.	.	.	.	.
10	0.722	-0.008	.	.	.	+	.	.	.	.	.	.
11	0.722	-0.008	.	.	.	+	.	.	.	.	.	.
12	0.722	-0.008	.	.	.	+	.	.	.	.	.	.
13	0.722	-0.008	.	.	.	+	.	.	.	.	.	.
14	0.722	-0.008	.	.	.	+	.	.	.	.	.	.
15	0.722	-0.008	.	.	.	+	.	.	.	.	.	.
16	0.722	-0.008	.	.	.	+	.	.	.	.	.	.
17	0.723	-0.007	.	.	.	+	.	.	.	.	.	.
18	0.726	-0.004	.	.	.	.	+	.	.	.	.	.
19	0.727	-0.003	.	.	.	.	+	.	.	.	.	.
20	0.727	-0.003	.	.	.	.	+	.	.	.	.	.
21	0.728	-0.002	.	.	.	.	+	.	.	.	.	.
22	0.729	-0.001	.	.	.	.	+	.	.	.	.	.
23	0.730	0.000	.	.	.	.	.	+	.	.	.	.
24	0.730	0.000	.	.	.	.	.	+	.	.	.	.
25	0.731	0.001	.	.	.	.	.	+	.	.	.	.
26	0.732	0.002	.	.	.	.	.	+	.	.	.	.
27	0.732	0.002	.	.	.	.	.	+	.	.	.	.
28	0.732	0.002	.	.	.	.	.	+	.	.	.	.
29	0.733	0.003	.	.	.	.	.	+	.	.	.	.
30	0.735	0.005	.	.	.	.	.	+	.	.	.	.
31	0.737	0.007	.	.	.	.	.	.	+	.	.	.
32	0.737	0.007	.	.	.	.	.	.	+	.	.	.
33	0.738	0.008	.	.	.	.	.	.	+	.	.	.
34	0.738	0.008	.	.	.	.	.	.	+	.	.	.
35	0.740	0.010	.	.	.	.	.	.	.	+	.	.
36	0.741	0.011	.	.	.	.	.	.	.	+	.	.
37	0.741	0.011	.	.	.	.	.	.	.	+	.	.
38	0.741	0.011	.	.	.	.	.	.	.	+	.	.
39	0.746	0.016	.	.	.	.	.	.	.	.	+	.
40	0.756	0.026	.	.	.	.	.	.	.	.	.	+
Count	40		2	3	3	9	5	8	4	4	1	1
Mean	0.728	-0.002										
S.D.	0.011	0.011										
High	0.756	0.026										
Low	0.702	-0.028										

Figure 14. Laboratory test report produced with spreadsheet software Lotus 1-2-3 and data acquisition program Lotus Measure. Report format is easily tailored to the application by the user.

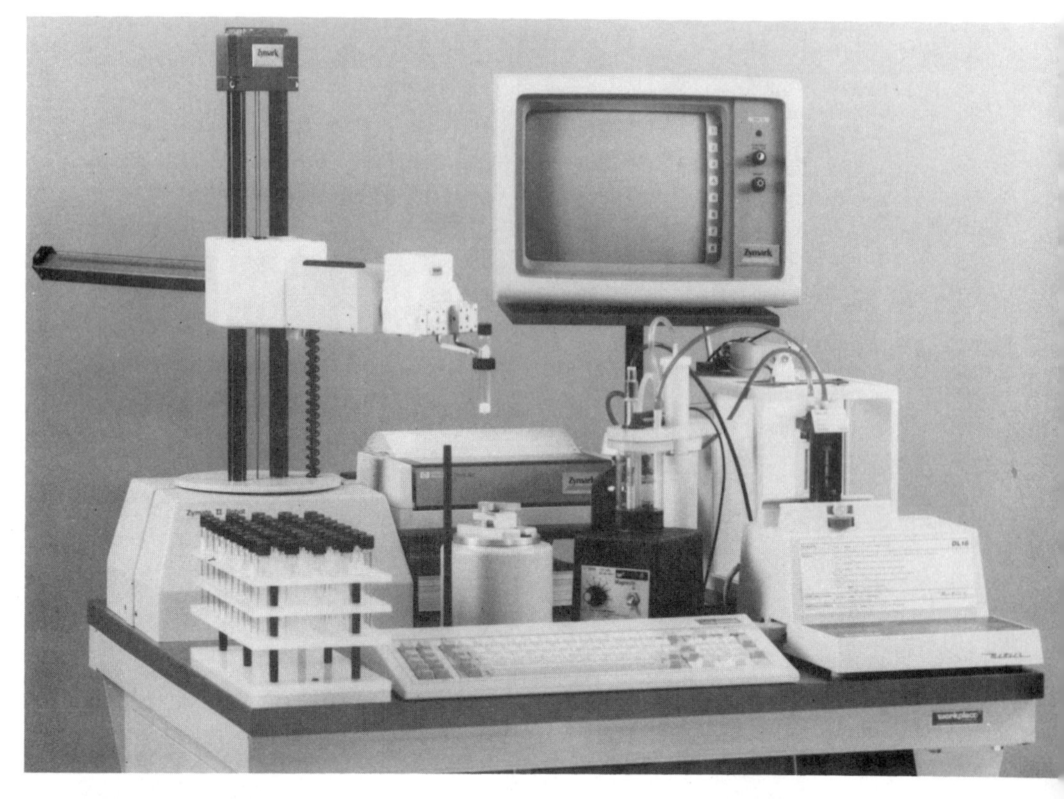

Figure 15. Automated titration system includes an analytical balance. Titration of an entire batch of samples is performed by robot. Robot manipulations on balance include loading and removing sample, opening and closing sliding door.

Lotus Measure, a new accessory program to Lotus 1-2-3, allows one to transfer data directly from a balance (or any other instrument with a data interface) into a Lotus spreadsheet (Figure 14).

Balances may also be connected to large centralized computers or integrated in laboratory robot systems (Figure 15), primarily for high-volume standardized laboratory, quality control, and production weighing procedures. Such systems are generally planned and installed by data processing and robot system specialists.

VII. BALANCES FOR SPECIAL APPLICATIONS

In a number of experimental processes, weighings are performed under special environmental conditions, for example, under vacuum, under high or low temperature, in special gas atmospheres, in magnetic or electrostatic fields, or with the sample immersed in a fluid. Dedicated systems and peripheral instruments are available for applications where a special environment around the weighing sample has to be maintained.

Appropriately designed weighing chambers are equipped with controlled heating or cooling units, vacuum pumps, or gas inlets, depending on the application. Experimental parameters such as pressure, temperature, atmospheric composition, and in some cases their rates of change are maintained with appropriate control instruments. The sample weight is measured by the balance as a function of the environmental parameters or as a function of time. The Mettler Thermoanalyzer in Figure 16 is a typical example of this type of equipment.

Weighings under special environmental conditions are generally done with samples ranging from milligrams to a few grams. The weight sensor is usually an electronic microbalance. Special problems arise from the fact

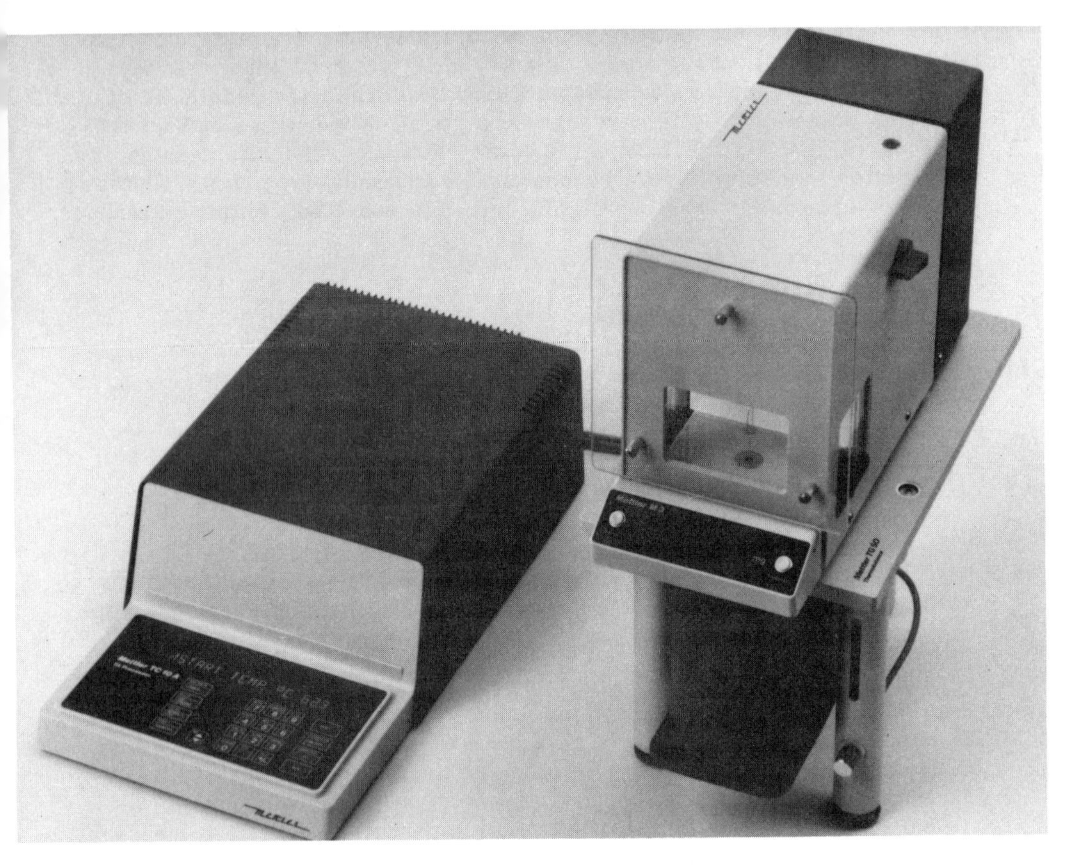

Figure 16. Thermogravimetric analysis system Mettler TA 3500, consisting of a microanalytical balance, furnace, and control instrument. Electronic weighing range 150 mg × 1 μg. Total load capacity 3000 mg, including tare. Sample weight changes are observed either at constant temperature or at controlled heating/cooling rate in the range between room temperature and 1000°C.

that the sample carrier is mechanically suspended from the balance beam into the weighing compartment. As a consequence, in the case of a vacuum balance, the entire balance assembly will have to be placed inside the vacuum compartment. For weighing under high and low temperatures and in gas atmospheres, there are special measures to protect the balance system. The compartment containing the balance may be separated from the sample compartment by baffles and labyrinth passages in order to keep the gas atmosphere and/or extreme temperature environment away from the sensitive microbalance mechanism. In addition, the balance compartment may be purged with an inert gas and the mechanism protected by a corrosion-resistant finish such as gold plating.

VIII. EXTRANEOUS FACTORS IN THE WEIGHING PROCESS

It is important to note that weighing accuracy is never mentioned in the technical specification sheet of a balance. The reason is that accuracy (i.e., the agreement between the weighing result and the true mass of an object) depends to a larger degree on operator technique and on the environment than on the balance. In most cases, a balance is capable of much better than the ±0.1% accuracy that is normally expected. Following are a few common sources of weighing errors caused by factors extraneous to the balance:

Buoyancy. This error is due to the buoyant force which the surrounding air exerts on the object being weighed. As it is relatively small, of the order of 0.1% of the weight of the object, this error is commonly disregarded except for high-accuracy mass determinations.

Changes in moisture or CO_2 content, or volatile samples. Some materials take up H_2O or CO_2 from the air during the weighing process, or they may be volatile and lose weight even at room temperature. For meaningful results such materials must be handled and weighed in closed containers.

Electrostatic forces. An object carrying a charge of static electricity is attracted to various parts of the balance, leading to erratic, nonrepeatable weighing results. This problem should be solved by keeping a normal humidity level in the laboratory.

Sample temperature. If an object is warm relative to the balance, convection currents cause the pan to be buoyed up. Weighing results will be unstable and nonrepeatable. Weighings should be carried out at room temperature, if possible.

Room climate. As a general rule, the finer the readability of a balance, the more critical is the choice of its location. Especially analytical balances should be kept away from air currents, such as doors, windows, and heat and air conditioning outlets. Exposure to strong radiant heat sources such as direct sunlight, ovens, and baseboard heaters should also be avoided.

Vibrations. Balances can be affected by vibrations, for example, from rotating machinery, elevators, or slamming doors. However, electronic balances are far less sensitive to vibrations than the mechanical balances used to be. It is seldom necessary to place an electronic balance on a stone table.

Radiofrequency and electromagnetic interference. The balance should be away from and not share the same powerline circuit with equipment that causes such interference, for example, through arcs or sparks in electric motors. A powerline filter may sometimes correct such a condition.

Except for buoyancy errors, which are systematic and can be mathematically corrected, extraneous weighing errors manifest themselves mostly through fluctuating, nonrepeatable balance readings. Solutions to such problems often require a diligent investigation into the nature and source of the errors.

IX. AIR BUOYANCY

The apparent mass (the weight as read directly from the balance) of a body with the same density as water is about 0.1% smaller than its true mass. The reason for this is that balances are calibrated with standard masses that have a density of 8.0 g/cm^3. Thus, a 1-kg mass standard displaces only 125 cm^3 of air volume, while 1 kg of water displaces 1000 cm^3. Consequently, the kilogram of water is subjected to a greater amount of air buoyancy and therefore appears lighter than the steel weight. A balance indicates the exact mass of an object only in the case where the object being weighed has the same density as the mass standard by which the balance was calibrated, and if the air density is the same as at the time of the calibration.

The following formula gives the correction that needs to be added to a weighing result in order to obtain the true mass of an object. The correction k is expressed as a fraction of sample weight: that is, if a sample weighs 35 g and the value for k from the formula is 0.00095, then the true mass of the sample is 35 × 1.00095 = 35.03325 g. Thus,

$$k = (1/d_2 - 1/d_1)/(1/d - 1/d_2)$$

where d_1 is the density of calibration standards (g/cm^3), d_2 is the density of the object to be weighed, (g/cm^3), and d is the air density in g/cm^3, depending on barometric pressure B (mm Hg) and temperature T (°C), as calculated from this formula for dry air (see air density tables [6] for moist air):

$$d = 0.001293(B/760)[273/(273+T)]$$

The following examples show the magnitude of the correction that would have to be added to the weights of objects within the range of densities encountered in most substances. Air density is assumed to be 0.0012 g/cm^3, and the density d_1 of the calibration standard is 8 g/cm^3.

d_2 =	0.5	1.0	2.0	5.0	8.0	20.0
k =	0.00226	0.00105	0.00045	0.00009	0	-0.00009

The correction for air buoyancy and the determination of true mass are extensively discussed in Ref. 7.

X. BALANCE SPECIFICATIONS AND WEIGHING ERRORS

The basic requirements of a balance are that the displayed weight be stable, repeatable and linear. As a general rule, weighing errors should not exceed the order of magnitude of the least significant decade of the display. Errors are usually a composite of several factors. The tolerances for them are given in the manufacturer's specifications under the following designations:

Reproducibility, also called repeatability or precision, is the standard deviation of the weighing results when making a series of weighings of the same test weight. The display is set to zero before each individual weighing.

Nonlinearity is the term used for errors at intermediate points in the weighing range in a balance that indicates correctly at no load and capacity load.

Temperature-related errors occur as the zero point and the calibration of a balance are affected by temperature changes. The tolerance values on temperature errors should be used as a reference for the degree of temperature control required for accurate weighing. Normal indoor conditions are in most cases sufficient. Since the balance electronics produce heat (10-20 W), it is recommended that balances be left constantly under power, in order to avoid warm-up drift at the beginning of every workphase. This is the reason why the on-off switch in many balances only turns off the display but keeps the power on for the rest of the system.

Error of the built-in calibration weight is found in balances that have this feature. In an analytical balance, the factory tolerance is of the order of 0.2 mg in a calibration weight of 100 g. This very small uncertainty is of no consequence in practical applications. However, the built-in weight may change over time from surface contamination. It should therefore be checked occasionally against a reliable mass standard.

Load shift errors are deviations of the weighing result when a test weight is first placed in the middle and then in any other position on the weighing pan. This kind of error is usually small enough to be unnoticeable as long as weighing loads are not deliberately placed off center.

The *accuracy* of a balance may be estimated by compounding all of these errors into an overall error band. For the purposes of legal metrology, test specifications and tolerances are found in ref. 3, which also contains the criteria by which a weighing instrument qualifies for one of the four OIML classes of accuracy.

A very practical approach to the question of weighing accuracy is found in the *United States Pharmacopeia* (U.S.P.) [8], where "accurate weighing" is defined as being within 0.1% of the true value (disregarding air buoyancy). To achieve this degree of accuracy, the U.S.P. recommends that the minimum sample weight be at least equal to 3000 times the standard deviation of the balance. While accuracy requirements may be more or less stringent in different fields, such a statement of the minimum load to be weighed on each balance is generally the best way of controlling the accuracy of weighings in a laboratory.

REFERENCES

1. B. Kisch, *Scales and Weights: A Historical Outline*, Yale University Press, New Haven, CT, 1965.

2. L. Bietry and M. Kochsiek, *Dictionary of Weighing Terms*, Mettler Instrumente AG, Greifensee, Switzerland.
3. Organisation Internationale de Metrologie Legale (OIML), *Recommandation Internationale No. 3, Reglementation Metrologique des Instruments de Pesage a Fonctionnement Non Automatique*, Bureau International de Metrologie Legale, 11 Rue Turgot, Paris IXe, France.
4. National Bureau of Standards, *NBS Handbook 44 (1987), Specifications, Tolerances, and Other Technical Requirements for Weighing and Measuring Devices*, U.S. Government Printing Office, Washington, D.C., 1987.
5. M. Kochsiek (ed.), *Handbuch des Waegens*, Vieweg, Braunschweig (West Germany), 1985.
6. *Handbook of Chemistry and Physics*, any edition/year, The Chemical Rubber Publishing Company, Cleveland, Ohio.
7. P. E. Pontius, *Mass and Mass Values*, National Bureau of Standards Monograph 133, U.S. Government Printing Office, Washington, D.C., 1974.
8. *The United States Pharmacopeia*, 21st ed., United States Pharmacopeial Convention, Inc., Rockville, MD, 1985.

3

Organic Elemental Analysis

T. S. MA* / Department of Chemistry, Brooklyn College of The City University of New York, Brooklyn, New York

I. GENERAL REMARKS

Organic elemental analysis is concerned with the determination of one or more chemical elements in an organic sample. The sample may consist of a single carbon compound [1], or may contain a mixture of chemical species [2]. In the case of pure carbon compounds, it is apparent that any element therein will be a significant part of the whole sample. Contrastingly, organic mixtures may possess some elements in large proportions (the major constituents), with other elements in small amounts (the minor or trace constituents). In this chapter, we discuss the instruments that are used to determine the elements in pure compounds or the major elemental constituents of organic mixtures. Whereas some of these instruments can also be employed to determine trace constituents [3], modification of the apparatus and/or the experimental procedure is often necessary.

Generally speaking, the methods for organic elemental analysis involve the transformation of carbon compounds into simple gaseous products and/or ionic species soluble in aqueous solutions. The instruments for such purposes must fulfill two requirements: (a) complete decomposition of the organic molecules irrespective of their chemical structures and physical consistencies, and (b) quantitative collection of the expected reaction products to be measured accurately by suitable means. When more than one kind of instrument is available, the choice will depend on a number of factors, such as the initial investment, maintenance cost, material consumption, availability of reagents, frequency of determinations, operational skill required, and so on.

At the present time, there are methods to determine any element that may form a part of an organic compound [1]. In the following sections, however, only those elements that are commonly determined will be mentioned.

II. CHN ANALYZERS

Practically speaking, all organic materials contain carbon and hydrogen. Apparatus for the determination of carbon and hydrogen only, however, is no longer available on the market. The instruments built for carbon

*Retired.

and hydrogen analysis now always include the determination of nitrogen; hence they are known as CHN analyzers.

In the CHN analyzer, organic compounds are oxidized at high temperatures to yield carbon dioxide, water, and oxides of nitrogen:

$$\text{CHN compound} \xrightarrow{O_2,\ 1000°C} CO_2 + H_2O + \text{N oxides}$$

The oxides of nitrogen are then converted to nitrogen gas in the presence of metallic copper:

$$\text{N oxides} \xrightarrow{\text{Cu},\ 600°C} N_2$$

Subsequently, CO_2, H_2O, and N_2 are quantitatively separated and individually measured. Other elements such as sulfur and halogens that may also be present in the organic material will produce gases that interfere with the determination. Therefore a suitable device is incorporated in the CHN analyzer to remove these interfering gases.

Commercial CHN analyzers may be classified in two categories. In one category, CO_2, H_2O, and N_2 are separated by gas chromatography, while in the other these three reaction products are separated by selective adsorption.

A. Instruments Based on Gas Chromatographic Separation

In the Perkin-Elmer PE 2400 CHN Elemental Analyzer [4], which was introducd in 1987, the organic sample is burned in a pure oxygen environment. The gaseous products are controlled to exact conditions of pressure, temperature, and volume in a gas mixing chamber, followed by separation of the bases using a high-sensitivity frontal chromatography technique. The N_2, CO_2, and H_2O are then allowed to depressurize through a column and are measured in a thermal conductivity cell. The schematic flow diagram of this system is shown in Figure 1. The combustion train has the same basic design as in the Perkin-Elmer model PE 240C (see Figure 2). Helium is normally used as the carrier gas. When helium is not readily available, it can be replaced by argon. The maximum analytical range for carbon is 3.6 mg; hydrogen, 1.0 mg; and nitrogen, 6.0 mg. The accuracy for all three elements is ±0.3% absolute using helium, and ±0.4% absolute using argon as the carrier gas. The PE 2400 CHN Elemental Analyzer is a self-contained instrument, being connected to a microprocessor that controls all system functions and calculations. The complete assembly is shown in Figure 3.

The Carlo Erba Elemental Analyzer 1106 (Figure 4) has two combustion chambers [5], one of which is for CHN determination while the other is for oxygen or sulfur determination. In the CHN chamber the organic substance is decomposed by flash combustion in an atmosphere enriched with oxygen. The resulting combustion gases then pass through a reduction furnace and are swept into the chromatographic column by helium. The N_2, CO_2, and H_2O are separated and recorded sequentially by the thermal conductivity detector. This instrument can be connected to an automatic sampler, which holds up to 196 samples.

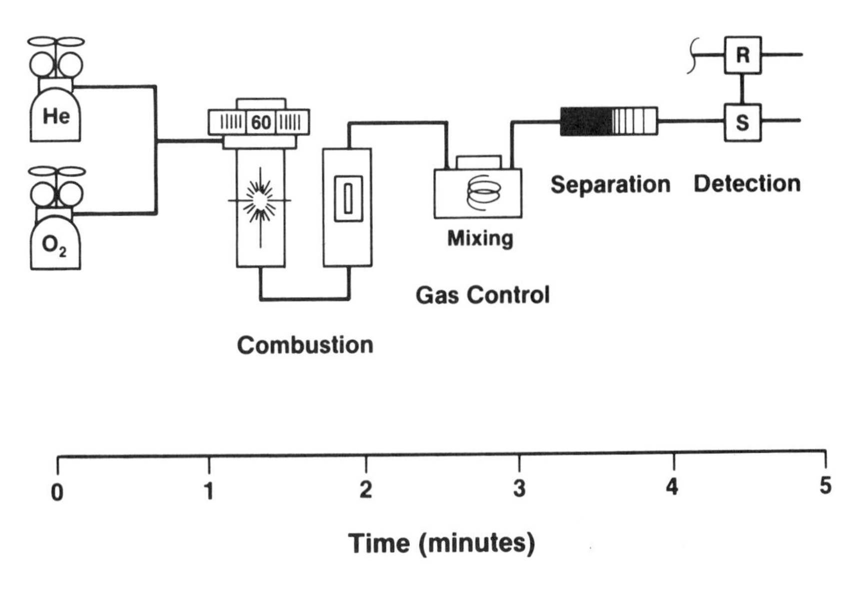

Figure 1. Schematic flow diagram of PE 2400 CHN Elemental Analyzer. (Courtesy of The Perkin-Elmer Corporation, Norwalk, CT.)

B. Instruments Based on Selective Adsorption

Simultaneous determination of carbon, hydrogen, and nitrogen by the Heraeus Elemental Analyzer CHN-O-Rapid (Figure 5) utilizes the Pregl combustion method to produce CO_2, H_2O, and N_2, followed by the separation of these three products by adsorption and desorption processes [6]. The functional diagram of this instrument is shown in Figure 6. The sample, sealed in a tin capsule, is injected from the sample feeder into the oxygen-filled quartz combustion tube, which is packed with cerium oxide and copper oxide to facilitate oxidation and with silver wool to remove halogens. Nitrogen oxides are reduced in a second combustion tube packed with copper wire, also containing lead chromate, which serves to remove sulfur oxides. Using helium as carrier and scavenging gas, CO_2, H_2O, and N_2 are introduced into a separation and measuring module. At first the gaseous mixture flows through a short silver column filled with silica gel, where H_2O is quantitatively retained. Then the gases flow through a long copper column also filled with silica gel. Due to the greater length and higher adsorptive capacity of the copper column, CO_2 is quantitatively retained. Thus N_2, which is unaffected by silica gel, enters the thermal conductivity cell together with helium. After the determination of the N_2 content, the copper column is heated to 85°C, whereby CO_2 is quickly desorbed and flows into the thermal conductivity cell. Once the measurement of CO_2 is completed, the silver column is heated to 250°C to liberate H_2O, which is carried by helium into the thermal conductivity cell. The working ranges of this system are as follows: C, 0.3-15 mg; H, 0.03-1.5 mg; N, 0.02-5 mg. For the analysis of extremely corrosive substances and organic materials containing alkalis or alkaline earths, a special combustion tube made of stainless steel is used in place of the quartz tube.

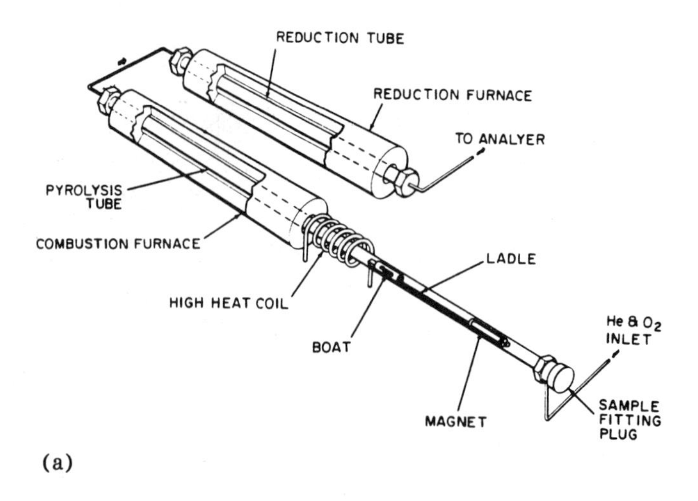

(a)

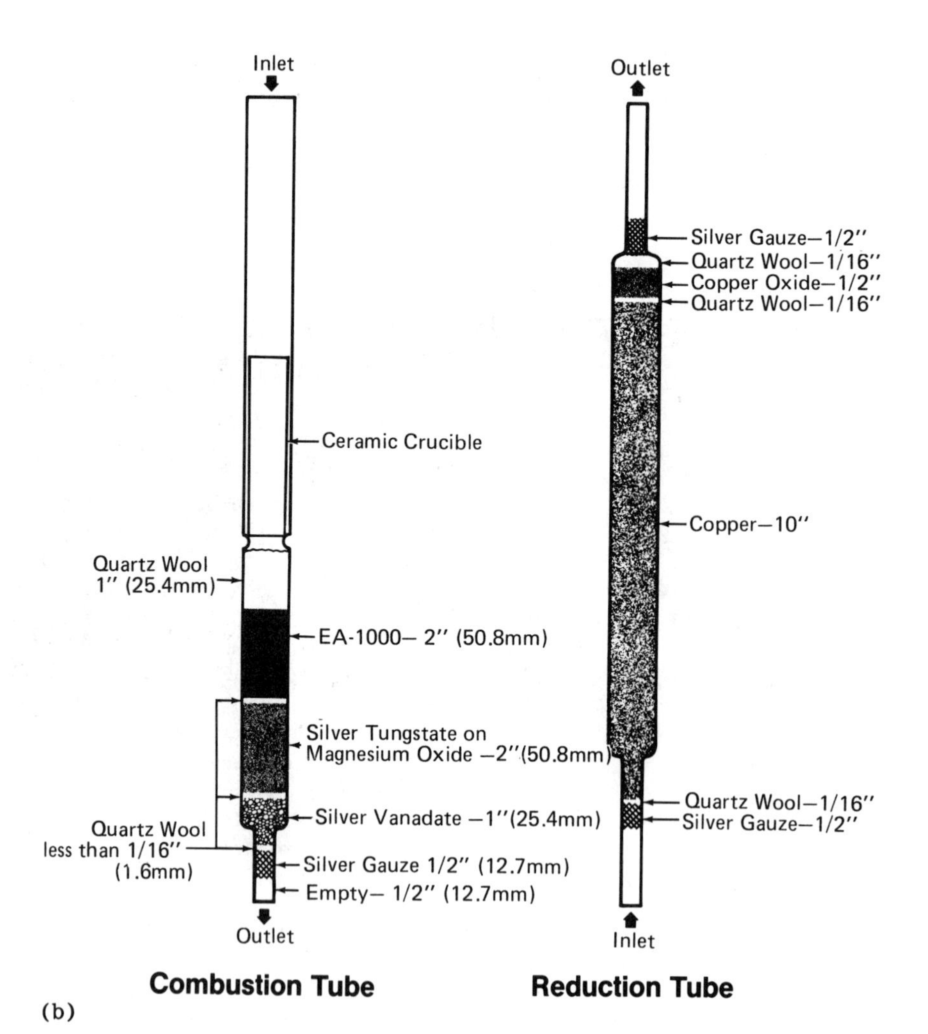

(b)

Figure 2. (a) The combustion train in PE 240C. (b) Packing of combustion tubes in PE 2400. (Courtesy of the Perkin-Elmer Corporation, Norwalk, CT.)

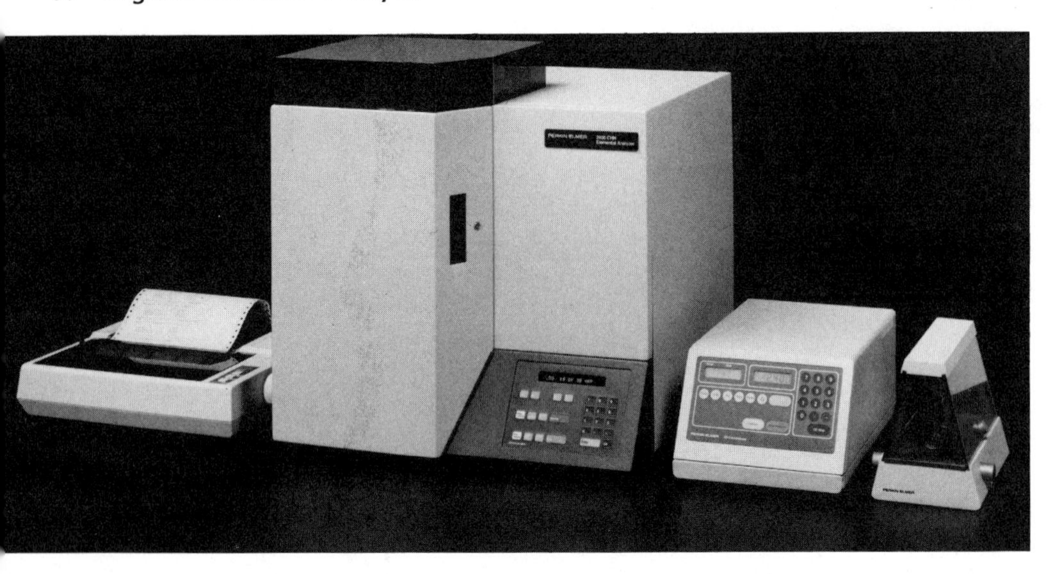

Figure 3. PE 2400 CHN Elemental Analyzer, with microbalance and microprocessor. (Courtesy of the Perkin-Elmer Corporation, Norwalk, CT.)

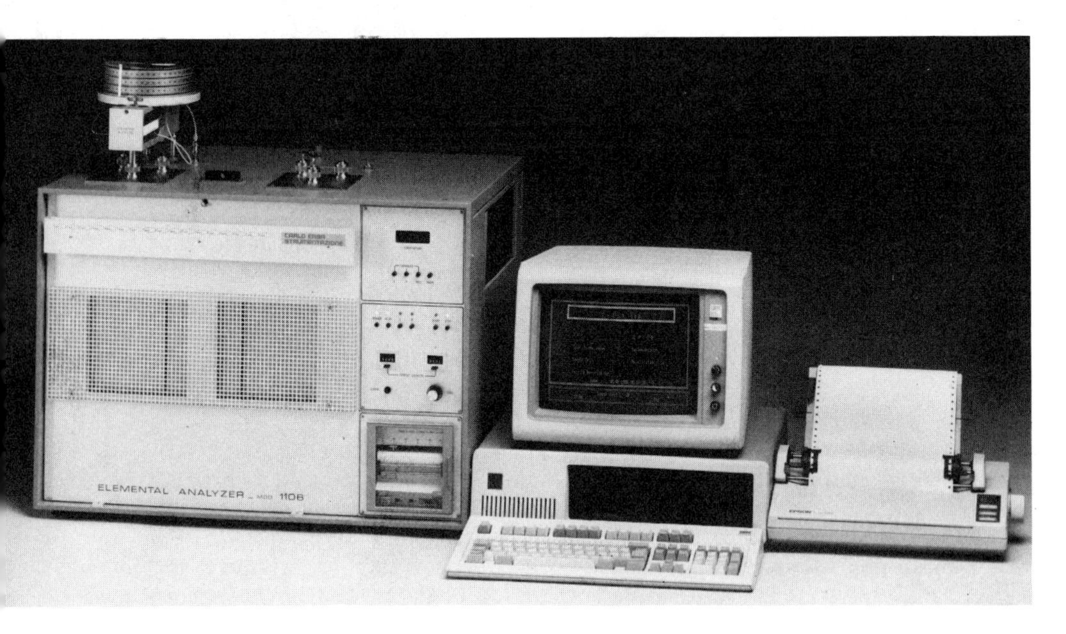

Figure 4. Carlo Erba Elemental Analyzer 1106. [Courtesy of Carlo Erba Strumentazion, Rodano (Milan), Italy.]

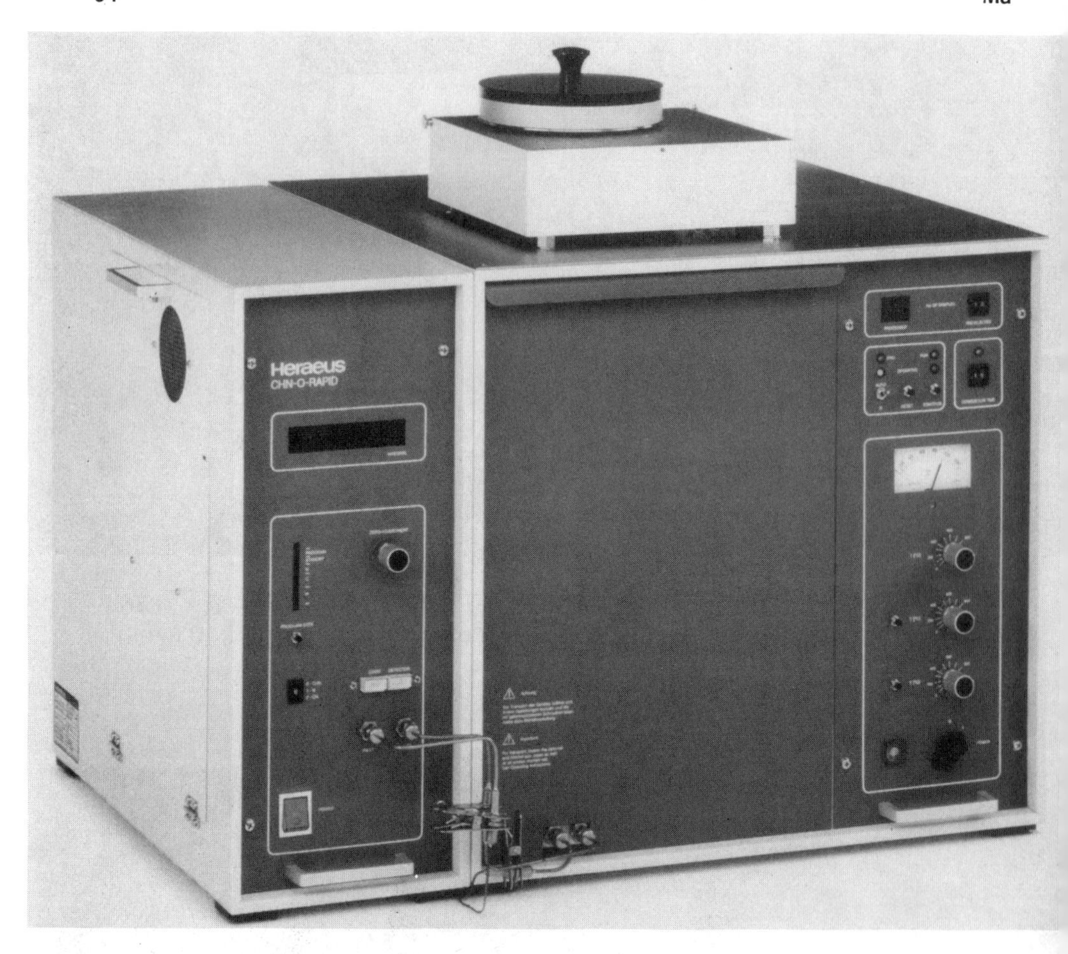

Figure 5. Heraeus Elemental Analyzer CHN-O-Rapid. (Courtesy of W. C. Heraeus GmbH, Hanau, Germany.)

In the Yanaco CHN Corder (Figure 7), a fixed volume 150 ml) of the mixture containing CO_2, H_2O, N_2, and helium is pumped through three pairs of thermal conductivity detectors connected in series, being joined to an H_2O absorption tube containing Anhydrone, a CO_2 absorption tube containing Ascarite, and a delay coil, respectively (see Figure 8). In the first pair, H_2O is removed, a difference is generated between the thermal conductivity detector at the inlet and outlet, and the H_2O content is determined. CO_2 is removed in the second pair, and a signal proportional to the CO_2 concentration is obtained. The delay coil of the third pair also has a capacity of 150 ml and is previously filled with helium gas. When pumping out starts, this helium gas flows to the outlet and a signal pro-

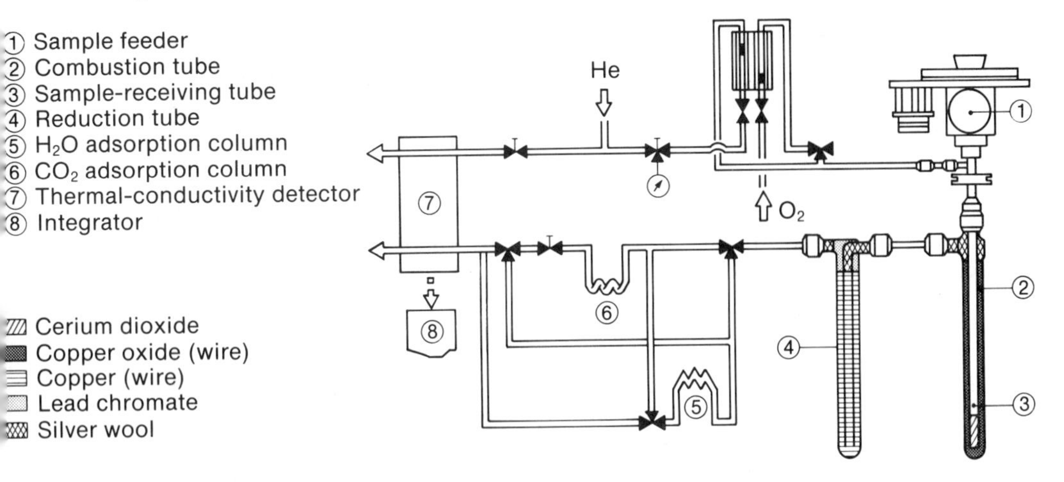

Figure 6. Functional diagram, CHN, of Heraeus Elemental Analyzer CHN-O-Rapid. (Courtesy of W. C. Heraeus GmbH, Hanau, Germany.)

Figure 7. Yanaco CHN Corder, model MT-3. (Courtesy of Yanaco New Science, Inc., Kyoto, Japan.)

portional to the N_2 concentration is obtained. It is claimed that this instrument consumes much less helium than the other systems [7].

The Leco CHN Elemental Analyzer uses a U-shape combustion tube (Figure 9). The organic sample is encapsulated in tin or copper and dropped into a reusable ceramic crucible centered in the primary hot zone of the combustion tube located in a resistance furnace. The sample is burned in oxygen at 950°C. Oxides of sulfur, if formed, are removed with a reagent in the secondary combustion zone, which also ensures complete combustion of all volatiles. Then the combustion products (CO_2, H_2O, N_2, NO, NO_2) together with oxygen are collected and mixed thoroughly in a glass tube under a sliding piston. The CO_2 and H_2O levels are constantly monitored during combustion by CO_2 and H_2O infrared detectors, and when they drop to a predetermined level, combustion is considered complete. An aliquot is taken and carried by helium through a reagent train for the removal of CO_2 and H_2O, and the reduction of nitrogen oxides to N_2. The content of N_2 is then measured by thermal conductivity. Simultaneously, the CO_2 and H_2O are measured by infrared detectors [8].

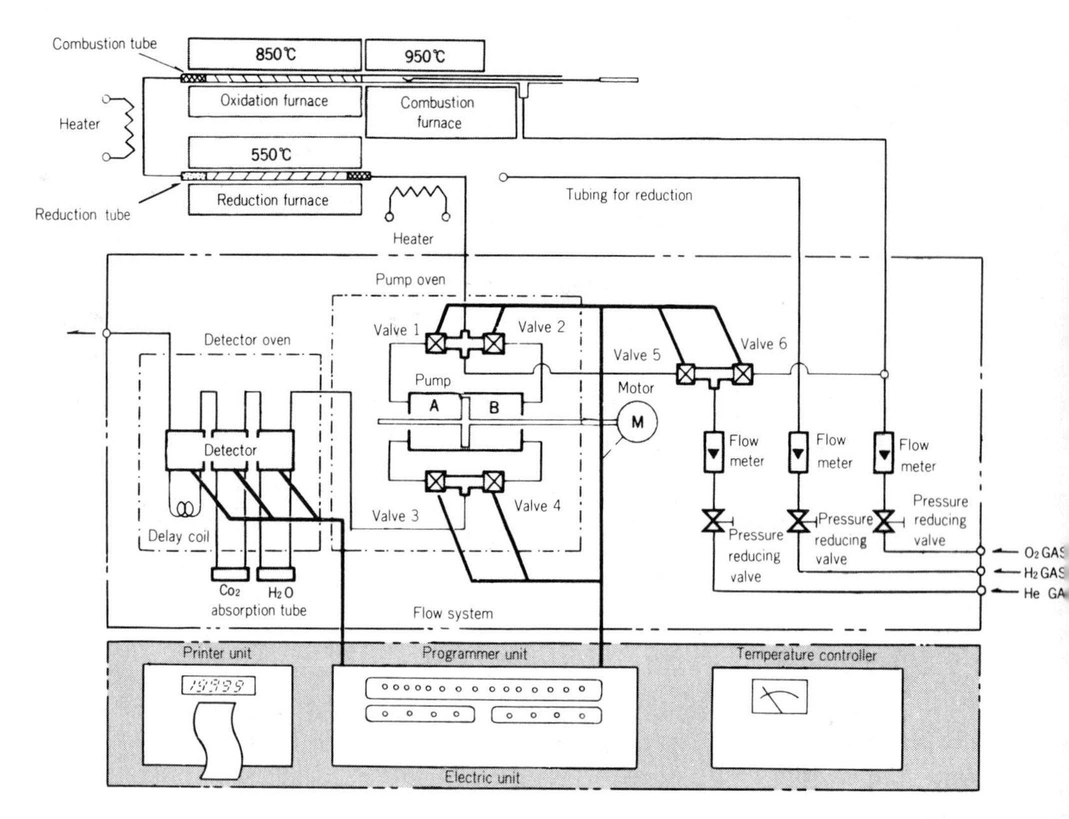

Figure 8. Gas flow diagram of Yanaco CHN Corder. (Courtesy of Yanaco New Science, Inc., Kyoto, Japan.)

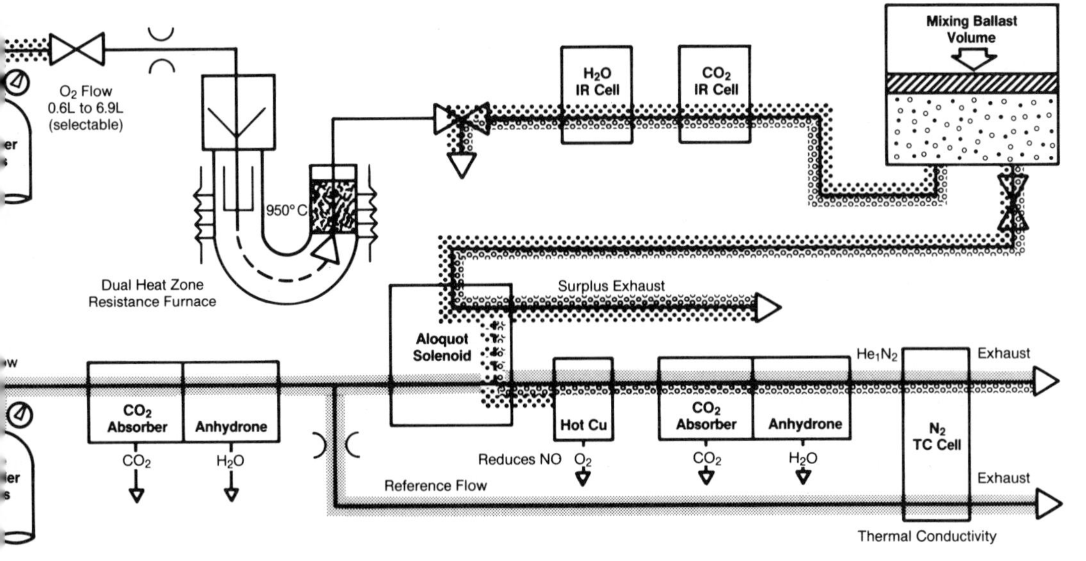

Figure 9. Gas flow diagram of Leco CHN Elemental Analyzer. (Courtesy of Leco Corporation, St. Joseph, Mich.)

III. APPARATUS FOR THE DETERMINATION OF NITROGEN

A. By the Kjeldahl Principle

When the nitrogen in an organic compound is of the amino type, it is preferable to use an apparatus for nitrogen determination that is based on the Kjeldahl principle. The sample is decomposed by heating in a glass vessel with sulfuric acid and suitable reagents, whereby organic nitrogen is converted to the ammonium ion:

$$\text{Amino compound} \xrightarrow{H_2SO_4,\ \text{catalysts}} NH_4^+$$

The nitrogen content of the resulting solution can be measured by means of the NH_4^+ ion-selective electrode [9]. The common practice, however, is to liberate ammonia from the solution by steam distillation after addition of alkali,

$$NH_4^+ \xrightarrow{NaOH} NH_3$$

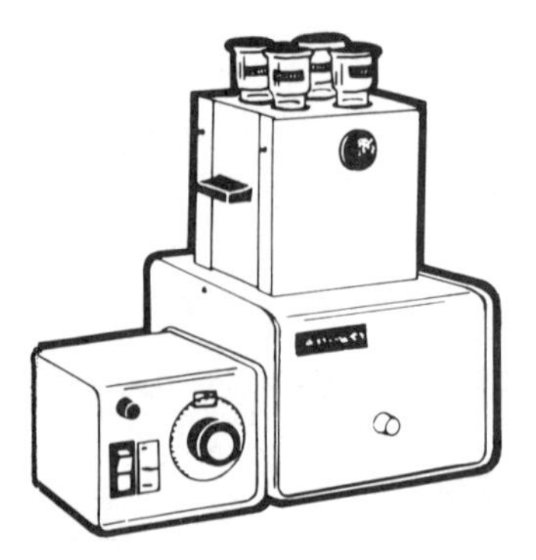

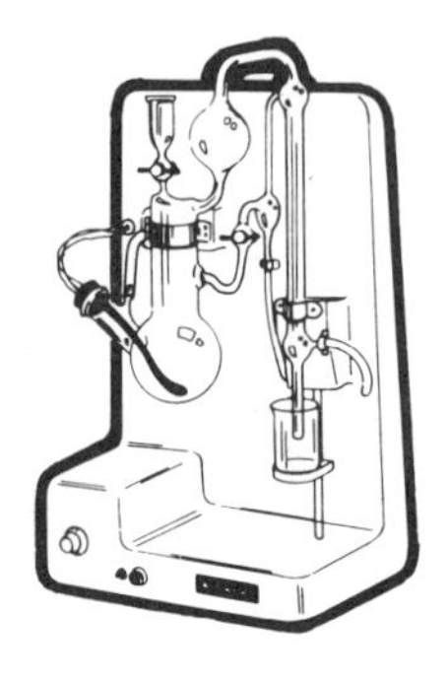

Figure 10. Digestion stand and distillation unit of Labconco Kjeldahl apparatus. (Courtesy of Labconco Corporation, Kansas City, Missouri.)

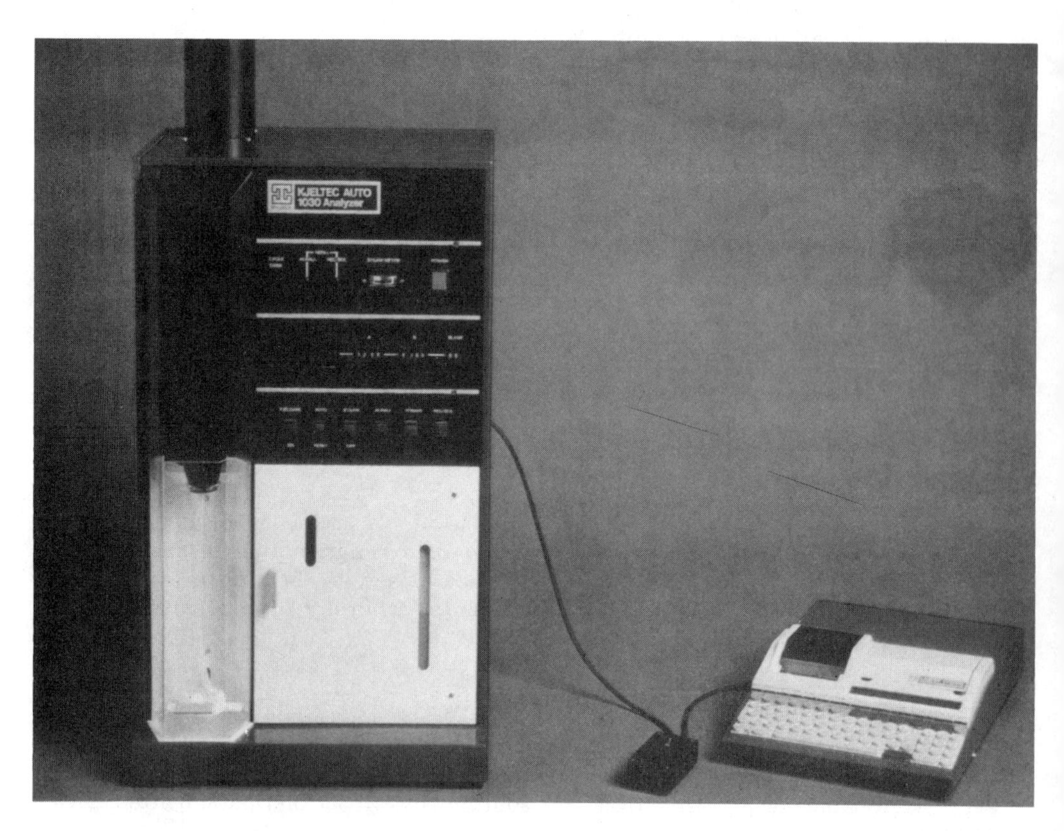

Figure 11. Kjeltec Auto System. (Courtesy of Tecator, Höganäs, Sweden.)

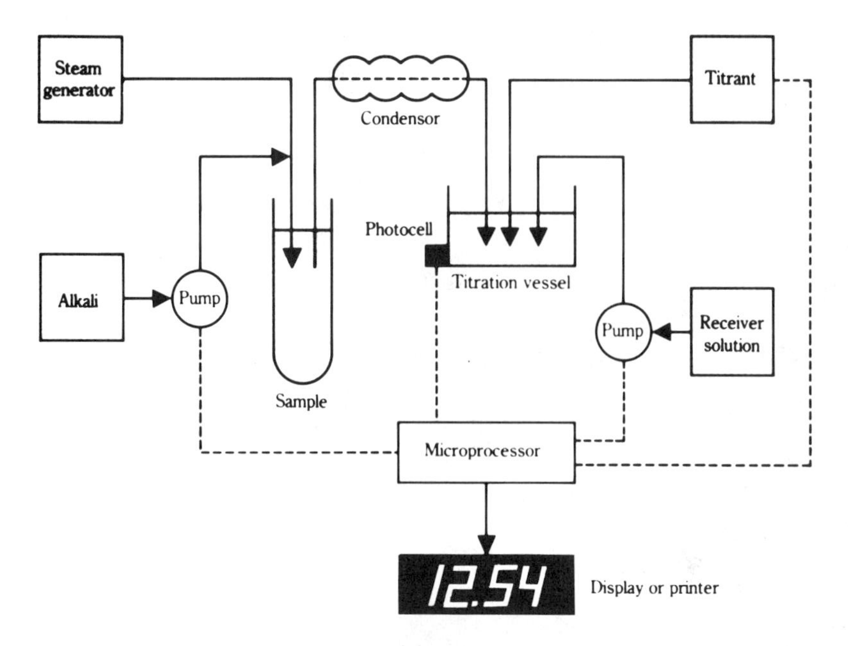

Figure 12. Flow diagram of Kjeltec Auto 1030 Analyzer. (Courtesy of Tecator, Höganäs, Sweden.)

followed by titration of the distillate. The apparatus can be easily constructed, and the experimental procedure is simple [10]. For the routine analysis of a large number of samples, there are available several commercial instruments [11-13] that emphasize speed and efficiency. Figure 10 shows the digestion stand and distillation unit of the Labconco Kjeldahl apparatus [12]. Figure 11 shows the Tecator Kjeldahl system, with the flow diagram depicted in Figure 12.

B. By the Dumas Method

The Leco Nitrogen Determinator (Figure 13) utilizes the Dumas principle to determine nitrogen by decomposing the organic compound in a combustion tube and converting all nitrogen into nitrogen gas to be measured. The sample is weighed into a tin capsule, which is then purged of any atmospheric gases and sealed. The capsule is automatically dropped into a crucible in the furnace and burned in pure oxygen at 1000°C. The gaseous combustion products (NO_2, NO, N_2, CO_2, H_2O) together with oxygen are collected in a "ballast volume" where the gases are homogenized at a pressure of 970 torr and a temperature of 52°C. When all the gases are collected, a 10-ml aliquot is forced out and carried by helium through a series of solid reagents, namely, (a) iron chips to remove dust particles, (b) hot copper to remove oxygen and to reduce NO_2 and NO to N_2, (c) Ascarite to remove CO_2, and (d) Anhydrone to absorb H_2O. The gas stream now contains only helium and N_2, which are passed on to the thermal conductivity cell [8].

Figure 13. Leco nitrogen determinator. (Courtesy of Leco Corporation, St. Joseph, Mich.)

IV. DETERMINATION OF OXYGEN

A. Instrument for Determining Oxygen Only

Determination of oxygen in organic materials is accomplished by pyrolysis of the sample with carbon in the presence of platinum as catalyst, thus converting all oxygen into carbon monoxide:

$$\text{CHO compound} \xrightarrow{\text{C}} \text{CO}$$

There are several ways to measure the CO produced [1]. In one procedure, the CO is carried by nitrogen into a tube containing iodine pentoxide at 125°C, where the following reaction takes place:

$$5\ CO + I_2O_5 \longrightarrow 5\ CO_2 + I_2$$

The iodine vapor is driven into the cathode chamber of an electrolysis cell, and I_2 is determined by electrolytic reduction. The assembly used in the Shanghai Institute of Organic Chemistry, Shanghai, China is shown in Figure 14. When the analysis is being carried out, the section for the electrometric determination of I_2 is covered with a hood in order to keep out the light [14].

B. Determination of Oxygen Using Commercial Elemental Analyzers

When commercial elemental analyzers are constructed to determine O in addition to CHN, the CO formed is measured as such without further chemical transformation in the Heraeus and Carlo Erba systems, but is reoxidized to CO_2 in the Perkin-Elmer system.

In the Heraeus CHN-O-Rapid Elemental Analyzer, pyrolysis takes place in a cracking tube filled with carbon black (Figure 15). A mixture consisting of 15% nitrogen and 5% hydrogen serves as carrier gas. The

Figure 14. Assembly for the determination of oxygen. (Courtesy of Prof. Z. Y. Hu, Shanghai Institute of Organic Chemistry, Academia Sinica, Shanghai, China.)

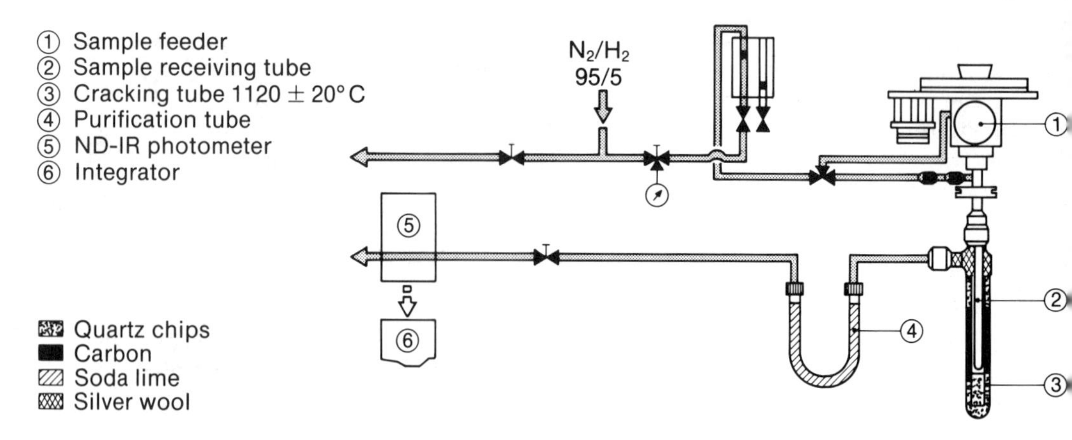

Figure 15. Operation diagram for oxygen determination in the Heraeus CHN-O-Rapid Elemental Analyzer. (Courtesy of W. C. Heraeus GmbH, Hanau, Germany.)

CO concentration is measured with a nondispersive infrared photometer. Since the detector has low cross-sensitivity toward nitrogen, hydrogen and methane, it is not necessary to remove these gases. On the other hand, acidic gases such as hydrogen chloride interfere seriously; they must be removed by incorporating a purification tube filled with soda lime.

For the determination of oxygen using the Carlo Erba Elemental Analyzer 1106 (Figure 4), the organic sample is weighed into a silver container and introduced into the second combustion tube packed with special carbon. The resulting CO is swept through the chromatographic column by helium gas and is measured in the thermal conductivity detector.

The Perkin-Elmer PE 2400 CHN Elemental Analyzer also has provision to adapt the instrument for the determination of oxygen [4]. The arrangements are similar to those for the model PE 240 [15].

V. DETERMINATION OF HALOGENS

A. Determination of Chlorine, Bromine, and Iodine

Currently the favorable apparatus for the determination of chlorine, bromine and iodine in organic materials is combustion flasks. Chloro, bromo, or iodo compounds are burned in the closed combustion flask in the presence of platinum as catalyst, in an atmosphere of pure oxygen. For the purpose of measuring the chlorine, bromine, or iodine content, however, it is necessary to convert these three elements to their lowest oxidation state, that is, to the halides [1]. This can be achieved by placing a reducing absorption solution in the combustion flask.

$$\text{Chloro compound} \xrightarrow{O_2} Cl_2,\ OCl^-,\ \text{etc.} \xrightarrow{H_2O_2,\ NaOH} Cl^-$$

$$\text{Bromo compound} \xrightarrow{O_2} Br_2,\ OBr^-,\ \text{etc.} \xrightarrow{H_2O_2,\ NaOH} Br^-$$

$$\text{Iodo compound} \xrightarrow{O_2} I_2,\ OI^-,\ \text{etc.} \xrightarrow{(NH_2)_2H_2SO_4,\ KOH} I^-$$

Two models of commercial combustion flasks [16] are shown in Figure 16. These thick-wall conical flasks are made of borosilicate glass. One model has bell-shape, flaring lip. The elongated ground-glass stopper is sealed to the platinum wire gauze sample carrier. The other model is fitted with a ball joint stopper, which extends to form a hook. The sample carrier, made of perforated platinum sheet, is detachable and rests on the hook during combustion. Both models come in two sizes: 500 ml and 1000 ml.

For the sake of safety, flask combustion can be carried out in a metal cabinet using an infrared heat source for ignition. The apparatus available commercially is shown in Figure 17. The cabinet has a baffled, polycarbonate door. The infrared lamp (120 V, 150 W) is mounted in the housing with a screw-crank elevating device. An angled platform tilts the combustion flask to bring the organic sample wrapped in special black paper close to the lamp. Alignment of sample in the infrared beam and ignition are done from outside the cabinet after the door is fastened.

Patrick et al. recently published a device that has been continuously used for many years [17]. The combustion flask and platinum gauze sample carrier are shown in Figure 18. The glassware is constructed with a 500-ml conical flask, screw cap, sealing ring, and borosilicate glass tube (80 mm long, 18 mm outside diameter) with a stud to hold the sample carrier. The flask is placed inside a metal cabinet which contains two quartz-halogen lamps (240 V, 800 W) mounted horizontally, one at each side (Figure 19). The lamps have parabolic reflectors made from polished aluminum. Ceramic lampholders are fitted to each reflector, and are arranged at such a distance that they produce a focused image halfway across the cabinet. Critical optics are not necessary due to the high-powered lamp. The lamps can only be operated when the cabinet door is closed, and they are controlled through an adjustable time switch to give a reproducible exposure. Ignition usually takes place after about 2 sec, and can be observed safely through a dark neutral-density filter in the door.

After the combustion, the halide in the absorption solution can be determined by titration [18] or by means of the halide ion-selective electrode [19]. When large numbers of samples are to be analyzed, the automated potentiostatic procedure of Scheidl and Toome [20] is recommended. The system is shown in Figure 20. A combustion furnace, a buret, and a titrator are modified and connected to the automated potentiostatic apparatus. A solid-state programmer using a monostable multivibrator circuit to generate various time delays completes a halogen determination in about 4 min. The organic sample is decomposed by heating in pure oxygen in a quartz tube. Aqueous sulfurous acid in 2 *M* perchloric acid mixed with isopropanol is used as the flush solution, while aqueous sodium bisulfite in the same mixture is used as the absorption solution to convert halogen to halide which is determined by titration with 0.01 *M* silver nitrate to the preset end point.

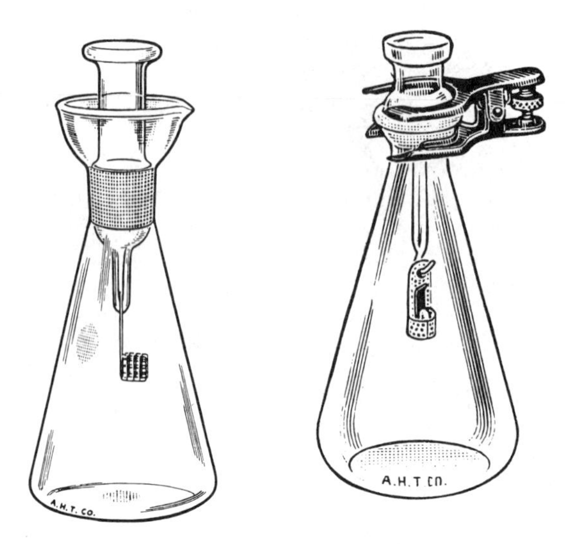

Figure 16. Combustion flasks. (Courtesy of Thomas Scientific, Swedesboro, NJ.)

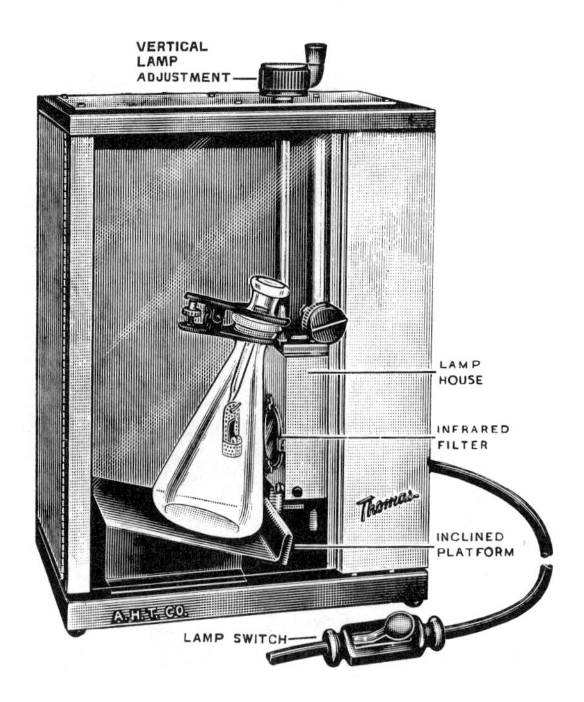

Figure 17. Metal cabinet for combustion-flask ignition. (Courtesy of Thomas Scientific, Swedesboro, NJ.)

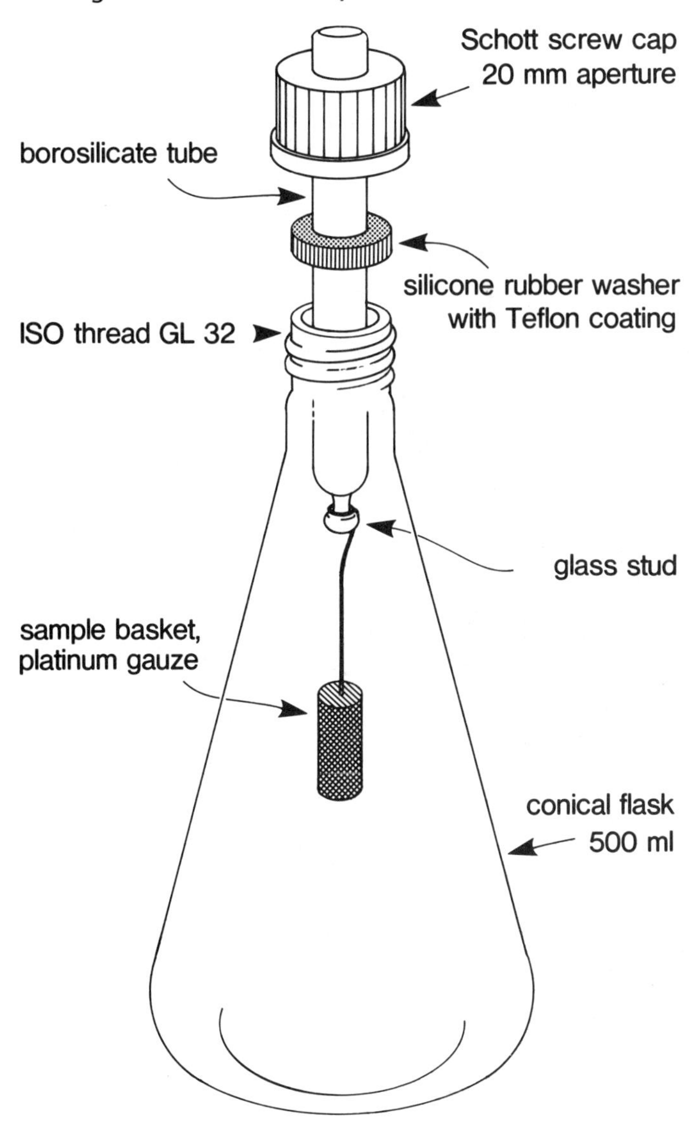

Figure 18. Combustion flask and sample carrier, used in Beecham Pharmaceuticals, Surrey, England. (Courtesy of Laboratory Practice.)

B. Determination of Fluorine

Because organic fluorine compounds are more difficult to decompose than the other halogenated compounds, it is better to carry out the decomposition by fusion with alkali metal [1], especially for perfluoro substances:

$$\text{Fluoro compound} \xrightarrow{\text{Na or K}} F^-$$

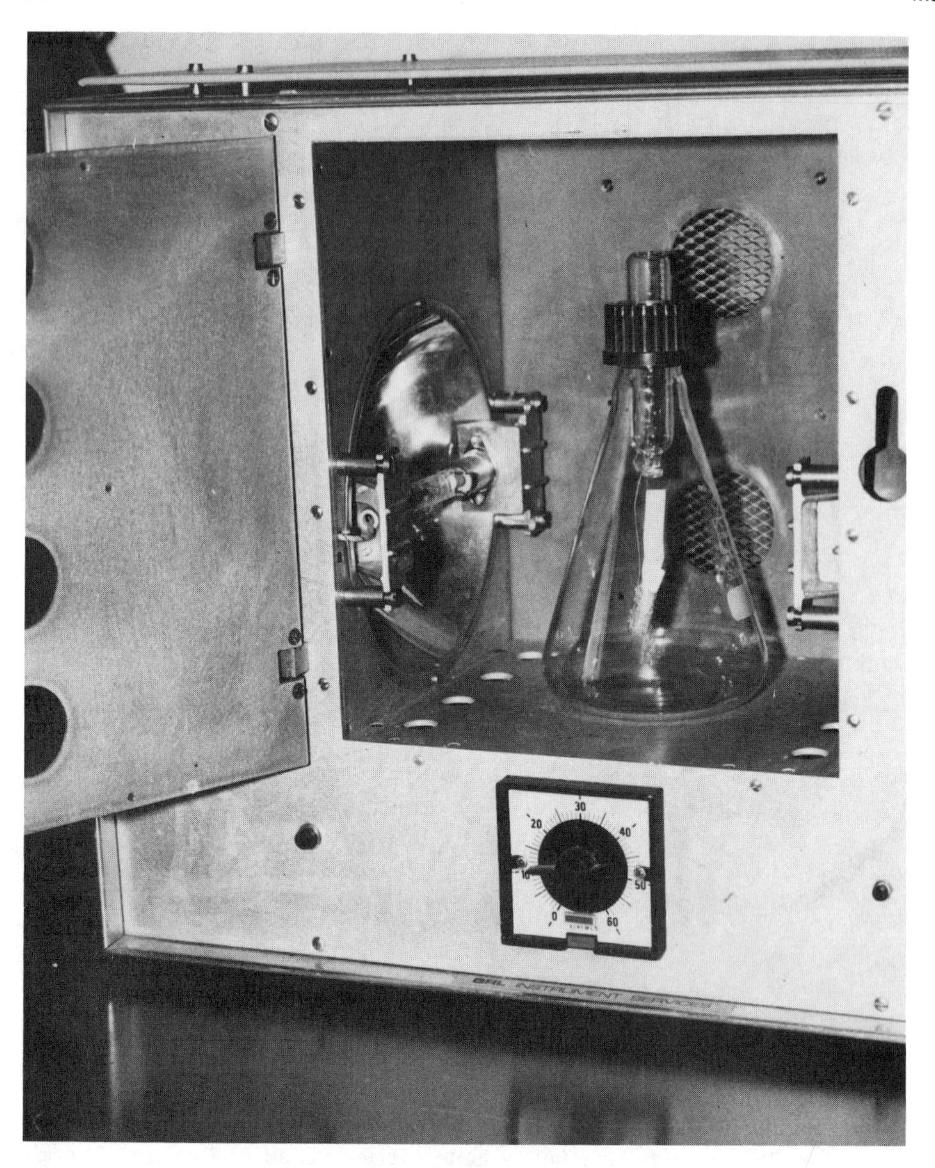

Figure 19. Cabinet for flask combustion, used in Beecham Pharmaceuticals, Surrey, England. (Courtesy of Laboratory Practice.)

The apparatus for such purpose is the Parr microbomb made of nickel [21]. Another method of decomposition involves the use of the oxyhydrogen flame, but this instrument is not commercially available. The oxyhydrogen flame is the device of choice for analyzing low-boiling fluoro compounds [22].

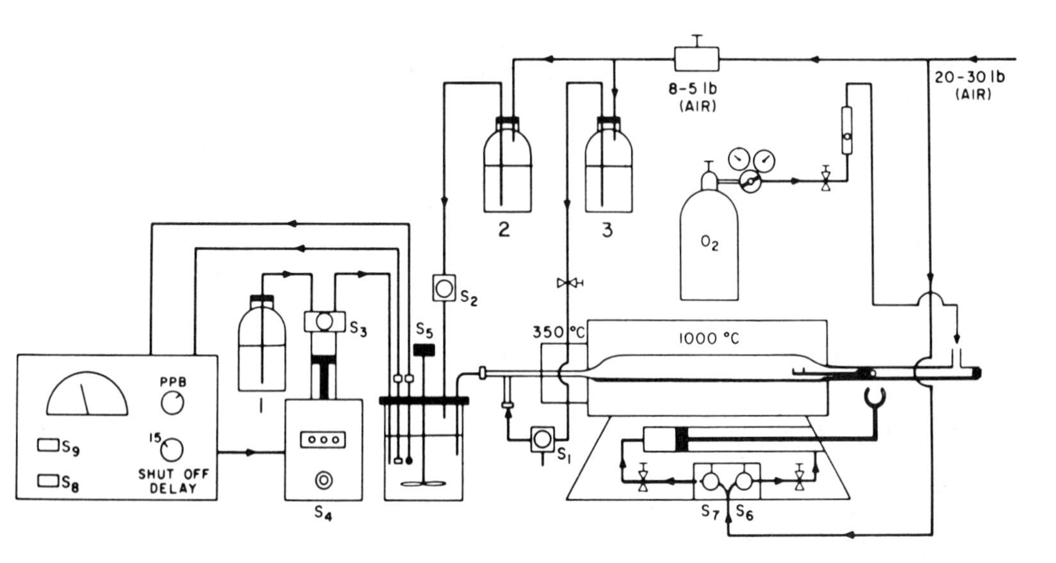

Figure 20. Assembly for automated determination of Cl, Br and I: (1) silver nitrate solution; (2) flush solution/ (3) absorption solution. Switches and solenoids operated by programmer (S_1 to S_9). (Courtesy of Microchemical Journal.)

The fluoride content obtained after decomposition of the organic sample can be determined with the F^- ion-selective electrode, or by one of the spectrophotometric methods [1]. In order to remove all interfering substances, the most reliable procedure is the separation of fluosilicic acid by steam distillation from perchloric acid solution [23].

VI. DETERMINATION OF SULFUR

A. Determination by Titrimetry

In contrast to the determination of halogens, decomposition of organic sulfur compounds is aimed at the conversion of sulfur to its highest oxidation state, sulfate. The closed-flask combustion technique (see Section V.A.) is commonly used, with water containing hydrogen peroxide as the absorbent:

$$\text{Sulfur compound} \xrightarrow{O_2} SO_2,\ SO_3 \xrightarrow{H_2O,\ H_2O_2} SO_4^{2-}$$

A second instrument suitable for the mineralization of organic sulfur is the Parr bomb [21]. The sample is mixed with sodium peroxide and a chemical accelerator such as potassium perchlorate and is placed in the metal bomb, which is then sealed and heated to ignite the charge:

$$\text{Sulfur compound} \xrightarrow{Na_2O_2,\ KClO_4} Na_2SO_4$$

Because the reactions inside the Parr sodium peroxide bomb tend to be violent, and because it is difficult to obtain high-quality sodium peroxide, this procedure is gradually being replaced by other techniques [24].

For the determination of the sulfate produced after decomposition of the organic sulfur compound, the common practice is titration with barium perchlorate. When the organic sample contains only carbon, hydrogen, oxygen, and sulfur, the product obtained after closed-flask combustion will be H_2SO_4, which can be conveniently measured by acidimetry using a standardized NaOH solution.

B. Determination by Gas Chromatography

In the Carlo Erba Elemental Analyzer (see Figure 4), the second combustion tube can be employed for the determination of sulfur. By using an oxidation catalyst different from that for CHN [5], sulfur in the organic sample is transformed exclusively to SO_2. The SO_2 is then separated by gas chromatography in a column different from that for CHN, and is measured by the thermal conductivity detector.

VII. DETERMINATION OF PHOSPHORUS AND ARSENIC

There are no special instruments for the determination of phosphorus or arsenic in organic substances. Acid digestion in open glass vessels such as the micro-Kjeldahl flask is generally employed for the decomposition of organic phosphorus or arsenic compounds:

$$\text{Phosphorus compound} \xrightarrow{HNO_3,\ H_2SO_4} PO_4^{3-}$$

$$\text{Arsenic compound} \xrightarrow{HNO_3,\ H_2SO_4} AsO_4^{3-}$$

The closed-flask combustion procedure (see Section V.A) can be employed for phosphorus compounds, but is not recommended for arsenic compounds because arsenic may ruin the platinum catalyst by forming an alloy.

The phosphate ion obtained after combustion is generally determined by a spectrophotometric method. In contrast, titrimetry is often the finishing step in the analysis of organic arsenicals [25].

VIII. DETERMINATION OF BORON AND SILICON

The combustion flask (see Section V.A) has been used for the decomposition of organic boron compounds. The Parr microbomb [21], however, is more reliable because practically all types of boron compounds can be decomposed in it. Both methods convert the boron in organic materials to borate:

$$\text{Boron compound} \xrightarrow{O_2,\ \text{or}\ Na_2O_2} BO_3^{3-}$$

The resulting borate is determined spectrophotometrically, or by a special titrimetric method involving the use of mannitol [26].

Decomposition of organic silicon compounds can be achieved by heating with oxidizing acids in a platinum crucible, or by the closed-flask combustion technique. The product in both cases is SiO_2, which can be weighed.

$$\text{Silicon compound} \xrightarrow{HNO_3 + H_2SO_4,\ \text{or}\ O_2} SiO_2$$

Another method for analyzing silicon compounds prescribes alkali fusion in a metal bomb to convert organic silicon into alkali silicate,

$$\text{Silicon compound} \xrightarrow{KOH} K_2SiO_3$$

which is determined titrimetrically or spectrophotometrically.

IX. DETERMINATION OF MERCURY

Analysis of organic mercury compounds can be carried out with two glass tubes, one of 10 mm diameter and the other of 6 mm diameter (Figure 21). The former (combustion tube) is packed with calcium oxide, while the latter (absorption tube) is packed with shredded gold leaves. The absorption tube fits loosely to the constricted end of the combustion tube. The organic sample is placed in a microboat, which is pushed into the combustion tube. Upon heating, the mercury compound is decomposed and mercury is liberated. A stream of air drives the mercury vapor through the calcium oxide into the absorption tube, where mercury is retained as gold amalgam. The increase in weight of the absorption tube gives the content of mercury in the sample [27]:

$$\text{Mercury compound} \xrightarrow{700°C} Hg$$

Organic mercury compounds can also be decomposed by digestion with oxidizing acids in the micro-Kjeldahl flask, producing mercuric ions:

$$\text{Mercury compound} \xrightarrow{HNO_3,\ H_2SO_4,\ KMnO_4} Hg^{2+}$$

The finishing step involves titration with standardized potassium thiocyanate solution using an automated second-derivative spectrophotometric titrator [28].

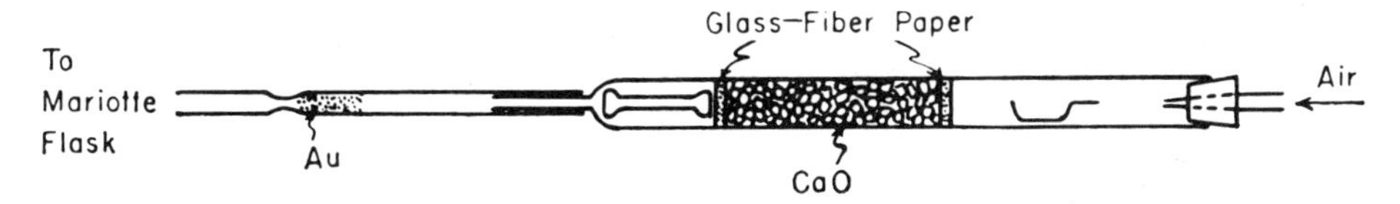

Figure 21. Apparatus for the determination of mercury. (Courtesy of Van Nostrand Reinhold Company.)

X. ANALYSIS OF ORGANOMETALLICS

There are no special instruments for the determination of metallic elements in organometallic compounds. When the organic substance submitted for analysis contains only one metallic element, the sample can be simply mineralized by heating it in a crucible or microboat, with the addition of small amount of oxidizing acids. If the sample does not contain phosphorus, silicon, or boron, the residue after combustion will be the free metal, an oxide, or a sulfate of the metallic element. The residue can be weighed or brought into solution in order to determine the metal cation by an appropriate method.

If the decomposition of the organometallic compound is carried out in a liquid medium, the reaction vessel may be a conical flask or micro-Kjeldahl flask. The conical flask is convenient because it can also serve as the titration flask for the determination of the metal cation in the resulting solution.

If the organic sample contains more than one metallic element, the solution obtained after digestion or solubilization of the combustion residue is then subject to quantitative inorganic analysis. The procedure will depend on the nature of the metallic elements present in the original organic material.

XI. MULTIELEMENTAL ANALYSIS

A. Using the Gas Chromatograph to Determine All Combustion Products

Upon total destruction of an organic compound, the nonmetallic elements are usually transformed into gaseous products. With the advancement of gas chromatography whereby columns and packing materials have been found that can handle corrosive gases, and reliable methods have been developed for detection and quantitation, a number of workers have investigated the application of gas chromatography to the multielemental analysis of organic samples. In contrast to the determination of organic functional groups by gas chromatography, which aims at converting only a portion of the organic molecule into a specific gaseous product [30], multielemental analysis of an organic compound involves the conversion of all elements present in the compound to various gases and accurate measurement of as many of these gases as possible.

In Table 1 are listed the combustion products obtained from nonmetallic elements that have been determined by gas chromatography. Three types of combustion processes have been employed: (a) oxidation in an atmosphere of pure oxygen, (b) catalytic hydrogenation with a stream of hydrogen, and (c) sulfurization by heating the organic sample with sulfur and sodium sulfite in a sealed tube.

It should be recognized that more than one gaseous product can be derived from a particular element during combustion. In each case, attempts are made to convert these products into one gaseous species to be measured, or to find an experimental condition under which only one kind of gas is produced from that element. For instance, Dugan [31] has observed that

Table 1. Combustion Products for Multielemental Analysis

	Combustion product		
Element	By oxidation [32,33]	By hydrogenation [34]	By sulfurization [35]
Carbon	CO_2	CH_4	CO, CO_2, COS, CS_2
Hydrogen	H_2O	—	H_2S
Nitrogen	N_2	NH_3	N_2
Oxygen	—	H_2O	CO, CO_2
Sulfur	SO_2	H_2S	
Chlorine	HCl	HCl	NaCl
Bromine	Br_2	Br_2	NaBr
Iodine	I_2	I_2	NaI
Phosphorus	—	PH_3	
Silicon	—	SiH_4	

in the combustion of sulfur compounds, sulfur is exclusively converted into SO_2 in the presence of metallic copper at 840°C, and he has constructed an instrument (in the Hercules, Inc., laboratories) to determine C, H, N, and S simultaneously by measuring N_2, CO_2, H_2O, and SO_2 using the gas chromatograph [32].

Grob and co-workers at Villanova University have proposed two schemes for multielemental analysis based on combustion and gas chromatography. In one scheme using oxidation reactions, C, H, S, Cl, Br, and I can be determined simultaneously [33]. The instrumentation consists of a combustion system and a gas chromatograph equipped for dual-column operation with thermal conductivity detection. The schematic diagram is shown in Figure 22. The organic sample is weighed into an aluminum capsule and covered with 50 mg of vanadium pentoxide. The sealed capsule is placed on a nichrome wire ignitor and introduced into the combustion tube. Combustion is initiated at the ignitor and completed in the oven, in an atmosphere of oxygen that also serves as carrier for gas chromatography. Two chromatographic columns are required to separate the six oxidation products, using the series-across-detector mode. HCl, Br_2, and I_2 are separated and quantitated when the combustion products pass through the first column packed with 20% SE-30 on Chromosorb W-AW and through one side of the thermal conductivity detector. The remaining components then pass through the second column packed with Chromosorb 102 and through the second side of the detector, whereby CO_2, SO_2, and H_2O are separated and quantitated. This method is applicable to the analysis of organic compounds with molecular weight below 300.

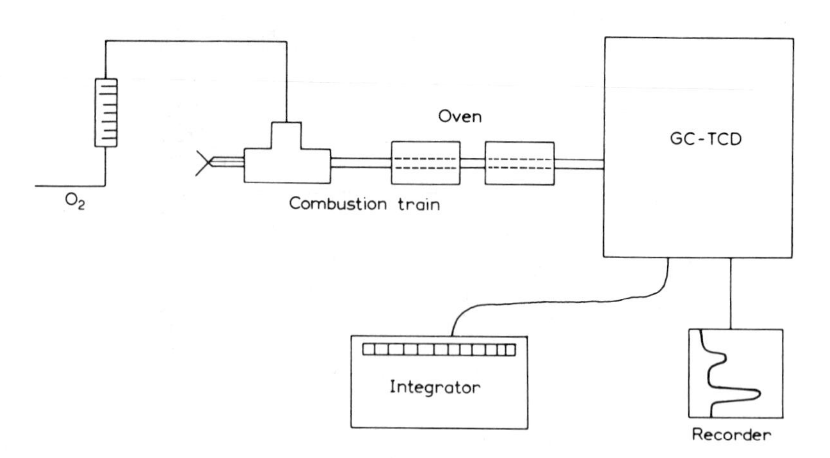

Figure 22. Multielemental analysis by oxidation and gas chromatography. From Ref. 33. (Courtesy of Journal of Chromatography.)

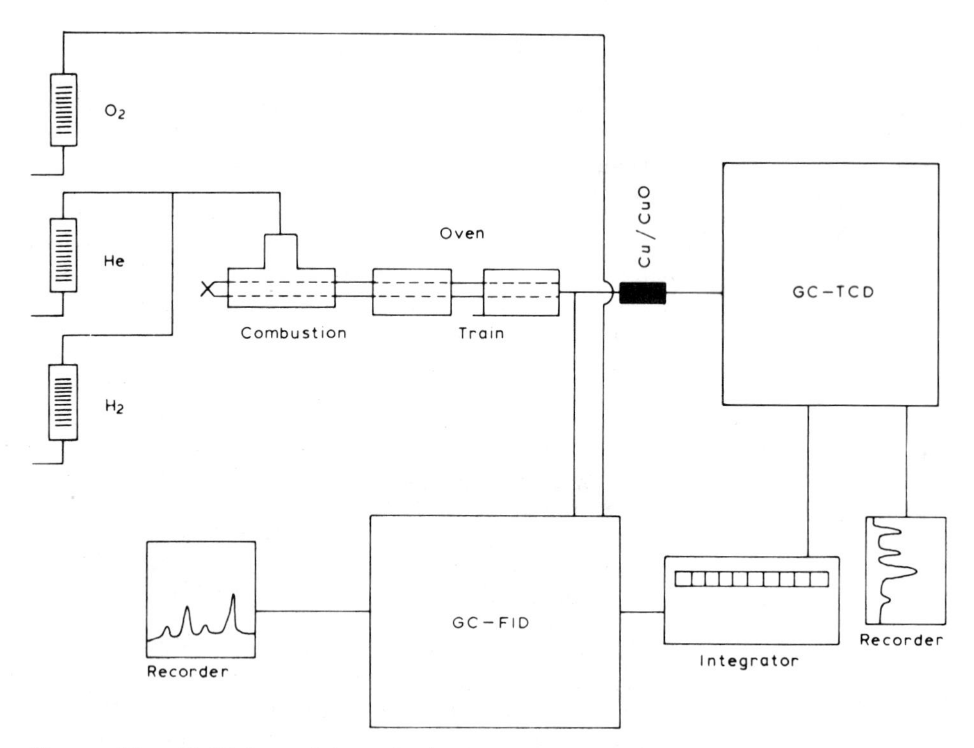

Figure 23. Multielemental analysis by hydrogenation and gas chromatography. From Ref. 34. (Courtesy of Journal of Chromatography.)

Another scheme developed by Sullivan and Grob [34], suitable for the simultaneous determination of C, N, O, S, Cl, Br, I, P, and Si, is based on catalytic hydrogenation. Two gas chromatographs are required, one for flame ionization detection and the other for thermal conductivity detection (see Figure 23). The organic sample is weighed into an aluminum capsule and covered with 50 mg of magnesium powder. The capsule is placed on a nichrome wire ignitor and introduced to the combustion tube containing the nickel catalyst. Combustion in an atmosphere of helium-hydrogen (80:20) converts the above mentioned elements to CH_4, NH_3, H_2O, H_2S, HCl, Br_2, I_2, PH_3, and SiH_4, respectively. One portion of the combustion products is passed into the gas chromatograph equipped with a flame ionization detector where H_2S, SiH_4, and CH_4-PH_3 combination are separated and quantitated. Another portion of the combustion products is conducted through a stainless steel tube filled with copper-copper oxide to convert NH_3 into N_2; the gaseous mixture then passes into the second gas chromatograph, which is equipped with a thermal conductivity detector and fitted with two columns aligned in the series-across-detector mode. One column packed with 20% SE-30 on Chromosorb W-AW serves to separate HCl, Br_2, and I_2, while the other column, packed with Carbosieve S, is used to separate and quantitate CH_4, N_2, and H_2O. Since PH_3 and CH_4 form a composite peak in the flame ionization chromatogram, the concentration of PH_3 is determined by the difference between the PH_3-CH_4 peak in the flame ionization chromatogram and the CH_4 peak in the thermal conductivity chromatogram. Like the oxidation scheme described in the previous paragraph, the hydrogenation scheme is also limited to organic compounds with molecular weight below 300. Furthermore, compounds containing a high oxygen content may form carbon monoxide, causing low results for both C and O.

Hara and co-workers at Doshisha University in Kyoto, Japan, have undertaken an extensive study of multielemental analysis based on sulfurization [35]. The organic sample is placed in a quartz ampoule together with 5 mg of sulfur. After displacing the air with helium, the quartz ampoule is sealed and heated at 600-950°C in an oven or in a high-frequency induction furnace. Then the quartz ampoule is inserted into a Teflon tube connected to a gas chromatograph. By driving a jack across the Teflon tube, the quartz ampoule is broken and the combustion products are introduced to the gas chromatograph using helium as carrier gas. If the organic sample contains C, H, O, and N, the chromatogram will exhibit six peaks, indicating N_2, CO, CO_2, COS, H_2S, and CS_2, respectively. The atomic ratios between C, H, O, and N in the organic sample is determined either by means of three calibration curves obtained by analyzing glycine, glutamic acid, citrulline and fumaric acid, or by using correction factors obtained by analyzing five standard compounds. If the organic sample contains Cl, Br, or I, heating it with sulfur and sodium sulfide transforms the halogens into sodium halides. Determination of Cl, Br, and I by gas chromatography was not successful because of the incomplete liberation of hydrogen halides from sodium halides. However, the sodium salts can be dissolved in water and the halide ions determined by ion chromatography.

B. Combining Gas Chromatography with Atomic Emission Spectrometry

A different way to utilize gas chromatography for multielemental analysis involves the linkage of the gas chromatograph with a spectrometer. Yu and co-workers at the Lanchow Institute of Chemical Physics, Lanchow, China, have determined the empirical formulas of volatile compounds by combining gas chromatography with microwave emission spectrometry. The instrumentation comprises three components, namely, a gas chromatograph fitted with high-resolution glass capillary column, microwave generator and plasma discharge tube, and polychromator and data recording device [36]. By using the helium plasma, an organic compound eluted from the gas chromatograph is fragmented completely to atoms. The signal recorded in the spectrometer for the specific element is nearly proportional to the quantity of the element; therefore it is possible to calculate the elemental ratios of the compound by referring to a standard. Compounds containing C, H, F, Cl, Br, I, S, and Hg have been successfully analyzed in this instrument.

Fry and co-workers at Kansas State University have investigated extensively the atomic emission of nonmetals in the red and near-infrared regions obtained with microwave-induced or inductively coupled plasmas [37,38]. A high-powered, atmospheric-pressure helium microwave induced plasma is employed as the interface between a gas chromatograph and Fourier transform near-infrared emission spectrometer using a silicon photodiode detector. By the collection of a series of time-resolved interferograms when organic compounds elute from the gas chromatograph into the plasma, a complete account of both spectral and chromatographic activity is recorded. Computer-generated, element-specific chromatographic reconstructions for each nonmetallic element are obtained from a single injection of the organic sample. Atomic emission intensity is plotted against chromatographic retention time for each of the frequencies chosen from the 15,700-8500 cm^{-1} region. As many as 11 elements (C, H, N, O, F, Cl, Br, I, S, P, and Si) can be determined simultaneously in this manner. When an argon inductively coupled plasma is used with a silicon photodiode detector, seven elements (C, H, N, O, Cl, Pr, and S) can be determined simultaneously.

C. Other Methods

Besides the above-mentioned techniques, there are many other methods for the simultaneous determination of several elements using one weighed sample. The reader is referred to the writer's book [1] and biennial reviews [39].

REFERENCES

1. T. S. Ma and R. C. Rittner, *Modern Organic Elemental Analysis*, Marcel Dekker, New York, 1979.
2. T. S. Ma and R. E. Lang, *Quantitative Analysis of Organic Mixtures*, Wiley, New York, 1979.

3. T. S. Ma and S. A. K. Hsieh, *Trace Element Determination in Organic Materials*, Academic Press, Orlando, FL, in preparation.
4. The Perkin-Elmer Corporation, Norwalk, CT, R. F. Culmo, private communication, March 1987.
5. See ref. 1, p. 65. B. Colombo, Carlo Erba Strumentazione, Rodano, Italy, private communication, February, 1987.
6. G. Weser, *Lab. Fach.*, October 1983; private communication, February 1987.
7. M. Matsumiya, Yanaco New Science, Inc., Kyoto, Japan, private communication, March 1987.
8. R. B. Fricioni, Leco Corporation, St. Joseph, MI, private communication, February 1987.
9. T. S. Ma and S. S. M. Hassan, *Organic Analysis Using Ion-Selective Electrodes*, Vol. 2, Academic Press, London, 1982.
10. See ref. 1, p. 100.
11. Kjel Foss Automatic, Foss Electric Ltd., York, England.
12. Labconco Kjeldahl Apparatus, Labconco Corporation, Kansas City, MO.
13. Kjeltec Auto Analyzer, Tecator, Höganäs, Sweden.
14. Z. Y. Hu, Shanghai Institute of Organic Chemistry, Academia Sinica, Shanghai, China, private communication, September 1984.
15. See ref. 1, p. 150.
16. B. Schlarp, Thomas Scientific, Swedesboro, NJ, private communication, February 1987.
17. L. Patrick, G. M. J. Powell, and P. Ridgway-Watt, Lab. Pract. *35*(11), 70 (1986); L. Patrick, private communication, May 1987.
18. See ref. 1, p. 178.
19. See ref. 9, p. 218.
20. F. Scheidl and V. Toome, Microchem. J. *18*, 42 (1978); F. Scheidl, private communication, May 1987.
21. Parr Instrument Company, Moline, IL.
22. T. S. Ma, in *Treatise on Analytical Chemistry* (I. M. Kolthoff and P. J. Elving, eds.), Part II, Vol. 12, Wiley, New York, p. 117, 1965.
23. See ref. 1, p. 189.
24. S. Hamel, Parr Instrument Company, Moline, IL, private communication, January 1987.
25. See ref. 1, p. 244.
26. See ref. 1, p. 270.
27. See ref. 1, p. 275.
28. See ref. 1, p. 277.
29. See ref. 1, p. 299.
30. T. S. Ma and A. S. Ladas, *Organic Functional Group Analysis by Gas Chromatography*, Academic Press, New York, 1976.
31. G. Dugan, Anal. Lett. *10*, 639 (1977).
32. See ref. 1, p. 338.
33. J. F. Sullivan, R. I. Grob, and P. W. Rulon, J. Chromatogr. *261*, 265 (1983).
34. J. F. Sullivan and R. L. Grob, J. Chromatogr. *268*, 219 (1983).
35. T. Hara et al., Bull. Chem. Soc. Jpn. *50*, 2292 (1977); *51*, 1110, 2951, 3079 (1978); *53*, 951, 1308 (1980); *54*, 2956 (1981); *55*, 329, 2127, 3450, 3800 (1982); *56*, 1378, 3615 (1983).

36. Q. I. Ou, K. H. Zeng, G. C. Wang, and W. L. Yu, in *Proceedings of the 4th International Symposium on Capillary Chromatography* (R. E. Kaiser, ed.), Hüthig, Heidelberg, p. 445, 1981.
37. J. M. Keane, D. C. Brown, and R. C. Fry, Anal. Chem., *57*, 2526 (1985); *58*, 790 (1986).
38. D. E. Pivonka, A. J. J. Schleisman, W. G. Fateley, and R. C. Fry, Appl. Spectrosc. *40*, 291, 464, 766 (1986).
39. T. S. Ma, C. Y. Wang, and M. Gutterson, Anal. Chem. *54*, 87R (1982); *56*, 88R (1984); *58*, 144R (1986).

II
SPECTROCHEMICAL INSTRUMENTATION

4

Instrumentation for Atomic Emission Spectroscopy

WILLARD A. HARELAND / Sandia National Laboratories, Albuquerque, New Mexico

I. INTRODUCTION

This chapter describes instrumentation for chemical analysis by atomic emission spectroscopy. Although atomic spectroscopy, in general, covers a broad range of disciplines, including absorption, fluorescence, and emission, this chapter is directed specifically toward analytical instrumentation for atomic emission spectroscopy using electrically generated energy to excite the atom. These include direct current (dc) and alternating current (ac) arcs, high voltage sparks, direct current, microwave, and inductively coupled plasmas, and electron and ion bombardment, such as glow discharge and hollow cathode sources. Instrumentation for the separation of electromagnetic radiation into discrete components and the measurement of spectra is presented.

II. PRINCIPLES OF ATOMIC EMISSION SPECTROSCOPY

When an atom is exposed to a high-temperature source, such as an electrical discharge, energy is transferred to the atom by collisions with energetic particles and by interaction with electromagnetic radiation. The excited atom decays to a lower energy level with emission of its own characteristic electromagnetic radiation. The frequency of the radiation is determined by the energy difference between the upper and lower energy states, ΔE, according to the equation

$$\Delta E = h\nu$$

where $h = 6.6254 \times 10^{-27}$ erg sec (Planck's constant) and ν is the frequency.

Spectrochemical analysis by atomic emission spectroscopy is based on the measurement of electromagnetic radiation emitted by electronically excited atoms and ions. The energy distribution of the radiation is unique to each element. By separating the electromagnetic radiation into its components and measuring the wavelengths, the element and sometimes the specific isotope can be identified. The radiant energy at a specific wavelength is a function of the concentration of the element. Detailed descriptions on the theory and principles of atomic spectroscopy can be found in a number of publications [1-7].

III. INSTRUMENTATION FOR ATOMIC EMISSION SPECTROSCOPY

An atomic emission spectrometer is designed for generating electronically excited atoms in an excitation source and separating and measuring the electromagnetic radiation emitted by the atoms. The instrumentation required for spectrochemical analyses can be described in terms of a simplified diagram in which a series of processes convert information concerning the elemental concentration and composition from the chemical domain to an observation domain. Figure 1 shows the information flow through a spectrometer and illustrates the processes that occur in atomic emission spectroscopy. The information, C, originates in the chemical domain in the form of the concentration of a particular element and is converted to an optical signal, I, by the sample excitation source. Electromagnetic radiation emitted by the excited sample is focused onto the entrance slit of a spectrometer. The incident radiation is resolved into discrete energy components by the spectrometer and is measured with a detector that converts the radiant energy, J, to an electrical signal, K. In the case of photographic detection, the spectral radiation interacts with a light-sensitive material to form a permanent image. The detector signal, K, is measured with a meter, computer, or other readout device in the observation domain.

A wide variety of instruments and techniques has been developed and applied for qualitative and quantitative analysis by atomic emission spectroscopy. Detailed descriptions of the various components and instruments for atomic emission spectroscopy are given next.

A. Generation of Electromagnetic Radiation

Arc Discharge

An arc discharge is a sensitive source for producing free atoms from a sample and electronically exciting the atoms to a high-energy state [4]. Samples are mixed with a conducting material, such as graphite, and packed into the crater of a carbon electrode. A dc arc discharge between the sample electrode and counter electrode vaporizes the sample, decomposes molecular species produced in the plasma, and atomizes the analyte. The atoms are excited to higher energy states by energetic particles in the arc plasma.

The dc arc is a unidirectional arc in which the sample is vaporized directly from an electrode. Power for the arc discharge is provided by a rectified power supply. The source unit consists of an isolation transformer and a rectifier circuit to convert the input ac voltage to dc voltage. Current is controlled in the circuit by means of a variable inductance in the supply side of the transformer. The rectifier produces a dc voltage with some ripple; however, the ripple can be reduced with an inductance circuit installed in series with the arc gap. The voltage drop across the arc gap ranges from 10 to 100 V, and the current can be varied from 1 to 30 A. The temperature in the dc arc discharge varies from 4000 to 8000 K, depending on the electrode material. The open circuit potential of 300 V is not sufficient to break down the gap; therefore, the arc discharge is initiated by striking the electrodes together momentarily or by applying a low-power spark across the arc gap. The arc discharge is

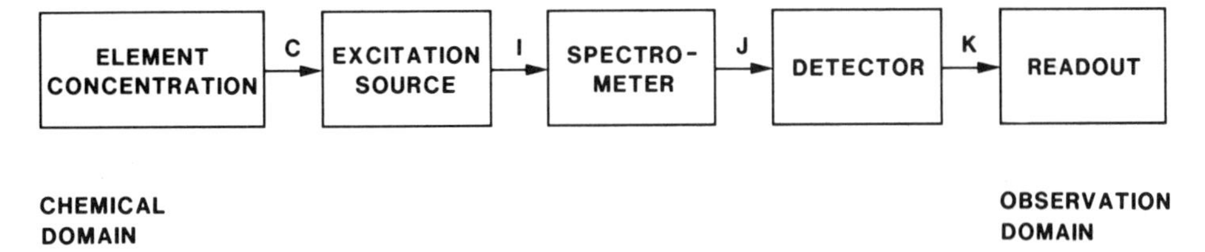

Figure 1. Conversion of information from a chemical domain to an observation domain by atomic emission spectroscopy: C, information; I, optical signal; J, electrical signal; K, detector signal.

maintained by charge carriers produced in the plasma. The dc arc provides a sensitive excitation source for the identification of trace elements because a large quantity of material is vaporized in the discharge. The minimum detectable limits for most elements range from 0.05 to 10 ppm. For quantitative analysis, however, the dc arc has a number of disadvantages over other sample excitation sources:

1. The arc tends to wander, resulting in poor analytical precision.
2. The arc temperature is determined by the species in the arc plasma; therefore, as volatile elements are vaporized, the composition and temperature of the arc change with time.
3. With selective volatilization, the more volatile elements distill from the sample electrode first, and the more refractory elements remain on the electrode.
4. The distribution of the elements in the arc plasma is not uniform throughout the length of the arc gap.
5. The arc temperature is low enough to permit molecular species, such as cyanogen, to exist in the arc plasma.

Many of the problems associated with dc arcs can be minimized or eliminated, however. The arc can be stabilized by using different electrode configurations to control the discharge point. Sample buffers are used to control the arc temperature and to maintain a smooth discharge [8]. Analytical errors resulting from selective volatilization can be reduced by burning the sample to completion. In some instances, selective volatilization can be used advantageously to improve the detection limits for certain elements. With graphite electrodes, the most serious interferences are due to molecular band emission from cyanogen, which is produced in the arc plasma by the combination of carbon from the graphite electrodes and nitrogen in the air. However, this interference is reduced by using an inert gas sheath around the arc to exclude nitrogen.

An ac arc provides more uniform sampling of the electrode than a corresponding dc arc. The ac arc operates at voltages ranging from 110 to 4400 V. The polarity of the discharge is reversed at each half-cycle, and the discharge is extinguished when the voltage drops to zero. Since the discharge is ignited during each half-cycle, sampling of the electrode is random, resulting in improved precision compared to dc arc excitation. The sensitivity of the ac arc source for the determination of trace elements is less than that of the corresponding dc arc.

High-Voltage (HV) Spark

An HV spark discharge provides one of the most accurate sources for spectrochemical analyses [2]. The circuit consists of an HV transformer, used to charge an oil-filled capacitor, and an auxiliary control gap to initiate and control the discharge. A potential of 15,000 to 40,000 V is developed across the analytical gap. The form and characteristics of the discharge are controlled by the capacitances, inductances, and resistances of the discharge circuit.

High-voltage spark sources are used for spectrochemical analysis of metals and alloys. A solid conducting sample is placed on a sample excitation stand of the type shown in Figure 2. An HV spark discharge between the metal sample and the counter electrode generates sufficient energy to vaporize the sample material and to excite the atoms. The emitted radiation is measured through an ultraviolet-transparent window mounted on the excitation stand. The stand has a flat plate on which to place the metal sample above a thoriated tungsten counter electrode. The source unit is designed such that the chamber is purged with argon gas during operation. To prevent deposition of the vaporized metal on the window, the gas inlet port is located in such a way that the gas sweeps over the window. With a controlled waveform source, the polarity of the discharge is not reversed.

Spark excitation sources are used in combination with direct-reading spectrometers to determine major, minor, and trace elements in metals and alloys. This type of instrument is used extensively in the metals industry to verify and control the composition of metal alloys during production. After sampling and surface preparation, the complete analysis for 15-20 elements can be done in less than a minute. Figure 3 is a standard calibration curve for silicon in low-alloy steels. The intensity of silicon is measured relative to the intensity of iron and plotted against the concentration ratio, [Si]/[Fe], in the standard. Thus, errors resulting from fluctuations in the spark excitation source are minimized. Typical detection limits for metals are 0.001 to 0.1 wt.%.

Glow Discharge Lamp

The glow discharge lamp is an extremely versatile radiation source for the atomic emission analysis of solid conducting samples [9-12]. A diagram of the glow discharge lamp designed by Grimm [13] is shown in Figure 4. The lamp consists of a flat sample cathode and cylindrical anode mounted in a sealed chamber containing an inert gas. When a potential of 600-1800 V is applied across the electrodes, the gas is ionized and accelerated to the cathode. Collisions of the energetic ions with the cathode surface cause vaporization of the sample material and excitation of the atomic vapor. The emitted electromagnetic radiation is viewed through a quartz window mounted in the glow discharge lamp.

The configuration of the anode, cathode, and vacuum port is very critical for the optimum operation of the glow discharge lamp. The cathode block, which is in direct contact with the sample, is water-cooled to remove excess heat generated in the electrical discharge. Argon or another inert gas is metered into the lamp through an inlet port located in the anode chamber, and the unit is continuously pumped through two evacuation

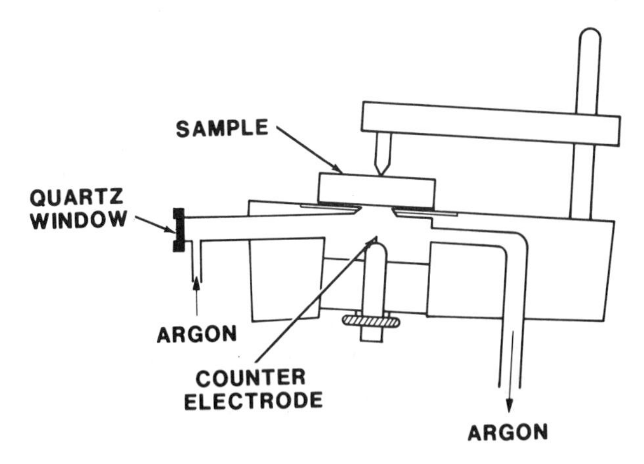

Figure 2. Spark excitation stand for point-to-plane analysis by atomic emission spectroscopy. (Adapted from Thermo Jarrell Ash, Waltham, MA, with permission.)

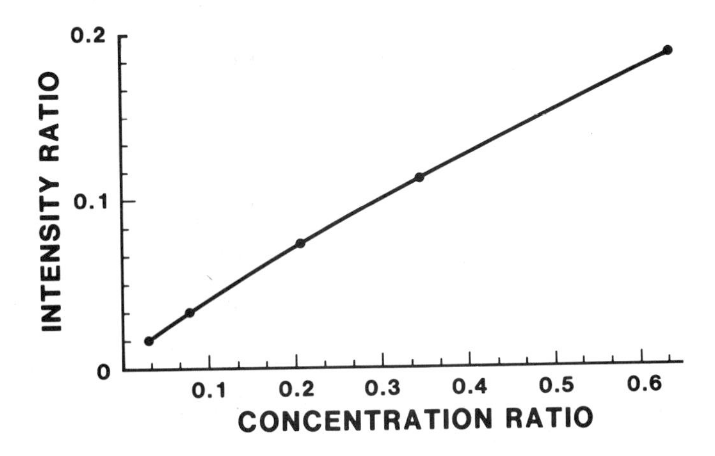

Figure 3. Standard calibration curve of intensity ratio I_{Si}/I_{Fe} versus concentration ratio [Si]/[Fe] for silicon in low-alloy steels with HV spark atomic emission spectroscopy.

ducts located near the sample cathode. A hollow cylinder, which projects from the anode chamber to within about 0.2 mm from the cathode surface, separates the two evacuation ducts. The purpose of the cylinder is to provide a barrier to control the gas pressure distribution across the cathode surface.

In the assembly and operation of the glow discharge lamp, a flat, electrically conductive sample is pressed against the open end of the

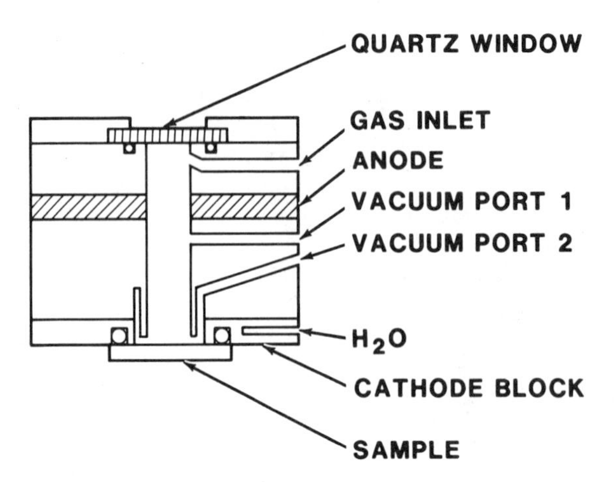

Figure 4. Glow discharge lamp designed by Grimm. (Reprinted with permission from Ref. 13.)

cathode body and held in place by vacuum. After evacuation of the chamber to remove traces of residual air and moisture, the chamber is backfilled with argon to a pressure of 5-15 torr. The pressure is maintained by metering the gas into the chamber at a constant rate while continuously pumping the chamber through the two evacuation ducts. The gas pressure distribution across the sample surface can be controlled in such a way that the sample is eroded in a uniform manner in the discharge. The body of the lamp can be fabricated from electrically nonconducting materials, which permits either the anode or cathode to be at ground potential [14].

The glow discharge lamp has been used to determine major, minor, and trace elements in a wide variety of iron-, copper-, gold-, and aluminum-based alloys. Its precision for spectrochemical analyses is equivalent to that of x-ray fluorescence spectroscopy and is typically 2% or less. Standard calibration curves for the determination of trace elements in pure materials are linear over three orders of magnitude in concentration. The atomic emission from the discharge provides a direct measure of the element concentration as a function of depth in the sample [15].

Hollow Cathode Discharge

Although the hollow cathode discharge has been used primarily as spectral line sources for atomic absorption spectroscopy, it is also an excellent excitation source for the determination of trace elements. The discharge tube consists of a hollow cylindrical cathode containing the sample material and an anode in a demountable chamber. The chamber is evacuated to remove residual gas and backfilled with an inert gas to a pressure of about 5 torr. When a potential of 600-1800 V is applied across the electrodes, the gas is ionized and accelerated to the cathode. Collisions with the cathode material cause atomization and excitation of the sample. The emitted electromagnetic radiation is measured through a quartz window mounted on the chamber.

Sampling in the hollow cathode discharge occurs by sputtering and vaporization caused by ion bombardment. The mechanism for sampling can be controlled by external cooling of the cathode. With a cold cathode, sampling occurs primarily by sputtering the surface. Without external cooling, the temperature of the cathode can exceed 2000 K, which is sufficient to melt many common metals and alloys. The volatile elements vaporize from the bulk of the sample. Because of fractional distillation from the hot cathode, the spectral radiation emitted by the cathode discharge must be monitored over an extended period [16].

Hollow cathode discharges have been used to determine trace elements in a wide variety of sample materials. Samples are prepared in either chip or powder form and are placed in the crater of a graphite electrode. Solutions can be analyzed by dispensing the sample solutions directly into a graphite electrode and evaporating the solution. Electrically nonconducting powders can be mixed with copper powder and prepared by briquetting the mixture in the form of a hollow cathode. The detection limits for 17 elements using the above sample preparation techniques have been reported [17], and they range from 0.06 to 10 parts per million (ppm) for solid samples and from 0.2 to 1.0 ppm for solutions.

Laser-Induced Microplasma

Pulsed lasers have been used in atomic emission analyses of extremely small samples and occlusions on the surfaces of metal alloys [18,19]. The laser beam is focused on the sample surface and the material is vaporized. The plasma plume formed above the target passes through an auxiliary electrode gap and is excited by a low-voltage spark discharge. The electromagnetic radiation is resolved with a spectrograph, and the spectra are recorded photographically or measured with an array detector.

A diagram of a laser-induced microplasma source is shown in Figure 5. The unit consists of a solid-state Nd glass laser, a low-voltage spark

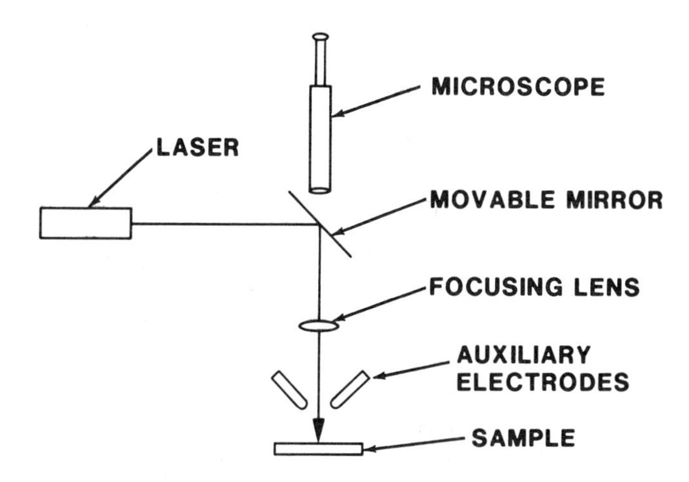

Figure 5. Laser-induced microplasma source.

source, and a mirror and focusing lens to direct the laser beam onto the surface of the sample. A laser pulse with an energy of 0.1-1.0 J and a duration of about 2 μsec produces a hemispherical crater with a diameter of 25-250 μm. The vaporized sample is excited by means of a spark discharge between two graphite electrodes positioned above the sample surface. The target area is identified visually with a microscope and is positioned under the laser beam by moving the sample, using a microscope stage and cross hairs for alignment. With the laser sampling source, as little as 10^{-9} to 10^{-11} g of an element can be detected [20].

Rotating Electrode

Spectrochemical analyses of liquids can be performed using spark excitation; however, special techniques and equipment are required for sample introduction. A rotating electrode source, which is shown in Figure 6, is used for the analysis of viscous materials, such as oils. The source unit consists of a cylindrical electrode and a pointed counter electrode in which a low-frequency spark discharge is applied across the electrode gap. The cylindrical electrode is mounted on a mandrel that is turned at a constant speed by a motor. The edge of the cylindrical electrode is in contact with a liquid sample contained in a small sample boat. As the electrode rotates, the solution is carried up into the spark gap, where it is vaporized and excited in the spark discharge. With the rotating electrode, the detection limits for trace metals in oils are comparable to those achieved by flame emission and atomic absorption spectroscopy [20A].

Plasma Sources

High-temperature plasmas are generated from electrical power sources. These differ from conventional flames in that the high temperatures result from collisions induced by electrical currents. Three general types of plasmas are commonly used in analytical atomic emission spectroscopy for the analysis of solid, liquid, and gaseous samples. They are direct-current plasmas (DCP), inductively coupled plasmas (ICP), and microwave-induced plasmas (MIP). The plasma sources can be distinguished by the characteristic frequencies and the mode of coupling.

A high-temperature plasma is produced in a dc discharge by constricting the arc through mechanical, hydrodynamic, or magnetic processes. A dc plasma source consists of a dc arc discharge between two electrodes in which a gas, such as argon, is introduced. Tangential flow of the gas into the arc causes the emergent hot gases to cool in such a way that the resultant gas jet is constricted. Constriction of the plasma causes an increase in the electron density with a concomitant increase in temperature. Self-induced magnetic fields are created as a result of the increased current density and further tend to constrict the arc. Figure 7 shows a diagram of a dc plasma source in which the gas exits from the discharge through a small orifice.

Direct-current plasmas used in atomic emission spectroscopy operate at a current of 5-30 A and gas flow rates up to 70 liters per minute, depending on the type of gas. Temperatures in the plasma range from

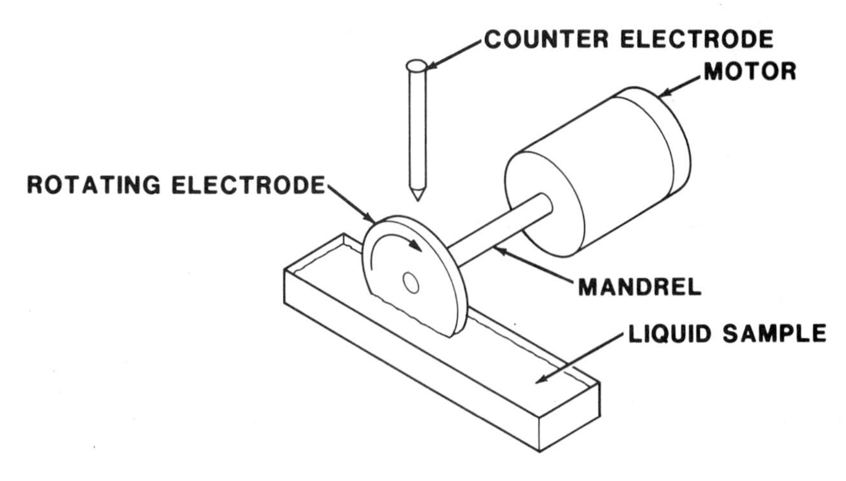

Figure 6. Rotating electrode source.

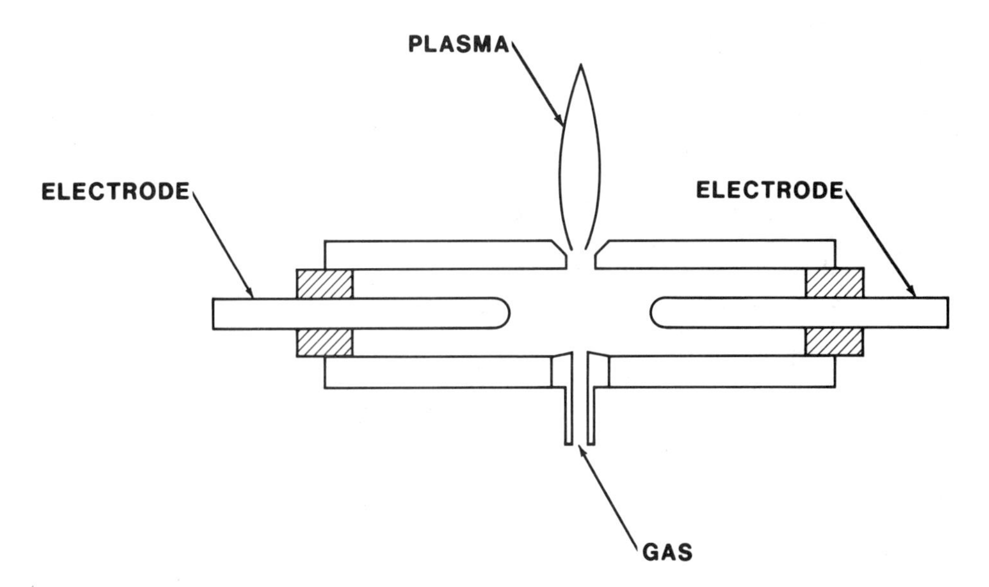

Figure 7. Direct current plasma source.

4700 K to over 11,000 K. With higher currents, however, temperatures of 16,000 K have been achieved.

A commercial three-electrode dc plasma source is shown in Figure 8 [21]. The unit consists of two pyrolytic carbon anodes and a single thoriated tungsten cathode. Argon gas is directed into the arc discharge through two ceramic sleeves that surround the anodes. The sample solution is introduced into the plasma as an aerosol by means of a nebulizer chamber

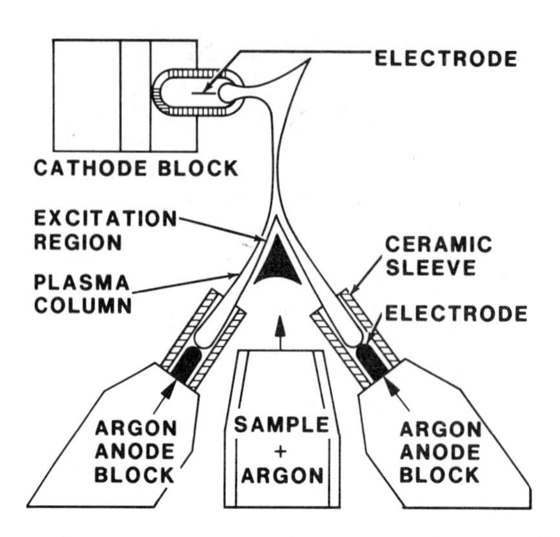

Figure 8. A commercial three-electrode direct-current plasma source. (Reprinted with permission from Ref. 21.)

positioned between the two anodes. The optimum measurement region is located near the base of the confluence, which is identified in Figure 8 as the shaded triangular region. In this way, the high background from the arc plasma is avoided in analytical work.

Direct-current plasma sources can be used for the analysis of a wide variety of materials. It is possible to analyze solutions containing up to 45% total dissolved solids. Under these conditions, however, the standards must be prepared in similar matrices to match the samples because the properties of the plasma and sample nebulizer vary with the sample matrix. The major advantage of the dc plasma over other plasma sources is that the plasma can be generated using a simple, low-cost dc power supply. However, continuous operation over long periods of time is not possible because the electrodes erode. Consequently, electrodeless plasma sources, such as the inductively coupled plasma, are more common in atomic emission spectroscopy.

The inductively coupled plasma is a high-temperature source in which energy to form the plasma is generated by a high-frequency magnetic field [22]. The operation of the plasma is analogous to that of a transformer in which the primary winding is an induction coil with an oscillatory current from a radiofrequency generator. The secondary winding, or load, is ionized argon gas, which is coupled inductively to the radiofrequency source. A diagram of an inductively coupled argon plasma is shown in Figure 9. The plasma torch consists of two concentric quartz tubes with argon gas flowing tangentially between the inner and outer tubes to provide cooling, and a water-cooled induction coil near the periphery of the outer tube. Argon gas flowing through the smaller tube interacts with the magnetic field when sufficient external energy is applied to ionize the gas. Initial ionization can be provided by a Tesla discharge, and the plasma is sustained by ionized argon formed in the high-temperature plasma. The sample is directed into the plasma through a capillary tube mounted in the center of the torch assembly. In commercial plasma torches, a quartz bonnet is inserted between the outer quartz cylinder and load coil to center the torch and to prevent direct contact between the coil and hot silica tube.

A radiofrequency generator with an output power of 1-5 kW at a frequency of 3-100 MHz will sustain an argon plasma. The temperature of the plasma varies from 5000 to 10,000 K, depending on the power, gas flow rate and composition, and coupling efficiency. The radiofrequency circuit is tuned to maximize the efficiency. The temperatures in the plasma vary over the length of the plume, with the hottest region within the annular region.

Some elements are completely (>99%) ionized in inductively coupled plasmas. Figures 10(a) and 10(b) are monochromatic images of an inductively coupled argon plasma showing the distribution of neutral and ionized calcium at 422.673 and 393.367 nm, respectively. Neutral calcium emission is at a maximum near the top of the load coil; however, the intensity decreases rapidly and is nearly constant up to a distance of 15-20 mm above the

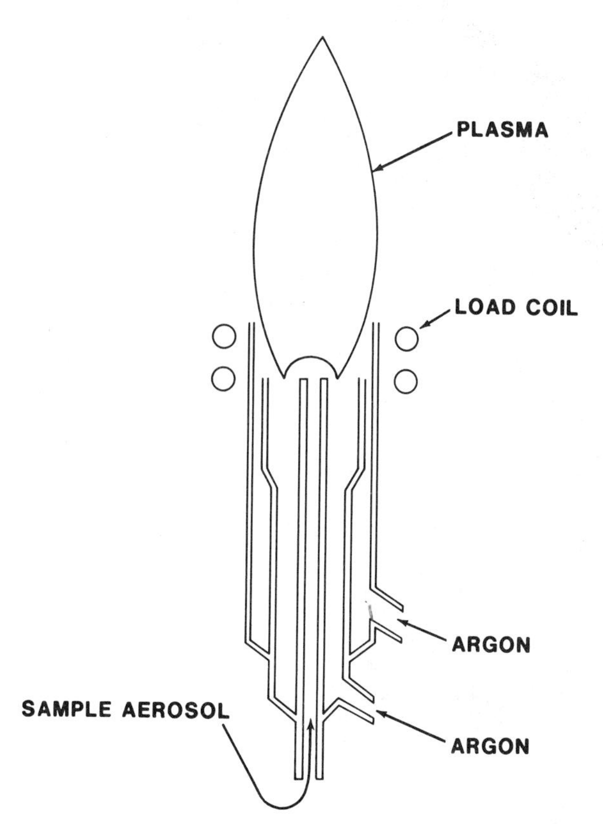

Figure 9. Inductively coupled argon plasma source.

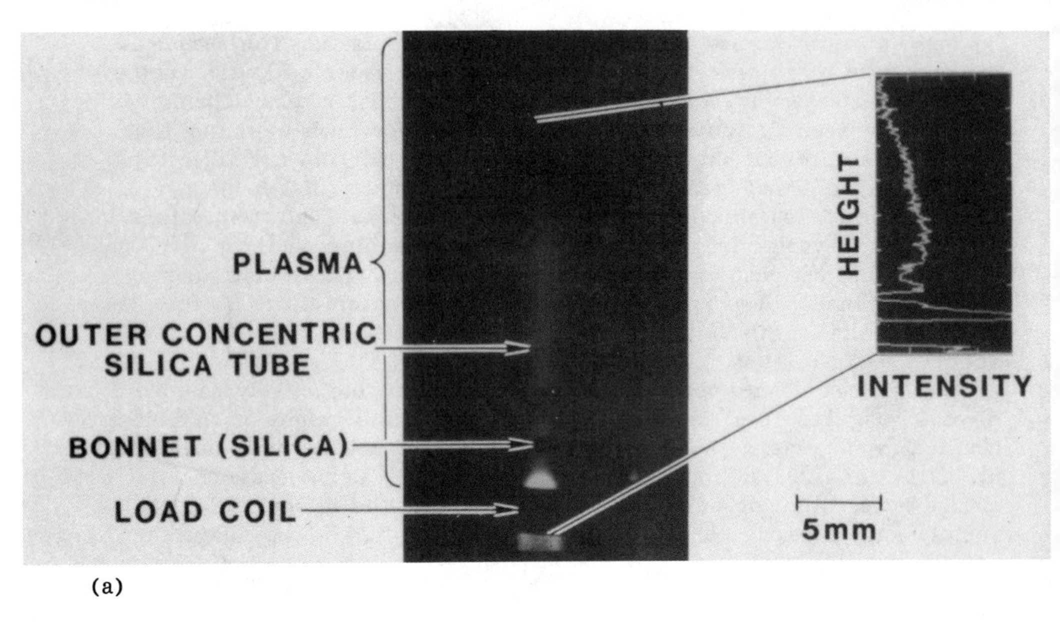

(a)

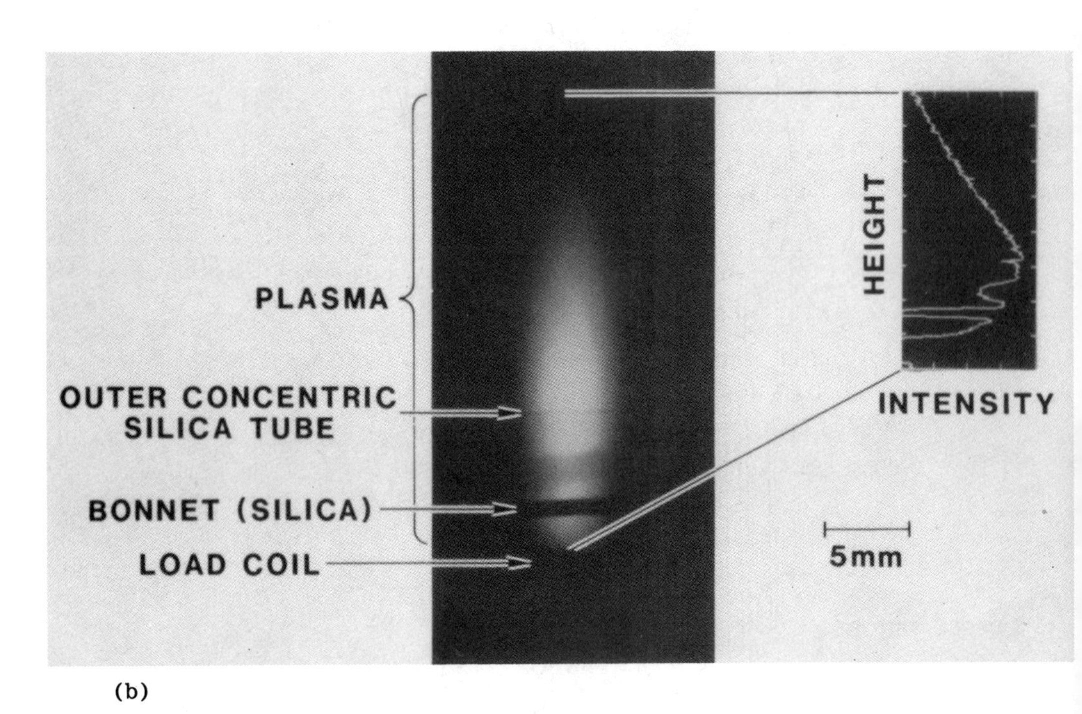

(b)

Figure 10. Monochromatic images of an inductively coupled argon plasma showing the relative distribution of neutral and ionized calcium in the plasma: (a) 422.673 nm; (b) 393.367 nm.

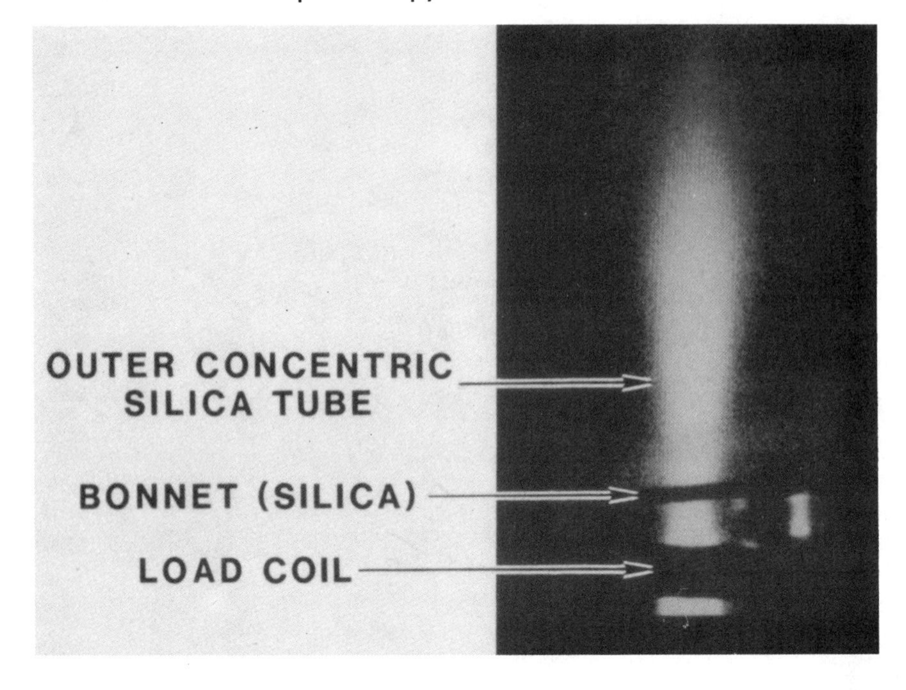

Figure 11. Computer-enhanced monochromatic image of the inductively coupled argon plasma in Figure 10(a) showing the relative distribution of neutral calcium in the plasma.

load coil. An analysis of the spatial distribution using a computer-enhanced image of Figure 10(a) shows that weak neutral calcium emission occurs in the center core of the plasma and in a region near the surface of the plasma plume, as shown in Figure 11. The latter appears as a result of the recombination of ionized calcium and the subsequent excitation of neutral calcium in the cooler regions of the plasma. Because ionized calcium emission intensity is nearly constant in a region from 3 to 10 mm above the load coil and decreases linearly with increased height in the plasma, choice of the optimum region for measuring the emitted electromagnetic radiation in the plasma is very critical. Optimum viewing heights vary with the analyte. In multielement ICP analyses, the plasma region is selected by compromise.

The inductively coupled plasma is an extremely versatile source for atomic emission spectroscopy. Sample solutions are nebulized into a chamber, and the aerosol is directed into the plasma using a carrier gas. By heating the chamber and passing the aerosol over a condenser, the solvent can be partially removed, increasing the efficiency of the sample introduction system. Standard calibration curves for most elements are linear over five to six orders of magnitude in concentration. Detection limits have been determined for many elements and range from 0.1 to about 200 parts per billion (ppb) in solution [23-27].

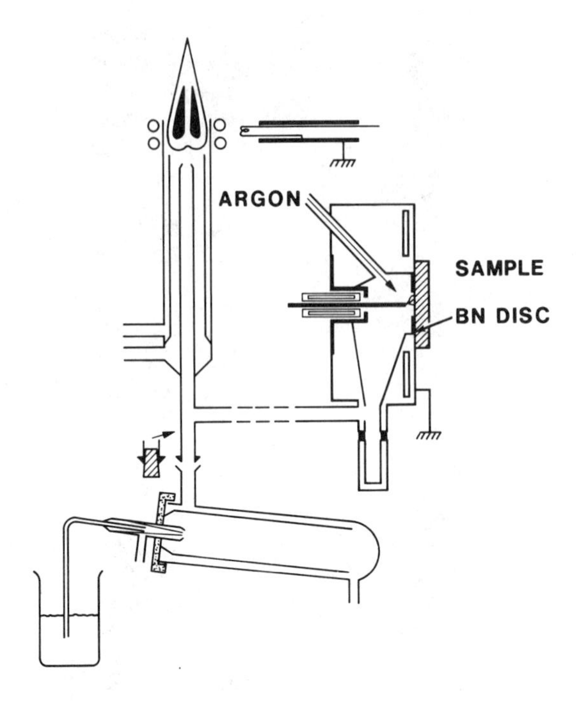

Figure 12. Spark ablation source coupled to an inductively coupled plasma. (Reprinted with permission from Ref. 28.)

Particulate materials produced by spark ablation can be directed into an inductively coupled plasma and analyzed [28]. Figure 12 is a diagram of a solid sampling device coupled to a plasma source. The sample is ablated using a low-voltage spark discharge, and the particulate material is transferred to the plasma by means of a carrier gas. This device has been used to determine minor elements in various types of aluminum alloys. To analyze nonconducting powder samples, pellets are prepared by mixing the samples with copper powder and briquetting the mixture. The detection limits for several common impurity elements in Al_2O_3 have been determined using the briquetting technique. These limits range from 0.6 to 265 ppm, comparable to the detection limits using conventional sample dissolution techniques and inductively coupled plasma analyses.

Microwave-induced plasmas (MIP) have been used as excitation sources for the analysis of gaseous samples and applied in many areas of atomic emission spectroscopy. The plasma source consists of a microwave resonator, or cavity, inductively coupled to a high-frequency generator by means of a coaxial waveguide. The microwave generator provides 200-500 W of power at a frequency of 2.45 GHz. Power is transferred to the microwave cavity by means of a coupling loop.

The TM010 microwave cavity designed by Beenakker [29] (Figure 13) is a 10-mm-high by 93-mm-diameter copper cylinder. The diameter of the cavity is critical for achieving a maximum electric field strength with a frequency of 2.45 GHz. In the Beenakker design, the field strength is a maximum at the center of the cavity. The discharge tube is mounted in the axial position. Plasma temperatures measured in an atmospheric pressure helium microwave plasma with 100 W of forward power ranged from 4420 to 5720 K [30].

In the operation of the microwave-induced discharge, a support gas, such as argon, helium, nitrogen, or air, is directed into the discharge tube. If the cavity is tuned properly, the plasma ignites spontaneously. Alternatively, the plasma may be ignited with a Tesla discharge. A concentric tube mounted in the discharge tube directs samples into the plasma.

Microwave-induced plasmas have been used to determine a wide variety of elements, including H, C, N, O, F, Cl, Br, and S. Detection limits for most elements range from 1 to 50 ppb, and calibration curves are linear over three to four orders of magnitude in concentration [31].

B. Separation of Electromagnetic Radiation

Electromagnetic radiation can be separated into its components by a variety of instruments and techniques. Dispersion techniques based on diffraction of electromagnetic radiation with a grating are commonly used in most commercial instruments. For high-resolution spectroscopy, the incident electromagnetic radiation can be separated in two dimensions by diffracting the beam with an Echelle grating and separating multiple orders with a prism [32]. Fourier transform spectroscopy has been developed that extends the wavelength range to include the ultraviolet region.

Grating Instruments

Two common types of grating spectrographs are based on concave and plane gratings. Concave grating instruments can be described using a

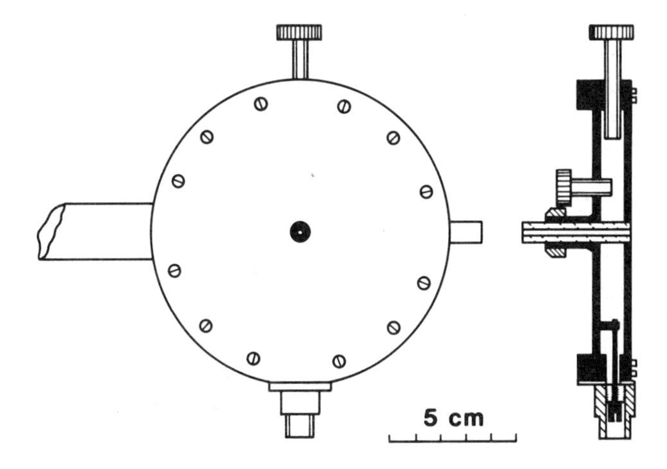

Figure 13. TM010 microwave cavity (designed by Beenakker). (Reprinted with permission from Ref. 29.)

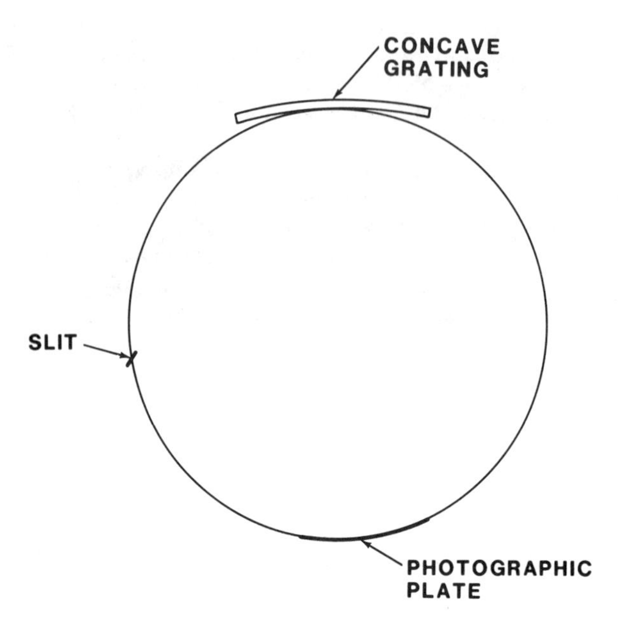

Figure 14. Diagram of the Rowland circle showing the relative position of the slit, concave grating, and detector.

Rowland circle. Using a grating ruled on a spherical lens with a radius of curvature R, Rowland observed that when a point source and a concave grating are positioned on a circle with a diameter equal to the radius of curvature, the diffracted light beam is in focus on the circle [2]. A diagram of the Rowland circle with a photographic plate as the detector is shown in Figure 14.

In the Rowland mounting the grating and detector are attached to a rigid bar that moves relative to a fixed entrance slit. The angle of diffraction remains constant, and the angle of incidence varies. The detector is always kept in a position normal to the grating. Under these conditions, the dispersion is linear over a broad wavelength range.

The properties of the Abney mounting are identical with those of the Rowland mounting. With the Abney mounting, however, the grating and detector are fixed and the slit moves. The slit is attached to a bar that pivots on the axis of the Rowland circle. In this way, the slit is always on the Rowland circle. The Abney mounting is very difficult and cumbersome to use because, to keep the slit parallel to the light beam, the slit must be rotated as the slit position is changed. In addition, light source and external optics must be repositioned when the angle of incidence is changed.

The Eagle mounting is a compact instrument with features similar to those of the larger Rowland mounting. In this mounting, however, the slit and detector are mounted on the same side of the grating normal to all components on the Rowland circle. The wavelength region is changed by rotating the grating and photographic plate and moving the grating in a direction parallel to the incident beam such that all components are positioned on the Rowland circle. A diagram of the Eagle mounting showing the relative positions of the grating and detector is given in Figure 15.

The Paschen-Runge mounting has all components mounted at fixed positions on a Rowland circle. The positions of the slit, grating, and detector are optimized to minimize astigmatism. The Paschen-Runge mounting is used in most commercial direct-reading spectrometers. In the detector module, the exit slits, photodetectors, and internal optics are mounted in fixed positions. For precise alignment of the exit slits with the diffracted spectral radiation, refractor plates may be mounted in front of each slit and used for fine adjustment to redirect the beam into the exit slit.

The Wadsworth mounting utilizes a spherical mirror to collimate the incident beam and to illuminate the grating with parallel light. When the wavelength region is changed, the detector-to-grating distance must be changed to maintain proper focus of the diffracted light.

The Ebert-Fastie and Czerny-Turner mountings utilize a plane grating in combination with a spherical mirror to collimate the incident beam and to focus the diffracted light on the exit slit or detector. The Ebert-Fastie mounting (Figure 16) has a single mirror and a plane grating positioned at an intermediate point between the mirror and slits. The incident beam passes over the grating and the diffracted light passes under the grating. The wavelength region is selected by rotating the grating. Since the detector is near normal to the grating plane, the dispersion is approximately linear. A major advantage of the Ebert-Fastie mounting is that a prism can be substituted for the grating.

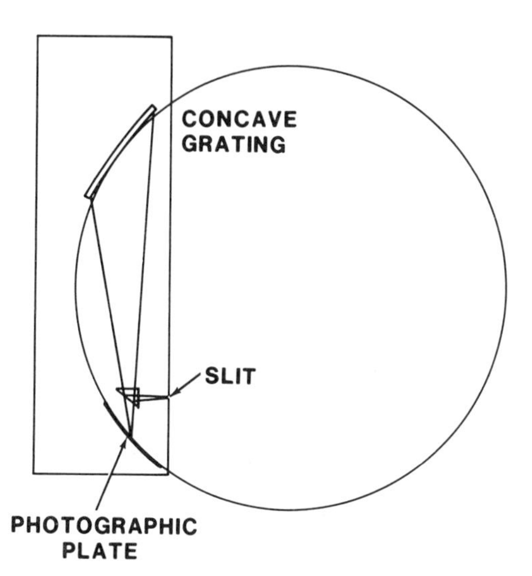

Figure 15. Diagram of the Eagle mounting showing the relative position of the concave grating, slit, and detector with respect to the Rowland circle.

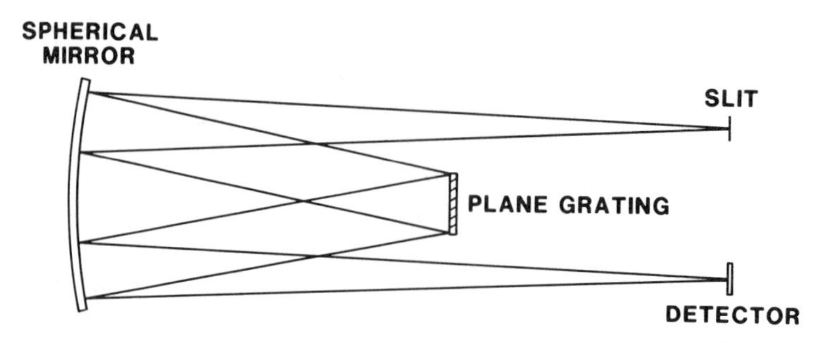

Figure 16. Diagram of the Ebert-Fastie mounting (side view).

The Czerny-Turner mounting differs from the Ebert-Fastie mounting in that the spectrometer has two spherical mirrors as shown in Figure 17. Compact Czerny-Turner polychromators are used in combination with array detectors for the simultaneous measurement of spectra over an extended wavelength range. In many research applications, array detectors have replaced photographic emulsions for the measurement of spectra.

Fourier Transform Spectrometer

A Fourier transform spectrometer separates incident electromagnetic radiation by encoding the optical signal, measuring the encoded signal with a detector, and decoding the detector output signal with a computer or other type of numerical processor. The Fourier transform spectrometer consists of an interferometer with a movable and fixed mirror assembly, a photomultiplier detector to convert electromagnetic radiation into an electrical signal, and a computer to digitize the interferogram and convert the interferogram to a spectrum using a fast Fourier transform (FFT) algorithm. The interferometers commonly used in Fourier transform spectroscopy are based on an original design by Michelson. Figure 18 is a diagram of an interferometer [33]. The instrument has a beam splitter in which approximately one-half the incident light is transmitted and the remaining is reflected from the surface, neglecting losses due to optical components. The transmitted beam is reflected back to the beam splitter by means of a fixed mirror where the beam is reflected to the photodetector. The incident beam that was initially reflected by the beam splitter is directed onto a second mirror mounted on a movable assembly. The reflected beam passes through the beam splitter and is combined with the beam from the fixed mirror. The photodetector measures the combined light beams. The position of the movable mirror is measured and monitored with a laser interferometer. The movable mirror assembly is supported by air bearings. A compensator, which is a transparent plate with a thickness equal to that of the beam splitter, equalizes the optical paths along both arms of the interferometer.

The function of the interferometer is to produce two distinct beams from a common source such that the distances traversed by one of the beams can be varied. The photodetector output signal represents the

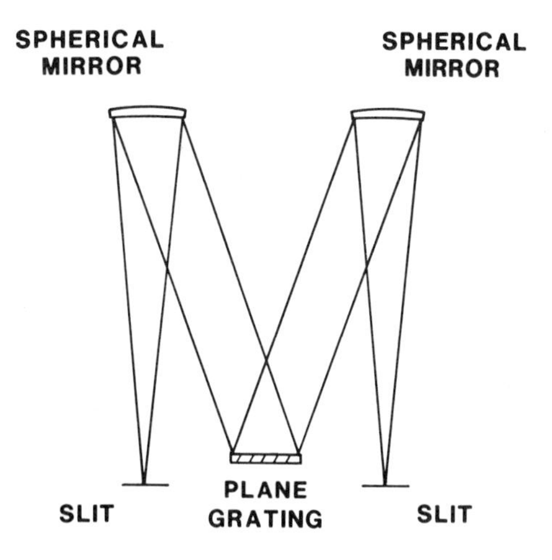

Figure 17. Diagram of the Czerny-Turner mounting.

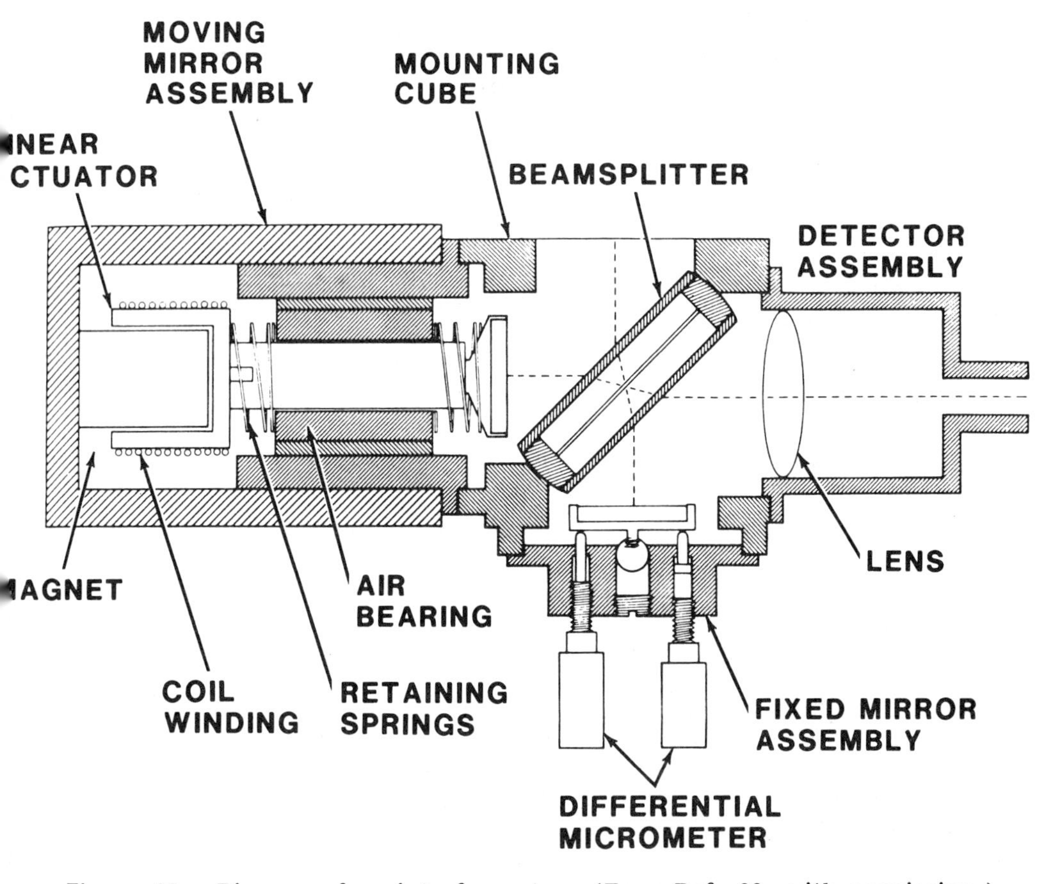

Figure 18. Diagram of an interferometer. (From Ref. 33, with permission.)

combination of the beams from the fixed and movable mirrors. When the optical path lengths along both arms of the interferometer are identical, all frequencies from the incident light source are in phase, and the photodetector output signal is a maximum. As the path length in one arm of the interferometer is changed, electromagnetic radiation of different frequencies change in intensity because of phase differences and produce a complex pattern representing a combination of cosine signals for each frequency component in the incident light source. In a simplified case using monochromatic radiation of wavelength λ and a mirror displacement of $\lambda/4$, the distance traversed by the beam differs by $\lambda/2$ and the two beams combine destructively. When the mirror is displaced by $\lambda/2$, the beams combine constructively. As the mirror is moved, the photodetector output signal is a cosine wave with a frequency equal to that of the incident monochromatic light source and can be represented by

$$I(X) = \frac{B(\nu)}{2} \cos 2\pi\nu X$$

where I(X) is the intensity at an optical path difference X, $B(\nu)$ is the intensity of the incident radiation, ν is the frequency of the source radiation in wavenumber units (cm^{-1}), and X is the optical path difference between the two arms of the interferometer. The intensity of the source radiation is reduced by a factor of 2 because of transmission and reflection of one-half the radiation back to the source. In addition, absorption and scattering of light by the beam splitter and optics contribute to light losses.

For broadband or continuum radiation, the intensity of light at the detector represents the sum of the intensities of all frequencies, as shown by

$$I(X) = \frac{1}{2} \int_{-\infty}^{\infty} B(\nu) \cos 2\pi\nu X \, d\nu$$

$$B(\nu) = \frac{1}{2} \int_{-\infty}^{\infty} I(X) \cos 2\pi\nu X \, dX$$

Since I(X) is an even function in which the interferogram is symmetrical around $\pm X$, the equation can be expressed as

$$B(\nu) = \int_{0}^{\infty} I(X) \cos 2\pi\nu X \, dX$$

For qualitative and quantitative analysis by Fourier transform spectrometry, the emission source must provide a stable signal over the entire measurement time, which is determined by the time required for the movable mirror to travel a minimum distance in the interferometer. This signal stability requirement excludes consideration of a number of sample excitation sources. The dc arc and hollow cathode discharge could not be used because of selective volatility of various elements in the plasma. Faires [34] and Stubley [35] have studied the applicability of Fourier transform spectrometry for atomic emission analysis using inductively coupled plasma sources. The capabilities of Fourier transform atomic emission spectroscopy are superior in many ways to those of conventional spectrochemical tech-

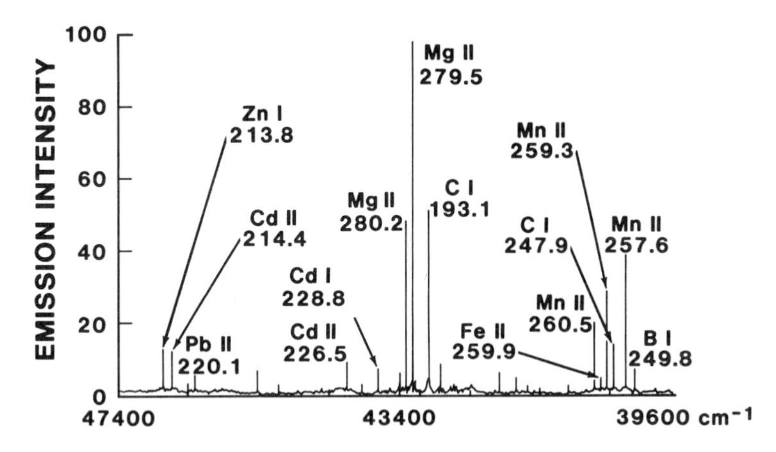

Figure 19. Emission spectrum of a 10-ppm multielement solution analyzed by inductively coupled plasma/Fourier transform spectrometry. (From Ref. 35, with permission.)

niques. An atomic emission spectrum can be measured over a broad wavelength region with a Fourier transform instrument. This offers the same advantages as an instrument with photographic detection (multiplex advantage). Since the electromagnetic radiation is measured with a photoelectric detector, the sensitivity and linear dynamic range are comparable to those of direct-reading spectrometers. However, the detection limits differ by an order of magnitude because of increased noise (multiplex disadvantage). Figure 19 is an emission spectrum obtained by nebulizing a solution containing 10 ppm of 16 elements into an inductively coupled plasma source. The 193.1-nm carbon line near the 279.5-nm Mg line is due to aliasing [35]. Aliasing is caused by undersampling of the interferogram [35A].

C. Detection of Electromagnetic Radiation

Photoelectric

The conversion of electromagnetic radiation into electrical energy is the most commonly used technique for the measurement of atomic spectra. Phototransducers measure radiation by converting the radiant energy into an electrical signal [36]. The photomultiplier tube consists of a photocathode, a series of dynodes, and an anode in an evacuated tube. The circuit is designed such that the photocathode and individual dynodes are at different potentials relative to the anode. A potential of 500-2000 V is applied between the cathode and anode. The dynodes form a divider network between the photocathode and anode, with each dynode at a more positive potential than the previous dynode.

The interaction of a photon with the photocathode surface causes the ejection of an electron. The electron is accelerated to a dynode by the electric field. Collisions between the energetic electrons and the dynode

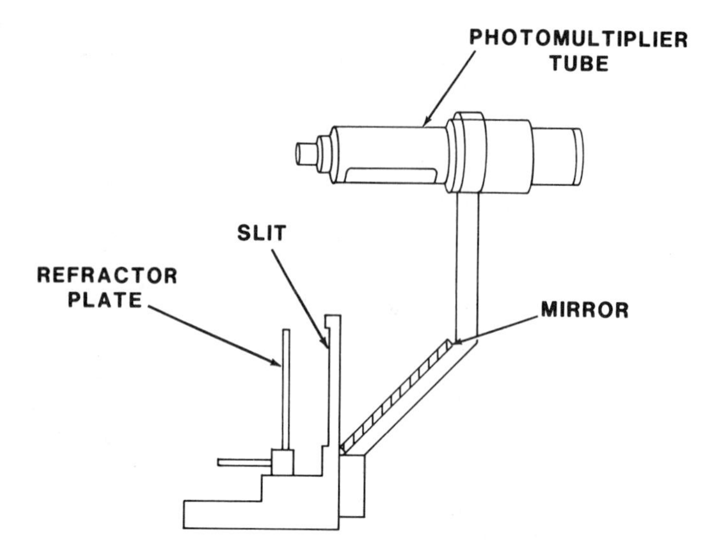

Figure 20. Diagram of a photodetector assembly for a commercial direct-reading atomic emission spectrometer. (Adapted from Thermo Jarrell Ash, Waltham, MA, with permission.)

surface eject several secondary electrons from the surface. The secondary electrons, in turn, are accelerated to the second dynode in the series. The net effect is the production of multiple electrons at each stage in the dynode chain. The amplification is typically one million and depends on the number of dynodes in the chain and the number of electrons emitted per incident electron. The latter is determined by the dynode material and the applied voltage in the photomultiplier tube. Spectral sensitivities for various photocathode materials used in photomultiplier tubes have been published [37].

In one direct-reading atomic emission spectrometer, a detector module of the type shown in Figure 20 is mounted on the focal curve at a position corresponding to a specific spectral line. The line image is focused on the slit. The light is directed to a photomultiplier detector and measured electronically. A refractor plate, which is a flat, optically transparent material, may be used to direct the spectral radiation onto the slit by rotating the plate. Some commercial instruments use a slit mask in place of individual exit slit assemblies. Direct-reading spectrometers with over 60 photodetectors are commercially available.

Photographic

Photographic detection is one of the most sensitive and versatile techniques for measuring and recording spectral radiation. A plastic film or glass plate with the photographic emulsion coated on the surface can be placed on the focal curve of the spectrometer and an image of the spectrum can be recorded. Long-term exposure of the spectrum to the photographic emulsion results in integration of the incident spectra. Variations in the intensity of the radiation that may be caused by instability in the sample excitation source are minimized.

D. Analysis of Photographically Recorded Spectra

A densitometer is an instrument used for measuring and identifying elements from photographically recorded atomic emission spectra. The instrument consists of a photoplate, or film holder assembly, and a light source and associated photodetector to measure the quantity of light passing through the photographic emulsion. The photoplate is moved in a direction normal to the light beam and can be positioned to measure any segment on the photoplate. An image of the spectrum is magnified and displayed on a screen. A reference spectrum with spectral lines of known elements is displayed adjacent to the sample spectrum.

Manual Analyses

For qualitative analyses, the sample and reference spectra are aligned such that known spectral lines coincide. Usually a known element, such as iron, is recorded on the photoplate to serve as a reference. Master reference spectra with the most sensitive lines of 70 elements are available, and these spectra can be matched to any linear dispersion spectrograph. If an unknown spectral line is not identified on the master plate, the wavelength of the line can be estimated and the element identified from standard wavelength tables [38-43].

For quantitative spectrochemical analyses, the densities of selected spectral lines are determined by measuring the intensity of light passing through the emulsion. The density of the line image is measured relative to an unexposed portion of the photoplate and compared with values obtained from standards analyzed under similar conditions. Since the photographic emulsion does not respond linearly with incident radiant energy, a correction factor is needed to compensate for the response of the photographic emulsion.

Emulsion calibration curves comparing optical density with intensity can be derived using a number of mathematical techniques. Many are based on measuring the densities of recorded images using standard sources. Probably the most common technique for general laboratory work is the two-step method [44,45]. With this method, an iron spectrum is recorded using a neutral-density filter to attenuate a segment of each spectral line. This produces a line spectrum with a segment of each line reduced in density by a known factor. The ratio of the densities is determined as a function of the incident radiant energy or intensity, and the measured values are used to construct preliminary and emulsion calibration curves. Since the photographic emulsion does not respond equally to electromagnetic radiation over all wavelengths, separate emulsion calibration curves covering selected wavelength regions are required for accurate spectrographic analyses.

Computer Analyses

The analysis of photographically recorded emission spectra is extremely time-consuming and laborious. To reduce the analysis time and improve the accuracy and efficiency for spectrographic analyses, a number of automated instruments and techniques have been developed. The first and probably the most extensive effort for high-speed spectrographic analyses occurred with the compilation of wavelength tables by Harrison and co-workers [38] at the Massachusetts Institute of Technology. The goal of this effort was to determine the wavelengths of all detectable spectral lines in spectra of known elements. These tables were compiled and published in the mid-1930s and represent the most complete and extensive tables of atomic spectra to date. Photographically recorded spectra were analyzed with an instrument that monitored the position of the photoplate while the spectrum was scanned and measured. When a peak was detected or noted by a change in slope of the photodetector output signal, the position corresponding to the peak was recorded.

Modern instruments for spectrographic analyses are based on similar concepts and techniques. With modern instruments, however, many of the functions necessary for high-speed spectrographic analyses can be performed in real time with a computer [46-54].

For computer analysis of photographically recorded emission spectra, the spectrum is digitized by measuring the transmittance of the photoplate at regular intervals and storing the data in computer-readable form. The measurement interval must be comparable to the half-width of a spectral line. A spectral line has a half-width of about 0.002-0.003 nm, and the line broadening is due primarily to Doppler effect. Using a spectrograph with a linear reciprocal linear dispersion of 0.5 nm/mm, the measurement interval should be about 0.004-0.006 mm per data point. The measured transmittance values are stored sequentially in memory, and the individual values can be accessed based on the memory address. A diagram of a computer-controlled microphotometer for analyzing photographically recorded spectra is shown in Figure 21. A high-resolution linear measurement transducer mounted directly on the plate-holder assembly is used to monitor the position of the photographic plate as the spectrum is scanned. The output signal from the linear measurement transducer gates the analog-to-digital converter and logs the transmittance reading directly into memory. Spectral lines are identified in the digitized spectrum by locating the memory address or position of each spectral line and calculating its wavelength based on the relative position of the line center with respect to a reference point, or known line, in the spectrum. A technique for determining the position of a peak is illustrated in Figure 22. Consecutive transmittance readings are compared with previous readings to determine the direction and magnitude of the changes in transmittance values. If a peak is detected, the transmittance values are fitted to a second-order polynomial. The first derivative of the polynomial, when set equal to zero, identifies the exact location corresponding to the peak. The equation for calculating the wavelength of a spectral line is determined from the dispersion of the spectrograph. For prism spectra, the wavelength can be calculated using the Hartmann dispersion formula,

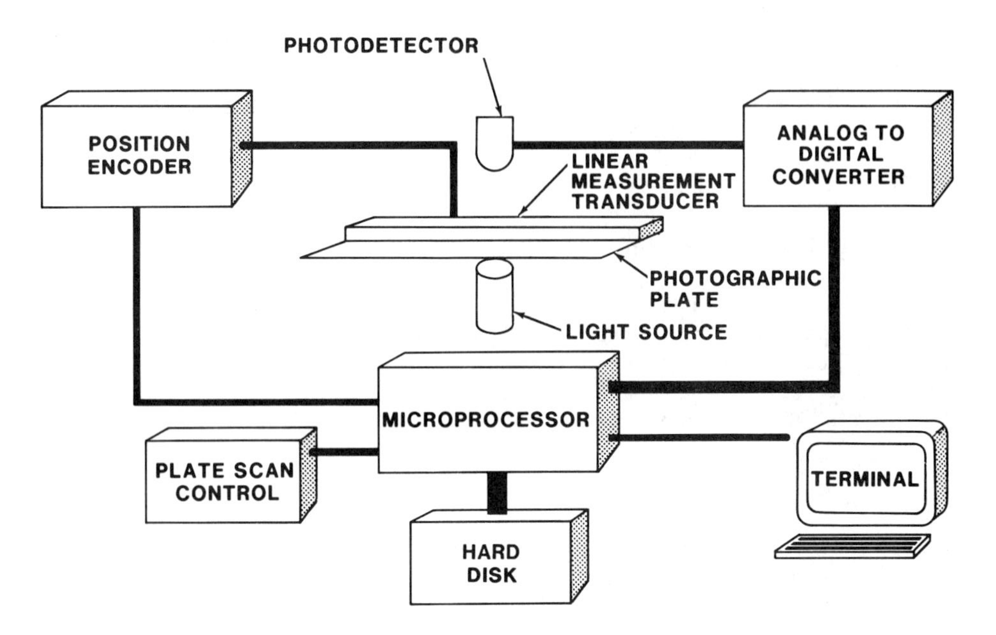

Figure 21. A computer-controlled microphotometer.

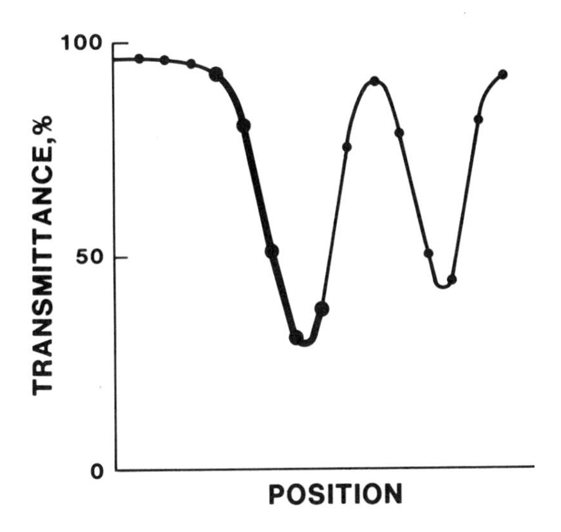

Figure 22. Measured transmittance values corresponding to a spectral line are fitted to a second-order polynomial. The first derivative of the polynomial is set equal to zero and solved to determine the position of the peak minimum.

$$\lambda = \lambda_0 + \frac{c}{d_0 - d}$$

where c, λ_0, and d_0 are constants. The angular dispersion of grating spectra varies only slightly with wavelength and can be described by

$$\frac{d\beta}{d\lambda} = \frac{n}{d \cos \beta}$$

where β is the angle of the diffracted beam and n is the order. For most monochromators used in spectrographic analyses, the angle of the diffracted beam is nearly normal to the grating, and β does not change significantly over the spectral region measured. Therefore, the dispersion is nearly linear. Figure 23 is a graph of wavelength versus position for an iron spectrum recorded with a 3.4-m Ebert spectrograph. The experimental data points were fitted to both linear and quadratic equations of the form

$$\lambda = a + bN$$

$$\lambda = a + bN + cN^2$$

where λ and N are the wavelength and position, respectively, and a, b, and c are constants. The wavelength differences are plotted as a function of wavelength in Figure 24. Using a linear equation, the wavelength difference can exceed 0.05 nm. However, the quadratic equation provides a more accurate representation of the dispersion for a grating spectrum and takes into account nonlinearity resulting from changes in the angle β. Using a second-order equation, the wavelength difference is less than 0.006 nm for spectral lines measured over a wavelength region from 250 to 350 nm.

Elements are identified by comparing the calculated wavelengths with reference values in wavelength tables. With the aid of computers, vast libraries of spectral data can be accessed within a very short time, enabling the identification of all detectable elements in a sample. Quantitative spectrochemical analysis can be performed with computers if the necessary calibration data are available in computer-readable form. Since emulsion and standard calibration curves can be described in terms of polynomial equations, the relationships between transmittance and relative intensity or concentration are readily obtained. An emulsion calibration curve relating transmittance with intensity is shown in Figure 25. The emulsion calibration curve does not fit a simple polynomial equation; however, various algorithms have been applied successfully to the conversion of transmittance or density to relative intensity [44]. Conversion of percent transmittance to Seidel function values simplifies the computer program significantly. For the calculation of element concentration, standard calibration curves of log concentration versus log intensity ratio are used. The intensity ratio is the intensity of the element spectral line divided by the intensity of an internal standard element line. The intensity values are corrected for background. With a computer-controlled microphotometer, semiquantitative analyses can be completed within minutes [52].

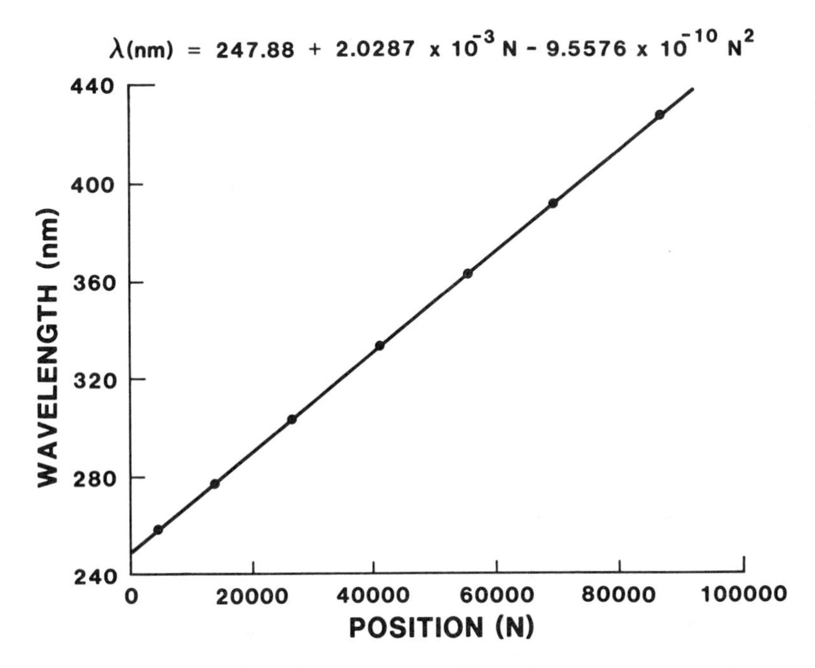

Figure 23. Wavelength versus position for an iron spectrum recorded with a 3.4-m Ebert spectrograph.

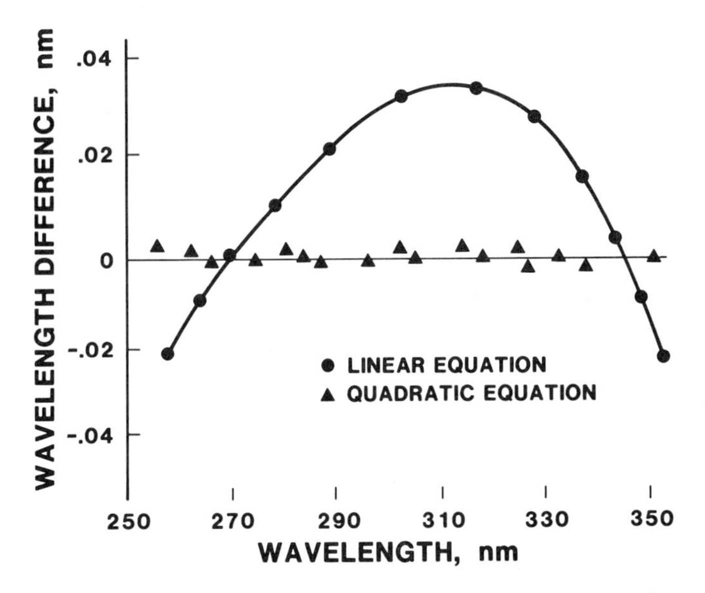

Figure 24. Wavelength difference as a function of wavelength.

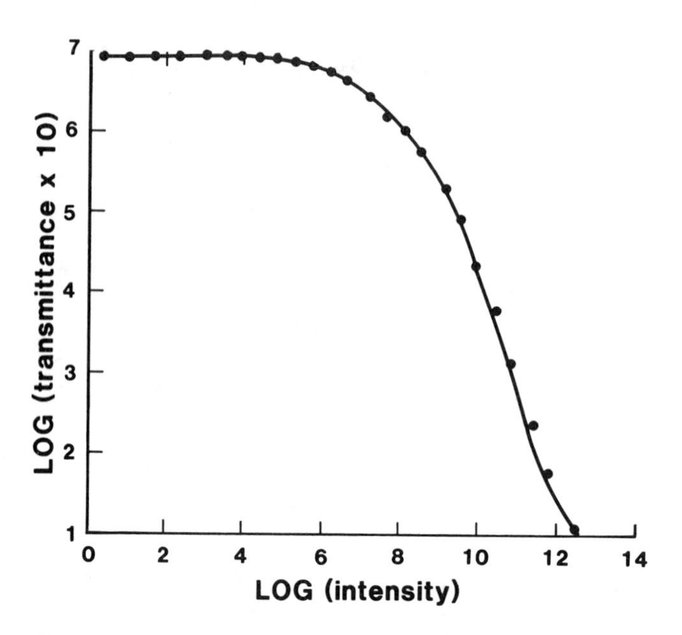

Figure 25. Emulsion calibration curve of transmittance versus intensity.

ACKNOWLEDGMENT

This work was performed at Sandia National Laboratories, supported by the U.S. Department of Energy under contract no. DE-AC04-DP00789.

REFERENCES

1. L. H. Ahrens, *Spectrochemical Analysis*, Addison-Wesley, Cambridge, MA, 1950.
2. N. H. Nachtrieb, *Principles and Practice of Spectrochemical Analysis*, McGraw-Hill, New York, 1950.
3. C. E. Harvey, *Spectrochemical Procedures*, Applied Research Laboratories, Glendale, CA, 1950.
4. L. H. Ahrens and S. R. Taylor, *Spectrochemical Analysis*, Addison-Wesley, Reading, MA, 1961.
5. R. A. Sawyer, *Experimental Spectroscopy*, 3rd ed., Dover, New York, 1963.
6. B. F. Schribner and M. Margoshes, in *Treatise on Analytical Chemistry, Part I, Theory and Practice* (I. M. Kolthoff and P. J. Elving, eds.), Wiley, New York, 1965.
7. P. B. Farnsworth, in *Metals Handbook*, Vol. 10, 9th ed. (R. E. Whan, ed.), American Society for Metals, Metals Park, OH, 1986.
8. I. G. Yudelevich, I. R. Shelpakova, S. B. Zayakina, and O. I. Scherbakova, Spectrochim. Acta *29B*, 353-363 (1974).
9. K. Wagatsuma and K. Hirokawa, Spectrochim. Acta *42B*, 523-531 (1987).

10. H. Bubert, Spectrochim. Acta *39B*, 1377-1387 (1984).
11. H. Jager and F. Blum, Spectrochim. Acta *29B*, 73-77 (1974).
12. K. Wagatsuma and K. Hirokawa, Anal. Chem. *56(6)*, 908-913 (1984).
13. W. Grimm, Spectrochim. Acta *23B*, 443-454 (1968).
14. J. B. Ko, Spectrochim. Acta *39B*, 1405-1423 (1984).
15. K. Wagatsuma and K. Hirokawa, Anal. Chem. *56(3)*, 412-416 (1984).
16. E. H. Daughtrey, Jr., D. L. Donohue, P. J. Slevin, and W. W. Harrison, Anal. Chem. *47(4)*, 683-688 (1975).
17. S. Caroli, O. Senofonte, N. Violante, F. Petrucci, and A. Alimonti, Spectrochim. Acta *39B*, 1425-1430 (1984).
18. L. Moenke-Blankenburg, Prog. Analyt. Spectrosc. *9*(3), 335-427 (1986).
19. G. Dimitrov and T. Zheleva, Spectrochim. Acta *39B*, 1209-1219 (1984).
20. R. H. Scott and A. Strasheim, in *Applied Atomic Spectroscopy*, Vol. 1 (E. L. Grove, ed.), Plenum Press, New York, 1978.
20A. B. E. Buell, in *Petroleum Industry Analytical Applications of Atomic Spectroscopy*, Vol. 2 (E. L. Grove, ed.), Plenum Press, New York, 1978.
21. A. T. Zander and M. H. Miller, Spectrochim. Acta *40B*, 1023-1037 (1985).
22. V. A. Fassel, Science *202*(13), 183-191 (1978).
23. P. W. J. M. Boumans and J. J. A. M. Vrakking, Spectrochim. Acta *42B*, 553-579 (1987).
24. R. K. Winge, V. J. Peterson, and V. A. Fassel, Appl. Spectrosc. *33*(3), 206-219 (1979).
25. R. M. Barnes, L. Fernando, L. S. Jing, and H. S. Mahanti, Appl. Spectrosc. *37*(4), 389-395 (1983).
26. K. E. LaFreniere, G. W. Rice, and V. A. Fassel, Spectrochim. Acta *40(B)*, 1495-1504 (1985).
27. P. W. J. M. Boumans, *Line Coincidence Tables for Inductively Coupled Plasma Atomic Emission Spectrometry*, Vols. 1 and 2, 2nd ed., Pergamon Press, New York, 1984.
28. A. Aziz, J. A. C. Broekaert, K. Laqua, and F. Leis, Spectrochim. Acta *39B*, 1091-1103 (1984).
29. C. I. M. Beenakker, Spectrochim. Acta *31B*, 483-486 (1976).
30. J. E. Freeman and G. M. Hieftje, Spectrochim. Acta *40B*, 475-492 (1985).
31. D. Kollotzek, P. Tschopel, and G. Tolg, Spectrochim. Acta *39B*, 625-636 (1984).
32. D. L. Wood, A. B. Dargis, and D. L. Nash, Appl. Spectrosc. *29(4)*, 310-315 (1975).
33. E. A. Stubley and G. Horlick, Appl. Spectrosc. *39(5)*, 800-804 (1985).
34. L. M. Faires, Anal. Chem. *58(9)*, 1023A-1034A (1986).
35. E. A. Stubley and G. Horlick, Appl. Spectrosc. *39(5)*, 805-810 (1985).
35A. P. R. Griffiths and J. A. deHaseth, *Fourier Transform Infrared Spectrometry*, Wiley, New York, 1986.
36. G. W. Ewing, in *Treatise on Analytical Chemistry, Part I*, Vol. 4 (P. J. Elving, V. G. Mossotti, and I. M. Kolthoff, eds.), Wiley, New York, 1984.
37. S. F. Jacobs, in *Handbook of Optics* (W. G. Driscoll and W. Vaughan, eds.), McGraw-Hill, New York, 1978.

38. G. R. Harrison, *Massachusetts Institute of Technology Wavelength Tables, with Intensities in Arc, Spark, or Discharge Tube*, Wiley, New York, 1939.
39. W. R. Brode, *Chemical Spectroscopy*, 2nd ed., Wiley, New York, 1943.
40. R. L. Kelly, *Table of Emission Lines in the Vacuum Ultraviolet for all Elements (6 Angstroms to 2000 Angstroms)*, U.S. Atomic Energy Commission, UCRL 5612, 1959.
41. C. Kerekes, *Tables for Emission Spectrographic Analysis of Rare Earth Elements* (L. Lang, ed.), Macmillan, New York, 1964.
42. F. M. Phelps III, *MIT Wavelength Tables*, Vol. 2, MIT Press, Cambridge, MA, 1982.
43. Qiu De-ren and Cheng Wan-xia, *Atlas for 2 A/mm and 4 A/mm Grating Spectrograph*, Fudan University, Shanghai, China, Shanghai Scientific and Technical Publishers, China, 1984.
44. J. W. Anderson, *Photographic Photometry, in Applied Atomic Spectroscopy*, Vol. 1 (E. L. Grove, ed.), Plenum Press, New York, 1978.
45. K. Zimmer, G. Heltai, and K. Florian, Prog. Anal. Atom. Spectrosc. *5*, 341-467 (1982).
46. A. W. Helz, F. G. Walthall, and S. Berman, Appl. Spectrosc. *23*, 508 (1969).
47. D. W. Steinhaus, K. J. Fisher, and R. Engleman, Jr., Chem. Instrum. *3*, 141 (1971).
48. R. Hoekstra and R. Slooten, Spectrochim. Acta *26B*, 341 (1971).
49. B. L. Taylor and F. T. Birks, Analyst *97*, 681 (1972).
50. F. G. Walthall, J. Res. U.S. Geolog. Surv. *2*, 61 (1974).
51. A. W. Witmer, J. A. J. Jansen, G. H. van Gool, and G. Brouwer, Philips Technol. Rev. *34*, 322 (1974).
52. C. P. Thomas, Appl. Spectrosc. *33*, 604 (1979).
53. M. E. Grandy, M. A. Sainz, and D. M. Coleman, Appl. Spectrosc. *36*, 643 (1982).
54. W. A. Hareland and D. M. Melgaard, in *Analytical Chemistry Instrumentation* (W. R. Laing, ed.), Lewis Publishers, Chelsea, Michigan, 1986.

5

Atomic Absorption and Flame Emission Spectrometry

M. L. PARSONS / Pacific Scientific Co., Duarte, California

I. INTRODUCTION

Atomic absorption and flame emission spectrometry have been mainstays in the area of trace elemental analysis for many years. In recent years, the extreme sensitivity of electrothermal or graphite furnace atomic absorption has placed it at the forefront of trace metal techniques. With a combination of these methods, over 60 elements are determined with detection limits of better than 1 ppm (microgram per gram of solution, or in aqueous solution microgram per milliliter) and in many cases better than 1 ppb (nanograms per gram or nanograms per milliliter) [1].

An International Union of Pure and Applied Chemistry (IUPAC) committee on spectroscopic nomenclature [2] has recommended the following abbreviations for these techniques:

AES atomic emission spectrometry
AAS atomic absorption spectrometry
FAES flame atomic emission spectrometry
FAAS flame atomic absorption spectrometry
EAAS electrothermal atomic absorption spectrometry

Other variations, such as cold vapor, hydride generation, etc., are not abbreviated but are spelled out, for example, cold-vapor AAS or hydride generation FAAS.

In very general terms, the elements that are most suitably determined by these methods are as follows:

FAES—the alkali metals and the alkaline earths
FAAS—the alkaline earths and the first-row transition metals
EAAS—most metals except the refractories
Cold-vapor AAS—Hg
Hydride generation FAAS—As, Se, Sb, Te, and Bi

These methods are usually thought of as single-element techniques; however, instruments with two simultaneous measurement channels and as many as 16 sequential channels have been sold commercially.

A. Measurement Phenomenon

The valence electrons associated with free atoms in the hot sample gases are involved in both AES and AAS measurements. The valence electrons

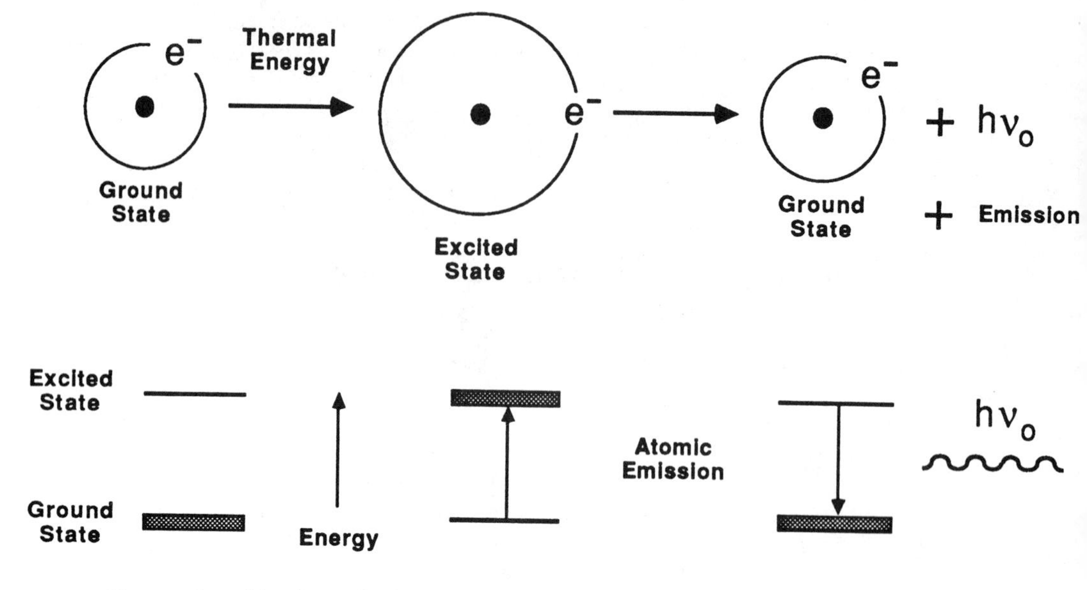

Figure 1. Atomic emission.

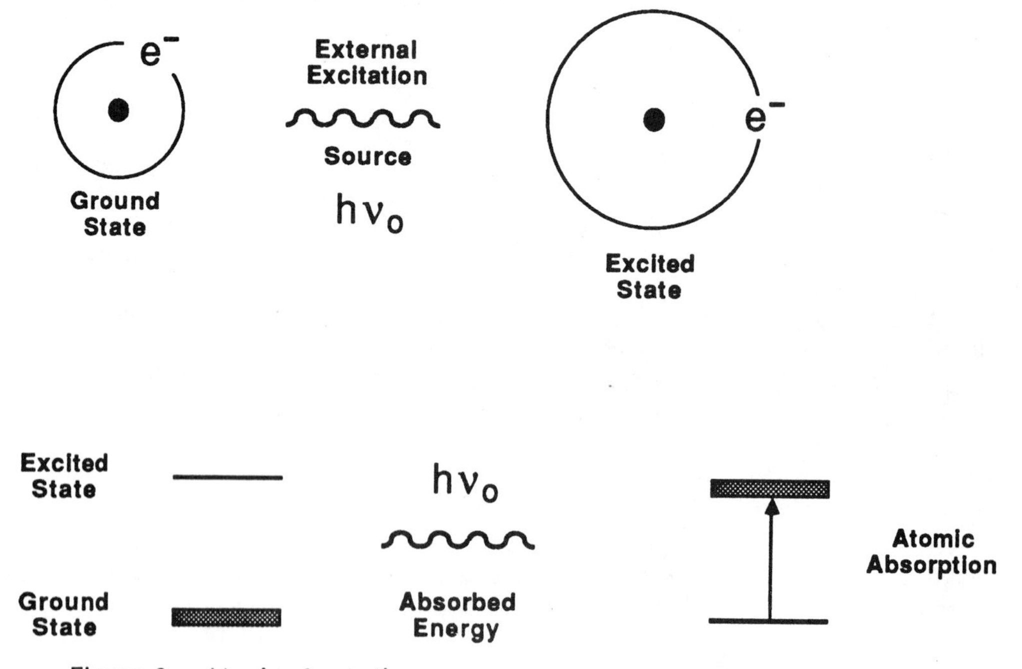

Figure 2. Atomic absorption.

are in the outermost orbitals in the electron cloud of the atom and are the ones involved in molecular bonding. They are the most easily excited electrons; therefore, they produce the most intense radiation in AES and have the most interaction with the excitation radiation in AAS. AES and AAS phenomena are illustrated in Figures 1 and 2, respectively.

In FAES the valence electrons are excited by means of collisions with flame-gas molecules with a resultant transfer of kinetic energy to the atom of interest. The excited state is very short lived, and upon relaxation back to the normal, or ground state, the atom emits a photon of energy that is equal to the energy difference between the ground and excited states. The intensity of emission is proportional to the number of atoms that are excited.

This energy is expressed by

$$E - h\nu_0 \tag{1}$$

where h is Planck's constant (6.626176×10^{-27} erg sec) and ν_0 is the frequency of the transition at the transition center in sec^{-1}. Throughout the electromagnetic spectrum, the relationship between frequency and wavelength, λ, is given by

$$c = \nu\lambda \tag{2}$$

where c is the velocity of radiation in appropriate units. The units of λ are in length. Because most of the transitions under consideration in AES and AAS are in the ultraviolet or visible range of the electromagnetic spectrum, the commonly used units are nanometer (nm), and the less preferred unit angstrom (Å), which is equal to one-tenth of a nanometer. The nanometer is the IUPAC recommended unit and will be used throughout the chapter.

The AAS measurement arises when a photon of the specific energy that corresponds to the difference between the lower and upper energy levels of a free atom in the hot sample gas is absorbed by the atom. The amount of radiation that is absorbed is related to the number of atoms contained in the ground level in the optical path of the excitation radiation.

The analytical functions are relative intensity of emission for AES and absorbance for AAS. The intensity of radiation emitted from an emission atom cell is related to a number of fundamental parameters associated with the specific atom under consideration and the specific transition being observed, in addition to the optical parameters of the instrument. In the final analysis, it is possible to relate the concentration of an element in solution to the intensity of a particular transition originating from the element in a flame or other appropriate excitation source. In the simplest form, the analytical relationship can be expressed as

$$I = KC \tag{3}$$

where I is the signal measured by the instrument, K is a proportionality constant, which includes both fundamental constants and instrumental constants, and C is the solution concentration of the analyte element. This expression assumes no background radiation, which would have to be subtracted from the total intensity. In most cases, a background signal is present.

For AAS the measurement is completely analogous to solution spectroscopy. The intensity of the excitation source is measured with and without the analyte element present in the sample cell. The logarithm of the ratio of these intensities is defined as the absorbance, A, as follows:

$$A = \log I_0/I = abC \quad (4)$$

where I_0 is the reference intensity (i.e., with no analyte element present in the sample cell), I is the transmitted light when the analyte is present, a is a proportionality constant incorporating most of the fundamental constants and optical constants as in AES, b is the optical path length of the sample cell, and C is the solution concentration of the analyte in the sample.

A valuable characteristic of most flames and other atom cells considered in this chapter is that they are either in thermodynamic equilibrium or local thermodynamic equilibrium. This means that the Maxwell law of velocity distribution, the Boltzmann law of internal energy distribution, the laws of mass action, and the laws of equilibrium can be applied to these cells.

An excellent treatise on flames and the fundamental nature of elemental interactions with flame gases and electromagnetic radiation has been written by Alkemade et al. [3]. Reading this treatise is a must for the serious practitioner of analytical atomic spectroscopy.

B. Overview of Basic Instrumental Requirements

The experimental arrangements for AES and AAS are illustrated in Figure 3. AES requires an atomic sample cell, which is usually a flame; entrance optics, which focus the emitted radiation onto the entrance slit of a monochromator; and a detection readout system. The addition of an external excitation source to provide radiation of the appropriate energy is needed for AAS. Because the optical components are essentially identical for both measurements, many manufacturers provide the opportunity for making both measurements with the same instrument.

Atom cells are usually flames or electrically heated graphite tubes. The most common flames are air/acetylene, nitrous oxide/acetylene, and natural gas, propane, or butane/air. The graphite tubes are often pyrolytically coated to decrease their porosity. Electrically heated metal filaments are sometimes used as well. The most common burner for flame work is a slot burner with a premix gas chamber. A variety of nebulizers are used to introduce sample solution into the burner; these will be described in detail in a later section.

Generally, entrance optics are rather short in focal length to keep the instruments compact and are made of quartz to ensure good light throughput in the ultraviolet spectral region.

The spectral output from most atom cells and elements is rather simple, consisting of line spectra, small-molecule spectra (usually only two or three atoms to a molecule), and, depending on the sample matrix, a low-level continuum. Because of the simple spectrum, the spectral resolution of the monochromator can be a moderate requirement. Most manufacturers employ medium resolution grating monochromators with focal lengths on the order of 0.33-0.5 m.

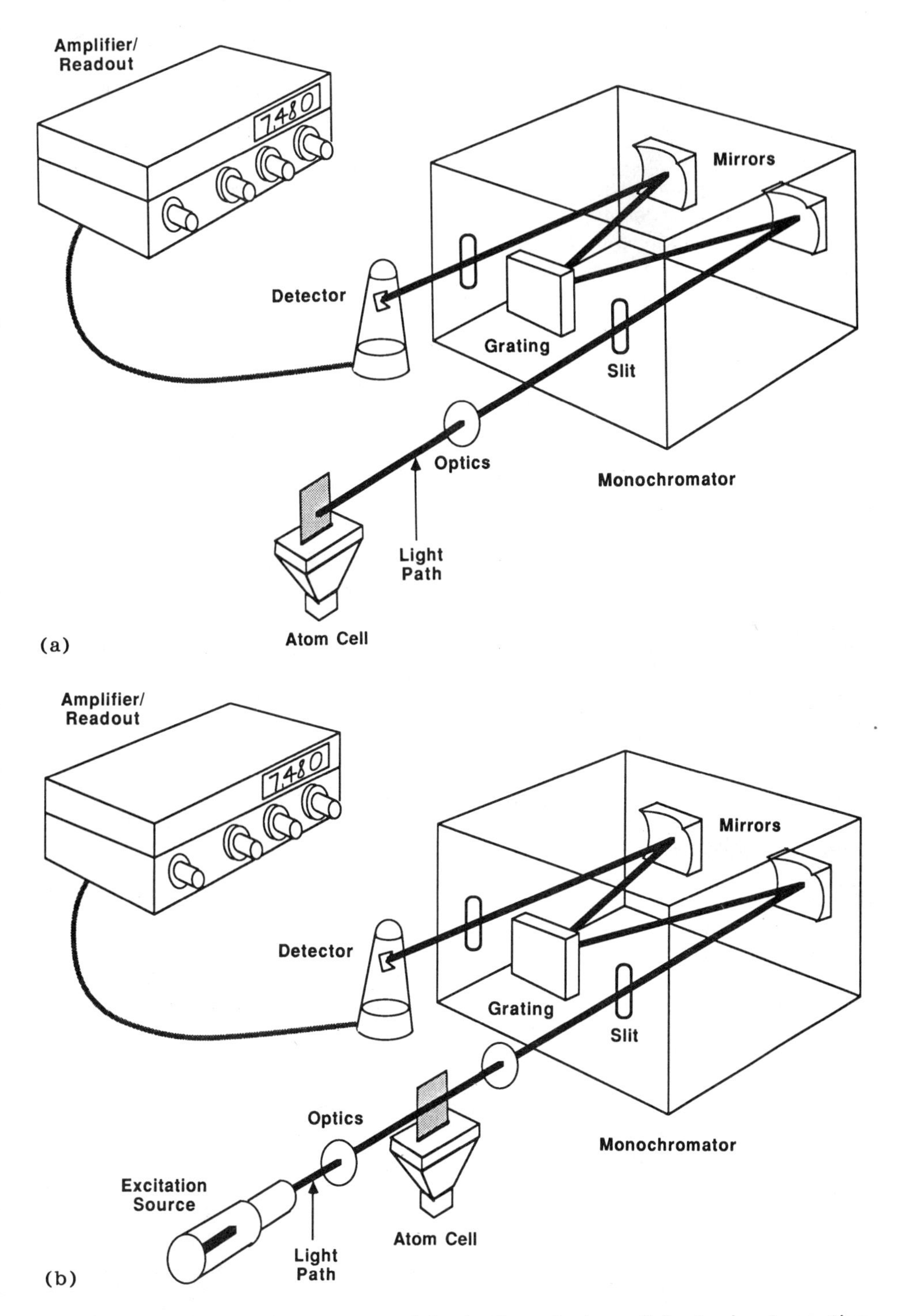

Figure 3. Experimental set-up: (a) atomic emission, (b) atomic absorption.

Photomultiplier-tube-based detection systems are by far the most common. However, experimentation with photodiode array and other types of solid-state detectors has been increasing. Most modern instruments use some form of computer for control or data reduction.

The most widely used external excitation sources in AAS are hollow cathode discharge lamps (HCLs) and electrodeless discharge lamps (EDLs). Both of these provide simple line spectra of the element of interest and the inert fill gas, usually Ar or Ne, have reasonable long- and short-term stability, and are simple to operate and replace.

C. History of "Flame" Methods

It is important to have a historical perspective with respect to the development of modern instrumentation. Breakthroughs in one area often feed another. For example, the development of microcomputer technology has provided major improvements for most analytical instruments, not only in the area of data reduction but also in parameter control, optimization of parameters, etc.

Glass was developed before recorded history; hence, the discovery of the lens and the prism are not well fixed in time. However, the development of the mirror is attributed to Euclid in about 300 B.C. Ptolemy wrote the first treatise on optics about 100 A.D. The scientific development of optical components was much more recent. The 1600s saw the first telescope by Galileo and the first spectrometer by Newton. Figure 4 provides an illustration of Newton's spectrometer as described in his article in the Philosophical Transactions of the Royal Chemical Society [4]. Newton also is responsible for the modern concepts of light and the entire electromagnetic spectrum, although he was only aware of the visible spectrum at that time.

The sun was the most important source of radiation for researchers in Newton's time, as well as for the next 100 years or so, and many important observations were made during this period. The first observation of atomic absorption was made by Wollaston, who described the dark lines in the sun's spectrum in 1802. In 1821 Fraunhofer measured some 700 of these dark lines, although he did not understand their origin.

The first experiments with flames involving the introduction of chemical salts were performed by Melville in the mid-1700s, but it was not until 1826-1830 that Talbot ascribed the resulting emission to atoms. He suggested that this method might be used for qualitative analysis. Soon after, in 1848, Foucault made the first observation of FAAS.

In spite of the fact that all scientific discoveries are built on previous studies, there are breakthroughs of major significance that cause the work of certain researchers to stand out from the rest. Such is the case of Bunsen and Kirchhoff. Bunsen was a chemist and the developer of the Bunsen burner, which is still in use in chemistry laboratories today. Kirchhoff was a physicist who collaborated with Bunsen in their spectroscopic studies and who explained AAS and AES in atomic terms in 1859. From 1860 to 1864, they discovered Cs, Rb, Tl, and In by FAES. They made measurements of both AAS and AES and put these techniques on a sound scientific basis. They also solved the riddle of the dark lines

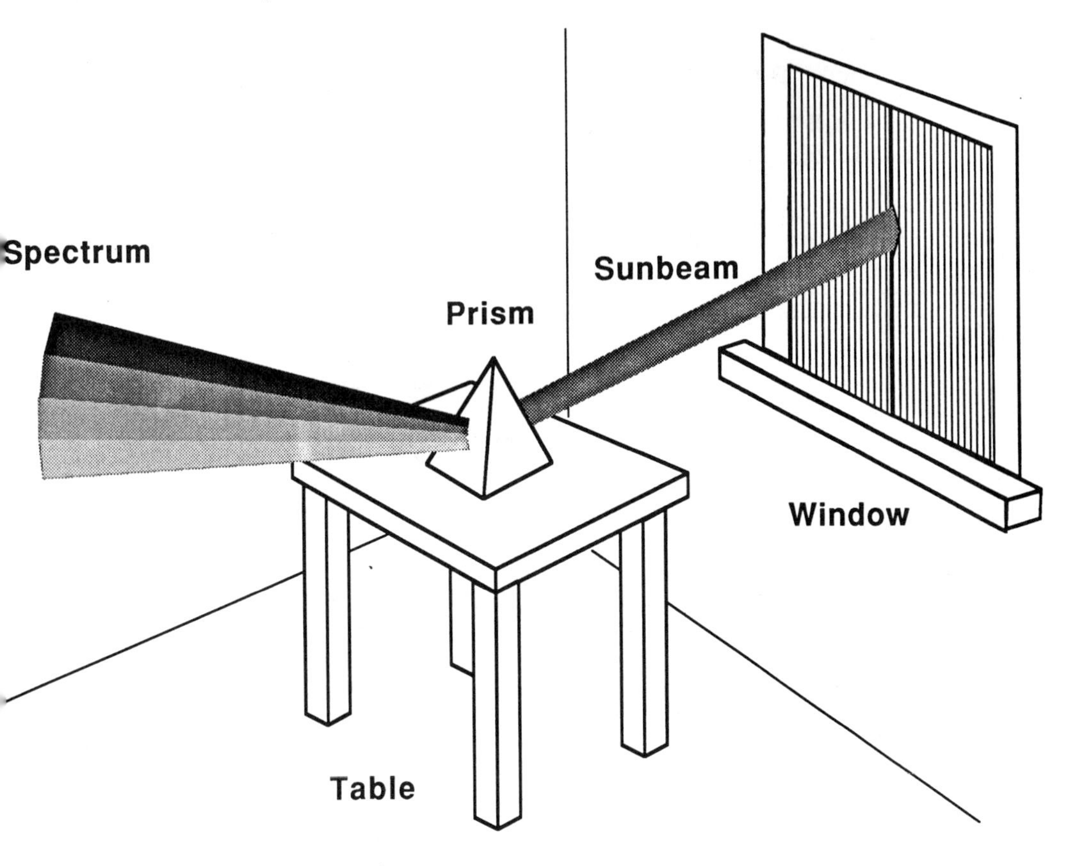

Figure 4. Newton's experiment.

in the solar spectrum, ascribing them to AAS of elements in the sun's atmosphere.

Many scientists studied flames in the last half of the nineteenth century and the field continued to develop; however, modern instrumentation required the development of modern electronics and detection devices. As is usually the case, wartime drives scientific development at a faster rate and World War II was no exception. The first flame-emission instrument was introduced by Beckman Instruments in 1948 in the form of the famous Beckman DU flame emission attachment.

In 1955, simultaneously and independently, Walsh in Australia [5] and Alkemade and Milatz in the Netherlands [6] published research that provided the fundamentals of modern AAS. Winefordner [7,8] placed both FAES and FAAS on a firm fundamental basis in the early 1960s. These methods became the most important trace metal techniques during the 1960s and 1970s and are still in major use in many laboratories throughout the world. A time line that presents some of the major milestones on the road to modern AES and AAS is presented in Figure 5.

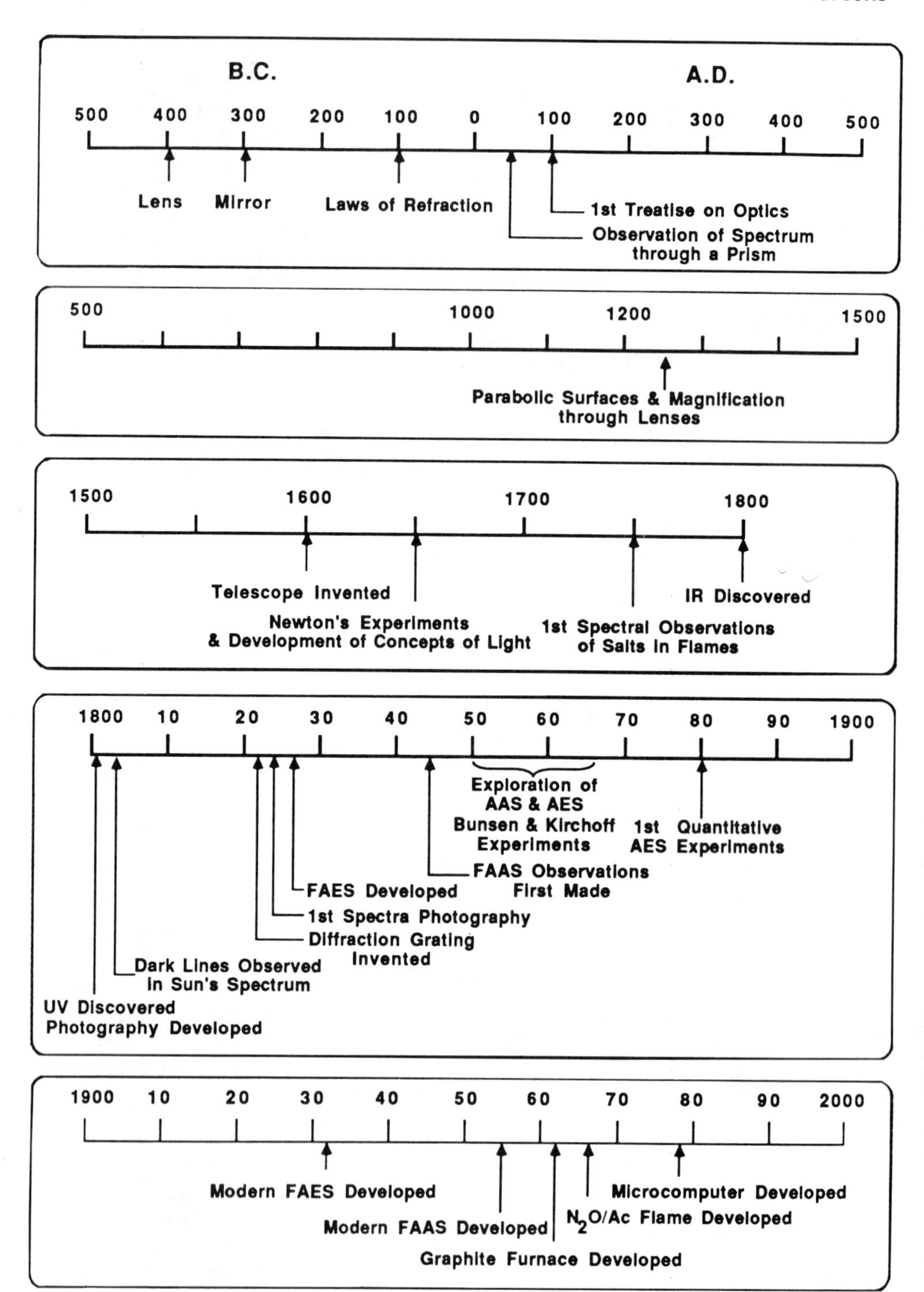

Figure 5. Historical time line.

D. Literature Sources

The primary literature sources for any area are in the peer-reviewed journals. Those in analytical chemistry that publish articles involving AES and AAS include Anal. Chem., Appl. Spectrosc., Spectrochim. Acta B, Analyst, Anal. Chim. Acta, Talanta, JAAS, Fresenius Z. Anal. Chem., Can. J. Spectrosc., Zh. Anal. Khim., and Prog. Anal. At. Spectrosc., and there are many lesser journals.

Perhaps of equal importance today are those journals that supply bibliographical and abstract information. The foremost of those is the Chemical Abstracts Selects titled Atomic Spectroscopy, published by the ACS Chemical Abstracts Services. Analytical Abstracts is another important abstracting journal. In addition, several journals provide monthly or at least periodic reviews or bibliographic information, including Anal. Chem., JAAS, and At. Spectrosc. In addition, it is now possible to obtain computer searches of *Chemical Abstracts* and many other databases on either a "pay as you go" or a subscription basis.

A rather modest number of books have been written on the subject; most of these are listed in the bibliography.

II. COMMERCIALLY AVAILABLE INSTRUMENTAL COMPONENTS

In this section, the instrumental components used in AES and AAS will be covered. Emphasis will be on those systems and subsystems that are available commercially and not on those of simple academic interest. The final section of the chapter is devoted to research directions in this field.

A. Atomic Sample Cells

The atom cell is the heart of the AES or AAS instrument. The measurement requires free, or unbonded, atoms, and the sample cell must provide the energy needed to break the molecular bonds and must create a cloud of atoms that are not bound to other atoms and are not ionized. So the cell must have a fair amount of, but not too much, energy. Most of the bonds in simple molecules require 1-6 eV to dissociate, whereas most ionization potentials are in the 5- to 10-eV range. Common flames generally provide the energy required to provide atoms in the free state. However, a variety of flames must be used to successfully determine the full range of elements. In AES, the atom cell must also provide the energy to excite the element of interest. The common flame does this very well if the resonance transitions (those originating from the ground state) are reasonably low in energy. Remembering that the energy of the transition is inversely proportional to its wavelength, transitions above a wavelength of 400 nm can usually be measured with greater sensitivity by AES than AAS. By the same reasoning, transitions below 300 nm are usually more sensitive by AAS, and the ones between 300 and 400 nm are about the same by either method. Graphite furnaces generally provide the energy to atomize a sample, but in only a few cases do they provide the energy to produce sensitive AES signals.

Flames

The most common flames in use today are air-acetylene, nitrous oxide-acetylene, and air-alkane. The characteristics of methane- (the main constituent of natural gas), propane-, and butane-air flames are very similar and can essentially be considered together. Other flames of historical importance or that have been used in academia, such as the air-hydrogen, oxy-hydrogen, and oxy-acetylene, will not be covered here. In addition, the design of burners and nebulizers is common to all flames and will be discussed in separate sections below.

a. Burner design

During the past two decades all commercial instrument companies have evolved very similar and somewhat standardized burner designs. A generic version is illustrated in Figure 6. This version contains a nebulizer that is operated by the oxidant gas, a mixing chamber into which both the oxidant and fuel gases flow, and a burner head. The mixing chamber contains an impact bead, a spoiler, or other device that provides a complicated mixing path to the burner head. The mixing chamber is usually fabricated from some type of chemically inert plastic. The chamber is shaped so that the larger droplets from the nebulizer will drain into a waste container through a trap. The trap prohibits the free flow of burner gases and therefore the possibility of a "flashback," which would damage the burner system. The mixing chamber causes the oxidant and fuel to mix before burning and creates a laminar-flow flame, which is reasonably quiet with respect to both sound and flicker, which would result in an electronically noisy signal.

The burner head is generally of the slot type (Figure 7) and is constructed of titanium; stainless steel; or other noncorrosive, high-temperature metal. The slot size depends on the oxidant-fuel combination. Early burners were sized by trial and error and flashbacks were common; however, several workers [3,9,10] made rather thorough experimental investigations of the geometry of burner slots, incorporating as experimental parameters the oxidant and fuel type, gas and burning flow rates, and geometry of the burner head, including total slot area, width and length, and burner head depth. From these studies, a safe burner head design has evolved with which it is virtually impossible to have a flashback.

Many other types of burner design have been employed experimentally, but few designs have been used in commercial instrumentation. Historically, much use was made of so-called diffusion burners, which, much like welding torches, caused the gases to mix at the end of concentric metal tubes. Oxygen was a common oxidant for these burners, and the aerosol generated by the nebulizer was inserted into the center tube. These burners were very noisy both electronically and with respect to sound and are no longer sold with commercial instrumentation. Variations of the Meeker burner design have been used. Of special interest for making fundamental measurements is the practice of using a flame sheath outside of a central flame using a Meeker type burner to simplify the temperature profile. Multiple slot and even circular slot designs have also been used. Burners have also been designed for specific experiments including low-pressure measure-

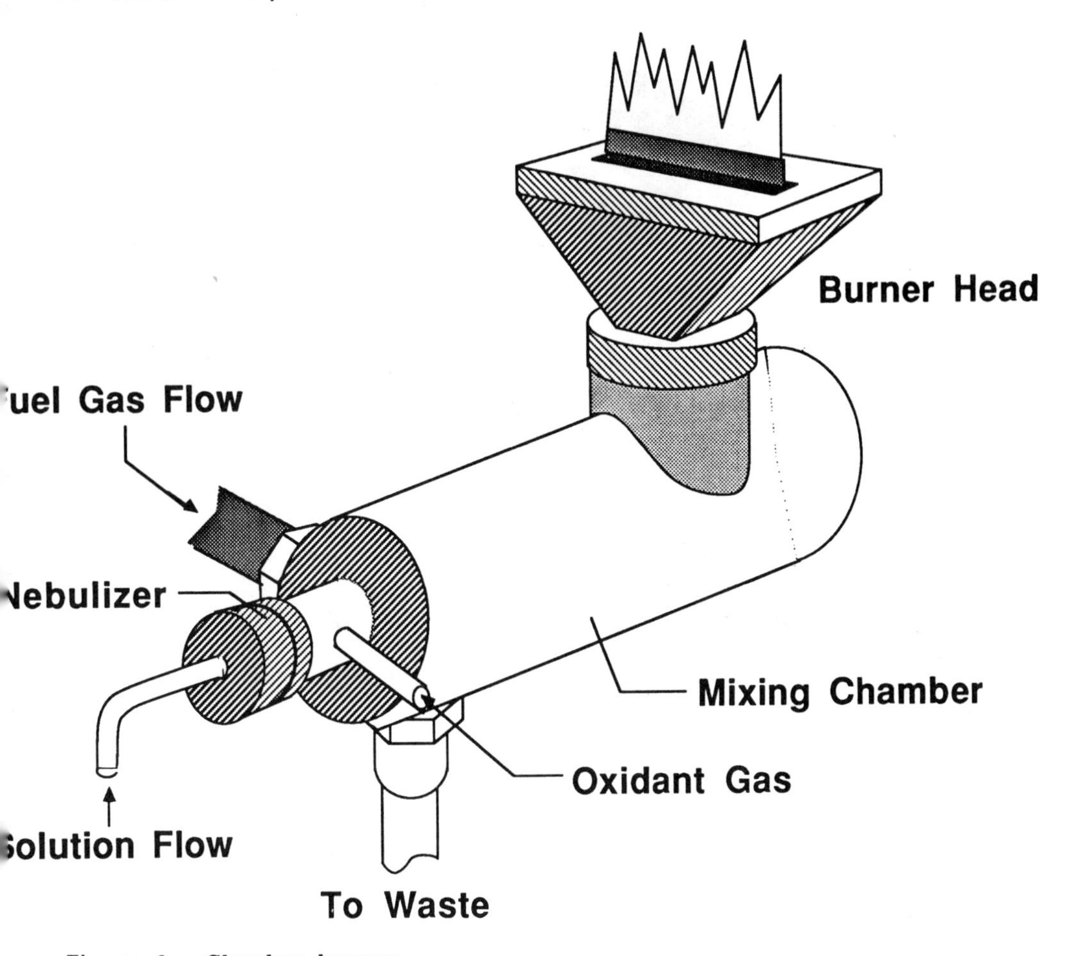

Figure 6. Chamber burner.

ments, burner interface to mass spectrometers, and the use of exotic oxidants (fluorine, chlorine, etc.) and fuels (CO, CN, etc.).

As stated, most commercial burners are of the type described in Figure 7; however, many manufacturers have offered specialized options. There are at least two types of multislot variations, one of which permits an increased flame length and another that permits an increased flame width. Absorption is directly proportional to the sample cell length, so the increase in length results in a concomitant increase in absorbance signal. The wider flame is purported to increase the system's tolerance to high-solids-containing solutions. In reality, the burner design was originally configured to take advantage of the rather poor optics used in early AAS instrumentation.

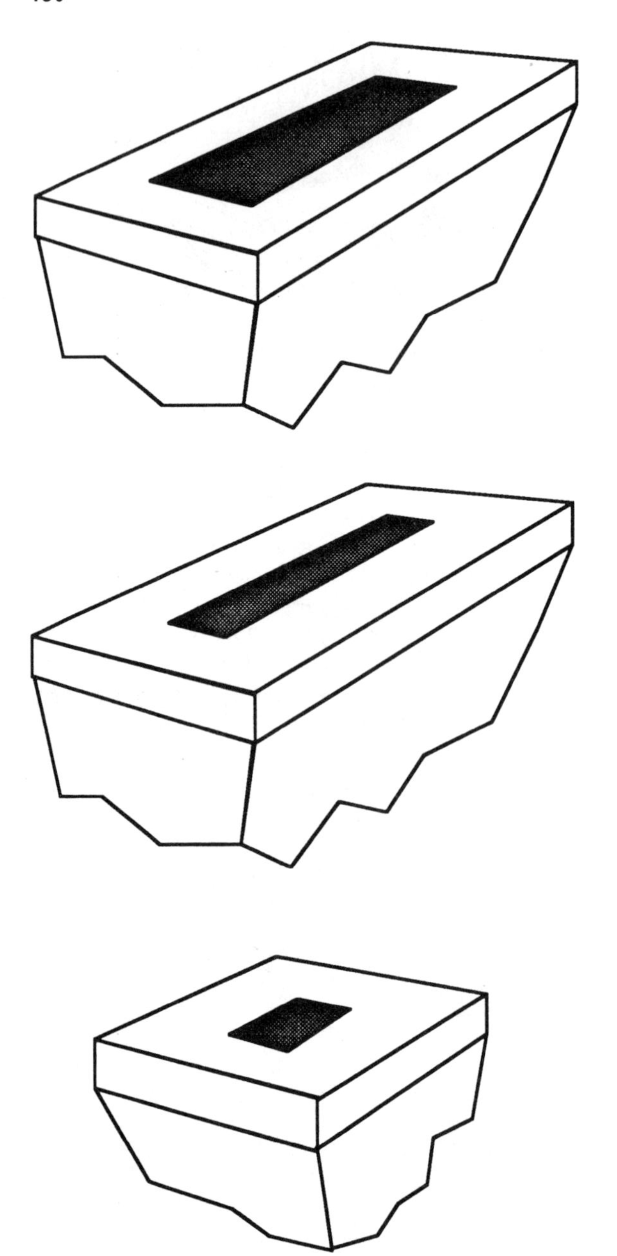

Figure 7. Burner head configuration.

Although several studies have been made in the last decade with respect to the various parameters affecting burner design, little research has been performed in the last 2-3 years. Most of the current research has been associated with the nebulizer components of burner systems.

b. Nebulizer systems

Introducing a solution containing an analyte into the burner has presented the analyst with the "weak link" in FAAS/FAES systems. There has been a great deal of research in this area, and many phenomena have been exploited in this endeavor. The nebulizer must provide transport to the burner system and, in some fashion, create an aerosol of fine droplets that can subsequently evaporate and react in the flame to generate the free atoms that constitute the species to be measured by either AAS or AES. By far the most common approach is by pneumatic means, for example, the utilization of gas pressure in conjunction with an orifice or multiple orifices to create an aerosol mist. Three of the most common approaches are illustrated in Figure 8. The cross-flow system is very similar to a perfume atomizer. The concentric tube approach is the most common commercial device; however, many companies are now offering a version of the Babbington nebulizer for use with highly dissolved solids solutions. Note that the Babbington nebulizer does not require the solution to flow through a small orifice. When a solution containing a high concentration of dissolved salts is forced through a small orifice, there is a tendency for the salts to precipitate and clog the orifice. This precipitation can create a rapidly changing flow rate of solution into the burner system and can cause errors in the analyses and downtime with the system.

All pneumatic systems are dependent on the physical nature of the solutions being introduced. The viscosity is of major importance. Most mineral acids in reasonably low concentrations will affect the viscosity of the solution. If samples differ in viscosity from the standards, errors will necessarily occur. Matching the physical and chemical matrix of samples and standards to achieve accurate results is very important. Many analysts use a force-feeding system, such as a peristaltic or syringe pump, to help ensure that a constant flow of solution is introduced into the nebulizer system. This method is effective over a reasonable range of viscosities; however, if the range is too large, errors can result because of the difference in efficiency of aerosol generation with too great a change in the physical nature of the solutions. It is the author's opinion that failure to recognize this problem is one of the greatest sources of error in FAAS and FAES methods. Several researchers have studied these problems in some detail [11-13].

One of the major drawbacks of the pneumatic nebulizers is the poor aspiration efficiency, that is, the percentage of the solution introduced into the burner system that actually reaches the flame. For most nebulizer systems, this percentage is only between 1% and 15%, but much research effort has been expended in trying to improve aspiration efficiency. Heating the chamber of the burner can help substantially; however, this usually causes an increase in signal memory from one sample to the next. The use of heated sample chambers is not commercially supported in today's instruments.

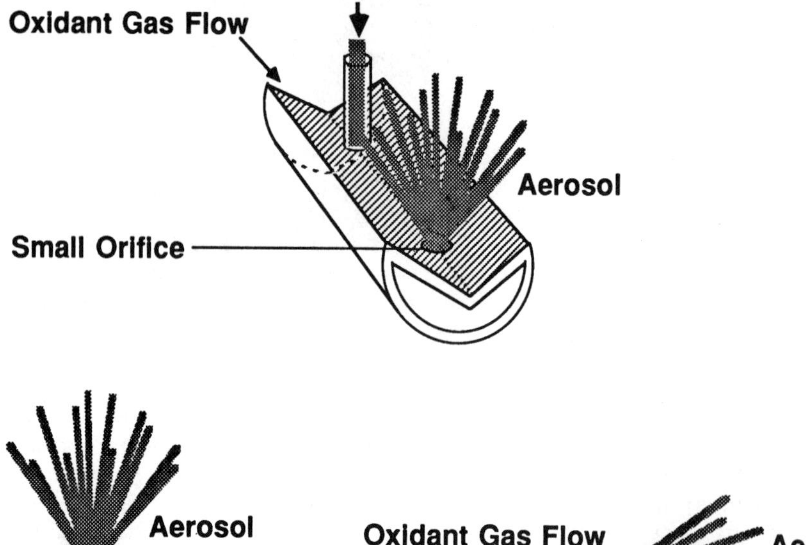

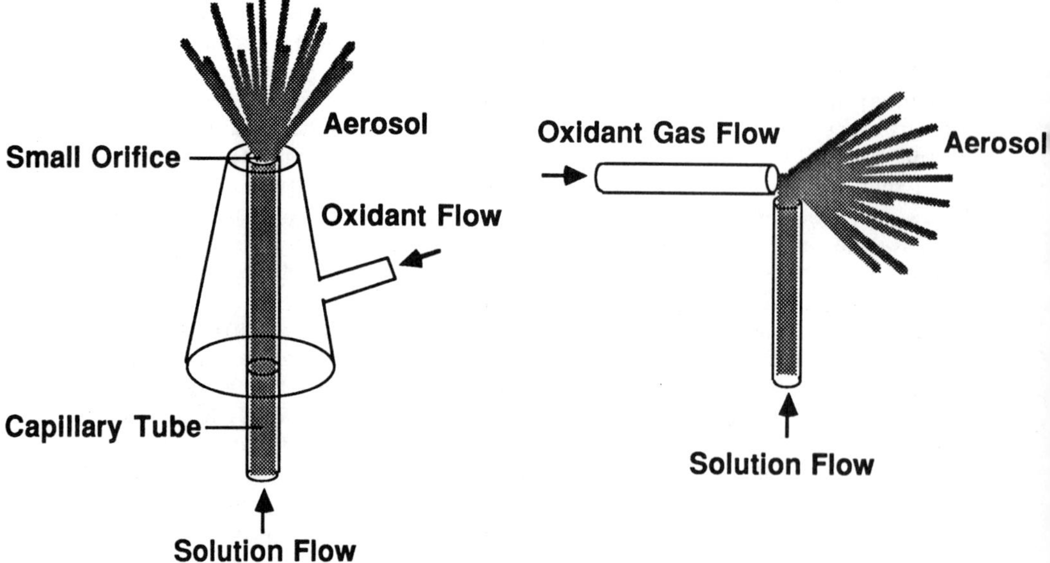

Figure 8. Common nebulizers: (top) modified Babbington, (left) concentric tubes, (right) cross-flow.

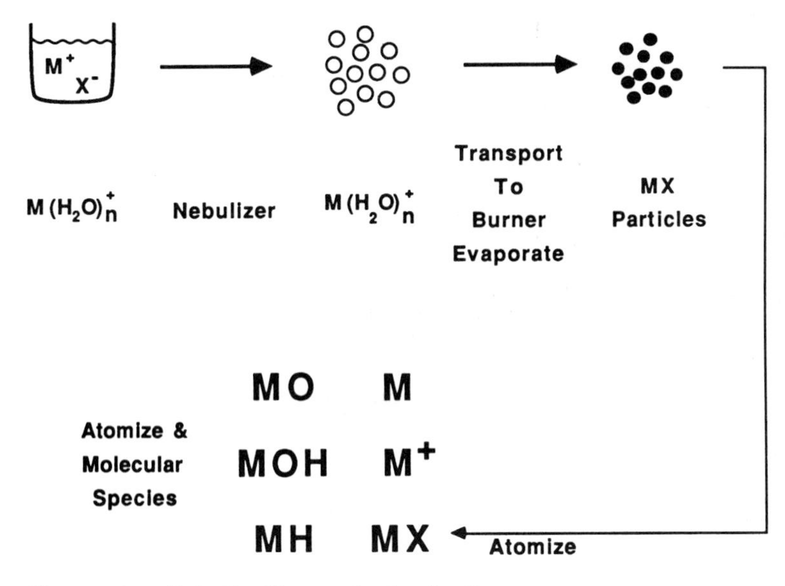

Figure 9. Nebulization and atomization process.

Ultrasonic and electrostatic nebulizers have also been studied; however, neither has had much commercial success. In many respects, the ultrasonic system has advantages over the pneumatic systems—for example, it provides a tighter droplet size distribution and a lower mean droplet diameter; however, the added electronics, cooling requirements, and maintenance problems have a negative economic effect.

c. *General considerations*

The overall process of solution introduction into a flame is shown schematically in Figure 9. As mentioned earlier, the solution flow rate is often determined independently with a pump. However, there is an inherent nebulizer-controlled solution flow rate that is dependent on the gas flow/pressure conditions and geometry for a specific nebulizer. If the external pump rate exceeds the nebulizer-controlled solution flow rate, there will be a degradation in the aspiration efficiency, and hence in the analytical results. The droplets small enough to travel completely through the burner system into the flame are completely evaporated or desolvated in the flame. At this point, vaporization and chemical reactions take place. Given a typical metal salt, MX, one of the following chemical reactions can occur:

$$MX = M + X \tag{5}$$

$$M = M^+ + e^- \tag{6}$$

$$M + O = MO \tag{7}$$

$$M + OH = MOH \tag{8}$$

$$M + H = MH \tag{9}$$

$$M + Y = MY, \text{ where Y is a different anion.} \tag{10}$$

Note that only reaction (5) increases the atomic population of M. All other reactions decrease M and provide a negative analytical result. Note also that all of the reactions are presented as equilibrium reactions. It is possible to use straightforward equilibrium theory to predict the atomic population of M, the analytical species, in a particular flame if the composition of the flame and the temperature are known. Such calculations have been performed by several analysts [3,14,15]. These calculations are possible because most flames used for analytical purposes are in localized thermodynamic equilibrium. This means that if an observation is made from a very small volume of the flame, thermodynamic equilibrium exists or is approached very closely; therefore, calculations based on a Boltzmann distribution can be used.

The fraction of atomic species with respect to all forms of the species (i.e., the sum of the atoms, ions, and all molecules containing the analyte species) can be calculated. This fraction is usually defined as the β factor. Knowledge of this factor in the various flames gives information as to which flame is most suited for each element. Table 1 presents the β factors for elements in the common flames.

d. Air-hydrocarbon flames

The light hydrocarbons methane, ethane, propane, and butane all burn in air with a flame having very similar properties. Many laboratories employ the air-natural gas flame, which is essentially an air-methane flame. Such flames can be burned in a stoichiometric manner to carbon dioxide and water as in equation (11), but it is more common to use a fuel-rich gas ratio that is closer to a carbon monoxide stoichiometry of equation (12), or somewhere in between.

$$2CH_4 + 4O_2 + N_2 \rightarrow N_2 + 4H_2O + 2CO_2 \tag{11}$$

$$2CH_4 + 2O_2 + N_2 \rightarrow N_2 + 2H_2O + 2CO \tag{12}$$

The nitrogen acts as a diluent only and participates in the combustion only to a minute degree. Similar reactions can be written for the other light hydrocarbons.

Simple molecular free radicals exist in such flames to a small degree and can be observed from the emission of the molecules, such as C_2, CH, OH, etc. These species contribute to the background of the flame but are generally unimportant analytically except for possible spectral overlap with an analytical transition. In addition, continuum emission is usually given off by recombination reactions in the flame. The continuum emission in most flames is minor and causes no real concern for either AE or AA measurements.

Temperatures of these flames are 1800-2200 K. In addition, the burning velocities are quite low, 30 to 60 cm/sec, and can be accommodated by rather large burner orifices or slots. These flames are most often used for AES or AAS measurements of easily ionized elements, such as the alkali metals or the alkaline earths, because the low flame temperature does not

Table 1. Beta Values for the Air-Acetylene and Nitrous Oxide-Acetylene Flames

Element	Symbol	Beta, A/AC flame	Beta, N/AC flame
Aluminum	Al	<0.0001	0.13
		<0.00005	0.29
		0.0005	0.97[a]
			0.5
Antimony	Sb	0.03	
Arsenic	As	0.0002	
Barium	Ba	0.0009	0.074
		0.002	0.074
		0.003	0.98
		0.0018	
Beryllium	Be	0.0004	0.095
		0.00006	0.98
			0.98
Bismuth	Bi	0.17	0.35
Boron	B	<0.0006	0.0035
		<0.000001	0.2
Cadmium	Cd	0.38	0.56
		0.50	0.60
		0.80	
Calcium	Ca	0.066	0.34
		0.14	0.52[a]
		0.05[a]	0.98
		0.018	
Cesium	Cs	0.02	0.0004
		0.0057	
Chromium	Cr	0.071	0.63
		0.13	1.02
		0.53	1.00
		0.042	
Cobalt	Co	0.023	0.11
		0.28	0.25
		0.41	
Copper	Cu	0.4	0.49
		0.82	0.66
		0.98	1.00[a]
Gallium	Ga	0.16	0.73
		0.16	

(continued)

Table 1 (Continued)

Element	Symbol	Beta, A/AC flame	Beta, N/AC flame
Germanium	Ge	0.001	
Gold	Au	0.21	0.16
		0.40	0.27
		0.63	
Indium	In	0.10	0.37
		0.67	0.93
		0.67	
Iridium	Ir	0.1	
Iron	Fe	0.38	0.83
		0.66	0.91
		0.84	1.00
		0.66	
Lead	Pb	0.44	0.84
		0.77	
Lithium	Li	0.21	0.34[a]
		0.26[a]	0.96[a]
		0.20[a]	0.041
		0.08	0.91[a]
Magnesium	Mg	0.59	0.88
		1.05	0.99
		0.62	0.92
			0.99[a]
Manganese	Mn	0.45	0.37
		0.93	0.77
		1.0	
Mercury	Hg	0.04	
Molybdenum	Mo	0.03	
Nickel	Ni	1	
Palladium	Pd	1	
Platinium	Pt	0.4	
Potassium	K	0.7[a]	0.12[a]
		0.25	0.0004
		0.45	0.17[a]
		0.59[a]	
Rhodium	Rh	1	
Rubidium	Rb	0.16	
Ruthenium	Ru	0.3	

Table 1 (Continued)

Element	Symbol	Beta, A/AC flame	Beta, N/AC flame
Selenium	Se	0.0001	
Silicon	Si	<0.001	0.55
		<0.0000001	0.12
			0.36
Silver	Ag	0.66	0.57
		0.70	
Sodium	Na	0.63	0.32
		1.00	0.97[a]
		1.00[a]	0.012
		0.56	0.80
Strontium	Sr	0.068	0.26
		0.10	0.57
		0.13	0.99
		0.021	
Tantalum	Ta		0.045
Thallium	Tl	0.36	0.55
		0.52	
Tin	Sn	<0.0001	0.35
		0.043	0.82
		0.078	
		0.061	
Titanium	Ti	<0.001	0.11
			0.33
			0.49
Tungsten	W	0.004	0.71
Vanadium	V	0.0004	0.32
		0.015	0.99
		0.000001	
Zinc	Zn	0.66	0.49
		0.45	

[a]Ionization has been suppressed for these measurements/calculations.
Source: From Refs. 15 and 16.

cause too much ionization and because the excitation energies are low. Some of the best detection limits for the alkali metals are measured in these flames. For the same reasons, the hydrocarbon-air flames are not suited to elements that tend to form oxides. These flames are neither energetic enough nor chemically reducing enough to break the oxide bonds and are of limited use for the majority of elements. Detection limits for FAAS using these flames are presented in Table 2.

Table 2. Limits of Detection for the Air-Hydrocarbon Flame

Element	Symbol	Wavelength (nm)	Type	LOD-AAS (ppb)
Antimony	Sb	217.581	I	100
		231.147	I	100
Bismuth	Bi	223.061	I	50
Calcium	Ca	422.673	I	2
Cesium	Cs	455.5276	I	600
		852.1122	I	50
Chromium	Cr	357.869	I	5
Cobalt	Co	240.725	I	5
Copper	Cu	324.754	I	50
		327.396	I	50
Gallium	Ga	287.424	I	70
Gold	Au	242.795	I	20
Indium	In	303.936	I	50
Iridium	Ir	208.882	I	15,000
		2639.71	I	2,000
Iron	Fe	248.3271	I	5
Lead	Pb	283.3053	I	10
Lithium	Li	670.776	I	5
Magnesium	Mg	285.213	I	0.3
Manganese	Mn	279.482	I	2
		403.076	I	2
Mercury	Hg	253.652	I	500
Molybdenum	Mo	313.259	I	30
Nickel	Ni	232.003	I	5
Osmium	Os	290.906	I	17,000
Palladium	Pd	244.791	I	2,000
		247.642		30
Platinum	Pt	265.945	I	100
Potassium	K	766.490	I	5
Rhodium	Rh	343.489	I	30
Rubidium	Rb	420.180	I	NA
		780.027	I	5
Ruthenium	Ru	349.894	I	300
		372.803	I	3,000

Table 2 (Continued)

Element	Symbol	Wavelength (nm)	Type	LOD-AAS (ppb)
Selenium	Se	196.09	I	100
		203.98	I	2,000
Silver	Ag	328.068	I	5
		338.2068	I	200
Sodium	Na	330.237	I	NA
		588.9950	I	2
		589.5924	I	2
Strontium	Sr	407.771	II	NA
		460.733	I	10
Tellurium	Te	214.281	I	100
Thallium	Tl	276.787	I	30
		377.572	I	2,400
Tin	Sn	224.605	I	30
Zinc	Zn	213.856	I	2

Source: From Ref. 16.

e. Air-acetylene flame

The air-acetylene flame is by far the most popular in use today for both FAES and FAAS. It is easy to use, quiet, and a good atomization cell for many elements. Some 30-40 elements can be determined in the parts per million range by this flame. Theoretically, burning acetylene to the stoichiometric products of CO_2 and H_2O should be possible; however, acetylene is normally burned to about equal amounts of CO and CO_2, as shown in equation (13). This is a slightly cooler flame, but it consists of chemically more reducing species. In cases where oxide formation persists, an even more reducing flame can be produced by burning the acetylene essentially completely to CO; see equation (14).

$$C_2H_2 + 2O_2 + N_2 \longrightarrow N_2 + H_2O + CO + CO_2 \tag{13}$$

$$2C_2H_2 + \frac{5}{2}O_2 + N_2 \longrightarrow N_2 + H_2O + 4CO + H_2 \tag{14}$$

Note that the fuel-rich burning mode produces two very good chemical reducing species in large quantities, namely, CO and H_2. Even when used in the normal mode, the air-acetylene flame produces CO. These reducing species help to account for the fact that it is a good atom cell for many elements even though it is only slightly hotter than the air-hydrocarbon flames. The air-acetylene flame usually is measured between 2360 and 2600 K and is often stated to be at 2450 K.

The most dominant feature in the background emission of the air-acetylene flame is due to the OH radical, which has a strong band peaking at 306 nm and degrades to the blue. However, small C_2, CH, and continuum emissions are also observed. All of these features are more prominent in the fuel-rich flame.

Table 3 gives the elements, transition, and detection limits for those elements normally determined by this flame in both FAAS and FAES. These limits were obtained with a premixed slot burner with a 10-cm flame length, the most common burner system used with the air-acetylene flame.

f. *Nitrous oxide-acetylene flame*

The most important development in flame spectroscopy since the commercialization of atomic absorption spectrometers was probably the development of the nitrous oxide-acetylene flame. It has chemical reducing capabilities well beyond the air-acetylene flame. This is clearly seen by comparing the data for the two flames in Tables 1 and 3. The nitrous oxide flame is capable of reducing highly refractory elements. The chemistry of this flame is not straightforward; however, it appears that the major reaction products are CO and H_2 rather than the customary CO_2 and H_2O. The chemistry is presented in equation (15).

$$C_2H_2 + 2N_2O \longrightarrow 2CO + 2H_2 + 2N_2 \quad (15)$$

Once again the major combustion products are very good chemical reducing species. In addition, probably about 0.05 mole fraction of free H radicals is formed in this flame. Hydrogen radicals are one of the most powerful chemical reducing species known. This flame may not be as close to chemical thermodynamic equilibrium as the others discussed, and the temperature may not be as meaningful. Temperature measurements range from 2830 to 3070 K, and a temperature of 2950 K is often stated.

A drawback of the nitrous oxide-acetylene flame is its increased spectral background features. The intensity of the OH and continuum emission is probably 10 times that in the air-acetylene flame, and there is a strong CN band emission that peaks at about 380 nm as well.

Detection limits for the nitrous oxide-acetylene flame are given in Table 4 for both FAAS and FAES.

g. *Limitations of flames*

The main limitation of the use of flames as atom cells for either AES or AAS is the very poor efficiency at transferring the sample from the solution to the hot gases at the site of the analytical measurement. As stated in the section on nebulization, the efficiency of transmitting the sample through the burner is nominally about 5-10%; however, this does not take into account the major dilution factor when the solution is vaporized into the flame gases. The dilution factor is 10^3 to 10^4, depending on the gas and solution flows. If a more efficient method of sample introduction could be employed, the same factors could be gained in sensitivity across the periodic table.

Table 3. Limits of Detection for the Air-Acetylene Flame[a]

Element	Symbol	Wavelength (nm)	Type	LOD-AES (ppb)	LOD-AAS (ppb)
Aluminum	Al	308.2153	I		700
		309.2710	I		500
		396.1520	I	NA	600
Antimony	Sb	206.833	I	NA	50
		217.581	I	NA	40
		231.147	I	3000	40
		259.805	I	NA	
Arsenic	As	193.759	I	10,000	140
Barium	Ba	455.403	II	NA	
		553.548	I	NA	
Bismuth	Bi	223.061	I	3000	25
Boron	B	249.677	I	NA	
Cadmium	Cd	228.8022	I	500	1
		326.1055	I	NA	NA
Calcium	Ca	393.366	II		5000
		396.847	II		5000
		422.673	I	0.5	0.5
Cesium	Cs	455.5276	I	NA	
		852.1122	I	NA	8
Chromium	Cr	357.869	I	NA	3
		425.435	I	NA	200
Cobalt	Co	240.725	I		4
		352.685	I	NA	125
Copper	Cu	324.754	I	NA	1
		327.396	I	NA	120
Gallium	Ga	287.424	I		50
		294.364	I	NA	50
		417.204	I	NA	1500
Germanium	Ge	265.1172	I	7000	
Gold	Au	242.795	I	NA	6
		267.595	I	NA	90
Indium	In	303.936	I	NA	30
		325.609	I	NA	20
		451.131	I	NA	200
Iodine	I	183.038	I		8000
		206.163	I	2,500,000	

(continued)

Table 3 (Continued)

Element	Symbol	Wavelength (nm)	Type	LOD-AES (ppb)	LOD-AAS (ppb)
Iridium	Ir	208.882	I		600
		263.971	I		2500
Iron	Fe	248.3271	I		5
		371.9935	I	NA	700
Lead	Pb	217.000	I		9
		283.3053	I	NA	240
		368.3462	I	NA	
Lithium	Li	670.776	I	NA	0.3
		451.857	I		NA
Magnesium	Mg	279.553	II		NA
		280.270	II		NA
		285.213	I	NA	NA
Manganese	Mn	279.482	I	NA	2
		403.076	I	NA	600
Mercury	Hg	253.652	I	NA	140
Molybdenum	Mo	313.259	I		20
		379.825	I	80,000	900
		390.296	I	100	1600
Nickel	Ni	232.003	I		2
		352.454	I	NA	350
Niobium	Nb	309.418	II		NA
Osmium	Os	290.906	I	NA	1200
Palladium	Pd	244.791	I		20
		247.642	I		20
		340.458	I	NA	660
		363.470	I	NA	300
Phosphorus	P	213.547	I		30,000
Platinum	Pt	214.423	I		350
		265.945	I	NA	50
Potassium	K	766.490	I	NA	1
Rhenium	Re	346.046	I		800
Rhodium	Rh	343.489	I	NA	2
		369.236	I	NA	70
Rubidium	Rb	420.180	I	NA	
		780.027	I	NA	0.3
Ruthenium	Ru	349.894	I	NA	400
		372.803	I	NA	250

Table 3 (Continued)

Element	Symbol	Wavelength (nm)	Type	LOD-AES (ppb)	LOD-AAS (ppb)
Selenium	Se	196.09	I	NA	50
		203.98	I	50,000	10,000
Silver	Ag	328.068	I	NA	1
		338.2068	I	NA	70
Sodium	Na	330.237	I		NA
		588.9950	I	NA	1
		589.5924	I	NA	0.2
Strontium	Sr	407.771	II	NA	400
		421.552	II		NA
		460.733	I	NA	2
Sulfur	S	180.7311	I		30,000
Tellurium	Te	214.281	I	500	30
		238.578	I		NA
Thallium	Tl	276.787	I	NA	30
		377.572	I	NA	1,200
		535.046	I	NA	12,000
Tin	Sn	224.605	I		10
		235.484	I	2,000	600
		283.999	I	NA	1,000
		326.234	I	NA	
Tungsten	W	255.135	I		3,000
		400.875	I	90,000	
Uranium	U	591.539	I	NA	
Vanadium	V	318.540	I	NA	
		437.924	I	300	
Ytterbium	Yb	398.799	I		80
Zinc	Zn	213.856	I	7,000	1
Zirconium	Zr	351.960	I		NA

[a]NA means that FAES or FAAS was observed but no detection limit was reported; a blank space indicates that no observation was made.
Source: From Refs. 16 and 17.

Table 4. Limits of Detection for the Nitrous Oxide-Acetylene Flame[a]

Element	Symbol	Wavelength (nm)	Type	LOD-AES (ppb)	LOD-AAS (ppb)
Aluminum	Al	308.2153	I	NA	
		309.2710	I	NA	20
		396.1520	I	3	900
Barium	Ba	553.548	I	1	8
Beryllium	Be	234.861	I	100	1
Boron	B	208.891	I		NA
		208.957	I		24,000
		249.677	I		700
		249.773	I		1,500
Cadmium	Cd	326.1055	I	800	
Calcium	Ca	422.673	I	0.1	1
Cesium	Cs	455.5276	I	600	
		851.1122	I	0.02	
Chromium	Cr	425.435	I	1	
Cobalt	Co	352.685	I	200	
Copper	Cu	324.754	I	30	
		327.396	I	3	
Dysprosium	Dy	353.170	II		800
		404.597	I	20	500
		421.172	I		50
Erbium	Er	337.271	II		100
		400.796	I	20	40
Europium	Eu	459.403	I	0.2	30
Gadolinium	Gd	368.413	I		2000
		440.186	I	1000	
Gallium	Ga	417.204	I	5	
Germanium	Ge	265.1172	I	400	50
Gold	Au	267.595	I	500	
Hafnium	Hf	307.288	I		2000
Holmium	Ho	345.600	II		3000
		405.393	I	10	400
		410.384	I		40
Indium	In	303.936	I		1000
		325.609	I		700
		451.131	I	1	3500
Iridium	Ir	208.882	I		500

Table 4 (Continued)

Element	Symbol	Wavelength (nm)	Type	LOD-AES (ppb)	LOD-AAS (ppb)
Iron	Fe	371.9935	I	10	
Lanthanum	La	408.672	II		7500
		550.134	I	4000	2000
Lead	Pb	368.3462	I	0.2	
Lithium	Li	670.776	I	0.001	
Lutetium	Lu	261.542	II		3000
		451.857	I	400	
Magnesium	Mg	285.213	I	1	
Manganese	Mn	403.076	I	1	
Mercury	Hg	253.652	I	10,000	
Molybdenum	Mo	313.259	I	10	25
		379.825	I	300	
		390.296	I	10	
Neodymium	Nd	463.424	I		600
		492.453	I	200	700
Nickel	Ni	352.454	I	20	
Niobium	Nb	334.906	I		1000
		405.894	I	60	5000
Osmium	Os	290.906	I		80
		442.047	I	2000	NA
Palladium	Pd	363.470	I	40	
Phosphorus	P	177.499	I		30,000
		213.547	I		29,000
Platinum	Pt	265.945	I	2000	2000
Potassium	K	766.490	I	0.01	
Praseodymium	Pr	495.137	I	500	2000
Rhenium	Re	346.046	I	200	200
Rhodium	Rh	343.489	I		700
		369.236	I	10	1400
Rubidium	Rb	780.027	I	8	
Ruthenium	Ru	372.803	I	300	
Samarium	Sm	429.674	I		500
		476.027	I	50	14,000

(continued)

Table 4 (Continued)

Element	Symbol	Wavelength (nm)	Type	LOD-AES (ppb)	LOD-AAS (ppb)
Scandium	Sc	391.181	I	10	20
Selenium	Se	196.09	I	100,000	
Silicon	Si	251.6113	I	3,000	20
		288.1579	I		NA
Silver	Ag	328.068	I	2	
Sodium	Na	588.9950	I	0.01	
		589.5924	I	0.01	
Strontium	Sr	460.733	I	0.1	50
Tantalum	Ta	271.467	I		800
		474.016	I	4,000	
Terbium	Tb	432.643	I	NA	600
Thallium	Tl	377.572	I	50	
		535.046	I	2	
Thorium	Th	324.4448	I		181,000
		491.9816	II	10,000	
Thulium	Tm	371.791	I	4	10
Tin	Sn	224.605	I		3,000
		235.484	I		90
		283.999	I	100	
Titanium	Ti	334.941	II		NA
		364.268	I	NA	10
		365.350	I	30	500
Tungsten	W	255.135	I		500
		400.875	I	200	7,500
Uranium	U	358.488	I		7,000
Vanadium	V	318.540	I	200	20
		437.924	I	7	100
Ytterbium	Yb	398.799	I	0.2	5
Yttrium	Y	410.238	I		50
Zinc	Zn	213.856	I	10,000	
Zirconium	Zr	351.960	I	1,200	
		360.119	I	3,000	1,000

[a]NA means that FAES or FAAS was observed but that no detection limit was reported; a blank space indicates that no observation was made.
Source: From Refs. 16 and 17.

Electrothermal Cells

a. Graphite furnaces

L'vov [18] and Woodriff [19] independently developed the field of EAAS during the 1960s. Both used rather large furnace cells held at constant temperature in conjunction with small sample size introduction for AAS measurements. These researchers saw the advantage of the furnace approach for avoiding the large dilution factors inherent in the use of flames as atom cells. Instrument manufacturers realized the tremendous advantage of increased sensitivity and developed and commercialized smaller versions of these devices during the 1970s under the name of carbon or graphite furnaces. The commercial devices consisted of a graphite tube that is resistively heated by means of an electrical current after a very small sample has been introduced. A typical cell is illustrated in Figure 10. An inert gas normally flows through the system to lengthen the life of the tube and to prevent unwanted oxide formation. Although there has been some academic evaluation of AES using the graphite furnace technique, most of the applications have been with AAS.

The key to the success of EAAS is to control the time-temperature behavior of the tube. Most instruments in use today are computer controlled with some type of temperature feedback. The ability to reproduce the time-temperature profile from sample to sample and from sample to standard is essential for achieving precision and accuracy of results. Samples are generally introduced volumetrically as liquids, most often as dilute aqueous solutions in 20-50 μl quantities. A typical time-temperature profile is illustrated in Figure 11. An advantage of this technique is that sample

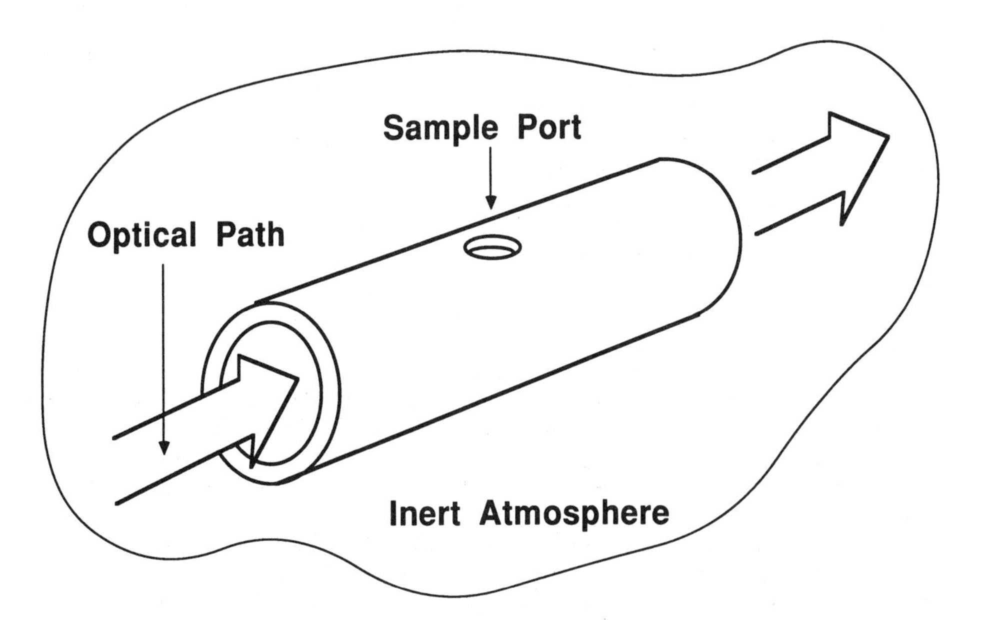

Figure 10. Typical graphite tube atom cell.

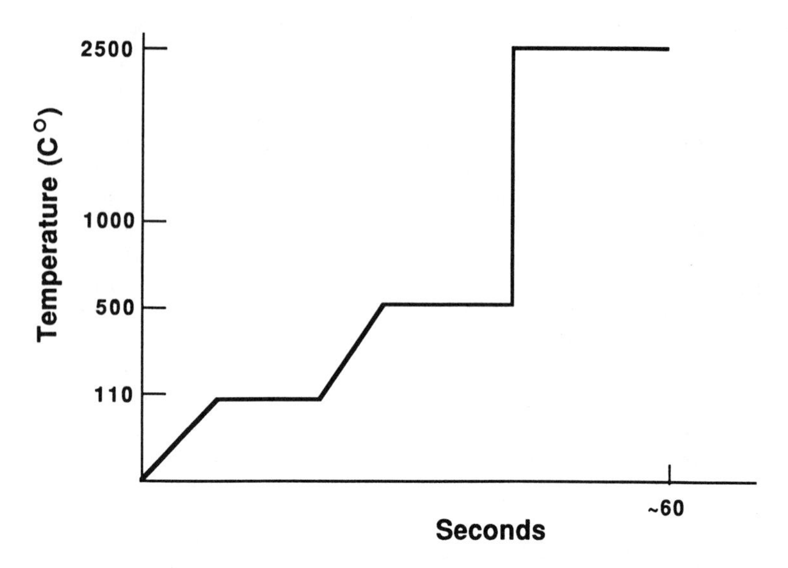

Figure 11. Typical time-temperature profile for a graphite furnace experiment.

pretreatment can be realized in the furnace tube. As a consequence, biological samples can be ashed or chemically pretreated as part of the analytical measurement procedure. The normal procedure is to dry the sample with a ramped temperature increase, to pretreat the sample (by ashing, charring, or chemical reaction) with another ramped temperature increase, and finally to atomize by means of a step function to the final temperature.

For many elements, EAAS is the most sensitive method of determination, in many cases, by more than an order of magnitude over any competing technique. Table 5 contains a list of 52 elements, with their transitions and detection limits, as determined by this technique.

On the negative side, the use of a graphite furnace generates a transient signal. This causes no problem with today's electronics, but the response of earlier systems was too slow to keep up with the rapid rise and fall of signal, and errors resulted. In addition, integrating electronics became more desirable. Current instrumentation handles these problems with little trouble.

However, not all problems have been overcome. The technique is simply not suitable for refractory elements, because sample introduction is not precise and time-temperature profiles are relatively nonrepeatable. The former is due to the very small sample solution volumes used, and the latter is due to the changes in electrical properties, caused by repeated heating and cooling in the graphite tube.

Further, the use of background correction is absolutely necessary to obtain accurate results. All modern instrumentation has background correction capabilities. This is more thoroughly discussed in Section II.D.

Table 5. Limits of Detection for Graphite Furnace AAS[a]

Element	Symbol	Wavelength (nm)	Type	LOD (ppb)
Aluminum	Al	308.2153	I	NA
		309.2710	I	0.01
		396.1520	I	600
Antimony	Sb	206.833	I	NA
		217.581	I	0.08
		231.147	I	NA
Arsenic	As	189.042	I	NA
		193.759	I	0.12
Barium	Ba	553.548	I	0.04
Beryllium	Be	234.861	I	0.003
Bismuth	Bi	223.061	I	0.01
Cadmium	Cd	228.8022	I	0.0002
Calcium	Ca	422.673	I	0.01
Chromium	Cr	357.869	I	0.004
Cobalt	Co	240.725	I	8
Copper	Cu	324.754	I	0.005
		327.396	I	NA
Erbium	Er	400.796	I	0.3
Gadolinium	Gd	440.186	I	0.3
Gallium	Ga	287.424	I	0.01
Germanium	Ge	265.1172	I	0.1
Gold	Au	242.795	I	0.01
Holmium	Ho	345.600	II	NA
		405.393	I	NA
Indium	In	303.936	I	0.02
Iodine	I	183.038	I	40,000
Iridium	Ir	208.882	I	0.5
Iron	Fe	248.3271	I	0.01
		371.9935	I	NA
Lanthanum	La	550.134	I	0.5
Lead	Pb	217.000	I	0.007
		283.3053	I	NA
Lithium	Li	670.776	I	0.01

(continued)

Table 5 (Continued)

Element	Symbol	Wavelength (nm)	Type	LOD (ppb)
Magnesium	Mg	285.213	I	0.0002
Manganese	Mn	279.482	I	0.0005
		403.076	I	NA
Mercury	Hg	253.652	I	0.2
Molybdenum	Mo	313.259	I	0.03
Nickel	Ni	232.003	I	0.05
Osmium	Os	290.906	I	2
Palladium	Pd	247.642	I	0.05
Phosphorus	P	177.499	I	NA
		213.547	I	20
		253.561	I	NA
Platinum	Pt	265.945	I	0.2
Potassium	K	766.490	I	0.004
Rhenium	Re	346.046	I	10
Rhodium	Rh	343.409	I	0.1
Rubidium	Rb	780.027	I	NA
Selenium	Se	196.09	I	0.05
Silicon	Si	251.6113	I	0.6
Silver	Ag	328.068	I	0.001
Sodium	Na	588.9950	I	0.004
Strontium	Sr	460.733	I	0.01
Sulfur	S	180.7311	I	NA
		182.0343	I	NA
		216.89		NA
Tellurium	Te	214.281	I	0.03
Thallium	Tl	276.787	I	0.01
Tin	Sn	235.484	I	0.03
		283.999	I	NA
Titanium	Ti	364.268	I	0.3
		365.350	I	NA
Uranium	U	358.488	I	30
Vanadium	V	318.540	I	0.4
Ytterbium	Yb	398.799	I	0.01
Yttrium	Y	410.238	I	10
Zinc	Zn	213.856	I	0.001

[a]NA means that EAAS was observed but that no detection limit was reported.

Source: From Refs. 16 and 17.

Finally, in EAAS the sample matrix causes more problems than with almost any other type of sample cell. The only form of energy available in EAAS to break molecular bonds is kinetic energy. Flames generally have higher kinetic energy and also utilize chemical energy in reducing oxides to the atomic state. Thus, if the element of interest is present in more than one chemical form, the results can be confusing and the signals nonrepeatable. Pretreating the samples is often necessary to ensure that all the analyte is in one chemical form. Hence, chemical "modifiers" are often added (similar to the "carriers" in classical dc-arc analysis) to aid in this requirement. The term matrix modification is used when chemicals are added. Because this is a sample-specific problem, literature searches are very important when a new sample type is to be analyzed. Slavin has published a major bibliography for EAAS [20].

Standard additions are often used to improve accuracy for these same reasons. Of course, several measurements are required for the development of one analytical result when standard additions are used. Therefore, both the effort and the cost per sample are increased.

Recently, a technique to delay the vaporization of the sample until the gases in the graphite tube are already hot has been used. This delay can be accomplished by changing the tube design or by using the "L'vov platform." The L'vov platform is simply a small piece of graphite that makes contact with a small part of the tube surface and therefore heats at a slower rate than the tube itself. The platform is illustrated in Figure 12, and the difference in the time-temperature profile is illustrated in Figure 13. When the sample vaporizes under these conditions, the gas temperature is already much hotter than the sample, which prevents unwanted chemical reactions with matrix species, provides a more constant temperature environment for the analytical measurement, and improves results.

b. Metal filament and other solid sampling cells

During the introduction of EAAS techniques in the 1970s, metal filaments were employed in an experimental set-up similar to graphite furnaces. The filament were heated by means of resistive heating using an electrical current. Temperature-time profiles could be produced in a similar fashion, and analytical procedures were developed in a completely analogous fashion to graphite furnace AAS. Unfortunately, the filament approach incorporates and exacerbates all of the problems associated with the graphite furnace technique. This method is no longer commercially available.

Recently, a new device for atomizing solid samples into a tube cell has been introduced [21]. This device utilizes a sputtering concept into a reduced atmosphere of argon. Figure 14 presents a schematic diagram of the cell. Note that the sample serves as the cathode for the electrical discharge and that atomization is accomplished by argon ion bombardment of the sample surface. The argon is introduced in six separate jets, which creates a sputter pattern about 6 mm in diameter. The obvious limitation of the procedure is that the sample must be conductive. However, for metal and alloy samples, this means less sample preparation than with other AAS cells. A further complication with respect to other EAAS cells is the necessity for a vacuum system. Because the evacuated volume is quite small, this is a rather simple requirement, however. The precision, sensitivity, and accuracy for metal samples probably compare with other AAS cells.

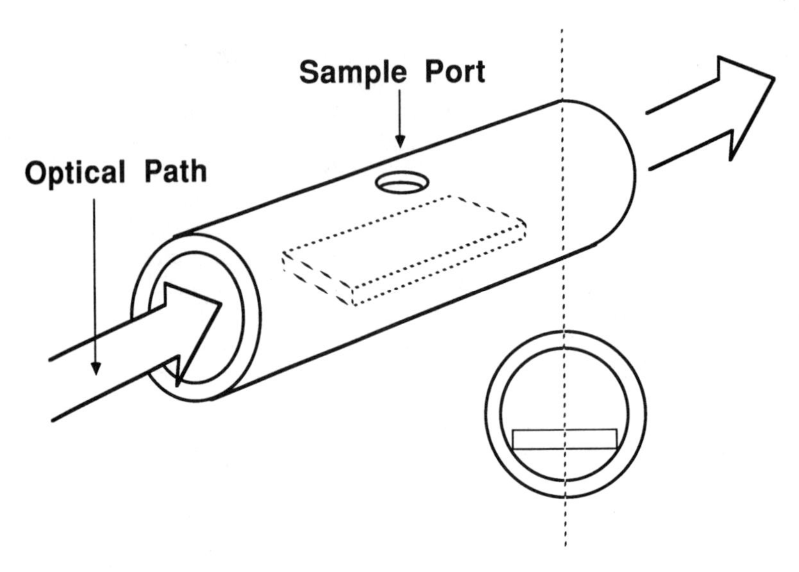

Figure 12. The L'vov platform.

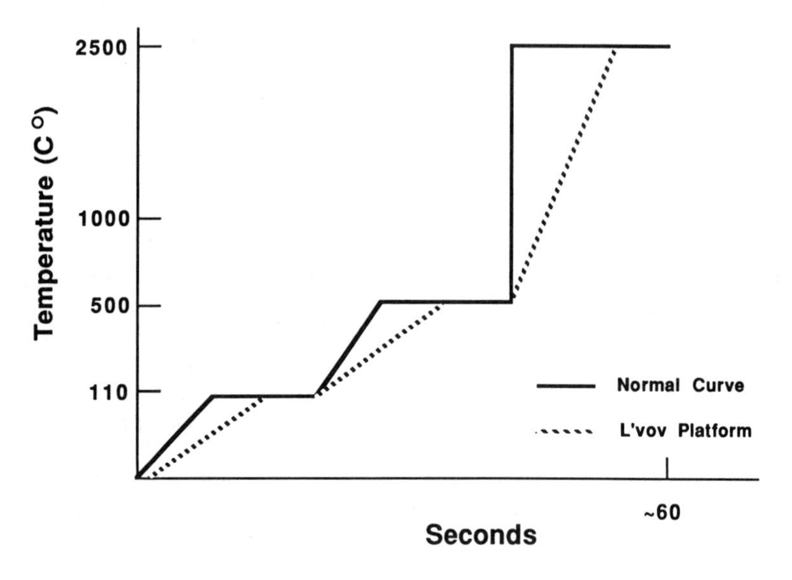

Figure 13. Time-temperature profile with the L'vov platform.

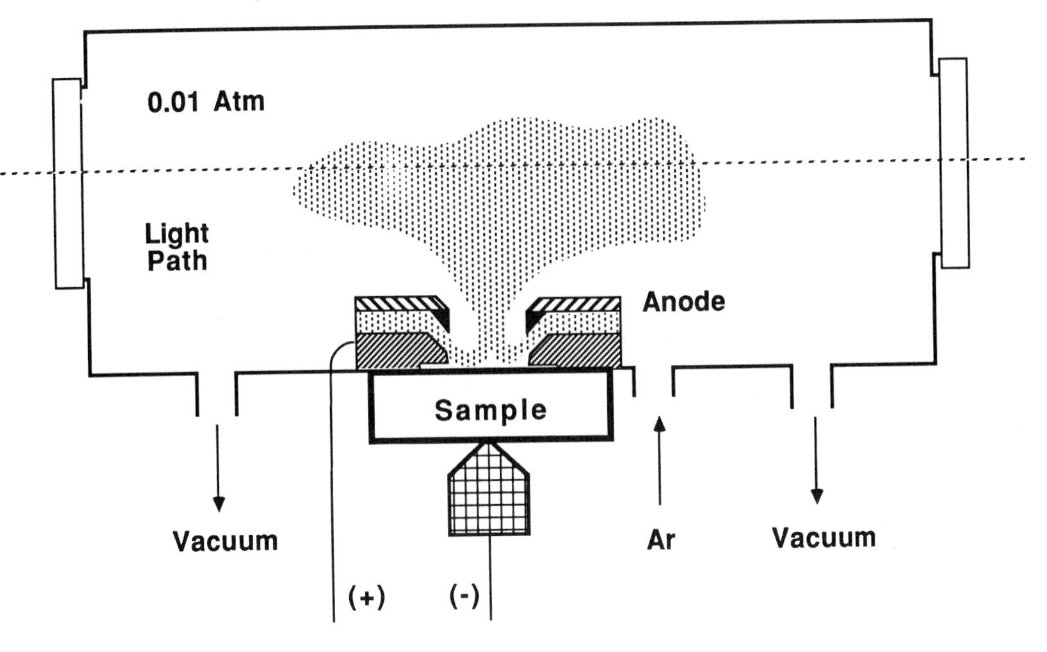

Figure 14. The Atomsource cell.

B. Radiation Sources

In AES, the sample cell also serves as the radiation source; therefore, the flames and other atom cells discussed in the previous sections also serve as discussion of the radiation sources for AES.

In AAS, the situation is quite different. The radiation source is external to the atom cell and must possess certain important radiational properties with respect to the atom cell. First and most important, the radiation source must emit radiation of the same frequency and therefore the same wavelength as the analyte atoms in the sample cell. Second, the width of the spectral lines emitted in the radiation source must be narrow with respect to the width of the analyte atomic absorption profile for a linear analytical curve to be achieved. Third, the intensity and stability of radiation from the source must be great enough for precise measurement by the detection system. Hollow cathode discharge lamps (HCLs) and electrodeless discharge lamps (EDLs) generally fulfill these requirements and are discussed in the section on line sources below.

An external radiation source also must measure a background or non-specific absorption in AAS, which is not part of the atomic absorption signal from the analyte species. Such an absorption signal, if not measured and subtracted from the analyte absorption, will result in an error in the determination of the concentration of the analyte species. Section II.D describes the various correction techniques for the background absorption

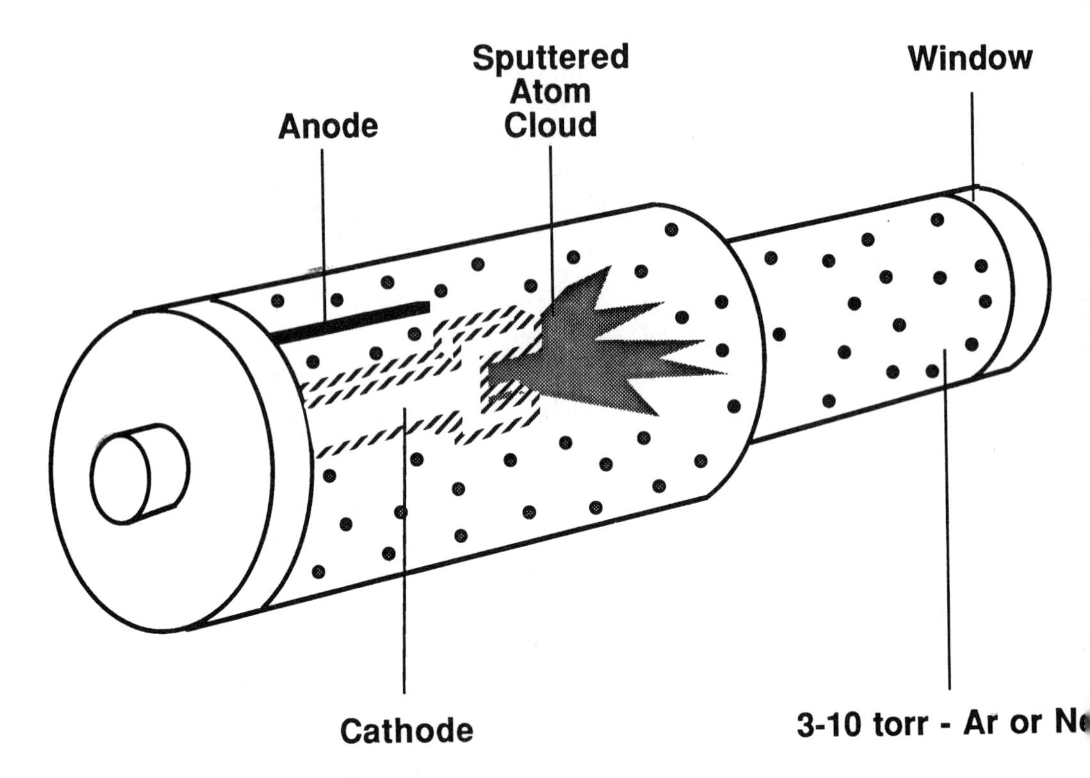

Figure 15. Schematic of a hollow-cathode lamp.

signals. One common procedure for background correction involves the use of an external continuum radiation source. Continuum sources are discussed in Section II.B.

Line Sources

The term "line source" refers to a source of radiation capable of producing a narrow band of radiation that is associated with a particular transition originating from a specific atomic species. If this type of emission were observed in the visible spectral region through a spectroscope or by means of the photographic output from a spectrographic instrument, the emission results would look like a line in the spectrum. In fact, the phenomenon is simple atomic emission as was described in Section I. For AAS, it is desirable to have such a source, specific for each atomic species required in the analysis.

a. Hollow cathode discharge lamps (HCLs)

Probably the simplest of all atomic line sources, the HCLs are available for most elements. They consist of a hollow cathode containing the element of interest in a reduced atmosphere (1-10 torr) of an inert noble gas, usually neon or argon, and a low-voltage power train, which can be operated in either a dc or ac mode. Normally, the lamps are powered at a few hundred volts and a few milliamperes. Under such conditions, a glow discharge exists and the atoms of the element of interest are sputtered from the surface of the hollow cathode and are excited by thermal collision with other atoms and ions in the glow discharge (Figure 15).

The spectral characteristics of HCLs are nearly ideal for AAS measurements. The intensity of the output radiation after a reasonably short warm-up time is quite stable with only about 1% short-term noise. For most elements, the intensity is sufficient for sensitive analytical measurement. Finally, the line widths are quite narrow. The only line broadening effect is Doppler broadening. The temperature of a typical HCL is about 50-100°C; this results in a Doppler width of only about 0.0005-0.005 nm. It is common to state the half width or full width at half maximum (FWHM), as illustrated in Figure 16. In addition, HCLs emit virtually no continuum radiation, very simple noble gas spectra, and few nonresonance analyte atom spectra. The spectral features for a typical HCL are presented in Figure 17.

Elements that are metallic and have good sputter characteristics produce intense resonance emission from HCLs and are most suitable for AAS measurements. Manufacturers have perfected many techniques to enhance the radiation from HCLs as well. Some elements do not emit intense resonance radiation with this technique, regardless of all efforts to improve their output. Table 6 provides a tabulation of the relative emission intensity for the resonance transitions for many elements based on a single measurement system. Most elements with intensities of less than 100 on this scale would probably not emit enough radiation to permit AAS measurements with sufficient precision to be deemed useful.

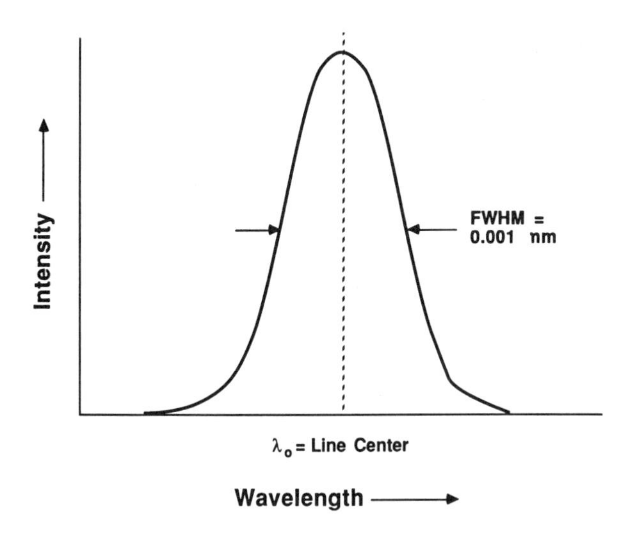

Figure 16. Line profile of a hollow-cathode lamp.

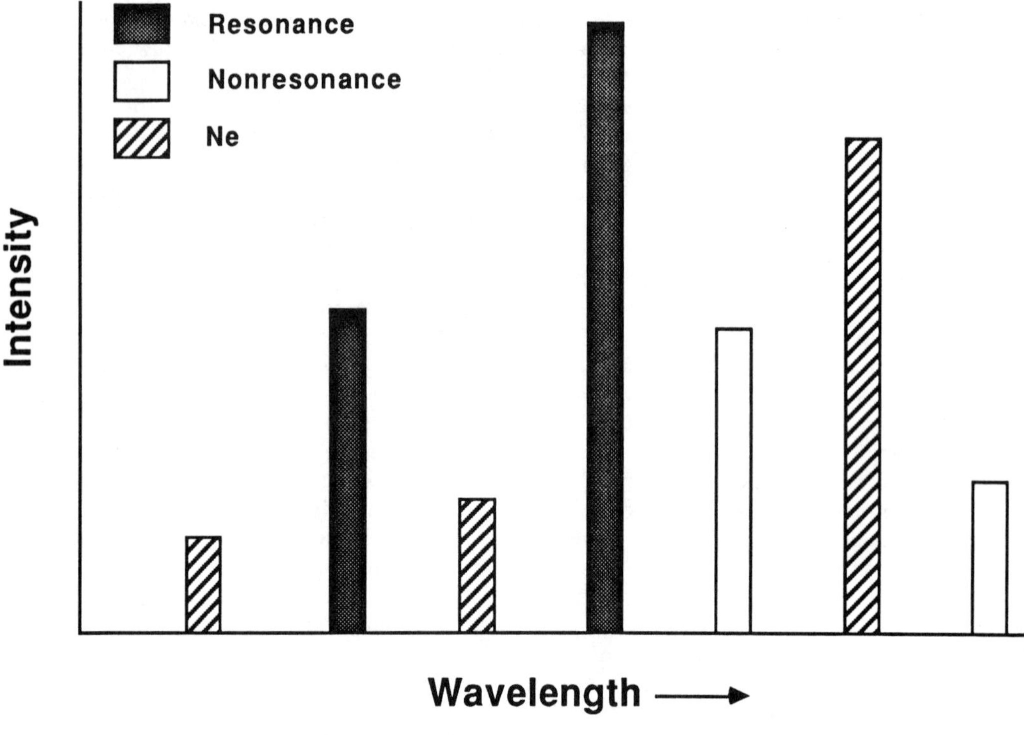

Figure 17. Spectral features of a typical hollow-cathode lamp.

b. Electrodeless discharge lamps (EDLs)

Another approach to atomic line emission is by inductively coupling microwave or radiowave energy into an ionized gas cloud. This process produces radiation very similar in characteristics to an HCL. An EDL is very simple: it consists of an evacuated quartz bulb into which a small amount of the element (or compound of the element) of interest and a few torr of a noble gas (argon or neon) have been introduced and sealed. The EDL is then placed into a coupling cavity and tuned to achieve maximum inductive coupling. Fortunately, many of the elements that provide poor HCLs produce excellent EDLs. Although no one has produced a comparison of relative intensities for EDLs, it is generally stated that, for those elements for which commercial EDLs are produced, there is at least an order of magnitude gain in intensity output over that of an HCL for the same element. Figure 18 shows the elements for which EDLs are available.

EDLs have spectral properties that can be described in an analogous way to HCLs. That is, EDLs produce simple, intense elemental resonance spectra that are very narrow in width. Only a few lines are produced

Table 6. Relative Intensities of Elemental Transitions from Hollow Cathode Lamps, All Measured with the Same Experimental System[a]

Element	Fill gas	Wavelength (nm)	Relative emission intensity[b]
Aluminum	Ne	309.271 }	1200
		309.284 }	
		396.152	800
Antimony	Ne	217.581	250
		231.147	250
Arsenic	Ar	193.696	125
		197.197	125
Barium	Ne	553.548	400
		350.111	200
Beryllium	Ne	234.861	2500
Bismuth	Ne	223.061	120
		306.773	400
Boron	Ar	249.772	400
Cadmium	Ne	228.802	2500
		326.106	5000
Calcium	Ne	422.673	1400
Cerium	Ne	520.012 }	8
		520.042 }	
		569.700	8
Chromium	Ne	357.869	6000
		425.435	5000
Cobalt	Ne	240.725	1000
		345.350	1500
		352.685	1300
Copper	Ne	324.754	7000
		327.396	6000
Dysprosium	Ne	404.599	2000
		418.678	2000
		421.172	2500
Erbium	Ne	400.797	1600
		386.282	1600
Europium	Ne	459.403	1000
		462.722	950
Gadolinium	Ne	368.413	350
		407.870	700
Gallium	Ne	287.424	400
		417.206	1100

(continued)

Table 6 (Continued)

Element	Fill gas	Wavelength (nm)	Relative emission intensity[b]
Germanium	Ne	265.118 }	500
		265.158 }	
		259.254	250
Gold	Ne	242.795	750
		267.595	1200
Hafnium	Ne	307.288	300
		286.637	200
Holmium	Ne	405.393	2000
		410.384	2200
Indium	Ne	303.936	500
		410.176	500
Iridium	Ne	284.972	400
		263.971	400
Iron	Ne	248.327	400
		371.994	2400
Lanthanum	Ne	550.134	120
		392.756	45
Lead	Ne	216.999	200
		283.306	1000
Lithium	Ne	670.784	700
Lutetium	Ar	335.956	30
		337.650	25
		356.784	15
Magnesium	Ne	285.213	6000
		202.582	130
Manganese	Ne	279.482	3000
		280.106	2200
		403.076	14,000
Mercury	Ar	253.652	1000
Molybdenum	Ne	313.259	1500
		317.035	800
Neodymium	Ne	463.424	300
		492.453	600
Nickel	Ne	232.003	1000
		341.476	2000
Niobium	Ne	405.894	400
		407.973	360
Osmium	Ar	290.906	400
		301.804	200

Table 6 (Continued)

Element	Fill gas	Wavelength (nm)	Relative emission intensity[b]
Palladium	Ne	244.791	400
		247.642	300
		240.458	3000
Phosphorus	Ne	213.547 }	30
		213.620 }	
		214.911	20
Platinum	Ne	265.945	1500
		299.797	1000
Potassium	Ne	766.491	6
		404.414	300
Praseodymium	Ne	495.136	100
		513.342	70
Rhenium	Ne	346.046	1200
		346.473	900
Rhodium	Ne	343.489	2500
		369.236	2000
		350.732	200
Rubidium	Ne	780.023	1.5
		420.185	80
Ruthenium	Ar	349.894	600
		392.592	300
Samarium	Ne	429.674	600
		476.027	800
Scandium	Ne	391.181	3000
		390.749	2500
		402.040	1800
		402.369	2100
Selenium	Ne	196.026	50
		203.985	50
Silicon	Ne	251.611	500
		288.160	500
Silver	Ar	328.068	3000
		338.289	3000
Sodium	Ne	588.995	2000
		330.232 }	40
		330.299 }	
Strontium	Ne	460.733	1000

Table 6 (Continued)

Element	Fill gas	Wavelength (nm)	Relative emission intensity[b]
Tantalum	Ar	271.467	150
		277.588	100
Tellurium	Ne	214.275	60
		238.576	50
Terbium	Ne	432.614 }	110
		432.647 }	
		431.885	90
		433.845	60
Thallium	Ne	276.787	600
		258.014	50
Thulium	Ne	371.792	40
		409.419	50
		410.584	70
Tin	Ne	224.605	100
		286.333	250
Titanium	Ne	364.268	600
		399.864	600
Tungsten	Ne	255.100 }	200
		255.135 }	
		400.875	1400
Uranium	Ne	358.488	300
		356.660	200
		351.461	200
		348.937	150
Vanadium	Ne	318.341 }	600
		318.398 }	
		385.537 }	200
		385.584 }	
Ytterbium	Ar	398.798	2000
		346.436	800
Yttrium	Ne	407.738	500
		410.238	600
		414.285	300
Zinc	Ne	213.856	2500
		307.590	2000

[a]These data were obtained using Westinghouse HCLs and a single experimental setup. No correction has been made for the spectral response of the monochromator/PMT system.

[b]The most intense line is the Mn 403.026 transition, with a relative intensity of 14,000.

Source: From Ref. 16.

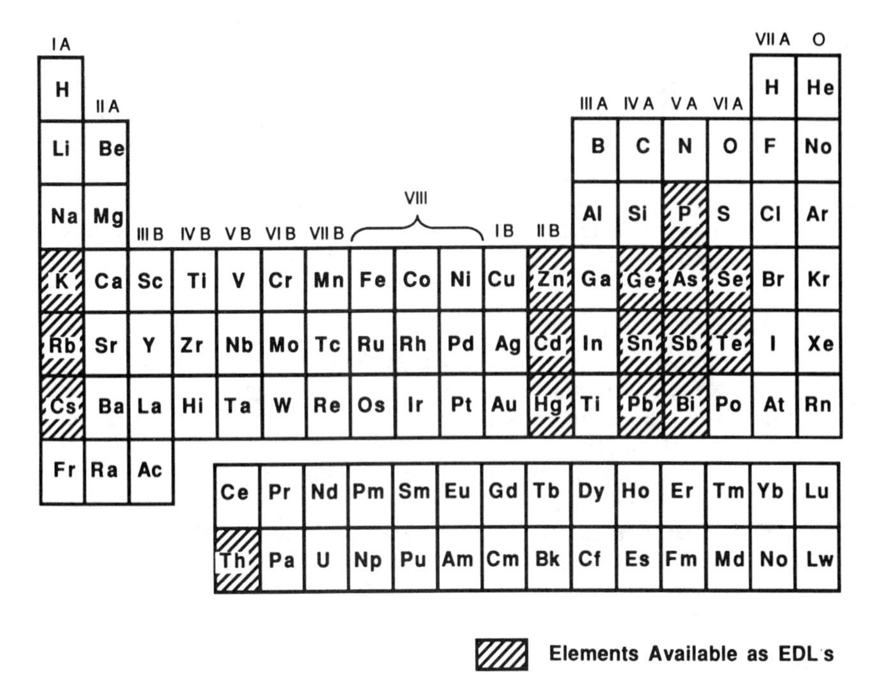

Figure 18. Elements for which EDLs are available.

by the noble diluent gas, as well as a few nonresonance lines. In general, the emission is stable and has little short-term noise, though generally more than HCLs.

Use of EDLs requires the purchase of an additional power supply; however, if the analyte elements include those for which HCLs cannot be used, the expense is certainly worth it.

Continuum Sources

As opposed to line sources, the purpose of a continuum source is to provide a broad spectral region of radiation so that no matter which wavelength is observed, some intensity of emission is found. The reason continuum sources are used in AAS is to be able to make a background correction. Background corrections are often required in FAAS and essentially always are required in EAAS. The concept of the background correction is very important in AAS, and part of Section II.D is devoted to the various ways to accomplish it. The most common continuum sources in the visible region are the tungsten-filament light bulbs used in everyday lighting applications. These would be suitable for AAS background correction if all of the analytically useful transitions were in the visible; however, most of the transitions important for AAS are found in the UV region of the spectrum (200-400 nm). The most common lamps for this region are discussed below.

a. Deuterium discharge lamp

The deuterium discharge lamp is probably the most widely used for AAS background correction because it can be made essentially like an HCL and operates with a similar power supply. Thus, it is easy to incorporate into the optical system of a commercial AAS instrument. This lamp consists of a dc discharge through a relatively low-pressure deuterium gas (e.g., 10 torr). Deuterium, rather than hydrogen, is used because it provides more intensity at lower wavelengths and because the lamps last longer.

b. High-pressure discharge lamps

Continuum lamps can also be made from a discharge through many gases at high pressure. Xenon and a mixture of xenon and mercury are commonly used at high pressure (several atmospheres) and rather high powers (e.g., 150-300 W). Although these types of continuum sources are widely used in molecular fluorescence spectrometers, they are not often used in AAS instrumentation because they require a different power supply and are not as easily incorporated into the optical system of most AAS instruments.

C. Optical Systems

Imaging Optics

The optical requirements for either AAS or AES are reasonably straightforward and rather simple. No stringent optical requirement is imposed on the system. This is especially true with flame systems in which the optical area of interest is 5-10 mm in diameter. A flame is quite large, and passing radiation from the excitation source through the flame is simple, often accomplished with a simple lens or a mirror/lens combination. Likewise, focusing radiation from the flame to the monochromator in FAES is accomplished with the same type of optics. Graphite furnace AAS imposes a tighter optical requirement of 0.5-1 mm in diameter; however, this is readily done with similar optics.

From a materials standpoint, quartz or fused silica lenses are required because of their UV transmission characteristics. Normal glass has a rather sharp UV cutoff at 350 nm. For the same reason, front-surfaced mirrors must be used.

All of the optical imaging is performed before the radiation enters the monochromator. The detector is normally mounted at the exit slit of the monochromator, and no imaging is done after the final mirror inside the monochromator. In systems that use two radiation sources, a line source and a continuum source for background correction, a beam splitter, chopper, or combination of both is usually employed.

Spectral Band Selection

The term monochromator is in reality a misnomer. Monochromatic radiation is radiation with only one wavelength or frequency. All spectral band selection devices pass a band of wavelengths or frequencies that may be larger or smaller according to the quality of the device. For instance, an optical filter may pass a spectral band of 50-100 nm, whereas a 1-m monochromator, depending on the grating and size of slits used, may pass a spectral band of 0.01-1 nm. However wrong, the term monochromator is with us to stay.

a. Optical filters

Colored glass filters have long been used to restrict the spectral region passing to a detector. Simple filters often will pass from 50 to 100 nm. However if two filters are used, one with an appropriate low-wavelength cutoff and the other with a high-wavelength cutoff, a variable and in many cases a very narrow spectral band can be selected with up to 50% transmission characteristics. Sets of filters with selected cutoffs are commercially available. These have been used in instruments that operate in the visible and have been incorporated into instruments that are specifically designed for elements that emit or absorb visible radiation. The most common of these are simple flame emission photometers designed for the alkali metals.

Another type of filter that has been commonly used for very narrow spectral band selection is the interference filter, which depends on the selective interference of radiation because of the thickness of dielectric separating two media of different refractive index. This phenomenon is similar to that used in the diffraction grating described below. The disadvantage of this filter is that one filter must be used for each wavelength of interest. Once again, this approach is satisfactory for a single-element instrument or for one that is devoted to a few elements, but it lacks flexibility. In addition, because of the geometric constraints, these filters are difficult to manufacture for use in the UV.

b. Monochromators

All monochromators have several features in common: entrance and exit slits, a collimating mirror, a grating, and a camera mirror, generally with the same focal length as the collimating mirror. The most commonly used optical arrangement for AES and AAS is the Czerny-Turner configuration illustrated in Figure 19. Although prisms were often used in early instruments, modern spectrometers use diffraction gratings exclusively. The switch to diffraction gratings probably is due to their increased light throughput in the UV and to the fact that a linear wavelength drive is easily accomplished.

Four characteristics can be used to describe any monochromator: free spectral range, light throughput, resolution, and spectral bandwidth.

The free spectral range of a monochromator depends on the type of configuration, the angle over which the grating can be rotated, and the number of grooves per unit length of grating. A spectral range of 200 to 800 or 1000 nm is common. Second- and third-order diffraction can be passed through a monochromator because of the laws of diffraction, and this can sometimes cause spectral interferences. For example, radiation at 200 nm will have a second-order diffraction feature at 400 nm and a third-order feature at 600 nm, etc. Because there is generally UV radiation associated with many elements, a UV cutoff filter is often used when the anlytical features are above 400 nm.

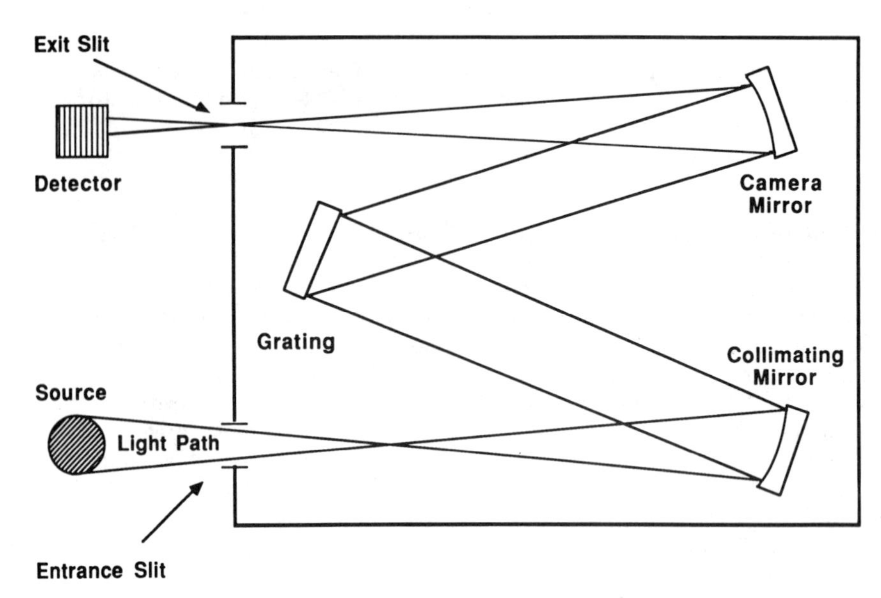

Figure 19. Monochromator using the Czerny-Turner mounted grating.

Light throughput depends on the solid angle that can be accepted into the monochromator and is usually dictated by the size of the mirrors or the size of the grating in combination with the slit width/height and the blaze of the grating. There is a tradeoff between light throughput and resolution. Because of the rather simple spectral features associated with flame measurements, many manufacturers opt for medium-resolution, high-light-throughput monochromators for AES/AAS instruments, which reduces the costs of these spectrometers.

Both resolution and spectral bandwidth depend on the number of grooves per unit length of the grating (which is illuminated by the collimating mirror), the focal length of the mirrors, and the slit width settings. More grooves, longer focal lengths, and narrower slit settings all produce higher resolution (and decrease light throughput). Common gratings have between 1000 and 3000 grooves/mm. The focal lengths are usually between 0.25 and 0.5 m, and slit widths are generally in the 10-500 μm range. Depending on all of these settings, a spectral bandpass between 0.05 and 10 nm is obtained.

Clearly, the selection of spectral bandpass depends on the particular element and matrix being measured. If the analyte is in a clear spectral region and if there are no interfering spectral features in the sample matrix, the slits can be opened up to increase the light throughput without degrading the measurement. On the other hand, if either the analyte or the matrix produces radiation in the spectral window that cannot be used in the analytical measurements, the result is stray light into the detector and a degradation of the signal-to-noise ratio of the signal being measured.

Detection Systems

A detection system includes the detector, associated power supplies and amplifiers, and readout devices. Historically, the readout devices have been analog meters, digital meters, or stripchart recorders; however, today computers are incorporated in all but the least expensive, bottom-of-the-line instruments. Most of the discussion of readout devices will center on computer systems and their capabilities.

a. Detectors

For the past two decades, the standard in detectors for radiation in the UV/visible spectral region has been the photomultiplier tube (PMT). This detector has the ability to transduce radiation into an electrical signal with very high gain (on the order of 10^6) and with very little dark current (on the order of 10^{-10} amps). Fluctuations in the atom cell dominate the noise in these systems, so the shot noise associated with PMTs is an insignificant noise source for the system. Further, the spectral response for modern PMTs is such that good sensitivity is available throughout the UV/visible. Some systems provide a red-sensitive PMT for wavelengths above about 600 nm. Very few elements other than the alkali metals have analytical transitions in that region. To date, there are no other detectors used in commercial instruments designed for AES or AAS.

However, photodiode arrays are being used in some UV/visible spectrometers for solution molecular absorption measurements. These systems seem to offer improved precision over PMT systems, which indicates that AES/AAS instrumentation may use arrays in the future. One advantage may be the possibility of simultaneous measurement of several or even many lines and, by suitable data reduction, it may be possible to obtain a better signal-to-noise ratio. Other types of solid-state photodetectors are also being evaluated in academic research organizations.

b. Readout electronics

The most common type of amplifier electronics used in AES/AAS is the lock-in amplifier. This type of device "locks in" on any signal of a specific modulation frequency, and therefore does not respond to signals at other frequencies. The radiation from the excitation source, either HCL or flame, is modulated at a fixed frequency that is synchronized with the lock-in amplifier. This means that only the signals of interest to the analyst are amplified and that the signal-to-noise ratio will be improved. The amplified signal can then be fed into a stripchart recorder or a meter for evaluation. In more modern instruments, it is converted into a digital signal and is fed into a computer for evaluation. A cleaner approach is to digitize the signal directly from the PMT and send the resulting signal directly into a computer that can be programmed to emulate a lock-in amplifier and can also perform many other operations on the data.

c. Computers

As stated, most commercial instruments being marketed today have a dedicated computer. This provides maximum flexibility for data evaluation and signal processing. In addition, a computer gives the analyst many other desirable features. A computer can store the information required for the instrumental parameters for many determinations. In more advanced instruments, it can provide control over instrument variables and can set the variables to optimum conditions for a particular determination. The option for several different analytical calculations, such as analytical curve fitting and concentration readout, standard additions, etc., is available. Of particular value are the statistical data treatment programs. For example, simple signal averaging has improved the typical precision from 3-5% to 0.5-1%.

As scientific programmers get smarter, the computer will be able to do even more for the analyst. Optimization programs, even more sophisticated background correction, and interference correction will undoubtedly be available. The ability to interface the instrument with a laboratory computer or LIMS system makes the transmission of sample reports by electronic rather than hard copy a realizable goal. The single largest source of errors results from transposition of numbers. Many laboratories have already realized the advantages of computer interface, and in a few years it will be the rule rather than the exception.

D. Miscellaneous Components/Concepts

Interferences

There are many interferences in AES/AAS; they can be caused by the nebulizer-atom cell system, the external excitation source (in AAS), or the sample matrix. Interferences can be spectral, chemical, or related to the physical properties of the sample solution.

a. Spectral interferences

There can be several types of spectral interferences. In FAAS and FAES there are spectral features caused by the hot gas molecules in the flame. The OH, CH, CN, and C_2 molecules all emit spectra in the UV/visible, and if the analyte emission or absorption falls in the same spectral region as the flame emission, the analytical signal can be degraded, causing misinterpretation of the resultant signals. The background signal should always be investigated. These signals are often easily corrected by making an appropriate blank measurement.

If there are elements in the sample matrix that emit very close to the analyte transition and if this emission is within the spectral bandpass of the monochromator, the emission will be observed by the detection system and will result in erroneous signal interpretation. This can be corrected by decreasing the spectral bandpass of the system (which also will decrease the light throughput). However, some of these spectral line interferences are too close to correct; these are called direct spectral overlaps. If two lines are too close for the monochromator to resolve, they will appear to be direct spectral overlaps for that particular system. Table 7 provides a list of observed and predicted direct spectral overlaps. If the sample has matrix elements that give a direct spectral overlap, the only alternative is to choose a different analyte transition for measurement. In any event,

Table 7. Spectral Overlaps

A. Observed Overlaps

Analyte element	Wavelength (nm)	Interferent element	Wavelength (nm)
Aluminum	308.215	Vanadium	308.211
Antimony	217.023	Lead	216.999
Antimony	231.147	Nickel	231.097
Cadmium	228.802	Arsenic	228.812
Calcium	422.673	Germanium	422.657
Cobalt	252.136	Indium	252.137
Copper	324.754	Europium	324.753
Gallium	403.298	Manganese	403.307
Iron	271.903	Platinum	271.904
Manganese	403.307	Gallium	403.298
Mercury	253.652	Cobalt	253.649
Silicon	250.690	Vanadium	250.690
Zinc	213.856	Iron	213.859

B. Predicted Overlaps

Analyte element	Wavelength (nm)	Interferent element	Wavelength (nm)
Boron	249.773	Germanium	249.796
Bismuth	202.121	Gold	202.138
Cobalt	227.449	Rhenium	227.462
Cobalt	242.493	Osmium	242.497
Cobalt	252.136	Tungsten	252.132
Cobalt	346.580	Iron	346.586
Cobalt	350.228	Rhodium	350.252
Cobalt	351.348	Iridium	351.364
Copper	216.509	Platinum	216.517
Gallium	294.418	Tungsten	294.220
Gold	242.795	Strontium	242.810
Hafnium	295.068	Niobium	295.088
Hafnium	302.053	Iron	302.064

(continued)

Table 7 (Continued)

Analyte element	Wavelength (nm)	Interferent element	Wavelength (nm)
Indium	303.936	Germanium	303.906
Iridium	208.882	Boron	208.884
Iridium	248.118	Tungsten	248.144
Iron	248.327	Tin	248.339
Lanthanum	370.454	Vanadium	370.470
Lead	261.365	Tungsten	261.382
Molybdenum	379.825	Niobium	379.812
Osmium	247.684	Nickel	247.687
Osmium	264.411	Titanium	264.426
Osmium	271.464	Tantalum	271.467
Osmium	285.076	Tantalum	285.098
Osmium	301.804	Hafnium	301.831
Palladium	363.470	Ruthenium	363.493
Platinum	227.438	Cobalt	227.449
Rhodium	350.252	Cobalt	350.262
Scandium	298.075	Hafnium	298.081
Scandium	298.895	Ruthenium	298.895
Scandium	393.338	Calcium	393.366
Silicon	252.411	Iron	252.429
Silver	328.068	Rhodium	328.060
Strontium	421.552	Rubidium	421.556
Tantalum	263.690	Osmium	263.713
Tantalum	266.189	Iridium	266.198
Tantalum	269.131	Germanium	269.134
Thallium	291.832	Hafnium	291.858
Thallium	377.572	Nickel	377.557
Tin	226.891	Aluminum	226.910
Tin	266.124	Tantalum	266.134
Tin	270.651	Scandium	270.677
Titanium	264.664	Platinum	264.689
Tungsten	265.654	Tantalum	265.661

Table 7 (Continued)

Analyte element	Wavelength (nm)	Interferent element	Wavelength (nm)
Tungsten	271.890	Iron	271.903
Vanadium	252.622	Tantalum	252.635
Zirconium	301.175	Nickel	301.200
Zirconium	386.387	Molybdenum	386.411
Zirconium	396.826	Calcium	396.847

Source: From Ref. 22.

the matrix should be evaluated for close spectral lines when a method is being developed.

In AAS, spectra from the HCL or EDL can also cause spectral interferences if the emission from the fill gas, usually neon or argon, falls within the spectral bandpass of the monochromator. Table 8 provides a list of neon lines that are close to analyte transitions in AAS, along with the resolution required by the system to resolve the lines and thus eliminate the problem.

In addition to spectral features related directly to atomic or molecular electronic transitions, there are spectral features that can be related to very small particles from sample solutions containing high concentrations of dissolved solids. These features can cause continuum emission or light-scattering problems. The results of either cause background problems that must be corrected or errors will be made. In fact, these problems are so pronounced in EAAS that background correction is always necessary. At least checking for background emission or absorption is important in FAAS and FAES as well. Background correction in AAS is discussed below.

b. Other interferences

The most important interferences other than spectral are those associated with the physical nature of the sample solutions. Particularly important are properties such as viscosity, total dissolved solids, surface tension, etc., which can affect the way in which a sample is nebulized. If samples vary substantially in any of these properties and if samples and standards vary from one another, errors will invariably result. Matching the standards and samples as closely as possible is essential but if it is not possible, techniques such as the use of internal standards or standard addition can often correct for these variations. When developing a method for a new sample matrix, it is often desirable to compare the analytical curve method to either the internal standard method or the method of standard additions. If the results are not comparable, it can be assumed that there is indeed variation between samples or between standards and samples, and the latter methods will yield more accurate results. It is always assumed that background problems have been checked as well.

Table 8. Neon Lines That Must Be Resolved

Analyte element	Wavelength (nm)	Neon line (nm)	Required resolution (nm)
Chromium	359.349	359.353	0.002
Chromium	357.860	357.464	0.020
Chromium	360.533	360.017	0.25
Copper	324.754	323.238	0.75
Dysprosium	404.599	404.264	0.16
Gadolinium	371.748	372.186	0.21
Gadolinium	371.357	370.964	0.19
Lithium	670.784	335.505 in 2nd order is 671.010	0.11
Lutetium	335.956	336.063	0.05
Niobium	405.894	404.264	0.81
Rhenium	346.046	346.053	0.003
Rhenium	346.473	346.658	0.09
Rhenium	345.188	345.419	0.11
Rhodium	343.489	344.770	0.64
Rhodium	369.236	369.420	0.09
Ruthenium	372.803	372.186	0.31
Scandium	402.369	404.264	0.94
Silver	338.289	337.828	0.23
Sodium	588.995	588.189	0.40
Sodium	589.592	588.189	0.70
Thulium	371.792	372.186	0.19
Titanium	365.350	366.411	0.53
Titanium	364.268	363.367	0.45
Titanium	337.145	336.981 and 336.991	0.077
Uranium	358.488	359.353	0.43
Uranium	356.660	356.853	0.09
Ytterbium	346.436	346.658	0.11
Zirconium	360.119	360.017	0.05
Zirconium	351.960	352.047	0.04

Source: From Ref. 22.

In the flame techniques, problems are often encountered because chemical reactions in the flame involve the analyte element. Any time the analyte is involved in chemical reactions with a matrix element, the result will be either a decrease or an increase in the analytical signal because the analytical signal depends on the concentration of the free atom concentration and because any chemical reaction will alter that concentration. A famous example of this problem is the decrease in calcium signal in AAS in the presence of phosphorus caused by the formation of a CaOP compound. If this type of problem is encountered, the standards must be matched to the matrix or the offending species removed from the samples before analyzing, or sometimes standard additions can be used. An internal standard will have no value unless it undergoes the same chemistry as the analyte element.

Background Correction

As can be noted from the discussion above, it is often necessary to perform some kind of background measurement and subsequent correction. In emission, this procedure is rather straightforward. One simply adjusts the wavelength of the monochromator off the line center of the analyte transition and obtains the background signal while aspirating the sample solution. It is best to observe the background signal on both sides of the analyte transition in case of a sloping background. This type of measurement can be programmed into modern instruments and should be a matter of routine practice.

In AAS, the measurement is much more complicated because the signal being observed is not from the sample but from the external excitation source. If the wavelength setting of the monochromator is changed, the signal from the HCL or EDL cannot be seen. Because there are several ways to accomplish the background correction in AAS and because different approaches are used in modern instrumentation, each will be discussed.

a. Two-line measurements

If the excitation source has a nonabsorbing emission line very close to the analyte transition, that line, such as a neon line or a line that originates from an excited state from the analyte species, is not absorbed by the analyte atoms. The absorbance signal from the measurement of the nonabsorber can be assumed to be caused by one of the phenomena, described in the previous section, that cause a background problem. Table 9 provides a list of lines close to analyte transitions that can be used for background correction by this technique. Two major problems with respect to this type of correction are of concern. First is the assumption that the background measured close to the analyte transition is the same as the background at the wavelength of the analyte transition. This is generally close, but probably never an exact assumption. Second is the necessity of making two separate measurements, each requiring an adjustment of the monochromator. The chance for misadjusting the system is possible. If a second excitation source is required, the possibility for misaligning the optics also exists.

Table 9. Close Lines for Background Correction

Element	Analysis line (nm)		Background line (nm)		Source
Aluminum	309.271	I	306.614	I	Al
Antimony	217.581	I	217.919	I	Sb
	231.147	I	231.398	I	Ni
Arsenic	193.759	I	191.294	II	As
Barium	553.548	I	540.0562	I	Ne
			553.305	I	Mo
			557.742	I	Y
Beryllium	234.861	I	235.484	I	Sn
Bismuth	223.061	I	226.502	II	Cd
	306.772	I	306.614	I	Al
Bromine	148.845	I	149.4675	I	N
Cadmium	228.8022	I	226.502	II	Cd
Calcium	422.673	I	421.9360	I	Fe
			423.5936	I	Fe
Cesium	852.1122	I	854.4696	I	Ne
Chromium	357.869	I	352.0472	I	Ne
			358.119	I	Fe
Cobalt	240.206	I	238.892	II	Co
			242.170	I	Sn
Copper	324.754	I	324.316	I	Cu
Dysprosium	421.172	I	421.645	II	Fe
			421.096	I	Ag
Erbium	400.796	I	394.442	I	Er
Europium	459.403	I	460.102	I	Cr
Gallium	287.424	I	283.999	I	Sn
			283.690	I	Cd
Gold	242.795	I	242.170	I	Sn
Indium	303.936	I	306.614	I	Al
Iodine	183.038	I	184.445	I	I
Iron	248.3271	I	249.215	I	Cu
Lanthanum	550.134	I	550.549	I	Mo
			548.334	I	Co
Lead	283.3053	I	280.1995	I	Pb
			283.6900	I	Cd
	217.000	I	220.3534	II	Pb

Element	Analysis line (nm)		Background line (nm)		Source
Lithium	670.791	I	671.7043	I	Ne
Magnesium	285.213	I	283.690	I	Cd
			283.999	I	Sn
Manganese	279.482	I	282.437	I	Cu
			280.1995	I	Pb
Mercury	253.652	I	249.215	I	Cu
Molybdenum	313.259	I	312.200	II	Mo
Nickel	232.003	I	232.138	I	Ni
Palladium	247.642	I	249.215	I	Cu
Phosphorus	213.618	I	213.856	I	Zn
Potassium	766.490	I	769.896	I	K
			767.209	I	Ca
Rhodium	343.489	I	350.732	I	Rh
			352.0472	I	Ne
Rubidium	780.027	I	778.048	I	Ba
Ruthenium	249.894	I	352.0472	I	Ne
Selenium	196.09	I	199.51	I	Se
Silicon	251.6113	I	249.215	I	Cu
Silver	328.068	I	332.374	II	Ne
			326.234	I	Sn
Sodium	588.9950	I	588.833	I	Mo
Strontium	460.733	I	460.500	I	Ni
Tellurium	214.281	I	213.856	I	Zn
			217.581	I	Sb
Thallium	276.787	I	280.1995	I	Pb
Tin	224.605	I	226.502	II	Cd
	286.332	I	283.999	I	Sn
Titanium	364.268	I	361.939	I	Ni
	365.350	I	361.939	I	Ni
Uranium	358.488	I	358.119	I	Fe
Vanadium	318.398	I	324.754	I	Cu
	318.540	I	324.754	I	Cu
Zinc	213.856	I	212.274	II	Zn

Source: From Ref. 23.

b. Continuum source measurement

The width of the line emitted from the excitation source is very narrow (0.0005-0.005 nm), in most situations an order of magnitude narrower than the width of the monochromator spectral band pass. Therefore, if a continuum source is passed through the sample instead of the HCL or EDL, the absorbed radiation from the analyte species is essentially negligible. However, if a molecular species absorbs or if small particulates in the atom cell cause light to scatter, the result will be to attenuate the entire continuum radiation that is being observed by the detector. This is illustrated in Figure 20.

This approach works quite well. The sample can be measured in a very short time with the analyte absorption and the background absorption being alternated at relatively high frequencies. In addition, the background subtraction can be made easily in the instrument, and the analyst can obtain a background-corrected number directly. Many instruments come with this feature as the standard method for background correction.

The correction is reasonably accurate and in most cases is satisfactory. However, in a few cases, this type of correction fails. If the analyte is present at high concentration, its absorption may actually represent a significant fraction of the background measurement and may result in overcorrection. For this reason, the correction is good up to an absorbance of about 0.5-0.7. Above those absorbances, a dilution is required for an accurate measurement. The correction is usually inaccurate if the background is highly structured as well. Further, this type of background measurement makes no correction for spectral interferences.

Experimentally, the light paths from two different excitation sources must be adjusted to precisely one optical beam through the atom cell and monochromator. If the beams become misaligned, errors will result. To cover the entire UV/visible spectral range, two continuum lamps must be used. Normally, a deuterium HCL is used for the UV and a tungsten filament lamp for the visible. Care must be taken to ensure that the background correction is indeed accurate.

c. Smith-Hieftje system

In 1983, Smith and Hieftje [24] published a paper describing a background measurement that requires the use of a single excitation source for both the analyte and the background absorption. Their system exploits the fact that at high current the HCL is generally self-absorbing because of the geometry of the HCL. There is a cloud of colder atoms at the mouth of the hollow cathode, and at higher currents, this cloud becomes larger because of the increased sputtering in the HCL. The self-absorption reduces the absorbance because of the analyte species and the absorbance caused by the background is enhanced. If the high current is applied on a periodic basis, the observed signal will alternate between analyte and background absorption, and the appropriate correction can be made in the instrument. At least one manufacturer is offering this background correction system with its instruments.

The Smith-Hieftje approach offers some advantages because the same optical path is used for both analyte and background measurement. This

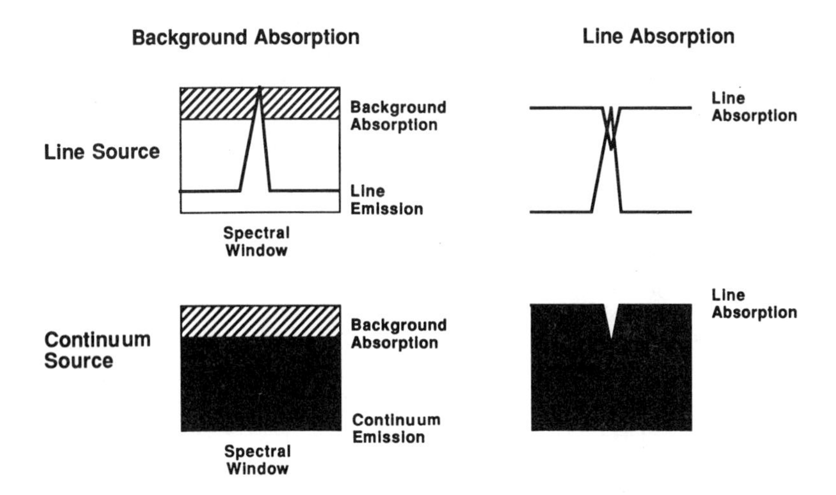

Figure 20. Background correction for continuum source.

approach can accurately correct for both spectral interferences and highly structured background absorbances. On the negative side, there is an inherent loss of sensitivity with the method because, even at high current, there is some absorption by the analyte species. This loss of sensitivity can be up to 50%. This approach works poorly for the refractory elements for which self-reversal is difficult to observe.

d. Zeeman splitting system

It has long been established that, in the presence of a magnetic field, an electronic energy level will be split into several, three or more, components. Further, the components have a polarized nature—that is, certain components are polarized parallel to the field and others are polarized perpendicular to the field. In general, the Zeeman splitting shifts the wavelength of the transition being observed. The simplest case is illustrated in Figure 21. Note that the π component remains at the same energy level and the σ components are shifted to both higher and lower energy. This transition of course results in a shift of wavelength to both lower and higher values and is observed through polarizers in the parallel and the perpendicular modes and in the absorbance observed at either the π or σ wavelengths. In the π measurement, the analyte signal and the background are measured, and in the σ measurement, the background alone is measured. Alternatively, the magnet can be turned off and a normal absorption measurement can be made. Then the background can be measured with the magnet on and with the polarizers in the σ position. Remember that this approach represents the ideal case. With more complicated splitting the evaluation is more complex. At any rate, the difference in measured signal, using both parallel and perpendicular polarized light, represents the background absorbance and can be used for correction. Alternatively, the difference between the normal absorbance measurement can be used with the magnet off or with it on in the σ position.

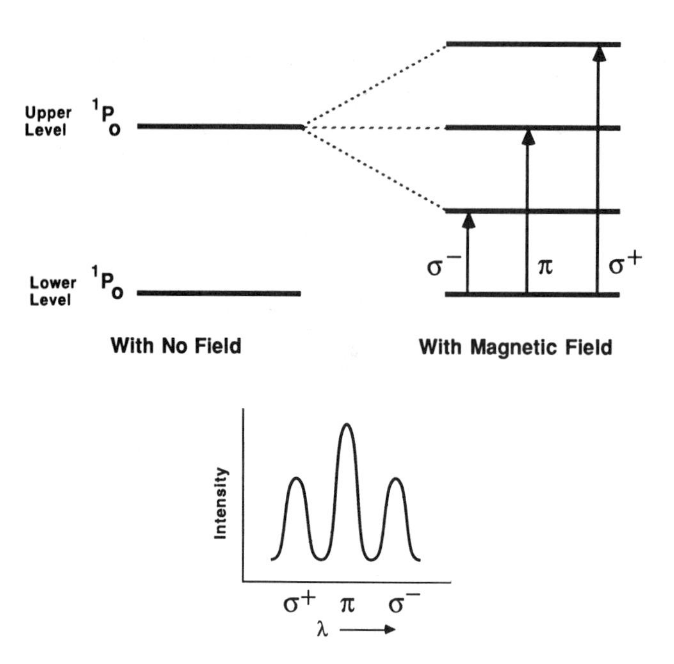

Figure 21. Zeeman splitting.

Commercial instrument manufacturers place the magnets either at the excitation source or at the atom cell. However, some elements cause problems when placing the magnet at the excitation source. Using the magnet in conjunction with a graphite furnace has proven convenient, and this is the most common configuration.

The most important negative with respect to the Zeeman background measurement is the requirement of additional equipment, namely, the magnet and associated electronics, and the polarizers. Polarizers reduce the light throughput of the optical system by a substantial amount, causing a loss of sensitivity by this technique; however, the loss is not as great as with the Smith-Hieftje system.

On the plus side, the system can correct for very high background absorbances and for backgrounds with highly structured features. It can accommodate very close, overlapping spectral features; however, no method can correct for a direct spectral overlap.

Less Common Atom Cells

a. Delves' cup

One of the limitations of FAAS is the requirement for reasonably large sample. Often a 10- to 20-ml sample is required for an accurate measurement. Certain sample types are simply too precious to permit use of large quantities for analysis. Examples are blood, biological tissues, and moon rocks. One would like to analyze for lead in blood in a drop of blood, not in several milliliters. This need has prompted researchers to look for different approaches to sample introduction. Of course, the search for additional sensitivity is always a driving force as well. One such approach was reported by Delves [25], who placed a few microliters, typically a hundred or so, into a metal cup and inserted the cup into the hot gases of an analytical flame. The absorbance was then measured just above the rim of the cup. This technique worked well for some of the more volatile elements of biological interest, such as lead, and led to the commercialization of the technique. The limitation predictably is that the analyte elements must be relatively volatile and nonrefractory.

Except for a very few specific applications, most of the uses of the Delves' cup technique probably have been replaced by the graphite furnace techniques.

b. *Cold mercury vapor cells*

Mercury has a significant atomic vapor pressure at room temperature and is the only element except the permanently gaseous elements to possess this property. Because of this, AAS signals can be measured in a cold, or room-temperature, cell. One of the first commercial AAS instruments was an Hg analyzer. Most companies sell a simple quartz cell, which fits into the optical path, for Hg. These systems provide for the reduction of Hg ions in solution to elemental Hg and provide a means of sweeping the elemental Hg into the cell for measurement.

This approach is extremely sensitive because *all* the Hg in a given sample can be concentrated in cell. Detection limits of 0.02 ppb can be achieved using 50 ml of sample solution. Preconcentration approaches can lower this detection limit to the parts per trillion level, which represents a measurement of a few tens of picomoles of Hg in the cell.

c. *Hydride generation systems*

Another approach has been used to concentrate those elements that form volatile hydrides. Chemical reaction with an excess of a suitable reductant, usually sodium borohydride ($NaBH_4$) or stannous chloride ($SnCl_2$) in acid solution, will quantitatively generate the appropriate gaseous metal or metalloid hydride compound, which is swept into the flame or a heated quartz cell for measurement by AAS. The elements that form volatile hydrides are As, Se, Sb, Bi, and Sn. The procedure is similar to that for Hg. The samples are dissolved and a suitable aliquot (10-100 ml) is placed in a reaction vessel, the reductant is added to the acidic solution, and the resultant gaseous hydride is flushed by an inert gas or one of the flame gases into the atom cell. This procedure concentrates all of the analyte element from the sample into the atom cell and increases the sensitivity dramatically.

The hydride bonds are very weak, so a heated quartz cell or an air-acetylene flame is adequate to essentially atomize 100% of the sample. Table 10 gives the detection limits for elements by this technique. The major problem with this type of analysis is dealing with one of these analytes in the presence of a large quantity of one of the other elements that can form a volatile hydride or other species that can react with the reductant.

Almost all of the instrument manufacturers produce a hydride generation system compatible with their AAS instruments.

Table 10. Detection Limits by Hydride Generation and Cold-Vapor AAS

Element	LOD (ppb)[a]
Antimony, Sb	0.1
Arsenic, As	0.02
Bismuth, Bi	0.02
Mercury, Hg	0.02
Selenium, Se	0.02
Tellurium, Te	0.02
Tin, Sn	0.5

[a]The detection limits are based on 50-ml solution volumes.
Source: From Ref. 26.

III. COMMERCIAL SYSTEMS

A. Typical Systems

Table 11 lists the current suppliers of AAS/AES instrumentation along with their addresses. Prices have been omitted because they are so variable. However, current instrumentation ranges from $10,000 to $30,000, depending on the options and the quality.

Most manufacturers package their instruments in one of two optical configurations described as "single-beam" or "double-beam" systems. The double-beam system is not a true double beam in the same concept as in molecular spectroscopy because there is no such thing as a "reference" flame that is identical to a "sample" flame. In reality, the reference beam does not pass through the flame at all and does not conpensate for any phenomena occurring in the flame. The double-beam configuration does compensate for variations in the excitation source. So if the system is a double-beam system, it is not necessary to wait for the excitation source to warm up, and if the excitation source drifts, the system can compensate for that as well.

B. General Considerations

Analytical Procedure

The analytical measurements in either AAS or AES are typical of most instrumental analytical methods. These are measurements in which the signal from the sample solution is compared with that of a standard. Even though the measurements can be related directly to a concentration, too many fundamental parameters are not known for this to be practical. In the best of cases, absolute measurements can only be made to an accuracy

Table 11. Current AAS/AES Instrument Suppliers

Supplier	Address
Alko Diagnostic Corp.	Alko Park, 333 Fiske St. Holliston, MA 01746
Analyte Corp.	611 Southeast L St. Grants Pass, OR 97526
Angstrom, Inc.	P. O. Box 248 Belleville, MI 48111
Applied Research Laboratories, ARL	9545 Wentworth St. Sunland, CA 91040
Baird-Atomic Ltd.	Warner Dr. Springwood Ind. Estate Braintree, Essex CM7 7YL, England
Baird Corp.	125 Middlesex Tnpk. Bedford, MA 01730
Buck Scientific	58 Fort Point St. East Norwalk, CT 06855
Chemplex Industries Inc.	160 Marbledale Rd. Tuckahoe, NY 10707
Ciba Corning Diagnostics Corp.	132 Artino St. Oberlin, OH 44074
Eppendorf Geratebau, Netheier & Hinz GmbH	P. O. Box 65 06 70 D-2000 Hamburg 65 West Germany
F. & J. Scientific	79 Far Horizon Dr. Monroe, CT 06468
Foss Electric (Ireland) Ltd.	Sandyford Ind. Estate Foxrock, Dublin 18 Ireland
GBC Scientific Equipment Pty. Ltd.	22 Brooklyn Ave Dandenone Victoria 3175 Australia
Grun Analysengerate GmbH	Industriestr. 27-31 D-6330 Wetziar West Germany
Hilger Analytical Ltd.	Westwood Estate Margate, Kent CT9 4JL, England

(continued)

Table 11 (Continued)

Supplier	Address
Instrumentation Laboratory (UK) Ltd.	Kelvin Close Birchwood Science Park Warrington, Chesire England
Instruments SA, Inc.	173 Essex Ave. Metuchen, NJ 08840
International Equipment Trading Ltd.	3005 Commercial Ave. Northbrook, IL 60062
Japan Spectroscopic Co. Ltd.	2967 Ishikawacho Hachioji, Tokyo Japan 192
Jenway Ltd.	Gransmore Green Feisted Dunmow Essex CM6 3LB England
Kontron Instruments GmbH	Oskar-v-Miller-Strasse 1 Eching bei Munchen D-8057, West Germany
Dr. Bruno Lange GmbH	P. O. B. 370 363 Koffnigsweg 10 D-1000 Berlin 37 West Germany
Lehr Siegler, Inc.	74 Inverness Dr. East Englewood, CO 80112
Micro Coatings	One Lyberty Way Westford, MA 01996
Nissel Sangyo America, Ltd.	Hitachi Scientific Instrument Div. 460 E. Middlefield Rd. Mt. View, CA 94043
Oxford Analytical Instruments Ltd.	20 Nuffield Way Abingdon, Oxon OX14 1TX, England
Perkin-Elmer & Co. GmbH	P. O. Box 1120 D-7770 Uberlingen West Germany
Perkin-Elmer Corp.	761 Main Ave. Norwalk, CT 06859
Perkin-Elmer Ltd.	Post Office Lane Beaconsfield Buckinghamshire HP91QA, England

Supplier	Address
Petracourt Ltd.	Gransmore Green Felsted Dunmow Essex CM6 3LB, England
Philips Analytical	Bldg. HKF NL-5600 MD Eindhover The Netherlands
Philips Export B.V.	Lelyweg 1, Almelo The Netherlands 7602EA
Photo Electric Instruments P. Ltd.	Diwan House Mehta Market Jodhpur (RAJ) 342001 India
Process & Instruments Corp.	1943 Broadway Brooklyn, NY 11207
Prolab Laboratory Construction Ltd.	P.O. Box 208 CH-3422 Kirchberg Switzerland
P.S. Analytical	P.O. Box 2387 Princeton, NJ 08540
PT Analytical	56 Jonspin Rd. Wilmington, MA 01887
Pye Unicam Ltd., A Scientific & Industrial Co. of Philips	York St. Cambridge CB1 2PX, Great Britain
Radiometer A/S	Emdrupvel 72 DK-2400 Copenhagen NV Denmark
Scientific & Medical Products Ltd.	Shirley Institute Didsbury, Manchester M&O 8RX, England
SEAC	Via C. del Prete 139 50127 Florence Italy
Shimadzu (Europa) GmbH	Ackerstrasse 111 4000 Dusseldorf 1 West Germany
Shimadzu Scientific Instruments, Inc.	7102 Riverwood Dr. Columbia, MD 21046

(continued)

Table 11 (Continued)

Supplier	Address
Spex Industries, Inc.	3880 Park Ave. Edison, NJ 08820
Strohlein GmbH & Co.	Girmeskreuzstrasse 55 D4044 Kaarst West Germany
Teledyne Analytical Instruments	16830 Chestnut St. City of Industry, CA 91749
Tennelec, Inc.	601 Oak Ridge Tpk. Oak Ridge, TN 37830
Thermo Jarrell Ash	590 Lincoln St. Waltham, MA 02254
Universal Biochemicals	6 Sathya Sayee Bagar Madurai Tamil Nadu 625 003, India
Varian Associates, Inc.	220 Humbolt Courtf Sunnyvale, CA 94089
Varian Associates	28 Manor Rd. Walton-on-Thames KT12 2QF, Great Britain
Varian Instrument Group	220 Humboldt Ct. Sunnyvale, CA 94089
Xerox Analytical Laboratories	800 Phillips Rd. W114/42D Webster, NY 14580

Source: From Refs. 27 and 28.

of about 10% and more often to 50% or worse. By the same token, with the proper use of standards, accuracy can approach 0.5% or even better.

The best situation occurs when the analyst can prepare a set of standards that is identical to the matrix of the samples. If this can be done, all chemical, physical, and spectral effects should be totally compensated, and the ultimate in accuracy should be accomplished. Unfortunately, preparing this identical set of standards is often not possible. Indeed, many sample types are inherently variable and quite inhomogeneous with respect to the matrix composition. Examples of this behavior include rocks, brines, the earth's atmosphere, and the ocean. Even if it is possible to match the major matrix components, the effects of varying minor and trace components can cause errors to arise.

If there is any doubt about the effects of sample matrix, the analyst should at the very minimum compare the results of two or more techniques in developing a method. For example, comparing the results obtained from the usual analytical curve with those obtained by the method of standard additions is often desirable. It is a good idea to explore the use of an internal standard as well. If the matrix affects the slope of the analyte signal as a function of analyte concentration, the standard additions approach should improve the accuracy. If the matrix affects the shape of the curve, for example, a changing slope with analyte concentration, the internal standard method may help. In either case, any background signal must be corrected independently. Remember that the choice of internal standard depends on the chemical/physical behavior of the internal standard species correlating with that of the analyte. A poor choice of internal standard will degrade rather than improve the accuracy of the measurement.

If the results by several of the methods discussed above agree, it is probably best to simply use the analytical curve method. If not, careful evaluation of the matrix should be made to determine the best analytical approach for the system in question.

Analytical Lines

In AAS/AES measurements for most elements, there is a single most sensitive transition to use. These are often the same in AAS and AES; however, they are often different as well. Tables 2-5 give the transitions and detection limits for the air-hydrocarbon flame, air-acetylene flame, nitrous oxide-acetylene flame, and graphite furnace system, respectively. Very little work has been done using the graphite furnace as an emission source, so only AAS limits of detection are provided.

Most of the AAS lines and many of the AES lines are resonance transitions—that is, they originate from the electronic ground level of the analyte atoms. Many of the AES lines tend to be of higher wavelength and thus require lower energies for excitation. All transitions that have been used by analysts in both AES and AAS are covered in detail in Parsons's *Handbook of Flame Spectroscopy* [16], along with a great deal of other useful information.

C. Useful Tips

Optimization

It may be apparent to the reader by now that AAS/AES measurements are not simply "push the button and out comes the number." The results are affected by a complex set of instrumental variables, in addition to complex chemistry in the flame or hot gases of the graphite furnace. To add to the confusion, essentially all of the variables that the analyst can control are interactive with most of the other variables. This means that an optimum signal *cannot* be obtained by adjusting each variable, one at a time, until a maximum signal is reached. Such approaches have been used time and again with little success and have caused many analysts a lot of grief.

When a set of interrelated variables is to be optimized, a systematic approach must be used in which all variables are adjusted concurrently to obtain the maximum signal. One such approach is called the modified simplex method, developed by Deming and co-workers [29,30]. In this method, the optimum settings for the four or five critical variables in AAS/AES measurements are generally found in 20-30 measurements and require only a few hours to perform. Of course, other experimental design methods are available for optimization [31], but the simplex method is ideally suited for the present purpose.

A word or two should be said about the variables to include in an optimization experiment and the function that should be optimized. Table 12 is a tabulation of the critical variables in FAAS, FAES, and EAAS. In the final analysis, a critical variable is one that affects the atom concentration of the analyte species. Factors that affect the sample solution flow rate into the system directly also affect the atom concentration, such as the gas flow that causes sample aspiration or the pump speed if the sampling system uses a peristaltic pump. However, other factors can cause very important effects in a more indirect way. The fuel-to-oxidant ratio in flames can drastically increase or decrease the analyte atom concentration because of chemical reactions in the analytical zone of the flame. The slit function of the monochromator can play a major role in the noise of the background as well as in the background signal. If the analyst needs to push the detection limit of the system to obtain the required information, attention must be paid to the variables that will be included in the optimization experiment. Further, it is the signal-to-noise ratio that should be optimized, not the signal alone. The noise of a measurement is the ultimate limiting factor in the detection limit of a system. By optimizing the signal-to-noise ratio, the noise is minimized and the signal is maximized. The result is the best measurement for which the system is capable.

Sample Preparation

Little will be said concerning sample preparation because of the infinite set of sample types and problems, each of which must be evaluated individually. However, there are several guidelines that should be used in the development of an analytical procedure.

First, it is probable that a similar sample has been addressed by someone, somewhere, and that they have written about their successes, but it is entirely possible that no written information will be available about failures. The literature can save the analyst untold time. Journals such as Anal. Chem., Appl. Spectrosc., Talanta, Anal. Chim. Acta, and Atomic Spectrosc. often have excellent articles concerning the preparation of a wide variety of samples, as well as experimental set-up information. These sources often have review articles about a particular sample type or method. In addition, often bibliographies are published on a regular basis and are indexed by element and sample type.

Next, it is important to remember that the more dissolved solids present in the sample solution, the less likely that the simple, dilute aqueous standards will be adequate for comparison. It is also likely that in samples with highly dissolved solids and/or high acid concentrations, physical properties (viscosity, surface tension) will be substantially different from water. On the other hand, any time that a chemical separation is required, it

Table 12. Critical of Optimization Parameters for AES/AAS Methods

Parameter	FAAS	EAAS	FAES
A. Independent parameters			
Excitation source power	Yes	Yes	NA[a]
Photomultiplier voltage[b]	Yes	Yes	Yes
Readout gain[c]	Yes	Yes	Yes
Noise suppression setting[d]	Yes	Yes	Yes
B. Dependent (interdependent) parameters			
Oxidant gas flow rate	Yes	NA	Yes
Fuel-to-oxidant ratio	Yes	NA	Yes
Sheath gas flow rate[e]	Yes	Yes	Yes
Solution flow rate[f]	Yes	NA	Yes
Sample size	NA	Yes	NA
Height of optical measurement	Yes	Yes	Yes
Monochromator slit setting	Yes	Yes	Yes
Burner variables[g]	Yes	NA	Yes
Furnace variables[b]	NA	Yes	NA

[a]NA stands for not applicable.
[b]The PMT voltage does not affect the SNR unless extreme voltages are used. It will specify the level of signal observed.
[c]The gain does not affect the SNR until electronic noise becomes important. It also specifies the level of signal observed.
[d]This specifies the frequency response of the system and is accompanied by a time requirement. More noise filtering requires a longer measurement.
[e]Most commercial burners do not use a sheath gas; however, there is always the possibility of a sheath gas in EAAS.
[f]This is important if the sample solution flow rate is controlled by a pump rather than the oxidant gas flow rate.
[g]Some burners have additional variables such as bead position, nebulizer position, etc.
[h]The timing cycle and temperature are always critical variables for the graphite furnaces.
Source: From Ref. 32.

is very likely that some analyte is lost in the process. In fact if separation is used for the samples, the best practice is to also treat the standards in the same fashion.

It has sometimes been suggested that the addition of certain chemical reagents reduces the interference of some matrix element on the analyte species. This is often true; however, care must be taken to ensure that this addition does not cause other problems, such as a change in the viscosity of the solution.

All of the above guidelines simply reinforce the concept of matching the matrix of the samples and standards if at all possible.

Finally, it should be remembered that there is more than one way to approach these problems. If the chemicals in the final sample solution cause terminal problems in an analysis, there is undoubtedly a different way to dissolve or prepare the sample. Don't forget that there are often different analyte lines that can be measured if the problems are spectral rather than chemical in nature.

Matrix Modification

The graphite furnace is by its nature a chemical reactor. The sample with its matrix is placed on a graphite platform and is heated to a very high temperature in some type of gaseous atmosphere. Reactions can take place in the solid, liquid, or gaseous phase, or multiple reactions can take place. Any compound formation involving the analyte degrades the analytical measurement. The reactive nature of the elements has generated an area in graphite furnace AAS called matrix modification. The concept is to add to the sample a chemical reagent that will cause desirable chemical reactions or inhibit undesirable reactions. A common type of matrix modification is to add a reagent that preferentially reacts with either the sample or the matrix elements, but not both, to create a volatile compound that is vaporized faster than the other. This causes a separation of the sample from its matrix in real time. The reverse approach is also used, that is, to add a chemical reagent that reacts with either the sample or the matrix elements, but not with both, to create a relatively nonvolatile compound that vaporizes slower than the other. Common reagents that have been recommended are NH_4NO_3, $Mg(NO_3)_2$, Ni, $(NH_4)_2HPO_4$, $(NH_4)H_2PO_4$, $(NH_4)_2SO_4$, $K_2Cr_2O_7$, ascorbic acid, and citric acid.

Much has been written about these effects [20,33]. The origin goes back to the "carrier distillation" techniques used in dc arc emission spectroscopy developed in the 1950s. The technique is very dependent on the composition of the sample matrix, and care must be taken not to extrapolate from one matrix to another. Experimental verification is the only valid approach for this technique.

D. Choices

Many of the choices in AAS/AES have already been discussed to some degree. What analytical method should be used? Should the AAS or AES be used? Should a single-beam or a double-beam instrument be bought? What type of background correction method is best? What is the best type of sample dilution for a particular sample matrix?

As stated, many of these questions have been addressed; however, in many cases, there is no one correct answer. In fact, there are invariably tradeoffs. For instance, the light throughput of a monochromator can always be improved, but at the expense of resolution. If the light output from the excitation source is weak, there is no choice. If the output is strong, the choice depends on the need to resolve close spectral features.

With respect to single versus double beam, several factors must be considered. Generally, the double-beam systems are more expensive. They also are more complicated because of additional moving parts, and there is usually a sacrifice of some light throughput. On the other hand, the double-beam systems require little or no warm-up time and compensate for long- and short-term noise originating from the excitation source. In fact, the precision of most measurements can probably be improved with the use of the double-beam system, all other factors being constant. All in all, other considerations, such as the type of background correction, are probably more important to most analysts.

For many applications, the use of continuum source background correction is quite satisfactory. However, if the samples often must contain high concentrations of dissolved solids or if the spectral region used for analysis contains a highly structured background, one of the other background techniques may be necessary to obtain accurate results. If most of the samples analyzed by a laboratory are of this type, the Smith-Hieftje or a Zeeman system should be carefully considered.

The question of AAS versus AES is simply answered. The best method is usually the one with the best detection limit, which relates to precision. The best detection limit almost always means the least noise and therefore the best precision. For this determination, the sample matrix of interest should be used, not dilute aqueous solutions. As stated, AES is often most sensitive for analytical transitions above 400 nm and AAS for transitions below 300 nm. Background structure observation and correction are easier by AES than by AAS, a point that is often overlooked.

Flame methods are almost always more convenient and simpler than EAAS. However, EAAS is almost always more sensitive than flame methods. If the limit of detection is the overriding factor, EAAS will always be the method of choice. If precision or sample throughput is more important, then flame methods will be better. Many analysts explore the ability of flame methods first, and if these prove inadequate, they look to EAAS.

In making the choice, the analyst should be guided by experimental measurement, not by preconceived concepts or by personal bias. The experimental results may prove surprising.

E. Research Directions

Because AAS/AES is a mature analytical technique, fewer academic researchers are devoting full time to fundamental studies. The advent of ICP-AES, DCP-AES, ICP-AFS, and ICP-MS, all with the capability of multielement measurements either with a simultaneous polychromator technique or with a fast slew-scanning monochromator technique, has damaged the AAS/AES market. However, the method is deeply entrenched in the field of analytical chemistry, and it will be years before it is replaced by the methods listed above.

Because all of the above-mentioned methods require dissolution of the sample, introduction of the sample into a nebulizer, and a free atomic cloud of analyte atoms for measurement, many of the processes being studied are common to all. Research into sample dissolution and preanalysis chemical treatment will undoubtedly continue to be a focus in the field of trace elemental analysis. In addition, the weak link in all of these techniques is the nebulizer system. Most of the current nebulizers have rather poor efficiency and constitute the largest source of flicker noise in the atom cells. This problem will be an area for research for the duration of the use of these techniques. Nebulizers with much higher efficiencies, tighter droplet distribution, and lower flicker noise will naturally result.

All of the methods competing with AAS/AES have multielement capabilities. If AAS/AES is to survive as a commercial product, multielement procedures must be developed. It is not uncommon for 20-50 elements to be simultaneously determined by the plasma techniques. One new approach that permits up to 16 elements to be determined by AAS has recently been introduced [34]. More work along these lines is very important.

Graphite furnace techniques, cold-vapor AAS, and hydride generation methods represent the most sensitivity for many elements. As long as this is the case, these methods will be used for those elements. Unfortunately, the techniques that utilize MS detection appear to be gaining rapidly in the quest for increased sensitivity. Both academic and industrial laboratories are devoting much research effort to EAAS techniques, and because of its extreme sensitivity, many more developments in this area are likely. Making the method into a multielement technique will help considerably.

The use of a continuum external excitation source, coupled with a high-resolution monochromator, is an approach with some promise. This approach has been spearheaded by O'Haver and Harnly [35]. A continuum source allows exploration of the entire spectral range capable of being detected by the PMT.

Acknowledgment: The author would like to recognize SAVANT Audio-visuals. Their excellent graphical presentations were the model for several of the figures in this chapter.

REFERENCES

1. Parsons, M. L., Major, Sandy, and Forster, Alan R., Appl. Spectrosc. *37*, 411 (1983).
2. Alkemade, C. T. J., Nomenclature, Symbols, Units and Their Usage in Spectrochemical Analysis III. Analytical Flame Spectroscopy and Associated Procedures, IUPAC, 1972.
3. Alkemade, C. T. J., Hollander, T., Snelleman, W., and Zeegers, P. J. T., *Metal Vapours in Flames*, Pergamon Press, New York, 1982.
4. Newton, I., *Opticks or a Treatise of the Reflections, Refractions, Inflections, and Colours of Light*, first edition, 1704; fourth edition, 1930; reissued by Dover, New York, 1952.
5. Walsh, A., Spectrochim. Acta, *7*, 108 (1955).
6. Alkemade, C. T. J. and Milatz, J. M. W., Appl. Sci. Res. B *4*, 289 (1955), and J. Opt. Soc. Am. *45*, 583 (1955).
7. Winefordner, J. D., and Vickers, T. J., Anal. Chem. *36*, 1939 (1964).
8. Winefordner, J. D., and Vickers, T. J., Anal. Chem. *36*, 1947 (1964).

9. Hieftje, G. M., Appl. Spectrosc. *25*, 653 (1971).
10. Suddendorf, R. F., and Denton, M. B., Appl. Spectrosc. *28*, 8 (1974).
11. Browner, R. F., and Boorn, A. W., Anal. Chem. *56*, 786A (1984).
12. Browner, R. F., and Boorn, A. W., Anal. Chem. *56*, 875A (1984).
13. Winefordner, J. D., Mansfield, C. T., and Vickers, T. J., Anal. Chem. *35*, 1607 (1963).
14. Harker, J. H., and Allen, D. A., J. Inst. Fuel *42*, 183 (1969).
15. Wittenberg, G. K., Haun, D. V., and Parsons, M. L., Appl. Spectrosc. *33*, 626 (1979).
16. Parsons, M. L., Smith, B. W., and Bentley, G. E., *Handbook of Flame Spectroscopy*, Plenum Press, New York, 1975.
17. Thermo Jarrell Ash, *Guide to Analytical Values for TJA Spectrometers*, Waltham, MA, 1987.
18. L'vov, B. V., Spectrochim. Acta *17*, 761 (1961).
19. Woodriff, R., Stone, R. W., and Held, A. M., Appl. Spectrosc. *22*, 408 (1968); Woodriff, R., and Ramelow, G., paper presented at the National SAS Meeting in Chicago, 1966.
20. Slavin, W., *Graphite Furnace Source Book*, Perkin-Elmer Corp., Ridgefield, CT, 1984.
21. *ATOMSOURCE*, Analyte Corp., Grants Pass, OR, 1987.
22. Lovett, R. J., Welch, D. L., and Parsons, M. L., Appl. Spectrosc. *29*, 470 (1975).
23. Sneddon, J., Spectroscopy *2*(5), 38 (1987).
24. Smith, S. B., and Hieftje, G. M., Appl. Spectrosc. *37*, 419 (1983).
25. Delves, H. T., Analyst *95*, 431 (1970).
26. Perkin-Elmer Corp., "Mercury/Hydride System," Norwalk, CT, No. 1876/6.79 (1987).
27. Anal. Chem. *59*, 1 (1987).
28. Am. Lab. *18*(19), Jan. (1987).
29. Deming, S. N., and Morgan, S. L., Anal. Chem. *45*, 278A (1973).
30. Parker, L. R., Jr., Morgan, S. L., and Deming, S. N., Appl. Spectrosc. *29*, 429 (1975).
31. Sharaf, M. A., Illman, D. L., and Kowalski, B. R., *Chemometrics*, Wiley, New York, 1986.
32. Parsons, M. L., and Winefordner, J. D., Appl. Spectrosc. *21*, 368 (1967).
33. Slavin, W., and Manning, D. C., Progress Anal. Atom. Spectrosc. *5*, 243 (1982).
34. "Analyte 16 Atomic Absorption Spectrometer," Analyte Corp., Grants Pass, OR, 1987.
35. O'Haver, T. C., Analyst *109*, 211 (1984).

BIBLIOGRAPHY

Alkemade, C. T. J., and Herrmann, R., *Fundamentals of Analytical Flame Spectroscopy*, A. Hilger, Bristol (U.K.), 1979.

Alkemade, C. T. J., Hollander, T., Snelleman, W., and Zeegers, P. J. T., *Metal Vapours in Flames*, Pergamon Press, New York, 1982.

Angino, E. E., and Billings, G. K., *Atomic Absorption Spectrometry in Geology*, 2nd ed., Elsevier, Amsterdam, 1972.

Bennett, P. A., and Rothery, E., *Introducing Atomic Absorption Analysis*, Varian Techron, Mulgrave, Australia, 1983.

Burriel-Marti, F., and Ramirez-Munoz, J., *Flame Photometry, A Manual of Methods and Applications*, 4th ed., Elsevier, Amsterdam, 1964

Christian, G. D., and Feldman, F. J., *Atomic Absorption Spectroscopy: Applications in Agriculture, Biology and Medicine*, Wiley-Interscience, New York, 1970.

Dean, J. A., *Flame Photometry*, McGraw-Hill, New York, 1960.

Dean, J. A., and Rains, T. C. (Eds.), *Flame Emission and Atomic Absorption Spectrometry, I: Theory*, Marcel Dekker, New York, 1969.

Dean, J. A., and Rains, T. C. (Eds.), *Flame Emission and Atomic Absorption Spectrometry, II: Components and Techniques*, Marcel Dekker, New York, 1971.

Dean, J. A., and Rains, T. C. (Eds.), *Flame Emission and Atomic Absorption Spectrometry, III: Elements and Matrices*, Marcel Dekker, New York, 1975.

Dvorak, J., Rubeska, I., and Rezac, Z., *Flame Photometry; Laboratory Practice*, Butterworths, London, 1971 (Transl. from the Czech).

Elwell, W. T., and Gidley, J. A. F., *Atomic Absorption Spectrophotometry*, 2nd ed., Pergamon, Oxford, 1966.

Hassan, S. S. M., *Organic Analysis Using Atomic Absorption Spectrometry*, Halsted Press, New York, 1984.

Herrmann, R., and Alkemade, C. Th. J., *Flammenphotometrie*, Springer Verlag, Heidelberg, 2nd ed., 1960; *Flame Photometry*, 2nd rev. ed., Interscience, New York, 1963 (translated by P. T. Gilbert, Jr.).

Hoda, K., and Hasegawa, T., *Atomic Absorption Spectroscopic Analysis*, Genshi Kyuko Bunseki, Kondanska, Tokyo, 1962.

Kirkbright, G. F., and Sargent, M., *Atomic Absorption and Fluorescence Spectroscopy*, Academic Press, London, 1974.

L'vov, B. V., *Atomic Absorption Spectrochemical Analysis*, A. Hilger, London, 1970 (translated from the Russian).

Mavrodineanu, R., and Boiteux, H., *Flame Spectroscopy*, Wiley, New York, 1965.

Mavrodineanu, R. (Ed.), *Analytical Flame Spectroscopy; Selected Topics*, Philips Technical Library, Macmillan, London, 1970.

Parsons, M. L., Smith, B. W., and Bentley, G. E., *Handbook of Flame Spectroscopy*, Plenum Press, New York, 1975.

Pietzka, G., *Flammenspektrometrie*, Ullmann's Enzyklopadie der Technischen Chemie, Band 2/1, Urban und Schwarzenberg, Müchen-Berlin, 1961.

Pinta, M., *Atomic Absorption Spectrometry, Vol. 2: Application to Chemical Analysis*, 2nd ed., Masson, Paris, 1980.

Pinta, M. (Ed.), *Spectrometrie d'Absorption Atomique, I: Problemes Generaux, and II: Application a l'Analyse Chimique*, Masson, O.R.S.T.O.M., Paris, 1971. English Transl., *Atomic Absorption Spectrometry*, A. Hilger, London, 1975.

Poluektov, N. S., *Techniques in Flame Photometric Analysis*, Consultants Bureau, New York, 1961 (translated from the Russian).

Price, W. J., *Spectrochemical Analysis by Atomic Absorption*, Heyden, London, 1979.

Pruvot, P., *Spectrophotometrie des Flammes*, Gauthier-Villars, Paris, 1972.

Pungor, E., *Flame Photometry Theory*, Van Nostrand, London, 1967 (translated from the Hungarian).
Ramirez-Munoz, J., *Atomic Absorption Spectroscopy and Analysis by Atomic Absorption Flame Photometry*, Elsevier, Amsterdam, 1968.
Reynolds, R. J., and Aldous, K., *Atomic Absorption Spectroscopy, A Practical Guide*, Griffin, London, 1970.
Rubeska, I., and Moldan, B., *Atomic Absorption Spectrophotometry*, Butterworths, London, 1969 (translated from the Czech).
Schuhknecht, W., *Die Flammenspektralanalyse*, Enke Verlag, Stuttgart, 1961.
Slavin, W., *Atomic Absorption Spectroscopy*, Wiley-Interscience, New York, 1968.
Slavin, W., *Graphite Furnace Source Book*, Perkin-Elmer Corp., Ridgefield, CT, 1984.
Sychra, V., Svoboda, V., and Rubeska, I., *Atomic Fluorescence Spectroscopy*, Van Nostrand Reinhold, London, 1975.
Tsalev, D. L., and Zaprianov, Z. K., *Atomic Absorption Spectrometry in Occupational and Environmental Health Practice, Vol. 1: Analytical Aspects and Health Significance*, CRC Press, Boca Raton, FL, 1983.
Tsalev, D. L., *Atomic Absorption Spectrometry in Occupational and Environmental Health Practice, Vol. 2: Determination of Individual Elements*, CRC Press, Boca Raton, FL, 1984.
Van Loon, J. C., *Analytical Atomic Absorption Spectroscopy: Selected Methods*, Academic Press, New York, 1980.
Welz, B., *Atom-Absorptions-Spektroskopie*, Verlag Chemie, Weinheim (Germany), 1972.
Winefordner, J. D. (Ed.), *Spectrochemical Methods of Analysis*, Wiley-Interscience, New York, 1971.

6

Ultraviolet and Visible Spectrophotometers

GALEN WOOD EWING* / Department of Chemistry, New Mexico Highlands University, Las Vegas, New Mexico

In this chapter will be described instruments for the determination of absorption spectra of condensed-phase samples, primarily in the ultraviolet and visible (UV-VIS) regions of the electromagnetic spectrum, but sometimes extending into the near infrared (NIR). In this context, the ultraviolet is taken to extend from about 175 to 350 nm, the visible from 350 to 750 nm, and the near infrared from 750 to 3200 nm. This is equivalent to a frequency range of about 10^{15} to 10^{14} Hz, or a wavenumber range of 5.7×10^{8} to 3000 cm^{-1}. In the present discussion, the independent variable will generally be expressed in nanometers.

I. INTRODUCTION

The determination of absorption spectra depend directly on the Beer-Lambert-Bouguer expression, commonly known merely as Beer's law:

$$A = abC$$

where A is the absorbance of the sample, a is the absorptivity, b is the path length through the absorbing sample, and C is the concentration of the absorbing substance in a transparent solvent. This relation is derived on the assumption of monochromatic radiation; both A and a are in general wavelength dependent. The absorbance is defined as the negative logarithm of the transmission, T:

$$A = -\log_{10} T = \log_{10}(P_0/P)$$

in which P is the power of the beam of radiation after passing through the sample, and P_0 is the power that would be seen if the sample concentration were reduced to zero.†

In most spectrophotometers the quantities P and P_0 are measured separately and the absorbance is calculated therefrom. The two measurements can be made sequentially on solution and blank, or simultaneously

---

*Retired

†The symbol I for intensity is sometimes used rather than P, but this is to be avoided, as the required quantity must have the dimensions of a power. The term "intensity" should properly be reserved to describe a source of radiation.

with the solution and blank in two identical cuvets. In either case, the blank must contain all components of the solution other than the analyte. In an alternative procedure, the method of standard addition, known increments of analyte can be added to the test solution to give a series of values of P from which the transmission of the sample can be deduced.

Beer's law is strictly valid only under certain conditions: (1) the beam of radiation must be monochromatic, and (2) the absorbing analyte must not enter into any chemical reaction with itself, or with any other component of the solution that is in any way dependent on the concentration. The first of these conditions must be met as nearly as possible in the design of the spectrophotometer, whereas the second is the responsibility of the operator.

II. DESIRED CHARACTERISTICS IN A SPECTROPHOTOMETER

The ideal UV-VIS-NIR spectrophotometer should be able to produce a beam of nearly monochromatic radiation anywhere within the wavelength range from well below 200 to about 3200 nm. The range that can actually be attained is determined by the transmission of optical materials, particularly glass and silica, the emission range of suitable lamps, and the response of detectors. The sensitivity should be such as to enable accurate measurement of samples with transmission from nearly 100 percent to as low as 10^{-3} or even 10^{-4} (i.e., absorbance between 0 and 3 or 4).

The limits of attainable absorbance are determined by the signal-to-noise (S/N) ratio. At high absorbances, as one approaches the low transmission limit, the signal, P, becomes vanishingly small; hence it is essential to eliminate noise as far as possible. The principal source of noise under these conditions is stray radiation, which reduces the spectral purity of the signal reaching the detector. In single monochromators with conventional diffraction gratings, stray radiation is typically about 0.1 percent of the total radiant energy at the detector, but this can be reduced to 0.001 percent by double monochromation or by double pass through a single monochromator, typically combined with modulation (chopping) of the beam of radiation. The stray radiation can be further decreased through the use of holographic gratings, to perhaps 0.0005 percent. The sources of stray radiation are discussed below.

At the low absorbance end of the useful range, the values of P and P_0 are so nearly equal that their ratio becomes lost in noise. The noise that is effective in this region is primarily the Johnson noise produced in the input resistors and shot noise from the photomultiplier detectors.

It is necessary to distinguish carefully between slit width and band width. The term slit width refers to the actual distance (in millimeters) between the slit jaws, whereas band width designates the range of wavelengths passed through the slit. The band width at any wavelength setting is determined by the angular dispersion produced by the diffraction grating, and by the width of the entrance and exit slits (these are normally equal).

The following considerations will help to clarify the throughput considerations for a monochromator. Suppose that the entrance slit is illuminated with approximately monochromatic radiation from a line source, such as

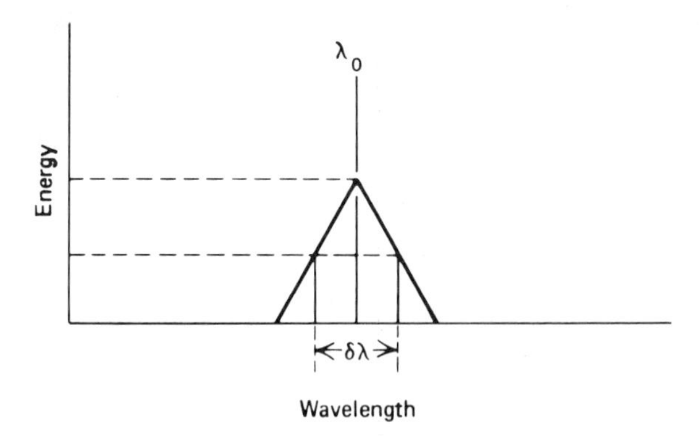

Figure 1. The slit function of a monochromator with equal width entrance and exit slits, centered at wavelength λ_0.

the green line (546.1 nm) from a mercury discharge lamp. If the wavelength control is swept slowly through the region from, say, 540 to 550 nm, the image of the entrance slit will move across the exit slit. Since the two slits are of equal width, the emergent energy will show a linear increase from zero to the maximum at exact congruence, followed by a linear decrease to zero again, the slit function (Figure 1). (If the slits were not equal, the function would be trapezoidal rather than triangular.) Clearly, if the illumination is broadband (i.e., white), radiation of wavelength λ_0 will be contaminated with radiation both of shorter and of longer wavelength. Thus a monochromator is not capable of isolating truly monochromatic radiation from a continuum.

The wavelength purity can be increased at the expense of power by decreasing the slit width. As the slits are narrowed, however, a limit is eventually reached at which the band width is limited by diffraction effects caused by the slit itself. The width of the slits when diffraction limited is of the same order of magnitude as the wavelength of the radiation, and is not ordinarily attained in the ultraviolet and visible regions.

The exit beam from a monochromator includes also some amount of unwanted stray radiation. This is partly caused by imperfections in the diffraction grating and other optical elements and partly by undesired reflections at optical surfaces. Much thought must be given to minimizing stray radiation in the design of spectrophotometers. Opaque baffles to trap the radiation should always be included, and the interior walls of the instrument should be painted dead black to absorb radiation.

III. MEASUREMENT OF STRAY RADIATION

Since the causes of stray radiation are so diverse and unpredictable, absorbance data found in the literature are seldom corrected for this source of error. For accurate results, it is imperative to use a spectrophotometer with stray radiation levels as low as possible. Hence it is desirable to have a method for measuring such levels.

Table 1. NBS Standard Reference Materials (SRMs) for Spectrophotometry[a]

SRM number	Type	Quantity checked	Wavelength range (nm)
930D	Glass filters	Transmittance	440-635
931d	Liquid filters	Absorbance	302-678
932	Quartz cuvet	Pathlength	—
935	Potassium dichromate	UV absorbance	235-350
936	Quinine sulfate dihydrate	Fluorescence	375-675
2009	Didymium-oxide glass	Wavelength	400-760
2031	Metal-on-quartz filters	Transmittance	250-635
2032	Potassium iodide	Stray light	240-280
2034	Holmium oxide solution	Wavelength	240-650

[a]More complete descriptions of these SRMs can be found in Ref. 11.

The usual method for the measurement of stray light in a spectrophotometer is to insert into the optical path a blocking filter that absorbs nearly completely at the wavelength of interest while passing radiation at some other wavelength band essentially unattenuated. A signal observed by the detector under these conditions is due solely to stray radiations scattered from the passed band. A 10-g/l solution of sodium iodide, for example, does not transmit appreciably below 259 nm, but is essentially completely transparent above 290 nm when observed in a 10-mm cuvet [1]. Hence if a spectrophotometer is set at a lower wavelength, say 250 nm, any signal that is observed must originate in stray radiation of wavelengths greater than about 290 nm. This would not serve to measure stray light of wavelengths between 259 and 290, where partial absorption occurs, unless elaborate calibration procedures were adopted. As indicated in Table 1, crystalline potassium iodide is certified by the NBS as a standard reference material for use as a blocking filter. Several other filter materials suited to various wavelength ranges are described by Poulson [1].

More elaborate mathematical procedures for the measurement of stray radiation have been described [2,3], but are rarely used in practice.

IV. THE ARCHITECTURE OF A SPECTROPHOTOMETER

A spectrophotometer must contain (1) a monochromator, (2) a source of radiation, (3) a detector, and (4) electronic means for presenting the acquired data in suitable form.

A. The Monochromator

The majority of instruments presently manufactured utilize diffraction gratings to effect wavelength dispersion. (Formerly silica prisms performed this function, but such designs are now obsolete.) Most widely used is a plane reflection grating in either a Littrow or modified Ebert (Czerny-Turner) mounting. A concave grating is more expensive to manufacture, but permits a design with fewer optical parts. Figure 2 shows the basic plan of these three grating mountings. Most modern plane-grating designs

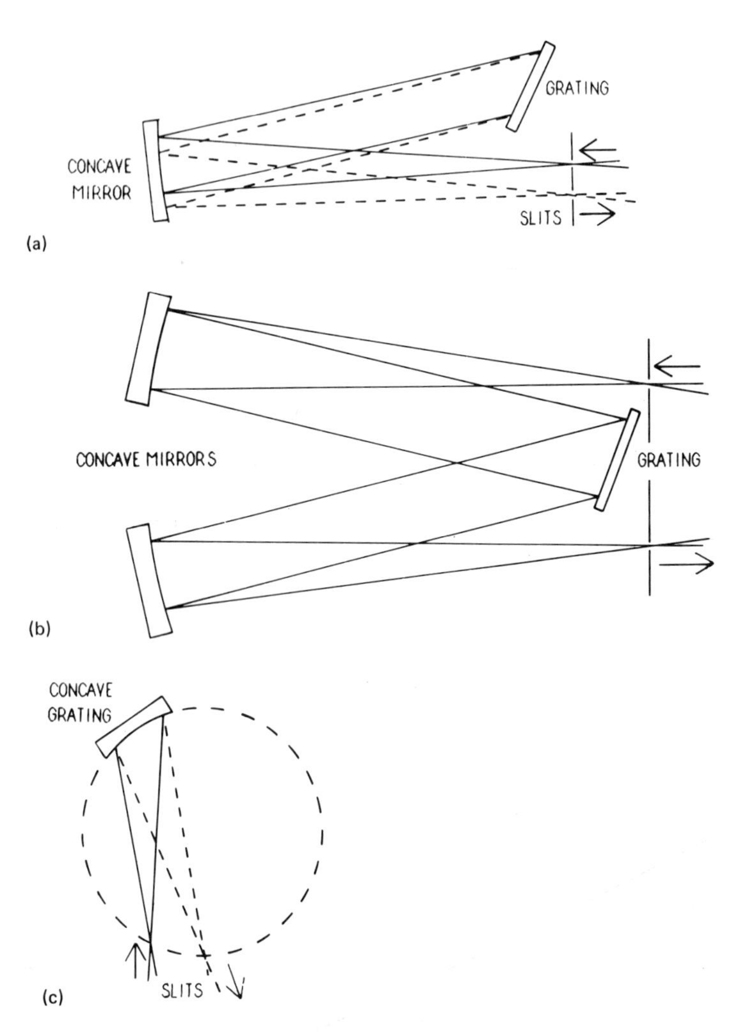

Figure 2. Basic designs of monochromators: (a) Littrow and (b) Czerny-Turner, both using plane reflection gratings, rotatable around an axis perpendicular to the plane of the paper; and (c) a design using a concave grating mounted in such a way that wavelengths can be selected by moving the exit slit along the circle.

use the Czerny-Turner layout, as its inherent symmetry reduces the optical aberrations significantly; the Littrow is somewhat more compact, however.

The monochromator is highly inefficient in its use of energy, since at any given position of the grating only a very small fraction of the total radiant energy passing through the entrance slit is able to reach the exit slit and be transmitted to the detector. Hence recording a complete spectrum can be relatively time-consuming, as each narrow wavelength band must be observed in turn. The latter restriction can be overcome by the use of a linear array of closely spaced detectors.

Both drawbacks can, in principle, be avoided by the use of an interferometer in place of the grating. Provision must be made for varying continuously the phase of the radiant beam in one leg of the interferometer, to produce at the output of the detector a record of the time dependence of the incoming signal, which is the Fourier transform of the conventional spectrum. This principle has rarely been used in the UV-VIS region; the only example known to this author is manufactured by Chelsea Instruments, Ltd., in Great Britain, available in the United States through Questron Corporation, Princeton, NJ [4]. The Fourier transform principle is widely used in infrared spectrophotometers, and a detailed description can be found in another chapter of this handbook.

B. Radiation Sources

An electric discharge in deuterium gas is almost universally used as a source of ultraviolet in spectrophotometers. An alternative is a high-powered xenon arc lamp, but this produces so much radiation and heat that it requires elaborate ventilation as well as ozone elimination. (It is widely used in fluorimetry, where the added intensity is an advantage.) In some special-purpose instruments, such as detectors for use in liquid chromatography, the continuous spectrum of the deuterium lamp is not required and a less expensive mercury lamp can be used instead. For the visible and NIR regions, an incandescent lamp is unexcelled. A tungsten-halogen type is preferred because of its higher output and better long-term stability.

Provision must be made for interchanging the deuterium and tungsten light sources. This is usually accomplished by a focusing mirror that swings between the two lamps, controlled either manually or automatically. The changeover wavelength is in the neighborhood of 320 nm. A slight glitch in the recorded spectrum is likely to appear as the wavelength scan passes the crossover point; hence in some models this point is made adjustable in wavelength, so that coincidence with some particularly vital point in the spectrum can be avoided. If an attempt is made to record the UV portion of a spectrum with the tungsten lamp or the visible region with a deuterium source, beyond about 10 nm from the crossover point the S/N ratio will be sufficiently degraded to render the spectrum useless.

Undoubtedly the most effective light source would be a laser, if one were available that could be tuned over a wide enough range. At the present state of the art, significant tuning is possible only in dye lasers, but even then the range (of the order of 100 nm) is not sufficient to enable recording of more than a very narrow portion of a spectrum. In certain

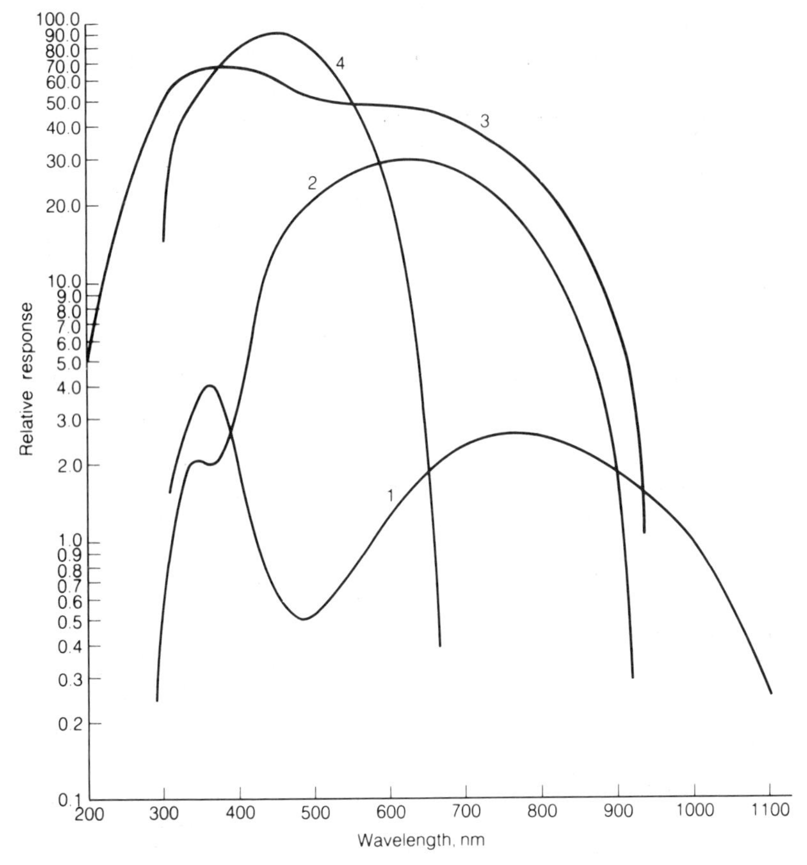

Figure 3. Spectral sensitivity curves for representative photomultiplier tubes. Curve 1: the cathode material consists of successive layers of silver, oxygen, and cesium, within an envelope of borosilicate glass. Curve 2: a multialkali cathode with a borosilicate glass envelope. Curve 3: the cathode is cesium-doped gallium arsenide, and the envelope is special UV-transmitting glass. Curve 4: the cathode is bialkali, the envelope borosilicate glass. (From literature of Hamamatsu Corporation, with permission.)

cases, a laser and detector combination serves admirably as an analyzer for a specific substance, where a single wavelength is all that is required.

C. Detectors

There are two classes of detectors in general use with UV-VIS spectrophotometers: vacuum photoemissive tubes, and solid-state photocells. The most widely used of the first type is the photomultiplier tube (PMT). This device utilizes a photosensitive cathode composed of a thin layer of a semiconductive material containing an alkali metal. Examples are Cs_3Sb, K_2CsSb, Na_2KSb with a trace of Cs, or successively deposited Ag, O, and Cs.

The active layer is supported on a metallic plate, or on the inner surface of a silica envelope. Different cathode materials show different spectral response curves, examples of which are shown in Figure 3.

The action of a PMT depends on the principle of secondary emission of electrons from a sensitive surface. The initial electrons liberated from the photocathode as a result of radiant energy are accelerated by a focused electric field and impinge on the first of a series of secondary emitters called dynodes. Each impacting electron causes the emission of n secondary electrons, with n between 3 and 5 depending on the applied voltage. Each of these electrons strikes the second dynode, where it releases n more electrons, and so on, cascading down the tube, which may contain 10-15 dynodes. The result is internal amplification of the photocurrent by 6 or 7 powers of 10. Typically, a single primary electron can result in a pulse lasting perhaps 5 nsec, corresponding to an average anode current of the order of 0.5 mA. For some applications, such as detection of faint fluorescence, a PMT is used in a pulse-counting mode, but in spectrophotometry the energy of the incident radiation is usually large enough that pulses from individual electrons are not resolved and only an average current is measured.

In some smaller instruments a less expensive vacuum diode is substituted for the PMT. This has the same type of photocathode as the PMT, and hence can have the same spectral response curve (Figure 3). Since no dynodes are present, amplification must be provided externally.

D. Semiconductor Photocells

There are several types of semiconductor photocells. The most generally useful is the p-i-n silicon diode, fabricated on a wafer of intrinsic (i.e., undoped) silicon with two heavily doped regions of opposite types, the radiation being incident on the p type. The p-i-n diode is preferable to the simpler p-n device in that it permits higher bias voltages to be applied without danger of breakdown, thus favoring a wide dynamic range. A transistor, either bipolar or field-effect (FET), can be fabricated on the same silicon chip with a photodiode, thus producing a phototransistor or photoFET, with higher sensitivity than the diode alone.

Arrays of closely spaced detecting elements can be employed to advantage if the dispersing optical system of the spectrophotometer is designed to focus the spectrum in a focal plane. Several manufacturers offer linear diode arrays, with many elements that can be sampled electronically at high speed. One widely used model (model RL1024S of EG&G Reticon) has 1024 diodes spaced 25 μm, center-to-center, thus extending over 25.6 mm in the focal plane [5]. To determine the extent of the spectral region observable at a single wavelength setting, this value must be multiplied by the reciprocal dispersion (nm/mm) of the spectrophotometer. For example, a quarter-meter Czerny-Turner monochromator with a 600-grooves/mm grating has a reciprocal dispersion of approximately 8 nm/mm, so that the array would cover about 200 nm.

This makes a very attractive detector assembly for spectrophotometers, in that a wide spectral region, broken into n small segments, can be observed simultaneously, n being the number of individual diodes. This device permits simplification of the monochromator, as the grating can be stationary

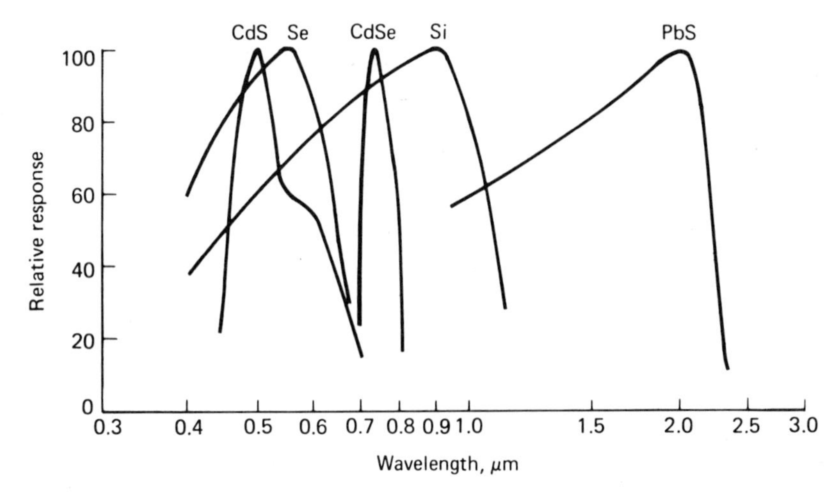

Figure 4. Wavelength response of typical photocells of various kinds. The CdS, CdSe, and PbS cells are photoconductive, whereas the Se and Si are photovoltaic.

rather than rotatable, but the electronic requirements are more severe, making a computer essential for the collection of data. The diodes are sampled sequentially and the voltage produced stored in memory. The chief advantage of an array detector is speed [6]. An entire spectrum can be recorded in a few seconds, depending on the length of time over which the signal is integrated. Hence it is useful in the study of kinetics of relatively fast reactions [7].

Photosensitive cells made of such binary compounds as CdS, CdSe, and PbS show photoconductive response. The only member of this series to be used extensively in spectrophotometry is the PbS cell, which is the detector of choice in the NIR region. Figure 4 shows the relative spectral response curves for a few representative photocells.

Photoacoustic cells can be considered to be general-purpose radiation detectors in both the UV-visible and infrared regions. They are discussed in a separate chapter in this handbook.

E. Photometric Devices

Under this heading are described several alternative methods for determining the power of the radiation at the selected wavelength. The simplest method, found in some small, manually operated instruments, calls for separate measurements of P and P_0, followed by calculation of their ratio. Commonly the instrument is adjusted to read 100 percent for a blank, and then the reading is taken on the desired solution, which gives percent T directly.

In automatically recording, single-beam spectrophotometers, this approach requires scanning the entire spectrum with the blank, capturing it in memory, then repeating the scan with the sample present. For this purpose the "memory" has traditionally been an analog stripchart recorder synchronized with the scan mechanism, so that both the blank and the analytical

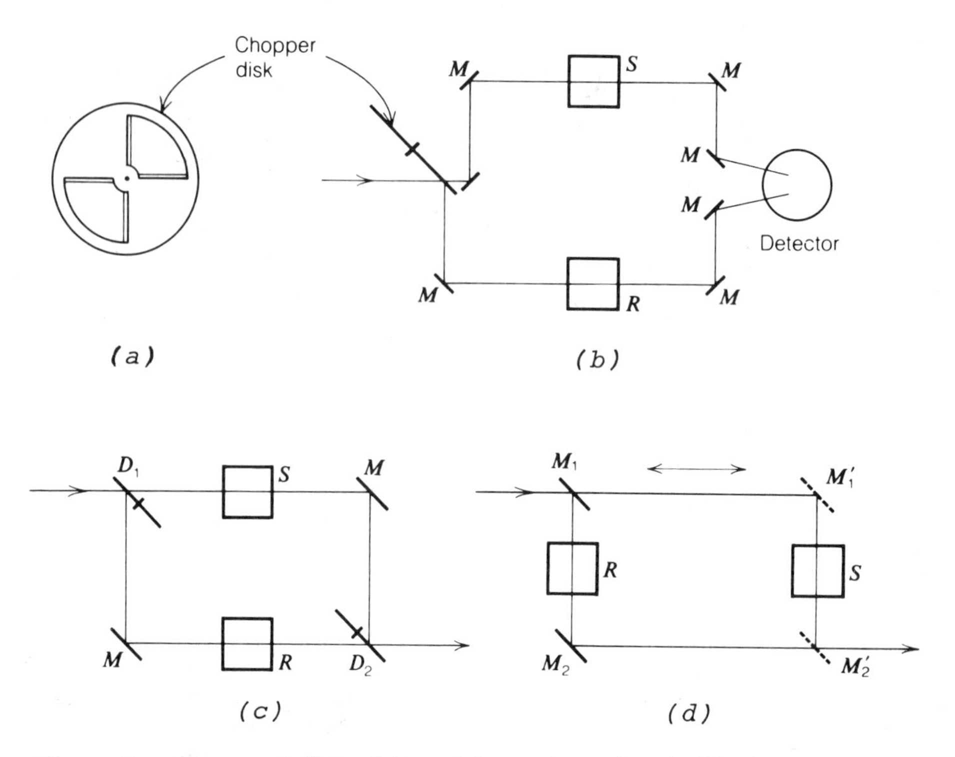

Figure 5. Representative photometric systems for double-beam spectrophotometers. (a) A chopper wheel, consisting of two mirror segments and two transparent segments, so that the light beam is alternately reflected and transmitted. (b) A photometer using a single detector alternately receiving radiation through the sample and reference cuvets. (c) A system using two identical chopper wheels (D_1 and D_2) and a single detector (not shown). (d) A photometer in which a pair of mirrors (M_1 and M_2) are mounted on a carriage that moves them back and forth between the sample and reference cuvets, a system used in the Beckman model DBG spectrophotometer. In each diagram, S refers to the sample and R to the reference.

spectra are recorded on the same chart for visual inspection. In more sophisticated instruments both spectra are digitized and stored in computer memory, so that ratioing or any other appropriate data processing manipulations can be carried out under software command. This is the only approach available for use with array detectors.

Another principle type of photometry requires the use of a double-beam configuration. A beam splitter is inserted in the optical path, generally following the exit slit of the monochromator, dividing the beam of radiation into two nominally equal parts. One of the half beams passes through the blank, and the other through the sample, contained in an identical cuvet. The beams are then either recombined to fall on the same detector, or measured by two separate but identical detectors. If a single detector is used, the beams must be separated in time rather than in space, whereas dual detectors receive signals simultaneously. Often the beam splitter itself consists of a mechanical chopper in the form of a sector disk with alternate sectors transparent and reflective. In any of these alternatives, the synchronized electronic system keeps track of which signal corresponds to which beam. Several schemes are shown in Figure 5. It is desirable for the sequence to include a time slot during which both beams are cut off, so that the detector will respond only to stray radiation that can be corrected for by the computer.

Note that the optical null as used in many infrared spectrophotometers is not used in the ultraviolet and visible regions because the residual imperfections in manufacture of the device would be intolerable in this region.

V. SPECIAL-PURPOSE SPECTROPHOTOMETERS

The features described above are required in all full-fledged spectrophotometers. However, some important instruments are designed for less general applications, which require either special designs or optional accessories.

Dual-wavelength spectrophotometers utilize a "polychromator" rather than the usual monochromator. The instrument is provided with two separate gratings, the beams from which are combined to strike the same detector. Figure 6 gives the optical layout for a typical instrument. This type of spectrophotometer can be operated in a number of distinct modes: (1) It can be used without scanning to monitor two components of the sample, recording their variations with time; the two wavelengths then correspond to absorption maxima of the two components. (2) One beam can be set at a fixed wavelength where the absorbance is constant (for example, at an isosbestic point), while the second beam is scanned. (3) A dual-wavelength instrument can be used to advantage in the examination of turbid samples, with one wavelength monitoring the absorbing solution, while the second is set at a point where absorption is nearly zero. The effective path length cannot be measured with certainty for this application, but can be assumed to be the same for both wavelengths. The sample cell should be placed as nearly as possible in contact with the face of the detector, so that scattered radiation over a relatively large solid angle will be intercepted. Generally dual-wavelength instruments can also be operated in a conventional single-wavelength mode.

Reflectance spectra can be obtained with many general-purpose spectrophotometers through the use of special accessories. These are of two types, for measuring specular and diffuse reflectance, respectively. A typical attachment for specular reflectance is shown in Figure 7a. It is constructed to fit into the sample compartment of a particular spectrophotometer. The beam of radiation is deflected by a system of mirrors to permit reflection from the sample without requiring any modification of either the monochromator or the detector assembly.

Diffuse reflectance in the VIS-NIR is best measured with the aid of an integrating sphere, as shown in Figure 7b. The two beams of a double-beam spectrophotometer enter the sphere through separate ports or windows,

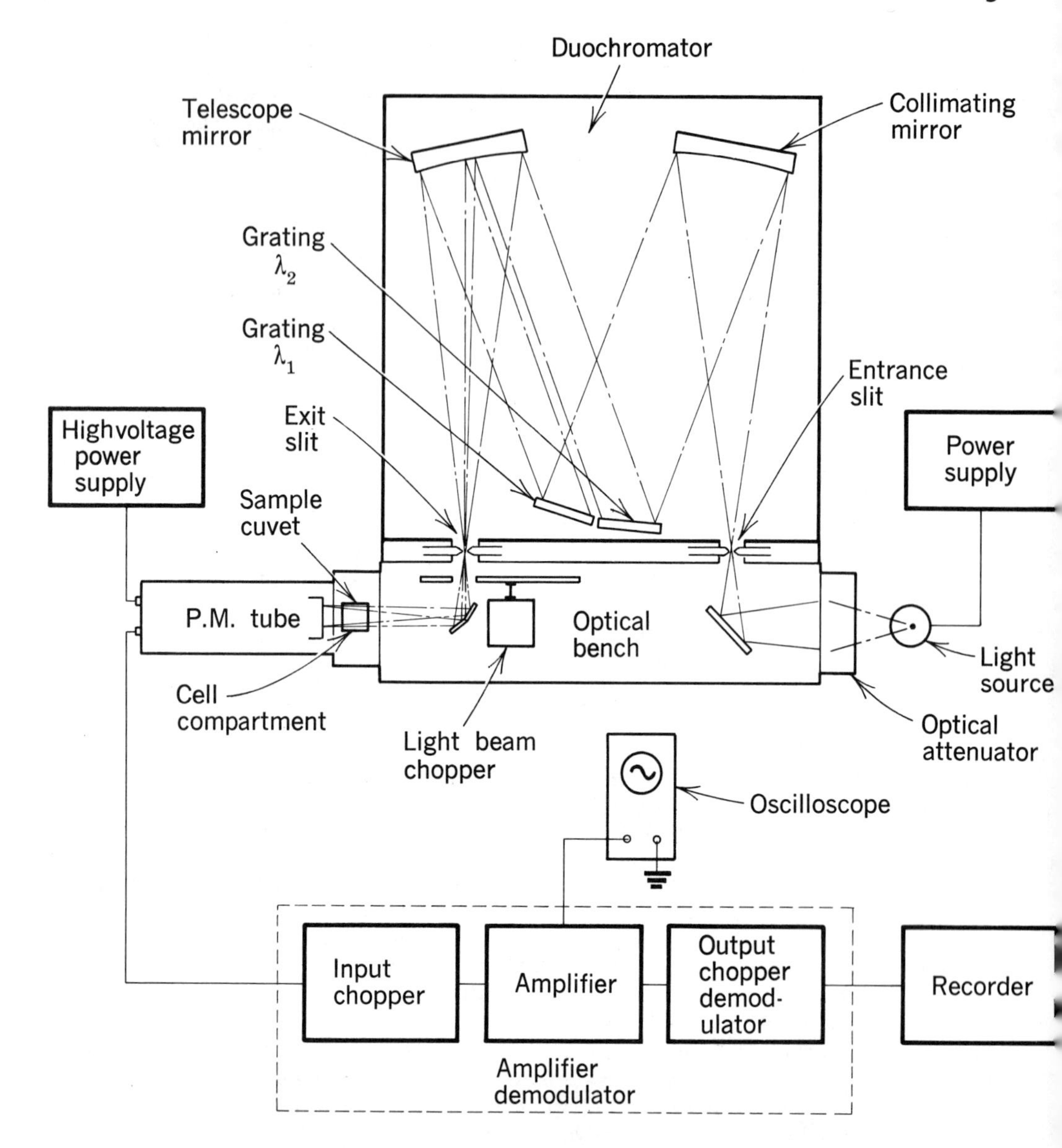

Figure 6. A dual-wavelength spectrophotometer. The radiation from a common source passes through a monochromator equipped with two gratings, separately adjustable. Each grating disperses half of the beam of radiation without intercepting the other. The two beams are recombined and passed through a single cuvet to the photomultiplier detector. A mechanical chopper passes the two beams alternately. (From SLM/Aminco, with permission.)

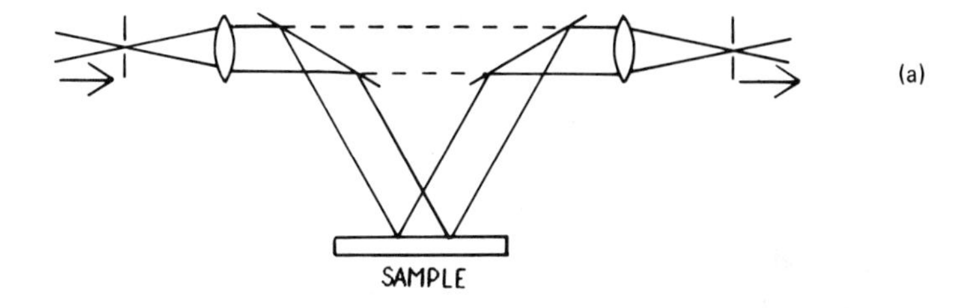

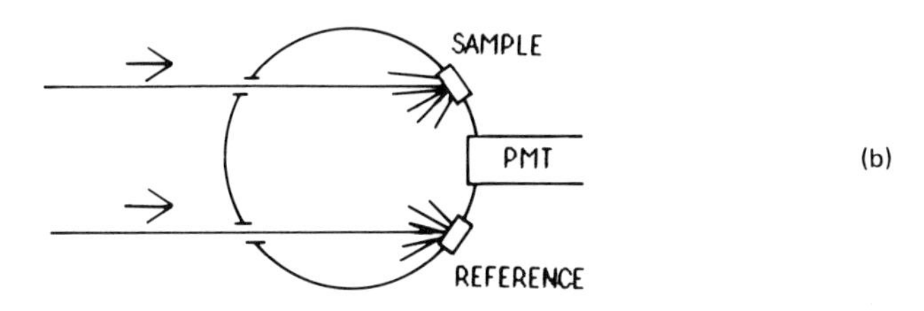

Figure 7. Reflectance attachments for a spectrophotometer. (a) A form designed for specular reflectance, consisting of two diagonal mirrors to reflect the beam from the sample surface; this unit can be placed in the sample compartment of a standard spectrophotometer. (b) An integrating sphere; the two beams from a double-beam spectrophotometer enter the sphere through a pair of ports, and strike panels bearing the sample and reference materials, respectively. The scattered radiation illuminating the photomultiplier detector measures the diffuse reflectance of the sample and reference materials alternately.

and impinge upon two specimen holders, one for the sample, the other for a blank. The blank is usually a rigid block made of pressed magnesium oxide or barium sulfate powder. These materials possess nearly perfect reflectance, over the entire visible and near-IR spectral region. The detector responds alternately to the two beams, as in any double-beam instrument. The sphere has the effect of averaging out local inequalities in the sample and reference beams.

Fluorescence attachments are available for many spectrophotometers. These can be inserted into the sample compartment, in a way similar to the specular reflectance accessory. They are described in Chapter 8.

Photometric detectors for high-performance liquid chromatography (HPLC) constitute a major application for spectrophotometry. The instrument must be specially designed to permit sufficient radiant energy to be passed through a narrow-bore section of effluent tubing. Figure 8 shows one variety of transmission cell. For many biochemical applications, a choice of two wavelengths, 254 and 280 nm, will give adequate selectivity; these are conveniently obtained from a mercury lamp with suitable filters, using the radiation directly for 254 nm and via a fluorescent converter for 280 nm. More details will be found in Chapter 22.

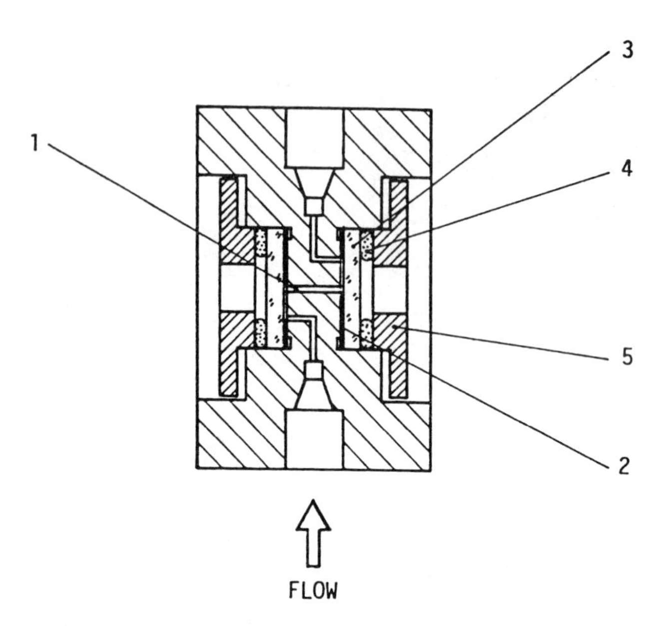

Figure 8. Schematic diagram of a cell for spectrophotometric monitoring of the effluent of a liquid chromatograph. The beam of radiation passes from right to left through the cell cavity (1), which is 5 mm in length with a diameter of 0.5 mm. The gasket material (2) has a slit cut for the solvent path, 3 is a quartz window, 4 is a soft metal gasket, and 5 is a compression member to hold the device together (From JASCO, Inc., with permission.)

The firm of E. Leitz, Inc., manufactures a "microspectrophotometer," which consists of a grating monochromator mounted on top of a research-quality microscope. The operator can focus the microscope on the exact small area desired, then activate the spectrophotometer and record the spectrum. Whatever is seen in the microscope, whether by transmitted or reflected light, will generate the spectrum. The spectral range covers the ultraviolet down to 220 nm, if the microscope is provided with quartz optics.

VI. SAMPLE HANDLING

The conventional sample cells (cuvets) are of square cross section, 10 mm in interior dimension, and about 4 cm in height. These can be had in borosilicate glass for the visible region, or in silica for the UV. Disposable plastic cuvets of similar size are also available for visible use. Similar cuvets provide various optical path lengths from about 0.1 to 100 mm. Special models are provided with different kinds of caps or stoppers, with temperature-regulating jackets, etc. Some are intended for on-line measure-

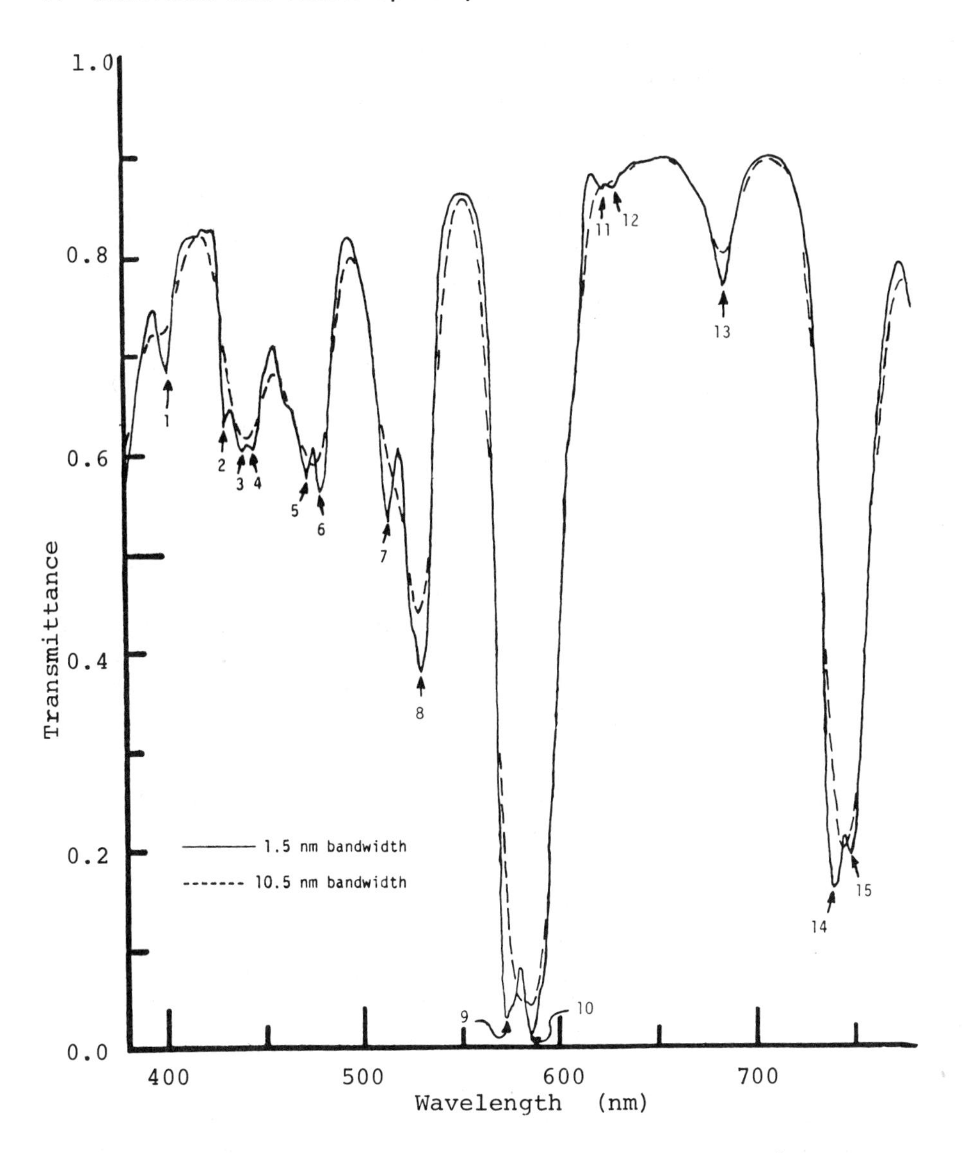

Figure 9. The spectral transmittance of a didymium glass filter over the wavelength range 380-780 nm. Note the effect of band width on the fine structure visible in several of the transmission minima. The wavelengths (in nanometers) corresponding to the numbered minima are as follows (for band width of 1.5 nm):

1:	402.36	6:	478.77	11:	623.62
2:	431.49	7:	513.44	12:	629.62
3:	440.17	8:	529.67	13:	684.70
4:	445.58	9:	572.67	14:	739.91
5:	472.71	10:	585.33	15:	748.34

(From Ref. 12.)

ments of flowing streams, either on an industrial scale or for applications in flow-injection analysis.

Stopped-flow accessories are available, providing for the rapid mixing of reagents followed by time-based measurements commencing immediately after the mixing is complete.

Some less expensive spectrophotometers with limited resolution use cylindrical cuvets (i.e., test tubes) rather than the square variety. For precise measurements, the refractive index of the liquid must be the same for samples and standards or blank, since the cylindrical shape modifies the optical focus of the system.

Special absorption cells are needed for measurements on gases, where the optical path must usually be rather long. Such cells can be provided with mirrors to reflect the radiation back and forth through a cell of reasonable dimensions, as needed to give the requisite path length.

VII. SIGNAL PROCESSING

Some of the smaller "abridged" spectrophotometers are operated manually, but all modern fully expanded instruments are equipped with computers. In the majority, the desired operating conditions—wavelength range, resolution, and so on—are specified by the operator through simple keyboard entry. In some, the computer stands alone, connected to the spectrophotometer proper by cables. In these, the computer is typically an IBM PC/AT or a Hewlett-Packard HP-1000, or equivalent, and the control information is introduced through a standard computer keyboard.

The computer has several concomitant functions to perform. It must control the preliminary operating settings, as indicated above, and in addition, in double-beam instruments, it must continuously adjust the width of the monochromator slits to maintain uniform energy in the reference beam. The computer must also be informed about the phasing of a beam chopper or its equivalent so that it will be able to steer the output of the detector to the portions of memory alotted respectively to the blank and sample. It must then be able, on command, to subtract the blank spectrum from that of the sample, and to display it on a screen. In instruments intended for research use, the computer should have access to a library of spectra, and be able to compare the current spectrum with the library for identification. It should then be able to determine quantitative concentration or other desired data by comparison with the library file or with standard samples.

Detailed study of a spectrum on the viewing screen is often possible. A cursor controlled by the operator can select significant portions of the spectrum, expand them, or manipulate them in other ways, thus increasing the ease with which useful information can be extracted.

An important ability of the computer is to take the first or higher derivative of the spectral curve, a process that emphasizes the significance of small deviations and inflection points in the original spectrum. (This was formerly done instrumentally, by wavelength modulating the beam within the monochromator, or by scanning both monochromators of a dual-wavelength spectrophotometer, with a small offset between them. These approaches have been rendered obsolete through the use of computers.)

VIII. CALIBRATION CHECKS

As with any precision laboratory instrument, occasional confirmation of calibration is desirable. In a spectrophotometer, this requires checking both the wavelength and photometric accuracy.

A. Wavelength Calibration

Undoubtedly the most accurate standards for checking the wavelength scale are the various types of lasers. An inexpensive, low-power, helium-neon laser will serve admirably for a check at its characteristic wavelength, 632.8 nm. For those spectrophotometers with a deuterium source, the lines at 486.6 and 656.1 nm can be used for calibration; note that this requires using the UV lamp in the visible region, not possible on some spectrophotometers. A filter of didymium glass has a number of sharp absorption peaks that can be used as secondary wavelength standards. The precision can be expected to be within about 0.5 nm. Figure 9 shows the absorption

Table 2. Absorptivities of $K_2Cr_2O_7$ in 0.001 *M* Perchloric Acid at 23.5°C (Absorptivity in kg g^{-1} cm^{-1})

$K_2Cr_2O_7$ (g kg^{-1})	235(1.2)[b] nm (minimum)	257(0.8) nm (maximum)	313(0.8) nm (minimum)	350(0.8) nm (maximum)	Uncertainty[c]
0.020[a]	12.243	14.248	4.797	10.661	0.034
0.040	12.291	14.308	4.804	10.674	0.022[d]
0.060	12.340	14.369	4.811	10.687	0.020[d]
0.080	12.388	14.430	4.818	10.701	0.020[d]
0.100	12.436	14.491	4.825	10.714	0.019[d]

[a]Nominal concentration; all weights corrected to vacuum.
[b]Wavelength and in parentheses spectral bandwidth.
[c]Includes estimated systematic errors and the 95 percent confidence interval for the mean.
[d]For wavelength of 313 nm, the uncertainty is reduced to half of these values for the $K_2Cr_2O_7$ concentrations marked.
Source: From Ref. 10, p. 124, with permission.

spectrum of such a filter, with wavelengths marked for the most useful maxima [8]. Holmium oxide, either as a glass or in solution, is also usable for wavelength checks; however, only the didymium data are given here, as this is the only solid wavelength calibration filter supplied by the National Bureau of Standards (see Table 1).

B. Photometric Accuracy

Solutions of potassium chromate and potassium dichromate have been widely used as reference standards for validating the photometric scale. Gibson [9], as early as 1949, gave tables for the transmittance of K_2CrO_4 in 0.05 *M* KOH solution. In more modern work, Burke and Mavrodineanu [Ref. 10, p. 121, also Ref. 11] recommend $K_2Cr_2O_7$ in $HClO_4$. Their data are summarized in Table 2.

The National Bureau of Standards issues standard reference materials for both wavelength and photometric calibration purposes [12]. These are listed in Table 1.

IX. SOURCES OF EQUIPMENT

Manufacturers of UV-VIS spectrophotometers and related equipment, distributed widely in the United States, are given in Table 3.

Table 3. Manufacturers of UV-VIS Spectrophotometers and Related Equipment Distributed Widely in the United States

Beckman Instruments, Inc., 2500 Harbor Blvd., Fullerton, CA 92643.
A complete line, both single and dual beam.

Buck Scientific, Inc., 58 Fort Point St., East Norwalk, CT 06855.
Imported instruments from Cecil Instruments, Ltd. (UK): also rebuilt instruments of various manufacture.

EG&G Gamma Scientific, 3777 Ruffin Rd., San Diego, CA 92123.
Concave grating monochromators; calibration sources.

Guided Wave, Inc., 5200 Golden Foothill Pkwy, El Dorado Hills, CA 95630.
Systems using fiber optics, array detectors, etc.

Hach Co., P. O. Box 389, Loveland, CO 80539.
A unique double-pass Czerny-Turner spectrophotometer.

Harrick Scientific Co., P.O. Box 1288, Ossining, NY 10562.
Accessories for diffuse reflection etc.

Hewlett Packard Co., 3000 Hanover St., Palo Alto, CA 94304.
Diode-array spectrophotometers.

Hitachi/NSA, 460 E. Middlefield Rd., Mountain View, CA 94043.
Complete line.

Instruments SA, Inc., 173 Essex Ave., Metuchen, NJ 08840.
A full line of monochromators.

JASCO, Inc., 314 Commerce Drive, Easton, MD 21601.
Czerny-Turner double-beam spectrophotometers.

Kontron Instruments, 9 Plymouth St., Everett, MA 02149.
Double-beam, concave grating spectrophotometers.

E. Leitz, Inc., Link Dr., Rockleigh, NJ 07647.
Microspectrophotometer.

Linear Instruments Corp., P.O. Box 12610, Reno, NV 89510.
Chromatographic detectors.

LKB Instruments, Inc., 9319 Gaither Rd., Gaithersburg, MD 20877.
Small computerized instruments.

Milton Roy Co., Anal. Products Div., 820 Linden Ave., Rochester, NY 14625.
"Spectronic" series single-beam spectrophotometers.

Perkin-Elmer Corp., Main Ave., Norwalk, CT 06856.
Complete line.

Princeton Instruments, Inc., 644 Newkirk Ave., Trenton, NJ 08610.
Diode-array spectrophotometers.

Shimadzu Scientific Instruments, Inc., 7102 Riverwood Dr., Columbia, MO 21046.
Complete line.

SLM Instruments, Inc., 810 W. Anthony Dr., Urbana, IL 61801.
Dual-wavelength spectrophotometer.

Spectra-Physics, Autolab Div., 3333 N. First St., San Jose, CA 95134.
Chromatographic detectors.

Spex Industries, Inc., 3880 Park Ave., Edison, NJ 08820.
Full line of monochromators.

Varian Instrument Group, 505 Julie Rivers Rd., Sugar Land, TX 77478.
The Cary series of double-beam, dual-pass, Czerny-Turner spectrophotometers.

REFERENCES

1. R. E. Poulson, Appl. Optics *3*, 99 (1964).
2. W. Kaye, Anal. Chem. *53*, 2201 (1981).
3. W. Kaye, Am. Lab. *15*(11), 18 (November 1983).
4. D. Snook and A. Grillo, Am. Lab. *18*(11), 28 (November 1986).
5. Y. Talmi and R. W. Simpson, Appl. Optics *19*, 1401 (1980).
6. J. Sedlmair, S. G. Ballard, and D. C. Mauzerall, Rev. Sci. Instrum. *57*, 2995 (1986).
7. Y. Talmi, Appl. Spectrosc. *36*, 1 (1982).
8. W. H. Venable, Jr., and K. L. Eckerle, *Didymium Glass Filters for Calibrating the Wavelength Scale of Spectrophotometers*, National Bureau of Standards Special Publication 260-66, Washington, DC, 1979.
9. K. S. Gibson, *Spectrophotometry*, National Bureau of Standards Circular 484, Washington, DC, 1949.
10. K. D. Mielenz, R. A. Velapoldi, and R. Mavrodineanu, *Standardization in Spectrophotometry and Luminescence Measurements*, National Bureau of Standards Special Publication 466, Washington, DC, 1977.
11. R. Mavrodineanu, J. I. Schultz, and O. Menis, *Accuracy in Spectrophotometry and Luminescence Measurements*, National Bureau of Standards Special Publication 378, Washington, DC, 1973.
12. R. W. Steward, Ed., *NBS Standard Reference Materials Catalog 1986-87*, National Bureau of Standards, Gaithersburg, MD, p. 100, 1986.

7

Instrumentation for Infrared Spectroscopy

JOHN COATES* / Spectra-Tech, Inc., Stamford, Connecticut

I. INTRODUCTION

This chapter presents an overview of infrared (IR) instrumentation with an emphasis on current trends but also with some perspective of the past. Infrared spectroscopy can be considered as a general technique because it provides methods for studying materials in all three physical states—gas, liquid, and solid—over what is loosely defined as the infrared region of the spectrum. From an analytical point of view, the generic term IR spectrum refers to the mid-infrared region, which traditionally covers a range from 4000 cm^{-1} (2.5 micrometers, μm) to between 400 cm^{-1} (25 μm) and 200 cm^{-1} (50 μm). These lower values are defined in terms of practical limits set by instrumentation rather than theoretical limits of the technique.

The other areas covered by the infrared spectrum, known as the near infrared and the far infrared, also have analytical importance. These overlap the mid infrared and nominally cover the ranges 15,000 cm^{-1} (0.67 μm) to 3000 cm^{-1} (3.33 μm) and 600 cm^{-1} (16.67 μm) to 10 cm^{-1} (1000 μm), respectively. The units of wavelength (microns or micrometers, μm) were used extensively in the past, but now wavenumbers (reciprocal centimeters, cm^{-1}) are the accepted units. A simple reciprocal relationship exists between wavelength, λ, and wavenumber ν:

$$\nu\,(cm^{-1}) = 10{,}000/\lambda\ (\mu m)$$

The most popular display format for the infrared data is known as the double-beam or ratioed spectrum. The vertical scale, or ordinate, of this format is traditionally recorded in transmittance (T or I/I_0), or more commonly in percent transmittance (%T or $100 \times I/I_0$), where I_0 is the intensity of the infrared beam incident on the sample, which is also the energy reaching the detector from the open beam of the instrument, and I is the intensity of the radiation reaching the detector after passage through the sample. The symbols I and I_0 have been used here to be consistent with the usual convention in infrared texts. Note that the symbols P and P_0 are used in a similar manner to represent the powers of the transmitted and incident radiation in other chapters of this book.

In this section a short synopsis of the history of analytical infrared spectroscopy is presented. It is by no means complete, but it does provide an insight into the critical changes and trends of the technology that are

**Current affiliation*: Nicolet Instrument Corporation, Madison, Wisconsin.

relevant to the analytical chemist. Infrared spectroscopy is now probably the most versatile and flexible of the molecular spectroscopy techniques. It can be applied to a broad range of materials of organic or inorganic origin. Most samples produce a unique spectrum that is a characteristic fingerprint of the molecular species involved. This is a generalization, but as a point of discussion, it only really breaks down in the area of simple, symmetrical diatomic molecules, and, in some people's eyes, in the area of macromolecules.

The information contained in the infrared spectrum originates from molecular vibrations. These are either fundamental modes that are associated with vibrations of specific functional groups, complex modes from composite motions of the total molecule, or vibrational overtone or summation modes of the fundamental vibrations. Infrared analysis may simply involve the characterization of a material with respect to the presence or absence of a specific group frequency associated with one or more fundamental modes of vibration. Or a problem may be solved by complex pattern recognition, which is the process used (unconsciously) by an experienced spectroscopist, or by a computer search-match algorithm, when an unknown spectrum is compared to a prerecorded reference data base. Alternatively, the spectral data is used to measure quantitatively one or several components in a simple or complex mixture. At this time (1989) infrared spectroscopy is underutilized (in this author's view) and the full potential is yet to be fully realized. From the point of view of instrumentation, the technique is ahead of its practical implementation—that is, most analyses are not limited by the performance of current instruments.

This rosy image of infrared spectroscopy has not always held true, and in the past the advancement of the technique has been constrained by lack of performance and flexibility. It has "developed" in an erratic manner, and this can be broadly summarized in terms of 10-year segments as follows:

1940s: Era for rapid growth of commercial instruments
1950s: Era of discovery and rapid growth in new analytical methods based on IR
1960s: The wax and wane of analytical IR
1970s: The rebirth of infrared spectroscopy aided by the introduction of Fourier transform infrared (FT-IR) spectrometers, the era of growth in new commercial instruments
1980s: Era for new methods of sample and data preparation.

By the start of the 1970s some observers were sounding the death knell of infrared spectroscopy with predictions of a takeover by newer emerging techniques. During the 1970s many of the older instrument companies either dropped out of the marketplace or were forced to update the technology of the older, dispersive instruments, changing to the interferometer-based FT-IR spectrometers. During the 1970s major improvements were made to the design of FT-IR instruments to improve the reliability and to increase the ease of use for less experienced operators. This led to the move from a research orientation to a general-purpose analytical tool. The reduction in size and cost of computers, especially with the introduction of microcomputer technology, also helped this changeover.

Existing dispersive instruments, which were redesigned to give many of the convenience features of FT-IR instruments, such as speed, graphic displays, and computational capabilities, for routine analysis, also benefited from these trends in technology.

From the start of the 1980s it was evident that FT-IR instruments would probably displace dispersive instruments for many routine analytical IR applications. Many new instrument companies were formed to take advantage of this newer technology, and the market has since become extremely competitive. This competition has been beneficial for the spectroscopist because it forced instrument companies to ensure that their instruments provide a favorable price-to-performance ratio and that the full versatility of the technique is exploited in new areas of application. This has included the development of combined techniques, involving other analytical methods such as gas chromatography (GC) and thermogravimetric analysis (TGA), giving rise to the techniques GC/IR and TGA/IR. New computer algorithms have also been developed for data manipulation, spectral searching and identification, and quantitative analysis. Competition has also reduced the price to the same level as most analytical dispersive instruments. Note, however, that the term used earlier was "displace" rather than "replace." There are still major application areas that are as well, if not better, served by dispersive instruments as by FT-IR technology. Some of these will be implied by discussions later in this chapter.

At one time infrared spectroscopy was considered to be an energy-starved technique and believed to be quite impotent for anything more than qualitative characterization and simple quantitative analysis. The situation has altered and although the fundamental energy of the process has not changed, instruments have been designed to handle the energy more efficiently. This has been accomplished by the use of higher-efficiency sources, more stable optics, increased detector sensitivity, and improved electronics. In fact, the technique is no longer limited by the instrument itself, but instead tends to be limited by the sample and the operator.

During recent years, it has become obvious that the traditional methods for sample preparation need to be reviewed and streamlined to meet the performance of the latest generation of instruments. Several factors must be considered. First, current performance is more than an order of magnitude better than with early instrumentation. Whereas at one time one could be satisfied with a signal-to-noise (S/N) performance of maybe 300:1 (peak to peak), we now enjoy a situation where it is possible to achieve an S/N of 100,000:1 or better (relative to the open beam) under certain experimental conditions, such as low-energy throughput situations with a high-sensitivity detector. It is possible to enjoy a performance that equates to between 1000:1 and 10,000:1 signal-to-noise performance, on a routine basis, and within 1 or 2 min.

Second, as a consequence of this improved performance, the time frame for acquiring an acceptable spectrum is much shorter. With an older instrument a scan time of 10-20 min is usually necessary to provide a spectrum of good quality. Therefore, 20 min spent in sample preparation is consistent with the scan time. Equivalent data can be produced on a modern spectrometer in less than 1 min, which means that a significant time penalty is experienced with traditional sampling methods. Furthermore, as spectroscopic software becomes more sophisticated, there is a need to produce data from a sample in a more consistent manner, and, wherever possible, more accurately. Recent developments in sampling methods now take advantage of the performance of both the instrument and the software.

II. INSTRUMENTATION

Modern (since late 1960s) commercial instruments fall into three categories: grating-dispersive, filter-dispersive, and Fourier transform (FT-IR) spectrometers. In the analytical laboratory the grating and FT-IR instruments are the most important for the current discussion. Whatever the mode of measurement, any infrared instrument, in common with spectrometers in other wavelength ranges, may be logically split into the following discrete components: the main optical system (the analyzer), the source and detector, the sample compartment, and the electronics and data handling. The importance of most of these components is self evident, but the sample and sample compartment are two important components that are often overlooked. In particular, the placement and interfacing of the sample relative to the other components are often critical to the success and accuracy of analytical measurements.

The term "analyzer" was used above as a generic description of the main spectrometer, or spectrophotometer, which forms the heart of the instrument. This unit is responsible for taking the broadband infrared radiation and ultimately splitting it into discrete frequencies or wavelengths with a given spectral resolution. This may either be performed directly via a monochromator in a dispersive instrument, or indirectly via an interferometer in a Fourier-transform instrument. In the latter case, the complete interferometer assembly is often referred to as the modulator, because the output from this unit is a modulated infrared beam that is decoded to produce the final infrared spectrum. It is difficult to discuss the "analyzer" component of a spectrometer in isolation from the other key components that make up the instrument. Therefore, in this section, the fundamental concepts of infrared instruments will be addressed relative to the operation of this analyzer component.

A. Monochromator Instruments

Monochromators can range from simple filter-based systems to high-resolution, double prism/grating or grating/grating systems. Two common designs of monochromator are used in modern optical spectrometers, the Littrow and the Czerny-Turner. Of these, the Littrow is the most important for dispersive infrared instruments. An example of a simple Littrow grating monochromator instrument is shown in Figure 1—the layout of the Perkin-Elmer model 700 series. Early instruments, such as the original workhorse, the Infracord (Perkin-Elmer model 137), were prism-based Littrow systems. These provided a simple wavelength scan in micrometers. Later versions of this instrument, and a vast array of subsequent generations, plus instruments from other manufacturers, such as Beckman and Pye-Unicam, incorporated similar Littrow designs but were equipped with one or more diffraction gratings.

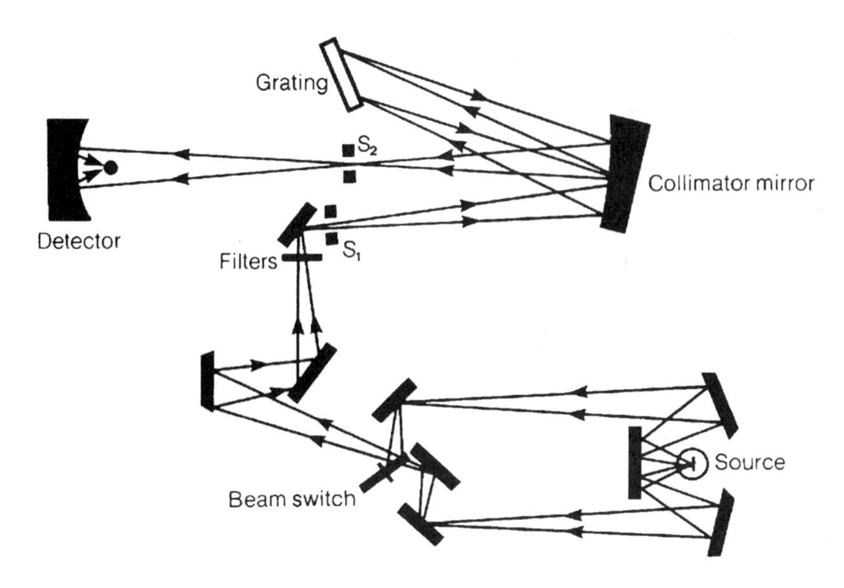

Figure 1. An example of a Littrow grating monochromator for an infrared dispersive spectrometer. Note that S_1 is the entrance slit and S_2 is the exit slit.

The principles of operation of monochromators are described elsewhere in this book and therefore only the features specific to infrared instrumentation will be discussed in this chapter. Instruments featuring gratings are designed to provide a linear output in wavenumbers, instead of wavelength. On older instruments this conversion is performed mechanically via a special cosecant cam. Modern instruments with microprocessor control feature direct-drive monochromators in which the grating is driven directly by a stepping motor. The stepping sequence of this motor is programmed in a nonlinear fashion to generate the equivalent of the cosecant relationship. Normally, the full range of a grating spectrometer, which may be anywhere between 4000 to 600 cm^{-1} and 5000 to 200 cm^{-1}, depending on price and performance, is necessarily not provided by a single order from the diffraction grating. Lower-priced instruments are often designed where the range from 4000 to 2000 cm^{-1} is obtained with the grating operated in the second order. The remaining part of the spectrum from 2000 cm^{-1} down is acquired in the first order. A series of cutoff and bandpass filters are placed between the monochromator and the detector in order to assure that the correct order of the grating is being used. The selection of the correct filter and filter type is synchronized with the scanning of the grating. In more expensive instruments, more than one grating is used, where each grating is utilized in the first order only. This usually provides optimum performance, because when used in the first order a grating produces its greatest overall output for a given spectral range. It is worth commenting that the instrument featured in Figure 1 used a unique grating that was ruled at two different blaze angles. This arrangement performed as two gratings in one, where each blaze angle was used in the first order. The wavenumber (or wavelength) precision and accuracy of early instruments was strongly dependent on the engineering of the cosecant cam. This is the basis of one of the most common competitive arguments in favor of FT-IR instruments. With modern dispersive instruments that feature direct grating drives this is a moot point because not only is the drive very reproducible (often better than 0.01 cm^{-1}), but it can also be precalibrated against a spectroscopic standard to give high accuracy. This provides precision and accuracy comparable to laser-referenced FT-IR spectrometers. Unfortunately, this has been overlooked in many of the emotive discussions comparing dispersive and FT-IR instruments.

Among the final key components of the monochromator system are the optical slits. On earlier instruments these were also cam driven and their operation was synchronized to the scanning of the grating either electronically or via direct mechanical linkage. Again, on modern instruments these are under microprocessor control and are directly driven by a stepper motor. It must be noted that with this type of control, the modern monochromator is no longer considered to be an analog-driven device.

B. Interferometer-Based Instruments

As noted earlier, FT-IR spectrometers are interferometric instruments, and almost without exception, commercial implementations are based on a single design: the Michelson interferometer. The optical layout of a typical interferometer is provided in Figure 2. The design is relatively simple, with the main components being the source, the modulator, the sample area, and the detector. In contrast to early instruments, most modern optical systems are designed with a minimum number of reflections and a relatively short optical path. The latter is important to help improve the overall stability of the instrument and to minimize the potential contact with atmospheric water vapor and carbon dioxide. In some of the latest systems this is less important because the main optical module is sealed, sometimes hermetically.

The heart of the FT-IR spectrometer is the modulator—the component that generates the optical interferogram. It is simple in basic design because there are only three major components—a fixed mirror, a moving mirror, and a beam-splitter assembly. The successful operation of the instrument requires very high mechanical and optical stability within the interferometer cavity, especially when operated at medium to high resolution, or used in the shorter wavelength regions. The infrared beam is first collimated (made parallel), with a parabolic mirror, and is directed onto the beam splitter, which is placed at an angle of around 45-60°, relative to the beam. The angle selected depends on the optical design selected by the manufacturer. Traditionally 45° has been used, but in some modern designs an angle close to 60° has been selected, primarily to reduce polarization effects. At the beam splitter a portion of the light is reflected and another portion is transmitted, forming the two arms of the interferometer. Under ideal conditions a beam splitter would transmit 50 percent and reflect 50 percent of the incident radiation. In practice, real beam splitters are not ideal, and they perform with less than full efficiency.

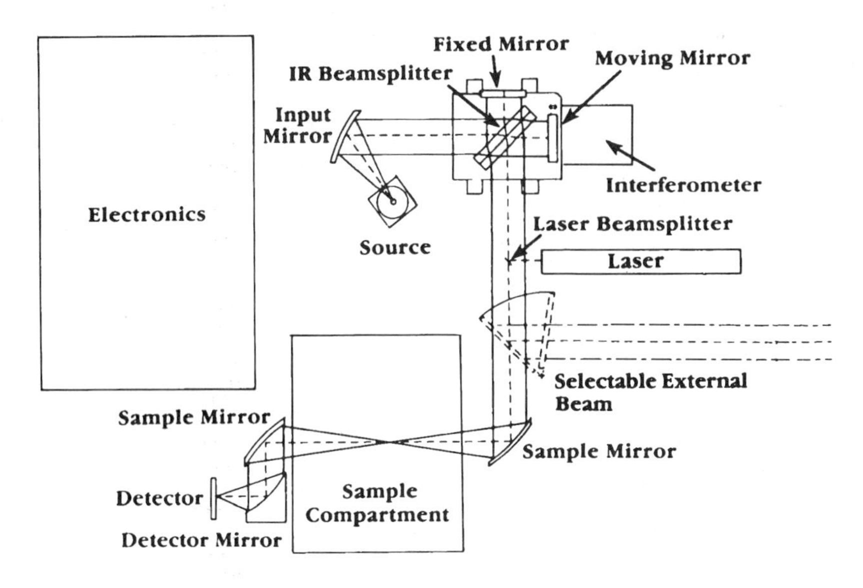

Figure 2. An example of a modern Michelson interferometer-based spectrometer.

Following passage through the beam splitter, the two beams are reflected back by the fixed and moving mirrors. At this point they recombine and interact to produce optical interference, caused by the beams being either in or out of phase depending on the position of the moving mirror relative to the fixed mirror.

To appreciate how the interferometer works and how it can be related to a spectroscopic measurement, it helps to discuss the results produced by this type of system when monochromatic light is used with wavelength λ, for example, from a laser. Figure 3 illustrates what happens when light from such a source is passed into the interferometer optics. At the position marked 1, the moving mirror and the fixed mirror are equidistant from the beam splitter. In this situation both light beams travel the same distance, and therefore when they recombine at the beam splitter they are mutually in phase and constructive interference of the beams occurs. This is observed as a maximum signal being passed through to the detector. For simplicity this is set to unity in the illustration. This unique location, where both optical paths are exactly the same, is called the ZPD, or zero path difference. As the mirror moves away from this position and passes through a position that is physically $\frac{1}{4}\lambda$ away from ZPD, or at an OPD (optical path difference) of $\frac{1}{2}\lambda$ ($2 \times \frac{1}{4}$ because it is doubled by reflection), the signal from the detector reaches a minimum value, or zero as represented in the illustration. This minimum is generated by the two beams now being completely out of phase, giving rise to destructive interference. As the mirror continues to move, it reaches a distance of $\frac{1}{2}\lambda$ from ZPD (OPD = 1λ), and the signal level again reaches a maximum value with constructive interference. This pattern continues where a series of maxima

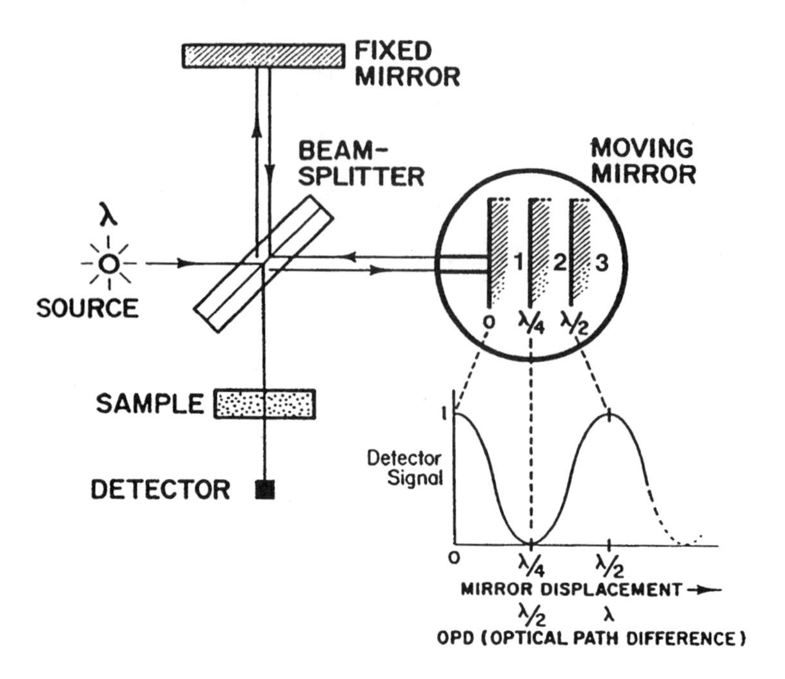

Figure 3. A simplified schematic illustrating the formation of an interferogram from a single-wavelength light source.

and minima are produced at integral numbers of $\frac{1}{4}\lambda$ and $\frac{1}{2}\lambda$ ($\frac{1}{2}\lambda$ and 1λ OPD) to yield an overall sine-wave pattern, or more accurately a cosine wave. The effect of moving the mirror in one arm of the interferometer, and generating an optical path difference, is termed "optical retardation," and the distance traveled by the mirror is often called the "retardation."

If a second wavelength is selected, a similar waveform is generated with a maximum at ZPD and maxima at integral wavelengths (OPD) from the ZPD, where the maxima are now separated by a distance equivalent to the new wavelength. Extending this discussion to a polychromatic source, where we have all wavelengths, we can expect a unique cosine wave to be generated for each of the component wavelengths. The observed signal at the detector is a summation of all of these cosine waves, which gives a maximum at the ZPD and rapidly decays to a complex overlapped signal, which continues to decay with increasing distance from the ZPD. Obviously if the component cosine waves can be resolved, then the contribution from the individual wavelengths can be observed, and a spectral output of the source could be constructed. This function is performed mathematically by a Fourier transformation, a technique that is used extensively in many branches of physical science and engineering for frequency analysis.

In practice, the measured interferometric signal is offset so that the signal is centered about zero on the intensity axis. This signal, now referred to as the "interferogram" (Figure 4), undergoes some preliminary mathematical processing—phase correction and apodization—prior to the

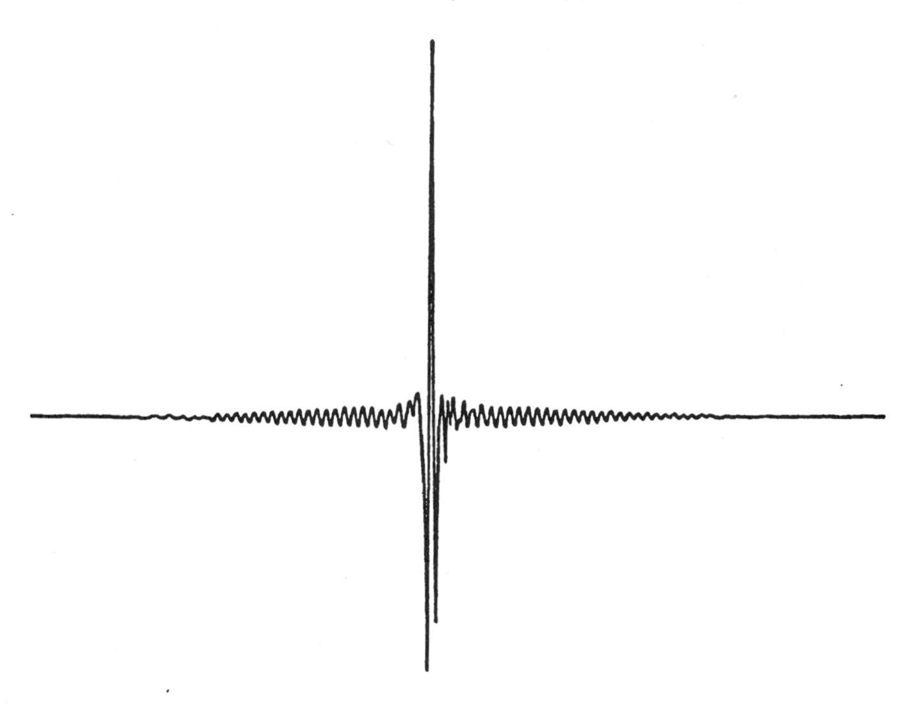

Figure 4. An idealized interferogram illustrating the center portion with the center burst and the region containing spectral information adjacent to the center burst.

Fourier transformation that generates the familiar spectral energy curve, as a function of wavenumber. The terms "phase correction" and "apodization" will be described briefly in a moment. In mathematical terms, the Fourier relationship is defined as a pair of integrals:

$$I(\delta) = \int_{-\infty}^{+\infty} B(\nu) \cos 2\pi\nu\delta \; d\nu$$

and

$$B(\nu) = \int_{-\infty}^{+\infty} I(\delta) \cos 2\pi\nu\delta \; d\delta$$

where $I(\delta)$ is the intensity of the interferogram as a function of retardation and $B(\nu)$ is the source intensity as a function of wavenumber.

From Figure 4 it can be seen that the interferogram has a unique appearance, where the central part, known as the center burst, corresponding to the ZPD, has a significantly larger intensity than the signal on either side. In an open-beam situation, this signal contains broadband information and has the greatest amount of information around the ZPD. As a sample is introduced, this distribution of information changes, dependent on the natural linewidths of the infrared spectrum of the material, and consequently more signal is observed out in the wings of the interferogram—the region beyond the center burst.

Spectral resolution is defined in a number of ways. It may be considered as the ability of an instrument to separate two spectral features, such as peaks, by a predefined criterion for separation, such as a trough between two peaks of defined percentage, or it may be considered as the ability of a system to observe the separation between a pair of discrete wavelengths. In the interferogram, this translates to being able to resolve, or to define uniquely, two cosine waves of different frequencies. For low resolution, where there is a relatively large difference between the two wavelengths to be resolved, the cosine waves in the interferogram can be differentiated at a relatively short distance (or retardation) from the ZPD point. At higher resolution, the differences between the wavelengths are smaller, and hence a much larger retardation is required before the individual cosine waves can be uniquely defined. This is a simplistic explanation but it does help to define the relationship between the collection of the interferogram and spectral resolution—that is, the higher the resolution in the final spectrum, the longer the interferogram must be to provide that resolution. In order for the two cosine waves to be individually identified they must go out of phase and back in phase at least once, and this requires that the interferogram is collected out to a distance of $1/\Delta\nu$ cm, where $\Delta\nu = \nu_2 - \nu_1$ (the separation between two bands expressed in cm^{-1}). Conversely, an instrument is normally setup to acquire data at a fixed retardation, and therefore the maximum resolution of the spectrometer is approximately defined as $1/\Delta_{max}$, where Δ_{max} is the distance that the mirror is moved. Once data are obtained at a specific resolution, a spectrum of lower resolution can always be artificially generated simply by using a subset of the data.

The actual resolution of a spectrum may be slightly lower than the predicted value because of a mathematical operation known as apodization. The Fourier relationships defined above included integrals with limits extended to infinity, which implies that data should be collected to infinite retardation. In practice this is obviously impossible and the data are acquired over practical distances with a resultant truncation of the interferogram. An apodizing function is applied to the interferogram to help reduce the effects produced by the premature truncation of the data. Without this function, any band that has a natural linewidth that is the same as or narrower than the resolution defined by the retardation is distorted in the final spectrum. The distortion is observed as negative lobes on the wings of the band. Apodization is applied by multiplying the interferogram by a mathematical function, such as a negative ramp (triangular apodization) or a raised cosine "bell," such as the Hanning or Happ-Genzel apodizations. This acts like a weighting function that changes the contributions of the cosine components relative to the center burst—that is, by reducing the contributions at the extremes. Since noise is evenly distributed across the entire interferogram and because its contribution relative to the signal is greater in the extremes of the interferogram, one side effect of the apodization is to reduce the noise level in the final spectrum.

The precision of the interferometric measurement is dependent on the accuracy in the formation and recording of the interferometric data. The signal produced is a phase-dependent phenomenon, and therefore any external factor that can change the phase will potentially cause errors in the measurement. Optical components that can cause dispersion, because of their refractive indices, such as the beam splitter, and electronic components that can change the phase of the signal, such as the detector, can and do introduce phase errors. These are numerically compensated by a process known as phase correction. Failure to apply adequate phase correction will result in photometric errors in the final spectrum. Both apodization and phase correction are normal operations performed in commercial spectrometers at the time of the Fourier transformation. Most high-performance instruments provide the operator with access to these functions. In low-cost instruments, however, the complete process is performed automatically and the numeric functions may be inaccessible to the user.

Some practical comments will now be made on the interferometer and the measurement of the interferogram. As indicated earlier, the critical component in the interferometer is the beam splitter. Beam splitters are made from a range of materials, and the actual choice is dependent on the spectral range to be measured. This text is focused on the mid infrared, and for the majority of measurements this range is adequately covered by a germanium beam splitter deposited on a KBr substrate. (The germanium film is too thin to be self-supporting.) Other materials are often also used as thin coatings to improve the performance at the extremes of the spectral range, and to reduce the susceptibility of the KBr to damage from moisture. With suitable coatings, the range of the Ge-KBr beam splitter can often be extended from around 6500 cm^{-1} to between 450 and 400 cm^{-1}. There is a potential shortfall because for many years dipersive instruments have offered reasonable performance down to 225 to 200 cm^{-1}. This can be handled by the use of a CsI beam splitter substrate in place of KBr. This has significant disadvantages when compared to KBr—notably a reduction in transmission (sometimes as much as 50 percent when compared to KBr), a greater propensity to attack by moisture, and a potential to distort when retained in a rigid mount. The latter point is debatable, because some manufacturers state that if the mounts are designed correctly then there is no problem. Ranges beyond the mid infrared are provided by materials such as Mylar, silicon, and metal meshes for the far infrared, and calcium fluoride and quartz for the near infrared and visible. It must be appreciated, however, that when working in regions outside the mid infrared it is normally necessary to change other components, such as source and detector, to obtain acceptable performance.

The instrument layout shown in Figure 2 illustrates additional components besides the modulator, source, and detector. One critical component not mentioned to this point is the helium-neon (He-Ne) reference laser. In the instrument shown, this component serves two important functions—it acts as a clocking device to synchronize the data collection of individual scans, and it provides standard reference information for the location of the center burst of each interferogram. In order to appreciate the importance of these two functions it should be realized that the interferogram must be digitized before any computation, such as the Fourier transformation, can be performed. Also, in most cases more than one coadded spectrum is collected from an FT-IR spectrometer in order to take full advantage of the sensitivity of these instruments. In a "perfect" spectrometer it is possible to take advantage of the square root law, which states that the random noise in a spectrum is reduced by the square root of the number of coadded scans. The laser is used to trigger the collection of data from the infrared interferogram, which is accomplished by passing the laser beam through a secondary visible interferometer linked to the main interferometer. The monochromatic light from the laser produces a cosine wave interferogram and zero crossing points—either each one, alternate ones, or, in some instruments, every third crossing is used as a trigger to collect data. There is a tradeoff: taking points more frequently makes the data arrays much larger, but it does extend the upper limit for data collection—for example, collecting at every crossing allows for collection to 15,804 cm^{-1} (the He-Ne line at 632.8 nm). Collecting data at every third zero crossing does produce a small array that provides benefits of storage and computing speed, but it also limits the upper spectral range, based on the Nyquist sampling criterion, to 5268 cm^{-1}. The He-Ne visible interferometer is incorporated into the infrared interferometer and the laser beam normally passes coaxially with the infrared beam through the center of the beam splitter and other key optics. It is accommodated by a small central spot in the beam splitter that does not have the germanium coating, and its entrance and exit through the optical system are made via small mirrors mounted in the center of the IR beam or small holes at the center of certain mirror optics. In the system illustrated, the laser circuitry serves a second function of absolute fringe counting to keep track of the location of the center burst of the infrared interferogram. It is essential that the location of the center burst is maintained at the exact same position, as failure to do this can result in a failure to comply with the square root law. The expected improvement in noise level is not realized unless the interferograms are in exact registration. The scheme mentioned here, where the registration is maintained by absolute counting of the laser fringes, may fail in situations where the infrared light level is low and there is any uncertainty in the exact location of the center burst. An alternate procedure that overcomes this problem is the use of a white light source. In older instrument designs this source was a separate tungsten filament lamp that illuminated a secondary interferometer located on the back or the side of the infrared interferometer. This produced an interferogram with a sharp center burst that was used as a trigger to initiate data collection, usually at a point before the center burst. These light sources were often unstable, and in instruments that feature white light triggering, the "white" light is taken from the infrared source itself.

The raw interferogram often shows some asymmetry caused by dispersion; this is particularly exaggerated on instruments that feature a moving optic as used in a refractively scanned interferometer. The interferogram is made symmetrical by phase correction and sometimes linearization by a dispersion correction. The symmetry can be utilized by acquiring a double-sided interferogram, which provides more efficient data collection because with less than twice the movement of the moving mirror the equivalent of two interferograms is acquired in less than the time of two scans. Further efficiency and speed can be gained from bidirectional collection, although not many instruments actually do this. In systems that do feature this mode of collection, the interferograms from each direction may not be coadded at the time of collection. Instead, they may be collected in two separate files and coadded, after some minor adjustments, at the end of the run.

Stability is of prime importance in the design of an interferometer. Since the spectral information encoded is of the order of wavelengths, the mechanical tolerances of the system must in most cases be of the same order or better. This means that a high-precision drive mechanism is required for the moving mirror, especially for high-resolution systems, because of the longer retardation distances involved. The location of the fixed mirror, the flatness and parallelism of the beam splitter, and its location are also important. In some commercial instruments, corner-cube optics are used for the moving and fixed mirrors to help overcome scan errors caused by irregularities in the mirror drive. High-pressure gas bearings are featured in many high-performance instruments to help produce smooth-running, frictionless drive systems. Other key factors in design that relate to operating environment are thermal stabilization and vibration isolation. Temperature changes within the interferometer can cause dimensional changes, and if these are too great and are not adequately compensated, then misalignment of the interferometer will occur with an accompanying loss of throughput, which is sometimes significant. Source placement and removal of heat from the source are also important design considerations for the same reasons. Vibration is another undesirable factor originating from both external and internal sources. Typical sources of vibration are the drive system, any cooling fans, and even components such as transformers that could transmit line-frequency vibrations at 60 Hz and higher harmonics. Systematic vibrations of this type produce overlapping frequencies in the interferogram that will transform to lines or spikes in the final spectrum. Intermittent or broad-frequency external vibrations will manifest themselves after transformation as regions of extraneous noise.

The interferometer is a single-channel instrument and consequently operates as a single-beam spectrophotometer. Therefore, in order to produce a conventional "double-beam" format spectrum it is necessary to obtain the open-beam spectrum and the sample spectrum independently and then calculate the ratio of the two spectra. When the acquisitions of the two spectra are separated by a long period of time, this can cause problems for two reasons. First, if there are medium- to long-term temperature drifts these will show up as changes in baselines and baseline slopes. Second, any variations in atmospheric water vapor or carbon dioxide will produce positive or negative absorptions in the ratioed spectrum. One way around this latter problem is to purge the optical bench and sample compartment with dry air or nitrogen. One precaution when purging is not to use a high flow rate of purge gas, as this can cause thermal gradients to occur because of localized cooling, giving rise to eddy currents that may lead to refractive index changes in the path of the infrared beam, again causing instability of the spectral background.

C. Sources

The ideal spectrometer would feature a point source from which each wavelength could be individually selected, and would have sufficient performance to provide many decades of linearity. This would require a line-selectable source such as a tunable laser, an option that some observers have suggested would make an ideal instrument. In fact, a commercial system based on tunable, solid-state lasers was developed as a high-resolution spectrometer, but it was never successful in the general analytical marketplace. For some applications it appeared to offer excellent characteristics—that is, in addition to providing very high resolution, the small diameter of the focussed laser beam made it ideal for microsampling, and it had a sufficiently high throughput to give an operating range out to around 12 absorbance units. The negatives, however, with the current state of the art in tunable laser technology, outweighed these positives. The usable scan range of these lasers is relatively narrow (often less than 100 cm^{-1}) and the individual cost of the lasers is high, making a system prohibitively expensive for an instrument covering a practical analytical range.

Most commercial infrared instruments use an extended (not a point) source having a continuous broadband output with an energy distribution close to that of a black-body radiator (Figure 5). Typical temperatures range from 1000 to 1500 K. It is readily appreciated from Figure 5 that although raising the source temperature does result in more power at the sample, the amount of usable infrared radiation is small compared to the increase in near-infrared and visible radiation. This is not unexpected, because visually there is a shift from red to yellow (to shorter wavelengths) as the source temperature increases.

Four different sources are commonly used in commercial instruments. These are the nichrome wire, the Nernst glower, the Opperman, and the Globar. The nichrome wire source is usually implemented in coil form, either unsupported or wrapped around a ceramic rod. It is used in many low to mid-priced instruments. Although it has the lowest cost, and is the simplest to maintain, it does have some drawbacks: it must be operated at relatively low temperatures to reduce oxide formation, and thermal stresses can result in deformation of the coil-form shape, leading to long-term changes in constancy and output. The Nernst and the Opperman sources have similar output characteristics and have been incorporated in many commercial systems. The Nernst glower is composed of a mixture of rare earth metal oxides that are heated by the passage of an electric current. This oxide mixture has a negative coefficient of electrical resistance—that is, its electrical conductivity increases with temperature, which is the opposite of most materials. This leads to certain inconvenient features—the source requires external heating to start the flow of electricity, and once started it has to be carefully regulated to prevent a "runaway" situation. The Opperman is a variant of this source, developed by Perkin-Elmer and used extensively in their instruments. Its construction features the rare earth oxides contained in a ceramic sheath, with a coaxial metal wire located in the center. The wire is electrically heated, and this in turn heats the oxide mixture. This has the advantage that it is self regulating and does not require external heating for startup. The last source, the Globar, is an electrically heated silicon carbide rod that is capable of being operated at high temperatures. In most cases water cooling is essential to help stabilize the source and to reduce the heat that is dissipated to the surrounding components in the spectrometer.

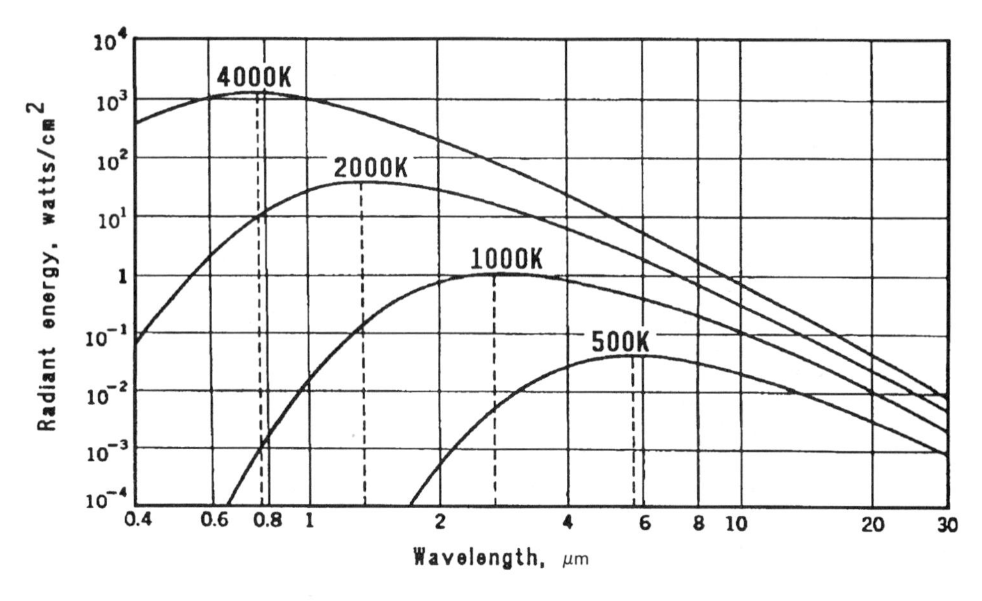

Figure 5. Radiant energy distribution curves for a blackbody source operating at different temperatures.

Most sources have a finite life span and at some point a replacement may be required. There are no hard and fast rules associated with the lifetime of a source, and there are no guarantees that a new source will last more than 6 months. Typical life spans should be 2-5 years, and in some isolated cases instruments have been used for 15-20 years without source replacement. The most common failure modes are a fracture, a hot spot, or loss of power caused by oxidation at the electrical connections. Symptoms of failure are either a loss of energy or an inconsistent distortion to the spectral background.

D. Detectors

There are two classes of detectors commonly used in infrared instrumentation: thermal detectors, and photon-sensitive or quantum detectors. Thermal detectors sense temperature changes by a change in some physical property, such as a dimensional change, the generation of a voltage, or a change in resistance. Photon-sensitive detectors operate by absorbing the energy of an incident photon, causing electrons in the detector material to become excited into a high-energy state. The speed of thermal detectors is influenced by the thermal or heat capacity of the detector element, and this results in such systems being slower in response when compared to photon-sensitive detectors. In the latter type the speed of movement of electrons in the detector material is significantly faster. The response of thermal detectors, however, tends to be relatively independent of the wavelength of the incident radiation, whereas photon-sensitive detectors are likely to have a limited operating range. The different types of detector within each class will be discussed in brief detail.

Thermal Detectors

Three different types of thermal detectors have been used, and at least two are still used, in infrared instruments—the thermocouple, the pneumatic or Golay detector, and bolometer detectors. The temperature changes detected by these systems are extremely small and they are generally of the order of 10^{-3} to 10^{-2} K. In all cases the radiation is modulated so that a "large" temperature change is monitored rather than a small percentage change on what is already a very small signal. The thermocouple and the Golay detectors have been used extensively in dispersive instruments. The Golay detector was also used in early, slow-scan, far-infrared interferometers. Pyroelectric bolometers are now used in both dispersive and FT-IR spectrometers.

A thermocouple is essentially a heat-sensitive junction formed when two dissimilar metals come into contact. When this junction is heated a voltage is produced that is proportional to the amount of heat energy falling on the detector. Detectors of this type are normally blackened to improve their response, which is more or less constant as a function of wavelength, except at longer wavelengths (below 400 cm^{-1}) where their efficiency drops off. The sensitivity of this detector increases inversely with the target area of the junction, which is therefore kept as small as possible. Smaller dimensions also reduce the total heat or thermal capacity of the detector reducing thermal lag and improving response. An increase in sensitivity is also achieved by housing the junction in a small evacuated enclosure, to reduce thermal noise from the environment. The radiation is focused through an infrared-transmitting window, usually made from cesium iodide or KRS-5 (thallium bromide/iodide eutectic), which is bonded with an adhesive to the enclosure to form a vacuum seal. The window is often made in the form of a lens to improve the focusing and to permit a further decrease in size of the junction. This lens/window is often mechanically the weakest part of the detector system. Detectors can fail by loss of vacuum either by leakage past the window, by outgasing of the adhesive, or by fogging of the lens (if made from the hygroscopic cesium iodide). In some modern instruments, produced in the past 10 years, the cesium iodide lens is protected against moisture with a thin layer of organic material.

Pneumatic or Golay detectors operate by the measurement of the thermal expansion of a nonabsorbing gas within a small, sealed, blackened chamber. The expansions or contractions of the gas within the chamber are detected by the movement of a flexible diaphragm, which is in turn linked to a movable mirror. Motions of the mirror are monitored by a light beam, which is directed to a photocell. These detectors were at one time more sensitive than thermocouples, but were more delicate and more susceptable to failure. They are, however, more responsive than thermocouples at longer wavelengths. It is worth noting that the photoacoustic detectors now used in FT-IR experiments are really a special case of the Golay cell, where the sample is actually placed within the detector. For a description of the photoacoustic effect, see Section V on sample handling, and Chapter 10.

Bolometers, which may be electrical conductors or semiconductors, provide a change in electrical resistance as a function of temperature. Today, pyroelectric bolometers in the form of DTGS (deuterated triglycine sulfate) detectors are the most commonly used. These are ferroelectric materials that show a strong temperature-electrical polarization dependency below their Curie point. A signal is produced at electrodes placed across the front surface of the material, as small polarization changes occur. In this way very small thermal changes can be detected. The response of this detector is inversely proportional to the modulation frequency of the incident radiation. The modulation rates over the normal wavelength range encountered in an interferometer scanning at 0.2 cm/sec are 1600 Hz at 4000 cm^{-1} and 160 Hz at 400 cm^{-1}. This implies that the response from a DTGS detector is noticeably less at 4000 cm^{-1} compared to 400 cm^{-1}. Also, if the moving mirror is slowed down by say a factor of 10, then a corresponding increase in performance can be obtained. One factor that must, however, be taken into account is that at these correspondingly lower modulation frequencies, interference from power-line frequencies (50 or 60 Hz, plus harmonics) can occur. Such interferences can be prevented by careful attention to the design of the electronics and by adequate vibration isolation to eliminate pickup from transformers, etc. The response of the DTGS detector is also temperature dependent, and its performance can be optimized by operating at just below the internal temperature of the spectrometer, regulated by a small thermo-electric device.

The thermal noise from DTGS detectors is very low, and most noise attributed to the detector usually originates from the amplifying electronics. DTGS detectors are used in modern dispersive instruments as well as FT-IR spectrometers. Higher scan speeds have been achieved with these detectors, relative to thermocouple detectors, in dispersive instruments. In these cases the modulation rates from the optical chopper are usually increased from 13-15 Hz for thermocouples to 50-60 Hz for the DTGS detector.

Another type of bolometer that is used for FT-IR measurements is the liquid helium-cooled doped germanium bolometer. These detectors are typically applied in the far-infrared region operating between 400 and 5 cm^{-1}. In theory they could be used in the mid-infrared, but thermal emission from surrounding surfaces constitutes a major interference in measurements with this detector and these must be eliminated.

Photon-Sensitive or Quantum Detectors

In a photon-sensitive detector the radiation is detected by the interaction of incident photons with a semiconductor device. If the energy of the photon is sufficiently high, electrons will be ejected from the surface of the material, and this forms the basis of operation of a conventional photocell or photomultiplier. In the infrared region the photons do not have sufficient energy to cause this effect. However, they have enough energy to raise an electron from a nonconducting energy state to a metastable conducting state. Therefore, under the right circumstances this class of material becomes photoconductive, assuming that the excited electrons can bridge the energy barrier between the two states. The height of this energy barrier increases with temperature, and therefore for these detectors to operate efficiently they must be cooled, sometimes down to liquid nitrogen temperatures. Materials commonly used for detectors in this class are silicon, lead sulfide, indium antimonide, and mercury cadmium telluride (MCT). The MCT detector is suitable for mid-infrared applications, and the other materials are used for detection in the near-infrared region.

Most photodetectors only operate over a limited range of frequencies (wavelengths). In the case of the MCT detector the operating range is dependent on the stoichiometry of mercury and cadmium in the telluride matrix. Variations in this composition give rise to detectors of different ranges, sometimes referred to as narrow-, medium-, and broadband detectors. The narrow-band detectors have the highest sensitivity but offer the narrowest operating range. The sacrifice is usually on the lower frequency end, often resulting in a cutoff (zero detection) around 700 cm^{-1}. The broadband detectors provide an operating range out to around 400 cm^{-1} but have about a fivefold reduction in detectivity. It is worth noting that some of the newer high sensitivity detectors do have an extended range of performance. The detector readily saturates at low modulation frequencies, and therefore fast scanning rates are used for optimum use of this detector. At high frequencies a plateau in the performance is achieved.

When operating near optimum conditions, the MCT detector offers significant gain in sensitivity over the DTGS pyroelectric detector. It is especially useful in low-energy applications such as microsampling with an infrared microscope, and for GC/IR applications where the high scan rates used for the detector are also important. Scan rates as high as 50 or 60 scans per second have been quoted for commercial instruments. Under these conditions the limiting step becomes the data transfer rate, that is, the time required to transfer data to a magnetic disk for storage. For routine sampling, however, the DTGS detector provides adequate sensitivity, it operates well over the full mid-infrared range, and it does not require the use of liquid nitrogen. For optimum performance from either detector it is desirable for the infrared image to fill the area of the detector. With normal macro sampling the detector element is typically 1 or 2 mm square. In the case of microsampling and GC/IR with small-bore lightpipes the best performance is achieved by matching the detector to the sample image. Therefore detectors as small as 250 or 100 μm square are used for these applications.

III. THE ELECTRONICS AND PRIMARY SIGNAL DATA HANDLING

For convenience, the discussion of the electronics will be separated into dispersive and FT-IR instrumentation, because the nature of the signal produced in the two instruments is different, even if the methods of detection are similar. The dispersive systems described will be restricted to double-beam systems because these cover the majority of instruments still in current use.

Two basic types of electronic systems have been used for dispersive instruments—optical null and ratio recording. The end result is essentially the same, double-beam, ratioed spectrum that is produced within the cycle time of the chopper system used on these instruments. Within a cycle of the chopper, the two beams, sample and reference, are sampled one or more times, and the light from each beam is passed along a common light path through the monochromator to the detector. In an optical null instrument the electronics is designed to ensure that the energy from each beam is maintained at the same level. Any imbalance between the beams is detected and a mechanical servo system is actuated to drive an optical wedge (or attenuator) into the reference beam channel to compensate optically for the imbalance. As the monochromator is sequentially scanned through the spectrum, sample absorptions result in an electrical imbalance of the detector signal between the two channels and this is immediately compensated by the optical wedge. In early instruments the optical wedge was mechanically coupled to the chart recorder pen. In this way, the spectrum was generated on the chart paper by the synchronized movement of the paper with the monochromator drive. This arrangement limited the possibility of ordinate (vertical) scale expansion, and this was restricted to fixed values by the use of gears on the wedge servo-recorder pen mechanism. On modern optical-null instruments this limitation is overcome by the use of a potentiometric recorder. With this arrangement, the wedge servo system drives a potentiometer, which in turn sends a voltage to the recorder, varying with the spectroscopic signal. Scale expansion is conveniently implemented by modifying the voltage range at the recorder. A general limitation experienced with optical null instruments is a loss of recorder response as band intensities approach zero transmittance. This is due to the fact that the optical wedge is fully extended in the reference beam so that there is zero signal from both sample and reference beams, resulting in a total lack of activity from the wedge servo system.

Ratio-recording instruments are now more popular, and these overcome the zero energy limitation indicated above. With ratio-recording electronics, the actual signals from the sample and reference channels are digitized by an analog-to-digital (A/D) converter. Once in digital form, the signals are numerically ratioed by a microprocessor to provide the double-beam spectrum on a point by point basis. The ratio is essentially performed on-the-fly as the monochromator is scanned. At fast scan speeds each point is individually processed. However, with slower scan speeds and high time constants, some instruments have sufficient time to produce a block average on each point. In this instance data are collected from several chopper cycles and a signal average is performed, over the period of the cycles, to help improve the signal-to-noise ratio of the data. One clear advantage of the ratio-recording method is that the reference channel always has full signal, unless a reference material is placed in the sample beam. In all cases the pen remains alive with a ratio-recording system, even when sample absorptions reach zero transmittance—this is sometimes referred to as a "live" zero. This retains an optimum pen response over the complete range of intensities, thereby ensuring an accurate representation of an absorption band profile.

Another feature adopted on some ratio-recording instruments is the technique of double chopping. This involves the use of a second chopper placed in front of the sample. The procedure is not unique to ratio-recording instruments, but it has been most successfully applied with this type of signal processing. In some systems the first chopper modulates the radiation falling on the sample at a different rate than the main chopping system. This provides a means for the electronics to differentiate infrared energy that originates from the source compared to energy that results by emission from the sample. An approach used in some instruments is to synchronize the two chopping systems and to generate four unique signals at the detector, corresponding to open and closed conditions for each beam. Hence a signal is obtained in each channel that is equivalent to the source switched on and switched off. In a simple system this can be accomplished by the use of a two-sector chopper in front of the sample and a four-sector chopper after the sample. The final signals for sample and reference channels are calculated by subtracting the signals obtained without source illumination from the signals produced with illumination. This effectively subtracts any signal that might originate from sample (or reference) emission.

Many modern dispersive instruments take full advantage of the use of an on-line microprocessor. In addition to the various signal processing manipulations described above, other numerical processes may be performed on-the-fly with the digital data that is produced, including "real-time" digital smoothing, baseline correction, and spectral subtraction. They also provide a real-time CRT display of the data, as well as a hard-copy output on a conventional chart recorder or digital plotter.

The signal produced on a dispersive instrument changes continually from an equivalent of 100% T to 0% T, and this is typically digitized with a 16-bit A/D converter. Note that on some older instruments 10- or 12-bit A/D converters were used. In a 16-bit system the signal is digitized to 1 part in 40,000 or better (depending on whether a bit is used for sign), and a zero offset is applied. For the interferometric measurement in an FT-IR spectrometer the situation is somewhat different because the signal varies dramatically from the high intensity of the center burst to the low intensity of most of the signal that extends from the wings of the center burst out to the extremes of the interferogram. This creates a dynamic range problem, since the noise in the interferogram must be adequately recorded from the bulk of the interferogram (2 bits required to describe the noise), while maintaining a faithful recording of the center burst. In the past, instrumental noise, mainly from the detector and the detector electronics, has been more or less adequately defined by the 16-bit A/D range. The current performance of detectors and detector electronics on modern instruments now demands that this range be extended. This is accomplished in two ways, either by using a wider A/D converter, such as an 18-bit device, or by using accurate gain switching. Some instruments use a combination of both approaches to produce the optimum performance for the detector systems. In gain switching, the gain is electronically switched at well defined points along the interferogram, around the region of the center burst. This is usually switched as binary values of the gain, that is, with 2×, 4×, 8×, etc. changes in gain level. The accuracy of this switching has to be critically controlled, so that no discontinuities will result in the final recorded interferogram. Failure to ensure this results in artifacts in the background of the transformed spectrum.

The digitized interferogram is stored in a special numerical register (accumulator) in the computer system of the interferometer. The digital precision of these registers is usually significantly higher than the primary digital data—typically 40, 64, or 80 bits wide (possibly more). The actual precision used is dependent on the instrument manufacturer, the type of computer used, and the resolution of the spectral data. In the past, working at high numerical precision provided certain overhead limitations in terms of speed and available space for storage. This is no longer a problem for modern computer systems. The accumulated data is eventually rescaled and passed on for further processing, such as phase correction, and possibly zero filling, leading to the final Fourier transformation into the normal spectral format. A single-beam spectrum is produced because the interferometer instruments are single-channel systems. The conventional double-beam format spectrum is generated by calculating the ratio of the sample spectrum to an independently generated background spectrum. It is normally the transformed spectrum that is ratioed, and not the interferogram. Many instruments provide a visual output on a CRT monitor while data is being collected. This is either the accumulated interferogram, or, on some systems, a transformed spectrum displayed in pseudo-real time.

IV. COMMERCIAL INSTRUMENTS AND ACCESSORIES

It is quite possible that a reader might be in the situation where he or she either inherits or is forced to use an older spectrometer. In order to remove any confusion and to appreciate some of the origins of the earlier infrared spectrometers, it is worthwhile to have a short discussion about commercial instrumentation. This section is intended to provide the reader with an insight into both the currently available IR instruments and any equipment from the past 10-20 years that is still potentially in active use. One unique feature of infrared instrumentation is that while it may become obsolete in terms of features, it can still provide usable information.

The first commercial instruments were developed out of collaborations with industry. The most notable of these were the Cyanamid/Perkin-Elmer and the Shell/Beckman liaisons in the development of the models 12 and IR-1 and -2 single-beam spectrometers, respectively. Developments from these early models led to double-beam instruments, such as the Perkin-Elmer model 21, which was one of the products that was responsible for the initial acceptance of infrared spectroscopy as a routine analytical technique. However, the first instrument to make a real impact on the analytical market was the Perkin-Elmer Infracord (model 137). This became a workhorse throughout industry, and its simplicity of operation helped to popularize infrared spectroscopy. It enjoyed a lifespan of 12 years, and even today, nearly 30 years after its introduction, some are still in use. Derivatives were made from this product based on its optical design concepts, leading to a host of new dispersive instruments that were introduced up until the early 1980s. Likewise, based on their early designs, companies such as Beckman, Hilger, and Pye-Unicam also produced many derivative products in the dispersive instrument market. Beckman developed an extensive series of double-beam instruments, many of them research quality, with an IRx designation. These continued to be produced until the late 1960s, early 1970s when a range of smaller instruments known as the Acculab series was introduced. The early Perkin-Elmer instruments featured drum-type recorders. The company's initial move away from this configuration was with the model 457, which featured a flat bed recorder and was also the first product to be developed at the firm's British factory. Drum recorders were still featured on the high performance x21 and x25 Perkin-Elmer series instruments until the early 1970s. The first products manufactured by Pye-Unicam (later to become a division of Philips) were the models SP100 and SP200 prism instruments, which were marketed in the late 1950s, early 1960s, mainly as competitive products to the Infracord. Grating versions of these instruments were developed and sold throughout the 1960s.

Interferometric instruments were developed at an early stage for far-infrared analysis, and a popular system was offered by Beckman-RIIC. Early versions of this instrument did not include an on-line computer, and instead the interferometric data was transferred to paper tape, which in turn was fed off-line into a mainframe computer for final processing. During the middle and late 1960s Block Engineering introduced two small, low-resolution interferometers, the models 200 and 600. These featured a data system that consisted of a coadder for the data collection, and a wave analyzer that performed an analog Fourier transform. The most significant breakthrough for analytical FT-IR came in 1969-1970 with the introduction of the FTS-14 by Block. The key feature of this product was the inclusion of the fast Fourier transform running on a dedicated, on-line minicomputer, a Data General Nova. It is interesting to note the configuration of one of these early instruments as offered in 1971: a 0.4-cm^{-1} resolution spectrometer, featuring a 4K (core memory) computer and a 128K fixed head disk drive, selling for just over $70,000. Two other products were soon introduced by Digilab (Block Engineering), the FTS-16 (far-IR) and the FTS-20. In 1973, EOCOM developed the "Black Beauty" interferometer, which became the basic building block of the Nicolet 7000 series FT-IR instruments. A few years later, in 1976, Nicolet introduced the model 7199, which was a fully integrated system including software running on the Nicolet 1180 computer. Both the early FTS products from Digilab and the 7199 became successful instruments in the research laboratory. It was not until 1979, however, with the introduction of the Nicolet MX-1, that FT-IR became an accepted technique for routine analysis. Two important features of the MX-1 caused this to happen: it was the first mid-priced FT-IR, selling around $40,000, and it featured a push-button control pad, somewhat analogous to the keyboard control panel of dispersive instruments of that time. This move away from the "computer" and "research" image of FT-IR was important in its acceptance. Most of the early commercialization of analytical FT-IR instrumentation originated in the United States, but it is worth noting that during the mid-1970s, JEOL introduced the JIR-03F, a 0.5-cm^{-1} resolution instrument, which was successful in the domestic Japanese market. Also, in West Germany, Bruker became a successful manufacturer. The early products were based on the EOCOM optics and featured a Nicolet 1080 data system. Later the company developed its own Aspect computer with a similar architecture to the EOCOM. Users of early FT-IR instruments would have noted some similarity in the appearance of software on the Digilab, Nicolet, and Bruker systems, in the use of three-letter nmemonics—partly due to a common engineering theme in the software development.

At the time of the introduction of the first FT-IR spectrometers there were many promises made about the performance of these systems. Much was made of their theoretical advantages compared to dispersive instruments. Many of the comparisons were, however, made with relatively early dispersive spectrometers, and by the mid 1970s, it was clear that the FT-IR manufacturers did not appreciate the state of the art in dispersive instrumentation. First, the early FT-IR instruments did not really meet all the performance characteristics that were claimed, and second, many of the so-called weaknesses of the dispersive instruments had been rectified on instruments that were current at that time. These comments are made in regard to the analytical user rather than the research spectroscopist, because from a research point of view the FT-IR instrumentation did provide advantages. However, one of the most important factors concerning the introduction of the FT-IR spectrometers was the impact of the dedicated computer. This caused instrument manufacturers, notably Beckman (United States), Perkin-Elmer (United States/UK), and Pye-Unicam (UK), to redesign dispersive instruments with built-in digital electronics and microprocessor control. This was to provide the necessary interfacing for a dedicated computer system. The first instrument to feature digital electronics was the Perkin-Elmer model 180, which was also offered with an on-line minicomputer option (Interdata 6/16). During the mid to late 1970s several mid-priced dispersive instruments were produced with built-in computer interfaces. Examples of these products, which were also marketed with optional data systems (noted in parentheses), were the Beckman 4200 series (DEC MINC), the Perkin-Elmer 283 and 580 (Interdata 6/16), and the Pye-Unicam SP2000/SP4000 and SP3 series (Hewlett Packard desktop system). Of these instruments, the model 283 must be given special reference because it was the first infrared product to incorporate an on-board microprocessor. Late in the 1970s, Beckman introduced the Microlab 600 and later, the 620. These were extensions of their earlier Acculab instruments and featured extensive built-in data processing, including some form of spectral searching and multicomponent quantitative analysis. The unique feature of these systems was that the software was loaded via plug-in ROM (read-only memory) cartridges. By 1980, most dispersive instruments were capable of providing the computer data processing facilities offered on FT-IR instrumentation. Probably the most significant data system at that time was the Perkin-Elmer Infrared Data Station (3500/3600)—a system that possibly outsold most other infrared data systems during that era, at least up to 1985.

During the latter part of the 1970s and the start of the 1980s new lower-cost FT-IR spectrometers were introduced, and these started to provide some of the performance that was promised, for the analytical chemist, on early instruments. This resulted in an increase in FT-IR instrument sales. Several new, and some traditional, infrared instrument companies entered the market to meet this demand—some of these being

Analect, Beckman, Bomem (Canada), IBM Instruments, Jeol (Japan), Mattson, and Perkin-Elmer. In the case of IBM and Perkin-Elmer, both companies entered the market with repackaged products manufactured by other companies—IBM offered a range of Bruker products and Perkin-Elmer produced the 1500, a version of one of the Analect refractively scanned spectrometers. After 2 years, Perkin-Elmer introduced their own products the 1700 Series (UK) and the 1800 (United States).

The list of companies involved with FT-IR is now different and there have been some realignments in the marketplace. Beckman and IBM instruments no longer manufacture or market FT-IR spectrometers. Some of the low-cost IBM technology was purchased by Nicolet early in 1987. This provided Nicolet with their first IBM Personal Computer (AT) based instrument, the model 5-PC. Some small companies have attempted to enter the low-cost, routine analytical market, such as Lloyd, Midac, and Janos, but they have made little impact at this point. In Japan, companies such as JEOL, JASCO, and Shimadzu have well-established domestic FT-IR markets but their products have not been fully introduced into Western markets at this time (1987). Pye-Unicam, in Europe, has entered the FT-IR market with a spectrometer manufactured by Nicolet. Also in Europe, Bruker has maintained a reasonable market presence in certain of the European Economic Community (EEC) countries with a broad-based product range. Dispersive instruments are still marketed by Perkin-Elmer, Pye-Unicam, and some of the Japanese manufacturers. It is unlikely that this market will completely disappear in the forseeable future. However, the market for these products will probably become more restricted as lower priced FT-IR products are purchased as replacement units.

Most of the emphasis of this chapter is on instrumentation used in the analytical and research laboratories. Infrared spectroscopy has been, for many years, an important technique for process and quality control analysis. In a quality control environment the instrument requirements are not too dissimilar from those of the routine analytical laboratory. However, there are higher demands on the reliability and ruggedness of the instrument hardware, and in addition, the instrument, software, and sampling equipment must be usable by non-technically qualified operators. One of the first companies to focus on these specific requirements was Wilks Scientific (now Foxboro) with their Miran series of instruments, especially the models 80 and 980 microprocessor-controlled instruments. In recent years some traditional instrument and accessory manufacturers, such as Analect, Bomem, Perkin-Elmer, Nicolet, and Spectra-Tech, have produced products focused on this market. Near-infrared instruments produced by Dickey-John, Bran & Luebe (Technicon), Pacific Scientific, Guided Wave, and LT Industries are also being successfully applied in this area by the application of statistical and correlation methods of quantitative analysis. Advantages of the near-IR products are in the simplicity of optical design and sample handling, plus the intrinsically high sensitivity of near-IR relative to mid-IR detection. The major disadvantage is in the lack of specificity that is provided by the mid-IR measurement. Both of these areas of infrared measurement are expected to grow in the quality control application in the next decade.

The requirements for process analysis, especially processes that require direct on-line coupling, are very different from those of laboratory-based instruments. First, the spectrometer must be rugged and housed in a waterproofed and sealed enclosure. It must conform with established electrical safety and hazard codes, and must be capable of operating in an alien environment—in terms of temperature, atmosphere, humidity, and mechanical vibration. Infrared is attractive for this type of analysis because it is nondestructive and it is sensitive to changes in composition. For many years, companies, and not necessarily traditional instrument companies, have incorporated infrared sensing devices (near-IR and mid-IR) in manufacturing lines for monitoring parameters such as moisture and thickness. Some dedicated infrared instruments have been produced by companies such as Foxboro and GAC (General Analytical Corporation), which monitor specific chemical components or functionalities. These are filter-based instruments where predefined optical filters are used to provide high-precision recording at specific wavelengths. The GAC LAN-II system is a typical example where two frequencies are used for the continuous monitoring of carbon dioxide and sugar, or artificial sweetener, in carbonated soft drinks. FT-IR spectrometers are starting to be used for process applications, and products manufactured by Analect, Bomem, and Nicolet do conform to the basic requirements demanded by a process environment. One of the advantages of the FT-IR approach is that multiwavelength determinations are possible, enabling a relatively large number of components to be monitored simultaneously. However, the simplicity, reliability, and sensitivity of filter-based instruments may outweigh the flexibility features of FT-IR systems for some manufacturing processes.

A. Display Formats

All modern spectrometers produce spectra with either wavelength or wavenumber units presented on a descending energy scale. The two spectra in Figures 6a and 6b help illustrate the visual differences in these two scales. A common scale convention on most commercial dispersive spectrometers features a 2:1 scale compression centered around the 2000 cm^{-1} point: that is, data above 2000 cm^{-1} are presented with half the scale of data below that point. Some modern FT-IR spectrometers offer this feature as an optional plotting format for visual comparison with archived dispersive data.

In a single-beam mode of operation the vertical scale is a profile of the energy from the source that is modified by the spectral contributions from the sample plus the characteristic background absorptions from components of the spectrometer, the response characteristics of the detector, atmospheric water vapor, and carbon dioxide. Data are seldom used in this format (Figure 7), but in very early instruments this was the only mode of operation.

Although the percent transmittance format is a measure of the energy transmitted through the sample relative to the original energy incident on the sample, it does not provide a direct measure of the absorption of energy by the sample. This quantity, known as *absorbance*, is determined from the logarithm (base 10) of (1/T) $\log_{10}$ (I_0/I). By convention,

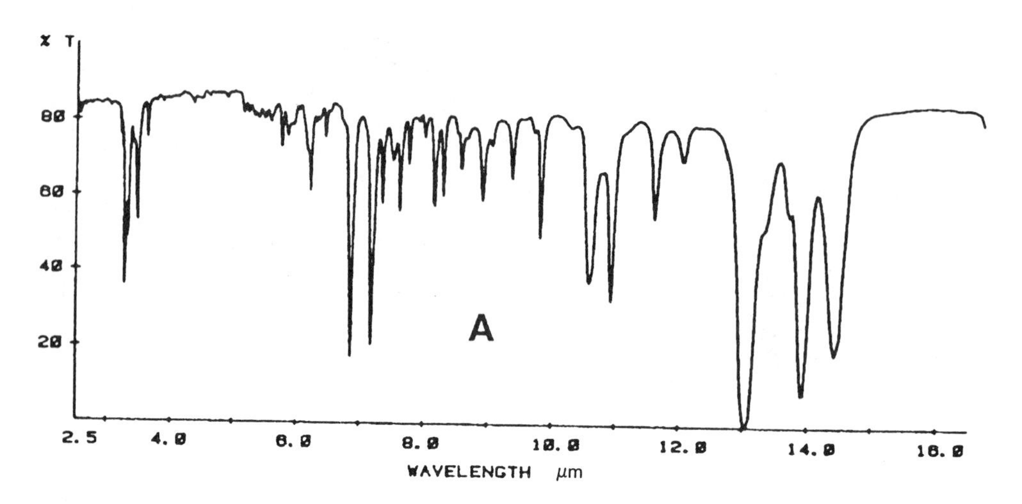

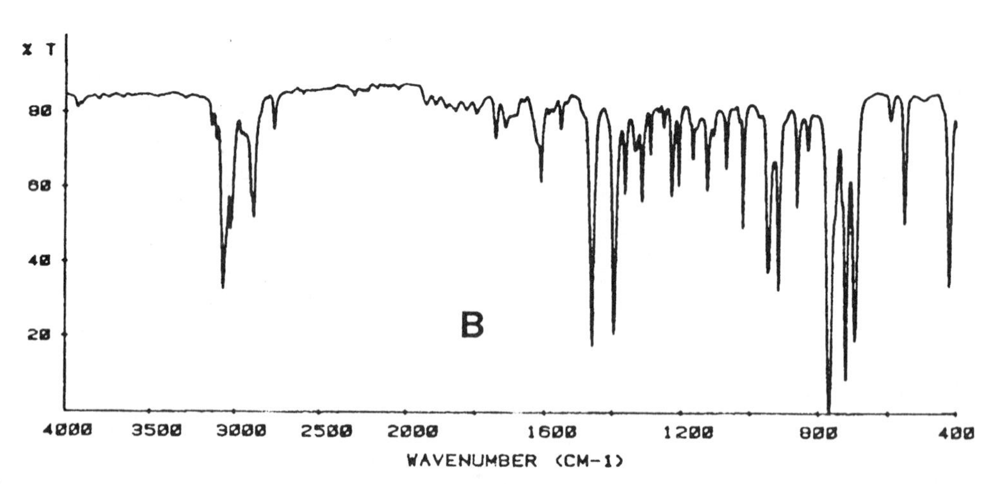

Figure 6. The infrared spectrum of indene (0.015 mm pathlength), (A) plotted in linear wavelength format and (B) plotted in linear wavenumber, with a 2:1 scale change at 2000 cm^{-1} (5 μm).

infrared spectra in this format are inverted with zero absorbance at the base of the axis and increasing absorbance presented up the scale, as illustrated in Figure 8.

On early instruments, the format was often fixed and defined by the primary method of measurement. All three formats shown on Figures 6a, 6b, and 8 are, however, often available on modern spectrometer systems. This facility is provided by computer-controlled output of the spectral data to a digital plotting device. This type of device provides additional flexibility in the form of scale expansion, because the spectral data may be fitted to a user-defined window over the entire spectral range or over

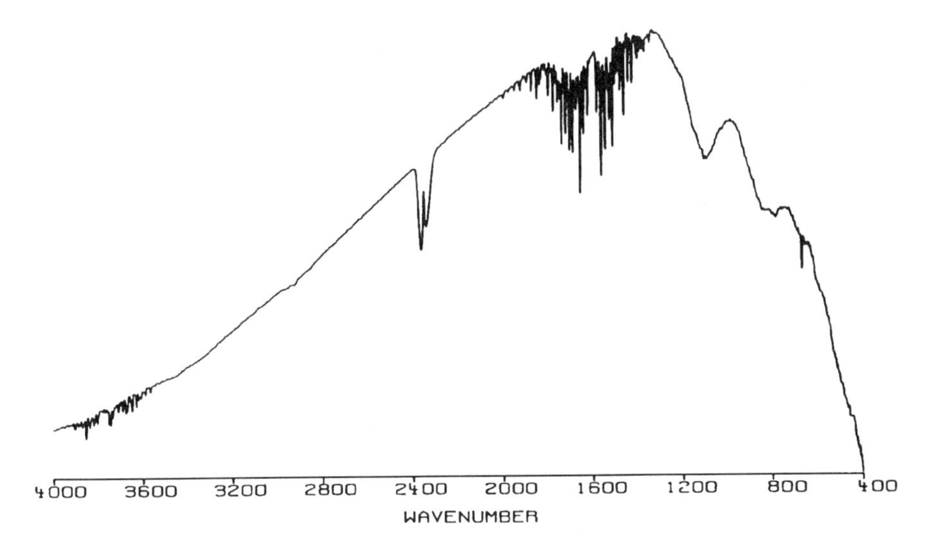

Figure 7. A typical single-beam background spectrum from an FT-IR spectrometer. Note that the profile of the background is a function of the source, the beam splitter, and the detector.

a selected portion of the data, for both axes. The choice of final format is at the preference of the user. This author is a traditional infrared spectroscopist, and prefers the percent transmittance (%T) format with the 2:1 scale compression for all qualitative evaluations—including spectral interpretation, identifying an unknown, and general problem solving. This 2:1 scaling provides the emphasis in the most important part of the spectrum, the fingerprint region, and the logarithmic output of the %T format emphasizes the weak spectral features and deemphasizes the stronger ones. Also, a trend with modern spectrometer systems is to produce the spectral output on plain paper, which is quite unsuitable for good spectral interpretation. Therefore, whenever practical, a gridded form of output is preferred for any spectral data that is to be subsequently used for sample identification.

For applications that require quantitative evaluations, whether by calculation or by visual comparison, the absorbance form, with either 1:1 or 2:1 scaling, is preferred. These may be personal preferences, but they are based on many years of practical experience with analytical infrared spectroscopy.

V. SAMPLE HANDLING

A. The Sample

Too often the sample is the last item considered in either the design or the performance of a spectrometer, yet it is undoubtedly the most important and most critical element. However, sample compartments of many instruments, especially those of older design, place a major physical constraint on the size of sample and sampling accessories that can be accommodated. Newer instruments, especially FT-IR spectrometers, now offer not only large sample compartments, but also the facility for additional sampling areas via the use of external beam options. These provide for the addition of one or more auxiliary sample compartments or other large external accessories, such as infrared microscopes, and interfaces to chromatographic and other hyphenated techniques.

Early dispersive infrared instruments had a single-beam design, and therefore only a single sampling position was required. The limitations of a single-beam instrument, in the absence of any data storage or data handling capabilities, were soon realized and double-beam systems were adopted. As an aside, it is worth noting that a magnetic recording system was used on a very early Beckman instrument in an attempt to perform a pseudo-real-time cancellation of the instrumental background. This proved to be unsuccessful because of the inability to obtain adequate synchronization of the two scans. In all later dispersive instruments the cancellation was obtained in virtually real time by alternate sampling of the analytical and reference beams, both directed through the sample compartment. For convenience, the sample beam is usually situated in the front part of the sample compartment, and the reference in the back. Most instruments provide a sample (and reference) slide mount where the sample, either in a cell or mounted in a card or other accessory, is held in place.

Another issue that influences both the design of accessories and the space potentially available for samples is how the infrared beam enters the sample compartment. In most spectrometers the beam is brought to a focus. The nature of this focus is strongly dependent on the type and design of the spectrometer. For a dispersive instrument this is usually an image of the slits and has a rectangular cross section. Typical dimensions of the image at the focus are about 3-4 mm by 10-12 mm. In an FT-IR spectrometer the image is essentially circular, although in practice it may be somewhat elliptical in shape. The dimensions can range from as small as 3-4 mm to 12-14 mm in nominal diameter. It is normally an image of the system aperture; when a smaller image is experienced this is indicative of a spectrometer that is specially designed for high spectral resolution. This image will change size if the size of aperture is changed in spectrometers that provide multiple apertures for throughput matching at different spectral resolutions. One notable exception is the Perkin-Elmer model 1800, where the image at the normal sampling position is a pupil image of the beam splitter—this image does not change in size when the system aperture is varied.

The nature of the sample and its physical and optical geometry are rarely considered. Similarly, many users seldom consider the interaction between the infrared beam of the spectrometer and a sampling accessory. Both samples and sampling accessories can significantly modify the shape and focus of the infrared beam, generating a distorted image. If a sampling accessory is designed correctly then distortions are usually minimized. Misuse or misalignment, however, will exaggerate any distortions irrespective of the original design considerations. It is essential that a user study both the instrument manufacturer's recommendations and the instructions of the accessory manufacturer before commencing a new sample handling technique. The sample itself, independent of a sampling accessory, must be considered to be an optical component of the spectrometer, and therefore care must be taken to ensure that the sample is correctly mounted in the sample slide with a suitable holder or cell. Lack of attention to the precautions mentioned above will produce several negative effects. If nothing else, the most significant result may be a loss in energy, which could be the deciding factor whether an experiment is a success or a failure. The worst case, however, may be that even though adequate energy is transmitted or transferred from the sample and the accessory, an undefined distortion nevertheless exists in the spectral information. This situation can and does lead to photometric errors, band distortions, or frequency shifts; one or more of these effects can occur with a given measurement if care is not exercised.

One additional sample-related factor that is not always considered is where the sample is placed relative to the analyzer, irrespective of whether the latter is a monochromator or an interferometer. Almost invariably in dispersive instruments the sample compartment, and hence the sample, is placed *before* the monochromator. However, nearly all FT-IR spectrometers, with only a few exceptions, have the sample compartment *after* the modulator, just in front of the detector. Although in the past a few dispersive (monochromator-based) instruments have provided access to a point just in front of the detector, none are used routinely in this mode. The reason for this is that the amount of nearly monochromatic light that exits the slits of a monochromator at any particular point of the spectrum is extremely low. Passage through a sample at this point would lead to significant energy losses and optical aberrations, and stray light would have exaggerated effects at these low light levels. In FT-IR instruments the sample is illuminated by all of the energy that emerges from the interferometer, and this contains light modulated by all wavelengths. Therefore, sample-related perturbations are relatively low compared to the signal level. Sample placement in these cases can have a noticeable influence on the quality and photometric precision of the final spectral data. Instruments that have the sample between the source and the analyzer allow the sample to become heated by the radiation from the source. One obvious factor to consider is that the sample may be thermally unstable and may decompose. Or, if it is a low-melting solid or a low-boiling liquid it may change state. Both of these are clearly undesirable situations. A less obvious problem is that the sample itself becomes a source (sample temperatures as high as 60°C have been measured) and reradiates infrared energy as an emission spectrum of the sample. Unless the instrument is specifically designed to handle this problem by modulating the energy source in some fashion, the effect will cause significant photometric errors throughout the spectrum, especially in regions of high sample absorption, where emissions from the sample are usually greatest. This phenomenon is experienced with dispersive instruments that do not have presample modulation, a feature that is commonly referred to as presample chopping or double chopping. The majority of FT-IR instruments, almost by definition, have presample modulation. Sample heating may cause physical or chemical modifications of the sample, or, if extreme, its effects can reach the detector causing

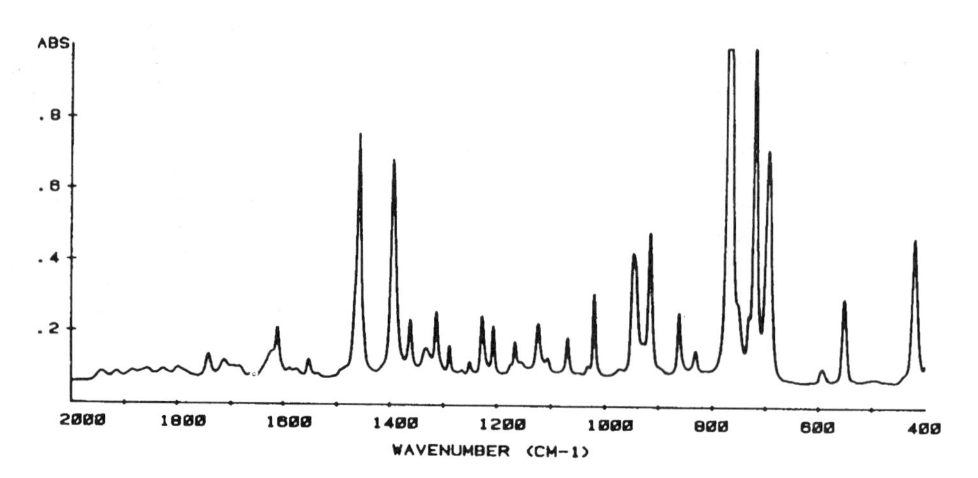

Figure 8. Partial plot of indene spectrum (0.015 mm pathlength) with a limited wavenumber range and with an absorbance format.

a loss of sensitivity from the production of a large DC offset. The effect in FT-IR instruments is not modulated and therefore can be discriminated from the interferometric signal. One example of sample emission causing sensitivity loss is in gas chromatographic-based applications (GC-IR), where emissions from the lightpipe (gas cell) may be high because of the 200-300°C operating temperature. In newer GC-IR interfaces manufacturers have taken steps to address this problem.

Sample emission can, as indicated above, cause practical problems in spectroscopic measurements. On occasions, however, an emission spectrum is required, and this may be obtained on either a dispersive or an FT-IR instrument. Many dispersive instruments, however, are not designed to make this measurement because the experiment requires only a single-beam configuration and these instruments are usually set up for double-beam operation. Some ratio-recording instruments introduced in the past 8-9 years have been adapted to make emission measurements. Not all FT-IR instruments are able to obtain the emission spectrum either, because it is necessary to place the sample at an equivalent position to the source. This necessitates beam switching of the source optics, a feature that is available on high quality analytical or research-grade instruments.

B. Sample Measurement

This section is only intended to provide an overview of sampling methods and does not include detailed descriptions of individual procedures. The reader is directed either to consult a text on practical infrared spectroscopy, or to view one or more of the audiovisual programs that are available on the subject (see the list of recommended reference material at the end of this chapter).

Sampling may be reduced to two principal approaches based on the size of the sample—macro, or bulk, sampling, and micro sampling. The underlying methods that are used for both are the same, and therefore the main part of this section will be devoted to the actual methods of measurement. Microsampling will be considered as a special case and will be discussed briefly in a separate section. In common with most forms of optical spectroscopy, the traditional method for sample measurement is via transmission through the sample, which can post certain problems depending on the physical state of the sample. All three normal physical states may be measured, but in certain cases physical size, shape, consistency, or optical-physical properties can influence the quality of the final spectrum. The selection of the correct sample preparation and handling method is therefore very important to ensure accurate collection of the data, free from artifacts or abberations. Most infrared transmission measurements require the sample to be either held in a cell fitted with infrared-transmitting windows, or supported in an infrared-transmitting medium. Exceptions are rigid samples, such as certain polymeric materials, that can be prepared as a self-supporting film. Table 1 provides a summary of common infrared-transmitting materials that are available for cell windows or as dispersing media. The two most common window-materials are sodium chloride, NaCl, and potassium bromide, KBr.

In many cases it is either impossible or inconvenient to analyze a sample by transmission, and therefore alternate methods utilizing reflection must be used. These techniques, based on front surface reflection (known as specular or external reflectance), bulk reflection (diffuse reflectance), or total internal reflection (ATR, attenuated total reflectance), are gaining popularity and are becoming accepted practices. One of the main attractions of reflection methods is that they typically require little or no sample preparation, and often provide a means for nondestructive testing.

Gases and vapors are relatively straightforward to sample, and the selection of the optimum cell is based on the anticipated concentration and pressure (or partial pressure) of the analyte. At atmospheric pressure and with moderate to high component concentration, relatively short-pathlength cells, typically 5 or 10 cm, are suitable. These provide adequate spectral intensities for systems containing down to fractional percentages for certain gases. In the most common form the cells consist of a glass body with removable IR-transmitting windows (usually NaCl or KBr). For some special applications, for example, with corrosive gases or vapors, metals such as stainless steel, nickel, or monel are used for cell bodies. Surface coatings of gold or Teflon have also been used to reduce surface activity and corrosion in cells. Measurements at the part per million (ppm) level, for trace analysis and environmental applications, are performed with cells with folded pathlengths ranging from 1 to 20, or even 40 m. The use of longer cells, even with effective pathlengths as long as 1 km, have been reported.

Condensed-phase materials may be studied in the vapor state with any of the cells described above when suitably equipped with a heating jacket. A special case of vapor-phase analysis is the on-line examination of gas-chromatographic fractions. The technique, commonly referred to as GC/IR, makes use of a narrow-bore lightpipe as the gas transmission

Table 1. Infrared-Transmissive Materials[a]

Material	Useful range (cm^{-1})	Refractive index at 1000 cm^{-1}	Water solubility (g/100, H_2O)
Sodium chloride, NaCl	40,000-590	1.49	35.7
Potassium bromide, KBr	40,000-340	1.52	65.2
Cesium iodide, CsI	40,000-200	1.74	88.4
Calcium fluoride, CaF_2	50,000-1140	1.39	Insoluble
Barium fluoride, BaF_2	50,000-840	1.42	Insoluble
Silver bromide, AgBr	20,000-300	2.2	Insoluble
Zinc sulfide, ZnS[b]	17,000-833	2.2	Insoluble
Zinc selenide, ZnSe[c]	20,000-454	2.4	Insoluble
Cadmium telluride, CdTe	20,000-320	2.67	Insoluble
AMTIR[d]	11,000-625	2.5	Insoluble
KRS-5[e]	20,000-250	2.37	0.05
Germanium, Ge	5,500-600	4.0	Insoluble
Silicon, Si	8,300-660	3.4	Insoluble
Sapphire	55,000-1780	1.74	Insoluble

[a]The wavenumber ranges quoted above refer to the useful transmission ranges. Shorter ranges are experienced when the materials are used as ATR crystals (elements) because of the increased pathlength.
[b]Also available in forms known as Cleartran and Irtran 2 (tradename of Eastman Kodak).
[c]Also available under the tradename Irtran 4 (Eastman Kodak).
[d]An IR glass made from germanium, arsenic, and selenium.
[e]Eutectic mixture of thalium bromide and iodide, Tl(BrI).

cell. Most GC/IR cells are constructed from glass capillary tubing, often as narrow as 1 mm in diameter, with the internal bore coated with gold. This provides a high internal reflectivity and ensures that the hot fractions, which are often maintained at 200°C or higher, are in contact with an inert surface.

Liquids can be sampled in a number of different ways depending on the sample characteristics, such as volatility, composition, and corrosivity (toward the cell materials), and the overall absorptivity. One of the simplest and most popular methods is to sandwich the liquid between two infrared windows, typically made from NaCl or KBr, thus producing a capillary film. This is certainly a rapid procedure, but it is inappropriate when precise measurements are required, for several reasons: it is prone to sample loss by evaporation, small air or vapor pockets can form producing an equivalent to stray light during measurement, and the cell is dimensionally unstable and has a nonreproducible pathlength. These problems are partially overcome by the use of a sealed, fixed-pathlength cell. Cells of this design are notoriously difficult to clean because of the relatively short pathlengths, which typically are 0.015-0.2 mm for "neat" (undiluted) materials. Sample absorptivity and viscosity are key issues in this matter. It is sometimes necessary to dilute the sample in a solvent to improve the measurement of moderate to intense absorptions at longer than ideal pathlengths and to reduce cell cleaning problems. This approach is limited in practice because of the unavailability of good solvents that do not themselves absorb infrared radiation. In recent years problems in liquid sampling have been reduced by the use of ATR or internal reflectance liquid cells, especially the cylindrical internal reflectance (CIRCLE) cell. This approach has improved the precision and eliminated the need to analyze samples in solution. It has also provided a method for studying aqueous-based systems, which tend to be difficult to handle on a routine basis in conventional transmission cells.

Solid samples pose a unique set of problems, and the final choice of sampling method is dependent on the physical nature of the sample, for example, whether it is a powder, an extended rigid solid, or a flexible material such as a plastic or rubber. Many solids either exist in powdered form or can be reduced to a powder by grinding. One physical limitation imposed by a powdered material, if examined by a transmission method, is the need to reduce the particle size to below the wavelength of the infrared radiation. Failure to do so results in excessive optical scattering with a resultant large loss in transmitted radiation. This is usually accompanied by a loss of spectral contrast, caused by inefficient energy interchange with the incident photons, and band shape distortions that result from a phenomenon known as the Christiansen effect. To help reduce energy loss and band distortions due to scattering, samples studied by transmission methods are normally dispersed in either an infrared transmitting solid medium, such as KBr, or an oil medium, such as a liquid hydrocarbon (Nujol) or a perhalogenated hydrocarbon. Samples prepared in admixture with KBr powder are compressed into self-supporting pellets in a high-pressure die under pressures of 10 tons or more, for a nominal 13-mm-diameter pellet. In the oil-based method or mull technique, the powdered sample is mixed with a small quantity of the oil to produce a paste with the consistency of petrolatum jelly (Vaseline). This is then squeezed between two IR-transmitting windows until acceptable band intensities are obtained for the sample. Both are accepted as traditional methods of solid sample preparation, but neither can be considered as ideal.

In recent years there has been a move away from these traditional methods for a number of reasons. For many people they are tedious and time-consuming, as well as being difficult to control in terms of both quality and quantitative integrity of the spectrum. Modern spectrometers have provided the opportunity to develop alternate methods of solid sample preparation. These include diffuse reflectance and photoacoustic measurement, as well as extensions to existing methods, such as internal reflectance.

Diffuse reflectance has become a popular alternative to the compressed KBr pellet technique for powdered samples. Powders may be examined either directly or in admixture with KBr or a similar dispersing medium. Particle size and sample homogeneity are probably the two major issues in the success of the application of this method. From a practical point of view the measurement involves focusing the infrared beam on the surface of the powdered specimen and collecting the diffusely scattered radiation with large curved mirrors placed around the surface. There are several commercial accessories available, and the major differences are in the design philosophies for the selective rejection of unwanted or spurious reflections. The presence of such reflections can cause anomalous band shapes and distortions, such as Reststrahlen bands—which are often encountered for intense absorption bands and are observed either as a strong negative lobe on the side of a peak or as an inversion at the apex of the peak. The user must be aware of such effects because if unaccounted for, they can give the effect of producing a spectrum that at first sight might appear to be acceptable, but with greater band structure than anticipated. In early applications of diffuse reflectance many spectral misinterpretations resulted from attempts to assign extra band structure caused by these artifacts. Anomalies of this type are most often observed with strong infrared absorbers, such as inorganic compounds and certain organics. With careful attention to sample preparation, and especially with the option of diluting the sample in a scattering, nonabsorbing matrix, the effects can be accounted for, and reduced or even eliminated.

Diffuse reflectance is not limited to the measurement of powders, and it can be applied to the analysis of extended solid surfaces, such as fibers and fabrics. Modifications to the sampling procedure are often necessary to accommodate these sample types. One novel approach to handling "nonideal" or intractable samples is to transfer a small quantity of the sample by abrasion to the surface of a fine-grade (240-400 grit) silicon carbide abrasive paper. The silicon carbide itself produces a relatively weak reflection spectrum that can be compensated for by ratioing the acquired sample spectrum to that of an unused piece of the abrasive paper.

"Nonideal" solids, including materials such as highly carbon-filled polymers, can be measured by photoacoustic spectroscopy (PAS). The measurement takes advantage of the modulated nature of the infrared beam of an FT-IR spectrometer. The sample is maintained in a small sealed chamber under a dry gas atmosphere, typically nitrogen, argon, or helium. The interaction of the modulated infrared radiation with the sample generates a modulated heat wave within the sample at the frequencies where the sample absorbs. The modulated wave is transferred to the gas atmosphere above the surface of the sample, which in turn interacts with the gas molecules and causes modulated expansions and contractions within the gas. These gas vibrations are in the audible frequency range and are in turn detected by an integral high-sensitivity microphone. This method of measurement is unique because it is the only method that provides a direct measure of infrared absorption by the sample. Photoacoustic detection is often described as the method of last resort because it has certain idiosyncrasies and it is limited to small samples. Problems with the technique essentially arise from the mode of measurement for the following reasons: (1) It is essential that the gas above the sample is dry, because water vapor, being a strong IR absorber, will dominate the spectrum (note that many powders will outgas surface-adsorbed moisture). (2) The magnitude of the signal is sample dependent because it is related to the thermal conductivity of the material and thermal transfer from the surface to the gas. (3) The sampling device must be vibrationally decoupled from the instrument and acoustically isolated from the operating environment—any pickup will result in signatures or noise spikes in the spectrum. In spite of these potential problems, PAS does offer a practical solution for certain, otherwise hard to handle, solid samples. It is discussed at greater depth in a separate chapter of this handbook.

Internal reflectance has long been accepted as a valuable technique for the study of extended solids, especially in sheet form. It is an accepted method for sampling polymeric films and surface coatings. The measurement involves the use of an optical element (sometimes referred to as the crystal, although it is not necessarily a crystal in the classical sense) made of a high-refractive-index, infrared-transmitting material. Examples of suitable materials are KRS-5 (a thalium bromide/iodide eutectic), zinc sulfide (IRTRAN 2) or selenide, or germanium. Samples are brought into intimate contact with one or two parallel faces of the crystal and infrared radiation is internally reflected from these faces throughout the length of the crystal as shown in Figure 9. Internal reflectance is a surface-related measurement, and typically only the first 1 or 2 μm of the surface are studied, depending on the relative refractive indices of the internal reflection element (crystal) and the sample, and the wavelength being studied (the depth of penetration is proportional to wavelength, inversely proportional to wavenumber). Some successful results have been reported for powders and liquids with traditional internal reflectance accessories. This could only be considered as a marginal success because of the vertical orientation of the crystal faces. Newer forms of internal reflectance accessories with horizontally oriented crystals provide a convenient method for sampling all forms of solid and liquid samples, including pastes, gels, and soft powders.

External or specular reflectance is another method of sample handling that has been used for many years but recently has gained renewed importance, in part because of the increased sensitivity of modern instruments and also because of the increased use of microreflectance with infrared microscopes. Various accessories are available for this method of sampling, and they mainly differ in the angle of incidence. In the simplest form, the analysis is performed at a fixed angle, which is either set as a low or a high value relative to normal incidence. Typical fixed angles are around 10°, 30-45°, and grazing angle, which is around 80-85°. The selection of angle is dependent on the nature of the sample and the type of information required. Many measurements made by this specular reflectance approach do not determine true reflectance. Instead they are hybrid reflection/absorption measurements in which the infrared beam passes through a surface coating and reflects back through the coating from a reflective back surface. A typical example is the study of a "thick" coating, such as a paint or resin, on a metal surface. The spectrum produced in this manner appears to be a normal absorption, or at least transmission, spectrum. Under certain circumstances, where the refractive index of the material is high, front surface reflection can occur. This is true of specular reflectance and is observed with the spectral bands inverted

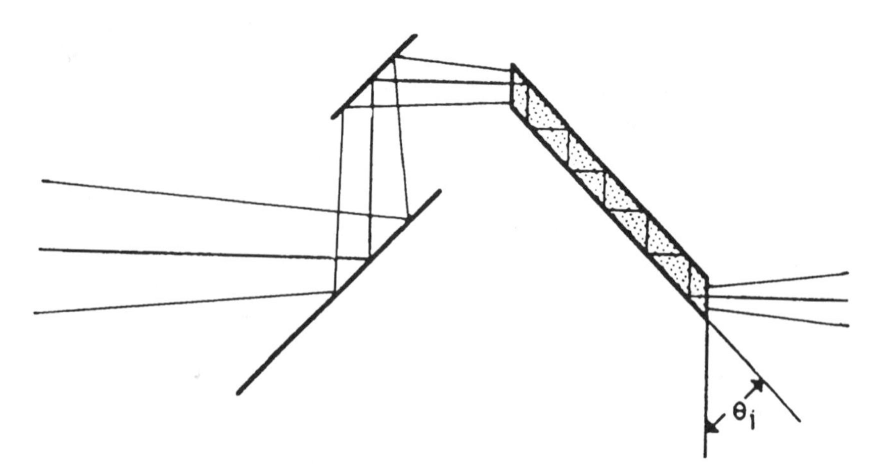

Figure 9. A typical optical layout of a modern ATR accessory featuring a multiple internal reflectance crystal (or element). Note that the angle indicated is the nominal angle of incidence, as well as the crystal angle.

relative to the normal absorption spectrum. In many cases, however, a far from ideal situation exists where a mixed mode of reflection occurs and both reflection and absorption are experienced at the surface of the sample. This produces a highly distorted spectrum that features negative lobes, in extreme cases resembling a first-derivative spectrum. Analytically this spectrum is not particularly useful because it is not as easily interpreted by the spectroscopist as is the normal absorption spectrum. It can be shown that the data can be separated into two components, the complex refractive index spectrum and the true absorption spectrum, by a mathematical computation, known as the Kramers-Kroenig transformation. Computer algorithms for this transform, which is computationally similar to a Fourier transform, are becoming available on commercial instruments.

C. Microsampling and Microspectrometry

Most of the discussion to this point is based on the assumption that there is sufficient sample for the sampling procedure used, which is frequently not the case. If the Beer-Lambert law is assumed to hold for the average sample and the units for the individual parameters are evaluated, then the observed absorbance is expressed as mass per unit area. This implies that a given spectrum intensity can be maintained for a smaller sample if the area is reduced by a proportional amount. This is the principle behind the concept of microsampling. Consequently for a given spectral intensity it is possible to reduce the area used for, say, a KBr pellet, from the normal 13 mm diameter, to either 3 mm or 1 mm (or even less). These correspond to approximately 19× and 169× reductions in sample size, respectively, while retaining the original intensity of the final spectrum—assuming constant sample thickness and pathlength. This reduction in sample size can reduce the amount of material required from a few milligrams to a few micrograms.

This decrease in sample size is not gained without some losses. The net loss in this case is a diminution in energy, since the average beam size at the sample point (focus) is on average 11 × 3 mm (dispersive) or 6-10 mm diameter (FT-IR)—note that these values vary depending on the instrument model and the manufacturer. Working with a sample size of 1 mm diameter or less will result in a significant reduction in the energy that reaches the detector. The throughput of energy may be improved by the use of a beam-condensing accessory, which effectively refocuses the beam and reduces its image size at the sample. The amount of beam condensation depends on the type of accessory used, and these may range from 4× to 8×. Micro devices for handling solids and micro cells for liquids are available for many of these accessories, enabling a wide variety of materials to be handled in the microgram range. Micro cups are also available for diffuse reflectance accessories. Careful application of dilute solutions of nonvolatile materials to a nonabsorbing diffusely reflecting surface, such as KBr powder, will extend detection levels down to nanogram quantities for certain compounds.

Microscopes for infrared spectroscopy have been in use since the 1950s. A commercial unit was produced at that time by Perkin-Elmer, but the technique was not widely accepted because of performance limitations of the infrared instrumentation available. In recent years there has been renewed interest in infrared microscopy, and the performance of modern instruments has made the technique into a powerful new analytical tool. Many new applications are being developed, including fiber and particle analysis on sample sizes down to 10 μm, and sometimes below. The technique has been extended to reflectance measurements, which has enabled infrared analyses to be made on surface deposits and contaminants, on a wide range of metals and other solid substrates. The applications of the technique are numerous and extend beyond the scope of this current chapter.

VI. DATA HANDLING AND COMPUTER APPLICATIONS

Infrared spectroscopy has, in the past, been applied to the general characterization of a wide range of materials. In recent years, the applications of the technique have been expanded by the increased performance of modern instruments, the availability of new sampling accessories, including infrared microscopes, and the development of new methods of data handling. This section addresses the role of data handling in the expansion of infrared spectroscopy as an analytical tool. Infrared spectral data processing can be separated into routine numerical manipulation of spectra, spectral identification by search methods, the determination of composition by quantitative analysis, and the measurement of time-dependent phenomena. It must be emphasized that these are all extremely broad areas of application, and the information provided below merely highlights the key areas of application.

A. Spectral Manipulation

One of the greatest frustrations of early infrared spectroscopists was the inability to extract information conveniently from the spectrum. Even simple operations like arithmetic manipulation of a spectrum can help to make more information available. The most useful functions are transmittance-absorbance interconversion, addition, subtraction, multiplication and division (by a factor or by another spectrum), digital smoothing, peak picking, baseline/background correction, and derivative taking (first or second order). In each case, care must be exercised in the use of a specific function. Once in digital format, the spectrum no longer exists as an analog function that can be expanded or compressed ad infinitum. There are limits defined by the numerical precision of the arithmetic functions used and the format selected for storing the data.

Conversions between transmittance and absorbance are the most common operations to be performed on an optical spectrum. However, unless the original software was written with attention to numeric precision, and unless the user is aware of the consequences of a transformation of this type, distortions may appear in the data. The problems arise from the logarithmic form of the conversion. Transforming from transmittance to absorbance results in a relative compression of weak bands and the noise on the baseline, whereas intense peaks and their accompanying noise are expanded, in some cases excessively. If the digital precision is not adequate then intense features may be truncated, and weak features may be lost in round-off errors. Therefore any conversion back to the original form would lead to an unacceptably modified spectrum. In principle, any reversible operation should return spectral data back to the original form, with any distortion being less than the noise level. If this is impossible within the digital limits of the software, then the user must be warned of the irreversible nature and consequences of the operation. Even a simple operation, such as a multiplication or a division, may lead to errors caused by exceeding truncation limits and data may be lost. Hence a multiplication followed by a division with the same number does not necessarily yield the same spectrum.

Other operations that result in band distortions, if not used correctly, are noise reduction procedures, including digital smoothing and signal averaging. In the first place, both of these operations must be performed in the transmittance form of the data because in this form the noise is constant throughout the full intensity range of a peak. In absorbance the noise varies as a logarithmic function, and this will result in a positive error on a peak, especially in situations where there is an intense absorption and a high noise content at the peak. In the design of an instrument, the accumulators used for signal averaging usually have sufficient precision to ensure that truncation does not occur. This must always be a concern in any homemade data system. Finally, digital smoothing requires the user to select a window for the moving average. Digital smoothing functions, of which the Savitzky-Golay algorithm is the most popular, are adjustable high-frequency filters by which noise is hopefully removed with minimal loss of spectral information. The window must be set to a width that is ideally less than the half bandwidth of the narrowest spectral feature, but presumably wider than the noise function. Dispersive data are usually oversampled and the noise is adequately separated from the spectral features. This is not necessarily the case in FT-IR data. If the linewidth of an absorption band is the same as the resolution used to record the spectrum, then smoothing cannot be used without the sacrifice of spectral data. This is because the width of the noise function is defined by the resolution of the spectrum.

Spectral subtraction is one of the most popular yet sometimes one of the most misused spectral manipulations. Many instrument software packages offer an interactive display for the subtraction, where one spectrum is automatically scaled and subtracted from a second spectrum and the result is placed on a CRT display in pseudo real time. There are two things that a user must be aware of: artifacts can be produced in a subtracted spectrum generated from a mixture (or solution), and in some chemical systems it is possible for apparent negative absorbances to be produced in the subtracted spectrum. In the first case, mutual interactions between components and refractive index changes are responsible for the formation of artifacts.

B. Spectral Searching

Computer-based spectral searches are one of the oldest forms of computer processing to be applied to infrared data, which is hardly surprising because of the fingerprint nature of the spectrum. Early searching programs were based on matching a subset of peaks between observed and archival spectra. The problem with this approach is deciding which peaks are the most important for a spectrum in the data base, and when faced with an unknown, which are then the most important peaks. Rules based on, say, the 10 most intense peaks worked relatively well on a controlled data base. However, the approach can break down, especially in the hands of a novice, for several reasons, one of the most significant being that early databases were formed from prism spectra. With the shift to grating and then to FT-IR data, considerable band intensity changes could be observed for spectra of the same compound because of the large differences in spectral resolution. Remember also that dispersive spectra approximate to a constant noise level with varying resolution, whereas FT-IR spectra have nearly constant resolution with a noise content that varies across the spectrum. Consequently the 10 most intense bands from a prism dispersive spectrum can be very different from those selected from an FT-IR spectrum.

If peaks are to be used in a searching scheme, then the most reliable approach is to set a minimum intensity threshold and to encode all peaks above that threshold. Also, the peaks should ideally be selected by a computer program rather than by the user. In that way the greatest consistency between the database and unknown spectra is maintained. In recent years there has been a trend toward the use of "full" spectra for the spectral matching. The term full often means a deresolved spectrum where the number of data points has been reduced to cut back on the storage overhead and to shorten the time required for the search. One of the greatest motivations for using this somewhat arbitrary approach has been the availability of the spectra for display and comparison with the unknown after the search is completed. The term arbitrary was used above because most commercial search systems use some form of a correlation metric or similarity fit, such as dot product, least-squares fit, root mean square (RMS) residues of the difference, or Euclidean distance as the criterion for goodness of match of the unknown to the data base spectra. These are purely numerical methods and the result of a search with these criteria does not necessarily yield a spectrum that has any chemical or spectroscopic resemblance to the unknown—unless the compound or very similar compounds are in the data base. The advantage of a peak-based search is that the emphasis is on the important regions of the spectrum—that is, the criterion can be based on both the presence and the absence of peaks, and peaks are therefore weighted accordingly.

The main problem with current approaches to spectral searching is that no one method is ideal. Peak-based searches generally fail to take the peak shape into account, and they can produce erroneous results when there is a high noise content in the data. Conversely, random noise does not cause a significant problem for full spectral searches but impurities and background artifacts can cause serious problems. Peak-based searches are also better for mixtures where a knowledge of the major components is important.

In the author's opinion, the ideal search should use a combination of peak data, possibly including second-derivative peak data, and full spectral fitting to provide an optimum match. There is a large amount of chemical and physical data inherent in the infrared spectrum, and this can be used to influence the outcome of a search. Currently one manufacturer, Perkin-Elmer, uses this concept with a pattern recognition approach in its search algorithm. Future search programs will probably utilize the intrinsic spectral data with a combination of comparison metrics plus a true expert system-based interpretation to weight the result.

C. Quantitative Analysis

The first analytical role of infrared spectroscopy was in the quantitative analysis of synthetic rubber compounds. In the early days of infrared spectroscopy, quantitative methods were popular but in most cases they failed to be transformed into acceptable routine methods. Instead, gas chromatography became a preferred technique for quantitative analysis in many of the applications that were served by infrared spectroscopy. The main limitations of the technique have already been covered in one form or another within this chapter—namely, the lack of suitable data processing, poor signal-to-noise performance on early instruments, lack of speed, and poor sample handling methods. This text has endeavored to show that these are no longer limitations, and therefore the role of quantitative infrared analysis must be reevaluated.

There are several numerical approaches for the extraction of quantitative data from the infrared spectrum. Many rely on a linear response to a changing concentration, such as the Beer-Lambert-Bouguer relationship, sometimes abbreviated to Beer's law. This is not, however, a prerequisite for quantitative analysis. If a function, either linear or nonlinear, relating concentration and some form of spectral response can be represented by a numeric expression, then it is possible to perform a computer-based analysis. The simplest form of an analysis involves the measurement of a single component within a chemical system by the measurement of a single spectral peak, expressed in absorbance, relative to a baseline reference point. The concentration of an unknown is derived by numerical calculation or by graphical means from a calibration series of samples of known concentration. Multiple-component systems may be handled by an extension of this method. These involve the production of either multiple calibrations for each component where the component absorption bands are not overlapped in the mixture spectrum, or a matrix solution for the individual component concentrations in situations that involve spectral band overlap. In this latter case at least as many standards, preferably known mixtures, as the number of components to be analyzed are required for a complete solution. Overdetermination—that is, the use of more than the minimum number of standards—is often desirable because this can help define systems that have a nonzero intercept, or a nonlinear response, especially where component interactions occur.

The above approach, where one or more discrete frequencies are selected for each component, works adequately for systems that are well defined and the contributions of individual components can be evaluated separately. The method for single or multiple components is not limited to just peak determinations. Other band parameters, such as derivatives (first and second order may be used where overlap is severe) and peak area, may be used as alternatives. The integration limits for peak area determinations must be carefully selected, and for some measurements an area slice may be preferable. Indiscriminate incorporation of too much information from the wings of an absorption band can result in a loss of precision, because of the proportionately higher noise-to-information content in the wings compared to the central region of the band.

In recent years there has been a move away from the use of discrete frequency measurements for multicomponent systems. Alternate methods involve the use of either a complete region or a number of segments of a spectrum. This provides a high level of redundancy in the determination and eliminates the need to select specific bands, which may not be the optimum bands, for each component. Spectral calibrations can range from least-squares fitting of the spectrum of the material to be analyzed against the calibration set, to the correlation or statistical evaluation methods, as used in the near-infrared. Popular approaches in these latter methods of calculation are principal component regression (PCR) and partial least-squares regression (PLS). In both methods the principal components are determined. These represent the major spectral contributions that change as a function of the concentrations of the components in a chemical system, and are not necessarily equivalent to the spectra of the individual ingredients. Once the components are mixed the dependent variables may be determined from the interaction products rather than from the individual component spectra. In this way interaction effects and physical and chemical properties may be correlated with infrared spectral data.

D. Time-Dependent Measurements

The high-speed data collection rates offered by FT-IR spectrometry provide a means for monitoring time-dependent phenomena. Probably the most familiar of these measurements are the chromatography/FT-IR interfaces, sometimes referred to as hyphenated or hybridized techniques. Of these techniques, the GC/IR method has been the most widely implemented. This is normally configured so that the effluent from a gas chromatograph is passed through a heated, narrow, gold-plated tube, which is sealed at the ends with infrared transmitting windows. This forms a gas flow cell, known as a lightpipe. The infrared beam from an FT-IR spectrometer is focused on one end of this cell while a high-sensitivity MCT detector is placed at the other end. The experiment requires some optimization, such as matching of the lightpipe dimensions to the column volume, matching of the detector element size to the focused image from the lightpipe, and providing adequate cooled baffles between the heated lightpipe and the detector to prevent the lightpipe from acting as a source and swamping the detector. With this setup, spectra are recorded and stored at regular intervals, as separated components from the column effluent pass through the lightpipe. After the chromatographic separation is complete, the spectra of the separated components are obtained from the stored data, and these may be used to reconstruct an infrared-based chromatogram.

The GC/IR measurement is not restricted to lightpipe technology. One alternate commercial approach involves the freezing out of separated components in a solidified gas matrix, where the spectrum is obtained by an infrared matrix isolation measurement. Other techniques that have benefited from this type of interfacing are thermogravimetric (TGA/IR) and supercritical fluid chromatography (SFC/IR).

The applications discussed so far involve time-dependent separations of materials. Chemical and physical processes may also be examined in this way using the gas, liquid, or solid physical state. The term "time-resolved measurement" is sometimes used, especially when high-speed processes are involved. This type of measurement requires a special temporal resolution of the individual points within an interferogram. In this way a snapshot interferogram of a reactant, an intermediate, or a reaction product can be reconstructed. It must be noted, however, that the traditional hyphenated techniques do not normally involve time-resolved computations in the generation of the spectra of separated components.

VII. COMMON SOURCES OF ERROR IN INFRARED MEASUREMENTS

There are many potential sources of error that can arise during the acquisition of an infrared spectrum. Frequently the instrument is blamed for many of the errors experienced, sometimes justifiably, but in most cases the errors are related to the practical implementation of the experiment—that is, they are operator dependent. Some predictable situations are noted in this section.

A. Sample and Sampling-Related Errors

In practice, errors may be attributed to the sample, the sampling methods, the instrument operation, and/or the data processing. The sample itself is often a major source of error, and this can be related to its physical state, its optical and other physical and chemical properties. Errors that arise from the sample are sometimes obvious and can be readily seen in the spectrum as recorded. In other instances they may be subtle and are only noticable after some form of data processing, such as spectral subtraction.

The optical pathlength of the sample is one of the most critical parameters that can influence the quality and usefulness of the final spectrum. For many of the operations that we perform on spectra we have to assume that the Beer-Lambert law is obeyed. The integrity of the data across the spectral range in use is essential if this is to be the case. The pathlength dependency (Lambert's law) in a practical measurement is such that the pathlength must be uniform across the sample and must provide absorption band intensities that lie within the linear operating range of the instrument. Both of these may seem to be obvious statements, but in practice they may be difficult to achieve. Liquids and self-supporting films, such as polymer films, will often cause problems in this area. When recording transmission spectra it is important to ensure that there are no bubbles or holes in the sample. If present, even as a microbubble or pinhole, these will give rise to an effect equivalent to stray light and will result in unpredictable photometric errors. If the pathlength is not uniform across the measured portion of the sample, then a nonlinear response will be experienced. This frequently occurs when polymer films are produced by casting or hot pressing procedures and when liquids are measured as capillary films. In such cases the spectra produced may be adequate for qualitative measurements but care should be taken if quantitative determinations are required. Any wedging of the sample in these measurements will lead to problems when spectral subtraction is used, resulting in incomplete cancellation of common bands and the occurrence of artifact residues.

The incorrect choice of pathlength is a common source of error for all forms of samples. Sometimes this is unavoidable because it may not be possible to adjust the sample thickness. When pathlengths are too long there is a dynamic range problem caused by the logarithmic relationship of the transmission format of the data. There is a point where the spectral information becomes unreliable, and this will vary from instrument to instrument. Some modern spectrophotometers are capable of producing useful data down to 0.1% transmittance (3 absorbance units), and possibly further, depending on the method of measurement. Many early instruments, however, were incapable of producing meaningful data below 10% transmittance (1 absorbance unit). In early dispersive instruments, the effect was often only appreciated when double-beam compensation experiments were performed. In such cases, either a loss of response was experienced with optical-null instruments, or an increase in noise level was obtained for ratio-recording instruments.

When handling spectral data with a computer, other problems can occur. Bands that approach zero transmittance experience a logarithmic increase in their noise content at the peak, and severe digitization errors can occur, caused by insufficient digital resolution close to the zero value. In this situation it is impossible to extract meaningful data by computer processing, and the only practical solution is to repeat the experiment with a shorter pathlength. Band distortions of this type can also lead to errors when performing quantitative measurements, including spectral subtraction. For example, attempts to subtract an absorption band close to zero may result in a computed spectrum that appears to contain spectral information in the region of this band. Invariably this is false information and it must be treated with extreme caution. In addition to the production of apparent spectral features, the subtraction will also produce an increase in the noise content in regions that originally contained intense bands. This is caused by the logarithmic expansion of the noise when the data is converted to the absorbance form. As bands approach zero transmission their absorbance increases toward infinity and any noise on these bands will correspondingly increase at the peak. After subtraction this noise is usually present as part of the residue. Note that in a computer data are always finite and there is no way to represent infinity. Consequently, bands at zero transmission will be truncated at some point in the absorbance scale.

There are similar limitations when there is insufficient sample and the pathlength is too short. In this circumstance spectral information may be lost in the instrumental noise. With computer-based instruments, spectral data may be recovered either by signal averaging to improve the signal-to-noise ratio of the spectrum, or by the careful application of digital smoothing techniques. On older dispersive instruments an analogous approach is to use a higher time constant and to scan for a longer period, dependent on the level of damping imposed by the increased time constant.

There are several other sample-related effects that can lead to the distortion of the measured spectrum. As noted earlier, some of these are readily observed in the final spectrum. A typical example is light scattering from solid samples in transmission measurements, as noted in the section on solid sampling. Concentration and mixing effects are subtle, but can produce anomalies and artifacts, especially when data processing methods are used.

In gas-phase experiments, concentration, controlled by gas pressures, is important. At increased pressures, band broadening is experienced, and this will change the profile of the absorption bands. Again, this is especially important when performing quantitative analysis and spectral subtraction. In the latter case, a mismatch in sample pressure may lead to second derivative-like residues in the final subtracted spectrum. Similar artifacts will also be produced if spectra of differing resolution are subtracted. In the liquid phase, chemical and physical interactions between components are important. As the concentration of the components increases, interaction effects become more pronounced, especially between polar materials. It must be realized that once a substance is dissolved in a solvent, its spectrum may change significantly from the spectrum in the free state—this can apply to the solvent as well as to the solute. Care must therefore be exercised when interpreting solution data. Experiments with spectral subtraction will cause difficulties when the spectrum of an isolated component is subtracted from the spectrum of a solution containing the component. Severe band shifts and distortions are often experienced, and these can lead to significant artifacts in the subtracted data.

B. Instrument and Data-Related Errors

The above discussion of sample-related errors was treated first because the sample has a profound influence on the quality of data produced from an instrument. Many instrument-related errors appear in the same manner as sample-related errors, and therefore it is important to remove the sample as a possible cause of a problem before attempting to isolate problems associated with instrument performance. It is also important to separate potential errors that are due to improper instrument operation from errors caused by instrument malfunction. In the past it was very easy to get into a situation where errors were compounded between instrument operation and sample preparation. On modern instruments this is less of a problem because the trade-off rules are better understood and in many cases these are incorporated into the operation of the instrument. Known error situations are anticipated and either the selection of incompatible parameters is prevented or an error condition is indicated. Also, built-in system diagnostics are available on many instruments, providing information about possible errors due to failed or failing components.

Common errors on older instruments are caused by failure to follow instrument manufacturers' guidelines on instrument settings for a given set of experimental conditions, or by failing to follow the trading rules for spectrum acquisition. Typically on optical-null instruments these result from incorrect setting of amplifier gain, electronic balance between the sample-beam and reference-beam channels, and time constants or instrument slit settings (resolution) that were incompatible with the scan speed. The spectral output of older instruments is usually recorded on precalibrated, gridded chart paper. Spectral calibration errors can occur due to misalignment of this chart paper, or due to failure to store the paper under the recommended conditions. Shrinkage or stretching of chart paper caused by humidity changes often accounts for frequency errors as high as 10 or 20 cm^{-1}. Obviously, mechanical factors can account for calibration errors, but often the common source of this type of error lies in the chart paper and the recording system.

While considering calibration issues, it is important to separate the concepts of wavelength reproducibility and accuracy. As indicated earlier, a common argument in favor of FT-IR instruments has been their intrinsic accuracy because of the use of an internal laser reference. This has at times been referred to as "the Connes advantage." The accuracy quoted, 0.01 cm^{-1}, is important, but only has real meaning when dealing with high-resolution spectra. This has little meaning for condensed-phase spectra, where resolutions of 2, 4, or even 8 cm^{-1} are commonly used. As a general guideline, the best accuracy that is achievable is one-tenth of a resolution element; that is, for a 4 cm^{-1} spectrum the best accuracy one can expect is 0.4 cm^{-1}. Calibration stability and wavelength reproducibility are more important parameters for everyday, practical applications. Most modern instruments, dispersive or FT-IR, are capable of providing a scan-to-scan reproducibility of 0.01 cm^{-1} or better, as long as care is taken in sample preparation and provided that the instrument is dimensionally stable. Variations in sample preparation and sample position within the instrument can certainly effect the wavelength precision. Also, dimensional changes, even in laser-referenced instruments, can cause frequency shifts as high as 1 or 2 cm^{-1}.

Instrument stability is another important issue, and this goes beyond the dimensional changes mentioned above. Thermal and mechanical effects do influence the quality of the data produced from a spectrometer. There is an obvious relationship between thermal effects and dimensional changes, and these can lead to calibration errors. Temperature can also have a profound effect on the sample, as mentioned earlier, and it can also influence the performance of critical components of a spectrometer, especially in interferometric instruments. Not all laboratories are air conditioned, and throughout the various seasons wild temperature swings may be experienced in the instrument's operating environment, sometimes as large as 10 or 20°C (possibly more). Under such conditions significant changes in instrument performance will be experienced, ranging from changes in background slope to the complete loss of energy. These usually result from the combination of changes in detector characteristics and dimensional changes within the interferometer, as mentioned in the section on instrumentation. While similar effects may occur in dispersive instruments, the outcome is sometimes less noticable because of the continuous cancellation of the background with the double-beam, sample/reference ratio.

Other environmental factors that can effect instrument performance are humidity and mechanical vibration. The presence of moisture in the air is often detected as perturbations or interfering absorption bands in a spectrum. Most instruments provide a purging facility, which aids the removal of this interference by the passage of dry air or nitrogen through the system. While this is not essential, it is a good practice because it also helps to keep critical, hygroscopic elements, such as beam splitters and detector windows, dry. Failure to do this in areas of high humidity can lead to loss of performance caused by fogging of these components. As mentioned earlier, in the section on FT-IR instrumentation, mechanical vibration will generate artifacts in spectral data. Obviously, it is undesirable to place a sensitive analytical instrument, whether it is a dispersive or FT-IR spectrometer, in a location that is prone to severe vibration, unless the instrument is environmentally hardened to withstand such treatment. If this is not the case, problems can certainly be anticipated, including the misalignment of optical components. Even mild vibrations can produce such artifacts as a baseline ripple (sine waves) or electronic spikes in a spectrum.

The electronics and the detector system are a common source of error if the instrument is used incorrectly or if it is used outside normal operating ranges. Most instruments provide a linear response within a defined, finite operating range. Severe deviations from linearity will occur when a system is used outside these ranges. Errors associated with spectral intensities have already been addressed in the section on sampling errors. When working with an FT-IR instrument it is very important that the detector is used within its linear range. Incorrect setting of the detector electronics will result in nonlinear response, and this is often detected by the occurrence of a false signal in cut-off regions of the spectrum, that is, in regions where zero transmittance is expected. For best results, the instrument gain should be set so that the signal fills the analog-to-digital (A/D) converter. Setting the gain too low will result in inadequate representation of the noise level, and the anticipated signal-to-noise improvement with signal averaging will not necessarily be achieved. Conversely, setting the gain too high will result in a saturation condition, which will lead to a nonlinear response and significant photometric errors.

One final issue that affects instrument performance is stray light, light that reaches the detector without passing through, or originating from, the sample, or radiation that originates from wavelengths beyond the normal measuring range. On dispersive instruments this can be an important factor because it can lead to unpredictable photometric errors. In the normal operation of an instrument the stray light is maintained at a minimum level by blackening the internal surfaces of the monochromator housing and critical components such as the slit jaws, and by the use of optical filters in front of the detector. Typically, this is kept to fractional percentages in most regions of the spectrum. The equivalent situation does not occur in an FT-IR spectrometer, although an analogous effect caused by a phenomenon known as aliasing can occur. Aliasing is the inclusion of higher frequencies from wavelengths outside of the normal operating range introduced into the spectrometer. This is normally kept to a minimum by optical or digital filtering of these higher frequencies.

VIII. RECOMMENDED REFERENCE SOURCES

Over the years there has accumulated a tremendous wealth of literature on the art and science of infrared instrumentation. If the reader has access to a good library then it is certainly worth checking back to literature published in the 1950s. Although any papers published at that time will now be grossly out of date, they do provide a tremendous source of ideas on applications, and many useful hints on sample handling, all of which are still perfectly valid. Today, it is very common to see the invention of many new wheels, partly caused by the lack of reference back to the older journals. The newer computer-based literature-search programs do not help in this matter because they only extend back about 15 years. Rather than quote an excessive amount of literature, the following titles are recommended for general information on instrumentation and applications.

1. A. L. Smith, *Applied Infrared Spectroscopy: Fundamentals, Techniques, and Analytical Problem Solving*, Chemical Analysis, Vol. 54, Wiley-Interscience, New York, 1979.
2. P. R. Griffiths and J. A. de Haseth, *Fourier Transform Infrared Spectrometry*, Chemical Analysis, Vol. 83, Wiley-Interscience, New York, 1986.
3. G. L. McClure, Ed., *Computerized Quantitative Infrared Analysis*, STP 934, American Society For Testing and Materials (ASTM), Philadelphia, 1987.
4. N. J. Harrick, *Internal Reflection Spectroscopy*, Harrick Scientific Corporation, Ossining, NY, 1979.
5. Audio-Visuals: SAVANT, P. O. Box 3670, Fullerton, CA.
6. C. D. Craver, Ed., *Coblentz Desk Book* [a reference source and collection of standard reference spectra], The Coblentz Society, P.O. Box 9952, Kirkwood, MO 63122.

8

Molecular Fluorescence and Phosphorescence

LINDA B. McGOWN and KASEM NITHIPATIKOM* / Department of Chemistry, P. M. Gross Chemical Laboratory, Duke University, Durham, North Carolina

I. INTRODUCTION

Molecular luminescence spectroscopy can be used for fundamental studies of molecular excited states as well as for selective and sensitive analysis of luminescent samples. Luminescence processes such as fluorescence and phosphorescence are emission processes, and luminescence techniques have dynamic ranges and detection limits that are several orders of magnitude better than molecular UV-visible absorption techniques for highly luminescent compounds. Selectivity can be derived from several sources. First of all, only certain groups of compounds that absorb UV-visible radiation are likely to undergo deexcitation by luminescence. Second, selectivity among luminescent molecules can often be accomplished on the basis of excitation and emission spectral characteristics and excited-state lifetimes. Polarization and selective quenching and enhancement techniques can be used to achieve further selectivity. Luminescence techniques can be extended to nonluminescent compounds by the use of indirect methods in which the analyte is involved in a reaction or interaction with luminescent reagents.

There are numerous sources of information in the literature on luminescence spectroscopy. Most instrumental analysis textbooks provide descriptions of components and configurations of luminescence instruments, as well as some background theory and applications. A number of books on fluorescence and/or phosphorescence analysis have been written [1-7], as well as a recent handbook chapter [8], and include discussions of theory, instrumentation, practical considerations, and applications. Several recent series and reviews have focused on state-of-the-art developments and current trends in luminescence instrumentation, techniques, and applications [9-13]. Biochemical applications of excited-state measurements and of steady-state and dynamic measurements of fluorescence polarization and anisotropy are emphasized in a recent text that also contains basic material on fluorescence spectroscopy and analysis [14]. Finally, the review of molecular luminescence in the biannual Fundamental Review issue of Analytical Chemistry provides an excellent guide to recent literature in the field. Other references on more specialized topics are cited throughout this chapter.

**Current affiliation*: Texas Tech University, Lubbock, Texas.

II. THEORY

The fundamental aspects of luminescence phenomena have been discussed in a number of books [15-19]. Berlman has provided a compilation of fluorescence spectra that also includes values for quantum yields, excited-state lifetimes, and other fundamental quantities [20]. An introduction to the theory of luminescence spectroscopy is presented in this section, including descriptions of fundamental processes and characteristics, spectra, and quantitative relationships between intensity and emitter concentration.

A. Excited-State Processes

The processes involved in molecular luminescence (photoluminescence) are illustrated in Figure 1. Molecular absorption of photons in the ultraviolet-visible (UV-vis) range causes the electronic transition from a lower energy level [typically the lowest vibrational level, v_0, of the ground electronic singlet state (S_0)] to an excited singlet state S^*. The molecule rapidly undergoes vibrational relaxation to the lowest vibrational level of the excited state, from which it may deexcite by one of several competitive pathways. Deexcitation directly back to the ground state can occur either by nonradiative processes or by emission of a photon. The latter process is referred to as fluorescence. Alternatively, the molecule may undergo a "forbidden" transition to an overlapping triplet state (T^*) by a process known as intersystem crossing. Once in the triplet state, the molecule may undergo vibrational relaxation to the lowest vibrational level of T^* or else return to S^* by intersystem crossing. In the latter case, deexcitation by photon emission results in delayed fluorescence. In the former case, deexcitation may occur by photon emission from the triplet state to the ground state, known as phosphorescence, or by nonradiative processes. Phosphorescence generally occurs at longer wavelengths than fluorescence, since the triplet state is lower in energy than the overlapping excited singlet state for molecules that phosphoresce.

B. Excited-State Lifetimes

Phosphorescence and fluorescence can be distinguished from one another on the basis of the lifetimes of their respective excited states. Both processes follow first-order kinetics,

$$I = I_0 e^{-kt} \tag{1}$$

where I is the intensity at time t and I_0 is the initial intensity. The rate constant k for deexcitation can be expressed in terms of the radiative and nonradiative deexcitation rate constants:

$$k = k_R + k_{NR} \tag{2}$$

The luminescence lifetime τ is the time required for the intensity to decay to 1/e of the initial intensity and is inversely proportional to k. Fluorescence lifetimes, τ_F, are on the order of nanoseconds to microseconds. Phosphorescence lifetimes, τ_P, are much longer due to the forbidden singlet-triplet-singlet transitions, and range from milliseconds to seconds.

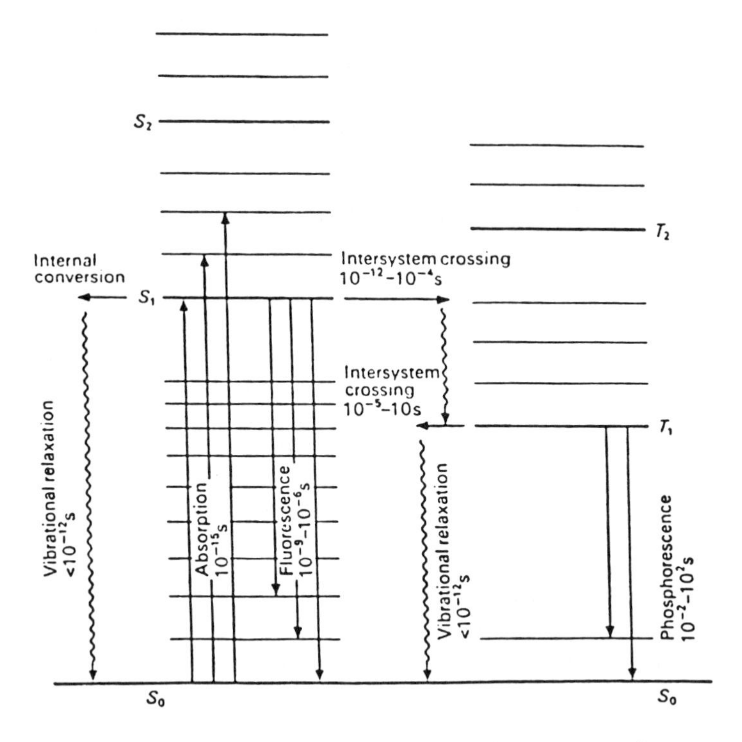

Figure 1. A Jablonski diagram showing electronic processes. (Adapted from Ref. 105, with permission.)

C. Quantum Yield

The quantum yields or efficiencies of the fluorescence and phosphorescence processes can be expressed as

$$\Phi_F = \frac{\text{Number of photons emitted as fluorescence}}{\text{Total number of photons absorbed}} \tag{3}$$

and

$$\Phi_P = \frac{\text{Number of photons emitted as phosphorescence}}{\text{Total number of photons absorbed}} \tag{4}$$

respectively. The quantum yield is related to the radiative rate constant k_R and the excited-state lifetime:

$$\Phi = k_R \tau \tag{5}$$

If the quantum yield equals unity, the observed lifetime τ is equal to $1/k_R$ and is called the intrinsic lifetime (also referred to as the natural, or radiative, lifetime). Various symbols including τ and τ_R have been used for the intrinsic lifetime.

D. Quenching

Quenching refers to any process that causes a reduction in the quantum yield of a given luminescence process. Quenching can be categorized as either collisional or complexation. In collisional quenching, the duration of the interaction between the luminescent molecule and the quencher is much shorter than the excited-state lifetime of the luminescent molecule, whereas complexation quenching involves the formation of a complex that is long-lived relative to the excited state. Collisional quenching is described by the Stern-Volmer equation,

$$I_0/I = 1 + k_q c_q \tau \qquad (6)$$

where c_q is the concentration of the quencher, k_q is the rate of collisional quenching, and τ is the observed lifetime for the excited state. Collisional quenching is evidenced by a linear decrease in the observed luminescence lifetime with increasing quencher concentration. Observed lifetime is independent of quencher concentration in complexation quenching.

E. Intensity and Concentration

The fluorescence intensity I_F is directly proportional to the intensity of the excitation beam (I_0), the absorbance (εbc, where ε is the molar absorptivity, b is the pathlength, and c is the concentration of the fluorescing species), and the quantum yield (Φ_F) of the species:

$$I_F = K_F I_0 \Phi_F \varepsilon bc \qquad (7)$$

The proportionality constant K_F is a function of instrumental response factors and other instrumental parameters.

Since I_F is proportional to I_0, sensitivity and detection limits can be improved by increasing the intensity of the source (limited by a maximum intensity above which photodecomposition of the sample occurs). This is in contrast to absorption spectroscopy in which an increase in source intensity does not directly effect sensitivity and detection limits. In addition, the directly measured quantity I_F is linearly proportional to concentration of the fluorescent molecule, unlike absorptiometric experiments in which concentration is linearly related to the log of the ratio of two measured intensities. The combination of these proportionality features results in sensitivities and detection limits for fluorescence measurements of strongly fluorescent molecules that are several orders of magnitude better than those for absorptiometric determinations of the same molecules. Fluorescence spectroscopy is also more selective than absorption spectroscopy, since not all molecules that absorb UV-vis radiation also fluoresce. The applicability of fluorimetric analysis is therefore limited to a select group of molecules, although many indirect methods have been described for nonfluorescent molecules.

An equation analogous to equation (7) relates phosphorescence intensity to incident intensity, quantum yield, and absorbance:

$$I_P = K_P I_0 \Phi_P \varepsilon bc \qquad (8)$$

The subset of absorbing molecules that exhibit phosphorescence is smaller than the group exhibiting fluorescence for several reasons, including the forbidden nature of the transitions involved and the longer excited-state lifetimes, which make phosphorescence much more susceptible to quenching. Phosphorescence is generally observed only at low temperatures (liquid nitrogen) or with special techniques, such as deposition of the sample on a solid surface or dissolution in micellar solutions to eliminate the quenching processes that occur in solution and reduce the rates of competitive deexcitation processes. Certain atoms and molecules, such as the "heavy atoms" (Br^-, I^-, Tl^+, Ag^+, etc.), enhance intersystem crossing and result in quenching of fluorescence and enhancement of phosphorescence.

Dissolved molecular oxygen in solution is probably the most troublesome and ubiquitous quencher of both fluorescence and phosphorescence. In order to maximize intensity, solutions are generally deoxygenated prior to analysis. Oxygen quenching causes decreases in both intensity and excited-state lifetime as the oxygen concentration is increased. The effect of oxygen is especially serious in the case of phosphorescence, which is generally not observed at all in solution even in the presence of only small amounts of dissolved oxygen.

F. Luminescence Spectra

Luminescence excitation spectra are acquired by scanning excitation wavelength at a constant emission wavelength. Emission spectra are similarly collected by scanning emission wavelength at a constant excitation wavelength. The excitation spectrum of a molecular species is analogous to its absorption spectrum, and is generally independent of the emission wavelength due to vibrational relaxation to the lowest vibrational level of S^* prior to photon emission (Figure 1). The emission spectrum is similarly independent of the excitation wavelength. The two independent dimensions of wavelength information can be more fully exploited by means of total luminescence spectroscopy, in which fluorescence intensity is plotted as a function of excitation wavelength on one axis and emission wavelength on the other (Figure 2); this is called an excitation-emission matrix (EEM).

Fluorescent molecules that have similar vibrational structure in the ground and excited singlet states will have excitation and emission spectra that are roughly mirror images of each other (Figure 3). Spectral resolution and detail increase as temperature decreases and as the rigidity of the sample matrix increases, and these effects can be exploited by low-temperature techniques such as matrix isolation and Shpol'skii spectroscopies (Section IV.E).

G. Polarization

A ray of light consists of a magnetic field and an electric field perpendicular to each other and to the direction of propagation of the ray. The electric field vector of the ray has a particular angular orientation with respect to a given coordinate system, and a beam composed of many rays can be characterized in terms of the overall angular distribution of the electric field vectors. The angular distribution of the rays may be random, resulting in a nonpolarized beam, or the rays may all have parallel vectors and be completely polarized (Figure 4). If a molecule is excited with completely

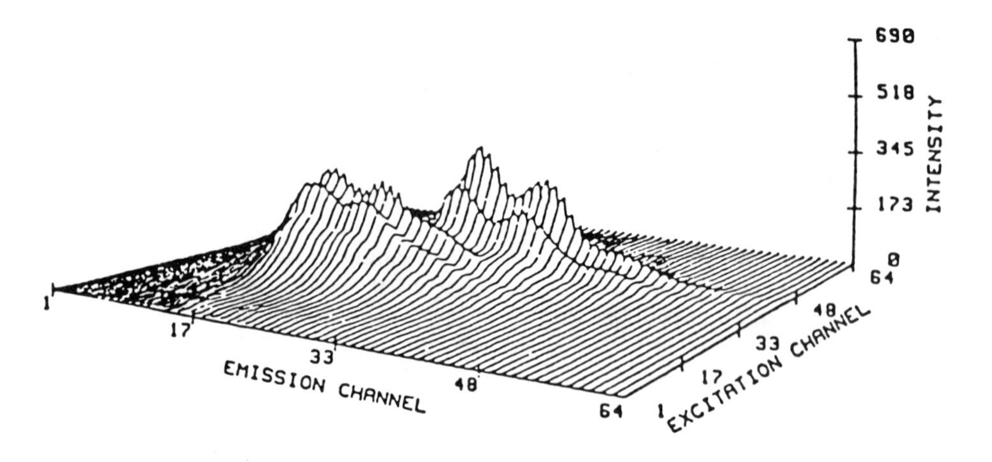

Figure 2. Total luminescence spectrum (EEM) of a perylene/anthracene mixture. (From Ref. 106, with permission.)

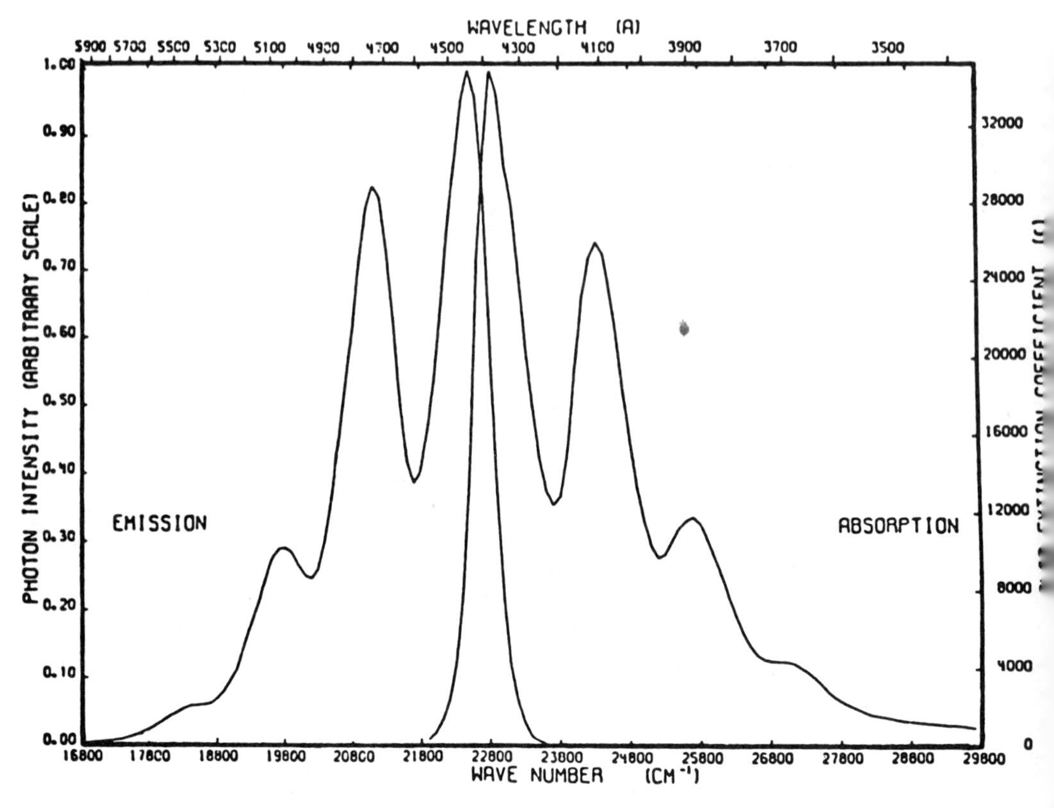

Figure 3. Absorption and fluorescence emission spectra of perylene in benzene. (Adapted from Ref. 20, with permission.)

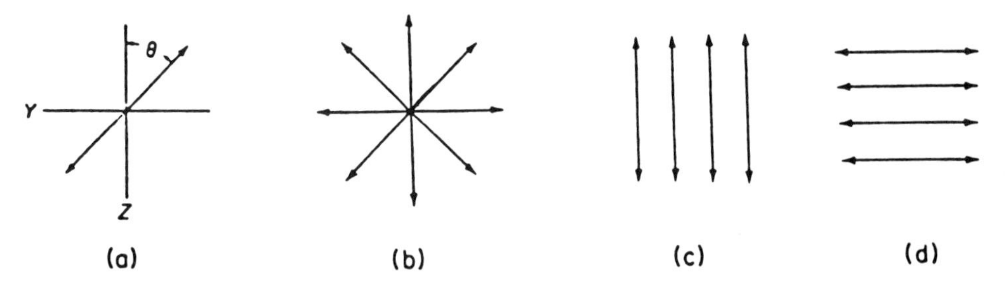

Figure 4. Conceptual diagram of polarization showing (a) coordinate system, (b) nonpolarized light, (c) vertically polarized light, and (d) horizontally polarized light. (From Ref. 7, with permission.)

polarized light, the emission will be somewhat depolarized due to effects from the absorption process, difference in orientations between the absorption and emission dipoles of the molecule, Brownian rotation of the molecule, and reabsorption of emitted light by a second molecule, which then emits it at a different orientation. Other effects may be observed due to scattered light and energy transfer. The polarization P is generally measured by exciting the sample with vertically polarized light and observing the vertically and horizontally polarized fluorescence intensities (I_V and I_H, respectively):

$$P = (I_V - I_H)/(I_V + I_H) \tag{9}$$

Fluorescence anisotropy r, where

$$r = (I_V - I_H)/(I_V + 2I_H) \tag{10}$$

is closely related to polarization and may be calculated from the same experimental data.

III. INSTRUMENTATION

A. Instrument Components

A general schematic diagram of a luminescence spectrometer is shown in Figure 5. Emission is usually measured at 90° to the excitation beam to avoid background from nonabsorbed radiation, although angles other than 90° are sometimes used for specific applications. The instrumental components include a light source with its own power supply, excitation wavelength selector, sample chamber, emission wavelength selector, detector, output, and, in some instruments, computer controller.

Light Sources

Luminescence intensity is directly proportional to the intensity of the light source [equations (7) and (8)], and high-intensity sources can therefore be used to increase sensitivity and lower detection limits for luminescence

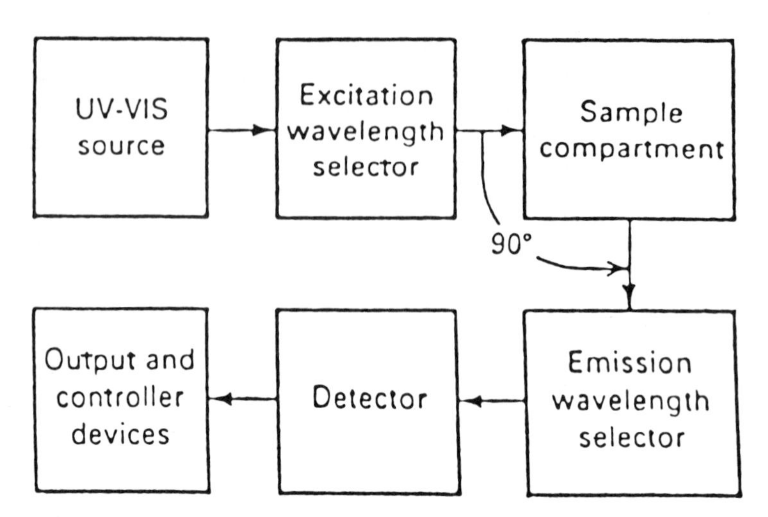

Figure 5. General schematic diagram of a single-beam fluorometer. (From Ref. 8, with permission.)

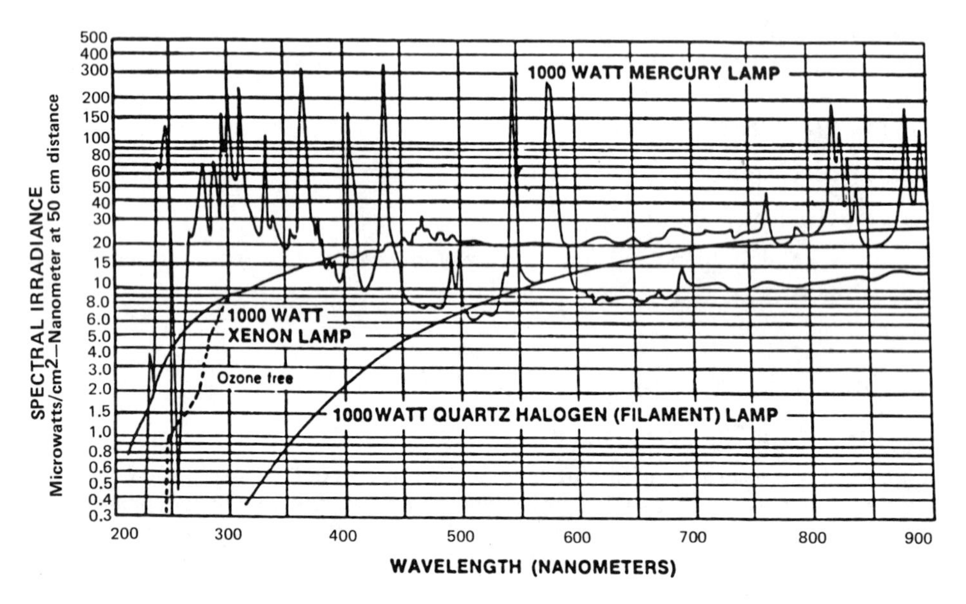

Figure 6. Output of several light sources commonly used for fluorescence excitation. (From Oriel, with permission.)

analysis. The xenon arc lamp is a commonly used source. The output is continuous over a broad wavelength range and is therefore well suited to spectral scanning. Another common source is the high-pressure mercury arc lamp. The output is a continuum with a line spectrum superimposed on it, making the mercury lamp better suited to nonscanning filter instruments. Other sources include halogen lamps and combination xenon-mercury lamps. The outputs of the sources used for luminescence measurements are shown in Figure 6. The intensity of the excitation beam can be increased by the use of mirrored backing in the source compartment to redirect emitted light in the direction of excitation.

Lasers are also used in luminescence experiments in which continuous scanning of excitation is not required. The properties and principles of lasers and laser light have been described [21,22]. Tunable lasers can be used to provide multiwavelength excitation capabilities. The principles and applications of dye lasers have been described [23], and the properties of a wide variety of dyes used in tunable dye lasers have been surveyed [24]. Improvements in detection limits can sometimes be obtained using lasers, although the increased emission signal is accompanied by an increase in light scatter and localized heating effects. Also, the excitation beam must often be greatly attenuated to avoid photodecomposition of the sample constitutents. Lasers are frequently used for time-resolved fluorimetry and other specialized applications. The use of lasers for chemical analysis has been discussed in recent books and articles [25-27].

Pulsed sources, including both lamps and lasers, are used for special applications such as for dynamic measurements of luminescence lifetimes and for time-resolved elimination of background signals (see Sections III.C, III.D, and IV.C).

Wavelength Selectors

Either filters or monochromators can be used for wavelength selection. Filters offer better detection limits but do not provide spectral scanning capabilities. Often, a filter will be used in the excitation beam along with a monochromator in the emission beam to allow emission spectra to be acquired. Full emission and excitation spectral information can be acquired only if monochromators or polychromators (monochromators with the slits removed, used with array detectors, Section III.D) are used in both the excitation and emission beams, so that two monochromators (or polychromators) are necessary, for techniques such as synchronous excitation and total luminescence spectroscopy.

Sample Compartment

Cuvettes for fluorescence measurements of solutions are usually rectangular with at least two adjacent faces that are transparent. The remaining two faces may also be transparent, or they may have reflective inner surfaces to direct more of the fluorescence emission to the detector. Quartz cuvettes are used for measurements in the UV-visible region, and less expensive glass cuvettes can be used for measurements in the visible region only.

Inexpensive disposable cuvettes made of polystyrene or polyethylene can also be used for work in the visible range with certain solvents. Disposable acrylic cuvettes can be used for the 275-350 nm wavelength range, and are useful for measurements of native protein fluorescence. Special cuvettes are available for microvolume work and for flow systems, such as flow injection analysis, fluorimetric detection for high-performance liquid chromatography (HPLC), and stopped-flow measurements. Mirrored backings on these cells are especially useful to maximize the detected emission intensities.

Low-temperature phosphorescence measurements are generally made on samples that are contained in special Dewar cells. The sample is contained in a quartz tube, which is surrounded by liquid nitrogen in the Dewar container. Samples must be cooled to a solid state very carefully to avoid cracking and "snow" formation. For room-temperature phosphorescence, samples may be deposited onto a strip of filter paper or other support material. A dry, inert gas is continuously passed over the sample to eliminate oxygen and moisture. Phosphorescence measurements of liquid samples and solutions may be made using conventional fluorescence cuvettes to contain the samples. Special cuvettes designed to facilitate oxygen removal can also be used.

Detectors

Photomultiplier tubes (PMTs) are the most commonly used detectors, and various types are available for different applications. In general, they are sensitive in the range from 200 to 600 nm, with maximum sensitivity obtained in the 300-500 nm range (Figure 7). Red-sensitive PMTs are also available for work beyond 600 nm. The PMT housings are sometimes cooled to temperatures as low as -40°C to minimize temperature-dependent noise.

Photomultipliers can be operated either as analog detectors, in which the generated current is proportional to the incident radiative intensity, or as photon counters, in which the current pulses generated by individual incident photons are counted. A preset threshold value allows noise and other low-energy events to be disregarded. Single-photon detection offers greatly improved detection limits and is used for the measurement of very-low-intensity emission signals. The linear response range is much narrower that that of the analog detector, being limited on the low end by signal-to-noise deterioration and on the high end by the response time of the counter.

The use of multichannel detectors [28-32] for the simultaneous acquisition of emission spectra or excitation-emission matrices has increased the range of applications of luminescence experiments to include real-time experiments, kinetic measurements, and on-line detection for chromatography and other flow systems. The ability to acquire complete spectral information essentially instantaneously has also greatly facilitated qualitative analysis by dramatically reducing the time required per analysis. Among the more commonly used detectors are diode arrays, vidicons, silicon intensified target vidicons, charge-coupled and charge-injection devices, and numerous other devices made possible by recent technological advances.

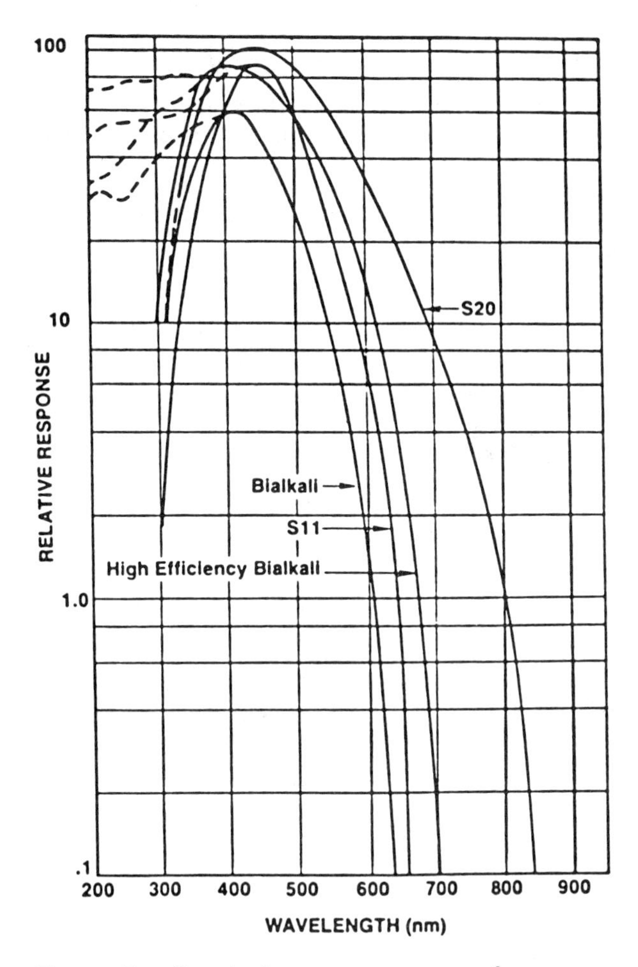

Figure 7. Spectral response curves for some common PMT detectors; — Pyrex window, --- quartz window. (From Hamamatsu, with permission.)

Output and Computers

Simple instruments such as those with filters for wavelength selection often have simple analog or digital readout devices. X-Y recorders can be used to record spectra with a spectrofluorometer. Computers are frequently interfaced to more sophisticated spectrofluorometers to control data acquisition, and for storage, manipulation, and analysis of spectral data. For large data arrays, such as those acquired for an EEM, and complex experiments such as those involving dynamic measurements of luminescence lifetimes, computerized data collection and analysis is almost indispensable. The use of on-line computers for fluorescence spectroscopy has been discussed [33].

B. Instrument Configurations for Conventional Fluorescence Measurements

Instruments designed for luminescence spectroscopy can usually be used for either fluorescence or phosphorescence measurements, with a few instrumental modifications generally required for the latter. In this section we will discuss the configurations used for fluorescence instruments. The following sections will address the specific instrumental requirements and some of the special techniques used for phosphorescence measurements.

Filter Fluorometers

Instruments in which wavelength selection is accomplished with filters are generally referred to as fluorometers, whereas monochromator-based instruments capable of spectral scanning are called spectrofluorometers. The simplest, least expensive fluorometers are single-beam filter instruments with halogen lamp sources, phototube or PMT detectors, and simple output devices. Ratiometric fluorometers are also available in which the ratio of the sample signal to a reference signal is used in order to compensate for fluctuations in line voltage, drift, etc. In some instruments, such as the one shown in Figure 8, the sample and reference light paths are taken from the same source and are alternately measured by the same detector by use of a mechanical cam. This configuration automatically corrects for fluctuations in both the light source output and the detector response. A configuration in which two detectors are used, one for the

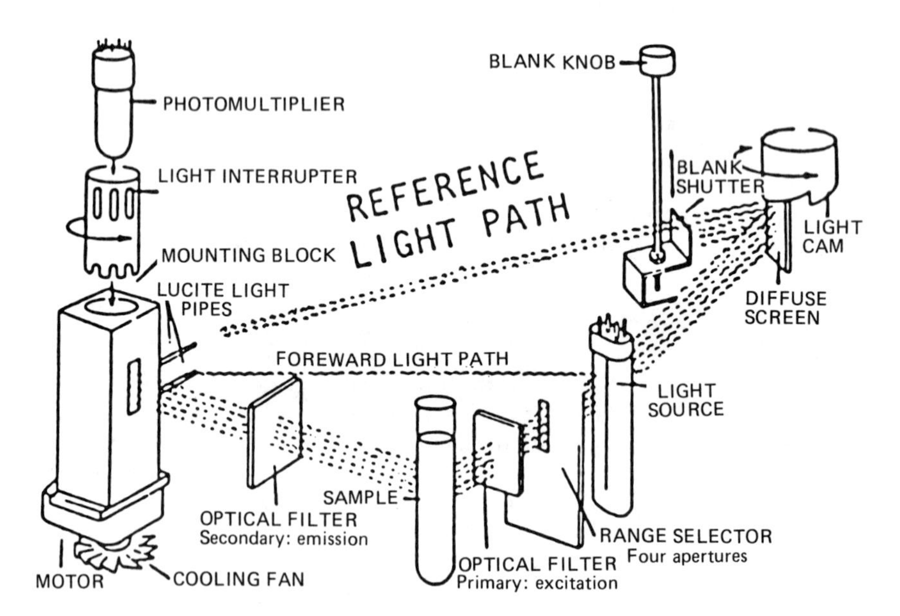

Figure 8. Schematic diagram of a double-beam (ratiometric) filter fluorometer. (From Sequia-Turner, with permission.)

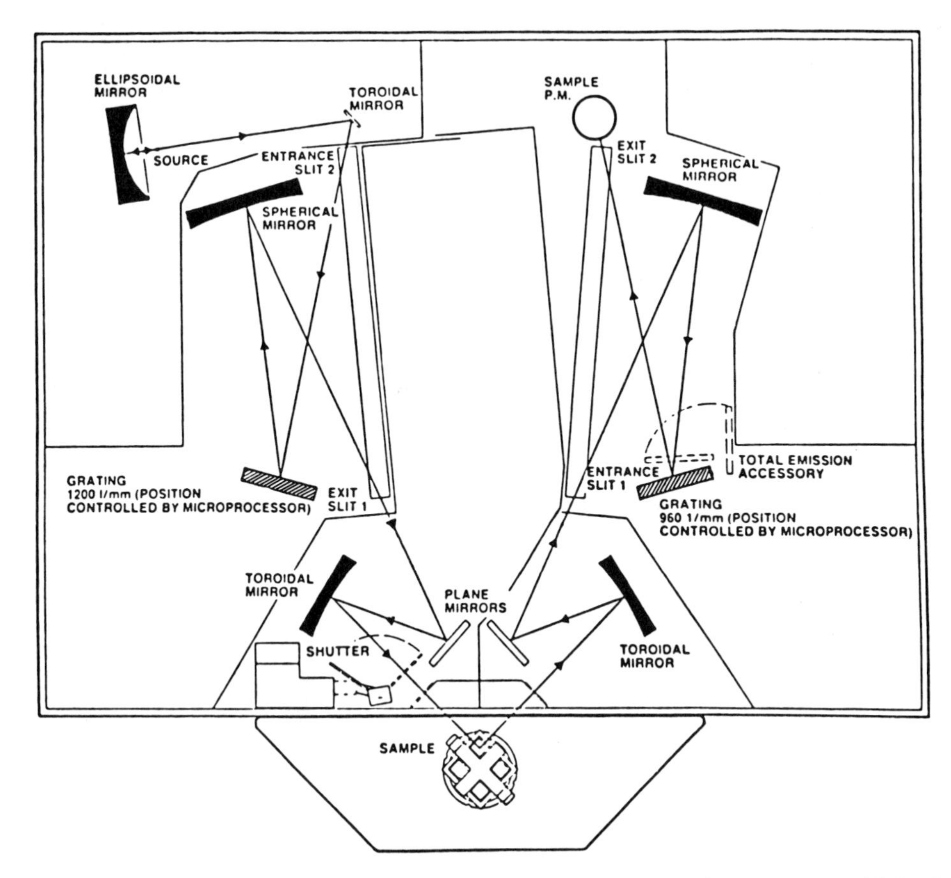

Figure 9. Schematic diagram of a single-beam spectrofluorometer. (Adapted from Perkin-Elmer, with permission.)

sample beam and one for the reference beam, will correct for source output fluctuations only.

Filter fluorometers are suitable for quantitative analysis applications in which spectral scanning and high resolution are not required. Filters transmit more light and cost less than monochromators, thereby providing better detection limits with less expensive instrumentation.

Spectrofluorometers

Spectrofluorometers are also available in both single beam and ratiometric configurations. A single-beam instrument is shown in Figure 9. A ratiometric spectrofluorometer configuration in which two separate detectors are used is shown in Figure 10. The reference PMT monitors the excitation beam after it has passed through the monochromator so that corrections are based on fluctuations in the wavelength of interest only. The reference

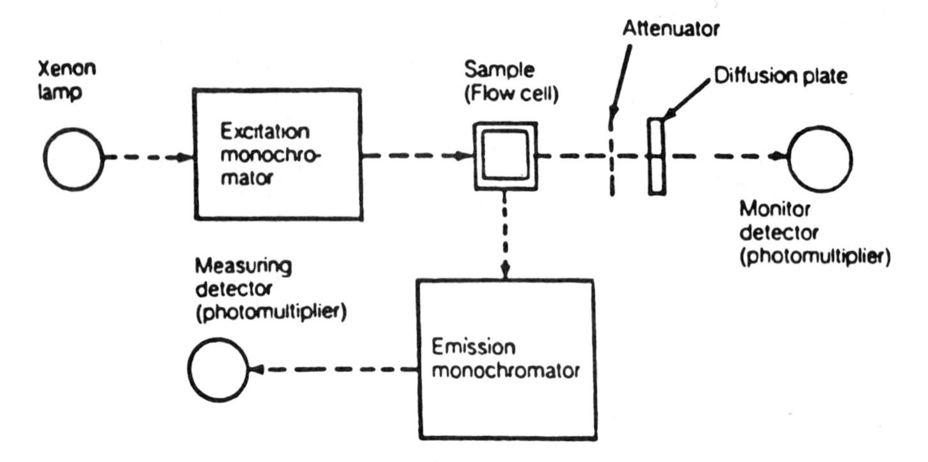

Figure 10. Schematic diagram of a ratiometric spectrofluorometer. (From Perkin-Elmer, with permission.)

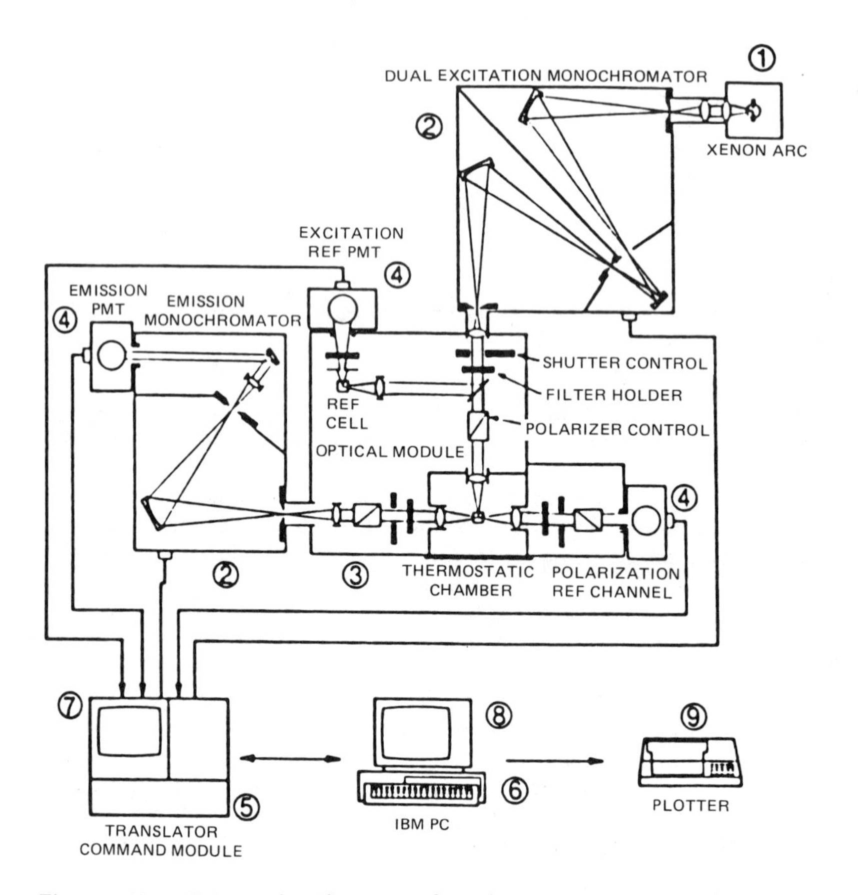

Figure 11. Schematic diagram of a single-photon counting, ratiometric spectrofluorometer. (From SLM Instruments, with permission.)

PMT monitors the excitation beam after it has passed through the monochromator so that corrections are based on fluctuations in the wavelength of interest only. The reference PMT is placed after the sample to monitor the transmitted beam, thereby allowing absorption measurements to be made on the sample.

A functional diagram of a single-photon counting spectrofluorometer is shown in Figure 11. Single-photon counting instruments are used to measure very low light levels and are designed for maximum signal-to-noise. In the instrument shown in Figure 11, a double monochromator is used in the excitation beam to minimize stray light. Double emission monochromators are used in some instruments in addition to or instead of double excitation monochromators. Cooled PMT housings may be used to minimize detector noise. Ratiometric detection uses a reference PMT to detect a portion of the excitation beam diverted to the detector with a beam splitter placed between the excitation monochromator and the sample. If an appropriate blank solution is placed in the cell compartment in the reference beam, absorption measurements can be made.

Spectrofluorometers offer the most flexibility for quantitative and qualitative applications. Many instruments are modular so that monochromators can be replaced with filters in both the emission or excitation beams if so desired.

C. Phosphorimeters

Phosphorescence can be measured with a conventional, continuous-source fluorescence instrument that is equipped with a phosphoroscope attachment. The phosphoroscope is used to eliminate fluorescence signals on the basis of the large difference in lifetimes between fluorescence and phosphorescence. Early phosphoroscopes used the rotating can assembly (Figure 12), but modern instruments most commonly employ mechanical choppers. The use of rotating mirrors (Figure 13) has also been described [34]. Pulsed-source instruments also can be used for phosphorescence measurements [35-37], in which case the intensity measurements are made after the fluorescence signal has completely decayed. The time-resolved approach

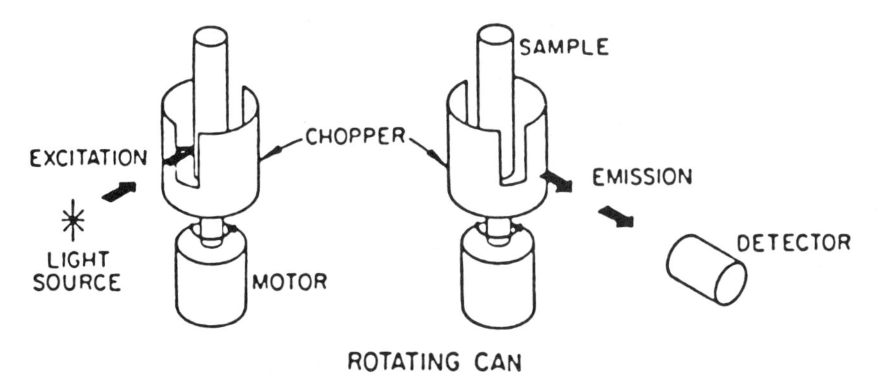

Figure 12. A rotating cylinder phosphoroscope. (From Ref. 39, with permission.)

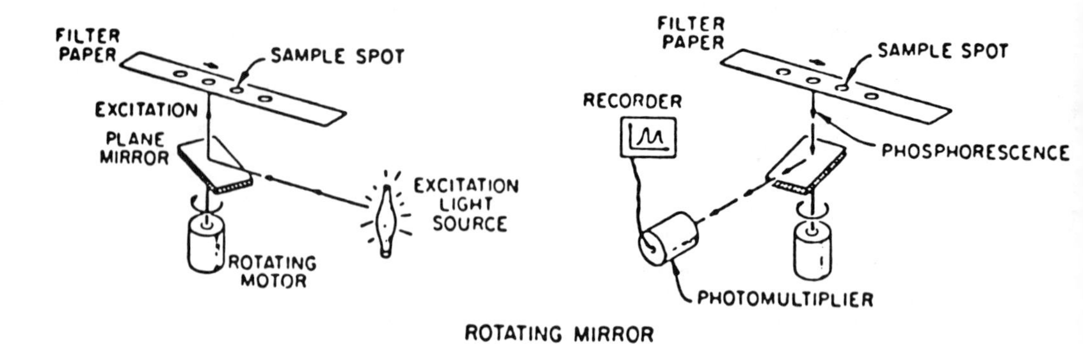

Figure 13. A rotating mirror phosphoroscope. (From Ref. 39, with permission.)

has the advantage of enabling the resolution of multiexponential phosphorescence decay signals in order to perform multicomponent determinations [37,38].

Room-temperature phosphorescence (RTP) spectroscopy [39] can be performed on conventional fluorescence instruments. In solid-surface techniques [40], modifications are necessary to measure the luminescence, which is either reflected or transmitted from the solid surface. Alternatively, commercial luminescence instruments with spectrodensitometers are available for solid-surface RTP and fluorescence analyses.

D. Instruments with Special Capabilities

Spectrofluorometers for Extended Spectral Acquisition

Spectrofluorometers can be used to generate excitation spectra, emission spectra, synchronous excitation spectra, and total luminescence (EEM) spectra. Conventional emission and excitation spectra simply require emission and excitation monochromators. Acquisition of synchronous excitation spectra [41-43] requires that the emission and excitation monochromators can be synchronously scanned. Total luminescence spectra [28,32,44-47] can be acquired on any instrument with both emission and excitation monochromators. As the name implies, total luminescence spectra may contain phosphorescence as well as fluorescence contributions, especially for solutions that contain micelles or other organized media to enhance phosphorescence.

Array detection can greatly reduce the time required for data acquisition. The instrument shown in Figure 14 has array detection of the emission beam so that the emission spectrum can be simultaneously measured at a given excitation wavelength. A spectrograph is used instead of an emission monochromator in order to disperse the emission beam and simultaneously detect all of the wavelengths in a given range. Total luminescence spectra can be acquired in much less time relative to conventional spectrofluorometers, since emission spectra can be simultaneously collected at the sequentially

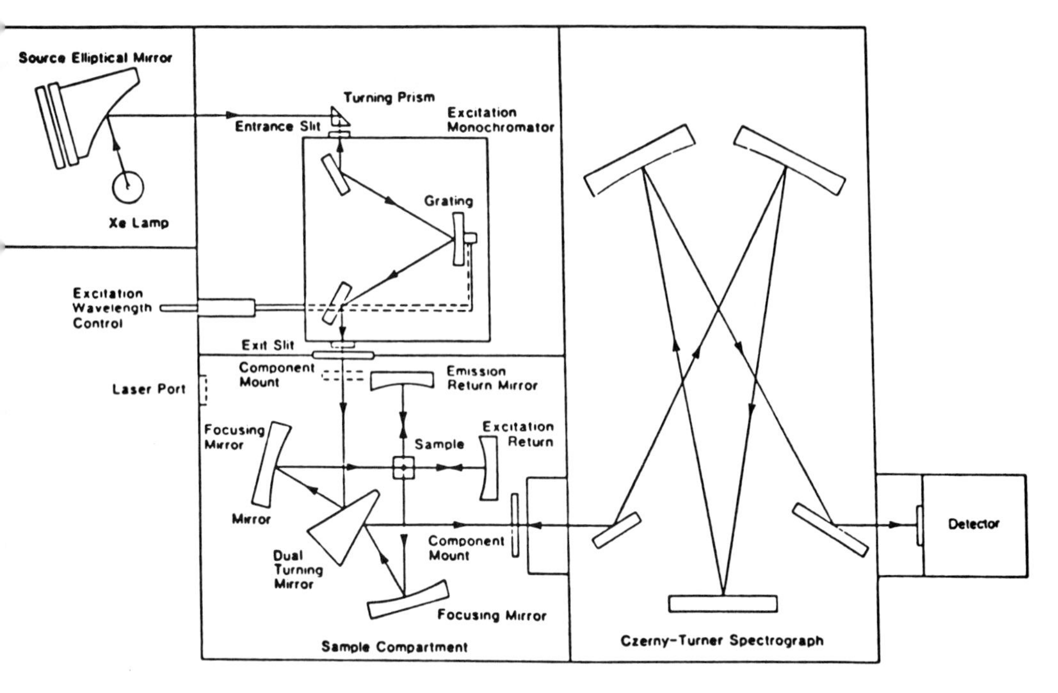

Figure 14. Schematic diagram of a fluorometer with array detection. (From Tracor-Northern, with permission.)

scanned excitation wavelengths. The use of two-dimensional array detectors can further facilitate the collection of total luminescence spectra. A rapid-scanning spectrofluorometer is shown in Figure 15 [48]. The exit slits are removed from the excitation and emission monochromators to form polychromators. The excitation wavelengths are spatially resolved along the vertical axis of the sample cuvette and therefore of the detector, and the emission wavelengths are spatially resolved along the horizontal axis of the detector (Figure 16). A silicon intensified target vidicon is used for the two-dimensional detection. Other imaging devices, including those listed in Section III.A, can also be used. In addition to the reduction in analysis time, advantages of array detection include multiwavelength kinetic studies that would be difficult or impossible with sequential scanning, and minimization of sample decomposition due to inherent instability or photochemical reactions.

Polarization Measurements

Fluorescence polarization can be measured with both fluorometers and spectrofluorometers. Polarizers are placed in the excitation and emission beams, each between the corresponding monochromator or filter and the sample. Two different formats can be used [14]. The "L" format (Figure 17)

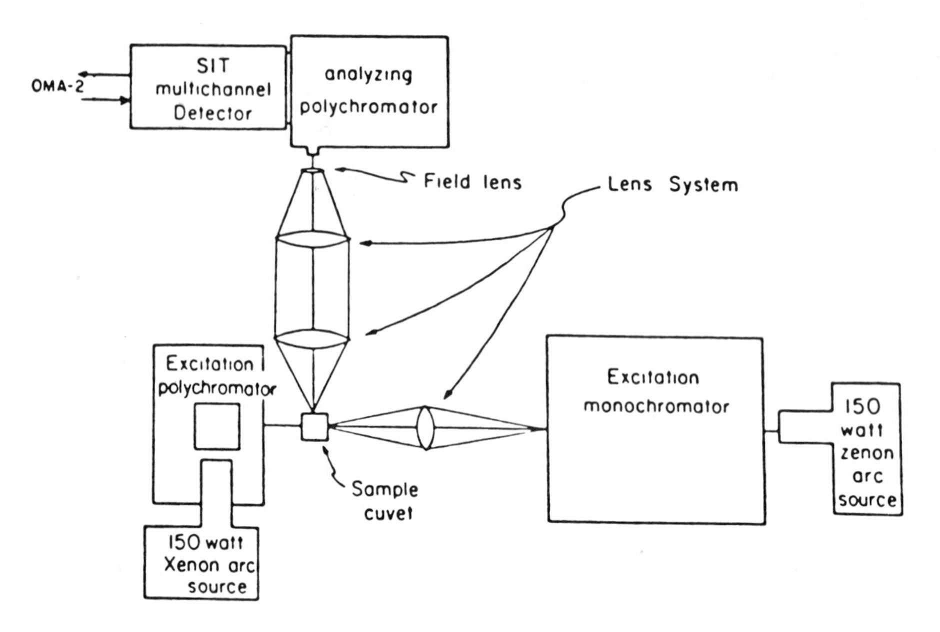

Figure 15. Schematic diagram of a rapid scanning fluorometer. (From Ref. 107, with permission.)

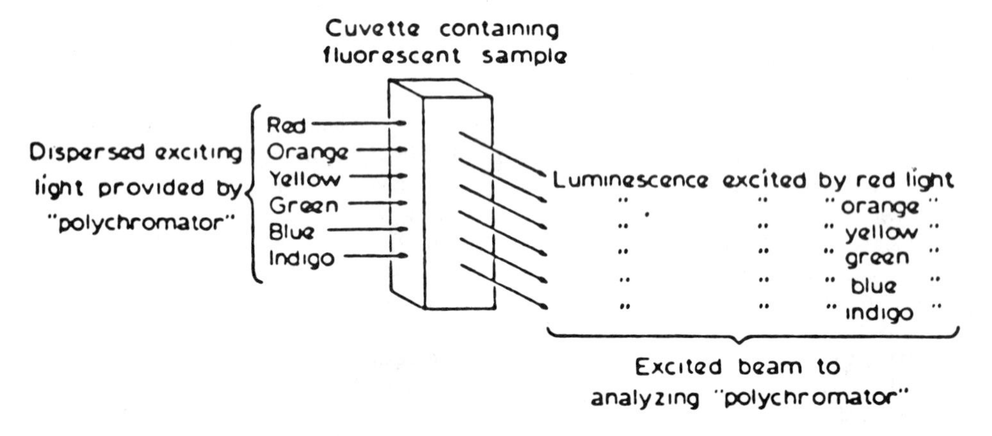

Figure 16. Illumination of a sample with polychromatic light. (From Ref. 46, with permission.)

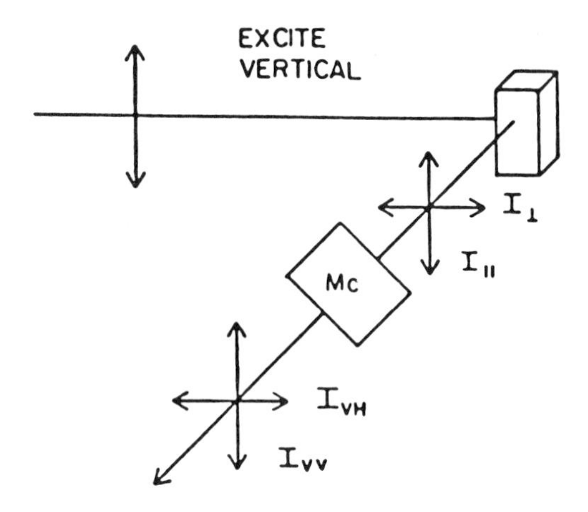

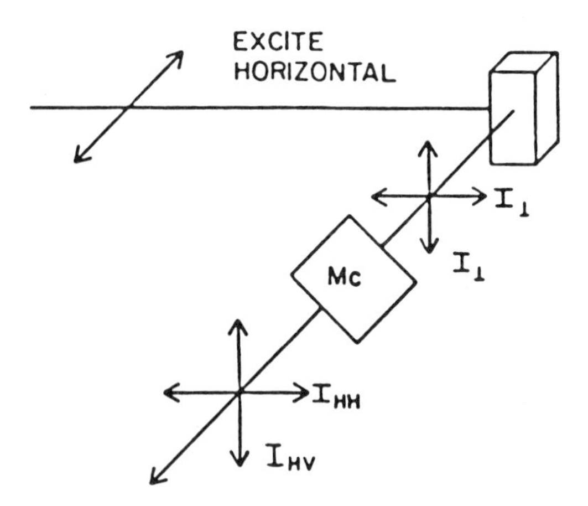

Figure 17. L Format for polarization measurements. MC, monochromator. (From Ref. 14, with permission.)

uses the conventional fluorometer configuration with a single excitation channel and a single emission channel. A correction factor must be used to account for partial polarization that occurs in monochromators and for the dependence of monochromator transmission efficiency on the polarizer angle. The "T" format (Figure 18) uses a single excitation channel and two emission channels, one for the vertical component and the other for the horizontal component. The two emission signals are measured ratiometrically to eliminate errors due to polarization that occurs in the excitation monochromator. A correction factor must still be used to account for the different sensitivities and polarizer settings of the two emission channels.

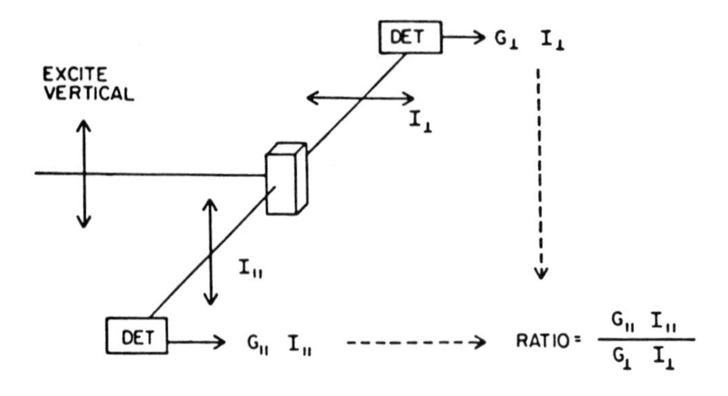

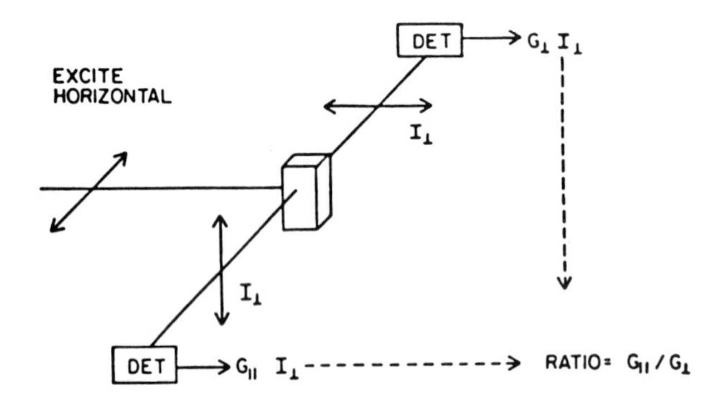

Figure 18. T Format for polarization measurements. DET, detector; G, gain. (From Ref. 14, with permission.)

Luminescence Lifetime Measurements and Related Techniques

Numerous books, chapters, and monographs have discussed techniques and instrumentation used for the determination of excited-state lifetimes and related techniques [14,49-54]. The most conceptually direct approach is the use of pulsed-source excitation and direct acquisition of the decay curve, which is analyzed to yield the exponential decay characteristic(s) of the emitting component(s). The performance of this technique depends on the width and reproducibility of the excitation pulses relative to the lifetime range under investigation as well as detector response, and is generally limited to lifetimes of several nanoseconds or longer. Measurements of subnanosecond fluorescence lifetimes with pulsed source excitation generally require more sophisticated techniques. The instrument shown in Figure 19 uses a flashlamp source with a high repetition rate and time-correlated single-photon counting (TCSPC) [55] to extend the range of lifetimes that can be determined into the subnanosecond range. Each pulse generates a start signal at a PMT, which triggers the voltage ramp of the time-to-amplitude converter (TAC) (Figure 20). The ramp is terminated when a photon emitted from the sample is detected. Therefore, the amplitudes of the TAC output pulses are proportional to the time between the start and stop signals, which in turn are statistically related to the luminescence lifetime. Lasers with high repetition rates can be used with fast PMT detection to extend the use of TCSPC well into the subnanosecond region. As with steady-state single-photon counting measurements, TCSPC is designed for the detection of very-low-level photon signals.

The long lifetimes associated with phosphorescence emission are readily measured by simple pulsed-source systems with direct acquisition of the decay curves. A delay time is imposed to allow fluorescence emission to completely decay before measurements are begun. Alternatively, a simple chopper device may be used to eliminate the fluorescence signal.

Phase-modulation luminescence spectroscopy [56-58] is an alternative to pulsed methods for luminescence lifetime determinations. The excitation beam from a continuous source (lamp or laser) is modulated at an appropriate frequency (megahertz range for fluorescence lifetimes, hertz range for phosphorescence lifetimes). As shown in Figure 21, the luminescence emission will be phase-shifted by angle ϕ and demodulated by a factor m relative to the excitation beam. The values of ϕ and m depend upon the modulation frequency and the luminescence lifetime of the sample, which can be independently calculated from ϕ and from m. A phase-modulation instrument for fluorescence lifetime determinations is shown in Figure 22.

Time-resolved methods, based on the pulsed-excitation approach, and phase-resolved spectroscopy [59-65], based on the phase-modulation technique, can provide selectivity for chemical analysis based on luminescence lifetime differences, as well as providing a means for eliminating background interferences (Section IV.C).

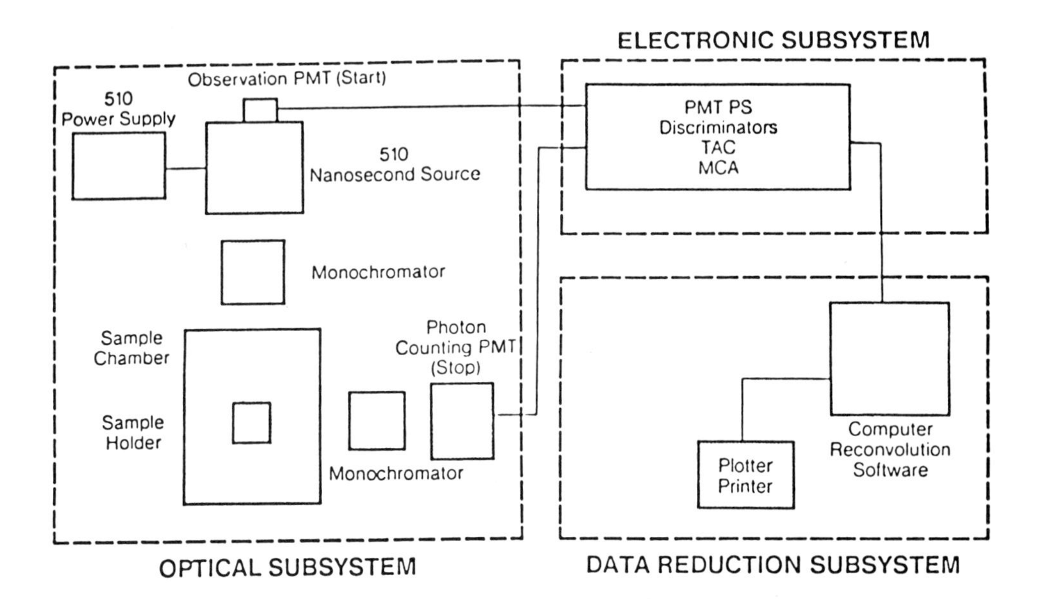

Figure 19. Time-correlated single-photon counting fluorescence lifetime instrument. (From PRA, with permission.)

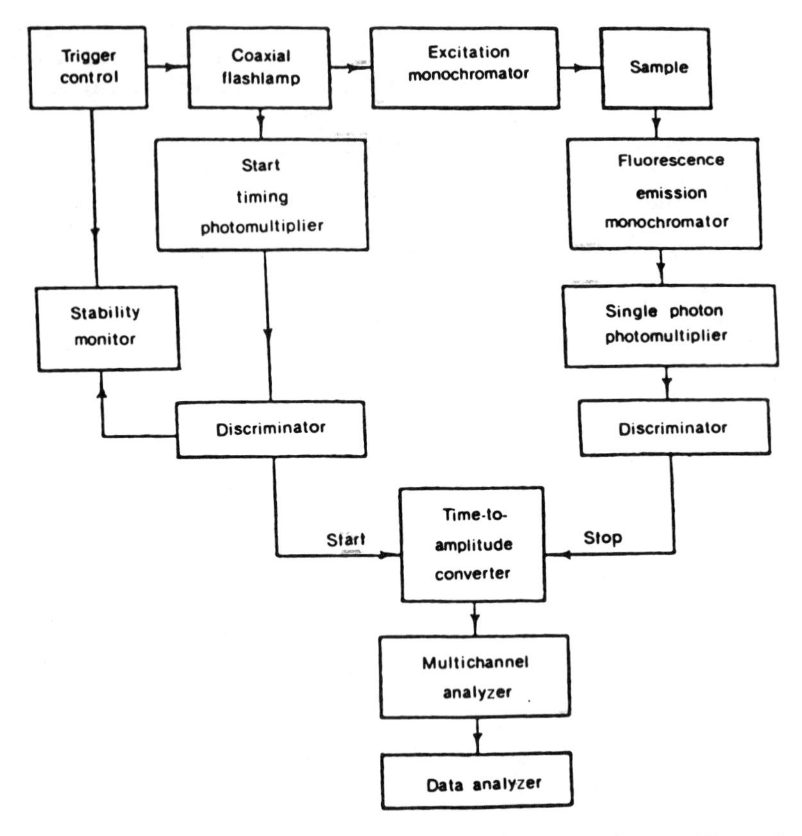

Figure 20. The single-photon correlation technique. (From Ref. 108, with permission.)

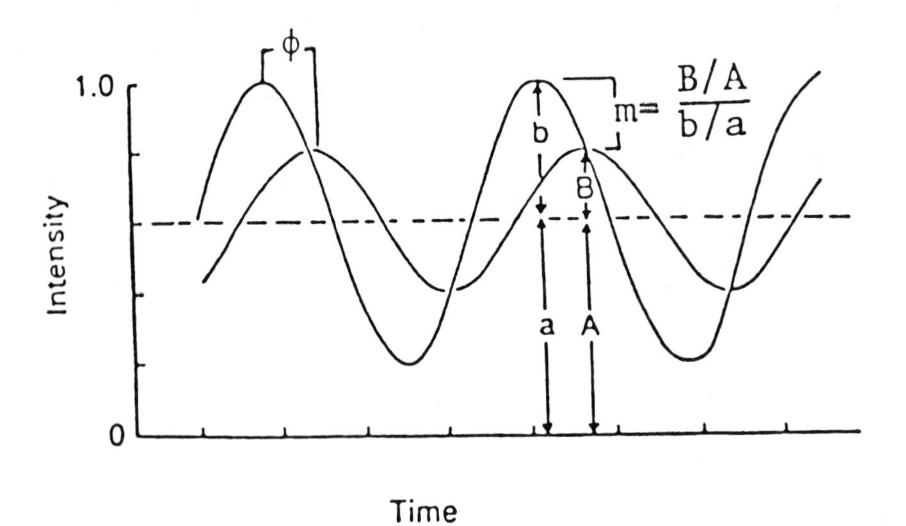

Figure 21. Depiction of the phase-modulation fluorescence lifetime technique. (Adapted from Ref. 14, with permission.)

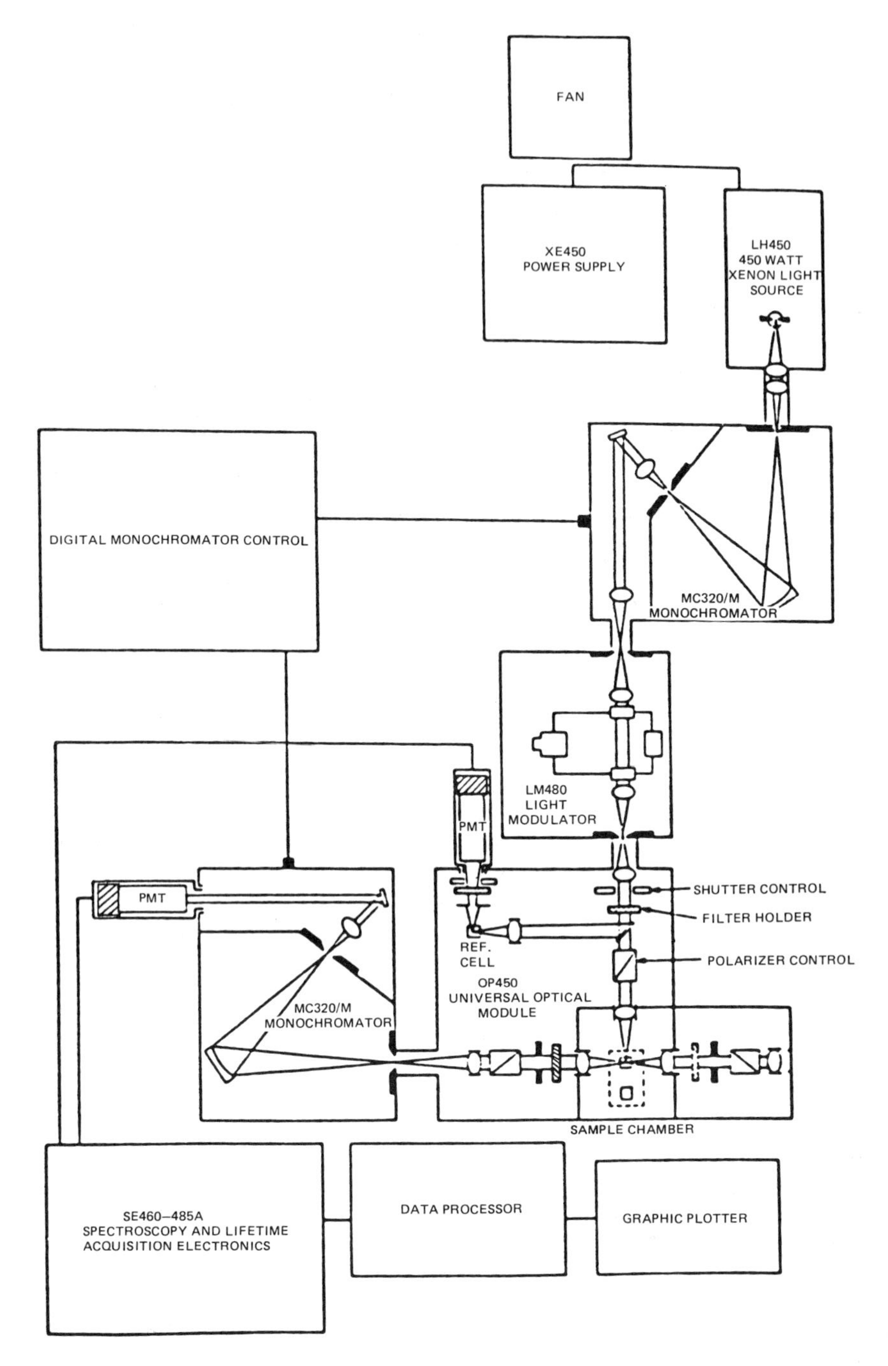

Figure 22. Phase-modulation spectrofluorometer. (From SLM Instruments, with permission.)

Fiber Optics

Optical fibers can be used to transport light to and from luminescent samples. Recent reviews have described the principles and use of fiber optics [66,67], which is a rapidly growing area of interest with a wide range of applications. Wavelengths ranging from 250 to 1300 nm can be transmitted, depending on the fiber material. Advantages of optical fibers include their low cost, the simplicity and flexibility of experimental configurations, and the ability to take in situ measurements rather than requiring that the sample be placed in a container located in the sample compartment of an instrument. Relatively inaccessible locations can be probed and analytes can be directly measured in the field [68], in process streams, and in vivo for biological studies [69]. Optical fiber fluoroprobes [70] that have a reagent phase immobilized on the probe can be used as chemical sensors, as an alternative to electrochemical sensors.

IV. PRACTICAL CONSIDERATIONS AND APPLICATIONS

A recent monograph has addressed the practical aspects of luminescence measurements, including calibration, spectral correction, data handling and other topics [71]. Discussions of practical considerations can also be found in the texts cited in Section I [1-8].

A. Environmental Effects

Luminescence measurements are very sensitive to experimental conditions due to the competition of the fluorescence process with other deexcitation processes, and the complexities of the vibronic structure of the molecular energy levels. The sensitivity to environmental effects can be a very valuable means for probing microenvironments of macromolecular structures [14] such as micelles [72], proteins [73], biological membranes [74], and cells, and for studying solvation processes [75], excited-state processes [76,77], molecular and macromolecular rotations [14], and many other processes. On the other hand, experimental conditions such as solvent composition and temperature must be very carefully controlled for qualitative and quantitative analyses in order to obtain accurate and reproducible results. For aqueous solutions, pH and ionic strength effects must be considered.

As discussed in Section II.E, dissolved oxygen quenches luminescence and decreases fluorescence lifetimes. Deoxygenation is desirable and often essential for many luminescence experiments, and can be accomplished by passing high-purity helium or nitrogen through the sample or by alternate methods such as rapid freeze-thaw cycles or molecular dialysis [78,79]. Other quenchers in the sample matrix must also be taken into account, as well as inner-filter effects, self-quenching and excited-state quenching processes.

B. Wavelength Calibration and Spectral Correction

Emission monochromators are commonly calibrated with a standard source such as a mercury arc lamp, which has a number of accurately known emission lines ranging from 253.65 to 809.32 nm. Excitation monochromators can then be calibrated by using a scattering solution to direct the calibrated excitation line through the emission monochromator to the detector.

Acquired spectra can be corrected for the wavelength dependencies of source output, monochromator efficiency, and detector response by various methods that include the use of thermopiles, quantum counters, calibrated phototubes with known spectral response curves, and solutions of standard fluorescence compounds.

C. Background and Interfering Signals

A common source of background intensity is scattered light resulting from particulate, Rayleigh, and Raman scattering phenomena. Fluorescence and phosphorescence may also be mutually interfering. Scattered light has an effective lifetime of zero on the nanosecond time scale of fluorescence, and phosphorescence lifetimes are on the millisecond time scale. Therefore, time-resolved and phase-resolved approaches can be used for the elimination of scattered light interferences in luminescence measurements [80,81], and of fluorescence interference in phosphorescence measurements [60]. Phase-resolved techniques can also be used to directly suppress longer-lived emission in order to measure the shorter-lived emission, such as measurements of Raman scattered light in the presence of fluorescence [80].

Background signals can also be corrected by blank subtraction, provided that the background signal is the same in the blank and the sample. Scattered light is usually easily identified in spectra and can often be avoided by appropriate choices of excitation and emission wavelength. Derivative and wavelength modulation techniques [82] can be used to improve selectivity and reduce background interferences.

D. Direct Quantitative and Qualitative Analysis

Luminescent molecules can be directly determined from measurements of luminescence intensity [equations (7) and (8), Section II.E]. Molecules that exhibit native luminescence include organic molecules with conjugated π-bond systems such as polycyclic and polynuclear aromatic hydrocarbons, fluorescent metal chelates, and certain rare earth elements and inorganic lanthanide and uranyl compounds. Simultaneous multicomponent determinations of luminescent compounds in mixtures can be performed if sufficient selectivity is available and the compounds absorb and emit independently in the absence of synergistic effects. Selectivity can be achieved through the exploitation of emission and excitation spectral characteristics, luminescence lifetimes, and selective enhancement and quenching [83]. Qualitative analyses are also based on one or more of these dimensions of information. For complex mixtures, techniques such as synchronous excitation and total luminescence (Section III.D) can be used for multicomponent determinations and qualitative analysis [41-47].

E. Low-Temperature Techniques

As discussed in Sections II.E and III.A, phosphorescence measurements are often made on solid materials at low temperatures. Low-temperature fluorescence techniques are also sometimes used [84]. In matrix isolation spectroscopy [85-87], the sample is vaporized, mixed with a large excess of a diluent gas, and deposited on a solid surface at a low temperature (10-15 K or lower) in order to minimize aggregation and quenching effects. Another technique employs the Shpol'skii effect [88-90] to obtain high resolution spectra in *n*-alkane or other solvents at low temperatures (77 K). High-resolution Shpol'skii spectra can be used for fingerprinting of fossil fuels and other organic samples. Resolution can be increased by using Shpol'skii media with the matrix isolation technique. These techniques can provide lower detection limits, wider linear response ranges, and better resolution and selectivity for many samples. However, they have not been widely used on a routine basis for chemical analysis because of the care and skill required for sample preparation.

F. Room-Temperature Phosphorescence

The application of phosphorescence spectroscopy to chemical analysis may be significantly increased by the use of room-temperature phosphorescence (RTP) [39] to avoid the need for low temperatures and frozen samples. Techniques include solid-surface RTP, micelle-stabilized RTP, and solution-sensitized RTP. Solid surfaces that have been used include cellulose, alumina, silica gel, sodium acetate, sucrose, chalk, inorganic mixtures, and polymer salt mixtures. Micelle stabilization involves the use of micelles to solubilize, stabilize, and enhance phosphorescent species in solutions. Heavy atoms are often used in both solid-surface and micelle techniques to enhance phosphorescence by increasing the rate of intersystem crossing from excited singlet to overlapping triplet states. Solution-sensitized RTP can be used to indirectly determine molecules that emit little or no phosphorescence. The analyte absorbs photons and transfers the energy via nonradiative excitation transfer to a phosphorescent acceptor molecule. The phosphorescence of the acceptor is measured and used to determine the analyte.

G. Other Techniques and Applications

A wide variety of approaches has been taken to the application of luminescence spectroscopy to chemical analysis and characterization. Indirect luminescence methods in which a nonluminescent analyte is coupled to a luminescent reagent can be used to extend the applicability of luminescence analysis. Kinetic methods [91] for both direct and indirect luminescence analysis have been described for a variety of analytes, including determinations of enzymes and their substrates. Immunochemical techniques [92,93] involving luminescent-labeled antigens, haptens, or antibodies have proven to be an important alternative to radioactive labeled reagents. Spectral fingerprinting with total luminescence spectra and synchronous excitation spectra have been used for the characterization of fossil fuels and other samples containing polycyclic aromatic hydrocarbons [94-96], and of human serum [97].

Fluorescence anisotropy and lifetime measurements have been widely used for the study of biological and biochemical systems (Section IV.A) [14]. Native protein fluorescence has been examined, in addition to the fluorescence of probe molecules that bind to sites on proteins. Fluorescence probes are also useful for studying transport across membranes. A variety of fluorescent probes has been described [14,98].

Luminescence detection for HPLC has become increasingly popular and includes the use of fluorescence detection [99,100] as well as micelle-stabilized RTP [101] and sensitized RTP [102] techniques. Detection for gas chromatography has also been discussed [99]. Rapid scanning detectors can be used for on-line acquisition of luminescence spectra of chromatographic peaks [32,103].

In the past decade there has been high activity and interest in many areas of luminescence instrumentation and applications. Phosphorescence analysis has greatly benefited by room-temperature and solid-surface techniques. The use of organized media such as micelles and cyclodextrins has great potential for increasing the selectivity, sensitivity, and applicability of luminescence analysis [104]. Advances in instrumentation for dynamic lifetime and time-resolved techniques have extended measurements well into the picosecond range. The availability of commercial instruments has made excited-state lifetime-based techniques accessible to a wider range of users. Multichannel detection and on-line computerization have greatly facilitated data acquisition and analysis. Future trends should include continued expansion of areas of application of conventional luminescence instruments and techniques for routine analyses and experiments, as well as increased accessibility of state-of-the-art instrumentation for both research and routine applications.

REFERENCES

1. G. G. Guilbault, *Practical Fluorescence: Theory, Methods and Techniques*, Marcel Dekker, New York, 1973.
2. S. G. Schulman, *Fluorescence and Phosphorescence Spectroscopy: Physicochemical Principles and Practice*, Pergamon, London, 1977.
3. M. Zander, *Phosphorimetry*, Academic Press, New York, 1968.
4. D. M. Hercules, Ed., *Fluorescence and Phosphorescence Analysis*, Wiley-Interscience, New York, 1965.
5. J. D. Winefordner, S. G. Schulman, and T. C. O'Haver, *Luminescence Spectrometry in Analytical Chemistry*, Wiley-Interscience, New York, 1972.
6. C. E. White and R. J. Argauer, *Fluorescence Analysis: A Practical Approach*, Marcel Dekker, New York, 1970.
7. A. Pesce, C-G. Rosen, and T. L. Pasby, Eds., *Fluorescence Spectroscopy: An Introduction for Biology and Medicine*, Marcel Dekker, New York, 1971.
8. L. B. McGown, in *Metals Handbook 9th Edition Vol. 10: Materials Characterization*, ASM Handbook Committee, American Society for Metals, Metals Park, OH, pp. 72-81, 1986.

9. E. L. Wehry, Ed., *Modern Fluorescence Spectroscopy*, Plenum Press, New York (Vols. 1 and 2, 1976, Vols. 3 and 4, 1981).
10. S. G. Schulman, Ed., *Molecular Luminescence Spectroscopy: Methods and Applications Part 1*, Wiley-Interscience, New York, 1985.
11. D. Eastwood, Ed., *New Directions in Molecular Luminescence*, STP 822, American Society for Testing and Materials, Philadelphia, 1983.
12. L. J. Cline Love and D. Eastwood, Eds., *Advances in Luminescence Spectroscopy*, STP 863, American Society for Testing and Materials, Philadelphia, 1985.
13. I. M. Warner and L. B. McGown, Recent Advances in Multicomponent Fluorescence Analysis, *Critical Reviews in Analytical Chemistry*, Vol. 13, CRC Press, Boca Raton, FL, pp. 155-222, 1982.
14. J. R. Lakowicz, *Principles of Fluorescence Spectroscopy*, Plenum Press, New York, 1983.
15. C. A. Parker, *Photoluminescence of Solutions*, Elsevier, Amsterdam, 1968.
16. J. B. Birks, *Photophysics of Aromatic Molecules*, Wiley-Interscience, London, 1970.
17. J. B. Birks, *Organic Molecular Photophysics*, Vol. 2, John Wiley and Sons, London, 1975.
18. S. P. McGlynn, T. Azumi, and M. Knoshita, *Molecular Spectroscopy of the Triplet State*, Prentice-Hall, Englewood Cliffs, NJ, 1969.
19. R. S. Becker, *Theory and Interpretation of Fluorescence and Phosphorescence*, Wiley-Interscience, New York, 1969.
20. I. B. Berlman, *Handbook of Fluorescence Spectra of Aromatic Molecules*, 2nd ed., Academic Press, New York, 1971.
21. W. F. Coleman, J. Chem. Educ. *59*, 441 (1982).
22. J. Wilson, in *Applications of Lasers to Chemical Problems* (T. R. Evans, ed.), John Wiley and Sons, New York, pp. 1-34, 1982.
23. H. W. Latz, in *Modern Fluorescence Spectroscopy*, Vol. 1 (E. L. Wehry, ed.), Plenum Press, New York, 1976.
24. M. Maeda, *Laser Dyes: Properties of Organic Compounds for Dye Lasers*, Academic Press, New York, 1984.
25. J. C. Wright, in *Laser Applications in Chemistry* (K. Kompa and J. Wanner, Eds.), Plenum Press, New York, pp. 67-73, 1984.
26. R. N. Zare, Science *226*, 298 (1984).
27. J. H. Richardson, in *Modern Fluorescence Spectroscopy*, Vol. 4 (E. L. Wehry, Ed.), Plenum Press, New York, pp. 1-24, 1981.
28. I. M. Warner, J. B. Callis, E. R. Davidson, M. Gouterman, and G. D. Christian, Anal. Lett. *8*, 665 (1975).
29. Y. Talmi, Anal. Chem. *47*, 658A and 699A (1975).
30. J. D. Ingle, Jr., and M. A. Ryan, in *Multichannel Image Detectors*, Vol. II (Y. Talmi, Ed.), ACS Symposium Series 236, American Chemical Society, Washington, DC, pp. 155-170, 1983.
31. D. W. Johnson, J. B. Callis, and G. D. Christian, in *Multichannel Image Detectors* (Y. Talmi, Ed.), ACS Symposium Series 102, American Chemical Society, Washington, DC, pp. 97-114, 1979.
32. G. D. Christian, J. B. Callis, and E. R. Davidson, in *Modern Fluorescence Spectroscopy*, Vol. 4 (E. L. Wehry, Ed.), Plenum Press, New York, pp. 111-165, 1981.

33. J. E. Wampler, in *Modern Fluorescence Spectroscopy*, Vol. 1 (E. L. Wehry, Ed.), Plenum Press, New York, pp. 1-43, 1976.
34. T. Vo-Dinh, G. L. Walden, and J. D. Winefordner, Anal. Chem. *49*, 1126 (1977).
35. T. Vo-Dinh, R. Paltzold, and U. P. Wild, Z. Phys. Chem. *251*, 395 (1972).
36. T. D. S. Hamilton and K. R. Naqvi, Anal. Chem. *45*, 1581 (1973).
37. G. D. Boutillier and J. D. Winefordner, Anal. Chem. *51*, 1384 (1979).
38. D. E. Goeringer and H. L. Pardue, Anal. Chem. *51*, 1054 (1979).
39. T. Vo-Dinh, *Room Temperature Phosphorimetry for Chemical Analysis*, Wiley-Interscience, New York, 1984.
40. R. J. Hurtubise, *Solid Surface Luminescence Analysis: Theory, Instrumentation, Applications*, Marcel Dekker, New York, 1981.
41. J. B. F. Lloyd, Nature (Lond.), Phys. Sci. *231*, 64 (1971).
42. T. Vo-Dinh, Anal. Chem *50*, 396 (1978).
43. T. Vo-Dinh, in *Modern Fluorescence Spectroscopy*, Vol. 4 (E. L. Wehry, Ed.), Plenum Press, New York, pp. 167-192, 1981.
44. G. Weber, Nature (Lond.) *190*, 27 (1961).
45. G. W. Suter, A. J. Kallir, and U. P. Wild, Chimia, *37*, 413 (1983).
46. I. M. Warner, J. B. Callis, E. R. Davidson, and G. D. Christian, Clin. Chem. *22*, 1483 (1976).
47. L. P. Giering and A. W. Hornig, Am. Lab. *9*, 113 (1977).
48. D. W. Johnson, J. A. Gladden, J. B. Callis, and G. D. Christian, Rev. Sci. Instrum. *50*, 118 (1979).
49. J. N. Demas, *Excited State Lifetime Measurements*, Academic Press, New York, 1983.
50. G. M. Hieftje and E. E. Vogelstein, in *Modern Fluorescence Spectroscopy*, Vol. 4 (E. L. Wehry, Ed.), Plenum Press, New York, pp. 25-50, 1981.
51. J. Yguerabide and E. E. Yguerabide, in *Optical Techniques in Biological Research* (D. L. Rousseau, Ed.), Academic Press, New York, pp. 181-290, 1984.
52. A. J. W. G. Visser, Ed., Anal. Instrum. *3 and 4*, 193-566 (1985).
53. R. B. Cundall and R. E. Dale, Eds., *Time-Resolved Fluorescence Spectroscopy in Biochemistry and Biology*, NATO ASI Series A: Life Sciences Vol. 69, Plenum Press, New York, 1983.
54. W. R. Ware, in *Creation and Detection of the Excited State*, Vol. IA (A. A. Lomola, Ed.), Marcel Dekker, New York, pp. 213-302, 1971.
55. D. V. O'Conner and D. Phillips, *Time-Correlated Single Photon Counting*, Academic Press, London, 1984.
56. Z. Gaviola, Z. Phys., *42*, 852 (1927).
57. J. B. Birks and D. J. Dyson, J. Sci. Instrum. *38*, 282 (1961).
58. R. D. Spencer and G. Weber, Ann. N.Y. Acad. Sci. *158*, 361 (1961).
59. T. V. Veselova, A. S. Cherkasov, and V. I. Shirokov, Opt. Spectrosc. *29*, 617 (1970).
60. J. J. Mousa and J. D. Winefordner, Anal. Chem. *46*, 1195 (1974).
61. E. Lue-Yen and J. D. Winefordner, Anal. Chem. *49*, 1262 (1977).
62. J. R. Lakowicz and H. Cherek, J. Biochem. Biophys. Methods *5*, 19 (1981).

63. J. R. Mattheis, G. W. Mitchell, and R. D. Spencer, in *New Directions in Molecular Luminescence* (D. Eastwood, Ed.), ASTM STP 822, American Society for Testing and Materials, Baltimore, MD, pp. 50-64 (1983).
64. L. B. McGown and F. V. Bright, Anal. Chem. *56*, 1400A (1984).
65. L. B. McGown and F. V. Bright, *Critical Reviews in Analytical Chemistry*, 18, CRC Press, Boca Raton, FL, pp. 245-298 (1987).
66. I. Chabay, Anal. Chem. *54*, 1071A (1982).
67. J. F. Rabek, *Experimental Methods in Photophysics, Part 1*, Wiley-Interscience, New York, pp. 272-286, 1982.
68. B. E. Jones, J. Phys. E. *18*, 770 (1985).
69. J. I. Peterson and G. G. Vurek, Science *224*, 123 (1985).
70. O. S. Wolfbeis, Trends Anal. Chem. *4*, 184 (1985).
71. K. D. Mielenz, *Measurement of Photoluminescence*, Academic Press, New York, 1982.
72. M. Gratzel and J. K. Thomas, in *Modern Fluorescence Spectroscopy*, Vol. 2 (E. L. Wehry, Ed.), Plenum Press, New York, pp. 169-216, 1976.
73. J. E. Churchich, in *Modern Fluorescence Spectroscopy*, Vol. 2 (E. L. Wehry, Ed.), Plenum Press, New York, pp. 217-237, 1976.
74. R. A. Badley, in *Modern Fluorescence Spectroscopy*, Vol. 2 (E. L. Wehry, Ed.), Plenum Press, New York, pp. 91-168, 1976.
75. T. C. Werner, in *Modern Fluorescence Spectroscopy*, Vol. 2 (E. L. Wehry, Ed.), Plenum Press, New York, pp. 277-317, 1976.
76. P. Froehlich and E. L. Wehry, in *Modern Fluorescence Spectroscopy*, Vol. 2 (E. L. Wehry, Ed.), Plenum Press, New York, pp. 319-438, 1976.
77. S. G. Schulman, in *Modern Fluorescence Spectroscopy*, Vol. 2 (E. L. Wehry, Ed.), Plenum Press, New York, pp. 239-275, 1976.
78. M. E. Rollie, G. Patonay, and I. M. Warner, Anal. Chem. *59*, 180 (1987).
79. M. E. Rollie, C-N. Ho, and I. M. Warner, Anal. Chem. *55*, 2445 (1983).
80. J. N. Demas and R. A. Keller, Anal. Chem. *57*, 538 (1985).
81. K. Nithipatikom and L. B. McGown, Anal. Chem. *58*, 3145 (1986).
82. T. C. O'Haver, in *Modern Fluorescence Spectroscopy*, Vol. 1 (E. L. Wehry, Ed.), Plenum Press, New York, pp. 65-81, 1976.
83. I. M. Warner, G. Patonay, and M. P. Thomas, Anal. Chem. *57*, 463A (1985).
84. E. L. Wehry and G. Mamantov, in *Modern Fluorescence Spectroscopy*, Vol. 4 (E. L. Wehry, Ed.), Plenum Press, New York, pp. 193-250, 1981.
85. J. S. Shirk and A. M. Bass, Anal. Chem. *41*, 103A (1969).
86. R. C. Stroupe, P. Tokousbalides, R. B. Dickinson, Jr., E. L. Wehry, and G. Mamantov, Anal. Chem. *49*, 701 (1977).
87. J. L. Metzger, B. E. Smith, and B. Meyer, Spectrochim. Acta *25A*, 1177 (1969).
88. E. V. Shpol'skii, A. A. Il'ina, and L. A. Klimova, Dokl. Akad. Nauk. SSSR *87*, 935 (1959).
89. E. V. Shpol'skii, Sovt. Phys. Uspekhi *2*, 378 (1959).
90. E. V. Shpol'skii, Sovt. Phys. Uspekhi *3*, 372 (1960).

91. J. D. Ingle, Jr., and M. A. Ryan, in *Modern Fluorescence Spectroscopy*, Vol. 3 (E. L. Wehry, Ed.), Plenum Press, New York, pp. 95-142, 1981.
92. D. S. Smith, M. Hassan, and R. D. Nargessi, in *Modern Fluorescence Spectroscopy*, Vol. 3 (E. L. Wehry, Ed.), Plenum Press, New York, pp. 143-191, 1981.
93. H. T. Karnes, J. S. O'Neal, and S. G. Schulman, in *Molecular Luminescence Spectroscopy: Methods and Applications: Part 1* (S. G. Schulman, Ed.), Wiley-Interscience, New York, pp. 717-779, 1985.
94. P. John and I. Soutar, Anal. Chem. *48*, 520 (1976).
95. J. A. Siegel, Anal. Chem. *57*, 934A (1985).
96. H. von der Dick and W. Kalkreuth, Fuel *63*, 1636 (1984).
97. O. S. Wolfbeis and M. Leiner, Anal. Chim. Acta *167*, 203 (1985).
98. G. S. Beddard and M. A. West, *Fluorescent Probes*, Academic Press, London, 1981.
99. P. Froehlich and E. L. Wehry, in *Modern Fluorescence Spectroscopy*, Vol. 3 (E. L. Wehry, Ed.), Plenum Press, New York, pp. 35-94, 1981.
100. D. C. Shelly and I. M. Warner, Chromatogr. Sci. *23*, 87 (1983).
101. R. Weinberger, P. Yarmchuk, and L. J. Cline Love, Anal. Chem. *54*, 1552 (1982).
102. J. J. Donkerbroek, N. J. R. Van Eikema Hommes, C. Gooijer, N. H. Velthorst, and R. W. Frei, Chromatographia *15*, 218 (1982).
103. J. R. Jadamec, W. A. Saner, and Y. Talmi, Anal. Chem. *49*, 1316 (1977).
104. L. J. Cline Love, J. G. Habarta, and J. G. Dorsey, Anal. Chem. *56*, 1132A (1984).
105. P. John and I. Soutar, Chem. Br. *17*, 278 (1981).
106. C.-P. Pau and I. M. Warner, Appl. Spectrosc. *41*, 496 (1987).
107. I. M. Warner, M. P. Fogarty, and D. C. Shelly, Anal. Chim. Acta *109*, 361 (1979).
108. D. J. S. Birch and R. E. Imhof, Eur. Spectrosc. News *48*, 31 (1983).

9

Instrumentation for Raman Spectroscopy

D. L. GERRARD and H. J. BOWLEY / BP Research Centre, Sunbury-on-Thames, Middlesex, England

I. INTRODUCTION

The choice of instrumentation for Raman spectroscopy depends, probably more than with any other spectroscopic technique, on the purpose for which that instrumentation is required. The principal items of hardware that need to be considered are the monochromator, the light source (laser), the detector, and the computing capability. No single combination of these items will be sufficient to fulfill all of the needs of a broad-based analytical Raman spectroscopy group. Each of the four requirements will be considered separately, but in the first instance it is worth looking briefly at the theory and history of the technique of Raman spectroscopy to see why its present status as an analytical spectroscopic method differs so much from other spectroscopic techniques, notably infrared, nuclear magnetic resonance, and mass spectrometry.

When a molecule is irradiated with light of a single fixed frequency, ν_0, then apart from the familiar effects of reflection, transmission, and absorption, there are also two molecular scattering effects. The first of these is elastic scattering and involves no exchange of energy between the molecule and the incident photons. This is the Rayleigh scatter, and appears at ν_0. The second effect is inelastic scattering and involves an exchange of energy between the incident photons and the vibrational energy levels of the molecule being irradiated. Some of the photons will lose energy and be shifted to lower frequencies (Stokes scatter), while others will gain energy and be shifted to higher frequencies (anti-Stokes scatter). The frequency shift will depend on the energy of the particular vibrational energy level concerned and will be independent of ν_0. Hence, since most molecules have more than one vibrational mode that is Raman active, there will be a range of frequencies in the scattered beam forming two mirror image spectra about ν_0 (Stokes and anti-Stokes), constituting the Raman spectrum of the molecule involved in the energy exchange. In most cases the majority of the molecules being studied will be in the ground state and so the Stokes spectrum will be considerably stronger than the anti-Stokes. This is illustrated in Figure 1, which shows the Raman spectrum (Stokes and anti-Stokes) of carbon tetrachloride. The large central peak that goes off scale is the Rayleigh scatter at ν_0, which is designated as 0 cm^{-1}. The Stokes and anti-Stokes spectra can be clearly seen to form mirror images, with the Stokes the more intense. Figures associated with

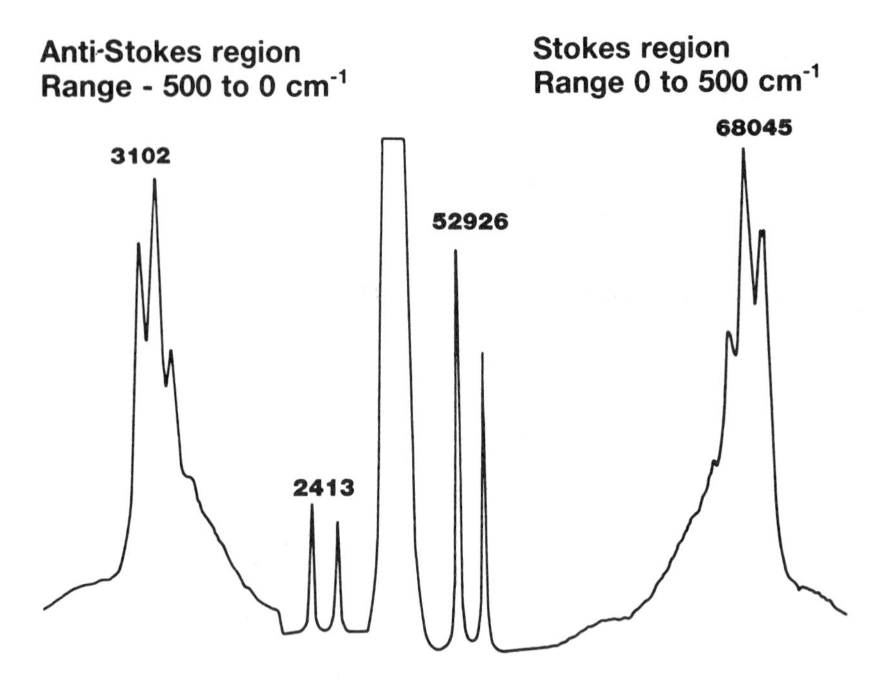

Figure 1. Stokes and anti-Stokes spectrum of carbon tetrachloride. Numerical values are the photon counts associated with the peak.

the peaks represent the photon counts. Hence, for most practical purposes Raman spectroscopy is only concerned with the Stokes scatter. There are some forms of Raman spectroscopy relating to nonlinear effects that use the anti-Stokes spectrum, but these are outside the scope of this chapter, and the reader is referred to the excellent book edited by Harvey [1], which covers all aspects of nonlinear Raman effects.

Although the Raman effect was demonstrated as long ago as 1928, which makes it very much a front runner in terms of spectroscopic techniques, its potential value as an analytical method is only just beginning to be realized. The vibrational energy levels involved in the effect are unique to any particular molecule and so, like infrared absorption spectra, which also involve vibrational modes, Raman spectra can be used for purposes of characterization and identification of unknown materials. However, the Raman spectrum is hardly ever used in this way, whereas the infrared technique is one of the most universal of all analytical methods. The reason for this lies in the very weak nature of the Raman effect. Of the incident photons from the monochromatic light source, typically only 1 in 10^8 exhibits inelastic scattering, with a resulting change in frequency or Raman shift. Hence, for many years after the effect was first demonstrated it could only be applied to the study of relatively simple, strongly scattering molecules, and the sensitivity was extremely poor. Even the most powerful light sources available, based on mercury vapor lamps, could not provide a sufficient photon flux to make the technique anything other than a source of complementary information to infrared spectroscopy, providing a limited amount of additional vibrational information about relatively few molecules.

This situation was changed by the development of continuous-wave gas ion lasers, notably the helium/neon, argon, and krypton lasers, in the late 1960s. These provided, for the first time, a sufficiently intense, monochromatic light source to make Raman spectroscopy a more broadly applicable technique with a sensitivity, in most cases, comparable with that of infrared or NMR. However, since the other spectroscopic methods had, in the interim period, established themselves as the preferred techniques for purposes of characterization and identification of unknown materials and for quantitative measurements, Raman spectroscopy found itself in the position of having to find applications in areas where these other techniques could not be readily applied. This gave rise to the growth of a number of Raman groups, each specializing in a comparatively narrow field and hence with specialized requirements in terms of instrumentation. It is only in the past few years that analytical groups have emerged that are applying the technique to a wide range of analytical problems and are taking advantage of the very high degree of versatility that the Raman effect offers. This has led to the commercial production of highly sophisticated spectrometers that incorporate, in one instrument, many different facilities. Despite these developments, however, the modern Raman spectroscopist still has some difficult choices to make, assuming he has a limited budget, to provide himself with the best system for his particular requirements.

II. THE LIGHT SOURCE

As mentioned above, the Raman effect is, in most circumstances, so weak that a very intense light source is required in order to obtain a detectable Raman signal, and this is now exclusively supplied by lasers. The most important of these initially was the helium-neon laser, particularly because of its relatively low price and its robust nature; it has now been largely superseded by the much more powerful argon ion and krypton ion lasers. These give a series of lines, which can each be used independently, and which are listed in Table 1. It should be noted that some of these lines, particularly those in the ultraviolet region, are only available with reasonable intensity from the most powerful versions of these lasers. On the whole the argon laser is of most value for studies requiring blue or green excitation, and the krypton laser for use in the yellow and red regions. Of the two, provided there is no particular wavelength requirement, the argon laser is probably more generally applicable, and if funds are limited it is normally the best option. This is because it can be obtained in higher power output versions than the krypton, is less vulnerable to pressure changes in the tube, and will give several lines without the need for changing the laser optics. All this goes to make it by far the easier of the two lasers for the nonexpert to use, and on the whole it has less down time than the krypton laser.

There are, however, other factors to be considered. For example, if the sample to be analyzed exhibits fluorescence, this may often be so

Table 1. Visible Wavelengths Emitted by Argon Ion and Krypton Ion Lasers

Laser	Wavelength (nm)
Ar^+	514.5
Ar^+	501.7
Ar^+	496.5
Ar^+	488.0
Ar^+	476.5
Ar^+	472.7
Ar^+	465.8
Ar^+	457.9
Ar^+	454.5
Kr^+	799.3
Kr^+	752.5
Kr^+	676.4
Kr^+	647.1
Kr^+	568.2
Kr^+	530.9
Kr^+	520.8
Kr^+	482.5
Kr^+	476.5
Kr^+	413.1

severe that it totally obscures the Raman signal. Fluorescence is normally considerably reduced when red excitation is used to obtain the Raman spectrum, and in this context the 752.5- and 799.3-nm lines available from the krypton laser can be particularly useful [2]. Similarly, it is often the case that samples that are vulnerable to photodecomposition or thermal damage due to absorption of the laser beam are much less prone to those effects with red excitation. If possible, therefore, a broadly based Raman analytical laboratory should be equipped with both argon and krypton lasers. If the analytical group can justify the use of more than one Raman spectrometer, then a 25-mW helium/neon laser is also a valuable asset because of its ease of use, reliability, and portability.

In general, the power of the laser purchased should be as high as possible, for several reasons. First, the power diminishes as the laser tube ages, and after 2-3 years in use it may well be down to only half of its original power. Second, if a high-power laser is available, then by careful planning of the location of the laser and the spectrometers and the use of a beam splitter it is possible to use one laser simultaneously on two spectrometers. Third, a high-power continuous-wave laser can be used in conjunction with a dye laser to give a high degree of wavelength tunability, if this is required. Reasons for requiring tunability of the laser source are considered in Section II.A. It must be remembered, however, that the laser tube only has a relatively short lifetime (typically 3-4 years) even when great care is taken to use very clean cooling water and to operate at power outputs well below the maximum. These tubes are expensive to replace, and the higher the output of the laser, the more expensive is the replacement tube.

As mentioned above, high-powered lasers can be used in conjunction with dye lasers to give a high degree of tunability. This is done by "pumping" a narrow jet of a solution of a dye in a suitable solvent (this varies with the particular dye being used) with a continuous-wave laser, usually argon. This produces a coherent broadband output, which can be tuned using an optical wedge over a narrow range. Again, the magnitude of the tunable wavelength range depends on the particular dye being used. The efficiency of these dyes is usually poor, with the result that the output at any particular wavelength is considerably less than the pump power. For this reason a high-power laser is essential. Also, it is only the high-power lasers that give a significant output in the ultraviolet region, which is required to pump dyes giving outputs in the blue region of the spectrum (<450 nm). The useful tunable range of these dyes is very variable, but is typically about 20 nm. Hence, in order to obtain tunability over a wide range several dyes have to be used. The number of dyes now available is considerable, and a very useful book by Maeda [3] is available that gives details of all the currently used compounds. It is possible to obtain complete tunability over the whole of the visible region of the spectrum by using a high-power argon ion laser and a suitable range of dyes. The main value of such tunability lies in the particular application of resonance-enhanced Raman spectroscopy and, to a lesser extent, surface-enhanced Raman spectroscopy. These are both methods of obtaining a Raman signal greatly increased over what is normally possible, by careful choice of the exciting line, and they will now be considered in some detail.

A. Resonance-Enhanced Raman Spectroscopy and Surface-Enhanced Raman Spectroscopy

When the laser wavelength being used to obtain the Raman spectrum lies close to or within the electronic absorption profile of a particular chromophore within a molecule, then the Raman signal due to that chromophore may be enhanced by several orders of magnitude compared with that of the rest of the molecule. The maximum enhancement occurs when the laser wavelength is at the point of maximum absorption. This effect is known as the resonance Raman effect, and for a full discussion of the theory the reader is referred to the article by Behringer [4]. Where this effect occurs the sensitivity of the technique is obviously much greater than normal and can be used to identify and quantify such chromophores at extremely low levels. This has been of particular value in studies of biologically important compounds [5,6], polymer degradation [7-9], and electroactive polymers [10-12]. In order to obtain maximum sensitivity in this type of work it is necessary to have a laser line as close to that of the appropriate electronic transition as possible, and so it is desirable to have the maximum degree of tunability of the laser system. With continuous-wave lasers, which even at their most powerful only give about 20 W (all lines), this tunability is really limited to the visible region of the spectrum. Although studies in this region have produced some extremely interesting and valuable results, it severely restricts the number of systems that can be studied. Since, for continuous-wave systems, the laser line being used to pump the dye laser is in the visible or just into the ultraviolet part of the spectrum, the dye laser output must necessarily be at longer wavelength (i.e., lower energy) than the pump laser, and this imposes the restriction. If the laser power is sufficiently high then the wavelength can be shortened by the use of frequency-doubling crystals, but outputs from continuous wave systems are not sufficiently great to make this a practical proposition. For tunability in the much more interesting ultraviolet region of the spectrum we have to resort to the use of pulsed lasers, and these are considered in Section II.B.

The second technique for producing a much enhanced Raman signal is known as surface-enhanced Raman spectroscopy (SERS). This is an effect whereby unusually high-intensity Raman signals are observed from molecules adsorbed on to electrochemically roughened metal electrode surfaces. The effect was first noted for the silver/pyridine system, for which it was thought initially to be specific. However, a large amount of subsequent work has shown it to be a more general phenomenon in terms of the characteristics of both the substrate and the adsorbate. In recent years many hundreds of papers have been published concerning both the experimental and theoretical aspects of SERS in an attempt to elucidate the exact nature of the effect [13]. Unlike the resonance effect, to which it must be closely related, SERS does not require a high degree of laser tunability, although the substrate dictates the exciting wavelength to be used. To date the only substrates that exhibit the effect for a useful range of materials are silver and gold and to a lesser extent copper and nickel. All of these metals can be used with standard outputs from argon or krypton ion lasers. However, the overall wavelength dependence of the enhancement remains a complex function of surface roughness and substrate dielectric properties. Other influencing factors include the extent of the surface coverage of the adsorbate and the working electrode potential. Although the SERS effect is still a relatively little exploited phenomenon, its potential value in electrochemical studies is considerable and is just beginning to make an impact in the literature.

B. Pulsed Systems

Because of the high powers involved, the principal use of pulsed laser systems to date has been in the area of nonlinear studies, and it is only relatively recently that they have been used in conventional Raman work. As mentioned above, their main value lies in the very high degree of tunability that they offer and hence the increased sensitivity that can be obtained by taking advantage of the resonance Raman effect. The group that has been outstanding in the application of tunable pulsed lasers for this purpose has been that of Asher [14-16]. He has applied a laser system based on a pulsed YAG laser to the study of a range of aromatic compounds.

The two commonly used systems are those based on either YAG or excimer lasers used in conjunction with dye lasers and frequency-doubling crystals. Such systems are now capable, by using a range of crystals and by frequency mixing as well as doubling and tripling, of giving complete tunability over the range of 190 nm to 4 μm. Their main disadvantage is their very low pulse rate (~10-300 Hz) and the short duration (~10 nsec) of the pulses. This means that the detector is only exposed to the Raman signal for a maximum of ~3 μsec in every second, but in the rest of the time it is still building up a noise signal. To avoid this situation the detector needs to be gated so that it is only operating during the duration of the laser pulse. The most successful Raman systems using pulsed lasers are based on gated intensified diode array detection, and the instrumentation described by Asher appears to be the best currently in use, although if considerable operator skill is available, the system described by Gustafson [17] should be seriously considered because of the quasi-continuous nature of his laser system.

Another feature of the excimer and YAG pulsed lasers is that they can be made to give output pulses of <1 psec in duration. Again, it requires considerable expertise to keep such systems operating and their limited value to an analytical Raman spectroscopy group means that they would normally only be considered for specialist use by a very well equipped group. One of the most important potential advantages of such short pulses is their potential application in the area of fluorescence rejection. The most serious limitation of Raman spectroscopy, which has prevented its application to a wide range of problems, particularly those relating to industrial systems, is that of sample fluorescence. It is often the case, particularly with oils and oil-based materials, commercial polymer samples, and catalysts, that the fluorescence produced by the laser beam is so great that it totally obscures the much weaker Raman signal. Various techniques have been used to overcome this problem, and it is worth considering the alternatives, briefly, because the particular technique chosen will depend very largely on the instrumentation available. The methods used fall into two categories: those that attempt to extract the Raman signal from the total signal (Raman + fluorescence) and those that prevent or reduce the fluorescence. The various alternatives are considered next.

Techniques That Prevent/Reduce Fluorescence

(a) "Burning out"

This is the most basic of the techniques available and is very limited in its application, being effective only for a relatively small number of thermally stable, solid inorganic compounds and unpigmented high-molecular-weight polymers. Basically the technique involves irradiating the fluorescing solid with a high-power laser source (usually > 1 W) of the same wavelength as that being used to obtain the Raman spectrum. In a limited number of cases this will gradually reduce the fluorescence and allow a Raman spectrum to be obtained. The disadvantages are:

1. It cannot be used for samples that exhibit a significant absorption at the wavelength being used unless they are highly thermally stable.
2. It cannot be used for samples that are liable to exhibit photodecomposition.
3. It may well decompose the particular compound being studied.
4. Since the area of sample "burned out" by the laser beam is very small, the sample must be in position for the spectrum to be taken before the process is carried out. This means that, for the duration of the "burning out," the spectrometer cannot be used. As the process may take several hours this can be inconvenient.

The mechanism of "burning out" is not understood, but one probable theory is that its effect is to decompose very low levels of fluorescing organic impurity.

(b) Use of laser lines in the ultraviolet and near infrared

Many fluorescence problems can be avoided by using laser wavelengths that lie beyond the range of those normally used in Raman spectroscopy. For many compounds these wavelengths will be outside the range that induces fluorescence, or will produce fluorescence that lies beyond the Raman signal in terms of wavelength. Of particular value in this context is the use of ultraviolet lasers, although the attendant problem of using a pulsed system has to be considered (see above). Some success has also been achieved using the far red lines (752.5 and 799.3 nm) of a krypton laser, which are easily obtained by the use of specially coated mirrors [2]. It is necessary when using these lines that the tracking range of the gratings in the monochromator is sufficient or that a second pair of gratings, suitable for operation up to about 1.06 μm, is available. The recently published work of Hirschfeld and Chase [18] has used the 1.064 μm output of a continuous-wave YAG laser and has produced good quality spectra of materials that normally exhibit strong fluorescence. In this case, instead of using the dispersive monochromator system normally used to obtain Raman spectra, they used the emission port of a Fourier transform infrared spectrometer. This system appears to have considerable potential for recording spectra of fluorescent samples but has the drawback that it is not possible to record signals close to the exciting line.

(c) Use of quenching agents

Another technique that can be of value in certain cases is the use of fluorescence quenching agents [19-21]. These are chemicals that form charge-transfer complexes, particularly with aromatic and polycyclic aromatic hydrocarbons, and considerably reduce the intensity of the fluorescence of such compounds. Compounds that can be used for this purpose include picric acid, nitrobenzene, butane 2,3-dione, and tetracyanoethylene. Their main disadvantage lies in the fact that they have to be used in high concentrations relative to the species being studied and may well interfere seriously with the spectrum of that species. They do, however, have the advantage, particularly in the case of tetracyanoethylene, that they produce an absorption in the visible region of the spectrum, which may be used to record the resonance enhanced spectrum of the compound being examined.

Techniques That Extract the Raman Spectrum from the Total Signal

(a) Use of short pulses and gated detectors

As mentioned above, one of the major drawbacks with pulsed systems is the low repetition rate and short duration of the pulses, which means that the detector is only recording noise for the majority of the time. If a photomultiplier tube is used as the detector, a boxcar system can be used to activate the tube for the duration of the pulse only. Of much more use with pulsed systems is the multichannel detector because it records a large portion of the spectrum simultaneously and hence has the potential to give good quality spectra of a particular region much more rapidly than the photomultiplier tube. The advantage of a gated diode array from the point of view of fluorescence rejection is that, for some compounds at least, the fluorescence is sufficiently slow compared with the Raman signal that by activating the detector for a few nanoseconds all of the Raman signal can be collected but, hopefully, only a small proportion of the fluorescence. Considerable success has been achieved in fluorescence rejection by this technique, but only for compounds where the rise time of the fluorescence signal is relatively slow.

(b) Mathematical techniques

These essentially involve the use of one or more mathematical functions to generate a simulated fluorescence background. Depending on the sophistication of the computer and software, this method can be the best technique available to most Raman groups for obtaining Raman spectra from compounds that exhibit a high degree of fluorescence. Even simple computer techniques, such as point-by-point manual simulation of the fluorescence, can prove extremely effective in this area. All commercial instrument manufacturers now supply software for simulating fluorescence as standard packages supplied with the spectrometer.

C. Quasi-CW Lasers

As mentioned above, the disadvantage of the pulsed systems such as the excimer and the YAG is their low repetition rate, which means that the detector is only exposed to the Raman signal for a very small proportion of the time taken to accumulate the spectrum. Hence, for most purposes the continuous-wave (CW) lasers are normally used. Another alternative to be considered if high powers are required is the metal vapor lasers, based on copper or gold, which give pulsed outputs at repetition rates of several kilohertz and so may be considered as quasi-CW lasers. Between them the copper and gold vapor lasers provide a range of lines in the visible region of the spectrum and give outputs equivalent to about 40 W from a CW laser. They can be used to pump dye lasers, but from the Raman spectroscopist's point of view their main disadvantage is their large beam diameter (-1 cm), which is not easy to focus down to a realistic diameter required for many Raman studies.

III. OPTICS AND MONOCHROMATORS

The main difficulty associated with the Raman effect is, in most cases, its very weak nature. This means that once the incident photons, supplied by the laser, have interacted with the sample being analyzed, the scattered photons must be collected as efficiently as possible, with minimum losses between the sample and the detector. The way in which this is done is to use a lens, normally a camera lens, to collect the scattered photons. The Raman signal is then brought to a focus at the entrance slit of a monochromator, which separates the light into its component frequencies. The simplest monochromator is the single, but because of its poor stray light rejection it is not capable of giving spectra in the low Δ cm^{-1} region. The most popular type of monochromator since the revival of Raman spectroscopy in the late 1960s has been the double, fitted with holographic gratings, and a diagram of such a system is shown in Figure 2. The gratings are available in a wide range of groove densities, and a compromise needs to be reached on the appropriate value for the type of work to be undertaken. The higher the groove density, the greater will be the resolution that can be achieved by the monochromator, but the narrower will be the operating range. For a spectrometer with a range extending through the visible region of the spectrum, but with an adequate resolution (~1 cm^{-1} in the middle of the operating range), a convenient compromise is 1800 grooves/mm.

Another important consideration is the mirror coatings. Since each reflection in the monochromator will result in losses it is necessary, provided no other consideration needs to be taken into account, to use the most efficient coating in terms of reflectivity, and these will normally be supplied by the spectrometer manufacturer. However, the range of the monochromator can be extended into the ultraviolet or near-infrared spectral regions by the use of alternative sets of gratings to those used in the visible region, in which case the mirrors in the monochromator must have suitable coatings. This will usually result in a lower throughput in the visible region in order to extend the overall operating range. The ultraviolet is normally a more useful region than the near-infrared because, other considerations being equal, for a given incident intensity the Raman signal is greater and also it is more likely that resonance enhancement will be observed in this region. An alternative method can be used to obtain Raman spectra in the near infrared region, and that is to use a Fourier transform infrared (FTIR) spectrometer fitted with a near-infrared detector. This technique was recently reported in the literature [18] and is attracting a great deal of attention at the present time. The main advantage is that it can be provided as an "add-on" to a conventional FTIR spectrometer, and the region covered may well prove particularly useful for the examination of materials that exhibit too much fluorescence to be studied in the visible or ultraviolet regions.

Another factor that must be considered when extending the normal operating range of the spectrometer is the detector response, and this is considered in Section IV. In general, although a reasonable compromise can be reached to extend the operating range of a conventional Raman spectrometer, if an appreciable amount of work is to be undertaken outside the visible range, serious consideration should be given to the use of an instrument specifically designed for the purpose.

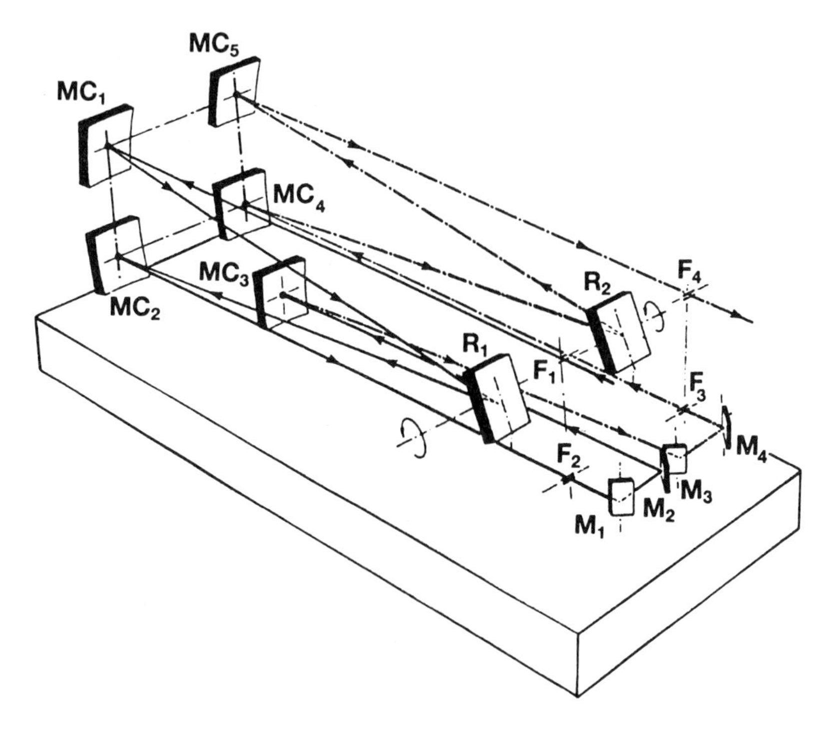

Figure 2. Diagramatic representation of a double monochromator: F, slit; M, plane mirror; MC, concave mirror; R, grating. (From Jobin-Yvon, with permission.)

The very best quality double monochromators, even with strongly scattering solid samples such as polyethylene, are capable of observing Raman signals down to 5 or 6 cm^{-1} before the tail of the exciting wavelength becomes too severe. It is possible to get even closer to the exciting line, although at the sacrifice of throughput, by using a triple monochromator. An alternative type of triple monochromator that is becoming increasingly widely used is one that comprises a double monochromator and a second element, a single monochromator. In such cases the double element is a spectrometer and the single a spectrograph. At this point it is worth briefly considering the difference between the two concepts. The spectrometer gives a much greater degree of dispersion than the spectrograph and uses very narrow entrance slits. In this way, the width of the signal (in terms of cm^{-1} or nm) that reaches the detector is extremely narrow and to a large extent dictates the resolution of the monochromator. In the case of the spectrograph the dispersion is not as great and the slit is much wider. In this case the detector normally used is a multichannel device and the resolution relates to the number of detector channels (normally up to 1024) and the width of the signal reaching the detector.

The most expensive part of the spectrometer hardware is the monochromator itself, and so it is reasonable to expect that the greatest attention should be paid to this aspect of the instrument. However, there is very little to choose between the monochromators supplied by the major manufacturers of Raman spectrometers, and so the most important choice that has to be made is what type of monochromator to opt for. In general a single-pass system can be ruled out because of its poor dispersion and poor stray light rejection. It may have some application where a very high throughput is required, as may be the case for working in the ultraviolet or near-infrared region, but for conventional Raman spectroscopy it is not suitable. The double monochromator has for many years been the standard for Raman spectroscopy, and the reduction in throughput is more than compensated for by the greater resolution and stray light rejection. On the whole, these have been used with photomultiplier tubes, and systems of this type have produced the bulk of the Raman work undertaken in the past 15 years or so.

The advent of the diode array detector and similar multichannel detector systems has now made the use of a triple monochromator system much more realistic. The true triple monochromator (i.e., a triple spectrometer) will give the best stray light rejection but at the expense of a considerable decrease in throughput, and would normally only be worth considering if most of the work to be carried out relates to spectra recorded within a few cm^{-1} of the exciting line. As mentioned above, other forms of the triple combine a double spectrometer and a single spectrograph. It is normal in such cases for the double element to act in a subtractive mode—that is, the first stage acts to disperse the light and the second to recombine it, with the scattered laser fundamental much reduced in intensity, before passing it into the spectrograph. In this way the stray light rejection, dispersion, and throughput of the whole system are only slightly worse than a conventional double spectrometer, but the spectrograph element gives the system the capability of operating with a diode array and giving a good spectral range. An instrument of this type combines the best features of spectrometer and spectrograph for a small sacrifice in performance. The great advantage of this type of triple configuration is that it can be conveniently fitted with both a photomultiplier tube and a diode array detector, the particular application dictating which one is used.

The basic choice of monochromators lies then between the double and triple systems. The triple is considerably more expensive but has greater versatility for the analyst in terms of the range of samples that can be conveniently analyzed, and would normally be the preferred instrument for a laboratory that is essentially analytically based. The double is slightly better in terms of performance, and for some laboratories this may be sufficiently important to opt for the double, particularly if it is to be used with a photomultiplier tube, when it operates at its best.

IV. DETECTORS

Detectors fall conveniently into two categories: single-channel and multichannel.

A. Single-Channel Detectors

The only single-channel detectors that are commonly used in Raman spectrometers are photomultiplier tubes. These may be sensitive to different regions of the spectrum, but for most purposes one such tube will cover the range of interest to Raman spectroscopists (~400-800 nm). Because the Raman signal is, in most cases, very weak, the best photomultiplier tube available should be used. Fortunately, even the best tubes are relatively inexpensive compared with the total cost of the system, and manufacturers will supply at slightly extra cost specially chosen tubes that have extremely low background noise. The performance of these tubes tends to fall off sharply in the red region, where the Raman signal in any case is itself weaker, and this should be taken into account when considering the type of work to be undertaken. In some cases separate UV/visible and red tubes may be an advantage, although in general sensitivity above about 750 nm is never as good as can be obtained at shorter wavelengths.

Another advantage of photomultiplier tubes is that they are relatively stable and robust and should give many years of service without problems. Despite all precautions on the part of the operator and the use of computerized safety systems, it is sometimes the case that reflected laser light will find its way into the detector. This can cause a drastic increase in the background, which can persist for several hours, but in most cases photomultiplier tubes will recover even from this type of mistreatment. If inexpert operators are going to use the system, this is another advantage of such detectors that does not apply to the more vulnerable multichannel systems.

Photomultiplier tubes, because of their ability to cope with high background levels, are the most suitable type of detector to use with materials that exhibit high fluorescence and for measurements at low frequency shifts (<~30 cm^{-1}). They also have the advantage of extremely high sensitivity, and so if very weak Raman scatterers are to be studied, such as is the case with many catalyst systems, metal oxides, or carbon, the photomultiplier tube is normally the preferred detector.

B. Multichannel Detectors

At first sight the advantages of a multichannel detector are so considerable that it would appear to be an obviously better choice of detector than a photomultiplier tube, despite the considerably greater cost. The photomultiplier tube, at any one time, monitors a very narrow signal, typically about 1 cm^{-1} wide, and so to obtain the whole spectrum the Raman scattered light must be scanned by the spectrometer until the whole of the range of interest has been covered. Despite the very good sensitivity of photomultiplier tubes, in order to maintain a reasonable resolution, this scanning operation normally takes at least 10-15 min and frequently considerably longer. Multichannel detectors, by contrast, comprise a large number, typically 512 or 1024, of photodiodes, each capable of monitoring a different portion of the spectrum simultaneously. These detectors will cover a wavelength range of about 26 nm for a 1024-element detector. If the exciting line used to obtain the Raman spectrum is, for example, the 488.0-nm line of an argon ion laser, this means that if the starting point of the spectrum is, say, 100 cm^{-1}, this is in absolute terms $(20,492 - 100)cm^{-1}$ = 20,392 cm^{-1} = 490.4 nm. If the coverage of the diode array is 26 nm, then the upper limit of the spectrum being acquired is (490.4 + 26)nm = 516.4 nm = 19,365 cm^{-1} or a Δ cm^{-1} from the starting point of 1027 cm^{-1}. For many analytical purposes this range will be adequate, and for 1024 elements covering 1027 cm^{-1} the ultimate resolution is ~1 cm^{-1}, perfectly adequate for most purposes. If, however, we consider the case where the exciting line being used is the 647.1-nm line of a krypton ion laser (15,454 cm^{-1}), at 100 cm^{-1} the starting point is 15,354 cm^{-1} = 651.3 nm, and the end of the range is (651.3 + 26)nm = 677.3 nm = 14,765 cm^{-1}. So the coverage at this wavelength is only 589 cm^{-1}. Obviously this will improve the resolution but will severely reduce the spectral range being observed at any one time.

Another restriction of the diode array detectors is that, because of their vulnerability to damage by high photon flux, it is not possible to use them at low Δ cm^{-1} values. For example, if it is required to study the longitudinal acoustic modes of polyethylene, which typically occur in the range 5-30 cm^{-1}, the diode array detector is not feasible. Furthermore, the software required to achieve a linear scale in Δ cm^{-1} across all of the elements of the detector is not trivial, and great care must also be taken to ensure that the readings at each end of the spectral range covered by the detector are as reliable in terms of intensity as in the middle of the range. Similarly, diode arrays need to be carefully checked for linearity of response—that is, to ensure that by doubling the photon flux on any one diode, the recorded intensity also doubles. For these reasons it is much more difficult to obtain reliable quantitative data with a diode array, although with care it can be achieved.

Despite the disadvantages just mentioned, there are enough advantages of multichannel detectors to make them worth serious consideration. Although the overall sensitivity at any one wavelength is, for most of the commonly used arrays, inferior to that of the photomultiplier tube, the fact that a large portion of the spectrum is being monitored simultaneously means that the time necessary to record a spectrum of any particular sample is always very much less than with the photomultiplier tube. Also, provided the background fluorescence is not too severe, sample alignment is usually very much easier with a multichannel detector. The reason for this is that it is necessary with a photomultiplier tube to find a Raman peak, usually by rapidly scanning the spectrum and observing the response of the tube on a meter. The spectrometer is then locked at the maximum of the peak and the various parameters such as sample position, slit width, and laser power are optimized. If the sample is of unknown chemical composition it is quite easy to miss the strongest Raman signal in a quick visual scan, and this may make any parameters set on a weaker band less than ideal. With a diode array, however, the whole of the portion of the spectrum to be recorded is displayed on a screen, and in modern instruments this can be redisplayed every few milliseconds; so as far as sample alignment is concerned the spectrum is displayed on what is essentially a "real-time" basis. In this way changes in the spectrum can be observed by eye while instrument parameters and sample position are optimized.

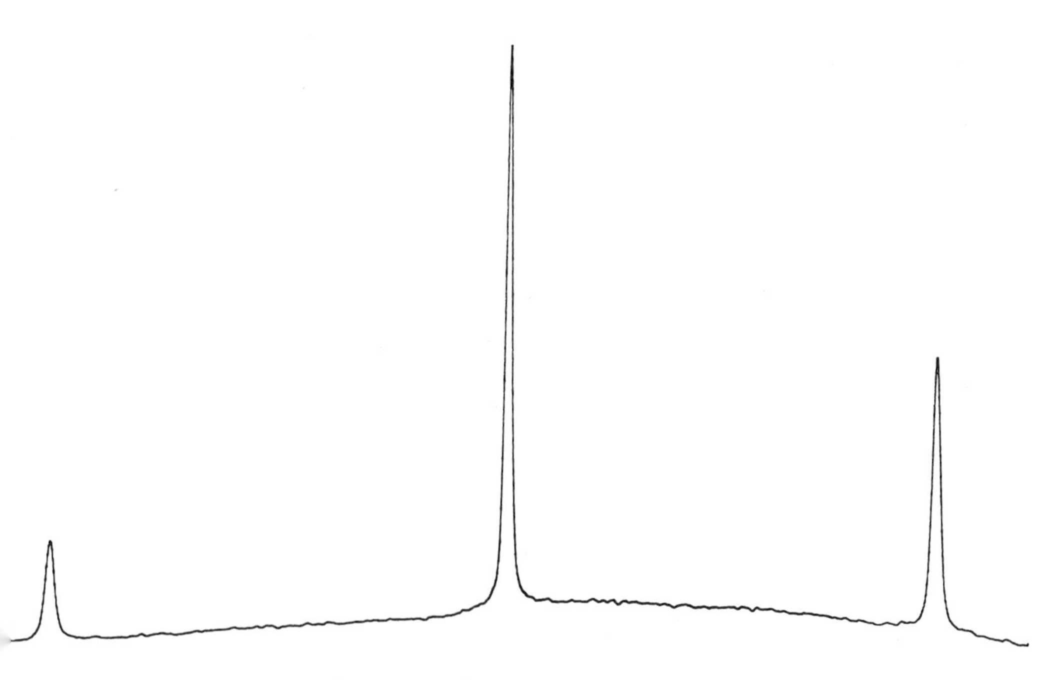

Figure 3. Response of a worn diode array. The spectrum has been recorded at three different positions on a 1024-channel array.

Another problem with diode arrays is that, particularly with the passage of time, the response of the individual elements of the array can differ. Hence, the observed intensity of a band can change depending on which elements of the array are used to record it. In some cases this effect can be so severe that the relative intensities of different bands in the spectrum can be reversed merely by using different elements of the array. This means that if reliable quantitative work is to be achieved a regular check must be made of the response of the various elements of the array. This can be done by using a strong Raman scatterer such as diamond, which gives a sharp, single-peak spectrum at 1332 cm^- , and, without altering the sample position or instrument parameters, noting the intensity of the peak at various positions on the diode array. Figure 3 shows the results for a worn array. In this case the peak for diamond has been recorded at three different positions and the change in the observed intensities as the peak moves across the array can be clearly seen. For an array in good condition and properly aligned there should be no more than a 1-2 percent difference in intensities across the array.

To a large extent, therefore, the choice of a photomultiplier tube or diode array depends, apart from financial constraints, on the nature of the work to be undertaken. If spectra need to be obtained rapidly—for example, if the technique is to be used to study reaction mechanisms and kinetics—then the diode array is essential. Also, the array can often be used to give a spectrum adequate for purposes of characterization and identification in only a few seconds. The most accurate quantitative work is always best carried out using a photomultiplier tube and, once calibrated for its response over the spectral range in which it is to be used, it will last for a long period of time without any significant deterioration. Also, if signals are to be observed close to the exciting line then a photomultiplier tube is essential.

If the Raman capability for a laboratory is to comprise only one spectrometer, then serious consideration should be given to an instrument fitted with both photomultiplier tube and diode array detectors. As already mentioned, commercial instruments of this type are available, and of particularly broad application are the triple monochromator systems based on a double spectrometer unit and a single spectrograph unit. This gives the overall capability in one instrument of virtually any application of conventional Raman spectroscopy and is an extremely powerful and versatile analytical tool.

V. COMPUTING

The use of computers is dealt with elsewhere in this handbook, so in this chapter we will limit ourselves to discussing the minimum basic requirements for the computer system and also the various desirable capabilities that the Raman spectroscopist finds useful. For many purposes it is possible to obtain perfectly adequate Raman spectra with only a minimal computing capability. The laser parameters are normally controlled by hand, which leaves the monochromator and detector parameters to be considered. It is normal on modern instruments to use a computer to control the most crucial monochromator parameter, the slit width, opening and closing of the slits being achieved by means of stepper motors. However, this is only a matter of convenience, and all of the instrumental parameters can be easily adjusted by hand. Indeed, perfectly satisfactory Raman spectra of a wide range of materials can be obtained without any computing capabilities except, in the case of a diode array, that required to operate the detector. Modern spectrometers are normally supplied with a dedicated computer and a software package to operate the spectrometer and to perform a range of operations on the acquired data. In this latter context the most important feature is the ability to accumulate spectra. The Raman spectra of many materials, particularly inorganic compounds, are extremely weak, and it is only by the accumulation of a large number of spectra that useful Raman data can be obtained. Also, as mentioned above, many materials exhibit fluorescence when exposed to the laser beam, and even when this is not so severe as to totally obscure the Raman signal, it will give a strong background. For reliable quantitative work to be carried out this background needs to be removed, and the spectrometer manufacturers approach this problem in a variety of ways.

The simplest approach is to use a point-by-point background simulation method. This involves moving a cursor over the background and using the computer to generate, from the cursor points, a simulated fluorescence background. More sophisticated methods use a combination of a range of mathematical functions to simulate the fluorescence signal. However, the nature of these fluorescence backgrounds is so variable that it is virtually impossible to match them even by a complex combination of such functions.

Another desirable feature of Raman software is the ability to undertake spectral subtraction. As with infrared spectroscopy, it is often the case that multicomponent systems are encountered and one or more components may need to be subtracted from the total spectrum in order to extract the spectrum of the required component. Such spectral subtraction routines are widely available from either the manufacturers of the Raman spectrometer or the manufacturers of FTIR spectrometers. Indeed, it is quite feasible, if Raman and FTIR instruments are available in close proximity to each other, to transfer Raman data to the computer of the infrared instrument and access the FTIR software, which at the present time is much more highly developed than the corresponding Raman software.

Although not essential for most purposes, the use of deconvolution techniques is becoming more widely used in Raman spectroscopy. It is of particular value in the study of reacting systems, where, for example, the Raman bands being monitored can overlap to such a great extent that, without deconvolution, they cannot be separately quantified. One of the most successful and widely used of these techniques is that of Fourier self-deconvolution [22], which is also widely used in infrared studies. This does not require particularly sophisticated or expensive computing capability and can be either incorporated into the software of the Raman instrument or used via an FTIR spectrometer where commercial software packages are readily available.

Although for complex time-resolved studies, particularly those involving chemically dynamic systems, relatively sophisticated computing is a distinct advantage, the technique of Raman spectroscopy can nonetheless be used very successfully for a wide range of applications with very little computing capability. In this sense, the Raman hardware still has many similarities with dispersive infrared instrumentation and the two techniques resemble each other very closely.

VI. ACCESSORIES

Because of the highly versatile nature of Raman spectroscopy, it is worth considering how to extend its capability by the use of accessories. The facts that we are using, in most cases, visible light to obtain the spectrum and that the Raman signal itself is also in the visible region of the spectrum mean that there is, for many purposes, a distinct advantage in using glass, which itself is a very weak Raman scatterer. This can be used as a medium for containing the sample and can be, if required, a sealed vessel or even a complete reaction rig. In addition, the highly transparent nature of glass to visible radiation means that Raman signals can be obtained for samples positioned behind very thick (1 cm) glass without any particular difficulty. This enables the construction of specialist cells for Raman studies at elevated temperatures and pressures. Another feature of glass is that its weak Raman signal can be simply removed from the overall signal by a suitable computer program, and in any event it appears in a region where most systems of interest (other than metal silicates or silica) are not observed. A further feature of glass that is becoming increasingly appreciated by the Raman spectroscopist is the value of fiber optics. This, together with other Raman accessories, is discussed next.

(a)

(b)

Figure 4. Specialist cells for Raman studies: (a) high temperature/pressure cell; (b) low temperature cell; (c) molten salt cell; and (d) mount and catalyst cell. (From Ventacon Limited, with permission.)

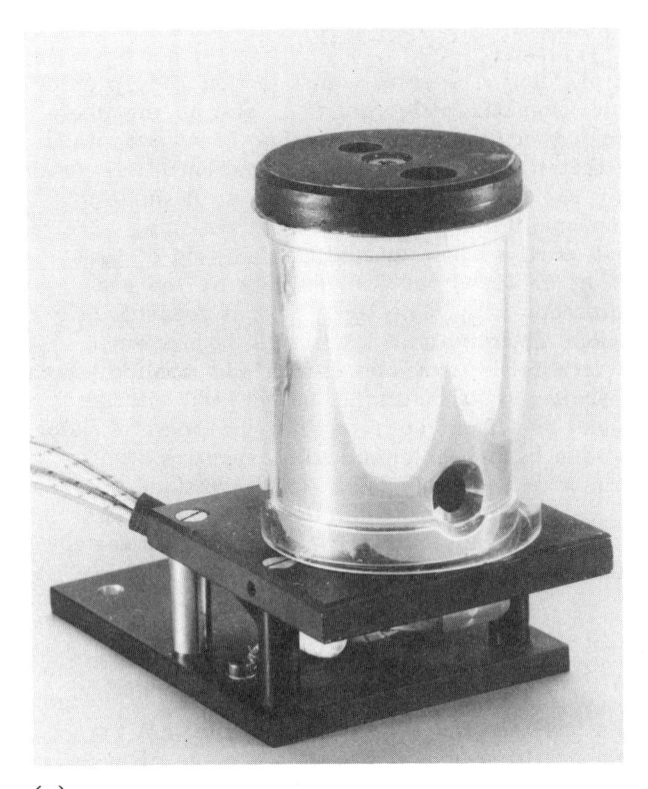

(c)

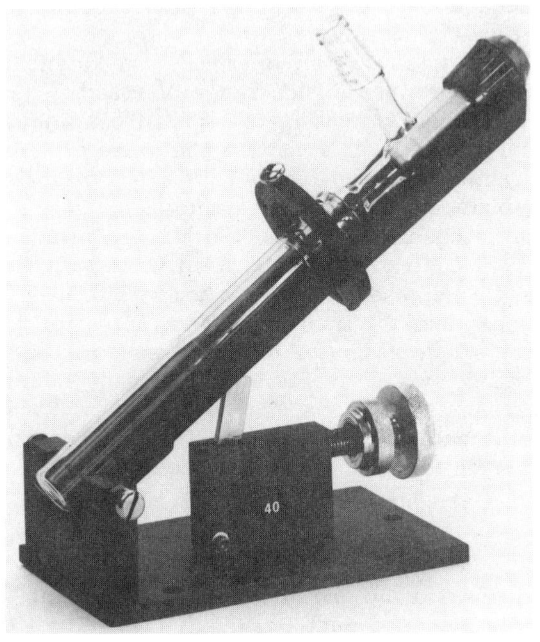

(d)

A. The Raman Microscope

The great value of lasers to the Raman spectroscopist is that they provide an intense source of coherent, monochromatic radiation, and on the whole we ask little more of them than that. However, the other great advantage of the laser is the potential that it offers for spatial resolution. It is possible to focus a laser beam, by use of a microscope objective, down to a spot size of ~1 μm. This then opens up the possibility of obtaining spectra of very small samples or small portions of a sample that may be of some particular interest. This can be achieved relatively simply by the use of a conventional optical microscope [23]. The laser light is focused on to the sample via the microscope objective and the sample is positioned using a glass screen or, preferably, a television camera and monitor attached to the microscope. The backscattered Raman signal is collected via the objective and passed, by means of a beam splitter, into the monochromator. In this way the Raman technique can be extended to microanalysis and has been used very successfully in the analysis of fluid inclusions in minerals [24], polymer studies [25], biological investigations [26], and a wide range of other applications. Manufacturers of Raman spectrometers all supply microscope attachments as "add-on" extras.

B. Specialist Cells

As mentioned above, the use of thick glass windows enables one to design for Raman spectrometers cells that are capable of operating over a very wide range of nonambient temperatures and pressures. A range of such cells is shown in Figure 4. The temperature range for in situ studies is from liquid helium temperature up to ~600°C, and the pressure range is from high vacuum to ~20 kbar. If higher pressures than these are required then diamond anvil cells can be used up to extremely high pressures, although their volume is necessarily very small.

Cells of the type shown in Figure 4 are of particular value in kinetic studies, where they can be used to reproduce a wide range of reaction conditions and are particularly valuable in simulating the sort of conditions prevailing in a large-scale chemical plant. In this context the value of Raman spectroscopy is only just beginning to be realized. It offers probably the simplest way for reproducing plant conditions in a spectrometer sample compartment and following reaction mechanisms and kinetics under "real" conditions. Sophisticated temperature controllers can be used in conjunction with these cells, and can be used to increase or decrease temperatures at a controlled rate and hold at a particular temperature for a preset period. In this way, and using a relatively simple computer program, spectra can be obtained automatically of a reacting system over a period of many hours under programmed temperature/pressure conditions. The technique of Raman spectroscopy can then be used for troubleshooting problems associated with the operation of the chemical plant, or for optimizing plant conditions.

C. Fiber Optics

The use of fiber optic probes will possibly be one of the most significant developments of Raman spectroscopy over the next few years [27]. The

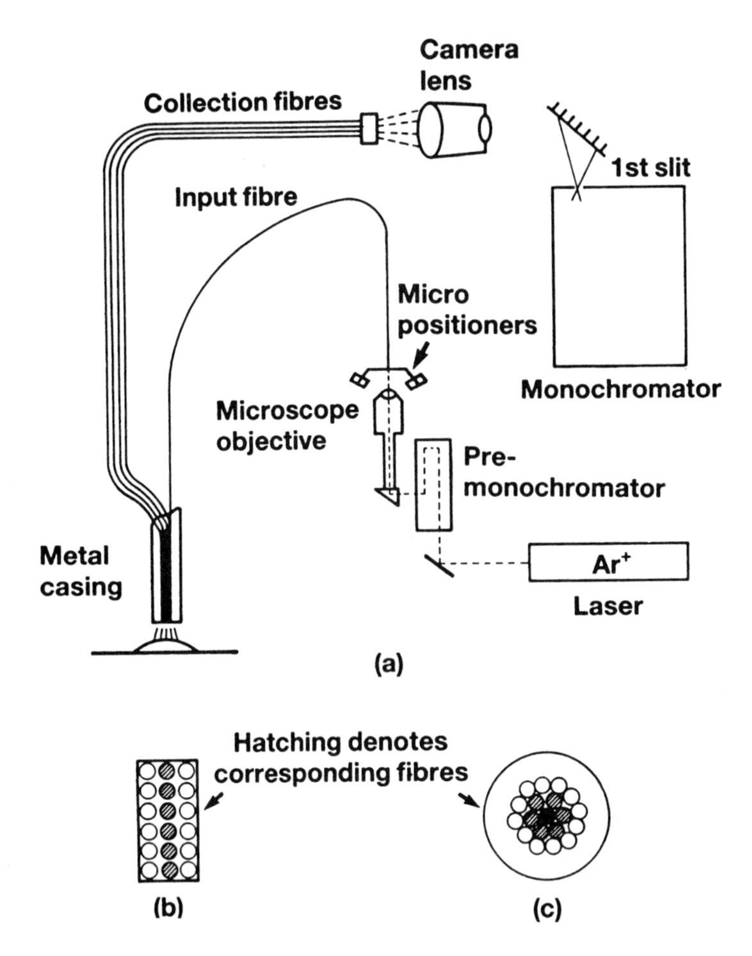

Figure 5. Diagram of fiber optic probe: (a) general layout of experiment, (b) arrangement of collection fibers in front of camera lens, and (c) probe end (dark center is input fiber).

laser beam being used to obtain the Raman spectrum can be focused, using a microscope objective, onto the end of a length of fiber optic. The other end of the fiber can then be placed close to the sample to be analyzed. In this way the sample can be situated remotely from the laser. Similarly, since the Raman signal is, in most cases, in the visible region of the spectrum, it is possible to collect the Raman scattered photons by means of one or more other fibers placed close to the sample and to take the signal, via these fibers, back to the spectrometer. Hence, the sample to be analyzed can be situated remotely from the spectrometer as well as from the laser. It is convenient to place the collection fibers round the fiber carrying the laser beam, and in this way a simple fiber optic probe can be constructed. A schematic diagram of such a probe is shown in Figure 5.

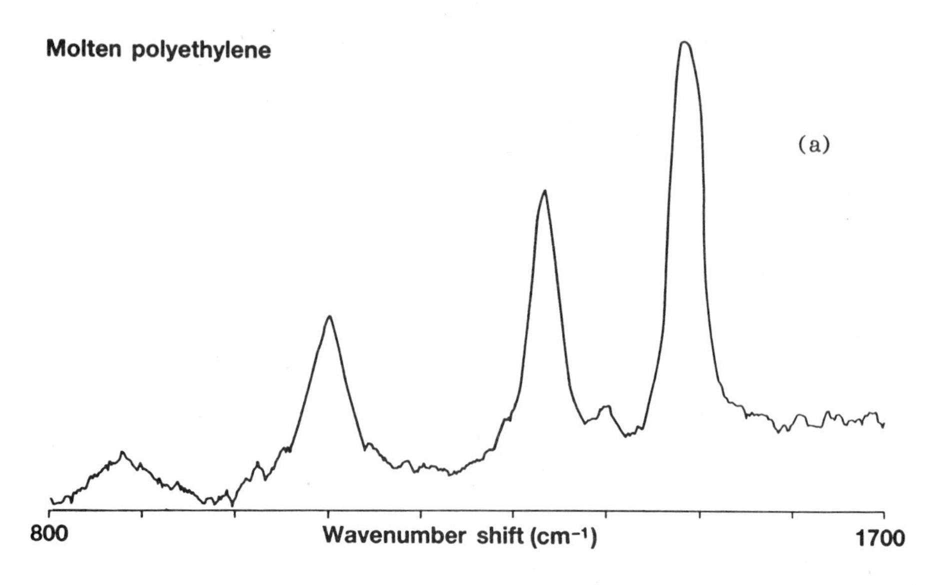

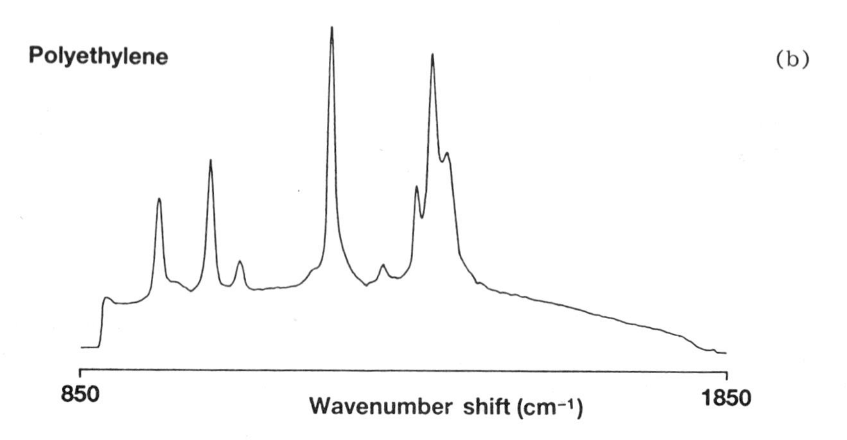

Figure 6. Raman spectra of polyethylene obtained using a fiber optic probe: (a) molten polymer and (b) solidified polymer.

The use of fiber optics open up the possibility of extending Raman studies to a wide range of systems that are not normally suitable for Raman spectroscopy. It can be used, for example, to study reactions occurring on, or in close proximity to, electrode surfaces. The probe end, sealed in a glass sheath, can be placed directly into the electrolyte and the reaction can be monitored at an appropriate location. Also, the fiber optic probe can be used to study samples in what may loosely be called hostile environments. This includes, for example, in situ studies in molten salts or molten polymers and in vivo biological studies. A spectrum of molten polyethylene obtained by placing the probe end directly into the molten polymer is shown in Figure 6. This contrasts with the spectrum of the solid polymer, which has a considerable degree of order, and is also shown in Figure 6. Studies of polymers crystallizing from the melt are of particular importance in the polymer processing industry. Other "hostile" environments could include highly corrosive, explosive, toxic, or radioactive systems or even large-scale chemical reactors—the potential is enormous.

VII. CONCLUSIONS

In conclusion it is fair to say that, although the technique of Raman spectroscopy has taken a long time to develop into a broad-based analytical tool, it is now in a position where it can be considered the equal of other spectroscopic techniques in terms of information value. Furthermore, if its intrinsic advantages are exploited to the full—and there is every sign that this is now beginning to happen—it is a technique that has such a high degree of versatility and adaptability that it should make a major contribution to both academic and industrial studies in the near future.

REFERENCES

1. A. B. Harvey, Ed., *Chemical Applications of Non-Linear Raman Spectroscopy*, Academic Press, New York, 1981.
2. K. P. J. Williams and D. L. Gerrard, Optics Laser Technol. *245* (1985).
3. M. Maeda, *Laser Dyes: Properties of Organic Compounds for Dye Lasers*, Academic Press, Tokyo, 1984.
4. J. Behringer, Observed Resonance Raman Spectra, in *Raman Spectroscopy Theory and Practice* (H. A. Szymanski, Ed.), Plenum Press, New York, 1967.
5. P. Carey, *Biochemical Applications of Raman and Resonance Raman Spectroscopies*, Academic Press, New York, 1982.
6. A. T. Tu, *Raman Spectroscopy in Biology: Principles and Applications*, John Wiley and Sons, New York, 1982.
7. I. S. Biggin, D. L. Gerrard, and G. E. Williams, J. Vinyl Technol. *4*, 150 (1982).
8. D. L. Gerrard, H. J. Bowley, and I. S. Biggin, J. Vinyl Technol. *9*, 43 (1987).
9. G. Martinez, C. Mijangos, J. L. Millan, D. L. Gerrard, and W. F. Maddams, Makromol. Chem. *80*, 2973 (1979).
10. K. P. J. Williams, D. L. Gerrard, D. C. Bott, and C. K. Chai, Mol. Cryst. Liq. Cryst. *117*, 23 (1985).
11. L. S. Lichtmann, A. Sarhangi, and D. B. Fitchen, Chem. Ser. *17*, 149 (1981).
12. H. Kuzmany, Chem. Ser. *17*, 155 (1981).
13. M. Fleischmann and I. R. Hill, Raman Spectrosc. Compr. Treatise Electrochem. *8*, 373 (1984).
14. S. A. Asher and C. H. Jones, Am. Chem. Soc. Div. Fuel Chem. preprint, *31*(1), 170 (1986).

15. C. R. Johnson, M. Ludwig, and S. A. Asher, J. Am. Chem. Soc. *108*(5), 905 (1986).
16. J. M. Dudik, C. R. Johnson, and S. A. Asher, J. Phys. Chem. *89*(18), 3805 (1985).
17. T. L. Gustafson and F. E. Lytle, Anal. Chem. *54*, 634 (1982).
18. T. Hirschfeld and B. Chase, Appl. Spectrosc. *40*, 133 (1986).
19. M. Kasha, J. Chem. Phys. *20*, 71 (1962).
20. C. H. J. Wells, *Introduction to Molecular Photochemistry*, Chapman & Hall, London, 75, 1972.
21. D. L. Gerrard and W. F. Maddams, Appl. Spectrosc. *30*(5), 554 (1976).
22. H. J. Bowley, S. M. H. Collin, D. L. Gerrard, D. I. James, W. F. Maddams, P. B. Tooke, and I. D. Wyatt, Appl. Spectrosc. *39*(6), 1004 (1985).
23. B. W. Cook and J. D. Louden, J. Raman Spectrosc. *8*(5), 249 (1979).
24. J. C. Touray, C. Beny-Bassez, J. Dubessy, and N. Guilhamou, Scanning Electron Microsc. *1*, 103 (1985).
25. F. Adar and H. Noether, Microbeam Anal. *20*, 41 (1985).
26. P. V. Huong, J. Pharm. Biomed. Anal. *4*(6), 811 (1986).
27. R. L. McCreery, M. Fleischmann, and P. J. Hendra, Anal. Chem. *55*, 146 (1983).

10

Photoacoustic Instrumentation

EDWARD M. EYRING and NELIA F. LEITE* / Department of Chemistry, University of Utah, Salt Lake City, Utah

STANISLAW J. KOMOROWSKI / Department of Calorimetry, Institute of Physical Chemistry of The Polish Academy of Sciences, Warsaw, Poland

TSUTOMU MASUJIMA / Institute of Pharmaceutical Sciences, Hiroshima University, Hiroshima, Japan

I. INTRODUCTION

Photoacoustic spectroscopy (PAS) has its origins in the research of Alexander Graham Bell and his contemporaries over 100 years ago [1]. PAS applied to gases is generally called optoacoustic spectroscopy (OAS) and was the principal analytical application of photoacoustic techniques for a long time. A still very useful survey of OAS was edited by Pao and published in 1977 [2]. PAS became an important analytical technique for the study of solids and liquids only after Rosencwaig and Gersho [3] developed a utilitarian mathematical treatment of microphonic PAS in 1975. The present chapter gives the greatest weight to the use of photoacoustic techniques for the investigation of solids and liquids because the authors have their expertise in these areas. The reader should not discount the possibility that the long-term most important analytical application by chemists of photoacoustic techniques may still be to the analysis of gas mixtures.

McDonald [4] has made a comparatively recent survey of the theory and concepts of photoacoustic (PA) and photothermal (PT) techniques. The common feature of all the experiments considered here is the excitation of a sample by pulsed or modulated energetic radiation with the absorbed energy producing heating of the sample as the excited states undergo radiationless decay processes. Since the incident radiation can either be periodic or single pulsed, the sample heating will give rise simultaneously to a thermal field and an acoustic field, both of which are correspondingly either periodic or pulsed. PT techniques measure the heat or acoustic wave that is created in a sample by energy absorption, so any type of incident energetic beam including electromagnetic radiation ranging from radiofrequency to x-ray wavelengths, electrons, protons, ions, ultrasound, etc. can be used to generate the PT signal.

The different PT techniques are named according to the mechanism used for detection of the thermal wave. Thus, we have the common PAS

*Current affiliation: Instituto de Pesquisa Espaciais, São Paolo, Brazil.

with microphonic detection [3], piezoelectric detection techniques [5], and photopyroelectric spectroscopy [6-8], all of which are contact techniques. Remote techniques are exemplified by photodisplacement spectroscopy [9,10], beam deflection spectroscopy (BDS) [11,12], photothermal radiometry (PTR) [13-15], electrothermal radiometry (ETR) [16], etc.

The PA signal intensity depends on the amount of heat generated in the sample (i.e., on the optical absorption coefficient and on the light-into-heat conversion efficiency) and on how the heat diffuses through the sample. Thus with PAS we can obtain absorption spectra, measure absorption coefficients (greater than 10^5 cm^{-1} measured for As_2S_3 [17] and MnO_2-impregnated polyethylene [18] and lower than 10^{-5} cm^{-1} for alkali fluoride crystals [18]), get information about thermal properties (thermal diffusivity and thermal conductivity measurements), do thermal imaging and investigate photoinduced energy conversion processes (monitoring of photovoltaic conversion efficiency of solar cells [19-20], study of nonradiative recombination processes in semiconductors [21], kinetics of quenching on a submicrosecond timescale [22], and determination of the quantum yields, energies, and lifetimes of metastable states [23,24]).

The PA signal is detected only for heat created within one thermal diffusion length $\mu = (\alpha/\pi f)^{1/2}$ beneath the sample interface. Here α denotes the thermal diffusivity. Thus, for high chopping frequencies f one obtains information about the sample surface, while at low frequencies one learns about deeper layers of the sample. If low resolution is required and one wishes to know how some properties change with sample thickness, the technique is called PA depth profiling [25,26] and one measures the PA signal amplitude and phase as a function of modulation frequency or sample position. If high lateral resolution is required, PA microscopy can be used and the PA signal is obtained when the focused pump laser beam is scanned over the sample [4].

The advantage of PT over conventional optical spectroscopies is that the PA signal is a direct measure of the energy absorbed by the material as a result of interaction with the energetic beam, while in conventional optical spectroscopy one measures the incident beam and compares it to the transmitted or reflected beam. It is clear that for transparent or weakly absorbing materials such as transparent gas mixtures or liquids containing minute quantities of an absorbing species or pollutant, or for highly light-scattering materials, such as powders, amorphous solids, gels, smears, and suspensions, it is not easy to use conventional spectroscopy, and PAS is a more suitable way to investigate these materials. Thus the considerable importance of the photoacoustic effect for analytical chemistry arises from the fact that one observes a signal against little background noise just as in fluorescence measurements and in contrast to the detection of small signals against a strong background in transmittance or absorbance measurements. A further important advantage of PAS is that only minimal or no sample preparation is often needed.

II. MICROPHONIC PHOTOACOUSTIC SPECTROSCOPY

The detection of a photoacoustic effect in a gaseous sample or in a solid sample in contact with a gas is customarily effected with a microphone.

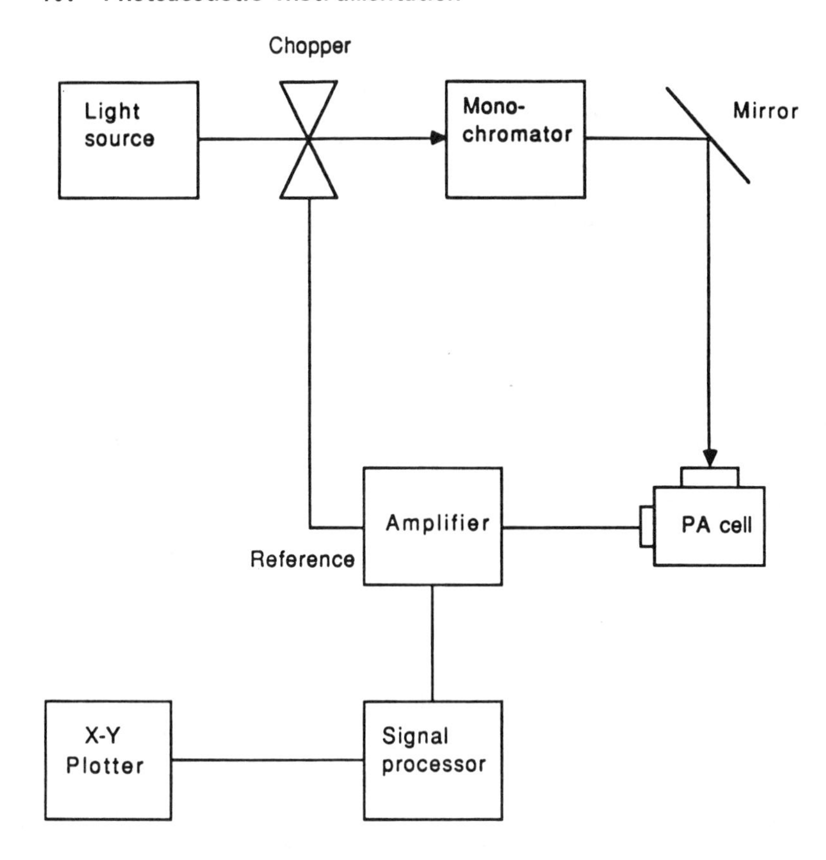

Figure 1. Block diagram of a typical photothermal spectrometer.

The standard model describing quantitatively the PA signal generation based on periodic heat flow from the sample to the ambient gas was proposed by Rosencwaig and Gersho [3] and is called the "thermal piston" model. McDonald and Wetsel [27] and McDonald [28] have shown that "a composite piston model"—that is, thermal diffusion and sample vibration effects both contributing to the observed PA signal—must be used in the equation that gives the PA signal when the acoustic wave produced in the sample causes the interface to vibrate. This effect was recently explored by Torres-Filho et al. [29] for monitoring epoxy adhesive curing.

The basic PA spectrometer (Figure 1) consists of a modulated or pulsed light source, the optical system (lenses, monochromator, etc.), the PA cell, and the signal recovery electronics. The PA spectrometer can measure the signal arising from the sample. The PA cell is a container for the sample and for the detector. Generally, a microphone is used in a gas cell for studying a gaseous sample or a solid sample bathed by a gas, while for liquid and solid samples piezoelectric transducers are often used. The microphone is typically a capacitor sensor and consists of a thin metal diaphragm or a metallized plastic dielectric and a rigid conducting back plate. Any change of pressure causes a deflection of the diaphragm in contact with the gas, changing the microphone capacitance and the output voltage. To eliminate the impact of any wavelength dependence of the source intensity, soot or carbon powder is usually used as a reference sample.

Many commercial microphones are available, ranging from the cheap electret type to the expensive, high-quality condenser microphone. The choice depends on the best compromise among such factors as sensitivity, bandwidth, size, noise, and cost. We will consider first the details of an experiment in which the sample of interest is a gas.

A. Microphonic Optoacoustic Spectroscopy of Gases (OAS)

PA techniques have been used to analyze the composition of gaseous samples, to obtain spectra of gases, and to study deexcitation and energy-transfer processes in gases. Gorelik [30] was the first to propose that measurements of the phase of the PA signal as a function of the modulation frequency in a gas system could be used to investigate the rate of transfer of energy between the vibrational and translational degrees of freedom of the gas molecules. This was put into practice for the first time by Slobodskaya [31] in 1948. Modern photoacoustic studies of gas-phase radiationless decay of excited states typically require a modulation of the incident laser beam, a microphone to detect the periodic pressure change (acoustic field) in the sample gas, and a lock-in amplifier to permit detection of the microphonic signal at the modulation frequency with a phase lag that depends on the time constant for radiationless decay of the excited vibrational states. A double-beam spectrometer with carbon black in the reference channel was used to study photochemical processes in NO_2 [32]. Intersystem crossing [33] was investigated by taking the in-phase and out-of-phase PA spectrum. Impurity quenching of the PA signal requires that the impurity have an excited level energetically close to the excited level of the absorbing molecule, and also that the lifetime of the impurity level be long with respect to f^{-1} [1]. Various designs for PA gas cells were discussed by Rosencwaig [1]. There is a single-pass nonresonant cell in which the optical beam passes through the cell once along the cylinder axis, and a resonant multipass system where the windows are internally reflective so that the optical beam goes back and forth between the reflective end windows. This increases the interaction length of the optical beam without increasing the cell volume. In another design the sample cell is placed within the laser cavity.

For most molecules, vibrational energy lies in the infrared region of the spectrum. Nearly all polyatomic gases absorb in 2-15 μm. There are several ways of obtaining continuously wavelength-tunable laser radiation in the mid-infrared. The most familiar possibility is the CO_2 laser that covers most wavelengths between 9.2 and 10.9 μm. Since many polyatomic gas molecules have strong, characteristic absorption bands within this wavelength range, CO_2-laser optoacoustic spectroscopy has been applied successfully [35] to the monitoring of the low (under 100 ppb) concentrations of toxic vapors present in gas mixtures. A second means of obtaining tunable infrared laser radiation is the use of a pulsed Nd:YAG pumped, tunable dye laser that is Stokes shifted in a multipass high-pressure H_2 gas cell to wavelengths at least as long as 12 μm [36]. A third possibility is the use of tunable infrared lead-salt diode lasers that cover the wavelength range 0.6-40 μm. The narrow line width of the lead-salt lasers (under 0.001 cm^{-1}) has the added potential advantage of resolving molecular species in very complex gas mixtures for which there is a great deal of vibrational band overlap.

For a gas system we have four principal sources of noise: acoustic noise in the background PA signal from light absorption in the cell window, ambient acoustic noise arising from building vibrations, absorption of scattered radiation by the cell walls, and noise from the electromechanical light chopping systems. Electrical noise is mainly attributable to noise sources in the amplifier connected to the microphone. Brownian motion noise arises from thermal fluctuations of the gas in the PA cell. Finally, there is also microphonic noise. Thus considerable attention must be given to minimizing the sample cell noise.

One can imagine a variety of circumstances in which an identification of molecular species present in a gas and their quantitation would be useful. We will focus here on the use of optoacoustic techniques to determine quantitatively traces of contaminant gases present in a continuously flowing gas stream at concentrations below 100 parts per billion (ppb). An FT-IR spectrometer is not well suited to this task, since the comparatively weak, broadband, infrared sources used in such a spectrometer do not permit high-sensitivity, rapid-response measurements. The far greater luminosity of a wavelength-tunable infrared laser addresses this problem nicely, with the added benefit of high spectral resolution characteristic of lasers.

Having replaced the Glo-Bar source and Michelson interferometer of an FT-IR spectrometer with a wavelength-tunable infrared laser, one must next select a more suitable detector than the triglycine sulfate (TGS) or liquid N_2 cooled Hg-Cd-Te sensor of the usual FT-IR spectrometer.

A resonant cell design in which the infrared laser is amplitude modulated at a frequency that corresponds to a standing acoustic wave frequency of the cell is most suitable for detection of trace gases in flowing ambient air using optoacoustic spectroscopy [34]. A schematic of such a Gerlach-Amer [36] cell is depicted in Figure 2. Such a cell accumulates the acoustic energy in a standing wave when the exciting infrared laser is amplitude modulated at a resonant acoustic frequency of the cell. Thus acoustic energy produced by laser pulses is accumulated from many successive laser pulses. Since background acoustic and electrical noise diminishes with increasing frequency, the cell dimensions are selected to resonate acoustically at moderately high laser modulation frequencies. Resonant acoustic frequencies will be of the order of c/l where c is the speed of sound in the gas ($\sim 3.5 \times 10^2$ m sec^{-1} for air at 298 K) and l is a characteristic dimension of the cell. For the cell of Figure 2 wherein radial modes are excited the resonant frequency is thus roughly $(3.5 \times 10^2/0.18)$ $sec^{-1} \approx 2 \times 10^3$ Hz, which is comfortably within the bandwidth of sensitive commercial microphones.

Background acoustic signals arising from gases adsorbed on the resonant sample cell windows are eliminated by positioning the windows at stationary nodal positions of the radial acoustic modes as depicted in Figure 2. Gas inlet and exit ports are similarly positioned at nodes to suppress noise arising from turbulence in the flowing gas. The generous dimensions of

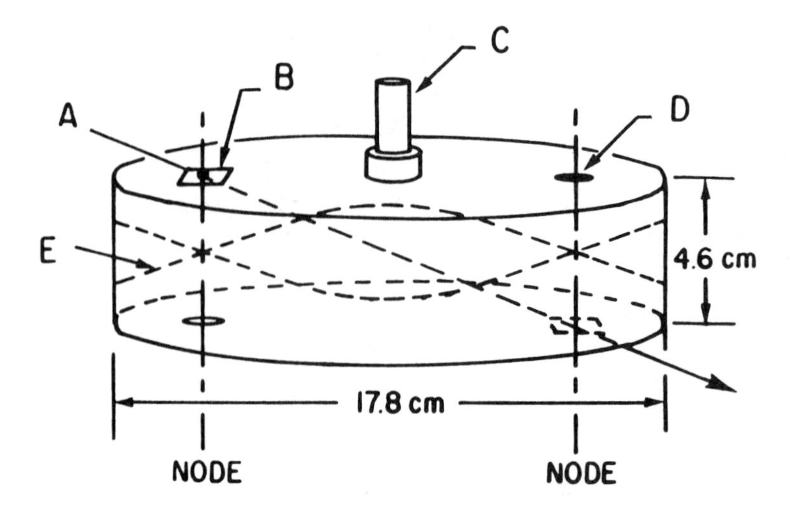

Figure 2. Resonant optoacoustic sample cell based on a design by Gerlach and Amer [7]. A, CO_2 laser beam; B, ZnSe window; C, microphone mount; D, gas port; and E, amplitude of first radial mode. (From Ref. 34, with permission.)

the resonant sample cell produce a low surface-to-volume ratio that reduces concentration inaccuracies arising from polar sample gas adsorption on the interior walls of the cell. The adsorption problem is further diminished by coating the interior surfaces of the stainless steel sample cell with tetrafluoroethylene (Teflon).

The resonant cell design poses the problem that the resonant frequency changes with gas temperature, pressure, and composition. This necessitates a continuous tuning of the laser modulation frequency to the acoustical resonance frequency of the cell. Loper et al. [34] automatically tracked the cell resonant frequency using the high-gain bandpass filter properties of the sample cell to detect the changes in the resonant acoustic frequency and to adjust the laser modulation frequency to preserve the peak optoacoustic signal.

Kamm [37] explored the increase in sensitivity achievable with a multiple-pass optical system involving mirrors that permits a large number of laser beam passes through the resonant optoacoustic sample cell.

B. Microphonic Photoacoustic Spectroscopy (PAS) of Solids

The Rosencwaig-Gersho (RG) one-dimensional model of the microphonic PA effect obtained with a solid sample is depicted in Figure 3. It consists of a solid sample with length l_s in a cell filled with a gas (usually air for visible wavelengths), with the sample at a distance l_g from a transparent window and supported by a backing material with length l_b.

A beam of electromagnetic radiation (x-ray [38], ultraviolet, visible, infrared, or microwave) is chopped at a low audio frequency, f, passes

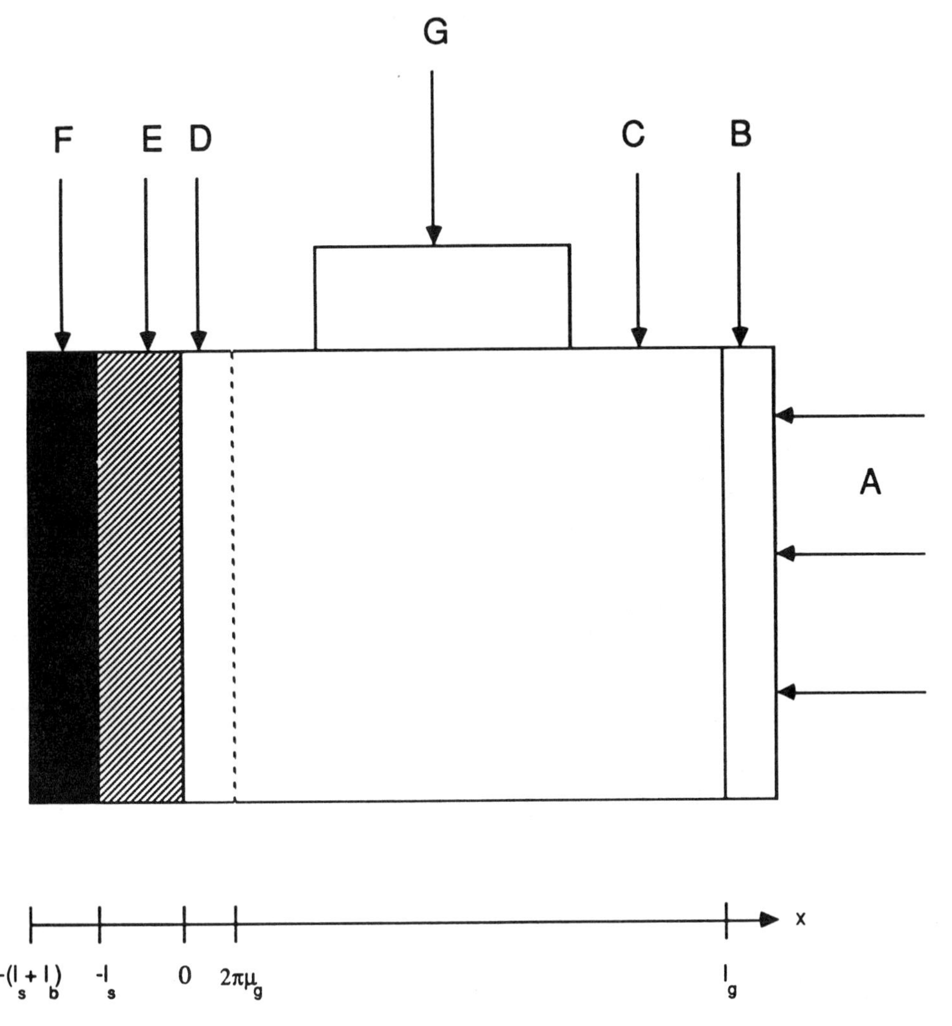

Figure 3. Schematic diagram of the one-dimensional model of the microphonic photoacoustic effect devised by Rosencwaig and Gersho [3]. A, Modulated light; B, transparent window; C, transducing gas; D, thermal piston; E, sample; F, backing material; and G, microphone. (From Ref. 3, with permission.)

through the window, and is incident on the surface of a solid sample. If the wavelength of the incident radiation is one that is absorbed by the sample material, the absorbed energy can be characterized by a wavelength-dependent optical absorption length $\mu_\beta(\lambda)$ given by

$$\mu_\beta(\lambda) = \frac{1}{\beta(\lambda)}$$

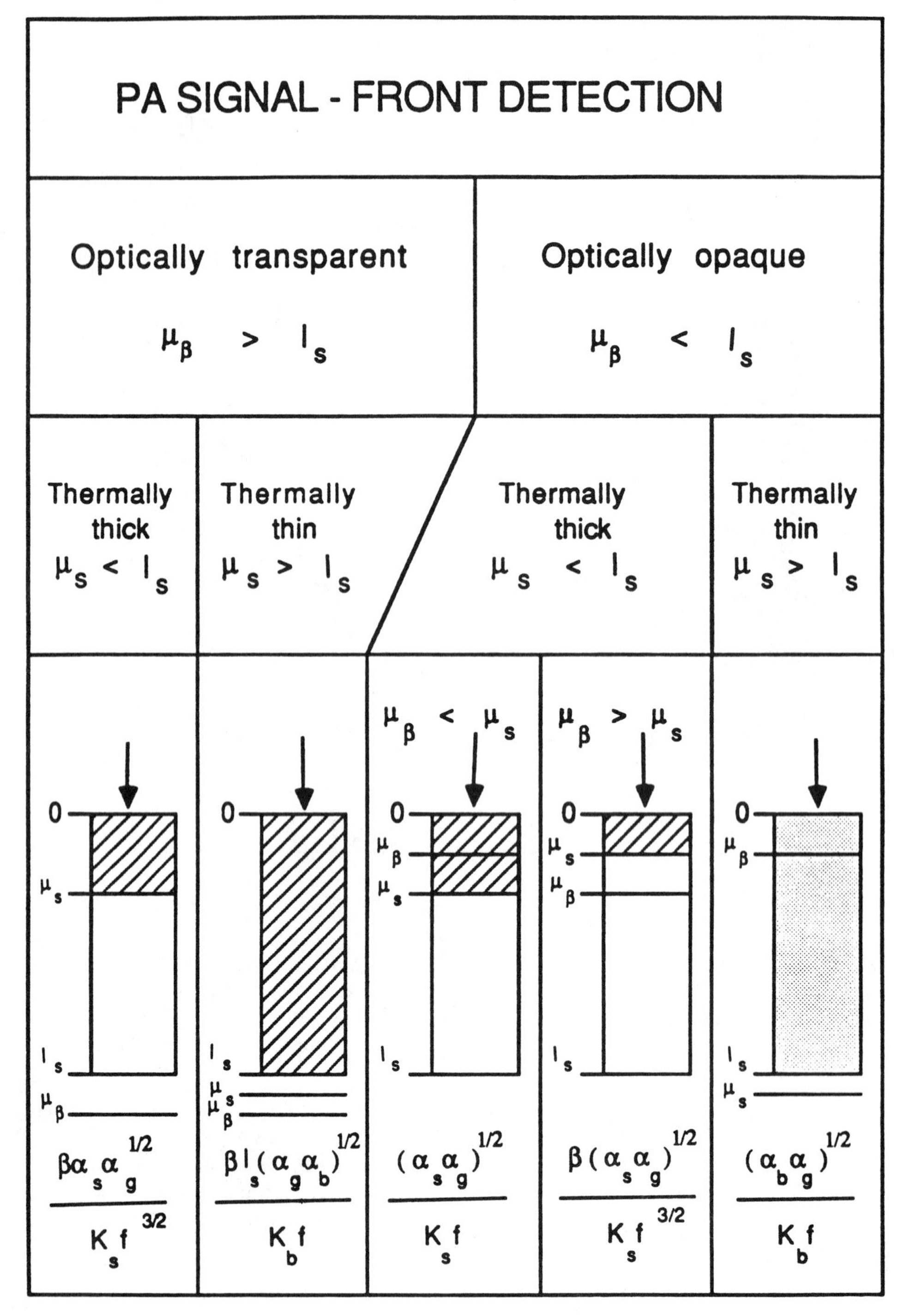

Figure 4. Schematic representation of the special cases for the RG photoacoustic theory of solids. l_s is the sample thickness, f is the modulation

with $\beta(\lambda)$ the usual optical absorption coefficient of the sample. This length μ_β represents the thickness of the sample within which light is absorbed. For optically transparent (thin) and opaque (thick) samples we have $\beta l_s \ll 1$ and $\beta l_s \gg 1$ respectively.

Part of the absorbed energy is converted by rapid radiationless transitions within the solid into heat, with the same time dependence of the incident energy. The resulting heat flow within the sample by thermal diffusion is characterized by the thermal diffusion length μ_s,

$$\mu_s = (\alpha_s/\pi f)^{1/2}$$

where α_s is the sample thermal diffusivity and f is the modulation frequency. Only heat generated within one thermal diffusion length of the sample surface will be able to reach this surface, and only a thin layer of gas adjacent to the sample surface, given by $\mu_g = (\alpha_g/\pi f)^{1/2}$, responds thermally to the periodic heat flow from the sample. This heated layer of gas acts as a thermal piston that sends a sound wave of frequency f through an isolating channel to a sensitive microphone. The signal is then transmitted to a lock-in amplifier and recorded [39].

According to its thermal properties, a sample can be thermally thin or thick depending on whether $\mu_s \gg l_s$ or $\mu_s \ll l_s$, respectively. In Figure 4 we summarize the dependence of the PA signal on the optical and thermal parameters of the sample, as well as its modulation frequency dependence as predicted by the RG model. We can see that for some cases the PA signal is proportional to β, which means that spectroscopy is feasible. Measurements of optical absorption coefficients using PAS with a microphone were first done by Wetsel and McDonald [40]. PAS has been found particularly useful for exploring the optical properties of very opaque [41] and highly transparent materials [42].

For optically opaque ($\beta l_s \gg 1$) and thermally thin ($\mu_s \gg l_s$) samples we have both optical and PA opacity and the PA signal is independent of the physical properties of the sample. An example is carbon black, which is used as a PA standard at visible wavelengths. In the case of a thermally thick but optically thin sample, the RG model yields the following proportionality (as shown in Figure 4):

$$S = C\beta\mu_s f^{-3/2}$$

where S is the PA signal, C is a constant of proportionality, and the other variables have been defined previously. One application of this relationship is to the photoacoustic signal obtained at infrared wavelengths from species chemisorbed on a heterogeneous catalyst. In such a case the PA signal arises from the chemisorbed species and the catalyst, with the optical properties of the chemisorbed monolayer dominating. From the above relationship it is clear that the PA signal intensity will drop off precipitously

---

frequency, μ_β is the optical absorption length, K_i is the thermal conductivity, β is the optical absorption coefficient of the sample and $\mu_i = (\alpha_i/\pi f)^{1/2}$ is the thermal diffusion length, where α_i is the thermal diffusivity and i can be s (sample), b (backing material) or g (gas). (Adapted with the permission of the author from A. C. Bento, Master of Physics thesis, I.F./UNICAMP, Brazil, 1987.)

with increasing beam-chopping frequency. It is also true that the PA signal reaching the microphone cannot come from a greater depth in the solid sample than the distance that heat can diffuse from within the solid to the solid surface between successive pulses of incident electromagnetic radiation. For low chopping frequencies ($f \simeq 50$ Hz) the sample depths probed are of the order of micrometers. Thus photoacoustic spectroscopy is a surface technique that complements other surface techniques, such as x-ray photoelectron spectroscopy (XPS or ESCA), Auger electron spectroscopy (AES), and secondary ion mass spectrometry (SIMS), that examine surfaces only to depths of the order of 10-40 Å. The nondestructive depth profiling of solid sample materials such as printed electronic surface boards [43] illustrates the unique capability of photoacoustics for studies of solid surfaces.

Thermal diffusivities of many materials can be measured by the PA technique. It was first done by Adams and Kirkbright [25]. Here we describe the phase-lag method used by Pessoa et al. [44]. It consists in measuring the relative phase lag $\Delta\phi = \phi_F - \phi_R$ at a single modulation frequency f between rear-surface illumination and front-surface illumination (Figure 5). This is an alternative to a method proposed by Yasa and Amer [45]. Using the RG thermal diffusion model for PA signal generation, we have

$$S_F/S_R = I_F/I_R \, [\cosh^2(l_s a_s) - \sin^2(l_s a_s)]^{1/2}$$

$$\tan(\Delta\phi) = \tanh(l_s a_s) \, \tan(l_s a_s)$$

where S_F/S_R is the ratio of the PA signal amplitude for front- and rear-surface illumination, I_F/I_R is the ratio of the absorbed light intensity for front and rear illumination, and $a_s = (\pi f/\alpha_s)^{1/2}$. The $\Delta\phi$ measurements for one frequency are sufficient to derive the thermal diffusivity if we know the sample thickness.

To get this relationship we must suppose that the sample is optically opaque to the incident light. For semiconductors this condition is easily achieved, but for transparent samples such as glasses and polymer foils, it is necessary to use aluminum foil (20 μm thick) attached to both sides of the sample with a thin layer of vacuum grease. The thermal diffusion time $\pi l_s^2/\alpha_s$ in this Al foil is the order of 13.6 μsec, so that the heat generated in the Al foil absorber may be assumed to be instantaneously diffused toward the sample. This method has been applied to semiconductors [44], glasses [46], and the polymer foils [47]. In the case of samples with large thermal expansion coefficients, such as polymers [47] and some glasses, the method is applicable only at low frequency ($f < 15$ Hz). This is true because the main contribution to the PA signal comes from the thermoelastic bending of the sample, as demonstrated by Leite et al. [47], so one must take account of this new contribution to the PA signal generation. In this case the thermal diffusion is obtained from the modulation frequency dependence of the front-phase signal:

$$\phi_F = -\frac{\pi}{2} + \arctan\left[\left(\frac{1}{l_s a_s}\right) - 1\right]$$

Measurements of thermal diffusivity were used by Leite et al. to study the kinetics of the doping process of iodine-impregnated polystyrene.

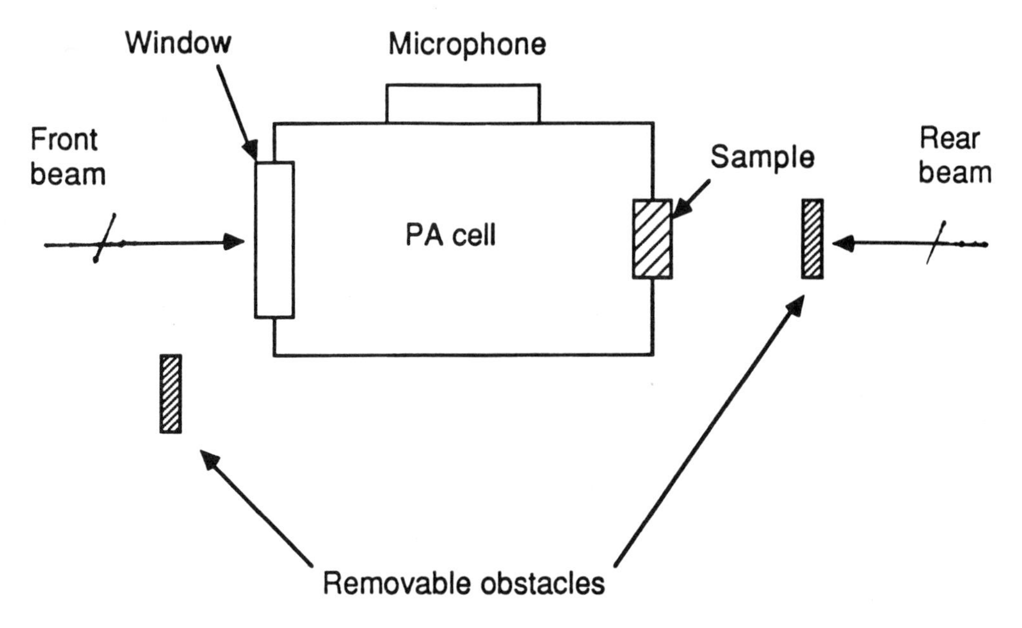

Figure 5. Schematic of experimental arrangement used in the determination of thermal diffusivity through a two-beam PA phase measurement, after Pessoa et al. [44].

The present authors believe that analytical chemists would benefit from viewing photoacoustics not as the basis for a new kind of spectroscopy but rather as the basis for alternative means of detection in various existing electromagnetic spectroscopies. Thus, for example, a microphonic, photoacoustic sample cell can be dropped into the sample compartment of a mid-infrared Fourier transform spectrometer where the PA cell replaces the usual triglycine sulfate (TGS) solid-state detector. This substitution of detectors proves especially advantageous in mid-infrared studies of opaque samples such as sulfided, aluminum oxide-supported, cobalt-molybdenum catalysts used in the hydrodesulfurization of petroleum fractions [48].

Successful mid-infrared studies of solids sometimes hinge only on the qualitative appearance and disappearance of characteristic peaks along the wavelength axis, with no need for quantitation [49]. (When quantitation is necessary, internal standards sometimes prove useful [50].) Thus many scientists with access to dispersive or FT-IR spectrometers would probably find applications for microphonic photoacoustic sample cells if they were less expensive and consequently more readily available. (Signal/noise ratio is so low in PA experiments that one greatly prefers an FT over a dispersive instrument.) A principal purpose of the present chapter is the description in detail of a cheap but sensitive microphonic PA sample cell and its associated electronics, which can be homemade and substituted for the TGS detector in commercial FT-IR spectrometers. A schematic diagram of the adaptation of a Digilab FT-IR spectrometer for photoacoustic measurements on solid samples is shown in Figure 6.

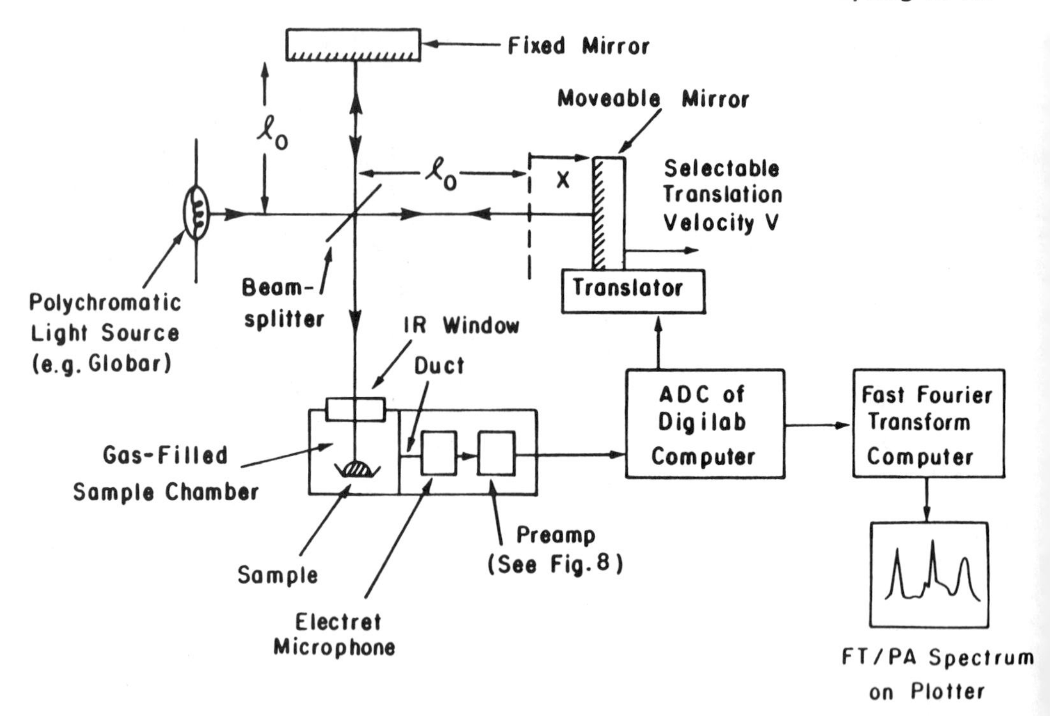

Figure 6. Schematic diagram of the adaptation of a Digilab FT-IR spectrometer for photoacoustic measurements on solid samples.

Figure 7 depicts a cheap microphonic PAS sample cell. The microphone is a commercially available electret type (e.g., Panasonic Omnidirectional Electret Condenser Microphone Cartridge P9930) costing about one U.S. dollar. High-quality but expensive condenser microphones (such as the Bruel & Kjaer model 4165) have at least twice as much sensitivity and a flatter frequency response curve. However, since one usually references the PA signal from the sample against that obtained from a black material, the flatness of the response curve is not critical as long as the microphone selected can detect low-frequency signals. Furthermore, the larger physical size of the B&K microphone precludes a compact sample cell design.

By drilling a 0.6-mm-diameter hole in the side surface of the P9930 microphone (about 2 mm from the front side edge), one front-vents the microphone. It can withstand pressure changes arising from evacuation of the cell required in the introduction of helium or nitrogen to serve as a noninfrared absorbing transducer gas. (By loading the sample cell in a controlled-atmosphere glove box one can circumvent this pressure change problem.)

Unanimity regarding the choice of the source for a mid-infrared reference spectrum is lacking. Fateley and co-workers [51] prefer Fisher Norit-A graphite, whereas Riseman and Eyring [52] recommended use of a deuterated triglycine sulfate (DTGS) detector.

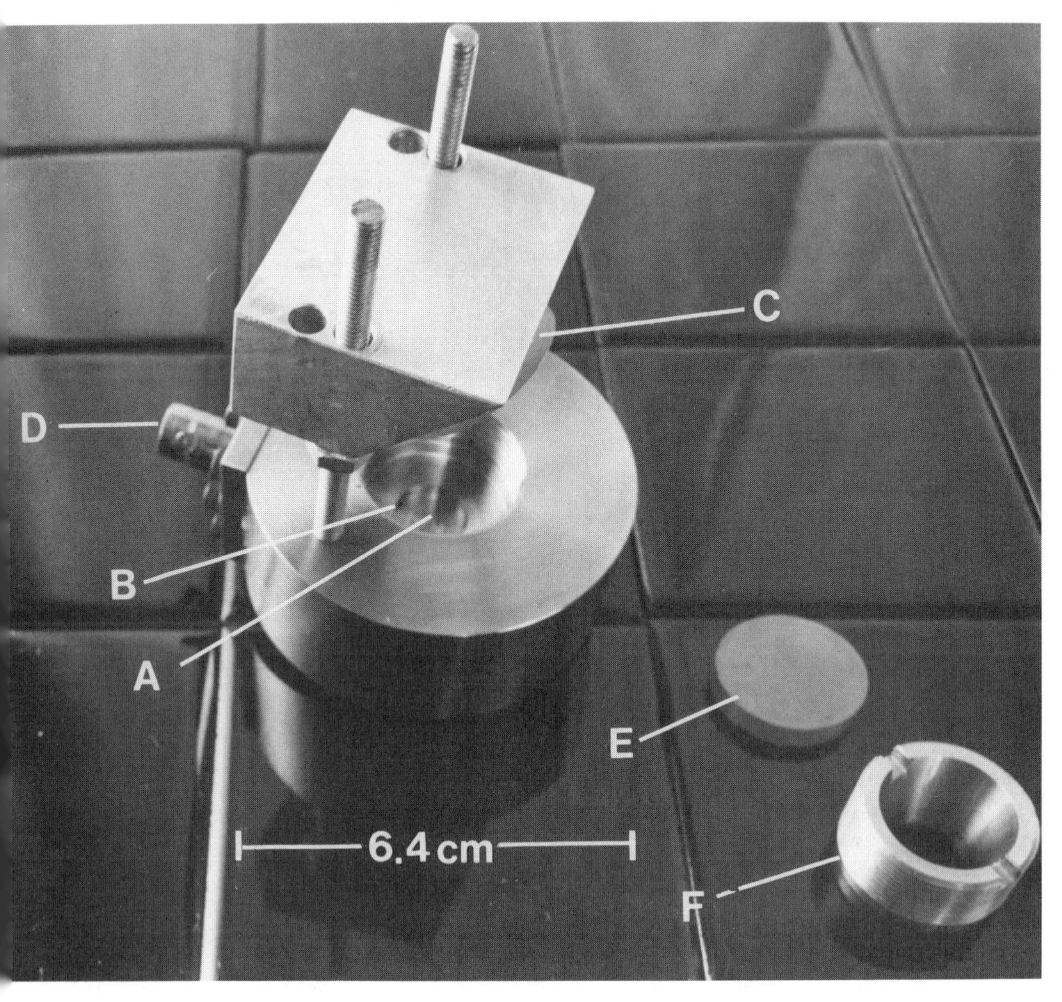

Figure 7. Photograph of a microphonic photoacoustic sample cell: A, shallow sample well; B, duct leading from sample well to the microphone; C, front-surface gold mirror for reflecting the infrared beam from the Michelson interferometer onto the sample; D, BNC connector leading from the microphone; E, KRS-5 infrared transmitting window; and F, screw-down clamping ring for holding the window in place.

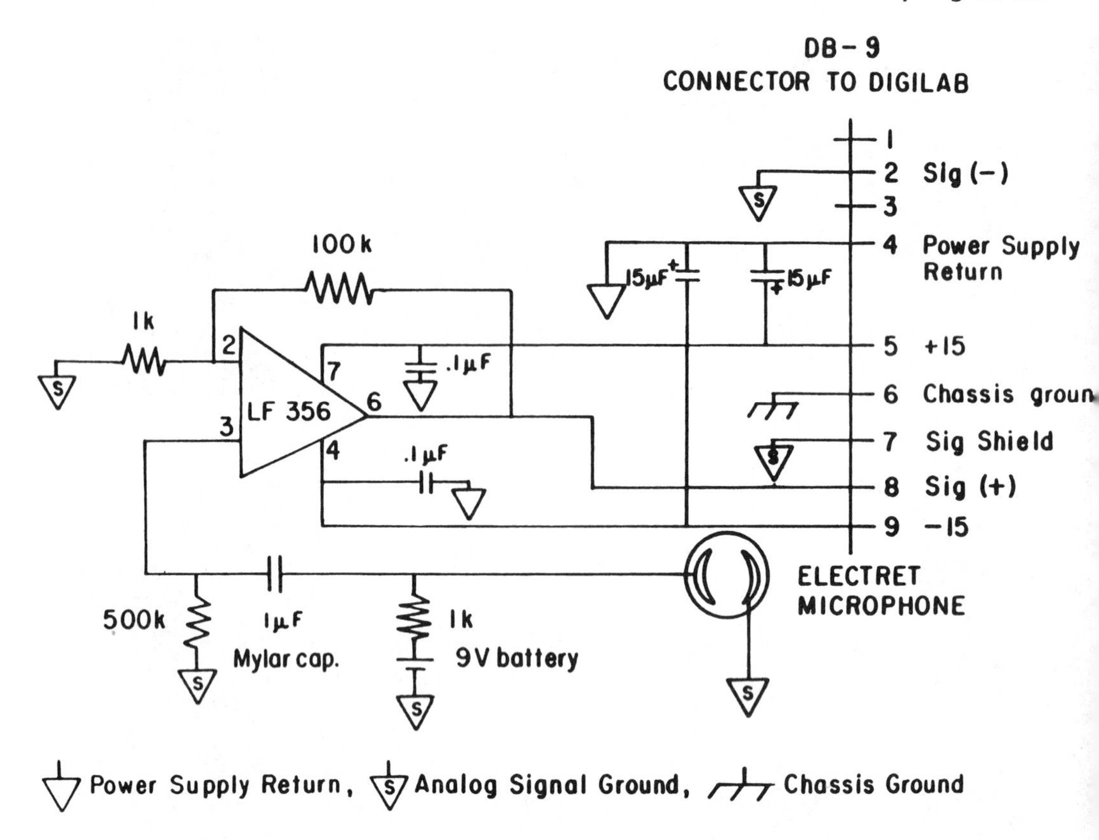

Figure 8. Electronics between the microphone and FT-IR spectrometer for the sample cell of Figure 7.

Figure 8 is a schematic of the electronic circuit servicing the electret microphone. A battery-powered preamplifier reduces noise. The PA signal emerging from this circuit is passed to a lock-in amplifier, in our case a Princeton Applied Research (PARC) model 124A.

For mid-infrared work a KRS-5 disk (available from Wilmad Glass Co., Buena, NJ) is a suitable window material. One changes the reference mode of the lock-in amplifier to internal (thus functioning no longer as a lock-in) and sets the bandpass filter precisely. The upper bound on the accepted frequency should be double or triple the highest estimated frequency to assure a multimodulated interferogram of the high-frequency signal. The final amplified photoacoustic signal is fed back to the commercial FT-IR spectrometer through the connector for the bypassed TGS detector. The motion of the moving mirror of the Michelson interferometer in a "rapid scan" FT-IR spectrometer modulates the PA signal in a manner analogous to the beam chopping described above. The equivalent modulation frequency f is given by

$$f = 2 \times \text{mirror scanning speed (cm/sec)} \times \text{wave number (cm}^{-1}\text{)}$$

Since the PA signal amplitude decreases with increasing frequency, a slow mirror scanning speed is desirable. However, one must make a trade-off against the need to accumulate data in a reasonable length of time.

Other considerations regarding frequency should be mentioned. From the above relationship it is clear that on the high-frequency (4000 cm^{-1}) end of the mid-infrared spectral range the "chopping" frequency of a rapid-scan FT-IR spectrometer is 10 times as high as it is at the low-frequency (400 cm^{-1}) end of the spectral range. Thus the photoacoustic signal measured at 4000 cm^{-1} is for chemical species nearer the sample surface than the species responsible for the photoacoustic signal at 400 cm^{-1}. A step and integrate motion of the moving mirror in an FT spectrometer [53] circumvents this problem, but the typical commercial FT-IR instrument is of the rapid scan variety. JASCO now manufactures an FT-IR spectrometer with a step and integrate mirror motion designed to facilitate photoacoustic depths profiling studies.

The total internal gas volume of the sample cell is made as small as possible for two reasons: First, and most important, the photoacoustic signal intensity is inversely proportional to cell volume. Second, the combination of the unoccupied sample cell volume and unoccupied microphone compartment volume with the connecting duct has the geometry of a Helmholtz resonator that enhances the PA signal at the unique Helmholtz frequency for that geometry [1]. Care is taken to select a geometry that will place this frequency outside the 4000 cm^{-1} to 400 cm^{-1} spectral range so that this resonance effect does not distort peak heights in a portion of the measured infrared spectrum.

The reason that the gas in the sample cell (which acts as a signal transducer between the sample surface and the microphone) must not be an absorber of infrared radiation is that the photoacoustic effect is much larger in absorbing gases than in solids. Thus even a trace of carbon dioxide gas, for example, in the PA sample cell will produce a spectrum that totally obscures the anticipated photoacoustic spectrum from the solid sample.

If one wishes to use the PA sample cell described above for measurements at higher frequencies (visible, ultraviolet, or x-ray), the sample should be placed on top of a second (transparent) window screwed into the bottom of the sample cell. (This eliminates reflection of the electromagnetic radiation from the stainless steel bottom of the sample well, which becomes a more serious problem at shorter wavelengths.)

The potential for IR studies of catalysis, corrosion, and other surface modification processes has been reported by Palmer and Smith [54]. Further examples of the use of FTIR/PAS in surface chemistry and catalysis studies can be found in Refs. 55 and 56.

III. PIEZOELECTRIC PHOTOACOUSTIC SPECTROSCOPY

When localized heating occurs in a material, the heat energy can flow to the surrounding matter through two mechanisms. Thermal diffusion is the first mechanism, in which the rate of energy transfer is determined by the thermal diffusivity ($\alpha = K/\rho c$) of the material and the distance of appreciable energy transfer is given by the thermal diffusion length. In the preceding expression K is the thermal conductivity, ρ is the sample density, and c is the sample heat capacity at constant pressure. This is the thermo-acoustic mode and is a dissipative process. The second mechanism is the thermoelastic process, which is generally nondissipative. Here the energy transfer is through a coupling of the local heat energy to the vibrational modes of the material, causing the sample to expand and contract, thus generating an elastic (acoustic) wave. The speed of energy transfer is determined by the speed of sound in the material, and the distance of appreciable energy transfer is limited by the sample dimensions.

The resulting expansion, caused by a combination of the normal thermal expansion and thermal elastic bending mechanism, produces a net displacement of the sample surface, which can be sensed by a piezoelectric device as a voltage output between the two surfaces of the transducer. Thus piezoelectric detection responds to absorption of radiation by the entire sample.

Many types of piezoelectric materials are available, such as lead zirconate titanate (PZT), lead metaniobate, lithium niobate, crystalline quartz, and polymer films such as polyvinylidene difluoride (PVF_2 or sometimes PVDF), Teflon, Mylar, etc., with PVDF being the most frequently used polymeric piezoelectric.

Piezoelectric ceramic transducers such as Pb-Zr-Ti PZT (Vernitron, Bedford, OH) are much less sensitive detectors of PA signals than are microphones. The typical sensitivity of a piezoelectric transducer is a few volts per atmosphere, compared to that of a sensitive microphone of a few thousand volts per atmosphere. However, advantages of piezoelectrics include a much faster signal detection risetime, that is, broad frequency bandwidth (as high as 1 GHz), and applicability to liquid and solid sample systems where a direct or indirect coupling can produce a good match of acoustic impedances. This opportunity was recognized [57,58] early in the development of interest in photoacoustic spectroscopy of solids, but it remained for Tam and Patel to demonstrate the overwhelming advantage of piezoelectric detectors for photoacoustic spectral studies of liquids [59,60].

Piezoelectric detection of the PA signal is also useful for a large sample that would be difficult to accommodate within a gas-microphone cell. It is also useful for a bulk sample with low optical absorption, since piezoelectric detection senses the total amount of light absorbed by the entire sample. A third important application is when the surrounding medium is a vacuum.

A. Piezoelectric Transducers with Liquid Samples

The piezoelectric technique is applicable to the study of weak absorptions in liquids in the UV, visible, or IR region of the electromagnetic spectrum [1]. Figure 9 depicts two examples of many possible piezoelectric sample cell configurations for photoacoustic spectroscopic studies of liquids. In Figure 9a cylindrical piezoelectric transducer (PZT) is incorporated into the body of the sample cell [61]. The most convenient mounting is shown in Figure 9b, where the transducer is in direct contact with the outside surface of a standard absorption cuvette [62]. Since the coupling of the piezoelectric transducer with the cuvette surface is critically important,

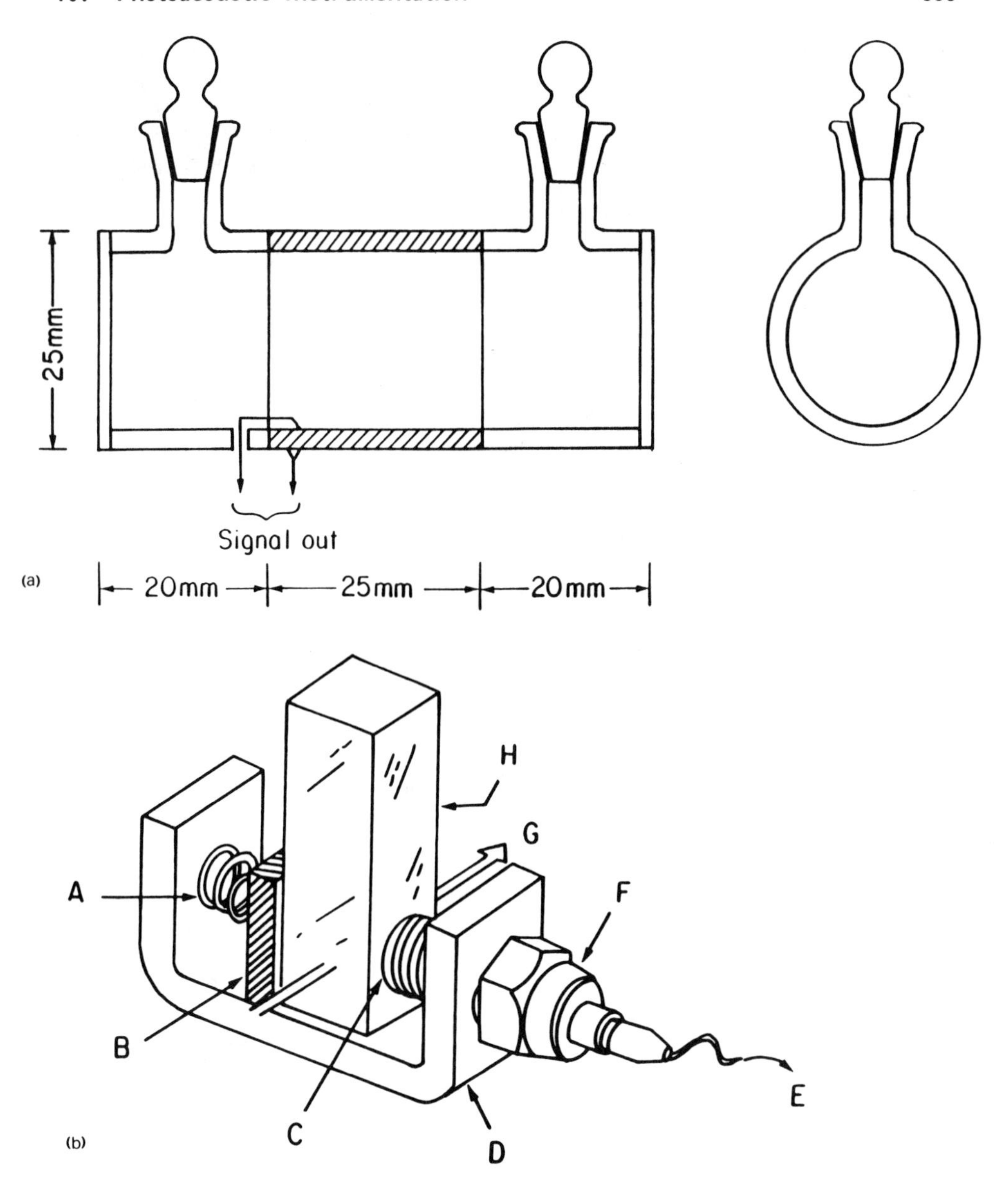

Figure 9. (a) Piezoelectric sample cell for liquids wherein the cylindrical transducer has become part of the body of the sample cell. (From Ref. 60, with permission.) (b) Photoacoustic sample cell with piezoelectric transducer making physical contact with the outside surface of the sample liquid cuvette: A, spring; B, rubber pad for mounting; C, thin layer of grease; D, steel clamp; E, to preamplifier; F, transducer; G, laser beam; and H, solid sample or cuvette with liquid sample. (Reprinted from Ref. 61 with permission.)

the surface must be flat and clean and a coupling material such as silicone grease or epoxy cement with similar acoustic impedance must be applied to achieve a high signal-to-noise ratio. In our experience with such sample cells, 50 mW of laser power may be needed to obtain adequate photoacoustic signals. Flowing liquid cells for photoacoustic measurements with piezoelectric detectors have also been constructed [63].

To verify that such a transducer is working properly one can blow directly on the piezoelectric. If it works, a signal will appear on a monitoring oscilloscope. It is also important to insure that the amplifier accepting the signal from the transducer has a sufficiently high input impedance. If it does not, a FET (field-effect transistor) operational amplifier may be inserted between the transducer and the amplifier.

While a piezoelectric transducer is insensitive to acoustic waves in surrounding air (because of the acoustic impedance mismatch), it is sensitive to mechanical vibrations arriving through the base of the apparatus. Thus a thin layer of foam rubber under the apparatus can be helpful.

A piezoelectric ceramic typically has a high electrical impedance and thus may function as a receiving antenna for ambient electromagnetic radiation arising from sources such as the capacitor bank of a laser system. A metal box (Faraday cage) surrounding the piezoelectric detection system can reduce this source of noise in the PA signal.

Laser light scattered onto the piezoelectric transducer can produce spurious photoacoustic signals. Thus special attention must be given to the elimination of bubbles and dust from a sample liquid in such a PA cell. This is a particularly important consideration in flow cells with small liquid volumes. The liquid may be degassed (using a standard laboratory ultrasonic generator or by stirring in vacuum), and a back pressure may be applied in flow systems to eliminate bubbles.

Because of its fast rise time a piezoelectric transducer is particularly well suited for pulsed PA experiments. During the past 10 years, many authors have applied the pulsed PA technique to the detection of small concentrations and/or weakly absorbing species in liquids [64-67].

The basis for study of photoinduced conversion processes with PA methods is that following optical absorption in a medium the nonradiative (heat) channels compete with several other deexcitation channels, such as fluorescence, photochemistry, photoelectricity, etc. If E_0 is the energy absorbed in a system, the heat released Q can be expressed in a general form as

$$Q = E_0\left(1 - \sum_i \gamma_i\right)$$

where the γ_i values are the conversion efficiencies of the several channels that decay without producing heat. Since the PA signal is proportional to the heat released, it follows that

$$S = S_0\left(1 - \sum_i \gamma_i\right)$$

where S_0 is the PA signal when all absorbed energy is converted into heat. Thus the PA method is complementary to other photophysical and photochemical methods.

Two experimental methods have been used to measure the heat released by radiationless relaxation processes. The first method employs a gas microphone and phase-sensitive detection of the acoustic wave and is sensitive to processes whose lifetimes are greater than 100 μsec, whereas the second method, called pulsed acoustic calorimetry, uses a piezoelectric detector for monitoring the acoustic wave and is sensitive to processes with lifetimes ranging from 50 nsec to 20 μsec. Thus the pulsed PA method with piezoelectric detection is the most attractive for direct measurement of fast nonradiative processes that follow photoexcitation and in the last few years has been used to monitor radiationless decay processes in liquids in a manner complementary to earlier luminescence photophysical studies [24,68-70].

When a ceramic PZT a few millimeters thick is employed in such studies the resonance frequency is usually below 1 MHz. This limits the time resolution achievable to the order of microseconds. This limitation has been overcome by using thin polymeric foils such as poly(vinylidene difluoride), denoted hereafter as PVF_2. Tam and co-workers [71-74] used PVF_2 foil to detect PA signals in solid samples. More recently this type of piezoelectric foil has been used in liquid-phase PA studies [75,76]. In our laboratory we have fabricated PVF_2 transducers for monitoring radiationless deactivation processes from excited states with lifetimes as short as a few nanoseconds [77]. These transducers can be immersed directly in the sample liquid and follow signal rise times of less than 10 nsec.

Figure 10a depicts schematically the way in which we mount a PVF_2 foil for liquid samples. Direct contact between the foil and the liquid considerably improves the sensitivity of the detector because of the similarity of the acoustic impedance of PVF_2 and that of the liquid. Our preference for Pennwalt KYNAR [78] is based on the sensitivity of this product and the durability of the aluminum metal film electrodeposited on the foil. For a chemically hostile environment one can use the gold or nickel electrodized foils. Figure 10b shows the typical shape of the photoacoustic pulse monitored with our PVF_2 transducer. In this instance, the ~7-nsec duration, 20 μJ, λ = 355 nm Nd:YAG laser pulse focused to a spot size of ~50 μm is absorbed by a $\sim 10^{-5}$ *M* solution of recrystallized ferrocene in methanol. The signals detected with a PVF_2 transducer are perfectly consistent with theoretically predicted [79,80] pulsed PA signal shapes. The signal shows a compression and a rarefaction of the liquid when the PA pulse propagates through the sample. The shape of the signal depends on the laser-beam-transducer geometry and also on the rate of the radiationless processes.

A detailed description of a PA calorimeter and the application of pulsed, time-resolved photoacoustic calorimetry with piezoelectric detection to a variety of problems in biochemistry and organic and organometallic chemistry has been reported by Peters and Snyder [81].

B. Piezoelectric Transducers with Solids Samples

In PA experiments with solid samples and piezoelectric detection, one can simply attach the piezoelectric to the sample with wax or a suitable cement [58]. Generally, for optically transparent samples the transducer is in the form of an annulus located on either side of the sample, while for

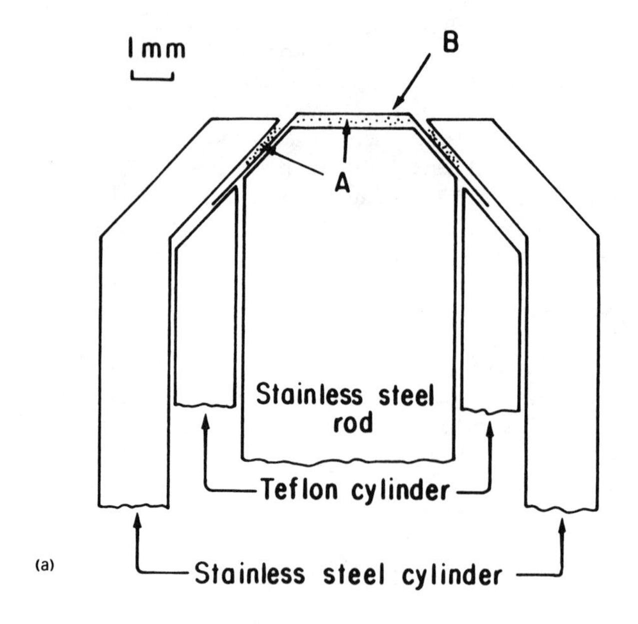

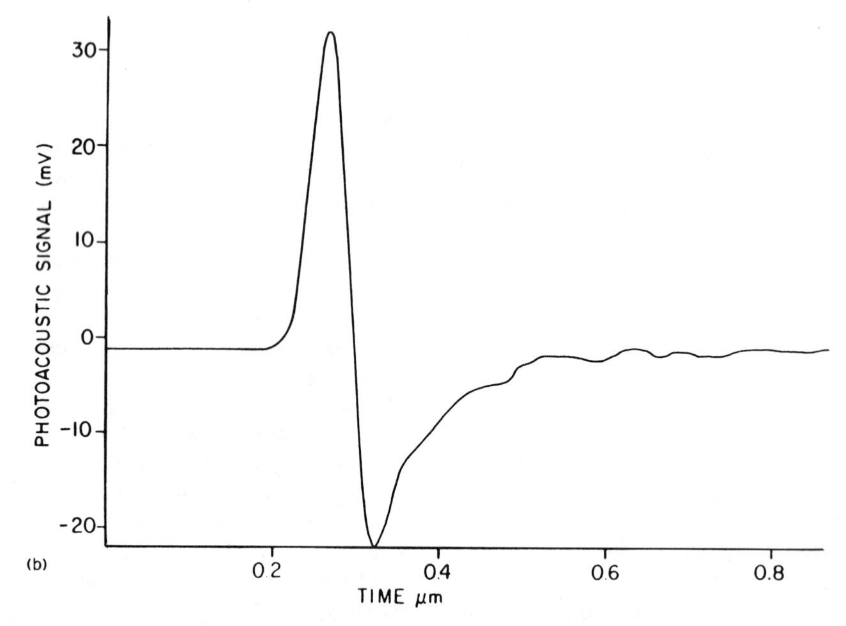

Figure 10. (a) Schematic diagram of the homemade cylinder mount for the small PVF_2 transducers used in fast radiationless decay experiments with liquid samples at Utah. A, conductive, epoxy adhesive; B, 28-μm-thick PVF_2 foil. (b) Voltage versus time display of a photoacoustic signal obtained with a ferrocene in methanol sample using the PVF_2 transducer of (a).

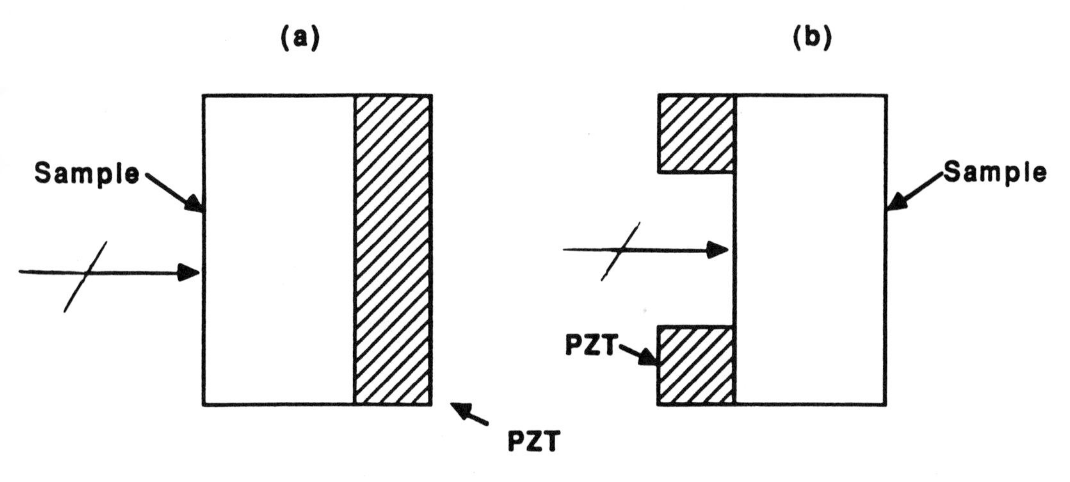

Figure 11. Transducer assembly for PZT detection of solid (a) opaque and (b) transparent samples. (From Ref. 82, with permission.)

optically opaque samples, the piezoelectric transducer is attached to the back surface of the sample [82] (Fig. 11). The cell of Figure 9b can also be used for solid samples. The coupling agent between the sample and the transducer must have a good acoustic impedence match with both sample and transducer, and the incident radiation must not reach the transducer.

In the first PA spectroscopic measurements on solids with piezoelectric detectors, Hordvik and Schlossberg [5] studied weak absorptions in high-power laser windows. In another experiment Hordvik and Skolnik [57] showed one can measure surface and bulk optical absorption in these materials, through the different phases of the piezoelectric signals. Studies of bulk solid and powder samples have also been reported by Farrow et al. [58].

Tam and Patel [83] measured absorption by thin films of powder and avoided the effect of scattered light by putting the sample and the transducer near the opposite ends of a substrate. Photoconductive quantum efficiency of a thin organic dye film (0.1-1 μm) was measured with piezoelectric detection [84] in an extension of Cahen's idea [19] for PA monitoring of photoelectricity to other systems.

Photothermal studies of electron paramagnetic resonance (EPR) of $CuSO_4$ single crystals cemented to a piezoelectric located inside the cavity of an X-band spectrometer have been reported by DuVarney et al. [85].

Photothermal techniques with piezoelectric detection have also been used to monitor transport properties and nonradiative processes in semiconductors. Goede et al. [86] studied single crystals of CdS:Se, CdS:Te, ZnS:Te, and ZnS:Mn and thin films of the last compound. They recommended the use of this technique for measurements of absorption tails down to 10^{-3} cm^{-1} using samples ~10^{-1} cm thick, for sensitive determination of isovalent impurities after suitable calibrations, and for measurements of quantum efficiencies of highly luminescent samples with impurity concentrations too low for conventional studies. Photothermal microscopy with piezoelectric detection has also been reported [71].

ACKNOWLEDGMENTS

The authors gratefully acknowledge financial assistance provided by the Department of Energy, Office of Basic Energy Sciences. N. F. Leite is grateful to the Brazilian agency FAPESP for partial financial support.

REFERENCES

1. A. Rosencwaig, *Photoacoustics and Photoacoustic Spectroscopy*, John Wiley and Sons, New York, 1980.
2. Y.-H. Pao, Ed., *Optoacoustic Spectroscopy and Detection*, Academic Press, New York, 1977.
3. A. Rosencwaig and A. Gersho, J. Appl. Phys. *47*, 64 (1976).
4. F. A. McDonald, Can. J. Phys. *64*, 1023 (1986).
5. A. Hordvik and H. Schlossberg, Appl. Opt. *16*, 101 (1977).
6. H. Coufal, Appl. Phys. Lett. *44*, 59 (1984).
7. A. Mandelis, Chem. Phys. Lett. *108*, 388 (1984).
8. A. Mandelis and N. Zwer, J. Appl. Phys. *57*, 4421 (1985).
9. S. Ameri, E. A. Ash, V. Neuman, and C. R. Petts, Electron. Lett. *17*, 337 (1981).
10. L. C. M. Miranda, Appl. Opt. *22*, 2882 (1983).
11. D. Fournier, A. C. Boccara, and J. Badoz, Appl. Phys. Lett. *32*, 640 (1978).
12. J. C. Murphy and L. C. Aamodt, Appl. Phys. Lett. *38*, 196 (1981).
13. P. E. Nordal and S. O. Kanstad, Phys. Scr. *20*, 659 (1979).
14. R. Santos and L. C. M. Miranda, J. Appl. Phys. *52*, 4194 (1981).
15. S. Pekker and E. M. Eyring, Appl. Spectrosc. *40*, 397 (1986).
16. S. Pekker and E. M. Eyring, Appl. Spectrosc. *41*, 260 (1987).
17. E. M. Monahan, Jr., and A. W. Nolle, J. Appl. Optics *48*, 927 (1977).
18. C. L. Cesar, C. A. S. Lima, N. F. Leite, H. Vargas, A. F. Rubira, and F. Galembeck, J. Appl. Phys. *57*, 4431 (1985).
19. D. Cahen, Appl. Phys. Lett. *33*, 810 (1978).
20. D. Cahen and S. D. Halle, Appl. Phys. Lett. *46*, 446 (1985).
21. C. C. Ghizoni and L. C. M. Miranda, in *IEEE Ultrasonics Symposium Proceedings* (B. R. McAvoy, Ed.), Vol. 2, IEEE, New York, 601, 1982.
22. S. J. Isak, S. J. Komorowski, C. N. Merrow, P. E. Poston, and E. M. Eyring, Appl. Spectrosc. *43*, 419 (1989).
23. T. A. Moore, Photochem. Photobiol. Rev. *7*, 187 (1983).
24. S. J. Komorowski, Z. R. Grabowski, and W. Zielenkiewicz, J. Photochem. *30*, 141 (1985).
25. M. J. Adams and G. F. Kirkbright, Analyst *102*, 678 (1977).
26. L. C. Aamodt and J. C. Murphy, Can. J. Phys. *64*, 1023 (1986).
27. F. A. McDonald and G. L. Wetsel, Jr., J. Appl. Phys. *49*, 2313 (1978).
28. F. A. McDonald, Appl. Optics *18*, 1363 (1979).
29. A. Torres-Filho, L. F. Perondi, and L. C. M. Miranda, J. Appl. Polymer Sci. *35*, 103 (1988).
30. G. Gorelik, Dokl. Akad. Nauk SSSR *54*, 779 (1946).
31. P. V. Slobodskaya, Izv. Akad. Nauk SSSR, Ser. Fiz. *12*, 656 (1948).
32. W. R. Harshbarger and M. B. Rosin, Acc. Chem. Res. *6*, 329 (1973).

33. K. Kaya, W. R. Harshbarger, and M. B. Robin, J. Chem. Phys. *60*, 4231 (1974).
34. G. L. Loper, J. A. Gelbwachs, and S. M. Beck, Can. J. Phys. *64*, 1124 (1986).
35. P. Rabinowitz, B. N. Perry, and N. Levinos, IEEE J. Quantum Electronics *QE-22*, 797 (1986).
36. R. Gerlach and N. M. Amer, Appl. Phys. *23*, 319 (1980).
37. R. D. Kamm, J. Appl. Phys. *47*, 3550 (1976).
38. T. Masujima, in *Synchrotron Radiation in Chemistry and Biology II, Topics in Current Chemistry*, Vol. 147 (E. Mandelkow, Ed.), Springer-Verlag, Berlin, 1988, pp. 145-157.
39. H. Coufal and J. F. McClelland, J. Mol. Struct. *173*, 129 (1988).
40. G. C. Wetsel, Jr., and F. A. McDonald, Appl. Phys. Lett. *30*, 252 (1977).
41. See, for example, M. G. Rockley and J. P. Devlin, Appl. Spectrosc. *34*, 407 (1980).
42. See, for example, A. C. Tam and C. K. N. Patel, Appl. Optics *18*, 3348 (1979).
43. A. Rosencwaig, in *VLSI Electronics; Microstructure Science*, Vol. 9 (N. G. Einspauch, Ed.), Academic Press, New York, 1985.
44. O. Pessoa, Jr., C. L. Cesar, N. A. Patel, H. Vargas, C. C. Ghizoni, and L. C. M. Miranda, J. Appl. Phys. *59*, 1316 (1986).
45. Z. Yasa and N. Amer, in "Topical Meeting on Photoacoustic Spectroscopy" (Ames, IA, 1979), paper WAS-1, unpublished.
46. A. C. Bento, H. Vargas, M. M. F. Aguiar, and L. C. M. Miranda, Phys. Chem. Glasses *28*, 127 (1987).
47. N. F. Leite, N. Cella, H. Vargas, and L. C. M. Miranda, J. Appl. Phys. *61*, 3025 (1987).
48. S. M. Riseman, S. Bandyopadhyay, F. E. Massoth, and E. M. Eyring, Appl. Catalysis *16*, 29 (1985).
49. W. P. McKenna, D. J. Gale, D. E. Rivett, and E. M. Eyring, Spectrosc. Lett. *18*, 115 (1985).
50. S. M. Riseman, F. E. Massoth, G. M. Dhar, and E. M. Eyring, J. Phys. Chem. *86*, 1760 (1982).
51. See, for example, C. Q. Yang and W. G. Fateley, J. Mol. Struct. *141*, 279 (1986).
52. S. M. Riseman and E. M. Eyring, Spectrosc. Lett. *14*, 163 (1981).
53. M. M. Farrow, R. K. Burnham, and E. M. Eyring, Appl. Phys. Lett. *33*, 735 (1978).
54. R. A. Palmer and M. J. Smith, Can. J. Phys. *64*, 1081 (1986).
55. S. B. Kinney, R. H. Staley, C. L. Reichel, and M. S. Wrighton, J. Am. Chem. Soc. *103*, 4273 (1981).
56. M. G. Rockley, D. M. Davies, and H. H. Richardson, Appl. Spectrosc. *35*, 185 (1981).
57. A. Hordvik and L. Skolnik, Appl. Optics *16*, 2919 (1977).
58. M. M. Farrow, R. K. Burnham, M. Auzanneau, S. L. Olsen, N. Purdie, and E. M. Eyring, Appl. Optics *17*, 1093 (1978).
59. C. K. N. Patel and A. C. Tam, Appl. Phys. Lett. *34*, 467 (1979).
60. C. K. N. Patel and A. C. Tam, Rev. Mod. Phys. *53*, 517 (1981).
61. S. Oda, T. Sawada, and H. Kamada, Anal. Chem. *50*, 865 (1978).

62. A. C. Tam and C. K. N. Patel, Optics Lett. *5*, 27 (1980).
63. T. Sawada and S. Oda, Anal. Chem. *53*, 471 (1981).
64. A. C. Tam and C. K. N. Patel, Appl. Optics *18*, 3348 (1979).
65. C. K. N. Patel and A. C. Tam, Chem. Phys. Lett. *62*, 511 (1979).
66. A. M. Bonch-Bruevich, T. K. Razumova, and J. O. Starbogatov, Opt. Spektrosk. *42*, 82 (1977).
67. E. A. Voigtman, A. Jurgensen, and J. Winefordner, Anal. Chem. *53*, 1921 (1981).
68. L. J. Rothberg, J. D. Simon, M. Bernstein, and K. S. Peters, J. Am. Chem. Soc. *105*, 3464 (1983).
69. S. E. Braslavsky, R. M. Ellul, R. G. Weiss, H. Al-Ekabi, and K. Schaffner, Tetrahedron *39*, 1909 (1983).
70. J. E. Rudzki, J. L. Goodman, and K. S. Peters, J. Am. Chem. Soc. *107*, 7849 (1985).
71. A. C. Tam and H. Coufal, Appl. Phys. Lett. *42*, 33 (1983).
72. W. Imain and A. C. Tam, Appl. Optics *22*, 1875 (1983).
73. A. C. Tam and W. P. Leung, Appl. Phys. Lett. *45*, 1040 (1984).
74. A. C. Tam and G. Ayers, Appl. Phys. Lett. *49*, 1420 (1986).
75. C. Y. Kuo, M. M. F. Vieira, and C. K. N. Patel, J. Appl. Phys. *55*, 3333 (1984).
76. K. Heihoff and S. E. Braslavsky, Chem. Phys. Lett. *131*, 183 (1986).
77. S. J. Komorowski and E. M. Eyring, J. Appl. Phys. *62*, 3066 (1987).
78. KYNAR (TM) Piezo Film Technical Manual, Pennwalt Corp., 900 First Ave., P. O. Box C, King of Prussia, PA 19406-0018.
79. H. M. Lai and K. Young, J. Acoust. Soc. Am. *72*, 2000 (1982).
80. J. M. Heritier, Optics Commun. *44*, 267 (1983).
81. K. S. Peters and G. J. Snyder, Science *241*, 1053 (1988).
82. H. Vargas and L. C. M. Miranda, Phys. Rep. *161*, 43 (1988).
83. C. K. N. Patel and A. C. Tam, Appl. Phys. Lett. *34*, 760 (1979).
84. A. C. Tam, Appl. Phys. Lett. *45*, 510 (1984).
85. R. C. DuVarney, A. K. Garrison, and G. Busse, Appl. Phys. Lett. *38*, 675 (1981).
86. O. Goede, W. Heimbrodt, and F. Sittel, Phys. Stat. Solid (a) *93*, 277 (1986).

11

Chiroptical Techniques

HARRY G. BRITTAIN / Squibb Institute for Medical Research,
New Brunswick, New Jersey

I. INTRODUCTION TO CHIROPTICAL PHENOMENA

Most forms of optical spectroscopy are usually concerned with measurement of the absorption or emission of electromagnetic radiation, the energy of which lies between wavenumbers 10 and 50,000 cm^{-1}. Any possible effects associated with polarization of the electric vectors of the radiant energy are not normally considered important to these experiments. For molecules lacking certain types of molecular symmetry, interactions with polarized radiation are important and can be utilized to study a wide variety of phenomena. Materials can therefore be classified as being either isotropic (incapable of influencing the polarization state of light) or anisotropic (having the ability to affect the polarization properties of transmitted light).

Molecules that are not superimposable on their mirror images are termed dissymmetric or chiral, and are capable of being resolved into enantiomers. The requirement for the existence of dissymmetry is that the molecule possess no improper axes of rotation, the minimal interpretation of which implies the lack of either a center of inversion or a reflection plane. The term dissymmetric should be used rather than asymmetric, since it is possible for a molecule to contain proper axes of rotation and still be capable of being resolved into its optical isomers.

Chiral molecules interact with electromagnetic radiation in exactly the same fashion as do achiral molecules: they will exhibit optical absorption, have a characteristic refractive index, and can scatter oncoming photons. Optically active compounds are also capable of interacting with light whose electric vectors are circularly polarized. This property manifests itself in an apparent rotation of the plane of linearly polarized light (polarimetry) or in a preferential absorption of either left- or right-circularly polarized light (circular dichroism). Circular dichroism can be observed in either electronic or vibrational bands. The raman scattering of a chiral compound can also reflect the optical activity of the molecule. Should the excited state of a compound be both luminescent and chiral, then the property of circularly polarized luminescence can be observed. The circular dichroism of an optically active molecule can be used in conjunction with fluorescence monitoring to provide a differential excitation spectrum (fluorescence-detected circular dichroism).

A variety of chiroptical techniques are available, and these can be useful in analytical work. The foremost problem is a determination of the absolute configuration of all dissymetric atoms within a molecule. Any chiroptical technique can be used to obtain the enantiomeric purity of a given sample, although certain methods are more useful than others. Questions of optical activity are of extreme importance to the pharmaceutical industry, where ever-increasing numbers of drugs are being synthesized that contain chiral substituents. Documentation of the properties of any and all dissymmetric centers is essential to the registration of such substances. The entire field of asymmetric chemical synthesis is supported by studies of molecular optical activity.

Each chiroptical technique is somewhat different in its instrumentation, experimental design, and in what information can ultimately be extracted from the measurements. These will be discussed in turn, and representative examples will be provided to illustrate the methods. A general introduction to optical activity has been provided by Charney [1], who has provided the most readable summary of the history and practice associated with chiroptical spectroscopy. A number of other monographs have been written that concern various applications of chiroptical spectroscopy, ranging from the very theoretical to the very practical [2-9], and these texts contain numerous references suitable for entrance into the field. The areas of circular dichroism and optical rotatory dispersion have naturally received the most extensive coverage, since these have been exhaustively studied since the effects were discovered. The newer methods are covered primarily in review articles, which will be cited when appropriate.

II. POLARIZATION PROPERTIES OF LIGHT

An understanding of the polarization properties of light is essential to any discussion of chiroptical measurements. It is the usual practice to consider only the behavior of the electric vector in describing the properties of polarized light, even though it is possible to use the magnetic vector equally well. Unpolarized light propagating along the Z axis will contain electric vectors whose directions span all possible angles within the X-Y plane.

Linearly polarized light represents the situation where all the transverse electric vectors are constrained to vibrate in a single plane. The simplest way to produce linearly polarized light is by dichroism, or passage of the incident light beam through a material that totally absorbs all electric vectors not lying along a particular plane. The other elements suitable for the production of linearly polarized light are crystalline materials that exhibit optical double refraction; these are exemplified by the Glan, Glan-Thompson, and Nicol prisms.

As with any vector quantity, the electric vector describing the polarization condition can be resolved into projections along the X and Y axes. For linearly polarized light, these will always remain in phase during the propagation process unless passage through another anisotropic element takes place. Attempted passage of linearly polarized light through another polarizer (referred to as the analyzer) results in the transmission of only the vector component that lies along the axis of the second polarizer. If the incident angle of polarization is orthogonal to the axis of transmission of the analyzer, then no light will be transmitted.

Certain crystalline optical elements have the property of being able to alter the phase relationships existing between the electric vector projections. When the vector projections are rendered 90° out of phase, the electric vector executes a helical motion through space and the light is now denoted as being circularly polarized. Since the helix can be either left- or right-handed, the light is referred to as being either left- or right-circularly polarized. It is preferable to visualize linearly polarized light as consisting of the resultant of left- and right-handed circularly polarized components, the electric vectors of which are always exactly in phase. When the phase angle between the two vector components in any light beam lies between 0 and 90°, the light is denoted as elliptically polarized. The production of a 90° phase shift is termed quarter-wave retardation, and an optical element that effects such a change is a quarter-wave plate. Passage of linearly polarized light through a quarter-wave plate produces a beam of circularly polarized light, the type of which depends on whether the phase angle has been advanced or retarded by 90°. The passage of circularly polarized light through a quarter-wave plate will produce linearly polarized light, whose angle is rotated by 90° with respect to the original plane of linear polarization. Circularly polarized light will pass through a linear polarizer without effect.

The chiroptical spectroscopic methods that have been developed take advantage of the fact that anisotropic materials are capable of producing the same effects in polarized light as do crystalline optical elements. Since these effects are determined by the molecular stereochemistry, the utility of chiroptical methods to the study of molecular properties is self-evident.

III. OPTICAL ROTATION AND OPTICAL ROTATORY DISPERSION

Charney [1] has provided an excellent summary of the history associated with chiroptical methods. The study of molecular optical activity can be considered as beginning with the work of Biot, and with the publication of his notes [10]. Biot demonstrated that the plane of linearly polarized light is rotated upon passage through an optically active medium, and designed a working polarimeter capable of quantitatively measuring the effect. The use of calcite prisms was introduced by Mitscherlich in 1844 [11], and the double-field method of detection was devised by Soleil in 1845 [12]. Since these early developments, many advancements in polarimetry have been made, and there is a large number of detection schemes now possible. An extensive summary of these methods has been provided by Heller [13].

In principle, measurement of optical rotation is extremely simple, as shown in the schematic diagram of Figure 1. The incident light is collimated and plane-polarized, and is allowed to pass through a medium. The plane of the incident light is specified, and then the angle of rotation is defined with respect to this original plane. This is carried out by first determining

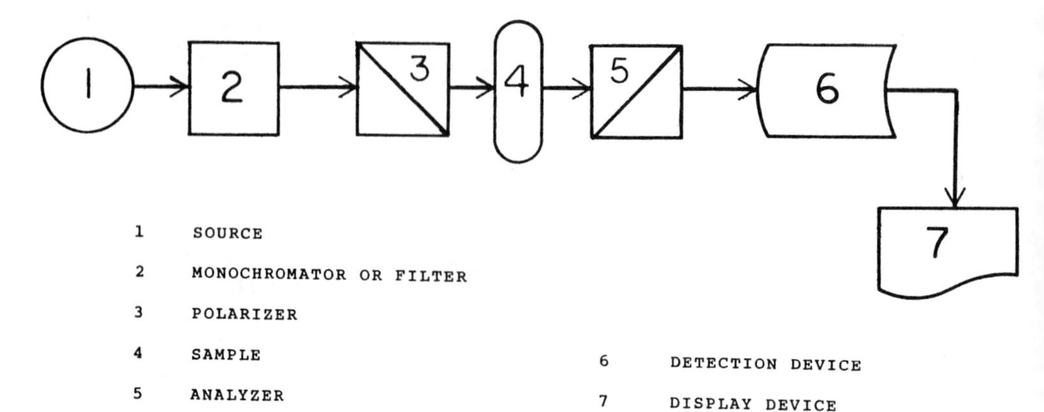

Figure 1. Block diagram of a simple polarimeter suitable for the measurement of optical rotation. For the measurement of optical rotatory dispersion, a tunable wavelength source would be used.

the orientation of polarizer and analyzer for which no light can be transmitted (the null position). The medium containing the optically active material is then introduced between the prisms, and the analyzer is rotated until a null position is again detected. The observed angle of rotation is taken as the difference between the two null angles.

The velocity of light (v) passing through a medium is determined by the index of refraction (n) of that medium:

$$v = c/n \tag{1}$$

where c is the velocity of light in vacuum. For an isotropic molecule, the refractive index is not affected by the polarization state of the light. For an anisotropic molecule, the refractive index associated with left-circularly polarized light will not normally equal the refractive index associated with right-circularly polarized light. It follows that the velocities of left- and right-circularly polarized light will differ on passage through a chiral medium. Since linearly polarized light can be resolved into two in-phase oppositely signed circularly polarized components, then the components will no longer be in phase once they pass through the chiral medium. Once the components are recombined, linearly polarized light is obtained whose plane is rotated (relative to the original plane) by an angle equal to half the phase angle difference of the circular components. Charney [1] has shown that this phase angle difference is given by

$$\beta = \frac{2\pi b'}{\lambda_0}(n_L - n_R) \tag{2}$$

In equation (2), β is the phase difference, b' is the medium path length (in centimeters), λ_0 is the vacuum wavelength of the light used, and n_L and n_R are the refractive indices for left- and right-circularly polarized light. The quantity $(n_L - n_R)$ defines the circular birefringence of the chiral medium, and is the origin of the effect. The observed rotation of the plane polarized light is given in radians by

$$\phi = \beta/2 \tag{3}$$

The usual practice is to express rotation in terms of degrees, and in that case equation (2) becomes

$$\alpha = \frac{1800b}{\lambda_0}(n_L - n_D) \tag{4}$$

In equation (4), b represents the medium path length in decimeters, which is the conventional unit. The optical rotation α of a chiral medium can be either positive or negative, depending on the sign of the circular birefringence directly, since the magnitude of $(n_L - n_D)$ is in the 10^{-8} to 10^{-9} range [13].

The optical rotation (α) exhibited by a chiral medium depends on the optical path length, the wavelength of the light used, the temperature of the system, and the concentration of dissymmetric molecules. If the solute concentration (c) is given in terms of grams per 100 ml of solution, then the observed rotation (an extrinsic quantity) can be converted into an intrinsic quantity, the specific rotation, by

$$[\alpha] = \frac{100\alpha}{bc} \tag{5}$$

The molar rotation, [M], is defined by

$$[M] = \frac{(FW)\alpha}{bc} \tag{6}$$

$$= \frac{(FW)[\alpha]}{100} \tag{7}$$

where FW is the formula weight of the dissymmetric solute. When the solute concentration is given in terms of molarity, then equation (6) becomes

$$[M] = \frac{10\alpha}{bM} \tag{8}$$

The temperature associated with a measurement of the specific or molar rotation of a given substance must be specified. Thermal volume changes or alterations in molecular structure (as induced by a temperature change) are capable of producing detectable changes in the observed rotations. In the situation where solute-solute interactions become important at high concentrations, it may be observed that the specific rotation of a solute is not independent of concentration. It is acceptable practice, therefore, to obtain polarimetry data at a variety of concentration values to verify that a true molecular parameter has been measured.

The most important parameter in determining the magnitude of optical rotation is the wavelength of the light used for the determination. Generally, the magnitude of the circular birefringence increases as the wavelength becomes shorter, and hence specific rotations increase in a regular manner at lower wavelengths. This behavior persists until the light is capable of being absorbed by the chiral substance, whereupon the refractive index exhibits anomalous behavior. The variation of specific or molar rotation with wavelength is termed optical rotatory dispersion, or ORD.

The anomalous dispersion observed in ORD spectra arises because the refractive index of a material is actually the sum of a real and imaginary part:

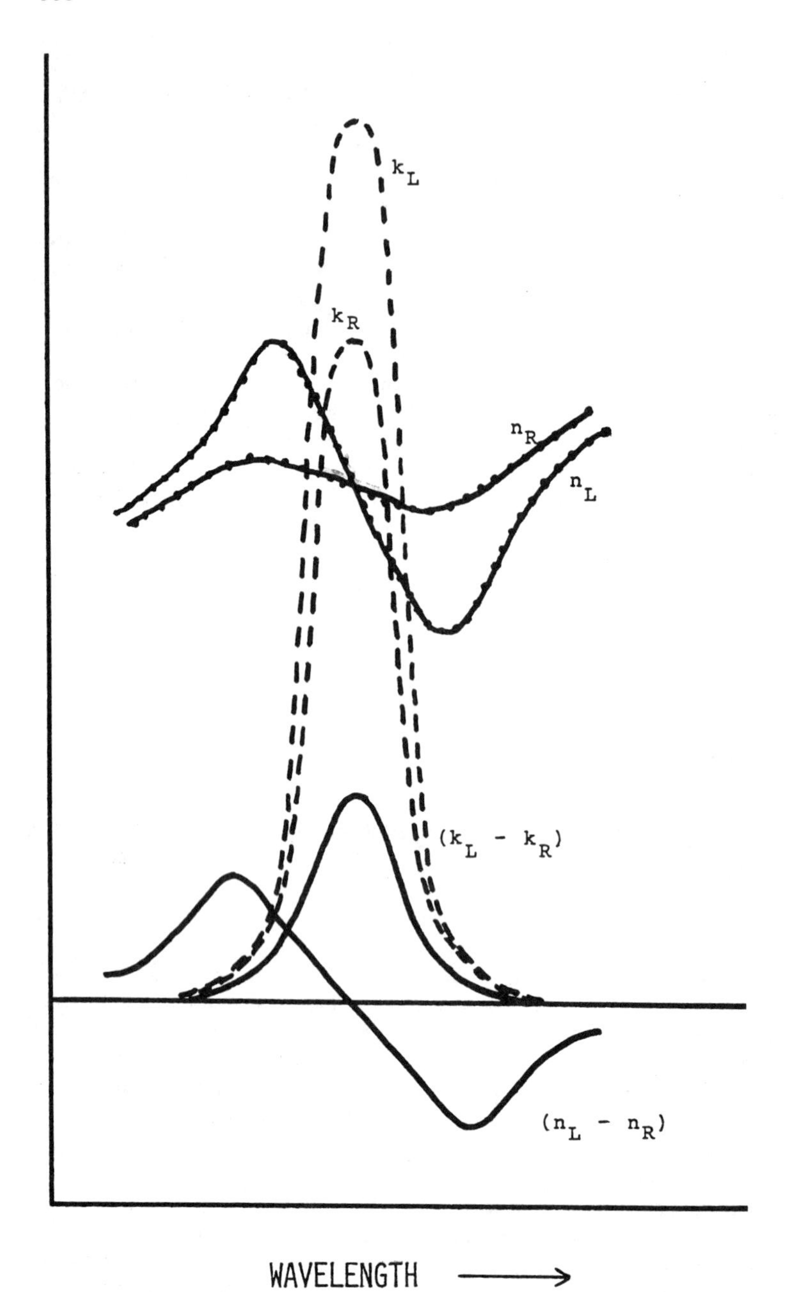

Figure 2. Schematic representation of the Cotton effect, illustrating the effects of circular birefringence and circular dichroism within an absorption band.

$$n = n_0 + ik \tag{9}$$

where n is the observed refractive index at some wavelength, n_0 is the refractive index at infinite wavelength, and k is the absorption coefficient of the substance. It follows that if $(n_L - n_D)$ does not equal zero, then $(k_L - k_D)$ will not equal zero. The relation between the various quantities was first conceived by Cotton [14], and is illustrated schematically in Figure 2.

The earliest work involving chiral organic molecules was entirely based on ORD methods, since little else was available at the time. One of the largest data sets collected to date concerns the cirality of ketone and aldehyde groups [15], which eventually resulted in the deduction of the octant rule [16]. The octant rule was an attempt to relate the absolute stereochemistry within the immediate environment of the chromophore with the sign and intensity of the ORD Cotton effects. To apply the rule, the circular dichroism (CD) within the $n-\pi^*$ transition around 300 nm is obtained, and its sign and intensity noted. The rule developed by Djerassi and coworkers states that the three nodal planes of the n and π^* orbitals of the carbonyl group divide the molecular environment into four front octants and four back octants. A group or atom situated in the upper left or lower right rear octant (relative to an observer looking at the molecule parallel to the C=O axis) induces a positive Cotton effect in the $n-\pi^*$ band. A negative Cotton effect would be produced by substitution within the upper right or lower left back octant. Although exceptions to the octant rule have been shown, the wide applicability of the rule has remained established. The ability to deduce molecular conformations in solution on the basis of ORD spectral data has proven to be extremely valuable to synthetic and physical organic chemists, and has enabled investigators of the time to develop their work without requiring more heroic methods.

IV. CIRCULAR DICHROISM

As just discussed, optical rotation and optical rotatory dispersion are easily interpretable outside regions of electronic absorption, but exhibit anomalous properties within absorption bands. This effect arises since the refractive index also contains a contribution related to molecular absorptivity, as described earlier in equation (9). Not only will the phase angle between the projections of the two circularly polarized components be altered by passage through the chiral medium, but their amplitudes will be modified by the degree of absorption experienced by each component. This differential absorption of left- and right-circularly polarized light is termed circular dichroism (CD), and is given by $(k_L - k_D)$.

If one circularly polarized component is more strongly absorbed than the other, when the light beams are recombined after leaving the chiral medium, they no longer produce plane-polarized light. Instead, the resulting components describe an ellipse whose major axis lies along the angle of rotation. The measure of the eccentricity of the ellipse that results from the differential absorption is termed the ellipticity, ψ. It is not difficult to show [1] that

$$\psi = \frac{\pi z}{\lambda} (k_L - k_R) \tag{10}$$

where z is the path length in centimeters and λ is the wavelength of the light. If C is the concentration of absorbing chiral solute in moles per liter, then mean molar absorptivity, a, is derived from the absorption index by

$$a = \frac{4\pi k}{2.303 \lambda C} \tag{11}$$

In that case, the ellipticity in radians becomes

$$\psi = \frac{2.303Cz}{4} (a_L - a_R) \tag{12}$$

The expression of ellipticity in radians is cumbersome, and consequently the quantity is converted into degrees by the relation

$$\theta = \psi(360/2\pi) \tag{13}$$

and then

$$\theta = (a_L - a_R) Cz (32.90) \tag{14}$$

The molar ellipticity is an intensive quantity, and is calculated from

$$[\theta] = \frac{\theta(FW)}{Lc'(100)} \tag{15}$$

where FW is the formula weight of the solute in question, L is the medium path length in decimeters, and c' is the solute concentration in grams per milliliter. The molar ellipticity is related to the differential absorption by

$$a_L - a_R - [\theta]/3298 \tag{16}$$

Most instrumentation suitable for measurement of circular dichroism is based on the design of Grosjean and Legrand [9], and a block diagram of the basic design is shown in Figure 3. Linearly polarized light is passed through a dynamic quarter-wave plate, which modulates it alternately into left- and right-circularly polarized light. The quarter-wave plate is a piece of isotropic material, which is rendered anisotropic through the external application of stress. The device can be a Pockels cell (in which stress is created in a crystal of ammonium dideuterium phosphate through the application of AC high voltage) or a photoelastic modulator (in which the stress is induced by the piezoelectric effect). The light leaving the cell is detected by a photomultiplier tube, whose current output is converted to voltage and split into two portions. One signal consists of an alternating signal proportional to the CD, and is due to the differential absorption of one component over the other. This signal is amplified by means of phase-sensitive detection. The other signal is averaged, and is related to the mean light absorption. The ratio of these signals varies linearly as a function of the CD amplitude, and is the recorded signal of interest.

It is impossible to summarize here the utility of CD spectroscopy for the study of molecular stereochemistry, since entire monographs have been written on the subject [4-9]. As a simple example of an application of CD

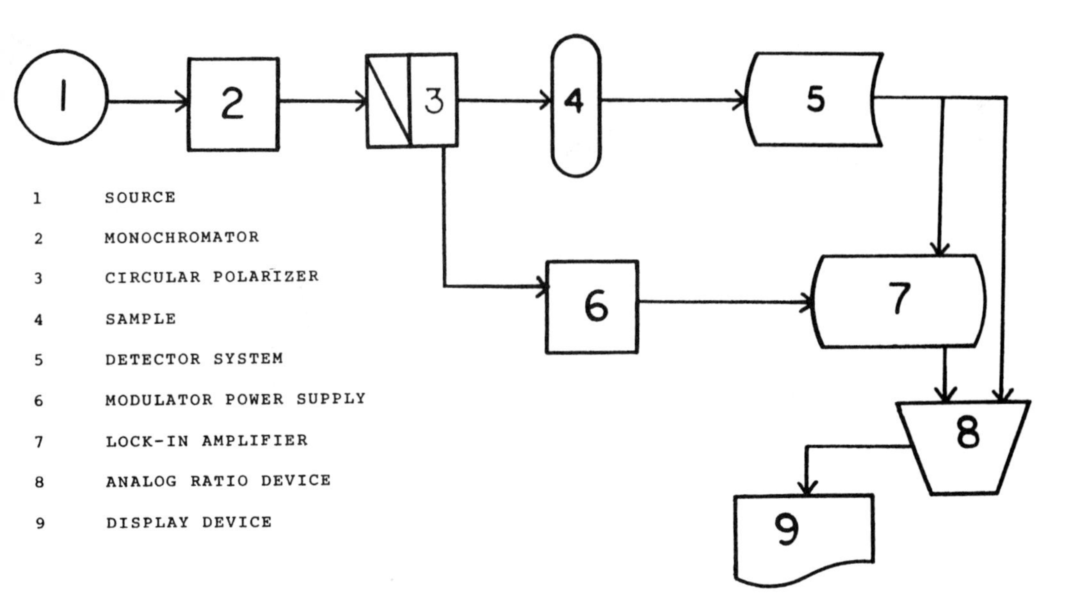

Figure 3. Block diagram of an analog apparatus suitable for the measurement of circular dichroism.

spectroscopy, the chiral properties of oximes formed upon reaction of a carbonyl group with hydroxylamine will be mentioned. Upon formation of the oxime chromophore, the UV absorption and CD spectra simplify, so that the sign and intensity of the Cotton effect accurately reflect the stereochemistry within the immediate vicinity [17]. The CD spectra of most oximes consist of one major peak (without any accompanying fine structure) around 240 nm. The CD spectrum of an oxime usually yields more reliable stereochemical information than does that of the parent ketone, since the single-signed CD peak is a reliable indicator of the absolute stereochemistry adjacent to the chromophore. Although such information would be available from considerations of the octant rule, confirmation is possible through formation of the oxime.

V. CIRCULARLY POLARIZED LUMINESCENCE SPECTROSCOPY

Optical rotation, optical rotatory dispersion, and circular dichroism methods yield information relating to the ground electronic state of the chiral molecule. Circularly polarized luminescence (CPL) spectroscopy involves a measurement of the spontaneous differential emission of left- or right-circularly polarized light by an optically active species. As with CD spectroscopy, the chirality can either be natural or induced by a magnetic field. CPL spectroscopy differs from CD spectroscopy in that it is a measure of the chirality of a luminescent excited state. If the geometry of the molecule remains invariant during the excitation process, then the same

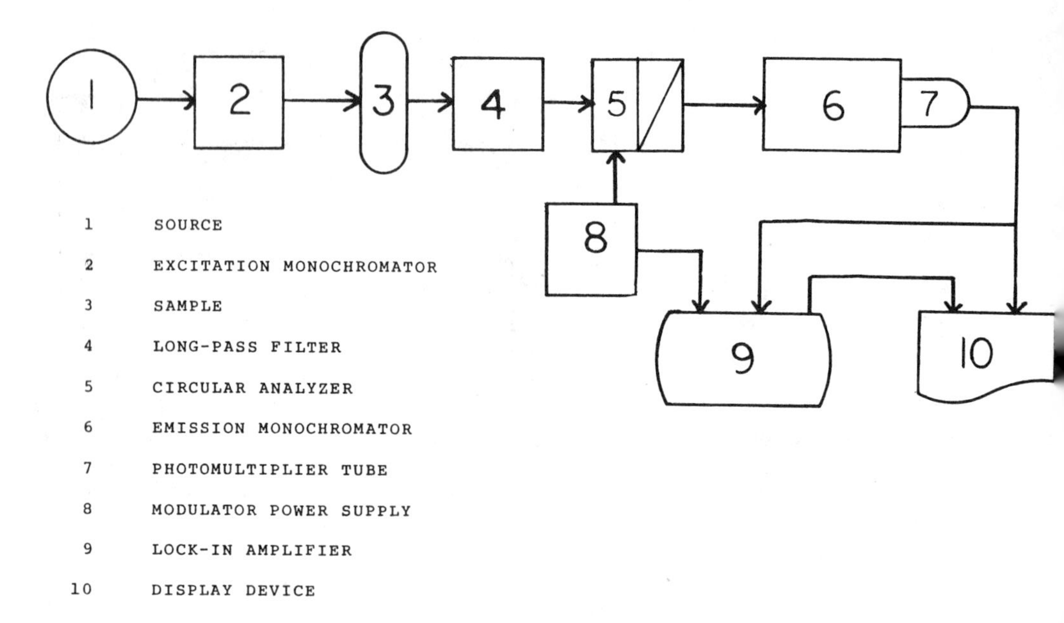

Figure 4. Block diagram of an analog apparatus suitable for the measurement of circularly polarized luminescence.

chiroptical information would be obtained using either CD or CPL methods. In the situations where geometrical changes are associated with excitation into higher electronic states, then comparison of CD and CPL results can be used to deduce the nature of these structural modifications. Reviews of CPL spectroscopy have been written by Richardson [18] and Brittain [19].

The major limitation associated with CPL spectroscopy is that it is confined to emissive molecules only. However, this restriction can be used to advantage in that it imparts selectivity to the technique. The CPL technique has been particularly useful in the study of chiral lanthanide complexes [20], since the absorptivity of these compounds is such that high-quality spectra can only be obtained under undesirable conditions. It is appropriate to consider CPL spectroscopy as a technique that combines the selectivity of CD with the sensitivity of luminescence.

Even though CPL spectroscopy has been known for some time, no commercial instrumentation has yet become available for its measurement. Instruments have been described that are based either on an analog design following the CD principles of Grossjean and Legrange [4], or on digital methods that employ photon counting electronics [21].

A block diagram describing the basic design of an analog CPL spectrometer is shown in Figure 4. The excitation source can be either a laser or an arc lamp, but it is important that the source of excitation be unpolarized to avoid possible photoselection artifacts. It Is best to collect the emitted light at 0° to the excitation beam ("head-on" geometry) so that spurious effects due to linear polarization in the emission will not find their way through the imperfect electronics used for the phase-sensitive detection. Extensive discussions of these effects have been provided by a number of investigators [18,19]. Any unabsorbed excitation energy is purged from the optical train through the use of a longpass filter.

The circular polarization in the emitted light is detected by the circular analyzer, since the insensitivity of photomultiplier tubes to light polarization states requires the use of a transducer. Through the use of a dynamically operated quarter-wave retardation element (a photoelastic modulator or a Pockels cell), the circularly polarized component of the emission is transformed into periodically interconverting orthogonal planes of linearly polarized light. This effect is accomplished by advancing and retarding the phase angle difference between the electric vectors by 90°. The linear polarizer following the modulator extinguishes one of the linear components, producing a modulated light intensity proportional to the CPL intensity in the emission.

The emission is analyzed in the usual manner by an emission monochromator, and detected by a photomultiplier tube. The current from the tube is converted to a voltage signal, and divided into two outputs. One analog output can be sent to a chart recorder or digitized for input into a laboratory computer. The other signal is fed into a lock-in amplifier, where the AC ripple is amplified by means of phase sensitive detection. The output of the amplifier is proportional to the CPL intensity, and becomes the other signal to be processed. It is useful to ratio the differential (CPL) and total (TL) luminescence signals, and to display this quantity also.

The CPL experiment thus produces two measurable quantities, which are obtained in arbitrary units and related to the circular polarization condition of the luminescence. The TL intensity is defined as

$$I = I_L + I_R \tag{17}$$

and the CPL intensity is defined as

$$dI = I_L - I_R \tag{18}$$

In equations (17) and (18), I_L and I_R represent the emitted intensities of left- and right-circularly polarized light, respectively. The ratio of these quantities must be dimensionless, and therefore free of unit dependence. In analogy to the Kuhn anisotropy factor defined for CD spectroscopy, the luminescence dissymmetry factor of CPL spectroscopy is defined as

$$g_{lum} = \frac{dI}{(1/2)I} \tag{19}$$

In the digital measurement method, the number of left-circularly polarized photons is counted separately from the number of right-circularly polarized photons [21]. The definitions for the various CPL quantities are still the same, however.

The calibration of CPL spectra is an important question as long as all CPL spectrometers are laboratory constructed. The best method reported to date is that of Steinberg and Gafni [22], who used the birefringence of a quartz plate to transform the empirically observed dissymmetry factors into absolute quantities.

As an example of a situation where CPL spectroscopy was used to unique advantage, the formation of lanthanide ternary complexes will be considered. It had been known that lanthanide complexes of ethylenediamine tetraacetate (EDTA) were coordinatively unsaturated, and that these could be used as aqueous shift reagents for the simplification of nuclear magnetic resonance spectra [23]. This property was based on the ability of these compounds to form ternary complexes:

$$Ln(EDTA) + S \rightleftarrows Ln(EDTA)(S) \qquad (20)$$

When the Ln ion was paramagnetic, then the induced shifts in the resonance lines of S could be used to deduce molecular conformations. The substitution of one methyl group on the ethylenediamine backbone of EDTA yields the PDTA ligand, but this ligand is optically active and can be resolved into its enantiomers. The Tb(PDTA) complex forms ternary complexes in exactly the same fashion as do the EDTA complexes, but now CPL spectroscopy can be used to study the possible stereochemical changes that accompany the formation of ternary complexes.

It was found that Tb(R-PDTA) could indeed form ternary complexes with a large variety of achiral substrate ligands [24]. In many cases, it was found that the CPL spectra of the ternary complexes were quite different from that of the parent Tb(PDTA) complex. This indicates that these ligands were capable of deforming the PDTA ring structure. When extremely flexible ligands were studied (e.g., succinic acid), it was found that the CPL spectra of the ternary complexes were not altered relative to the parent complex. However, for this particular ring system another study [25] demonstrated that the succinate-type structure was forced to adopt a new conformation upon complexation with a Tb(EDTA) complex. These works indicate that although the lanthanide EDTA complexes function as aqueous shift reagents, steric interactions between the EDTA ligand and the substrate are significant. The NMR data may be used to deduce solution-phase conformations of the substrate, but this stereochemical information represents only the situation for the perturbed ligand. It is probably not possible to deduce the stereochemistry of the free ligand in solution from data obtained using Ln(EDTA) complexes as shift reagents. This information could only have been obtained from a chiroptical study, and CD spectroscopy would not have been the method of choice for such work.

VI. VIBRATIONAL OPTICAL ACTIVITY

Since all chiral molecules exhibit strong absorption in the infrared region of the spectrum, extensive investigation into the optical activity of vibrational transitions has been carried out. The IR bands of a small molecule can easily be assigned with the performance of a normal coordinate analysis, and these can usually be well resolved. The optical activity observed within a vibrational transition is determined by the symmetry properties of its group vibration, and consequently can represent a probe of local chirality. When coupled with other methods capable of yielding information on solution phase structure, vibrational optical activity can be an extremely useful method.

One of the problems associated with vibrational optical activity is the weakness of the effect. The rotational strengths of vibrational bands are much smaller than those of electronic bands, since the magnetic dipole contribution is significantly smaller. In addition, instrumental limitations of infrared sources and detectors create additional experimental constraints on the signal-to-noise ratios. In spite of these problems, two methods suitable for the study of vibrational optical activity have been developed. One of these is vibrational circular dichroism (VCD), in which vibrational optical activity is observed in the classic method of Grosjean and Legrand. The other method is Raman optical activity (ROA), in which chirality is studied through Raman spectroscopy.

A. Vibrational Circular Dichroism

Vibrational CD (VCD) was first measured in the liquid phase by Holzwarth and co-workers in 1974 [26,26A], and was soon confirmed by Nafie and his group [27]. These and other studies carried out during this period demonstrated that VCD could be measured at good signal-to-noise levels, and that the effect was quite sensitive to details of molecular stereochemistry and confirmation [28].

An infrared CD spectrometer can be constructed using the same detection scheme as described for a UV/VIS CD spectrometer, using optical components suitable for infrared work (Figure 5). Reflective optics optimized for the infrared are normally used to eliminate problems associated with the limited transmission of infrared energy by lenses. The light source can be a tungsten lamp, glowbar, or carbon rod source, depending on the particular spectral range to be covered. The monochromator used to select the analyzing wavelengths should contain a grating blazed for first-order work in the infrared.

The circular polarization in the infrared source is obtained by first linearly polarizing the light, and then passing it through a photoelastic modulator. The linear polarizer is a CaF_2 crystal overlaid either by a wire grid (producing the linear polarization through dichroism) or by an $LiIO_3$ crystal polarizer. The optical element within the photoelastic modulator is made out of either CaF_2 or ZnSe, chosen on the basis of the desired spectral range. The detector is usually a photovoltaic device capable of responding to infrared energy. The AC component within the detector output is discriminated and amplified by phase-sensitive detection, and the mean detector output is obtained to normalize the VCD signal. The ratio of the differential absorbance to the total absorbance is

$$g_{abs} = \frac{a_L - a_R}{1/2(a_L + a_R)} \tag{21}$$

where a_L and a_R represent the absorptivities for left- and right-circularly polarized light, respectively.

Early VCD measurements were complicated by artifacts arising from imperfections in the optical components. These effects were detected by running the VCD spectrum of the racemic compound, and corrected spectra were obtained as the difference between the two spectra. In most instances, careful alignment of the optical components is able to eliminate most such effects.

An alternate method for the detection of VCD was introduced by Nafie, who was able to show that a modified Fourier transform infrared (FTIR)

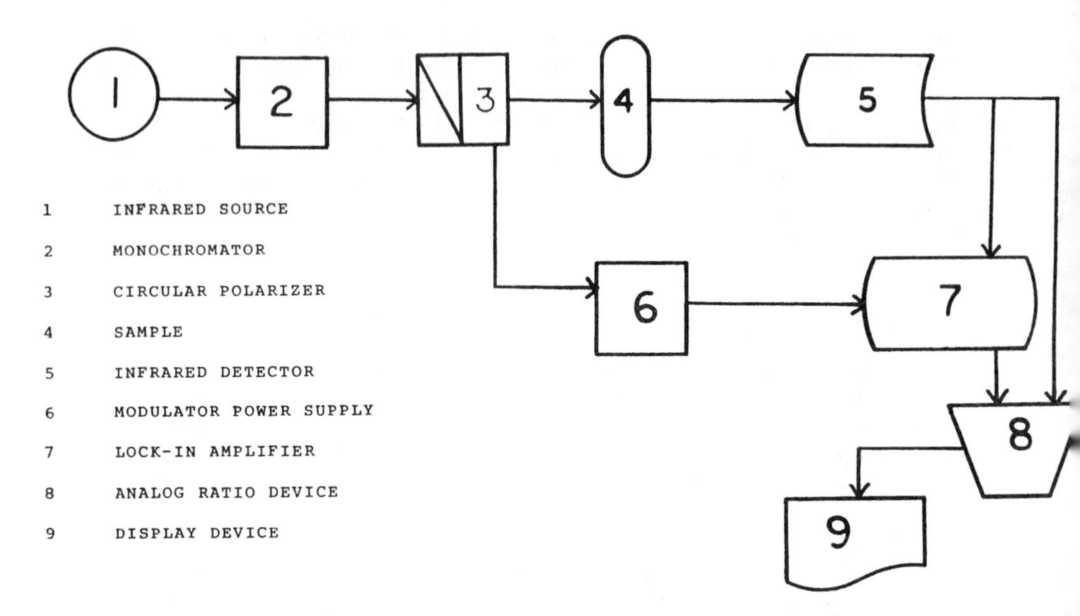

Figure 5. Block diagram of an analog apparatus suitable for the measurement of vibrational circular dichroism.

spectrometer could be used [29]. The instrumental advantages of the FTIR method have provided significant enhancements in signal-to-noise ratios, and thus reduce the time required to obtain a VCD spectrum.

As an example of the utility of VCD spectroscopy, one work of Nafie and co-workers will be discussed. The VCD within the carbon-hydrogen stretching region of 21 amino acids, five bis(amino acid) copper (II) complexes, and the tris(alaninato) cobalt(III) complex was obtained in deuterium oxide solvent [30]. For the free amino acids, a positive VCD band at 2970 cm^{-1} was noted, corresponding to the methine -CH stretching mode. The intensity of this VCD was associated with the presence of an intramolecular hydrogen bond existing between the ND^{3+} and COO^- functionalities. Upon complexation with transition metal ions, the VCD intensity of this band increases even further, due to increased bonding strength between the amine and carboxylate groups through the bonds of the Cu(II) or Co(III) ion. The mechanism of this effect was proposed to originate in a ring current that was closed either by the intramolecular hydrogen bond or by the metal-ligand bonds.

B. Raman Optical Activity

A method complementary to VCD spectroscopy is that of Raman optical activity (ROA). The ROA effect is the differential scattering of left- or right-circularly polarized light by a chiral substrate. ROA is particularly useful within the 50-1600 cm^{-1} spectral region, and is currently the only method suitable for obtaining optical activity within vibrational bands having energies less than 600 cm^{-1} (spectral limitations of VCD optical components preclude work below 600 cm^{-1}). The ROA effect was first demonstrated in 1973 by Barron and Buckingham [31,32], who observed differential scattering effects in the raman spectra of α-phenethylamine and α-pinene. Development of the technique required substantial experimental effort, since small imperfections in the optical elements could yield artifacts substantially larger in magnitude than a genuine ROA effect. Barron has provided both a general introduction [33] and a detailed review [34] of the field.

ROA spectrometers have been described by McCaffery [35,36] and Diem [37]. With the introduction of a special optical system, Hug was able to drastically reduce ROA artifacts [38], and this design is summarized in the block diagram of Figure 6. The laser output is passed through a small monochromator to remove unwanted plasma lines, and is totally polarized by a linear polarizer. The beam is circularly polarized by a quarter-wave plate and is allowed to impinge on the sample cell. The scattered light is passed through a photoelastic modulator, which is set to advance and retard phase angles by 90°. Since the ROA effect employs visible light scattering, no special modulator materials are required. The modulated light beam is passed through a linear polarizer, thus providing an AC signal in the scattered light that is proportional to the ROA intensity. The scattered light is collected over a 20° cone with respect to the laser [35], analyzed by a high-resolution double grating monochromator, and detected by a photomultiplier (PMT). The PMT output is split, with one

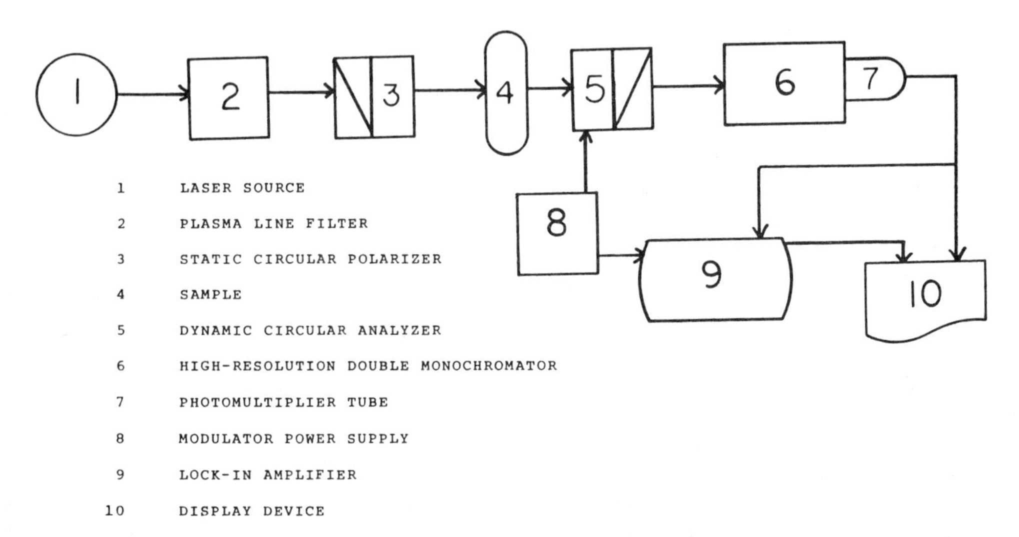

Figure 6. Block diagram of an analog apparatus suitable for the measurement of Raman optical activity.

signal being amplified by a DC device and the other discriminated through phase-sensitive detection. The two quantities can be either displayed directly or ratioed before being displayed. Diem [37] has described the design of a diode array ROA spectrometer, but the basic detection is comparable to that just described.

The amplified DC signal is proportional to the mean scattered light intensity,

$$I = I_L + I_R \tag{22}$$

while the AC signal (after amplification by means of phase-sensitive detection) is proportional to the differential scattering intensity:

$$dI = I_L - I_R \tag{23}$$

Taking the ratio of these quantities removes any unit dependence, producing a quantity that is termed the circular intensity differential (CID):

$$CID = dI/I \tag{24}$$

As with VCD spectroscopy, these CID magnitudes are quite small, with most values ranging between 10^{-3} and 10^{-4}.

One of the most complete studies involving ROA was the work of Nafie and co-workers [39] involving studies on (+)-(3*R*)-methylcyclohexanone. The Raman scattering observed between 100 and 600 cm^{-1} was found to contain strong ROA in all the major bands. The assignment of these bands to 11 skeletal motions was made on the basis of a full normal coordinate analysis, and through studies on several deuterated compounds. The ROA bands were found to occur as couplets, an effect that was explained in terms of chiral vibrational perturbations due to the presence of the methyl group mixing the A' and A'' skeletal modes of cyclohexanone.

VII. FLUORESCENCE-DETECTED CIRCULAR DICHROISM

Fluorescence-detected circular dichroism (FDCD) is a chiroptical technique first developed by Tinoco and co-workers. A CD spectrum is obtained by measuring the difference in total luminescence obtained after the sample is excited by left- and right-circularly polarized light. For the FDCD spectrum of a given molecular species to match its CD spectrum, the luminescence excitation spectrum must be identical to the absorption spectrum.

A block diagram of a typical FDCD spectrometer is shown in Figure 7. A tunable UV source, whose output is relatively independent of wavelength, is required. A high-powered xenon lamp (whose output is selected by a suitable grating monochromator) is appropriate for this purpose. The spectral output is linearly polarized by a Glan or Glan-Thompson polarizer, selected for its UV transmission. A modulated circular polarization is induced in the beam through the use of a dynamic quarter-wave plate. This circularly polarized excitation beam is allowed to enter the sample cell and to produce the required luminescence. Up to this point, the optical train is identical to that of a conventional CD spectrometer.

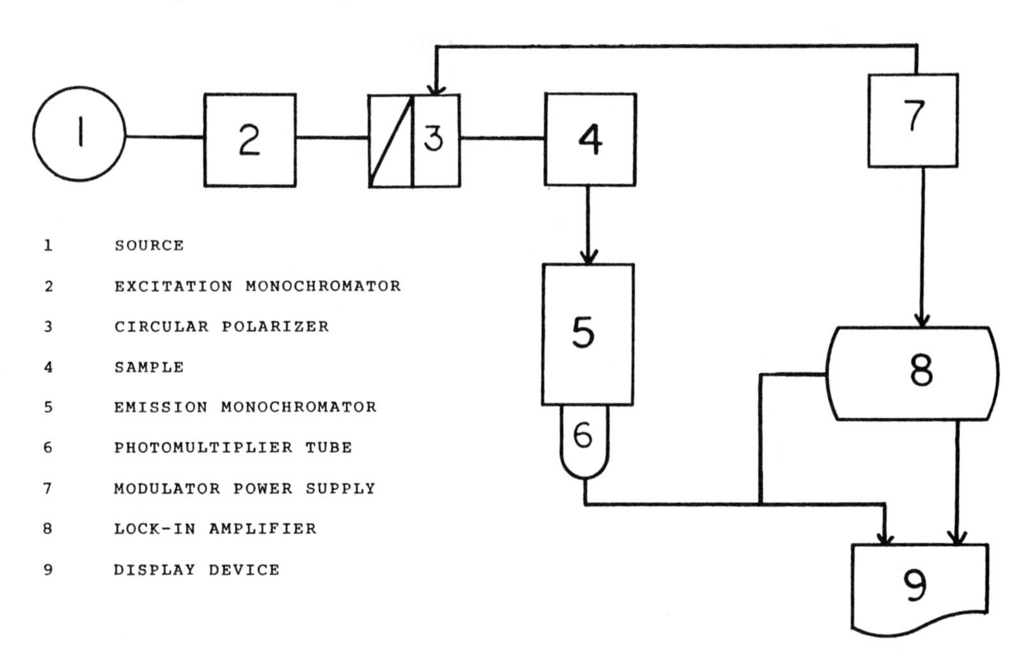

Figure 7. Block diagram of an analog apparatus suitable for the measurement of fluorescence-detected circular dichroism.

The emitted light is detected at right angles to the excitation beam by a photomultiplier tube. As was described for CPL spectroscopy, the tube output will contain an AC signal equal to $(I_L - I_R)$ and proportional to the differential absorption spectrum. A measurement of the mean DC output of the tube is equal to $(I_L + I_R)$ and is proportional to the mean absorption spectrum. The ratio of these two signals yields the FDCD signal:

$$\mathrm{FDCD} = \frac{I_L - I_R}{I_L + I_R} \tag{25}$$

This FDCD signal can be either displayed on a chart recorder or digitized for introduction into a computer system. As in the case of the CPL experiment, accurate results can only be obtained if the two measurements are made simultaneously.

A general theory relating to the FDCD of isolated chiral fluorescent compounds has been presented, and the effect of solute concentration has been considered [40]. FDCD is potentially extremely suitable for the study of solute-solute interactions, since the added selectivity of fluroescence in either species can be very useful. Three such situations have been identified and analyzed for the possible effects on the FDCD spectra [41]. These are when (a) the optically active material is nonfluorescent, but can be coupled with an optically inactive fluorescent species, (b) a fluorescent optically active material is coupled to a nonfluorescent achiral compound, and (c) an optically active fluorescent species is coupled to a chiral fluorescent compound.

The FDCD experiment can easily yield instrumental artifacts when photoselection effects are possible [42,43]. When the fluorescent molecule is situated in a medium where molecular motion is restricted (vicous or rigid phases), then the excitation energy may not be absorbed equally by all molecular orientations. These effects would be most pronounced for planar aromatic hydrocarbons, whose transition dipoles exhibit a strong dependence on the molecular orientation with respect to the incident energy. When these photoselection effects occur, the excitation spectrum can be quite different from the absorption spectrum and the FDCD spectrum will not match the CD spectrum. Turner and co-workers have described a dual photomultiplier tube detection system in which the effects due to photoselection can be minimized [44].

The power of FDCD spectroscopy was recently illustrated in studies of the CD and FDCD associated with poly-L-tryptophan [45]. It had been inferred from ORD and CD studies that this polypeptide adopted a right-handed α-helical conformation in 2-methoxyethanol, but the actual conformation of the peptide backbone could not be studied directly due to effects associated with the indole side chains [46,47]. While the CD spectrum would reflect the optical activity of both the backbone and side chains, the FDCD spectrum would only reflect the chirality experienced by the indole side chains. In studies of the CD and FDCD of this polymer system, it was noted that both similarities and differences were observed. Through a series of assumptions about how the two effects were observed, these workers were able to use the CD and FDCD data to produce separate CD spectra corresponding to the side chains and to the backbone. The backbone CD spectrum was found to closely resemble the genuine CD spectra normally observed for α-helical compounds [48]. This result was taken as confirmation that poly-L-tryptophan does indeed adopt the right-handed α-helical conformation in 2-methoxyethanol.

VIII. CONCLUDING REMARKS

The range of techniques suitable for the study of molecular optical activity is extensive, and sophisticated instrumentation for each has been developed. Each method is associated with its own particular selectivity, and each has its advantages and disadvantages. The chirality within electronic transitions can be studied by optical rotatory dispersion (ORD), circular dichroism (CD), circularly polarized luminescence (CPL), or by fluorescence-detected circular dichroism (FDCD). Chirality within vibrational bands can be studied by either vibrational circular dichroism (VCD) or Raman optical activity (ROA). Optical activity within rotational bands has been shown to be theoretically feasible [49], and undoubtedly a method will eventually be developed for this effect. The development of asymmetric chemistry is currently an extremely active research area, and the development of chiroptical methods will proceed at the same pace.

REFERENCES

1. E. Charney, *The Molecular Basis of Optical Activity*, John Wiley & Sons, New York, 1979.

2. C. Djerassi, *Optical Rotatory Dispersion: Applications to Organic Chemistry*, McGraw-Hill, New York, 1960.
3. P. Crabbe, *ORD and CD in Chemistry and Biochemistry*, Academic Press, New York, 1972.
4. L. Velluz, M. Legrand, and M. Grosjean, *Optical Circular Dichroism: Principles, Measurements, and Applications*, Verlag Chemie, Weinheim, 1965.
5. G. Snatzke, Ed., *Optical Rotatory Dispersion and Circular Dichroism in Organic Chemistry*, Heyden & Sons, London, 1967.
6. K. Nakanishi, *Circular Dichroic Spectroscopy: Exciton Coupling in Organic Stereochemistry*, University Science Books, Mill Valley, CA, 1983.
7. D. J. Caldwell and H. Eyring, *The Theory of Optical Activity*, Wiley-Interscience, New York, 1971.
8. L. Barron, *Molecular Light Scattering and Optical Activity*, Cambridge University Press, Cambridge, 1982.
9. S. Mason, *Molecular Optical Activity and the Chiral Discriminations*, Cambridge University Press, Cambridge, 1982.
10. J. B. Biot, Mem. Inst. de France *50*, 1 (1812).
11. R. Mitscherlich, *Lehrbuch der Chemische*, 4th ed., 1844.
12. H. Soleil, Compt. Rend. *20*, 1805 (1845); *24*, 973 (1847); *26*, 162 (1848).
13. W. Heller, Optical Rotation—Experimental Techniques and Physical Optics, in *Physical Methods of Chemistry*, Vol. I (A. Weissberger and B. W. Rossiter, Eds.), John Wiley & Sons, New York, 1972.
14. A. Cotton, Compt. Rend. *120*, 989, 1044 (1895).
15. D. N. Kirk, Tetrahedron *42*, 777 (1986).
16. W. Moffitt, R. B. Woodward, A. Moscowitz, W. Klyne, and C. Djerassi, J. Am. Chem. Soc. *83*, 4013 (1961).
17. P. Crabbe and L. Pinelo, Chem. Ind. 158 (1966).
18. F. S. Richardson and J. P. Riehl, Chem. Rev. *77*, 773 (1977); *86*, 1 (1986).
19. H. G. Brittain, Excited-State Optical Activity, in *Molecular Luminescence Spectroscopy: Methods and Applications—Part 1*, Wiley-Interscience, New York, 1985.
20. H. G. Brittain, Coord. Chem. Rev. *48*, 243 (1983).
21. P. H. Schippers, A. van den Beukel, and H. P. J. M. Dekkers, J. Phys. E *15*, 945 (1982).
22. I. Z. Steinberg and A. Gafni, Rev. Sci. Instr. *43*, 409 (1972).
23. G. A. Elgavish and J. Reuben, J. Am. Chem. Soc. *98*, 4755 (1976).
24. L. Spaulding, H. G. Brittain, L. H. O'Connor, and K. H. Pearson, Inorg. Chem. *25*, 188 (1986).
25. H. G. Brittain, Inorg. Chim. Acta *70*, 91 (1983).
26. E. C. Hsu and G. HOlzwarth, J. Chem. Phys. *59*, 4678 (1973).

26A. G. Holzwarth, E. C. Hsu, H. S. Mosher, T. R. Faulkner, and A. Moscowitz, J. Am. Chem. Soc. , 251 (1974).

27. L. A. Nafie, J. C. Cheng, and P. J. Stephens, J. Am. Chem. Soc. *97*, 3842 (1975).
28. T. A. Keiderling, Appl. Spectrosc. Rev. *17*, 189 (1981).
29. L. A. Nafie, M. Diem, and D. W. Vidrine, J. Am. Chem. Soc. *101*, 496 (1979).

30. M. R. Oboodi, B. B. Lal, D. A. Young, T. B. Freedman, and L. A. Nafie, J. Am. Chem. Soc. *107*, 1547 (1985).
31. L. D. Barron and A. D. Buckingham, J. Chem. Soc. Chem. Commun. 152 (1973).
32. L. D. Barron, M. P. Boggard, and A. D. Buckingham, J. Am. Chem. Soc. *95*, 603 (1973).
33. L. D. Barron, Am. Lab. *12*, 64 (1980).
34. L. D. Barron, Acc. Chem. Res. *13*, 90 (1980).
35. A. J. McCaffery and R. A. Shatwell, Rev. Sci. Instrum. *47*, 247 (1976).
36. L. I. Horvath and A. J. McCaffery, J. Chem. Soc. Faraday II *73*, 562 (1977).
37. M. R. Oboodi, M. A. Davies, U. Gunnia, M. B. Blackburn, and M. Diem, J. Raman Spectrosc. *16*, 366 (1985).
38. W. Hug, Appl. Spectrosc. *35*, 115 (1981).
39. T. B. Freedman, J. Kallmerten, C. G. Zimba, W. M. Zuk, and L. A. Nafie, J. Am. Chem. Soc. *106*, 1244 (1984).
40. I. Tinoco, Jr., and D. H. Turner, J. Am. Chem. Soc. *98*, 6453 (1976).
41. T. G. White, Y.-H. Pao, and M. M. Tang, J. Am. Chem. Soc. *97*, 4751 (1975).
42. B. Ehrenberg and I. Z. Steinberg, J. Am. Chem. Soc. *98*, 1293 (1976).
43. I. Tinoco, Jr., B. Ehrenberg, and I. Z. Steinberg, J. Chem. Phys. *66*, 916 (1977).
44. E. W. Lobenstein and D. H. Turner, J. Am. Chem. Soc. *102*, 7786 (1980).
45. K. Muto, H. Mochizuki, R. Yoshida, T. Ishii, and T. Handa, J. Am. Chem. Soc. *108*, 6416 (1986).
46. A. Cosani, E. Peggion, A. S. Verdini, and M. Terbojevich, Biopolymers *6*, 963 (1968).
47. E. Peggion, A. Cosani, A. S. Verdini, A. Del Pra, and M. Mammi, Biopolymers *6*, 1477 (1968).
48. R. J. Woody, J. Polymer Sci. Macromol. Rev. *12*, 181 (1977).
49. P. L. Polavarpu, J. Chem. Phys. *86*, 1136 (1987).

12

Lasers in Chemical Instrumentation

FREDERICK L. YARGER / Department of Physics, New Mexico Highlands University, Las Vegas, New Mexico

I. INTRODUCTION

One of the most important advances in the field of optical physics in the past 50 years is the development of the laser. Since T. H. Maiman built the first laser system in 1960, the stimulus to the optical sciences has made it one of the fastest growing fields in science and technology. The word laser is an acronym for light amplification by stimulated emission of radiation. The phenomenon involves the generation and amplification of electromagnetic radiation by stimulated emission of photons by atoms or molecules.

The possibility of stimulated emission of radiation was predicted as early as 1918 by A. Einstein. However, the first application of the principle was not made until 1954, when C. H. Townes and co-workers achieved stimulated emission of microwave radiation in a ruby crystal. N. Bloembergen introduced the concept of the three-level solid-state maser (the microwave version of the laser) in 1956, and built one using chromium-doped potassium cobalt cyanide. A Schalow and C. H. Townes adapted the principle to light in the visible region, and in 1960 T. Maiman constructed the first laser using a ruby crystal for the active medium, producing light at a wavelength of 694.3 nm. Shortly afterward A. Javan and associates developed the first laser using a gas as the medium, with mixed helium and neon, emitting laser light at 1.15 μm and 632.8 nm. Other lasers, with different lasing media and wavelengths, soon followed. The availability of these intense coherent light sources opened up the fields of nonlinear optics, second-harmonic generation, holography, and high-resolution laser spectroscopy. The optical sciences and the optics industry have been revolutionized by the laser, with new electrooptic systems and applications appearing frequently.

II. FUNDAMENTALS OF ATOMIC TRANSITIONS AND STIMULATED EMISSION

The theory introduced by Einstein explained the interaction of radiation with matter in terms of three basic processes: absorption, spontaneous emission, and stimulated emission of electromagnetic energy by atoms (or molecules). The energy states of atoms are characterized by the quantum numbers (1,2,3,...) assigned to the quantized energy levels (E_0, E_1, E_2,...). These energy levels have populations (number of atoms per unit volume) N_0, N_1, N_2,..., etc. The lowest energy level is called the ground state, and the higher levels, the excited states. If the system is in equilibrium with thermal radiation at a given temperature T, then the relative populations of any two levels are given by Boltzmann's equation:

$$\frac{N_2}{N_1} = \frac{\exp(-E_2/kT)}{\exp(-E_1/kT)} = \exp[-(E_2 - E_1)/kT] = \exp(-h\nu/kT)$$

where k is the Boltzmann constant. Note that if $E_2 > E_1$, then $N_2 < N_1$. The total photon number is unchanged, since the emission and absorption processes add and remove photons to and from the radiation field at a constant rate. The proportionality constants, known as the Einstein coefficients, A_{21}, B_{21}, and B_{12}, express the transition probability per unit time for each of the basic processes. For a two-level system in equilibrium, this condition of balance is illustrated in Figure 1. Instead of considering the number of photons, it is more convenient to use the photon energy density per unit frequency, $u(\nu)$, since they are proportional. Thus for the spontaneous emission process we have

$$\text{Number of photons emitted/unit volume/unit time} = A_{21}N_2$$

The absorption process depends on the strength of the radiation field and is also a stimulated process. Thus,

$$\text{Number of photons absorbed/unit volume/unit time} = B_{12}u(\nu_{12})N_1$$

A. Stimulated Emission

Stimulated emission is the process whereby electromagnetic waves of the radiation field drive the atomic oscillators that generate the emission. The number of stimulated photons emitted is proportional to the photon energy density, the density of electrons in the upper state, and the proportionality constant B_{21}. Thus we can write

$$\text{Number of stimulated photons emitted/unit volume/unit time} = B_{21}u(\nu)N_2$$

If the atoms are in thermodynamic equilibrium with the radiation field, with N_1 and N_2 constant in time, then the net rate of downward transitions must equal that of upward transitions:

$$N_2A_{21} + N_2B_{21}u(\nu) = N_1B_{12}u(\nu)$$

and

$$u(\nu) = N_2A_{21}/(N_1B_{12} - N_2B_{21})$$

Then, using the Boltzmann equation, we can write

$$u(\nu) = (A_{21}/B_{21}) \times 1/[(B_{12}/B_{21}) \exp(h_\nu/kt) - 1]$$

The Planck radiation formula requires that the Einstein coefficients satisfy the relations

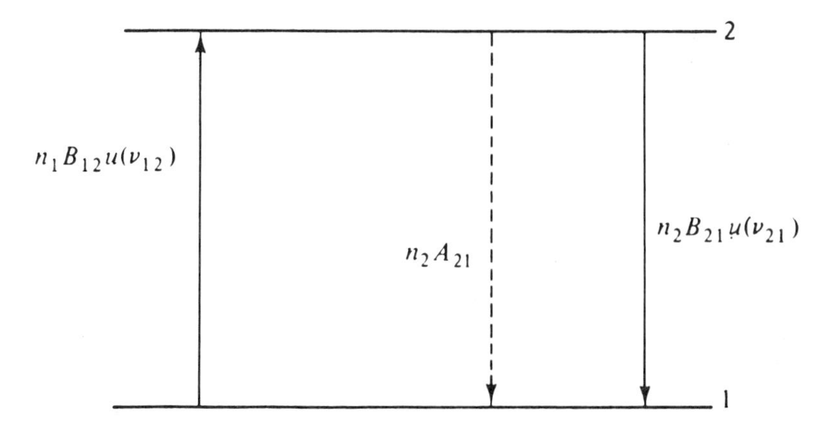

Figure 1. A two-level system, showing absorption (up arrow), spontaneous emission (---), and stimulated emission (——).

$$B_{12} = B_{21}$$

and

$$A_{21}/B_{21} = 8h\pi^3/c\nu^3$$

The ratio of the stimulated emission rate to the spontaneous rate for atoms in equilibrium with thermal radiation is then given by the equation

$$\frac{\text{Stimulated emission}}{\text{Spontaneous emission}} = \frac{B_{21}u(\nu)}{A_{21}} = \frac{1}{\exp(h\nu/kT) - 1}$$

The above result tells us that the rate of stimulated emission is very small in the visible region of the spectrum for ordinary optical soruces ($T = 1000$ K). For these sources, most of the radiation is emitted in a random way by spontaneous transitions (incoherent radiation). Also, the above ratio is smaller at shorter wavelengths, and it is thus easier to achieve lasing at longer wavelengths, everything else being equal.

The emission ratio must be large for lasing to occur, and we see from the above ratio that this happens when the photon energy density is high. Consider an optical medium through which the radiation is passing, whose atoms are in various energy levels, such as E_1, $E_2, \ldots$, where $E_2 > E_1$. From the above relations, we see that the rate of stimulated downward transitions will exceed that of upward transitions if the population of the upper state is greater than that of the lower state, that is, if $N_2 > N_1$. In other words, the ratio of stimulated emission to absorption must have a value greater than unity for lasing to occur. Such a condition is contrary to the thermal equilibrium distribution given by the Boltzmann formula. This condition is known as population inversion (Figure 2). When this occurs, the photon energy density of the radiation is amplified when it passes through the medium. This induced emission is propagated in the same direction as the incident radiation, and in phase with it (coherent radiation).

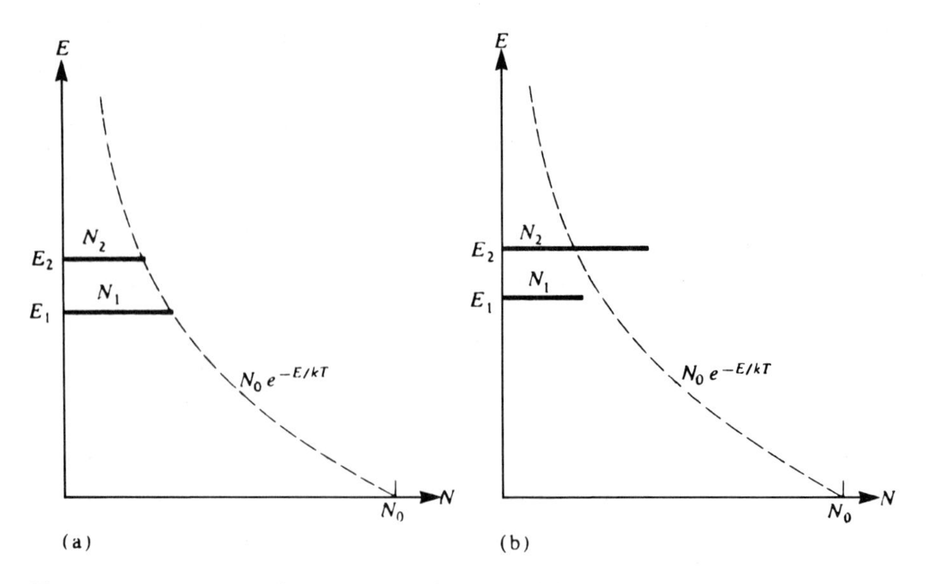

Figure 2. Graphs of the population densities of two energy levels of an atom: (a) normal Boltzmann distribution and (b) inverted distribution.

B. Excitation Methods and Population Inversion

The population inversions required for optical amplification to take place in the active medium of a laser may be produced by several methods. Some of these are optical pumping or photon excitation, electron excitation, and inelastic interatomic collisions.

For optical pumping, an external light source (such as a flashlamp or another laser) is used to produce a high population in a particular

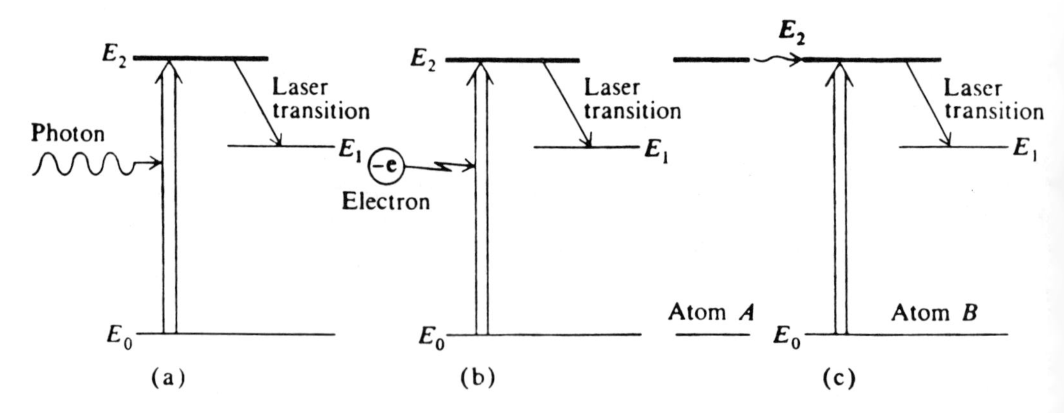

Figure 3. Diagrams showing the three processes for producing a population inversion: (a) optical pumping, (b) direct electron excitation, and (c) inelastic interatomic collision.

energy level in the active medium by selective optical absorption. This is shown in Figure 3a, and is the method used in the solid-state lasers such as the ruby or the Nd:YAG (neodymium-doped yttrium aluminum garnet).

Direct electron excitation in a gaseous discharge, as shown in Figure 3b, is used in some of the gaseous ion lasers (the argon-ion laser, for example) in which the active medium carries the discharge current. With appropriate values of gas pressure and current, the discharge electrons directly excite the atoms to the upper lasing energy level, which has the longer lifetime.

The third method also makes use of an electrical discharge, but in a suitable combination of two gases. In this case the two kinds of atoms each have excited states that exactly or nearly coincide. An excited atom of type A gas then transfers its excitation energy to an atom of type B, by the process known as collisional excitation. This method is used in the helium-neon (He-Ne) laser, where helium plays the role of type A, and neon that of type B atoms. The actual laser transitions occur in the neon atoms.

III. LASER SYSTEMS

A laser consists of four basic elements: the active medium, the excitation mechanism, the optical cavity, and the output coupler. Since the output coupler is part of the optical cavity, the laser may also be described as having three basic elements. Each element is a necessary part of the optical oscillator that emits a highly collimated beam of intense coherent radiation. This applies to normally radiating atomic or molecular systems. In addition we have some superradiant systems with such high gain that feedback mechanisms are not necessary to achieve stimulated emission.

A. The Active Medium

Many lasers are named for the amplifying (active) medium that is required to support the population inversion between two energy levels. The laser medium may be a gas (He-Ne, CO, argon, etc.), a liquid (dye solutions), or a solid (ruby, Nd:YAG, etc.). Lasing has been observed in over half of the elements and in many compounds, such as HF, KrF, CO, CO_2, and HCN, among others.

The active medium may consist of two parts, the host medium and the lasing substance. This is the case for the ruby (sapphire doped with Cr^{3+}), and the Nd:YAG lasers. This distinction is usually not made in gas lasers even though they may consist of two or more gases. The most important characteristic of the amplifying medium is its ability to maintain a population inversion between the two energy levels that define the lasing transition. The procedure for maintaining inversion is called pumping (exciting) the atoms or molecules to the desired higher energy level, rather than exciting *all* the energy levels as in an ordinary discharge.

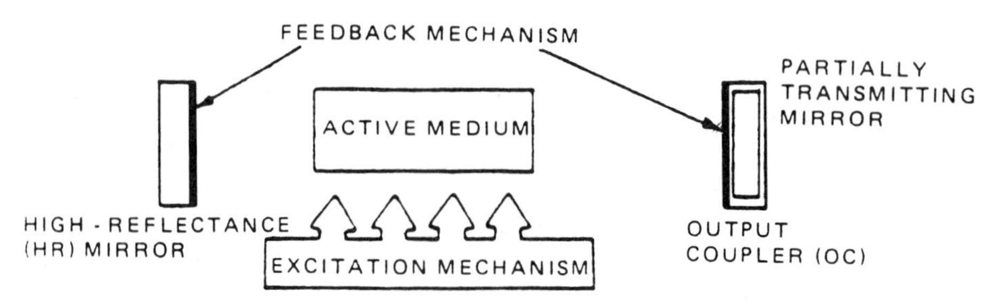

Figure 4. Elements of a laser.

B. The Excitation Mechanism

The excitation of the active medium to the upper energy level to attain population inversion is produced by an external energy source, which may be optical, chemical, electrical, or thermal. For gas lasers an electrical discharge is usually employed. In some, the free electrons in the discharge collide with and excite the atoms, ions, or molecules directly, while in others the excitation is produced by inelastic atom-atom or molecule-molecule collisions. For example, the He-Ne laser employs electron excitation of the helium atoms, which in turn excite the neon atoms through collisions.

Solid-state lasers, such as the ruby and Nd:YAG, make use of xenon flashlamps, either continuous or pulsed, to provide photon excitation of the chromium or neodymium ions embedded in the crystal lattice. This is referred to as optical pumping, which classification also includes the use of another laser in place of the flashlamp. This is the only practicable method for excitation of liquid or solid active media.

C. The Optical Cavity

The third basic element of a laser is the optical cavity or resonator, which consists of two mirrors that direct the photons back and forth through the active (amplifying) medium located between them. The mirrors (shown in Figure 4) may be plane or concave, and must be carefully aligned along the optical axis of the laser. One of the mirrors is as nearly 100 percent reflective as possible, while the second, called the output coupler, has a reflectivity in the range of 50-99 percent. This allows part of the amplified radiation to escape, forming the output beam.

IV. CHARACTERISTICS OF LASER LIGHT

The light emitted by a laser is distinctly different from that emitted by other sources. These differences are described in terms of monochromaticity, directionality, and coherence.

A. Monochromaticity

The radiation emitted by a laser is much more nearly monochromatic than that from other light sources used in spectroscopy. That is to say, the linewidth of laser radiation is extremely narrow. For example, the band of wavelengths emitted from a He-Ne laser is about one ten-millionth that from an ordinary discharge lamp (see Table 1).

B. Directionality

The laser is unique as a light source with high directionality and minimum angular spread, both difficult to achieve with ordinary light sources. These properties result from the geometry of the laser cavity and from the monochromatic and coherent nature of the radiation within the cavity.

C. Coherence

Coherence is a measure of the degree of phase correlation in a radiation field at different times and locations. The laser was the first truly coherent light source to be discovered, and this property distinguishes it from any other sources. To achieve this high degree of coherence, a laser must oscillate in single axial and transverse modes in the optical cavity. A multimode laser may be no more coherent than a well-filtered thermal source. A radiating source, in order to emit highly coherent light, must be small in extent (in the limit, a single atom) and must emit in a narrow bandwidth (in the limit, equal to zero). In reality many atoms are involved, and in a laser their oscillations and emissions are coordinated to as high a degree

Table 1. Line Widths of Light Sources

Source	Center wavelength (nm)	FWHM[a] line width (nm)
Mercury discharge lamp	546.1	0.217
Sodium discharge lamp	589.6	0.1
Cadmium low-pressure lamp	643.8	0.013
Helium-neon laser	632.8	8×10^{-8}
Diode-pumped Nd:YAG laser	1064	5×10^{-5}
Color-center laser	1500	1×10^{-5}
Dye laser	(Variable)	0.0036
Lead-salt diode laser	5000	0.0003
Excimer laser	300	0.001

[a]FWHM, full width at half maximum height.

Table 2. Lasers for Spectroscopic Applications

Type	Wavelength (nm)	Power	Mode	Excitation
A. Solid media				
1. Nd:YAG	1064,533	0.04-600 W 21 W	CW[a] Pulsed	Flashlamp Diode laser
2. Co:MgF_2	1510-2450	170 mJ 4.3 W	Pulsed CW	Nd:YAG
3. Ti:Al_2O_3	400-700	100 mJ 1.6 W	Pulsed CW	Nd:YAG Argon ion
4. Cr:$BeAl_2O_3$	710-820	> 1 J	Pulsed	Flashlamp
5. Diode	780-1500	< 200 mW	CW or pulsed	Diode current
6. Lead-salt diode	2000-32,000	mW-μW	CW	Diode current
7. Color center	800-4000	< 300 mJ	5-20 psec[b]	Dye or Kr laser
B. Liquid media				
1. Dye lasers	300-1000	0.05-15 W	3-50 nsec	N_2, Nd:YAG
2. Dye lasers	248	20 mJ	30 psec	Excimer
3. Dye lasers	340-940	0.25-50 W	0.2-4 μsec	Flashlamp
4. Dye lasers	400-1000	< 2 W	CW or pulsed	Ion laser
C. Gas media				
1. Excimer laser, ArF	193	> 50 W	Pulsed	Electric discharge or electron beam
Excimer laser, KrF	248	> 100 W	Pulsed	
Excimer laser, XeCl	308	> 150 W	Pulsed	
Excimer laser, XeF	351	> 30 W	Pulsed	
2. He-Cd laser He-Cd laser	325 442	2-50 mW 1.5-10 mW	CW CW	Electric discharge Electric discharge
3. He-Ne laser	632.8	0.5-100 mW	CW	Electric discharge
4. Argon ion laser	488,514.5	2 mW-20 W	CW	Electric discharge
5. Nitrogen laser	337	1-330 mW	Pulsed	Electric discharge

[a]CW, continuous wave.
[b]Pulse duration.

as possible through the mechanism provided by stimulated emission. Laser light is described as being both temporally and spatially coherent.

V. LASERS IN INSTRUMENTATION

The application of the laser to problems in instrumentation has, in recent years, provided researchers with the tools to carry out investigations previously either very difficult or impossible. When coupled with parallel advances in electronics and computers, the range of applications and research appears almost limitless. In Table 2 are listed the principal lasers used in spectroscopic instrumentation, together with some of their most important parameters.

The lasers listed in Table 2 are widely used in the investigation of photochemical processes, as probes or preparation agents, using absorption, fluorescence, and Raman techniques, as well as laser-induced dynamic processes. Since the methods of laser involvement vary widely, each of these areas will be discussed separately.

A. Absorption Spectroscopy

One of the early applications of lasers in instrumentation was in absorption spectroscopy. Since both atoms and molecules absorb radiation at discrete frequencies when excited by electromagnetic waves of appropriate energy, their energy states—electronic, vibrational, and rotational—can be determined by a study of their absorption spectra. Such studies may involve single-photon or multiple-photon excitation, as well as photoionization, photodissociation, and photoisomerization, as discussed by Ready and Erf [11]. Many of the papers in the September 1987 issue of Applied Optics [12] also discuss these applications, as does the book of Priov et al. [13]. Szappanos and Truett [14] have provided a recent update of laser spectroscopy with many references to the original literature.

The early use of the Nd:YAG laser (A.1 in Table 2) was as a frequency-doubled pump source for dye lasers in both continuous and pulsed modes. The Nd:YAG has usually been flashlamp pumped, but recent developments in diode lasers provide an alternative pump source. Much of the new technology is described in references 15-17. Several diode lasers are required to cover the tuning range listed in Table 2. Arrays of diode lasers are currently being developed for commercial applications. Moulton [18] has described advances in the use of cobalt magnesium fluoride (A.2 in Table 2), titanium sapphire ($Ti:Al_2O_3$; A.3 and A.4), and alexandrite ($Cr:BeAl_2O_3$). The progress in pumping techniques has provided more tools for the spectroscopist. In some near-infrared applications requiring a broadly tunable source, the titanium sapphire could replace the dye laser.

The lead-salt diode lasers listed in A.6 of Table 2 have been in use in spectroscopy for some time; Eng et al. [19] have presented a review describing their applications. Several diodes are required to cover the range of wavelengths listed.

The color-center (alkali halide) lasers (A.7 of Table 2) emit in the near-infrared, but at present the usable ranges for spectroscopic applications are approximately 1400-1700 and 2200-3500 nm. German and Slambrouk [20] describe the current state of the art.

B. Raman Spectroscopy

Fourier-transform (FT) infrared spectrometers have been in use for many years for measurement of absorption. Recently the combination of Raman spectroscopy with the FT systems has greatly reduced the problem of background fluorescence, which often tends to mask the Raman signal. Several instrumental techniques, such as time-resolved detection, excitation outside the absorption region, and coherent anti-Stokes Raman spectroscopy (CARS) have been used in efforts to solve the fluorescence problem. The continuous Nd:YAG laser represents a considerable advance over the argon and krypton lasers previously in use. Good discussions are given by Buijs [21] and by Hallmark et al. [22].

Raman spectroscopy has been used in the study of supported metal oxide catalysts by the use of an argon ion laser [23]. The methods used in the study of surfaces have been greatly improved by the use of lasers.

The use of He-Cd and Nd:YAG lasers in resonance Raman spectroscopy has been reported recently. Armstrong et al. [24] found it possible to reduce the fluorescence problem using continuous radiation from the He-Cd laser. Friedman and Campbell [25] have utilized the time-resolved capabilities of a pulsed Nd:YAG laser in a study of the protein dynamics of hemoglobin.

During the last few years both coherent and noncoherent Raman spectroscopic techniques have been used to study molecular gases in combustion processes and the structure and energy transfer in molecular liquids at high pressure. Stimulated Raman scattering, CARS, and Raman-induced Kerr-effect scattering have been used by Schmidt et al. [26] and by Moore et al. [27] in measurements on shocked liquids using a dye laser pumped by a 6-nsec Nd:YAG laser pulse. The development of commercial subpicosecond dye lasers [28] should enhance research in these areas.

The investigation of the fluorescence and chemiluminescence spectra of molecules, discussed by Schulman [29] in his first volume, should be a useful reference.

C. Other Applications

An application of laser instrumentation to mass spectrometry is discussed by McMahon and McMahon [30] in their paper on laser desorption as applied to the study of biochemical compounds.

Research in molecular dynamics has also been discussed by Ready and Erf [11] in their laser applications book. They show that it is not necessary for photons to be in resonance with the energy-level separations of the atoms or molecules of gases. Also discussed are several approaches to the investigation of the behavior of molecules in liquid form, such as chemical reactivity, light scattering, vibrational activity, etc. A discussion of surface spectroscopy of solids is also included.

The journals [33-36] mentioned in the bibliography are very useful for keeping abreast of laser applications. The books [37-44] provide a source for delving deeper into specific areas of investigation.

REFERENCES

1. Fowles, G. R., *Introduction to Modern Optics*, Holt, Rinehardt and Winston, New York (1967).
2. Hecht, E., and A. Zajac, *Optics*, Addison-Wesley, Reading, Mass. (1976).
3. Jones, K. A., *Introduction to Optical Electronics*, Harper and Row, New York (1987).
4. Klein, M., *Optics*, Wiley, New York (1970).
5. Lengyel, B. A., *Lasers*, 2nd ed., Wiley, New York (1971).
6. Levi, L., *Applied Optics, a Guide to Optical System Design*, Vol. 1, Wiley, New York (1967).
7. Pedrotti, F. L., and L. S. Pedrotti, *Introduction to Optics*, Prentice-Hall, Englewood Cliffs, NJ (1987).
8. Siegmann, A. E., *An Introduction to Lasers and Masers*, McGraw-Hill, New York (1971).
9. Young, M., *Optics and Lasers*, Springer-Verlag, New York (1986).
10. Kompa, K. L., and J. Waner (Eds.), *Laser Applications in Chemistry*, NATO ASI Series, Vol. 105, Plenum Press, New York (1984).
11. Ready, J. F., and R. K. Erf (Eds.), *Laser Applications*, Vol. 5, Academic Press, New York (1984).
12. "Laser Applications in Chemical Analysis" (a series of articles), Applied Optics, *26*, 3495ff (1987).
13. Priov, Y., A. Ben-Reuven, and M. Rosenbluth (Eds.), *Methods for Laser Spectroscopy*, Plenum Press, New York (1986).
14. Szappanos, J., and W. L. Truett, "Laser Spectroscopy: An Update," Photonics Spectra, *21(3)*, 61 (1978).
15. Hunsperger, R. G., *Integrated Optics: Theory and Technology*, 2nd ed., Springer-Verlag, New York (1985).
16. Holmes, L. M., "Tunable Crystals and Diodes Lase in the Infrared," Laser Focus/EO Technology, *22(4)*, 70 (1986).
17. DeShazer, L. G., "Advances in Tunable Solid State Lasers," Laser Focus/EO Technology, *23(2)*, 54 (1987).
18. Moulton, P. F., "Tunable Solid State Lasers," Laser Focus/EO Technology, *23(8)*, 56 (1987).
19. Eng, R. S., J. F. Butler, and K. J. Linden, "Tunable Diode Laser Spectroscopy," Opt. Eng., *19(6)*, 945 (1980).
20. German, K. R., and T. V. Slambrouk, "Applications of Color-Center Lasers," Spectroscopy, *1(19)*, 42 (1986).
21. Buijs, H., "Fourier-Transform Raman Spectroscopy—An Overview," Spectroscopy, *1(8)*, 14 (1986).
22. Hallmark, V. M., C. G. Zimba, J. D. Swalen, and J. F. Rabolt, "Fourier-Transform Raman Spectroscopy: Scattering in the Near Infrared," Spectroscopy, *2(6)*, 40 (1987).
23. Wachs, I. E., F. D. Hardcastle, and S. S. Chan, "Raman Spectroscopy of Supported Metal Oxide Catalysts," Spectroscopy *1(8)*, 30 (1986).
24. Armstrong, D. W., L. A. Spino, T. Vo-Dinh, and A. Alak, "Resonance Raman Analysis of Dilute Solutions of Fluorescent Molecules," Spectroscopy, *2(4)*, 54 (1987).

25. Friedman, J. M., and B. F. Campbell, "Time-Resolved Raman Scattering as a Probe of Protein Dynamics in Hemoglobin," Spectroscopy, *1(11)*, 34 (1986).
26. Schmidt, S. C., D. S. Moore, D. Schiferl, M. Chatlet, T. P. Turner, J. W. Shaner, D. L. Shampine, and W. T. Holt, "Coherent and Spontaneous Raman Spectroscopy in Shocked and Unshocked Liquids," in *Advances in Chemical Reaction Dynamics* (P. M. Rentzepis and C. Capellos, Eds.), D. Reidel Publishing Co., New York (1986).
27. Moore, D. S., S. C. Schmidt, J. W. Shaner, D. L. Shampine, and W. T. Holt, "CARS in Benzene and Nitromethane Shock-Compressed to 11 Gpa," in *Shock Waves in Condensed Matter* (Y. M. Gupta, Ed.), Plenum Press, New York (1986).
28. Forrest, G. T., "Commercial Subpicosecond Dye Lasers," Laser Focus/EO Technology, *23(5)*, 40 (1987).
29. Schulman, S. G. (Ed.), *Molecular Luminescence Spectroscopy: Methods and Applications*, Part 1, Wiley, New York (1986).
30. McMahan, S. H., and J. M. McMahan, "Laser Desorption Mass Spectrometry," Spectroscopy, *1(9)*, 42 (1986).
31. Brau, C. A., "Recent Developments in Free Electron Lasers," Laser Focus/EO Technology, *23(2)*, 40 (1987).
32. Anon, "An Interview with Jim Hopkins: Diode Pumping the Nd:YAG Laser," Lasers and Optronics, *6(6)*, 57 (1987).

Journals

33. "Lasers in the Life Sciences," Harwood Academic Publishers, New York.
34. "Laser Chemistry," Harwood Academic Publishers, New York.
35. "Chinese Physics—Lasers," American Institute of Physics, New York.
36. Journal of Raman Spectroscopy, Wiley, New York.

Books

37. Bagratashvili, V. N., V. S. Letokhov, A. A. Makarov, and E. A. Ryabov (Eds.), *Multiple Photon Infrared Laser Photophysics and Photochemistry*, Institute of Spectroscopy, USSR Academy of Sciences, Harwood Academic Publishers, New York (1985).
38. Zewail, A. (Ed.), *Photochemistry and Photobiology*, Harwood Academic Publishers, New York (1984).
39. Letokhov, V. S., *Laser Picosecond Spectroscopy and Photochemistry of Biomolecules*, Taylor & Francis, Philadelphia (1987).
40. Martellucci, S., and A. N. Chester (Eds.), *Analytical Laser Spectroscopy*, NATO ASI Series, Vol. 119, Plenum Press, New York (1985).
41. Haensch, T. W., and Y.-R. Shen (Eds.), *Laser Spectroscopy VII*, Series in Optical Sciences, Vol. 49, Springer-Verlag, New York (1987).
42. Letokhov, V. S., *Laser Photoionization Spectroscopy*, Academic Press, New York (1987).
43. Kliger, D. S., *Ultrasensitive Laser Spectroscopy*, Academic Press, New York (1983).
44. Levenson, M. D., and S. Kano, *Introduction to Nonlinear Laser Spectroscopy*, Academic Press, New York (1988).

13
Nuclear Magnetic Resonance

MATTHEW PETERSHEIM / Department of Chemistry, Seton Hall University, South Orange, New Jersey

I. INTRODUCTION

Nuclear magnetic resonance (nmr) spectroscopy has kept pace with the explosion of instrumental techniques available for physical and analytical experimentation. In virtually every area of chemical research it has become an essential tool in the identification of substances and in the study of their structural and dynamic characteristics. Consequently, the breadth of nmr as a field is too great to be dealt with in a single volume if any detail is to be presented. This chapter will deal with the instrumentation, theory, and practical aspects of high-resolution nmr experiments with liquid solutions, since a preponderance of the chemical applications are of this type. Likewise, the discussion is restricted to Fourier transform instruments, although there are less versatile alternatives. By no means does this selection encompass all of the exciting advances being made in nmr spectroscopy. Two primary omissions are high-resolution spectra from solid samples [1-4] and whole-body imaging [5,6]. Even within the confines of high-resolution Fourier transform nmr spectroscopy of liquid solutions, this chapter only outlines the methods and potential applications of this technique.

The versatility of nmr spectroscopy derives from the atomic nature of the information it provides. In principle, each chemically distinct atom within a compound will have a unique resonance, provided that the nucleus of interest has a magnetic dipole moment (Figure 1a). When the sample is placed in a strong, stable, homogeneous magnetic field, B_0, the states for the nuclear magnetic dipoles become quantized in a manner that is macroscopically measurable. This static field induces the dipoles of the nmr "active" isotopes to precess with moments parallel and antiparallel to the static field lines (Figure 1b). The precession frequency depends primarily on the applied field strength and an intrinsic property of the nucleus called the gyromagnetic ratio, γ, which has a unique value for each "active" isotope. The electronic environment of the nucleus and proximity of other nmr-active nuclei also affect the precession frequency. Defining the static field axis as the z direction, the dipole states are not quantized in the x and y directions; thus, at equilibrium, the dipoles precess at a well-defined angle with respect to the z axis but with arbitrary phase. In the case of a system with only two allowed states, this can be

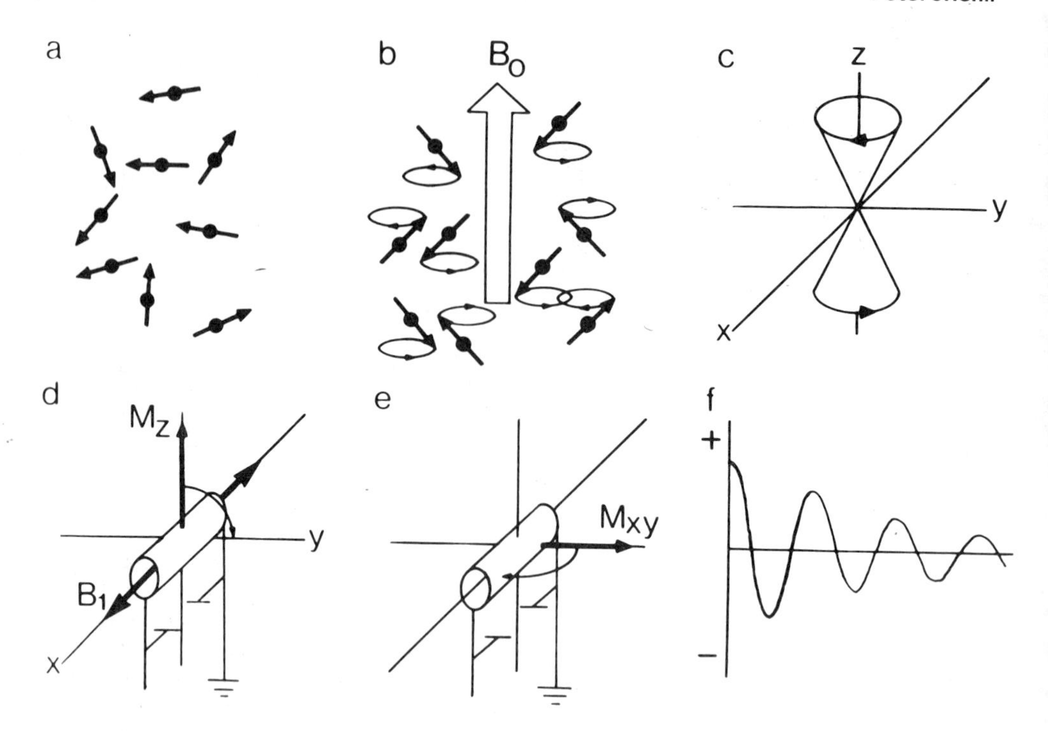

Figure 1. The pulsed nmr experiment. (a) NMR active nuclei have magnetic dipole moments. (b) In the presence of an external static field, B_0, the dipoles precess about the field axis, each with an average component either parallel or antiparallel to the field. (c) In this case there are two spin populations precessing in opposite directions represented by the two cones. At equilibrium, the top cone, that is, the dipoles with components parallel to the field, has a slightly greater population. (d) The difference between the two populations is represented by the net magnetization, M_z. The nmr signal derives from M_z, which can be observed by rotating it into the plane of an rf coil using a resonant rf field, B_1. (e) B_1 is usually gated on just long enough to rotate M_z into the x-y plane. After B_1 is gated off, the net magnetization, M_{xy}, begins to rotate about the z axis at a frequency equal to the difference in precession frequencies for the two populations. (f) This rotating M_{xy} induces an emf in the rf coil along the x axis. The signal appears as a cosine wave decaying in the time following the pulse, and is referred to as the free induction decay (fid).

depicted macroscopically as two hollow cones of precession (Figure 1c). The population of nuclei with precession moments parallel to the applied field is very slightly more abundant than the antiparallel population, giving rise to a net macroscopic magnetization, M_z, parallel to the applied field (Figure 1d).

The Fourier transform nmr experiment is initiated by a short pulse of a strong, radiofrequency (rf) alternating current through an inductive coil arranged about the sample, producing an rf magnetic field, B_1, perpendicular to the static field (Figure 1d). When the frequency of B_1 matches the difference in precession frequency for the two populations of nuclei, M_z rotates about the B_1 axis (Figure 1d). The rate of this rotation depends on the gyromagnetic ratio of the nuclei and the strength of the rf field. As a matter of convention, the z axis is parallel to the static field and the x axis is parallel to the rf field.

In the simplest experiment, the duration and power of the rf pulse are chosen so that when B_1 is gated off the net magnetization lies along the y axis (Figure 1e). Following the pulse, this magnetization:

1. rotates about the z axis at the original precession frequency of the nuclear dipoles,
2. dephases or "spreads out" in the x-y plane, and
3. relaxes back to the z axis.

Rotation of the *net magnetization* about the z axis provides the observed signal by inducing an rf current in the inductive coil used for the rf pulse (Figure 1e). The signal amplitude depends on the magnitude of the net magnetization in the x-y plane, which decays by way of the dephasing and relaxation processes. This decaying rf signal (Figure 1f) is referred to as a free induction decay, or fid. It is ultimately Fourier transformed to yield a spectrum with a peak at the precession frequency and a peak width defined by the rate of decay.

Pulsing the rf field serves two purposes in the experiment: it provides a rapid adiabatic rotation of the net magnetization for the initiation of the fid, and it introduces uncertainty in the rf frequency. The latter permits the use of a monochromatic rf source to deliver a range of frequencies centered about the source frequency. The power distribution is fairly constant over a frequency range roughly equal to the inverse of the pulse width. Thus, a 10-μsec rf pulse would provide substantial rf power at frequencies nearly 100 kHz removed from the source rf. With a high-field instrument, this is not a wide enough distribution for simultaneous matching of the precession frequencies for two different isotopes but is usually adequate for excitation of all chemically distinct species of a given isotope. That is, a single pulse yields the entire spectrum for the chosen isotope. Simultaneous detection of two elements using a single rf source would require pulses of nanosecond duration, since in high-field instruments resonance frequencies vary by megahertz with the identity of the isotope.

In the pulsed experiment, then, a high-power monochromatic rf source close to the precessional frequencies for a particular isotope is gated on for a few microseconds. This causes rotation of all components of the z magnetization associated with precession frequencies within several kilohertz of the carrier rf. Consequently, the fid is a superposition of signals, each with a characteristic rf and decay rate. The Fourier transform of this signal is the nmr spectrum for that isotope.

This brief discussion reflects most routine nmr experiments. There is, however, a growing number of techniques that are not readily described in terms of these simple vector diagrams. This chapter expands on the concepts presented in this introduction. The serious nmr spectroscopist will have to confront the more rigorous texts on this subject in order to understand and properly interpret the more exotic and exciting experiments available with this technique.

II. INSTRUMENTAL COMPONENTS

Although the quality and versatility of Fourier transform nmr instrumentation have improved significantly over the past decades, the basic designs remain the same [7,8]. A general description of the principal components will be presented here. Fukushima and Roeder [9] provide a more comprehensive introduction to nmr instrumentation; however, those interested in more advanced discussions of current improvements are referred to primary sources such as the *Journal of Magnetic Resonance*.

A. Magnet and Assembly

The first instrumental requirement mentioned in the introduction was a strong, stable magnet with a homogeneous field. The low end of "strong" is usually a field of 1.4 tesla, used mostly for 60-MHz ^{1}H experiments. The high end for commercial instruments is currently 11.7 tesla (500-MHz ^{1}H) but is rapidly approaching 14.1 tesla (600-MHz ^{1}H). For a great many applications, the field strength defines the sensitivity and the spectral resolution of the instrument. Whether higher field is "better" depends on the particular experiment.

There are four classes of magnets employed in nmr instruments: permanent, ferrous core electromagnets, persistent superconducting solenoids, and nonpersistent superconducting solenoids. The permanent magnets are very stable, having virtually no operational expense associated with them. Unfortunately, they are relegated to the low end of strong. Electromagnets with ferrous cores have been used to attain what are now considered intermediate field strengths. These magnets consume a great deal of electrical power and require bulk heat exchange as a result of the resistive heating in the magnet coil. At the low field end they continue to be cost-effective. The commercial intermediate and high-field instruments almost invariably employ persistent superconducting solenoid magnets. Although these magnets are very stable and consume no electrical power directly, they do require a weekly supply of liquid nitrogen and a monthly supply of liquid helium. The cost effectiveness in operating a conventional electromagnet at intermediate field strength versus an equivalent superconducting magnet obviously depends on the local electricity rates and the cost and availability of the cryogenic liquids. A major breakthrough in superconducting materials occurred during the writing of this chapter [10]. Obviously, if persistent superconducting magnets can be operated at room temperature or even liquid nitrogen temperature, magnet operational costs will become negligible. The highest fields used in nmr experiments at present are produced by nonpersistent superconducting solenoids. These magnets require a continuous supply of current and consume much larger quantities of cryogenic liquids. They are not used with commercial instruments.

By far the majority of new instruments employ superconducting solenoids. Figure 2 shows the basic components of this class. The magnet consists of a main-field superconducting coil with a dozen or so smaller coils wound in such a way as to provide control over field gradients in various directions with respect to the main field. The latter are referred to as the shim coils and are used to improve the homogeneity of the field. These

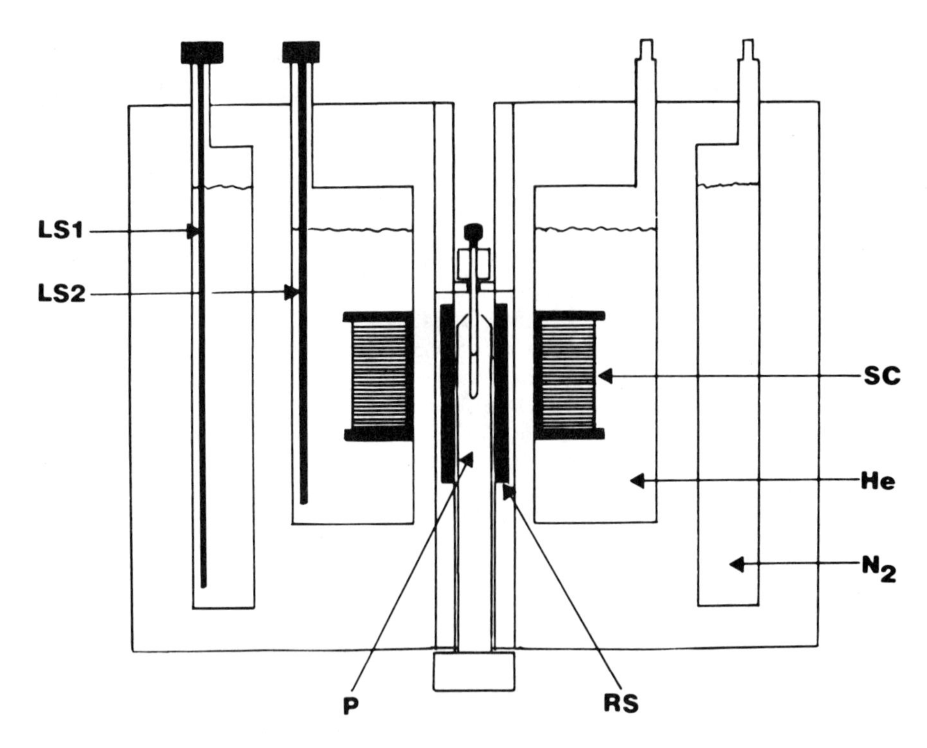

Figure 2. Magnet and probe assembly. The superconducting coils (SC), both the primary solenoid and the superconducting shim coils, are submerged in a liquid helium reservoir (He), which is suspended in an evacuated chamber. A liquid nitrogen reservoir (N) is used to shield the He reservoir from external thermal radiation in order to reduce the rate of He loss. Level sensors (LS1 and LS2) are used to electronically monitor the reservoir levels. The room-temperature shim coils (RS) and rf probe (P) are mounted in the bore of the magnet dewar and the sample is introduced pneumatically at the top of the bore.

superconducting coils must remain submerged in liquid helium during operation, with the current in each established during installation. The liquid nitrogen shield reduces heat transfer to the helium reservoir, decreasing the boil-off rate of the more expensive cryogen. Liquid helium level sensors consist of a superconducting wire extending vertically into the reservoir. Periodically, the resistance of the wire is measured, and it is directly proportional to the length of the wire not submerged in the liquid. Liquid nitrogen sensors can consist of a long coaxial capacitor placed vertically into the reservoir. Liquid and gaseous nitrogen fill the capacitor gap, yielding a change in capacitance with liquid level.

The room-temperature shim stack is a set of coils mounted inside the bore of the magnet dewar. These dozen or so coils are used to provide the final adjustments to the field homogeneity, and their contributions must be optimized regularly. Minor adjustments of the room-temperature shim coils are required for changes in sample composition, volume, and temperature; more time-consuming adjustments are required if instrumental components are moved within the bore. Consequently, instruments of the current generation have the room-temperature shim coil currents under computer control using individual digital-to-analog converters. The shim currents can then be optimized either manually or through the use of a multivariable optimization procedure.

The upper stack is an insert in the dewar bore that provides conduits for air streams used to pneumatically change samples and spin the sample once it is in place. Liquid samples are usually spun at several revolutions per second to average residual field inhomogeneities that cannot be removed by shimming. High-resolution experiments with solid samples generally require spinning at kilohertz frequencies and at the "magic angle" with respect to the main field. This is done in order to average chemical shielding, dipolar coupling, and quadrupolar coupling, which are responsible for the broad and complex resonance patterns observed with solids. The hardware near the sample for experiments with solids is bulkier than that for conventional liquid samples, requiring larger-bore magnets for the former.

The "size" of the magnet is usually discussed in terms of the field strength, expressed as kilogauss, tesla, or the precessional frequency for protons in the field, and the usable bore of the solenoid. The size of the bore defines the space available for the instrumental components that must be in the field, and the size of the sample, as mentioned above. The extreme in bore size is found with the whole-body imaging instruments being developed for medicine. More conventional spectrometers are classified as narrow, medium, or wide bore, with bore diameters ranging from roughly 5 to 10 cm. Field homogeneity within the sample volume is generally better with the narrower-bore magnets, and cryogen consumption is lower for a given field strength. Some experiments, however, require the larger bores, either because of sample size or, as in the case of magic-angle spinning of solids, the additional space needed for the instrumental components inside the bore. Thus, two groups of instruments have evolved, those intended for high-resolution experiments with liquids and those for solid samples or broad-line experiments. The bore size is one of the few major distinctions between the two groups of instruments. The terms narrow, medium, and wide also apply to the gap between poles on permanent and solid-core electromagnets with the same implications.

B. Probes

The sample "chamber" in nmr spectrometers is referred to as the probe. A general probe design for solenoid magnets is presented in Figure 3a. The most essential component is the rf transmitting and receiving coil discussed in the introduction. This is a coil of wire arranged so that an rf current generates an rf magnetic field perpendicular to the main field axis. With solenoid magnets the rf coil design is complicated by the need for ease in introducing the sample. Pneumatic sample changing necessitates the use of saddle or Helmholtz coils, shown in Figure3b. Horizontal solenoid coils are used when more efficient power transfer is needed; however, they require removing and dismantling of the probe for sample changes. The coil is actually part of a circuit constructed in the probe body, tuned to resonate at the precession frequency and impedence matched with the rf source. A simple circuit that accomplishes this is given in Figure 3c, where the series capacitor controls the resonant frequency and the parallel capacitor is used for impedence matching. For an excellent and thorough discussion of probe design, refer to Hoult [11]. Although his review preceeds this chapter by a decade, the information contained reflects the design used in most commercial probes. A more elementary presentation is given by Fukushima and Roeder [9]. For specialized advances in design the reader should peruse current issues of the *Journal of Magnetic Resonance*.

There are several properties of the rf circuit of interest to the experimentalist:

1. sample volume within the coil
2. rf power delivery to the sample and sensitivity to the signal magnetization
3. homogeneity of the rf field generated during transmission
4. decay time of the resonant circuit
5. versatility in the resonance properties of the circuit

The sample volume has an immediate effect on the sensitivity of the probe, since the signal is directly proportional to the number of a particular atomic species within the coil. The magnet bore and the space required for other components of the probe put an upper limit on the sample diameter. The total active volume of the coil is restricted by the need for the strong rf field to be homogeneous and for the circuit to have a high quality factor [11]. The quality factor defines the efficiency of power delivery and signal detection. Inhomogeneities in the rf field result in dephasing of the net magnetization during the rotation period, which increases the observed widths of the resonances and generates spectral artifacts in some multipulse experiments. The decay time of the circuit determines the dead time between the rf pulse and the detection period of the experiment. It is defined by the effective RC component of the circuit, which usually translates to a time constant of microseconds. Given that the power of the initial rf pulse is nine orders of magnitude greater than that of the signal, a 1-μsec time constant would result in a dead time greater than 20 μsec in order to dissipate the pulse power before detection. In high-resolution experiments this dead time is negligible since the signal decay times are on the order of milliseconds or longer. Broad-line experiments may require shorter dead times, usually at the expense of the quality factor for the circuit.

There is a growing demand for nmr experiments with elements other than hydrogen and for double-resonance experiments. This demand is often on a daily basis and has given rise to three different probe designs in order to minimize time lost to exchanging probes, retuning, and reshimming. The simplest in terms of circuitry is to have concentric coils each with an independent resonant circuit and rf source. This is a common approach in introducing decoupling and lock capabilities, where the primary signal coil would be innermost, a decoupler coil next, and then an outer

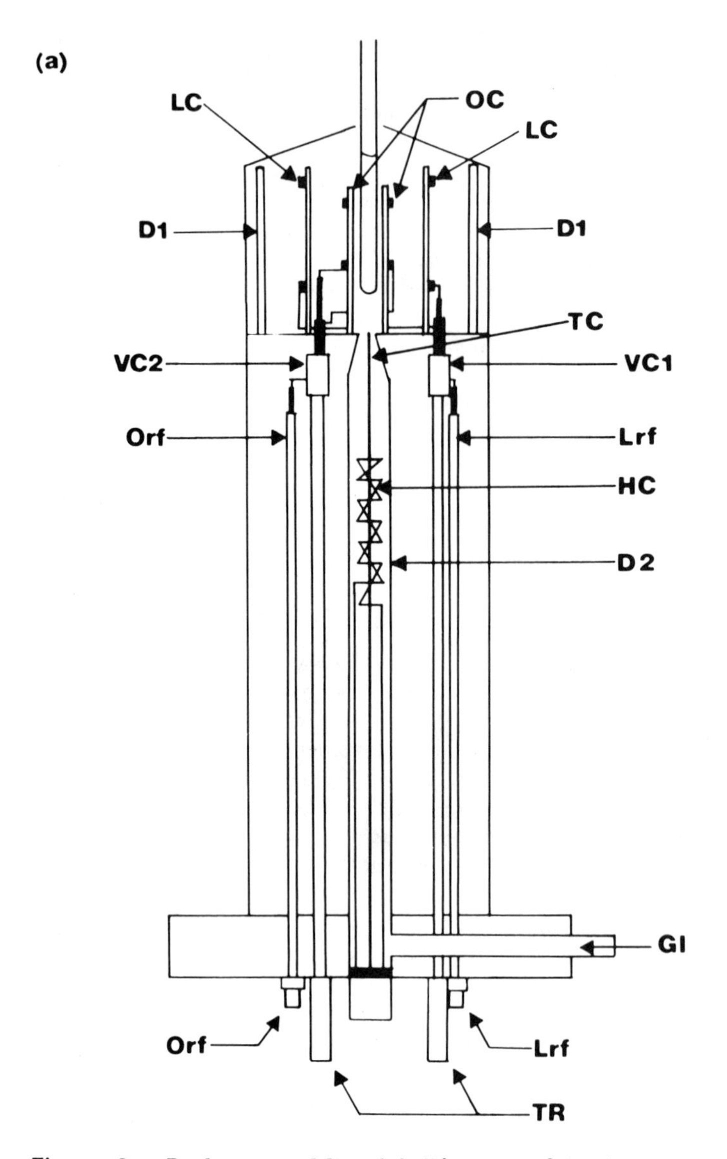

Figure 3. Probe assembly. (a) Diagram of probe components. The rf coil for the observed nucleus in the experiment (OC) is mounted on a glass insert closest to the sample volume; a second rf coil for the lock signal (LC) is shown mounted on a larger glass insert. Only part of the rf circuitry in the probe is shown here. VC1 is a variable capacitor intended for tuning the lock circuit to be in resonance with the lock rf and VC2 is the same for the observe rf circuit. The observe (Orf) and lock (Lrf) transmission lines terminate in connectors at the base of the probe housing. The two variable capacitors have tuning rods (TF), which are also accessed from the base when the probe is mounted in the magnet. The temperature

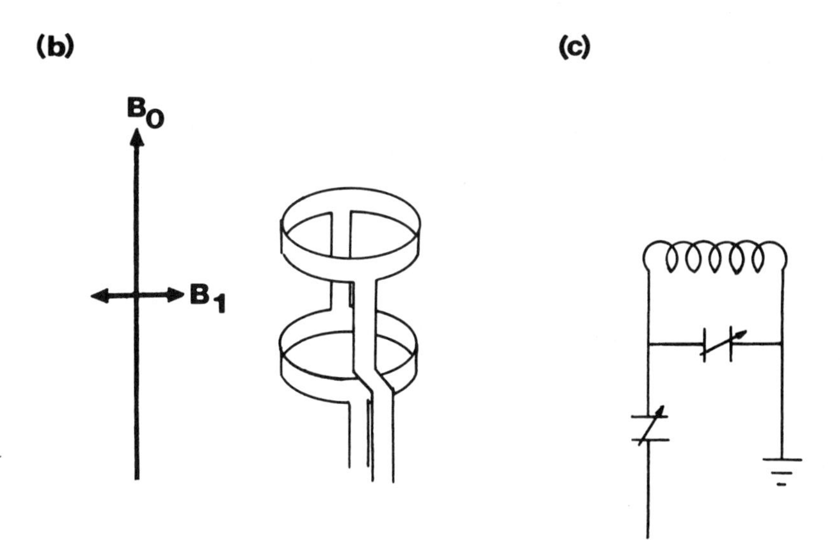

in the sample volume can be monitored with a thermocouple (TC) just outside of the rf coil volume and regulated with a flow of gas passing through a heating coil (HC). The gas inlet (GI) is also in the base of the probe. The sample volume is insulated with a glass dewar (DI) for greater thermal stability, and the conduit for the flowing gas is also a dewar (D2) insulating the other probe components. (b) Basic configuration of the rf coils with respect to the static (B_0) and rf (B_1) magnetic fields. (c) RF circuit with tunable capacitors for rf resonance and impedence matching.

lock coil (Figure 3a). This order reflects power and homogeneity demands for the three coils. Efficiency of power delivery decreases with the distance from the coil surface to the sample, and interposed coils introduce inhomogeneities in the rf field from outer coils. The primary, or "observe," coil demands the greatest power efficiency and field homogeneity, and the lock coil demands the least. By convention, decoupler circuits operate at the proton resonance frequency and lock circuits at the deuterium frequency. The decoupler is used either to provide rf power over a very narrow frequency range for selective double-resonance experiments or to irradiate the entire proton spectrum in order to collapse scalar coupling contributions from protons on the spectra of other nuclei. The lock circuit is used to monitor changes in position of a particular resonance, usually a single deuterium peak arising from a deuterated solvent. It is generally assumed that the deuterium will be present at molar concentrations, so there is a low demand for sensitivity in the lock coil. The resonance position changes as a result of fluctuations or drift in the effective field strength at the sample. The lock signal is used to control the current in one of the room-temperature shim coils, which will add or subtract with the main field in order to compensate for the drift.

A second approach in providing probe versatility is to make the resonant frequency of the observation circuit adjustable over a range of several megahertz. This is usually accomplished by allowing the capacitance to be adjusted with the probe mounted in the magnet. Screw-driven variable capacitors and plug-in fixed capacitors can be used in combination to cover a range of over 100 MHz. Probes of this type are referred to as broadband. Their primary disadvantage is that the circuit quality factor is a function of both the inductance and capacitance. Since the inductance is fixed, the quality factor, or sensitivity, of the circuit varies over the entire tuning range. When maximum sensitivity and rf power are needed it is best to work with a fixed-frequency probe optimized for a narrow frequency range. The third approach is to use doubly or multiply tuned circuits with a common coil. The advantage in using this type of probe is that no adjustments are required to switch among the frequencies for which it is tuned. This approach is widely used in proton-carbon combination probes and allows for automatic switching between the two nuclei in batch sample processing. As is the case with the broadband probes, the quality factor for the circuit is lower than it would be for equivalent fixed-frequency systems.

Besides the rf circuitry, most commercial probes are equipped with variable temperature capabilities. This usually consists of a conduit up the center of the probe, which directs a stream of gas onto a thermocouple and then the sample tube. Inside the conduit is a resistive heating filament used to warm the gas for high temperatures. Low temperatures are obtained by first passing the gas through a coil submerged in an ice, dry ice, or liquid nitrogen bath and then heating to the desired temperature. The rf coils are surrounded by a glass dewar for better temperature control at the sample (Figure 3a).

C. RF Source and Detection

The Fourier transform nmr experiment is initiated by a pulse of rf power on the order of 100 W, while the signal of interest is on the order of microwatts. Isolation and preservation of this very weak signal is an electrical engineering challenge discussed in detail by Hoult [11] and Fukushima and Roeder [9]. The general experimental design is diagrammed in Figure 4. The rf synthesis is accomplished using either fixed-frequency sources or a broadband frequency synthesizer. In both cases, the rf originates with a crystal oscillator, the output of which is amplified and mixed to the desired frequency. The mixing process generates harmonics in the rf that must be removed with a bandpass filter before amplification to the high powers needed. Broadband instruments, which are used to perform experiments at discrete frequencies over a range of up to 500 MHz, would require a very large collection of narrow-band filters for this purpose. Fortunately, a more general solution is available through the use of a common intermediate frequency. The synthesizer produces a second rf tailored to mix with the rf of interest to yield the intermediate frequency. This is passed through a fixed, narrow bandpass filter and then mixed back down to the rf needed. A fixed low-pass filter can be used to remove leakage of the intermediate frequency. This filtered rf is essentially monochromatic; however, the objective in a standard pulsed experiment is to provide even irradiation over the entire spectral range for a given isotope. Pulsing

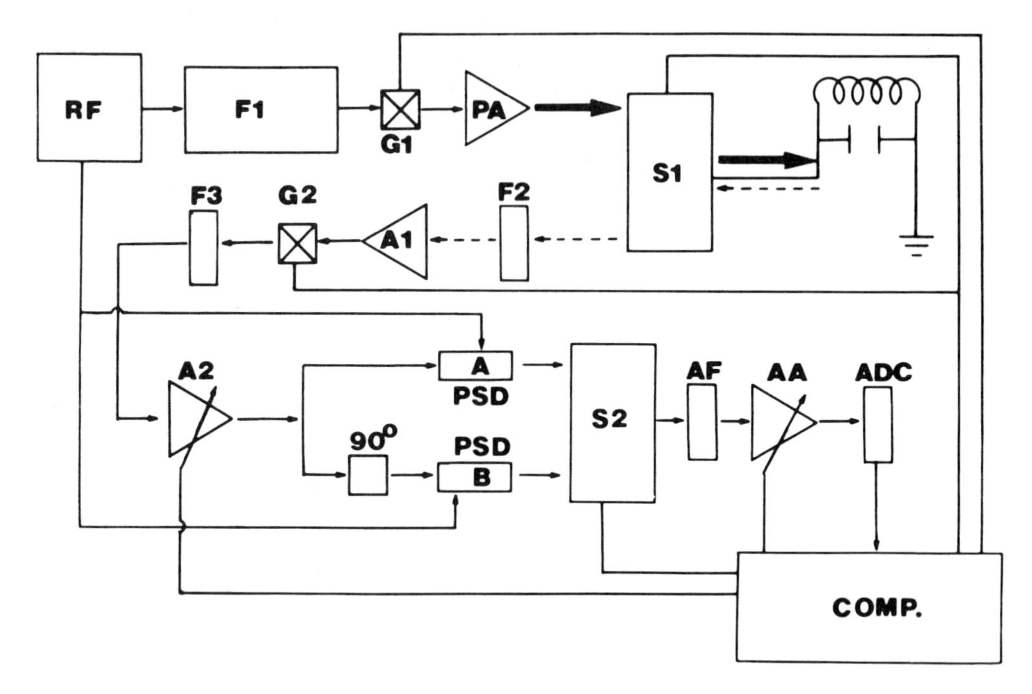

Figure 4. Block diagram for pulsed nmr experiment: RF = rf synthesizer; F1 = rf bandpass filter; G1 = pulse forming fast gate under pulse programmer control; PA = power amplifier for the pulse; S1 = primary switch for isolating the detector from the strong rf pulse; F2 = first bandpass filter for the signal; A1 = signal preamplifier; G2 = fast gate for secondary isolation of detector; F3 = rf bandpass filter; A2 = variable gain amplifier; 90° = rf phase shifter; PSD = phase-sensitive detector; S2 = switch for selecting the phase-sensitive detector to be sampled; AF = audio bandpass filter; AA = audio amplifier, variable; ADC = analog-to-digital converter; COMP. = computer.

produces distortion in the rf, which results in the AC potential being distributed about the original rf according to

$$V(f) = \text{sinc}(2\pi f t_p) = \frac{\sin(2\pi f t_p)}{2\pi f t_p} \tag{1}$$

where f is a frequency offset from the monochromatic rf in hertz and t_p is the pulse duration. A plot of this function is provided in Figure 5 for a 10-μsec pulse of 500 MHz rf. In this case V(f) is constant to within 0.5% over a range of ±2500 Hz about the central rf. This represents a 10-ppm range for a 500-MHz ^1H experiment.

The rf pulse is formed by a fast gate using a pulse programmer under computer control. A variable-gain power amplifier is used to convert this milliwatt pulse to a hundred-watt pulse. This is directed to a switching circuit used as the first stage of isolation for the detector. Switching

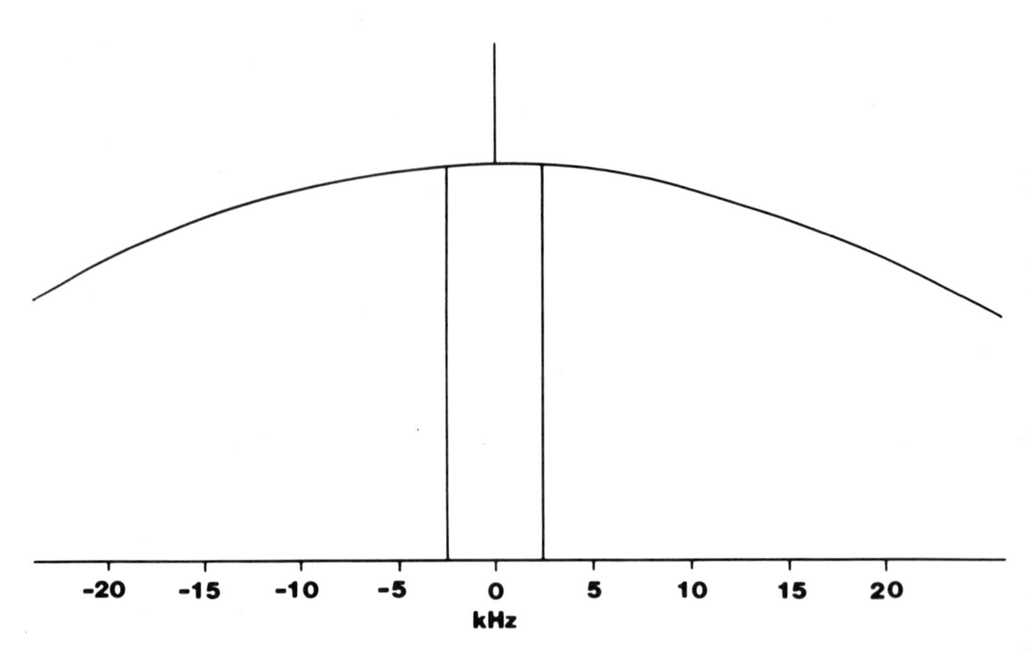

Figure 5. Frequency distribution of pulsed rf voltage about the original monochromatic rf. A 10-μsec pulse of 500 MHz rf was assumed for this calculation.

can be accomplished passively, using crossed diodes or box isolators [11], and with diode switches triggered by the pulse gate [9]. As discussed in the preceding section, the probe is impedence and frequency matched for maximum transmission of the pulse power.

It generally takes on the order of 20 μsec for the pulse power to dissipate to a level below the desired signal. Shorter recovery periods can be imposed on the circuit, but at the expense of probe sensitivity. The first switching circuit generally opens to the detector before the pulse power has decayed to a level lower than the signal, necessitating secondary gates in the detection circuit. These secondary gates are under pulse programmer control. Immediately following the first switch, the signal is filtered to remove contributions from the lock circuit and the decoupler if the observed signal nucleus is other than the proton. This very weak signal is then amplified. In broadband instruments, a broadband preamplifier is generally employed for the lower half of the rf operating range, that is, from ^{31}P resonance and down, and narrow-band components for ^{1}H and ^{19}F at the high end. Similarly, a low-pass filter with a trap for the lock signal would be used before the preamp at the low-frequency end and narrow bandpass filters for ^{1}H and ^{19}F.

Additional filtering with the broadband systems is accomplished by conversion to an intermediate frequency, as was done with the original rf synthesis. Again, filtering is performed using a narrow bandpass about the intermediate frequency and the resultant is mixed back down. A final stage of rf amplification is applied, usually with a variable-gain amplifier under computer control. This signal is then processed with a phase-sensitive detector. Since the spectral information for a given isotope generally falls within a range of several kilohertz, the reference frequency for the phase-sensitive detection is subtracted to produce an audio-range signal more amenable to digitization. Sharp cut-off variable range filters are used to remove noise contributions from outside the useful spectral region, and a final stage of amplification can be applied before analog-to-digital (A/D) conversion. The signal at this stage is of the form

$$V(t) = \sum_i V_i \cos[2\pi(f_i - f_r)t] \exp(-t/T_{2i}^*) \tag{2}$$

where the sum is taken over all resonances in the spectrum, V_i is the maximum signal amplitude for resonance i, f_i is the resonance frequency of i, f_r is the reference frequency for the phase-sensitive detection, t is the delay time following the pulse, and $T\ _i^*$ is the relaxation time for the signal component i. This signal is refered to as the free induction decay (fid). The spectrum is obtained by a Fourier transformation, which for this signal has the form

$$V(f) = \sum_i \frac{V_i}{2} \left[\frac{T_{2i}^*}{[T_{2i}^* (f' - f)]^2 + 1} + \frac{T_{2i}^*}{[T_{2i}^* (f' + f)]^2 + 1} \right] \tag{3}$$

$$f' = f_i - f_r$$

where f is an arbitrary point in the spectrum. There are two Lorentzian terms for each resonance frequency, f_i, as a consequence of the method of detection. The cosine of the difference frequencies in equation (2) cannot provide the sign of the difference. This uncertainty is transformed as mirror-image resonances at frequencies on either side of the reference. If f_r happens to be placed in the midst of a set of resonances, the two spectral images are folded onto one another, resulting in ambiguity in resonance positions. An example of this is given in Figure 6.

There are two general solutions to this problem. The f_r can be selected to be on one side of the spectrum. This approach doubles the audio frequency range that must be digitized. Not only does this double the A/D rate required to acquire the data, but the signal-to-noise ratio decreases by a factor of $2^{1/2}$ as a result of the noise from the empty half of the spectrum "folding" into the region of interest. Commercial instruments supply a second, more frequently applied, option referred to as quadrature detection. The rf signal is split and one component is phase shifted by 90°. The two signals are then processed with separate phase-sensitive detectors and digitized with either two independent A/D converters (ADC) or alternately sampled by a single ADC (Figure 4). The phase-shifted signal behaves as a sine wave relative to the original signal, providing the sign on the audio frequency when the two signals are compared. This permits f_r to be placed in the center of the spectrum without foldover. Therefore, a narrower spectral range can be studied and the audio filters can be used to reject noise outside the region of interest. If two independent ADCs are used, the A/D rate required is actually one-fourth that for the first solution, since the maximum audio frequency is lower by one-half

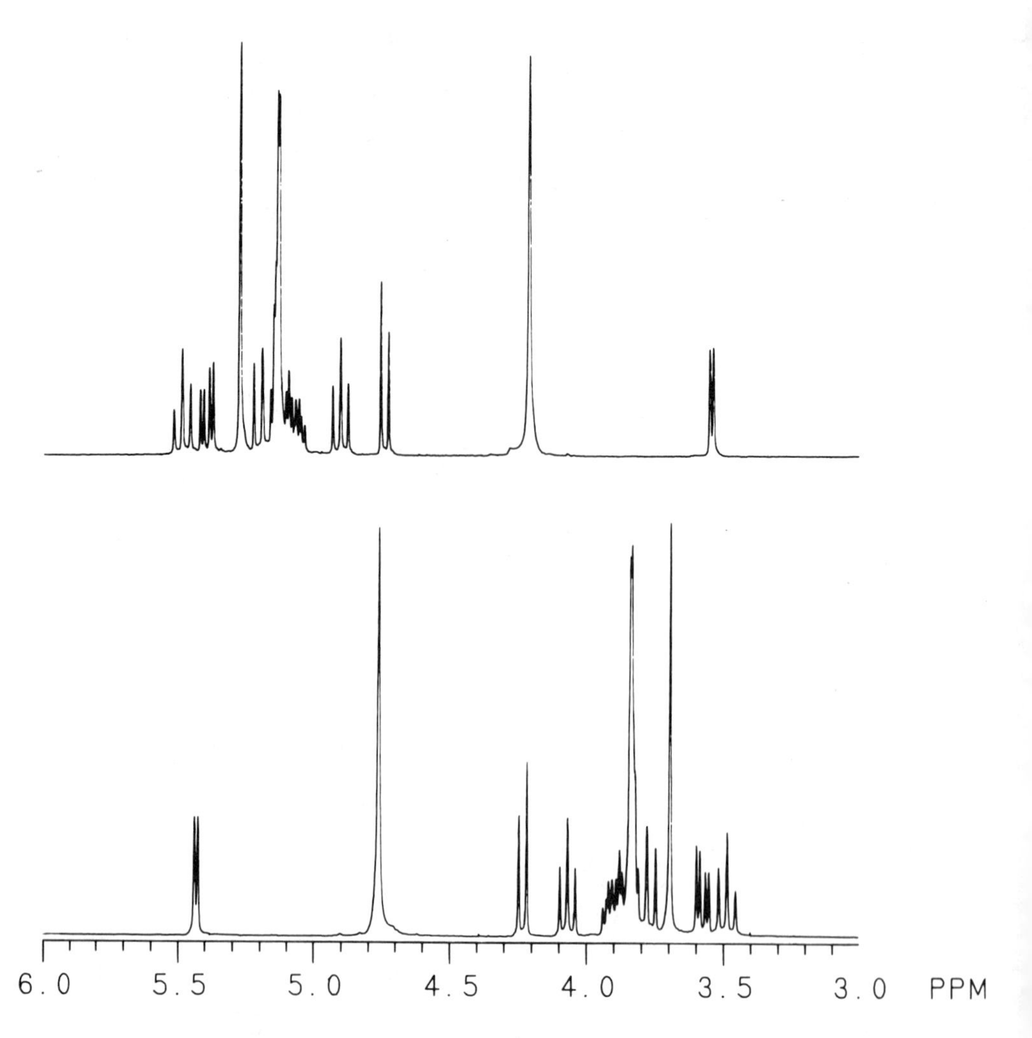

Figure 6. Spectral fold-over. The top trace is a spectrum of sucrose in D_2O collected without quadrature phase detection and the bottom trace is with it. In both cases the source frequency was centered in the spectrum.

and the 90° phase shift results in each of the signals satisfying half of the Nyquist criterion of two points per wave cycle. When a single ADC is used for both signals, the second factor of 2 is lost.

The main disadvantage in quadrature detection is the need for matched detection circuits. This is alleviated by a process referred to as phase-cycled detection. This can be accomplished by shifting the phase of f_r 90° with each successive acquisition. On alternate acquisitions the two signals then interchange the sine and cosine character. By routing the data so that "sine" and "cosine" signals are added separately, the detector mismatch is averaged.

D. Pulse Programmer

In modern commercial instruments the pulse programmer is a logic circuit having a memory stack with each address in memory consisting of two word components: a control register and an interval value. The control register uses TTL outputs that will enable instrument components such as:

1. the pulse-forming gate on the rf transmitter
2. the switches that isolate the transmitter and receiver
3. the A/D trigger
4. phase shifters for the rf transmitters and the phase-sensitive detectors
5. decoupler on/off switch
6. "high" and "low" power toggles for transmitters

or any other hardware that need be triggered during the pulsing cycle. The interval value is loaded into a counter, which is used to define the duration for the selected register settings. This period may vary from microseconds up to minutes or hours. Thus a very large dynamic range is required for both the interval word and the associated counter. Submicrosecond precision in timing is generally required with phase locking of transmitter and detector gates to the rf source oscillator. This can be accomplished by using one of the source mixing frequencies as a megahertz clock for the interval counter.

At the end of the defined interval the counter signals move to the next address in the memory stack. The size of this stack defines the limit in complexity of the pulse sequence that can be implemented on the instrument. This will be discussed further in the section on multipulse experiments.

E. The Computer and Peripherals

Speed, multitasking, available RAM (random access memory), word size, and A/D and D/A capabilities are all important aspects of modern instruments. A single data set will range from 2K to 64K points with conventional transform processing times now on the order of seconds. In two-dimensional nmr experiments there can be over one thousand of these data files, requiring megabytes of storage per experiment. A fast array processor with a bulk storage disk drive or a link with a main-frame computer is almost essential for routine use of two-dimensional techniques. Multitasking is standard on new commercial instruments, which is a boon to people running long experiments by permitting processing of existing data sets without recourse to a second computer. This, of course, requires substantial blocks of RAM given the size of individual data sets. The word size of the computer defines the ultimate dynamic range available, although the number of bits provided by the A/D converter determines the immediate dynamic range of the signal. This can be a crucial factor with dilute solutions that have substantial solvent peaks in the spectrum. The current trend is toward 16-bit A/D converters with memory word sizes greater than 20 bits. The A/D converters must operate at about 100 kHz to satisfy the Nyquist criterion in some of the high-field experiments. D/A converters have proven very useful in automation of bulk sample processing. They can be used to control the sample eject air flow, the spinning rate of the sample, and the currents in the room-temperature shim coils. As discussed before, the shim coils are used to improve the homogeneity of the magnetic field, which can change appreciably from sample to sample due to differences in the magnetic permeability of the solvents, imperfections in the sample tubes, or even differences in sample volume for a given size of tube. Digital control of the shim currents permits the use of multivariable optimization routines to adjust the field homogeneity for each sample. Fortunately, it is usually sufficient to adjust only three or four of the dozen or so shim coils with each sample change.

The development of two-dimensional nmr experiments over the past decade is largely responsible for the high-speed graphics systems now being used on most commercial instruments. As stated above, these experiments generate megabytes of data. Once processed, the data require substantial manipulation before final plots are generated, and often several different plots may be needed. At the beginning of the 1980s, this stage required days of effort. With the newer systems this processing takes minutes and can be automated to some extent. Some of the speed comes from the use of array processors; however, fast raster screens greatly facilitate manipulating the view of the data and fast digital plotters make it possible to generate extensive documentation.

III. PHYSICAL PHENOMENA IMPORTANT IN NMR

There are several good texts that present the theory of nuclear magnetic resonance from various perspectives. The rudiments on an elementary level are given by Yoder and Schaeffer [12], Fukushima and Roeder [9], Becker [13], Farrar and Becker [14], and many other specialized publications. Intermediate levels of theory along with various applications beyond compound identification are presented in the texts edited by Harris and Mann [15] and Laszlo [16]. According to Fukushima and Roeder [9], *The Principles of Nuclear Magnetism* by Abragam [17] "is the premier reference work in nuclear magnetic resonance." *Principles of Nuclear Magnetism in One and Two Dimensions* by Ernst, Bodenhausen, and Wokaun [6] is in the same league. Corio [18] and Slichter [19] also provide a rigorous background in the theory of nmr, and Harris [20] provides a good treatment of the formalisms useful in understanding some of the more exotic experiments being performed. The approach taken here is introductory.

A. Contributions to Spectral Features

The resonance features in most nmr spectra derive from three fundamental phenomena: the nuclear Zeeman effect, chemical shielding, and scalar coupling. Experiments in which the sample has macroscopic molecular order, as in crystals, field-oriented liquid crystals, and field-oriented molecules, direct dipolar coupling and electric quadrupolar coupling will give rise to resonance features not observed for the same material in isotropic samples [21]. All of these phenomena require the nucleus of interest to have a nonzero spin quantum number—that is, it must have spin angular momentum that generates a magnetic dipole moment centered at the nucleus. The magnitude of the moment is determined by the gyromagnetic ratio, γ, for the particular isotope.

The *Zeeman effect* is the interaction of the nuclear dipoles with a strong static magnetic field and provides the macroscopic definition of the energy levels required for the resonance experiment, as depicted in Figure 1. In the presence of the strong field the nuclear magnetic dipoles are driven to precess about the field as a consequence of their associated spin angular momentum (Figure 1). There are 2I + 1 allowed spin states for a given nucleus, where I is the total nuclear spin quantum number. This is zero for nuclei having even numbers of protons and neutrons, an integer value for nuclei with an even number of protons only, and I is a half integer if the number of protons is odd [22]. The energy of interaction of the field with the dipole is given as

$$E(m_I) = -\gamma h m_I B_0 / 2\pi \tag{4}$$

$$m_I = \pm(I - 1), \ . \ . \ . \tag{5}$$

where m_I is the magnetic spin quantum number, h is Planck's constant and B_0 is the strength of the applied magnetic field. The experimentally observable quantity is the bulk magnetization induced in the sample by the Zeeman interaction:

$$M_0 = \sum_{m_I} E(m_I) N(m_I) / V \tag{6}$$

where $N(m_I)$ is the number of nuclei with magnetic quantum number m_I and V is the volume of the sample within the rf coil. The term $N(m_I)$ is defined by a Boltzmann distribution in terms of $E(m_I)$:

$$N(m_I) = N \frac{\exp[E(m_I)/RT]}{\Sigma \exp[E(m_I')/RT]} \tag{7}$$

where N is the total number of these nuclei in the coil volume and the sum is over m_I'. In a 14-tesla (140-kG) field, the Zeeman energy for protons is $\pm$ 0.75 J/mol. This is less than RT for attainable temperatures in nmr experiments. Thus, the exponential functions can be expanded in m_I and truncated after the first-order term. Doing this and substituting equations (4) and (7) into equation (6) yields

$$M_0 = \frac{N\gamma^2 h^2 I(I + 1) B_0^2}{(2\pi)^2 \, 3VkT} \tag{8}$$

where k is Boltzmann's constant. Note that the magnetization depends on the square of the applied field, which has been one of the motivations in developing higher-field instruments. The signal in a Fourier transform experiment derives from the time dependence of this magnetization after it has been rotated into the x-y plane (Figure 1). It will be shown that this signal has an additional factor of γB_0, making it cubic in this term [23].

The magnetic dipoles of the $m_I = 1/2$ population have components parallel to the applied field, and those of the $m_I = -1/2$ population have components antiparallel to the field. Thus the sum in equation (6) yields a net magnetization that is parallel to the applied field. Assigning the z axis as parallel to the field, there is no order imposed in the x or y directions. Consequently, the $N(m_I)$ magnetic dipoles in each spin population can be described as establishing conical distributions about B_0 at equilibrium (Figure 1c).

Each of the elements in the first six rows of the periodic table has at least one stable isotope with a nonzero spin quantum number, except argon, technetium, cerium, promethium, bismuth, and polonium, which have only radioactive isotopes with spin greater than zero. Each element has a unique gyromagnetic ratio that effectively provides complete resolution of the elements in the nmr spectrum. The corresponding precession frequency for the isotope is equal to $\gamma B_0/2\pi$.

The *chemical shift* is probably the most important perturbation of the Zeeman interaction in terms of the evolution of nmr as a tool for identification and structural studies of compounds. This phenomenon also goes by the names chemical shielding and nuclear screening. It derives from magnetic fields generated in the vicinity of the nucleus as a result of coherence in the motion of electrons induced by the applied field. These local fields add to or subtract from the applied field, altering the spin precession frequencies. The particular induced field depends on the electronic, or chemical, environment of the nucleus. In general, these local fields result in a separate spectral band for each chemically distinct atom in a given compound or mixture. The field experienced by a given nucleus is

$$B = B_0(1 - \sigma) \tag{9}$$

where σ is the chemical shielding parameter. This shielding parameter depends on the orientation of the molecule with respect to the applied field unless the electronic environment of the atom is spherically symmetric. In liquid solutions, isotropic tumbling averages the effect, reducing the shielding to a scalar quantity. The convention in reporting nmr spectra is to use the chemical shift expressed in ppm rather than the shielding parameter. The relation between the two is

$$\delta(\text{ppm}) = -\sigma \times 10^6 \tag{10}$$

Table 1 contains a list of isotopes and the usual range over which δ varies with the chemical environment of that element. Chemical shifts are not expressed as absolute quantities because the applied field is altered macroscopically by the bulk magnetic susceptibility, or permeability, of the sample, which varies with sample composition. Consequently, a standard, or set of standard compounds, is chosen for each element as a reference for comparing δ values obtained under diverse conditions. The standard is either dissolved directly in the solution (internal) or contained in a coaxial tube suspended in the sample (external). Although there is some tendency to search for a standard with close to the maximum shielding expected for the particular isotope, the usual criteria are more practical. A standard preferably has a single, narrow resonance that does not overlap with peaks of interest. It should be a readily available substance, have minimal hazard associated with it, and, if it is to be used as an internal standard, should have minimal chemical interactions with the component of interest. Table 1 lists accepted chemical shift standards for some of the more commonly observed isotopes.

Table 1. Chemical Shift Ranges

Element	Range (ppm)[a]	Standard[b]	Isotope	Spin	Frequency (MHz)[b]
H	20	TMS, TSP[e]	1	1/2	100.00
			2	1	15.35
			3	1/2	106.66
Li	10	Li^+ (aq)	6	1	14.72
			7	3/2	38.86
B	200	$Et_2O \cdot BF_3$[c]	10	3	10.75
			11	3/2	32.08
C	650	TMS	13	1/2	25.15
N	1,000	NO_3^- (aq)[c]	14	1	7.22
		CH_3NO_2	15	1/2	10.14
O	1,500	H_2O[d]	17	5/2	13.56
F	800	CCl_3F	19	1/2	94.09
Na	20	Na^+ (aq)	23	3/2	26.45
Al	500	$Al(H_2O)_6^{3+}$	27	5/2	26.06
Si	400	TMS	29	1/2	19.87
P	700	85% H_3PO_4	31	1/2	40.48
S	650	CS_2[d]	33	3/2	7.67
Cl	1,000	Cl^- (aq)	35	3/2	9.80
			37	3/2	8.16
K	50	K^+ (aq)	39	3/2	4.67
			41	3/2	2.56
Ca	20	Ca^{2+} (aq)	43	7/2	6.73
Cd	800	$Cd(CH_3)_2$	113	1/2	22.19
Pt	13,000	$Pt(CN)_6^{2-}$	195	1/2	21.41

[a]Webb, G. A., Chapter 4 in Ref. 16.
[b]Harris, R. K., Chapter 1 in Ref. 15.
[c]Mann, B. E., Chapter 4 in Ref. 15.
[d]Rodger, C., N. Sheppard, H. C. E. McFarlane, and W. McFarlane, Chapter 11 in Ref. 15.
[e]TMS, tetramethylsilane; TSP, trimethylsilylpropane sulfonate.

Often, the chemical shift of a nucleus is defined by the nature of its bonds. Two special environmental contributions, however, involve more remote electron coherences: ring current shifts and pseudo-contact paramagnetic shifts. The term "ring current" refers to the facile electron coherences that can be induced in aromatic rings. The resulting magnetic field is usually stronger than that induced locally about single atoms and is great enough to perturb nuclei several angstroms removed, whether or not they are covalently attached to the ring. Likewise, strong magnetic fields are generated by the coherences induced in the unpaired electrons of paramagnetic centers. These fields can have very pronounced effects on chemical shifts of atoms many angstroms removed, again, whether or not the atoms are covalently connected to the paramagnetic center [24]. The term "pseudo-contact" derives from the similarity in manifestation of this effect to the Fermi contact interaction. The latter involves through-bond interactions between the unpaired electron(s) of the paramagnetic center and the nucleus of interest. Fermi contact interactions are much stronger than the pseudo-contact, in general.

Probably the second most important perturbation of the Zeeman interaction is scalar coupling, which arises from mixing of the spin states for nuclei separated by one or more bonds. The magnetic dipole of one nucleus introduces perturbations in the orbital and spin angular momentum of the bonding electrons, which in turn perturbs the magnetic environment of neighboring nuclei. The fields inducing these perturbations are all local, arising from the nuclei themselves. Thus, scalar coupling is independent of the applied field strength. The magnitude of the interaction depends on the number of the bonds separating the nuclei, the relative geometry of the intervening bonding orbitals, the s character of the orbitals for the immediate bonds to the atoms of interest, and the product of the gyromagnetic ratios for the coupled nuclei [15,17,18]. Scalar couplings are manifested as a division of the resonance into two or more new resonances, with the resonance splittings described in terms of a scalar coupling constant, J. The units of J are hertz by convention, and the value of J is independent of field strength. Corio [18] provides a comprehensive treatment of the often complex coupling patterns that can be encountered. The simplest case consists of two spin 1/2 nuclei, A and B, having resonances with a frequency difference, $f_A - f_B$, greater than the scalar coupling constant. This is referred to as an AB system with weak coupling, or an AX system if the two nuclei are of different isotopes. The A and B resonances are each split into symmetric doublets, as shown in Figure 7. The new set of transitions results from mixing the A and B spin wave functions. The symmetry of these doublets changes drastically if $f_A - f_B$ is of the same order as J or smaller, as shown in Figure 7. This is referred to as a strong coupling condition. The transition from strong to weak coupling is often observed in comparing spectra obtained on low- and high-field instruments. If A and/or B has other than a spin of 1/2, the weak coupling case would exhibit $2I_B + 1$ resonances for the A cluster and $2I_A + 1$ resonances for the B cluster. In the strong coupling domain, the spectrum could be very complex.

The next order problem in coupling of only spin 1/2 nuclei is the A_2B system. The subscript 2 indicates two chemically equivalent A atoms capable of identical coupling with B. The transition from weak to strong coupling as a consequence of changing the applied field strength is depicted in Figure 8. As Corio [18] demonstrates very clearly, higher-order couplings can become almost intractable for even simple molecules, especially at low field strengths. Thus, higher-field instruments not only provide cubic improvement in sensitivity but can also greatly simplify coupling patterns by increasing the magnitude of $f_A - f_B$.

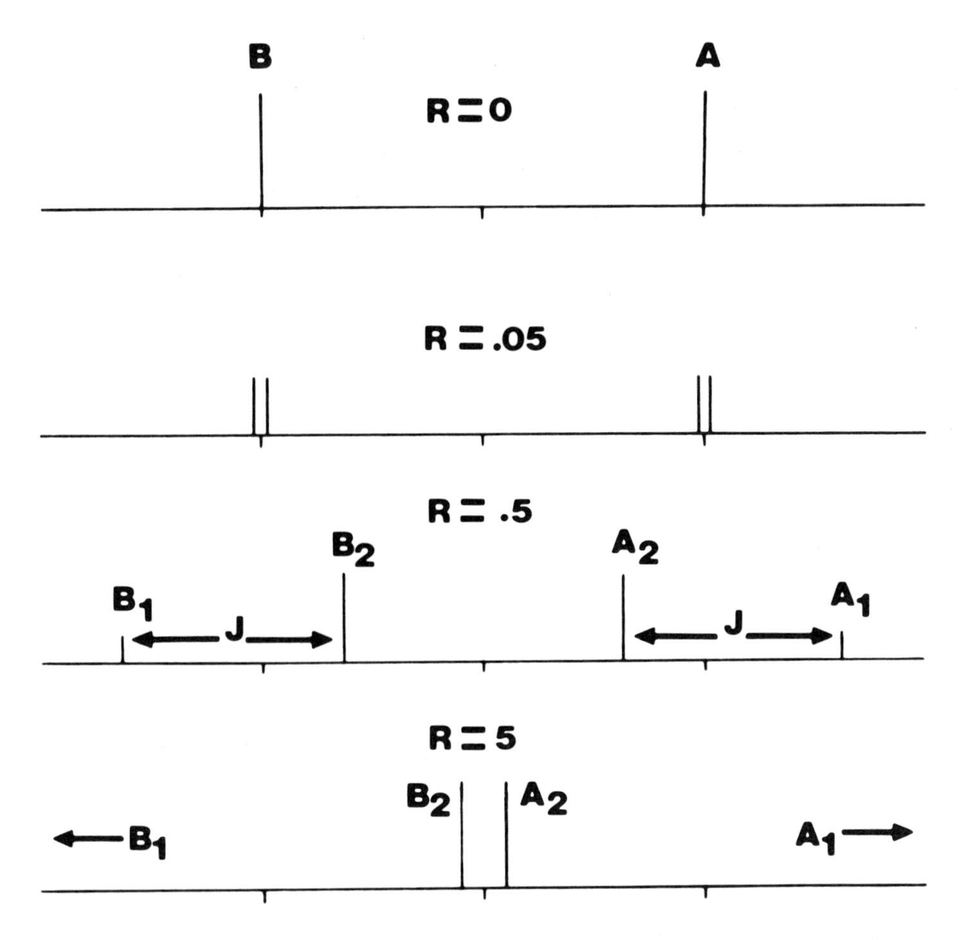

Figure 7. AB scalar coupling patterns: R is the ratio of the coupling constant, J, to the difference in resonance frequency for the two nuclei, $f_B - f_A$. The A_1 and B_1 wings for R = 5 are off the frequency scale and are roughly 2% of the A_2 and B_2 intensities [18].

The value of J depends on several parameters, as mentioned previously; however, there are two cases in which J can be interpreted in terms of fundamental properties of the molecule. One-bond couplings, ^1J, are the strongest and can be on the order of 100 Hz. The magnitude can usually be interpreted in terms of the amount of s character in the bond, with ^1J increasing with the s character. Examples of ^1J values are given in Table 2. Three-bond couplings, ^3J, are usually the most prevalent in ^1H spectra and are sensitive to the torsion angle, ϕ, between the coupled pair of nuclei (Figure 9). The relation between ^3J and ϕ is defined empirically; for example, the Karplus equation has the form [15]

$$^3J = A \cos 2\phi + B \cos \phi + C \tag{11}$$

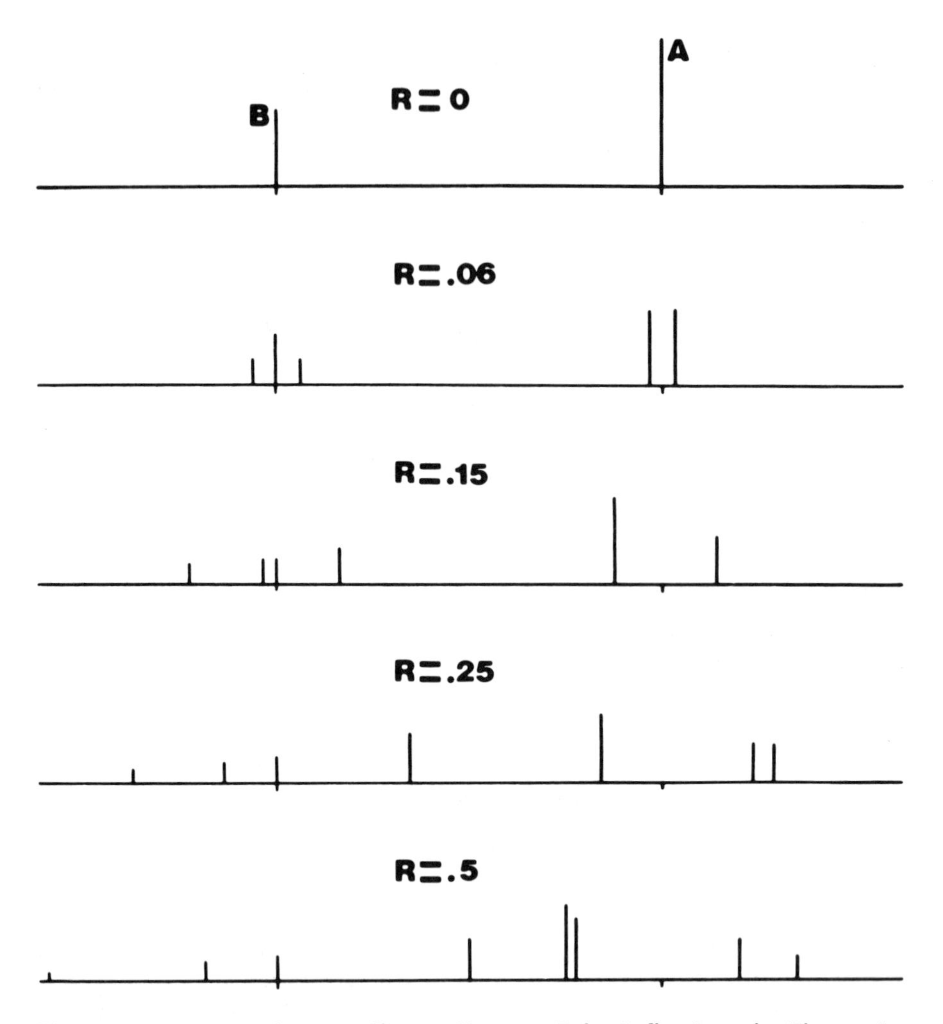

Figure 8. ABX scalar coupling patterns; R is defined as in Figure 7.

where the parameters A, B, and C are determined from measured values of ^3J for systems with known values of ϕ. In general, the scalar couplings are as rich in structural information as the chemical shifts of the resonances.

Direct dipolar coupling between two nuclei is rarely observed as a spectral feature with liquid samples because the coupling depends on the orientation of the molecule with respect to the magnetic field [17,21]:

$$D_{ij} = \frac{h^2 \gamma_i \gamma_j}{2r^3} (1 - 3 \cos^2 \theta) \tag{12}$$

where the gyromagnetic ratios are for the two coupled nuclei, r is the length of the position vector connecting the two nuclei, and θ is the angle

Table 2. One-Bond Scalar Coupling (^1J)

Nuclei	^1J (Hz)	Description
$^1H-^1H$	287[a]	H_2
$^1H-^{11}B$	100-190[a]	terminal H
	< 80[a]	H bridge
$^1H-^{13}C$	96-320[a]	
	125[a]	ethane
	156.2[a]	ethylene
	248.7[a]	acetylene
	500 αc [a]	hydrocarbons, αc = s-character
$^1H-^{15}N$	61.2[a]	ammonia
	73.3[a]	ammonia
	89[a]	urea
$^1H-^{17}O$	79[c]	$H_2O(g)$
	76; 85.5[c]	methanol in acetone
$^1H-^{19}F$	530[a]	HF
$^1H-^{29}Si$	-147 to -382[b]	
$^1H-^{31}P$	~200[a]	P(III) compounds
	400-1100[a]	P(IV) compounds
$^7Li-^{13}C$	15[a]	$[LiCH_3]_4$
$^{11}B-^{11}B$	14-28[a]	difficult to resolve
$^{11}B-^{13}C$	18-76[a]	
$^{11}B-^{19}F$	1.4-189[a]	dependent on B substituents
$^{11}B-^{31}P$	30-100[a]	
$^{13}C-^{13}C$	34.6[a]	ethane
	67.2[a]	ethylene
	170.6[a]	acetylene
$^{13}C-^{15}N$	5.8[a]	tetramethylammonium
	5.9[a]	cyanide
$^{13}C-^{19}F$	158[a]	CFH_3
	372[a]	$CFBr_3$
$^{13}C-^{29}Si$	-37 to -113[b]	
$^{13}C-^{31}P$	0 to -25[a]	P(III) compounds
	45-300[a]	P(IV) compounds
$^{15}N-^{15}N$	19[a]	$(PhCH_2)_2N-NO$
$^{15}N-^{29}Si$	6-12[b]	

(continued)

Table 2 (continued)

Nuclei	^{1}J (Hz)	Description
$^{17}O-^{31}P$	188[c]	OPF_3
	203[c]	$OPCl_3$
	195[c]	$OPBr_3$
	165[c]	$\underline{O}P(OCH_3)_3$
	120[c]	$\underline{O}P(CH_3)_3$
	90[c]	$OP(\underline{O}CH_3)_3$
$^{19}F-^{31}P$	800-1500[a]	
$^{19}F-^{17}O$	424[c]	F_2O_2
$^{19}F-^{29}Si$	167-488[b]	
$^{29}Si-^{29}Si$	53-186[b]	
$^{29}Si-^{31}P$	7-50[b]	
$^{29}Si-^{195}Pt$	-1600[b]	*trans*-$PtCl(SiCH_2Cl)(PEt_3)_2$
$^{31}P-^{31}P$	-443[a]	$(CH_3)_3P{=}PCF_3$
	465[a]	$HO_2P-PO_3^{3-}$

[a]Mann, B. E., Chapter 4 in Ref. 15.
[b]Harris, R. K., J. D. Kennedy, and W. McFarlane, Chapter 10 in Ref. 15.
[c]Rodger, C., N. Sheppard, H. C. E. McFarlane, and W. McFarlane, Chapter 11 in Ref. 15.

of the position vector with respect to the applied field (Figure 10a). This form assumes two spin 1/2 nuclei and neglects additional couplings. In most cases molecules in liquids tumble isotropically, resulting in D_{ij} being averaged to zero. Studies in which there is macroscopic order in the sample can yield resonance splittings as great as kilohertz, depending on the orientation of the coupled nuclei with respect to the macroscopic director and their separation. Solution order is most often accomplished using field-oriented liquid crystals [21]; however, there is increasing work being done with the very high-field instruments currently available in which molecules having anisotropic magnetic susceptibilities demonstrate a measurable deviation from isotropic tumbling [25]. There has also been some work using static electric fields to partially orient molecules in the magnetic field [26-28]. The essential point of this discussion is that dipolar coupling is generally not apparent as a spectral feature but can be an important source of structural information in ordered systems where it is not averaged to zero.

Electric quadrupolar coupling is also averaged to zero for molecules that tumble isotropically. It arises from a coulombic interaction between the nucleus and the surrounding electrons, providing that

1. the nucleus has a nonspherical proton distribution and
2. the electrons have a nonspherical distribution about the nucleus.

Nuclei that have a total spin quantum number of 1 or greater are referred to as quadrupolar and satisfy the first criterion. The second criterion is necessary for an observed quadrupolar coupling, since it derives from electrostatically preferred orientations of the nucleus with respect to the field generated by the electrons. There are no preferred orientations if the electrons are spherically distributed. When macroscopic order is present, as discussed with direct dipolar coupling, each resonance for a quadrupolar nucleus can split into 2I peaks. The magnitude of the splitting is directly proportional to the magnitudes of the electric field gradients generated by both the nucleus and the local electrons. The combined terms are

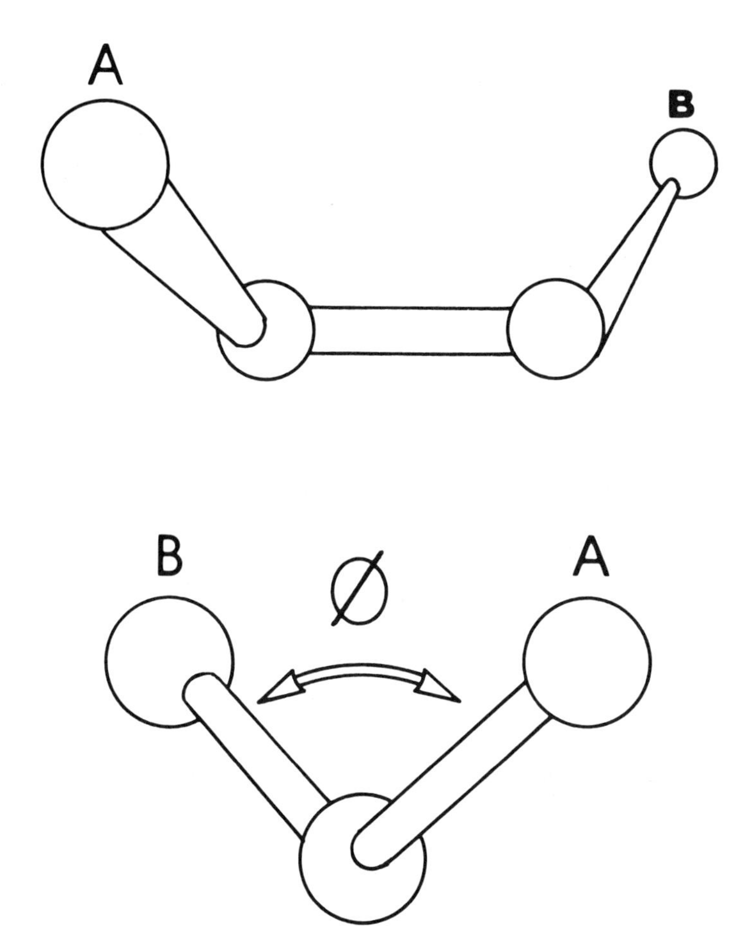

Figure 9. Definition of the torsion angle, ϕ, pertinent to three-bond scalar couplings and the Karplus equation.

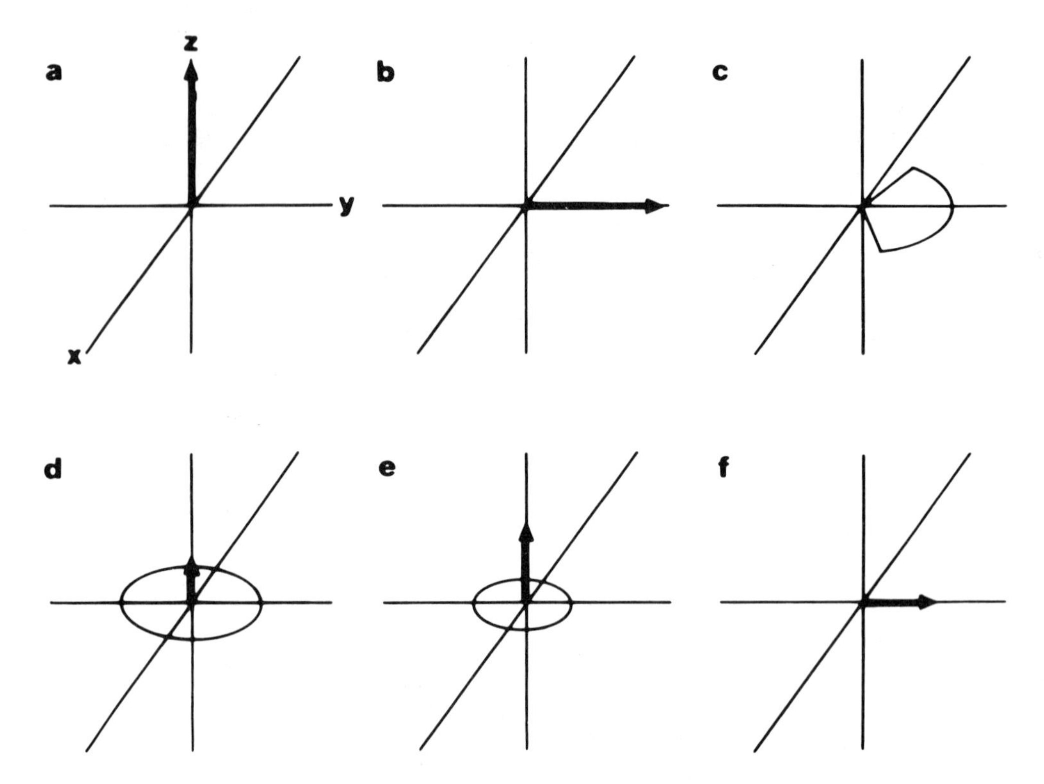

Figure 10. Pulsed experiments and the behavior of the macroscopic nuclear magnetization. (a) At equilibrium, the order imposed on the microscopic magnetic dipoles gives rise to a net magnetic moment along the z axis, that is, parallel to the strong static field. (b) The magnetization is rotated into the y axis by a rf magnetic field along the x axis at the nuclear Larmor frequency. In order to simplify the diagrams, it is assumed that there is only one resonance in the spectrum and that the coordinate system rotates about the z axis at the Larmor frequency. This results in the rotating magnetization appearing stationary. (c) Variation in the local field results in the original magnetization "fanning out" in the x-y plane. (d) After a period roughly five times T_2^* there is no net magnetization remaining in the x-y plane even though it has not all returned to the z axis. (e) and (f) If a second rf pulse is applied before the magnetization returns to equilibrium, the signal can be substantially reduced. An interval of five times T_1 between pulses is usually considered adequate for recovery of most of the magnetization.

referred to as the electric quadrupolar coupling coefficient; the component from the electrons varies significantly with the nature of the bonds to the atom. The magnitude of the splitting also depends on the orientation of the molecule with respect to the applied field. The functional form is generally more complex than that for direct dipolar coupling [17,21]. The requisite ordering of the molecules restricts both dipolar and quadrupolar coupling measurements to a very small range of applications.

B. Spin Relaxation and Signal Decay

As mentioned in the introduction, following a 90° pulse the resultant magnetization in the x-y plane rotates about the z axis, gradually dephases within the x-y plane, and more slowly returns to the equilibrium state. The rotation frequency is normally defined by the Zeeman interactions with perturbations from chemical shielding and scalar coupling, as discussed in the preceding section. Dephasing of the magnetization and its return to the z axis are both involved in the exponential decay of the signal after the 90° pulse [equation (2)]. The apparent lifetime for the decay can usually be represented by four primary terms:

$$\frac{1}{T_2^*} = \frac{1}{T_2} + \frac{1}{T_{fh}} + \frac{1}{T_{vf}} + \frac{1}{T_1} \tag{13}$$

The first three terms pertain to the dephasing process and the last defines the return to equilibrium.

The term T_2 is referred to as the spin-spin or transverse relaxation time, and it is often the dominant term in this sum. It derives from changes in spin state within the x-y plane. As a result of a 90° pulse, the 2I + 1 spin states defined by the static field are transformed into a new set of 2I + 1 states described in terms of their behavior in the x-y plane. For a spin 1/2 nucleus, there are two such states, with one having the slightly greater population established at equilibrium. Transverse relaxation results from a net conversion of nuclei from the dominant population to the minor spin state. The signal decays to zero as the two states become equally populated.

The second and third terms in equation (13) derive from variation in the local field experienced by the nucleus. The term T_{fh} represents the signal decay due to aberrations in the field homogeneity, which arise from inherent imperfections in the static field, distortions in the field imposed by the probe components, imperfections in the sample tube, and samples with phase heterogeneity, such as liquid solutions with suspended particulates. The term T_{vf} reflects variations in the local field due to fluctuations in the packing and motion of the molecules or simply structural heterogeneity. Both the T_{fh} and the T_{vf} processes effect a decay in the signal by generating a distribution of Larmor frequencies for each resonance in the spectrum. As shown in Figure 10c, the initial magnetization along the y axis fans out as it rotates about the z axis because of this variation in the local field. The contribution from T_{fh} can usually be minimized by adjusting the shim coils and spinning the sample at several hertz. The term T_{vf}, however, is an intrinsic property of the sample under a given set of conditions and cannot be dealt with instrumentally. Fortunately, it is usually a negligible term for liquid solutions of chemically inert substances. Neither term is truly a spin relaxation mechanism.

The term T_1 is referred to as the spin-lattice or longitudinal relaxation time, and it governs the return to equilibrium following a 90° pulse, or any other perturbation of the spin population. It involves a true change in spin state, as does transverse relaxation, and it is on the same order as T_2 or longer. Although T_2 generally determines the rate of decay of the signal, T_1 defines the frequency with which the signal can be sampled. Figure 10 depicts the result of applying two sampling pulses with an insufficient delay between them. The amplitude of the signal after the second pulse is substantially less than that of the first, as a result of incomplete longitudinal relaxation. In signal averaging experiments it is desirable to have a period of at least five times T_1 between pulses in order to preserve the full amplitude of the signal. This is especially important when resonance integrals are to be compared, since each resonance is likely to have a different T_1 value.

Both transverse and longitudinal relaxation arise from an exchange of angular momentum with some aspect of the nuclear environment by way of fluctuations in the magnetic and electronic interactions with the nucleus. In diamagnetic liquid solutions the interactions can be discussed in terms of the four principal perturbations of the Zeeman effect: chemical shielding, scalar coupling, direct dipolar coupling, and electric quadrupolar coupling. The theory relating measured relaxation times to these interactions and their fluctuations is quite involved and is treated rigorously by Abragam [17], Redfield [7], Schlichter [19], and several other texts on nmr. The essential point is that *nmr relaxation studies provide a versatile source of information about molecular dynamics*. Because the information is atomic in nature, it has great potential for detailed descriptions of molecular motions. However, caution must be taken in interpreting relaxation studies. Often the dynamic models are greatly simplified in making them tractable and there may be contributions to the relaxation pathway from several different interactions with the environment [29]. Here, only a brief statement will be given of the relaxation mechanisms important in liquid solutions. The reader is advised to pursue details within a particular area of application before seriously dealing with relaxation experiments.

Working in reverse order, *electric quadrupolar coupling* usually dominates the relaxation mechanism of nuclei with $I > 1/2$, unless the distribution of electrons about the nucleus is close to spherically symmetric. The quadrupolar nucleus, by definition, has an asymmetric distribution of protons represented by the nuclear quadrupole moment, Q. Because of this nonspherical nuclear charge distribution, the nucleus has preferred orientations within the asymmetric electrostatic field generated by the surrounding electrons. The magnitude of this interaction between the nucleus and the electrons is given by the electric quadrupole coupling constant, e^2qQ, where e is the charge on an electron and q is the electronic field gradient at the nucleus. As the molecule tumbles or reorients in the applied magnetic field, this electrostatic interaction generates torque on the nuclear magnetic dipole. When the angular velocity of the motion matches the resonance frequency for an allowed nuclear magnetic transition, the torque can result in a change in nuclear spin state. The functional forms for $T_1{}^q$ and $T_2{}^q$ can be very complex [30,31]. However, there are two conditions that yield fairly simple equations [17]: fast motion limit, all values of I:

$$\frac{1}{T_1{}^q} = \frac{1}{T_2{}^q} = \frac{3\ (2I + 3)}{40\ I^2\ (2I - 1)}\ K_q t_c \tag{14a}$$

$$K_q = \left[1 + \frac{n^2}{3}\right]\left[\frac{2\pi e^2\ qQ}{h}\right]^2 \tag{14b}$$

t_c = rotational correlation time, assuming simple isotropic motion

n = asymmetry parameter; deviation from axial symmetry in the electronic field gradient

and I = 1 only, all values of t_c:

$$\frac{1}{T_1{}^q} = \frac{3K_q}{80}\ [J(1) + 4J(2)] \tag{15a}$$

$$\frac{1}{T_2{}^q} = \frac{K_q}{160}\ [9J(0) + 15J(1) + 6J(2)] \tag{15b}$$

$$J(0) = 2t_c \tag{15c}$$

$$J(f_I) = \frac{2t_c}{1 + (2\pi f_I t_c)^2} \tag{15d}$$

$$J(2f_I) = \frac{2t_c}{1 + (2\pi 2f_I t_c)^2} \tag{15e}$$

Note that all of the information about molecular motion is contained in the spectral density functions, J(0), $J(f_I)$, and $J(2f_I)$. The terms $J(f_I)$ and $J(2f_I)$ can be thought of as the probabilities of having a reorientation angular velocity matching the resonance frequency for a single-quantum, f_I, and a double-quantum, $2f_I$, transition, respectively. The term $T_2{}^q$ has J(0) in addition to $J(f_I)$ and $J(2f_I)$, which arises from nonresonant effects of the torque. This zero-quantum term, as it is usually labeled, reflects a dephasing of the transverse spin states with no discrete transition. This is reminiscent of the T_{fh} and T_{vf} contributions to signal decay. These equations are only valid for the rotational motion of a sphere. More complex motion would require modification of these functions [17,29] and a greater number of unknown parameters describing the motion.

Direct dipolar coupling involves the through-space interaction of two or more magnetic dipoles. It is often the dominant mechanism for relaxation of spin 1/2 nuclei, provided there are other magnetic dipoles within roughly 0.5 nm either on the same molecule or by way of collisions. The coupling of two dipoles can be described in terms of an interaction vector (Figure 11a). Rotation of this interaction vector with respect to the applied field results in each of the dipoles exerting torque on the other. Again, when the angular velocity of the interaction vector matches a resonance frequency, the torque can induce a transition. Figure 11b shows the allowed transitions for a coupled pair of spin 1/2 nuclei of the same isotope. The zero- and

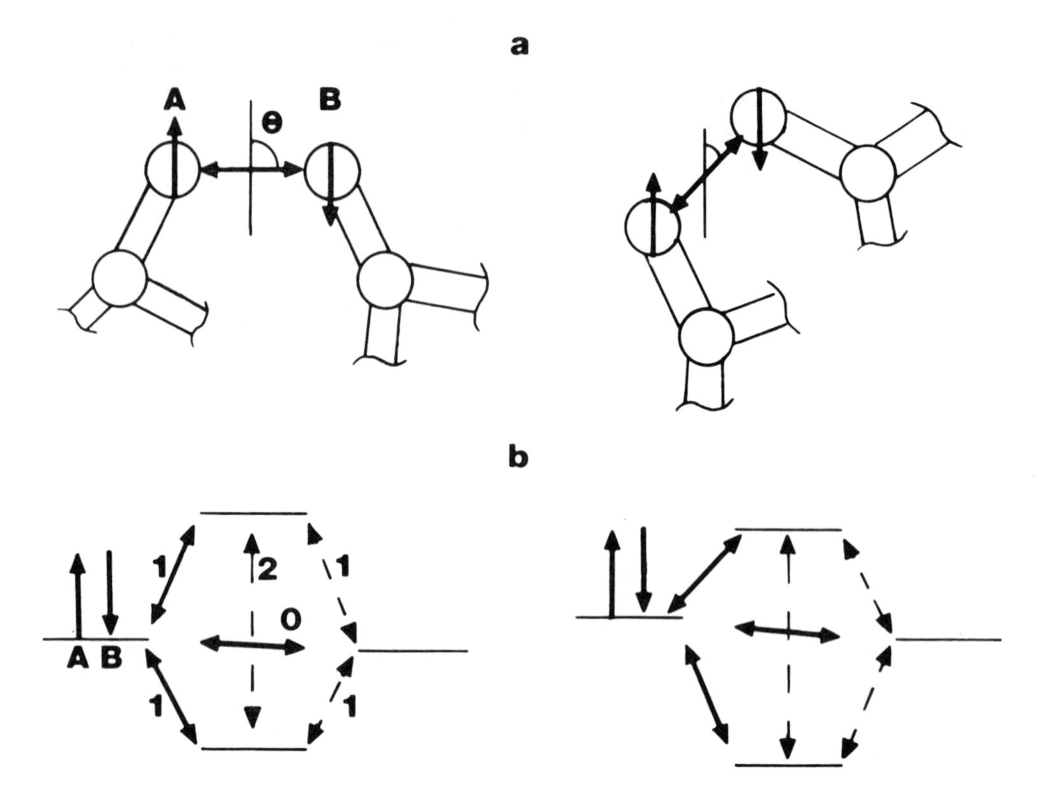

Figure 11. Dipolar coupling of two spin 1/2 nuclei. (a) Diagrams of the interaction vector for the coupled magnetic dipoles in two orientations with respect to the applied field. (b) Energy levels for this two-spin system. The relative positions of the levels change with the orientation of the interaction vector with respect to the static field. The double-headed arrows connecting levels indicate the possible spin transitions for this system. With the two dipoles initially antiparallel, only two types of transitions are immediately operative: zero-quantum (0) involving an exchange of spin between A and B, and single-quantum (1) with either A or B independently exchanging angular momentum with some other aspect of the environment. If the two dipoles are initially parallel, a double-quantum (2) transition is allowed in place of the zero-quantum. This involves both A and B simultaneously exchanging angular momentum with the environment.

double-quantum transitions, labeled 0 and 2, respectively, involve both spin states changing. The double- and single-quantum transitions involve a net change in the spin angular momentum for the coupled pair that corresponds to a change in the rotational angular momentum in the molecule.

For two dipolar coupled nuclei of the same isotope and chemical shifts much smaller than the Larmor frequency, the relaxation times are given [17] as

$$\frac{1}{T_1{}^d} = \frac{\gamma^4 h^2}{8\pi^2 r^6}\left(\frac{2}{5}\right) I(I + 1)\ [J(f_I) + 4J(2f_I)] \tag{16a}$$

$$\frac{1}{T_2{}^d} = \frac{\gamma^4 h^2}{8\pi^2 r^6}\left(\frac{1}{5}\right) I(I + 1)\ [3J(0) + 5J(f_I) + 2J(2f_I)] \tag{16b}$$

where r is the separation of the two nuclei and the spectral density functions have the same form as for the quadrupolar relaxation if the tumbling is isotropic. If several nuclei are dipolar coupled to a given nucleus, equations (16) would consist of sums over all pairs of interactions. Dipolar coupling of two different isotopes or two nuclei with chemical shifts differing on the order of the Zeeman effect yields relaxation times of the form

$$\frac{1}{T_1{}^d} = \frac{\gamma_I{}^2 \gamma_S{}^2 h^2}{30\pi^2 r^6} S(S + 1)\ [J(f_I - f_S) + 3J(f_I) + 6J(f_I + f_S)] \tag{17a}$$

$$\frac{1}{T_2{}^d} = \frac{\gamma_I{}^2 \gamma_S{}^2 h^2}{120\pi^2 r^6}$$

$$\times\ S(S + 1)[4J(0) + J(f_I - f_S) + 3J(f_I) + 6J(f_S) + 6J(f_I + f_S) \tag{17b}$$

$$J(f_I - f_S) = \frac{2t_c}{1 + [2\pi(f_I - f_S)t_c]^2} \tag{17c}$$

$$J(f_I + f_S) = \frac{2t_c}{1 + [2\pi(f_I + f_S)t_c]^2} \tag{17d}$$

where I is the spin quantum quantum number for the observed nucleus and S is the quantum number for the second spin population in the coupled pair. The relaxation of I will be exponential with these decay times only if the S population is always close to its equilibrium state. If both populations are significantly perturbed from equilibrium, data analysis involves solving coupled rate laws [17]. Again, these spectral density functions are valid only for simple isotropic tumbling, and if several nuclei are dipolar coupled, equations (17a) and (17b) would consist of sums over all pairs of interactions.

Unlike electric quadrupolar and direct dipolar coupling, scalar coupling does not vary with the orientation of the molecule in the applied field. Consequently, molecular tumbling does not generate a fluctuation in the scalar coupling of nuclei that would lead to spin transitions. Two other sources of fluctuations, however, can result in relaxation by way of scalar coupling. *Scalar relaxation of the first kind* requires that one of the coupled nuclei undergoes chemical exchange with an exchange lifetime shorter than the inverse of the scalar coupling constant. The chemical exchange can be in the form of breaking covalent bonds or a conformational change which alters the scalar coupling constant. *Scalar relaxation of the second kind* is more commonly encountered. It requires that one of the coupled nuclei has a longitudinal relaxation time shorter than the inverse of the scalar coupling constant. The broad resonances of protons bonded to nitrogen result from this mechanism. The protons are scalar coupled to the nitrogen, which is 99.63% ^{14}N. Rapid quadrupolar relaxation of this spin 1 nucleus induces the relaxation of attached protons. Equations for the two types of scalar relaxation have the same form:

$$\frac{1}{T_{I_1}{}^{sc}} = \left(\frac{8\pi^2 J^2}{3}\right) \frac{S(S + 1)T_{S2}}{1 + [2\pi(f_I - f_S)T_{S2}]^2} \tag{18a}$$

$$\frac{1}{T_{I_2}{}^{sc}} = \frac{4\pi^2 J^2}{3} S(S + 1) \frac{T_{S2}}{1 + [2\pi(f_I - f_S)T_{S2}]^2} + T_{S1} \tag{18a}$$

where I is the spin population being observed, S is the quantum number for the spin population undergoing chemical exchange or rapid relaxation, and J is the scalar coupling constant. For *scalar coupling relaxation of the first kind*, T_{S2} and T_{S1} are both equal to the chemical exchange lifetime of the S nuclei. In *relaxation of the second kind*, they are the transverse and longitudinal relaxation times for S.

Chemical shift relaxation, the fourth mechanism, is important in spin 1/2 nuclei that have *anisotropic chemical shielding*. In this case the nuclear magnetic dipole interacts with the magnetic moment generated by the electronic coherences induced in the presence of the strong static field. For an atom having a spherically symmetric electron distribution, the induced magnetic moment does not change with the motion of the molecule, and hence the interaction cannot give rise to spin transitions. As the electron distribution about the atom deviates from spherical symmetry, the magnitude of the induced magnetic moment becomes more dependent on the orientation of the molecule in the field. The nucleus, then, experiences a fluctuating magnetic field as the molecule tumbles in the static field. When the frequency of the fluctuation matches a resonance frequency, a spin transition can occur with a corresponding change in the rotational angular momentum of the molecule. Equations for $T_1{}^{csa}$ and $T_2{}^{csa}$ are

$$\frac{1}{T_1{}^{csa}} = \frac{[\sigma_z \gamma B_0]^2}{40} \left(1 + \frac{n^2}{3}\right) 6J(f_I) \tag{19a}$$

$$\frac{1}{T_2{}^{csa}} = \frac{[\sigma_z \gamma B_0]^2}{40} \left(1 + \frac{n^2}{3}\right) [3J(f_I) + 4J(0)] \tag{19b}$$

where σ_z is the largest deviation from the average shielding parameter and n is the asymmetry parameter, which is a measure of deviation from axial symmetry in the shielding parameter. In general, relaxation by this mechanism is important when there are only weak dipolar couplings to nearby nuclei or in the slow motion regime. However, the $B_0{}^2$ dependence has made anisotropic chemical shift relaxation of greater importance with the increasing field strengths of the new instruments.

A fifth, less common, mechanism is referred to *spin-rotation relaxation*. At a glance, the concept resembles that of the anisotropic chemical shielding mechanism in that the interaction is between the nucleus and a magnetic moment induced in the molecule. They differ in the source of the induced magnetic moment. In the chemical shielding effect, the magnetic moment arises from coherences imposed on orbital motion of local electrons by the applied magnetic field. With spin rotation, the magnetic moment arises from rotational angular momentum of the molecule and the fact that the molecule is a collection of charged particles. This magnetic moment is independent of the presence of an external field. As the angular velocity of the molecule fluctuates, the magnitude of the magnetic moment will fluctuate and thereby induce nuclear spin transitions. This pathway in spin relaxation is significant only when the other mechanisms are unusually inefficient.

Paramagnetic centers can induce nuclear relaxation by way of direct dipolar coupling (pseudo-contact) or, with covalently attached nuclei, scalar coupling (Fermi contact) between the nuclear magnetic dipole and the magnetic moments of the unpaired electrons. The equations resemble those for the equivalent nuclear-nuclear mechanisms [24].

C. Dynamic Contributions to Spectral Features

Two types of dynamic effects on nmr spectra will be discussed here: those arising from chemical exchange and those from spin relaxation. The term chemical exchange pertains to any structural change giving rise to a new chemical shift for a particular group of atoms. This can be a simple conformational change or it can involve bond breaking. If the chemical exchange frequency is lower than the change in resonance frequency for the nucleus of interest, the process is referred to as *slow exchange* and there will be a discrete set of resonances for each chemical state. In the *fast exchange* limit the exchange frequency is greater than the change in spectral frequency and there is a single average resonance for each unique nucleus.

Intermediate exchange processes have received a great deal of attention because the resonance shapes and positions can be used to estimate the chemical lifetimes of the reactants and products. For a thorough discussion of these processes and data analysis refer to Kaplan and Fraenkel [32], Sandstrom [33], or Oki [34].

When the chemical reaction involves formation of radical intermediates there is the potential for *chemically induced dynamic nuclear polarization* (CIDNP) [35-37]. Recombination of the radicals requires that the unpaired electrons revert to a singlet state. This process can occur by an exchange of spin angular momentum between unpaired electrons and scalar, or hyperfine, coupled nuclei. When this occurs, the nmr spectrum of the recombination products can exhibit orders of magnitude changes in the amplitude of the resonances for hyperfine coupled nuclei and an altered phase with respect to other resonances.

Nuclei with a spin quantum number greater than 1 have an additional contribution to line shape referred to as a *dynamic chemical shift*. The resonance line shape deviates markedly from a Lorentzian when nuclear relaxation is in the slow motion regime [30,31,38]. Although this phenomenon is only observed with some of the more unusual nuclei, it can be an important consideration in either quantitative or qualitative interpretation of resonance line shape analysis.

IV. NMR EXPERIMENTS

A. RF Pulses and Phase Manipulations

In most commercial probes there is a single rf coil that is used to perturb the equilibrium magnetization and to detect the resultant x-y magnetization, as depicted in Figure 1. Figuratively speaking, all modern pulsed experiments require rf transmission and detection in the ±x and ±y directions, not just the single physical axis of this coil. This versatility is accomplished with one coil by controlling the transmitter and detector phase relative to some initial rf pulse in the experiment. The most common example of this operation is quadrature detection used to provide real and imaginary signals, or cosine and sine components of the Fourier transform. The phase cycling for quadrature detection not only removes the spectral "fold-over" problem discussed previously, but also compensates for differences in the two phase-sensitive detectors used and cancels coherent noise from rf sources other than the intended pulse.

Two phase-sensitive detectors, A and B, are used in these experiments with the rf signal split and phase shifted by 90° for the detector referred to as B (Figure 4). The first 90° pulse in a quadrature sequence will be defined as rotating the z magnetization directly onto the positive y axis (Figure 12a) and is assigned a phase of zero. This magnetization then begins precessing into the positive x axis, generating the positive cosine fid for detector A and a positive sine fid for detector B (Figure 12a). The signals from these two detectors will be referred to as A(1) and B(1), where the arguments are the pulse number. After the sample returns to equilibrium (e.g., $5 \times T_1$), a second 90° pulse is generated but using rf that is phase shifted by 90° relative to the continuous rf source. This has the same effect as having a coil oriented along the y axis with the rf magnetic field being defined initially along the negative y axis. The z magnetization rotates into the positive x axis (Figure 12b), and its precession results in a negative sine fid in detector A and a positive cosine fid in detector B (Figure 12b), both of which are locked to the original synthesizer rf. Since the "real" signal should be a positive cosine and the "imaginary" signal a positive sine, as defined by the first pulse, the data from the second pulse are combined with the data of the first pulse as

$$\text{Real data} = A(1) + \varepsilon + B(2) + \delta + \varepsilon$$

$$= A(1) + B(2) + 2\varepsilon + \delta$$

$$\text{Imaginary} = B(1) + \varepsilon + \delta - A(2) - \varepsilon$$

$$= B(1) - A(2) + \delta$$

where ε contains all contributions to the signal before phase-sensitive detection that are not dependent on the rf phase and δ is any distortion in the signal introduced by the 90° phase shifter for detector B. The rf from the source is then phase shifted by 180° relative to that used for the first pulse. It is used to generate a third 90° pulse, which rotates the z magnetization into the negative y axis (Figure 12c). The A(3) signal is, then, a negative cosine fid and the B(3) signal is a negative sine fid (Figure 12c):

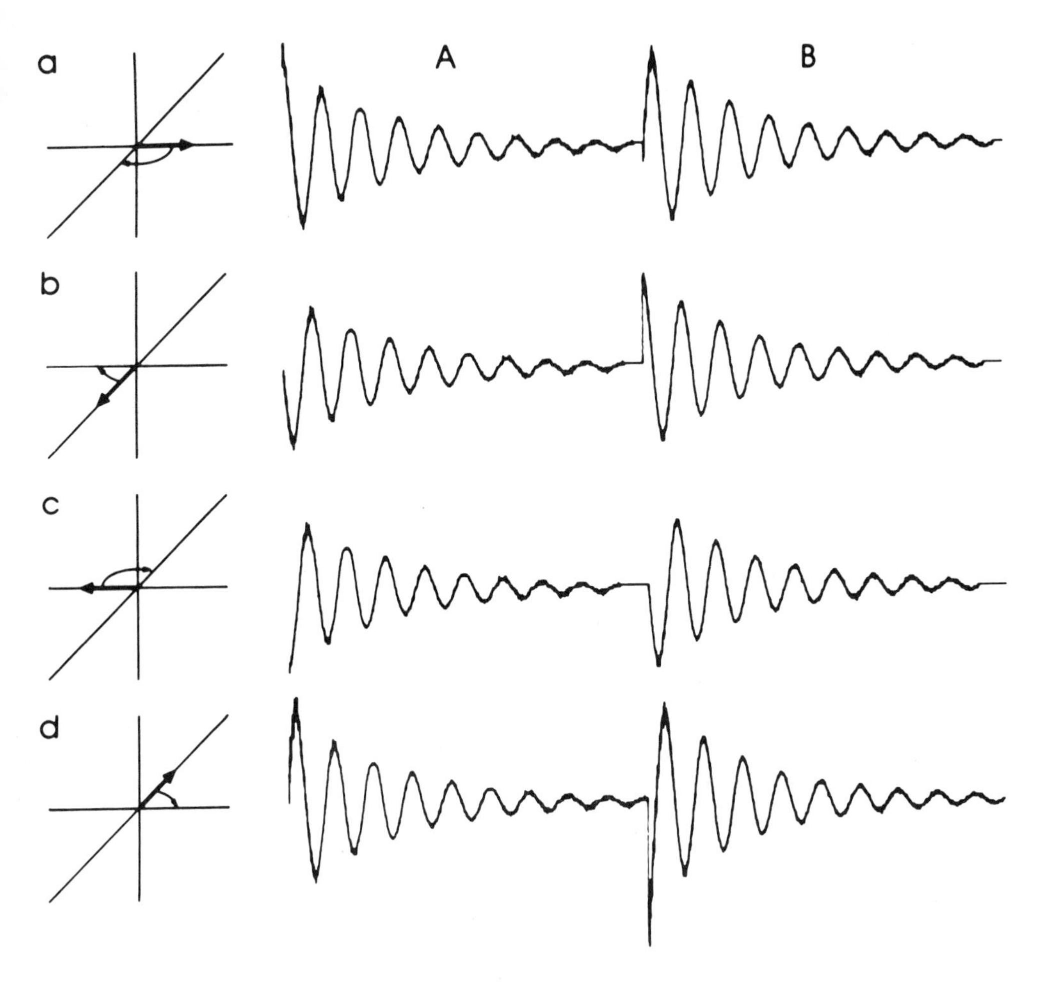

Figure 12. Signal cycle in quadrature detection. (a) The first 90° pulse is along the x axis and is assigned a phase of 0°. The initial y magnetization begins to rotate into the x axis, generating an emf in the rf coil that results in a positive cosine fid at the A detector and a positive sine fid at the B detector. (b) The second 90° pulse is timed to be 90° out of phase with respect to the first pulse and is shown here as having rotated the z magnetization into the x axis. The x-magnetization begins to rotate into the -y axis, resulting in a negative sine fid at A and a positive cosine fid at B. (c) the Third pulse is 180° out of phase with respect to the first and results in a negative cosine fid at A and a negative sine fid at B. (d) The final pulse of the cycle is 270° out of phase and results in a positive sine for A and a negative cosine for B.

$$\text{Real data} = A(1) + B(2) + 2\varepsilon + \delta - A(3) - \varepsilon$$

$$= A(1) + B(2) - A(3) + \delta + \varepsilon$$

$$\text{Imaginary} = B(1) - A(2) + \delta - B(3) - \delta - \varepsilon$$

$$= B(1) - A(2) - B(3) - \varepsilon$$

This cycle is completed with a fourth pulse having the rf shifted 270° relative to the first pulse. The resultant magnetization is along the negative x axis (Figure 12d), and its precession leads to a positive sine fid from detector A and a negative cosine for detector B (Figure 12d):

$$\text{Real data} = A(1) + B(2) - A(3) + \delta + \varepsilon - B(4) - \delta - \varepsilon$$

$$= A(1) + B(2) - A(3) - B(4)$$

$$\text{Imaginary} = B(1) - A(2) - B(3) - \varepsilon + A(4) + \varepsilon$$

$$= B(1) - A(2) - B(3) + A(4)$$

Thus, at the end of the cycle the δ and ε contributions to the original signal have been removed and any differences in gain for detectors A and B have been averaged in both data sets. Signal averaging beyond four acquisitions would involve repeating this cycle from the beginning, with the resultant data sets being

$$\text{Real data} = A(1) + B(2) - A(3) - B(4) + A(5) + B(6) - A(7) - B(8) + \cdots \tag{20a}$$

$$\text{Imaginary} = B(1) - A(2) - B(3) + A(4) + B(5) - A(6) - B(7) + A(8) + \cdots \tag{20b}$$

Many nmr experiments involve two or more pulses preparatory to a single acquisition, often with additional phase considerations. Implementation of quadrature detection may involve more elaborate phase cycles in order to preserve the desired component of the signal.

B. Relaxation Studies

The usual objective in relaxation experiments is to determine the inherent transverse and longitudinal relaxation times for each spectral component. The *transverse relaxation time* is often estimated from resonance line widths, which, for the Lorentzian peak shapes of equation (3), follow the relation

$$\text{FWHM} = (2\pi T_2^*)^{-1} \tag{21}$$

where FWHM is the full width in hertz at half the maximum amplitude. In using this relation to determine T_2, it is generally assumed that all other terms in equation (13) are negligible. This is a good assumption in high-resolution experiments when T_2^* is on the order hundreds of milliseconds or less and it is much less than T_1. For longer-lived signals, field inhomogeneities become more significant and longitudinal relaxation cannot be neglected.

The effect of T_{fh} on the T_2 measurement can be removed by performing a *spin echo experiment*. The simplest form of the echo experiment involves a pulse sequence (PS) composed of two rf pulses:

$90°-t_1-180°-t_1$-detector on (PS1)

where t_1 is a variable delay, which is given values ranging from some fraction of the expected T_2 to roughly $3T_2$. During t_1 the components of the x-y magnetization corresponding to the individual spectral resonances precess at their Larmor frequencies and decay according to their particular values of T_2^* (Figures 13a, b, and c). The 180° pulse has twice the duration of the 90° pulse and causes each component of the magnetization to rotate 180° about the axis of the rf coil, for example, the x axis as shown in Figure 13d. Note that the x components of the microscopic magnetization vectors are unaltered by an rf pulse along the x axis, while the signs on the y components are reversed, and that the angular velocities of the vectors are unaltered. This results in a reversal of the dephasing process dictated by field inhomogeneities (Figures 13d, e and f). The rephasing is complete at a time t_1 after the 180° pulse, yielding an echo of the fid generated by the first 90° pulse. Figure 14 shows the t_1 dependence of the echo and transformed spectra for a spin echo experiment of water. The principal assumption in using this experiment to remove the T_{fh} contribution from the signal decay is that the field gradients within the sample are small enough that the molecules do not diffuse to a radically new field density during one cycle of the pulse sequence. In high-resolution experiments this assumption is valid unless the field homogeneity is intentionally degraded. Ideally, the echo amplitude follows an exponential decay of the form

$$M(2t_1) = -M_0 \exp(-2t_1/T_2') \tag{22}$$

where $M(2t_1)$ is the echo amplitude for delay t_1, M_0 is the equilibrium magnetization (assuming a perfect 90° pulse), and T_2' is the effective transverse relaxation time with T_{fh} removed.

Quadrature detection of this echo signal could be accomplished by phase cycling (QPC) the 90° pulse and keeping the phase of the 180° pulse fixed:

90° (QPC)-t_1-180°(+x)-t_1-detector on (PS2)

The effect of this fixed 180° is to reverse all of the signs on the signal components in equations (21). Another source of artifacts in multipulse experiments is imperfections in the 90° and 180° pulses, such as inhomogeneities in the rf field and incorrect definition of the pulse duration. Field inhomogeneities are always present with rf coils of finite length and diameter, no matter how perfectly the coil is made. As with the static field, the inhomogeneities result in distributions of precession frequencies, which vary across the sample. Even in the absence of these inhomogeneities, definition of a 90° pulse is not an absolute process, since it involves finding the rf pulse that yields the greatest signal, or lowest signal in the case of a 180° pulse. The accuracy of the pulses depends to a large extent on the patience of the operator and the homogeneity of the static field. This process is made more tedious by the fact that the pulse accuracy varies with sample composition, especially with changes in solvent or ionic strength. In this experiment, a constant error in the pulses has no effect on the relaxation time provided the maximum amplitude of the echo is treated as a variable parameter in the data analysis.

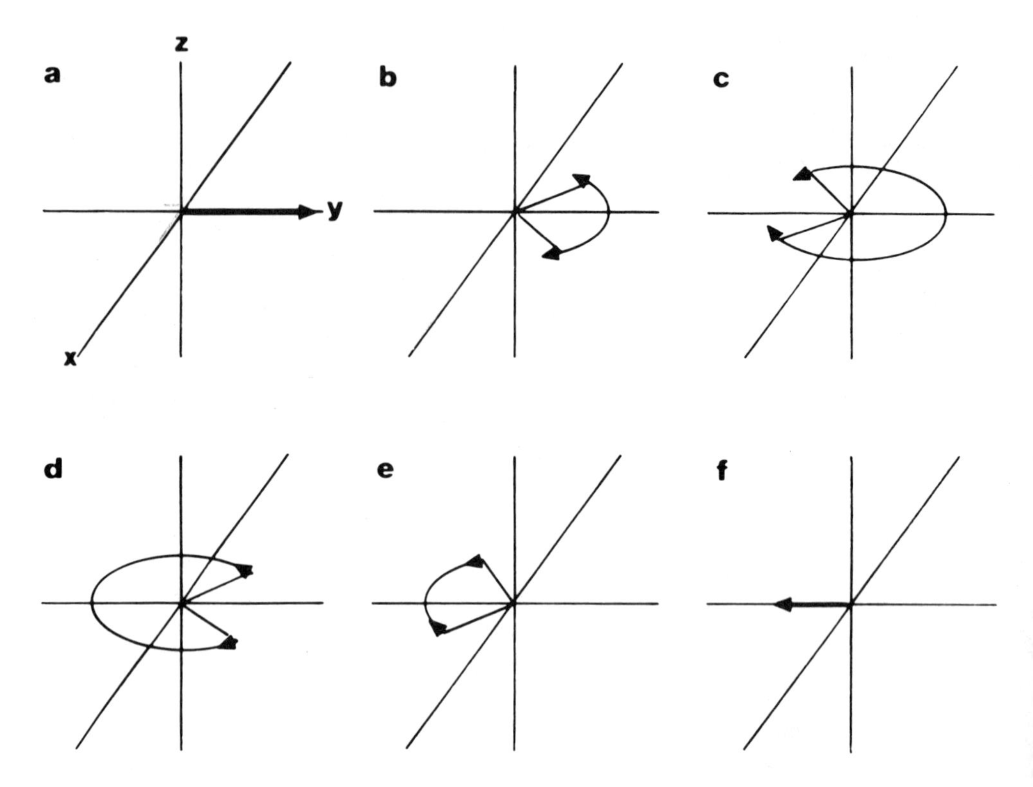

Figure 13. Vector diagram of the spin echo experiment. (a) The 90°(x) pulse rotates the z magnetization into the y axis. Assuming a single resonance, this magnetization can be treated as stationary if a coordinate system is used which is rotating about the z axis at the Larmor frequency. (b) and (c) Although there is a single resonance, small variations in the local field throughout the sample generate a distribution of Larmor frequencies. This results in the net magnetization gradually "fanning out" in the x-y plane of the rotating frame. (d) In this echo experiment, a 180°(x) pulse is applied after some evolution period, t_1. This has no effect on the x components, but all y components of the magnetization undergo a change in sign. Note that the individual components of the magnetization retain their original direction of rotation. (e) and (f) Following this second pulse, much of the "fanning" process is reversed until after a second t_1 interval a maximum -y component exists. The detector is gated on at this moment and a negative signal would be detected. This magnetization has a smaller amplitude than the original as a result of true transverse and longitudinal relaxation during the two t_1 periods.

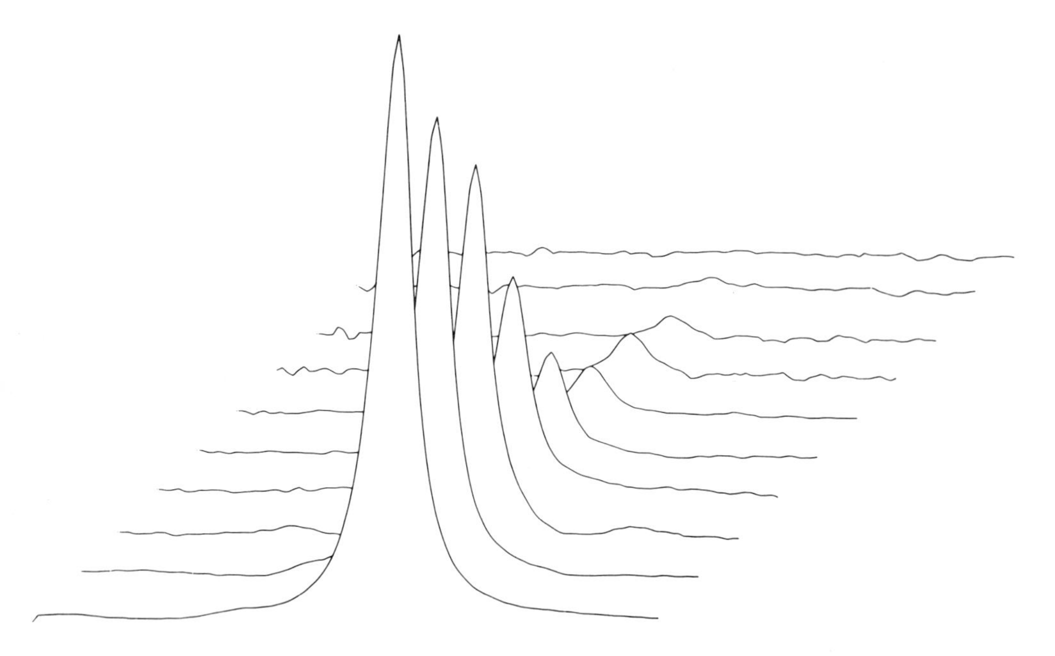

Figure 14. Stacked plot of a 300-MHz ^1H echo experiment. The residual HDO resonance is shown for 99.6% D_2O with increasing values of t_1 in the 90° - t_1 - 180° - t_1 - acquire echo sequence. The resonances were phased positive for display and t_1 is 1 msec for the first spectrum and was incremented by 50 msec in successive experiments.

The echo experiment is usually successful in canceling the T_{fh} contribution to the signal decay, and T_{vf} is negligible for homogeneous liquid solutions of inert substances. This still leaves the T_1 component to be accounted for in the echo decay before T_2 can be determined. A second experiment is required to separate these two relaxation times. A common approach is to measure the longitudinal relaxation by inversion recovery. This experiment involves a pulse sequence that is a simple rearrangement of PS3:

$$90°(x)\text{-}180°(y)\text{-}90°(x)\text{-}t_1\text{-}90°\ (QPC)\text{-detector on} \qquad (PS4)$$

The composite 180° pulse minimizes the effect of errors in pulse duration in rotating the equilibrium magnetization into the negative z axis (Figures 15a, b, and c). This negative z magnetization has only T_1 relaxation pathways available to it; however, it is invisible to the detector. The phase-cycled 90° is used to sample this z magnetization after a variable decay period, t_1 (Figures 15d, e, and f). An example of an inversion recovery experiment is given in Figure 16. Ideally, the amplitude of the signal follows an exponential decay of the form

$$M(t_1) = M_0\ [1 - 2\ \exp(-t_1/T_1)] \qquad (23)$$

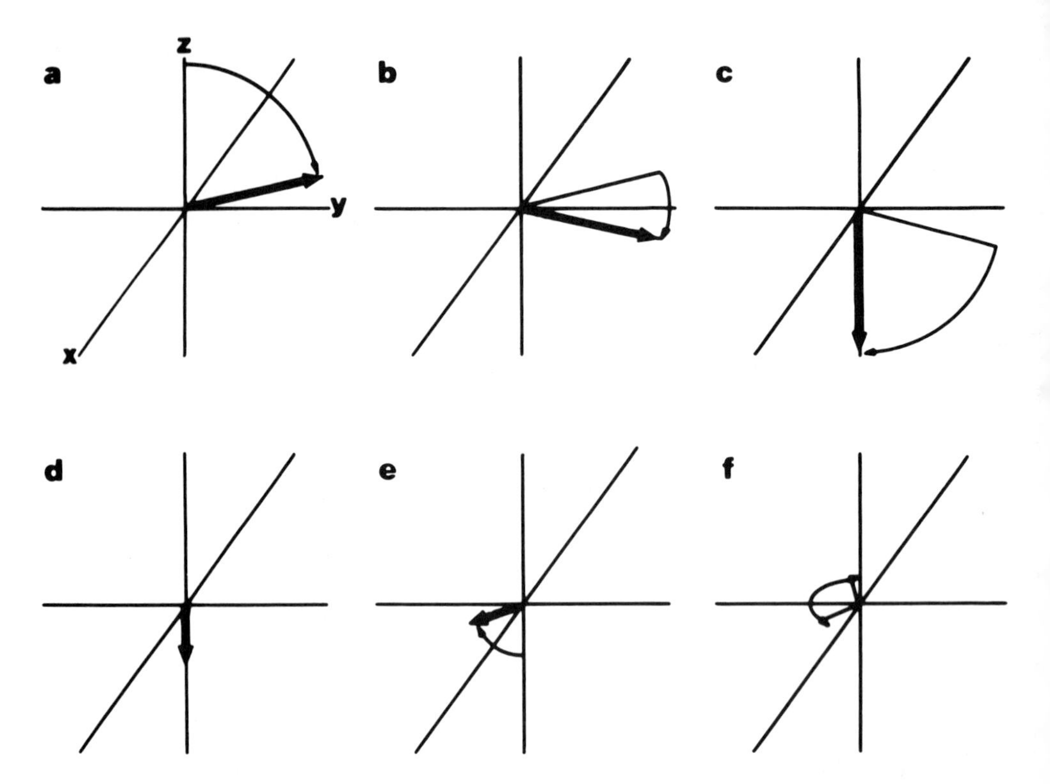

Figure 15. Vector diagram of an inversion recovery experiment using a composite 180° pulse. The 90° and 180° pulses are assumed to have an unintentional error, which can be canceled by using a composite 180° inverting pulse. (a) This composite pulse begins with a 90°(x), which is shown falling short of fully rotating the magnetization into the y axis. (b) The first pulse is followed immediately with a 180°(y) pulse, which changes the sign on any residual z magnetization. (c) A second 90°(x) pulse is applied, completing the inversion process with the error inherent in the two 90° pulses being canceled to a large extent by the effect of the 180°. (d) During the evolution period, t_1, the z magnetization gradually returns to its equilibrium state by way of longitudinal relaxation. (e) and (f) The extent to which the magnetization has recovered after a particular t_1 interval is measured by a final 90° pulse. This rotates the current net z magnetization into the plane of the rf coil for detection.

Log plots or nonlinear regression of the echo and inversion recovery data yield T_2' and T_1, respectively. The intrinsic transverse relaxation is then

$$\frac{1}{T_2} = \frac{1}{T_2'} - \frac{1}{T_1} \tag{24}$$

assuming T_{vf} is negligible and the echo has effectively removed contributions from the field inhomogeneities.

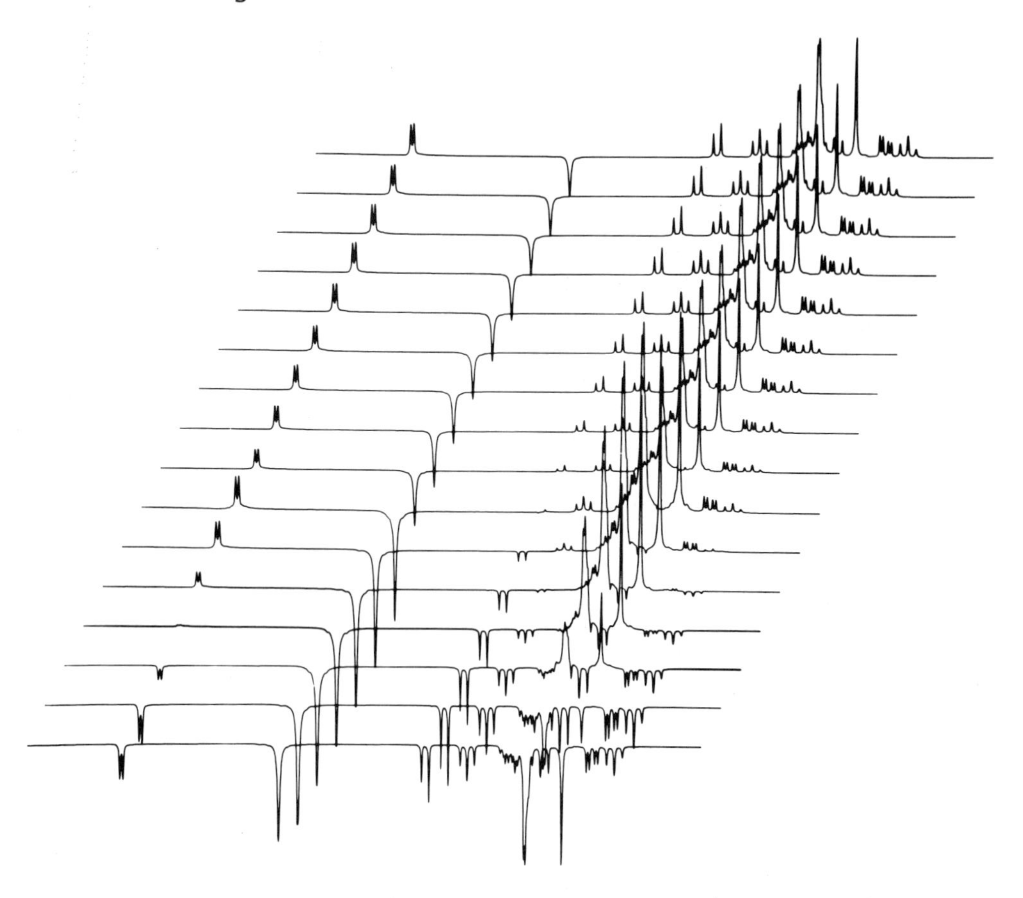

Figure 16. Inversion recovery experiment: 300-MHz ^1H spectra of 0.1 *M* sucrose in 99.6% D_2O. The negative spectrum at the front of the stack was collected with a 1-msec delay (t_1) between the inverting 180° and the sampling 90° pulses and t_1 was incremented by 0.2 sec successively in the remaining experiments. Notice that virtually every resonance has a unique longitudinal relaxation time as is evident by their different null points. The HDO resonance close to the center of the spectrum has the longest T_1 by far, having not reached its null point after 3 sec. By contrast, Figure 14 shows that T_2 for HDO is a factor of 10 shorter—that is, the transverse magnetization decays within 0.4 sec.

It remains to translate these experimental relaxation times into a description of molecular dynamics. Their relations to the microscopic terms given in equations (14-19) are

$$\frac{1}{T_1} = \frac{1}{T_1{}^q} + \frac{1}{T_1{}^{csa}} + \frac{1}{T_1{}^{sr}} + \sum_j \frac{1}{T_{1j}{}^d} + \sum_k \frac{1}{T_{1k}{}^{sc}} \tag{25}$$

$$\frac{1}{T_2} = \frac{1}{T_2{}^q} + \frac{1}{T_2{}^{csa}} + \frac{1}{T_2{}^{sr}} + \sum_j \frac{1}{T_{2j}{}^d} + \sum_k \frac{1}{T_{2k}{}^{sc}} \tag{26}$$

where the first summation is over all dipolar couplings to the atom of interest and the second is over all scalar couplings to that atom that satisfy one of the two conditions for scalar relaxation. Assuming the molecule tumbles as a simple sphere, there is a single motional parameter, the rotational correlation time. Besides this, the quadrupolar [equations (14) and (15)] anisotropic shielding [equations (19)], and spin rotation terms each require two intrinsic parameters. The dipolar term requires the set of distances between the atom of interest and all dipolar coupled atoms [equations (16) and (17)], and the scalar coupling term requires the set of scalar coupling constants and associated kinetic parameter [equations (18)]. Rarely are enough of these parameters known so that only two measured relaxation times can be used to determine the rotational correlation time. Fortunately, equations (25) and (26) are often dominated by one mechanism.

Nuclei with spin greater than 1/2 usually relax by way of quadrupolar coupling, unless the surrounding electronic environment is spherically symmetric. This reduces the parameter list to the rotational correlation time, the quadrupolar coupling constant, and the coupling asymmetry parameter, which is one more than can be defined from two measurements. In some cases, the electron density has axial symmetry, which results in a zero asymmetry parameter. The normal single-bonded state of deuterium falls into this category: hence its relaxation is one of the simplest to represent microscopically, that is, T_1 and T_2 measurements are adequate for simple spherical rotation. Quadrupolar nuclei with more than one bond may also have axial symmetry; if not, at least one other type of experiment must be performed to define all three parameters. Nuclear quadrupolar resonance (nqr) on solid samples provides a direct measure of the quadrupolar coupling constant and the asymmetry parameter. Given this information, only one relaxation study is required to determine the rotational correlation time for the spherical model. If this information is not available, these three parameters can be determined by performing the relaxation experiments at a second field strength, which changes the contributions from the three spectral density functions.

Relaxation of spin 1/2 nuclei is usually dominated by dipolar couplings if there are other nmr-active nuclei within a few angstroms. This can be a very useful source of structural as well as dynamic information, since the couplings decrease rapidly with distance. The multiplicity of couplings, however, can also cause the model to be intractable. Directly bonded nuclei will dominate the relaxation, reducing the dipolar sum to only one or two terms for a given resonance. For example, ^{13}C relaxation can often be described in terms of the number of hydrogens directly bonded to the carbon. In this case the dipolar distances are simply the bond lengths, and as the number of couplings is known, only the correlation time remains to be determined.

Chemical shielding anisotropy becomes important as a relaxation mechanism as the electron distribution about the nucleus of interest deviates from spherical symmetry. It can be more significant than dipolar relaxation when the nucleus of interest has only spin zero nuclei directly bonded, such as the phosphorus of a phosphate with only ^{16}O atoms or a ^{13}C atom in a carbonyl position with ^{16}O and ^{12}C atoms bonded to it. This is particularly true as the motion of the molecule becomes slower than the resonance frequency of the nucleus. If the electron density has axial symmetry, then there is a single shielding parameter and a single motional parameter, both being determined by the two relaxation measurements. Otherwise, there is also an asymmetry parameter. This third parameter can be measured by solid-state nmr experiments, similar to the nqr experiments discussed with quadrupolar relaxation. Very often, when chemical shielding anisotropy is important in relaxation there is a contribution from dipolar couplings as well. These dipolar couplings may not be to directly bonded atoms, thereby complicating the model. The dipolar and chemical shielding anisotropy contributions are best separated by performing relaxation experiments at two or more field strengths.

The motions of large molecules can be much more complex than the rotation of a simple sphere [29]. As mentioned in the theory section, the spectral density functions will have additional terms for motion deviating from spherical rotation. Studies of this type of system are usually restricted to the atoms within the molecule that have simple and well-defined relaxation pathways. The motional parameters can be defined by performing relaxation experiments at several field strengths and by combining the results from several different atoms within the molecule. In general, it is difficult to justify more than a three-parameter model for the motion.

C. Double Resonance

The phrase "double resonance" refers to experiments in which two different rf sources are employed. Examples of this class are homo- and heteronuclear decoupling, nuclear Overhauser experiments, spin locking, attached proton tests (APT), and insensitive nuclei enhancement by polarization transfer (INEPT). There are also several two-dimensional experiments in this category and related multipulse experiments, some of which are listed in Table 3. The second source of rf in the double resonance is customarily referred to as the decoupler because of the most common purpose for which it is used, that is, to remove scalar coupling effects. As discussed in the instrumental section, modern commercial instruments have as standard equipment a separate ^{1}H rf source and a separate coil designated as the decoupler within most probes. The choice of ^{1}H as standard is a consequence of prevalent couplings with this nucleus, such as ^{1}H-^{1}H, ^{1}H-^{13}C, ^{1}H-^{31}P, and ^{1}H-^{15}N couplings. Double-resonance experiments with other combinations of isotopes require optional synthesizers and probes. Instruments of the current generation have the decoupler power, duration, and phase under control of the pulse programmer, which is required for some of the two dimensional experiments.

The *decoupling* experiments are intended to collapse scalar coupling multiplets by generating rapid fluctuations in the spin state of one of the coupled nuclei. The fluctuations result in the unperturbed nucleus experiencing an averaged spin state for the perturbed nucleus, causing the effective scalar coupling to average to zero. *Homonuclear decoupling* usually involves selective irradiation of one resonance in the spectrum with collapse of only those resonances to which it is coupled. This can be a useful tool in making resonance assignments or even simplifying crowded spectral regions. Examples are given in Figure 17. Since homonuclear decoupling

Table 3. Multipulse and Two-Dimensional NMR Experiments

Description	Reference
Multipulse	
APT Attached proton test, used to distinguish even and odd numbers of protons one-bond scalar coupled to another isotope, usually ^{13}C. The spectrum usually shows all resonances but even numbers of bound protons yield positive peaks and odd numbers yield negative peaks in most examples.	41
DEPT Distortionless enhancement by polarization transfer, used to define the number of bound protons, as with APT. The pulse conditions are chosen so that only those nuclei with the same number of bound protons will have enhanced resonances. Thus four experiments must be performed in ^{13}C studies to unambiguously define all resonances, although it is more definitive than APT.	49
INADEQUATE Incredible natural abundance double quantum transfer experiment, most commonly used to observe natural abundance ^{13}C-^{13}C, or scalar couplings among any other low natural abundance nuclei. This is "incredible" given that only one carbon in 10^4 can be ^{13}C bonded directly to another ^{13}C.	50,51
INEPT Insensitive nuclei enhancement by polarization transfer, an earlier version of DEPT. Both can be used in determining numbers of bound protons and instead of NOE enhancement in simple spectra, for example, nuclei other than ^{13}C that have less diversity in numbers of bound protons.	52
RELAY Relayed correlation spectroscopy, used to establish connectivities among nuclei separated by several bonds. Although the two nuclei need not be directly scalar coupled, there must be an unbroken chain of intervening couplings.	53
1331 A solvent suppression sequence used as a 90° pulse for all but a single resonance at the central frequency.	54
Two-Dimensional Homonuclear Scalar Coupling	
COSY Correlated spectroscopy, provides connectivities of scalar coupled resonances by way of polarization transfer, chemical shift vs. chemical shift.	55
HDCOSY Homodecoupled correlated spectroscopy, a variation on COSY, which provides a homodecoupled spectrum along the f_1 axis, chemical shift vs. chemical shift.	56
SECSY Spin-echo correlated spectroscopy, essentially the same information as COSY but the data are displayed differently, chemical shift vs. chemical shift.	57

(Table 3 continued)

Description	Reference
MQF-COSY Multiple quantum filter COSY can be used to suppress singlet diagonal peaks and to selectively observe p^{-th} order couplings, chemical shift vs. chemical shift.	57
FOCSY Fold-over corrected spectroscopy, a variation on COSY, which provides better contour resolution, chemical shift vs. chemical shift.	58
CCCP Carbon-carbon connectivity plot, two-dimensional equivalent of INADEQUATE, the contours indicate natural abundance ^{13}C-^{13}C scalar couplings, ^{13}C chemical shift vs. ^{13}C chemical shift.	50,51
J-Resolved spectroscopy provides scalar coupling constants but not connectivities, scalar coupling constant vs. chemical shift.	48
Heteronuclear Scalar Coupling	
Heteronuclear chemical shift correlated experiment, usually ^{1}H-X double resonance with observation of X rf, contours connect ^{1}H resonances with scalar coupled X resonances, ^{1}H chemical shift vs. X chemical shift.	59
Relayed coherence transfer spectroscopy, two-dimensional equivalent of INEPTR, provides connectivities of ^{13}C resonances with those of ^{1}H directly bonded and ^{1}H's scalar coupled to the directly bonded ^{1}H's, ^{13}C chemical shift vs. ^{1}H chemical shift.	60
Nuclear Overhauser Effect (NOE)	
NOESY provides resonance connectivities based on cross relaxation, the dipolar coupled nuclei must be within roughly 0.4 nm for first-order couplings, chemical shift vs. chemical shift.	47
Small flip angle NOESY provides the means to determine the zero-, single-, and double- quantum contributions to the dipolar relaxation of scalar coupled nuclei even where the cross relaxation (NOE) is zero as a result of $2\pi f t_c$ being near unity; chemical shift vs. chemical shift.	61
Chemical Exchange	
ACCORDIAN spectroscopy, essentially a variation on the NOESY experiment, rate constant vs. chemical shift.	62
zz-Exchange spectroscopy, distinguishes between chemical exchange and NOE, chemical shift vs. chemical shift.	63

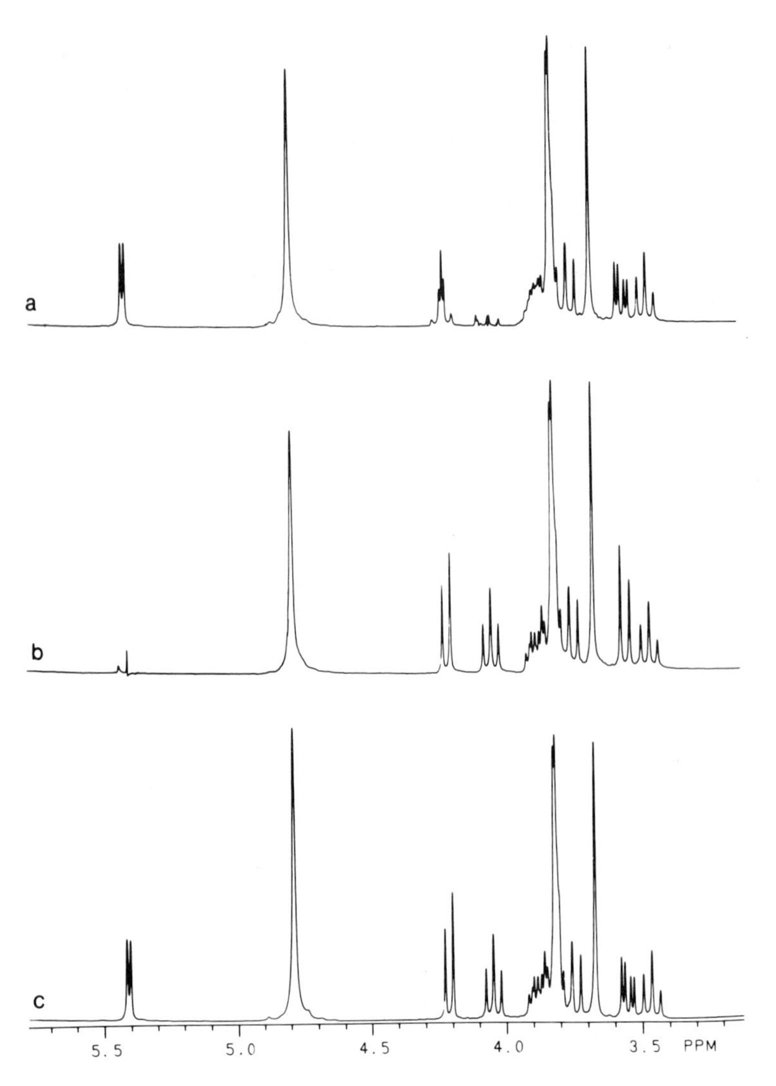

Figure 17. Homonuclear decoupling of the 300-MHz ^1H spectrum of sucrose in D_2O. (a) Selective saturation of the triplet at 4.05 ppm partially collapses the doublet at 4.22 ppm. (b) Saturation of the doublet at 5.41 ppm collapses the doublet of doublets at 3.55 ppm, leaving a doublet. (c) Fully coupled spectrum. The singlet at 4.8 ppm is residual HDO.

must be performed during data acquisition, the principal problem with this type of experiment is separating decoupling rf from the signal. This is accomplished by using a fast sample-and-hold-circuit for the A/D converter and gating the decoupler on between samplings. Technically the decoupler should be on during data sampling in order to completely decouple the resonances. However, the evolution of the couplings in the unperturbed resonance while the decoupler is gated off is usually insignificant compared with the resonance line widths, thus obscuring the residual coupling.

Heteronuclear decoupling experiments can involve either selective irradiation of, say, a particular ^1H resonance with observation of multiplet collapse in the heteroatom (X) spectrum, or irradiation of the entire ^1H spectrum and collapse of all ^1H-X couplings. The selective experiment can be used for resonance assignments in either spectrum, and the total decoupling provides a simpler heteroatom spectrum with an increase in sensitivity by compression of multiplets to single resonances (Figure 18). Neither approach presents a problem in decoupling during acquisition since the two rfs differ by several megahertz. The principal problem is having adequate decoupling power over the entire ^1H spectrum with essentially continuous operation. This is accomplished by either amplitude or phase modulation of a continuous monochromatic rf source of moderate power (several watts). Modulation has the same effect on monochromatic rf as pulsing, in that it introduces a frequency distribution of the power about the original rf. The shape of the distribution depends on the particular type of modulation, and the power requirements depend on the magnitude of the scalar couplings and the spectral width of the irradiated species. Two very successful phase modulation techniques are MLEV-16 and WALTZ-16 [39]:

$$\text{MLEV-16:}\quad RR\underline{RR}\ \underline{R}RR\underline{R}\ \underline{RR}RR\ R\underline{RR}R \tag{PS4}$$

$$R = 90°(x)\text{-}270°(y)\text{-}90°(x)$$

$$\underline{R} = 90°(-x)\text{-}270°(-y)\text{-}90°(-x)$$

$$\text{WALTZ-16:}\quad \underline{3}4\underline{2}3\underline{1}2\underline{4}2\underline{3}\ 3\underline{4}2\underline{3}1\underline{2}4\underline{2}3\ 3\underline{4}2\underline{3}1\underline{2}4\underline{2}3\ \underline{3}4\underline{2}3\underline{1}2\underline{4}2\underline{3} \tag{PS5}$$

where each numeral in the WALTZ-16 sequence represents that number of 90°(x) pulses from the decoupler and underscored numerals correspond to pulses with a -x phase relation. The MLEV-16 pulse interposes a 270° decoupler pulse.

The *nuclear Overhauser effect* (NOE) derives from spin relaxation by way of direct dipolar coupling and is also referred to by the term *cross polarization* [40]. The zero- and double-quantum transitions for a dipolar coupled pair of nuclei provide the means by which a transition in one nucleus induces a transition in the second nucleus (Figure 11b). The conventional NOE experiments involve selective irradiation of one spin population for a time on the order of T_1 and observing changes in the second spin population. In homonuclear experiments a single resonance is irradiated, or saturated, with the decoupler hardware before the 90° pulse. Saturation of a resonance involves irradiation with enough power to maintain equal populations in the spin states associated with that resonance. An example of this type of experiment is given in Figure 19. The fractional change in a nonirradiated resonance, B, as a result of saturation of resonance A is given as

$$\frac{B(t)}{B(0)} = \frac{W_2 - W_0}{2W_1 + W_2 + W_0}\{1 - \exp[(2W_1 + W_2 + W_0)/t]\} \tag{27a}$$

$$W_0 = \frac{(\gamma_A \gamma_B)^2}{10 \quad r^6} \quad \frac{t_c}{1 + (f_A - f_B)^2 \ t_c^2} \qquad (27b)$$

$$W_1 = \frac{3 \ (\gamma_A \gamma_B)^2}{20 \quad r^6} \quad \frac{t_c}{1 + (f_B t_c)^2} \qquad (27c)$$

$$W_2 = \frac{3 \ (\gamma_A \gamma_B)^2}{5 \quad r^6} \quad \frac{t_c}{1 + (f_A + f_B)^2 \ t_c^2} \qquad (27d)$$

where t is the period during which the A resonance is saturated, B(t) is the area of the B resonance at the end of the saturation period, B(0) is the area when A is not saturated, γ_A and γ_B are the gyromagnetic

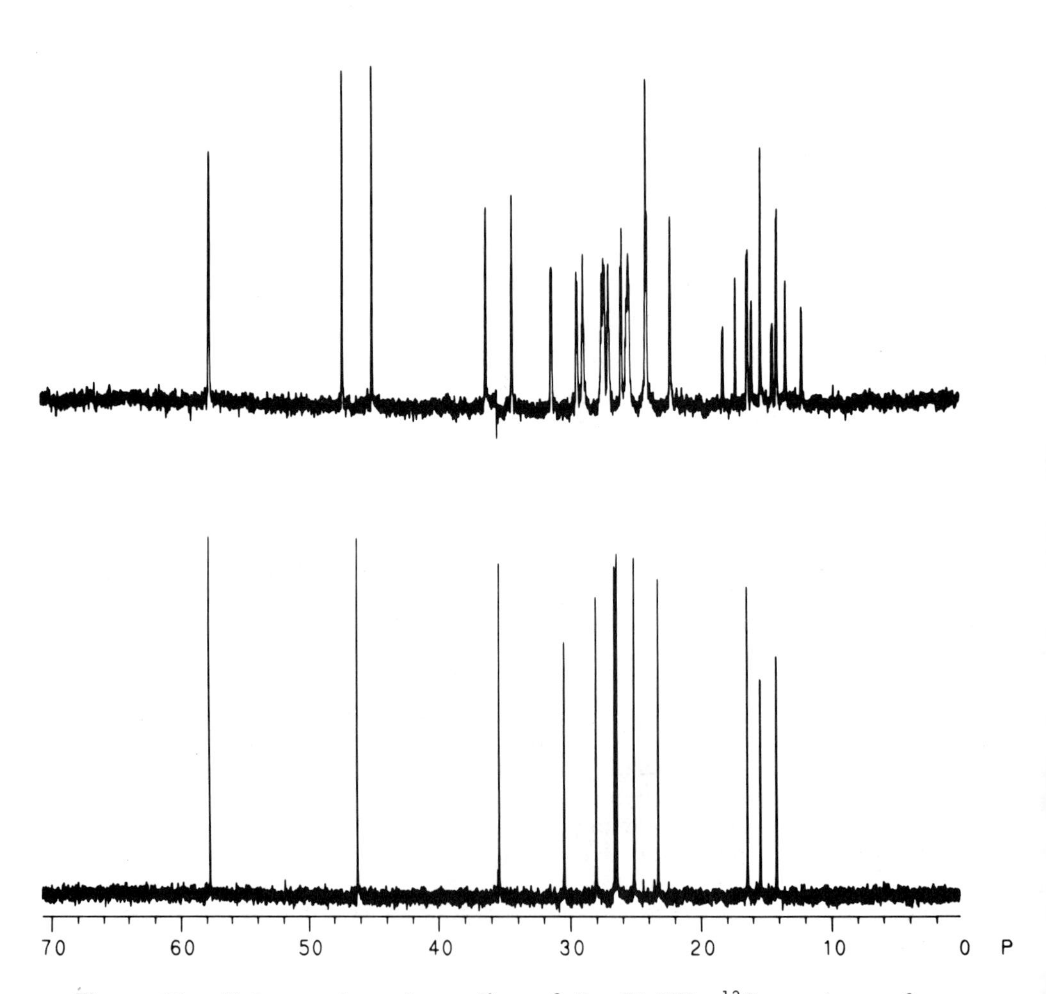

Figure 18. Heteronuclear decoupling of the 75-MHz ^{13}C spectrum of sucrose: broadband ^1H decoupled spectrum (bottom) and fully coupled spectrum (top). Both were collected with NOE enhancement.

ratios, r is the distance between the two nuclei, f_A and f_B are the resonance frequencies, and t_c is the rotational correlation time. These equations assume spin 1/2 nuclei and spherical rotation. Note that there is no NOE when the zero- and double-quantum transitions have equal probability, that is, $W_2 = W_0$. In homonuclear experiments where f_A and f_B are approximately equal, this occurs when

$$f = \frac{5^{1/2}}{2t_c} \qquad (28)$$

Thus, for a given field strength certain motions will not yield an NOE or for a given correlation time there is a field at which the NOE is zero. In addition, the change in area of B is positive for "fast" motions and/or at high field strengths (large f), whereas it is negative for slow motions and/or low field strengths.

There are two practical applications of the nuclear Overhauser effect. In heteronuclear experiments the NOE is used to increase the signal amplitude of low-sensitivity nuclei by cross-polarization with ^1H when the sign on the change is positive. This is used very often in ^{13}C experiments where directly bonded protons are strongly coupled. Saturation of the proton resonances can be accomplished by continuous decoupling using modulated rf. Under ideal conditions the NOE can yield a factor of 2 increase in the ^{13}C peak intensities, which is a factor of 2 increase in the signal-to-noise ratio or a factor of 4 decrease in the time required to collect a usable spectrum (Figure 20). In general, the maximum NOE between any two isotopes is given as

$$\text{NOE} = \frac{\gamma\ (\text{observed})}{2\gamma\ (\text{saturated})} + 1 \qquad (29)$$

where the NOE is defined as the original resonance area plus the fractional change in area.

The second general application deals with measuring distances between nuclei. According to equation (27a), $W_2 - W_0$ and $2W_1 + W_2 + W_0$ can be determined by measuring the peak area of B as a function of saturation time. Data reduction requires a nonlinear regression. These two experimental parameters are sufficient for determining both t_c and r, provided that equations (27b)-(27d) hold. If the motion is more complex than that of a rotating sphere, the additional terms required to describe the motion are determined by combining results from other relaxation studies.

Spin locking experiments can provide a fourth type of relaxation experiment. They involve a 90° pulse along, say, the x axis followed by a very long pulse of moderate power along the y axis. Recall that the axis is chosen by the relative phase of the pulses and not by two orthogonal coils. The long pulse "locks" the magnetization along the y axis for its duration. This magnetization decays to zero by a mechanism referred to as longitudinal relaxation in the rotating frame. This relaxation is exponential and, assuming a homogeneous rf field, has a decay time T_{1r} that is very sensitive to slow motions. Thus, T_{1r} measurements are useful in supplementing T_1 and T_2 data for molecules with complex motion such as polymers.

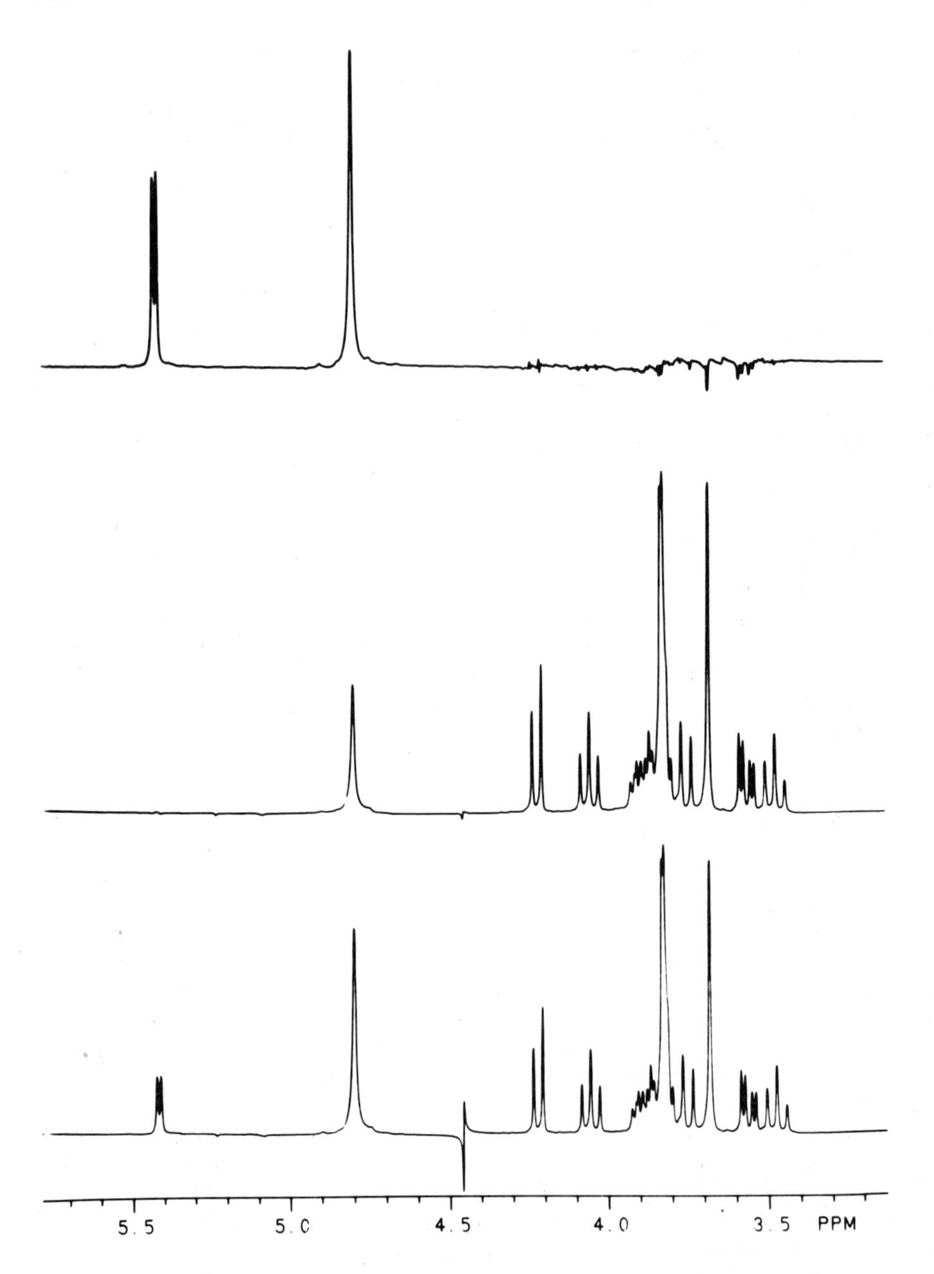

Figure 19. The 300-MHz ^1H homonuclear NOE experiment with sucrose in D_2O. The bottom spectrum is that of sucrose with the decoupler frequency set to the center of the spectrum where there are no resonances. For the middle spectrum, the decoupler frequency was set to coincide with the anomeric resonance at 5.41 ppm and the decoupler power was applied for 0.5 sec just prior to the sampling 90° pulse. In this case the anomeric peaks are not present because the populations of the spin states for these

A second application of spin locking is in optimizing the nuclear Overhauser effect, or cross-polarization, for signal enhancement in solids. In these experiments, the objective is to satisfy the Hartmann-Hahn condition for two different isotopes, such as ^1H and ^{13}C, which has the form

$$\gamma(^1H)B_{rf}(^1H) = \gamma(^{13}C)B_{rf}(^{13}C) \tag{30}$$

where the γ terms are the gyromagnetic ratios and the B_{rf} terms are the applied rf field strengths for the respective nuclei. Experimentally, this condition is met by a 90° pulse for the ^1H followed immediately by the spin-locking pulse of field strength $B_{rf}(^1H)$, all performed with the decoupler hardware. During the spin lock period, a ^{13}C rf pulse is applied through the primary rf hardware with a field strength satisfying equation (30). The Hartmann-Hahn condition provides a maximum rate of cross-polarization, which in this experiment results in rotation of ^{13}C magnetization into the x-y plane with the full NOE between ^1H and ^{13}C. At the end of the ^{13}C pulse the ^{13}C fid is collected. If T_{1r} for the ^1H atoms is longer than this acquisition period, then several ^{13}C pulse-fid cycles can be performed during a single ^1H spin-locking period. This process provides a substantial reduction in signal averaging time required for solid-state experiments, but the demands on the ^1H decoupler power are usually too great for the standard decoupler hardware. The experiments are usually performed with a special high-power decoupler source and probes designed to accommodate it.

The *attached proton test* (APT) provides an alternative to selective heteronuclear decoupling experiments in making resonance assignments [41]. This experiment combines broadband proton decoupling with an echo-type pulse sequence for the heteronucleus:

^1H: [BB-NOE]-[off]-[BB-decoupling]

^{13}C: D1-90°-D2-180°-D2-D3-180°-D3-detector on (PS6)

where D1 is a relaxation delay for carbon during which broadband proton decoupling is used to develop NOE enhancement. During the first D2 interval, the decoupler is gated off and the components of the ^{13}C magnetization evolve with coupling multiplicities. The decoupler is then gated on for the remainder of the pulse cycle. This "freezes" the relative phase of coupled magnetization components generated during D2 and yields a final spectrum that is proton decoupled. The first 180° pulse is used to

---

nuclei have been forced to be equal by irradiation with the decoupler. During the 0.5-sec irradiation period, cross-relaxation occurs between the anomeric protons and any protons within about 0.4 nm. This results in a change in area for the corresponding resonances, which is seen more readily by subtracting the middle spectrum from the bottom spectrum. The difference is shown at the top. There are very small, negative NOEs to the doublet of doublets at 3.55 ppm and the singlet at 3.68 ppm. These are confirmed in a two-dimensional experiment (Figure 24). There are several other small features in this difference spectrum that are most likely a result of fluctuations in temperature during the experiment with corresponding small shifts in the resonance positions.

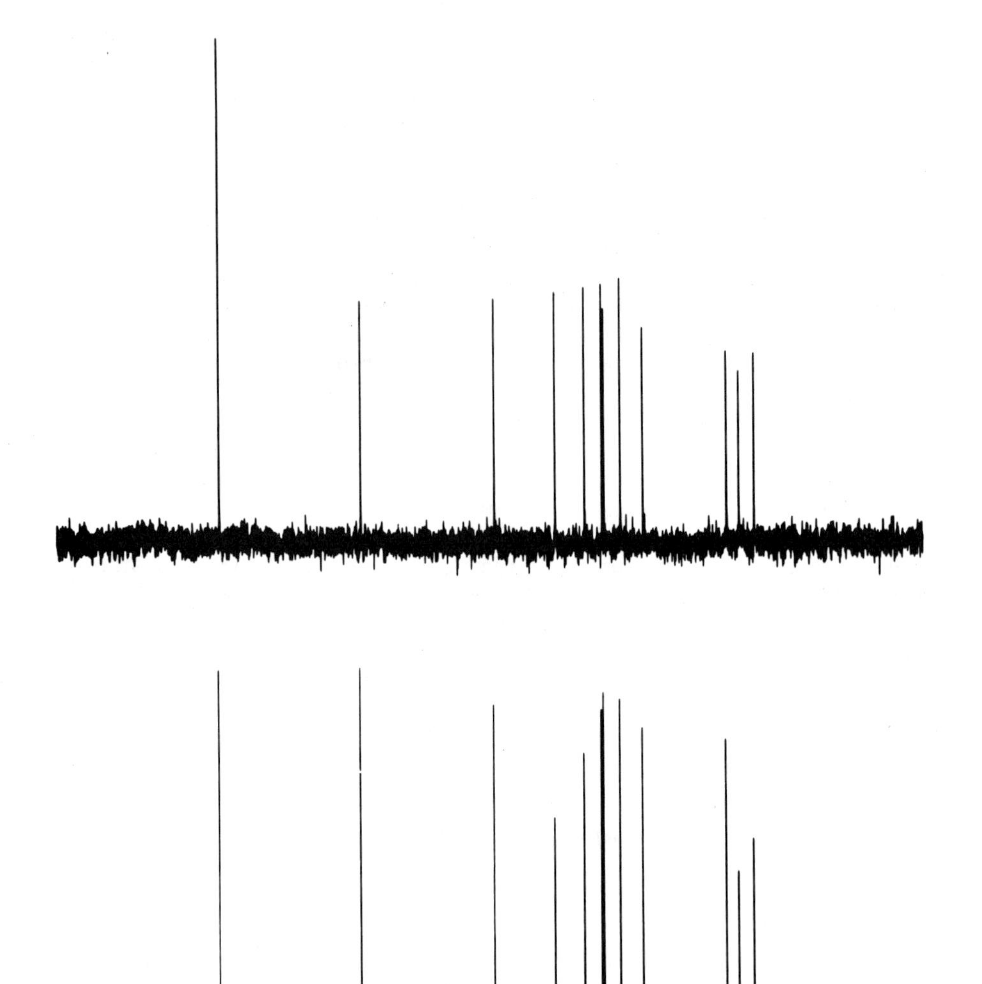

Figure 20. The 75-MHz ^{13}C spectrum of sucrose. For the bottom spectrum the ^{1}H decoupler was on during the entire experiment in order to generate the maximum nuclear Overhauser effect and to decouple the ^{1}H-^{13}C scalar interactions. The decoupler was on only during acquisition for the top spectrum with minimal enhancement. In both cases a 10-sec delay between pulses was used.

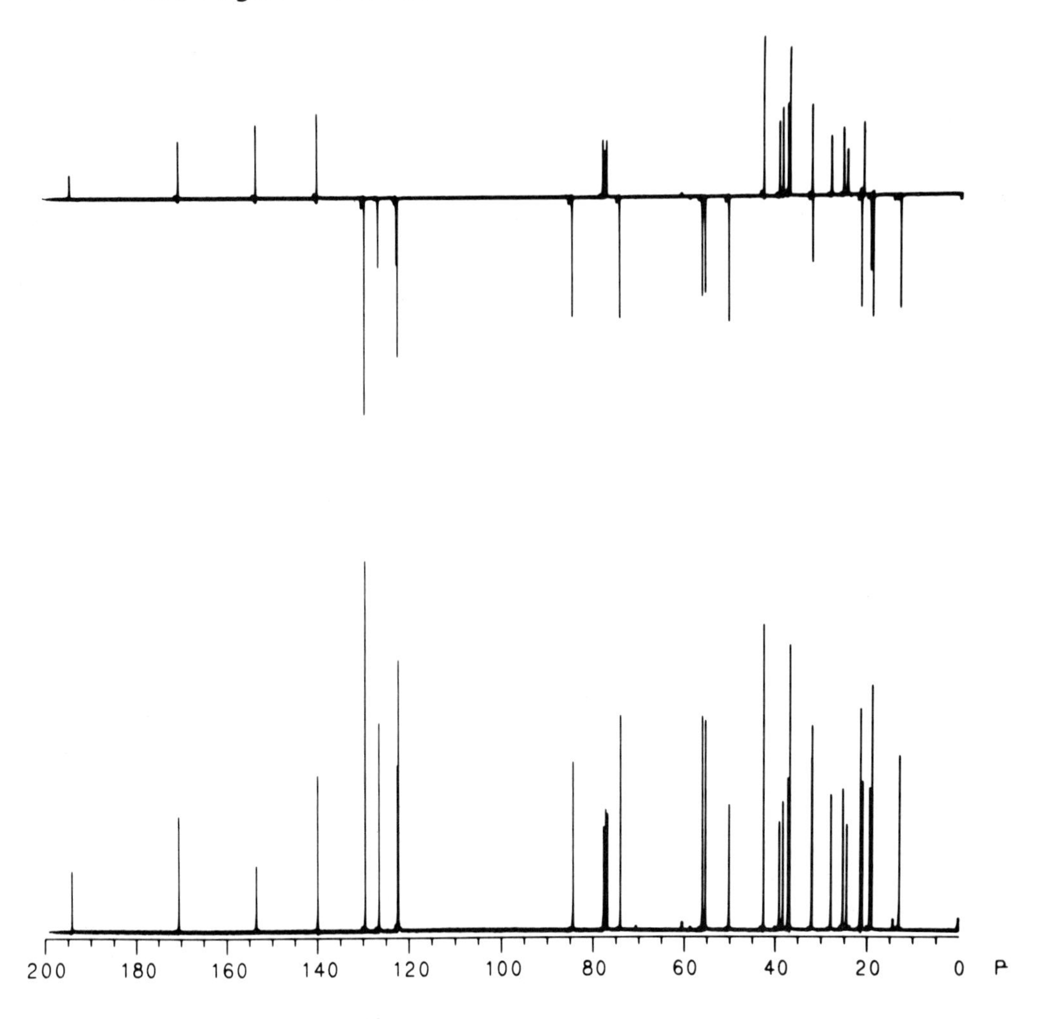

Figure 21. The 75-MHz ^{13}C APT experiment. The top spectrum of 3α-acetoxy-5'-pregnen-20-phenylthiocarbamate was collected using the APT pulse sequence with D1 = 7.5 msec (PS6). Positive resonances correspond to carbons with zero or two protons bound, and negative resonances are carbons with one or three protons. The normal decoupled spectrum is beneath it.

refocus the centers of the magnetization components generated during the coupled period and the second 180° pulse re-establishes the phase of the entire spectrum. An example of an APT experiment is given in Figure 21 where the positive ^{13}C resonances correspond to carbons with zero or two bound protons and the negative peaks are for one or three bound protons. This experiment can be very useful in assigning complex ^{13}C spectra.

The last type of double resonance experiment to be discussed here is exemplified by INEPT. This is one of a class of multipulse experiments that can be understood only with careful attention to spin dynamics. The INEPT experiment is similar to the nuclear Overhauser effect except that the polarization transfer occurs by way of scalar coupling rather than dipolar relaxation. The following is one version of the pulse sequence in which the decoupler hardware is used for the ^{1}H rf and ^{13}C is the insensitive nucleus:

^{1}H: 90° (x)-D1-180°(y)-D1-90°(±y)-D2-180°(y)-D2-90°(x)

^{13}C: 180°(y) 90°(x) 180°(y) detector on

D1 = 1/(4J) (PS7)

where J is the scalar coupling constant for which polarization transfer is to be optimized and D2 is a multiple of D1 depending on the number of protons scalar coupled to the ^{13}C atom of interest. INEPT has been shown to be more effective in enhancing resonances than NOE in cases where the T_1 of the heteroatom is long, since the recycle time of the INEPT experiment can be shorter than T_1.

D. Two-Dimensional Experiments

The fundamental concepts behind two-dimensional nmr experiments were presented by Jeener in 1971 [42] and implemented by Ernst in 1974 [43]. Details of the very powerful techniques that have evolved since then are presented by Ernst, Bodenhausen, and Wokaun [6], Bax [44,45], and Turner [46]. There are three general categories of these experiments described in terms of the two experimental axes:

1. chemical shift versus time
2. chemical shift versus chemical shift
3. coupling constant versus chemical shift

The relaxation experiments presented in Section IV.B are representative of the first category. Discussion in this section will be restricted to the second and third classes. Most of these experiments rely on some type of coupling of one spectral feature with another. This can be in the form of scalar coupling, the nuclear Overhauser effect, or even slow exchange between two chemical states. Table 3 contains an abridged compilation of two-dimensional experiments listed according to the phenomenon giving rise to resonance coupling. In order to give the reader a general feeling for these very powerful techniques, two experiments will be discussed in detail: nuclear Overhauser effect spectroscopy (NOESY) and J-resolved spectroscopy.

NOESY is a two-dimensional nuclear Overhauser experiment developed by Macura and Ernst [47]. Unlike the conventional NOE experiment, it does not involve double resonance. Instead, "selective" perturbation of spin populations is accomplished by the timing of three 90° pulses:

$90°_1-t_1-90°_2-t_m-90°_3$-detector on ($t_2$) (PS8)

where quadrature phasing has been ignored in the labels but is normally performed. This is a great boon in NOE studies of crowded spectra, since the resonances need not be well resolved, as they must be in order to obtain unambiguous results with the double-resonance experiment. The two-dimensional experiment involves collecting signal-averaged fids for on the order of a thousand different values of t_1, where the maximum value of t_1 is defined by the desired spectral resolution. Cross-polarization occurs during the mixing period, t_m, which is generally less than the T_1 times of the resonances of interest and is held constant for the entire set of experiments. The first step in data processing involves Fourier transforming the fid collected for each value of t_1; this is referred to as the transform from the t_2 to the f_2 domain. At this stage, there is the equivalent of a conventional nmr spectrum of the sample for each value of t_1; however, the amplitudes of the resonances oscillate as t_1 is varied. The second transform is performed by selecting a particular value of f_2 and transforming as a function of t_1 to yield the f_1 domain.

In order to demonstrate the manner in which this pulse sequence works, a hypothetical spectrum will be considered consisting of two resonances, A and B. Resonance A is given a frequency, f_A, of 10 Hz relative to the phase-sensitive detector reference and B is at 40 Hz (f_B). First it will be assumed that the A and B spin populations are *not dipolar coupled*. Following the first 90° pulse the B magnetization leads the A magnetization as they precess about the z axis (Figure 22). Assuming all pulses have +x phase, $90°_2$ rotates y components of the A and B magnetization into +z axes, leaving the residual x-y magnetization along +x. Figure 22 shows the effect of $90°_2$ after a t_1 value of 0.01 sec. The general expressions for the z magnetizations following this pulse are

$$A_z(t_1,2) = -A_0 \cos(2\pi f_A t_1) \exp(-t_1/T_2^*) \tag{31a}$$

$$B_z(t_1,2) = -B_0 \cos(2\pi f_B t_1) \exp(-t_1/T_2^*) \tag{31b}$$

where the 2 within the parentheses indicates the state immediately after $90°_2$, A_0 and B_0 are the equilibrium amplitudes of the magnetizations, the cosine argument is the angle through which the x-y magnetization has rotated in the time t_1, and it was assumed that A and B have the same transverse relaxation times. For the sake of simplicity, the residual x components at this stage will be neglected from here onward. During t_m, the z magnetization decays to equilibrium according to

$$A_z(t_1,2,t_m) = A_0 - [A_0 - A_z(t_1,2)] \exp(-t_m/T_1) \tag{32a}$$

$$B_z(t_1,2,t_m) = B_0 - [B_0 - B_z(t_1,2)] \exp(-t_m/T_1) \tag{32b}$$

where it is assumed that both spin populations have the same T_1 in order to simplify the expressions once dipolar coupling is included. At the end of the mixing period the third 90° pulse rotates the existing z magnetization into the ± y axis. The signal after $90°_3$ is

$$S(t_1,t_m,t_2) \approx A_y(t_1,2,t_m,3)f_A^0 \cos(2\pi f_A t_2) \exp(-t_2/T_2^*) + B_y(t_1,2,t_m,3)f_B^0 \cos(2\pi f_B t_2) \exp(-t_2/T_2^*) \tag{33}$$

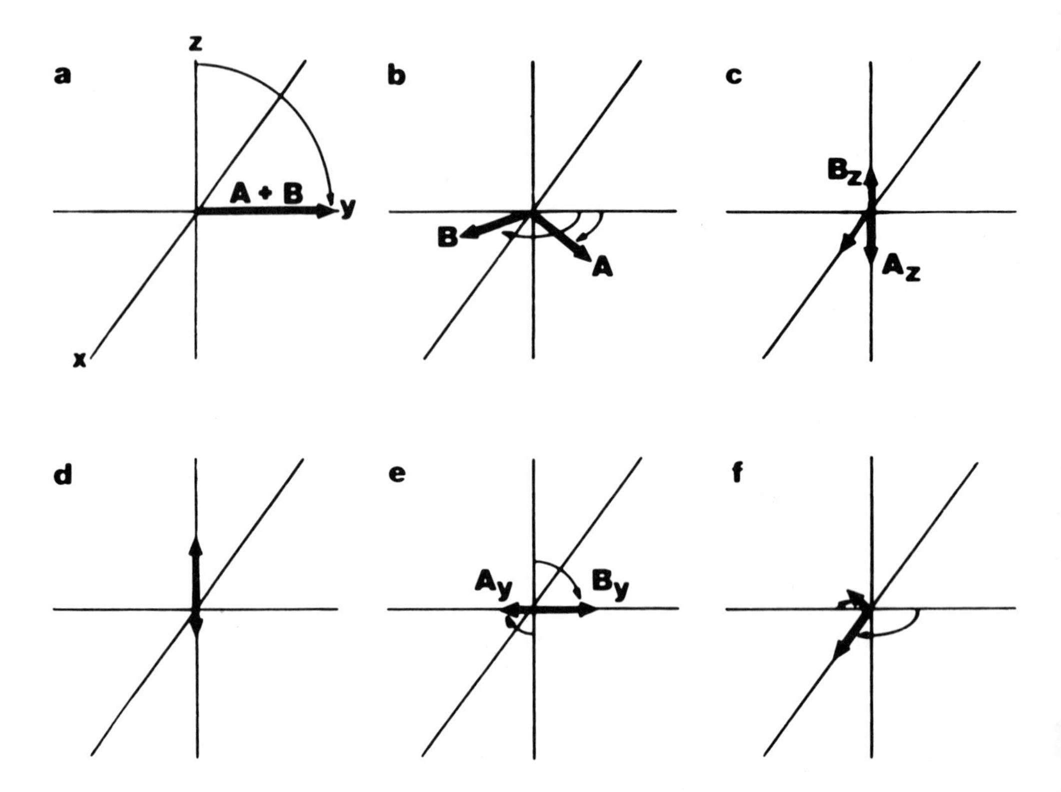

Figure 22. Vector diagram of the NOESY experiment for two dipolar coupled nuclei, A and B. (a) The first 90° pulse rotates the A and B magnetization into the y axis. (b) During t_1 the A and B components of the magnetization separate according to their relative Larmor frequencies. Here it is assumed that B has a greater resonance frequency. (c) The second 90° pulse, applied at the end of t_1, rotates y components into the z axis, and x components are unaffected. In this case, the A_x and B_x components are both in the positive x direction. (d) Ignoring the x-y magnetization during t_m, A_z and B_z undergo longitudinal relaxation. If the two are dipolar coupled, then cross-relaxation occurs by way of the zero- and double-quantum pathways. (e) At the end of t_m, the current z magnetization is measured by applying a third 90° pulse, rotating it into the plane of the detector. (f) These A_y and B_y components then undergo free induction decay, generating signals proportional to their amplitudes at the beginning of t_2.

where the A_y and B_y terms are the amplitudes on the y axis following this pulse and $f_A{}^0$ and $f_B{}^0$ are the resonance frequencies before phase-sensitive detection. Fourier transformation of the data in the t_2 domain yields

$$S(t_1,t_m,f_2) \approx A_y(t_1,2,t_m,3)f_A{}^0\ L(f_A - f_2) + B_y(t_1,2,t_m,3)f_B{}^0\ L(f_B - f_2) \tag{34a}$$

$$L(f_n - f_m) = 1/[T_2{}^* (f_n - f_m)^2 + 1/T_2{}^*] \tag{34b}$$

Thus, the spectrum for each value of t_1 consists of two Lorentzians with the amplitudes defined by the values of t_1 and t_m. Transformation in the t_1 domain yields

$$S(f_1,t_m,f_2) \approx A_0 f_A{}^0 L(f_A - f_1)L(f_A - f_2) + B_0 f_B{}^0 L(f_B - f_1)L(f_B - f_2) \tag{35}$$

where the Lorentzians in f_1 come from the decaying cosines of equations (31). Figure 23a is a contour diagram of this function, a common form for two dimensional plots.

If A and B are dipolar coupled, equations (32) have the form [6]

$$A_z(t_1,2,t_m) = A_0 - [A_0 - A_z(t_1,2)][1 + \exp(-R_c t_m)][\exp(-R_L t_m)/2] + Q[B_0 - B_z(t_1,2)][1 - \exp(-R_c t_m)][\exp(-R_L t_m)/2] \tag{36a}$$

$$B_z(t_1,2,t_m) = B_0 - [B_0 - B_z(t_1,2)][1 + \exp(-R_c t_m)][\exp(-R_L t_m)/2] + Q[A_0 - A_z(t_1,2)][1 - \exp(-R_c t_m)][\exp(-R_L t_m)/2] \tag{36b}$$

$$R_c = 2|W_2 - W_0|$$

$$R_L = 2W_1 + W_0 + W_2 - |W_2 - W_0| + R_{ext}$$

$$Q = [W_0 - W_2]/\{|W_0 - W_2|\}$$

where W_0, W_1, and W_2 are the zero-, single-, and double-quantum transition probabilities given in equations (27), and R_{ext} is the contribution to the relaxation constant from all pathways other than those involving A-B coupling. The essential point is that the A_z magnetization now has B_z dependence and likewise for the B_z magnetization. The signal after $90°_3$ still has the form of equation (32), and the transform in t_2 yields equation (33). Substituting equations (36) into (33) and transforming in the t_1 domain yields

$$S(f_1,t_m,f_2) \approx A_0 f_A{}^0 E_+(t_m)L(f_A - f_1)L(f_A - f_2) + B_0 f_A{}^0 E_-(t_m)L(f_B - f_1)L(f_A - f_2) + B_0 f_B{}^0 E_+(t_m)L(f_B - f_1)L(f_B - f_2) + A_0 f_B{}^0 E_-(t_m)L(f_A - f_1)L(f_B - f_2) \tag{37a}$$

$$E_+(t_m) = [1 + \exp(-R_c t_m)][\exp(-R_L t_m)/2] \quad (37b)$$

$$E_-(t_m) = Q[1 - \exp(-R_c t_m)][\exp(-R_L t_m)/2] \quad (37c)$$

Notice that the $E_-(t_m)$ terms define cross-relaxation between A and B during t_m and they result in a second set of double Lorentzians involving both f_A and f_B. These are the cross peaks demonstrating correlation between two resonances in the spectrum and are shown in Figure 23b for this hypothetical system. Similar cross peaks are observed in all chemical shift versus chemical shift experiments, although the mechanism by which they arise may be scalar coupling (COSY, SECSY) or chemical exchange (ACCORDIAN) rather than the NOE. Figure 24 presents profile and contour plots of a real NOESY experiment. Figure 25 is a COSY contour plot showing scalar couplings among the resonances.

J-Resolved spectroscopy falls in the third category, in which one axis is the scalar coupling constant in hertz and the second axis is the chemical shift. Aue, Karhan, and Ernst [48] demonstrated that the spin-echo experiment used for T_2 measurements could provide a two-dimensional data array that distinguished resonances according to their scalar coupling. Using pulse sequence PS1,

$$90°\text{-}t_1\text{-}180°\text{-}t_1\text{-detector on } (t_2) \quad (PS1)$$

the two-dimensional experiment involves varying t_1 over a range covering $1/J_{max}$ to $1/J_{min}$, where J_{max} is the maximum scalar coupling constant expected and J_{min} is the minimum coupling constant to be measured. The observed echo at time t_1 after the 180° pulse has the form [48]

$$M(t_1,t_2) = \sum_i M_i(0,0) \cos(2\pi J_i m_i t_1 + f_i t_2)[\exp(-t_1/T_2 - t_2/T_2^*)] \quad (38)$$

where t_2 is a particular time during the acquisition period, m_i is the magnetic quantum number associated with resonance i, J_i is the scalar coupling constant associated with resonance i, f_i if the precession frequency of resonance i in radians per second, T_2 is the effective transverse relaxation time, and T_2^* has the contribution from field inhomogeneities. The scalar coupling argument in the cosine term arises from the fact that the two components of magnetization for a given spin population continue to separate in the x-y plane even after the refocusing 180° pulse. Their contribution to the echo is maximal when the leading component catches the trailing component. If more than a single coupling is involved, J_i is replaced by a sum over the set of couplings reflected in a given resonance [48].

A full data set will consist of a hundred or so different values of t_1. As with NOESY, the first transform involves taking the fid for each value of t_1 and transform in the t_2 domain. Using the trigonometric identity

$$\cos(\theta + \phi) = \cos(\theta) \cos(\phi) + \sin(\theta) \sin(\phi) \quad (39)$$

and assuming quadrature detection, the resultant form of the real part of the data is

$$M(t_1,f_2) = \sum_i M_i(0,0) \cos(2\pi J_i m_i t_1) \exp(-t_1/T_2) L(f_i - f_2) \quad (40)$$

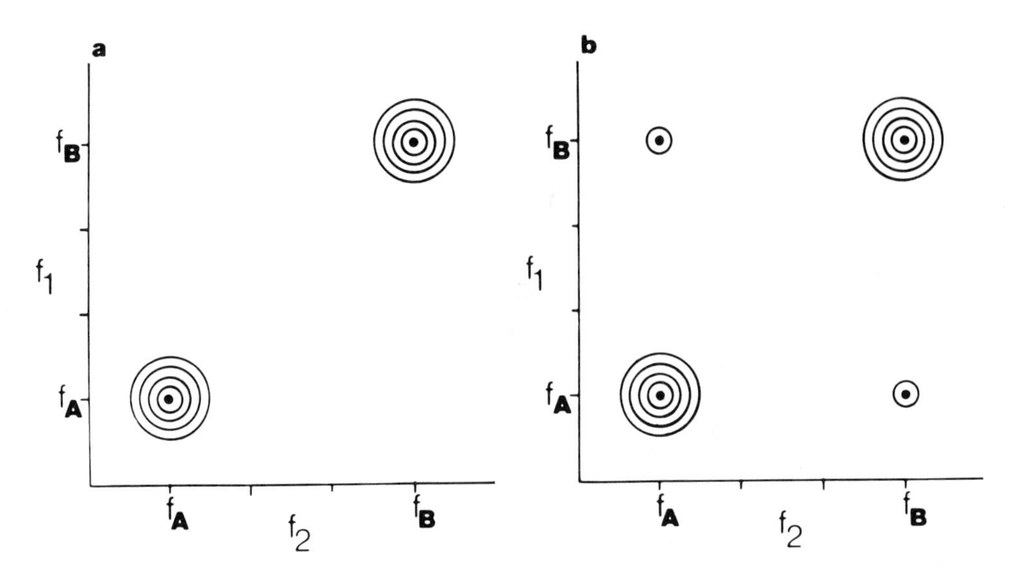

Figure 23. Diagram of NOESY contour plots. (a) Nuclei A and B are not dipolar coupled, yielding only the $L(f_2 - f_A)L(f_1 - f_A)$ and $L(f_2 - f_B)L(f_1 - f_B)$ contours along the diagonal. (b) The two nuclei are dipolar coupled, yielding the lower-amplitude off-diagonal contours, $L(f_2 - f_A)L(f_1 - f_B)$ and $L(f_2 - f_B)L(f_1 - f_A)$.

where the Lorentzian function has the form of equation (34b). Figure 26 is a stacked plot of a data set after this first transform. The second transform is, again, in the t_1 domain, yielding

$$M(f_1, f_2) = \sum_i M_i(0,0)L(2\pi J_i m_i - f_1)L(f_i - f_2) \tag{41}$$

where the Lorentzian in J_i has T_2 rather than $T_2{}^*$. Figure 27 is a profile plot of this second transform of the spectra in Figure 26. Instead of cross peaks connecting one spectral component with another, the positions of the peaks define the scalar coupling constant and chemical shift of a particular resonance in the conventional spectrum.

E. Data Collection

There are several parameters that must be considered in routine collection of spectra. Most of these have been alluded to throughout this chapter, but will be summarized here in the form of a glossary of experimental terms.

Spectrometer Frequency

This is monochromatic rf that will be used to form the primary, or observation, pulses. When the spectrometer is operating with quadrature detection,

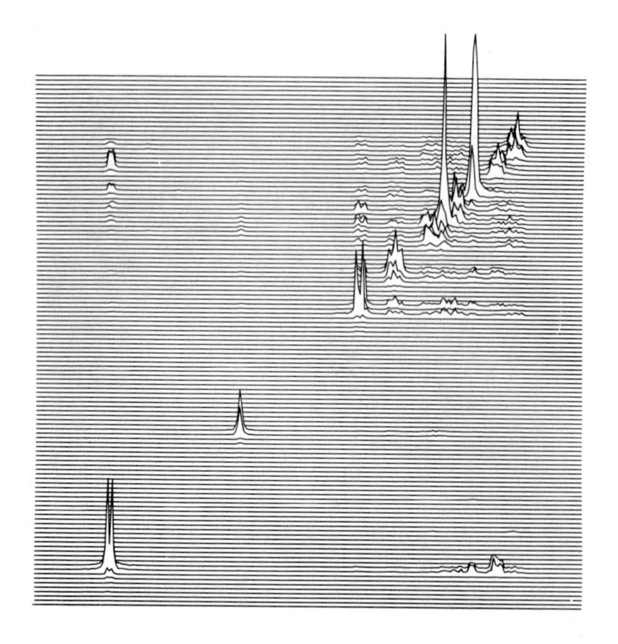

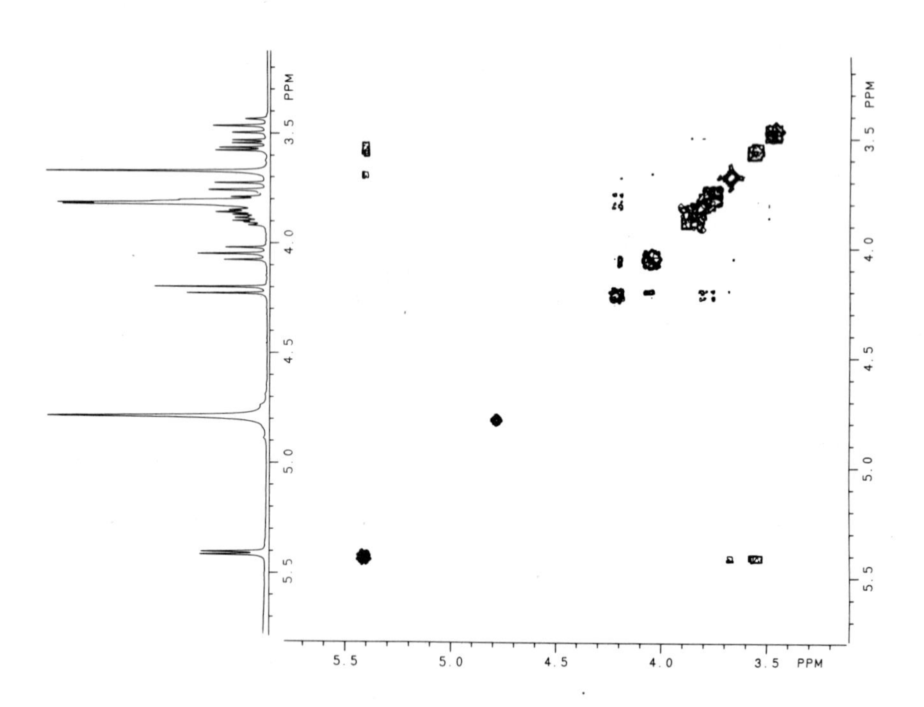
5.5
5.0
4.5
4.0
3.5
PPM

this is usually set close to the center of the spectrum but removed from resonances of interest. Despite the extensive precautions in attenuating contaminating rf, there is usually a small "spike" in the center of the spectrum due to minute leakage of the source to the detecter circuit.

Transmitter Gain

The rf pulse power is under software control by the pulse programmer in modern instruments. It is usually variable from roughly 100 watts down. This power setting is used to establish the pulse duration required for a 90° rotation of the magnetization. The pulse shortens with increasing power, yielding a wider usable spectral region receiving nearly equal power.

Pulse Width

There are usually several software parameters dealing with pulse widths in order to accommodate some of the very complex multipulse sequences that have evolved. Modern instruments provide for changes in pulse width and power within a complex cycle. In routine experiments the pulse width is on the order of 10 μsec or les, providing even rf power over several kilohertz.

Spectral Width

Most commercial high-resolution instruments have a maximum frequency range of ±50 kHz about the central frequency, which is adjustable downward. The analog-to-digital conversion rate is usually the limiting factor in defining this maximum, given the Nyquist criterion. The narrow bandpass filters used by the detector are usually automatically set by the software to be just beyond the spectral width chosen for a given experiment.

Block Size

The number of words in memory devoted to the collection of the current data set. Once the A/D conversion rate, or the spectral width, is chosen, the number of points taken defines the maximum spectral resolution. That is, the digital resolution is approximately the spectral width divided by the number of data points.

Acquisition Time

The notion of resolution is also indicated in the total time during which data is collected in a single pulse cycle. This is, effectively, the block

---

Figure 24. NOESY profile and contour spectra of sucrose in D_2O. The data were collected using a mixing time, t_m, of 0.4 sec, and a 180° pulse during this period to suppress scalar coupling cross peaks. Notice that the anomeric resonance demonstrates the same NOEs as observed in the presaturation experiment (Figure 19).

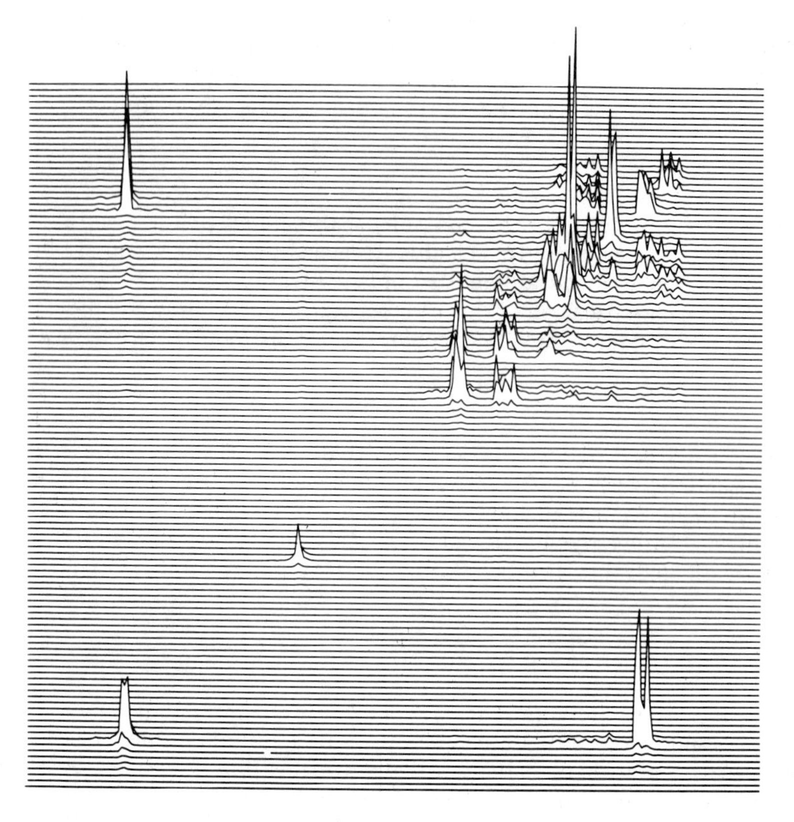

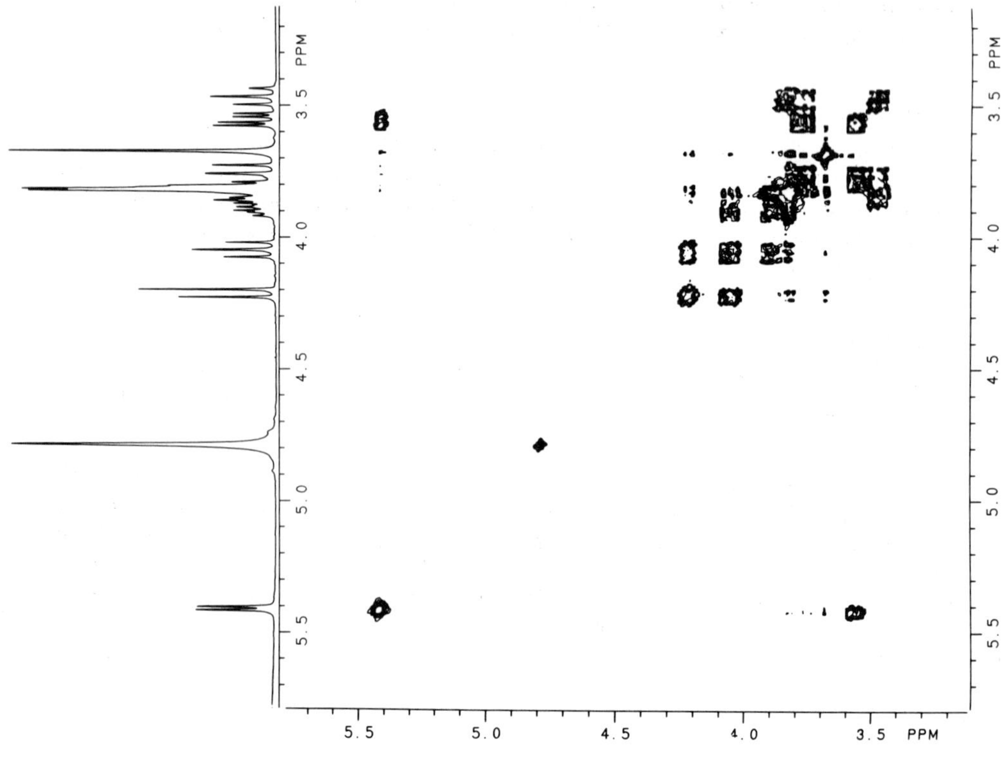
5.5
5.0
4.5
4.0
3.5
PPM

size divided by the A/D conversion rate. If the acquisition time is much shorter than four or five times the T_2^*, the resonances will exhibit "ringing" in their tails as a result of truncating the decay. This is often a problem with very wide spectral widths and will be discussed in the section on data processing.

Preacquisition Delay

The detector dead time during which the transmitted power is dissipated is usually several microseconds. This delay is responsible for some of the phase distortion observed in the transformed data. For rapidly decaying signals, as with many quadrupolar nuclei, this delay can lead to substantial loss of signal amplitude.

Recovery Delay

As stated in the section on spin relaxation, the free induction decay is virtually complete long before the system has returned to equilibrium. Since the acquisition time is usually chosen to encompass most of the fid but not include the noise after the signal has decayed, a delay is inserted between pulse cycles to allow for recovery of the equilibrium magnetization.

Receiver Gain

Besides the rf preamplifier immediately after the pulse/signal switching circuit, there is a variable-gain rf amplifier that is used to control the signal amplitude entering the phase-sensitive detector. This is generally set to ensure an optimum signal within the dynamic range of the A/D converter. Modern instruments have this as both a manual and an automated adjustment.

Audio Gain

Following phase-sensitive detection, the audio signal can be further amplified before conversion. Combined use of the rf and audio gains in improving signal quality is a matter of trial and error.

Number of Acquisitions and Overflow

Most data processing is performed with fixed decimal point operations for speed. This sets an upper limit on the number of acquisitions that can

---

Figure 25. COSY profile and contour spectra of sucrose in D_2O. Here, the off-diagonal peaks demonstrate scalar coupling among the resonances. Notice that there is a strong contour connecting the diagonal peak for the anomeric proton at 5.41 ppm with the doublet of doublets at the 3.55 ppm diagonal position. These two resonances were shown to be coupled in the homonuclear decoupling experiments of Figure 17. Likewise, the triplet of 4.05 ppm has very strong off-diagonal contours connecting it with the doublet at 4.22 ppm, as demonstrated in the decoupling experiment (Figure 17).

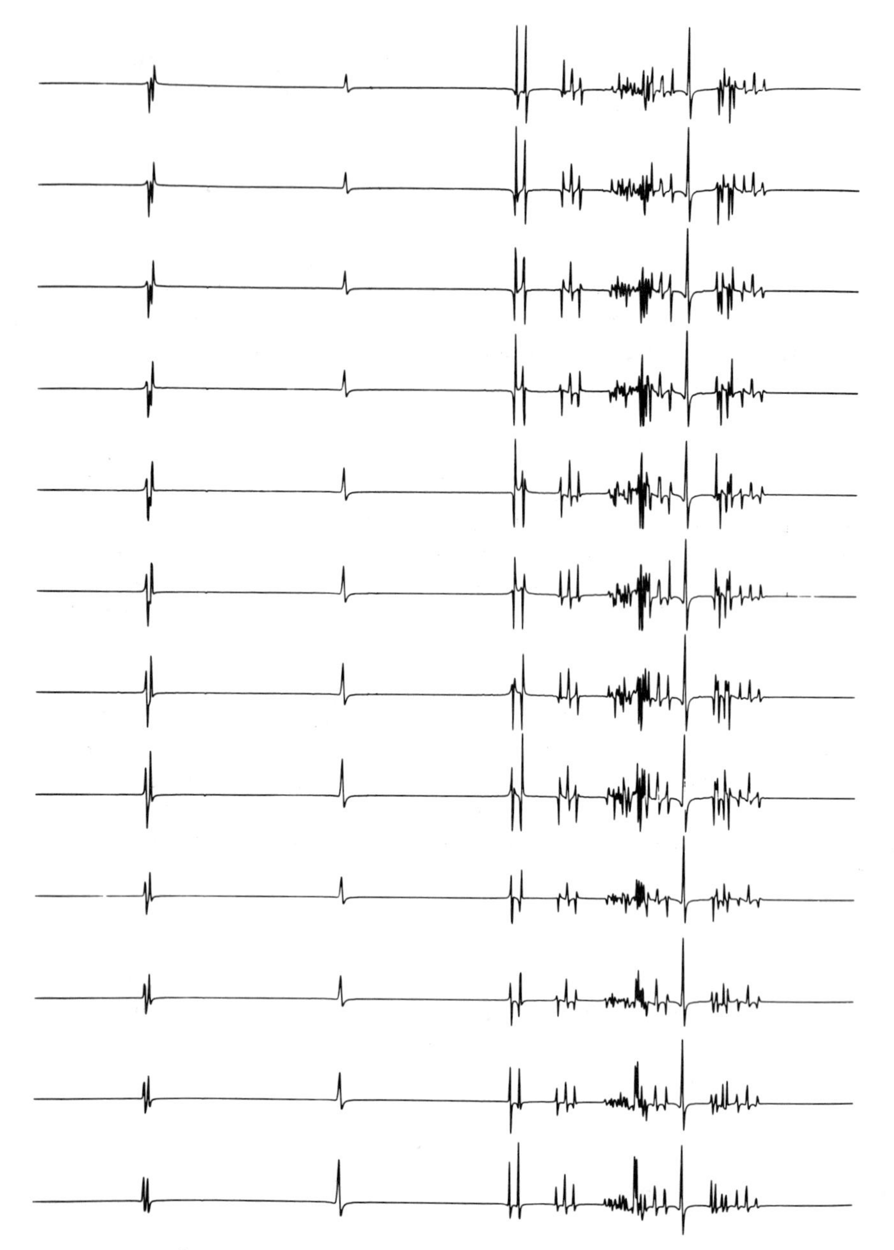

Figure 26. ^{1}H J-Resolved spectroscopy: transform in f_2 only. The first 12 spectra, from the bottom up, are of a J-resolved experiment with sucrose in D_2O. The data were apodized in the t_2 domain with a sine-bell (first half of a sine) function before transforming; t_1 was incremented by 10 msec with each successive experiment, with a total of 128 experiments. Note that all of the scalar coupled resonances oscillate in the t_1 domain, whereas the HDO singlet simply undergoes transverse decay.

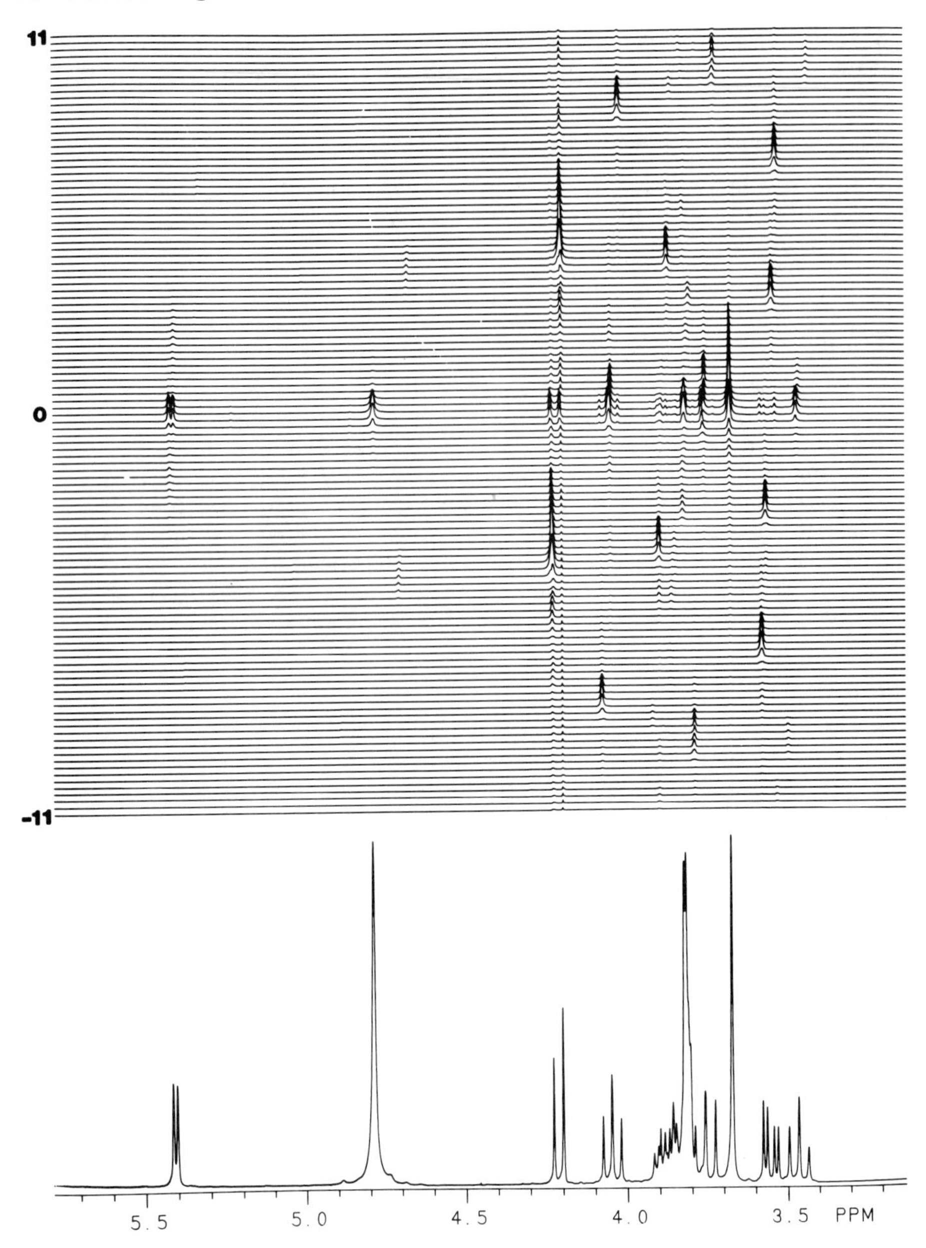

Figure 27. Profile plot of a J-resolved spectrum. The vertical frequency axis is the scalar coupling constant, which ranges from +11 Hz to -11 Hz in this plot. Ideally, only resonances consisting of odd numbers of peaks would have peaks occurring on the 0 Hz axis; however, in this experiment most of the resonances had residual amplitude on this axis. Each scalar coupled resonance has matching peaks on either side of zero, defining the coupling constants. Note that the anomeric doublet at 5.41 ppm and the triplet at 3.45 ppm have very-low-amplitude peaks on either side of zero. These resonances happen to have significantly shorter T_2 times than the other resonances.

be coadded before the data word size is exceeded. Assuming the signal amplitude is just within the dynamic range of the A/D converter, this limit is

$$\text{Maximum number of acquisitions} = 2^n/2^m$$

where n is the number of bits per data word and m is the number of bits resolution in the A/D converter. Word overflow results in baseline distortions in the transformed spectrum.

F. Routine Data Processing

Data processing usually involves apodization or selective weighting of the digital fid, Fourier transformation, viewing expanded regions of the spectrum, spectral phase correction, peak area determination, and, of course, plotting. Fast Fourier transform programs are now very fast with the modern coprocessors available on new instruments. The transform is the least time-consuming operation in the experiment. Likewise, there is little to be said about plotting the spectra other than that the trend is toward fast digital plotters with continuous paper feed, which greatly simplify all aspects of routine, automated, and two-dimensional plotting.

Apodization

The principal objective of this class of operations is either to artificially improve the signal to noise ratio or to improve resolution in the transformed data. Both are accomplished by multiplying the signal-averaged fid with a weighting function designed to optimize one of these two parameters. The most commonly applied weighting functions are:

Exponential	$W(t_2) = \exp(-kt_2)$
Gaussian	$W(t_2) = \exp(-kt_2{}^2)$
Double exponential	$W(t_2) = \exp(At_2 - Bt_2{}^2)$
Half-cycle of sine (sine bell)	$W(t_2) = \sin(\pi t_2/t_f)$

where $W(t_2)$ is the normalized weighting factor for the fid amplitude at time t_2, k is a user-defined constant and is π times the change in line width imposed by the weighting operation, A and B define the exponential and Gaussian contributions in the double exponential, and t_f is the final value of t_2 in the digital fid. Positive values of k result in a weighted fid that decays more rapidly than the original (Figure 28b). This has the undesirable effect of causing the resonances to broaden; however, it also forces the end of the fid, which is usually mostly noise, to decay to zero. Thus a *positive value of k improves the signal-to-noise-ratio, but at the expense of resolution* (Figure 28b). Often a few tenths of a hertz line broadening is negligible compared with the original line width but can yield a substantial noise reduction. The converse holds for negative values of k, that is, the fid is forced to be more gradual, yielding narrower resonances, but the noise at the end of the fid is weighted more heavily

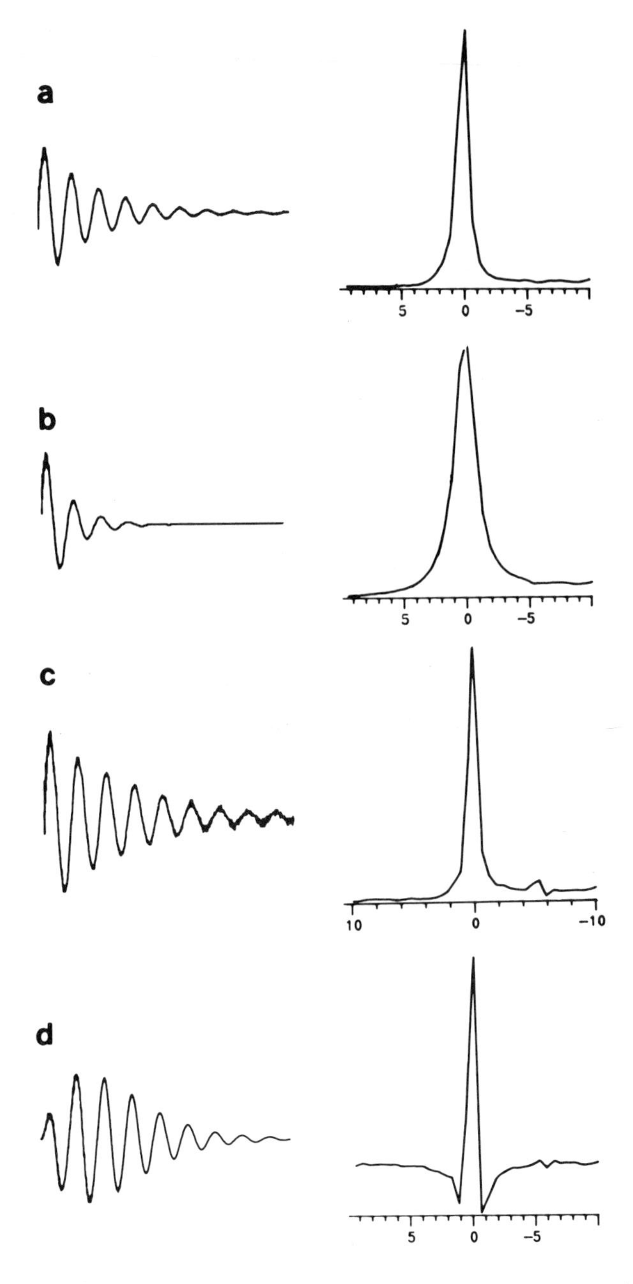

Figure 28. Apodization of fids for noise reduction or resolution enhancement. (a) Water fid and transformed spectrum with no apodization; the unit for the frequency axis is hertz. (b) The same using an exponential apodization function with a line-broadening factor of 0.3 Hz. (c) The same using an exponential apodization function and a line-broadening factor of -0.3 Hz. (d) The same using the first half of a sine wave (sine bell) as the apodization function.

(Figure 28c). The exponential function reinforces a Lorentzian line shape in the transformed data, whereas the Gaussian function imposes Gaussian character on the resonances.

The double-exponential and sine functions are preferred for *resolution enhancement* over the exponential and Gaussian weighting, since they both make the initial part of the fid more gradual and force the noise at the end of the fid to approach zero. The double multiplication is more tedious to optimize, but the sine function always introduces oscillations in the baseline about each resonance (Figure 28d).

There are far more effective methods for resolution enhancement, such as "optimal matched filters" [6,23,64] and Fourier self-deconvolution [65], which involve more exotic weighting operations. These techniques are not yet routine on commercial instruments.

Phase Correction

The resonances in the transformed spectrum are usually distorted from a pure "absorption" band shape (Figure 29a). This can arise from inaccuracies in the preacquisition delay and pulse and detector phase definition. These have the effect of introducing a "dispersive" character to the "absorption" band. A common solution to this problem is to maintain "real" and "imaginary" data sets after the transform, corresponding to the cosine and sine transform components, respectively. Most of the phase distortion can be corrected by a simple frequency-dependent weighting in the combination of these two data sets. A manual adjustment of the phase involves viewing the spectrum as the weighting parameters are varied under knob control. The final solution is a matter of esthetics (Figure 29b). The newer instruments have automated phase correction routines that minimize the dispersive components. These work well for spectra with phase distortion close to that expected for the preacquisition delay. However, several multipulse experiments generate phase relations among the resonances that are purposefully or inadvertently more complex. These effects must be dealt with individually.

Peak Area Determination

Provided that the combined acquisition time and recovery delay is roughly five times the longest T_1 of the sample, the spectral peak areas are proportional to their relative concentrations. For resonances originating from the same molecule, the relative areas define the number of atoms of each type within the molecule, which can be useful information in identifying unknown compounds. Modern instruments have simple and even automated integration routines that provide compensation for curved baselines about the resonances. These provide accurate estimates for peak areas if the resonances are well resolved.

Area determination of overlapping peaks requires some type of peak simulation. This is not as routine as integration, since a function for the resonance band shape must be defined and the resonances are rarely the pure Lorentzians expected for exponential fid. Often it is possible to represent a resonance with a weighted sum of a Lorentzian and a Gaussian peak.

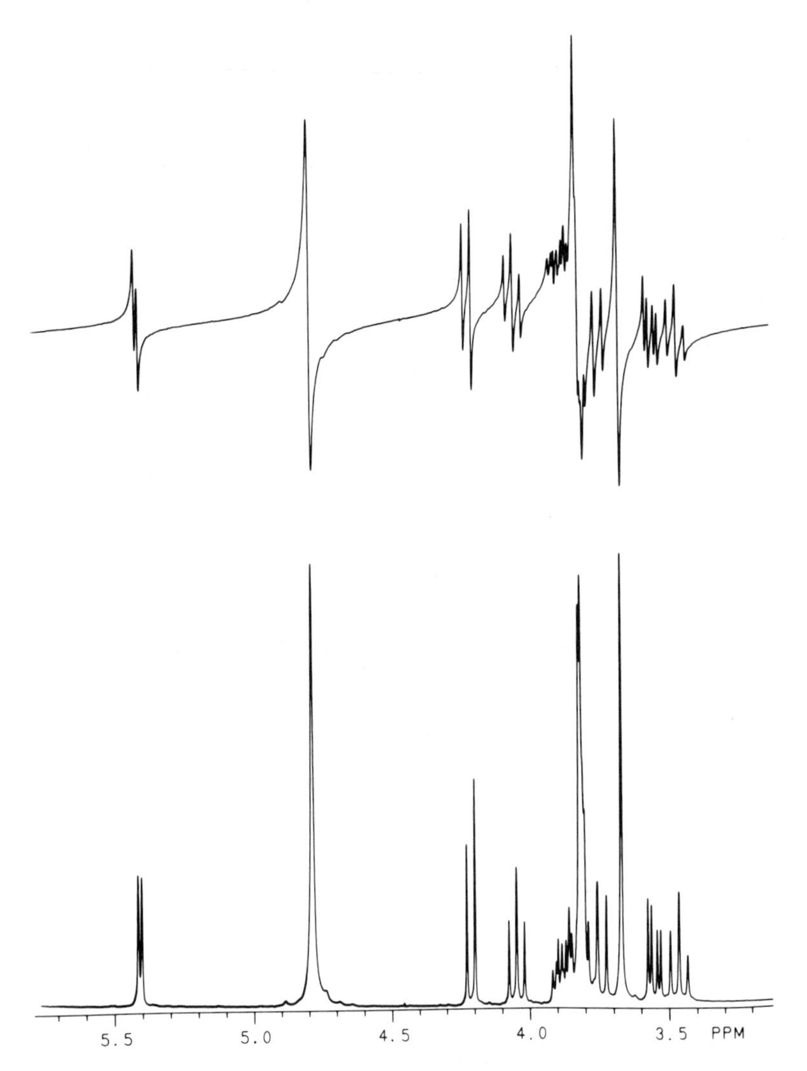

Figure 29. Phase adjustments in Fourier transform spectra. At the top is the spectrum of sucrose with no phase adjustment, and at the bottom is the same with a standard linear phase adjustment.

The simulation may be performed manually, visually comparing real and calculated peaks, or using a nonlinear regression algorithm.

G. Computer-Assisted Analysis

The current trend in multiuser nmr facilities is toward computer networks for data processing [66] and more flexible software for data manipulation [67,68]. This is essential if there is to be a major advance in nmr data analysis, since the instrument's dedicated computer is involved in increasing multitasking as a matter of course in the execution of the experiments. Two major areas of nmr data analysis that will benefit from expanded access to the raw data are spectral data base searches and direct spectral interpretation.

Computer searches are for the most part restricted to natural abundance ^{13}C nmr spectra [69-72] for two main reasons: the ^{13}C chemical shifts are relatively insensitive to solvent unless a conformational change is induced, and complications from homonuclear coupling are essentially nonexistent. The most prevalent alternative, ^{1}H, has greater variation in chemical shift with solvent, and the observed homonuclear scalar coupling patterns depend strongly on the field strength [18].

Pattern recognition has been used to correlate chemical structure with spectral features [73] and in identifying related cross peaks in two-dimensional spectra [6,74]. In the same vein, nmr data is being incorporated in energy minimization routines in protein structure determination [75].

V. PURCHASING AN NMR SPECTROMETER

Modern nmr spectrometers are major investments by most standards, and intelligent selection of an instrument requires familiarity with a wide range of concepts, preferably by way of direct experience. Reading this chapter cannot substitute for experience, but it will introduce the novice to the hardware, terminology, and some practical aspects of nmr experiments. With this general knowledge it is advised that an outline be made of long-term goals for the instrument, and then one should contact someone who operates a nmr spectrometer for a living to discuss needs and budget constraints. Some general considerations in comparing instruments are:

1. relative quality of the probes in terms of resonance band shape, resolution, sensitivity, and duration of a 90° pulse
2. quality of an "unsymmetrized" two-dimensional data set, such as COSY or NOESY
3. maximum spectral resolution in two-dimensional data sets
4. quality of ^{1}H spectrum with multinuclear probes
5. data storage capacity and multiuser capabilities
6. automation software and hardware

The manufacturer usually guarantees probe pulse widths, spectral band shape, resolution, and sensitivity with respect to a set of specified standards. It is up to the users to decide whether these are adequate for their needs. In comparing probes it would be useful to collect a spectrum using a 180° pulse. Ideally, there should be no features in the spectrum, but both static and rf field inhomogeneities can result in residual broad peaks where the narrower component of the resonance is nulled. Although these artifacts may be negligible for most routine experiments, they can complicate relaxation studies and yield false peaks in many two-dimensional experiments. The symmetrization procedure mentioned above removes noise and artifacts from two-dimensional data sets by assuming it is necessarily diagonally symmetric. Artifacts survive this process if there is a corresponding artifact fortuitously occuring in the second dimension. The likelihood of this happening increases with the severity of the band shape distortions. Thus, an "unsymmetrized" two-dimensional spectrum from each instrument being considered would be very useful for making comparisons. The quality of the ^1H spectrum from multinuclear probes can be very poor if it is necessary to use the decoupler coil for acquisition. Often it is difficult to obtain a usable proton spectrum through the decoupler coil, effectively making routine ^1H experiments unavailable while the multinuclear probe is installed. Since it is undesirable to be frequently changing probes, this issue may be important for multiuser systems.

The last two issues mentioned are more simply a matter of the intended use of the instrument. If it will be used for routine bulk sample processing, then automated features are essential; if large data sets are to be generated and manipulated, then data storage and retrieval systems must be considered carefully. These are obvious statements and are best expanded through discussion with an nmr operator.

REFERENCES

1. Haeberlen, U., "High Resolution NMR in Solids," *Adv. Magn. Reson.*, Supp. 1, Academic Press, New York, 1976.
2. Maciel, G. E., *Science, 226*, 282 (1984).
3. Oldfield, E., and R. J. Kirkpatrick, *Science, 272*, 1537 (1985).
4. Belton, P. S., S. F. Tanner, and K. M. Wright, *Nucl. Magn. Reson., 13*, 133-173 (1984).
5. Mansfield, P., and P. G. Morris, "NMR Imaging in Biomedicine," *Adv. Magn. Reson.*, Supp. 2, Academic Press, New York, 1982.
6. Ernst, R. R., G. Bodenhausen, and A. Wokaun, "Principles of Nuclear Magnetic Resonance in One and Two Dimensions," Oxford University Press, Clarendon, 1987.
7. Redfield, A. G., and R. K. Gupta, "Pulsed Fourier Transform Nuclear Magnetic Resonance Spectrometer," *Adv. Magn. Reson.*, Vol. 5, Academic Press, New York, 1971.
8. Ellet, J. D., Jr., M. G. Gibby, U. Haeberlen, L. M. Huber, M. Mehring, A. Pines, and J. S. Waugh, "Spectrometers for Multi-Purpose NMR," *Adv. Magn. Reson.*, Vol. 5, Academic Press, New York, 1971.
9. Fukushima, E., and S. B. W. Roeder, "Experimental Pulse NMR, A Nuts and Bolts Approach," Addison-Wesley, Reading, MA, 1981.
10. Nelson, D. L., M. Stanley, and T. F. George, "High-Temperature Superconductors," ACS Symposium Series No. 351, American Chemical Society, Washington, DC, 1987.
11. Hoult, D. I., "The NMR Receiver: A Description and Analysis of Design," *Prog. Nucl. Magn. Reson. Spec., 12* (1978).
12. Yoder, C. H., and C. D. Schaeffer, Jr., "Introduction to Multi-Nuclear NMR," Benjamin/Cummings, Menlo Park, 1987.
13. Becker, E. D., "High-Resolution NMR: Theory and Chemical Applications" (2nd ed.), Academic Press, New York, 1980.
14. Farrar, T. C., and E. D. Becker, "Pulse and Fourier Transform NMR," Academic Press, New York, 1971.

15. Harris, R. K., and B. E. Mann (Eds.), "NMR and the Periodic Table," Academic Press, London, 1978.
16. Laszlo, P. (ed.), "NMR of Newly Accessible Nuclei," Vols. 1 and 2, Academic Press, New York, 1983.
17. Abragam, A., "The Principles of Nuclear Magnetism," Oxford University Press, Clarendon, 1978.
18. Corio, P. L., "Structure of High-Resolution NMR Spectra," Academic Press, New York, 1966.
19. Schlicter, C. P., "Principles of Magnetic Resonance" (2nd ed.), Springer-Verlag, New York, 1978.
20. Harris, R. K., "Nuclear Magnetic Resonance Spectroscopy," Longman, Essex, 1987.
21. Emsley, J. W., and J. C. Lindon, "NMR Spectroscopy Using Liquid Crystal Solvents," Pergamon, New York, 1975.
22. Davis, J. C., Jr., "Advanced Physical Chemistry: Molecules, Structure, and Spectra," Wiley, New York, 1965.
23. Ernst, R. R., *Adv. Magn. Reson.*, *2*, 1 (1966).
24. La Mar, G. N., W. DeW. Horrocks, Jr., and R. H. Holm (Eds.), "NMR of Paramagnetic Molecules," Academic Press, New York, 1973.
25. Lohman, J. A. B., and C. MacLean, *J. Magn. Reson.*, *42*, 5 (1981).
26. Buckingham, A. D., and J. A. Pople, *Trans. Faraday Soc.*, *59*, 2421 (1963).
27. Hilbers, C. W., and C. MacLean, *Mol. Phys.*, *16*, 275 (1969).
28. Plantenga, T. M., P. C. M. Van Zijl, and C. MacLean, *Chem. Phys.*, *66*, 1 (1982).
29. Heathly, F., "Nuclear Magnetic Relaxation and Models for Backbone Motions of Macromolecules in Solution," *Annu. Rep. NMR Spectrosc.*, *17* (1986).
30. Werbelow, L. G., and G. Pouzard, *J. Phys. Chem.*, *85*, 3887 (1981).
31. Westlund, P.-O., and H. Wennerstrom, *J. Mag. Resonance*, *50*, 451 (1982).
32. Kaplan, J. I., and G. Fraenkel, "NMR of Chemically Exchanging Systems," Academic Press, New York, 1980.
33. Sandstrom, J., "Dynamic NMR Spectroscopy," Academic Press, New York, 1982.
34. Oki, M., "Applications of Dynamic NMR Spectroscopy to Organic Chemistry," Academic Press, New York, 1985.
35. Bargon, J., H. Fischer, and U. Johnsen, *Z. Naturforsch.*, *22*, 1551 (1967).
36. Ward, H. R., and R. G. Lawler, *J. Am. Chem. Soc.*, *89*, 5518 (1967).
37. Closs, G. L., "Chemically Induced Dynamic Nuclear Polarization," *Adv. Magn. Reson.*, Vol. 7, Academic Press, New York, 1974.
38. Neurohr, K. J., T. Drakenberg, and S. Forsen, "NMR of Newly Accessible Nuclei" (P. Laszlo, Ed.), Vol. 2, Chapter 8, Academic Press, New York, 1983.
39. McFarlane, W., and D. S. Rycroft, "Multiple Resonance," *Annu. Rep. NMR Spectrosc.*, *16* (1985).
40. Noggle, J. H., and R. E. Schirmer, "The Nuclear Overhauser Effect: Chemical Applications," Academic Press, New York, 1971.
41. Pratt, S. L., and J. N. Shoolery, *J. Magn. Resonance*, *46*, 535 (1982).

42. Jeener, J., Ampere International Summer School, Basko Polje, Yugoslavia, 1971.
43. Ernst, R. R., Sixth International Conference on Magnetic Resonance in Biological Systems, Kandersteg, Switzerland, 1974.
44. Bax, A., "Two-Dimensional NMR in Liquids," Reidel, Dordrecht, 1982.
45. Bax, A., "A Simple Description of Two-Dimensional NMR Spectroscopy," *Bull. Magn. Reson.*, *7* (1985).
46. Turner, D. L., "Basic Two-Dimensional NMR," *Prog. Nucl. Magn. Resonance Spectrosc.*, *17* (1985).
47. Macura, S., and R. R. Ernst, *Mol. Phys.*, *41*, 95 (1980).
48. Aue, W. P., J. Karhan, and R. R. Ernst, *J. Chem. Phys.*, *64*, 4226 (1976).
49. Doddrell, D. M., D. T. Pegg, and M. R. Bendall, *J. Magn. Resonance*, *48*, 323 (1982).
50. Bax, A., R. Freeman, and S. P. Kempsell, *J. Am. Chem. Soc.*, *102*, 4849 (1980).
51. Bax, A., R. Freeman, and S. P. Kempsell, *J. Magn. Resonance*, *41*, 349 (1980).
52. Morris, G. A., and R. A. Freeman, *J. Am. Chem. Soc.*, *39*, 163 (1978).
53. Eich, G., G. Bodenhausen, and R. R. Ernst, *J. Am. Chem. Soc.*, *104*, 3731 (1982).
54. Hore, P. J., *J. Magn. Resonance*, *55*, 283 (1983).
55. Bax, A., Freeman, R., and G. A. Morris, *J. Magn. Resonance*, *42*, 164 (1981).
56. Rance, M., O. W. Sorensen, G. Bodenhausen, G. Wagner, R. R. Ernst, and K. Wuthrich, *Biochem. Biophys. Res. Commun.*, *117*, 479 (1983).
57. Piantini, U., O. W. Sorensen, and R. R. Ernst, *J. Am. Chem. Soc.*, *104*, 6800 (1982).
58. Nagayama, K., A. Kumar, K. Wuthrich, and R. R. Ernst, *J. Magn. Resonance*, *40*, 321 (1980).
59. Bodenhausen, G., and R. Freeman, *J. Am. Chem. Soc.*, *100*, 320 (1978).
60. Bolton, P. H., *J. Magn. Resonance*, *48*, 336 (1982).
61. Oschikinat, H., A. Pastore, and G. Bodenhausen, *J. Am. Chem. Soc.*, *109*, 4110 (1987).
62. Bodenhausen, G., and R. R. Ernst, *J. Am. Chem. Soc.*, *104*, 1304 (1982).
63. Wagner, G., G. Bodenhausen, N. Muller, M. Rance, O. W. Sorensen, R. R. Ernst, and K. Wuthrich, *J. Am. Chem. Soc.*, *107*, 6440 (1985).
64. Wittbold, W. M., Jr., A. J. Fischman, C. Ogle, and D. Cowburn, *J. Magn. Resonance*, *39*, 127 (1980).
65. Ni, F., G. C. Levy, and H. A. Scheraga, *J. Magn. Resonance*, *66*, 385 (1968).
66. Levy, G. C., and J. Begemann, *J. Chem. Inf. Computer Sci.*, *25*, 350 (1985).
67. Dumoulin, C. L., and J. E. Harrington, *Computer Enhanced Spectrosc.*, *2*, 113 (1984).
68. Levy, G. C., F. Delaglio, A. Macur, and J. Begemann, *Computer Enhanced Spectrosc.*, *3*, 1 (1986).

69. Hiraishi, J., *JETI, 32*, 54 (1984).
70. Heller, S. R., *J. Chem. Inf. Computer Sci., 25*, 244 (1985).
71. Novie, M., and J. Zupan, *Anal. Chim. Acta, 177*, 23 (1985).
72. Bremser, W., and W. Fachinger, *Magn. Res. Chem., 24*, 183 (1986).
73. Beerwinkle, K. R., R. C. Beier, and B. P. Mundy, *Comput. Chem., 10*, 3 (1986).
74. Pfandler, P., G. Bodenhausen, B. U. Meier, and R. R. Ernst, *Anal. Chem. 57*, 2510 (1985).
75. Havel, T., and K. Wuthrich, *Bull. Math. Biol., 46*, 673 (1984).

14

Electron Paramagnetic Resonance

GARETH R. EATON / Department of Chemistry, University of Denver, Denver, Colorado

SANDRA S. EATON / Department of Chemistry, University of Colorado-Denver, Denver, Colorado

This chapter focuses on the instrumental and experimental aspects of EPR. Throughout the chapter, maximal use will be made of references to other readily available sources for theoretical background, tabulations of spectral parameters, alternative spectrometer configurations, etc. The chapter will attempt to be a fairly comprehensive guide to most of what the novice will have to learn eventually, but wherever good discussions already exist reference will be made to them rather than repeating the details. The best introductory general text on EPR is the one by Wertz and Bolton [1]. Introductory treatments are also provided in Refs. 2-6. For additional information on EPR instrumentation, see the books by Alger [7], Poole [8], and Wilmshurst [9]. After the sample has been properly prepared, and the spectra have been competently obtained, the data have to be interpreted. For this aspect of EPR, reference should be made to Refs. 10-15 and the references cited in Section VIII. Specific applications to biological systems are discussed in Refs. 16-20.

I. INTRODUCTION TO EPR

Electron paramagnetic resonance (EPR)* is a technique for measuring the absorption of electromagnetic radiation by an electron spin system. Usually the sample is placed in a magnetic field, and the transitions monitored are between Zeeman levels. The most common experiment uses 9.0-9.5 GHz microwaves (in the X band), for which the free-electron resonance occurs at about 3200-3400 G.†

---

*EPR is also called electron spin resonance (ESR). EPR is a more general term, since it encompasses both the spin and orbital contributions to the electron magnetic moment.

†SI units are rarely used in the magnetic resonance literature: 10,000 gauss (G) = 1 tesla (T), 1 ampere-turn/m = $4\pi \times 10^{-3}$ oersted. In most cases the assumption is made that the magnetic field inside the sample is the same as the external magnetic field; that is, gauss and oersted are not usually distinguished. Most experimentalists use units of gauss.

Common applications of EPR include studies of electronic structure, reaction kinetics and mechanism, radiation damage, defect centers of semiconductors, organic triplet states, transition metal ions, and spin labels or spin probes attached to biological molecules. The information that can be obtained from EPR spectra is discussed in Sections II and VIII.

A variety of paramagnetic centers can be studied by EPR, including organic free radicals, radiation-induced radicals, paramagnetic transition metal complexes, triplet species, defect centers in solids, and conduction electrons in metals. The samples can be in the solid, liquid, or gas phase. The sample is discussed in Section IV.

There are fundamental similarities in the physics of NMR (nuclear magnetic resonance) and EPR. However, several characteristics of EPR are different from NMR, which makes separate study important:

1. The sign of the electron moment is the opposite of that of the proton moment.
2. EPR spectra commonly extend over a wider range of magnetic fields than do NMR spectra.
3. EPR relaxation times are much shorter than NMR relaxation times, making multiple-pulse and Fourier transform techniques less generally applicable in EPR than in nmr.
4. The wavelengths used in EPR are much shorter than those used in NMR. For example, the X-band wavelength is about 3 cm, whereas the wavelength for 200-MHz NMR is 150 cm. Some lumped circuit designs commonly used in NMR cannot readily be carried over into EPR.
5. The EPR sample interacts with the EPR spectrometer to a greater extent than samples generally interact with most spectrometers, including NMR.
6. EPR has enjoyed less widespread use than NMR. The manufacturers of EPR spectrometers have commonly assumed that the user was more sophisticated about the use of the spectrometers than the users of NMR spectrometers. Consequently, the EPR user has to know more about the way the spectrometer works than do users of other types of commercial spectrometers.

A. Precis of the EPR Experiment

An electron has quantized angular momentum. Although both L and S contribute, the fundamentals of EPR are best illustrated by considering first just the spin angular momentum, $S = 1/2$. When an electron is placed in a magnetic field, the projections of the magnetic moment on the axis defined by the magnetic field (conventionally called the z axis) take on two discrete values, +1/2 and −1/2. The energy for the $m_S = +1/2$ level is greater than the energy for the $m_S = -1/2$. The separation between these levels is linearly proportional to magnetic field strength, B (see Figure 1). When an electromagnetic field irradiates the sample with energy equal to the separation between the spin energy levels, transitions between the spin states occur. The transitions are magnetic-dipole allowed, and are effected by the component of the microwave magnetic field vector, B_1, perpendicular to the magnetic field created by the external electromagnet. The sample is placed in a resonator that maximizes the perpendicular

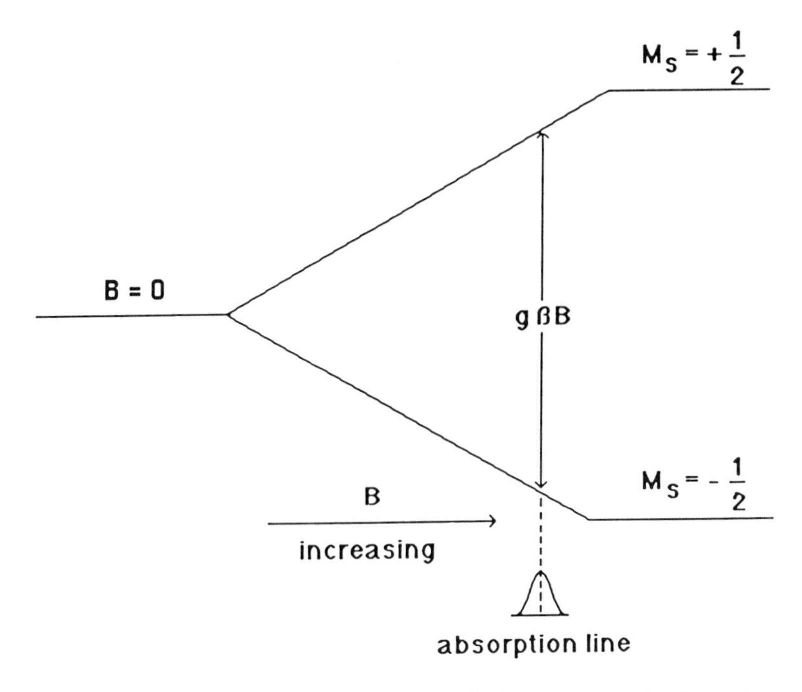

Figure 1. Zeeman energy-level splitting for an electron in a magnetic field. The energy separation is linearly proportional to magnetic field strength, B. Transitions between the two electron energy levels are stimulated by microwave radiation when $h\nu = g\beta B$. The microwave magnetic field, B_1, is perpendicular to B. If the line shape is due to relaxation, it is Lorentzian. The customary display in EPR spectroscopy is the derivative of the absorption line shown here.

B_1 field at the sample. The match of $h\nu$ with $g\beta B$ is called resonance—hence the name electron paramagnetic resonance. The term β is a fundamental constant, the Bohr magneton, whereas g is a characteristic value for a particular sample. The magnetic field of neighboring nuclei or electrons can add to or subtract from the external magnetic field, resulting in spin-spin splitting analogous to the nuclear spin-spin splitting in NMR. This is illustrated in Figure 2.

The instrumentation needed for the EPR experiment consists of four somewhat separate subsystems: (a) microwave system including source, cavity or other resonator, and detector, (b) magnet, power supply, field controller, (c) modulation and phase-sensitive signal detection system, and (d) data display and manipulation system including oscilloscope, analog recorder, A/D converter, computer. A typical CW EPR system is sketched in Figure 3. The subsystems are discussed in detail in Section III.

To help grasp the approximate magnitudes of some numerical values associated with the EPR experiment, the following results are presented. For electrons,

$$\Delta E = g\beta B$$

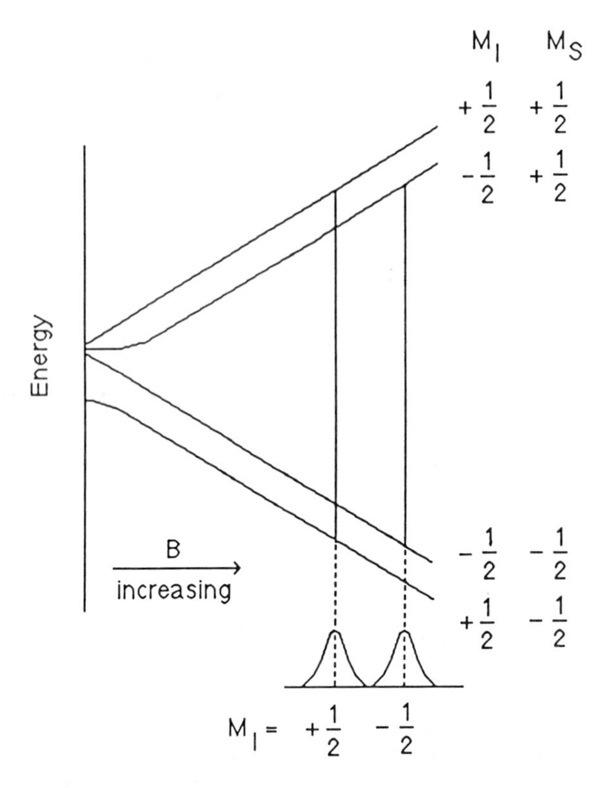

Figure 2. The coupling of nuclear spins to the electron spin adds additional energy levels to those shown in Figure 1. The nuclear hyperfine coupling, and the selection rule that nuclear spins do not flip when electron spins flip, result in 2nI + 1 absorption lines in the EPR spectrum. In the usual EPR experiment the microwave frequency is held constant and the magnetic field is swept. Resonance absorption of microwave energy occurs at each magnetic field for which the nuclear-hyperfine-shifted energy levels are separated by the energy equivalent of the microwave quantum, $h\nu$.

For an organic radical at the usual X-band EPR magnetic field of about 3400 G, this becomes

$$\Delta E = (2.0023)(0.92731 \times 10^{-20} \text{ erg/G})(3400 \text{ G}) = 6.3129 \times 10^{-17} \text{ erg}$$

Then $\Delta E/kT = 1.52 \times 10^{-3}$ and $e^{\Delta E/kT} = 0.998477$. Thus, the difference between the energy levels is a very small fraction of thermal energies at room temperature and the population difference of the levels is only about 0.2%. Conversion of ΔE to frequency units (1 erg = 1.5094×10^{26} sec^{-1}) yields 9.53 GHz. In wave numbers, this is 0.3 cm^{-1}.

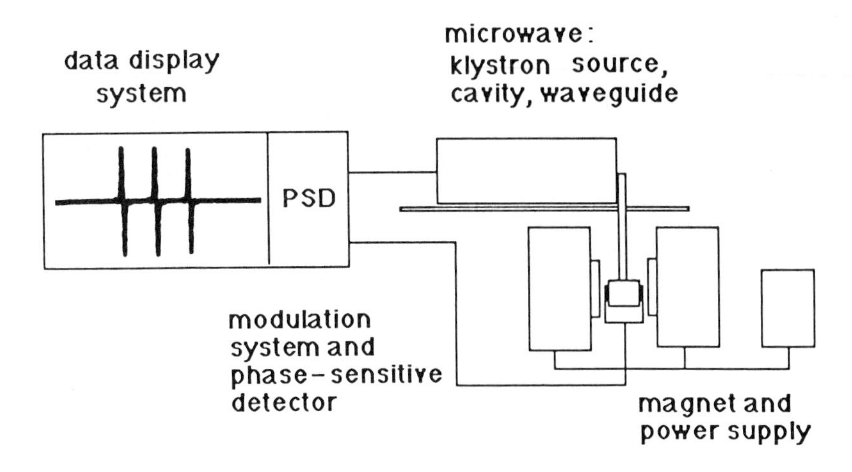

Figure 3. The fundamental modules of a continuous wave (CW) EPR spectrometer include (a) the microwave system including source, cavity or other resonator, and detector, (b) the magnet system including power supply and field controller, (c) the modulation and phase-sensitive detection system, and (d) the data display and manipulation system including oscilloscope, analog recorder, A/D converter, and computer. Each is described in detail in the text. It is important to recognize that each of these modules can be largely independent of the others. In modern spectrometers there is a trend toward increasing integration of these modules via interfaces to a computer. The computer is becoming the controller of the spectrometer operation as well as being the data display and manipulation system.

B. The Use of Magnetic Field Modulation

Since an understanding of the process by which the signal is detected is fundamental to the rest of the information in this chapter, the topic is discussed here, rather than in Section III.

As a first step it is necessary to recognize that the spectrometer measures absorption of microwave energy by the sample. Absorption of energy by the sample changes the microwave energy in the standing wave pattern in the cavity (cavity Q), which modifies the match between the cavity and the waveguide, thereby changing the amount of microwave energy reflected back from the cavity to a detector. If the magnetic field is scanned slowly through resonance, the absorption of energy for a nitroxyl radical looks like Figure 4. The problem with this method of detecting the EPR signal is that the signal-to-noise level (S/N) is poor. Throughout science, a common method of improving S/N is the use of phase-sensitive detection. Commercial spectrometers use magnetic field modulation and phase-sensitive detection. The physical principles are, briefly:

1. Modulation—A lower-frequency variation (called modulation) is imposed on a high-frequency signal. In EPR the high frequency is that of the microwaves, and the lower frequency is magnetic field oscillation, usually at 100 kHz.

2. Phase-sensitive detection—The low-frequency signal imposed on the detected microwave signal is compared with the wave form that generated the low-frequency modulation. Only the signals that have a particular phase relative to that of the original generating signal pass through the circuit. Since the noise is not in phase, most of the noise is rejected and the S/N is improved. The output of a modulation and phase-sensitive detection system is a signal that is the first derivative of the absorption signal.

It is important to note that although the modulation is at a low frequency relative to the microwave frequency (10^5 versus 10^{10} Hz), it is a very high frequency relative to the magnetic field scan rate ($\sim 4 \times 10^{-3}$ Hz for a 4-min scan; 0.4 Hz if the line occupies only 1% of the scan width). The modulation of the signal is achieved by varying the magnetic field. The 100-kHz signal is a sinusoidal voltage that is applied to coils on the walls of the cavity, which creates a magnetic field. The shape of the coils is

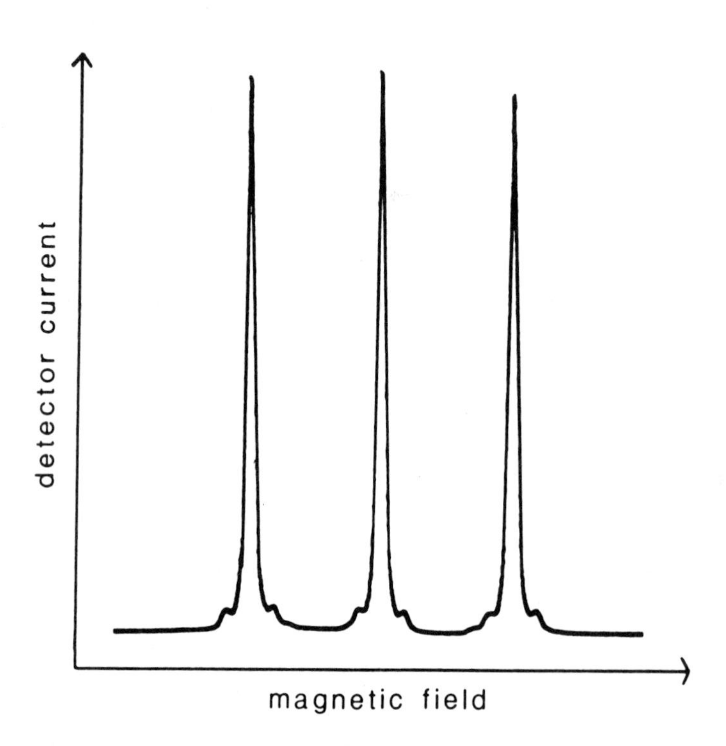

Figure 4. Extension of the ideas in Figures 1 and 2 to the case of electron coupling to a ^{14}N nucleus (I = 1) predicts a three-line EPR spectrum. The spectrum shown here is the absorption spectrum (integral of the usual first derivative spectrum) of a nitroxyl radical. At each magnetic field for which there is resonant microwave energy absorption, there is a change in cavity Q, which results in a reflection of microwave power from the cavity, and hence an increase in detector crystal current.

such that the magnetic field generated adds to or subtracts from the magnetic field produced by the large electromagnet. Since the magnetic field of the electromagnet is changed slowly, one can consider it to be constant during a few cycles of the modulation field. Thus, as illustrated in Figure 5, the 100-kHz modulation field causes the magnetic field seen by the sample to vary between B_{m1} and B_{m2} 100,000 times per second. The difference $B_{m2} - B_{m1}$ is equal to the "modulation amplitude" setting on the spectrometer console. If the modulation amplitude is set to 2 G, the magnetic field seen by the sample oscillates from 1 G below the magnetic field established by the electromagnet to 1 G above the field established by the electromagnet.

As shown in Figure 4, the detector current changes as the magnetic field is scanned through resonance. Consider one of the absorption peaks in more detail, as shown in Figure 5. When the 100-kHz modulation causes the magnetic field to be at B_{m1} the detector current is i_1. As the field increases sinusoidally from B_{m1} to B_{m2} the detector current increases sinusoidally from i_1 to i_2. The amplitude of the 100-kHz modulation of the detector output is determined by the slope of the absorption signal between B_{m1} and B_{m2}. Note that the phase of the detected signal for the rising portion of the absorption signal is the inverse of that for the falling portion. Thus the output of the phase-sensitive detector gives the value of the derivative (slope) of the absorption signal. However, the signal will be a true derivative only if the absorption curve is a straight line between B_{m1} and B_{m2}. This is never exactly true, but an adequately close approximation is achieved if $B_{m2} - B_{m1}$ is less than one-tenth of the distance between the inflection points of the absorption curve. These inflection points correspond to the peaks in the first-derivative curve.

C. Dispersion Effects on EPR Spectra

The microwave magnetic susceptibility has real and imaginary components. EPR spectrometers are usually tuned to detect the absorption signal, which is the imaginary part of the microwave susceptibility. In any region of the electromagnetic spectrum, including EPR, absorption is accompanied by dispersion, which is the real part of the microwave magnetic susceptibility. Dispersion manifests itself as a shift in the resonant frequency of the cavity and a change in line shape. Spectrometers can be designed to detect the dispersion signal, by using two detector crystals. There are experiments for which this is important [8]. The fundamental paper describing dispersion line shapes is Ref. 21.

Even though the spectrometer is designed to detect the absorption signal, it is possible for the observed line shape to be a mixture of absorption and dispersion. For example, the conductivity and dielectric absorption of some solutions can provide sufficient mixing of the real and imaginary components of the microwave magnetic susceptibility to cause asymmetry in the spectral lines. Published examples of such effects include spectra of 75 m*M* aqueous vanadyl solutions [22] and 0.05-0.06 *M* aqueous solutions of Cr^{3+} and Fe^{3+} [23]. Related work on the EPR of paramagnetic ions in metal was reported [24]. References 22-24 and Poole ([8], 2nd ed., p. 490) contain several diagrams illustrating mixtures of absorption and dispersion line shapes.

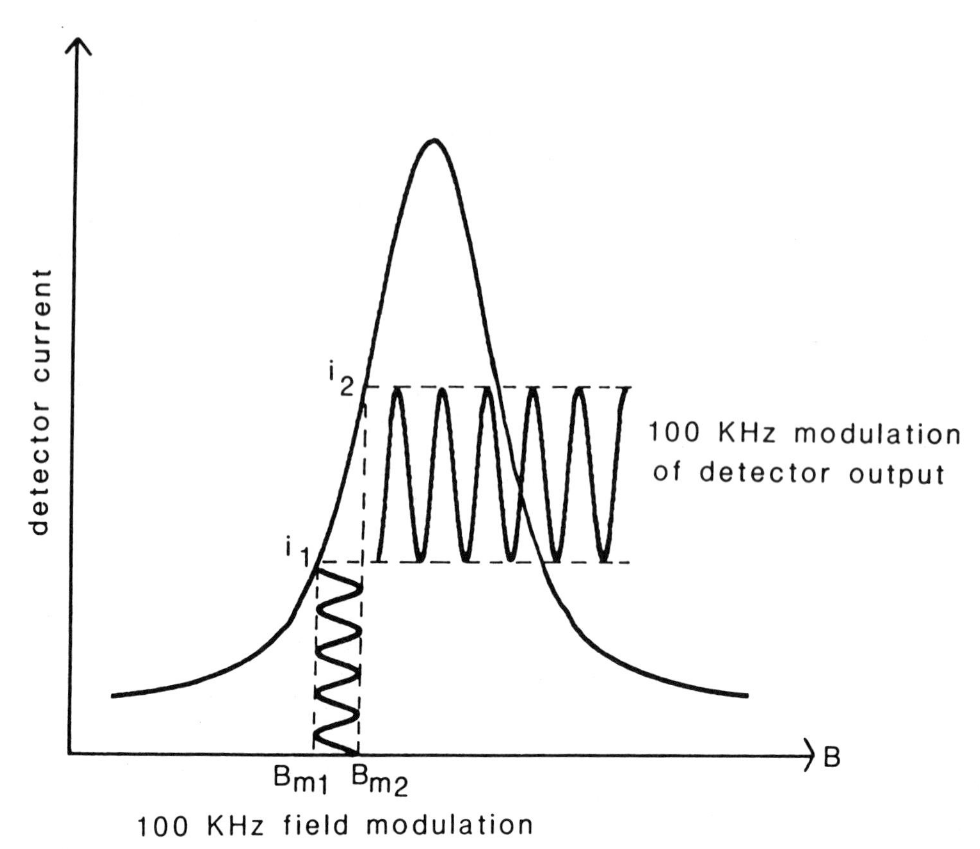

Figure 5. This figure displays how the detector current changes as a function of magnetic field modulation (here assumed to be at 100 kHz). Relative to the modulation frequency, the sweep of the magnetic field generated by the electromagnet can be considered to be stationary. The 100-kHz modulation of the current in the coils associated with the cavity results in a small magnetic field that oscillates at 100 kHz between values that symmetrically add to and subtract from the main field. As a result, the magnetic field oscillates at 100 kHz between B_{m1} and B_{m2}. Since the absorption of microwave energy depends on magnetic field, the detector crystal current oscillates at 100 kHz also, between i_1 at B_{m1} and i_2 at B_{m2}. If $B_{m1} - B_{m2}$ is small enough relative to the line width of the EPR signal, $i_2 - i_1$ is proportional to the slope of the absorption signal. The output of the detector crystal is amplified and compared with the output of the 100-kHz oscillator in the phase-sensitive detector, resulting in a first-derivative display. For a more detailed picture, see Ref. 8, 1st ed., Fig. 10-3.

D. Passage Criteria

The usual treatments of EPR spectroscopy assume slow-passage conditions, unless stated otherwise. In practice, routine experiments may not actually satisfy slow-passage criteria. Hence, it is important to understand passage criteria at least to the point of knowing whether failure to achieve slow-passage conditions could be affecting the experimental spectrum.

Physically, the terms "slow," "rapid," and "fast" passage describe the relationship of the spectral scan rate ("passage" through the spectrum) to relaxation times (see Table 2-1 in Alger's book [7]).

1. Slow passage—the spin system is always in thermal equilibrium.
2. Rapid passage—the modulation field sweeps through a spin packet in a time shorter than the relaxation time of the spin packet.
3. Fast passage—the electron spins do not have time to relax between modulation sweep cycles.

With organic radicals, or with some metals at low temperature, it is difficult to achieve slow-passage conditions. Often, the use of modulation frequencies less than 100 kHz is needed. Failure to achieve slow-passage conditions invalidates the use of continuous wave (CW) saturation methods for estimating relaxation times.

II. WHAT ONE CAN LEARN FROM AN EPR MEASUREMENT

The resonant absorption of energy by an electron spin system provides information about the environment of unpaired electrons in the material under study. Some of the parameters that are commonly examined are enumerated below.

Often the question is asked, "how long does it take to obtain an EPR spectrum?" Even ignoring signal-to-noise questions (and attendant signal-averaging time requirements), the time required depends on what one wants to learn. For example, to answer the question of whether a particular column chromatography fraction contains a nitroxyl radical would take only a few seconds, but to determine the relaxation characteristics of a spin system might take several days of experiments. The importance of knowing what one wants to learn from an EPR measurement cannot be stressed too strongly. The care with which various parameters are adjusted in obtaining a spectrum is a function of the information that one wishes to obtain from the spectrum.

Any of the parameters discussed below may be studied as a function of a variety of factors, including solvent, temperature, sample phase (solid solution, liquid solution, frozen solution, single crystal, powder, etc.), other species in the sample, sample size, sample orientation, or irradiation (frequencies from acoustic to ionizing, including resonant frequencies for electrons or nuclei).

A. g Value

Although the g value is commonly reported as if it were a scalar, it is really a dyadic with components

$$\begin{matrix} g_{xx} & g_{xy} & g_{xz} \\ g_{yx} & g_{yy} & g_{yz} \\ g_{zx} & g_{zy} & g_{zz} \end{matrix}$$

For example, g_{yx} is the g value along the y axis when the magnetic field is applied along x. The mathematics of g values is discussed in section 15-8 of Ref. 15. The matrix product of g with its transpose is the g^2 tensor, which can be diagonalized. Some of the literature is a bit careless about the nomenclature with regard to g values.

B. Spin-Spin Coupling

Multiple peaks in EPR spectra can result from many things. One should always suspect multiple chemical species, and even multiple isomers or conformers of a single chemical species. However, there are in fact very few chemical species whose EPR spectra consist of only a single line, since most species have nuclei with spins that are in the vicinity of the unpaired electron.

Electron-Nuclear Coupling

The electron-nuclear coupling constant, generally denoted a_n, is independent of magnetic field. The observed splittings, measured on a magnetic field plot, do depend on magnetic field, due to some second-order corrections (see Ref. 1 for further discussion). However, most electron-nuclear couplings are small enough that the second-order corrections are sufficiently small at X band that quartets, sextets, etc., are clearly recognizable. Many nuclear couplings are observable in the standard CW EPR spectrum. Some of those that are too small to resolve in the standard CW spectrum can be obtained from ENDOR, TRIPLE (see Section VII), or ESEEM spectra [25].

Electron-Electron Coupling

Recently, it has been recognized that electron-electron spin-spin coupling can lead to resolved splitting in EPR spectra just as can electron-nuclear coupling in EPR and nuclear-nuclear coupling in NMR [26,27].

C. Line Shape

If the line shape is due entirely to relaxation phenomena, it is Lorentzian. The EPR spectrum of Fremy's salt (peroxylamine disulfonate) in fluid solution is nearly perfectly Lorentzian. However, most samples exhibit some unresolved hyperfine structure, which makes the line shape more complicated. If there is a lot of unresolved hyperfine structure, the line shape will approach Gaussian. (See Refs. 1 and 8 for extensive discussions of the details of Lorentzian and Gaussian lines.) One key thing to remember is that the Lorentzian line has a large fraction of its area far from the center of the line. Hence, it is necessary to integrate rather wide scans to obtain proper integrations of EPR spectra [28].

It is important to recognize that the derivative display emphasizes sharp features at the expense of broad features. This is the reason that derivative presentations are being used increasingly in other types of spectroscopy. It is easy to be fooled into ignoring broad signals that might in fact be the major species in the sample. A full-scale display of the sharpest feature in the spectrum might reveal only a percent or so of the total signal. A simple example illustrates this point. The EPR spectrum of high-spin Fe(III) porphyrins in frozen solution is commonly described as having two EPR lines, at g = 6 and at g = 2. In fact, there is signal extending all the way from g = 6 to g = 2, but since the derivative display is used, the small slope except at the turning points of the powder pattern results in an almost zero derivative signal over a large portion of the spectrum (see Figure 7-10 in Ref. 1).

The EPR line shape can reveal kinetic information such as the slow tumbling of nitroxyl spin labels [17,18,29,30] and rates of chemical reactions in equilibrium mixtures [31].

D. Intensity

The intensity of the EPR signal is a measure of the concentration of species present, as long as the nature of the spin system is known. In this regard, nomenclature is poor. Many reports do not distinguish between the amplitude of the spectrum in the usual derivative display and the intensity of the signal as revealed by the double integration of the derivative signal. Note that if the line shapes are the same, regardless of the details of the shape [32], the amplitude of the signal (height of the spectral display) is inversely proportional to the square of the width of the peak. For two signals with the same area, if one is twice as wide as the second, its amplitude will be only one-fourth that of the second. Section V provides a detailed discussion of aspects of quantitative EPR.

E. Saturation Behavior

The microwave power saturation behavior of a sample reveals much about the nature of the spin system. CW power saturation yields information proportional to the product of the relaxation times, T_1T_2 [8,33,34] (see the following section for a discussion of relaxation times). One convenient way to display power saturation behavior is to plot spectral amplitude versus the square root of microwave power (Figure 6). The spectral amplitude can be defined by the difference between the signal intensities at the magnetic fields that correspond to the maxima and minima of the unsaturated signal (a plot at constant field) or by the difference between signal maxima and minima at each power. Since the lines broaden as the signal is saturated, these two definitions will give different plots. A plot at constant magnetic field is more sensitive to changes in relaxation times and easier to simulate using a computer. A linear plot indicates no saturation. The lower the microwave power at which the plot deviates from linearity, the more readily the signal saturates.

The ease of saturation could reveal, for example, how well the sample is degassed, whether aggregation is occuring, whether a good glass has

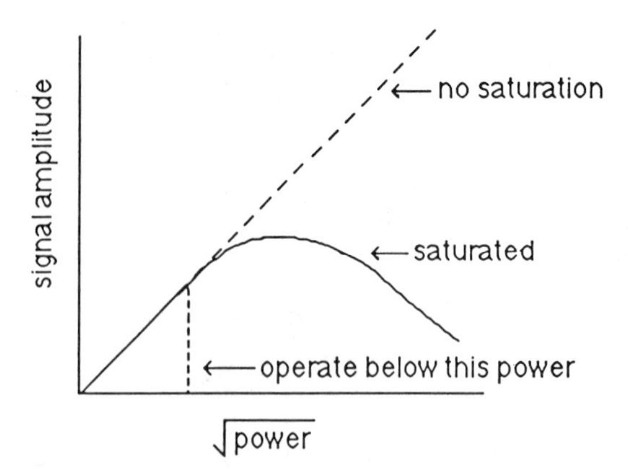

Figure 6. A plot of signal amplitude as a function of the square root of the microwave power incident on the cavity is called a power saturation curve. The plot would be linear if no saturation of the electron spin system occurred. To do quantitative EPR, it is necessary to operate at power levels below the power at which the curve becomes nonlinear. Curvature of the plot can be interpreted in terms of relaxation of the spin system, as discussed in Section VI.D.

formed upon freezing, whether unsuspected paramagnetic impurities are present, or whether the species being studies is in close proximity to another paramagnetic species.

F. Relaxation Times

A physical system that is moved away from equilibrium by a change in one or more variables of state will return to equilibrium by relaxation processes. The study of relaxation is a more general view of kinetics than the usual reaction rate measurement. Relaxation times are, effectively, a different dimension in which to view chemistry. Relaxation times are parameters characteristic of an ensemble of spins, and have no direct meaning for isolated chemical species. This is in contrast to g values and spin-spin splittings, which are characteristic of the isolated molecule (though they change depending on the environment). The measurement of relaxation times is important because the relaxation times (a) affect the choice of experimental parameters (see Section VI.B) and (b) provide chemical information not available by other techniques.

Saturation and relaxation measurements are more important in magnetic resonance than in higher-frequency spectroscopies because spontaneous emission from an excited state is proportional to the square of the frequency. There are numerous contributions to the relaxation processes. The measured relaxation times have to be interpreted in terms of all of the possible contributions, so a list of these contributions serves to guide the experimental design. Relaxation times may depend on the species being studied, intermolecular interactions including solvent, collisions with like molecules or other paramagnetic species, magnetic field, temperature, diffusion, spin diffusion, viscosity, cross-relaxation, phonon bottleneck, and electric field. The interpretation of relaxation times in terms of various contributions can represent a major effort, and is beyond the scope of this chapter. Discussion of several of these contributions can be found in Refs. 12 and 35-37.

If an electron spin system is displaced from equilibrium by the application of a microwave magnetic field, B_1, perpendicular to the magnetic field of the electromagnet, the spin system will attain a new macroscopic state that has (a) higher energy and (b) lower entropy (nonzero M_x, M_y requires some degree of phase coherence, and hence has lower entropy). Removal of B_1 allows the spin system to relax toward equilibrium. Achieving thermal equilibrium is slower than the loss of phase coherence. This is easily seen to be true, since there can be absence of phase coherence without thermal equilibrium, but there cannot be thermal equilibrium while there is still phase coherence [38].

The relaxation time for attaining thermal equilibrium of the M_z magnetization is called T_1, the longitudinal or spin-lattice relaxation time. The T_1 processes change electron spin states. This means that the total energy of the spin system is changed by T_1 processes. Also, T_1 can contribute to the line width due to the finite lifetime of the spin states. A small T_1 leads to a smearing out of energy levels and therefore a broadening of the EPR lines. The magnitude of this effect can be estimated from the Heisenberg uncertainty principle.

The relaxation time for loss of phase coherence (i.e., achieving $M_x = M_y = 0$) is called T_2, the transverse or spin-spin relaxation time. The T_2 processes are adiabatic—that is, the total energy of the spin system is not changed by T_2 processes. For example, a mutual spin flip

$$\downarrow\uparrow \rightleftharpoons \uparrow\downarrow$$

between one spin in the upper state and one in the lower state does not change the total number of spins in each state.

The terms T_1 and T_2 are related to the line width by the following formula:

$$1/T_2{}^* = 1/T_2 + 1/2T_1$$

where T_2 is the relaxation time described above, and $T_2{}^*$ is a line-width parameter. Sometimes the notation is changed to

$$1/T_2 = 1/T_2{}' + 1/2T_1$$

where $T_2{}'$ is the transverse relaxation time described above, and T_2 is a line-width parameter.

If $T_1 >> T_2$, then $T_2 = T_2{}^*$. Since this is commonly the case in solids, many texts drop the distinction between T_2 and $T_2{}^*$. In general,

$$T_2{}^* < T_2 \leq T_1$$

III. THE EPR SPECTROMETER

The user, or potential user, of EPR spectroscopy should learn about the instrumentation because (a) commercial spectrometers are so modular that one has to be well informed about the function of each component to be an intelligent customer, (b) the sample interacts with the spectrometer to a greater extent than in most techniques, and (c) some important experiments require home-built equipment.

There are two general types of EPR spectrometers. Continuous wave (CW) spectrometers are the "normal" spectrometers. These can have accessories for multiple resonance including ENDOR, TRIPLE, and ELDOR (see Section VII). The second type is pulsed spectrometers, including ones designed for saturation recovery and spin-echo experiments. The following paragraphs discuss features and components that are common to most EPR spectrometers.

A. The Microwave System

Diagrams of the microwave circuits are given in Figures 7, 8, and 9 for CW, saturation recovery, and spin-echo EPR spectrometers, respectively. In each case there are attenuators to set power levels, phase shifters to establish the correct phase relationships for the two or three arms of the bridge, circulators to direct the flow of the microwaves, and isolators to prevent unwanted interaction of components. In the pulsed spectrometers PIN-diode switches are used to shape the pulses and protect the detector. Limiters are also used to protect sensitive components. Information about microwave components is provided in the first edition of Poole's treatise [8] and in Refs. 43-45. The novice may also find it useful to look at one of the books by Laverghetta [46,47]. The important advance of the Varian E-Line and Bruker ER420 and ER200 spectrometers over the earlier (e.g., Varian V4500) systems was the introduction of the reference arm and the use of a circulator in place of the magic T. The reference arm permits biasing the crystal at critical coupling of the cavity, so that signal amplitudes as a function of incident power truly reflect the power response of the spin system. The microwave power incident on a magic T is divided equally with half going to each of two arms of the T. Thus, half of the klystron power goes to the cavity, and half of the reflected power goes to the detector. The circulator directs the microwave power without dividing it, and hence gives a factor of 4 greater signal power at the detector relative to the magic T. The new Bruker ESP300 spectrometer continues the evolution of EPR to the incorporation of the computer as a central component in the design of the spectrometer, instead of just as a data acquisition device.

Choice of Frequency

Considerable care needs to be taken with regard to choice of microwave frequency. As has been emphasized repeatedly in this chapter, the crucial decision is what one wants to learn from the sample. Most EPR spectra have been obtained at X band, and specifically in the range 9.0-10.0 GHz. If a second microwave frequency was available in a lab, it usually was 34-35 GHz (normally called Q band among EPR spectroscopists, but Ka in the IEEE notation). These frequency ranges were chosen largely because

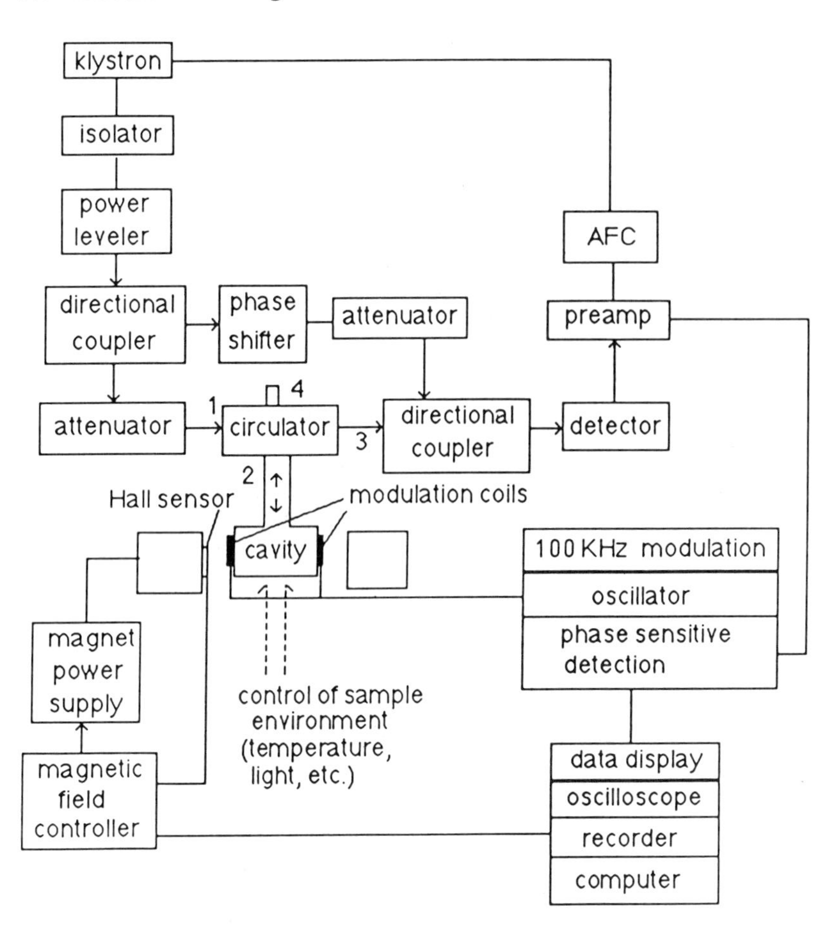

Figure 7. CW EPR spectrometer block diagram. This figure extends the information in Figure 3 to the component level, with particular emphasis on the microwave system. Pictured here is the fundamental reference-arm bridge, with four-port circulator, which is the basis for all modern CW EPR spectrometers. The key features are the use of directional couplers (unequal power dividers) to pick off part of the klystron output for the reference arm, where its phase relative to the sample arm is established, and then recombine the two arms prior to the detector crystal. Hence, the detector crystal can be biased by using power from the detector arm, and the cavity can be used critically coupled. The use of a circulator instead of a magic T as in earlier designs uses power and signal more efficiently, increasing signal/noise. Additional components are added to this basic system for dispersion operation (two detector crystals are needed), bimodal cavity operation (two reference arms are needed), etc.

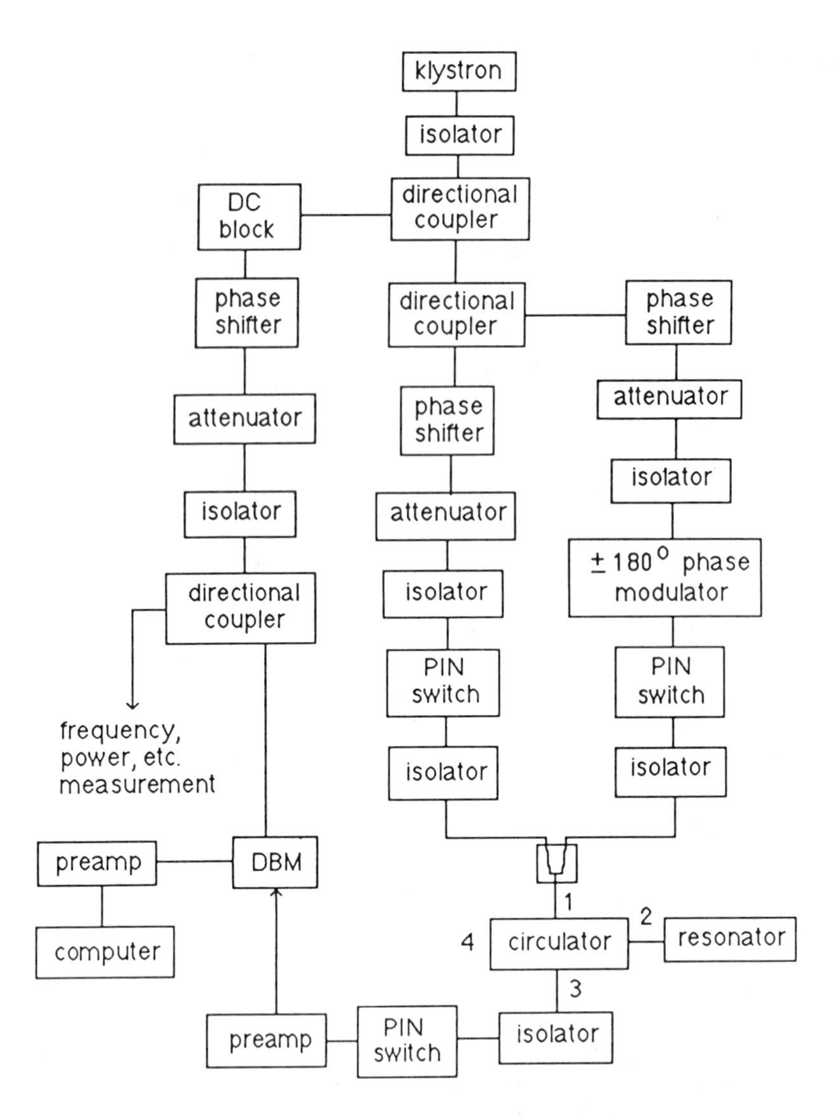

Figure 8. Saturation recovery spectrometer block diagram. Only the microwave aspects of the spectrometer are shown. The magnet and data display aspects are similar to the CW spectrometer in Figure 7, except that the analog-to-digital converter has to be faster, and the computer does more control of the system timing. To achieve pulsed saturation recovery (or stepped-power EPR) operation, a third arm is added to the microwave bridge. In the third arm are components to establish the phase and the power level of the saturating pulse. PIN diode switches and phase-shift modulators are bounded by isolators (three-port circulators with one port terminated in a load, so that microwaves pass in only one direction) to prevent switching transients from getting back to the klystron or into the signal. It is common in pulsed spectrometers to compare the reference arm and the signal arm with a "double-balanced-mixer" (DBM), which acts in this application as a phase-sensitive detector of microwaves. The "DC block" is a device that does not have electrical continuity for DC signals, but does for microwave signals. It prevents ground loops in the microwave system. The microwave system shown here for use with a reflection cavity (or other reflection resonator) can be modified to use a bimodal cavity—see Hyde's chapter in Ref. 35. Additional details are presented in Ref. 39.

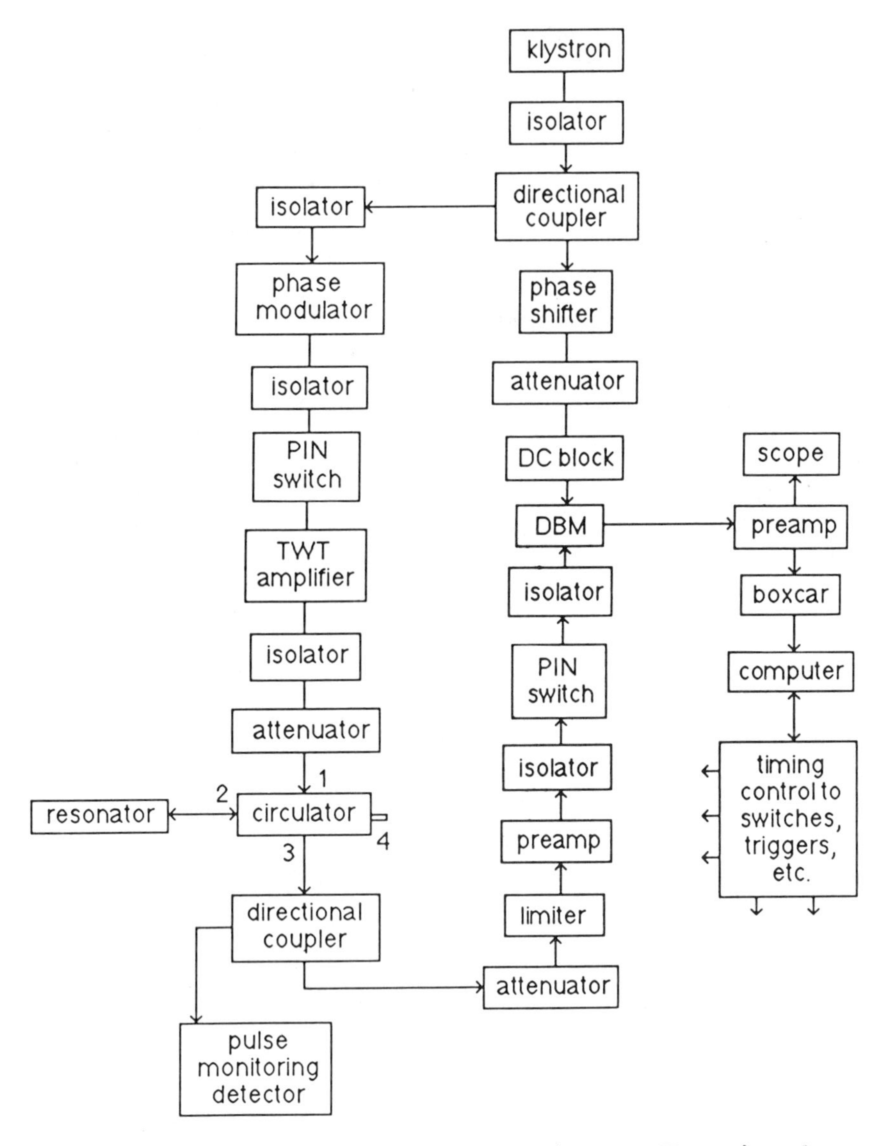

Figure 9. Spin-echo spectrometer block diagram. The spin-echo spectrometer requires only two arms in the microwave bridge, but requires much more sophisticated timing control for pulse shaping and signal detection. Usually the ESE spectrometer uses a higher-power pulse than does the saturation recovery spectrometer, so a traveling wave tube (TWT) amplifier is used to amplify the klystron output. Because of the high pulse power, special care is needed in the microwave echo detection components. Some spectrometers use a low-noise TWT amplifier to amplify the echo prior to the DBM. GaAsFET amplifiers are available with lower noise figures; however, they need to be protected with limiters on their input. The limiter is a device that clips high-level signals and only passes low-level signals. Details on spin-echo spectrometers are given in Refs. 40-42.

of the development of microwave technology in this range during World War II and the subsequent availability of components. The X band has proven to be a convenient compromise, and remains the frequency of choice if one is to have only a single spectrometer. No strong argument could be made for 9 versus 8 or 10 GHz on the basis of spectral resolution. The Varian spectrometers operate at slightly lower frequency within this range than the Bruker/IBM spectrometers.

Bigger is not always better. Higher microwave frequency yields greater spectral dispersion and higher sensitivity (for a fixed sample size), but increasingly inconveniently small sample sizes. There is also increasing problem with sample lossiness at higher frequencies. Recently it has been recognized that some important characterization information is more readily obtained at lower frequencies than at higher frequencies [48]. Consequently, there has been intense activity in developing the S-band (2-4 GHz) region for EPR. The effort to obtain EPR spectra in living organisms has driven spectroscopy to regions of the spectrum in which water is less lossy than at X band. Consequently there has been some activity at even lower frequencies, 1 GHz and below. Resonant cavities at these frequencies are unmanageably large for many of the experiments, and have very poor filling factors [49]. Consequently, other types of resonators are used at these lower frequencies. There are 1-GHz and 4-GHz bridges available from Bruker.

Microwave Source

Klystrons have been the most common microwave source in EPR spectrometers, and have dominated the X- and Q-band systems. A klystron locked via an automatic frequency control (AFC) circuit to a high-Q cavity has been an effective arrangement. For CW EPR a klystron is still the best choice of microwave source at most frequencies. Adequate power (~200 mW), with low noise (amplitude, frequency, and phase noise) is readily available. These systems are also constructed almost exclusively with waveguide components. Recent improvements in solid-state microwave sources, as well as their longer life, smaller size, and lower power supply requirements, have caused them to be used in recent EPR systems. Many X-band pulsed EPR systems, and all recent low-frequency CW EPR spectrometers, have used solid-state microwave sources. Most such sources are noisier than klystrons. Standard high-Q cavities demodulate too much of this noise, resulting in low signal to noise in the spectrum. Thus, a lower Q resonator is beneficial, so long as the B_1 can be made high enough. At frequencies below X band a solid-state source is probably the best choice. For example, several S-band spectrometers have been constructed using the Engelmann CC-24 mechanically tuned transistor cavity oscillator. Bruker uses a klystron in the S-band bridge, but uses a transistor oscillator in the L-band bridge.

The frequency readouts on commercial spectrometers are not direct measurements of the frequency. On the Varian E-Line bridge the frequency readout is a metal tape that is connected via gearing to the mechanical adjustment of the klystron. [Note that on the Varian bridge, the frequency indicated on the metal tape is not valid unless the AFC output is zero (centered).] On the IBM/Bruker ER-200, there is a digital readout, but it also is not actually a measured frequency. To be useful, these must be calibrated. For most purposes it would be adequate to calibrate relative to an HPX532 cavity wavemeter, which has an "overall accuracy" specification of 0.080% (~7 MHz at 9 GHz). More precise, and usually more accurate, measurement can be made with a digital frequency counter. A sample of the klystron output is available at an SMA connector on the back of newer commercial bridges to facilitate such measurements. If a frequency counter is not routinely available, it is worthwhile to rent one to determine the calibration correction for the frequency readout if g-value measurements are important to the EPR study. The calibration will usually be nonlinear, with corrections on the order of a percent or so.

Waveguide versus Coaxial Cable

Wherever it is mechanically convenient, waveguide is used because of its lower loss and higher shielding properties compared to coaxial cable. However, waveguide becomes unmanageably large at frequencies below X band, and even at X band it is awkward to build a microwave bridge assembly using waveguide components. In addition, many of the newer high-performance microwave components are becoming available only in coaxial units. Most microwave bridges are assembled using semirigid coaxial cable. The use of flexible coaxial cable should be avoided except for prototype modules, since it has lower performance than semirigid coaxial cable.

New fabrication techniques make it possible to design most of the microwave system of an EPR spectrometer on a printed circuit board. Thus, there is a trend toward nonwaveguide components in EPR spectrometers. In assembling systems, and especially in attaching the resonator, coupling schemes that are mechanically suitable for waveguide have to be replaced by coupling schemes suitable for coax, stripline, etc. The one component for which it is still essential to use waveguide is the rotary vane attenuator, which is a very useful component since it does not introduce phase shifts.

Resonator

In the 40-year history of EPR spectroscopy, with very few exceptions, almost all EPR measurements have been done using resonant cavities [9]. Indeed, for much of this period those who applied EPR to interesting problems in chemistry, physics, biology, medicine, and materials science have primarily used commercially available spectrometers with TE_{102} resonant cavities. Many articles have been published that provide detailed characterization of the performance of the Varian E-231 TE_{102} cavity [50-52 and references therein].

Several developments have spurred the interest in improved resonators.

1. The very small filling factor of the standard sample geometry in a resonant cavity is a major limitation in the application of electron paramagnetic resonance (EPR) spectroscopy.
2. The large size of cavities at low frequencies is inconvenient.
3. New sources do not have as much power as klystrons so it is important to maximize B_1 for a given power.

4. The urgency to obtain larger B_1 at the sample per watt of incident microwave power in time-domain (pulsed) EPR has also driven development of alternate resonator designs. The larger B_1 and the lower Q, the better for pulsed EPR. Low Q with high B_1 is also desirable for ELDOR (electron double-resonance) and dispersion EPR studies.

Over the past few years several researchers have explored alternate resonator designs, including slotted tube resonators, loop gap resonators, folded half-wave resonators, and dielectric resonators [39,40,53-63]. The figure of merit for a new resonator has been the increase in B_1 at the sample relative to the TE_{102} cavity, and the increased signal-to-noise obtainable with the new resonant structure relative to a cavity. In many cases a report of this aspect of a new resonator has been published even though the device was not characterized even with regard to the B_1 distribution.

Several dramatic improvements over the standard cavity for CW EPR have been reported at L-, S-, X-, and Q-band frequencies. Most of these devices have been generated by "cut and try" methods. Except for those designs that essentially copy Whitehead's and Hyde's papers and patents [53-56], the guiding principle has been, roughly, that something with dimensions of approximately 1/4 or 1/2 wavelength will resonate if brought near a small loop or probe on the end of a semirigid coaxial cable. The geometry has to be selected to maximize the B_1 field at the sample.

The fundamental problem for the design and use of these new resonant structures is to achieve the high B_1 fields simultaneous with a mechanically convenient coupling design that is consistent with the experimental realities of the variety of types of samples that chemists, biologists, and materials scientists want to study.

Some resonators described in the literature are fine for some types of sample, but not practical for other types of samples. The resonator developed by Mims [64,65] was optimized for using liquid samples that were then to be frozen and studied at liquid helium temperature. It has not been used for the more common sample in dilute liquid solution at room temperature, and probably would not function well under those conditions. Varian and Bruker produced a cavity (E-238 and ER/4103TM, respectively) designed specifically to improve the S/N of lossy (especially aqueous) samples at or near room temperature. The new loop gap resonator reported by Hyde [66] for X-band use is designed to give improved S/N for sample-limited aqueous biological samples. The sample holder intended for use with this resonator is now available from Wilmad. It is made of a plastic that is highly permeable to gases, so that a sample can be deoxygenated by passing nitrogen over the outer surface.

The most important features of a resonator are:

1. Background signal—This can be due to contamination, or it can be due to impurities in the materials from which the resonator is fabricated. For example, see the discussion by Buckmaster [67] of impurities in brass cavities.
2. Mechanical stability—Microphonics in the resonator may be the limiting feature in the signal-to-noise ratio of home-built systems.
3. Filling factor—The filling factor is the fraction of the microwave field in the resonator that is filled with the sample to be studied. The filling

factor of cavity resonators is usually very small. One of the main advantages of some of the newer types of resonators is their much larger filling factor relative to a cavity. Signal amplitude is proportional to filling factor.

4. Q—The signal amplitude is proportional to resonator Q. The Q is strongly influenced by the sample, and must be known or controlled to do quantitative EPR.

The Q of a Resonator

a. *Significance of Q*

The applications of resonant circuits exploit the voltage magnification that occurs at the resonant frequency. A definition of Q (sometimes called the quality factor) of a resonator is

$$Q = (\text{frequency})(\text{energy stored})/(\text{energy dissipated per cycle})$$

It can also be shown that Q = (frequency)/(Δfrequency) where Δfrequency is the difference in frequency at the half-power points, that is, the bandwidth. The Q is also a measure of the response time to a perturbation—the ringing time of the cavity after a pulse. In both CW and pulsed EPR the signal amplitude increases with cavity (or resonator) Q. The various contributions to Q are discussed in Ref. 51. In this chapter Q is used to refer to the loaded Q of the cavity in the usual experimental configuration. The microwave magnetic field, B_1, increases as the square root of Q. Consequently, it is important to know Q for quantitative EPR and for measurement of relaxation times. The AFC stability depends on Q. If the Q is too small, the frequency dependence of the cavity response is too shallow and the AFC cannot maintain constant frequency. If the Q is too large, the AFC circuit will not be stable enough to hold the frequency. Note that the Q depends on the surface conductivity of the resonator, which increases rapidly with decreasing temperature. Thus, if the variable temperature system cools the entire resonator and not just the sample, the Q will change with temperature of the sample. At liquid helium temperatures a copper cavity may have too high a conductivity to have a usable Q. The minimum microwave pulse that can be introduced into a resonator depends on the Q. Mims [68] showed that the minimum pulse time, t_p, assuming a Gaussian shape, is

$$t_p = 4Q/\omega$$

The Q-determined ringing time after a pulse limits how soon an echo, FID, or recovery signal can be measured following a pulse. For example, to detect an echo, the cavity has to decay by about 140 dB. With a Q of 300 at X band, this takes about 170 nsec.

b. *Measurement of Q*

On the test bench, for instance, during resonator development, the best method is a swept frequency display of impedance and reactance. This would not be available in many laboratories.

On a pulsed spectrometer, a measurement of cavity ringing time would be the most convenient. The formula to use [69] is

$$i(t) = i_0 \, e^{-\omega t/2Q}$$

For example, if the Q is 300, and the frequency, ω, is $2\pi \times 9 \times 10^9$, the time to decay by 1/e is about 10 nsec.

On a CW spectrometer, the most convenient method is the measurement of the resonator band pass. Details for performing this measurement on the Varian E-line spectrometer are presented in Ref. 51.

An experimental definition of Q for a critically coupled cavity is

$$Q = \nu_0/(\nu_2 - \nu_1)$$

where ν_0 is the resonant frequency of the cavity, and ν_2 and ν_1 are the frequencies at half power (Figure 10). The width of the "dip" at the half-power points is $\nu_2 - \nu_1$. Therefore, it is necessary to measure ν_0, ν_1, and ν_2. The ν_0 is the frequency read on the frequency counter when the spectrometer is in operate mode. Since you want to measure the frequencies at the half-power points, you first need to determine the detector current readings that correspond to half-power. You then adjust the frequency (with the AFC off) until the detector current meter indicates the half-power values. This occurs once on the high-frequency side of the dip and once on the low-frequency side. Since the spectrometer was not designed with this measurement in mind, it is difficult to do. The frequency is changed via mechanical linkages, and neither the detector current meter nor the frequency scale has high resolution. Greater precision can be achieved by using an external frequency meter and by wiring a high-impedance voltmeter across the leads of the detector current meter. With nonlossy solvents the value of Q is about 3500 for the Varian E-231 cavity. Lossy solvents cause a reduction in Q.

B. The Magnet System

Magnet Homogeneity

The magnetic field homogeneity requirements are not as demanding for EPR as for NMR, since the EPR line widths are usually rather large in comparison with NMR line widths. Indeed, inherent in the use of 100-kHz magnetic field modulation is the assumption that the approximately ±35 mG sidebands thereby created will contribute negligibly to the overall line widths. Important cases with narrower lines are known, but unless these are the focus of the study, it is rarely necessary to use shim coils on an electromagnet intended for EPR studies. The commercial systems have magnetic field homogeneity specifications of about ±15 mG over a volume 2.5 cm in diameter and 1.3 cm long. It is important to confirm the homogeneity of the field, especially after a magnet has been moved. The easiest way to do this is to use an oscilloscope display of an NMR signal.

Size of Magnet

For many EPR experiments a 3- or 4-in diameter pole face is adequate. Certainly the Varian E-3, E-4, and Bruker ER-100 spectrometers have been used for many high-quality studies. However, a larger magnet makes possible a wider variety of studies. There are three factors that should

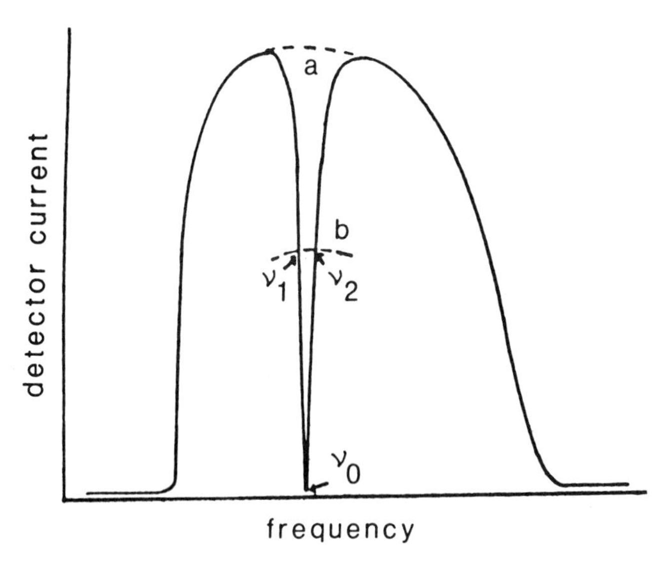

Figure 10. Klystron power mode, showing the "dip" due to cavity absorption. The vertical axis is power (detector current), and the horizontal axis is frequency; a is the detector current off resonance, b is the detector current at half-power, ν_0 is the resonant frequency of the cavity, and $\nu_2 - \nu_1$ is the half-power bandwidth of the cavity. The width of the dip is a measure of cavity Q. The dip should be in the moddle of the power mode to get maximum power from the spectrometer. When the reference arm power is turned on, the display increases in amplitude and the dip does not touch the baseline, since in this case the power from the reference arm adds to the power reflected from the cavity. When the reference arm is on, the symmetry of the power mode with respect to the dip is a measure of the relative phases of the reference arm and the signal arm.

be considered in selection of the size of the magnet—maximum field required, volume of region with homogeneous field, and gap size.

For most X-band studies the maximum magnetic field needed seldom exceeds ~10,000 G and the magnetic field strength needed does not usually dictate the size of magnet. For Q band (35 GHz), the center field needed is 12,500 G. Higher-field EPR becomes so demanding of the magnet that the primary focus shifts from the microwave system to the magnet.

The volume of space needed for the experiment places the most stringent requirements on the size of magnet. The magnet pole face diameter needed is determined primarily by the homogeneous volume needed for use of a dual cavity. For example, with two TE_{102} cavities combined, as in the Varian E-232 dual-cavity assembly, the two samples are 43 mm apart. Such cavities would most commonly be used for comparative g-value measurements with highly resolved spectra, so the magnet homogeneity needed is maximum for this case. The larger the pole face diameter, the easier it is to produce homogeneous fields. Usually, ~9-in diameter is adequate, but 10-in diameter is needed for the Bruker variable-temperature Q-band accessory.

The magnet gap (the distance between the pole faces) needed has to be at least 27 mm (Varian) to 35 mm (Bruker/IBM) to accommodate the width of the standard X-band EPR cavities. Less space could be adequate for some of the newer resonant structures, but even here a shield of about these dimensions is commonly used. The mechanical housings of some common commercial variable temperature apparatus are wider than the cavities with which they are to be used. Care needs to be taken to avoid size incompatibilities. For example, some of the closed-cycle cryogenic systems will not reach to the center of a 12-in magnet unless the gap is very large. Experiments involving external coils, such as imaging studies, require additional gap space.

If resources are available, a wider gap is preferable, since it facilitates future experiments. The tradeoffs are that magnet weight and cost increase with pole face diameter and gap size for a given field strength. Also, the magnet power supply size needed increases with gap size for a given field strength.

Magnetic Field Measurement and Control

The accuracy of magnetic field measurement and control determines the accuracy with which one can measure g and a values, compare areas under spectra, and subtract background spectra. The most common field measurement devices incorporated in EPR spectrometers for feedback control are Hall probes and rotating-coil gaussmeters. Most magnetic field controllers have gain and offset controls that permit calibration of the field scan. The portable Hall-probe gaussmeters that have recently become available are very handy, but are not sufficiently accurate for calibration of EPR spectrometers. The temperature-compensated Hall probes built into the commercial EPR spectrometers are more accurate than the test equipment. The new Bruker/IBM ER 032M Hall probe has especially impressive specifications: absolute field accuracy better than 500 mG over the full 20 kG range, 10 mG resetability, and sweep linearity ±5 mG or $\pm 1 \times 10^{-3}$%. For calibration of the magnetic field it is necessary to use an NMR gaussmeter, or a sample whose spectral features are well documented.

Since the magnetogyric ratio for the electron is about 658 times that for the proton, the corresponding NMR frequency at the magnetic field used for X-band EPR is about 14.5 MHz. The magnetogyric ratio for the proton is known with good accuracy, so an NMR system makes a good gaussmeter for calibration of the magnetic field. Many laboratories have the old, but still utilitarian, Harvey-Wells G-502 NMR gaussmeter. Users of this system should note the detailed calibration of the standard "yellow probe" that has been published [70,71]. It is also important to recognize that the field at the sample is not the same as the field at the gaussmeter probe. A special "in-cavity" probe is available from Bruker for this measurement. For very precise measurements, it has been shown that the limiting error in the measurement of the magnetic field at the sample is the reproducibility of the field inside the cavity for a given field outside the cavity [70].

If an NMR gaussmeter is not available, secondary standards can be used with accuracy adequate for most studies. Near g = 2, Fremy's salt is useful (see Ref. 8, 2nd ed., p. 445). Over wider ranges some Mn(II) and Fe(III) samples can be used [52].

Limiting the accuracy of field measurements is the method of varying the magnetic field. Early spectrometers used a motor-driven potentiometer (e.g., Varian Fieldial). Later, stepping motors with 9600 steps per scan were used on the Varian E-line spectrometers. The evolution toward computer-controlled spectrometers has actually decreased the precision of magnetic field control, since D/A resolution and computer memory limitations have often restricted data collection to 1024 points per spectrum. Modern systems now collect 4096 points per spectrum routinely, and 8192 points per spectrum is possible. However, now that memory costs are decreasing rapidly, it seems reasonable to expect a return to the precision of the mid-1960s. This precision is not useful unless the accuracy of the field is calibrated and appropriate time constants are used in recording the spectrum.

C. Signal Detection, Amplification, and Display

The field-modulation and phase-sensitive detection aspects of EPR spectrometers were discussed in Section I.B. Numerous microwave detection schemes have been used for EPR, but most spectrometers use crystal detectors (point-contact, Schottky barrier, tunnel or backward diode). The crystal rectifies the microwaves, producing a voltage that is subsequently amplified. This voltage carries the various modulation information, such as the 70-kHz AFC and 100-kHz (or lower) magnetic field modulation, used in the phase-sensitive detectors in later stages of the detection system. Varian and Bruker/IBM have used different types of crystals as microwave detectors, so the visual display available to the operator is different in these two major types of spectrometers. The Varian V4500 and E-Line spectrometers use a tunnel-diode crystal. During tuning, the operator makes adjustments to bias the crystal with about 200 μA detector current to put it in a region in which the output voltage is linear with microwave power. The oscilloscope display during the "tune" mode of the spectrometer is a plot of microwave power reflected from the cavity (vertical) versus microwave frequency (horizontal). The power output from a klystron is a nonlinear function of frequency, so the display is a broad hump. When the frequency is tuned to the resonant frequency of the cavity (or other resonator), the energy incident on the cavity is absorbed in the cavity, and there is a "dip" in the power reflected from the cavity. The relative frequency width of this dip reflects the Q of the cavity (see Section III.A.5 for details). The dip touches the baseline (zero reflected power) when the cavity is critically coupled to the source. The Bruker/IBM ER200 spectrometer uses a Schottky barrier diode crystal. This crystal produces output only for microwave powers greater than a certain minimum value. In "operate" mode, the crystal is biased into this region by the reference arm. However, in "tune" mode, the crystal is not biased in some versions of the spectrometer, and the diode turns off at the low microwave power at the bottom of the dip. The result is that the dip is broad and flat-bottomed for all except the lowest Q or grossly mismatched cavity. Consequently, the operator cannot make a quantitative measurement of cavity Q on the ER200. No provision was made for measuring Q on the Bruker/IBM spectrometers, because of the necessity of eliminating DC coupling to the detector preamplifier. This is a serious problem for quantitative EPR (see Section V).

In the earlier Varian V4500 EPR the operator had sufficient control over the klystron control voltages that several klystron power modes could be seen, and the proper mode had to be selected. In the newer systems the control voltages are designed to keep the operator on the power mode of interest.

High-quality double-balanced mixers (DBM) are now available in the microwave region, making their incorporation into EPR detection systems attractive. Most pulsed EPR spectrometers use DMBs, with the reference arm to the LO side and the detected signal to the RF side. Since these inputs have the same frequency, the DBM works as a phase-sensitive detector, and the output is a voltage that carries the modulation information, if any.

Since the first amplifier usually determines the noise level of the detection system, and the EPR signal is a very low-level signal, the use of a low-noise (e.g., NF = 2.5) microwave amplifier is an attractive concept for improved S/N in EPR spectroscopy. For electron spin-echo systems either a low-noise traveling-wave tube amplifier or a GaAsFET amplifier is a standard component of a detection system. In CW EPR spectrometers the use of such an amplifier must be considered carefully. In the ideal case, with a perfectly critically coupled resonator, a microwave preamplifier would provide a significant S/N advantage over the usual detectors. To take optimum advantage of a high-gain low-noise preamplifier, very good coupling to the resonant structure is essential. For example, if the coupling is only 40 dB (1 μW reflected when the incident power is 10 mW) a preamplifier gain of more than 20-30 dB will result in saturation of the output of the preamplifier (this usually occurs at about ±10 dBm). Coupling of 40 dB is as good as is usually achieved in the waveguide systems with iris and tuning-screw coupling. The older systems, for which these coupling schemes were first developed, were purposefully run mismatched to provide reflected power to bias the crystal detector. The transition to modern detection systems has had only limited success so far because the resonator coupling mechanism has not been as carefully designed as the rest of the system. Consequently, GaAsFET preamplifiers will be useful for very low-level signals, such as are used for easily saturable signals and for saturation recovery experiments, but will not soon be useful for high-power studies of metals etc.

Another portion of the detection system is the oscilloscope that displays the output of the phase-sensitive detector. When the magnetic field is swept by the recorder or a computer, the oscilloscope displays the signal for a particular point in the spectrum. To obtain a display of a spectrum on the oscilloscope, the voltage from the field controller is held constant and the magnetic field is swept by another voltage applied to the modulation coils (or to auxiliary rapid scan coils). This is why the oscilloscope can display only a narrow segment of the spectrum (up to 40 G wide on the Varian E-9, and 50 G wide on the ER200; 200 G with the rapid scan coils installed). In this mode ("oscilloscope field sweep" on the Varian, "rapid"

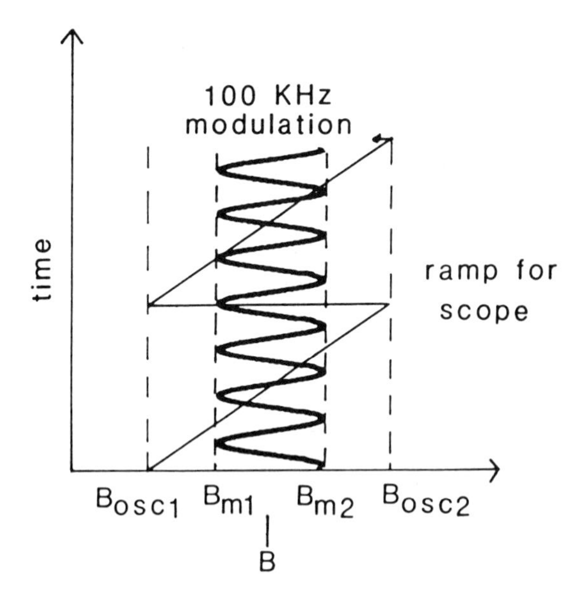

Figure 11. For display on the spectrometer oscilloscope, the magnetic field is swept synchronously with the oscilloscope display. To accomplish this, a second, lower-frequency "sawtooth" modulation of the magnetic field is superimposed on the 100-kHz field modulation. This sawtooth is at 60 Hz in the V4500 series, 34 Hz in the E-Line and Century series, and selectable from 50 Hz to 0.10 Hz in the ER-200 and ESP300 spectrometers.

on the ER200) there are two separate signals applied to the cavity modulation coils: (1) the 100-kHz modulation discussed in Section I.B and (2) a lower-frequency (34 Hz on the E-line, 60 Hz on the V4500, selectable from 50 to 0.10 Hz on the ER200) linear sawtooth modulation as illustrated in Figure 11. The oscilloscope is synchronized with the sawtooth so that the entire signal between B_{osc1} and B_{osc2} is displayed on the oscilloscope. The field scan is not linear in this mode. Because of the trigger rate, it is not possible to get an oscilloscope display when very-low-frequency magnetic field modulation is used. The ER200 has an additional set of modulation coils mounted on the magnet pole faces, which can be used to obtain 100- and 200-G wide displays on the oscilloscope. These coils can also be used for rapid-scan averaging for signals up to 200 G wide. The Varian E-271 rapid-scan accessory adds a pair of coils for rapid scans up to 100 G wide.

D. Data System—Capabilities Needed

About 15 years ago even computer manufacturers were skeptical that there was a valid need for use of computers in EPR. Pioneering work by Klopfenstein [72] and Goldberg [73] showed the path, which has now led to such elaborate manifestations as the elegant computer-controlled ENDOR studies of Spaeth [74]. A few years ago the field was reviewed comprehensively by VanCamp and Heiss [75].

Data collection by computer is discussed in another chapter in this volume and in several reviews [72-77]. The minimal features that should characterize a computer data acquisition and analysis system for EPR include (1) 12-bit (plus sign) A/D for signal intensity, (2) 12-bit (plus sign) D/A for magnetic field control, (3) memory to store several spectra simultaneously with easy access, (4) software for baseline correction, subtraction, integration, smoothing, and control of as many functions as are accessible on the spectrometer, (5) ability to transfer spectra to other computers, and (6) high-level language programming (e.g., FORTRAN, Pascal). The speed required for the A/D capabilities is strongly dependent on the experiment to be done. Normal slow-passage EPR spectra with narrow lines rarely are scanned in less than 2-4 min, using a filter time constant in the detection/display system of ~0.05-0.1 sec. If each step in the magnetic field is collected as a single data point in the array, a very slow A/D can be used. Actually, higher-frequency noise comes through the system than is implied by the filter time constant, so it is worthwhile to clean up the spectrum by sampling the A/D at each field step as many times as the speed of the A/D and the processor speed permits. As a practical compromise, something of the order of 2000 samples/sec gives adequate signal averaging [78]. In the new Bruker/IBM 023M module, there is an A/D whose resolution and sampling rate depend on the magnetic field scan rate, in order to optimize digitization.

Time-domain EPR experiments make greater demands on the digitizer. Indeed, components are not yet available to directly digitize signals as fast as is needed for spin-echo experiments. The usual response to this problem is to use a boxcar integrator to sample many echoes and digitize the much slower output from the boxcar. Saturation recovery and rapid response CW spectrometers can take advantage of the new digital storage oscilloscopes, such as the LeCroy 9400, which can digitize directly at 100 MHz, and as fast as 5 GHz with the so-called random interleaved sampling technique on repetitive signals.

E. Commercial EPR Spectrometers

Early commercial EPR spectrometers were listed in table 3-4 of Alger's book [7]. Over the past two decades, Varian, JEOL, and Bruker/IBM have been the major commercial suppliers of EPR spectrometers. The total market is fairly small, relative to the market for other types of analytical instrumentation, so marketing decisions by large corporations have caused major fluctuations in availability of various models. For example, JEOL stopped marketing EPR in the United States in about 1972, though JEOL EPR spectrometers continue to be available elsewhere. Varian, after defining the field with, in succession, the V4500 series, the E-3, and the E-Line series (E-4, E-9, etc.), and finally the Century Series (E-109, etc.), stopped marketing EPR spectrometers in 1984. Very few parts are still available for Varian spectrometers. When IBM entered the analytical instrumentation market in 1981, Bruker EPR spectrometers, enhanced by IBM design criteria and support, became more widely marketed in the United States. When Varian left the market, the ER200 became the successor to the E-109 as the state-of-the-art spectrometer. The next generation EPR, the ESP300, was marketed in Europe starting in 1986, and is now available in the USA. This spectrometer was announced as the ER300 by IBM just before IBM left the EPR market in 1987. Many of the modules are the same as in the ER200, but there is a new computer system integrated into the design of the spectrometer, rather than attached to it. Also during 1987, Micro-Now Inc. began marketing a routine research EPR spectrometer in the spirit of the Varian E-3 and E-4 (for many years they have marketed a "teaching-grade" apparatus for demonstrating EPR, and upgrades for Varian V4500 series bridges).

Bruker recently introduced the ESP380 pulsed EPR accessory for the ESP300 series spectrometer. The ESP380 is capable of a wide range of pulse sequences for saturation recovery, spin echo, echo envelope modulation, etc., with microwave phase rotation. The bridge has four dual-channel microwave pulse forming units, controlled by a pulse programmer with 2 nsec resolution.

There is very little competition in the commercial EPR market. One must either accept the features of the one or two spectrometers available, modify one to fit one's own needs, or build one's own.

IV. THE SAMPLE

The chemical and physical nature of the sample, as well as expected magnetic behavior, relates to the requirements for and limitations placed on the spectrometer. For example, the effect of the loss factor of the solvent, whether the sample is limited in size, whether it is stable, etc., determine the type of EPR experiment that can be performed. Some other features of the interaction of instrumentation with the information you want to get from a sample were mentioned in Section III. Some of the sample conditions to be considered in the design of EPR experiments follow.

A. Spin System

EPR spectra are more readily observed for systems with an odd number of unpaired electrons [1]. For systems with even numbers of unpaired electrons EPR spectra are frequently not observed. For example, it would be difficult to prove whether paramagnetic Ni(II) is present by taking an EPR spectrum, except in very special circumstances. More details on the spectra expected for various species are given in Section VIII.

B. Temperature

Selection of Temperature for Operation

The first question to consider is the temperature range in which prior work would lead one to expect to be able to observe the spectrum of interest. Some guidance is given in Section VIII. It is important to know how sensitive T_1 and T_2 might be to temperature. In general, the lower temperature, the more likely it is that an EPR signal can be observed, both because of relaxation times becoming longer and because the Boltzman population difference becomes more favorable. However, if the relaxation time becomes too long relative to the modulation frequency in CW EPR, passage effects can affect the signal shape.

Temperature Control

Temperature control in magnetic resonance has never been engineered as carefully as the microwave and electronics aspects. Usually, the nature of the sample and the interaction of the cavity or resonator with anything put into it makes it most convenient to change the sample temperature by passing heated or cooled gas over the sample. If the resonator is small, as in Q-band measurements, or some of the newer S- and X-band resonators, it is mechanically more convenient to cool the resonator and the sample together. This changes the mechanical dimensions of the resonator, and its electrical conductivity, so both frequency and Q change with temperature. Mechanical instabilities can also be a problem. In gas flow systems there is usually a temperature gradient over the sample. This gradient might be more than a degree Kelvin. Consequently, at very low temperatures, where the spectrum changes rapidly with temperature, or at room temperature for some biological samples (such as membrane melting or protein unfolding studies), the gradient might be larger than the temperature range of interest. Special efforts should be taken to map the temperature gradient if temperature is a critical variable. For precise temperature control near room temperature a recirculating liquid (remember it must be nonlossy, such as a silicone fluid, or a hydrocarbon or fluorocarbon) system is probably better than a gas stream. This would have to be locally constructed, since no commercial unit is available. Reference 79 reviews both high- and low-temperature experimental techniques.

C. Amount of Sample

The amount of sample to be used in an experiment depends on the information sought. However, in general, the smaller the amount of sample the better, so long as adequate S/N can be obtained. Too large a sample can cause problems. For the spectrometer response to be linear the Q change due to the microwave resonant energy absorption by the sample must be a small perturbation of the cavity Q. Goldberg has described the limits in detail [80]. With the standard TE_{102} cavity at X band the number of spins must be kept below about 10^{18}, corresponding to less than about a milligram of compound, in order to avoid excessive change in Q at resonance [80]. Commercial X-band EPR spectrometers can detect roughly 10^{10} spins of a material having a 1-G-wide line. In practice, with saturable organic radicals in solution, one would use about 100 μl of solution, and expect to be able to observe radicals at micromolar concentration. Concentrations between 10^{-4} and 10^{-3} molar are good practice.

When positioning the sample in the cavity, one must be aware of B_1 distribution in the cavity and be aware of the effect of sample on B_1.

Position is more important for small samples than for samples that extend the full length of the cavity. This is especially important in efforts to compare two samples, as in spin concentration determinations, and in efforts to measure relaxation properties.

D. Phase of Sample

Intermolecular electron-electron spin interaction causes loss of resolution of EPR spectra. Thus spectra are usually obtained on samples that are magnetically dilute. Magnetic dilution can be achieved by dissolving a sample in a diamagnetic solvent or doping into a diamagnetic solid. It is much harder to interpret EPR spectra of magnetically concentrated samples. Unless the goal is to characterize the magnetically concentrated material, it is better to dissolve the solid (if it is stable to the solvation).

Solid State

Solid-state samples can be single crystals, powders, or frozen solutions. If frozen solutions are used, solvents should be selected that give glasses. Techniques for single-crystal EPR are discussed in Refs. 81 and 82.

Fluid Solution

a. Concentration

In NMR one usually strives for the highest sample concentration that does not lead to viscosity broadening of the spectrum. In contrast, in EPR one strives for the lowest concentration for which adequate S/N can be obtained. The difference is the effect of intermolecular interactions on spin relaxation, and hence on broadening of the spectrum. How careful one must be about this depends on the type of information one seeks from the sample. If the question is simply, "is square-planar Cu(II) present?" almost any sample that is about 1-5 mM in Cu(II) will be adequate to answer the question. If the question is "how much Cu(II) is present?" or "is the Cu(II) in the proximity of another paramagnetic species?" then much more careful sample preparation is required. In general known, low concentrations should be used. For some purposes it will be necessary to extrapolate to infinite dilution, and to check for solvent effects, including aggregation.

b. Solvent

Much of the discussion of solvents in this chapter emphasizes the fact that lossy solvents make it more difficult to obtain quantitative EPR spectra. Note, in this regard, that the solvent loss tangent can be very temperature dependent. One dramatic case is CH_2Cl_2—small changes in temperature near 0°C can dramatically change the apparent concentration of species dissolved in CH_2Cl_2 unless care is taken to account for changes in solvent loss tangent (and hence in Q).

The mere fact that a solute dissolves in a solvent means that there are significant solvent-solute interactions. This chemistry can affect spectra.

Experiments in which relaxation times are important can be affected by nuclear spins in the solvent, so the use of deuterated solvents can be important in special cases.

c. Oxygen

Most EPR spectra are sensitive to the presence of oxygen or other paramagnetic gases. The longer the relaxation time, the more sensitive the spectrum is to the presence of oxygen. Degassing usually means getting rid of oxygen. Bubbling argon or nitrogen through the sample, or evacuating with a few cycles of freeze-pump-thaw, will reduce the oxygen concentration to the point that it will not broaden the CW EPR spectrum. However, for relaxation-time measurements, or for measurements such as ENDOR and ELDOR, which are very sensitive to relaxation times, one has to continually improve degassing techniques. The longest relaxation time measured is the closest to the correct value. NMR relaxation times are particularly sensitive to removal of air, and the lack of effectiveness of some of the common oxygen-removal techniques has been documented (p. 161 in Ref. 83; [84]). The best advice in this aspect of experimental technique is never trust your result, and keep trying.

d. Other paramagnetic impurities

Other paramagnetic impurities can affect EPR spectra and relaxation times. Beyond the presence of unsuspected components of the sample (e.g., it was not known that it contained iron), the most likely problem is dirty sample tubes. Also be alert that attempts to clean tubes may introduce paramagnetic impurities, ranging from rust particles to Cr(III) from cleaning solutions.

Gaseous Samples

EPR has only limited applicability to gaseous samples. Most of the species that have been studied are very small molecules. See Ref. 85 for a discussion of gas-phase EPR.

E. Impurities, Overlapping Spectra

Always check the "background" signal. Even high-quality synthetic quartz sample tubes yield EPR signals. Pyrex, or other borosilicate glasses, commonly used because they are so much cheaper and are easier to use in fabricating special apparatus, yield strong EPR signals. The signal due to the tube has been reported in the literature and attributed to the sample. The cavity and the variable-temperature Dewar insert also produce signals. Just as stopcock grease keeps showing up in NMR spectra, iron, copper, biphenyl ion, and adventitious nitroxyl radicals recur in EPR spectra and often are attributed to other species.

Be quantitative—you might be looking at only a small part of the sample. EPR spectra can provide detailed information about chemical equilibria and purity. Conversely, if the chemistry of the sample is not taken into consideration, superposition of spectra due to chemical equilibria, decomposition products, or impurities in the sample may be misinterpreted as features of the spectrum of the species the experimenter thinks (or hopes) is present. Remember that the special benefits of derivative spectra for highlighting sharp spectra also can be a liability when sharp and broad spectra are superimposed. Even less than 1% of a sharp spectrum, when recorded "full scale" in the usual way, can mislead the unwary into ignoring the rest of the signal, which may be the signal of the species of interest. Many examples of this problem can be found in the literature.

F. Other Environmental Effects

Varying other environmental conditions, such as pressure, controlled atmosphere, light, or other radiation, may be central to the information one wants from the EPR spectrum. The standard commercial X-band cavities are designed with slots in the end of the cavity for irradiation of the sample. Irradiation is harder to do with the Q-band resonator and some of the newer resonators. However, the loop-gap and slotted-tube resonators can be designed with adequate slots for irradiation. Special sample arrangements for high-pressure studies have been described [86,87].

V. QUANTITATIVE MEASUREMENTS OF SPIN DENSITY

The older literature (e.g., Alger's book [7]) suggests that EPR is inherently nonquantitative. This impression is erroneous. Both improvements in instrumentation and improvements in awareness of relevant experimental parameters make it reasonable to aspire to about 1% accuracy in signal area measurements. However, it is necessary to be careful about many facets of the experiment to obtain reliable quantitative EPR spectra. This section provides guidance. Several reviews provide details and full documentation [52,88,89].

The key to quantitative EPR is attention to detail, and awareness both of the sample and of the spectrometer. Special attention must be paid to the effect of solvent, either by measurement of cavity Q or by keeping all sample parameters constant, and to the selection of the reference material (including knowing its behavior as a function of temperature).

The fundamental article on EPR sensitivity is by Feher [90]. Most textbook treatments are based on Feher's article. A summary of Feher's results is provided in the Varian E-Line Century Series EPR instruction manual. Reading these various sources can be confusing, even after differences in notation are sorted out, because so many of the variables are functionally related to other variables, permitting many quite different looking "correct" expressions. Furthermore, many treatments seek expressions for the dependence of minimum detectable number of spin on, for example, frequency. These calculations can be done for many different situations. For example, the Varian manual summarizes the dependence of signal amplitude on frequency for eight combinations of whether the signal can be saturated, whether it is limited in size, and whether it is lossy. An emphasis on minimum detectable number of spins does not provide formulas directly applicable to the usual case of a sample in a lossy solvent contained in a cylindrical sample tube. Several articles present relevant material [51,80,89,91,92]. The most applicable summary is in Ref. 51.

The general principles in the following discussions apply to most spectrometers that use a reflection cavity and magnetic field modulation, but the details relate specifically to the Varian E-3 and E-Line spectrometers, since most relevant data in the literature are for these spectrometers. The reference-arm bridge was a major contribution toward making routine quantitative measurements of signal area possible.

A. Spectrometer Calibration

Goldberg provided detailed consideration of linearity in Varian V-4500, E-3, and E-Line spectrometers [93]. The important message is that investigators should check the linearity of modulation amplitude, microwave power, receiver amplification, and magnetic field scan of their spectrometers.

In addition to checking linearity of modulation amplitude settings, it is important to calibrate the absolute magnitude of the modulation amplitude. This is best done with a small sample that has a narrow line. A speck of DPPH is adequate. One measures the broadening of the line due to excessive modulation amplitude. Reference to Figure 5, which shows how the derivative line shape is obtained, reveals that when the modulation amplitude exceeds the line width, a signal such as that given in Figure 12 will be obtained. The splitting between the positive and negative excursions, corrected for the original nonbroadened line width, gives the magnitude of the modulation amplitude. Details of the procedure and tables necessary for the task are given in Poole [8, 2nd ed., p. 242]. The modulation coils are resonated at 100 kHz on commercial spectrometers. The modulation amplitude at the sample depends on the modulation frequency, and has to be separately calibrated and adjusted for each frequency and for each cavity.

Note also that the shape of the modulation field depends on the cavity design. The modulation field in the Varian E-231 cavity has a roughly cosine shape relative to the center of the cavity, decreasing toward the top and bottom. This must be accounted for if samples of different length are compared. In the Varian E266 Q-band cavity assembly the modulation coils are mounted on the outside of the variable-temperature Dewar, and are larger than the cavity, so the modulation is nearly constant over the cavity. The details of these considerations are discussed in Ref. 50.

B. Calibration of Sample Tubes

Unless one buys the most expensive EPR tubes, there will be a range of sizes (e.g., 3.2-4.4 mm for nominal 4.0-mm OD tubes) in each batch. Therefore, one should select similar tubes (measured with a micrometer) to minimize tube corrections. Internal diameters should be calibrated gravimetrically using standard procedures for volumetric glassware (e.g., weighing water in tubes). The simplest way to perform routine measurements of signal area is to calibrate a few tubes with aliquots of the same sample. Such tubes, used with the same type of sample and the same

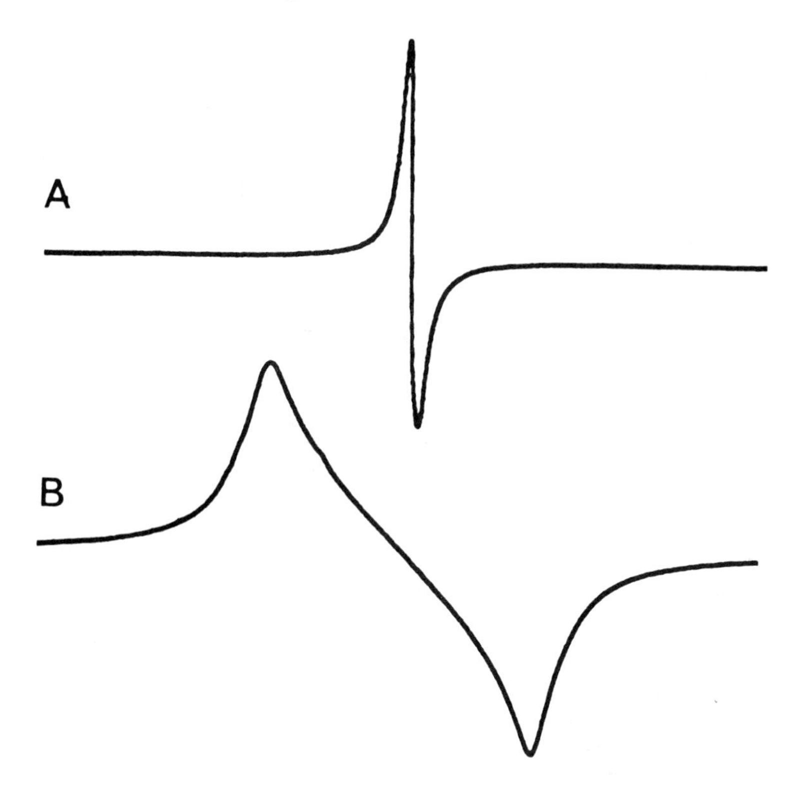

Figure 12. (A) A spectrum of solid DPPH, obtained with modulation amplitude $\ll$ line width. (B) Spectrum of the same sample obtained with modulation amplitude $\gg$ line width, which demonstrates the distortion of the spectrum due to overmodulation. The distortion follows directly from the picture in Figure 5. For example, if the field is centered on resonance, and the modulation amplitude is large relative to line width, i_1 and i_2 would both be approximately zero. The separation between the two "peaks" in this display depends on the modulation amplitude and on the linewidth of the undistorted EPR signal. The use of this distortion to calibrate the modulation amplitude is described in detail in Ref. 8.

solvent, temperature, etc., carefully positioned in the cavity, can yield results accurate to within 5%, and within 2-3% with care. In favorable cases, with good S/N, one may reasonably aspire to 1% accuracy. If a small amount of solid sample is used it must be positioned carefully at the center of the cavity.

C. Consideration of Cavity Q

In measurements of spin concentrations it is necessary to determine cavity Q or keep all conditions affecting Q rigorously constant. Several papers in the analytical chemistry literature have described sample tube/holder assemblies intended to keep the experimental conditions (sample tube diameter and positioning) as constant as possible [94,95]. There are many experimental situations in which such an approach is not very convenient—for example, samples that have to be prepared on a vacuum line.

At constant incident power the signal amplitude is linearly proportional to cavity Q. Insertion of any material into the EPR cavity changes the resonant frequency of the cavity, the Q, and the distribution of electric and magnetic fields in the cavity. This effect is presented pictorially in Figure 3-8 of Refs. 89 and 96. Particularly important manifestations of this effect include the distortions caused by the quartz variable-temperature Dewar commonly used in EPR, and the distortions caused by the sample tube and solvent themselves. The presence of dielectric in the cavity shifts the resonance frequency to lower frequency (a conductor has the opposite effect). The range of sample tubes and solvents commonly used in chemical laboratories can affect Q by much more than a factor of 2 even when sizes, sample amounts, etc. are chosen to make it easy to tune the spectrometer. The most detailed examination of this matter is given in Ref. 96, to which reference should be made for derivations of the relevant formulas.

In view of the importance of cavity Q, a discussion of the measurement of Q on the widely used Varian E-Line spectrometers is included in Section III.A.

D. Reference Samples for Quantitative EPR

One always does relative measurements of EPR spectral areas [52,88,97]. Absolute measurements are so difficult that in practice one would seldom have reason to attempt such measurements. The more similar the reference standard is to the unknown, the more accurate the quantitation of the EPR spectrum will be. Some especially egregious errors have been made in sincere attempts to achieve as accurate a result as possible. For example, researchers have gone to great effort to obtain carefully hydrated single crystals of copper sulfate to use as a reference for the quantitation of the EPR signal of an aqueous solution of an enzyme. The difference in the effects on cavity Q of a small single crystal and of an aqueous solution presents one of the worst possible cases for accurate comparison. Wherever possible, use a sample of the same spin state, and hence the same spectral extent, at about the same concentration in the same solvent, and in the same tube, or one that has been calibrated relative to the tube used for the unknown.

E. Scaling Results for Quantitative Comparisons

Unless the spectra to be compared are obtained under exactly the same conditions, it will be necessary to scale the area obtained for the unknown to compare it with the standard. This involves:

1. Subtract the background spectrum.
2. Correct for modulation amplitude. Area scales linearly with modulation amplitude, even when spectra are overmodulated.

3. Correct for gain. Area scales linearly with gain settings of the detector amplifier.
4. Correct for microwave power. Be sure to obtain spectra at power levels below saturation. Under these conditions, area scales as the square root of the incident microwave power.
5. Correct for Q differences. Area scales linearly with Q.
6. Correct for temperature differences. Note that EPR is a measurement of bulk spin magnetic susceptibility. Thus, the area scales with the population difference of the ground and excited states. For isolated $S = 1/2$, the temperature dependence obeys a simple Boltzmann behavior. Interacting spin systems, and cases of $S > 1/2$, can result in complicated temperature dependences. In this regard, note that most solid reference materials are not magnetically dilute, so their temperature dependence does not obey a simple Curie law [88,98,99]. A paper on chromic oxide is illustrative of the efforts needed in these types of cases [100]. Also note that the density of solutions, and hence the number of spins at a given position in the B_1 field, varies with temperature. In addition, solvent loss tangent varies with temperature.
7. Correct for spin differences. The transition probability, and hence the area, scales as $S(S + 1)$. For example, comparing $S = 1/2$ and $S = 3/2$, assuming all transitions are observed, the relative areas are 3/4 to 15/4 from this effect. If one observes only part of the transitions, which is common for $S > 1/2$, it is also necessary to correct for the relative multiplicities, which go as $(2S + 1)$.
8. Correct for g-value differences. Area scales as g. Note that papers and books published prior to Aasa and Vanngard's paper state incorrectly that area scales as g^2 [101].
9. Correct for field scan width. For first-derivative spectra, the scan width correction factor is $(1/\text{sweep width})^2$.

   Finally, note that if a computer is being used, the accuracy of the comparison may be limited by the digitization accuracy or resolution of the A/D converters or memory word length. For example, a large amount of the true area of a wide Lorentzian line could be lost in the least significant bit of an 8-bit A/D.

VI. GUIDANCE ON EXPERIMENTAL TECHNIQUE

A. Words of Caution

Unlike many modern instruments, EPR spectrometers, having been built in a tradition of use only by specialists, are not designed to be foolproof. Considerable care is required in using an EPR spectrometer to (a) prevent costly damage and (b) obtain useful results.

Cooling Water

Water cooling is needed for the klystron and for the electromagnet. Unlike the recent Bruker spectrometers, the Varian systems do not have interlocks to prevent damage to the klystron if you forget to turn the cooling water on.

Cleanliness

Most dirt, and tobacco smoke(!), has an EPR signal. The cavity must be protected from dirt. It is good practice to prohibit smoking in the spectrometer room, and to clean the outside of all sample tubes just prior to inserting them into the cavity. A tissue such as Kimwipe and a cleaner such as a residue-free electronic contact spray cleaner are handy for this purpose. Fingerprints will transfer dirt to the cavity.

Inevitably, the cavity will become contaminated, possibly from a broken sample tube, causing significant background signals. The mildest cleaning that will remove the known or suspected contaminant should be used. Check the manufacturer's guidance on solvent compatibility. Varian suggests chloroform and acetone, while IBM and Bruker suggest ethanol, hexanes, toluene, and/or 0.1 M EDTA followed by methanol. Some people use an ultrasonic cleaner, but this is not wise—it may loosen critically torqued parts of the Varian cavity. After cleaning, purge with clean dry nitrogen gas to remove the last traces of solvent.

Changing Samples

Always attenuate the power to ~40 dB below full power before changing samples, to avoid an unnecessary spike of power to the detector crystal.

Detector Current

The output of the crystal detector depends on the magnitude of the bias current. Each spectrometer should be checked to determine the range of detector current values within which the signal amplitude is independent of detector current. If the detector current drifts, as can happen with lossy solvents, or when the temperature is changed, significant errors in signal amplitude can result; S/N is degraded, and quantitative measurements are prevented.

B. Selection of Operating Conditions

While initially searching for a signal in a sample whose spectroscopic properties are not known, one can use relatively high spectrometer settings, such as 10 mW power, 1 G modulation amplitude, and a fast scan. This is likely to be adequate to at least detect a signal, if it is present, with reasonable S/N. Recognize, though, that such a cursory scan could miss samples at two extremes: (a) a signal with such a long relaxation time and narrow line that it is saturated or filtered out, and (b) a broad signal in the presence of a more obvious sharp signal. Always look for spectra you do not expect.

To obtain a quantitatively correct spectrum requires adjustment of microwave power, phase, modulation amplitude, gain, scan rate, and filter time constant. The criteria for selection of these settings are discussed in the following paragraphs.

Microwave Power

To obtain quantitative CW EPR spectra and to obtain undistorted EPR spectra, it is necessary to obtain spectra at microwave powers below those that cause significant saturation of the EPR spectrum. Most EPR samples can be saturated with the power levels available in commercial spectrometers. Thus, it is always important to check for saturation.

The incident microwave power should be set to a value below that at which the power saturation curve deviates from linearity (see Section II.E and Figure 6). A quick way to check that you are operating in a range in which the signal intensity varies linearly with the square root of microwave power is to decrease the attenuation by 6 dB (a factor of 4 increase in power). The spectral amplitude should change by a factor of 2; if it doesn't, reduce the power and try again.

Phase

There are two phase settings in the normal CW spectrometer, the reference-arm phase and the phase of the phase-sensitive detector working at the field modulation frequency (usually 100 kHz). If the reference-arm phase is wrong the signal amplitude will be low, and dispersion will be mixed with absorption. This phase setting should be adjusted to obtain the maximum detector current. If the 100-kHz detector phase is wrong the signal amplitude will be low. Since a null is easier to see than a maximum, the best approach is to set the phase for null signal, and then change phase by 90°. For saturation-transfer spectroscopy this setting is very critical, and a substantial literature has been devoted to it [102-104].

Modulation Amplitude

The experimental criterion is that the modulation amplitude (B_{m2} - B_{m1} in Figure 5) should be kept less than one-tenth of the derivative peak-to-peak line width. If the modulation amplitude is too large, the EPR signal will be distorted. In fact, it is possible to obscure hyperfine splitting when the modulation amplitude is too large. In practice you should always scan the spectrum with a very small modulation amplitude, determine the narrowest line width, and set the modulation amplitude to one-tenth of that line width. If you were using too large a modulation amplitude during the initial scan, repeat the measurement of the line width, and adjust the modulation amplitude again if necessary. For small modulation amplitudes the S/N increases linearly with modulation amplitude. It may happen that the S/N is too low to obtain a spectrum with the modulation amplitude meeting the above criteria. In this case, one must go back to the basic question—what information do you want from the sample? If line shape is the crucial information, then signal averaging will be needed to improve the S/N. If subtleties of line shape are of less significance, it may be acceptable to increase the modulation amplitude up to roughly half of the line width (see section 6H in [8], 2nd ed., for details on degree of line distortion). If the area under the peak is the information desired, overmodulation is acceptable, since the area is linearly proportional to modulation amplitude, even when the modulation is so large as to cause distortion of the line shape.

Gain

The gain is adjusted to give the desired size of display. Note, however, that the gain should not be so high as to cause the amplifier to saturate. On the Varian E-Line the PSD output becomes nonlinear when the "receiver level" meter is at the high end of the scale.

Scan Rate and Filter Time Constant

These two parameters are related to each other and to the line width by the following formula. The inequality must be satisfied to obtain undistorted lines.

$$\frac{\text{(Spectrum width in mm)}}{\text{(Line width in mm)}} \times \frac{\text{(time constant in sec)}}{\text{(sweep time in sec)}} < 0.1$$

A faster sweep or longer time constant does not give the system enough time to respond to changes in signal amplitude as the line is traversed. When using a computer you also need to know the response time of the D/A.

The front-panel readings of filter time constants do not have the same meanings on all EPR spectrometers. The Varian E-201 100-kHz module and the E-204 low-frequency module use a single-stage RC filter network. The newer E-207 high-frequency module employs a double-stage RC network, which drops off more quickly at higher frequencies. On the Bruker ER200 the filter system in the 022 module is such that the effective time constant is about a factor of 2-2.5 times less than for the same setting on the E-207. The differences in filtering need to be considered in comparing S/N on spectrometers with nominally the same settings. The Bruker 023M module filtering is a pure RC filter network followed by a digital integrating A/D converter.

The time-constant setting on the E-207 required to maintain the noise amplitude at a constant level is nominally proportional to the square of the receiver gain setting. Thus, if the gain is increased by a factor of 10, the time constant must be increased by a factor of 100 to maintain constant noise amplitude. This is seldom practical because of scan time limitations, so in practice noise usually increases with gain.

Signal-to-noise ratio can be improved by using a longer time constant and a slower scan. The Varian E-Line console provides for a 16-h scan and the Bruker ER200 provides for a 10,000-sec scan for this purpose. Why then, would one want to use an expensive computer system for S/N improvement, when the spectrometer is designed for extensive analog filtering? The answer is that with a perfectly stable sample and stable instrument, roughly equal time is involved in either method of S/N improvement. The problem is that perfect stability is not achieved, and the filtering discussion focuses on high-frequency noise. Long-term spectrometer drift due to air temperature changes, drafts, vibration, line voltage fluctuations, etc. limit the practical lengths of a scan. Signal averaging will tend to average out baseline drift problems just as it averages out high-frequency noise. Drifts in the magnetic field magnitude are not averaged out by any process, and they always increase apparent line width. Ultimately, the resultant line broadening limits the spectral improvement possible with any averaging or filtering technique.

If the sample decays with time, a separate set of problems emerges. Assume, for example, that you want to compare line shapes of two peaks in a noisy nitroxyl EPR spectrum, and that the amplitude of the spectrum is changing with time due to chemical reaction (shifting equilibria, decay, oxygen consumption or diffusion, etc.). In this case one wants to minimize the time spent scanning between points of interest. It would be wise to scan the narrowest portion of the spectrum that will give the information of interest. Then a numerical correction for the measured rate of change in the spectrum is the best way to handle the problem. The impact of the time dependence can be minimized more effectively by averaging rapid scans than by filtering a slow scan.

Choice of Modulation Frequency

As a tradeoff between low-frequency noise and distortion by modulation sidebands, most EPR spectra are obtained using 100-kHz modulation. ST-EPR spectra are usually obtained with modulation at a lower frequency and detection at the second harmonic, for example, 50 and 100 kHz.

For slow-passage EPR, it is necessary to have the reciprocal of the modulation frequency much greater than T_1:

$$\omega_m^{-1} >> T_1$$

This criterion is not met as often as it is assumed to be. Some samples at liquid nitrogen temperature, and many samples at liquid helium temperature, have T_1 too long to use 100-kHz modulation. Passage effects (see Section I.D), recognizable as distortions of line shapes, or even inversion of signals upon reversal of the field scan direction, alert you to the need to use a lower modulation frequency.

C. Second-Derivative Operation

With magnetic field modulation and phase-sensitive detection, the voltage at the detector can be expressed in a Fourier series [8,105-107]. The ω_m term is a measure of the first derivative, and the component detected at $2\omega_m$ is the second derivative. Note that whereas the amplitude of the first derivative EPR signal is proportional to the modulation amplitude, A, the amplitude of the second derivative EPR signal is proportional to the square of the modulation amplitude. If you integrate a first-derivative spectrum twice (I_1) and a second-derivative spectrum three times (I_2), the results are related by $I_2 = I_1 A/4$ if the actual absorption signal areas are identical [105].

It is also possible to obtain a second derivative of the EPR signal by using two modulation frequencies and two phase detectors. This follows directly from the discussion of how the first-derivative line shape is obtained (Section I.B), since everything stated there would remain true if the original signal were the first derivative. Thus, modulation at 100 kHz and 1 kHz, followed by phase detection at 100-kHz, yields a first-derivative spectrum with a 1-kHz modulation signal on it, and then phase detection at 1 kHz yields the second-derivative spectrum. If you go one step further and use the second harmonic of the high-frequency modulation plus a low-frequency modulation, you get the third-derivative display.

Derivatives can also be generated with computer manipulation of digitally stored data. However, the discrete nature of the data array causes such a "noisy" looking derivative that it is not very useful unless the original data were virtually noise-free or unless extensive multipoint averaging is used.

D. CW Saturation

Even as pulsed EPR spectrometers become more common, CW continuous saturation characterization of spin relaxation will remain important. Faster relaxation times can be characterized by CW methods than by pulse methods. In addition, the methods are complementary in the information they provide.

Data Presentation

In the literature $P_{1/2}$ is commonly used as a saturation parameter [50]. The term $P_{1/2}$ is the incident microwave power at which the EPR signal has half the amplitude it would have in the absence of saturation. The value of $P_{1/2}$ can be read off a plot of spectral amplitude as a function of $\sqrt{\text{power}}$ (Figure 6), which would be linear if there were no saturation. If $P = P_{1/2}$, $T_1 T_2 = 1/\gamma^2 B_1{}^2$ where γ is the electron magnetogyric ratio. Thus if B_1 is determined for a particular sample geometry as discussed in Section VI.F, the $T_1 T_2$ product can be calculated from $P_{1/2}$.

Another common presentation of saturation behavior is the Beinert/Orme-Johnson plot (also called the Vanngard plot) [108]. The plot is generated as follows: (a) Normalize the signal amplitude by dividing the measured amplitude by the receiver gain and modulation amplitude. The normalized signal amplitude is called S. (b) Divide S by $\sqrt{P}$ (P = power) and take the log. (c) Plot log $(S/\sqrt{P})$ versus log $(\sqrt{P})$. The result is a straight line parallel to the abscissa ($\sqrt{P}$ axis) when no saturation occurs. The lines bend down toward the abscissa when saturation occurs (Figure 13). For a purely homogeneously broadened line (no unresolved hyperfine structure), the slope of the line for $P > P_{1/2}$ is -4, and for a purely inhomogeneously broadened line (line shape dominated by unresolved hyperfine) the slope is -1 [109]. The advantages of this kind of plot are that (a) if all samples followed the same saturation behavior, all curves would be identical in shape, and would differ only in scaling factors, and (b) curvature of the descending branch of the saturation curve indicates that saturation curves with different values of $P_{1/2}$ are superimposed. The converse of (b) is not true. The absence of a significant deviation from the theoretical saturation behavior for a one-electron system does not necessarily imply that only a single species is present.

The most complete analysis of CW saturation data is to plot the saturation curve as in Figure 6 and simulate it using measured values of Q and B_1, for various guesses of T_1 and T_2, until a best fit is obtained [33,34]. The problem is to have a reasonable estimate of one of the relaxation times. If the hyperfine couplings are known (from simulation, ENDOR, etc.), then simulation of the spectrum in the absence of saturation effects yields

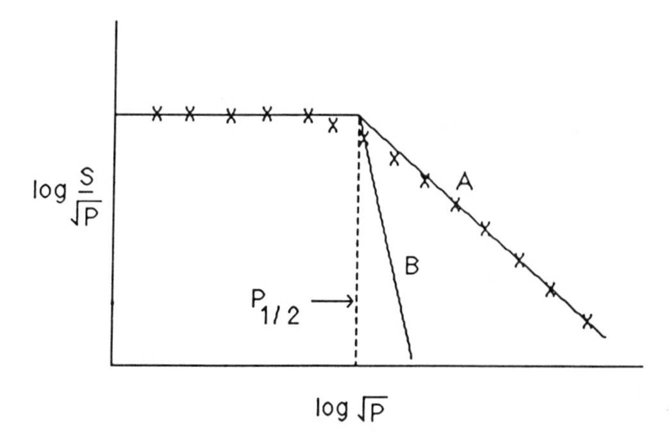

Figure 13. Beinert/Orme-Johnson plot of CW power saturation behavior. The construction of this plot is described in the text. Lines A and B are for inhomogeneously and homogeneously broadened signals, respectively.

T_2, and only T_1 remains to be varied to obtain a fit to the saturation curve. If the hyperfine structure is not known, then the problem is much more difficult and the result much less certain [34].

Line versus Point Samples

Since both modulation amplitude and microwave B_1 vary over the height of the cavity in the standard TE_{102} cavity [50,110,111], the experimentalist has a difficult trade-off. A stronger signal will be obtained, for a constant sample concentration, if a "line" sample is used that extends the entire height of the cavity. However, for a constant number of spins, without saturation, a point sample at the center of a TE_{102} rectangular cavity gives an EPR spectrum amplitude 2.3 times that of the same number of spins distributed along a line sample that exactly matches the cavity length. Interpretation of saturation behavior and relaxation times is more difficult for a line sample than for a point sample, since each portion of the line sample experiences a different B_1. If a point sample is used, the placement of the sample is critical. Several papers have discussed these problems in detail [50,51,112].

This is less of a problem in Q-band measurements and in some of the newer resonators (loop-gap, etc.). The power distribution in the TE_{011} Q-band cavity has the same variation as in the X-band cavity, but the modulation coils commonly used at Q band are large relative to the cavity dimensions, so the modulation amplitude is roughly constant over the sample. Many of the new resonators, such as the loop-gap resonators, are similarly small relative to the modulation coils. In addition, in the loop-gap and related resonators B_1 varies less over the sample length than in cavities. However, less detail has been reported for these cases than for the Varian E-231 cavity.

Computer simulation of power saturation curves or saturated spectra must take account of the B_1 and modulation distribution over the actual spatial extent of the sample [50]. To emphasize this point, consider a sample whose relaxation is such that at 200 mW the portion at the center of the cavity yields a signal amplitude that is about 2.6% of the maximum aplitude in the power saturation curve (this could happen if the relaxation times were approximately $T_1 = T_2 = 4 \times 10^{-7}$ sec/rad). Then consider an increment of sample 65% of the way to the top of the cavity that is experiencing a B_1 as if the incident power were 50 mW instead of 200 mW. The amplitude of its signal would be 6-7 times that of the portion at the center, if the modulation amplitude were uniform. Introducing the nonuniform modulation cuts its amplitude to about 1.8 times that at the center. This example shows that for a sample with a long relaxation time, the high-power portion of the saturation curve can be dramatically affected by portions of the sample experiencing lower B_1, due to the cosine dependence of B_1, in spite of the cosine-squared "selectivity" of modulation.

E. Methods of Measuring Relaxation Times

Both CW and pulsed EPR methods can be used to measure relaxation times (ch. 3 and 4 in Ref. 12). Some details are given in the following paragraphs. For each method, it is necessary to know the microwave magnetic field at the sample. Hence, it is necessary to know how to measure cavity Q (see Section III.A.5) and B_1 (see Section VI.F). Each method of measuring relaxation times may give different results, because it measures different contributions to the relaxation. See, for example, the comparison of CW progressive saturation and saturation recovery in Hyde [113].

No Unresolved Hyperfine Structure

When the line width is determined by relaxation, and not by unresolved hyperfine or by magnetic field inhomogeneity, then

$$(\Delta B_{pp})^2 = [4/(3\gamma^2 T_2^2)](1 + \gamma^2 B_1^2 T_1 T_2)$$

where ΔB_{pp} is the peak-to-peak line width and γ is the electron magnetogyric ratio [114,115]. If B_1 is small enough that the $B_1^2 T_1 T_2$ product term is $\ll 1$, the signal is unsaturated and the line shape depends only on T_2

$$T_2 = 2/(\sqrt{3}\,\gamma\,\Delta B_{pp})$$

$$= 6.5578 \times 10^{-8}/\Delta B_{pp} \text{ sec/rad when } \Delta B_{pp} \text{ is in gauss}$$

Usually there are other contributions to line width, so this formula gives a lower limit to T_2. The shape of a saturated line depends on both T_1 and T_2, so a plot of ΔB_{pp}^2 versus P (B_1^2 is proportional to P) can be used to estimate T_1 and T_2.

Saturation Recovery

Conceptually, one of the simplest relaxation time measurements is the saturation recovery method [35]. This is sometimes heuristically called "stepped EPR." In the saturation recovery method one saturates the spin system with a high microwave power, then reduces the power and observes the recovery of the spin system toward equilibrium by performing a CW EPR measurement at very low power level. This is a measure of T_1. The capabilities of current instrumentation limit these measurements to values of T_1 longer than about 1 μsec.

Fourier Transform EPR

Following a single microwave pulse, the spin system is displaced from equilibrium and recovers to equilibrium with a time constant T_2^* (recall that this depends on field inhomogeneity, see Section II.F). The signal so observed is called the free induction decay (FID). Fourier transform of the FID yields the frequency-domain spectrum. This is the basis of high-resolution FT-NMR. The same experiment can be done in EPR in favorable cases. The problem in applying this technique extensively in EPR is that it is difficult to obtain a microwave magnetic field at the sample with amplitude larger than the magnetic field width of the EPR signal. So far, FT-EPR has been demonstrated for a few organic radicals [116,117].

Spin Echo

The Hahn spin echo technique, first developed for NMR, can also be used in EPR. Any two microwave pulses will generate an echo. Detailed diagrams of the magnetization following the microwave pulses can be found [8,19,35, 118,119]. The amplitude of the echo following a two-pulse sequence decays with a time constant usually denoted as T_m, the phase memory decay time. Under certain circumstances $2T_m$ can be identified as the transverse relaxation time T_2. The value does not depend on magnetic field inhomogeneity, or on hyperfine coupling (in the absence of spin diffusion), so it is T_2, not T_2^*. This is a better measure of relaxation times than is line width, except for the case of the purely homogeneously broadened line, for which T_2 can be obtained accurately from line width. With modern instrumentation it is possible to measure T_m times as short as ~0.1 μsec.

Estimates of T_1 can be made with various pulse sequences on a spin-echo spectrometer, just as in NMR. If the two-pulse sequence that gives a spin echo is repeated faster than about once every five times the T_1 relaxation time, the spin system will not return to equilibrium between pulse sequences. In this case, the z magnetization is decreased, less magnetization is available to project into the x-y plane, and the echo amplitude is decreased. The value of T_1 can be determined from the dependence of echo amplitude on pulse repetition rate.

Another spin echo approach is to first invert the spin system with a 180° pulse, and then sample this inverted magnetization with a normal two-pulse spin echo. During the time between the inverting pulse and the sampling pulse, the spin system relaxes by T_1 processes. Hence, a plot of echo amplitude versus delay time after the inverting pulse yields T_1.

Many pulse sequences are described in texts on NMR [83,120-124] and details of the application to EPR are discussed [19,35].

F. Measurement of B_1

One of the most uncertain parameters in EPR measurements is the magnitude of the microwave magnetic field, B_1, at the sample. Since all materials in the cavity, including the sample tube and the sample itself, affect the distribution of the microwave magnetic and electric fields, it is virtually impossible to know the value of B_1 at the sample exactly. An extensive discussion of the measurement of B_1 is given by Bales and Kevan [125]. Some of the most practical ways to estimate B_1 are summarized next.

Calculation from Q and Cavity Dimensions

The value of B_1 can be calculated with the following equation:

$$B_1^2 = \frac{8\,Q\,P\,(1 \pm |\Gamma_0|)}{\nu_0 V_c[1+d^2/(n^2a^2)]}$$

where Q is the cavity Q, Γ_0 is the voltage reflection coefficient for the cavity at microwave resonance but off magnetic resonance [126], P is the power incident on the cavity in watts, ν_0 is the microwave frequency, V_c is the cavity volume, d is the length of the cavity, a is the breadth of the cavity, n is the number of wavelengths along d, and the plus and minus signs refer to overcoupled and undercoupled cavity systems, respectively. Normally, the cavity is critically coupled ($\Gamma_0 = 0$) for CW EPR and overcoupled for pulsed EPR.

If you measure the loaded Q, as described in Section III.A.5, you can calculate B_1 fairly accurately. For example, for the Varian E-231 cavity, $a = 2.286$ cm, $b = 1.016$ cm, $d = 4.318$ cm, $V_c = 10.029 + 1.53 = 11.56$ cm^3 (there is a correction for the fact that the side walls are dished out a bit [126]). If $\nu_0 = 9.5$ GHz, then

$$2B_1 = 3.92 \times 10^{-2}\,(QP)^{1/2}.$$

The value of B_1 calculated with this formula is not valid if there is substantial dielectric in the cavity to distort the B_1 distribution. The quartz Dewar used for variable temperature studies can increase the B_1 at the sample by nearly a factor of 2, depending upon the thickness of the quartz [127,128]. Each Dewar has to be calibrated. Similarly, the lens effect of the sample tube and solvent itself needs to be calibrated; it has been found to increase the integrated signal area by up to 55% [51].

Method of "Perturbing Spheres"

The fundamental reference is Ginzton's book [129, see also 130]. One measures the frequency change due to the presence of a conducting sphere. The formulas to use are

$$B_0^2 = \frac{\nu^2 - \nu_0^2}{\nu_0^2}\left(\frac{1}{2\pi a^3}\right) \qquad B_1^2 = \frac{B_0^2}{4\pi\nu_0}\,(Q\,P)$$

where a is the diameter of the steel ball (1/16-in balls from Small Parts, Inc., are satisfactory; $a = 0.079$ cm), ν is the cavity frequency with the ball, and ν_0 is the cavity frequency without the ball. The best experimental setup to use for this measurement is one that closely replicates the conditions under which the spectra will be obtained on a sample [131].

Line Broadening

If the relaxation times are known for a sample, it is possible to use that sample as a standard to determine B_1. Fremy's salt has been advocated for this purpose [132]. Other well-characterized samples, such as irradiated sugar and irradiated glycylglycine [133,134], can be used also. The line width of a signal, ΔB_e, is related to the line width in the absence of saturation, ΔB_0, and the relaxation times by the equation

$$\Delta B_e^2 = \Delta B_0^2 + 4T_1(B_1)^2/3T_2$$

Since $B_1^2 = KP$, where K is a proportionality constant, a plot of ΔB_e^2 versus P has a slope of $4T_1K/3T_2$ and the intercept ΔB_0^2. The value of K is then determined from the slope of the line and the known values of T_1 and T_2. This technique can be used to measure K for various experimental arrangements such as Dewar inserts and flat cells in the cavity.

Pulsed EPR Measurement of B_1

In pulsed EPR B_1 is related to the flip angle in radians, θ, and the length of the pulse in seconds, t_p, by the equation

$$B_1 = \theta/[(1.76 \times 10^7 \text{ rad s}^{-1} \text{ G}^{-1})(t_p)]$$

The shape of the echo for a standard sample, such as a coal sample, can be compared with the echo shapes calculated by Mims [135] to determine t_p for 90° and 180° pulses. The values of θ and t_p can then be used to calculate B_1.

Other Methods

Peric et al. [136] recently reported a method for obtaining B_1 by measuring the splittings between sidebands as a function of modulation frequency, using modulation frequencies from 100 kHz to 1.5 MHz. Recent reports discuss the utility of DPPH (diphenylpicrylhydrazyl) as a secondary standard whose absolute signal intensity measures B_1 [137], and the use of the magnetization hysteresis spectrum and power-dependent line width of a small crystal of the TCNQ (tetracyanoquinone) salt of methyl phenazine [138].

VII. LESS COMMON MEASUREMENTS WITH EPR SPECTROMETERS

A. Multiple-Resonance Methods

Irradiation with a second frequency during an EPR experiment can provide much more detailed information about the spin system than can the usual single-frequency measurement. The most common of these multiple-resonance methods are ENDOR, ELDOR, and TRIPLE. In order to keep this chapter brief, only references to other sources of information are given here. Figure 14 shows the types of transitions observed in these experiments.

In ENDOR one adjusts the microwave frequency and magnetic field to a resonance of interest, and then sweeps a radiofrequency signal through the range of nuclear resonant frequencies. It is necessary to operate at

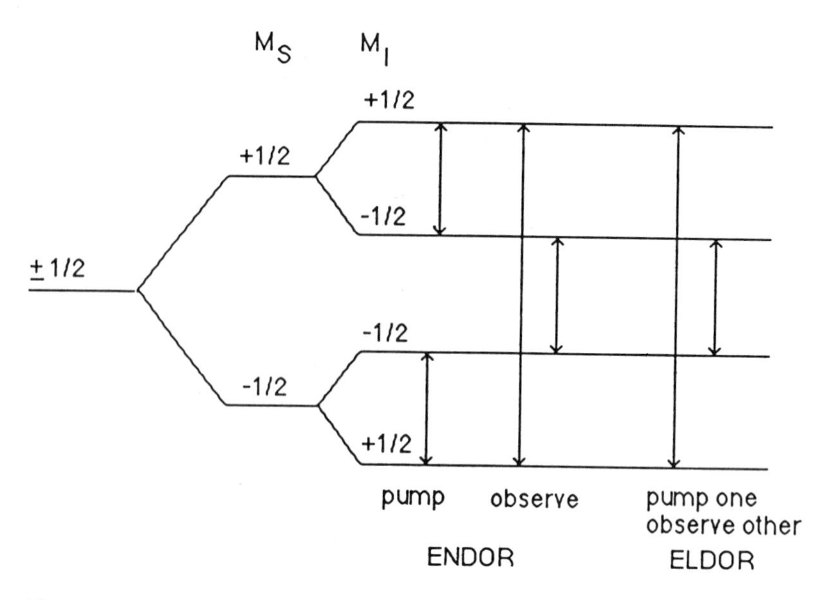

Figure 14. ENDOR and ELDOR transitions. This energy-level diagram is sketched for the case of $S = 1/2$, $I = 1/2$, $A > 0$, $g_n > 0$, and $A/2 > \nu_n$. The allowed EPR transitions have $\Delta m_S = \pm 1$ and $\Delta m_I = 0$. Forbidden EPR transitions with $\Delta m_I = \pm 1$ sometimes have observable intensities. NMR transitions have $\Delta m_I = \pm 1$ and $\Delta m_S = 0$. Observation of a change in the intensity of one of the EPR transitions when one of the NMR transitions is irradiated is the ENDOR experiment. Observation of a change in the intensity of one of the EPR transitions when another EPR transition is irradiated is the ELDOR experiment.

a microwave power level that partially saturates the EPR transition (i.e., at microwave power greater than the linear portion of the curve in Figure 6), and one must be able to saturate the NMR transition. The radiofrequency field at the nuclear Larmor frequency induces transitions between states such that the overall effect is to relieve the degree of saturation of the electron spin system. This causes an increase in the amplitude of the EPR signal. Thus, there is a "peak" corresponding to each nuclear transition that is coupled to the electron spin. Electron-nuclear couplings are observed via ENDOR with an effective resolution that is much higher than in conventional EPR. In addition, quadrupolar couplings can be observed by ENDOR, whereas the quadrupole couplings are not observable, to first order, in EPR transitions. The ENDOR experiment can be extended to the use of two simultaneous radiofrequency fields (TRIPLE). While the ENDOR experiment provides only the magnitudes of the nuclear coupling constant, this TRIPLE experiment provides the signs of the coupling constants.

The ELDOR experiment involves two microwave frequencies. One EPR transition is saturated, and a second EPR transition is observed with a nonsaturating microwave field from a second source. Typically, the microwave frequencies differ by about 350 MHz. When one transition is saturated, various relaxation mechanisms make the energy levels for the other (non-saturated) transition more nearly equally populated. The result is a reduction in the intensity of the second transition.

Note that in both the ENDOR and ELDOR experiments the experimental observable is a change in the intensity of the EPR line upon irradiation with a second frequency. Thus, the S/N is poorer than for the conventional EPR experiment for the same species. Each of the measurements depends on having relaxation times in an appropriate range, so not every compound can be studied by these techniques over the range of concentrations, temperatures, etc. that might be of interest. This strong dependence on relaxation times translates into a powerful tool for studying relaxation behavior.

Introductions to ENDOR, TRIPLE, and ELDOR are provided [8,139-143]. A convenient table comparing the techniques is in Ref. 8, 2nd ed., p. 650. The conditions needed to observe ENDOR of various nuclei in organic radicals in solution is discussed in detail [144]. Recent illustrations of TRIPLE provide leading references to the literature [145,146]. A wide range of variations on the basic theme of multiple resonance has been developed, with CW and pulsed methods, field sweep and frequency sweep, and various combinations. Poole [8] also describes double-resonance experiments in which in addition to the microwave field, there is optical irradiation (optically detected magnetic resonance, ODMR), pulse radiolysis (dynamic electron polarization, DEP), an electric field [147], and acoustic or ultrasonic paramagnetic resonance (UPR) [148].

B. Saturation-Transfer Spectroscopy

The technique of saturation-transfer spectroscopy can be used to measure rates of molecular motion that are a little faster than electron spin relaxation rates [104,149]. This time scale is particularly applicable to the study of biological systems labeled with nitroxyl spin labels. Standard conditions for the measurement of saturation transfer EPR spectra have been delineated [150]. For the most common ST-EPR technique, the spectrometer must be capable of phase-sensitive detection at twice the magnetic field modulation frequency. The Varian Century Series and IBM/Bruker ER200 spectrometers have this capability.

C. Electrical Conductivity

Since electrical conductivity of a sample affects the magnitude of the EPR signal observed from the sample, it is possible to measure the microwave electrical conductivity of a sample with an EPR spectrometer [151]. The conductivity of the walls of a resonant cavity affect the Q of the cavity. This effect has been used to measure surface resistance of nonferromagnetic metals at X band [152]. Electron spin diffusion rates in conducting crystals have been measured with electron spin echoes in a magnetic field gradient [153]. EPR can be used to monitor aspects of superconductivity in new materials [154]. A novel combination of EPR and the ac Josephson effect provided a new type of spectrometer [155].

D. Static Magnetization

The dc magnetization of a sample can be measured by using the EPR of a standard sample attached to the sample whose magnetization is to be measured. The EPR probes the magnetic field outside the sample [156].

E. EPR Imaging

With suitable magnetic field gradients, electron paramagnetic resonance (EPR) can provide pictures of spin concentration as a function of the three spatial dimensions x, y, and z. Most of the early work in EPR imaging has emphasized making pictures of objects. In this effort there has been an implied concept that the dimensions of interest were the three cartesian dimensions of the laboratory coordinate system [157-159]. However, the properties of the spin systems provide access to several additional dimensions, which may, in some cases, provide more insight into the nature of a sample than the spatial dimensions alone. We urge an expanded view that considers as dimensions of the imaging problem several additional features of EPR spectroscopy, including g values, hyperfine splitting, relaxation times T_1 and T_2, microwave field distribution, spin flip angles, and chemical kinetics (for example, formation and decay of radicals or diffusion). Combinations of these dimensions yield numerous possible multidimensional imaging experiments. Recent papers have emphasized relaxation times and spin flip angles as imaging dimensions [160-162]. There is a mathematical isomorphism between a two-dimensional spatial imaging experiment and the spectral-spatial problem [159,163-168]. Several examples of spectral-spatial imaging have been published recently [164,166-169]. The experimental aspects of EPR imaging have been reviewed [159], as have the rapidly accumulating experimental results [157-159,162].

VIII. SPECTRAL PROPERTIES OF COMMONLY OBSERVED SPECIES

EPR transitions can occur over the entire range of field/frequency relationships accessible to current spectrometers. At X band, most of the signals occur within the 0-10,000 G range of the standard 9- to 12-in magnets. However, transitions can be predicted that are not accessible at X band. For example, species with large zero-field splittings have transitions that require frequencies much higher than the X-band frequency. Even for a "simple" organic diradical system, full characterization of the spectrum requires examination of the half-field region of the spectrum. Recent examples of multiple-spin organic compounds emphasize this point [170]. Only for very limited purposes can one restrict the field of view to a few hundred gauss around $g = 2$.

The references given in this section are intended as a guide to introductory and review literature. There are many nuances of the observability and interpretation of the EPR of each species. No attempt is made to provide comprehensive coverage of the many fascinating details of the dependence of EPR spectra on, for example, zero-field splittings, g anisotropy, viscosity, and temperature.

A. Transition Metals

The EPR spectra of spin-1/2 species can be observed at room temperature in many cases. Lower temperatures are usually required to observe spectra for species with S > 1/2. Since most of the transition metals have at least one isotope with a nuclear spin, fairly distinctive patterns occur in the EPR spectra. Cu(II) (I = 3/2), Mn(II) (I = 5/2), and V(IV) (I = 7/2) (as VO^{2+}) are the most extensively studied. General introductions to EPR of transition metals have been given [6,15,171,172]. Comprehensive referencing of the literature is provided [173-184]. Some sample spectra are shown in Ref. 6. Reviews and selected examples for specific metals are listed in Table 1. The references cited were chosen to provide reviews or specific cases that are illustrated with pictures of spectra.

EPR of paramagnetic ions in metals permits the study of a wide range of metallic phenomena [236].

B. Main-Group Radical Species

The characteristic spectra of doublet organic radicals occur largely in a small range around g = 2 and are usually observable at room temperature. Spin-spin interaction can result in spectra at other field positions. EPR is very useful for identification of organic radicals, since the spectra tend to exhibit well-resolved hyperfine structure characteristic of the nuclei in the molecule. References for various characteristic classes of compounds are given in Table 2. Fairly comprehensive referencing of the literature is provided [173,174,237,238], as well as some sample spectra [6,238,239].

C. Triplets

Molecules that contain two unpaired electrons are called biradicals if the zero-field splitting (ZFS) is small enough that singlet-triplet state transitions are observable in the EPR, and triplets if the ZFS is so large that only transitions within the triplet state are accessible. Early efforts to observe the EPR of triplet species were unsuccessful until it was recognized that ZFS anisotropy was broadening the spectra. Hutchison made the first observation of triplet EPR in 1958. Some of the large number of triplet EPR spectra that have been observed are reviewed in Ref. 3, ch. 5. Recently, very-high-spin organic radicals have been synthesized [170,199]. Quintet [267], septet [268], and nonet [269] ground-state EPR spectra have been reported. Biradicals have exhibited a rich variety of spin-spin interactions. The literature of metal-radical interactions has been reviewed [26,27].

D. Fossil Fuels

Since most fossil fuels contain paramagnetic species, EPR is useful for the characterization of the fuels and the changes they undergo upon processing. Examples include oil shale [34], tar sands, coal, and petroleum [270]. Extensive studies of the EPR of carbonaceous materials have been conducted by Singer [271].

Table 1. EPR of Transition Metals

Metal	Oxidation state	Reference
General		142,172,177-195
Metalloproteins		83,118,196,197
Organometallic		198
First row		
Ti	+2	199
	+3	200
V	0	201
	+4	202,203,204
Cr	+3	205,206
	+5	207,208
Mn	+2	209,210
Fe		193,211-213
Co		214-216
Cu		217-219
Ni	+1	220,221
	+2	222
	+3	221,223
Second and third rows		
Nb, Ta		224
Mo	+3	225
	+5	226,227
Ru	+3	228,229
Lanthanides		177-185,230,231
Gd	+3	232,233
Actinides		177-184,234,235

Table 2. EPR of Main-Group Radical Species

Radical Type	Reference
General	186,188,191,195,237-244
Quinones	146,245
Nitroxyls	20,26,27,113,246,247
Spin-trapped species	248,249
Polymers	250
Species generated by irradiation	251-255
Radiation-damaged biological materials	141,256,257
General biological	19,258,259
S	260
Ge, Sn	261

Table 2 (Continued)

Radical type	Reference
Ge, Sn	261
Photosynthesis	262
Enzymatic systems	263
Semiconductors	264,265
Blood	266

IX. REPORTING RESULTS

This section is a summary guide to nomenclature, and to what parameters should be reported to communicate the results of the measurement. It is assumed that the chemistry of the problem is adequately described, including source of the sample, concentration, temperature, etc., so only the spectroscopic aspects of the results are covered in this section.

A. Reporting Experimental Spectra

Spectra should be reported with magnetic field increasing to the right. First-derivative spectra should be reported such that the low-field rise of the first line has a positive excursion.

The figure or its caption or the "methods" section of a paper should include the following information: (a) magnetic field at some point in the spectrum, or g factor marker, (b) magnetic field scale and scan direction, (c) microwave frequency, (d) microwave power level, (e) magnetic field scan rate, (f) modulation frequency and amplitude, (g) whether a Dewar or other cavity insert was used, (h) whether these parameters were calibrated or "read from the spectrometer," and (i) what standards were used. If multiple derivatives, saturation-transfer EPR, or other such techniques were used, the above list should be expanded accordingly. For pulsed EPR, or use of other home-built components such as new resonators, details of design and construction should be included or referenced.

B. Reporting Derived Values

The field uses a chaotic mix of units. The following notation and units are recommended.

EPR—Call the spectroscopy EPR, not ESR.

g—Unless particular attention is paid in the paper to the mathematical properties of g, use a noncommittal term such as g factor or g value. Thus, one would report, for example, $g = 2.0031$ as the isotropic g factor, and $g_z = 2.31$ as the z component of the g factor.

B—The magnetic field generated by the electromagnet should be denoted B, not H, and the units should be gauss or tesla.

B_1—The microwave magnetic field should be denoted B_1. The units should be gauss or tesla. It should be clearly stated whether the B_1 reported is the maximum value or the average value, and whether it is the amplitude of the linearly polarized microwave magnetic field or the circularly rotating microwave component.

a, A—All electron-nuclear interactions should be called hyperfine interactions. The term superhyperfine interactions does not serve a theoretically useful role and should be discarded. When a peak splitting is measured on a magnetic field scan it should be denoted by a, given in units of gauss or tesla, and called a "hyperfine splitting." When the peak splitting is obtained by computer fitting of the spectrum including second-order corrections, it should be denoted A, given in energy units (cm^{-1}), and called a "hyperfine coupling."

T_1, T_2, T_m—The electron spin relaxation times should be given in units appropriate to their experimental definition. In most cases this will be sec rad^{-1}, not sec as is usually reported.

τ_c, τ_r—The correlation time for molecular motion of the species whose EPR signal is being studied should be given in units appropriate to the experimental definition. When molecular rotation is the motion being considered the units are sec rad^{-1}, not sec as is usually reported.

Exchange/dipolar interactions—This is one of the most chaotic nomenclature areas in magnetic resonance—for a detailed discussion see Ref. 272. With careful attention to the mathematics, one of the following terms should be used: isotropic exchange, antisymmetric exchange (no unambiguous examples are known), symmetric anisotropic exchange (don't call it pseudo-dipolar), isotropic dipolar contribution that arises from g anisotropy, antisymmetric dipolar contribution that arises from g anisotropy, and dipolar (the major portion of the dipolar interaction). Use of other terms should be discontinued. Terms such as "fast" or "slow" should not be used to qualitatively describe magnitudes of J. The isotropic exchange should be given with sign and magnitude appropriate for an energy separation of -2Jh between the ground and first excited states.

Acknowledgments

This chapter builds on the contributions of many EPR spectroscopists, who have, via their papers or conversations, taught us what we have attempted to communicate here. We are especially grateful to Dr. James S. Hyde and Dr. Michael K. Bowman for their patient tutorials. Dr. Ira Goldberg and Dr. Arthur H. Heiss provided detailed comments on a draft of this chapter.

REFERENCES

1. J. E. Wertz and J. R. Bolton, *Electron Spin Resonance*, McGraw-Hill, New York, 1972. Reissued by Chapman and Hall, New York, 1986.
2. A. Carrington and A. D. McLachlan, *Introduction to Magnetic Resonance*, Harper & Row, New York, 1967.

3. N. M. Atherton, *Electron Spin Resonance: Theory and Applications*, Wiley, New York, 1973.
4. C. P. Slichter, *Principles of Magnetic Resonance*, 2nd ed., Springer-Verlag, Berlin, 1978.
5. G. E. Pake and T. L. Estle, *The Physical Principles of Electron Paramagnetic Resonance*, 2nd ed., W. A. Benjamin, Reading, MA, 1973.
6. R. S. Drago, *Physical Methods in Chemistry*, W. B. Saunders, Philadelphia, 1977, Ch. 9 and 13.
7. R. S. Alger, *Electron Paramagnetic Resonance: Techniques and Applications*, Wiley-Interscience, New York, 1968.
8. C. P. Poole, Jr., *Electron Spin Resonance: A Comprehensive Treatise on Experimental Techniques*, Wiley-Interscience, New York, 1967. Second edition, Wiley, 1983.
9. T. H. Wilmshurst, *Electron Spin Resonance Spectrometers*, Plenum Press, New York, 1968, see especially Ch. 4.
10. P. H. Rieger, in *Techniques of Chemistry* (A. Weissberger and B. W. Rossiter, eds.), Wiley-Interscience, New York, Vol. 1, Part III A, pp. 499-598, 1972.
11. W. Gordy, *Theory and Applications of Electron Spin Resonance: Techniques of Chemistry*, Vol. XV, Wiley, New York, 1980.
12. C. P. Poole, Jr., and H. A. Farach, *Relaxation in Magnetic Resonance*, Academic Press, New York, 1971.
13. J. E. Harriman, *Theoretical Foundations of Electron Spin Resonance*, Academic Press, New York, 1978.
14. C. P. Poole, Jr., and H. A. Farach, *The Theory of Magnetic Resonance*, Wiley, New York, 1972.
15. A. Abragam and B. Bleaney, *Electron Paramagnetic Resonance of Transition Ions*, Oxford University Press, Oxford, 1970.
16. H. M. Swartz, J. R. Bolton, and D. C. Borg, eds., *Biological Applications of Electron Spin Resonance*, Wiley, New York, 1972.
17. L. J. Berliner, ed., *Spin Labeling: Theory and Applications*, Academic Press, New York, 1976.
18. L. J. Berliner, ed., *Spin Labeling II*, Academic Press, New York, 1979.
19. L. R. Dalton, ed., *EPR and Advanced EPR Studies of Biological Systems*, CRC Press, Boca Raton, FL, 1985.
20. G. I. Likhtenshtein, *Spin Labeling Methods in Molecular Biology*, Wiley, New York, 1976.
21. G. E. Pake and E. M. Purcell, *Phys. Rev.*, 74:1184 (1948).
22. R. N. Rogers and G. E. Pake, *J. Chem. Phys.*, 33:1107 (1960).
23. H. Levanon, S. Charbinsky, and Z. Luz, *J. Chem. Phys.*, 53:3056 (1970).
24. M. Peter, D. Shaltiel, J. H. Wernick, H. J. Williams, J. B. Mock, and R. C. Sherwood, *Phys. Rev.*, 126:1395 (1962).
25. P. A. Narayana and L. Kevan, *Magn. Resonance Rev.*, 7:239 (1983).
26. S. S. Eaton and G. R. Eaton, *Coord. Chem. Rev.*, 26:207 (1978).
27. S. S. Eaton and G. R. Eaton, *Coord. Chem. Rev.*, 83:29 (1987).
28. S. S. Eaton, M. L. Law, J. Peterson, G. R. Eaton, and D. J. Greenslade, *J. Magn. Resonance*, 33:135 (1979).
29. H. M. McConnell and B. G. McFarland, *Q. Rev. Biophys.*, 3:91 (1970).

30. P. Jost and O. H. Griffith, in *Methods in Pharmacology*, Vol. II (C. Chignell, ed.), Appleton-Century-Crofts, New York, 1972, Ch. 7.
31. P. D. Sullivan and J. R. Bolton, *Adv. Magn. Resonance*, 4:39 (1970).
32. D. B. Chesnut, *J. Magn. Resonance*, 25:373 (1977).
33. K. M. More, G. R. Eaton, and S. S. Eaton, *Inorg. Chem.*, 24:3820 (1985).
34. G. R. Eaton and S. S. Eaton, *J. Magn. Resonance*, 61:81 (1985).
35. L. Kevan and R. N. Schwartz, eds., *Time Domain Electron Spin Resonance*, Wiley, New York, 1979.
36. K. J. Standley and R. A. Vaughan, *Electron Spin Relaxation Phenomena in Solids*, Plenum Press, New York, 1969.
37. L. T. Muus and P. W. Atkins, eds., *Electron Spin Relaxation in Liquids*, Plenum Press, New York, 1972.
38. J. Reisse, in *The Multinuclear Approach to NMR Spectroscopy* (J. B. Lambert and F. G. Riddell, eds.), Reidel, Hingham, MA, 1983, p. 63.
39. C. Mailer, J. D. S. Danielson, and B. H. Robinson, *Rev. Sci. Instrum.*, 56:1917 (1985).
40. J. R. Norris, M. C. Thurnauer, and M. K. Bowman, *Adv. Biol. Med. Phys.*, 17:365 (1980).
41. S. W. Tan, J. S. Waugh, and W. H. Orme-Johnson, *J. Chem. Phys.*, 81:576 (1984).
42. R. W. Quine, S. S. Eaton, and G. R. Eaton, *Rev. Sci. Instrum.*, 58:1709 (1987).
43. J. C. Slater, *Rev. Modern Phys.*, 18:441 (1946).
44. H. J. Reich, P. F. Ordung, H. L. Krauss, and J. G. Skalnik, *Microwave Theory and Techniques*, D. Van Nostrand, Princeton, NJ, 1953.
45. *MIT Radiation Laboratory Series*, Published by McGraw-Hill, New York, especially the following volumes: Vol. 8, C. G. Montgomery, R. H. Dicke, and E. M. Purcell, *Principles of Microwave Circuits*, 1948; Vol. 9, G. L. Ragan, *Microwave Transmission Circuits*, 1948; Vol. 10, N. Marcuvitz, *Waveguide Handbook*, 1951; Vol. 11, C. G. Montgomery, *Technique of Microwave Measurements*, 1947; Vol. 14, L. D. Smullin and C. G. Montgomery, *Microwave Duplexers*, 1948; and Vol. 17, J. F. Blackburn, *Components Handbook*, 1949.
46. T. S. Laverghetta, *Handbook of Microwave Testing*, Artech House, Dedham, MA, 1981.
47. T. S. Laverghetta, *Microwave Measurements and Techniques*, Artech House, Dedham, MA, 1976.
48. J. S. Hyde and W. Froncisz, *Annu. Rev. Biophys. Bioeng.*, 11:391 (1982).
49. M. J. Hill and S. J. Wyard, *J. Sci. Instrum.*, 44:433 (1967).
50. C. Mailer, T. Sarna, H. M. Swartz, and J. S. Hyde, *J. Magn. Resonance*, 25:205 (1977).
51. D. P. Dalal, S. S. Eaton, and G. R. Eaton, *J. Magn. Resonance*, 44:415 (1981).
52. S. S. Eaton and G. R. Eaton, *Bull. Magn. Resonance*, 1:130 (1980).
53. J. S. Hyde and W. Froncisz, *Specialist Periodical Reports on Electron Spin Resonance*, 10:175, 1986.
54. M. Mehdizadeh, T. Koryushi, J. S. Hyde, and W. Froncisz, *IEEE Trans. Microwave Theory Tech.*, 31:1059 (1983).

55. W. Froncisz, A. Jesmanowicz, and J. S. Hyde, *J. Magn. Resonance*, 66:135 (1986).
56. W. N. Hardy and L. A. Whitehead, *Rev. Sci. Instrum.*, 52:213 (1981).
57. H. J. Schneider and P. Dullenkopf, *Rev. Sci. Instrum.*, 48:68 (1977).
58. M. Mehring and F. Freysoldt, *J. Phys. E.*, 13:894 (1980).
59. C. P. Lin, M. K. Bowman, and J. R. Norris, *J. Magn. Resonance*, 65:369 (1985).
60. W. M. Walsh, Jr., and L. W. Rupp, Jr., *Rev. Sci. Instrum.*, 57:2278 (1986).
61. J. S. Hyde, J.-J. Yin, W. Froncisz, and J. B. Feix, *J. Magn. Resonance*, 63:142 (1985).
62. J. P. Hornak and J. H. Freed, *J. Magn. Resonance*, 62:311 (1985).
63. R. W. Dykstra and G. D. Markham, *J. Magn. Resonance*, 69:350 (1986).
64. W. B. Mims and J. Peisach, *Biochem.*, 15:3863 (1976).
65. W. B. Mims, *Rev. Sci. Instrum.*, 45:1583 (1974).
66. W. Froncisz, C.-S. Lai, and J. S. Hyde, *Proc. Natl. Acad. Sci. USA*, 82:411 (1985).
67. H. A. Buckmaster, C. Hansen, V. M. Malhotra, J. C. Dering, A. L. Gray, and Y. H. Shing, *J. Magn. Resonance*, 42:322 (1981).
68. W. B. Mims, *Rev. Sci. Instrum.*, 36:1472 (1965).
69. A. F. Harvey, *Microwave Engineering*, Academic Press, New York, 1963, p. 204.
70. B. G. Segal, M. Kaplan, and G. K. Fraenkel, *J. Chem. Phys.*, 43:4191 (1965).
71. R. D. Allendoerfer, *J. Chem. Phys.*, 55:3615 (1971).
72. C. Klopfenstein, P. Jost, and O. H. Griffith, *Computers in Chemical and Biochemical Research*, 1:175 (1972).
73. I. B. Goldberg, Rockwell International Science Center Report SC549.31FR, Computer Controlled ESR, April 1979, and *J. Magn. Resonance*, 18:84 (1975).
74. J.-M. Spaeth, in *Electronic Magnetic Resonance in the Solid State*, Canadian Institute of Chemistry Symposium Series, Vol. 1 (J. A. Weil, M. K. Bowman, J. R. Preston, and K. F. Preston, eds.), Canadian Society for Chemistry, Ottawa, Ontario, Canada, 1987.
75. H. L. VanCamp and A. H. Heiss, *Magn. Resonance Rev.*, 7:1 (1981).
76. J. W. Cooper, *The Minicomputer in the Laboratory*, Wiley, New York, 1977.
77. P. R. Rony, in *Physical Methods of Chemistry*, 2nd ed., B. W. Rossiter and J. F. Hamilton, eds., Wiley-Interscience, 1986, Vol. 1, Ch. 6 and 7.
78. R. W. Quine, G. R. Eaton, and S. S. Eaton, *J. Magn. Resonance*, 66:164 (1986).
79. W. Berlinger, *Magn. Resonance Rev.*, 10:45 (1985).
80. I. B. Goldberg and H. R. Crowe, *Anal. Chem.*, 49:1353 (1977).
81. J. R. Morton and K. F. Preston, *J. Magn. Resonance*, 52:457 (1983).
82. J. C. W. Chien and L. C. Dickenson, *Biol. Magn. Resonance*, 3:155 (1981).
83. E. Fukushima and S. B. W. Roeder, *Experimental Pulse NMR: A Nuts and Bolts Approach*, Addison-Wesley, Reading, MA, 1981.
84. J. Homer, A. R. Dudley, and W. R. McWhinnie, *J. Chem. Soc. Chem. Commun.*, 893 (1973).

85. A. A. Westenberg, *Prog. React. Kin.*, 7:23 (1975).
85A. A. Carrington, *Microwave Spectroscopy of Free Radicals*, Academic Press, New York, 1974.
86. D. W. Grandy and L. Petrakis, *J. Magn. Resonance*, 41:367 (1980).
87. I. B. Goldberg, *Rev. Sci. Instrum.*, 55:1104 (1984).
88. I. B. Goldberg and A. J. Bard, in *Treatise on Analytical Chemistry*, Part I, Vol. 10, Section I, Chapter 3, 2nd ed. (P. J. Elving, M. M. Bursey, and I. M. Kolthoff, eds.), Wiley-Interscience, New York, 1983.
89. M. L. Randolph, in *Biological Applications of Electron Spin Resonance* (H. M. Swartz, J. R. Bolton, and D. C. Borg, eds.), Wiley, New York, 1972, Ch. 3.
90. G. Feher, *Bell System Technical Journal*, 36:449 (1957).
91. D. C. Warren and J. M. Fitzgerald, *Anal. Chem.*, 49:250 (1977), and references therein.
92. D. C. Warren and J. M. Fitzgerald, *Anal. Chem.*, 49:1840 (1977).
93. I. B. Goldberg, *J. Magn. Resonance*, 32:233 (1978).
94. R. Chang, *Anal. Chem.*, 46:1360 (1974).
95. K. Nakano, H. Tadano, and S. Takahashi, *Anal. Chem.*, 54:1850 (1982).
96. G. Casteleijn, J. J. TenBosch, and J. Smidt, *J. Appl. Phys.*, 39:4375 (1968).
97. T.-T. Chang, *Magn. Resonance Rev.*, 9:65 (1984).
98. Yu. N. Molin, V. M. Chibrikin, V. A. Shabalkin, and V. F. Shuvalov, *Zavod. Lab.*, 32:933 (1966) (p. 1150 in transl.).
99. H. J. M. Slangen, *J. Phys. E.*, 3:775 (1970).
100. I. B. Goldberg, H. R. Crowe, and W. M. Robertson, *Anal. Chem.*, 49:962 (1977).
101. R. Aasa and T. Vanngard, *J. Magn. Resonance*, 19:308 (1975).
102. J. S. Hyde, *Methods in Enzymology*, Academic Press, New York, Vol. XLIX, Part G, 480, 1978.
103. T. Watanabe, T. Sasaki, and S. Fujiwara, *Appl. Spectrosc.*, 36:174 (1982).
104. J. S. Hyde and D. D. Thomas, *Annu. Rev. Phys. Chem.*, 31:293 (1980).
105. G. V. H. Wilson, *J. Appl. Phys.*, 34:3276 (1963).
106. G. A. Noble and J. J. Markham, *J. Chem. Phys.*, 36:1340 (1962).
107. A. M. Russell and D. A. Torchia, *Rev. Sci. Instrum.*, 33:442 (1962).
108. H. Beinert and W. H. Orme-Johnson, in *Magnetic Resonance* in *Biological Systems* (A. Ehrenberg et al., eds.), Pergamon Press, New York, 1967, p. 221.
109. J. A. Hamilton, Y. Tamao, R. L. Blakley, and R. E. Coffman, *Biochemistry*, 11:3696 (1972).
110. J. W. H. Schreurs, G. E. Blomgren, and G. K. Fraenkel, *J. Chem. Phys.*, 32:1861 (1960).
111. R. G. Kooser, W. V. Volland, and J. H. Freed, *J. Chem. Phys.*, 50:5243 (1969).
112. P. Fajer and D. Marsh, *J. Magn. Resonance*, 49:212 (1982).
113. J. S. Hyde and T. Sarna, *J. Chem. Phys.*, 68:4439 (1978).
114. J. W. H. Schreurs and G. K. Fraenkel, *J. Chem. Phys.*, 34:756 (1961).

115. M. P. Eastman, R. G. Kooser, M. R. Das, and J. H. Freed, *J. Chem. Phys.*, 51:2690 (1969).
116. R. H. Crepeau, A. Dulcic, J. Gorchester, T. R. Saarinen, and J. H. Freed, *J. Magn. Resonance*, 84:184 (1989).
117. A. Angerhofer, M. Toporowicz, M. K. Bowman, J. R. Norris, and H. Levanon, *J. Phys. Chem.*, 92:7164 (1988).
118. W. B. Mims and J. Peisach, *Biol. Magn. Resonance*, 3:213 (1981).
119. S. Geschwind, ed., *Electron Paramagnetic Resonance*, Plenum Press, New York, 1972.
120. T. C. Farrar and E. D. Becker, *Pulse and Fourier Transform NMR*, Academic Press, New York, 1971.
121. R. K. Harris, *Nuclear Magnetic Resonance Spectroscopy*, Pitman Books, Marshfield, MA, 1983.
122. J. W. Akitt, *NMR and Chemistry: An Introduction to the Fourier Transform-Multinuclear Era*, 2nd ed., Chapman and Hall, London, 1983.
123. M. L. Martin, G. J. Martin, and J.-J. Delpuech, *Practical NMR Spectroscopy*, Heyden, London, 1980.
124. B. C. Gerstein and C. R. Dybowski, *Transient Techniques in NMR of Solids: An Introduction to Theory and Practice*, Academic Press, New York, 1985.
125. B. L. Bales and L. Kevan, *J. Chem. Phys.*, 52:4644 (1970).
126. R. D. Rataiczak and M. T. Jones, *J. Chem. Phys.*, 56:3898 (1972).
127. P. Wardman and W. A. Seddon, *Can. J. Chem.*, 47:2155 (1969).
128. S. J. Wyard and J. B. Cook, in *Solid State Biophysics* (S. J. Wyard, ed.), McGraw-Hill, New York, 1969, p. 67.
129. E. L. Ginzton, *Microwave Measurements*, McGraw-Hill, New York, 1957, p. 492.
130. J. H. Freed, D. Leniart, and J. S. Hyde, *J. Chem. Phys.*, 47:2762 (1967).
131. D. D. Thomas, L. R. Dalton, and J. S. Hyde, *J. Chem. Phys.*, 65:3006 (1976).
132. A. Beth, K. Balasubramanian, B. H. Robinson, L. R. Dalton, S. K. Venkataram, and J. H. Park, *J. Phys. Chem.*, 87:359 (1983).
133. C. Mottley, L. D. Kispert, and P. S. Wang, *J. Phys. Chem.*, 80:1885 (1976).
134. E. S. Copeland, *Rev. Sci. Instrum.*, 44:437 (1973).
135. W. B. Mims, in *Electron Paramagnetic Resonance* (S. Geschwind, ed.), Plenum Press, New York, 1972, Ch. 2.
136. M. Peric, B. Rakvin, and A. Dulcic, *J. Magn. Resonance*, 65:215 (1985).
137. M. A. Hemminga, F. A. M. Leermakers, and P. A. de Jager, *J. Magn. Resonance*, 59:137 (1984).
138. A. I. Vistnes and L. R. Dalton, *J. Magn. Resonance*, 54:78 (1983).
139. M. M. Dorio and J. H. Freed, eds., *Multiple Electron Resonance Spectroscopy*, Plenum Press, New York, 1979.
140. L. Kevan and L. D. Kispert, *Electron Spin Double Resonance Spectroscopy*, Wiley, New York, 1976.
141. H. C. Box, *Radiation Effects: ESR and ENDOR Analysis*, Academic Press, New York, 1977.

142. A. Schweiger, *Electron Nuclear Double Resonance of Transition Metal Complexes with Organic Ligands*, Structure and Bonding 51, Springer-Verlag, New York, 1982.
143. R. S. Eachus and M. T. Olm, *Science*, 230:268 (1985).
144. M. Plato, W. Lubitz, and K. Mobius, *J. Phys. Chem.*, 85:1202 (1981).
145. H. Kurreck, M. Bock, N. Bretz, M. Elsner, W. Lubitz, F. Muller, J. Geissler, and P. M. H. Kroneck, *J. Am. Chem. Soc.*, 106:737 (1984).
146. B. Kirste, W. Harrer, and H. Kurreck, *J. Am. Chem. Soc.*, 107:20 (1985).
147. W. B. Mims, *The Linear Electric Field Effect in Paramagnetic Resonance*, Oxford University Press, Oxford, 1976.
148. S. D. Devine and W. H. Robinson, *Adv. Magn. Resonance*, 10:53 (1982).
149. L. R. Dalton, B. H. Robinson, L. A. Dalton, and P. Coffey, *Adv. Magn. Resonance*, 8:149 (1976).
150. M. A. Hemminga, P. A. deJager, D. Marsh, and P. Fajer, *J. Magn. Resonance*, 59:160 (1984).
151. M. Setaka, K. M. Sancier, and T. Kwan, *J. Catal.*, 16:44 (1970).
152. A. Hernandez, E. Martin, J. Margineda, and J. M. Zamarro, *J. Phys. E*, 19:222 (1986).
153. G. G. Maresch, M. Mehring, J. U. von Schutz, and H. C. Wolf, *Chem. Phys.*, 85:333 (1984).
154. See, for example, T. J. Emge, H. H. Wang, M. A. Beno, P. C. W. Leung, M. A. Firestone, H. C. Jenkins, J. D. Cook, K. D. Carlson, J. M. Williams, E. L. Venturini, L. J. Azevedo, and J. E. Schirber, *Inorg. Chem.*, 24:1736 (1985), and numerous recent papers in *Solid State Communications*, including D. Shaltiel, J. Genossar, A. Grayevsky, Z. H. Kalman, B. Fisher, and N. Kaplan, *Solid State Commun.*, 63, 987 (1987).
155. K. Baberschke, K. D. Bures, and S. E. Barnes, *Phys. Rev. Lett.*, 53:98 (1984).
156. S. Schultz and E. M. Gullikson, *Rev. Sci. Instrum.*, 54:1383 (1983).
157. K. Ohno, *Appl. Spectrosc. Rev.*, 22:1 (1986).
158. K. Ohno, *Magn. Resonance Rev.*, 11:275 (1987).
159. S. S. Eaton and G. R. Eaton, *Spectroscopy*, 1:32 (1986).
160. S. S. Eaton and G. R. Eaton, *J. Magn. Resonance*, 67:73 (1986).
161. G. R. Eaton and S. S. Eaton, *J. Magn. Resonance*, 67:561 (1986).
162. G. R. Eaton and S. S. Eaton, in *Electronic Magnetic Resonance in the Solid State*, Canadian Institute of Chemistry Symposium Series, Vol. 1 (J. Weil, ed.), Canadian Society for Chemistry, Ottawa, Ontario, 1987.
163. M. M. Maltempo, *J. Magn. Resonance*, 69:156 (1986).
164. A. E. Stillman, D. N. Levin, D. B. Yang, R. B. Marr, and P. C. Lauterbur, *J. Magn. Resonance*, 69:168 (1986).
165. P. C. Lauterbur, D. N. Levin, and R. B. Marr, *J. Magn. Resonance*, 59:536 (1984).
166. M. L. Bernardo, Jr., P. C. Lauterbur, and L. K. Hedges, *J. Magn. Resonance*, 61:168 (1985).
167. M. M. Maltempo, S. S. Eaton, and G. R. Eaton, *J. Magn. Resonance*, 72:449 (1987).

168. M. M. Maltempo, S. S. Eaton, and G. R. Eaton, *J. Magn. Resonance*, 77:75 (1988).
169. U. Ewert and T. Herrling, *Chem. Phys. Lett.*, 129:516 (1986).
170. H. Iwamura, *Pure Appl. Chem.*, 58:187 (1986).
171. G. F. Kokoszka and G. Gordon, in *Technique of Inorganic Chemistry* (H. B. Jonassen and A. Weissberger, eds.), Vol. VII, Wiley-Interscience, New York, 1968, p. 151.
172. H. A. Kuska and M. T. Rogers, in *Coordination Chemistry*, Vol. 1 (A. E. Martell, ed.), ACS Monograph 168, Van Nostrand Reinhold, New York, 1971, Ch. 4.
173. *Electron Spin Resonance Specialist Periodical Reports*, Royal Society of Chemistry, London, Vols. 1-10, 1973-1986.
174. Landoldt-Bornstein Tables, Neue Series II, *Magnetic Properties of Coordination and Organometallic Transition Metal Complexes*, Vol. 8, 1976; Vol. 10, 1979; Vols. 11 and 12b, 1984; and *Magnetic Properties of Free Radicals*, Vols. 9a,b, 1977; Vol. 9c, 1979; and Vol. 9d, 1980.
175. J. W. Orton, *Electron Paramagnetic Resonance: An Introduction to Transition Group Ions in Crystals*, Gordon and Breach, New York, 1968.
176. S. A. Al'tschuler and B. M. Kozyrev, *Electron Paramagnetic Resonance in Compounds of Transition Elements*, 2nd ed., Wiley, New York, 1974.
177. H. A. Buckmaster, *Magn. Resonance Rev.*, 2:273 (1973).
178. H. A. Buckmaster and D. B. Delay, *Magn. Resonance Rev.*, 3:127 (1974).
179. H. A. Buckmaster and D. B. Delay, *Magn. Resonance Rev.*, 4:63 (1976).
180. H. A. Buckmaster and D. B. Delay, *Magn. Resonance Rev.*, 5:25 (1979).
181. H. A. Buckmaster and D. B. Delay, *Magn. Resonance Rev.*, 5:121 (1979).
182. H. A. Buckmaster and D. B. Delay, *Magn. Resonance Rev.*, 6:85 (1979).
183. H. A. Buckmaster and D. B. Delay, *Magn. Resonance Rev.*, 6:139 (1979).
184. H. A. Buckmaster, *Magn. Resonance Rev.*, 8:283 (1983).
185. J. O. Artman, *Magn. Resonance Rev.*, 1:169 (1972).
186. J. R. Bolton, *Magn. Resonance Rev.*, 1:195 (1972).
187. P. D. Sullivan, *Magn. Resonance Rev.*, 2:35 (1973).
188. P. D. Sullivan, *Magn. Resonance Rev.*, 2:315 (1973).
189. C. P. Poole, Jr., and H. A. Farach, *Magn. Resonance Rev.*, 4:137 (1977).
190. C. P. Poole, Jr., H. A. Farach, and T. P. Bishop, *Magn. Resonance Rev.*, 4:137 (1978).
191. P. D. Sullivan, *Magn. Resonance Rev.*, 4:197 (1978).
192 T. F. Yen, ed., *Electron Spin Resonance of Metal Complexes*, Plenum Press, New York, 1969.
193. A. Bencini and D. Gatteschi, *Transition Metal Chemistry*, 8:1 (1982).
194. H. A. Kuska and M. T. Rogers, in *Radical Ions* (E. T. Kaiser and L. Kevan, eds.), Interscience, New York, 1968, Ch. 13.
195. P. D. Sullivan, *Magn. Resonance Rev.*, 3:250 (1974).

196. N. R. Orme-Johnson and W. H. Orme-Johnson, *Meth. Enzymol.*, 52C:252 (1978).
197. J. F. Boas, in *Copper Proteins and Copper Enzymes*, Vol. I (R. Lonte, ed.), CRC Press, Boca Raton, FL, 1984, Ch. 2.
198. M. F. Lappert and P. W. Lednor, *Adv. Organomet. Chem.*, 14:345 (1976).
199. W. Weltner, *Magnetic Atoms and Molecules*, Van Nostrand Reinhold, New York, 1983.
200. T. C. DeVore and W. Weltner, Jr., *J. Am. Chem. Soc.*, 99:4700 (1977).
201. S. W. Bratt, A. Kassyk, R. N. Perutz, and M. C. R. Symons, *J. Am. Chem. Soc.*, 104:490 (1982).
202. N. F. Albanese and N. D. Chasteen, *J. Phys. Chem.*, 82:910 (1978).
203. N. D. Chasteen, *Biol. Magn. Resonance*, 3:53 (1981).
204. R. P. Kohin, *Magn. Resonance Rev.*, 5:75 (1979).
205. L. S. Singer, *J. Chem. Phys.*, 23:379 (1955).
206. W. L. Klotz and M. K. DeArmond, *Inorg. Chem.*, 14:3125 (1975).
207. V. Srinivasan and J. Rocek, *J. Am. Chem. Soc.*, 96:127 (1974).
208. E. G. Derrouane and T. Ouhadi, *Chem. Phys. Lett.*, 31:70 (1975).
209. G. H. Reed and G. D. Markham, *Biol. Magn. Resonance*, 6:73 (1984).
210. E. Meirovitch and A. Lanir, *Chem. Phys. Lett.*, 53:530 (1978).
211. G. Palmer, *Iron Porphyrins*, Part Two (A. B. P. Lever and H. B. Gray, eds.), Addison-Wesley, Reading, MA, 1983, Ch. 2.
212. T. D. Smith and J. R. Pillbrow, *Biol. Magn. Resonance*, 2:85 (1980).
213. E. Konig, in *The Organic Chemistry of Iron*, 1:257 (1978).
214. C. Daul, C. W. Schlapfer, A. von Zelewsky, *Structure and Bonding*, 36:129 (1979).
215. R. S. Drago, J. S. Stahlbush, D. J. Kitko, and J. Breese, *J. Am. Chem. Soc.*, 102:1884 (1980).
216. F. A. Walker, *J. Magn. Resonance*, 15:201 (1974).
217. J. F. Boas, J. R. Pilbrow, and T. D. Smith, *Biol. Magn. Resonance*, 1:277 (1978).
218. B. J. Hathaway and D. E. Billing, *Coord. Chem. Rev.*, 5:143 (1970).
219. B. J. Hathaway, *Struct. Bond.*, 57:55 (1984).
220. R. R. Gagne and D. M. Ingle, *Inorg. Chem.*, 20:420 (1981).
221. F. V. Lovecchio, E. S. Gore, and D. H. Busch, *J. Am. Chem. Soc.*, 96:3109 (1974).
222. R. S. Rubins and S. K. Jani, *J. Chem. Phys.*, 66:3297 (1977).
223. A. G. Lappin, C. K. Murray, and D. W. Margerum, *Inorg. Chem.*, 17:1630 (1978).
224. G. Labauze, E. Samuel, and J. Livage, *Inorg. Chem.*, 19:1384 (1980).
225. B. A. Averill and W. H. Orme-Johnson, *Inorg. Chem.*, 19:1702 (1980).
226. R. C. Bray, *Biol. Magn. Resonance*, 2:45 (1980).
227. G. R. Hanson, G. L. Wilson, T. D. Bailey, J. R. Pilbrow, and A. G. Wedd, *J. Am. Chem. Soc.*, 109:2609 (1987).
228. P. Bernhard, A. Stebler, and A. Ludi, *Inorg. Chem.*, 23:2151 (1984).
229. R. E. DeSimone, *J. Am. Chem. Soc.*, 95:6238 (1973).
230. L. A. Sorin and M. V. Vlasova, *Electron Spin Resonance of Paramagnetic Crystals*, Plenum Press, New York, 1973.
231. L. E. Iton and J. Turkevich, *J. Phys. Chem.*, 81:435 (1977).

232. S. K. Misra and G. C. Upreti, *Magn. Resonance Rev.*, 10:333 (1986).
233. V. M. Malhotra and H. D. Bist, *Chem. Phys. Lett.*, 48:334 (1977).
234. I. Ursu and V. Lupei, *Magn. Resonance Rev.*, 10:253 (1986).
235. L. A. Boatner and M. M. Abraham, *Rep. Prog. Phys.*, 41:87 (1978).
236. S. E. Barnes, *Adv. Phys.*, 30:801 (1981).
237. K. W. Bowers, *Adv. Magn. Resonance*, 1:317 (1965).
238. B. H. J. Bielski and J. M. Gebicki, *Atlas of Electron Spin Resonance Spectra*, Academic Press, New York, 1967.
239. F. Gerson, *High Resolution E.S.R. Spectroscopy*, Wiley, New York, 1970.
240. W. M. Gulick, Jr., *Magn. Resonance Rev.*, 8:33 (1983).
241. H. Fischer, in *Free Radicals*, Vol. II (J. K. Kochi, ed.), Wiley, 1973, ch. 19.
242. G. R. Stevenson, *Magn. Resonance Rev.*, 6:209 (1980).
243. R. D. Allendoerfer, *Magn. Resonance Rev.*, 5:175 (1980).
244. P. D. Sullivan and E. M. Menger, *Adv. Magn. Resonance*, 9:1 (1977).
245. J. A. Pederson, *CRC Handbook of EPR Spectra from Natural and Synthetic Quinones and Quinols*, CRC Press, Boca Raton, FL, 1985.
246. E. G. Rozantsev, *Free Nitroxyl Radicals*, Plenum Press, New York, 1970.
247. L. B. Volodarsky, I. A. Grigorev, and R. Z. Sagdeev, *Biol. Magn. Resonance*, 2:169 (1980).
248. E. G. Janzen, C. A. Evans, and E. R. Davis, *ACS Symp. Ser.*, 69:433 (1978).
249. E. G. Janzen, in *Free Radicals in Biology*, Vol. IV (W. A. Pryor, ed.), 1980, Ch. 4.
250. B. Ranby and J. F. Rabek, *ESR Spectroscopy in Polymer Research*, Springer-Verlag, Berlin, 1977.
251. A. B. Denison, *Magn. Resonance Rev.*, 2:1 (1973).
252. H. Shields, *Magn. Resonance Rev.*, 3:375 (1974).
253. D. Griller, *Magn. Resonance Rev.*, 5:1 (1979).
254. A. D. Trifunac, *Magn. Resonance Rev.*, 7:147 (1982).
255. H. A. Farach and C. P. Poole, Jr., *Adv. Magn. Resonance*, 5:229 (1971).
256. J. H. Hadley, Jr., *Magn. Resonance Rev.*, 6:59 (1980).
257. J. W. Wells, *Magn. Resonance Rev.*, 8:117 (1980).
258. H. Thomann, L. A. Dalton, and L. R. Dalton, *Biol. Magn. Resonance*, 6:143 (1984).
259. D. J. E. Ingram, *Biological and Biochemical Applications of Electron Spin Resonance*, Plenum Press, New York, 1969.
260. H. S. Low and R. A. Beaudet, *J. Am. Chem. Soc.*, 98:3849 (1976).
261. J. D. Cotton, C. S. Cundy, D. H. Harris, A. Hudson, M. F. Lappert, and P. W. Lednor, *J. Chem. Soc. Chem. Commun.*, 651 (1974).
262. J. T. Warden, *Biol. Magn. Resonance*, 1:239 (1978).
263. D. E. Edmondson, *Biol. Magn. Resonance*, 1:205 (1978).
264. T. A. Kennedy, *Magn. Resonance Rev.*, 7:41 (1981).
265. G. Lancaster, *Electron Spin Resonance in Semiconductors*, Plenum Press, New York, 1967.
266. G. B. Friedmann, *Magn. Resonance Rev.*, 8:243 (1983).
267. D. E. Seeger and J. A. Berson, *J. Am. Chem. Soc.*, 105:5144 (1983).

268. T. Takui and K. Itoh, *Chem. Phys. Lett.*, 19:120 (1973).
269. Y. Teki, T. Takui, K. Itoh, H. Iwamura, and K. Kobayashi, *J. Am. Chem. Soc.*, 105:3722 (1983).
270. L. Petrakis and D. W. Grandy, *Free Radicals in Coals and Synthetic Fuels* (Coal Science and Technology 5), Elsevier, New York, 1983.
271. I. C. Lewis and L. S. Singer, *Chemistry and Physics of Carbon*, 17:1 (1981).
272. S. S. Eaton and G. R. Eaton, in *Biol. Magn. Resonance 8: Spin Labeling III* (L. J. Berliner and J. Reuben, eds.), 340, 1989.

15

X-Ray Photoelectron and Auger Electron Spectroscopy

NOEL H. TURNER / Naval Research Laboratory, Washington, DC

I. INTRODUCTION

The ability to analyze the composition of the outermost few atomic layers of solid surfaces has been expanded greatly over the past 20 years. Two analytical techniques have been in the forefront of the methods for such determinations: X-ray photoelectron spectroscopy (XPS) and Auger electron spectroscopy (AES). XPS is known also as electron spectroscopy for chemical analysis (ESCA). Both XPS and AES have been used in numerous areas of important technology; examples have included catalysis, corrosion, lubrication, electrodes, electronic devices, magnetic storage media, adhesion, adsorbents, biological surfaces, and polymers. In addition, much basic information has been supplied by XPS and AES to help understand reactions at gas-solid interfaces, the atomic environment of solids at and near the surface, and differences between the surface region and the bulk. Every year, a few thousand papers are published in which one or both of these techniques have been used, and there does not appear that there will be any decrease in the interest and use of these techniques in the near future. While the rapid advancement in both equipment and technique that characterized the development of these methods 10-15 years ago has declined, many more subtle improvements are still being made. These advances will continue to increase the usefulness of these methods for years to come.

II. BASIC PRINCIPLES

An extensive discussion of the basic principles involved with XPS and AES is beyond the scope of this chapter; other sources give more detailed explanations [1-15]. However, many of the important factors will be considered as they relate to available commercial instrumentation.

The most basic question that has to be answered is: Why are these techniques sensitive to the surface region of solids? Both XPS and AES make use of the fact that electrons with kinetic energies from about 30 to 2000 eV that are ejected from an atom in a solid will have an inelastic mean free path (IMFP or λ) of about 0.5 nm to 2-3 nm. The IMFP is defined as the average distance that an electron will travel before it undergoes an inelastic collision with another matrix constituent atom. Therefore, very few electrons from depths greater than two to three times the IMFP will

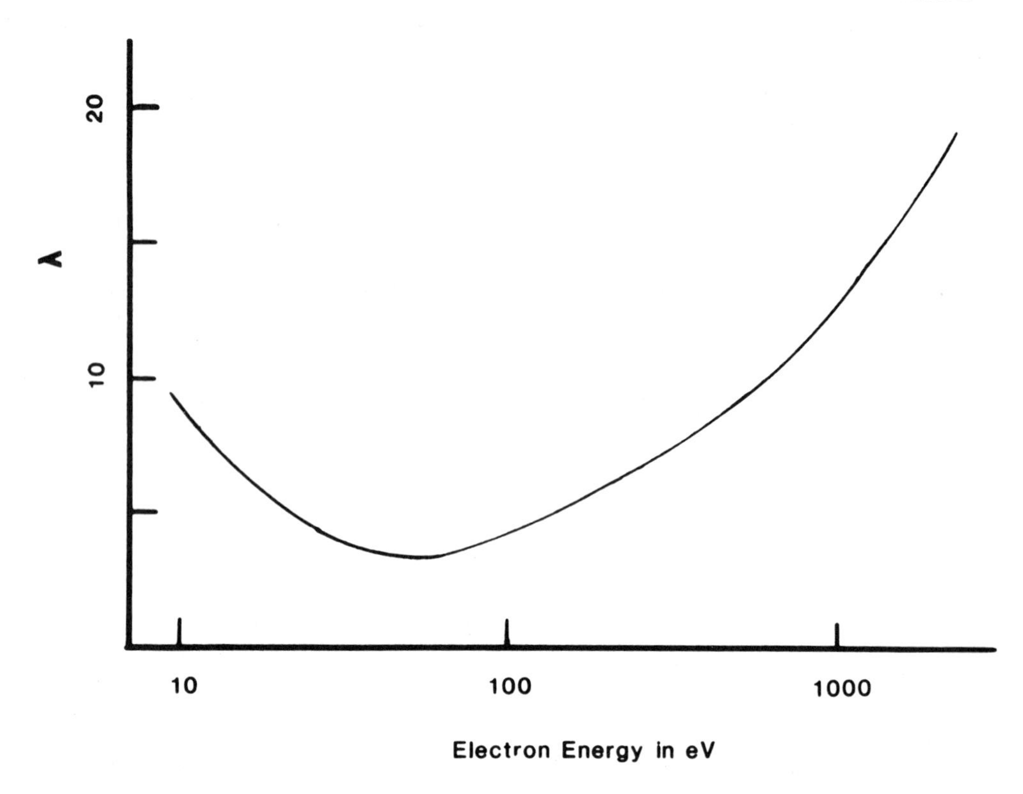

Figure 1. A representation of the inelastic mean free path (IMFP) versus the kinetic energy of an ejected electron computed for iron from the relationship given in Ref. 18.

leave the bulk with the characteristic photoelectron or Auger electron energy at which they were created. The depth dependence of the IMFP is proportional to 1/e, that is, 37 percent. In some cases it has been suggested that the IMFP for organic compounds and films is somewhat larger than those for metals and inorganic layers, that is, up to 8-12 nm [16,17]. An idealized representation of the IMFP of electrons as a function of kinetic energy is given in Figure 1. Note that there is a minimum value for the IMFP of about 0.5 nm for kinetic energies in the 100-eV region and then it slowly increases as the kinetic energy becomes larger. The data on which such a curve as this is based have a large amount of scatter. IMFPs have been subdivided into three material-dependent classes: elemental, inorganic, and organic [18]. Even with these subdivisions, there is still considerable scatter in the data. It also has been suggested from this study that the IMFP follows the relationship

$$\lambda = aE^b \tag{1}$$

where a is a material-dependent constant, b is an exponent with a value of 0.5, and E is the kinetic energy of the ejected electron. Others have proposed that b has a value near 0.75 [19-22]. Regardless of the various suggested functional dependences of the IMFP, the actual experimentally determined values are within the region shown above. With XPS the use of grazing takeoff angles combined with angular resolution detection is used to increase surface sensitivity; this topic will be treated in detail later.

A. XPS

XPS is based on the photoelectron effect in which a valence or core electron is ejected when irradiated with soft X-rays. The Einstein relationship governs the process:

$$KE = h\nu - BE - \phi_{sp} \tag{2}$$

where KE is the measured kinetic energy of the ejected electron, $h\nu$ is the photon energy (for XPS this is usually in excess of 1 keV), BE is the binding energy of the ejected electron, and ϕ_{sp} is the work function of the spectrometer. This latter factor is constant for a particular instrument and should vary little during its lifetime. With a simple rearrangement of equation (2), the binding energy can be evaluated from an experimental spectrum. All of the elements except hydrogen and helium can be detected by XPS. Each element will have its own unique set of binding energies, so that identification of the element or elements that contribute to an XPS spectrum can be made. In a few cases there is an overlap of lines from different elements. However, there are other lines that will be observable (except when the signal is due to a very small amount of an individual element) to differentiate elements in these overlapping situations.

The energy level of the ejected photoelectron is denoted by atomic notation. For example, in a typical XPS spectrum of Cu, as shown in Figure 2, peaks due to the 2p, 3s, 3p, and 3d levels of Cu would be observed. The 2s level, not shown in the figure, also is excited with the usual X-ray sources (i.e., Mg or Al) used in XPS. However, the 1s level could not be observed, since its binding energy is greater than the characteristic photon energies of these X-ray sources (1253.6 and 1486.6 eV for Mg and Al, respectively). With high-resolution spectra, the $3p_{1/2}$ and $3p_{3/2}$ lines for Cu could be partially distinguished. In addition to the photoelectron peaks, several Auger transitions are found, as indicated in the spectrum. These lines are due to the holes created in the 2p levels in the photoemission process; details about the Auger mechanism will be given below. Binding energy values for all of the elements have been tabulated [23].

In addition to the identification of the elements in the surface region, information about the chemical environment of these elements is usually available from XPS spectra. For example, there is a shift of about 5 eV between the $2p_{3/2}$ peak of metallic iron (707 eV) and Fe_2O_3 (712 eV) [24]. For sulfur, there is about an 8 eV difference in the 2p level between S^{-2} and SO_4^{-2} (161 and 169 eV, respectively) [25]. For some elements such as Cu, the shifts are smaller, and other procedures, that is, the use of Auger lines and "shake-up" peaks, have to be used for gaining chemical information [26]. Shake-up peaks are due to a discrete energy loss of a

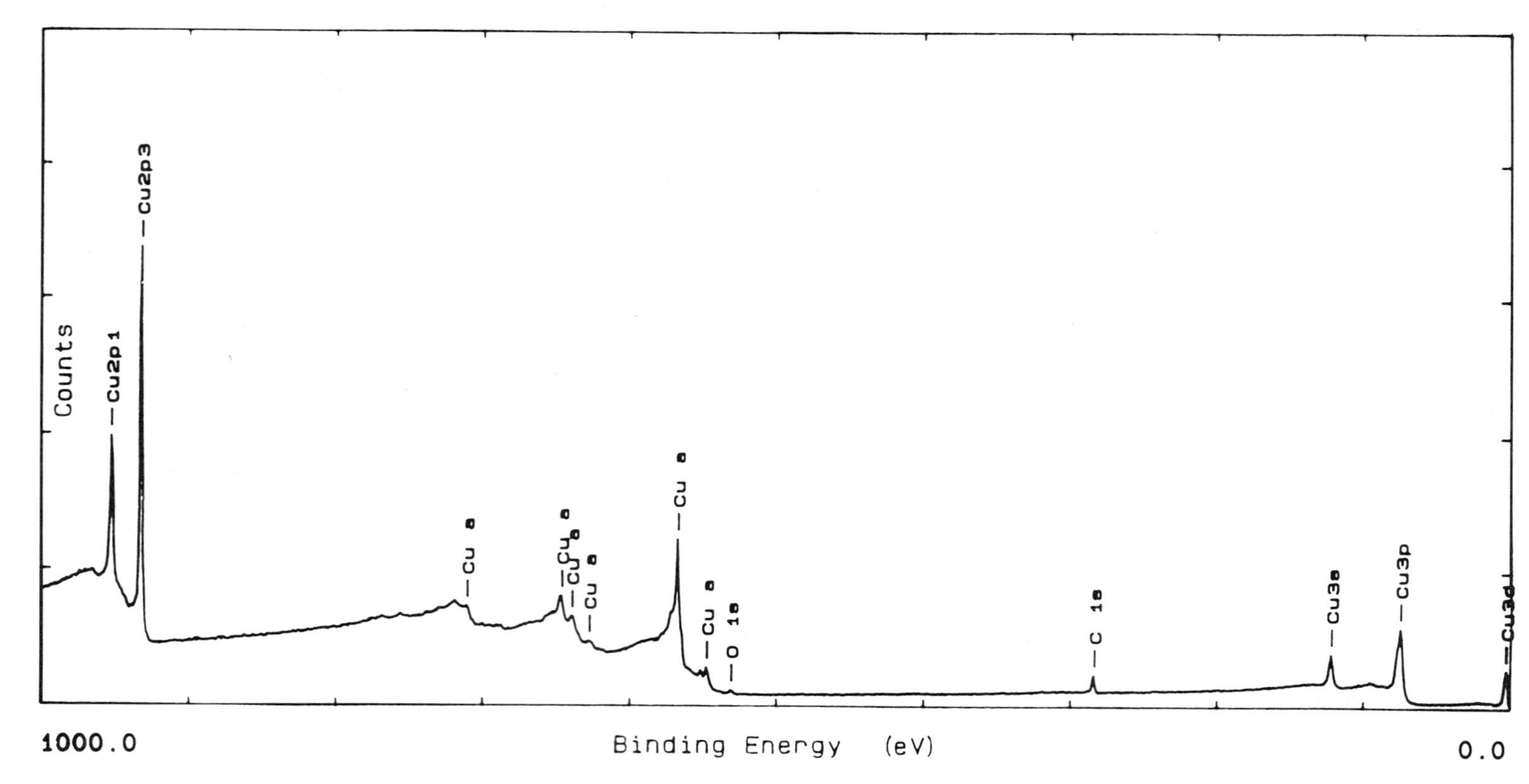

Figure 2. An XPS survey scan of copper with the energy of the photo electron transitions in terms of binding energy. Cu2p1 and Cu2p3 refer to the Cu $2p_{1/2}$ and $2p_{3/2}$ lines, respectively. The Cu a notation refers to X-ray-excited Auger transitions. Note also the small peaks due to carbon and oxygen. This figure courtesy of R. Siordia of Surface Science Industries.

core electron when a valence-band electron is ejected into an unoccupied conduction band during the photoemission event.

The surface sensitivity of XPS can be enhanced by analyzing only those electrons that leave the surface at a glancing angle, and this technique is known as angular resolved XPS (ARXPS). A simple model of a thin uniform layer of B at a depth d on a substrate A indicates this effect. The intensity of the substrate signal has been approximated by the following equation:

$$I_A = I_A^0 \exp(-d/\lambda \sin \theta) \tag{3}$$

where I_A is the measured intensity of the substrate signal, I_A^0 would be the signal of A without B, and θ is the angle at which the detected electrons leave the surface relative to the surface normal. The relationship for the signal intensity of the overlayer is given in equation (4).

$$I_B = I_B^0[1 - \exp(-d/\lambda \sin \theta)] \tag{4}$$

where I_B is the intensity of the overlayer signal and I_B^0 is intensity for an overlayer of several λ. Details about the actual experimental procedures will be given later.

B. AES

When a core hole in an atom is created, there are two ways for the atom to relax. The first way is for an X-ray to be emitted when an electron from a higher orbital falls into the hole (X-ray fluorescence). The second process is the creation of an Auger electron. In this case, an electron from a higher level falls into the empty hole while another electron is ejected from the same or even higher level of the atom. The energy relationship for the Auger process is given by

$$KE = W - X - Y \tag{5}$$

where W is the binding energy of the electron ejected during the initial ionization event, X is the binding energy of the electron that falls into the empty level, and Y is the binding energy of the ejected electron. Note that the energy of the Y level is going to be different from the initial ground-state atom, since there is now a net positive charge on the atom. There are more complex decay mechanism that contribute to the overall Auger transitions beyond the description given above. These effects usually do not alter the energy of the observed Auger transition to the extent that an elemental identification cannot be made. Details on these points have been reviewed previously [27].

Auger electrons can be created in a number of different ways. However, for surface analysis the two most predominant methods are from an electron beam that has a kinetic energy of about 2-10 keV, or from an X-ray source. The electron beam has to have at least two to three times the kinetic energy of the Auger transition of interest in order to observe the transition with reasonable signal strength. Beam currents range from tens of microamperes to about a nanoampere. In certain Auger spectrometers, the beam diameter can be adjusted to be less than 50 nm. Compilations of Auger spectra have been made for routine elemental identification [28-30]. Auger transitions are observed in XPS spectra and often provide additional information with this type of analysis [23,31-34].

AES lines are identified with X-ray notation. AES spectra are displayed in two ways: derivative and nonderivative. Details about the choice of presentation method will be given later. For example, if a hole is created in the 1s level of Si, and a $2p_{3/2}$ electron falls into the empty hole and another $2p_{3/2}$ electron is ejected from the atom, the transition would have the notation KL_3L_3. In actual fact the energy difference between the $2p_{1/2}$ and $2p_{3/2}$ levels in Si is small compared to the observable width of this Auger transition. Thus, the observed peak is a combination of these levels, and its notation is $KL_{2,3}L_{2,3}$. If valence electrons are involved in the Auger transition, often they are denoted by V. Examples of this are illustrated in Figures 3 and 4, which show the AES spectrum of Li_2SO_4 in both the integral and derivative display modes. In this case a hole in either the $2p_{1/2}$ or $2p_{3/2}$ level is filled with an electron falling from a valence level and ejecting another valence electron, and the notation would be $L_{2,3}VV$. Chemical effects often are observed in AES. In some cases there are shifts in the position of the observed peaks; for example, the $KL_{2,3}L_{2,3}$ peak for Si is observed at 1606 eV while this peak is at

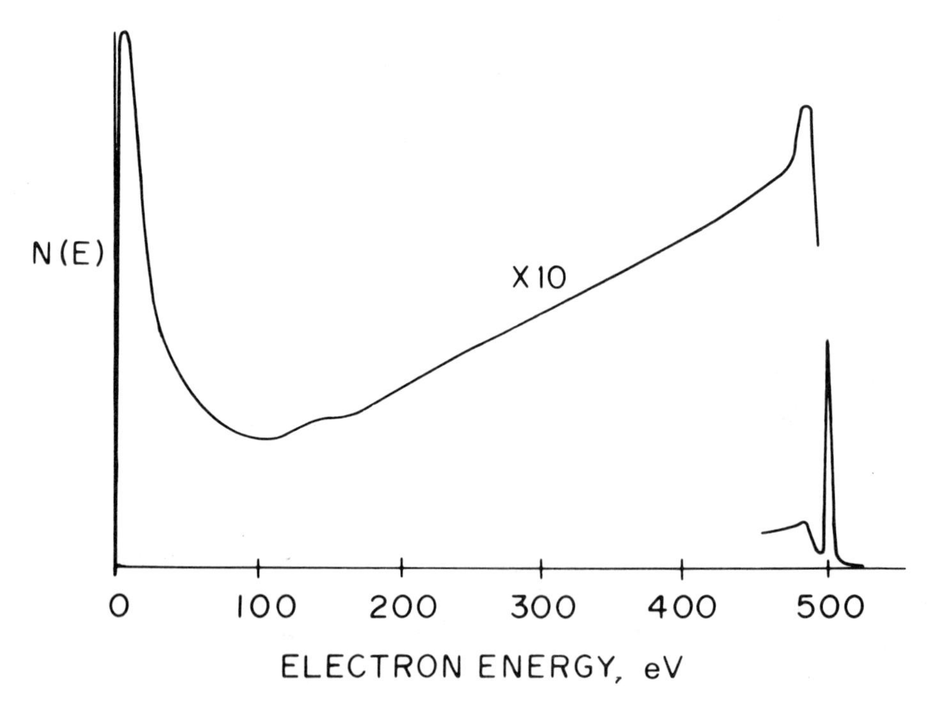

Figure 3. The electron excited AES spectrum of Li_2SO_4 observed with an electron beam excitation of 500 eV in the nonderivative display mode (from Ref. 81). The small bump at about 140 eV is due to the sulfur LVV transition. The strong peak at 500 eV is due to the reflected primary electron beam.

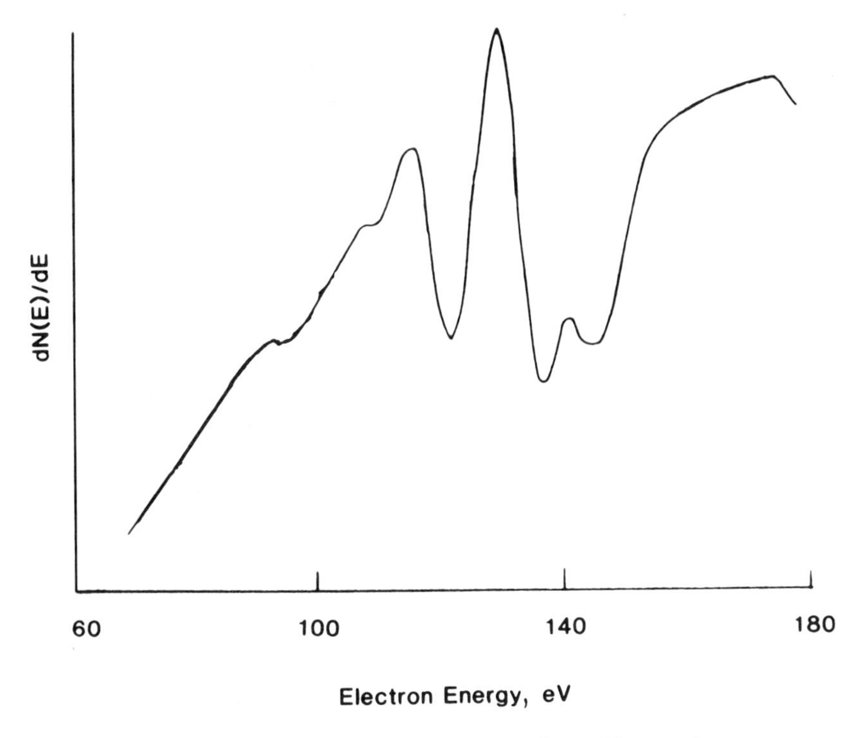

Figure 4. The derivative AES spectrum from Figure 3.

1611 eV for SiO_2 [35]. In other instances there are both an energy shift and a change in the peak shape. An example of this is with the S $L_{2,3}VV$ peaks of sulfides (150 eV) and sulfates (126 and 140 eV) [25]. The peak at 150 eV in Figure 4 is due to reduction of a sulfide-type species by the electron beam. Most of the peak shape differences that are observed involve transitions with valence electrons.

III. INSTRUMENTATION

There are several components that make up an XPS or AES spectrometer, and several commercially available systems that are capable of performing both XPS and electron-excited AES. There are, of course, some compromises with such systems. The various portions of XPS and AES spectrometers that will be considered are the vacuum system, sample handling and introduction, production of the photoelectrons or Auger electrons, the detection of these electrons, and the data collection and handling procedures.

A. Vacuum Systems

Both XPS and AES require vacuums in the 10^{-3} pascal (Pa) region (1 pascal is 1 newton/m^2 and is equal to about 7.5×10^{-3} torr). With pressures higher than this value, a significant number of the electrons to be analyzed will undergo collisions with the residual gas in the system and will not be detected. Now, almost all commercially available systems are capable of achieving pressures in the ultra-high-vacuum regime (about 10^{-8} to 10^{-9} Pa). It should be noted that it will take several hours for a surface to be covered (assuming that every molecule sticks to the surface) at these pressures. This means that once a clean surface has been produced, it should be possible under normal operating conditions to completely characterize by either or both XPS and AES most materials, without significant contamination from the vacuum system. Since not all combinations of clean surfaces and gases will have a sticking coefficient of one (i.e., upon each collision with a surface a gas molecule will be adsorbed), this reduces these problems in many cases.

The achievement of ultra-high vacuum can be accomplished in many ways. Consideration has to be given to the materials used in both the construction of a vacuum chamber and the components that are employed for analytical purposes. Fortunately, the technology to produce an ultra-high vacuum (UHV) has been available for many years. Thus, only routine considerations now have to be made for such systems.

Surface analytical instruments available now are almost entirely metal in construction. The metals used must be compatible with the requirements of obtaining an ultra-high vacuum, that is, a low outgassing rate per unit area, the ability to be heated to above room temperature (and in some uses for certain metals to over 1000°C), and ease of construction. Stainless steel is used for the main body of the vacuum chamber as it fulfills these requirements. Some metals such as brass have an appreciable vapor pressure in the ultra-high vacuum regime at room temperature, and therefore cannot be employed for such uses. In addition, the flanges on the various ports of an ultra-high vacuum system are constructed of stainless steel. On these ports are attached the various devices needed for a surface analysis system. The flanges, which almost always are circular, have a variety of trade names such as Con-Flat, Vac-U-Flange, etc. A circular recess with a mating knife edge is used to seal flanges with a gasket. Usually the gasket is made of copper, but other soft metals can be used. Copper is the usual material of choice, since it is relatively inexpensive compared to other soft metals such as gold, and can withstand higher temperatures than indium. For certain special flanges, a soft copper wire is used as the gasket. The flanges are held in position with a series of bolts. Almost all of the circular flanges are made in standard sizes and designs. In the United States, the flange size is referred to by the outside diameter; the range of sizes is from 1.33 to 10 in. This permits attaching flanges from different manufacturers to each other. Thus, easy modification to existing systems usually can be made. However, there is a tendency for some manufacturers of surface analysis systems to have only a minimum of extra flanges on their chambers. This reduces the capability to make needed alterations to existing systems when experimental circumstances demand change.

Certain nonmetallic materials are found also in an ultra-high-vacuum (UHV) surface analysis chamber. Electrical feedthroughs (usually on flanges) from the outside of the chamber to the interior of the system require ceramic insulators. Ceramic materials also are used as insulators for wires inside the vacuum chamber, especially when there is the chance of shorting electrical leads. Windows are almost always present on surface analysis systems so that the specimens can be visually observed. Often small amounts of polymeric material are present in UHV systems. Their most common use is for gaskets on certain valves (e.g., Viton). This material can be heated to about 150°C without significant outgassing, if it is not under compression. The surface area that these gaskets expose to the vacuum system is usually relatively small compared to the total area in such systems. In addition, polymeric materials such as Teflon sometimes are used to insulate wires inside an UHV system. Such usage should be kept to a minimum, since these materials will have a high outgassing rate during bakeout of the analysis chamber.

Several different types of vacuum pumps are capable of producing the necessary vacuum for XPS and/or AES spectrometers. Each type has its own set of advantages and disadvantages. In many systems combinations of pumps are used, either when operating in the UHV environment or in pumping the system down from atmospheric pressure. When it is necessary to bring a chamber up to atmospheric pressure, a nonreactive, dry gas such as nitrogen is admitted.

Ion pumps are used quite extensively for UHV systems. Electrons produced from cold cathodes (usually titanium) ionize gas molecules. The cathodes and anodes are in a strong magnetic field to increase the electron path length. The positive ions then strike the cathode and this releases neutral Ti atoms that coat the surrounding surfaces. These fresh Ti atoms will react with all gases except the inert gases, producing the pumping mechanism. Inert gases are buried under succeeding layers of Ti, and the efficiency for pumping these gases is less than with more reactive species. These pumps have a long lifetime, especially when operated in the UHV regime, since the amount of titanium removed is proportional to the pressure. If the load of the inert gases is too large, then small amounts of these gases are observed for extended periods. In addition, if the pressure in the system is raised—for example, for sputtering (to be discussed later) or to react a surface with a gas—ion pumps will take a long time to recover to the base pressure. The pumping speed for surface analysis systems that use ion pumps is usually about 200 liter/sec, but higher-speed pumps are available. Smaller ion pumps are used sometimes to evacuate special regions where the conductance to the main pump of a system is poor.

Often built into the main ion pumps is a separate Ti sublimination pump. This type of pump operates by the deposition of active layers of Ti onto a special metal piece within the vacuum system. These layers are produced by heating Ti filaments for a short period of time. Until the surface is covered by gas molecules, this type of pump is very effective for removing active gases. Titanium sublimination is often used during the pumpdown of a system after it has been taken up to atmospheric pressure. Also, this type of pump is used when sputtering with an inert gas if a turbomolecular pump (to be described later) is not available. In this case, the ion pump usually is turned off. The effective pumping speed of the Ti sublimation pumps can be increased if the surface on which the Ti is deposited is cooled with liquid nitrogen [36].

An ion pump cannot be operated if the pressure is greater than about 10^{-2} Pa. The time for which these pumps can be used at this pressure should be brief. Evacuation of a chamber from atmospheric pressure to a point at which the ion pump can be turned on has to be done with a roughing pump. One method is the use of a sorption pump. The sorption pump is a relatively small, valved vessel that contains molecular sieves. When these sieves are cooled with liquid nitrogen, they can adsorb a large amount of gas relative to their volume due to their large specific area. Sorption pumps can achieve pressures in the 10^{-3} Pa region, but in normal usage the pressure is one to two orders of magnitude higher. If the volume of the system that needs to be evacuated is relatively large, then more than one sorption pump may be required to bring the pressure down to a level at which an ion pump may be started.

The use of a rotary mechanical pump for the initial pump down of a system is not recommended. The possibility of oil contamination to a UHV system is too large. An oil trap in the vacuum foreline will reduce this risk, but there are other pumping methods that can achieve the same results (i.e., sorption pumps) at about the same cost.

Turbomolecular pumps also can be used for surface analysis chambers. These pumps are somewhat similar to turbine engines. A rotating set of blades compresses gas molecules against a set of stationary blades. The gas is then forced to a following set of rotating and stationary blades. This process is repeated several times until the gas is expelled from the turbine area. When operating at full speed the turbines rotate above 20,000 rpm. In order to achieve UHV, the turbopump is backed up with a mechanical pump. For protection of the vacuum system from the oil of the mechanical pump, and oil trap should be installed between the mechanical and the turbomolecular pump. This is especially important in the event of a power failure. Most turbomolecular pumps are capable of evacuating a chamber from atmospheric pressure without the use of a separate roughing pump.

Turbomolecular pumps are capable of pumping all gases. However, the pumping speed of lighter gases such as H_2 and He is much less than that for higher-molecular-weight gases. This type of pump is capable of handling gas loads that are much greater than those of ion pumps. This can be of use when sputtering or exposing a surface to a gas or gases in a controlled fashion. Cooling of turbomolecular pumps is achieved with water, or in some cases by air. Care must be taken with these pumps to eliminate vibrations in the analysis chamber. This is especially important when analyzing small areas on a sample. Also, with some pumps, measures may be needed to reduce the high-pitched noise that many of these pumps produce, such as baffles or enclosing the pump. Turbomolecular pumps require service more often than ion pumps, given equal service time. Turbomolecular pumps are available in many sizes, and in some cases systems will have both ion pumps and turbomolecular pumps, so as to take advantage of the strengths of each pump design.

Cryopumps can be used in surface analysis chambers also, but their use is not as great as either ion pumps or turbomolecular pumps. These pumps employ a high-specific-area adsorbent material, usually activated carbon, that has been cooled to a temperature near that of liquid helium. These pumps are vibration free, except that a compressor is needed to liquefy the helium. This compressor can be away from the main vacuum chamber. Cryopumps have the capacity to adsorb large quantities of gas, that is, 1000 liters at STP. In addition they can have very high pumping speeds for most gases. However, when the gas load capacity has been reached, the pump must be warmed to release the adsorbed gas. Provision for this is made so that the evacuated chamber does not have to be exposed to the released gases via proper valving. By careful operation of the system, and the use of a roughing pump, the time between warming cycles for a cryopump can be quite long. The ultimate pressure that can be achieved with a cryopumped system is in the 10^{-7} to 10^{-8} Pa region.

The venerable oil diffusion pump is capable of achieving the vacuum required for a surface analysis system, and it is offered by some manufacturers. This type of pump offers high pumping speed with relatively low cost. When used with a UHV system, a trap that is usually cooled with liquid nitrogen has to be employed. This is to prevent oil from backstreaming into the vacuum system. Such traps reduce the pumping speed of the oil diffusion pump by about 50 percent. This reduces the pumping speed advantage that oil diffusion pumps have over the other pumps discussed. In addition, this trap must be in operation at all times to prevent creeping by the diffusion pump oil. Most oil diffusion pumps require water cooling.

As stated above, each of these pumps has its own set of strengths and weaknesses. Some suppliers offer a choice of a vacuum pump or pumps for their instruments, while others have a fixed design that does not allow a choice. For these latter systems, it is possible to have modifications made to allow for an additional pump. With certain types of applications this can be very important. While there are some differences in the cost between the various types of pumping systems, they are not large. This is especially true when considering the overall cost of a surface analysis system.

Most surface analysis systems provide for heating the analysis chamber to temperatures of about 150-200°C. This "baking out" of the system results in the attainment of the base pressure of the system much more rapidly than if the system is not heated after it has been brought up to atmospheric pressure. Water vapor is the most usual contaminant in a vacuum system that impedes the achievement of the base pressure. In some cases a vacuum chamber will never reach the lowest pressure that it can attain without being baked out. Most systems are provided with either ovens that enclose the system, or insulated blankets that may contain heating elements. In addition, there may be heating elements in the base of the instruments. When baking out a system, certain heat-sensitive components have to be removed, such as micrometer drives, certain electrical connections, etc. Another approach to baking out a system is the use of a quartz heater that is placed permanently in the vacuum chamber. This method of heating is not as effective as the use of an oven or heating blankets, but it reduces the need for the removal or protection of components that are heat sensitive. This method also probably is not as efficient as baking, but again does not require that certain heat-sensitive components be removed from the system.

B. Sample Introduction and Handling

Early XPS and AES spectrometers required most of the vacuum system to be brought up to atmospheric pressure in order to change samples. The samples were placed upon a carousel. With modifications to the carousel it was often possible to handle a sample that was irregular in shape and/or bulky. Also, this arrangement allowed the introduction of a number of specimens at one time (provided that they were not very bulky), but much time was lost in reestablishing adequate vacuum conditions for surface analysis on a routine basis. In certain cases such an approach is still used when the samples require heating and/or cooling beyond the capability of sample holders designed for routine analysis.

In place of the older approach, many manufacturers now use a prechamber for the placement of the samples onto a holder. This prechamber usually has a small volume that can be evacuated quickly to a pressure of about 10^{-4} Pa. A small turbomolecular pump is usually appropriate for such systems. Then a gate valve on the main chamber is opened, and the sample or samples are transferred into the main chamber. The pressure in the main chamber usually will not rise by more than one to two orders of magnitude, and it will then fall to a value close to the original pressure in a matter of minutes. (Note that this presumes that the sample or samples do not outgas excessively.) Also, it is possible to design a glove box arrangement around the prechamber such that a sample does not have to be exposed to the atmosphere. Other types of sample treatment are possible, such as plasma etching. It is possible to have a second chamber between the ultra-high surface analysis vacuum system and the atmospheric pressure introduction assembly. This is done to lessen the pressure rise in the main analysis chamber, but it increases the complexity of the introduction system.

A number of different introduction approaches have been used for the placement of specimens into surface analysis chambers. Some designs allow only one sample holder to be introduced at a time into the analysis chamber. In some cases a parking arrangement is used to hold several individual sample holders in the analysis chamber. This feature is useful when it is desired to pump on the samples for an extended period of time, such as overnight, before analysis. Other designs allow a number of individual sample holders to be placed on a larger holder that is then introduced into the main chamber. Recently, introduction systems have been constructed that allow for the precooling of a sample in a prechamber before the holder is placed in the analysis chamber on a cooled sample stage.

There are several different mechanisms available for the actual transfer of the sample holder assembly into an analysis chamber. One approach is the use of a rack and pinion drive in the evacuated prechamber. The sample holder is placed on the end of the drive where it can be positioned in the main chamber via simple rotation of an exterior mechanical feedthrough. Another design for the transfer assembly involves a rod that is sealed with elastomeric gaskets. The sample holder is positioned on the end of the rod. The transfer is accomplished by manually moving the rod into the main analysis chamber. These gaskets require replacement on a regular basis, since this design requires a very tight fit between the rod and the gaskets. A magnet-controlled rod inside the UHV chamber that in turn is attached to the sample holder is another possible introduction approach. Trolley arrangements have been employed also for the transfer of sample holders. With this design the holder is positioned upon the trolley, which then transfers the sample into the main chamber. A manual rod feedthrough is used to remove the sample holder from the trolley and place it onto the main sample holder. Another approach is the use of a pneumatically controlled sample holder that is transferred into the analysis chamber. Each of these designs has its own advocates. Many of the transfer systems are highly automated and reduce or eliminate the possibility of operator error, such as an unintentional exposure of the main chamber to atmospheric pressure. This type of feature is particularly useful if a system will have a number of operators that do not use the equipment regularly.

Many of the regularly available sample holders only allow specimens of limited size, that is, 1-2 cm on a side and less than 1 cm thick. In addition, the sample entry port to the analysis chamber may not allow the passage of anything much larger than the standard sample holder. Unfortunately, many real-world samples cannot easily be placed on such holders, due to their size or shape, and they cannot be cut to dimensions that would allow them to be placed on such holders. Sample holders can be designed for bigger samples. In addition, larger sample entry ports can be placed on most surface analysis systems at the time of system construction. Some systems have been designed so that larger samples can be studied; for example, silicon wafers 15-20 cm in size can be introduced into certain XPS spectrometers. Unless only a certain size of specimen will always be analyzed, it is prudent to allow for as much flexibility in the sample introduction design as is possible.

Samples are placed on a stage that allows optimal positioning for the energy analyzer. This is especially important for irregularly shaped samples and/or for analyses where high spatial resolution is required. Positioning should be relatively straightforward and allow movement in all directions. In some instances this is achieved with motorized micrometers that can be under computer control. For scanning Auger microscopy (SAM—to be discussed later), a completely eucentric specimen stage should be used. and the ability to rotate and tilt the sample is required.

In addition to having the capability to accommodate samples of irregular geometry and size, sample manipulators should possess a number of other features if any nonstandard analyses may be encountered. For example, in some instances it is useful to be able to heat or cool samples relative to the ambient temperature of the analysis chamber. Heating can be done by direct or indirect electron bombardment, resistive heating (if the sample is conductive), or with special heating elements. Heating is useful in certain cases to allow minor contaminants to diffuse into the bulk of the sample under investigation or to come to the surface where sputtering (to be discussed later) is used for their removal. Temperatures of over 1000°C can be achieved with certain heatable sample manipulators. Cooling is useful if a sample has a relatively high vapor pressure or if a low-temperature process is being investigated. Cooling is usually done with liquid nitrogen, either through flexible tubing connected to a small reservoir, or with flexible braids. With certain designs it is possible to achieve temperatures close to 20 K, but temperatures near liquid nitrogen (77 K) are more usual. Obviously, provisions must be made to measure the temperature of the sample with a thermocouple. A manipulator should also have the capability of electrically isolating a sample. This is useful when measuring the current from an ion gun or an electron beam.

It is very important that the samples to be analyzed have not been touched with bare hands. Disposable plastic gloves (powder free) or lint-free cloth gloves should be used at all times when handling anything that goes into a UHV system. If the samples have been handled with bare hands, they have to be cleaned with a solvent or a series of solvents. Acetone, methanol, or various Freons are often used, and an ultrasonic cleaner will speed the cleaning process. In some instances certain solvents may not be used due to incompatibilities with components in analysis systems. Sample holders and tools used with the sample and sample holders (i.e., small screw drivers, tweezers, etc.) also have to be cleaned in a similar manner. A laminar flow hood is often used as a work station for the placement of samples onto holders, and for other operations for equipment that will be placed into an UHV system. The basic rule is to keep everything that is in contact with anything that goes into the vacuum system as clean as possible.

C. Excitation Sources

As was noted above, the electrons that are analyzed in XPS are produced by an X-ray source. Auger electrons are observed also when an X-ray source is used. However, with AES a beam of electrons is most often employed.

X-Ray Sources

The most common X-rays used in XPS are the Kα lines from Mg or Al anodes, that have energies of 1253.6 and 1486.6 eV, respectively. The anode material is usually a thin coating of the metal on a copper substrate. X-rays are produced when the anode is bombarded by electrons with a potential of about 10-15 keV with respect to the anode. In order to produce sufficient X-ray flux, the emission current from the electron source filament can be up to tens of milliamps. With normal X-ray sources a thin window of Al or Be is used to prevent stray electrons from striking the sample being analyzed. Also, adequate cooling of the anode is required, since up to several hundred watts have to be dissipated. Usually a closed-loop pressurized water system is used for cooling purposes. If the cooling is not adequate, a reduced lifetime of the anode is experienced. Designs where the anode is at ground potential or where the electron filament is at ground potential have been produced. When the latter arrangement is used, adequate precautions have to be made so that there is no electrical leakage from the cooling fluid to the rest of the system. If much higher potentials are required, additional safety measures would be needed.

There are several reasons that Mg and Al usually have been chosen as X-ray sources. The X-ray linewidth for the K$\alpha_{1,2}$ lines of Mg and Al is about 0.6 and 0.7 eV, respectively. Therefore, these sources will not excessively broaden or distort most XPS peaks. The $K\alpha_{1,2}$ lines are actually two closely spaced lines that are separated by approximately 0.3 and 0.4 eV for Mg and Al, respectively. The $K\alpha_1$ line is twice as intense as that of the $K\alpha_2$. All of the elements that can be observed by XPS have transitions that are accessed with X-rays produced with either of these elements as anodes. The kinetic energy of the XPS peaks from either of these X-ray sources is such that the IMFP is only about 1-2 nm.

There is no clear-cut advantage for either of these anodes. While Mg has a somewhat narrower X-ray line, a somewhat higher power level can be used with Al. Also, Al sources have a longer lifetime. In some cases, there is an overlap of an Auger transition with an XPS line when employing one particular anode. Since the kinetic energy of an Auger transition is constant and the kinetic energy of an XPS peak is dependent on the energy of the X-ray source [see equations (2) and (5)], changing anode materials will resolve the overlap that might be present. In order to overcome this potential problem, anode designs have been produced that have both Mg and Al sources in one assembly. (Assemblies with four different anodes have been made, but they have not been used widely). Each anode has its own filament, and can be used independently or together with the other anode. An example of such a design has been given [37]. The design has to be made so that there will not be any "cross-talk" between the two anodes [38]. For some designs, modifications have to be made to existing products to eliminate this problem. In some instances both anodes have been made of the same material. This allows a higher X-ray power to be used for an analysis.

While Mg and Al have been the most widely employed X-ray sources, other elements have been used. Silicon (1739 eV), Ag (2984 eV), and Ti (4510 eV) are available. Some of these materials are difficult to produce as anodes, and others have heat conduction problems. In addition, the X-ray line widths are somewhat larger with these elements than those for Mg or Al. Since the kinetic energy of the electrons excited by these anodes is higher than those from Mg or Al for a given XPS peak, it is possible to probe somewhat more deeply into the bulk with these anodes. This has proven useful in certain studies of polyemrs [39]. These other elements do offer the possibility of exciting higher energy Auger transitions than Mg and Al. This can extend the range of use of the Auger parameter for certain elements. The Auger parameter is defined as the sum of the binding energy and kinetic energy of the most intense (usually) X-ray photo emission line and X-ray induced Auger line. Also, it is possible to use one of these other anode materials with the usual Mg or Al anodes in a dual anode assembly.

Monochromatic X-ray Sources

It was noted above that the usual Mg or Al X-ray sources have two major X-ray lines, that is, $K\alpha_{1,2}$. In addition, there are satellite peaks about 10 eV higher in kinetic energy, the $K\alpha_{3,4}$. These lines have about 10 percent of the intensity of the main $K\alpha_{1,2}$ lines. Thus, small additional photo emission peaks will appear in the XPS spectra when a normal X-ray source is used. In order to circumvent this problem, X-ray monochromators for

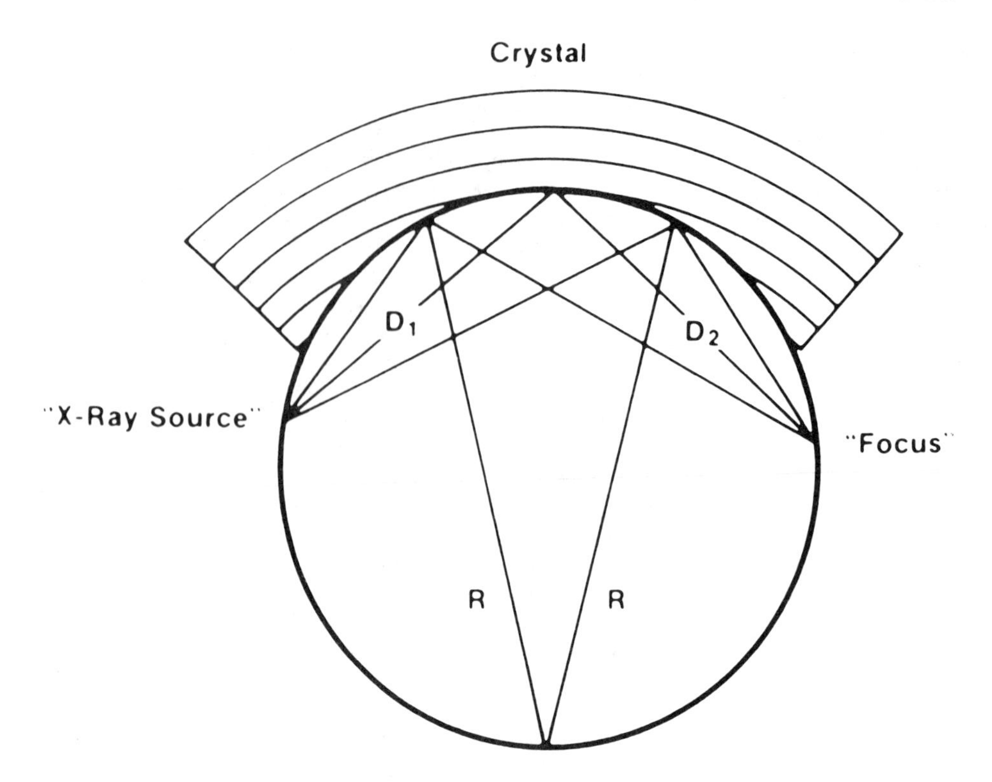

Figure 5. A schematic illustration of the bent crystal arrangement for the production of monochromatic X-rays. (From Ref. 41, courtesy of John Wiley and Sons.)

use in XPS spectrometers have been developed. The X-rays produced from the anode irradiate a bent crystal as shown in Figure 5. Due to the crystal lattice spacing, X-rays of only one given energy will be reflected, as given by the Bragg equation. Due to crystal spacings, Mg cannot be used with a monochromator; however, Al and elements that have a primary X-ray line at an energy greater than Al can be used. It has been shown that the monochromator used for Al can also be used with only slight modifications for Ag, since this latter element has a major X-ray line that is almost exactly double the energy of the Al $K\alpha_1$ line [40].

There are both advantages and disadvantages with monochromatic X-ray sources. With only one component of the main X-ray line, that is, the $K\alpha_1$ for Al, it is possible to obtain sharper XPS peaks in many cases. This can allow the resolution of the XPS line shape of an element in more than one elemental state much more easily than with normal X-ray sources. Also, damage to samples that are sensitive to X-ray radiation is reduced with a monochromatic source. Bremmstrahlung radiation that gives rise to a higher background with the normal X-ray sources is absent. One company has developed a focused X-ray generator with an Al monochromatic source that can irradiate an area of 150 μm [41]. Other designs irradiate much larger areas.

The X-ray intensity that a sample receives with a monochromatic source is very much less than from a normal source. This requires much longer data acquisition times or advanced electron detection systems in order to obtain a spectrum in a reasonable amount of time. Charging, that is, a shifting of XPS peaks that is observed with nonconducting samples due to a deficit of electrons on the sample surface from the photo emission process, is observed much more often with monochromatic X-ray sources. With normal X-ray sources the thin film in the assembly between the anode and sample releases secondary electrons as a result of the X-ray irradiation, and that often compensates for the electrons leaving the surface [42]. It has been suggested that the electrons from the thin film window could cause damage to sensitive samples [41]. Several approaches are employed to reduce or eliminate the effects of charging. A flood gun that irradiates the sample with electrons with only a few electron-volts of energy is often used. The placement of a fine metal screen above a charging sample in conjunction with a flood gun has been reported recently to be better than a flood gun alone [41]. Often, the use of a known photoline (usually the C1s line from contamination) has been used to correct for the effects of charging. In this case, a standard value for this line is used, and all of the peaks are shifted by a like amount [43], but this method has been criticized [44]. Argon, implanted from sputtering, has been suggested also as a binding energy standard [45], but again relaxation effects may make this line shift in energy for different samples. The Auger parameter can be used as an identification procedure in many instances. Bremmstrahlung can be used to advantage in certain cases with a conventional X-ray source. Certain Auger lines that are above the primary X-ray energy level have been observed, such as Si $KL_{2,3}L_{2,3}$ with nonmonochromatic Al and Mg sources [35]. The ability to remove the effect of the $K\alpha_{3,4}$ induced photolines is possible with fairly simple computer data-handling routines. In addition, deconvolution procedures have been suggested to remove the $K\alpha_2$ contribution to photopeaks [46-48]. Some of the results with these methods have been fairly good, but their usefulness in all cases is still questionable.

Electron Sources

While Auger transitions can be observed with X-ray excitation (and also by ion beams [49-51], most AES spectra in surface analysis are produced by electron excitation. The source of the electrons is a heated filament that is from 2-10 keV below the sample, which is at ground potential. It is possible to obtain AES spectra when the beam voltage is less than 2 keV, but many of the transitions used for routine analysis will be observed weakly or not at all. There are two major types of electron sources available. Early Auger electron spectrometers were equipped with tungsten filaments, enclosed in a holder that is held at a potential of around 70-100 V less than the filament, to draw off the electrons through a small hole. The electron beam was then focused with an Einsel lens arrangement. A diagram of an Einsel lens for an electron gun is shown in Figure 6. These filaments can produce electron beams with a diameter of a few micrometers. The electron beam can be deflected after it passes through the Einsel lens by opposing electric fields, either plates or segmented cylindrical pieces.

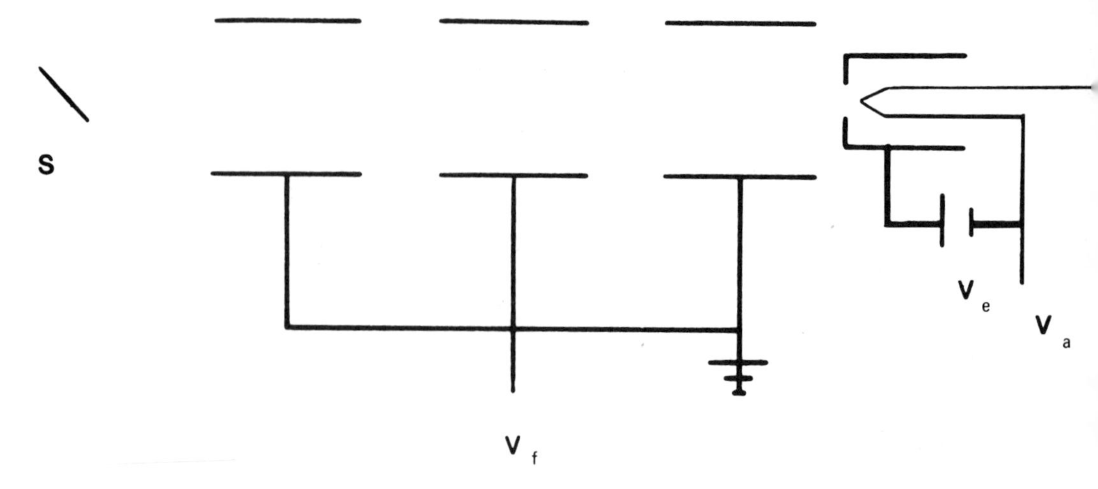

Figure 6. A schematic cross section of an electron gun with an Einsel lens focusing arrangement: S is the sample, V_f is the focus voltage, V_e is the emission voltage, and V_a is the acceleration voltage.

Since electron beams can be easily deflected, the ability to spatially resolve the surface composition of materials by AES was developed only a few years after the initial AES spectrometers were commercially available. Thus, the technique of scanning Auger microscopy (SAM) was developed. Here the AES signal as a function of position on the sample was displayed on a storage oscilloscope where it could be photographed. Such an approach had only a limited dynamic range of signal observation. With the addition of a secondary electron detector, a scanning electron microscope (SEM) that could employ the resolution of the electron beam was a readily available attachment. Absorbed current images could also be produced on the storage oscilloscope. Even today, systems are available with this spatial capability, usually in combined XPS-AES systems. The imaging and display capacity have been improved greatly, and will be discussed later.

With the interest in analyzing even smaller areas, LaB_6 electron gun sources have been developed for use in AES. As was noted in the introduction, beam diameters of under 50 nm for AES are now available commercially from several manufacturers. In the SEM mode of operation, 25-nm resolution is possible. With these systems the beam voltages can be raised to 25 or 30 keV. The design for these electron sources is much different from those that use a tungsten filament, with the electron gun arrangement very similar to that used for SEM. The focusing of the electron beam is accomplished with electromagnetic condenser and objective lenses. In addition, there are steering plates, stigmators, and deflectors. Beam blanking plates are placed to deflect the electron path, if desired. The spatial resolution of these systems is not as great as with most dedicated SEM instruments. A discussion of the location of the electron source in relation to the electron energy analyzer will be given below.

D. Electron Energy Analyzers

After a photoelectron or an Auger electron has been ejected from an atom and has left the solid with its characteristic kinetic energy, it has to be energy analyzed for surface analysis. There are several different types of energy analyzers, and each has its own set of strong and weak points. This is due to the requirements for XPS and AES being somewhat different. Some compromises have had to be made for systems where both techniques have been incorporated. While there have been a number of different designs for energy analyzers, there are only four major types commercially available. These are the cylindrical mirror analyzer (CMA), the concentric hemispheric analyzer (CHA), the retarding field analyzer (RFA), and a semi-imaging analyzer that combines elements of the CMA and RFA. The RFA most often used is a four-grid low-energy electron diffraction (LEED) spectrometer equipped so that AES spectra can be obtained. (LEED is a technique that determines the atomic structure in the near surface region and is applicable to single crystals.) The performance of this type of analyzer is not as good as with the other designs. However, many investigators who study single crystals use this type of analyzer for obtaining AES spectra, since there is not a large additional expense. Because the RFA is not used for routine surface analysis, it will not be discussed further. Detailed presentations of this analyzer are available [52].

A schematic diagram of a CMA is shown in Figure 7. For the case where simple AES spectra are obtained, the inner cylinder is held at ground potential. As an electron enters the region between the cylinders, it is deflected by the outer cylinder, which has a negative potential proportional to the electron's kinetic energy. Only those electrons that are in a solid angle cone $42 \pm 3°$ with respect to the center axis of the cylinders will be analyzed. (If the surface from which electrons are being ejected is perpendicular to this axis, the angle with respect to the surface is 48°.) Between the inner and outer cylinders are installed resistively coated ceramic pieces that have a high resistance (usually about 1 megohm). These ceramic pieces ensure a constant potential between the inner and outer cylinders. The effects of fringe fields at the ends of the cylinders are reduced drastically by these ceramic pieces. The sample surface has to be at the focal point of the analyzer, usually about 1 cm from the end of the CMA. This position usually is determined by analyzing the reflected peak at the energy of the incident electron beam (for AES). The peak intensity should be the greatest at this point [52], but this approach has been questioned [53]. For a regular CMA analyzer the electrons are deflected back to a focal point (denoted by C in Figure 7) where an electron multiplier is placed. By ramping the voltage on the outer cylinder, an electron energy spectrum can be obtained. AES spectra are displayed in one of two ways. In the first way (the integral mode) the number of electrons as a function of energy is plotted. In the first several years of AES, the detectors and electronics could not easily display AES spectra in this fashion. Usually Auger peaks are fairly small, and often hard to measure when displayed in this manner, since there is a large background. Improvements in detectors and electronics now make this mode of display much more readily possible. It was found by Harris that by taking the derivative of an AES spectrum, it was much easier to determine the energy

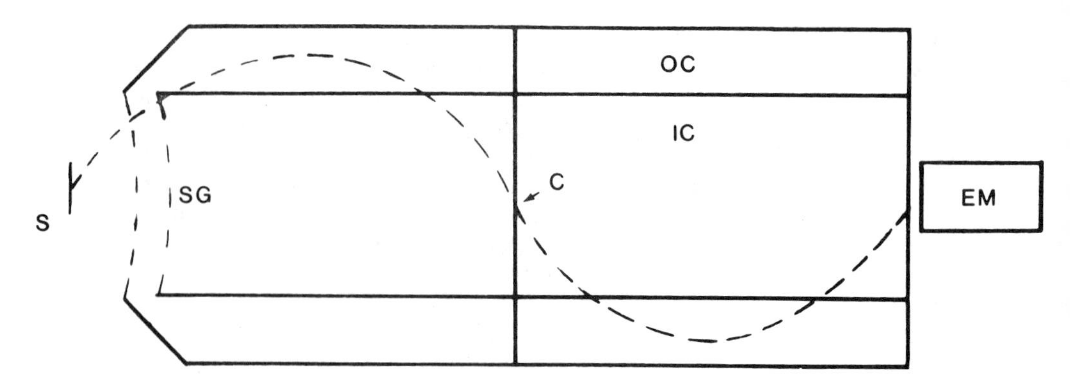

Figure 7. A schematic cross section of a double-pass CMA analyzer: S is the sample, the dotted line represents the path of the electrons from the sample to the electron multiplier EM, SG are spherical grids, IC and OC are the inner and outer cylinders, respectively, and C is the crossover point between the first and second stages of the analyzer. In a single-pass CMA the electron multiplier would be located just behind point C.

at which the peaks occurred [55]. One method of obtaining AES derivative spectra is to impose a small, high-frequency sine-wave potential on the ramping voltage applied to the outer cylinder. The magnitude of the high-frequency voltage varies from about 1 to 10 V, and the frequency is usually around 10 kHz. The phase-sensitive signal component from the electron multiplier is then measured with a lock-in amplifier. More recently, another approach has been developed in which the electron beam is blanked at a high frequency [56]. Again, the phase-sensitive signal is measured with a lock-in amplifier. Other methods of obtaining a derivative AES spectrum will be considered later.

Some analyzers similar to that described above can achieve an energy resolution of up to 0.3 percent. Certain designs allow for variable energy resolution up to 2 percent. While this performance is acceptable for AES, this design cannot be used for XPS. For example, 0.3 percent resolution will only give a resolution of 3 eV for an electron with a kinetic energy of 1000 eV, which is clearly inadequate for XPS. In order to use a CMA for XPS, the design has been altered. Instead of having the electrons describe only a single ellipse, the design has been changed so that another elliptical path is added to the first one. This type of CMA is known as a double-pass analyzer. Also, screens are placed before the entrance of the CMA. These screens reduce the energy of the incident electrons, which allows better energy resolution. The potential difference between the inner and outer cylinders is called the pass energy, and can be varied to produce an effective retardation of 5-200 V. Note that the inner cylinder is not held at ground potential in this mode of operation. This design differs from the single-pass analyzer in that a variable aperture is placed at the first cross point for the electrons and at a point near the exit plane to the electron multiplier. For high-resolution work (i.e., usually XPS) the aperture is about 2 mm, while for AES it is about 0.5 mm. The large aperture is needed to increase the signal, since much of the incoming signal is lost due to the retardation. Further details about this type of analyzer are available [57].

In order to use a CMA for angle-resolved XPS, another modification has been made to this design. A notched cylinder has been placed after the first cross-over point of the electron path. This cylinder is attached to a mechanical feedthrough that allows for both rotation and movement along the axis of the CMA. The notches on the cylinder are cut in such a way that a relatively small amount of the signal will be allowed to pass to the electron multiplier. If there is enough movement allowed along the axis of the CMA for the cylinder, then different sized notches are possible [58]. In order for this approach to be used, the sample surface cannot be normal to the axis of the CMA, but must be at about 45°. Rotation of the cylinder then allows the detection of those electrons that leave the surface at a glancing angle. It should be noted that this approach reduces greatly the number of electrons that can be detected. It is also possible to pull the notched cylinder back so that it does not interfere with any of the electrons passing through the CMA.

There are two general types of electron detectors that have been used with the CMA. The first is the Cu-Be detector, which is used only with CMAs that are used for AES. When a high-energy electron strikes the first stage, it releases numerous secondary electrons that are attracted to a second stage. The process is repeated several times until enough electrons are released so that a measurable current is produced. This type of electron multiplier requires an initial current of about 1 μA. The Cu-Be electron multiplier is rugged and can be used for many years. However, it cannot be exposed to the atmosphere for an extensive period of time, that is, on the order of hours, due to deterioration caused by water vapor. It is possible to rejuvenate Cu-Be detectors. For XPS (and low-current AES measurements) the Chanelltron type electron multiplier is used. This type of detector uses a metal coating on separated glass pieces that have a high potential between them. These detectors are air stable, but have somewhat shorter lifetimes. They can be destroyed very quickly if the initial current reaching them is too high.

An electron gun can be placed coaxially inside the inner cylinder of a CMA, as indicated in Figure 7. The electron multiplier is located behind the electron gun. In certain SAM systems the LaB_6 filament and much of the associated focusing lenses etc. have been positioned behind the electron multiplier. This requires that the electron detector must be placed so that it will not be in the path of the electron beam. Electron sources can be separate from the CMA, since it is possible to focus the electron beam at the focal point of a CMA.

Another widely used analyzer design is the concentric hemispherical analyzer (CHA). There are three main parts to the CHA: a retardation assembly, the hemispheres, and the electron detector. A simple schematic diagram of the CHA is given in Figure 8. Before the electrons enter the analyzer, their energy is reduced to a relatively low value. This allows for much better energy resolution with a reasonable sized analyzer. The retardation is usually accomplished with a lens assembly that focuses the electrons onto the entrance slit of the CHA. The heart of this type of

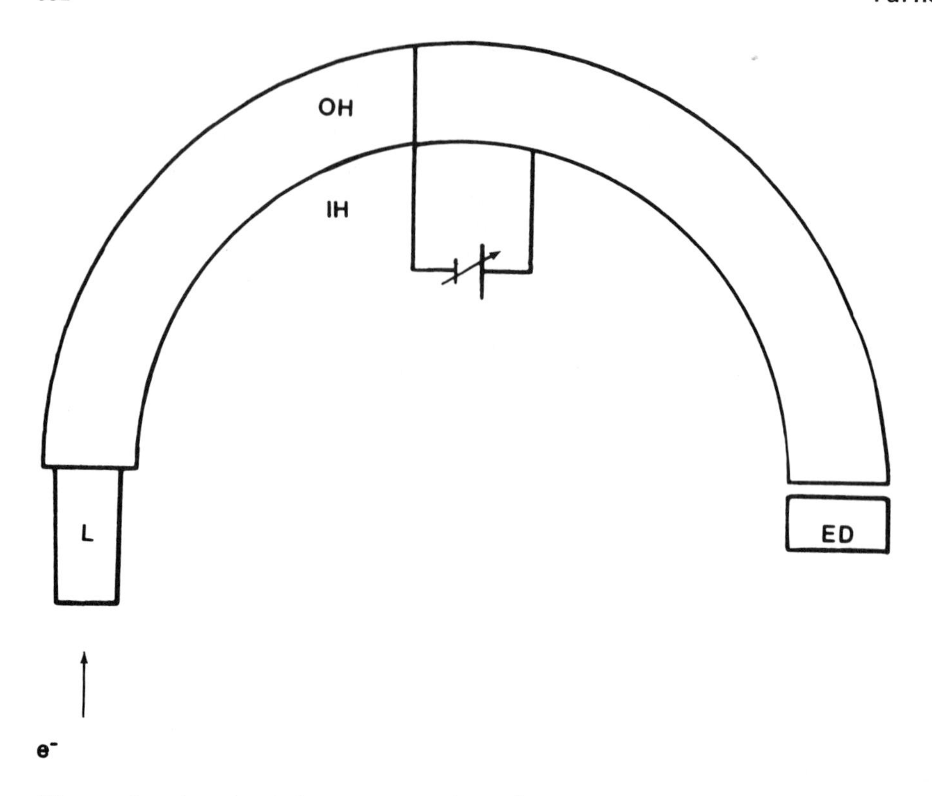

Figure 8. A schematic cross section of a CHA analyzer: L is the electron focussing lens, IH and OH are the inner and outer hemispherical sections, respectively, and ED is the electron detector.

analyzer is the set of concentric hemispheres. The inner and outer hemispheres have a potential difference applied such that electrons with only a small energy difference will reach the other end of the hemispheres. The potential difference is then ramped so that an energy spectrum of the electrons ejected from the surface is produced. At the end of the analyzer is the electron detector. With some instruments the electron detector is designed so that electrons of only one energy are measured. However, electrons of slightly different energies fall across the exit plane of the CHA. Over the past several years this fact has lead to the use of position-sensitive detectors that allow for the simultaneous detection of electrons with slightly different energies. This allows for much more rapid data collection, and, in fact, makes the use of certain low-intensity X-ray sources possible.

There are two possible modes of operation with the CHA. One method is known as fixed-analyzer transmission (FAT) or constant-analyzer transmission (CAT). Here the absolute resolution is independent of the kinetic energy of the incoming electrons; this allows easier quantification of electron spectroscopy data. The other mode of operation involves a fixed retardation ratio (FRR) or constant relative ratio (CRR). Peaks with low kinetic energies are easier to detect with this method due to their better signal-to-noise transmission characteristics.

The CHA is used quite extensively for XPS, due to its better energy resolution compared to the CMA. The CHA also is used for AES with an electron gun source. Due to the design of the CHA, the electron gun cannot be coaxial with the entrance area of the CHA. The throughput of the CHA is much less than that of the CMA, so that it is not as efficient in the collection of electrons. With the CHA it is easier than with the CMA to perform angular resolved XPS investigations.

A third type of electron energy analyzer, the semiimaging design, is commercially available for XPS and AES, and is shown in Figure 9. The first stage is used as a broadband energy filter in which the electrons are refocused onto an aperture A. This field is created by a cylinder that is ramped with a negative potential. The electrons are then retarded with a cylindrical electric field that is created with two grids. These grids also serve as the energy analyzer when the potential on the second grid is ramped. After the electrons pass through the second grid they are dispersed by a cylindrical field and refocused through an exit slit onto an electron detector. This type of energy analyzer operates in the constant

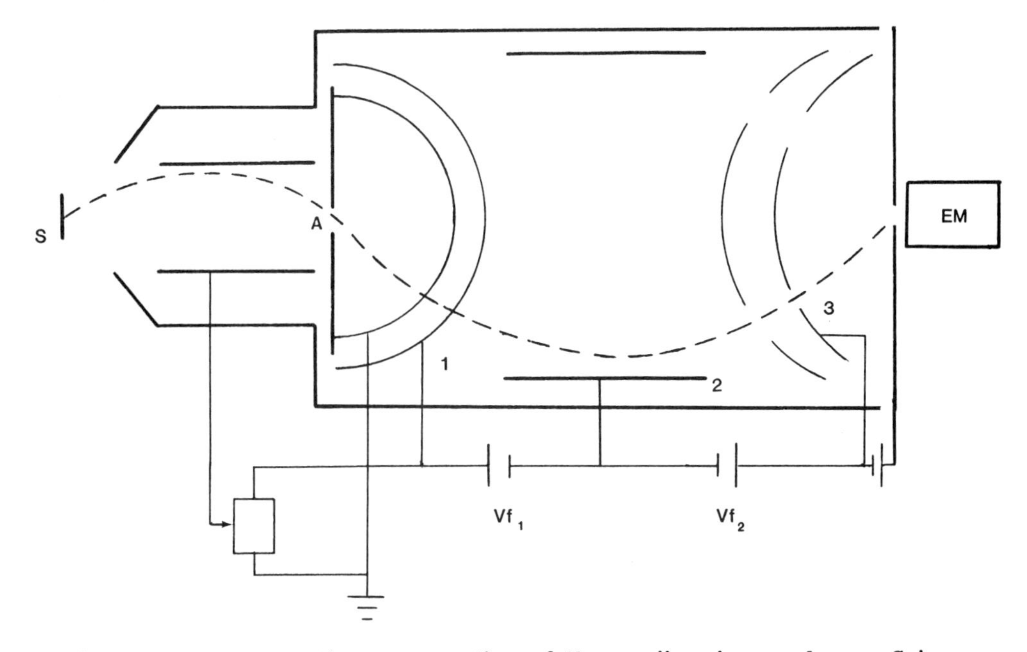

Figure 9. A schematic cross section of the semiimaging analyzer: S is the sample, the dotted line represents the path of the electrons where they first pass through a broadband filter, A is an aperture, 1 is a retarding grid, 2 is a cylindrical dispersion field, and 3 is an exit slit that refocuses the electrons into the electron multiplier EM.

energy resolution mode. The energy resolution is adjusted electrically from 0.3 to 4 eV. It is claimed that this analyzer (available from only one manufacturer) has both relatively high transmission and acceptance angle. With the current design of this analyzer it is impossible to obtain angle-resolved spectra. In addition, an electron gun has not been placed inside the analyzer. Further details about the design have been presented elsewhere [59].

It was mentioned previously that XPS analyses can be made with a beam as small as 150 μm. This is accomplished by focusing an electron beam on an X-ray anode. Another approach to analyze small areas by XPS has been to place an aperture in a CHA such that only the ejected electrons from a small area are detected. With this method, areas of 250 μm can be analyzed [60]. Of prime concern is to know exactly where the analysis is occurring. With the small area X-ray beam approach, an X-ray-sensitive phosphor or a film that changes color upon irradiation can be used to determine where the beam is striking. Then an optical microscope is adjusted such that the cross hairs coincide with the irradiated area. In systems where the analyzer determines the area being sampled, different methods have to be used. One approach is to maximize the XPS signal from a small deposition on a substrate, such as gold on copper, of the dimensions of the area to be analyzed. Again, an optical microscope is adjusted such that the cross hairs are aligned with the deposition. With either of the alignment procedures the microscope adjustments should be checked on regular basis. Also, the deposited material should have a strong photopeak. Another method can be used with systems that have SAM capability. The SEM image can be used for alignment with a small deposition of a substrate [60]. Then the position can be reproduced on the sample of interest with the same settings as employed with the standard.

E. Ion Guns

Most surface analysis systems are purchased with an ion gun as an accessory, and it has several different uses. It operates by striking the surface with energetic ions (usually fron an inert gas with energies of a few hundred to a few thousand electron volts) in order to remove or sputter atoms from the surface. Often, this is done to clean a surface of the almost inevitable layer of contamination present on a sample. Usually this can be done quickly, well within a minute. Another common use is to help obtain a depth profile of a sample by either XPS or AES. This is useful to determine the thickness and the composition of a layer that has been formed by some process, such as an oxide layer formed on a corrosion specimen or a sample from a deposition procedure. Another use is to study a material that has been implanted with highly energetic ions to change physical or chemical properties. Also, it is possible to analyze the species that are removed from the surface. The most common method of analysis of the sputtered species is with a mass spectrometer; this technique is called secondary ion mass spectrometry (SIMS). It is possible to have this type of analysis capability on many XPS or AES systems, but much of the work with SIMS is done on instruments designed solely for this method of analysis.

There are several components in an ion gun: the ionization, acceleration, focusing, and deflection regions. A heated filament boils off electrons that are attracted by a positive potential of about 70 V in an ionization chamber containing the sputtering gas. The ions are accelerated to the desired sputtering energy by a potential placed on the ionization chamber with respect to ground. Positive ions are usually employed. The ions are then focused by an Einsel lens arrangement. Most presently available ion guns have deflection plates that steer the beam after the ions have passed through the focussing region. It has been suggested that the sputtered area should be several times larger than the area analyzed [61]. This is due to the fact that the ion beam, which is about 0.1-1 mm wide with a well-designed ion gun, will have a Gaussian distribution. Thus, there will be edge effects, that is, a depth gradient in the crater produced by the ion beam. The ideal case would be a flat-bottomed crater. If only a small area (about 0.1 mm^2) is being analyzed, this would not present much of a problem. This assumes that the ion gun is focused on the same area that is being analyzed. Since the area of analysis in XPS is usually much larger than 0.1 mm^2, the sputtered area will have to be correspondingly larger. It is better to raster (scan) the ion beam rather than defocus it. With a defocused beam the edge effects of the crater will be larger. The same reasons apply if an AES analysis is performed by rastering the electron beam.

Most ion guns operate in a dynamic or differentially pumped mode—that is, the source gas is continuously being admitted into the ionization source. Some guns function by backfilling the vacuum chamber to a set pressure, usually in the 10^{-3} Pa region, with the main pumps turned off. In this latter mode of operation contaminants will build up in the vacuum chamber. The use of a simple Ti sublimination pump will help to reduce the levels of these contaminants. With the dynamic mode of operation the ionization chamber of the ion gun is usually pumped by a separate pump, most often with a small turbomolecular pump. This allows the ionization source to be at a pressure in the region used in the backfilled mode of operation, while the main chamber is continuously pumped. Thus, the pressure increase in the main chamber will be about one to three orders of magnitude above the base pressure. It should be pointed out that with a vacuum system that uses ion pumps, there is often residual sputtering gas even after the gas inlet has been closed. This is due to the memory effect with ion pumps, and is more pronounced with inert gases.

Usually ion guns will operate with a current density of a few tens of microamperes per square centimeter. For many materials this current density results in a removal rate of 10 nm/min or more. In many instances this rate is too great. The sputter rate can be reduced by a number of means. The pressure in the ionization chamber can be reduced, or the ionization filament current can be lowered. Also, the sputtering rate can be reduced by rastering the ion beam over a larger area than is needed. The sputter rate can be lessened by using ions with lower kinetic energy, that is, only a few hundred electron-volts. Relative sputtering rates can be measured with standard materials, that is, a known thickness of an oxide layer on Si or Ta, or with a standard sandwich material of Ni and Cr [62].

Sputtering is a destructive process. Even the removal of surface contaminants will alter the surface by disturbing surface crystallinity or changing the chemistry of the surface species. For example, reduction of many metal oxides has been observed with inert gas sputtering [63]. Also, different sputtering rates occur between elements, and in some cases with the same element in different oxidation states [64].

F. Data Collection and Handling

The signal from the electron multiplier must be further processed before useful information about the surface can be obtained. When the signal level is low, pulse counting techniques are used. This is the case for XPS and when a low electron beam current is employed for AES. When the beam current is higher, the current generated by the electron multiplier is sufficient to be amplified directly. This amplified current then can be fed into a lock-in amplifier as indicated earlier. Another approach has been to use an isolation amplifier directly. This results in obtaining an integral AES spectrum directly [65]. When pulse counting techniques are used, a rate meter can then produce an output voltage in the range of 0-10 V. The output from a lock-in amplifier can produce similar voltages. With an X-Y plotter a spectrum can be recorded, when a voltage proportional to the energy ramp for the electron energy analyzer is used for the x-axis input and the output from the electron detector system for the y-axis input. Many early XPS and AES instruments operated in this fashion, and even some instruments available today can obtain spectra in this manner. However, as is the case for many scientific instruments currently produced, computer control and data collection for XPS and AES spectrometers is the norm.

Computer systems for XPS and AES systems vary widely in size and capability. Microcomputers and personal computers are able to directly control simple data collection and do some data manipulation [66]. While most of these systems have been used to upgrade older equipment, they could be used with some currently available instruments or components. In the simplest case, the computer will provide a voltage via a digital-to-analog converter (D/A) to operate the energy ramp electronics. The voltage produced by the D/A usually will be in the range of 0-10 V. This voltage is then amplified by the analyzer control electronics to the level needed by the energy analyzer. The D/A has to have sufficient resolution so that the energy resolution of the electron analyzer is not compromised. For example, a 16-bit D/A can produce 65,536 different voltages over its output range. If it is desired to analyze the kinetic energy of electrons from 0 to 2000 eV, then the smallest energy difference that could be sampled would be just over 0.03 eV. This, of course, assumes that there are no other sources of error in the electronics and analyzer systems. Such an energy increment should be sufficient for all AES spectra and almost all XPS measurements. Kinetic energy increments will be in multiples of 0.03 eV in this example. Note that this means that an exact difference between points of 0.10 eV separation, for example, is not possible. The computer is then programmed to select the input for the D/A, which in turn provides the required output voltages. In some instances where a rather large increment is being used, such as 1 eV, the computer may be programmed such that the output to the D/A will provide the values closest to the desired output energy increment. Errors could build up if a set increment were used in such a case. The computer also will provide the dwell time during which the voltage will be at each step.

The signal from the electron multiplier can be stored in computer memory. For example, if the signal is being processed by a lock-in amplifier, its output voltage has to be fed into an analog-to-digital converter (A/D). This device operates exactly opposite to the method of a D/A. The A/D converts an analog signal, that is, a time-varying, continuous voltage, into a digital code that can be stored by the computer. At each energy interval, a separate electron intensity will be recorded. The computer can repeat energy sweeps and sum them to improve the signal-to-noise ratio of the spectrum. If the signal is being processed by a rate meter and the output is a voltage, the electron intensity can be treated in exactly the same way as that from the lock-in amplifier. The low levels encountered when pulse-counting techniques are used can be treated in a more direct fashion. Each individual pulse can be counted by separate counting circuitry, which then transfers the total count after a preset time to the computer. Such an approach must be used with position-sensitive detectors, since multiple signals are being processed at the same time. If the signal levels are about one million counts per second, or greater, most counting circuits are not able to function properly, as one pulse has not been processed before another has arrived. In this case, the signal can be converted into the frequency domain for signal processing [67].

After the spectrum has been recorded and stored by the computer, it has to be displayed in order to make use of the collected information. The spectrum can be displayed on a cathode display tube (CRT) for a rapid visual inspection by an operator. Permanent plots can be obtained from a dot matrix or laser printer. Higher-resolution output can be obtained also with a conventional pen plotter. This type of plotter also offers the possibility of having different colors when multiple spectra are being displayed. Additionally, overhead transparencies can be produced directly. Spectra also can be recorded for archival purposes on small magnetic disks. This has to be done even with computers that have large-capacity hard disks, since these latter systems will eventually be filled.

Computers can do much more than just collect a spectrum and then display and store it. Even the most basic systems will allow for the recording of other necessary experimental information, such as system pressure, electron multiplier settings, operator, date, etc. The abilities to scan only those regions of interest and to enhance low-level signals or take spectra at high resolution compared to survey experiments are additional advantages of computer data collection. Also, there is a reduction in the time required for an operator to be involved directly. In some instances extended or even overnight data collection is routine. With the rapid increase in computer capability coupled with the reduction in cost of desk-top computers, much more automated data collection is now available compared to the capabilities of even a few years ago. It is not possible to list here all of the various features for each commercially available system, but some important areas will be covered.

Perhaps one of the most useful features available on advanced systems is the capability to take data while processing previously acquired results. This is commonly referred to as foreground/background operation, that is, data collection in the foreground and the processing of previously acquired information in the background. (Some define foreground/background in the opposite manner.) With some systems, data manipulation procedures being done in the background mode will reduce the rate of data collection. Conversely, at some times during data collection, background operations cannot be done. Many of the operational settings on XPS and AES spectrometers are now under computer control, such as resolution controls on the electron energy analyzer and the X-ray power level in XPS, or the electron energy analyzer and the X-ray power level in XPS, or the electron gun control settings in AES. The control of accessories, such as ion gun operation, can be under computer command. In this instance this allows for either continuous (if not prohibited due to other experimental conditions) or interrupted depth profiles. With certain XPS systems the sample manipulator can be under computer control so that different samples can be analyzed in succession without operator intervention. System checks can be monitored with a computer. In the case of a failure, system diagnostics may be incorporated in the operating software that can help to pinpoint the problem.

Advanced data handling with system computers is now available with most currently available XPS and AES instruments. The display of only a small region of a spectrum is often useful. On occasion a point will be quite far removed from its neighbors due to a noise spike or some other temporary malfunction, and its repositioning can be easily accomplished without jeopardizing the value of the experiment. The ability to overlay two or more spectra, or to subtract one spectrum from another for comparative purposes, is available. Adding a fixed quantity to each point in a spectrum has advantages for certain display situations. Smoothing of a spectrum via mathematical procedures is easily accomplished [70,71]. Taking the derivative of a spectrum is done often, usually for AES spectra. Also, it has been suggested that taking the second derivative of XPS spectra can indicate the presence of more than one component in what appears to be the envelope of only one peak [72]. In addition, procedures utilizing deconvolution methods have been suggested as an alternative to the use of a monochromatic X-ray source [47,48].

Many of the above procedures are displayed on the CRT of the computer. It should be kept in mind that some of the data-handling procedures mentioned above do have drawbacks. For example, if smoothing is done over too wide a region of the spectrum of interest, some minor features can be lost. There are several different ways in which a derivative can be obtained, and each method has its own set of advantages and disadvantages [73]. The correction of a data point due to a spike is arbitrary. The use of the various methods to remove the X-ray doublet line width contribution to XPS spectra has not been as successful as hoped, but this point has been debated [47,48].

One of the most potentially useful data reduction procedures is that of semiquantitative treatment of XPS or AES data. However, much care and understanding are required with most of the currently available procedures. The accuracy of many of the methods has been questioned, and some have suggested that the results may have an accuracy of no better than 50 percent in certain cases. For both XPS and AES it should be remembered that the mean free path of the ejected electrons depends on the kinetic energy of the transition being analyzed. Therefore, a different volume will be analyzed for each element in the surface region.

In XPS the usual method for quantitative analysis is to determine the area of each peak. Another approach has been to measure the peak height of selected transitions [74]. This latter method is most often done manually, but it can be used as a rough indication of the composition of the surface region. Usually the most intense peak of each element is used in either procedure. There are several ways that the area of a peak can be measured. Before the area of a peak can be determined, a baseline has to be established. One procedure is to select two points on either side of the peak of interest and assume that a straight line (defined by the two points) will represent the baseline for the peak. This method will work, at least on a level that is visually satisfying for many peaks, but not for many of the major peaks of the transition metals. Another baseline selection procedure is the use of a parabolic curve that splits between the two halves of the peak [75]. This represents, on an empirical basis, the effect of electron energy losses in the primary peak. The area under the curve then can be determined by the use of a simple integration calculation.

Another approach has been the use of nonlinear least-squares methods. Here the baseline can be defined by either of the two methods mentioned above. The peak can be fitted with one of several types of curves. One of the simplest curves is the Gaussian, which is defined by three separate parameters: peak height, width, and position. In many instances this is not a bad approximation, especially with peaks that are fairly broad, for example, those with peak widths of about 1.5 eV or larger. However, with narrow peaks the Gaussian representation does not coincide very well with the experimental peak shape due to the Lorentzian contribution of the excitation source. Some programs allow for the curve to be represented by a combined Gaussian-Lorentzian function. The amount of the Lorentzian contribution is fixed or can be a variable, depending upon the program. These fitting programs can take into account the presence of more than one component in a complex curve. The intensity can then be normalized with the use of sensitivity factors [76]. These theoretical cross sections have been shown to be in reasonable agreement with experimental results in comparative studies [77]. When the relative areas of all of the curves for all of the detected elements have been determined, an elemental composition of the surface region can be calculated.

There are still several shortcomings to the above methods that are often encountered in commercial software. First, the choice of the baseline model is often done with a simple model, that is, a straight line or an empirical curve to attempt to correct for electron energy loss effects. In addition, the limits usually chosen for the baseline may be too narrow. Recently, it has been suggested that losses due to shake-off effects in XPS spectra will extend about 30-40 eV below the main peak, and expressions to compute the baseline to these energy levels have been proposed [78]. Second, even a combined Gaussian-Lorentzian curve often cannot correctly describe a narrow XPS peak. The Donack-Sumjic lineshape is probably a better representation [79], but this curve has not been incorporated into many curve analysis programs. Third, angular effects also have to be considered when using the theoretical cross-sections. When s-level peaks are being used, a correction due to the electron ejection angle is not needed. However, when other energy levels are being analyzed, that is, p, d, and f, then corrections must be made. For some analyzers such calculations will yield only an approximation to the correction needed. Also consideration has to be taken of shake-up effects when trying to measure the relative intensity of an XPS peak. As can be seen from the preceding discussion, the correct determination of the intensity of an XPS peak is complex, and often none of these factors is considered and included in commercial systems. Other factors, such as detector response and analyzer transmission as a function of kinetic energy of the incident electrons, must be taken into account. Some of these variables will cancel when making relative measurements.

Unfortunately, the case is much the same for the use of AES spectra for quantitative analysis purposes. There have been a number of different methods suggested for quantitative AES. The currently available procedures that are included with most commercial software use only the simplest approaches. These methods include the use of peak-to-peak heights in a derivative AES spectrum, the peak height of a N(E) AES spectrum, or, with a slight modification, the negative excursion of a derivative AES spectrum. All of these simple methods suffer from the fact that changes in the chemical state between the standard used and the material under investigation may alter significantly the response between the two different elemental environments. Also, these methods will not be usable if the signal results from two different elements. There are numerous examples where the peaks will be at different energies between two different chemical states. Even the area of an AES peak can be different with elements in two different chemical states [80]. The measurement of the area of an AES peak is not simple, due to redistributed primary and secondary electrons [81]. The ability to have user-supplied standards that may more nearly represent the problem at hand reduces some of the most glaring problems in the use of these simple analytical procedures.

With some SAM systems the SEM image can be collected digitally. This is in contrast to earlier SEM instruments that displayed the image and required a photo, if wanted, to be made in real time if a permanent record was desired. It should be pointed out that the storage of an SEM image requires a large amount of space on a magnetic disk. Therefore, it may not be desired to record images of every sample.

Other useful displays with SAM are elemental line scans and maps. A line scan is a series of elemental analyses at set intervals as horizontal or vertical lines across the area of the SEM image. This mode of analysis is quite useful for sandwich materials, thick coatings, etc. However, the method used to determine the intensity, that is, peak-to-peak heights, or the measurement of the peak area of the analyzed elements, again will be subject to all of the shortcomings noted above. The ability to detect minor components usually will be limited due to time requirements. For example, suppose it is desired to analyze a line with 100 points over a distance of 50 μm. If a minute is needed to sweep for a minor constituent at each point in order to achieve a signal with a good signal-to-noise ratio and another half minute is needed for all of the expected major components, then the line scan would require 150 min. The storage of the spectra obtained for the line scan usually is not done due to the large amount of data storage that would be required.

Elemental maps are AES determinations of the composition across the area of a SEM micrograph. This type of display can be useful to determine the elemental distribution of a catalyst on a support, the area of wear on a lubricated surface, or the uniformity of a coating. The same types of problems in data collection and evaluation that were indicated for line scans also apply to elemental maps. Display of the data can be quite sophisticated, if the data is stored in a digital format. In this instance the resolution will be determined by the number of pixels employed in the analysis. Different elements can be displayed by different colors. Overlays of the analyses can be made with the SEM display. Variations on an extended gray scale with the determined elemental intensity are possible also. In addition, image enhancement procedures can be used to improve the elemental representations. All of these post data collection procedures require advanced computer capability.

There are a number of other factors that are important in the software that is available for the collection and further treatment of data in either XPS or AES. Of considerable importance are the language and availability of the programs. If programs are written in a low-level language (i.e., machine language or micro instructions) or a high-level program language that is not available to the user, then checking or modifying the program is impossible. Errors are not unknown in computer programs in either data collection or data reduction. The inability to alter the data collection methods will forestall the use of new methods that may be developed. The same is true in the area of data reduction and presentation. Also, standard software may not be applicable to the needs of a specific laboratory. With many instruments, updates to the software are made, but the timing and direction that the revisions take may not be where an individual laboratory's interests lie. Another very important consideration in the use of data collection and reduction programs is the ease of use: is it "user friendly" or "user antagonistic"? If a program is difficult to learn and use, much time and effort is lost when operator mistakes are made. It is important to know how much time is required if the program must be restarted or rebooted. In addition, if a program has to be restarted, much operational information may need to be reentered. These factors are especially important if there are a number of operators of an instrument that are only occasional users, hence who will not have a great familiarity with the system.

Important also is the ability of a spectrometer's computer to communicate to other, usually more powerful, computers. The most direct use of such capability is the transfer of spectral files. This allows the use of programs that have advanced data handling programs that are not available on instrument computers. These programs can include background corrections, comparisons between theoretical models and experimental results, and advanced display capability, that is, plotting in ways of which the instrument computer is not capable. Most of these programs are available in standard, high-level languages, such as FORTRAN or BASIC, such that they can be transferred easily between laboratories. Some laboratories will find the ability to transfer tables and other information to spreadsheet and/or word processing programs useful.

IV. CONCLUSIONS

The purchase of an XPS and/or AES spectrometer involves a high initial cost and many technical factors. All currently available instruments have strengths and weaknesses, so that any choice is going to involve some compromises. In the evaluation of surface analysis instruments, a demonstration test with a variety of samples of both current and possible future interest should be made. While in some laboratories an instrument will be analyzing only a certain type of specimen, the availability of surface analysis equipment often will generate interest and use beyond that originally intended. Therefore, flexibility should always be kept in mind with the evaluation of this type of instrumentation. The use of computers undoubtedly will become more advanced in the future, but judgment will still be required in the interpretation of experimental results.

Acknowledgment

The author would like to thank Drs. J. H. Wandass and A. Moorish for their careful reading and comments for improvements of this chapter.

Note added in proof: Since this chapter was written major advances have been made in the analysis of small areas with XPS by several manufacturers. Resolution of features of about 10 μm has been achieved by one supplier, and under 100 μm by another with changes to the input lenses on CHA-based instruments. With these spectrometers a small energy window is fixed for the element of interest, and a two dimensional elemental image is obtained. In the other case resolution of about 30 μm in one dimension has been claimed by changes in both the design of the x-ray source and the input lens. With this instrument a line image is obtained with a somewhat larger energy window than with the other designs. In some cases it is possible to analyze for more than one element in an individual scan. These advances will be among the most important of the early 1990s in XPS.

REFERENCES

1. D. Briggs and J. C. Riviere, in "Practical Surface Analysis by Auger and X-ray Photoelectron Spectroscopy," edited by D. Briggs and M. P. Seah, John Wiley & Sons, New York, 1983, pp. 87-139.
2. M. Thompson, M. D. Baker, A. Christie, and J. F. Tyson, "Chemical Analysis, Vol. 74: Auger Electron Spectroscopy," John Wiley & Sons, New York, 1985.

3. M. Hayes, *Surface Technol.* 1983, *20*, 3.
4. J. C.Riviere, *Analyst (Lond.)*, 1983, *108*, 649.
5. D. S. Urch, *Annu. Rep. C*, 1983, 285; *Annu. Rep. C*, 1983, 327.
6. D. Spanjaard, C. Guillot, M.-C. Desjonqueres, G. Tregila, and J. Lecante, *Surface Sci. Rep.*, 1985, *5*, 1.
7. S. B. Hagstrom, M. O. Krause, and S. T. Manson, *Pure Appl. Phys.*, 1983, *43(Appl. At. Phys., Vol. 4)*, *450*.
8. M. W. Roberts, *Sci. Prog. (Oxford)*, 1982, *68*, 65.
9. H. Windawi and C. D. Wagner, *Chem. Anal. (NY)*, 1982, *63(Appl. Electron Spectrosc. Chem. Anal.)*, 191.
10. D. M. Hercules and J. C. Klein, *Chem. Anal. (NY)*, 1982, *63(Appl. Electron Spectrosc. Chem. Anal.)*, 147.
11. J. T. Grant, *Appl. Surface Sci.*, 1982, *13*, 35.
12. G. C. Allen and R. K. Wild, *Can. J. Spectrosc.*, 1983, *28*, 35.
13. C. J. Powell, *Chem. Anal. (NY)*, 1982, *63 (Appl. Electron Spectrosc. Chem. Anal.)*, 19.
14. C. R. Brundle, *Anal. Chem. Symp.Ser.*, 1982, *199 (Ind. Appl. Surf. Anal.)*, 13.
15. M. G. Lagally in "Advances in the Technology of Characterizing Microstructures," edited by F. W. Wiffer and J. A. Spitznagal, Metallurgical Society, AIME, Warrendale, PA, 1982, p. 213.
16. D. T. Clark, Y. C. T. Fok, and G. G. Roberts, *J. Electron Spectrosc. Related Phenom.*, 1981, *22*, 173.
17. B. Hupfer, H. Schupp, J. D. Andrade, and H. Ringsdof, *J. Electron Spectrosc. Related Phenom.*, 1981, *23*, 103.
18. M. P. Seah and W. A. Dench, *Surface Interface Anal.*, 1979, *1*, 1.
19. J. Szajman, J. Liesegang, J. G. Jenkin, and R. C. G. Lockey, *J. Electron Spectrosc. Related Phenom.*, 1981, *23*, 97.
20. C. D. Wagner, L. E. Davis, and W. M. Riggs, *Surface Interface Anal.*, 1980, *2*, 53.
21. H. Gant and W. Monch, *Surface Sci.*, 1981, *105*, 217.
22. J. C. Ashley and C. J. Tung, *Surface Interface Anal.*, 1982, *4*, 52.
23. C. D. Wagner, W. M. Riggs, L. E. Davis, J. F. Moulder, and (ed.) G. E. Muileberg, "Handbook of X-Ray Photoelectron Spectroscopy," Perkin-Elmer Corp., Eden Prairie, MN, 1979.
24. N. S. McIntyre and D. G. Zetaruk, *Anal. Chem.*, 1977, *49*, 1521.
25. N. H. Turner, J. S. Murday, and D. E. Ramaker, *Anal. Chem.*, 1980, *52*, 84.
26. S. Larrson, *Chem. Phys. Lett.*, 1976, *40*, 362.
27. D. E. Ramaker, *Appl. Surface Sci.*, 1985, *21*, 243.
28. L. E. Davis, N. C. MacDonald, P. W. Palmberg, G. E. Riach, and R. E. Weber, "Handbook of Auger Electron Spectroscopy," 2nd ed., Physical Electronics Industries, Eden Prairie, MN, 1976.
29. G. E. McGuire, "Auger Electron Spectroscopy Reference Manual," Plenum Press, New York, 1979.
30. T. Sekine, Y. Nagasawa, M. Kudoh, Y. Sakai, A. S. Parks, J. D. Geller, A. Mogami, and K. Hirata, "Handbook of Auger Electron Spectroscopy," JEOL Ltd., Tokyo, 1982.
31. C. D. Wagner, *Faraday Disc. Chem. Soc.*, 1975, *60*, 291.
32. C. D. Wagner, *Anal. Chem. Symp. Ser.*, 1981, *162 (Photon, Electron,*

Ion, Probes Polym. Struct. Prop.), 203.
33. C. D. Wagner in "Practical Surface Analysis," edited by D. Briggs and M. P. Seah, John Wiley & Sons, New York, 1983, pp. 477-483.
34. C. D. Wagner, D. A. Zatko, and R. H. Raymond, *Anal. Chem.*, 1980, *52*, 1445.
35. J. E. Castle and R. H. West, *J. Electron Spectrosc. Related Phenom.*, 1980, *18*, 355.
36. A. Roth, "Vacuum Technology," 2nd ed., North Holland, New York, 1982, p. 270.
37. O. Ganschow and P. Steffens, *J. Vac. Sci. Technol.*, 1982, *21*, 845.
38. D. D. Hawn, *Rev. Sci. Instrum.*, 1983, *54*, 767.
39. D. T. Clark, M. M. Abu-Shbak, and W. J. Brennan, *J. Electron Spectrosc. Related Phenom.*, 1982, *28*, 11.
40. K. Yates and R. H. West, *Surface Interface Anal.*, 1983, *5*, 133.
41. R. Chaney, *Surface Interface Anal.*, 1987, *10*, 36.
42. C. D. Wagner, *J. Electron Spectrosc. Related Phenom.*, 1980, *18*, 345.
43. P. Swift, *Surf. Interface Anal.*, 1982, *4*, 47.
44. H. Jaegle, A. Kalt, G. Nanse, and J. C. Peruchetti, *Analysis*, 1980, *9*, 252.
45. S. Kohki, T. Ohmura, and K. Kusao, *J. Electron Spectrosc. Related Phenom.*, 1983, *28*, 229; *31*, 85.
46. M. F. Koenig and J. T. Grant, *J. Electron Spectrosc. Related Phenom.*, 1985, *36*, 213.
47. M. F. Koenig and J. T. Grant, *J. Electron Spectrosc. Related Phenom.*, 1984, *33*, 9.
48. G. K. Wertheim, *J. Electron Spectrosc. Related Phenom.*, 1975, *6*, 239.
49. E. W. Thomas, *Pure Appl. Phys.*, 1983, *43(Appl. At. Collision Phys., Vol. 4)*, 299.
50. E. W. Thomas, *Vacuum*, 1984, *34*, 1031.
51. R. A. Baragiola, *Rad. Effects*, 1982, *61*, 47.
52. E. N. Sickafus and H. P. Bonzel in "Progress in Surface and Membrane Science," edited by J. F. Danielli, M. D. Rosenberg, and D. A. Cadenhead, Vol. 4, Academic Press, New York, 1971, pp. 164-167.
53. E. Sickafus and P. H. Holloway, *Surface Sci.*, 1975, *51*, 131.
54. C. LeGressus, J. D. Geller, and A. LeMoel, *Scanning Electron Microsc.*, 1983(2), 553.
55. L. A. Harris, *J. Appl. Phys.*, 1968, *39*, 1419.
56. A. Mogami and T. Sekine, *Proc. 6th European Cong. Electron Spectroscopy*, Tal International, Jerusalem, 1976, p. 322.
57. P. Staib and U. Dinklage, *J. Phys.*, 1977, E10, 914.
58. P. W. Palmberg, *J. Vac. Sci. Technol.*, 1975, *12*, 379.
59. P. Staib, *J. Phys.*, 1972, *E5*, 484; 1977, *E10*, 914.
60. J. A. Knapp, G. J. Lapeyre, N. V. Smith, and M. M. Traum, *Rev. Sci. Instrum.*, 1982, *53*, 781.
61. K. Yates and R. H. West, *Surface Interface Anal.*, 1983, *5*, 217.
62. H. J. Mathieu and D. Landolt, *Surface Interface Anal.*, 1983, *5*, 77.
63. C. P. Hunt, M. T. Anthony, and M. P. Seah, *Surface Interface Anal.*, 1984, *6*, 92.
64. B. Navinsek, P. Panjan, A. Zabkar, and J. Fine, *J. Vac. Sci. Tech-*

nol. A, 1985, *3*, 671.
65. K. S. Kim, W. E. Baitinger, J. W. Amy, and N. Winograd, *J. Electron Spectrosc. Related Phenom.*, 1974, *5*, 351.
66. H. H. Anderson, in "Sputtering by Particle Bombardment I," edited by R. Behrisch, Springer-Verlag, New York, 1981, p. 72.
67. M. C. Burrell, R. S. Kaler, and N. R. Armstrong, *Anal. Chem.*, 1982, *54*, 2511.
68. D. P. Griffith, W. S. Woodward, and R. W. Linton, *Anal. Chem.*, 1981, *53*, 2377.
69. H. V. Malmsted, C. G. Enke, and S. R. Crouch, "Electronics and Instrumentation for Scientists," Benjamin/Cummings, Menlo Park, CA, 1981, p. 433.
70. A. Savitsky and M. J. E. Golay, *Anal. Chem.*, 1964, *36*, 1627.
71. G. Horlick, *Anal. Chem.*, 1972, *44*, 943.
72. A. Proctor and P. M. A. Sherwood, *Anal. Chem.*, 1982, *54*, 13.
73. M. P. Seah, M. T. Anthony, and W. A. Dench, *J. Phys. E*, 1983, *16*, 848.
74. C. D. Wagner, L. E. Davis, M. V. Zeller, J. A. Taylor, R. H. Raymond, and L. H. Gale, *Surface Interface Anal.*, 1981, *3*, 211.
75. D. A. Shirley, *Phys. Rev.*, 1972, *B5*, 4709.
76. J. H. Scofield, *J. Electron Spectrosc. Related Phenom.*, 1976, *8*, 129.
77. J. Szajman, J. C. Jenkin, R. C. G. Leckey, and J. Liesengang, *J. Electron Spectrosc. Related Phenom.*, 1981, *19*, 393.
78. S. Tougaard, *J. Vac. Sci. Technol.*, 1987, *A5*, 1230, 1275.
79. S. Doniach and M. Sunjic, *J. Phys.*, 1970, *C3*, 285.
80. N. H. Turner and D. E. Ramaker, *J. Vac. Sci. Technol.*, 1983, *A1*, 1229.
81. D. E. Ramaker, J. S. Murday, and N. H. Turner, *J. Electron Spectrosc. Related Phenom.*, 1979, *17*, 45.

III

ELECTROCHEMICAL INSTRUMENTATION

16

Potentiometry: pH and Ion-Selective Electrodes

TRUMAN S. LIGHT* / The Foxboro Company, Foxboro, Massachusetts

I. INTRODUCTION

The study of electrochemistry and its associated theory and analytical applications may be conveniently grouped into three categories:

1. Cells that are a source of electrical energy and are called *voltaic or galvanic* cells (example: a battery) and from which electrical current is withdrawn.
2. Cells that obtain current from an external potential source and are called *electrolytic cells* (example: cells for the electroplating of metals) and result in the liberation of chemical species.
3. Cells to which is applied an external potential source that is *exactly equal but opposite* to that of the galvanic cell and are called *potentiometric cells*. These are characterized by zero or negligible current flow and are widely used in chemical analysis to derive quantitative information concerning concentration or activity of chemical species in solution in accordance with the Nernst equation (example: measurements of pH, pAg, pCl, pF).

A. Electrochemical Cell

Figure 1 illustrates a zinc-copper electrochemical cell sometimes called the Daniell cell. It consists of a piece of zinc metal immersed in a zinc ion solution of concentration $C_{Zn^{2+}}$ and copper metal immersed in a cupric ion solution of concentration $C_{Cu^{2+}}$. The two solutions are connected by means of a nonmixing, chemically inert, salt bridge, such as a gel-immobilized potassium chloride solution in an inverted U-tube. If each of the solutions is at a concentration called unit activity (see below for the difference between activity and concentration), and the current is zero microamperes, then the emf reading on the voltmeter will be +1.10 V. The plus sign indicates the polarity of the cell, namely, that the copper electrode is positive and the zinc electrode is negative. This polarity is invariant, and it is interesting to note that if the cell is permitted to operate as a galvanic (spontaneous) cell, then the observed emf will decrease slightly in magnitude but will not change in sign. Similarly, if the cell is permitted to operate electrolytically by driving it with an external source of emf, the

---

*Retired

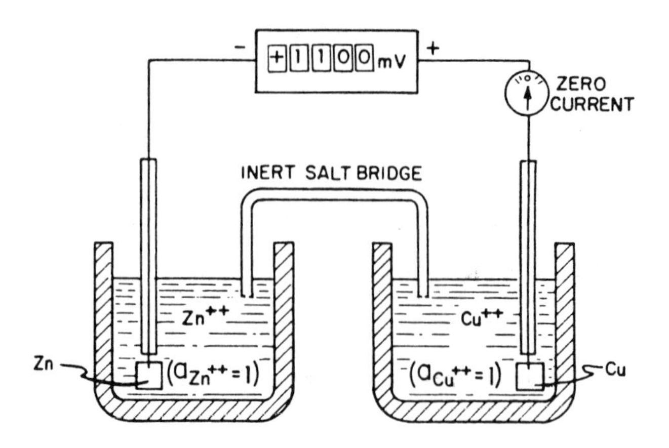

Figure 1. Zinc-copper electrochemical cell illustrating the principle of a potentiometric cell. The current or electron flow is essentially zero. The polarity, plus and minus, is invariant.

observed emf will increase slightly from the equilibrium value, but the polarity will not change.

The convention for designation of electrochemical cells starts with designation of the *half-cells*:

$$Cu^{2+} + 2e^- \quad Cu \rightleftharpoons \quad E^0 = +0.337 \text{ V} \tag{1}$$

$$Zn^{2+} + 2e^- \quad Zn \rightleftharpoons \quad E^0 = -0.763 \text{ V} \tag{2}$$

Both half-cells are written as reductions to keep the analytical or electrochemical cell convention of invariant signs in harmony with the thermodynamic convention of bivariant signs [1-3].

By subtracting equation (2) from equation (1) so as to cancel the electrons, the *whole-cell* reaction may be written:

$$Cu^{2+} + Zn \rightleftharpoons Cu + Zn^{2+} \qquad E^0 = +1.100 \text{ V} \tag{3}$$

A simplified method of writing electrochemical cells is conventionally used. The cell of Figure 1 is written as

$$Zn \,|\, Zn^{2+}\ (a=1) \,||\, Cu^{2+}\ (a=1) \,|\, Cu \tag{4}$$

where the single vertical bar indicates a phase boundary across which a potential is developed and the double vertical bar indicates a liquid junction region—such as the salt bridge in Figure 1 that separates two liquid phases of different concentrations. The concentrations indicated by brackets [], or activities, indicated by a_x terms (see below), of these liquid phases are indicated within parentheses.

Although we are discussing potentiometry (measurements at which no current is drawn through the cell), cell (4) is also conventionally used to mean a galvanic cell in which electrons are flowing externally from left to right if a wire is connected from the zinc to the copper terminal. In this case, the zinc terminal at which oxidation occurs is called the anode and the copper terminal at which reduction occurs is called the cathode.

If the emf of the cell is positive, the reaction is spontaneous in the forward direction of equation (3).

The emf, E, of cell (4) may also be written, and calculated from

$$E = E_{right} - E_{left} + E_j \quad (5)$$

where E_{right} is the potential of the right half-cell, equation (1), E_{left} is the potential of the left half-cell, equation (2), and E_j is the liquid junction potential at the salt bridge separating the two solutions.

B. Activity and Activity Coefficients

Equation (3) assumes that all concentrations are at unit activity. Other activities (or concentrations) change the emf of a cell. The relationship between activity and emf is called the *Nernst equation*. First, however, the connection between activity and concentration needs to be defined. Because of interionic effects, denoted by the *ionic strength*, the effective concentration of dilute solutions is often less than the actual concentration and is called the *activity*. The equations may be written

$$a_i = f_i C_i \quad (6)$$

$$I = \tfrac{1}{2}\Sigma C_i Z_i^2 \quad (7)$$

where a_i is the activity of ion i, C_i is the molar or molal concentration of ion i, Z_i is the charge of ion i, and I is the ionic strength of the solution.

The activity coefficient may be determined experimentally by one of several physicochemical methods. It is also calculated theoretically by equations using the Debye-Hückel theory. For very dilute solutions, where the ionic strength is <0.01, the Debye-Hückel limiting equation may be used:

$$-\log f_i = 0.512 Z_i^2 \sqrt{I} \qquad (\text{at } 25°\text{C}) \quad (8)$$

For solutions of up to tenfold higher ionic strength, the somewhat more rigorous form of the Debye-Hückel equation is often used:

$$-\log f_i = (0.512 Z_i^2 \sqrt{I})/(1 + \sqrt{I}) \qquad (\text{at } 25°\text{C}) \quad (9)$$

Some conventions concerning activities are worth noting. Solids, such as zinc or copper metal or undissolved crystals, have unit activity and thus drop out of the Nernst equation. Partial pressures of gases and molar concentrations of uncharged molecules are used directly because the activity coefficient of an uncharged species ($Z = 0$) is unity in accordance with equation (8). As an approximation, especially in dilute solutions of univalent ions, the concentration is sometimes substituted for the activity. For more quantitative results, activity coefficients may be calculated from the Debye-Hückel equations, (8) or (9). An even better approximation for activity coefficients may be obtained from the tables of Kielland [4], which take into account the nature of common ion species including their ion size and charge. Table 1 gives Kielland's listing of cation and anion activity coefficients based on the extended Debye-Hückel equation. More extensive discussion of the Debye-Hückel equation is available [5,7].

Table 1. Single-Ion Activity Coefficients Calculated from the Extended Debye-Hückel Equation [4]

Ion	Ion size (Å)	Ionic strength 0.001	0.005	0.01	0.05	0.10
H^+	9	0.967	0.933	0.914	0.86	0.83
Li^+	6	0.965	0.930	0.909	0.845	0.81
Na^+, IO_3^-, HCO_3^-, HSO_3^-, $H_2PO_4^-$, $H_2AsO_4^-$	4	0.964	0.927	0.901	0.815	0.77
K^+, Rb^+, Cs^+, Tl^+, Ag^+, NH_4^+, OH^-, F^-, SCN^-, HS^-, ClO_3^-, ClO_4^-, BrO_3^-, IO_4^-, MnO_4^-, Cl^-, Br^-, I^-, CN^-, NO_3^-	3	0.964	0.925	0.899	0.805	0.755
Mg^{2+}, Be^{2+}	8	0.872	0.755	0.69	0.52	0.45
Ca^{2+}, Cu^{2+}, Zn^{2+}, Sn^{2+}, Mn^{2+}, Fe^{2+}, Ni^{2+}, Co^{2+}	6	0.870	0.749	0.675	0.485	0.405
Sr^{2+}, Ba^{2+}, Ra^{2+}, Cd^{2+}, Pb^{2+}, Hg^{2+}, S^{2-}, CO_3^{2-}, SO_3^{2-}	5	0.868	0.744	0.67	0.465	0.38
Hg_2^{2+}, SO_4^{2-}, $S_2O_3^{2-}$, CrO_4^{2-}, HPO_4^{2-}	4	0.867	0.740	0.660	0.445	0.355
Al^{3+}, Fe^{3+}, Cr^{3+}, Ce^{3+}, La^{3+}	9	0.738	0.54	0.445	0.245	0.18
PO_4^{3-}, $Fe(CN)_6^{3-}$	4	0.725	0.505	0.395	0.16	0.095
Th^{4+}, Zr^{4+}, Ce^{4+}, Sn^{4+}	11	0.588	0.35	0.255	0.10	0.065
$Fe(CN)_6^{4-}$	5	0.57	0.31	0.20	0.048	0.021

The above principles may be illustrated by calculating the pH of a solution that is a mixture 0.01 *N* HCl and 0.09 *N* KCl. Several different approaches may be used involving different degrees of approximation:

1. A simple calculation would consider the hydrochloric acid as the only source of acidity and neglect activity coefficients;

$$pC_H = -\log C_{H^+} = -\log(0.01)$$

$$pC_H = 2.00$$

2. A better approach would be to first calculate the ionic strength, I, of the solution using equation (7), and then calculate the activity coefficient from the limiting Debye-Hückel equation using equation (9):

$$I = \tfrac{1}{2}\Sigma C_i Z_i^2 = \tfrac{1}{2}(0.01 \times 1 + 0.01 \times 1 + 0.09 \times 1 + 0.09 \times 1)$$

$$= 0.10 \text{ M}$$

$$\log f_{H^+} = -(0.512)(1)(0.10)^{\frac{1}{2}}/[1 + (0.1)^{\frac{1}{2}}] = -0.162/1.316$$
$$= -0.123$$
$$f_{H^+} = 0.753$$

$$paH = -\log a_{H^+} = -\log(f_{H^+})(C_{H^+}) = -\log(0.753)(0.01)$$

$$paH = 2.12$$

3. Or, we may select the activity coefficient from Table 1, and for a solution of ionic strength of 0.10, $f_{H^+} = 0.83$, and

$$paH = -\log a_{H^+} = -\log(f_{H^+})(C_{H^+}) = -\log(0.83)(0.01)$$

$$paH = 2.08$$

The latter value, paH = 2.08, is the best value for the solution and agrees with that obtained with a pH meter. Note the distinction between the terms "pC_H" ($-\log C_{H+}$) and "paH" ($-\log a_{H+}$). Activity, paH, is the "pH" quantity measured by a properly standardized pH meter.

For solutions more concentrated than an ionic strength of 0.1, the activity coefficients do not necessarily continue to decrease, as might be inferred solely from Table 1. Activity coefficients may increase to values much larger than unity for concentrated solutions, and advanced treatments of the Debye-Hückel equation discuss these cases [5].

C. Nernst Equation

With the concept of activity of an ion established, the Nernst equation, which relates the concentration or activity to the potential of a cell, may be discussed. For the generalized half-cell reaction involving the reduction of an oxidizing ion, Ox (example, Fe^{+3}), by electrons to a reducing ion, Red (example, Fe^{+2}),

$$Ox + ne^- \rightleftharpoons Red \tag{10}$$

the potential of the half-cell, E, is given by the Nernst equation:

$$E = E^0 + (RT/nF) \ln(a_{Ox})/(a_{Red})$$
$$= E^0 + (2.303RT/nF) \log(a_{Ox})/(a_{Red}) \tag{11}$$

where E is the electrode (half-)cell potential in volts, E^0 is the standard electrode potential in volts, R is the molar gas constant (8.31441 $J \cdot K^{-1} \cdot mol^{-1}$), T is the absolute temperature in Kelvin, F is the Faraday constant (96,487.0 coulombs $equivalent^{-1}$), n is the number of electrons involved in the reaction, a_{Ox} is the activity of the oxidized form, and a_{Red} is the activity of the reduced form. The term 2.303RT/F is called the Nernst coefficient and has the value of 59.16 mV at 25°C. Values at other temperatures are given in Table 2.

D. Standard and Formal Potentials

The standard electrode potential, E^0, as illustrated in equations (1) and (2) for zinc and copper and in equation (11) for the general form of the

Table 2. Values of the Nernst Coefficient, 2.303RT/F

Temperature (°C)	Nernst coefficient (mV)
0	54.20
5	55.19
10	56.18
15	57.17
20	58.17
25	59.16
30	60.15
35	61.14
40	62.13
45	63.13
50	64.12
55	65.11
60	66.10
65	67.09
70	68.09
75	69.08
80	70.07
85	71.06
90	72.05
95	73.05
100	74.04

Nernst equation, are available for many half-cell reactions [6]. Table 3 gives a selection of standard potentials [6]. A more useful form is called the *formal potential*, and is denoted by $E^{0\prime}$. Formal potentials use the actual analytical concentration* (rather than the activity) of the species involved in the reaction and specify the medium in which the reaction is occurring. An example is that of the ferrous-ferric couple:

For the reaction at unit activities:

$$Fe^{3+} + e^- = Fe^{2+} \qquad E^0 = +0.771\ V$$

For the reaction in 1 *F* HCl*

$$Fe^{3+} + e^- = Fe^{2+} \quad (1F\ HCl) \qquad E^{0\prime} = +0.70\ V$$

For the reaction in 1 *F* H_2SO_4

$$Fe^{3+} + e^- = Fe^{2+} \quad (1F\ H_2SO_4) \qquad E^{0\prime} = +0.68\ V$$

*The concentration notation "1*F* HCl" stands for one formula weight HCl/liter instead of one molecular weight. From this is derived the name "formal potential."

Table 3. Standard Electrode Potentials, Selected Values in Volts Versus SHE[a]

Half-reaction	E^0
$F_2(g) + 2H^+ + 2e^- = 2HF$	3.053
$O_3 + 2H^+ + 2e^- = O_2 + H_2O$	2.075
$S_2O_8^{2-} + 2e^- = 2SO_4^{2-}$	2.01
$Ag^{2+} + e^- = Ag^+$	1.980
$Co^{3+} + e^- = Co^{2+}$	1.92
$H_2O_2 + 2H^+ + 2e^- = 2H_2O$	1.763
$MnO_4^- + 4H^+ + 3e^- = MnO_2(s) + 2H_2O$	1.70
$Ce^{4+} + e^- = Ce^{3+}$	1.72
$H_5IO_6 + H^+ + 2e^- = IO_3^- + 3H_2O$	1.603
$MnO_4 + 8H^+ + 5e^- = Mn^{2+} + 4H_2O$	1.51
$2BrO_3^- + 12H^+ + 10e^- = Br_2 + 6H_2O$	1.478
$PbO_2 + 4H^+ + 2e^- = Pb^{2+} + 2H_2O$	1.468
$Cr_2O_7^{2-} + 14H^+ + 6e^- = 2Cr^{3+} + 7H_2O$	1.36
$Cl_2 + 2e^- = 2Cl^-$	1.3583
$Tl^{3+} + 2e^- = Tl^+$	1.25
$MnO_2(s) + 4H^+ + 2e^- = Mn^{2+} + 2H_2O$	1.23
$O_2(g) + 4H^+ + 4e^- = 2H_2O$	1.229
$2IO_3^- + 12H^+ + 10e^- = I_2 + 6H_2O$	1.195
$Br_2(l) + 2e^- = 2Br^-$	1.065
$2ICl_2 + 2e^- = I_2 + 4Cl^-$	1.07
$HNO_2 + H^+ + e^- = NO(g) + H_2O$	0.996
$NO_3^- + 3H^+ + 2e^- = HNO_2 + H_2O$	0.94
$2Hg^{2+} + 2e^- = Hg_2^{2+}$	0.9110
$Cu^{2+} + I^- + e^- = CuI$	0.861
$Ag^+ + e^- = Ag$	0.7991
$Hg_2^{2+} + 2e^- = 2Hg$	0.7960
$Fe^{3+} + e^- = Fe^{2+}$	0.771
$C_6H_4O_2(\text{quinone}) + 2H^+ + 2e^- = C_6H_4(OH)_2$	0.699
$O_2(g) + 2H^+ + 2e^- = H_2O_2$	0.695
$Hg_2SO_4(s) + 2e^- = 2Hg + SO_4^{2-}$	0.613

(continued)

(Table 3, continued)

Half-reaction	E^0
$Sb_2O_5 + 6H^+ + 4e^- = 2SbO^+ + 3H_2O$	0.605
$H_3AsO_4 + 2H^+ + 2e^- = HAsO_2 + 2H_2O$	0.560
$I_3^- + 2e^- = 3I^-$	0.536
$I_2 + 2e^- = 2I^-$	0.5355
$Cu^+ + e^- = Cu$	0.520
$Fe(CN)_6^{3-} + e^- = Fe(CN)_6^{4-}$	0.3610
$Cu^{2+} + 2e^- = Cu$	0.340
$UO_2^+ + 4H^+ + e^- = U^{4+} + 2H_2O$	0.38
$VO^{2+} + 2H^+ + e^- = V^{3+} + H_2O$	0.337
$Hg_2Cl_2(s) + 2e^- = 2Hg + 2Cl^-$	0.26816
$AgCl(s) + e^- = Ag + Cl^-$	0.2223
$Cu^{2+} + e^- = Cu^+$	0.159
$SO_4^{2-} + 4H^+ + 2e^- = H_2SO_3 + H_2O$	0.158
$Sn^{4+} + 2e^- = Sn^{2+}$	0.15
$S + 2H^+ + 2e^- = H_2S(g)$	0.144
$TiO^{2+} + 2H^+ + e^- = Ti^{3+} + H_2O$	0.100
$AgBr(s) + e^- = Ag + Br^-$	0.0711
$2H^+ + 2e^- = H_2(g)$	0.0000
$Pb^{2+} + 2e^- = Pb$	-0.1251
$Sn^{2+} + 2e^- = Sn$	-0.136
$AgI(s) + e^- = Ag + I^-$	-0.1522
$V^{3+} + e^- = V^{2+}$	-0.255
$Ni^{2+} + 2e^- = Ni$	-0.257
$Co^{2+} + 2e^- = Co$	-0.277
$PbSO_4(s) + 2e^- = Pb + SO_4^{2-}$	-0.3505
$Cd^{2+} + 2e^- = Cd$	-0.4025
$Cr^{3+} + e^- = Cr^{2+}$	-0.424
$Fe^{2+} + 2e^- = Fe$	-0.44
$H_3PO_3 + 2H^+ + 2e^- = H_3PO_2 + H_2O$	-0.499
$U^{4+} + e^- = U^{3+}$	-0.52
$Zn^{2+} + 2e^- = Zn$	-0.7626

(Table 3, continued)

Half-reaction	E^0
$Mn^{2+} + 2e^- = Mn$	-1.18
$Zr^{4+} + 4e^- = Zr$	-1.70
$Al^{3+} + 3e^- = Al$	-1.67
$Th^{4+} + 4e^- = Th$	-1.83
$Mg^{2+} + 2e^- = Mg$	-2.356
$La^{3+} + 3e^- = La$	-2.37
$Na^+ + e^- = Na$	-2.714
$Ca^{2+} + 2e^- = Ca$	-2.84
$Sr^{2+} + 2e^- = Sr$	-2.89
$K^+ + e^- = K$	-2.925
$Li^+ + e^- = Li$	-3.045

[a]Corrected to data given in Ref. 23.

The Nernst equation, equation (11), containing the standard electrode potential and ion activities, E^0, may be rewritten in terms of the formal electrode potential, $E^{0'}$, and concentrations:

$$E = E^0 + (RT/nF)\ \ln(f_{Ox}[Ox])/(f_{Red}[Red])$$

$$= E^0 + (RT/nF)\ \ln(f_{Ox})/(f_{Red}) + (RT/nF)\ \ln[Ox]/[Red]$$

$$= E^{0'} + (RT/nF)\ \ln\ [Ox]/[Red] \tag{12}$$

where $E^{0'} = E^0 + (RT/nF)\ \ln(f_{Ox})/(f_{Red})$, commonly called the *formal electrode potential.*

E. Liquid Junction Potentials

Liquid junction potentials arise when two solutions of different compositions are brought into contact. For example, a liquid junction potential occurs at each of the two tips of the salt bridge of Figure 1. The potential arises from the interdiffusion of the ions at the junction of two solutions, and is designated by E_j. Detailed discussion of liquid junction potentials are given by Lingane [2] and Bates [7].

Ideally, the liquid junction potential would be zero so that the Nernst equation could be applied directly for determination of concentration. Most electrochemical cells contain at least a small liquid junction potential that is not conveniently measured or calculated. Experimentally, this potential is minimized by using a salt solution, immobilized in a salt bridge, between dissimilar solutions. A saturated potassium chloride salt solution, which has nearly equal mobilities of potassium and chloride ions, is a favored choice. The largest potentials arise when the junction involves extremely mobile hydrogen or hydroxyl ions. Data of Table 4 reflect the magnitude of typical liquid junction potentials at the tips of reference electrodes in various "real" solutions.

The effect of liquid junction potentials may be illustrated by an example. If a pH electrode system were standardized in biphthalate buffer at pH 4.01 ($E_j = -2.6$ mV, Table 4) and then used to measure a 1 *M* hydrochloric acid solution at pH of approximately 0 ($E_j = -14.1$ mV, Table 4), an error of -11.5 mV (-0.19 pH unit) would have been introduced even though the electrode system was functioning properly.

Sometimes potassium chloride solutions cannot be used in the salt bridge or reference electrode for chemical reasons. This might be the case when a chloride ion-selective electrode is used to measure chloride ions. Then substitution of less ideal salts, such as sodium nitrate, potassium sulfate, or lithium trichloroacetate, might be used, with creation of a somewhat larger liquid-junction potential.

Table 4 is based on a theoretical estimation of liquid junction potentials. Another, less well recognized, source of liquid junction potential errors lies in the construction of the junction. The most reproducible values are obtained when the junction is formed at the interface of two solutions, called a flowing junction [2]. Common junctions are a porous ceramic plug, asbestos wick, agar-gel plug, ground glass sleeve, or capillary tip. Extraneous and spurious potentials may be created by mechanical strain or chemical contamination of this junction. The flow at the junction tip may be fast or slow, and precipitation or crystallization may occur in or on it.

The velocity of flow out of the junction should be sufficient to overcome the backflow of the sample into the junction. Otherwise, there would be a steady buildup of sample ions inside the junction, which might contaminate it, plug it, and eventually contaminate the internal electrolyte if the junction is part of a reference electrode. In that respect, reference electrodes that are described as fully sealed and never in need of refilling have a limited lifetime. For optimum accuracy, refillable reference electrode junctions are preferable, although less convenient. Liquid-junction potentials may sometimes be eliminated by using an ion-selective electrode as the reference electrode. This method is illustrated below in the discussion under reference electrodes.

F. Temperature Coefficients

In the usual *isothermal* operation of an electrochemical cell, the two electrodes are at the same temperature and the temperature coefficient of the system is the difference between the coefficients of the individual electrodes. *Thermal* operation of a cell refers to the case when the two electrodes are at different temperatures.

Both the measuring and the reference electrodes of a potentiometric cell are characterized by the Nernst equation [equation (11) rewritten to emphasize the temperature term]:

$$E = E^0 + (0.19841T/n) \log a_i \tag{13}$$

Table 4. Liquid-Junction Potentials (mV) [7][a]

Description of solution X	Sat'd KCl	0.1 *M* KCl
HCl, 1 *M* (pH = 0)	-14.1	-26.8
HCl, 0.1 *M* (pH = 1)	-4.6	-9.1
HCl, 0.01 *M* (pH = 2)	-3.0	
HCl, 0.01 *M*; KCl, 0.09 *M* (pH = 2)	-2.1	
KH phthalate, 0.05 *M* (pH = 4.01)	-2.6	
KCl, 0.1 *M* (pH neutral)	-1.8	
KCl, 1.0 *M* (pH neutral)	-0.7	0.0
KCl, 4.0 *M* (pH neutral)	-0.1	
KCl, sat'd (pH neutral)	0.0	
KH_2PO_4, 0.025 *M*; Na_2HPO_4, 0.025 *M* (pH = 6.87)	-1.9	
$NaHCO_3$, 0.025 *M*; Na_2CO_3, 0.025 *M* (pH = 10)	-1.8	
NaOH, 0.01 *M* (pH = 12)	-2.3	
NaOH, 0.10 *M* (pH = 13)	+0.4	+4.4
NaOH, 1 *M* (pH = 14)	+8.6	+19.1

[a]The voltages given are the junction potential (E_j) at 25°C between solution X and the noted concentration of KCl.

and the cell temperature coefficient is given by

$$dE/dT = dE^0/dT + (0.19841/n) \log a_i + (0.19841T/n)\, d(\log a_i)/dT \quad (14)$$

This equation indicates that there are three characteristic terms influencing the temperature coefficient of any given electrode:

1. A constant (dE^0/dT) that is characteristic of the cell and is not normally zero, since the standard potential of the cell, E^0 is a constant only at a given temperature.
2. A term $(0.19841/n) \log a_i$ that varies with the activity of the ion in which the electrode is immersed. For a reference electrode with its internal solution in a captive solution of a strong electrolyte, it is a constant; for a measuring electrode, it depends on the activity of the test solution.
3. A term that varies with the temperature coefficient of the solution in which the electrode is immersed, $(0.19841T/n)\, d(\log a_i)/dT$. This includes the temperature variation of the activity coefficient and the concentration. For a strong electrolyte as in the reference electrode, this term is mainly due to the volume expansion of the solution with

Table 5. Thermal Temperature Coefficients of Common Reference Electrodes (mV/Kelvin) [8]

Reference electrode	Thermal temp. coeff.[a]
Hg, Hg_2Cl_2, KCl (sat'd, 4.16 *M*) (SCE)	+0.17
Ag, AgCl, KCl (1 *M*)	+0.25
Ag, AgCl, KCl (4 *M*)	+0.09
Ag, AgCl, KCl (sat'd, 4.16 *M*)	+0.1[b]
Pt, H_2 (p = 1), H^+ (a = 1), KCl (sat'd) (SHE)	+0.87

[a]A positive coefficient means that the warmer electrode is the positive terminal in a thermal cell. Isothermal coefficients can be computed from the thermal coefficients by subtracting 0.871 mV/Kelvin.
[b]Estimated at about 25°C.

temperature and may be neglected; for the measuring electrode, it may be a complex function of the chemistry of the solution, especially if weak electrolytes are involved.

Potentiometric instruments for measuring pH and ion activity have provisions for manual or automatic temperature compensation. From the above discussion, it should be understood that instrument temperature compensation refers only to correction for the slope term of the Nernst coefficient (0.19841T) and that the instrument cannot know the temperature coefficients of the particular electrodes or solutions being used. Proper application of these instruments indicates that the same temperature must be used for both standardization and measurement of unknown solutions. Temperature coefficients for the common reference electrodes are given in Table 5. More detailed discussion of temperature coefficients is available [8,9].

II. REFERENCE ELECTRODES

Electrical potentials cannot be measured with one wire! Only potential differences can be measured. The potential of a single electrode, which indicates a concentration or activity in accordance with the Nernst equation, cannot be measured without use of a second or reference electrode. A *reference electrode* ideally will not change potential as a function of the composition or concentration of the solution in which it is immersed, nor will its potential change as a function of the small amounts of current that might be drawn through it in the process of measuring the potential.

A. Standard Hydrogen Electrode

The universally adopted primary standard reference electrode is the *standard hydrogen electrode* (SHE) shown in Figure 2, represented and defined by

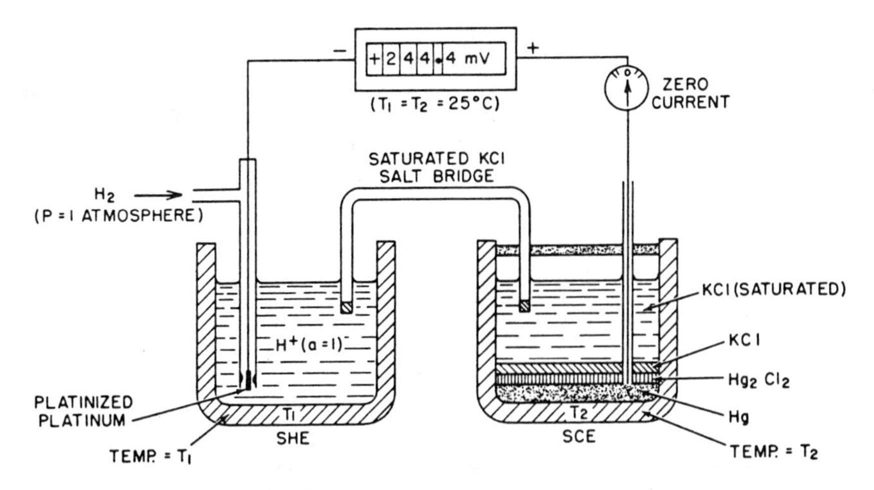

Figure 2. Reference electrodes: saturated hydrogen electrode (SHE) and saturated calomel electrode (SCE).

$$2H^+ + 2e^- = H_2 \qquad E^0 = 0.0000 \text{ V (at all temp.)} \tag{15}$$

It consists of platinum wire or foil, coated with platinum black, immersed in a hypothetical solution of unit hydrogen ion activity and in equilibrium with hydrogen gas at unit partial pressure. The potential of this standard electrode by convention is defined as zero millivolts at all temperatures. deBethune et al. [8] have shown that the standard hydrogen electrode actually has a "thermal" temperature coefficient of +0.871 mV/K when measured against the SHE at 25°C. However, when measured against another SHE at the same temperature, the "isothermal" temperature coefficient would be zero by definition. The SHE is not convenient to use, and several secondary electrodes have been devised for use in working practice.

B. Saturated Calomel Electrode

The *saturated calomel electrode* (SCE) shown in Figure 2 is a widely used secondary reference electrode made from mercury, mercurous chloride (calomel), and a saturated solution of potassium chloride that is also saturated with calomel. It has a potential of 244 mV measured relative to the SHE, as shown in Figure 2, and is represented by

$$Hg|Hg_2Cl_2, \text{ KCl(sat'd)} \tag{16}$$

The half-cell reaction is

$$Hg_2Cl_2 + 2e^- = 2Hg + 2Cl^- \qquad E^0 = 0.244 \text{ V } (25°C) \tag{17}$$

C. Silver, Silver Chloride Electrode

Another and more commonly used reference electrode for potentiometric measurements is the silver, silver chloride electrode. This may be prepared by anodizing (electroplating as the positive terminal) a metallic silver wire or foil in a chloride solution, thus depositing a layer of silver chloride on the silver. The electrode is immersed in a saturated solution of potassium chloride (or one of known concentration), which is in contact with solid silver chloride and potassium chloride; this assures saturation at all temperatures and prevents the dissolving of silver chloride from the silver metal surface.

The half-cell may be written as

$$Ag|AgCl \text{ (sat'd)}, \; KCl \; (xM) \tag{18}$$

and the half-cell reaction as

$$AgCl + e^- = Ag + Cl^- \qquad E^0 = +0.2222 \text{ V } (25°C) \tag{19}$$

A commercial version of this electrode is shown in Figure 3a and may consist of a tube 5-15 cm in length and 0.5-1.0 cm in diameter. An anodized silver wire surrounded by a relatively thick layer of silver chloride is contained in an inner tube connected to the potassium chloride filling solution in the outer tube by a porous junction of fiberglass. The liquid junction to the solution under test is created by a ceramic plug, a ground glass sleeve, an asbestos wick, or similar low-flow junction. The potassium chloride filling solution in the outer tube may diffuse into the test solution at a typical rate of 0.01-0.1 ml per day, and a fill hole is provided for replenishment. The concentration of the fill solution, as specified by the manufacturer, must be observed and the fill solution should be saturated with respect to silver chloride. Many reference electrodes are indicated as never requiring refilling and may be constructed with a very restricted junction and by immobilizing the fill solution in a gel. These electrodes have low flow rates, <0.001 ml per day and no provision for refilling. They are convenient but in time are prone to having the junction area contaminated by test solutions resulting in drifting and uncertain lifetime. A "no-flow" reference electrode is shown in Figure 3b.

D. Comparison of Calomel and Silver Chloride Reference Electrodes

Saturated KCl calomel and silver chloride reference electrodes (saturated KCl solution at 25°C is 4.17 *M*) offer ease of preparation but will have larger temperature hysteresis effects due to the change of saturation concentration with temperature. An additional problem with the silver chloride-saturated KCl electrode is the large increase in solubility of AgCl in saturated KCl solutions. For example at 25°C, the solubility of AgCl (1 g/liter) is 100 times as large in 4.17 *M* KCl (saturated) solution as in 1 *M* KCl [10]. A variety of KCl filling solution concentrations are offered in commercially available reference electrodes, including 0.1 *M*, 1.0 *M*, 3.5 *M*, 4.0 *M*, and saturated. Manufacturer's filling solution instructions for reference electrodes should be heeded. Table 6 lists some reference electrode potentials vs. the standard hydrogen electrode at typical concentrations and temperatures. Figure 4 shows a reference electrode "tree" whereby the potentials of several commonly used reference electrodes may be compared with each other and with the standard hydrogen electrode [10a,10b].

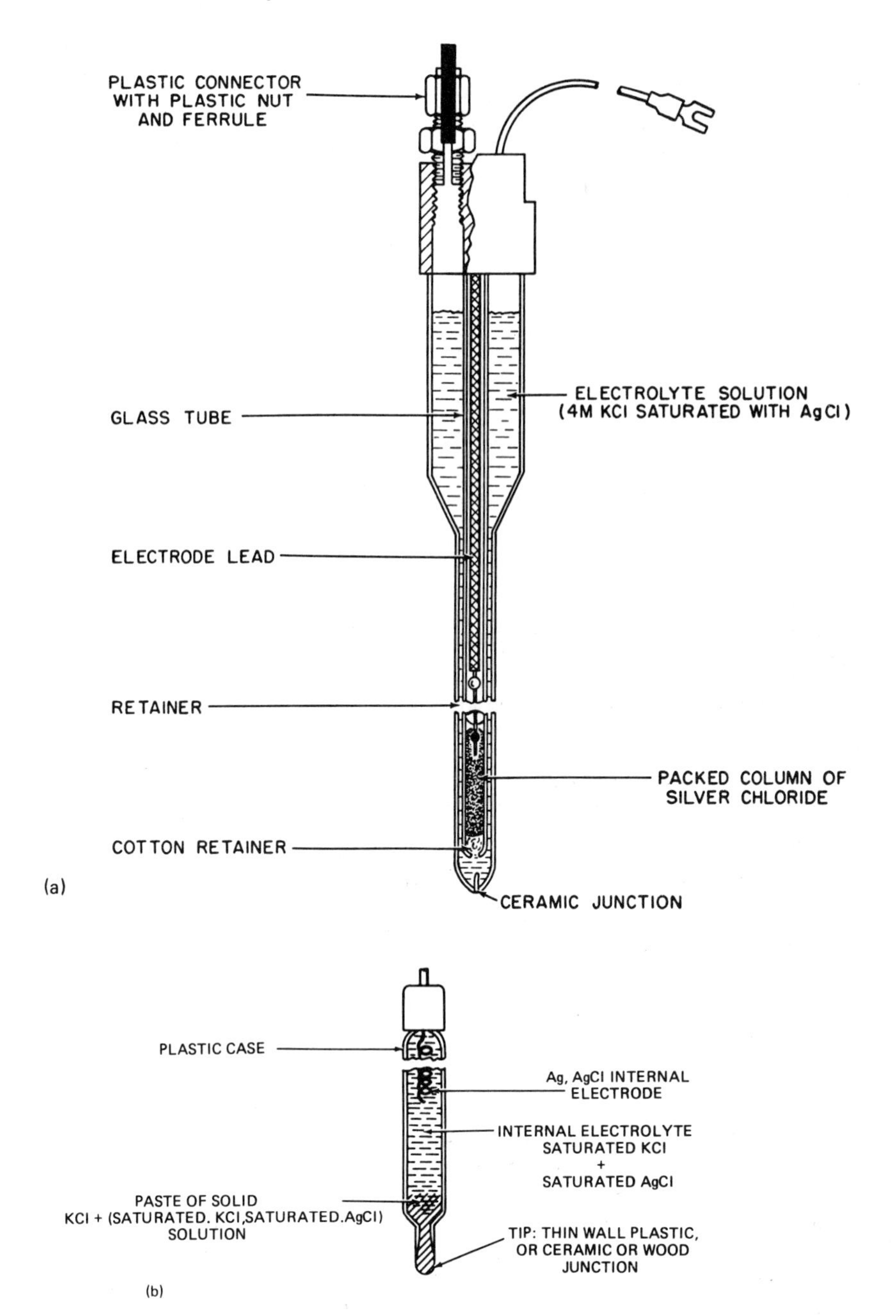

Figure 3. (a) Silver, silver chloride (Ag,AgCl) reference electrode. (b) Sealed "no-flow" reference electrode.

Table 6. Standard Potentials (mV) of Common Calomel and Silver Chloride Reference Electrodes [7]

Temperature (°C)	Calomel		Ag/AgCl	
	3.0 *M* KCl	Sat'd KCl	3.5 *M* KCl	Sat'd KCl
10	260.2	254.3	215.2	213.8
20	256.9	247.9	208.2	204.0
25	254.9	244.4	204.6	198.9
30	253.0	241.1	200.9	193.9
40	248.7	234.0	193.3	183.5

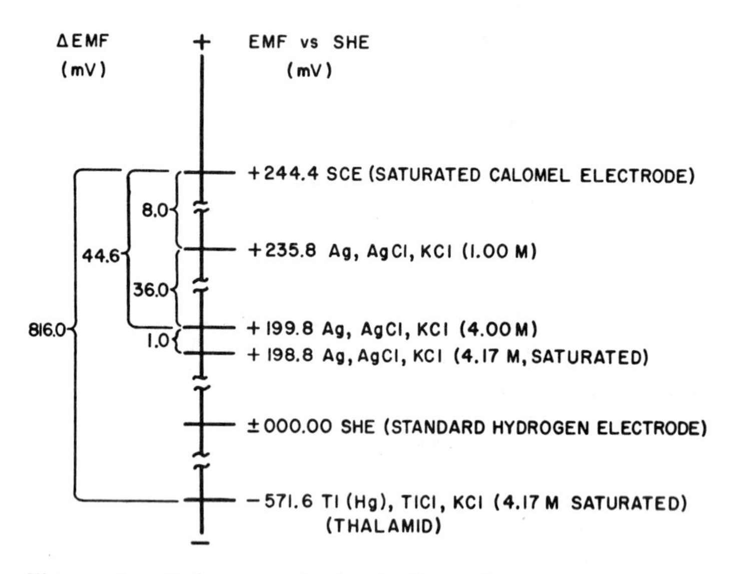

Figure 4. Reference electrode "tree."

There are some pros and cons concerning the use of calomel versus silver chloride reference electrodes, both of which are in common commercial use. In order to minimize temperature sensitivity and compensation problems, a glass pH electrode with an internal reference electrode of silver chloride must be matched with an external reference electrode containing the same internal silver chloride construction and concentration of potassium chloride. The silver chloride electrode is generally believed to have better thermal stability and less hysteresis than the saturated calomel electrode. The toxicity and environmental problems with consequent cleanup and disposal difficulties in the event of broken mercury-calomel electrodes do not exist with silver chloride electrodes. A comparison of properties of 1 *M* and 4 *M* KCl filled Ag, AgCl reference electrodes has been made [10].

E. Reference Electrode Practice

Sluggish or drifting response experienced with pH or ion-selective electrode measurements are often traceable to the reference electrode. One cause of this has been discussed in Section I.E (Liquid Junction Potentials) with the suggestion that faster-flowing liquid junctions are more desirable, albeit less convenient. Another possible problem arises when there is a chemical reaction between the solution under test and the reference electrode filling solution. An example of this might occur when the test solution contains sulfide ion and the saturated potassium chloride reference electrode solution is also saturated with silver chloride. The tip of the reference electrode could become fouled with insoluble silver sulfide. The remedy for this would be a "double junction" reference electrode containing two inner compartments. The second internal section would contain an electrolyte compatible with the test solution—in the case of the present example this could be potassium chloride without the dissolved silver. In other cases, such as those where chloride interference is undesirable, the second electrolyte might be potassium sulfate or nitrate.

A reference electrode without liquid junction potential may be used in special cases to eliminate the reference electrode "tip" problem. In practice this could be an ion-selective reference electrode that is sensitive to some ion that is constant in the solution. An example would be the measurement for fluoride using the fluoride ion-selective electrode. Usually, the test solution is at a constant buffered pH value that gives an optimum condition for this measurement. A pH glass electrode may now be used as the reference electrode, with the caveat that the measuring instrument must have a dual high-impedance input.

III. pH MEASURING ELECTRODES

The *hydrogen electrode*, previously described, is the ultimate standard to which the pH standard solutions of Tables 8 and 9 are referred. However, it is not a practical electrode to use. Not only is it cumbersome and requires purified hydrogen gas and frequent replatinization of the electrode, but it is also subject to much oxidation-reduction interference and can only be used in limited cases.

The *glass pH electrode* is one of the most selective tools of analytical chemistry. It is responsible for the most frequently measured analytical parameter—the hydrogen ion concentration or activity—in laboratory, field, and continuous industrial process applications. In spite of its excellent attributes, there are some limitations and errors, which are discussed below.

The construction of a glass electrode is illustrated in Figure 5. It consists of a thin membrane of specially formulated glass sealed to a stem glass. The stem glass is an electrical insulator and matches the temperature expansion coefficient of the membrane but is not response to hydrogen ions. The internal electrode is usually silver or calomel, that is, Ag/AgCl or Hg/Hg_2Cl_2, immersed in a buffered solution that also contains KCl saturated with AgCl or Hg_2Cl_2. The concentration of the KCl in the glass electrode internal solution is selected to match that of the reference electrode

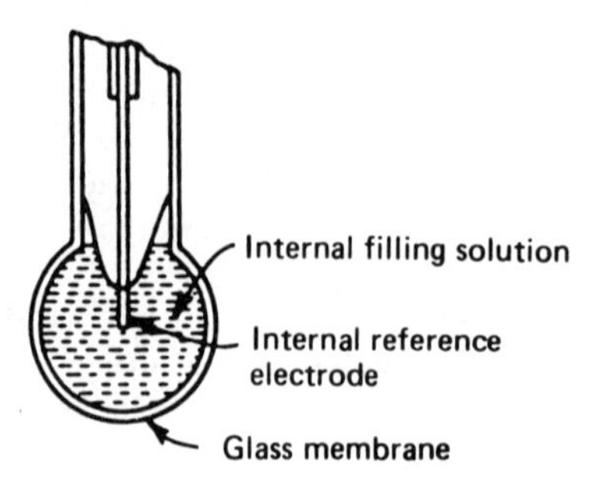

Figure 5. Glass pH electrode.

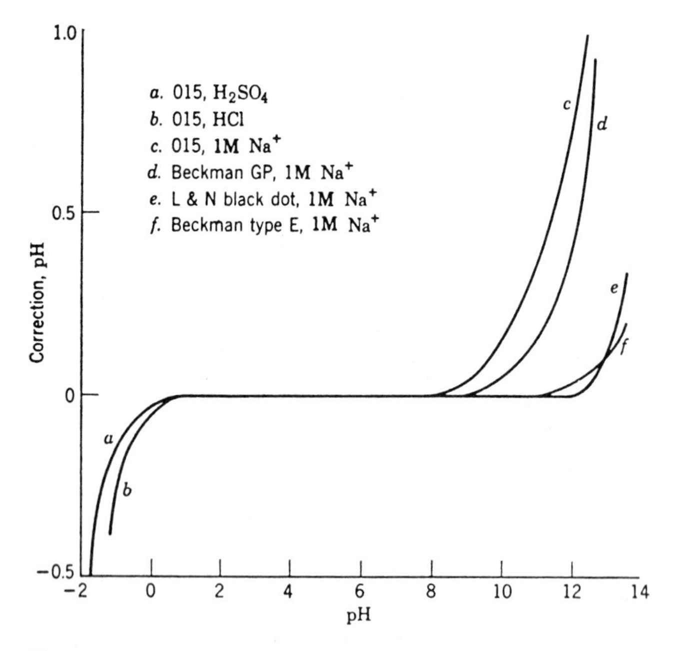

Figure 6. Glass electrode corrections in acid and alkaline solutions at 25°C [7].

already described in Figure 3. Thus the internal electrode of the glass electrode and the external reference electrode are chemically matched, an advantage that leads to minimal temperature coefficient for the cell.

The glass electrode cell with internal and external reference electrodes of Ag, AgCl, 4 M KCl might then be represented by

$$\underset{\text{(external ref.)}}{\text{Ag, AgCl, 4 } M \text{ KCl}} \;\Big\|\; \underset{\text{solution}}{\text{Test}} \;\Big|\; \underset{\text{membrane}}{\text{Glass}} \;\Big|\; \underset{\text{(internal reference)}}{\text{pH 7 buffer, 4 } M \text{ KCl, AgCl, Ag}} \qquad (20)$$

When immersed in a test solution of pH 7, the above cell would be a "symmetrical cell." At 25°C the potential would ideally be 0.0 mV and practically would be within a few millivolts of this value because of small liquid junction potentials and imperfections or strains in the glass membrane.

The equation for the above cell would then be

$$E = E^0 - (0.19842T)(pH) \tag{21a}$$

or

$$pH = (E^0 - E)/(0.19842T) \tag{21b}$$

where E^0 is determined by measuring the emf of the cell when immersed in a pH 7 buffer. If the resulting emf is 0.0 mV, as predicted theoretically, then the resulting standard potential for a glass electrode pH cell constructed as above would be

$$E^0 = 414.1 \text{ mV} \quad (\text{at } 25°\text{C}) \tag{22}$$

At other temperatures, E^0 would be a different number, influenced both by the temperature change of the Nernst equation (see Table 2) and the pH change with temperature of the standard buffer (see Tables 8-10).

Glass electrodes, in spite of being known as "almost specific for pH" measurement, are known to have several interferences and limitations. When the hydrogen ion concentration becomes very small, which occurs in alkaline solutions, the glass electrode will respond to other univalent alkali metal cations. This alkaline error is frequently referred to as the "sodium ion error" because Na^+ is the most frequently used cation in alkaline solutions, although K^+ and Li^+ also may interfere. Other common univalent cations such as Ag^+ and $NH_4{}^+$ are insoluble or slightly dissociated in alkaline solutions and do not exist in significant quantities. The sodium ion error may occur at pH 9 or greater and is very much a function of the glass composition. The extent of the error with different glass electrodes is shown in Figure 6. Correction tables need to be obtained from the manufacturer because of the variety in glass compositions. The sodium ion error also increases with temperature. At elevated temperatures in sodium hydroxide solutions the glass pH electrode is mostly a sodium ion electrode. Hot alkaline and hydrofluoric acid solutions attack the glass electrode and may dissolve enough to destroy it. There is also an "acid error" associated with glass electrodes as illustrated in Figure 6. This occurs in strong acid solutions where the utility of the glass electrode is already severely limited by the large liquid junction potentials [7].

Glass electrodes are commercially available in a variety of configurations. These include micro sizes, which fit into hypodermic needles for medical and biological research, submersible assemblies for mixing tanks and pressurized deep ocean research, immersible electrode assemblies for continuous process monitoring and control in chemical and waste-treatment plants, and many others.

Electrodes for the measurement of pH other than hydrogen and glass are known. Two examples of these are the *quinhydrone* and the *antimony* electrodes [7]. Quinhydrone, which is an equimolar compound of quinone and hydroquinone, supports a reversible oxidation-reduction reaction:

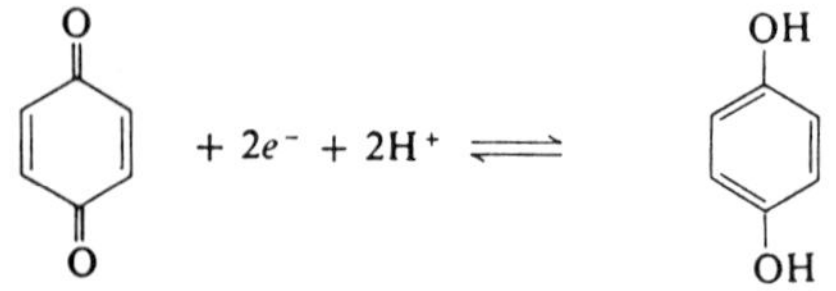

$$Q + 2e^- + 2H^+ \rightleftharpoons H_2Q \quad (23)$$

Quinone, Q *Hydroquinone*, H_2Q

The above reaction, which involves the H^+, determines the potential of a redox electrode such as gold or platinum wire. The measurement is very simple to make because a solution is quickly saturated with a little sparingly soluble quinhydrone and the electrical measurement between the wire and the reference electrode may be made by a low-impedance voltmeter. The disadvantages are due to the unstable nature of quinhydrone, its susceptibility to oxidation-reduction reactions that disturb the equilibrium, and its decomposition above pH 8.

The antimony electrode, which is composed of metallic antimony covered with a thin layer of antimonous oxide (Sb_2O_3), is represented by the reaction

$$Sb_2O_3 + 6H^+ + 6e^- = 2Sb + 3H_2O \quad (24)$$

This electrode also has the advantage of using low-impedance circuitry and is reputed to be more useful in hydrofluoric acid solutions than the glass electrode. In practice it is found not to be very accurate and is also susceptible to measuring other oxidation-reduction potentials in the solution [7,11].

The antimony-antimony oxide electrode is one of a class of electrodes known as metal-metal oxide electrodes [11] that have been reported for pH measurements but found to be wanting in comparison with the glass electrode. Two such electrodes recently studied are the *zirconium electrode* [12] and the *iridium electrode* [13]. These electrodes, which may have some advantages in higher-temperature regions where the glass electrode deteriorates, remain controversial for practical usage.

A. Definitions of pH

In the aqueous solutions that constitute the largest part of the earth's crust, the concentrations of the hydrogen ion, H^+, and hydroxyl ion, OH^-, are among the most significant variables. The relationship between these two ions and water is fixed by

$$H_2O = H^+ + OH^- \quad (25)$$

and the equilibrium constant expressed by the well-known water ionization constant, K_W:

$$K_W = a_{H^+}a_{OH^-} \quad (26)$$

For pure water, or when solutions are very dilute, activity coefficients are assumed to be unity and equation (26) is often simplified to

$$K_W = [H^+][OH^-] \quad (27)$$

The ionization constant of water, K_W, is very temperature dependent. At 25°C, $K_W = 1.012 \times 10^{-14}$, and $pK_W = 13.995$. A neutral solution is defined as one in which the hydrogen and hydroxyl ion concentrations are equal:

$$[H^+]_{neut} = [OH^-] = \sqrt{K_W} = 1.006 \times 10^{-7} \tag{28}$$

$$pH_{neut} = -\log[H^+]_{neut} = \tfrac{1}{2}pK_W = 6.997 \tag{29}$$

The pH of a neutral solution at 25°C is 7.00, as shown in equation (29). However, the neutral solution pH is temperature dependent and varies from pH 7.47 to pH 6.13 as water goes from the melting to the boiling point. The temperature variation of K_W is given in Table 7.

One of the earliest definitions of pH was given by Sørensen [14], and is easily remembered as "powers of 10":

$$[H^+] = 10^{-pH} = 1/10^{pH} \tag{30a}$$

$$pH \equiv -\log[H^+] \tag{30b}$$

Since the concentration of hydrogen and hydroxyl ion does not usually exceed 10 M, the practical range of the pH scale is -1 to 15 at ambient temperatures. Under special circumstances, a pH of as much as 17.6 has been described in potassium hydroxide solutions [15]. Thus the statements that the "pH of a neutral solution is 7" and that the "pH of all solutions are between 0 and 14" are fallacies that should not be perpetuated!

When the thermodynamic concept of activity (see earlier discussion) was introduced and observed to be more accurately correlated with the theoretical derivations and practical measurements of hydrogen ion concentration, the modern definition of pH was adopted:

$$pH \equiv paH \equiv -\log a_{H^+} = -\log[H^+]f_{H^+} \tag{31}$$

The brackets, [], are used to denote molar concentration. Molality, a temperature-independent concentration unit, is also used for concentration, and the definition permits either use, as long as the concentration units are clearly indicated.

B. Operational Definition of pH and Standards

The evaluation of activities involves theoretical calculations of the activity coefficient [the f term in equation (31)]. An experimental definition of pH based on potentiometric measurements with hydrogen or glass electrodes and standard solutions for pH (denoted by pH_S) is in modern use and is called the operational definition of pH:

$$pH_X = pH_S - (E_X - E_S)F/(RT \ln 10) \tag{32}$$

This equation corresponds to the usual laboratory operation of first standardizing a pH meter and then reading the unknown pH, and is for the electrochemical cell

Reference electrode	Salt bridge	Solution of pH(X) or pH(S)	Glass or hydrogen electrode

(33)

Table 7. Values of the Ionization Product of Water, K_w, and the pH of Neutral Water at Various Temperatures[a]

Temp. (°C)	$K_w \times 10^{14}$	$pK_w = -\log K_w$	pH_{neut}
0	0.1153	14.938	7.469
10	0.2965	14.528	7.264
20	0.6871	14.163	7.081
25	1.012	13.995	6.998
30	1.459	13.836	6.918
40	2.871	13.542	6.771
50	5.309	13.275	6.638
60	9.247	13.034	6.517
70	15.35	12.814	6.407
80	24.38	12.613	6.307
90	37.07	12.431	6.215
100	54.45	12.264	6.132

[a]From Ref. 24.

where pH_x is the pH of the unknown solution, pH_s is the pH of the standard solution, E_x is the electromotive force of the cell with the unknown pH solution, E_s is the electromotive force of the cell with the standard pH solution, R is the gas constant (8.3143 J K^{-1} mol^{-1}), and T is the temperature in Kelvin = t(°C) + 273.15. The standard solutions required in equation (32) have been carefully developed over many years principally by the U.S. National Bureau of Standards (NBS). The seven primary pH standard solutions are listed in Table 8 over a range of temperatures.

Authoritative discussion of the concepts of activity, the operational definition of pH, and standards for pH is given by Bates [7].

C. Secondary pH Standards

The seven primary pH standards of the NBS listed in Table 8 are for intermediate values of pH where the errors of the liquid junction potential that creep into the operational definition of pH [equation (32)] are minimal. The NBS has also designated acid and alkaline solutions as secondary standards, and these are listed in Table 9. Commercial sources for pH standards have found it convenient to formulate round-number pH values as standard pH buffers by addition of acid or alkali to NBS buffers. While these are without NBS sanction, values for two of them, pH 7 and pH 10, with their temperature variations are listed in Table 9.

Table 8. pH_s of NBS Primary Standard Buffer Solutions [7]

Temperature (°C)	KH tartrate (sat'd. at 25°C)	KH_2 citrate (m = 0.05)[a]	KH phthalate (m = 0.05)	KH_2PO_4 (m = 0.025) Na_2HPO_4 (m = 0.025)	KH_2PO_4 (m = 0.008695), Na_2HPO_4 (m = 0.03043)	Borax (m = 0.01)	$NaHCO_3$ (m = 0.025), Na_2CO_3 (m = 0.025)
0	—	3.863	4.003	6.984	7.534	9.464	10.317
5	—	3.840	3.999	6.951	7.500	9.395	10.245
10	—	3.820	3.998	6.923	7.472	9.332	10.179
15	—	3.802	3.999	6.900	7.448	9.276	10.118
20	—	3.788	4.002	6.881	7.429	9.225	10.062
25	3.557	3.776	4.008	6.865	7.413	9.180	10.012
30	3.552	3.766	4.015	6.853	7.400	9.139	9.966
35	3.549	3.759	4.024	6.844	7.389	9.102	9.925
38	3.548	3.755	4.030	6.840	7.384	9.081	9.903
40	3.547	3.753	4.035	6.838	7.380	9.068	9.889
45	3.547	3.750	4.047	6.834	7.373	9.038	9.856
50	3.549	3.749	4.060	6.833	7.367	9.011	9.828
55	3.554	—	4.075	6.834	—	8.985	—
60	3.560	—	4.091	6.836	—	8.962	—
70	3.580	—	4.126	6.845	—	8.921	—
80	3.609	—	4.164	6.859	—	8.885	—
90	3.650	—	4.205	6.877	—	8.850	—
95	3.674	—	4.227	6.886	—	8.833	—

[a] m = molality (mol kg^{-1}).

Table 9. pH_S of Secondary and Commercial Standard Buffers

Temp. (°C)	K tetroxalate[a] pH 1.68	Phosphate[b] pH 7	CO_3^{2-} [b] pH 10[d]	$Ca(OH)_2$ [c] pH 12.45[d]
0	1.666	7.12	10.32	13.423
10	1.670	7.06	10.25	13.003
20	1.675	7.02	10.18	12.627
25	1.679	7.00	10.01	12.454
30	1.683	6.99	9.97	12.289
40	1.694	6.98	9.89	11.984
50	1.707	6.97	9.83	11.705
60	1.723	6.98	—	11.449
70	1.743	6.99	—	—
80	1.766	7.00	—	—
90	1.792	7.02	—	—
95	1.806	7.03	—	—

[a] 0.05 *m* $KH_3(C_2O_4)_2 \cdot 2H_2O$; NBS Certificate 189 [7].
[b] Beckman Instruments, Fullerton, CA 92634.
[c] Saturated at 25°C [7].
[d] Caution: Absorption of atmospheric CO_2 by open containers will lead to errors in alkaline buffers.

Table 10. Glass Ion-Selective Electrodes [3,16]

Electrode for	Range (*M*)	Glass composition	Interferences
H^+	$1\text{-}10^{-14}$	$Li_2O\text{-}BaO\text{-}La_2O_3\text{-}SiO_2$	$H^+ \gg Li^+$, $Na^+ > K^+$
Na^+	$1\text{-}10^{-6}$	$Na_2O\text{-}Al_2O_3\text{-}SiO_2$	$Ag^+ > H^+ > Na^+ \gg Li^+$, K^+, NH_4^+
Cation (univalent)	$1\text{-}10^{-5}$	$Na_2O\text{-}Al_2O_3\text{-}SiO_2$	$K^+ > NH_4^+ > Na^+$, H^+, Li^+

D. pH Meters

Ordinary laboratory voltmeters cannot be used for the measurement of the emf of glass electrode cells because of the high electrical resistance of glass electrodes, typically 10-200 megohms. Special high-impedance voltmeter circuits are required, which draw 10^{-12} amperes (picoamperes) or less from the cell circuit. Modern electronic techniques permit the production of simplified pH meters for less than $50 that measure pH with an accuracy of ±0.1 pH units. For over $1000 there are available microprocessor-equipped pH meters that measure pH to ±0.001 pH units and emf to ±0.1 mV. These may include a temperature probe to display temperature and permit automatic temperature compensation, a memory for the pH values of the standard buffers (some of which are listed in Tables 8-9), a waiting period to allow for drift before taking pH readings, and built-in diagnostics to alert for electronic malfunctions or defective electrodes. Meters that are also used with ion-selective electrodes may have the capacity to calculate concentration from the Nernst equation and may also contain dual high-impedance amplifiers permitting the use of ion-selective electrodes as reference electrodes instead of the conventional low-impedance silver or calomel reference electrodes.

IV. ION-SELECTIVE ELECTRODES

Ion-selective electrodes (ISEs) are potentiometric electrodes that respond to individual cations or anions, with the glass pH electrode being the best known of the ISEs. All the ISE electrodes respond to a potential generated across a membrane characteristic of the electrode. The potential difference is measured with respect to a reference electrode and obeys the Nernst equation. The earlier observations concerning activities, activity coefficients, reference electrodes, liquid junction potentials, and temperature effects are all equally applicable.

For an ISE system based on the potentiometric cell

Reference electrode	‖	Test solution	|	ISE	(34)

the cell emf is

$$E = E^0 + (2.303RT/nF) \log a_i \quad (35)$$

for the ion i of charge n. Note that the charge n includes the sign, and therefore the second term of equation (35) is positive for cations and negative for anions.

The relative concentration error when using the Nernst equation is given by [18]

$$RE = (3.89n) \Delta E \quad (36)$$

where RE is the percent relative error in concentration, $RE \equiv (100 \Delta C/C)$. When using ISEs, the relative error is independent of concentration and is approximately 4n% per millivolt uncertainty in the measurement. If the uncertainty is ± 1 mV, then the relative error for a univalent ion would correspond to ±4%, and for a divalent ion ±8%. This uncertainty is for direct potentiometry; if the ISE is used for end-point determination in a potentiometric titration, the classical analytical titration accuracy of ±0.1% may be achieved.

A. Interferences and Selectivity Coefficients

The sodium ion error that exists with the glass pH electrode has been described. This interference is expressed in a general equation for use with ion-selective electrodes:

$$E = E^0 + (2.303RT/nF) \log (a_i + k_{ij}a_j^{n/z} + \cdots) \quad (37)$$

where a_i is the activity of the ion i with charge n, a_j is the activity of the interfering ion of charge z, and k_{ij} is the selectivity coefficient.

When the selectivity coefficient $k_j \ll 1$, little interference from ion j is predicted; when $k_{ij} = 1$, the ISE is equally responsive to the interferent ion j and to the ion i for which it is intended; when $k_{ij} \gg 1$, then the ISE responds better to the interferent j than to the ion i. Although selectivity coefficients are given in the literature, they should not be taken as quantitatively reliable quantities, for they will vary with the experimental conditions.

B. Classes of Ion-Selective Electrodes

ISEs may be characterized by subdivision into four classes [16-22]. Glass electrodes are responsive to univalent cations, and selectivity for these cations is obtained by varying the composition of the glass membrane. Table 11 shows the properties of three commercially available glass electrodes, selective for hydrogen ion (pH), sodium ion, and ammonium or other univalent cations.

As noted above and in Table 10, the pH glass electrode is almost specific for the hydrogen ion, with only the alkali metals such as sodium, potassium, or lithium ions causing significant errors in alkaline solutions. The sodium-ion glass electrode is more selective for two other cations (Ag^+ and H^+). The sodium ion electrode is 100 times more responsive to H^+, on a molar basis, and its usefulness in acid solutions is restricted. For a solution with $[H^+] = 10^{-5}$ (i.e., pH = 5), the sodium-ion electrode will give the same response as a solution with $Na^+ = 10^{-2}$. Alkaline or buffer-controlled solutions are used with glass ISEs to limit response to the H^+. These buffered solutions, sometimes called "ionic strength adjustors" (ISA solutions), are also used to fix the ionic strength of standards and solutions under test to keep activity coefficients constant and permit interpretation in terms of concentration rather than activity. Similarly, the univalent cation electrode is generally used as an ammonium-ion electrode although K^+ and H^+ are interferents; its usefulness is limited to solutions where the ammonium ion predominates.

Solid-state electrodes have membranes of a single crystal or of a homogeneous pressed pellet electrode. Ten of the commercially available solid state electrodes are listed in Table 11 and the construction is depicted

Table 11. Solid-State Ion-Selective Electrodes [16]

Electrode	Useful range (M)	Interferences (approx. conc. M)
F^-	$1\text{-}10^{-6}$	$OH^- < 0.1F^-$
Ag^+ or S^{2-}	$1\text{-}10^{-7}$	$Hg^{2+} < 10^{-7}$
Cl^-	$1\text{-}5 \times 10^{-5}$	10^{-7} CN^- and I^-; 10^{-3} Br^-; 10^{-2} $S_2O_3^-$; 0.1 NH^3; OH^-
Br^-	$1\text{-}5 \times 10^{-6}$	10^{-5} CN^-; 10^{-4} I^-; 400 Cl^-; 2 NH_3;
I^-	$1\text{-}5 \times 10^{-8}$	CN^-, 500 Br^-
SCN^-	$1\text{-}5 \times 10^{-6}$	10^{-6} I^-; 10^{-3} Br^- and CN^-; 20 Cl^-; 100 OH^-
Cd^{2+}	$10^{-1}\text{-}10^{-7}$	Ag^+, Hg^{2+}, Cu^{2+}
Cu^{2+}	$10^{-1}\text{-}10^{-8}$	Ag^+, Hg^{2+}, Cu^{2+}
Pb^{2+}	$10^{-1}\text{-}10^{-6}$	Ag^+, Hg^{2+}, Cu^{2+}

in Figure 7a. One of them, the fluoride ion electrode, rivals the glass pH electrode as a sensor with freedom from interferences and rapid and reproducible response. The membrane sensor is a single crystal of europium-doped LaF_3. This crystal is sealed into the end of a plastic tube and typically contains an internal Ag,AgCl reference electrode and internal solution with ions common to the membrane and internal reference electrode, namely NaF and NaCl. Figure 7b shows the utility of the fluoride ion electrode in analysis of public water supplies for fluoride in the 1 mg/liter (1 ppm) range.

The other solid-state electrodes, listed in Table 11 with their interferences, are usually formed from a homogeneous mixture of silver sulfide and an insoluble salt containing the ion to be determined. For example, the chloride-sensitive electrode may be a mixture of silver sulfide and silver chloride. Not surprisingly, silver salts that are more insoluble than silver chloride, such as the bromide, iodide, cyanide, and sulfide, will interfere with the chloride electrode. A pure silver sulfide pellet will respond to silver ions or sulfide ions, and since silver sulfide is among the most insoluble of salts there are few interferences to the Ag^+/S^{2-} electrode. The silver sulfide electrode responds to the silver ion activity at the electrode/solution interface, which in turn is governed by the equilibrium

$$AgX(\text{insoluble}) = Ag^+ + X^- \tag{38}$$

where X^- might be an anion such as Cl^-, Br^-, I^- or S^{2-}. The solubility product constant K_{sp} is given by

$$K_{sp} = a_{Ag^+}a_{X^-} \tag{39}$$

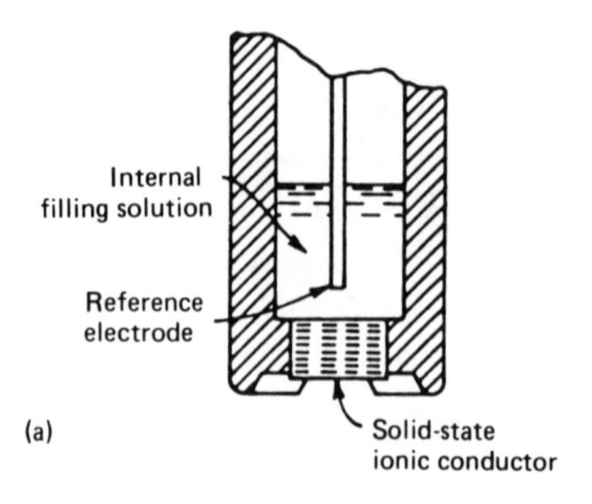

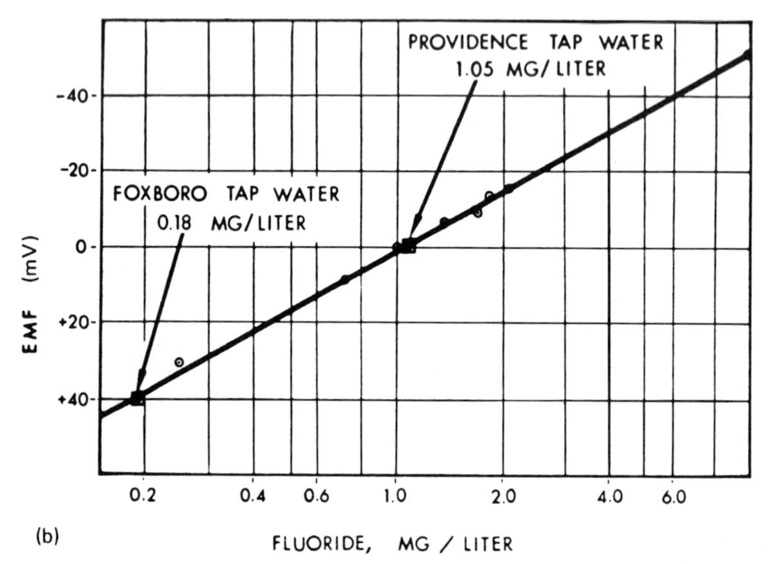

Figure 7. (a) Solid-state ion-selective electrode. (b) Fluoride electrode use in water analysis [18].

Substituting into the Nernst equation, equation (11), for ion-selective electrodes,

$$E = E^0 + (RT/F) \ln a_{Ag^+} = E^0 + (RT/F) \ln K_{sp}/a_{X^-}$$

$$E = E^{0\prime} - (RT/F) \ln a_{X^-} \tag{40}$$

which is the equation for the univalent anion-selective ISE.

A similar expression may be obtained for the divalent metal cation ISEs for Cd^{2+}, Cu^{2+}, and Pb^{2+} listed in Table 11:

$$E = E^0 + (RT/2F) \ln a_{X^{2+}} \tag{41}$$

An earlier type of the *liquid-membrane ion-exchange ion-selective electrode* is shown in Figure 8. It is constructed of two concentric plastic tubes with an inert porous hydrophobic membrane mounted to contact both the inner and outer volumes of the concentric tubes. A charged or neutral ion exchanger dissolved in an organic solvent is in the outer tube and soaked into the membrane. The inner tube contains the usual internal reference electrode of Ag,AgCl with the internal aqueous solution of KCl plus a salt of the ion to which the electrode is responsive. The difference in activity of the inner and outer solutions surrounding the membrane generates an electrode potential that obeys the familiar Nernst equation.

For convenience in renewing the ion-exchange liquid, which has a lifetime of the order of months, a commercial version of this electrode utilizes a replaceable cartridge that screws into the electrode body.

The selectivity, range, and sensitivity of seven of the liquid-membrane ion-exchange electrodes are listed in Table 12. These properties are determined by the organic ion exchanger and the solvent in which it is dissolved. Phosphate diesters, $(RO)_2PO_2^-$ with R groups from C_8 to C_{16}, dissolved in a nonpolar solvent such as dioctylphenylphosphonate, respond to Ca^{2+} in the presence of other univalent and divalent cations. Less polar solvents such as decanol yield electrodes that give almost equal response to alkaline earth metals. These latter electrodes are dubbed "water hardness" electrodes, since the parameter water hardness is the sum of Ca^{2+}, Mg^{2+}, and other divalent ions.

The liquid ion-exchange electrode for Cl^- is composed of a dimethyldistearyl ammonium ion exchanger, R_4N^+. This is a different electrode than the solid-state ISE for Cl^-, with some differences in selectivity (refer to Tables 11 and 12).

The operating temperature range of liquid-membrane electrodes is generally 0-50°C, which is more limited than the range for glass and solid-state electrodes. Liquid-membrane electrodes are restricted to aqueous media, since many organic solvents react with the ion exchanger or increase its solubility.

Gas-sensing electrodes are still another class of ion-selective electrodes that are sensitive to the concentration of dissolved gases such as CO_2

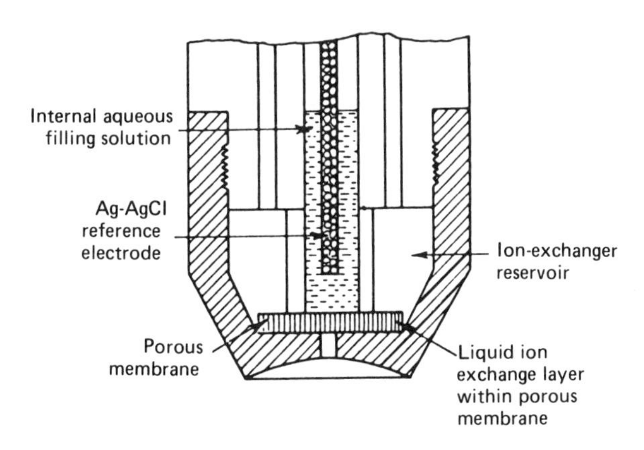

Figure 8. Liquid-membrane ion-selective electrode.

Table 12. Liquid-Membrane Ion-Selective Electrodes[a]

Electrode	Concentration range (M)
Ca^{2+}	$1\text{-}5 \times 10^{-7}$
Cl^-	$1\text{-}5 \times 10^{-6}$
BF_4^-	$1\text{-}7 \times 10^{-6}$
NO_3^-	$1\text{-}7 \times 10^{-6}$
ClO_4^-	$1\text{-}7 \times 10^{-6}$
K^+	$1\text{-}10^{-6}$
Water hardness	$1\text{-}6 \times 10^{-6}$

[a]From Ref. 16b; refer to this for interference information.

(from carbonates), NH_3 (from ammonium ions), NO_2 (from nitrites), and SO_2 (from sulfites). These electrodes utilize a membrane, permeable only to gas, between the solution under test and the sensor electrode, as illustrated in Figure 9.

For example, the carbon dioxide electrode functions when an acid solution liberates CO_2 from carbonates:

$$HCO_3^- + H^+ = CO_2 + H_2O \tag{42}$$

The sensing electrode in this case is a conventional pH glass electrode, surrounded by an internal solution that is a buffered bicarbonate solution. The CO_2 diffusing through the membrane changes the pH in proportion to the pCO_2 concentration on the outer surface of the membrane. Possible gas-sensing equilibria are listed in Table 13 [17]. Those known to be commercially available are marked with an asterisk [16].

C. Enzyme Ion-Selective Electrodes

These are special electrodes, presently more in the research than in the commercial realm, which are similar in construction to the gas-sensing electrodes in that a coating is applied over the membrane of an ISE. Unlike the coatings that are gas permeable, the enzyme electrodes have an enzyme coating that converts the substrate to an ion that is detected by an ISE [22]. An example of the enzyme electrode is the urea (NH_2CONH_2) electrode:

$$\text{Urea} + H_2O \xrightarrow{\text{urease}} NH_4^+ + HCO_3^- \tag{43}$$

The substrated urease covers the cation-sensitive glass electrode, which is used for the determination of NH_4^+ concentration (Figure 10) and may be used for the detection of urea by the above reaction. Since the substrate is consumed by the electrode reaction, enzyme electrodes require

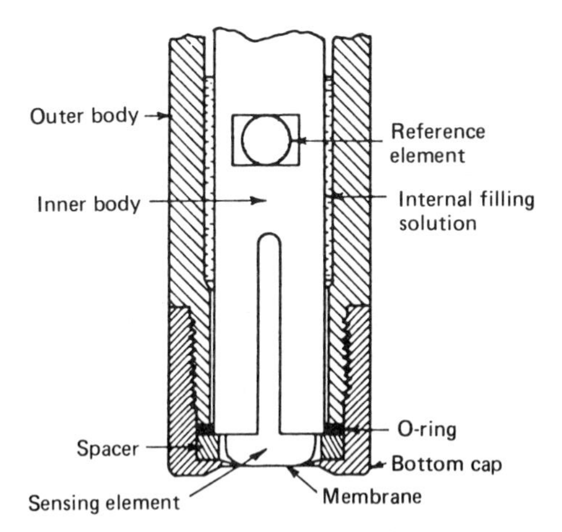

Figure 9. Gas-sensing ion-selective electrode.

Table 13. Possible Equilibria Associated with Gas-Sensing Electrodes [17]

Diffusing species	Equilibria	Sensing electrode
NH_3 [a]	$NH_3 + H_2O \rightleftharpoons NH_4^+ + OH^-$	H^+
	$xNH_3 + M^{n+} \rightleftharpoons M(NH_3)_x^{n+}$	$M = Ag^+, Cd^{2+}, Cu^{2+}$
SO_2	$SO_2 + H_2O \rightleftharpoons H^+ + HSO_3^-$	H^+
NO_2 [a]	$2NO_2 + H_2O \rightleftharpoons NO_3^- + NO_2^- + 2H^+$	H^+, NO_3^-
H_2S	$H_2S + H_2O \rightleftharpoons HS^- + H^+$	S^{2-}
HCN	$Ag(CN)_2 \rightleftharpoons Ag^+ + 2CN^-$	Ag^+
HF	$HF \rightleftharpoons H^+ + F^-$	F^-
	$FeF_x^{2-x} \rightleftharpoons FeF_y^{3-y} + (x - y)F^-$	Pt (redox)
HOAc	$HOAc \rightleftharpoons H^+ + OAc^-$	H^+
Cl_2	$Cl_2 + H_2O \rightleftharpoons 2H^+ + ClO^- + Cl^-$	H^+, Cl^-
CO_2 [a]	$CO_2 + H_2O \rightleftharpoons H^+ + HCO_3^-$	H^+
X_2	$X_2 + H_2O \rightleftharpoons 2H^+ + XO^- + X^-$	$X = Br^-, I^-$

[a]Commercially available.

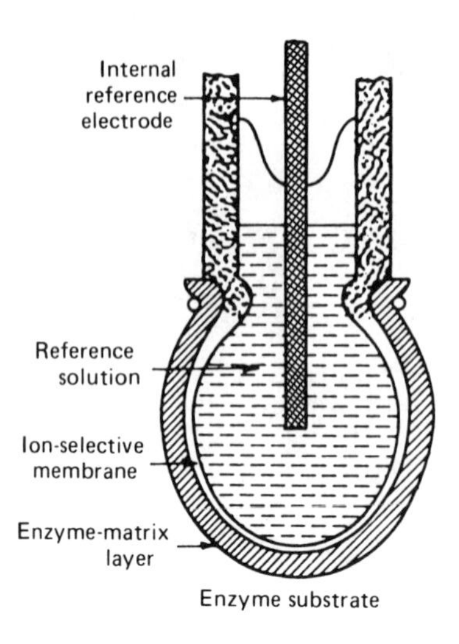

Figure 10. Enzyme ion-selective electrode.

periodic renewal of the membrane layer. There are many enzyme ISEs described in the literature [22], such as for urea, glucose, amino acids, lactic acid, alcohols, penicillin, amygdalin, and cholesterol.

REFERENCES

1. T. S. Licht and A. J. deBethune, J. Chem. Educ. *34*, 433-440 (1957).
2. J. J. Lingane, *Electroanalytical Chemistry*, 2nd ed., Interscience, New York, 1958.
3. G. D. Christian and J. E. O'Reilley, *Instrumental Analysis*, 2nd ed., Allyn and Bacon, Boston, 1986.
4. J. Kielland, J. Am. Chem. Soc. *59*, 1675-1678 (1937).
5. D. G. Peters, J. M. Hayes, and G. M. Hieftje, *Chemical Separations and Measurements*, W. B. Saunders, Philadelphia, 1974.
6. L. Meites, Ed., *Handbook of Analytical Chemistry*, McGraw-Hill, New York, 1965.
7. R. G. Bates, *Determination of pH*, 2nd ed., Wiley, New York, 1973.
8. A. J. DeBethune, T. S. Licht, and N. Swendeman, J. Electrochem. Soc. *106*, 616 (1959); and in *Encyclopedia of Electrochemistry* (C. A. Hempel, Ed.), pp. 412-434, Reinhold, New York, 1964.
9. L. E. Negus and T. S. Light, Instrum. Technol. *19*, 23-26 (1972).
10. T. S. Light, An Improved Reference Electrode for Process Measurement, in *Analysis Instrumentation*, Vol. 8, Instrument Society of America, Pittsburgh, 1970.

10a. T. S. Light, New Developments in pH Measurement and Reference Electrodes, in *Instrument Engineers Handbook* (B. G. Liptak, Ed.), Supplement 1, Chilton Book Co., Philadelphia, 1972.

10b. T. S. Light, pH Measurement, in *Environmental Engineers Handbook* (B. G. Liptak, Ed.), Vol. 1, Chilton Book Co., Radnor, PA, 1974.

11. D. J. G. Ives and G. J. Janz, *Reference Electrodes*, Academic Press, New York, 1961.
12. T. S. Light and K. S. Fletcher, Anal. Chim. Acta *175*, 117-126 (1985).
13. T. Katsube, I. Lauks, and J. Zemel, Sensors and Actuators *2*, 399-410 (1982).
14. S. P. L. Sørensen, Biochem. Z. *21*, 131, 201 (1909).
15. S. Licht, Anal. Chem. *57*, 514-519 (1985).
16a. *Handbook of Electrode Technology*, Orion Research, Boston, 1982.
16b. *Concise Guide to Ion Analysis*, Orion Research, Boston, 1983.
17. J. W. Ross, J. H. Riseman, and J. A. Krueger, Potentiometric Gas Sensing Electrodes, in *IUPAC International Symposium on Selective Ion-Sensitive Electrodes* (G. J. Moody, Ed.), Cranke, Russak, New York, 1973.
18. T. S. Light, Industrial Analysis and Control with Ion-Selective Electrodes, in *Ion-Selective Electrodes* (R. Durst, Ed.), NBS Special Publication 314, U.S. Government Printing Office, 1969.
19. G. J. Moody and J. D. R. Thomas, *Selective Ion-Sensitive Electrodes*, Merrow, Watford, U.K., 1971.
20. H. Freiser, Ed., *Ion-Selective Electrodes in Analytical Chemistry*, Vol. 1, Plenum, New York, 1978; Vol. 2, 1980.
21. J. Koryta, *Ion-Selective Electrodes*, Cambridge University Press, London, 1975.
22. G. G. Guilbault, *Ion-Selective Electrode Rev. 4*, 187-231 (1982).
23. A. J. Bard, R. Parsons, and J. Jordan, Eds., *Standard Potentials in Aqueous Solutions* (Copyright, International Union of Pure and Applied Chemistry), Marcel Dekker, New York, 1985.
24. T. S. Light and S. L. Licht, Anal. Chem. *59*, 2327 (1987).

17

Voltammetry

ANDREW F. PALUS / EG&G Princeton Applied Research, Princeton, New Jersey

I. INTRODUCTION

The choice of instrumentation for electroanalytical chemistry in general and for voltammetry in particular may be difficult for those chemists who are not well versed in the techniques. Confusion over conventions and definitions contributes to clouding the process of choosing instrumentation in this area.

Almost any commercially available voltammetric instrument will make the desired measurements, but some are easier to use or more versatile than others. These are among the features that must be considered when moving into the electroanalytical field. In order to facilitate such choices, a few definitions and explanations are needed.

Voltammetry is a technique in which a potential is applied to a polarizable electrode (against a suitable reference electrode) and the resulting current is measured. *Polarography* is a subset of this technique in which the active electrode consists of a series of small droplets of mercury falling from the tip of a fine glass capillary, the *dropping mercury electrode* (DME). Thus a polarographic method is a particular type of voltammetric procedure.

Electroanalytical chemistry is unique in that a single instrument can often be used in various modes of operation to implement many different analytical techniques. This is in contrast to spectroscopy, where instruments are for the most part less versatile. Thus a single voltammetric instrument can be used in such varied techniques as classical DC polarography, normal pulsed polarography, differential pulsed polarography, AC polarography, square-wave polarography (and their counterparts with other electrodes than the DME), cyclic voltammetry, and anodic stripping analysis. All of these methods are based on the behavior of an electrolyte when an excitation potential causes a response current.

There exists a large literature of applications of voltammetric techniques, including methods for the determination of metals, anions, and organic substances [1-15]. The limits of detection of such methods are typically of the order of 100 ppb or even less. In fact by anodic stripping analysis (treated in another chapter in this handbook), many researchers have reported parts per trillion determinations [16-18].

Instrumentation commercially available today runs the gamut from analog instruments capable only of taking data, to elaborate digital instruments acting under computer control. At 1987 prices, these instruments run from about $3000 to $25,000.

A. Basics

Historically, voltammetry originated with the polarographic work of Heyrovsky and his students in the 1920s [19]. This constituted a major breakthrough in rapid, automatically recording, instrumental techniques. In the early days, most chemists interested in the technique constructed their own apparatus. It is still possible to assemble a homemade instrument, but this is seldom worth the time and effort, in view of the excellent commercial implementations.

Voltammetry can be divided into many different techniques, the distinction between them lying in the waveform of the exciting potential. A typical modern voltammetric instrument must include a signal generator capable of producing a variety of waveforms as needed to implement the various techniques. The signal produced is impressed upon a potentiostat, which in turn forces the working electrode to follow the desired programmed potential. The resulting current is measured by suitable electronic circuitry, and converted to a convenient form to yield the desired analytical information. The basic phenomenon being observed in all cases is the oxidation or reduction of the desired species in solution.

B. The Electrochemical Cell

The simplest voltammetric cells consist of two electrodes dipping into the solution contained in a small flask with an inlet tube for nitrogen flushing. One of these electrodes (the DME in polarographic work), called the *working electrode*, is where the reaction of interest takes place. The second electrode, the reference electrode, must be designed to have as low a resistance as possible since the cell current must pass through it without appreciably modifying its inherent potential. Such cells were used universally for many years, but are now for the most part displaced by three-electrode cells in which the current is carried by the third electrode, called the *counter* or *auxiliary electrode*. The reference electrode is thus relieved of the necessity of carrying current, and can therefore be of higher resistance, simplifying its design.

The working electrode can take many forms. It can be a dropping mercury electrode, a hanging mercury drop, a platinum disk, or a specially prepared carbon rod, among others. The reference electrode forms a stable benchmark against which the potential of the working electrode is established. The reference electrodes most used in voltammetry are the saturated calomel electrode (SCE) and the silver/silver chloride electrode. The auxiliary electrode can be simply a platinum wire, or in some cases a glassy carbon electrode.

II. INSTRUMENTATION

A. Potentiostats

The heart of any voltammetric instrument is the potentiostat, which controls the applied potential and provides for measurement of the cell current. A simple potentiostat can be constructed from an operational amplifier with its power source (Figure 1). The more sophisticated instrumentation now available makes use of multiple operational amplifiers with feedback loops (Figure 2). In the present context, the important parameters for a potentiostat are low noise, high compliance voltage, broad bandwidth, iR compensation, and the ability to handle low currents and a wide range of applied potentials.

By compliance is meant the maximum potential that the instrument can apply to the counter electrode in order to produce the desired potential at the working electrode. The larger the compliance voltage, the greater is the ability of the potentiostat to exert adequate control over the cell. It is particularly important for work in nonaqueous or other low conductivity solutions. Most commercial instruments have compliance in the range of 10-100 V. Generally, the higher the compliance the better, but there are tradeoffs: high-compliance potentiostats often generate more noise than those with lower maximum voltages.

Two types of electrical noise can be present in a potentiostat. Potential noise, which may run from about 2 to 100 μV, is mostly power-line related. It is important because it translates directly into current fluctuations that are difficult to distinguish from true electrochemical effects. Potential noise is inherent in the design of a given instrument, and cannot be reduced through shielding.

It may appear that 100 μV of potential noise is unimportant if, say, 500 mV is being applied to an electrode. However, the noise is actually an AC signal riding on the DC potential, and this can give rise to a serious noise level in the current measurement. The effect is easily calculated. The capacitance of a typical mercury drop is of the order of 1 μF, which corresponds to an impedance of about 1300 ohms at 120 Hz (twice the power line frequency). The 100 μV of noise acting through this impedance will produce about 75 nA of AC current. This should be compared to the DC current, which might well be in the 100-1000 nA range.

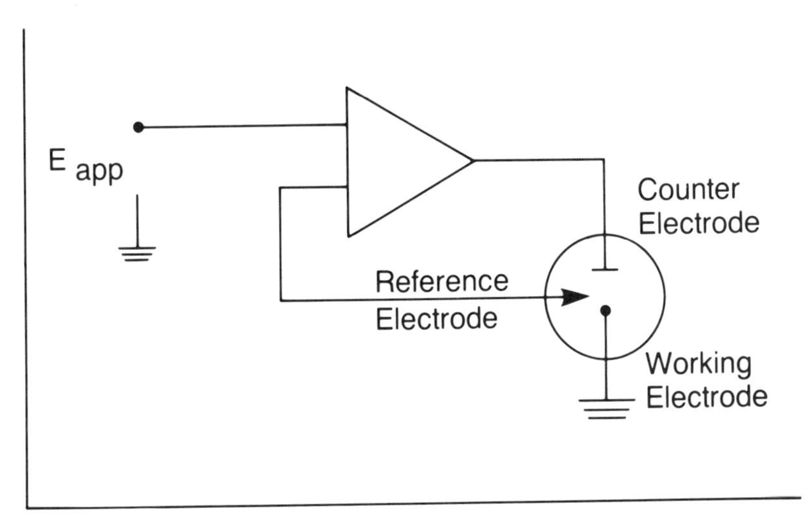

Figure 1. Simple potentiostat circuit.

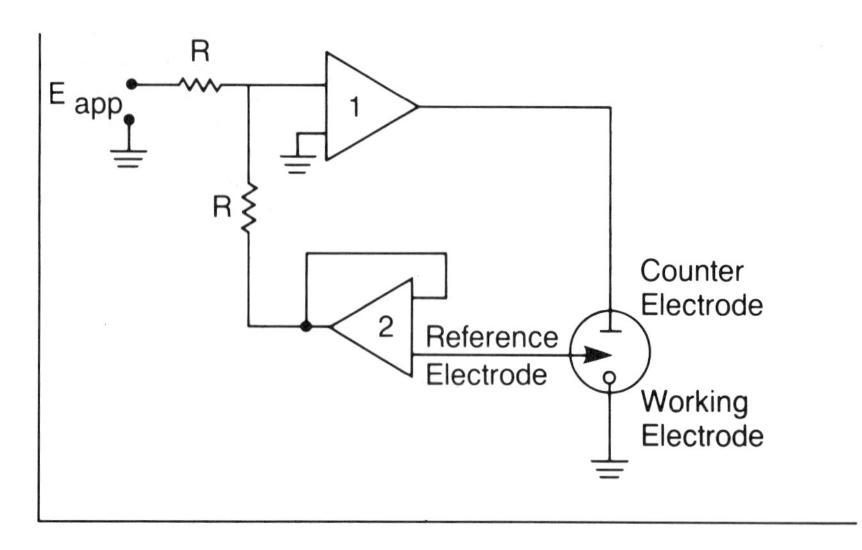

Figure 2. Potentiostat circuit with a feedback loop.

The second major type of noise of importance in voltammetry is current noise, generally due to inductive or capacitive pick-up from external sources. It is independent of electrode size, and can usually be reduced through shielding, as by the use of a Faraday cage. Current noise can also be reduced through electronic filtering.

Both types of noise, to the extent that they originate with stray effects at the power-line frequency or its harmonics, can be reduced markedly by integrating current measurements over exactly one period of the AC wave, namely, 16.67 msec (for 60 Hz power), or a small multiple thereof. This has the effect of averaging out that component of the noise that varies at this frequency.

Another parameter of importance is the *bandwidth*. This tells the user something about the speed of response of the potentiostat. For rapid scans, a broad bandwidth is essential. The bandwidth can be viewed in terms of the rise time of the potentiostat or the frequency response of the current follower. The trade-off for speed is stability: a faster potentiostat is more prone to oscillation. Speed is worthless if the potential of the working electrode is oscillating out of control.

Comparison of potentiostats with respect to their rise times may be difficult, in that not all manufacturers measure the rise time in the same way. The figures in a specification sheet are useless if the conditions under which they are determined are not given. Typically the rise time is defined as the length of time taken for a potential pulse to rise from 10% to 90% of its peak value, but this time can be affected by resistive or capacitive components associated with the working electrode. Most commercial instruments have a rise time on the order of 1 to 100 μsec, which translates to a bandwidth of 100 kHz or less, fully adequate for most voltammetric experiments. For instance, this would permit running a cyclic voltammogram at a rate as high as 1000 V/sec. Unfortunately, the potential control is not the limiting feature: the cell current must be measured as well, for which we must consider the operating speed of the current-to-voltage (I/E) converter.

The speed of an I/E converter is usually expressed in terms of its bandwidth, reported as the frequency at which the current measured for a sinusoidal potential excitation is only 70.6% of the expected value (i.e., down by -3 dB). While manufacturers can produce instruments with high bandwidth, it is usually at the expense of sensitivity. An I/E converter capable of measuring nanoamperes or picoamperes of current is designed by choosing stable, low-noise components, but such components tend to be slow. The trade-off is speed against sensitivity.

Another feature to consider in a potentiostat is compensation for the iR drop within the cell, where R is the resistance of the solution and of the wires and connections, and i is the cell current. The iR product is, by Ohm's law, a voltage, and if it is uncompensated will subtract from the potential signal given to the potentiostat.

As an example, suppose that a cell with 500 ohms resistance passes a current of 20 μA. The iR drop will be 10 mV. If the potentiostat is set up to apply 500 mV to the cell, the potential actually appearing at the working electrode will be only 490 mV. This 2 percent error may or may not be significant, depending on the application. Electronic compensation can be designed to reduce the error to negligible proportions.

Some commercial instruments do not provide for iR compensation. In this case the error can usually be overcome by using a highly conducting supporting electrolyte. This may not be possible with nonaqueous solutions.

The remaining specifications to consider are the voltage and current ranges available. The potential range required in voltammetry is seldom greater than ± 2V, hence is not likely to be a limiting factor unless the instrument is to be used for special purposes or nonaqueous solutions. The lowest detectable level of current alone is not a satisfactory measure of sensitivity; the above-mentioned noise factors must also be taken into account. It is useless to purchase an instrument with good low-current response if the noise present makes that current level unusable. Typical voltammetric instruments should be capable of measuring 1 nA or lower.

B. Electrodes

The dropping mercury electrode is the classical "front end" of a voltammetric (here polarographic) instrument. This electrode is versatile, easily renewable, and is backed up by some 50 years of articles, textbooks, and application information for analytical and research uses.

A number of other working electrodes are in use. These include the glassy carbon electrode (GCE), platinum or gold electrodes, thin-film electrodes (TFE), wax-impregnated graphite (WIG), and the rotating disk electrode (RDE). A thin-film electrode is typically prepared by electrodepositing a small amount of mercury on one of the solid electrodes mentioned above.

Solid electrodes of a noble metal or carbon have the advantage of being usable for oxidation studies. Mercury is of limited use for oxidations, since at potentials more positive than about +0.4 V (against the SCE), mercury itself is subject to oxidation.

The rotating disk electrode consists of a solid electrode (usually platinum) fitted into a stand that allows the electrode to be rotated in its own plane at a precisely controlled speed. In an analytical context this is often mercury-plated (a rotating TFE) for use in stripping voltammetry.

Still the most commonly used electrode for analytical voltammetry is mercury. From 1927 until about 1970 the electrode of choice was the classic DME, consisting of a section of very fine-bore, but thick-walled, glass capillary, of the type known as "marine barometer tubing," connected through a length of flexible tubing to an elevated reservoir of mercury. Electrical connection to the mercury is made by either a wire dipping into the reservoir or a section of metal tubing inserted into the delivery tube. The drop time (the length of time during which each drop remains attached to the mercury thread in the capillary) is controlled by adjusting the height of the mercury column. It is difficult to get reproducible drop times less than 3 or 4 sec.

A considerable improvement is achieved by adding to the above-described assembly a mechanical drop knocker. This is a device that taps the capillary at precisely equal time intervals, thereby causing the drop to fall. This permits drop times as small as 1/2 sec. The availability of mechanical drop knockers together with modern electronics is what led to a renaissance in polarography.

Most modern voltammetric systems (Figure 3) use a newer type of DME, called the SMDE (static mercury drop electrode). This device was introduced as a commercial product by EG&G Princeton Applied Research Corporation (Figure 4), and more recently other companies (such as Metrohm) have developed similar products. The SMDE can be used to perform the same experiments as the DME, but with improved signal-to-noise character-

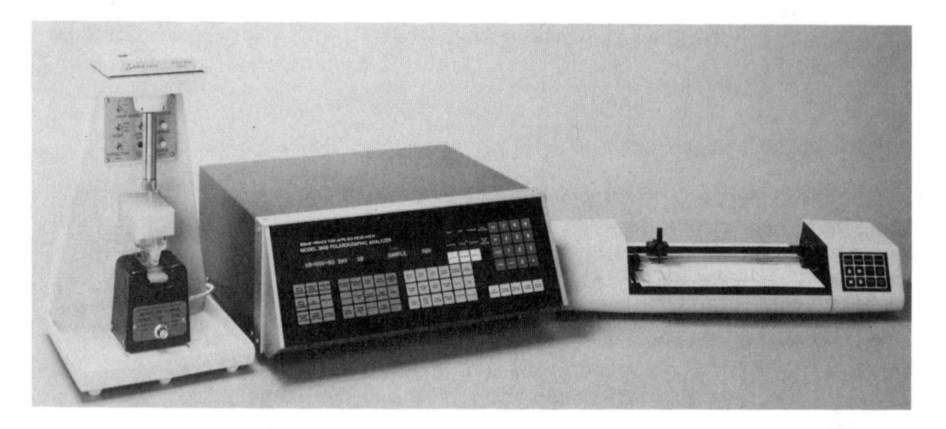

Figure 3. A modern voltammetric analyzer system. (Courtesy of EG&G Princeton Applied Research.)

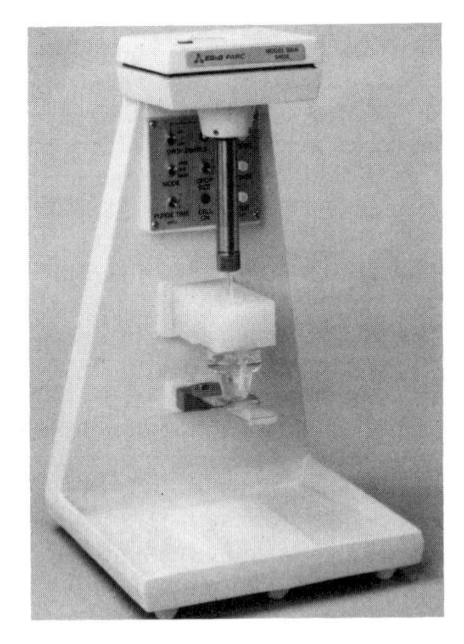

Figure 4. Static mercury drop electrode. (Courtesy of EG&G Princeton Applied Research.)

istics [20]. In addition, it is capable of acting as a hanging mercury drop electrode (HMDE). In this mode it is useful for single drop experiments, including cyclic voltammetry and Osteryoung square-wave voltammetry [21].

The principle behind the SMDE is to allow a mercury drop to grow rapidly and then be held hanging from the capillary at constant size, until discarded. This contrasts with the DME, in which each mercury drop grows continually during its drop life, until it is knocked off or falls of its own weight. Since the cell current is sensitive to changes in the area of the electrode, the curves obtainable with the SMDE are much superior to those from the DME with respect to the signal-to-noise ratio.

The current passing through a voltammetric cell consists of two components: the faradaic current, which is the quantity to be determined, and the capacitive, or nonfaradaic, current, which is undesirable and tends to obscure the faradaic current. The capacitive, or charging, current goes to maintaining the ionic double layer at the electrode surface, and must flow whenever the surface area is increasing. This interference, which is actually a form of noise, is greatly reduced by the use of the SMDE, because the drop grows rapidly only in the first 50-200 msec of the drop life. If the time window during which the current is measured occurs later than this, the effect of the capacitive current is suppressed.

Another advantage of the SMDE is its ability to function as an HMDE. Prior to the introduction of the SMDE, a separate electrode assembly was needed for single-drop experiments (Figure 4). Although this type of electrode works well, it is much less convenient.

Commercially available SMDEs take two different approaches in producing the mercury drop. One method is to use a large-bore capillary (0.5 mm versus the conventional 0.01 mm), with a low-pressure head. A solenoid-actuated valve dispenses the mercury. The drop size is controlled by the time during which the valve is open, to have a reproducible area within the range of about 1-2.5 mm^2. Since the current is proportional to the electrode area, the larger the electrode, the more sensitive the determination. If classical DME operation is desired, a smaller-bore capillary and a pressurized system is required.

A drawback to this type of SMDE is a phenomenon known as the Barker effect. The large bore of the capillary allows some solution to migrate up the capillary between the mercury and the glass. This can cause interruption of the thread of mercury in the capillary. The effect is most severe in basic solutions and at very negative potentials. For the great majority of applications this type of SMDE (commercialized by EG&G Princeton Applied Research) has proven to be a reliable, useful electrode.

The second type of SMDE uses a more conventional bore capillary. Here the mercury is enclosed in a small reservoir (only a few milliliters, as opposed to several pounds in the electrode described above), and the pressure is developed by tank nitrogen gas. The electrode can function as a DME, an SMDE, or an HMDE. The drop area can be selected to be within the range 0.25-0.5 mm^2. The small-bore capillary is not prone to the Barker effect. This electrode assembly (manufactured by Metrohm and imported by Brinkman Instruments, Inc.) performs nicely in basic and nonaqueous media as well as in more conventional environments.

Either of the commercial SMDE assemblies allows the user to replace the mercury capillary with a variety of solid electrodes. In addition, the Metrohm unit has provision for a rotating disk electrode.

Recently Yarnitzky [22] has introduced an electrode assembly that combines the SMDE with a rapid deoxygenation step. His electrode allows a sample to be freed of oxygen and transferred into the cell in 30-45 sec. By comparison, work with a conventional mercury electrode requires the removal of oxygen by bubbling (sparging) with an inert gas for 5 min or more. This is necessary because molecular oxygen is itself reducible at the mercury electrode, hence interfering with the majority of analytical reductions. The Yarnitzky electrode lends itself well to the automation of analytical methods, especially when combined with an automatic sample changer, all under computer control [23].

III. TECHNIQUES

In the 1950s Barker [24] did much excellent work on polarographic methods. The drop knocker was available to allow precise control of drop times, and adequate electronics existed to permit the sampling of current after precisely defined intervals. Combining these two features allowed the introduction of new potential waveforms for excitation, and the measuring of current at the most favorable time in the life of the drop. Barker's work became the foundation of many of the presently used pulsed techniques, including normal pulse polarography (NPP), differential pulse polarography (DPP), and square-wave polarography and voltammetry (SWP and SWV).

The differences between these techniques are only in their excitation waveforms and their current sampling regimens. As in all voltammetry, the underlying phenomena do not change. The large body of literature available for classical polarography is still applicable to the more modern pulse techniques.

When considering purchase of voltammetric instrumentation, it is important to notice which of the many electroanalytical methods the equipment can support and how the techniques are implemented. Some commercial instruments use exclusively digitally generated waveforms. Other instruments may not implement a given technique in the fashion the user expects. In order to be able to decide what is best for a given application, the user must first be familiar with each of the techniques.

Normal pulse polarography is a more sensitive technique than classical DC polarography, but still generates a sigmoid-shaped polarogram. Figure 5 shows a typical excitation waveform for NPP. The potential pulse is applied to the electrode near the end of the drop lifetime. The pulse width is on the order of 70 msec, while the lifetime of the drop may be from 0.5 to 5 sec. The current is sampled once, near the end of each potential pulse. The advantage of the method is that the electrode stays at the low initial potential most of the time, thus ensuring that the layer of solution surrounding the electrode is not changed in concentration from the bulk of the solution at the instant when the reading is taken. The signal/noise ratio is improved over the classical method because the charging current has a chance to die out before the current is measured. NPP is particularly useful for solid electrodes. In this case, keeping the electrode at the low initial potential minimizes the chances of fouling the electrode with reaction products.

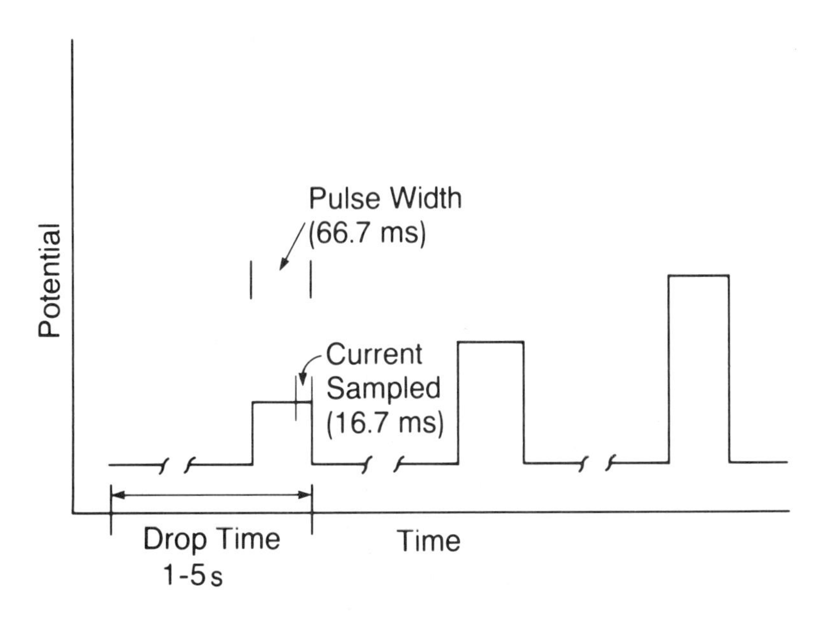

Figure 5. Waveforms for normal pulse polarography.

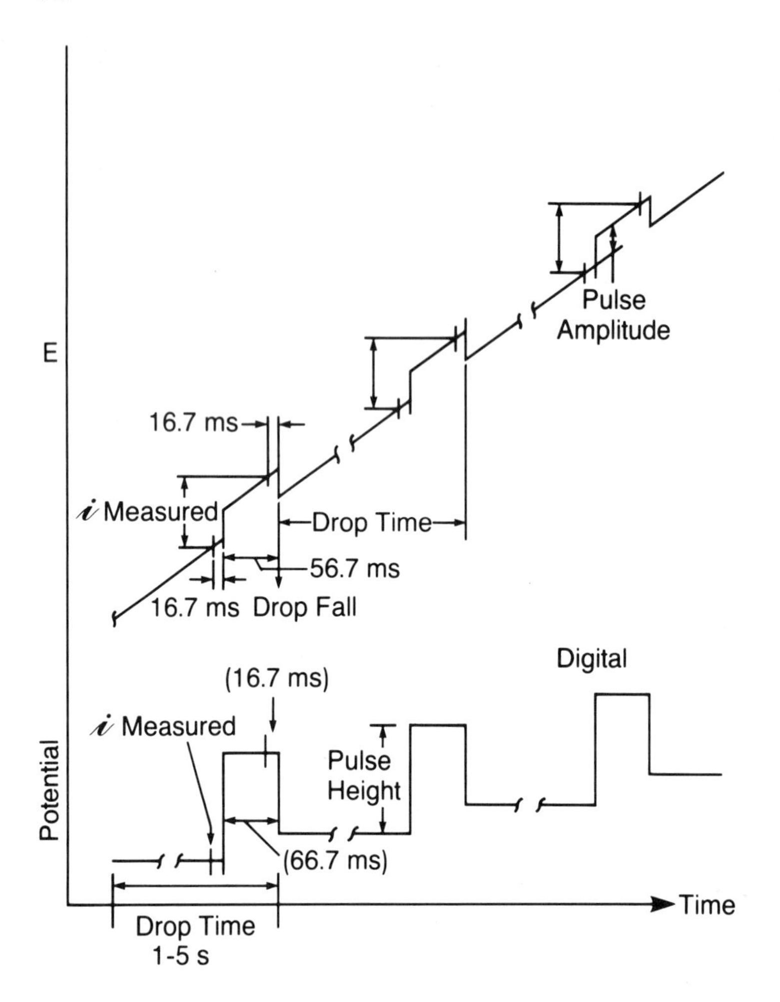

Figure 6. Waveforms for differential pulse polarography. The top curve is the analog version of the waveform; the bottom curve is the digital or "staircase" equivalent.

Differential pulse polarography has been the analytical method of choice for many years [25]. The waveform used is illustrated in Figure 6. The current is sampled twice, just before the start of the pulse, and again just before the end of the pulse. The first reading is subtracted from the second (giving rise to the term "differential"), and the value is plotted as the derivative, di/dt. The two different waveforms (analog and digital) produce similar looking data, but the digital waveform has some advantages. In the analog wave-form, the noise (capacitive current) is higher because of the sloping base line. In the digital waveform, the drop is held at constant potential for most of its lifetime. Since the capacitive current decreases with time, the first current reading is smaller than it would be if the potential continually increased with time.

In both of these pulsed techniques, the adjustable instrumental parameters are the scan rate, the drop time, and the pulse height (also called the modulation amplitude). In general, the more choices for these parameters, the more versatile the instrument. In practice drop times of 1 or 2 sec are adequate. Similarly, for good, well-resolved polarograms, scan rates of 2-5 mV/sec are sufficient.

The pulse height (in DPP) is a parameter for which many choices are important. The pulse height will affect sensitivity and resolution more than any other parameter. The rule of thumb is that a larger pulse height gives better sensitivity, but at the expense of resolution. A typical pulse height is 20 mV, with many instruments giving a range of 5-250 mV.

Square-wave polarography is the technique with the greatest variation in implementation. The two most important can be designated as Barker square wave and Osteryoung square wave. The methods differ in the details of the potential waveform and the sampling program. The frequency is also of importance.

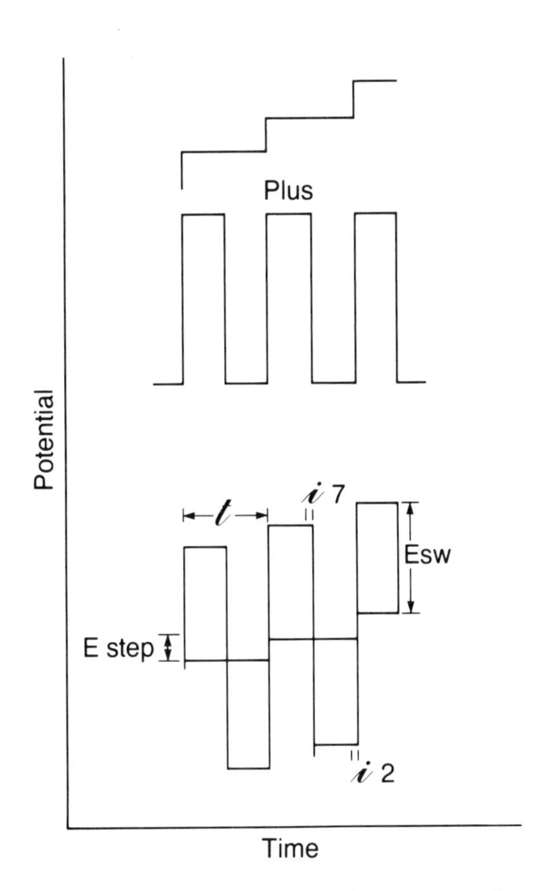

Figure 7. The digital staircase waveform (top curve) added to the square wave beneath it results in the lower curve, the waveform for Osteryoung square-wave voltammetry.

The Osteryoung square-wave technique merges a staircase potential waveform with a square-wave oscillation (Figure 7) [21]. The experiment is performed on a single hanging mercury drop. The current is sampled near the end of the forward pulse and again near the end of the reverse pulse. The method has two great advantages over DPP. The first advantage is speed: while DPP data is acquired at scan rates of 2-5 mV/sec, the effective scan rate in the square wave technique is 100 times faster. The second advantage is its sensitivity: SWV can be up to 10 times more sensitive than DPP. The sensitivity is controlled, to a first approximation, by the frequency, and so an instrument capable of a wide range of frequencies is desirable. For most analytical work, a frequency limitation to about 120 Hz is permissible, but for kinetic studies, a wider range is needed.

The original development of square-wave voltammetry was carried out by Barker, who worked exclusively with the DME. He used simply a square wave superimposed on an increasing potential ramp. This technique, while still offered on some commercial equipment, does not offer as great advantages as the Osteryoung modification.

Another variation of the Barker technique [26] is illustrated in Figure 8. Here, after a pulse, five cycles of a square-wave oscillation are applied to the electrode. The current is sampled at the forward and reverse pulses. The current for the last three cycles is averaged and reported as a single value. This technique typically has a fixed frequency of oscillation.

Cyclic voltammetry is a technique that can also be implemented with either an analog or a digital waveform. The important considerations in the digital version, called cyclic staircase voltammetry, are the size of the potential step and the time when the current is sampled [27]. In some instruments the sampling time is user adjustable. This can be useful in

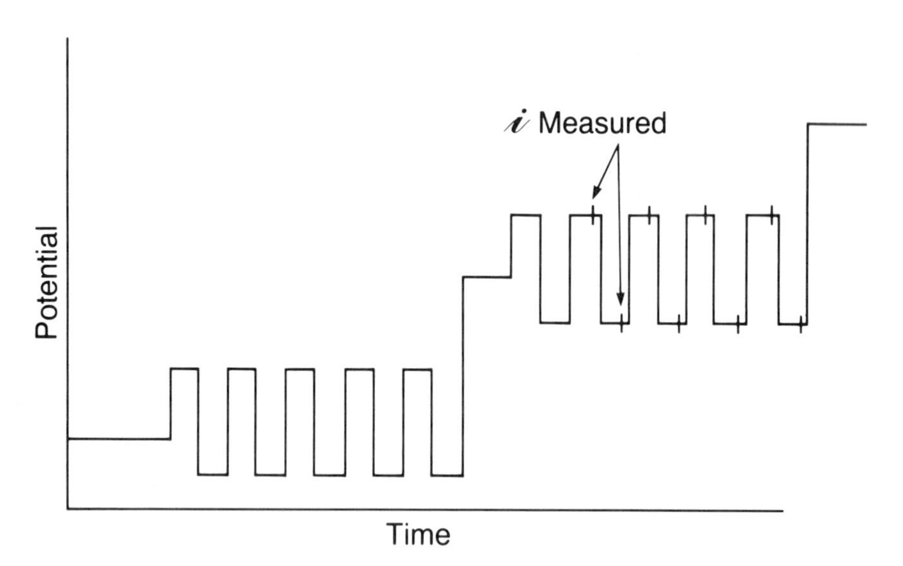

Figure 8. The waveform used by Metrohm in their square-wave technique. (Courtesy Metrohm, Ltd.)

studying reversible and irreversible systems [28]. Typical scan rates are 1 mV/sec to 1V/sec, but faster scans, up to 10 kV/sec, have been used for special purposes.

There are other waveform variations that give rise to corresponding techniques but that are seldom used in analytical work. Prominent among these is *AC polarography*, which uses a sine-wave voltage superimposed on the DC ramp. *Tast polarography* is a modification of the classical DC method in which the signal/noise ratio is improved by a sampling technique.

IV. THE USER INTERFACE

While analog instruments are still available, most modern instrumentation is digital in nature and under the control of either an internal microprocessor or an external computer. These may differ significantly in "friendliness" or ease of use. Instruments with a built-in microprocessor are capable of doing calculations, storing results, and recognizing peaks and other features of interest. The output available and its interaction with an external computer (a mainframe, local network, or personal computer) are also of importance.

There are two approaches toward creating a user interface: a complete keyboard and monitor can be included, or commands can be introduced through special-purpose keys only. Each approach has its advantages and disadvantages. The keyboard and monitor can be designed in several ways. The user may be required to remember specific commands, or the interface may be menu-driven or interactive, involving questions and answers.

The specific key approach has the advantage of allowing entry of commands in any sequence. The keys are labeled with abbreviations or mnemonics to assist the user. Techniques can be changed or parameters adjusted by pressing a key and entering a value. In many cases the keystrokes can be put together in a program to run specific techniques. The keys are all visible, and choosing the proper key is not difficult. The disadvantage of this approach is the physical limitation to the number of keys that can be placed on the front panel. If new techniques or new parameters are added, the instrument cannot be easily updated.

The menu-driven and interactive approaches allow the display of only those parameters that are needed in a particular application. For example, when running a DPP experiment, the menu would not show a choice of frequencies. This can make set-up easier for the unsophisticated user, who is not overwhelmed by numerous different buttons that do not apply to the technique he is using. Since most of the operating system is written in software or firmware, new revisions can add new menus or features with little difficulty. A disadvantage is that changing parameters can require multiple entries to reach the appropriate menu or question that needs to be changed.

V. DATA HANDLING

All commercial instruments are capable of acquiring the data from the voltammetric experiment. What it can do with that data is of more interest.

Some instruments are designed to draw calibration curves and to calculate results. If this is the case, one must note how the standards are to be run and the results calculated. For instance, one might ask whether there is a choice between a linear least-squares data fit and the more general polynomial fit. Is the instrument able to perform the calculations needed for calibration through the standard addition procedure? If a blank is run, can the instrument subtract that blank from both standards and sample? Is the instrument capable of storing the data, the calibration curve, and the pertinent parameter settings? All these questions are important to the user.

One of the differences in instrumentation is how peaks are recognized and their heights measured. The measurement can be made in one of three ways. The simplest is to measure the height of the peak above zero, but this does not always give correct answers because of the nonzero background level. The situation can be improved by means of an estimated baseline drawn by fitting a tangent to adjacent valleys on each side of the peak. This can still give erroneous results if the peaks are not well resolved. The third method assumes a symmetrical peak shape, but can be modified to take account of asymmetry due to overlapping peaks. Since not all voltammetric peak shapes follow clear mathematical equations, this method may also give rise to incorrect peak heights.

Another relevant consideration is in what form the analytical information is made available. Is there a hard copy output to a printer or plotter? The resolution in a printer is often limited by the resolution available on the monitor screen, whereas a digital plotter can have considerably greater resolution. If a printer is used, does it utilize commonly available paper? Thermal paper, in particular, may be expensive, and tends to lose contrast with time.

One should also inquire whether the data can be made available to an external computer. If so, the data format, the transfer rate (baud rate), and hand-shaking information must be available. If on-stream process control applications are planned, it is important that all features of the instrument can be controlled by a remote computer.

The ultimate choice of a particular instrument may be difficult. The nature of the measurements to be made dictates the requirements of the instrumentation. One can safely assume that any commercial instrument will be able to perform the great majority of analytical applications. The purchaser should pay particular attention to special features, ease of use, and reliability. It is always difficult to assess instruments solely by studying the specification sheets. The potential user should operate the instrument and try running some typical experiments. Only then can he make an intelligent choice of which instrument will best fit his needs.

VI. APPLICATIONS

The following references will provide entrance to the applications literature. First are listed references to metal determination, then to organics.

Cr^{VI}: C. Locatelli and F. Fagioli, Mikrochim. Acta *3*, 269 (1987).

Cd, Cu, Pb: J. H. Opperman et al., S. Afr. J. Chem. *41* 26 (1988).

Cd, Cu, Pb, Zn: M. Lugowska et al., J. Electroanal. Chem. Interfacial Electrochem. *226*, 263 (1987).
Pb: N. Ivicic and M. Blanusa, Fresenius Z. Anal.Chem. *330*, 643 (1988).
Pt: P. Shearan and M. R. Smyth, Analyst (Lond.) *113*, 609 (1988).
G. M. Schmid and D. R.Atherton, Anal. Chem. *58*, 1956 (1986).
Se: P. Breyer and B. P. Gilbert, Anal. Chim. Acta *201*, 23 (1987).
Sn: W. Holak and J. J. Specchio, J. Assoc. Off. Anal. Chemists *71*, 857-895 (1988).
Metals: P. Hoffmann and K. H. Lieser, Sci. Total Environ. *64*, 1 (1987).
Acetaldehyde: R. M. Ianniello, J. Assoc. Off. Anal. Chemists *70*, 566 (1987).
Anilines: N. E. Zoulis and C. E. Efstathiou, Anal. Chim. Acta *204*, 201 (1988).
Biologicals: W. F. Smyth, CRC Crit. Rev. Anal. Chem. *18*, 155 (1987).
Bromazepam: J. L. Valdeon et al., Analyst (Lond.) *112*, 1365 (1987).
Caffeine: A. Bazzi et al., Analyst (Lond.) *113*, 121 (1988).
Carcinogens: J. Barek et al., Collect. Chem. Commun. *51*, 2083 (1986); *52*, 867 (1987); *52*, 2149 (1987).
Cephalothin: D. Peled et al., Analyst (Lond.) *112*, 959 (1987).
Chlorpromazine: N. Zimova et al., Talanta *33*, 467 (1986).
Daminozide: R. M. Ianniello, Anal. Chim. Acta *193*, 81 (1987).
1,2-Dibromoethane: R. Tokoro et al., Anal. Chem. *58*, 1964 (1986).
Gallic Acid: S. Azhar Ali et al., J. Nat. Sci. Math. *28*, 187 (1988).
Indole: J. M. Pingarron Carrazon et al., J. Electroanal. Chem. Interfacial Electrochem. *234*, 175 (1987).
Methyl salicylate: O. W. Lau et al., Analyst (Lond.) *113*, 865 (1988).
Nitrobenzene: E. Lorenzo et al., Fresenius Z. Anal. Chem. *330*, 139 (1988).
Pentachlorophenol: E. C. Guijarro et al., Analyst (Lond.) *113*, 625 (1988).
Saframycins: P. M. Bersier and H. B. Jenny, Analyst (Lond.) *113*, 721 (1988).
Sulfadiazine: J. M. Pingarron Carrazon et al., Electrochim. Acta *32*, 1573 (1987).
Theaflavin: A. R. Fernando and J. A. Plambeck, Analyst (Lond.) *113*, 479 (1988).

REFERENCES

1. R. N. Adams, *Electrochemistry at Solid Electrodes*, Marcel Dekker, New York, 1969.
2. A. Bard, Ed., *Electroanalytical Chemistry*, Vol. 14, Marcel Dekker, New York, 1986.
3. A. Bard, Ed., *Electroanalytical Chemistry*, Vol. 1, Marcel Dekker, New York, 1966.
4. A. Bard and L. R. Faulkner, *Electrochemical Methods*, Wiley, New York, 1980.
5. Anonymous, *Bibliography of Polarographic Literature, 1922 to 1967*, Sargent-Welch Scientific Company, Skokie, IL, 1969.
6. A. M. Bond, *Modern Polarographic Methods in Analytical Chemistry*, Marcel Dekker, New York, 1980.

7. J. Heyrovsky and P. Zuman, *Practical Polarography*, Academic Press, New York, 1968.
8. P. T. Kissinger and W. R. Heineman, *Laboratory Techniques in Electroanalytical Chemistry*, Marcel Dekker, New York, 1984.
9. I. M. Kolthoff and J. L. Lingane, *Polarography*, Wiley-Interscience, New York, 1952.
10. L. Meites, *Polarographic Techniques*, 2nd ed., Wiley-Interscience, New York, 1965.
11. H. W. Nurnberg, Ed., *Electroanalytical Chemistry*, Wiley-Interscience, New York, 1974.
12. H. Siegerman, in *Techniques of Electro-Organic Synthesis* (N. Weinken, Ed.), Wiley, New York, 1975.
13. P. Zuman, *Organic Polarographic Analysis*, Macmillan, New York, 1964.
14. P. Zuman, *Topics in Organic Polarography*, Plenum Press, New York, 1970.
15. P. Zuman, *Substituent Effects in Organic Polarography*, Plenum Press, New York, 1967.
16. K. Z. Brainina, *Stripping Voltammetry in Chemical Analysis*, Wiley, New York, 1976.
17. F. Vydra, K. Stulik, and E. Julakova, *Electrochemical Stripping Analysis*, Wiley, New York, 1976.
18. J. Wang, *Stripping Analysis*, VCH Publishers, New York, 1985.
19. J. Heyrovsky, Trans. Faraday Soc. *19*, 785 (1924).
20. W. M. Peterson, Am. Lab. *11*, 69 (1979).
21. J. A. Turner, J. H. Christie, M. Vukovic, and R. A. Osteryoung, Anal. Chem. *49*, 1904 (1977).
22. C. Yarnitzky and E. Ouziel, Anal. Chem. *48*, 2024 (1976).
23. P. D. Carpe, R. M. Jackson, and A. F. Palus, Am. Lab. *19*, 73 (1987).
24. G. C. Barker and A. W. Gardner, Z. Anal. Chem. *173*, 79 (1960).
25. J. B. Flato, Anal. Chem. *44* (11), 75A (1972).
26. Metrohm, Ltd., product literature on the Metrohm 646VA Processor.
27. J. J. Zipper and S. P. Perone, Anal. Chem. *45*, 452 (1973).
28. R. Bilewicz, R. A. Osteryoung, and J. Osteryoung, Anal. Chem. *58*, 2761 (1986).

18

Instrumentation for Stripping Analysis

JOSEPH WANG / Department of Chemistry, New Mexico State University, Las Cruces, New Mexico

I. INTRODUCTION

In recent years stripping analysis has become one of the most widely used methods for trace metal analysis. Interest in stripping analysis has been sparked by its ability to measure simultaneously four to six trace metals at concentration levels down to 10^{-11} *M*, utilizing relatively inexpensive instrumentation. The major advantage of the method, compared to direct voltammetric analysis of the test solution, is the preconcentration of the analyte onto the working electrode (by factors of 100 to more than 1000). Because of the "built-in" preconcentration step, extremely high sensitivity can be obtained. Besides its inherent sensitivity, the technique offers a wide linear range, and is uniquely suited to the direct study of trace metal speciation in natural waters or to in situ field measurements. It is fair to say that there is no technique for trace metal analysis that can compete with stripping analysis on the basis of sensitivity per dollar investment. Hence, the technique is finding an increasing use for monitoring trace metals in environmental, clinical, food, or pharmaceutical samples. For example, stripping analysis appears to be the best analytical tool for the direct, and simultaneous, determination of four metals of prime environmental concern (lead, cadmium, zinc, and copper) in seawater [1]. Other metals conveniently measured by stripping analysis include bismuth, indium, tin, thallium, gallium, mercury, silver, selenium, manganese, arsenic, and gold. New stripping strategies, particularly the metal-chelate adsorption approach, permit quantitation of additional trace metals, including uranium, nickel, vanadium, cobalt, titanium, thorium, iron, and aluminum. Besides trace elements, low levels of various organic compounds of clinical or environmental interest can be detected using the adsorption approach or cathodic stripping voltammetry.

This chapter will briefly present the fundamentals of various stripping procedures. Then the instrumentation requirements, including the selection of the voltammetric analyzer, working electrode, cell and other variables, will be discussed. The literature of stripping analysis is reviewed briefly to help potential users in understanding and choosing among several variants of the technique. A recent monograph [2] and review articles [3-5] give a comprehensive outline of modern stripping analysis, embracing new strategies, associated problems, and numerous applications.

II. PRINCIPLES

Stripping analysis is a two-step electroanalytical technique (Figure 1). The first step consists of an electrolytic deposition of the metal onto the working electrode. In the most common versions of stripping analysis, anodic stripping voltammetry (ASV) and potentiometric stripping analysis (PSA), this preconcentration step involves reduction of a metal ion to the metal, which then dissolves in the mercury electrode:

$$M^{n+} + ne^- + Hg \rightarrow M(Hg) \quad (1)$$

For this purpose, a deposition potential E_d is applied to the working electrode; this potential should be in the limiting current region of the metal ion(s) being deposited. Under these conditions, the deposition step is considered as being controlled by mass transport, and hence can be facilitated by convective transport of the metal ion(s) to the electrode surface. This is usually accomplished by stirring or flowing the solution, or rotating the working electrode. The concentration of the reduced metal in the mercury, C_{Hg}, is given by Faraday's law:

$$C_{Hg} = \frac{i_L t_d}{nFV_{Hg}} \quad (2)$$

where i_L is the limiting current for the deposition of the metal, t_d the length of the deposition period, and V_{Hg} the volume of the mercury electrode.

Preconcentration is followed by a stripping (dissolution) step—the measurement step—in which the metal is oxidized electrolytically (or chemically in PSA) from the working electrode. In ASV the stripping step consists of scanning the potential in the positive (anodic) direction, linearly, or in other potential-time waveform. When the potential reaches the standard potential of a metal-metal ion couple, that particular metal is reoxidized back into solution and a current is flowing:

$$M(Hg) \rightarrow M^{n+} + ne^- + Hg \quad (3)$$

The current-potential voltammogram, recorded during the stripping step (shown also in Figure 1), provides the desired analytical information. The peak potential (position), E_p, of each metal is a characteristic of that metal and is related to the standard potential of its redox couple. Thus, it can be used for qualitative identification. The peak current (height), i_p, is proportional to the concentration of the corresponding metal ion in the test solution. The concentration is determined by a standard addition or a calibration curve. Because only a small fraction of the metal ions is deposited, it is essential that all experimental parameters (deposition time, convection rate, etc.) be reproducible as closely as possible during a series of measurements.

Most stripping measurements require an addition of appropriate supporting electrolyte and removal of dissolved oxygen. The former is needed to decrease the resistance of the solution and to ensure that the metal ions of interest are transported toward the electrode by diffusion and not by electrical migration. Contamination of the sample by metal impurities in the reagents used for preparing the supporting electrolyte is a serious problem. Dissolved oxygen severely hampers the quantitation (except in

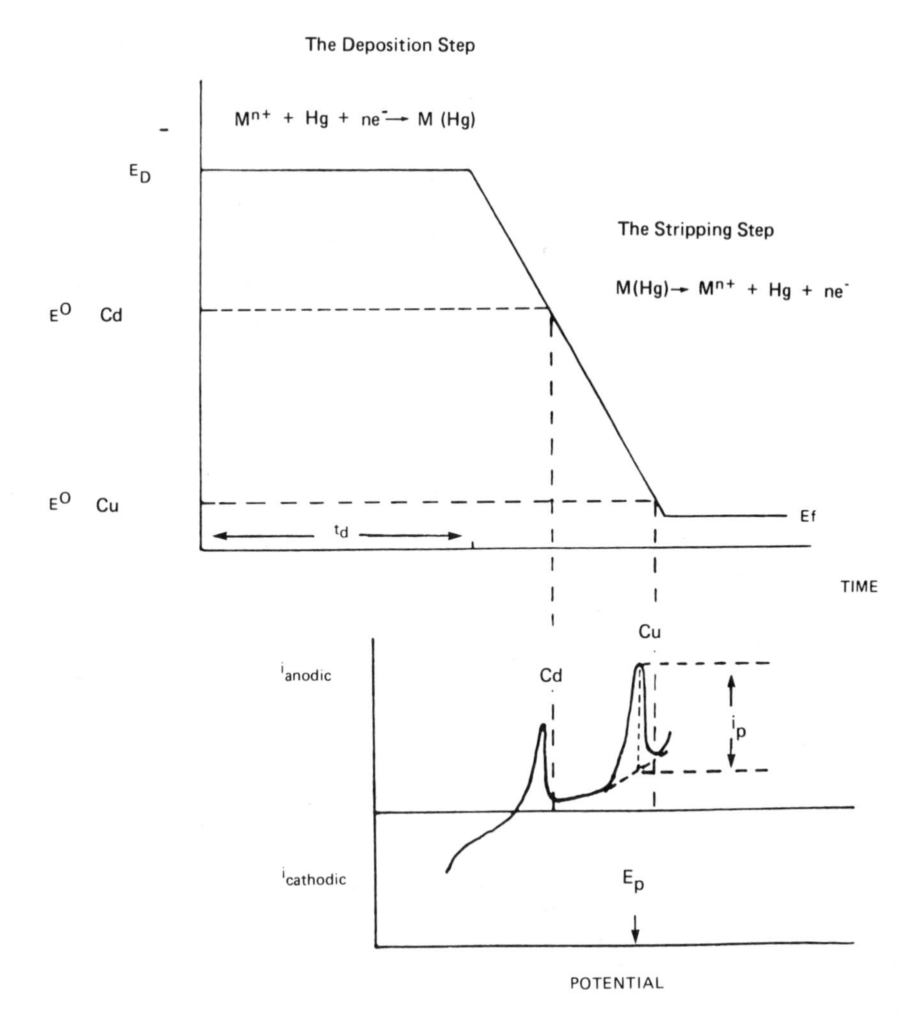

Figure 1. The potential-time sequence used in ASV along with the resulting stripping voltammogram.

PSA) and must be removed, usually by purging the sample with high-purity nitrogen.

The main types of interferences in stripping analysis are overlapping stripping signals, the adsorption of organic surfactants on the electrode surface, and the formation of intermetallic compounds by metals (e.g., copper and zinc) codeposited in the mercury electrode. Overlapping signals cause problems in the simultaneous determination of metals with similar redox potentials (e.g., lead and tin). Intermetallic-compound formation and surfactant adsorption cause a depression of the stripping response; shifts of the signal may also be observed. Various chemical, instrumental, or mathematical strategies, available for the elimination or minimization of these interfering effects, have been described [2].

III. STRIPPING STRATEGIES

A. Voltammetric Stripping Techniques

A variety of potential-time waveforms has been used during the anodic scan to oxidize the metals out of the electrode. Theories for the stripping peaks produced at the mercury film and hanging mercury drop electrodes using such waveforms are available [2].

The instrumental simplicity and speed of the linear scan dc stripping mode has resulted in its use for numerous stripping studies (particularly those aimed at the 10^{-6}-10^{-8} M level of the analyte). Unfortunately, the continuous change in potential generates a relatively large charging-background current. Nevertheless, with the commonly used mercury film electrode, the linear scan is only slightly (three- to fivefold) less sensitive than the more complex and slower differential pulse waveform [6]. Linear scan rates of typically 50-100 mV/sec are employed and offer the desired compromise between the relative magnitudes of the analytical and charging currents, resolution of adjacent peaks, and speed.

Other potential-time waveforms used in the stripping process are designed to discriminate against the charging current. Hence, increased signal-to-background characteristics are observed, particularly when using the hanging mercury drop electrode. At present, the most commonly used modulation waveforms in stripping analysis are differential pulse and square wave (Figure 2). The deposition step for these is the same as that employed in linear-scan stripping voltammetry (and most other stripping strategies).

The commercial availability of reliable pulse polarographs of modest price resulted in a plethora of applications of the differential pulse waveform for ultratrace stripping measurements. Another feature of the differential pulse waveform (besides the effective compensation of the charging current) is that some of the material stripping from electrode during the pulse is redeposited into the electrode during the waiting period between pulses. Therefore, the same material is "seen" many times, resulting in enhanced sensitivity. A disadvantage of differential pulse anodic stripping voltammetry is that the potential scan must be quite slow (5 mV/sec), hence limiting the analysis rate. Pulse amplitudes of 25 or 50 mV are commonly employed. An increase in the scan rate or pulse amplitude will result in peak broadening and impaired selectivity. Valenta et al. [7] demonstrated that minor changes in the timing circuitry of commercial instruments are required to achieve optimum performance. In particular, by shortening the pulse duration from 56 to 29 msec, the replating effect is much more pronounced.

The square-wave waveform has the advantage over the differential pulse mode that much faster scan rates (up to 1 V/sec) can be used. Hence, a complete stripping voltammogram can be obtained in about 1 sec. The analytical time can be further reduced, as removal of oxygen is not necessary to obtain analytically useful stripping voltammograms [8]. Use of the rapid square-wave waveform in stripping analysis is likely to increase substantially in the near future, as more commercial instruments containing this waveform become available.

In addition to the differential pulse and square-wave waveforms, other voltammetric modes capable of minimizing various background-current contributions during the stripping step have been attempted [2]. These include

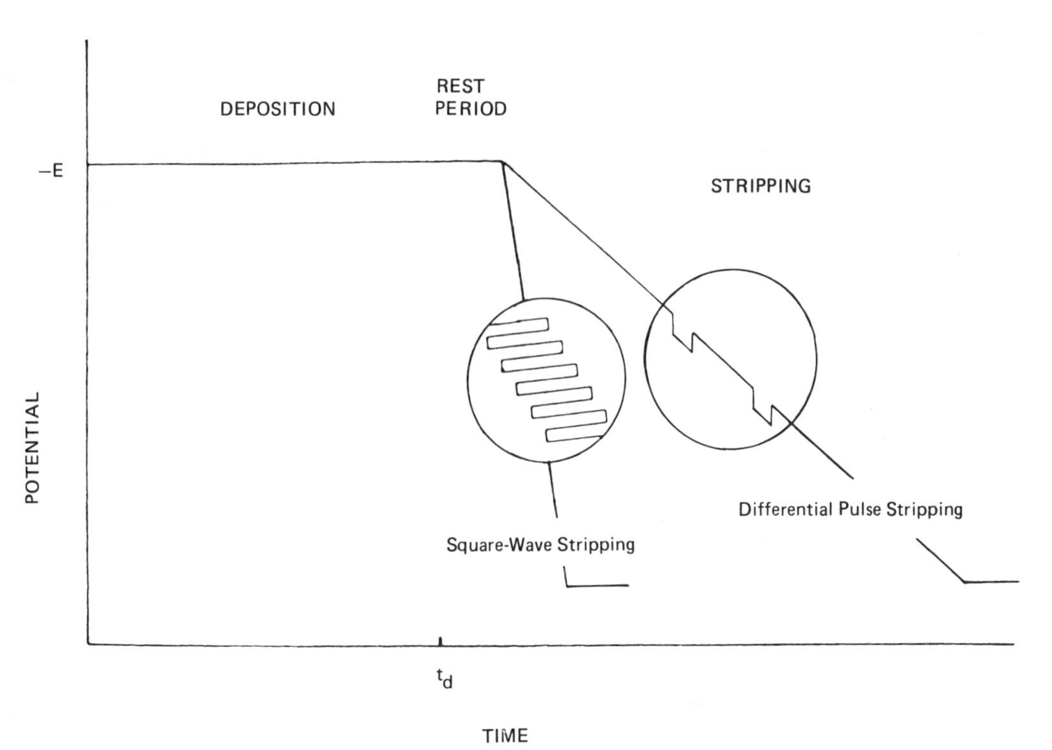

Figure 2. Waveforms for differential-pulse and square-wave stripping experiments.

modulation waveforms, such as alternating current or staircase, as well as twin-electrode strategies like subtractive stripping voltammetry or anodic stripping with collection. A recent study [9] critically compared the performance of these stripping procedures. While these offer detection limits comparable with those obtained using the differential pulse mode, their utility in stripping analysis has been hampered by the complexity or unavailability of commercial instruments.

Another promising stripping strategy, aimed at enhancing the stripping response, is the multiple-scanning approach [10]. Here the deposition potential is reapplied at the end of a fast linear scan; after a 1-sec delay a new scan is initiated. Because of the short stripping times, significant fraction of the oxidized metal ions can be redeposited. By integrating 20 such repetitive scans, significant enhancement of the signal is observed. Presently, commercial multiscan instruments are unavailable.

B. Potentiometric Stripping Analysis

Potentiometric stripping analysis (PSA) has attracted increasing attention during the last decade as an effective alternative to stripping voltammetry. PSA resembles stripping voltammetry in that preconcentration of the analyte is accomplished by electrodeposition on a mercury-coated glassy-carbon electrode; however, it differs from the voltammetric variant in that the metal amalgam is subsequently oxidized ("stripped") chemically, rather than electrochemically. For this purpose, the potentiostatic circuitry is disconnected after preconcentration, permitting amalgamated metals to be oxidized according to

$$M(Hg) + \text{oxidant} \rightarrow M^{+n} \qquad (4)$$

Typical oxidants used for the reoxidation are O_2, Hg(II), and Cr(VI). During the oxidation process, the variation of potential of the working electrode is recorded as a function of time, and a stripping curve, like the one shown in Figure 3, is obtained. A sudden change in potential occurs when all the metal deposited in the electrode has been oxidized. The transition time needed for the oxidation of a given metal, t_M, is a quantitative measure of the metal ion concentration in the sample:

$$t_M \propto ([M^{+n}]t_{dep}) \qquad (5)$$

As predicted by the Nernst equation, the potential at which the reoxidation takes place serves for qualitative identification of the different metals.

While the sensitivity of PSA approaches that of stripping voltammetry, it requires simpler instrumentation and is less prone to interferences (e.g., dissolved oxygen or other electroactive solutes, organic surfactants). Indeed, the use of non-deaerated samples is the main reason for the growing interest in this approach. Hence, PSA has proven useful for the rapid and accurate determination of several trace metals in a wide variety of media and materials. These applications, as well as the theory of PSA, have been reviewed by Jagner [12].

C. Cathodic Stripping Voltammetry

Cathodic stripping voltammetry (CSV) is the "mirror image" of anodic stripping voltammetry. It involves anodic deposition of the analyte, followed by stripping in a negative-going potential scan. Quantification is accomplished by measuring the height of the resulting reduction peak. Unlike anodic stripping voltammetry, the deposited analyte forms an insoluble film of the mercury salt on the mercury surface:

$$A^{n-} + Hg \underset{\text{stripping}}{\overset{\text{deposition}}{\rightleftarrows}} HgA + ne^- \qquad (6)$$

As a result, it is not unusual for calibration curves to display nonlinearity at high concentrations. A wide range of inorganic and organic analytes, capable of forming insoluble salts with mercury, has been measured by CSV. Among these are ions such as halide, cyanide, sulfide, or selenide, as well as various thiols, peptides, or penicillins. Most of the reported applications of CSV have been carried out at hanging mercury drop or mercury pool electrodes. A rotating silver disk electrode offers various advantages for the determination of anions that form insoluble silver salts. In this case, the preconcentration and stripping steps involve the following reaction:

$$Ag + X^- \rightleftarrows AgX + e^- \qquad X^- = Cl^-, Br^- \qquad (7)$$

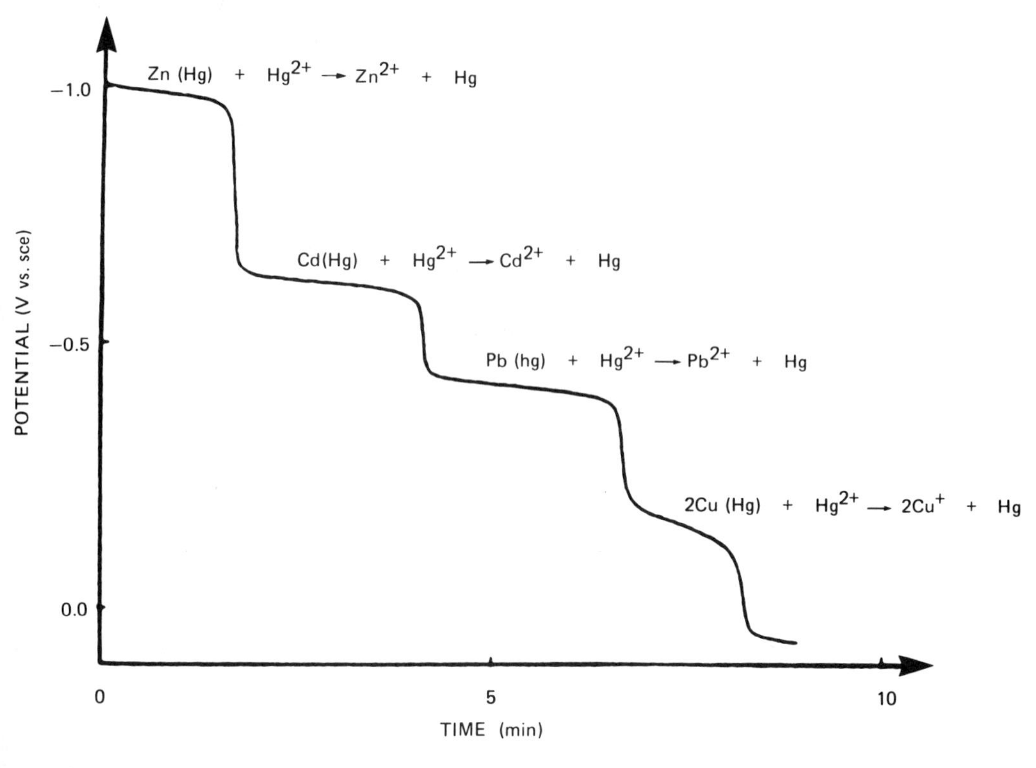

Figure 3. Potentiometric stripping analysis of a solution containing 1.5×10^{-6} *M* of Zn(II), Cd(II), Pb(II), and Cu(II). Three-minute deposition at -1.25 V. (From Ref. 11, with permission.)

Copper-based mercury electrodes may also be advantageous in some situations. Other versions of the cathodic approach, based on potentiometric stripping or flow analysis, have been reported.

D. Adsorptive Stripping Voltammetry

Adsorptive stripping voltammetry is a realtively new approach aimed at extending the scope of stripping analysis toward numerous analytes. It is similar to conventional stripping analysis in that the analyte is preconcentrated onto the working electrode prior to its voltammetric measurement, but it differs from the conventional scheme in the way preconcentration is accomplished. Rather than using electrolytic deposition, adsorptive stripping voltammetry utilizes controlled adsorptive accumulation for preconcentration. Hence, for a wide range of surface-active organic and inorganic species—that cannot be preconcentrated electrolytically—the adsorption approach serves as an effective alternative. The analyte is allowed to interfacially accumulate on the electrode surface for a specified length of time, under conditions of maximum adsorption (electrolyte, pH, potential, mass-transport, etc.); then the adsorbed material is determined by applying a negative- or positive-going potential scan (for reducible or oxidizable species, respectively). The adsorptive stripping response reflects the corresponding adsorption isotherm, as the amount of analyte adsorbed is proportional to the bulk concentration of the analyte. Hence, a linear response is expected, based on the Langmuir adsorption isotherm, provided that full coverage of the electrode area is avoided.

Typically the hanging mercury drop electrode is used for measuring reducible species, while carbon paste electrodes are used for oxidizable ones. The other instrumental requirements (voltammetric analyzer, cell, etc.) are as those used in conventional stripping analysis. With relatively short preconcentration times, very low concentrations (10^{-9}-10^{-11} *M*) can be determined. The method has been applied for measuring trace levels of compounds of biological and pharmaceutical significance, such as riboflavin [13], digoxin [14], codeine [15], bilirubin [16], diazepam [17], or adriamycin [18]. Typical voltammograms for solutions of increasing levels of the anticancer agent mitomycin C are shown in Figure 4.

Adsorptive collection of metals as their surface-active complexes on the hanging mercury drop electrode has been shown to be extremely useful for ultratrace measurements of metals, such as Ti, Al, Ni, Mo, V, U, Pd, Pt, Fe, or Th, that cannot be conveniently measured by conventional stripping analysis. This strategy offers also effective alternative procedures for metals, for example, Cu, Ga, Mn, or Sn, that are measurable with certain difficulty by conventional ASV. The formation of an appropriate surface-active complex of the metal, based on adding a suitable chelator (dimethylglyoxime for Ni, catechol for V, or solochrome violet RS for Ti), precedes the preconcentration step. The adsorbed complexes are then reduced by scanning the potential in the negative direction. In addition to voltammetric quantification of the surface-bound chelate, a potentiometric approach based on applying a constant reducing current and monitoring the potential-time behavior has been suggested recently [20]. Overall, reliable trace metal analysis can be performed using samples of environmental and biological origin. Because of the nature of the preconcentration step, interfering surfactants (that complete on adsorption sites) should be destroyed. For more details of adsorptive stripping voltammetry the reader is referred to a recent review [21].

E. Stripping Analysis of Flowing Streams

Highly sensitive and versatile automated methods for stripping analysis can be developed by coupling with flow systems, particularly those based on flow injection. The interest in stripping analysis/flow systems has increased considerably in recent years [22,23]. Besides the obvious advantages of on-line monitoring and possible automation, such coupling offers several advantages over stripping analysis performed in batch (beaker) systems. These include reduced analysis time, analysis of small-volume (100-500 μl) samples, improved reproducibility, and minimization of errors (due to contamination or adsorption losses). Flow systems offer also a simple and convenient way to perform the medium-exchange procedure, where the sample solution—after preconcentration—is replaced with a more

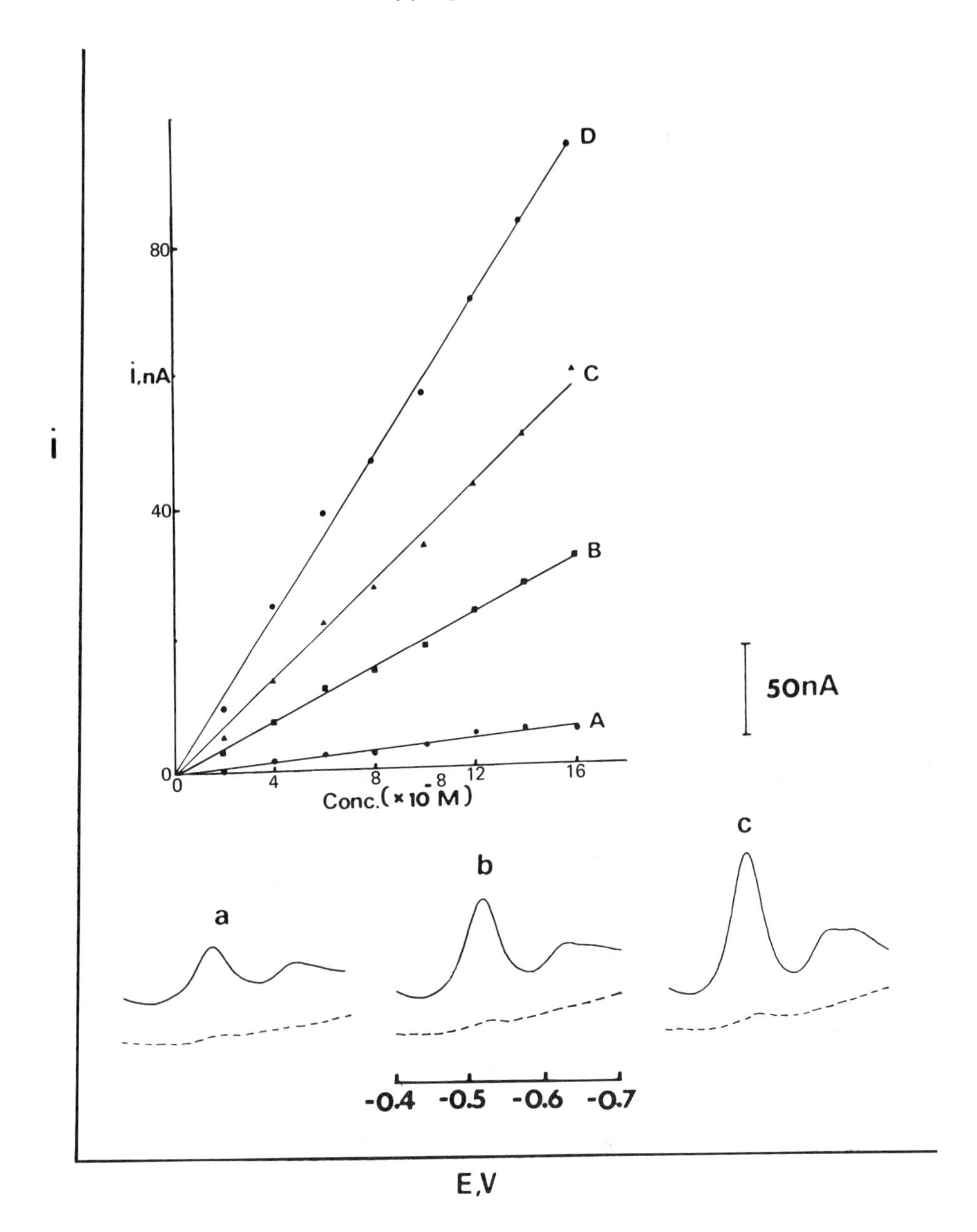

Figure 4. Adsorptive stripping for mitomycin C solutions of increasing concentration: (a) 4×10^{-8} M, (b) 8×10^{-8} *M* and (c) 1.2×10^{-7} *M*. Two-minute preconcentration. The dotted line represents the response without accumulation. Also shown are calibration plots after 0 (A), 30 (B), 60 (C), and 120 (D) preconcentration. (From Ref. 19, with permission.)

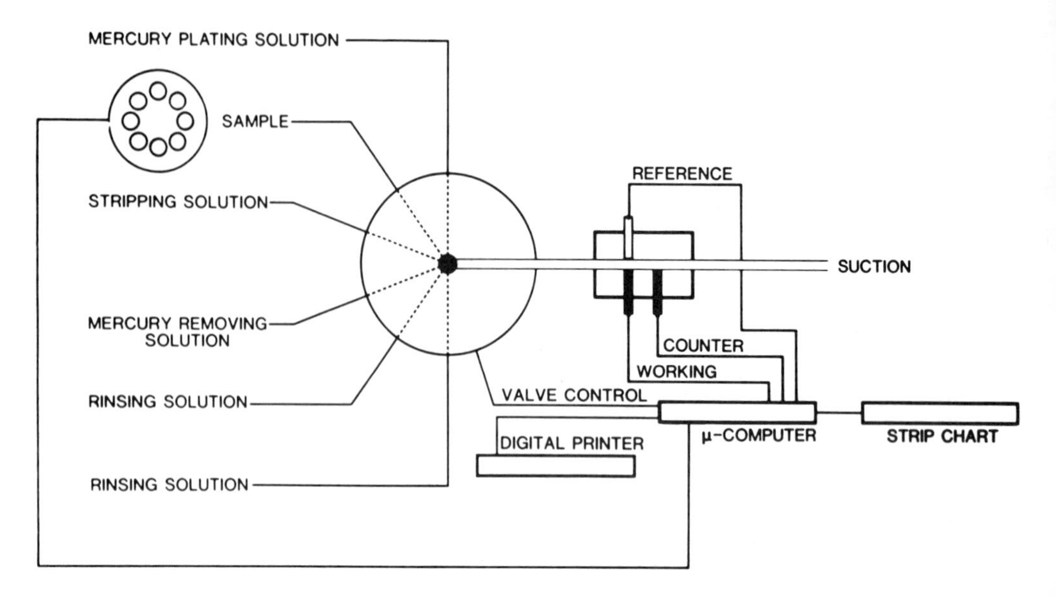

Figure 5. Schematic diagram of an automated flow system for potentiometric stripping analysis. (From Ref. 23, with permission.)

ideal supporting electrolyte in which stripping is carried out. The manifold of flow injection systems is particularly useful for this purpose; after the sample plug passes through the detector, stripping can occur in the carrier environment. As a result, stripping analysis can be substantially improved with respect to resolution between overlapping peaks or sample deoxygenation. Improved sensitivity can be obtained using subtractive stripping measurements in which the carrier (blank) response is subtracted from that of the sample plug [24]. Figure 5 shows a schematic of a computerized flow system for automated potentiometric stripping analysis possessing several of the above advantages. Besides flow-injection/anodic-stripping systems, automated flow systems have been applied to shipboard stripping analysis of trace metals [25]. Such system is controlled by a specially designed programmer, which controls the potentiostat, pumps, valves, and recorder. A variety of commercial and homemade flow cells, suitable for on-line stripping analysis, is described in Section IV. Specially designed cells, such as dual coulometric-voltammetric cells, can offer additional advantages, such as reduced intermetallic or oxygen interferences [26].

Table 1 lists detection limits reported for the various stripping strategies discussed in previous sections. It is emphasized that these limits can be obtained by operators with expertise in stripping analysis and ultratrace chemistry. In particular, the ability to obtain such low values strongly depends on the degree to which contamination can be minimized. Clearly, stripping analysis offers extremely low detection limits that compare favorably with those obtained at other (nonelectrochemical) "trace" techniques.

Table 1. Relative Sensitivity of Various Stripping Strategies

Stripping technique	Limit of detection (*M*)	Analyte	Working electrode	Ref.
Differential pulse	2×10^{-10}	Lead	Mercury drop	27
Square wave	1×10^{-10}	Lead	Mercury drop	27
Linear scan	5×10^{-11}	Lead	Mercury film	27
Differential pulse	1×10^{-11}	Lead	Mercury film	27
Square wave	5×10^{-12}	Lead	Mercury film	27
PSA	1.5×10^{-10}	Lead	Mercury film	28
Chelate adsorption	1×10^{-10}	Nickel	Mercury drop	29
Analyte adsorption	2.5×10^{-11}	Riboflavin	Mercury drop	13
CSV	2×10^{-10}	Penicillin	Mercury drop	30

Table 2. Representative Applications of Stripping Analysis

Analyte measured and matrix	Stripping strategy	Working electrode	Ref.
Pb, Cd, Se, Co, Ni, and Mn in rain water	Differential pulse, chelate adsorption, CSV	Mercury drop	31
Bi, Cd, Cu, Pb, Sb, and Zn in seawater and marine samples	Linear scan	Mercury film	32
Cu and Zn in pharmaceutical tablets	Differential pulse	Mercury film	33
Ni in nail	Chelate adsorption	Mercury drop	34
Cd, Pb, and Tl in fly-ash	PSA	Mercury drop	35
Sb, Cu, and Pb in gunshot residue	Linear scan	Mercury film	36
Pb and Sn in fruit juices	PSA	Mercury film	37
Cd, Cu, and Pb in urine	Differential pulse	Mercury drop	38
Chlorpromazine in urine	Analyte adsorption	Carbon paste	39
As in blood and urine	Staircase	Mercury film	40
Thioamide drugs in plasma and urine	CSV	Mercury drop	41

The high sensitivity of stripping analysis has led to its application to a large number of analytical problems. The representative applications given in Table 2 can be used to indicate the general utility of the technique.

IV. INSTRUMENTATION

A. Cells

Many designs of cells for stripping measurements have been reported in the literature or made available commercially. In most cases, the performance requirements for the cell are minimal, so that it may be designed for convenience of use (cleaning, purging, stirring, size, etc.). Conventional voltammetric cells, that is, covered beakers (10-100 ml volume), are commonly employed. The specific shape depends primarily on the working electrode used. For stripping work, aimed at measurements at the 10^{-7}-10^{-9} *M* level, precleaned glass or quartz cells can be used. For quantitation of lower levels (10^{-10}-10^{-11} *M*) Teflon cells are desired, because of possible release of heavy metals from glass or quartz cells. The cell contains the three electrodes (working, reference, auxiliary), which are immersed in the sample solution. These electrodes, as well as a tube for the purging gas, are supported in four holes in the cover. The cover is usually made of Teflon or Plexiglas. Most manufacturers of voltammetric analyzers make electrochemical cells suitable for stripping measurements. Other special-purpose cell configurations (micro, flow) are also available for work in small volumes or on-line measurements. These include miniature cells with stationary [42] or rotating [43] mercury film electrodes, or cells in which the solution flows through a thin-layer channel [44], onto a stationary disk in a wall-jet design [26], or through an open-tubular electrode [45]. Specially designed cells, aimed at minimizing or eliminating the time for oxygen removal, including a rotating cell assembly, a large-volume wall-jet cell, or a multicell system (with separated deaeration and measurement racks), have also been described [46-48].

B. Electrodes

Unlike polarography, the dropping mercury drop electrode is not used in stripping analysis. The working electrode must be stationary and have a favorable redox behavior of the analyte, reproducible area, and low background current over a wide potential range. The most widely used form of stripping analysis involves a micromercury electrode; by reducing the volume of the mercury, one can enhance the concentration of deposited metals [equation (2)]. Most commonly used electrodes that fulfill the above requirements are the mercury film electrode (MFE) and the hanging mercury drop electrode (HMDE). Thin mercury film electrodes offer improved sensitivity (larger area-to-volume ratio) and excellent resolution of neighboring peaks compared to the HMDE (Figure 6). The latter offers reproducible renewal of the surface, reduced intermetallic interferences, and high hydrogen overvoltage. The HMDE is preferred for adsorptive stripping and cathodic stripping work, while the MFE is preferred in potentiometric stripping analysis and field work.

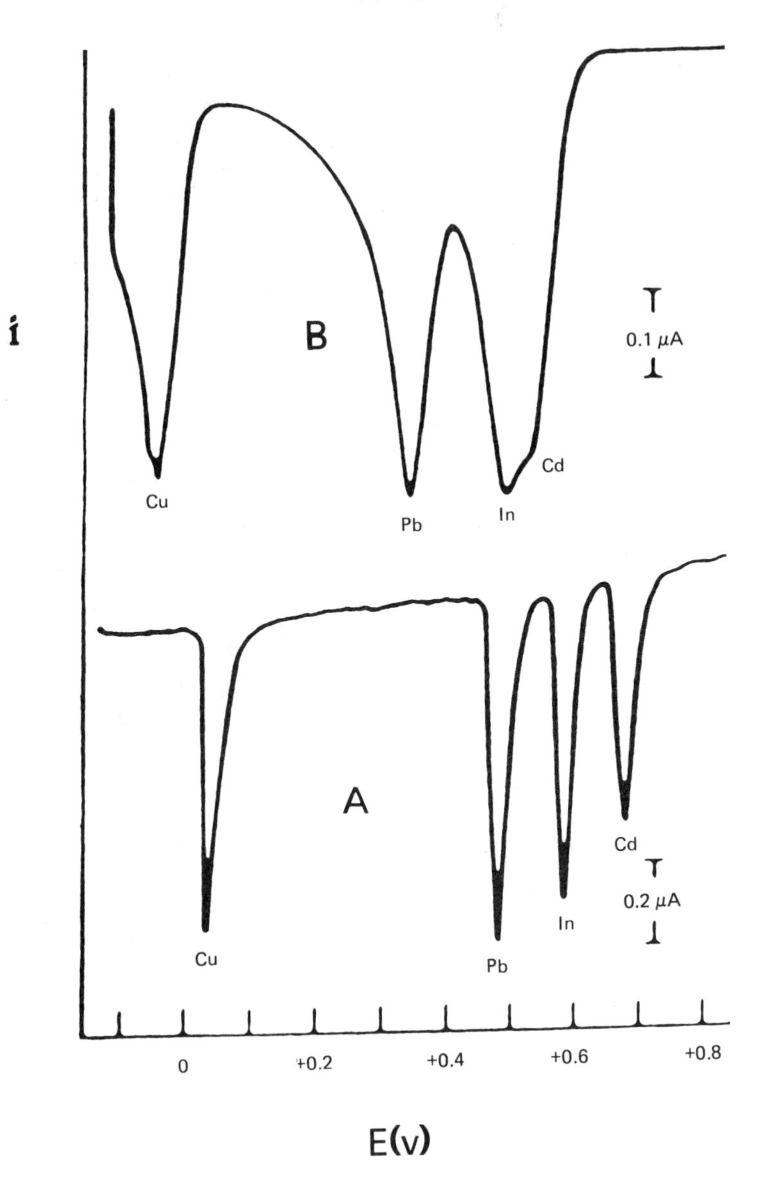

Figure 6. Anodic stripping voltammograms for a solution containing 2×10^{-7} *M* Cd(II), In(III), Pb(II), and Cu(II). (A) Mercury film electrode; (B) hanging mercury drop electrode. Deposition for 5 (A) and 30 (B) min. Scan rate of the linear scan stripping, 0.3 V/min. (Adapted from Ref. 49, with permission.)

Mercury Film Electrodes

The MFE consists of a thin (1-100 μm) film of mercury on a solid support. The mercury film can be applied to the substrate surface prior to the voltammetric measurement (preplated MFE) or during the measurement (in situ plated MFE). The second scheme, originated by Florence [49], requires a spike of Hg^{2+} ions to the sample (~5 × 10^{-5} *M*); mercury and analytes are then codeposited. Among the materials examined, carbon is well suited as support for the mercury film. In particular, glassy carbon carefully polished to a mirror finish has been shown as the superior substrate. The resulting MFE is not homogeneous, but composed of small, individual, and discontinuous mercury droplets. Best results are obtained when a rotating glassy carbon electrode is coupled with an in situ plated mercury film. Hence, the wide accessible potential range and low background current of mercury-coated glassy carbon are coupled with high plating efficiency. A rotating-disk MFE is thus finding increasing use in numerous practical situations. The use of stirred solution and a stationary MFE is simpler and often adequate [50]. A more complex, but effective, mode of transporting the metal ions toward the surface of the glassy-carbon MFE is obtained at the jetstream configuration [51]; this utilizes a flat disk, with a hole (positioned under the center of the MFE), that vibrates at high frequency in a vertical plane. Following completion of the experiment, the mercury film is removed from the surface by wiping with filter paper (and not by anodic polarization that may degrade the surface if chloride is present). Rotating or stationary glassy-carbon disk electrodes are available commercially from manufacturers such as Bioanalytical Systems, EG&G PAR, Pine Instruments, Metrohm, Tacussel, or Mee Instruments. "Homemade" glassy-carbon disks can be constructed from rods available from Tokai (Japan) or Le Carbone-Lorraine (France) by epoxy-sealing the glassy carbon tips into Plexiglas tubes.

Other carbon materials have been used successfully as substrates for the MFE. Among these are wax-impregnated graphite [52], Kelgraf [53], graphite-epoxy [54], or carbon fiber [55]. The latter offers unique properties, such as high plating efficiency (in the absence of forced-convection), reduced ohmic drop or sample volume, and low cost. These characteristics result in major simplification of the instrumentation and operation for stripping analysis, as stirring or rotating devices are eliminated and non-potentiostatic two-electrode systems can be used. Such ultramicroelectrodes have recently become available from commercial sources, such as Bioanalytical Systems or Tacussel. Edmonds [56] has reviewed the physical properties and voltammetric behavior of carbon fiber microelectrodes.

While most stripping applications of MFEs are based on a disk-shaped geometry, other electrode configurations may be employed for specific applications. These include MFE-based flow cells (for on-line monitoring) or dual-electrode designs, such as ring-disk (for anodic stripping with collection [57]) and split-disk electrode (for subtractive stripping voltammetry [58]).

Hanging Mercury Drop Electrode

The HMDE is widely used in stripping analysis. With the popular screw-type (Kemula-type) HMDE the mercury drop is formed by extrusion of

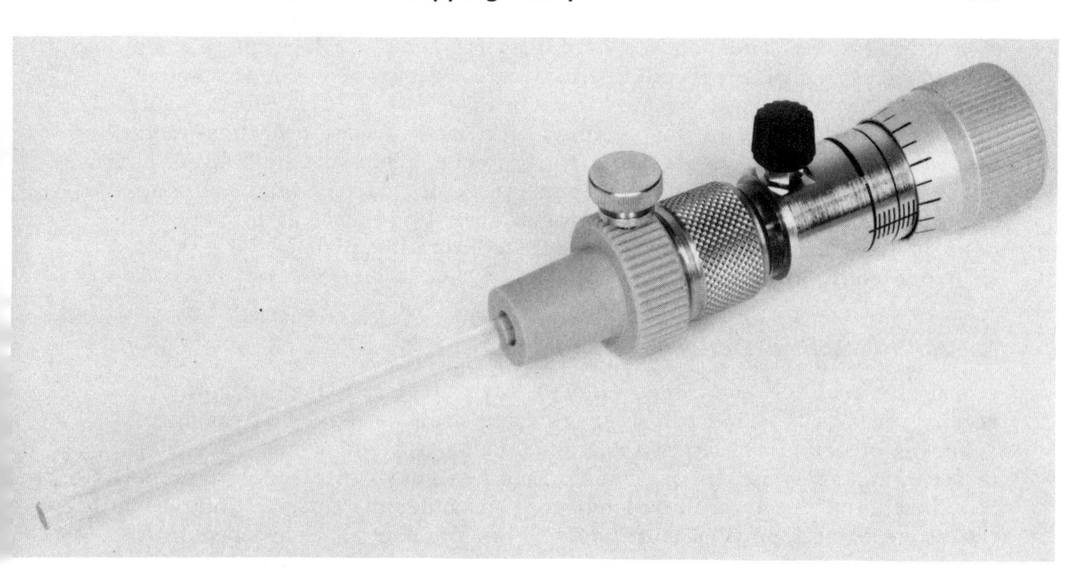

Figure 7. Kemula-type hanging mercury drop electrode. (Courtesy of Brinkmann Instruments Co., Westbury, NY.)

mercury from a capillary, fed from a reservoir by a micrometer-driven syringe (Figure 7). The calibrated micrometer permits reproducible delivery of drops of known area. To ensure proper operation, the capillary should be cleaned, and there must be no entrapped air in the mercury reservoir and the capillary. Electrodes based on this principle are available from Metrohm, Amel, or Tacussel.

The use of a static mercury drop electrode (SMDE), developed by EG&G PAR, instead of the Kemula-type HMDE, improves the reliability and convenience of stripping measurements at the HMDE. This is due to the automatic dispension of highly reproducible drops that hang at the capillary tip. A built-in valve is utilized for this purpose, allowing the mercury flow to be stopped to produce the stationary electrode. The SMDE also improves the ability to hold the drop, over long deposition periods, in a stirred solution. SMDEs are now available also from Metrohm, Amel, Tacussel, and Laboratorni Prestroje (Prague).

Solid Electrodes

The determination of metals, such as Au, Ag, Se, As, Re, Te, or Pt, possessing oxidation potentials positive of that of mercury or low solubility in mercury requires use of bare solid electrodes. Similar considerations are applied to trace measurements of mercury. Gold and glassy-carbon electrodes are especially suited for this purpose, with the metal of interest being determined following its preconcentration as metallic layer.

Such measurements yield low sensitivity and reproducibility compared to analogous measurements at mercury electrodes. In addition, solid-electrode stripping measurements may suffer from difficulties such as the appearance of multiple peaks (and hence a nonlinear response) or incomplete stripping. Nevertheless, reliable procedures for trace measurements of the above elements in various real samples have been reported [2]. These require precise electrode cleaning, polishing, and pretreatment procedures; the nature of these steps depends on the material involved. Substantial improvements in the quantification of selenium, arsenic, or tellurium are observed after codeposition with either gold or copper [59]. Most stripping applications of bare solid electrodes have utilized rotating-disk or flow-through electrode configurations.

Chemically Modified Electrodes

Chemically modified electrodes (CMEs) have attracted much attention in recent years [60]. This active area of research is rapidly advancing in different directions. Properly designed CMEs can add a new dimension to stripping analysis. In particular, modification of electrode surfaces can provide alternative approaches for accumulation of ions, thus extending the scope of stripping measurements. Most popular schemes used to trap metal species on the electrode surface are based on complexation and electrostatic attraction. The modifying agent (ligand, ion exchanger) can be introduced to the surface by functionalizing an appropriate polymer coating or directly into the matrix of a carbon paste electrode. Hence, preconcentration is accomplished by a purely nonelectrolytic deposition step; the collected analyte is subsequently measured during a voltammetric scan. For example, Baldwin et al. [61] reported on the determination of traces of nickel based on dimethylglyoxime-containing carbon paste electrodes. A poly(vinylpyridine)-coated platinum electrode can be used for measuring low levels of Cr(VI) [62]. Trace measurements of uranium can be obtained at a trioctylphosphine oxide-coated glassy-carbon electrode [63]. Nafion- [64] or amine- [65] modified electrodes can be used to preconcentrate organic analytes; in the later case, the preconcentration is based on covalent attachment. Another avenue, the use of permselective electrode coatings, offers other advantages for stripping analysis. Wang and Hutchins [66] used a cellulose acetate layer to cover the MFE and found that electrode fouling by surfactant adsorption was greatly minimized. It is expected that the utility of CMEs for stripping analysis will increase rapidly in the near future.

C. Stripping Analyzers

The popularity of stripping analysis depends on the availability of suitable instrumentation. The basic instrumentation required for stripping analysis is relatively inexpensive and readily available commercially. The modern microprocessor-based voltammetric analyzer, with an electronic potentiostat and associated signal generator, provides the desired potential control (during preconcentration) and potential programming (during stripping). An X-Y plotter or stripchart recorder are commonly used to display the stripping voltammogram.

A relatively inexpensive, but very reliable, instrument is the EG&G PAR model 264 stripping analyzer. This microprocessor-controlled device repeats the stripping cycle three times, including control of the purge time, of the static mercury drop electrode, and of the magnetic stirring and the recorder; both linear scan and differential pulse stripping modes can be employed. For a similar price the same company offers the model 174 polarographic analyzer. This versatile instrument has been widely employed in many of the developments of stripping voltammetry, especially those involving the differential pulse waveform. Wide current and scan-rate ranges (0.02 μA-10 mA and 0.1-500 mV/sec, respectively) are available.

More powerful (and expensive) instruments are available from various sources. Besides the basic requirements of potential control and programming, such instruments provide numerous advantages, as applied to flexibility, data treatment and storage, and sample processing. For example, Figure 8 shows the Metrohm model 646 voltammetric analyzer, coupled to the model 647 electrode stand. This instrument features multiple scans, utilizing different electrodes and/or measuring techniques on a single sample, without operator intervention. Both square-wave and differential pulse stripping modes can be performed in conjunction to the hanging mercury drop or rotating mercury-film electrodes. Automatic standard additions and sample changing are also available. Very advanced microprocessor-based instruments are also the Bioanalytical Systems model 100 and EG&G PAR model 384B. These instruments offer a menu of various voltammetric waveforms (including rapid square-wave), and are capable of data manipulation such as background subtraction, smoothing, or averaging; peak currents and potentials are displayed directly in numerical form.

Other multimode voltammetric analyzers, suitable for stripping analysis, include the Tacussel models PRG4 and PRG5 (France), Sargent-Welch model 7000 (United States), Amel model 472 (Italy), IBM models EC/220 or EC/225 (United States), Mitsubishi model AS-01 (Japan), Bioanalytical Systems

Figure 8. Metrohm model 646 voltammetric analyzer and model 647 electrode stand. (Courtesy of Brinkmann Instruments Co., Westbury, NY.)

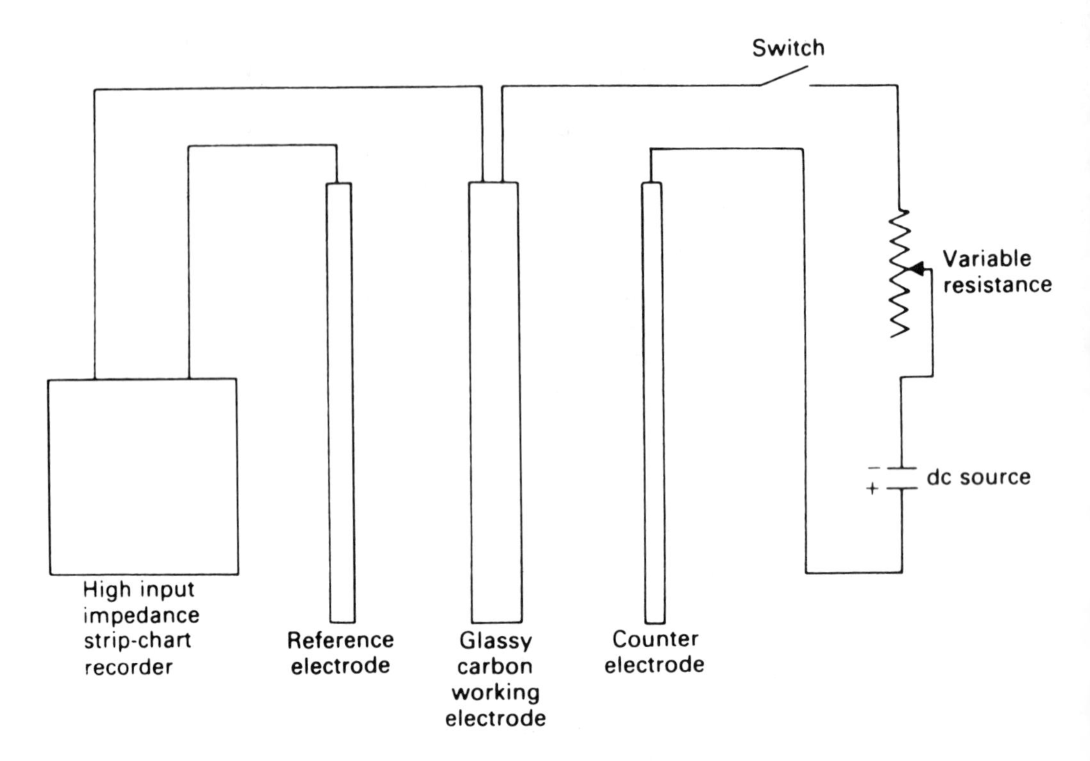

Figure 9. Schematic diagram showing the main components of a potentiometric stripping analyzer. (From Ref. 12, with permission.)

model CV-27 (United States), Metrohm models 506 and 626 (Swiss), Tesla Lab model PA3 (Czechoslovakia), Yanaka model P-1000 (Japan), or Bruker models E100 and E310 (Belgium).

Figure 9 shows the basic instrumentation required for potentiometric stripping analysis. This includes a potentiostatic circuitry for controlling the working-electrode potential during the deposition, a timing switching control, and an input impedance for measuring the potential during the stripping period; the latter must be larger than 10^{12} Ω, to prevent electrochemical oxidation of the amalgamated metals. A microcomputer is desirable for rapid data acquisition, experimental control, and background subtraction. For example, the microcomputer can transform the basic response to a differential (peak-shaped) potentiogram, as well as be used for registering rapid stripping events prior to their display at a stripchart recorder. (For concentrations of metals below the 10^{-8} *M* level, the stripping step normally takes 50-200 msec and the microcomputer becomes a necessity.) Instruments for performing potentiometric stripping analysis are available from Radiometer-Denmark (ISS-820 ion scanning unit) and Tecator-Sweden (Striptec system).

Some stripping experiments require simultaneous control of two working electrodes. Bipotentiostats for such dual-electrode operation are available commercially from Pine Instrument Company. Heineman's group [67] described the circuitry and stripping operation of a microprocessor-based bipotentiostat, capable of applying the differential pulse waveform to two working electrodes.

Computer-based technology has recently offered other advances in instrumentation and methodology for stripping analysis. For example, Bond et al. [68] described a compact microprocessor-based voltammetric analyzer for stripping measurements in harsh environments (e.g., hazardous or radiation laboratories) as well as clean laboratories for ultratrace measurements. The microprocessor systems were interfaced to a microcomputer system, which is external to the laboratories. The same group described also a computerized multi-time-domain method for differential-pulse or square-wave anodic stripping voltammetry [69]. The system can detect possible matrix effects (e.g., organic surfactants) and accordingly "makes decisions" regarding the need for a pretreatment procedure or the method of quantification. Valenta et al. [70,71] described a fully automated system for differential pulse anodic stripping voltammetry that greatly simplifies routine trace metal analysis. Apart from controlling the voltammetric measurement and processing the data, all stages of the sample pretreatment (filtration, acidification, UV irradiation) can be automated. Other versatile computer-operated systems, capable of making all types of stripping measurements, have been described [72,73]. Computer-assisted systems for generating pseudopolarograms, commonly used (in speciation analysis) for determining the complexation capacity of natural waters, have been reported [74]. As was described in Section III.E, computerization has the further advantage that it greatly simplifies the use of stripping flow systems.

Overall, a complete and reliable stripping system—including the stripping analyzer, cell, electrodes, and recorder—would generally cost between $10,000 and $20,000, depending primarily on the desired degree of versatility and automation (i.e., the cost of the stripping analyzer). This price range is considerably lower than that of instrumentation for nonelectrochemical techniques used for trace metal analysis.

V. CONCLUSIONS

Stripping analysis has been shown to be a very sensitive, rapid, and reproducible technique for measurements at the trace and ultratrace levels. Because the preconcentration is done electrochemically in the same cell as the final measurement, contamination risks are greatly reduced. Recent developments, particularly the development of new procedures based on chelate adsorption or chemically modified electrodes for measuring a great number of metals and the introduction of low-cost sophisticated ("push-button, do-all") instrumentation, has stimulated further interest in the field. Certainly, stripping analysis is a technique that supplements and complements the analytical capabilities of any laboratory.

REFERENCES

1. G. E. Batley, *Mar. Chem.*, *12*:107 (1983).
2. J. Wang, *Stripping Analysis: Principles, Instrumentation and Applications*, VCH Publishers, Deerfield Beach, FL (1985).

3. T. R. Copeland and R. K. Skogerboe, *Anal. Chem.*, *46*:1257A (1979).
4. J. Wang, *Environ. Sci. Technol.*, *16*:104 (1982).
5. T. M. Florence, *J. Electroanal. Chem.*, *168*:207 (1984).
6. T. M. Florence, *Anal. Chim. Acta*, *119*:217 (1980).
7. P. Valenta, L. Mart, and H. Rutzel, *J. Electroanal. Chem.*, *82*:327 (1977).
8. M. Wojciechowski, G. Winston, and J. Osteryoung, *Anal. Chem.*, *57*:155 (1985).
9. D. Turner, S. Robinson, and M. Whitfield, *Anal. Chem.*, *56*:2387 (1984).
10. L. Kryger and D. Jagner, *Anal. Chim. Acta*, *78*:250 (1975).
11. D. Jagner and A. Graneli, *Anal. Chim. Acta*, *83*:19 (1976).
12. D. Jagner, *Analyst*, *107*:593 (1982).
13. J. Wang, D. B. Luo, P. A. M. Farias, and J. S. Mahmoud, *Anal. Chem.*, *57*:158 (1985).
14. J. Wang, J. S. Mahmoud, and P. A. M. Farias, *Analyst*, *110*:885 (1985).
15. R. Kalvoda, *Anal. Chim. Acta*, *138*:11 (1982).
16. J. Wang, D. B. Luo, and P. A. M. Farias, *J. Electroanal. Chem.*, *185*:61 (1985).
17. R. Kalvoda, *Anal. Chim. Acta*, *162*:197 (1984).
18. E. N. Chaney and R. P. Baldwin, *Anal. Chem.*, *54*:2556 (1982).
19. J. Wang, M. S. Lin, and V. Villa, *Anal. Lett.*, 19 (1986) 2293.
20. H. Eskilsson, C. Haraldsson, and D. Jagner, *Anal. Chim. Acta*, *175*:79 (1985).
21. J. Wang, *Am. Lab. (Fairfield, Conn.)*, *17(5)*:41 (1985).
22. J. Wang, *Am. Lab. (Fairfield, Conn.)*, *15(7)*:14 (1983).
23. D. Jagner, *Trends Anal. Chem.*, *2*:53 (1983).
24. J. Wang and H. D. Dewald, *Anal. Chem.*, *56*:156 (1984).
25. A. Zirino, S. H. Lieberman, and C. Clavell, *Environ. Sci. Technol.*, *12*:73 (1978).
26. J. Wang and H. D. Dewald, *Anal. Chem.*, *55*:933 (1983).
27. T. M. Florence, *Analyst*, *11*:489 (1986).
28. A. Graneli, D. Jagner, and M. Josefson, *Anal. Chem.*, *52*:2220 (1980).
29. K. Torrance and C. Gatford, *Talanta*, *32*:273 (1985).
30. U. Forsman, *Anal. Chim. Acta*, *146*:71 (1983).
31. L. Vos, Z. Komy, G. Reggers, E. Roekens, and R. Van Grieken, *Anal. Chim. Acta*, *184*:271 (1986).
32. T. M. Florence, *J. Electroanal. Chem.*, *35*:237 (1972).
33. J. Wang and H. Dewald, *Anal. Lett.*, *16*:925 (1983).
34. B. Gammelgoard and J. Andersen, *Analyst*, *110*:1197 (1985).
35. J. Christensen, L. Kryger, and N. Pind, *Anal. Chim. Acta*, *141*:131 (1982).
36. R. Briner, S. Chouchoiy, R. Webster, and R. Popham, *Anal. Chim. Acta*, *172*:31 (1985).
37. S. Mannino, *Analyst*, *107*:1466 (1982).
38. W. Lund and R. Eriksen, *Anal. Chim. Acta*, *107*:37 (1979).
39. J. Wang and B. Freiha, *Anal. Chem.*, *55*:1285 (1983).
40. P. Davis, F. Berlandi, G. Dulude, R. Griffin, and W. Matson, *Am. Ind. Hyd. Assoc. J.*, *6*:480 (1978).
41. I. Davison and F. Smyth, *Anal. Chem.*, *51*:2127 (1979).

42. T. P. DeAngelis, R. E. Bond, E. D. Brooks, and W. R. Heineman, *Anal. Chem., 49*:1792 (1977).
43. R. Eggli, *Anal. Chim. Acta, 91*:129 (1977).
44. L. Anderson, D. Jagner, and M. Josefson, *Anal. Chem., 54*:1371 (1982).
45. S. Lieberman and A. Zirino, *Anal. Chem., 46*:20 (1974).
46. R. Clem, G. Litton, and L. Ornelas, *Anal. Chem., 45*:1306 (1973).
47. J. Wang and B. Frieha, *Anal. Chem., 57*:1776 (1985).
48. L. Mart, H. Nürnberg, and P. Valenta, *Fresenius Z. Anal. Chem., 300*:350 (1980).
49. T. M. Florence, *J. Electoranal. Chem., 27*:273 (1970).
50. J. Wang, *Talanta, 29*:125 (1982).
51. T. Magjer and M. Branica, *Croat. Chem. Acta, 49*:L1 (1977).
52. W. Matson, D. Roe, and D. Carrit, *Anal. Chem., 37*:1594 (1965).
53. J. Anderson and D. Tallman, *Anal. Chem., 48*:209 (1976).
54. J. Wang, *Anal. Chem., 53*:2280 (1981).
55. A. Baranski and H. Quon, *Anal. Chem., 58*:407 (1986).
56. T. Edmonds, *Anal. Chim. Acta, 175*:1 (1985).
57. D. Laser and M. Ariel, *J. Electroanal. Chem., 49*:123 (1974).
58. L. Sipos, P. Valenta, H. Nürnberg, and M. Branica, *J. Electroanal. Chem., 77*:263 (1977).
59. R. W. Andrews, *Anal. Chim. Acta, 119*:47 (1980).
60. R. W. Murray, Chemically Modified Electrodes, in *Electranalytical Chemistry* (A. J. Bard, Ed.), Marcel Dekker, New York, 1983, Vol. 13, p. 191.
61. R. P. Baldwin, J. K. Christensen, and L. Kryger, *Anal. Chem., 58*:1790 (1986).
62. J. Cox and P. Kulesza, *Anal. Chim. Acta, 154*:71 (1983).
63. K. Izutsu, T. Nakamura, R. Takizawa, and H. Hanawa, *Anal. Chim. Acta, 149*:147 (1983).
64. M. Szentirmay and C. Martin, *Anal. Chem., 56*:1898 (1984).
65. J. F. Price and R. P. Baldwin, *Anal. Chem., 52*:1940 (1980).
66. J. Wang and L. Hutchins, *Anal. Chem., 58*:402 (1986).
67. D. W. Paul, T. H. Ridgway, and W. R. Heineman, *Anal. Chim. Acta, 146*:125 (1983).
68. A. M. Bond, H. B. Greenhill, I. D. Heritage, and J. B. Reust, *Anal. Chim. Acta, 165*:209 (1984).
69. A. M. Bond, I. D. Heritage, and W. Thormann, *Anal. Chem., 58*:1063 (1986).
70. P. Valenta, L. Sipos, I. Kramer, P. Krumpen, and H. Rützel, *Fresenius Z. Anal. Chem., 312*:101 (1982).
71. W. Dorten, P. Valenta, and H. W. Nürnberg, *Fresenius Z. Anal. Chem., 317*:264 (1984).
72. R. Neeb and D. Saur, *Fresenius Z. Anal. Chem., 222*:200 (1966).
73. L. Kryger, D. Jagner, and H. Skov, *Anal. Chim. Acta, 78*:241 (1975).
74. S. Brown and B. Kowalski, *Anal. Chem., 51*:2133 (1979).

19

Measurement of Electrolytic Conductance

TRUMAN S. LIGHT* / The Foxboro Company, Foxboro, Massachusetts

GALEN WOOD EWING* / Department of Chemistry, New Mexico Highlands University, Las Vegas, New Mexico

I. INTRODUCTION

This chapter concerns a measurement technique that provides information about the *total* ionic content of a solution, usually aqueous. It is nonspecific, and this can be perceived as a drawback. However, there are many situations where this "weakness" can be converted into a strength. One obvious case is in characterizing the purity of potable waters and monitoring the effectiveness of water demineralizers. Conductance can be used effectively to measure the concentration of acids and bases, since it gives a monotonic, though not always linear, conductance-concentration plot over wide concentration ranges. Since electrolytes vary with respect to their ability to conduct current, conductance measurements provide a valuable method of analyzing binary mixtures of electrolytes. Conductometric titrimetry is a useful method for following the course of reactions involving electrolytes.

II. DEFINITION OF TERMS AND SYMBOLS

The electrical resistance of a material may be denoted by the symbol R, specified in ohms (Ω). The electrical conductance is the reciprocal of the resistance, usually given the symbol L, with the units of siemens (S) or reciprocal ohms (mhos, Ω^{-1}). The relation is

$$L = 1/R \tag{1}$$

The resistance of a sample is a property of its geometry—its cross-sectional area A and its length l. In the special case of liquid samples, our chief area of interest in this chapter, the area A usually is taken as the area of each of a pair of parallel electrodes, and the length l of the sample is replaced by d, the distance between them. The specific resistance, also called the resistivity, denoted by the symbol ρ, has the units of ohm-centimeters and is defined by

$$\rho = R(A/d) \tag{2}$$

---

*Retired

Table 1. Reference Solutions for Calibration of Cell Constants

Approx. molarity	Method of preparation	Temp. (°C)	L (μS/cm)
1.0	74.2460 g KCl per 1 liter solution at 20°C	0	65,176
		18	97,838
		25	111,342
0.1	7.4365 g KCl per 1 liter solution at 20°C	0	7,138
		18	11,167
		25	12,856
0.01	0.7440 g KCl per 1 liter solution at 20°C	0	773.6
		18	1,220.5
		25	1,408.8
0.001	Dilute 100 ml of 0.01 M to 1 liter at 20°C	25	146.93

Source: From Ref. 25.

Similarly, the specific conductance, or conductivity, is denoted by the symbol κ with the units of siemens per centimeter, so that

$$\kappa = 1/\rho = (1/R)(d/A) = L(d/A) = L\theta \tag{3}$$

where θ is equal to d/A in cm^{-1}. Since θ is a function only of the geometry of the measuring electrodes, it is useful in the characterization of cells, and hence is called the *cell constant*. If the electrodes are other than plane and parallel, the cell constant can best be determined by measurement of solutions of known specific conductance, $\kappa = L\theta$. Reference solutions for this purpose have been characterized with great care in cells of known geometry. A number of such solutions are listed in Table 1.

III. ELECTROLYTE THEORY

Since the conductance is determined by the totality of ions present in a solution, largely acting independently of each other, it can be expressed as a summation:

$$L = 10^{-3}\,\theta^{-1} \sum_i z_i C_i \lambda_i \tag{4}$$

in which C_i represents the molar concentration of the i-th ion, z_i its charge, and λ_i its equivalent ionic conductivity, to be defined below. The summation covers all ions of both signs. For some purposes it is desirable to define a *molar conductivity*, for which the usual symbol is Λ. The units are $S \cdot cm^2 \cdot mol^{-1}$. It is given by the relation

$$\Lambda = \lambda_+ + \lambda_- = 1000\kappa/C \tag{5}$$

The quantity λ is a property of ions (either positive or negative) that gives quantitative information about their relative contributions to

Table 2. Equivalent Ionic Conductivity of Selected Ions at Infinite Dilution at 25°C ($S \cdot cm^2 \cdot mol^{-1}$)

Cations[a]	λ_0	Tempco[b]	Anions[a]	λ_0	Tempco[b]
H^+	349.8	0.0139	OH^-	198.6	0.018
$Co(NH_3)_6{}^{3+}$	102.3		$Fe(CN)_6{}^{4-}$	110.5	0.02
K^+	73.5	0.0193	$Fe(CN)_6{}^{3-}$	101.0	
$NH_4{}^+$	73.5	0.019	$Co(CN)_6{}^{3-}$	98.9	
Pb^{2+}	69.46	0.02	$SO_4{}^{2-}$	80.0	0.022
La^{3+}	69.6	0.023	Br^-	78.14	0.0198
Fe^{3+}	68.0		I^-	76.8	0.197
Ba^{2+}	63.64	0.023	Cl^-	76.4	0.0202
Ag^+	61.9	0.021	$NO_3{}^-$	71.42	0.020
Ca^{2+}	59.50	0.0230	$CO_3{}^{2-}$	69.3	0.02
Cu^{2+}	53.6	0.02	$C_2O_4{}^{2-}$	74.2	0.02
Fe^{2+}	54.0		$ClO_4{}^-$	67.3	0.020
Mg^{2+}	53.06	0.022	$HCO_3{}^-$	44.5	
Zn^{2+}	52.8	0.02	$CH_3CO_2{}^-$	40.9	0.022
Na^+	50.11	0.0220	$HC_2O_4{}^-$	40.2	
Li^+	38.69	0.0236	$C_6H_5CO_2{}^-$	32.4	0.023
$(n\text{-}Bu)_4N^+$	19.5	0.02	Picrate$^-$	30.4	0.025

[a]For ions of charge z, the figures given are on an equivalent basis, that is, they apply to the fraction (1/z) of a mole.
[b]The temperature coefficient, when known, is given as $(1/\lambda_0)(d\lambda_0/dT)$, with the units K^{-1}.
Source: Ref. 26.

the conductance of a solution. Its value is to some extent dependent on the total ionic concentration of the solution, increasing with increasing dilution. It is convenient to tabulate numerical values of λ_0, the limiting value of λ as the concentration approaches zero (infinite dilution). Representative values are given in Table 2.

Much information pertaining to ionic equilibria can be obtained from conductometric data, particularly in situations where ions tend to be removed from solution by an equilibrium process. This applies, for example, to the combination of anions with hydrogen cations to form partially dissociated molecular acids, to the formation of complexes between metallic cations and various ligands, and to the formation of sparingly soluble salts. Thus the measurement of conductance can lead to the establishment of acid and basic dissociation constants, stability constants, and solubility product constants.

Further details of theory would be out of place here; they can be found in any modern text on physical chemistry (see, for example, Levine [1]).

IV. INSTRUMENTATION

There are two general types of conductance measuring devices. The first, and most widely used, employs a pair of electrodes, usually platinum, immersed in the test liquid, whereas the second depends on inductive or capacitive effects to measure the conductance. The remainder of this chapter is divided into two parts, treating these two, quite different, instrumental types.

A. Immersed-Electrode Measurements

Cells

Figure 1 illustrates some types of conductivity cells that are commercially available. As detailed in the caption, certain cells are primarily intended for precision physicochemical measurements, while others are more convenient for routine use and for titrimetry. For precise work, cells should be held at a constant temperature, since the conductance of most electrolytic solutions increases with temperature at the rate of about 2 percent per kelvin (cf. Table 2). Cells are made with various cell constants, of which 1.0 and 0.1 are the most widely useful. Table 3 indicates the cell constants appropriate for various ranges of conductance.

The electrodes are usually fabricated from platinum, though graphite, titanium, and tungsten are occasionally used. Platinum electrodes are best coated with a finely divided form of the metal, known as platinum black. This coating can be produced by a few minutes' electrolysis in a solution of chloroplatinic acid containing a small amount of lead acetate. The electrolysis should be repeated with reversed polarity. After platinizing, the electrodes should be stored in distilled water.

Excitation

The conductivity cell can be represented schematically by the equivalent circuit shown in Figure 2 [2]. In this diagram, R_{L1} and R_{L2} represent the resistances of the connecting wires (usually negligible), C_{DL1} and C_{DL2} are the double-layer capacitances at the electrode-solution boundaries, C_P is the interelectrode capacitance (in parallel with the cell), and R_{Sol} is the ohmic resistance of the solution between the electrodes. The components marked R_{F1} and R_{F2} denote the faradaic resistances at the two electrodes, that is, the electrical equivalent of any possible electrode reactions. If a constant (DC) potential is impressed on this network, no current will flow, except for a brief transient, provided that the voltage is small enough that no electrochemical process can occur. If the voltage is higher, current will flow through the R_F components and R_{Sol}.

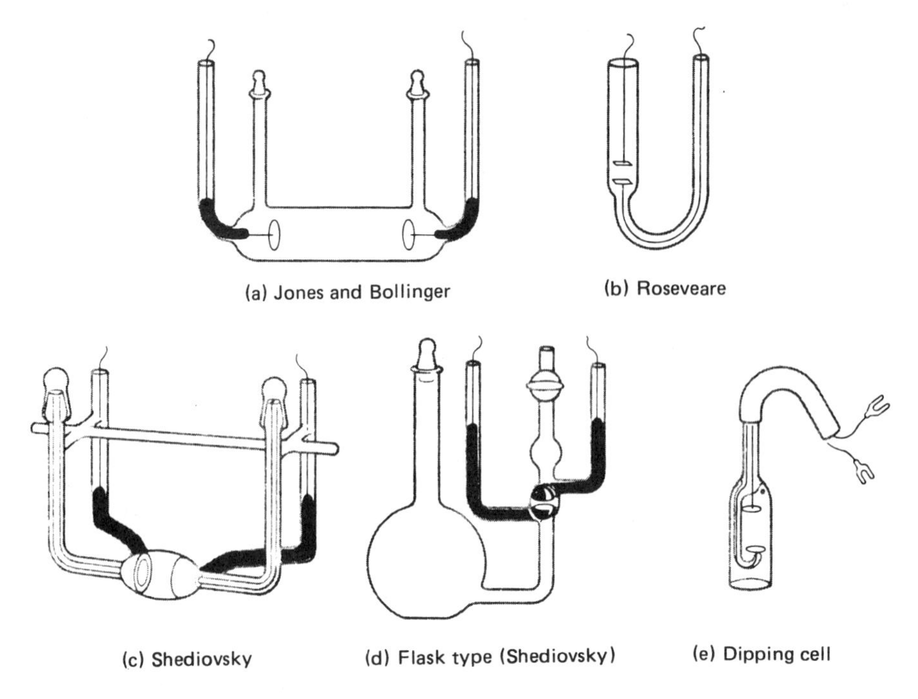

Figure 1. A selection of cells for measurement of conductance. (a) The form most used for exacting physico-chemical measurements. (b) The Roseveare cell is easy to make, with the corners of square pieces of platinum sealed into the glass walls. (c) This form has conical electrodes, useful for liquids that have a tendency to foam. (d) A flask form, convenient for preparing and measuring very dilute solutions. (e) The dipping cell, widely used for routine measurements. (From Ref. 27, with permission.)

Table 3. Recommended Cell Constants for Various Conductance Ranges

Conductance range (μS/cm)	Cell constant (cm^{-1})
0.05-20	0.01
1-200	0.1
10-2000	1.0
100-20,000	10.0
1000-200,000	50.0

Source: From Ref. 25.

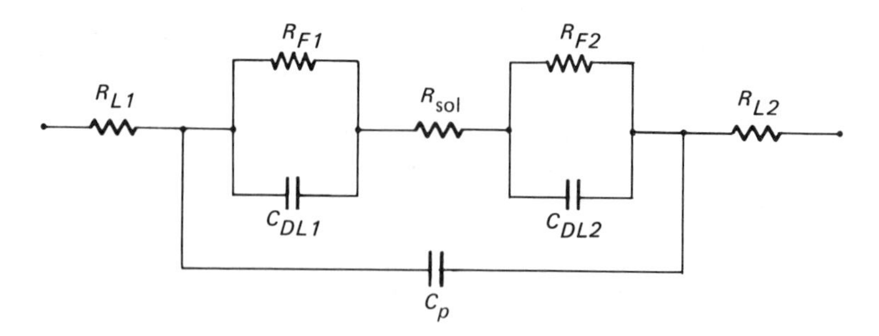

Figure 2. Electronic circuit equivalent to a conductance cell with immersed electrodes. (From Ref. 2, with permission.)

On the other hand, if an AC potential is applied, alternating current will flow through the C_{DL} terms and R_{Sol}, and at the same time through C_P. Each C_{DL} provides a low-impedance path for alternating current so that voltage cannot build up across the corresponding R_F sufficient to permit faradaic current to flow. Hence if C_P can be kept negligibly small, and the C_{DL} values large, the effect of R_{Sol} can be studied by itself. In practice the double-layer capacitances can be increased substantially by platinization which greatly increases the effective surface area of the electrodes. Then C_P becomes significant only with solutions of high resistance (low conductance), when large electrodes close together must be selected in order to keep the measurement within range.

Hence most commercial instruments for measuring electrolytic conductance operate on alternating current. Some use the 50- or 60-Hz power-line frequency for convenience; however, higher frequencies favor low impedances for the double-layer capacitances, so many instruments include built-in oscillators to provide excitation at (typically) 1 or 2 kHz at 5 V (RMS) or less.

Circuitry

The traditional circuit for measuring conductance is the Wheatstone bridge (Figure 3). This circuit is well adapted to static measurements, but since the bridge must be balanced to give a reading, it is not readily applicable to continuous measurements. The two-ganged, multiple-point switch shown selects between several ranges (only three are shown, but there may well be several more). At balance, when the meter shows a null, it is easily demonstrated that the resistance of the test cell, R_X, is given by

$$R_X = R_1R_3/R_2 \tag{6}$$

or the conductance, L_X, by

$$L_X = R_2/R_1R_3 \tag{7}$$

Note that R_3 is a calibrated variable resistor or bank of decade resistors, so that its reading at balance, multiplied by the R_1/R_2 ratio, gives the resistance of the sample directly. In some bridges, a switch is included to interchange the positions of R_X and R_2, which gives a dial reading (on R_3) directly proportional to conductance rather than resistance:

$$L_X = R_3/R_1R_2 \tag{8}$$

For precise results, it is necessary to include a small variable capacitor to cancel the effects of the cell capacitance, as without this the balance point of the bridge will not be sufficiently sharp. Reference 3 includes a detailed discussion of the capacitive effect.

A number of electronic circuits have been developed to give an output voltage proportional to the conductance of a sample without the need for balancing a bridge. Such an instrument is usable for continuous monitoring of flowing streams, including, for example, the effluent from a liquid chromatograph. It is also useful for conductometric titrimetry with constant inflow of reagent, the voltage signal being displayed against time on a strip-chart recorder [4-6]. An example of a self-balancing bridge controlled by a microcomputer is given in Ref. 7.

Commercial Instruments

The majority of instruments in today's market for measurement of electrolytic conductance are self-contained electronic units provided with digital readout. Top-of-the-line instruments include a temperature-sensing probe and either automatic or manual temperature compensation to correct for a range of values of the temperature coefficient. They may also have an

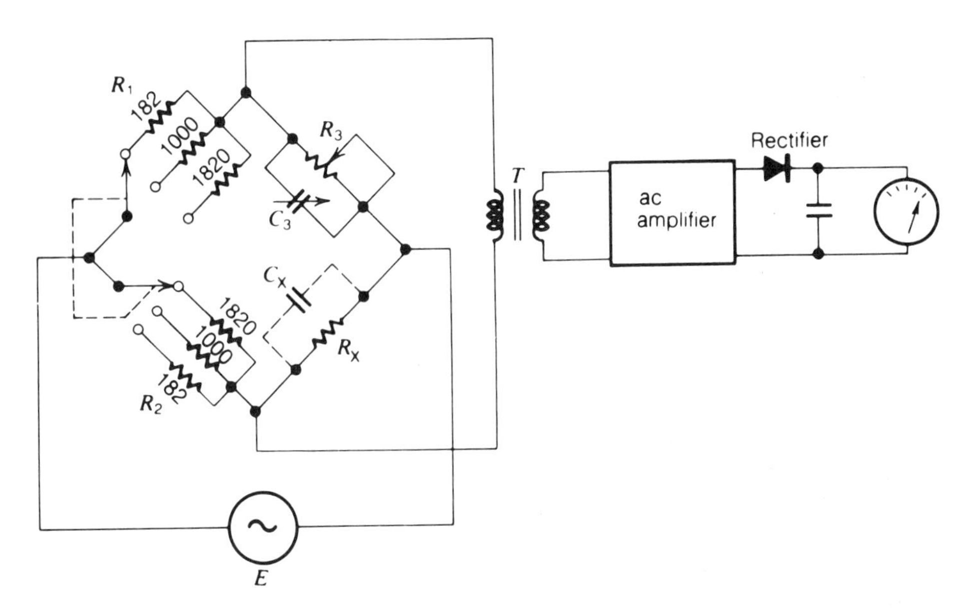

Figure 3. An AC Wheatstone bridge for conductance measurements. The two-pole, three-throw switch permits choice of 0.1, 1.0, or 10 as ratios. (From Ref. 28, with permission.)

option of displaying the temperature of the solution. Provision is usually made for entering the cell constant appropriate to the cell in use, so that the displayed readings are directly expressible in terms of siemens per centimeter.

Some instruments offer a selection of operating frequencies, with lower frequency for use with solutions of low conductivity, where cells with low cell constant (0.1 cm^{-1}, for example) and hence higher interelectrode capacitance must be used. The Radiometer model CDM80, for example, uses 165 Hz for the 20- and 200-mS/cm ranges, and 3 kHz for the 2-, 20-, and 200-mS/cm ranges. Others provide only a single frequency, usually 1 kHz, for all ranges.

One instrument that is unusual in that it uses a digitally controlled excitation waveform is model 32 of the Yellow Springs Instrument Company. This unit uses a modification of the waveform and circuitry described by Daum and Nelson [4], and needs no compensation for the capacitance of the cell.

Figure 4 shows a representative commercial instrument, Radiometer model CDM83.

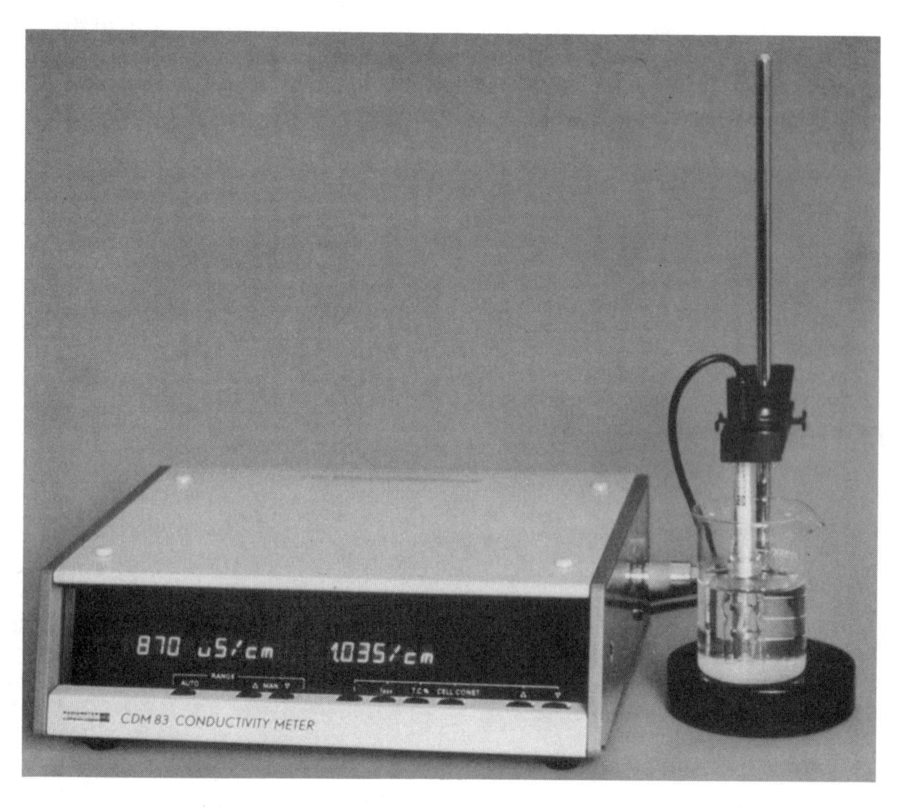

Figure 4. The Radiometer model CDM83 conductivity meter. (Courtesy of Radiometer Copenhagen.)

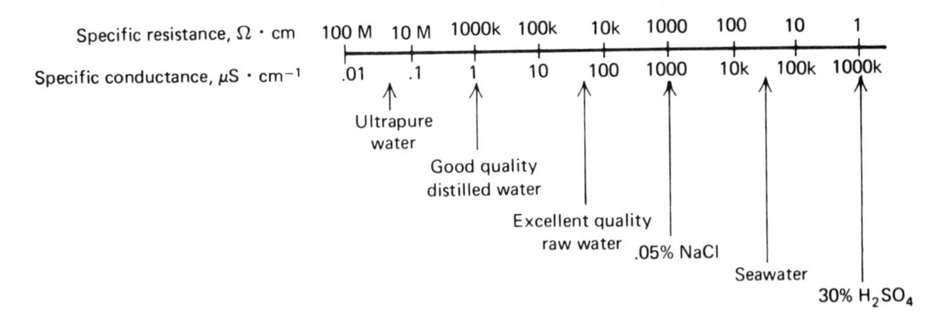

Figure 5. Specific resistance and conductance ranges for some typical materials. (From Ref. 28, with permission.)

Applications

Figure 5 shows specific conductances and resistances for a number of materials. Water itself is a very poor conductor; its specific conductance due to dissociation into H_3O^+ and OH^- ions is approximately 0.0551 μS/cm at 25°C [9]. Water of this theoretical purity is produced using commercially available, nuclear-grade, ion-exchange resins. It is used extensively in the semiconductor, power, and pharmaceutical industries. The conductivity measurement is extremely sensitive to traces of ionic impurities. The presence of 1 ppb (1 μg/liter) of sodium chloride will raise the conductivity to 0.0571 μS/cm at 25°C [9]. Ordinary distilled or deionized water falls far short of this purity.

Water purification equipment is often provided with conductance monitors, which can be configured to shut down a still or initiate regeneration of the ion-exchange bed of a demineralizer if the conductance becomes too high.

Solutions of strong electrolytes show a nearly linear increase of conductance with concentration up to about 10-20 percent by weight. At higher concentrations the conductance decreases again, due to such interactions as complexation reactions, formation of dimers or higher polymers, and increased viscosity. Figure 6 shows the relation between conductance and concentration for a few representative substances.

Conductance measurements can be useful in following the kinetics of reactions that involve a change in ionic content or mobility. See, for example, a report by Queen and Shabaga [10] on the kinetics of a series of solvolytic reactions. Kiggen and Laumen [7] have used similar measurements to observe diffusion processes in solution.

Conductometric titrimetry is widely applicable for titration reactions involving ions. Figure 7 indicates the kind of curves that result in the titration of acids of various dissociation constants by sodium hydroxide. Gaal et al. have reported good results with catalytic titrations [11]. (See also Chapter 26 of this handbook.)

Conductance detectors are routinely used with ion-exchange and similar types of chromatography [12].

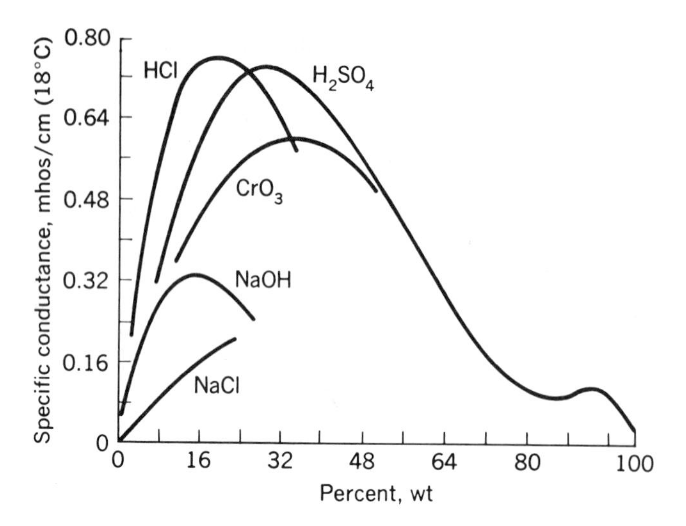

Figure 6. Conductivity-concentration curves for selected electrolytes. (From Ref. 28, with permission.)

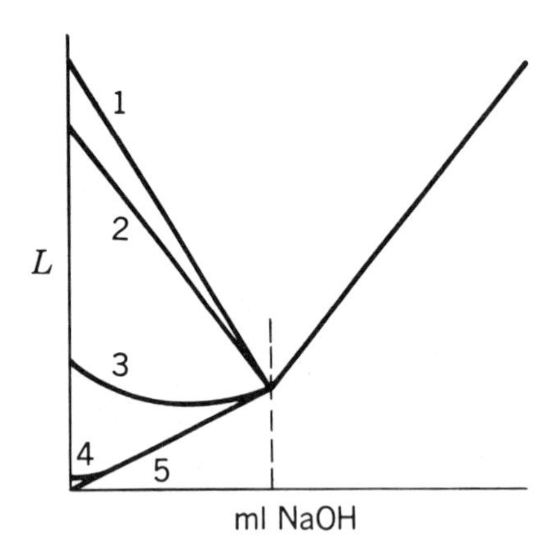

Figure 7. Conductometric titration curves of various acids by sodium hydroxide. Curve 1 represents a strong acid, and curve 5 an extremely weak one, while the others are intermediate. The acids are (1) hydrochloric acid, (2) dichloroacetic acid, (3) monochloroacetic acid, (4) acetic acid, and (5) boric acid. (From Ref. 28, with permission.)

Table 4. Compounds Amenable to Ion Chromatographic Analysis

Major classes of compounds		Primary detectors
Inorganics	Sulfate sulfite, thiosulfate, sulfide, halides, nitrite, nitrate, borate, phosphate, phosphite, hypophosphite, pyrophosphate, tripolyphosphate, trimetaphosphate, selenate, selenite, arsenite, arsenate, chlorite, perchlorate, chlorate, hypochlorite, carbonate, cyanide	Conductivity Amperometry
Metals	Alkali metals, alkaline earths, gold (I and III), platinum, silver, palladium, iron (II and III), copper, nickel, tin, lead, cobalt, manganese, cadmium, chromium, aluminum	Conductivity Visible color
Organics	Organic acids and amines, carbohydrates, alcohols, amino acids, saccharin, sugars/sugar alcohols, surfactants (anionic/cationic)	Conductivity Amperometry Fluorescence
Amino acids	Amino acids	Fluorescence Visible color Amperometry

Source: From Ref. 12b.

The ion chromatograph, principally using a conductivity detector, has been developed in the last decade and has rapidly established itself as one of the major instrumental analytical tools. Ion chromatography filled a long-standing gap in analytical methods for anions and was able to rapidly and reliably determine the inorganic anions such as chloride, bromide, fluoride, nitrate, nitrite, sulfate, and phosphate using a conductivity detector. Subsequent development broadened the application of ion chromatography beyond simple anion analysis. A modern definition is given by McNair and Polite [12a]: "Ion chromatography is the separation of either weakly or strongly ionic compounds on ion exchange columns by both chemical and electronic suppression using a variety of detectors." Table 4 lists the four major classes of compounds amenable to ion chromatographic analysis: inorganics, metals, organics, and amino acids [12b].

Ion chromatography is a form of liquid chromatography (refer to Chapter 22), and modern liquid chromatographic instruments may be used with proper accessories. Ion chromatography was first reported in 1975 by Small, Stevens, and Bauman [12c], who defined an ion chromatograph

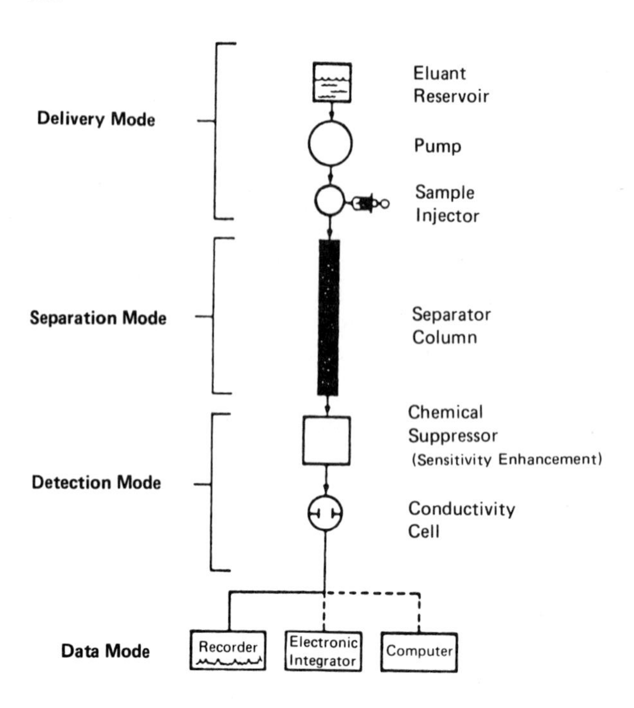

Figure 8. Typical ion chromatograph. (From Ref. 12b, with permission.)

as a device consisting of two ion-exchange columns connected in series and followed by a conductivity detector as shown in Figure 8 [12b,12d]. Their unique contributions were the use of the second "suppressor" column to react with the ions of the eluent, and the use of a sensitive small-volume conductivity cell detector to measure only the conducting analyte ions in the sample. Conductivity is a nonspecific measurement and cannot distinguish between the eluent and sample ions. The suppressor column diminishes or eliminates the eluent ions but not the sample ions; sensitivities in the microgram per liter (parts per billion) range may be achieved.

For example, a sample containing anions is injected at the head of an anion-exchange resin bed, which is the separator column (refer to Figure 8). If dilute sodium salicylate solution, which is the eluent, is passed through the anion column, separation of all anions is achieved and these are eluted in the form of the sodium salts. This effluent then passes through the suppressor column, which is a cation exchange resin that exchanges hydrogen ions for the sodium and other cations in the eluent. The sodium salicylate becomes salicylic acid, which is only slightly ionized. The anions of interest, which might be salts of stronger acids, are completely ionized and are readily detected by the conductivity cell. A typical anion chromatogram is shown in Figure 9.

Cation determinations are carried out in a similar manner by using a cation-exchange resin in the separator column. The sample cations are eluted with a strong acid, which in turn is eliminated in the suppressor column by using an anion-exchange resin in the hydroxyl form.

Single-column methods also may now be employed for ion chromatography. These may make use of hollow-fiber cation- or anion-exchange membranes, which permit an outside and counterflowing stream of acid or alkali to effectively neutralize the eluent in what now appears to be a single column [12d].

Conductivity is the most commonly used detector for ion chromatography. A host of improvements have been made in cell design and detector circuitry driven by the needs of this new analytical method. These include minimization of cell volume, improved electronic suppression and stabilization techniques, and better temperature control and regulation. Discussion of these improvements and others pertaining to ion chromatography is given in several review papers [12a,12e,12f,12g,12h].

B. Electrodeless (Noncontacting) Measurements

There are two ways in which conductance measurements can be made without electrodes in contact with the solution. The first is a high-frequency method, which uses frequencies in the megahertz region. The electrodes take the form of a pair of metal sheets or bands on the *outside* of the sample cell, which is made of an insulating material such as glass. The glass plays the part of the dielectric in the capacitors C_{DL} (Figure 2). The impedance of a capacitor is so low at this high frequency that even with a glass dielectric the alternating current passes freely into the sample. At the same time, both R_F terms become essentially infinite (i.e., open

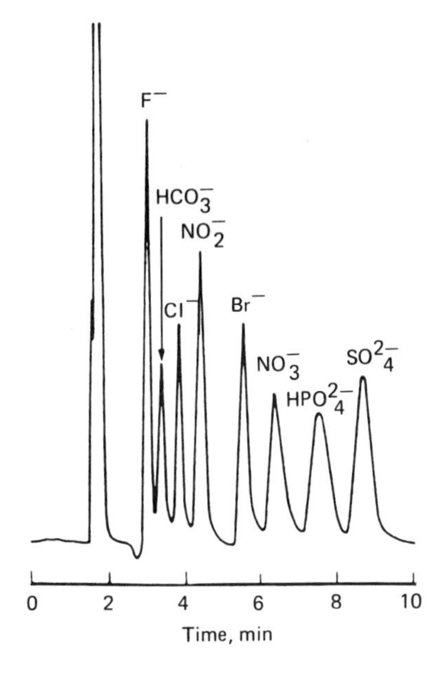

Figure 9. Typical ion-chromatogram for anion analysis. (From Ref. 12d, with permission.)

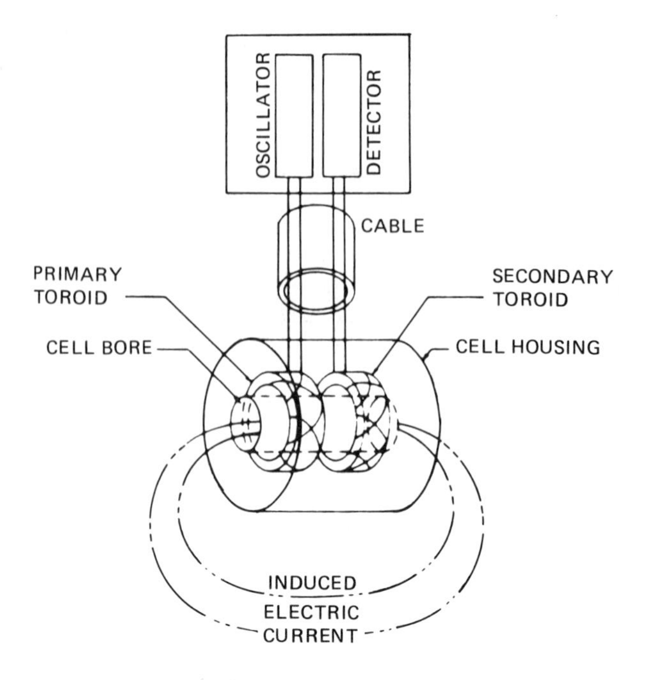

Figure 10. Principle of the electrodeless conductivity cell and instrument.

circuit). In this way the resistance of the cell becomes the desired R_{Sol}, paralleled by C_P. The cell capacitance C_P can be made very small by suitable placement of the electrodes.

High-frequency conductometry, sometimes called "oscillometry," has been treated extensively by Pungor [13] and is seldom used now. The most recent paper on the subject known to the authors appeared in 1981 [8]. We will not discuss this method further.

Inductive Methods

Because of the drawbacks of electrode fouling and polarization, a second method of measuring conductance without the use of contacting electrodes has become popular, especially in chemical process and industrial solution applications. Usually referred to simply as "electrodeless conductivity," it has also been called "inductive conductivity." This system utilizes a probe consisting of two encapsulated toroids in close proximity to each other, as shown in Figure 10. One toroid generates an electric field in the solution, while the other acts as receiver to pick up the small alternating current induced in the electrolytic solution, as shown in Figure 11. The equivalent electrical circuit may be compared to a transformer with the toroids forming the primary and secondary windings, and the core replaced by a loop in the conducting solution.

Several configurations are possible. One form, designed for immersion, is shown in Figure 12. The toroids are covered with a chemically resistant

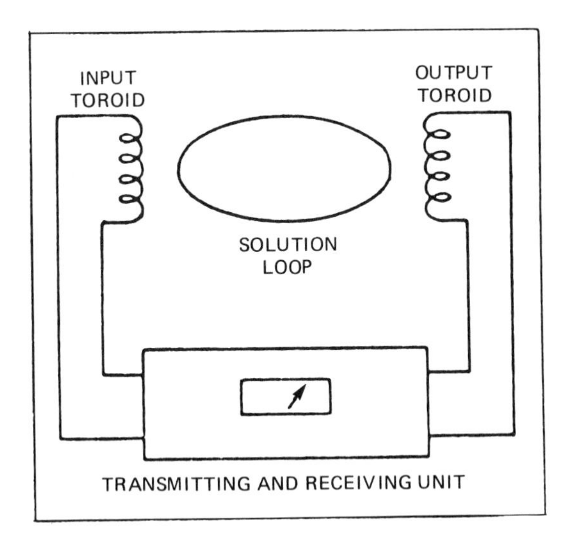

Figure 11. Simple representation of an electrodeless conductivity measuring circuit.

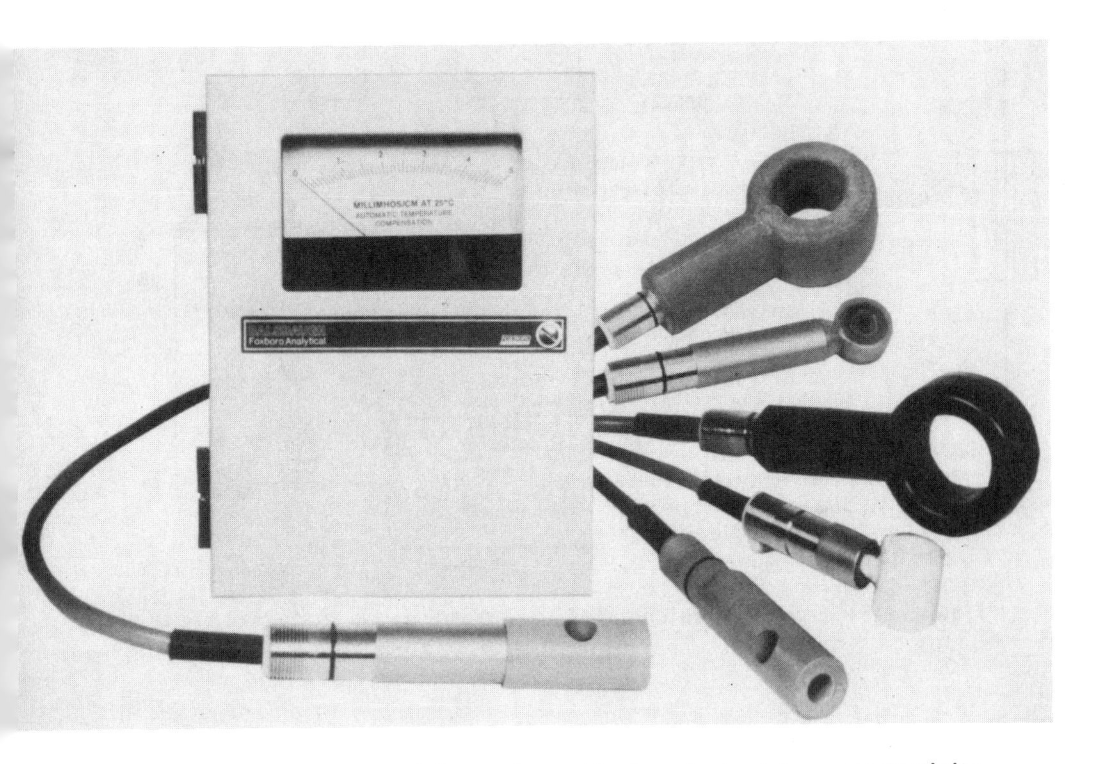

Figure 12. Various types of immersion cells for electrodeless conductivity, with an associated instrument. (Courtesy of The Foxboro Company.)

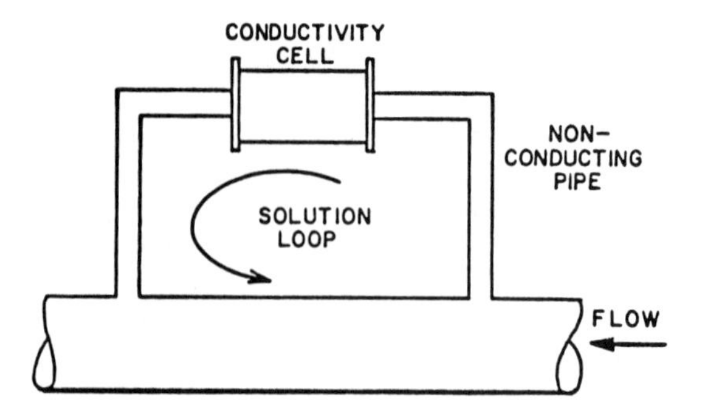

Figure 13. An electrodeless conductivity cell mounted externally to a segment of glass or plastic pipe forming a solution loop with an industrial flow line.

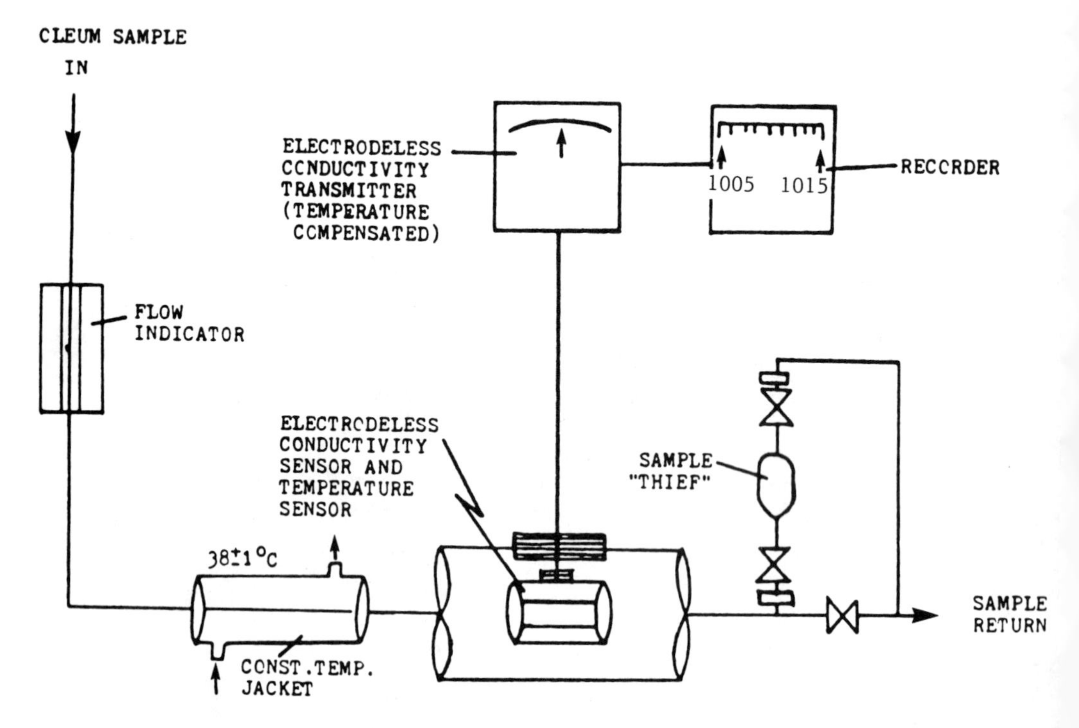

Figure 14. A continuous analyzer for oleum, using an electrodeless conductivity sensor [15].

fluorocarbon or other high-temperature resistant thermoplastic material. Any precipitates or coatings adhering to this probe generally have little or no effect on the measured conductance. A configuration in which the probe does not come in contact with the solution is illustrated in Figure 13. The toroid unit is installed around a section of nonconductive pipe containing the test solution. A complete loop must exist for this arrangement to work.

In either case, the generating toroid is energized from a stable audio frequency source, typically 20-50 kHz. The pickup toroid is connected to a receiver, which measures the current through this secondary winding. The current is then amplified and displayed on a meter or an analog strip-chart recorder. The output is a direct function of the conductance of the solution in the loop, in a manner completely analogous to the traditional measurement with contacting electrodes.

The useful range of commercially available instruments extends from 0-100 μS/cm to 0-2 S/cm, with relative accuracy of a few tenths of a percent of full-scale, after temperature compensation. A temperature sensor is incorporated in the toroid probe, and a compensation circuit corrects all readings to the standard reference temperature of 25°C.

The electrodeless conductivity technique using low-frequency inductive cells has been known since 1951 [14]. It is manufactured commercially for analysis and control in the chemical process industries and in other continuous monitoring applications. Although its stability, freedom from maintenance, and accuracy are superior to contacting conductivity techniques, lack of bench models of this type has hindered its laboratory use and application to date.

The smallest electrodeless probe is about 3.6 cm in diameter, has an equivalent cell constant of 2.5 cm^{-1}, and requires a minimum solution volume of several hundred milliliters to ensure a complete solution loop without wall effects which distort the apparent cell constant. For the lower conductance ranges, which require a smaller cell constant (cf. Table 3), the diameter of the probe must increase, and accurate measurement below about 10 μS/cm is not practical.

Numerous applications of electrodeless conductivity sensors have been published for the chemical, pulp and paper, aluminum, mining, and food industries [16-23]. Similar instrumentation has been used for in-situ measurements of the salinity of seawater [24]. An instrument for continuous analysis of oleum in the range of 100-102 percent equivalent sulfuric acid, with an accuracy of 0.01 percent, is illustrated in Figure 14 [15]. A historical review of electrodeless conductivity has recently been published [29].

REFERENCES

1. I. N. Levine, *Physical Chemistry*, 2nd ed., McGraw-Hill, New York, 1983, p. 473 et seq.
2. G. W. Ewing, *Instrumental Methods of Chemical Analysis*, 5th ed., McGraw-Hill, New York, 1985, p. 332 et seq.
3. J. Braunstein and G. D. Robbins, *J. Chem. Educ.*, *48*, 52 (1971).
4. P. H. Daum and D. F. Nelson, *Anal. Chem.*, *45*, 463 (1973).
5. G. M. Muha, *J. Chem. Educ.*, *54*, 677 (1977).

6. M. Ahmon, *Electronics*, Sept. 15, 132 (1977).
7. H. J. Kiggen and H. Laumen, *Rev. Sci. Instrum.*, *52*, 1761 (1981).
8. A. Scher and C. Yarnitzky, *Anal. Chem.*, *53*, 356 (1981).
9. T. S. Light and S. L. Licht, *Anal. Chem.*, *59*, 2327 (1987).
10. A. Queen and R. Shabaga, *Rev. Sci. Instrum.*, *44*, 494 (1973).
11. F. F. Gaal, B. F. Abramovic, and R. I. Cservenak, *Microchem. J.*, *34*, 295 (1986).
12a. H. M. McNair and L. N. Polite, *Am. Lab.*, *20*(10), 116-121 (1988).
12b. G. O. Franklin, *Am. Lab.*, *17*(6), 65-80 (1985).
12c. H. Small, T. S. Stevens, and W. C. Bauman, *Anal. Chem.*, *47*, 1801-1809 (1975).
12d. H. H. Willard, L. L. Merritt, Jr., J. A. Dean, and F. A. Settle, Jr., *Instrumental Methods of Analysis*, 7th ed., Wadsworth, Belmont, CA, 1988, pp. 641-644.
12e. G. Horvai, F. Pal, Z. Niegreisz, K. Toth, and E. Pungor, *LC-GC*, *6*(12), 1058-1064 (1988).
12f. K. Harrison, W. C. Beckham, Jr., T. Yates, and C. D. Carr, *Am. Lab.*, *17*(5), 114-121 (1985).
12g. T. Jupille, *Am. Lab.*, *18*(5), 114-126 (1986).
12h. T. E. Miller, Jr., *Am. Lab.*, *18*(5), 49-56 (1986).
13. E. Pungor, *Oscillometry and Conductometry*, Pergamon, Oxford, 1965.
14. M. J. Relis, "Method and Apparatus for Measuring the Electrical Conductivity of a Solution," U.S. Patent 2,542,057, Feb. 20, 1951.
15. R. Shaw and T. S. Light, *ISA Transactions*, *21*, 63 (1982).
16. W. A. Gow, H. H. McCreedy, and F. J. Kelly, *Can. Mining Metallurg. Bull.*, July, 1966.
17a. A. R. Timm, E. M. Liebenberg, G. T. W. Ormrod, and S. L. Lombaard, *Nat. Inst. Metallurg.*, Report No. 2003, Johannesburg, Feb. 28, 1979.
17b. G. T. W. Ormrod, "Electrodeless conductivity meters in the measurement and control of the amount of lime in alkaline slurries," paper presented at the NIM-SAIMC Symposium on Metallurgical Process Instrumentation, *Nat. Inst. Metallurg.*, Johannesburg, 1978.
18. R. Calvert, J. A. Cornelius, V. S. Griffiths, and D. I. Stock, *J. Phys. Chem.*, *62*, 47 (1958).
19. K. M. Queeney and J. E. Downey, *Adv. Instrum.*, *41* (part 1), 339-352 (1986).
20. G. D. Fulford, "Use of Conductivity Techniques to Follow Al_2O_3 Extraction at Short Digestion Times," in *Light Metals* (H. O. Bohner, Ed.), The Metallurgical Society of America, AIME, Warrendale, PA, pp. 265-278, 1985.
to slaked lime addition control," Canadian Pulp and Paper Assoc., Montreal, 1968.
22. W. Musow and A. Bolland, "Toroidal conductivity sensor technology applied to cyanidation of flotation tailings circuits," Instrument Society of America, 12th Annual Mining and Metallurgy Industries Symposium, Vancouver, 1984.

23. W. Musow and J. Montgomery, "Recausticizing control utilizing toroidal magnetic sensor technology and controller with artificial intelligence," Pulping Conference/TAPPI, Atlanta, 1985.
24. H. Hinkelmann, *Z. Angew. Phys. Einschl. Nukleonic, 9*, 505 (1957).
25. ASTM D1125-82, *Standard Test Methods for Electrical Conductivity and Resistivity of Water*, 1983 Annual Book of ASTM Standards, Vol. 11.01, American Society for Testing and Materials, Philadelphia, pp. 149-156.
26. R. P. Frankenthal, in *Handbook of Analytical Chemistry* (L. Meites, Ed.), McGraw-Hill, New York, 1963, pp. 5-30 et seq.
27. F. Daniels, R. A. Alberty, J. W. Williams, C. D. Cornwell, P. Bender, and J. E. Harriman, *Experimental Physical Chemistry*, 7th ed., McGraw-Hill, New York, 1970.
28. G. W. Ewing, *Instrumental Methods of Chemical Analysis*, 3rd ed., McGraw-Hill, New York, 1969.
29. T. S. Light, "Electrodeless Conductivity," Chap. 29, pp. 429-441, in *Electrochemistry, Past and Present*, ed. by J. T. Stock and M. V. Orna, ACS Symposium Series 390, American Chemical Society, Washington, D.C., 1989.

20

Coulometry

ANDREW F. PALUS / EG&G Princeton Applied Research, Princeton, New Jersey

I. INTRODUCTION

All coulometric methods are based on Faraday's law. This states that the extent of a reaction that occurs because of the passage of an electric current is directly proportional to the charge Q that is passed. This is true at any point in the electrochemical reaction. For analytical purposes, coulometry is concerned with experiments in which the electrochemical reaction is carried to completion. In this case the charge is given by

$$Q = nFVC_0$$

where n is the number of electrons involved in the reaction, F is the Faraday constant (very nearly 96,487 coulombs per equivalent), V is the volume of the solution in liters, and C_0 is the initial concentration in moles per liter.

The quantity of the charge (which is just the time integral of the current) is measured by a coulometer. It is apparent from the above equation that in a coulometric method the charge passed is a direct measurement of concentration, with no need for chemical standards or standardization. Furthermore, the charge is not influenced by the physical character and temperature of the solution.

Since no chemical standards are needed for coulometry, the precision and accuracy of the method are comparable to the precision and accuracy of gravimetric methods. In fact, measuring charge is merely a means of counting the number of electrons used in a reaction. This means that in coulometry, just as in gravimetry, the quantity measured is a fundamental physical unit.

Coulometric methods may be divided into two types, designated as primary and secondary. In primary coulometry, the species of interest undergoes electrochemical oxidation or reduction directly. The number of electrons required for the reaction is a measure of the amount of material reacted. In secondary techniques, a reagent, generated electrochemically, undergoes a stoichiometric reaction with the material of interest. The number of electrons used to generate the reagent can be related to the number of moles of reagent produced. Knowing the number of moles of reagent generated allows the calculation of the concentration of the analyte. This secondary technique allows titration without requiring standardization of a titrant.

Both primary and secondary methods can be operated under either controlled current or controlled potential conditions. In practice the primary methods are usually carried out with controlled potential, because primary controlled-current experiments are apt to suffer reduced current efficiency as the concentration of the analyte decreases, giving rise to errors in the determination. Most coulometric titrations make use of the constant-current technique. Here the concentration of the material being oxidized or reduced by the direct action of the current is not decreased greatly as the experiment is run. The advantage of the constant-current technique is that the measurement of charge is straightforward, being merely the product of the current and the length of time it is applied. Controlled-potential experiments require the use of a coulometer to measure the charge [1-5].

In coulometry, the chemist must also have a way to tell when the electrolysis is completed. This end point may be detected in many different ways. In controlled-potential experiments the electrolysis is complete when the current drops to zero [6]. This point can be established by following the current with a strip-chart recorder. In controlled-current methods the end point is detected by more conventional techniques, involving, for example, colorimetric or electrometric observations.

Colorimetric techniques are straightforward in that the same indicators may be used as those familiar in conventional volumetric titrations. The commonly used electrometric techniques are potentiometric and amperometric. The potentiometric method requires a separate circuit (e.g., a pH meter) to monitor the potential. An abrupt change in the cell potential indicates the end point for constant-current electrolysis. The amperometric technique uses twin, polarized microelectrodes, which only pass current when an excess of reagent is present in the cell [7].

Many chemists consider all coulometric techniques to be coulometric titrations. Their rationale considers that a primary coulometric method involves titrating with electrons, just as one titrates with a standard solution in a volumetric technique. For the present chapter, we will restrict the term "coulometric titrations" to those techniques in which a titrant is generated coulometrically, usually under controlled-current conditions [8].

II. INSTRUMENTATION

The instrumentation for coulometry consists of a constant-potential or constant-current source, a coulometer, and a cell. The potential or current source is either a three-electrode potentiostat or a galvanostat. The coulometer can take many forms; it can be an analog or digital coulometer, or it may consist of the simple integration of the current against time. The cell should have a working electrode with large surface area, and provision for isolation of the counter electrode from the rest of the solution. Let us consider each part separately.

A. Controlled-Potential Coulometry

Potentiostat

Controlled-potential coulometry (CPC) places more of a burden on the coulometer than on the potentiostat. The potentiostat must be able to control the potential accurately, to handle relatively large currents (hundreds of milliamperes), and its current noise must be low (for a discussion of noise in a potentiostat, see Chapter 17 and Ref. 9). Accurate control of the potential is usually guaranteed by using a three-electrode cell and employing a potentiostat with adequate compliance voltage. Compliance is a measure of the ability of the potentiostat to generate the required current at the working electrode, and should not be confused with the voltage range that the potentiostat can control. A potentiostat may have a compliance of 100 V, but only be capable of controlling potential over a ±10 V range. Typical compliances range from 10 to 100 V.

Coulometer

The specifications of the coulometer in CPC need to be more stringent, as results of the experiment will only be as accurate and precise as the coulometer, assuming all other experimental conditions are held constant. To better understand the coulometer, one must consider three approaches to the charge measurement problem.

Charge is just the time integral of the current. A simple current integrator circuit is shown in Figure 1. The voltage across the capacitor can be monitored by connecting a voltmeter from the output to ground. The voltage V across a capacitor C is proportional to the charge Q according to the relation $V = QC$, so the voltage built up in the coulometer is a measure of the charge accumulated. This circuit thus forms a charge meter or coulometer [1,10].

The analog coulometer has a fast response, and is usable for CPC, but does suffer from shortcomings. The total charge must be kept small or the size of the capacitor becomes prohibitive. Capacitors cannot be obtained with the same tolerance precision as resistors; the capacitor may be "leaky" or in other ways change with time. Analog coulometers tend to drift with time, due partly to the capacitor problem and partly to changes in the operational amplifier. The charge on an analog coulometer cannot be held for a long period of time. If the input is disconnected, the charge on the capacitor will begin to bleed off with time.

A digital coulometer is a device that takes an analog signal (the current) and converts it to a digital form (the number of coulombs). This is usually accomplished by converting the current to a proportional voltage, then sending the voltage to a voltage-to-frequency converter. This is a device that produces a train of pulses at a frequency determined by the applied voltage. The pulse train is then taken to a counter that is incremented by each pulse. The counter thus can display the number of coulombs that have been used.

The advantage of the digital coulometer is that it uses in the current follower a resistor instead of a capacitor. The resistor is a much more precise element in electronic circuits. The digital coulometer can be designed to be drift free and have a precision better than ±0.1 percent. Digital coulometers allow the user to sample anodic (oxidative) current, cathodic (reductive) current, or both. This gives added selectivity, since the analyst can choose to integrate only the current due to a reduction. The digital coulometer is also capable of remembering the accumulated charge even after the current input is disconnected. When the current is turned off the counter does not begin to drift to a smaller number, but holds its value indefinitely.

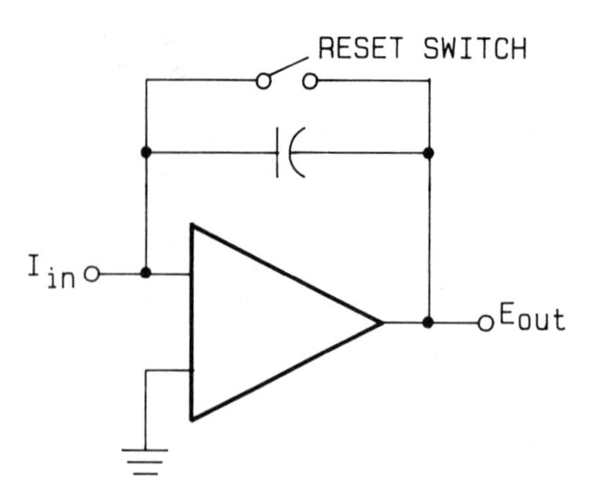

Figure 1. A simple circuit for an analog coulometer.

Both analog and digital coulometers should be provided with a means for subtracting a background current. In all practical experiments the coulometric cell will show a residual current flowing, even though there is no material to be oxidized or reduced. This residual current will cause an error in the measurement. Experiments are usually performed over a minimum time period (perhaps 10 min) by proper selection of the cell configuration and other experimental parameters, in an effort to reduce errors.

The third approach to integrating the current is to make a numeric approximation of the integral of current with time by adding up discrete current measurements. This can be done easily with computer-controlled instrumentation. Here the current is sampled at successive times and stored in the computer. The computer can then be used to good advantage to sum all of the currents during the experiment. The more frequently the current is sampled (i.e., the more data points that are taken), the more accurate will be the result. The limiting factor in this type of integration is the number and linearity of bits in the analog-to-digital converter (ADC). Most ADCs have some residual nonlinearity, and this gives rise to a small constant error that is propagated through the measurement.

Commercial instrumentation for CPC usually comes in separate pieces, one of which is typically a digital coulometer. The coulometer may be as simple as a unipolar, single-current range unit (e.g., model 640 of The Electrosynthesis Company) to a multirange, bipolar instrument (e.g., model 379 from EG&G Princeton Applied Research). The difference between the different classes, other than cost, is their versatility. A unipolar coulometer is only capable of integrating cathodic or anodic current, whereas a bipolar coulometer can integrate cathodic current, anodic current, or both. Most coulometers can be teamed with any good potentiostat.

Cell

The cell for CPC consists of a working electrode, a counter (auxiliary) electrode, a reference electrode, and a means for deoxygenating the sample.

The reference electrode can be any well-behaved half-cell. Typical reference electrodes are saturated calomel (SCE), Ag/AgCl, and Hg/Hg_2SO_4 (for more details, see Chapter 16). The working and counter electrodes require special consideration.

The counter electrode is usually a piece of noble metal such as a platinum wire or flag. Its size is not important. What is important is to separate physically the counter electrode from the analytical solution so that any reaction occurring at the counter electrode will not interfere with the desired determination. The usual way of isolating the counter electrode is to place it in a glass tube equipped with a porous plug (a bridge tube). The tube is then filled either with the same electrolyte used in the analytical compartment or with some other suitable conductive solution.

The working electrode is fabricated from a conductive material at which the reaction of interest can take place. Typical working electrodes are a platinum or silver basket, a mercury pool, or a graphite rod. In all cases the area of the working electrode should be large relative to the volume of the solution. In order for efficient electrolysis to occur in short times, as much solution as possible must come in contact with the working electrode.

The final consideration in the cell design is the incorporation of a good stirrer. This is needed to ensure that all of the reactant will come in contact with the working electrode. Most coulometric cells incorporate a stirrer and working electrode combination that allows the experiment to proceed to completion in 10 min or less [4]. See Ref. 11 for a description of a complete system for CPC, including potentiostat, coulometer, and cell.

Applications

Controlled-potential coulometry has been used in many areas of analysis. The pharmaceutical industry uses CPC for assays of drugs [12-14]. The mineral processing industry has made use of the technique for precious metal assays [15], uranium determinations [16-19], transuranium element determination [19,20], and as an alloy composition check [21-23]. Clinical chemists have used CPC to determine certain carcinogens [24,25]. Other analysts have used CPC for the determination of organic compounds [26] and metals [27,28]. The relatively slow speed of analysis is offset by the excellent precision and the "absolute" nature of the technique.

B. Controlled-Current Coulometry

Galvanostat

The galvanostat is the heart of the controlled-current coulometry (CCC) technique. It requires only two electrodes, though many commercial instruments provide a reference electrode input as a means to monitor the cell potential. The accuracy and precision of a controlled-current experiment is only as good as the current control. Commercial instrumentation is typically capable of a ±0.2 percent accuracy in the current control circuit.

Coulometer

In most cases a coulometer is not needed in CCC, since the charge can be calculated from the known current and the length of time it is applied.

Most commercial coulometers can be used to monitor the number of coulombs used in a constant current experiment if desired. See the earlier section on coulometers for more information.

Cell

The cell in CCC is similar to that described above for CPC. Unless the potential is being monitored, there is no need for a reference electrode. Usually additional electrodes are provided for end-point detection. In all cases the two generator electrodes must be isolated from one another by a porous frit or sintered glass barrier, to minimize contamination and interference. Adequate stirring is essential to ensure that the two species involved in the chemical reaction come in contact with each other.

General Applications

As noted above, little primary coulometry is done under constant-current conditions. The development of constant-current, potential-limited, coulometry in 1962 has never replaced the constant-potential approach [29]. The principal analytical applications of CCC are the generation of titrant in coulometric titrations. The reader is referred to the compilation by Farrington [30] for a comprehensive listing of methods for the coulometric titration of many substances.

C. Coulometric Titration Techniques

Coulometric titrations, as mentioned earlier, are almost always done under constant-current conditions, greatly simplifying measurement of the charge. While coulometric titrations could be done with the general-purpose instrumentation discussed above, the majority of commercially available coulometric titrators are specific-use devices, many of which are simply called analyzers. An example is the Dohrmann total organic halide analyzer. This device consists of a sample introduction system and a microcoulometer. Perhaps the best-known coulometric titration instruments are those optimized for the generation of iodine in the Karl Fischer method for the determination of water, as exemplified by the Aquatest IV of Photovolt.

The coulometric titration portions of these special-purpose instruments are usually designed solely for the titrations specified. These dedicated titrators cannot readily be adapted to other applications, especially since their readouts are in terms of specific concentration units rather than simply in coulombs.

Among these analyzers are a carbon dioxide titrator (from Coulometrics, Inc.), a sulfur analyzer, and a halide analyzer. The determination of halides in hydrocarbons [31] is frequently carried out by CCC. In this procedure samples are pyrolyzed and the resulting gases are absorbed into a titration cell where the halides react with electrolytically generated Ag^+ ion.

The CO_2 coulometric titrator can be used for the determination of either carbon [32,33] or oxygen [34]. This versatility is not due to any change in the coulometric step, but rather to a modification of the sample introduction system. The sample is always converted to CO_2, which is then swept into the reaction cell where titration by electrogenerated OH^- ion takes place to a photometric end point. Electronic circuitry is provided to allow the readout to be presented in terms of carbon, oxygen, or carbon dioxide, as desired.

Sulfur in petroleum products [35,36], and in environmental samples [37], has been determined coulometrically. The titrators (available, e.g., from Coulometrics and Dohrmann) are based on a redox principle. The solution in the cell contains an excess of free iodine, which reacts with the SO_2 produced by pyrolysis of the sample. The iodine is continuously replaced by electrolytic oxidation of iodide. The titration is susceptible to the same sources of interferences that may be expected in the corresponding volumetric titration [38].

A likely problem with coulometric titrators of the types just described is the possibility of overshooting the end point. This may occur because of slow reaction time, slow end-point detector response, or delay in the electronic circuit that makes the stop-titrating decision. The manufacturers attack this problem by reducing the current level as the end point is approached. This is comparable to using small increments of solution in a volumetric titration when nearing the end point.

III. CONCLUSIONS

Coulometry is a useful analytical technique. The chief advantage is that it provides good precision and does not rely on chemical standards. The method is based on the measurement of an absolute unit. General instrumentation is available for constant-potential coulometry, but it requires a knowledgeable user.

Coulometric titrators are available for specific determinations. These instruments are easy to run, can be automated, and are well accepted as analytical tools, but are not easily adapted for general-purpose applications. For arguments in favor of coulometric titration as a general analytical method, see Ref. 39.

A chemist wishing to introduce coulometry should decide whether the equipment will be used for a specific determination on a routine basis or for various determinations at different times. A generalized instrument can be put to other electrochemical uses in the laboratory. Purchase of a general system, consisting of a coulometer combined with a potentiostat, makes more sense for the researcher or nonroutine analyst. Instruments dedicated to specific reactions are excellent for a routine quality-control laboratory. Most manufacturers can provide application notes for the use of their instruments for various types of samples.

IV. APPLICATIONS

The following references will provide entrance to the applications literature.

As: K. Kuramoto, T. Sato, and K. Isida, *J. Electrochem. Soc.*, *134*, 1286 (1987).

Cu: M. Van Eenoo, C. Labar, and L. Lamberts, *Anal.*, *19*, 2063 (1986).

Cu, In, Se: M. H. Yang, M. L. Lee, Y. M. Lin, and H. L. Hwang, *Thin Solid Films*, *155*, 317 (1987).

Fe, steel: H. J. Boniface, *39th Proc. Chem. Conf.*, 111 (1986).
R. H. Jenkins, *Steel Res.*, *57*, 491 (1986).

Organics: J. Barek, A. Berka, and J. Jima, *Microchem. J.*, *34*, 166 (1986).

Pu, U: R. C. Sharma, P. K. Kalsi, L. R. Sawant, S. Vaidyanathan, and R. H. Iyer, *J. Radioanal. Nucl. Chem.*, *126*, 1 (1988).
H. Aoyagi, Z. Yoshida, and S. Kihara, *Anal. Chem.*, *59*, 400 (1987).
B. F. Myasoedov, *Inorg. Chim. Acta*, *140*, 231 (1987).

SO_2: Kh. Sheitanov, E. Pancheva, and M. Neshkova, *Izv. Khim.*, *20*, 421 (1987).

Ti: C. K. Baskare and K. N. Ganage, *Indian J. Chem.*, *Sect. A*, *27A*, 350 (1988).

V: V. Stuzka and J. Urbanek, *Acta Univ. Palacki. Olomuc.*, *Fac. Rerum Nat.*, *88* (Chem 26), 97, (1987).

REFERENCES

1. H. V. Malmstadt and C. G. Enke, *Electronics for Scientists*, Benjamin, New York, 1962.
2. G. A. Rechnitz, *Controlled-Potential Analysis*, Macmillan, New York, 1963.
3. J. A. Plambeck, *Electroanalytical Chemistry*, Wiley, New York, 1982.
4. J. E. Harrar, in *Electroanalytical Chemistry* (J. A. Bard, Ed.), Vol. 8, Marcel Dekker, New York, 1975.
5. W. R. Heineman and P. T. Kissinger, in *Laboratory Techniques in Electroanalytical Chemistry* (P. T. Kissinger and W. R. Heineman, Eds.), Marcel Dekker, New York, 1984.
6. F. B. Stephens, F. Jakob, L. P. Rigadon, and J. E. Harrar, *Anal. Chem.*, *42*, 764 (1970).
7. H. A. Strobel, *Chemical Instrumentation*, Addison-Wesley, Reading, Massachusetts, 1973.
8. D. J. Curran, in *Laboratory Techniques in Electroanalytical Chemistry* (P. T. Kissinger and W. R. Heineman, Eds.), Marcel Dekker, New York, 1984.
9. J. A. von Fraunhofer and C. H. Banks, *The Potentiostat and Its Applications*, Davey, Hartford, Connecticut, 1972.
10. B. H. Vassos and G. W. Ewing, *Electroanalytical Chemistry*, Wiley, New York, 1983.
11. H. Sigerman, J. Chang, and J. Thompson, *Am. Lab.*, *7*, 72, February (1975).
12. M. A. Brooks, in *Laboratory Techniques in Electroanalytical Chemistry* (P. T. Kissinger and W. R. Heineman, Eds.), Marcel Dekker, New York, 1984.
13. K. I. Nikolic and K. R. Velasevic, *Acta Pol. Pharm.*, *42*, 209 (1985).
14. K. I. Nikolic and K. R. Velasevic, *Acta Pharm. Jugosl.*, *35*, 41 (1985).

15. L. P. Rigdon and J. E. Harrar, *Anal. Chem.*, *46*, 696 (1974).
16. H. Takeishi, H. Muto, H. Aoyagi, T. Adachi, K. Izawa, Z. Yoshida, H. Kawamura, and S. Kihara, *Anal. Chem.*, *56*, 458 (1986).
17. N. M. Saponara and D. D. Jackson, *Anal. Chem. Symp. Ser.*, *19*, 345 (1984).
18. G. R. Relan et al., *J. Radioanal. Nucl. Chem.*, *83*, 239 (1984).
19. T. L. Frazzini, *IEEE Trans. Nucl. Sci.*, *NS29*, 866 (1982).
20. M. K. Holland and K. Lewis, *Anal. Chim. Acta*, *149*, 167 (1983).
21. B. Alfonsi, *Anal. Chim. Acta*, *20*, 277 (1959).
22. B. Alfonsi, *Anal. Chim. Acta*, *23*, 375 (1960).
23. B. Alfonsi, *Anal. Chim. Acta*, *19*, 569 (1958).
24. J. Barek et al., *Collect. Czech. Chem. Commun.*, *50*, 2853 (1985).
25. J. Barek et al., *Collect. Czech. Chem. Commun.*, *50*, 1819 (1985).
26. J. Barek, A. Berka, and J. Zima, *Microchem. J.*, *34*, 166 (1986).
27. R. L. Altman, *Anal. Chim. Acta*, *63*, 129 (1973).
28. Y. Su and D. E. Campbell, *Anal. Chim. Acta*, *47*, 261 (1969).
29. I. J. McColm et al., *Talanta*, *10*, 387 (1963).
30. P. S. Farrington, in *Handbook of Analytical Chemistry* (L. Meites, Ed.), McGraw-Hill, New York, 1963, pp. 5-187 et seq.
31. Y. Takahashi et al., *Chem. Water Reuse*, *2*, 127 (1981).
32. C. C. Y. Chan, *Geostand. Newsl.*, *10*, 131 (1986).
33. E. E. Engleman et al., *Chem. Geol.*, *53*, 125 (1985).
34. W. H. Benner, *Anal. Chem.*, *56*, 2871 (1984).
35. S. M. Farroha, A. E. Habboush, and M. N. Michael, *Anal. Chem.*, *56*, 1182 (1984).
36. I. J. Oita, *Anal. Chem.*, *55*, 2434 (1983).
37. E. Canelli and L. Husain, *Atmos. Environ.*, *16*, 954 (1982).
38. F. Z. Grandi et al., *Analyst*, *107*, 17 (1982).
39. G. W. Ewing, *Am. Lab.*, *13*(6), 16 (June 1981).

IV
CHROMATOGRAPHIC METHODS

21

Modern Gas Chromatrographic Instrumentation

SHERRY O. FARWELL / Department of Chemistry, University of Idaho, Moscow, Idaho

I. INTRODUCTION

Since the initial description of a gas-liquid chromatographic apparatus in 1952 [1], the number of gas chromatographs in use has grown to approximately 200,000. In fact, gas chromatography (GC) remains the most widely used separation method in analytical chemistry despite recent advances in liquid, supercritical fluid, and ion chromatography. Given the above, it is not surprising that there are at least 52 different commercial manufacturers of gas chromatographs, most of whom offer several models. Column A of Table 1 lists the companies manufacturing GCs in 1986-1987. Thus, the tremendous growth in GC research and applications during the last several years has been matched by a similar increase in the types of GC instrumentation. The major purpose of this chapter is to assist future GC users in selecting the instrumentation best suited to their applications from this chromatographic equipment plethora. No one GC is best for all analysis problems, and effective problem solving and purchasing involves selecting the tool most appropriate for the solution of the problem, that is, the proper analytical approach to problem-solving.

A. Analytical Approach to Instrument Selection

The decision to purchase a chromatographic system is usually based on the need to obtain measurement data for various analytes in multicomponent samples. The guidelines presented in the general introduction to this handbook should be used to assess the nature of the problem and to determine whether chromatography is the correct method for providing key information for solving the problem.

Chromatographic Choices

Now let's assume the application of the general criteria has advised you to consider chromatographic methods in general. How do you decide whether this is an analytical problem whose solution is appropriate for gas chromatography, as opposed to high-performance liquid (HPLC), thin-layer (TLC), or supercritical fluid (SFC) chromatography?

Table 1. Firms Offering Gas Chromatographs and Accessories[a]

Company	A	B	C	D	E	F	G	H	I	J	K	L	M	N	O	P	Q	R	S	T	U	V
Accuspec Inst. Corporation	x		x									x	x				x					
Ace Scientific Supply Co., Inc.																					x	
Aldrich Chemical Co., Inc.		x																				
Alltech Associates	x	x	x				x						x			x	x	x			x	x
ALPHA Applied Research																					x	
Analytical Inst. Dev., Inc.	x							x	x				x			x	x					
Analytical Parameters																				x		
Ancal, Inc.												x	x				x	x				
The Anspec Co.	x	x	x	x														x			x	x
Antek Instruments, Inc.	x	x	x																			x
Applied Analytical Industries																					x	
Applied Automation, Inc.	x			x					x				x			x	x					
Applied Chromatography Systems, Ltd.																					x	
Arnel Laboratory Services, Inc.	x			x																		
Axxiom Chromatography																					x	
Baseline Industries, Inc.	x							x	x				x			x	x					
Beckman Instruments, Inc.																					x	
Beckman Insts., Inc., Altex Div.		x																			x	
Benson Polymeric		x																				
Bid Three Industries			x																			
Bio-Rad Laboratories		x		x																		
The Boc Group, Inc.	x																					
Bristol Babcock, Inc.																						x
Buck Scientific, Inc.	x			x				x	x			x	x				x					x
Cal-Glass for Research, Inc.		x																				
Carlo Erba Strumentazione SpA	x	x	x	x	x			x	x		x		x				x				x	
Chemcon, Inc.			x																			
Chemical Data Systems			x																			
Chemical Research Supplies	x	x	x	x													x					x
Chromatography Services, Ltd.				x																		
Chrompack, Inc.		x	x																			

Table 1 (Continued)

Company	A	B	C	D	E	F	G	H	I	J	K	L	M	N	O	P	Q	R	S	T	U	V
Chrompack International BY		x	x																		x	
Chrompack-Packard	x						x	x	x				x		x	x	x	x			x	
Cole-Parmer Instrument Co.																						x
Cole Scientific, Inc.																						x
Corning Glass Works		x																				
Dani SpA	x			x				x	x				x			x	x	x				
Data Translation, Inc.																					x	
Delsi Instruments	x							x	x				x	x			x					
DET, Inc.																		x				
Duryea Associates, Inc.																					x	
Dynamic Solutions																						x
Dynatech Precision Sampling Corp.			x	x																		
Envirochem, Inc.	x		x																			x
Erba Instruments, Inc.		x	x																			
EM Science																					x	
ES Industries	x			x			x	x	x				x	x		x	x			x		
Fenwal Electronics	x																x		x			
FiAtron Laboratory Systems			x																			
Field Analytical UK, Ltd.								x		x			x				x					
Finnigan-MAT														x								
Fisher Scientific Co.	x																					
Foss Electric, Ltd.	x			x			x	x	x			x	x				x	x				
Foxboro Co.	x	x											x									
Foxboro Analabs		x																				
Galbraith Laboratories, Inc.		x																				
Galileo Electro-Optics Corp.		x																				
Gallenkamp		x																				
GAT Gamma Analysentechnik Gmbh																					x	
Gilson Medical Electronics SA																					x	
Gow-Mac Instrument Co.	x									x			x									
Haake Buchler Instruments	x			x			x	x	x		x		x			x	x	x				

(continued)

Table 1 (Continued)

Company	A	B	C	D	E	F	G	H	I	J	K	L	M	N	O	P	Q	R	S	T	U	V
Hach Co./Carle Chromatography	x												x				x					
Hamilton Co.			x																			
Handy & Harmon Tube Co.		x																				
Herrmann-Moritz	x																					
Hewlett-Packard Co.	x	x	x	x			x	x	x			x	x	x			x	x			x	x
HNU Systems, Inc.	x																					
Houston Atlas, Inc.																				x		
HPLC Technology, Ltd.				x																		x
IBM Instruments, Inc.																	x			x	x	
ICI Australia Operations Pty., Ltd.																					x	
Inacom Instruments BV				x																		
Informetrix, Inc.																					x	
Instrumentos Cientificos	x							x	x				x			x	x					
Interactive Microware, Inc.																					x	x
International Equipment Trading, Ltd.	x																					
IR&D, Inc.				x																		
Isolab, Inc.		x																				
J & S Scientific, Inc.		x																				
J & W Scientific, Inc.		x	x																			
Japan Spectroscopic Co., Ltd.																					x	
Jasco, Inc.		x																				
Jones Chromatography, Ltd.																					x	
Joyce-Loebl																					x	
Kimble Div.		x																				
Konik Instruments SA	x			x			x	x		x			x	x		x	x	x			x	
Kontes Glass		x																				
Kontron AG																					x	
Kratos Analytical																					x	
Lab-Crest Scientific		x																				
Lab Glass, Inc.		x	x																			
Laboratorium Prof. Dr. Berthold																					x	
Labtronix, Inc.				x																		

Table 1 (Continued)

Company	A	B	C	D	E	F	G	H	I	J	K	L	M	N	O	P	Q	R	S	T	U	V
Lachat Instruments																					x	
LDC/Milton Roy																					x	
Lee Scientific		x											x							x		
Linseis, Inc.																						x
LKB Instruments, Inc.		x																				
LKB Produkter AB																					x	x
Magnus Scientific Instruments	x			x				x	x				x	x			x					
Mark Instrument Co.		x																				
Markson Science													x									
MCRA Applied Technologies, Inc.																					x	
Micromeritics Instrument Corp.		x																				
Microsensor Technology, Inc.	x	x	x														x					
Mikroalb Aarhus A/S	x												x				x					
Nalge Co.		x																				
Nelson Analytical, Inc																					x	
Nordion Instruments Oy, Ltd.																					x	
Nuclear Sources & Services					x			x			x				x							
Nuclide Corp.	x																					
OI Corporation							x															
On-Line Instrument Systems, Inc.																					x	
Packard Instruments Co, Inc.			x																			
PCP, Inc.	x																			x		
Perkin-Elmer Corp.	x	x	x	x			x	x	x			x	x	x		x	x	x			x	x
Perkin-Elmer & Co. Gmbh																					x	
Perkin-Elmer, Ltd.			x																			
Pharmacia, Inc.		x																			x	
Pharmacia AB																					x	
Phase Separations, Ltd.		x																				
Phenomenex		x																				
Philips Analytical	x			x				x	x				x				x	x				
Philips Export BY			x																			x
Photovac, Inc.	x															x						

(continued)

Table 1 (Continued)

Company	A	B	C	D	E	F	G	H	I	J	K	L	M	N	O	P	Q	R	S	T	U	V
P. J. Cobert Associates, Inc.	x									x			x				x	x				
Process Analyzers, Inc.	x					x			x	x			x			x	x					
Prolabo		x																				
Pye Unicam, Ltd.	x	x	x																		x	x
Quadrant Scientific, Ltd.																					x	
Quadrex Corporation		x																				
Rainin Instrument Co., Inc.																					x	
Reliance Glass Works, Inc.		x	x																			
Restek Corporation		x																				
Sargent-Welch Scientific Co.																						x
Sarstedt, Inc.		x																				
Scientific Glass & Instruments, Inc.		x																				
Scientific Glass Engineering, Inc.		x	x																			
Scientific Marketing	x			x			x	x	x				x			x	x		x			
Scott Environmental Technology, Inc.			x																			
S-Cubed																					x	
Sentex Sensing Technology, Inc.	x				x			x								x						
Seragen, Inc.		x																				
Shimadzu Corp.	x		x	x				x	x				x			x	x					x
Shimadzu (Europa) Gmbh																					x	
Shimadzu Scientific Instrument, Inc.		x	x	x				x	x				x	x			x	x			x	
Siemens Energy & Automation	x			x					x				x				x					
Sievers Research, Inc.																				x		
Signal Instrument Co., Ltd.													x									
Soltec Corporation																					x	
SOTA Chromatography, Inc.																					x	
Spectra-Physics																					x	x
Spectra-Physics Gmbh																					x	
Spectra-Physics, Autolab Div.		x																			x	
Spectrum Scientific, Inc.		x																				
Spectrum Medical Industries, Inc.																					x	
Spiral Sarl	x							x					x	x			x					
SRI	x		x					x					x				x	x				

Table 1 (Continued)

Company	A	B	C	D	E	F	G	H	I	J	K	L	M	N	O	P	Q	R	S	T	U	V
Sun Brokers, Inc.			x																			
Supelco, Inc.		x		x																		
Suprex Corporation		x																				
Sycopel Scientific, Inc.	x					x			x													
System Instruments Corp.																					x	
Techmation, Ltd.	x																					
Technical Associates															x							
Tegal Scientific, Inc.	x															x						
Tekmar Co.			x	x																		
Thermedics																				x		
Tracor Instruments	x			x			x	x	x				x			x	x	x	x		x	x
Trivector Systems Internat., Ltd.																					x	
Tudor Scientific Glass Co.		x																				
Ultra-Scientific		x																				
U-Microcomputers, Ltd.																					x	
Unimetrics Corporation		x																				
Universal Scientific, Inc.		x	x																		x	x
Vacumetrics, Inc.														x								
Valco Instruments Co., Inc.	x		x	x				x			x											
Varex Corporation	x		x	x																	x	
Varian Associates	x			x	x		x	x	x				x			x	x	x			x	
Varian Instrument Group		x	x																			x
VG Instruments														x								
VWR Scientific		x																				
Waters Div., Millipore Corp.																					x	x
Wheaton Scientific		x																				
VICI Metronics, Inc., Condyne Div.			x																			
VICI AG/Valco Europe								x			x											
Wilmad Glass Co., Inc.		x	x																			
YMC, Inc.		x																				
Young Laboratories		x																				
Zymark Corporation				x																		

(continued)

Table 1 (Continued)

[a]Identifier for Columns of Table 1:

	Column	Sources
A	Gas chromatographs	*LC-GC*, *4*:824 (1986); *Anal. Chem.*, *58*:48 (1986); *Am. Lab.*, *18*:44
B	Open tubular columns	*Anal. Chem.*, *58*:47 (1986)
C	Injection systems	*Anal. Chem.*, *58*:50 (1986)
D	Automatic samplers	*LC-GC*, *4*:835 (1986)
E	Argon ionization detectors	*LC-GC*, *4*:830-831 (1986)
F	Coulometric detectors	"
G	Electrolytic conductivity detectors	"
H	Electron-capture detectors	"
I	Flame photometric detectors	"
J	Gas density balance detectors	"
K	Helium ionization detectors	"
L	Infrared detectors	"
M	Flame ionization detectors	"
N	Mass spectrometer detectors	"
O	Nuclear GC detectors	"
P	Photoionization detectors	"
Q	Thermal conductivity detectors	"
R	Thermionic ionization detectors	"
S	Ultrasonic detectors	"
T	Other detectors (atomic emissions, etc.)	"
U	Data systems	*Am. Lab.*, *18*:44 (1987)
V	Recording integrators	*Anal. Chem.*, *58*:50 (1986)

When maximum resolution, detectability, and simplicity are important considerations, then GC should be evaluated for its compatibility with the sample and the desired analytes. The major limitations of GC are established by three factors: (a) the volatilities of the actual analytes, or their chemical analogs produced by simple derivatization reactions, (b) the thermal stabilities of the analytes, or their chemical derivatives, at the temperatures required for volatilization, and (c) the thermal stabilities of the column coatings and the stationary phases. Until 1986, GC was restricted to an upper temperature of approximately 375°C due to column degradations at higher temperatures. This upper temperature limit meant a compound had to have an atmospheric boiling point of less than about 500°C and a molecular weight of less than about 1000 amu. Recent developments [2,3] in the fabrication of aluminum-clad, fused-silica open tubular columns with bonded and cross-linked stationary phases have extended the upper column temperature limit to the 440°C range, which in turn makes it feasible to use GC for compounds with boiling points up to about 750°C. This widened boiling point range and the number of simple derivatization reactions available for introducing sufficient volatility [4] are continuing to increase the practical versatility of modern GC. The broad application boundaries of GC make it a logical first choice for consideration in solving analytical problems that require analyte separation, identification, and quantification.

Future Problems and Needs

It is also a worthwhile idea to consider potential uses in the future for the GC, especially if the instrumentation being considered is so-called "top-of-the-line" models with corresponding prices. For example, does the instrument have the necessary features to handle perceived uses in the future? If not, can such features be added to the existing unit, and what are the costs of adding these options now versus adding them as retrofits? If anticipated use will be high for this GC in one type of application, will such additional features on this one instrument really be practical compared to the acquisition of several specifically equipped GCs?

Identifying practical rather than imagined needs is a job that can be done in cooperation with competent sales personnel, but the potential customer should control such interactive decision-making rather than vice versa. Defining one's future requirements, if any, is an essential step to obtaining what your laboratory actually needs and not what you think they need. The salesperson will appreciate your understanding of the instrumental requirements, since he or she now has a better chance of making you a satisfied customer and thereby enhancing the opportunity for repeat business.

Training and Service

Depending on the sophistication of the GC system and the experience of the users, manufacturer training may be either required or desirable. In addition to basic training, if any, when the GC system is delivered and installed, what other training programs are available? Are these further training programs offered at convenient times and places, and at reasonable cost?

Eventually your instrument will require service. What facilities and services are available to support both new application problems and any necessary maintenance repairs? What is the availability of spare parts and how many field engineers are there in your service area? What are the types and costs of service contracts? Opinions of other possible users in your area are especially valuable with respect to judging the instrument's performance record and the likely quality of after-sales support and service. If possible, contact users you know in addition to those names provided on a user list from the manufacturer.

II. SCOPE OF MODERN GAS CHROMATOGRAPHY

A. General Column Classification

The practice of gas chromatography falls into two general categories based on the types of columns: packed or unpacked [5-8]. Packed columns consist of metal or glass tubing filled with solid powder, either uncoated adsorbents (gas-solid chromatography) or a solid support coated with a stationary phase (gas-liquid chromatography). Unpacked columns are the result of Golay's [9,10] recognition that open tubular columns could provide vast improvements in chromatographic resolution because of their "openness." The low gas flow resistance in such open columns compared to packed columns permits the use of longer columns without requiring extremely high inlet pressures. Since resolution (R) is expressed by

$$R = \frac{\sqrt{n}}{4} \left(\frac{\alpha - 1}{\alpha}\right)\left(\frac{k_2}{k_2 + 1}\right) \tag{1}$$

where α is the relative retention defined by the ratio of the two adjusted retention times t'_{R2}/t'_{R1}, k_2 is the partition ratio defined by t'_{R2}/t_M where t_M is the column holdup time, and n is the number of theoretical plates defined by L/h where L is the column length and h is height equivalent to a theoretical plate. Substitution of n = L/h into equation (1) yields

$$R = \frac{1}{4} \sqrt{L/h} \left(\frac{\alpha - 1}{\alpha}\right)\left(\frac{k_2}{k_2 + 1}\right) \tag{2}$$

which indicates that resolution is directly proportional to the square root of column length. Thus, increasing the column length is one way to improve resolution, and open tubular columns permit us to take advantage of the $R \propto \sqrt{L}$ relationship.

There are three main types of open tubular columns: wall-coated open tubular (WCOT), support-coated open tubular (SCOT), and porous-layer open tubular (PLOT). Such open tubular columns are also commonly referred to as capillary columns because of their small internal diameters, but this capillary nomenclature is not recommended due to its lack of specificity [11]. The WCOT columns are prepared by coating the inner column walls with the liquid stationary phase. Current technology for the preparation of WCOT columns also usually includes deactivation of the fused silica surfaces prior to coating with the stationary phase and immobilization of this stationary phase by surface bonding and/or cross-linking [12]. SCOT columns are dissimilar to WCOT columns in that their stationary phases are coated onto a layer of fine porous support material attached to the inside column walls [13,14]. PLOT columns are similar to SCOT columns in that a porous material is deposited on the inner column walls. For example, new PLOT columns coated with aluminum oxide, porous polymers, or molecular sieves are useful in the separation of C_1-C_{10} hydrocarbons, mixtures of polar and apolar volatiles, and permanent gases, respectively [15,16]. Because of continuing advances in WCOT and PLOT technology, these columns are preferred over SCOT columns for most applications.

B. Wide-Bore WCOT Columns versus Packed Columns

Although packed columns are still being used for determinations requiring the routine separation of a limited number of components, the reasons for their use are becoming more difficult to justify. This conclusion is due to the introduction of fused-silica open tubular (FSOT) columns with larger inside diameters of 0.53 and 0.75 mm (i.e., wide-bore or megabore capillary columns) and variable film thicknesses of stationary phase [17-21]. As shown in Tables 2 and 3, these wide-bore open tubular columns complement those FSOT columns with smaller internal diameters of 0.10 mm (ultra narrow-bore), 0.25 mm (narrow-bore), and 0.32 mm (medium-bore) [8,13, 14,18,19,22,23]. For example, the wide-bore FSOT columns with relatively

Table 2. General Characteristics of FSOT Columns by Internal Diameter[a]

Characteristic	0.10 mm	0.25 mm	0.32 mm	0.53 mm	0.75 mm
Sample capacity (per component on column)	10-50 ng	50-250 ng	100-300 ng	0.5-2 μg	10-15 μg
Direct/on-column injection volume limit (μl)	0.5	1	2	5	6
Efficiency (n/m)	10,000	4000	3200	1900	1100
Optimum carrier-gas velocity (He, ml/min)	0.2	0.7	1.4	2.5	5.0

[a]The "generalized" values given in this table will also vary with the thickness of the stationary film, the column length, and the types of stationary phases and analytes. Values listed in the table are representative of columns with the following internal diameters and film thicknesses: 0.10 mm ID, 0.2-μm film; 0.25 mm ID, 1.0-μm film; 0.32 mm ID, 1.0-μm film; 0.53 mm ID, 5.0-μm film; 0.75 mm ID, 5.0-μm film.

Table 3. General Relationships among FSOT Column Internal Diameter, Stationary-Phase Film Thickness, and Maximum Sample Capacity

Column ID (mm)	Stationary-phase film thickness			
	0.1 μm	0.2 μm	1.0 μm	5.0 μm
0.10	10-20 ng	20-50 ng	—	—
0.25	25 ng	50 ng	250 ng	1 μg
0.32	30 ng	60 ng	300 ng	1.5 μg
0.53	50 ng	100 ng	500 ng	2.5 μg

thick films of stationary phase have sample and injection volume capacities that are comparable to those of conventional 2-mm ID packed columns, that is, a sample capacity of 10-20 μg and a maximum injection volume of 5-10 μl. The other major obstacle to the replacement of packed columns by FSOT columns has been the technical requirements of GC instrumentation for efficient WCOT operation. Since wide-bore FSOT columns and packed columns can be operated at similar 10-30 ml/min flow rates, the necessary conversions required to use wide-bore FSOT columns on packed-column gas chromatographs are both simple and easy [20,21]. The conversion consists of installing two injector and detector adapters, which are available as complete conversion kits from a number of the manufacturers of FSOT

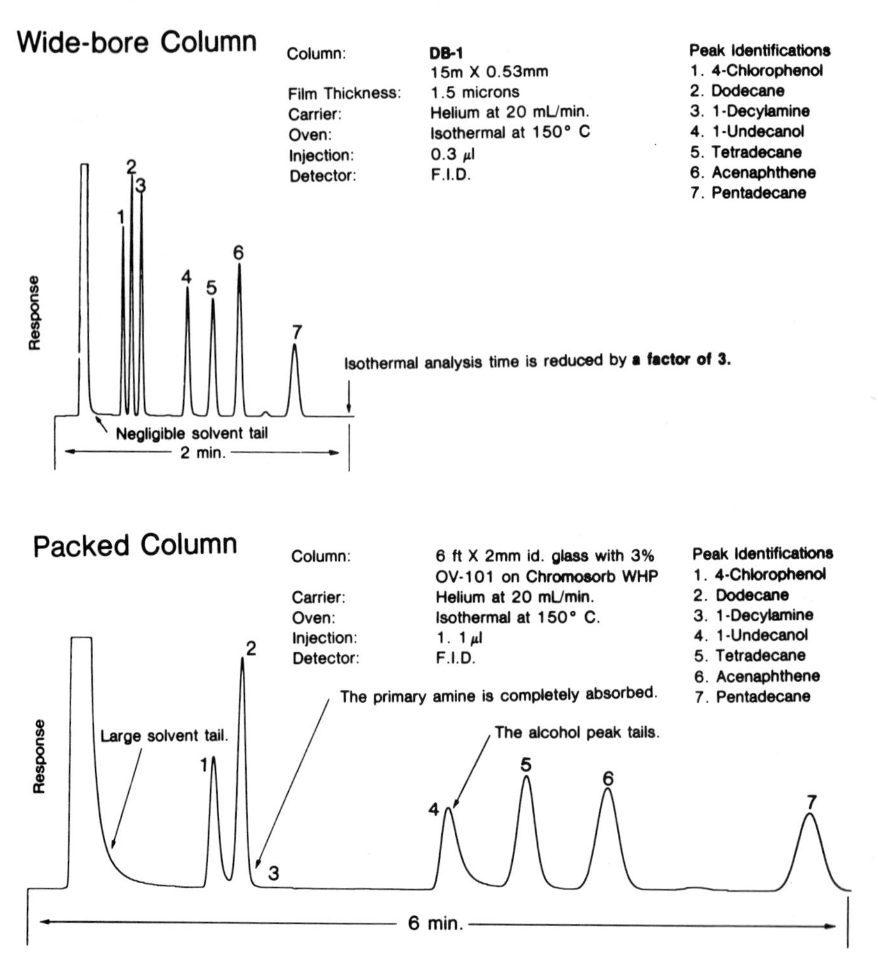

Figure 1. Performance comparison of a 0.53-mm ID × 15 m wide-bore WCOT column to a conventional 2-mm ID × 6 ft packed glass column.

columns listed in Column B of Table 1. The advantages of wide-bore FSOT columns over packed columns include (a) greater column inertness with fewer adsorption problems, (b) higher resolution as demonstrated in Figure 1 (c) analysis time reduced by approximately 33% for equivalent resolutions, (d) lower column bleed at high oven temperatures, (e) enhanced column-to-column reproducibility of retention times, (f) greater solvent compatibility, (g) fewer types of stationary phases required for adequate separation of diverse analytes, and (h) lower detectability because of higher column efficiencies and less adsorptive losses to column surfaces. In the vast majority of those applications where packed analytical columns are still being used, the only remaining arguments for their use must be based on lower initial column costs, durability of metal columns, satisfactory performance characteristics with a nonoptimal chromatographic system, and the common human trait to resist changes in routine activities. These new thick-film, wide-bore WCOT fused-silica columns with their less restrictive operational requirements both on the gas chromatograph and its operator will continue to lure packed-column users into trying current GC column technology.

Because of the author's perception of the diminished importance of conventional packed columns for even simple mixture analysis, the remaining portions of the chapter will focus almost entirely on the instrumental requirements and equipment for modern open tubular column gas chromatography.

III. FUSED-SILICA OPEN TUBULAR COLUMNS

A. General Technology Trends

The recognition of the chromatographic characteristics of open tubular columns has prompted numerous research efforts to improve column technology. Various materials have been used to construct open tubular columns, such as metals, plastics, Teflon, and glasses [11,13]. Presently, almost all open tubular columns are prepared from either soda-lime, borosilicate, or fused silica glasses [22,24]. Because of their straightness, flexibility, and relative surface inertness, thin-walled fused silica columns have rapidly become the most popular since being introduced in 1979 [25]. The current art of preparing efficient fused silica columns usually includes [a] surface treatments to deactivate active silanol (Si-OH) and siloxane groups (Si-O-Si), (b) surface treatments to enhance surface wettability, (c) stationary-phase coating, and (d) stationary-phase immobilization [8,12,22,24]. These methods of column preparation are adequately described in the preceding references and will not be repeated in this chapter.

B. Guidelines for Stationary-Phase Selection

The stationary phase can be a high-boiling liquid, a cross-linked polymer, an adsorbent, or a combination of these three. Fortunately, selecting the stationary phase suitable for a particular application in open tubular gas chromatography is less difficult than in packed-column gas chromatography because of the greater efficiency and resolution of open tubular columns. The magnitude of this reduction in potential complexity for the selection of liquid stationary phases is apparent when one considers that the list of common phases used in packed columns includes about 300 types. Additional confusion can often arise since certain popular types have a multiplicity of different commercial names for the same chemical phase: for example, SE-30, DC-11, DC-200, DC-410, SP-2100, OV-1, OV-101, HP-1, DB-1, and SF-96 are all methylsilicone phases.

Four basic criteria are important in selecting WCOT columns with a suitable stationary phase for a specific separation: (a) the polarity of the analytes, (b) the polarity of the stationary phase, (c) the selectivity of the stationary phase for the types of analyte molecules, and (d) stationary-phase stability. Criteria (a) and (b) are used in combination with the general rule that "like dissolves like," that is, for separating alcohols use a polyglycol stationary phase, or for separating nonpolar hydrocarbons use a nonpolar phase such as a methylsilicone. However, certain polar analytes can often be adequately separated on nonpolar stationary phases provided that the column surfaces have been properly deactivated. For example, polar sulfur-containing gases such as H_2S, CH_3SH, and SO_2 have been resolved from relatively nonpolar sulfur-containing compounds, like CS_2 and CH_3SCH_3, on nonpolar dimethylpolysiloxane WCOT columns [26,27]. In such cases, the resolution of mixtures containing polar/nonpolar compounds is usually better on nonpolar phase WCOT columns because of their uniform films of methylpolysiloxane phase and their concomitant higher column efficiency. For this reason, these nonpolar dimethylpolysiloxane WCOT columns (e.g., DB-1, HP-1, BP-1, CPSil-5CB, GB-1, ULTRA-1, RSL-150, OV-1, OV-101, SE-30, etc.) are a good all-purpose column for many types of samples. Thus, every chromatographic laboratory should have at least one cross-linked dimethylpolysiloxane WCOT column. Only when sufficient separations cannot be obtained on this general-purpose column will it be necessary to consider other choices for the stationary phase.

Two further exceptions to the above "like-for-like" guideline are in the cases of the light C_1-C_5 hydrocarbons and the permanent gases such as He, Ne, Ar, H_2, O_2, N_2, CO, etc. These gases have limited solubilities and small partition ratios in common liquid stationary phases and are analyzed preferably on molecular-sieve PLOT columns [16].

The general polarity and selectivity of liquid stationary phases can be determined by reference to McReynolds constants [28]. Almost all the commercial liquid stationary phases have been classified using this system, which employs a set of liquid phase constants (X', Y', Z', U', S', H', J', K', L', and M') for a series of test compounds selected to characterize the major interactions responsible for retention and different groups of compounds. However, McReynolds constants are often reported for just the first five symbols, X' through S'. Table 4 summarizes these general molecular interactions and groups for the complete McReynolds system. The values of the McReynolds constants increase with increasing polarity of the stationary phase. For example, the X', Y', Z', U', and S' values for the nonpolar OV-101 dimethylsilicone phase are 17, 57, 45, 67, and 43, respectively; the corresponding values for the more polar OV-225 cyanopropylmethylphenylmethylsilicone are 228, 369, 338, 492, and 386.

Table 5 lists the most widely used stationary phases in open tubular column GC along with their approximate temperature ranges and polarity ranges as estimated from adding the McReynolds numbers for the corresponding five values for X' through S'.

Specific difficult separations can often be obtained by combining different stationary phases. This can be performed either by coating one column with the appropriate phase mixture or by series coupling of two or more WCOT columns with the different phases [29-31]. Coupling of the columns via zero dead volume butt column connectors or conical-shaped glass connectors eliminates potential interaction problems that arise when the phases are actually mixed. The choice of column lengths and stationary phases may be determined empirically; however, an increasingly popular technique for determining the best phase mixture for specific separations is the window diagram method [6,22,32-35].

Table 4. General Molecular Interactions and Component Groups of the McReynolds System

Symbol	Test compound	Basic molecular interactions	Analogous compound groups
X'	Benzene	Dispersion with some weak proton acceptor properties	Aromatics, olefins
Y'	Butanol	Orientation properties with both proton donor and acceptor capabilities	Alcohols, nitriles, acids
Z'	2-Pentanone	Orientation properties with proton acceptor capabilities	Ketones, ethers, aldehydes, esters, epoxides, dimethylamino derivatives
U'	Nitropropane	Dipole orientation properties	Nitro and nitrile derivatives
S'	Pyridine	Weak dipole orientation with strong proton acceptor capability	Aromatic bases
H'	2-Methyl-2-pentanol		Branched alcohols
L'	1,4-Dioxane	Orientation properties with proton acceptor capability	
M'	*cis*-Hydrindane		

As previously noted, one of the significant advantages of present WCOT column technology is the immobilization of the stationary phase, that is, cross-linked or bonded phases [12,22]. The three most common immobilization methods are (a) free-radical-induced reactions between alkyl and alkenyl substituents on the polymer chains [36], (b) ^{60}Co gamma-ray-induced radical cross-linking of methylpolysiloxanes [37], and (c) condensation of polysiloxane chains [38]. Regardless of the specific method, the product is a stationary film that is both relatively nonextractable and thermally stable. Such immobilization allows the injection of samples in polar solvents such as water and alcohols. In addition, contaminated WCOT columns can be rejuvenated by flushing the impurities from the column with pure solvents. The upper temperature limit for a particular stationary phase is also extended via immobilization, which in turn means less column bleed, more stable baselines during temperature programs, and greater column lifetimes.

Table 5. Most Common WCOT Stationary Phases

Chemical phase	Common names	Temperature range (°C)	McReynolds polarity[a]
Methylsilicone (gum and oil)	DB-1, HP-1, BP-1, CPSil-5CB, GB-1, ULTRA-1, RSL-150, OV-1, OV-101, SE-30, SP-2100, DC-200, SF-96, 007-1, Rt_x-1	-60 to 280 (oil) -60 to 325 (gum)	220 to 229
5% Phenyl-methylsilicone	DB-5, BP-5, SPB-5, GC-5, CPSil-8CB, RSL-200, SE-52, OV-73, 007-2, Rt_x-5, HP-5	-60 to 325	334 to 337
50% Phenyl-methylsilicone	DB-17, HP-17, OV-17, GC-17, SP-2250, PE-17, RSL-300,007-17, Rt_x-50	40 to 280	884
50% Trifluoropro-pylmethylsilicone	DB-210, OV-210, SP-2401, QF-1	40 to 240	1520 to 1550
25% Cyanopropyl-methyl, 25% phenylmethyl, 50% methylsilicone	DB-225, OV-225, HP-225, SP-2300, CPSil-43CB, RSL-500	40 to 240	1813
Polyethylene glycol gum	DB-WAX, HP-20M, SP-1000, CPWAX-52CB, Superox-20M, Carbowax-20M, Stabilwax-DB	50 to 220	2301 to 2309

[a]Values determined by adding the McReynolds numbers for X', Y', Z', U', and S'.

The introduction of "Chirasil-Val" and other similar stationary phases prepared by chiral substitution into alkylpolysiloxane backbones [22,39] has allowed WCOT columns to be successfully used for the GC separation of many types of chiral molecules, including amino acids, carbohydrates, and alcohols [22]. The cross-linking of chiral phases for WCOT applications has also been reported [40]. The thermal stability of these phases for GC enantiomeric separations is limited to approximately 220-240°C, which unfortunately restricts their application to many biological and pharmaceutical chiral molecules.

Stationary-Film Thickness

Three general thicknesses of stationary films are available for the most popular WCOT columns: (a) thin film (0.1-0.2 μm), (b) medium film (0.5-1.5 μm), and (c) thick film (3-5 μm). Besides influencing the sample capacity of a WCOT column as previously discussed, the film thickness affects separation efficiency and resolution, surface inertness, and analyte retention. For maximum efficiency (i.e., peak narrowness), shorter analysis times [41], and minimum bleeding, film thicknesses of 0.1 and 0.2 μm are recommended. On the other hand, thick films offer significant benefits in the analysis of compounds that have very low boiling points or are quite polar and adsorptive. The WCOT columns with film thicknesses of 0.5-1.5 μm are most often used because of their "overall" performance characteristics.

C. Column Dimension Considerations

Length

Commercial WCOT columns are available in several standard lengths, such as 10, 25, 30, 50, and 60 m. For most applications either the 25- or 30-m column is a good initial choice. As implied earlier by equation (2), one way to improve chromatographic resolution is to employ longer columns. However, it is also possible to optimize the separation such that the shortest column length is used that yields the required resolution and the fastest analysis time. As shown by Ettre [42], the retention time of the last eluting compound, t_{RZ}, is given by

$$t_{RZ} = \frac{L}{\bar{\mu}}(1 + k_Z) \tag{3}$$

where L is the column length, $\bar{\mu}$ is the average linear carrier gas velocity, and k_Z is the partition, or capacity, ratio of the last eluting analyte Z. Thus, if potential resolution can be reduced to only the resolution actually required for a desired separation, shorter analysis times from columns with similar internal diameters can be realized by either shortening the column, increasing the carrier gas velocity, or decreasing the partition ratio. Figure 2 illustrates a high speed GC separation of a mixture of hydrocarbons on a short 3-m fused-silica WCOT column.

Internal Diameter

The two most common internal diameters currently used in FSOT chromatography are 0.25 mm for higher resolution applications and 0.53 mm for an attractive alternative to packed columns. However, Table 2 revealed that five different internal diameters are commercially available. Each of these five types of column diameters have their specific advantages and limitations; hence, the choice of column diameter depends on the required efficiency, the amount of sample to be injected, the desired analysis time, and the sophistication of the GC instrumentation with respect to extra column effects and response times.

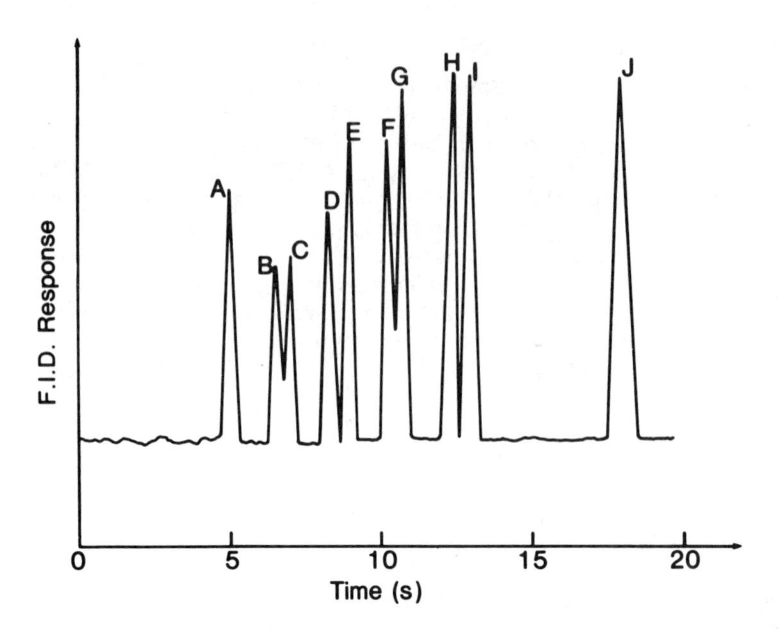

Figure 2. Example of a high-speed separation of hydrocarbons on a short 3 m × 0.25 mm ID fused-silica WCOT column. Peak components: A = *n*-pentane, B = 4-methylpentane, C = 2,3-dimethylbutane, D = 2-methylpent-1-ene, E = *n*-hexane, F = methylcyclopentane, G = 2,4-dimethylpentane, H = benzene, I = cyclohexane, and J = nitronaphthalene.

Interest in ultra narrow-bore WCOT fused-silica columns has been around for some time, but has recently increased as a result of their commercialization for capillary supercritical fluid chromatography [43]. The major advantages of the 0.10-mm ID columns in GC are increased efficiency and shorter analysis times. From the modified Golay relationship that describes band broadening in open tubular columns [13,14,22,23], one can derive the relationship in equation (4) by assuming thin films of stationary phase and negligible resistance to mass transfer in the stationary phase. The minimum height equivalent to a theoretical plate, h_{min}, that corresponds to the maximum column efficiency can be written as

$$h_{min} = r\left[\frac{1 + 6k + 11k^2}{3(1 + k)^2}\right]^{1/2} \tag{4}$$

where r is the column internal radius and k is the partition ratio or capacity factor. Equation (4) shows that for any given partition ratio, the WCOT column efficiency increases (i.e., h_{min} decreases) as the internal column diameter becomes smaller. Thus, ultra narrow-bore WCOT columns have efficiencies approaching 10,000 theoretical plates per meter when operated at optimum linear velocities of 40 to 50 cm/sec of hydrogen. For high-speed GC work, 5- to 10-m lengths of these 0.10-mm ID WCOT columns are used. Figure 3 illustrates the combined high resolution and analysis speed of a

10 m × 0.10 mm column coated with a 0.17-μm film of cross-linked 5% phenyl-, 95% methylsilicone. Additional examples of high-speed GC are presented in Ref. 41.

D. Future Directions in GC Columns

Many revolutionary advances have occurred in column technology during the past 10 years [11,22]. Nevertheless, certain problems of varying severity still exist and ongoing research is directed toward further progress in these areas.

One such area concerns the continuing development of open tubular columns whose performance has been optimized to provide efficient separation of specific sets of compounds in common samples, for example, classes of drugs in urine, volatile priority pollutants desorbed from water, volatile pesticides in air, sulfur-containing gases in air, certain types of chiral molecules, etc. The further development of these so-called specialty columns (e.g., DB-608, DB-624, DB-Dioxin, CPSil-88, Rt_X-Volatiles, etc.) will

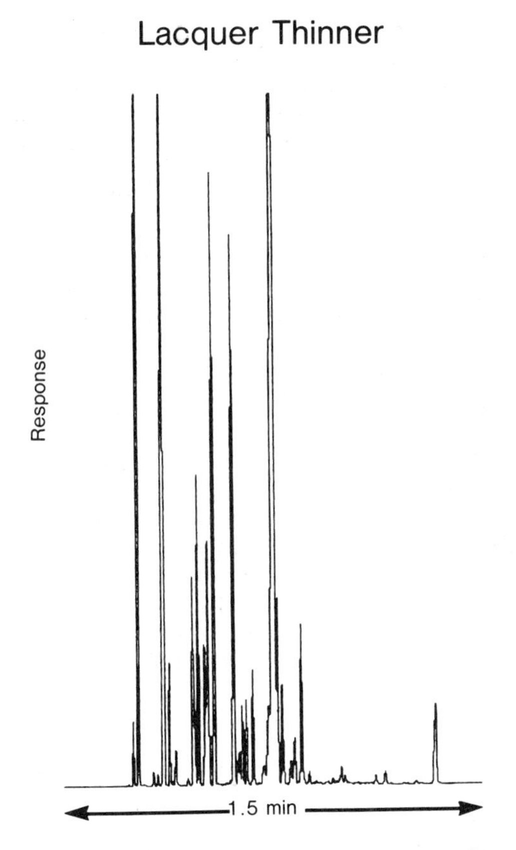

Figure 3. Example chromatogram of the high resolution and speed capabilities of 10 m × 0.10 mm ID fused-silica WCOT columns (n ≃ 100,000 plates).

require concomitant progress in superselective stationary phases via selectivity tuning [35] and in more column inertness via either new construction materials or better surface deactivation.

Although aluminum-clad, fused-silica WCOT columns with bonded phases have extended the upper temperature range of GC [2,3], additional improvements in overall thermal stability are needed for the analysis of even higher-molecular-weight compounds. Novel uses of certain ionic stationary phases such as molten organic salts [44-46] may also expand GC applications into molecules characterized by both high molecular weight and polar functional groups.

Anyone who has injected a sample into a GC and impatiently waited for the completion of the chromatographic separation has experienced the desire for faster analysis times. Continued improvements in the ability to coat microbore columns with very thin, yet uniform, films of many stationary phases will enable chromatographers to perform more runs in less time for a broader spectrum of sample types.

The need to inject relatively large amounts of sample onto columns that have both significant sample capacity and ultra-high resolution is desirable for several reasons, including detectability. One possible way to achieve this novel capability is through the use of a large number of capillary tubes of extremely small internal diameter that are essentially bundled together and externally sealed to form one larger tube with a multitude of equivalent, internal WCOT paths. Modern micromachining technology [47] for silicon could be used to etch these uniform, multiple microcolumn grooves onto an appropriate wafer. A silicon-chip, single-column gas chromatograph has already been constructed with micromachining techniques [48].

Subambient temperature gas chromatography, which uses relatively expensive liquefied gas cryogens from bulky Dewar containers to operate column ovens at temperatures from -100 to 0°C, is often employed for the analysis of volatile and/or polar compounds. However, WCOT columns with thick 3- to 5-μm films of stationary phase have demonstrated a greater retention, and therefore an increased elution temperature, for low-boiling compounds. Such thick-film WCOT columns have recently been used to separate certain classes of volatile and/or polar compounds without subambient oven temperatures. For example, Figure 4a shows the baseline separation of methane (b.p. = -164°C), ethane (b.p. = -89°C), propane (b.p. = -42°C), butane (b.p. = -0.5°C), and pentane (b.p. = 36°C) with a 5-μm film thickness, wide-bore WCOT column held at an isothermal oven temperature of 40°C. Figure 4b shows the baseline separation of seven sulfur-containing gases, without the use of subambient oven temperatures, on series-coupled, wide-bore WCOT columns of DB-1 and DB-WAX. Further, such developments with coupled columns containing thick, cross-linked films may eliminate the need for cryogenic GC and its associated problems [27,49] and costs.

E. Fundamental Relationship between Column and GC Technology

To repeat an old chromatographic adage, the column is the heart of the gas chromatograph. Hence, it should come as no great surprise that new

Column: **DB-1**
30m x 0.53 I.D.
5 micron film thickness
Carrier: Hydrogen @ 6mL/min.
500 mL (gas) injection
Split 1:5
Oven: 40° C Isothermal

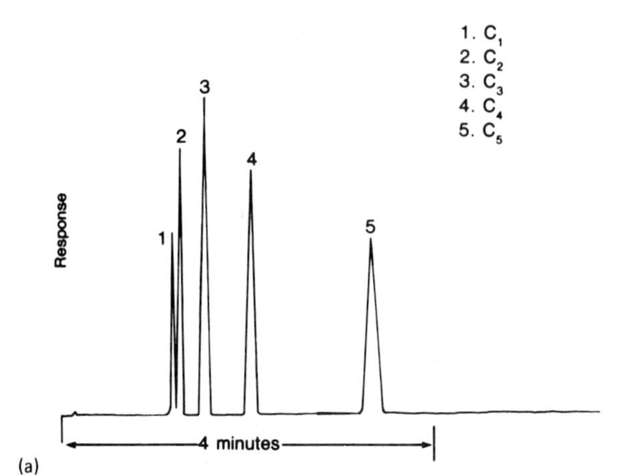

Combined DB-1 (30m) + DB-WAX (3m) 0.53μm FSOT Columns
(30° C for 1.2 min, then 30°C/min)

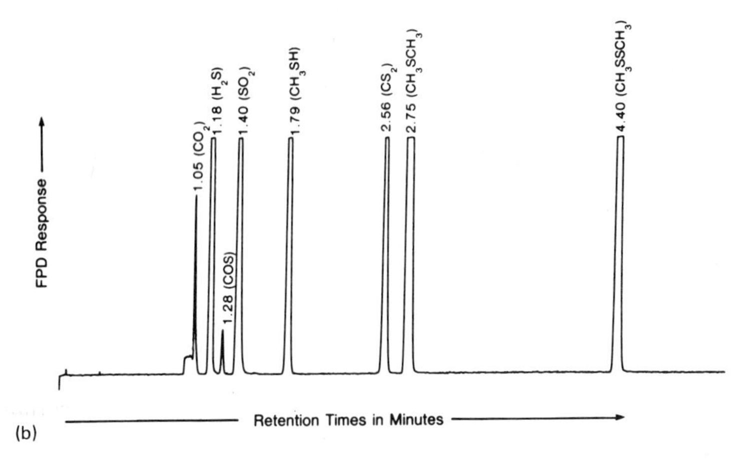

Figure 4. Noncryogenic GC separation of (a) volatile C_1-C_5 hydrocarbons with a 5-μm DB-1 WCOT column and (b) volatile sulfur gases with series-coupled DB-1 and DB-WAX WCOT columns.

developments in column technology always serve as the primary impetus for corresponding changes in the other essential parts of a state-of-the-art gas chromatograph. Such interactive development in GC instrumentation is mandatory in order to realize the full potential of existing columns. The remaining discussions in this chapter will examine the other major components of a modern GC system with respect to their design and operational requirements for optimum performance compatibility with current open tubular columns.

IV. SAMPLE INJECTION SYSTEMS

The injector on a GC fulfills three basic purposes: (a) to introduce the sample onto the column with minimal contribution to band spreading, (b) to connect the column to the injector, and (c) to present the carrier gas to the head of the column. The specific design and performance of the injection systems for task (a) in open tubular column GC are extremely important to the resultant peak shape (hence, both efficiency and resolution) and to quantitative precision and accuracy. Ultra narrow-, narrow-, and medium-bore WCOT columns are especially susceptible to difficulties in sample introduction due to both their low flow rates of carrier gas and limited sample capacities. In other words, the proper selection of the type of injector is essential to successful analyses via open tubular column GC. Unfortunately, different applications, samples, and columns require different injection systems, since no universal injector has yet been constructed to handle every sample type. Although the design and testing of new injector designs are on-going research efforts, there are currently four popular injection systems for liquid samples in addition to several commercial gas sampling inlet devices. Table 6 lists the most common injection systems used with different diameters of WCOT columns. In this section the basic operation and design for each of these injection systems will be briefly described, and special emphasis will be directed toward their specific uses and performance requirements.

Column C of Table 1 lists the companies that sell various sample injection systems for gas chromatography.

A. Injectors for Liquid Samples

Regardless of the injector type, most liquid samples containing both solvent and solutes are introduced into the injector by means of a microliter syringe. After injection, the liquid sample must be rapidly vaporized and the solutes must be subsequently transferred without molecular discrimination or decomposition to a narrow band at the head of the column. Significant loss in chromatographic efficiency will occur if the solutes are not introduced in a very narrow injection profile, for example, <500 msec for 0.32- and 0.25-mm ID columns.

Flash Vaporizer

Because wide-bore 0.53-mm and 0.75-mm ID WCOT columns can be operated at carrier flow rates similar to packed columns, the popular flash vaporization injector can be easily converted for use with these state-of-the-art wide-

Table 6. Common Injection Techniques for Different Diameters of WCOT Columns

Column ID (mm)	Possible injection techniques
0.75 and 0.53	Direct flash vaporization, on-column (hot or cold), gas valves, cold traps
0.32 and 0.25	Split (hot or cold), splitless (hot or cold), on-column (hot or cold), gas valves, cold traps
0.10	Split (hot or cold), splitless (hot or cold), gas valves, cold traps

bore WCOT columns. Figure 5 is a cross section of a typical flash vaporization injector, and Figure 6 shows the two general types of glass adapter inserts available for use with wide-bore WCOT columns. One type, the liner for direct flash injection, is used to vaporize the 0.5- to 5-μl volume of injected liquid in the upper glass chamber and then sweep the sample vapors into the WCOT column, whose inlet is positioned at the lower tapered restriction. The tapered section at the upper end of this glass adapter minimizes sample vaporization flash-back against the septum, while the lower tapered section aligns and seals the column inlet.

The other liner in Figure 6 is used for on-column injection. A standard 26-gauge (0.48 mm ID) syringe needle will fit through the upper restriction, which serves to guide the needle inside the wide-bore column where the sample is directly injected. One of the biggest problems with on-column injection is the possible scratching of the stationary phase and fused silica tubing by the tight-fitting syringe needle. Such scratching can damage the phase coated on the column wall, expose active surfaces, and cause the column to break. Because of the limited volume within the column, liquid injection volumes should be less than 0.5 μl at normal flash injector temperatures, that is, 20-30°C higher than the maximum column temperature. However, a temperature-programmable injector [50-52] is ideal for such on-column injections, since the sample can be injected under nonvaporizing cold temperatures and then vaporized at a controlled rate as the injector is heated to higher temperatures. In this situation it is possible to inject sample volumes up to about 3 μl using the on-column mode and wide-bore columns. Regardless of whether the injection mode is on-column or direct, optimization of sample injection rate, injector temperature, and sample solvent will improve chromatographic efficiency and resolution [53]. Optimization of these three parameters is especially important when wide-bore WCOT columns are operated at $\leq$ 5 ml/min carrier-gas flow rates. An autosampler that provides variable injection rates is extremely useful for these initial optimization experiments and the following routine analyses.

Cold On-Column Injectors

On-column injection is also possible with 0.32- and 0.25-mm ID WCOT columns when the syringe needle is made of small enough OD (e.g., 32 gauge) so that it will fit into the ID of the column. Replaceable, fragile

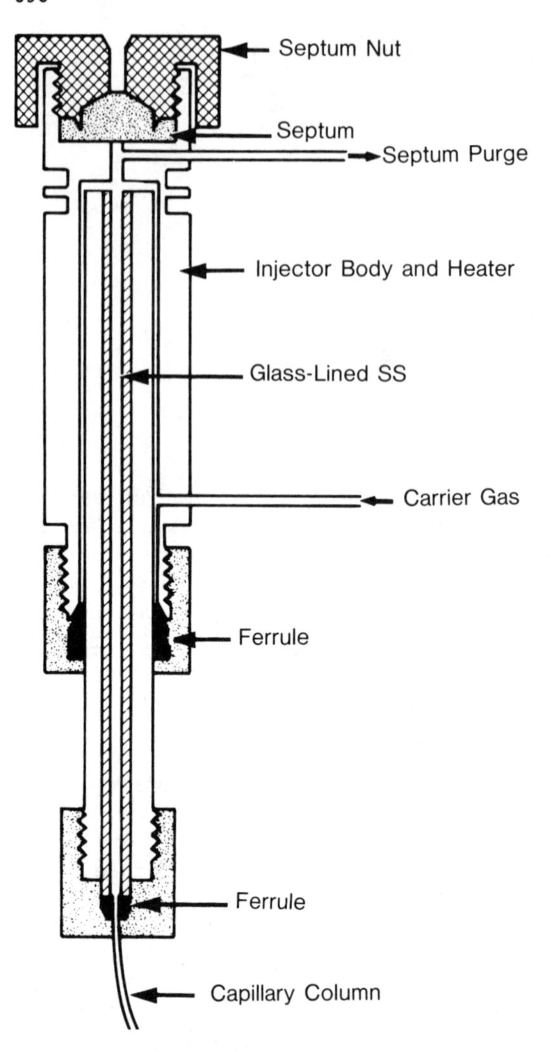

Figure 5. Schematic cross section of a typical flash vaporization injector.

syringe needles constructed from sufficient lengths of either fused silica or stainless steel capillary tubing are commonly used in this version of on-column injection into medium- and narrow-bore WCOT columns. Because all of the liquid sample is injected directly on the column without prior vaporization, the possibility of analyte discrimination during vaporization and transfer to the column is eliminated. Thus, cold on-column injection has the advantages of accurate and precise quantification and the minimization of sample decomposition due to combined thermal/catalytic effects. It is often the injection method of choice for the reliable quantification of sample analytes with wide molecular weight ranges, trace concentrations, and/or thermal instability [6,14,22,23,51]. A potential disadvantage of

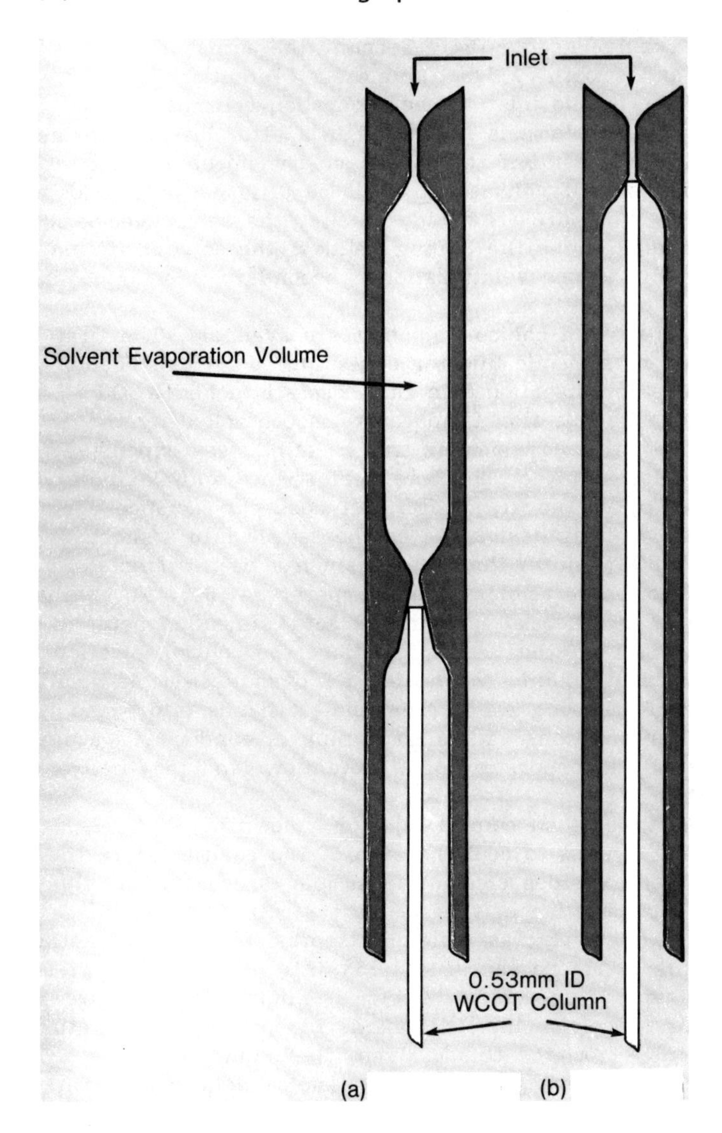

Figure 6. Two injection port liners for use with wide-bore WCOT columns. Design (a) is for direct flash injection and (b) is for on-column injection. (From Ref. 20, with permission.)

on-column injection is the deposition of relatively nonvolatile sample components in the column, which can produce a loss in both column efficiency and inertness.

For successful cold on-column injections, four criteria are important. First, the injection area must be below the boiling point of the sample solvent to prevent molecular discrimination associated with selective volatilization from the syringe needle [54]. Second, the actual injection must be made rapidly and smoothly. Third, unless a "retention gap" (a length of uncoated column at the inlet to the separation column) [55] is used, the maximum injection volume is restricted to 2 μl or less. Distorted peaks with large widths will result if the sample volume injected is too large. However, the injection of volumes up to 30-50 μl is possible if long retention gaps (i.e., 10-20 m) are employed. Finally, the column temperature must be rapidly raised above the boiling point of the solvent after sample injection in order to cause volatilization if secondary injector cooling is not employed.

Three general designs of on-column injectors are available. The first type essentially consists of a simple injector containing a syringe needle guide to align the needle for insertion into the closely attached WCOT column. In such designs the injection temperature is controlled by the column oven temperature. Although simpler and somewhat less expensive, this particular injector design often sacrifices performance and flexibility. However, one such simple, yet novel, design that does not sacrifice performance is shown in Figure 7. This design places the sample into a short section of the flexible FSOT column that previously has been retracted from the GC oven and cooled to ambient room temperature [56]. No auxiliary cooling is required since the thermal mass at the actual injection point is essentially that of the FSOT column. After injection, the column injection point is automatically reinserted into the heated GC oven for sample volatilization. The possibility of smearing the narrow analyte bands into broad bands because of their interactions with a large plug of condensed sample solvent can be reduced by incorporation of a "retention gap" [55] in front of the analytical column [57].

The second general type of on-column injector, shown in Figure 8a, uses a secondary cooling system to cool the area of the column where the sample is actually injected. Besides eliminating sample discrimination, this design also minimizes back ejection of the sample [58]. One version of this design uses a "duckbill" valve as a flexible septum [23], while another uses a stop valve and is therefore septumless [58,59].

The third general type of on-column injector is illustrated in Figure 8b. This general design for on-column injectors allows independent temperature control of the column inlet from the column oven. Depending upon the specific design, such temperature-programmable sample inlets can be controlled and programmed from approximately -100 to 400°C [51,52,60-63]. Some commercial models of these so-called programmed temperature vaporizers (PTV) use flow controllers instead of pressure regulators for the carrier-gas flow rate. This feature is desirable since it allows a constant carrier-gas flow rate during sample injection and temperature programs. The advantage of using PTV systems for on-column injection is that nonvolatile sample components will remain in the inlet and will not contaminate the analytical column. Because certain PTV injectors can be easily used in either the cold on-column, hot on-column, split, or splitless modes with good precision and accuracy [51,61,62], they are attractive universal liquid injection systems.

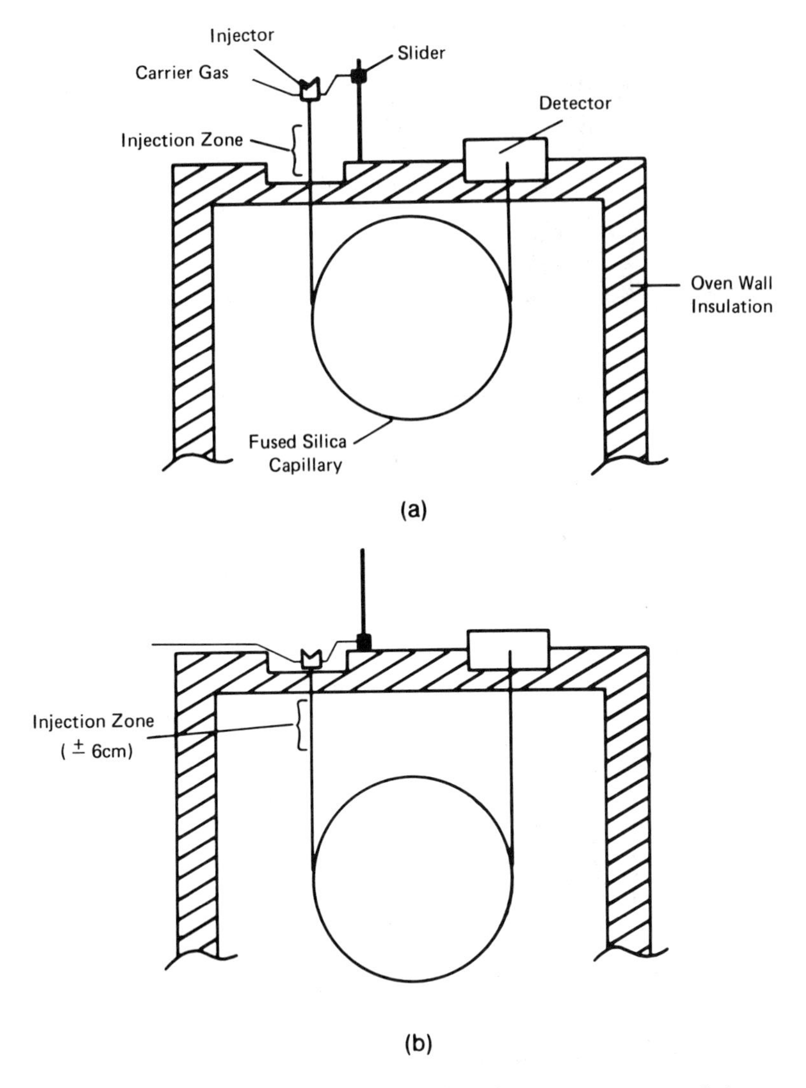

Figure 7. Schematic operation of a simple on-column injector. (a): Before and during sample injection. (b): After sample injection to end of run.

Split Injectors

Dynamic splitting was developed as the first successful type of sample introduction system for open tubular GC columns by Desty et al. [64]. Such injectors rapidly vaporize the liquid sample (approximately 0.5-2 μl) and allow only a small fraction to actually enter the column. The remainder is vented. Homogeneous mixing of the sample vapor and the carrier gas is crucial prior to splitting if the sample portion that enters the column is to reflect accurately the composition of the injected sample, that is,

(a)

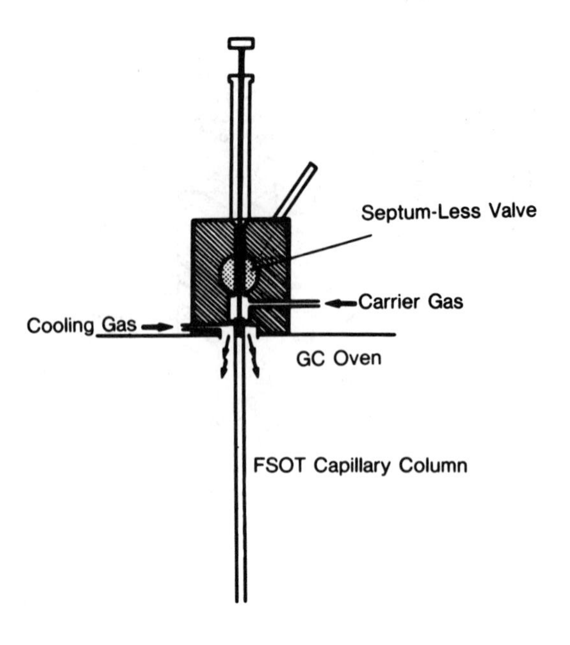

(b)

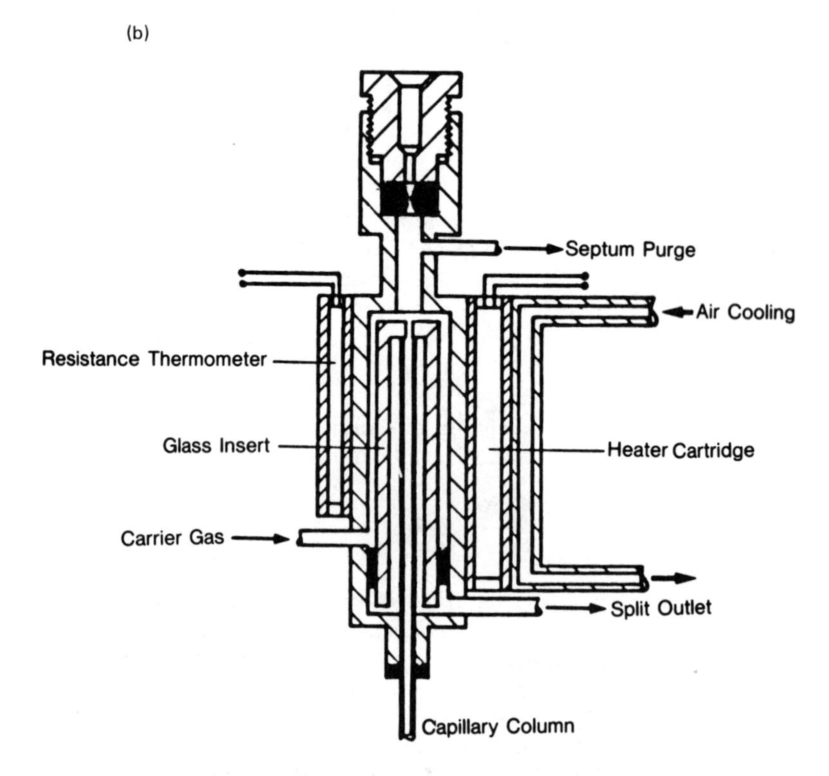

Figure 8. Two designs of on-column injectors: (a) cold and (b) temperature-programmable.

be a linear split [14]. The splitter not only reduces the amount of the sample that enters the column, but also permits a high flow rate of carrier gas through the injection chamber so that exponential dilution of the vaporized sample into the column does not occur.

Figure 9 illustrates two common pneumatic methods for controlling the split ratio, which is calculated by ratioing the column flow rate to the vent flow rate. Typical split ratios range from 1:10 to 1:1000. High split ratios are used with narrow-bore WCOT columns, and lower split ratios are used with the larger-bore columns, which have greater sample capacity. The column inlet is the split point. Pneumatic configurations like the one in Figure 9b have the advantage of being able to set the split ratio without affecting the column inlet pressure. Once the column inlet pressure has been set with the proper configuration, only the inlet flow, and not the column flow, will vary as the split ratio is changed. Furthermore, the pressure regulator is not in line with the column flow, so any contamination from the regulator diaphragm is vented. The configuration in Figure 9b also prevents accidental large flows of carrier gas into the oven in case of a system leak and thereby decreases safety concerns about the use of hydrogen as a carrier gas.

Other design requirements for a split injector include (a) an internal surface of glass or fused silica rather than metal for greater inertness, (b) a wide range of possible split ratios, (c) a minimum extra column volume, and (d) interchangeable glass or fused-silica liners. Several liner designs are available that differ in their internal flow patterns and heat capacity [6,14,23]. The optimum liner choice is dependent on the relative volatilities of the sample analytes and is essential to minimize analyte discrimination during the split operation. The descriptions of liner configuration and their selection as provided in Ref. 23 should be consulted for additional details.

Split inlets have two basic limitations. First, reproducible and linear splitting of samples containing analyte components with a wide range of volatilities is difficult to achieve in practice. Chapters 4 and 5 in Ref. 22 provide valuable guidelines in selecting experimental split sampling conditions for minimizing the degree of discrimination. Second, most of the injected sample is vented and only a small fraction goes into the column. Thus, the split technique cannot be used to analyze sample components present at trace concentrations. However, if the analytes are present at concentrations exceeding approximately 10^{-3} percent, if they show reasonable thermal stabilities, and if their volatilities are approximated by the range of the C-6 to C-19 *n*-alkanes, then split injection can be used quite successfully.

Splitless Injectors

Like cold, on-column injection, splitless injection is used for trace analysis. Most of the injected sample goes into the column and there are fewer sample discrimination problems with splitless as opposed to split streams, since as the name implies, no sample splitting occurs within the injector. Splitless injection should be considered for dilute analyte solutions whose volatilities are approximated by the volatility range of the C-10 to C-26 *n*-alkanes.

(a)

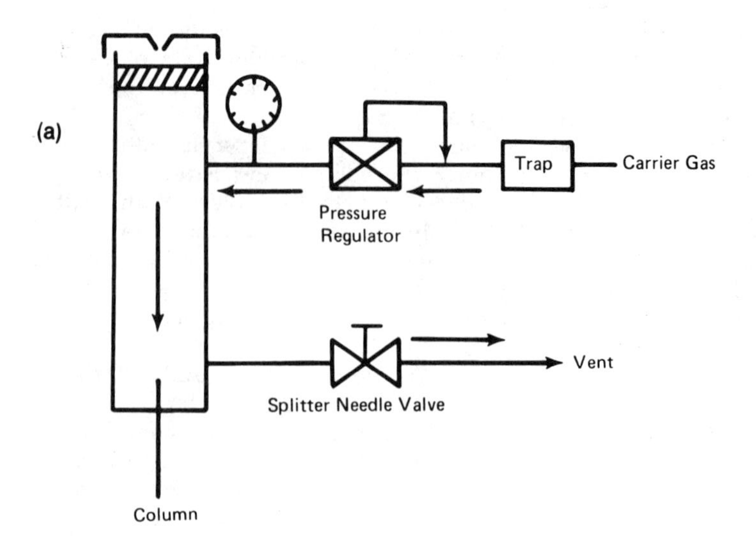

(b)

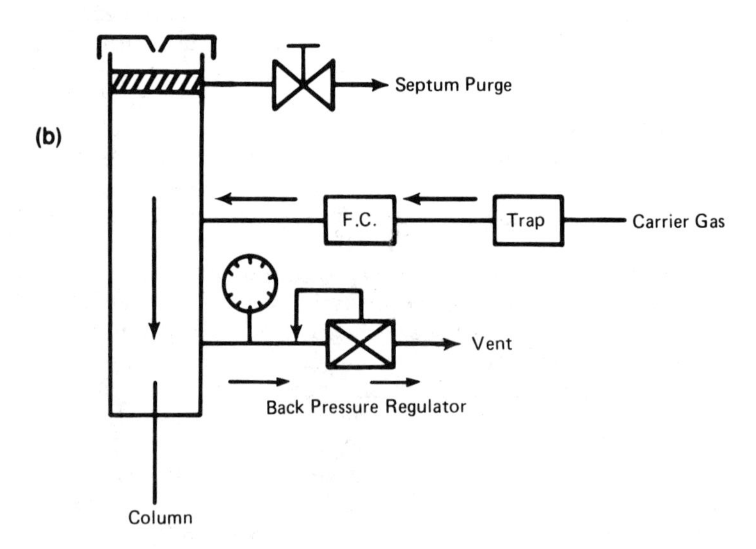

Figure 9. Two pneumatic methods for controlling split ratios in split injectors: (a) fixed split restrictor using a pressure regulator and a variable needle valve vent restrictor, (b) constant column inlet flow using a flow controller (F.C.) and a back-pressure regulator.

The design of splitless injectors differs from most split injector designs only in the size of the sample vaporization chamber and in the pneumatic configuration. Figure 10 shows a typical splitless inlet system with its different pneumatic configurations prior to injection, during injection, and after the injection and purge period. During the actual sample injection period it is necessary to stop the purge gas flow through the injector; however, the column flow is maintained by keeping the column inlet pressurized. The time period before the purge flow is reactivated depends on the sample solvent, the column flow rate, and the volatilities of the sample analytes. Typical purge-off times range from 20 to 60 sec. That is, the injector purge flow is only activated after 80-90 percent of the injected sample has entered the WCOT column. At this point, the purge flow is reactivated to rapidly vent the injector volume and thereby reduce the solvent tailing [14,65-67].

The splitless technique depends on a reconcentration of the vaporized analyte components at the head of the column. Two different reconcentration methods are used: (a) the "solvent effect" [67] or (b) cold trapping [14]. In the solvent effect case, the sample solvent forms a small band at the head of the column that reconcentrates the analyte vapors. This solvent effect can be explained with the aid of the following equation:

$$K = \beta k \tag{5}$$

where K is the distribution constant, β is the phase ratio (the volume of the column occupied by the gas phase relative to the volume occupied by the stationary phase), and k is the partition or capacity ratio. Thus, the small solvent plug at the inlet to the column creates a zone of very small phase ratio. This decrease in phase ratio causes a corresponding increase in the partition ratio, which in turn accounts for the strong retention of the analyte vapors at the column inlet. Because the analyte bands are passed through an area of decreasing phase ratio within the head of the column, the fronts of the analyte bands undergo greater interactive retention than do the rears of the bands—that is, the analyte bands are narrowed during the reconcentration.

The steps to choosing proper splitless operating conditions are summarized in Appendix A of Ref. 23 and in Chapter 3 of Ref. 6. Desirable design features for splitless injectors include (a) internal gas flow streamlining and no unswept areas since low inlet flows are used, (b) a purged septum to minimize septum bleed contaminants from entering the column, and (c) replaceable fused-silica liners that can be chemically deactivated if necessary. Because the basic injector designs for splitless and split are very similar, they are often constructed as a split/splitless unit that can be operated in one mode or the other.

B. Injectors for Gas Samples

Gas Sampling Loops and Valves

Although gas-tight syringes with volumes from about 1 to 100 μl are available for direct injection of gas samples [5], better injection reproducibility and accuracy are possible with a combination of a gas sampling valve and

(a)

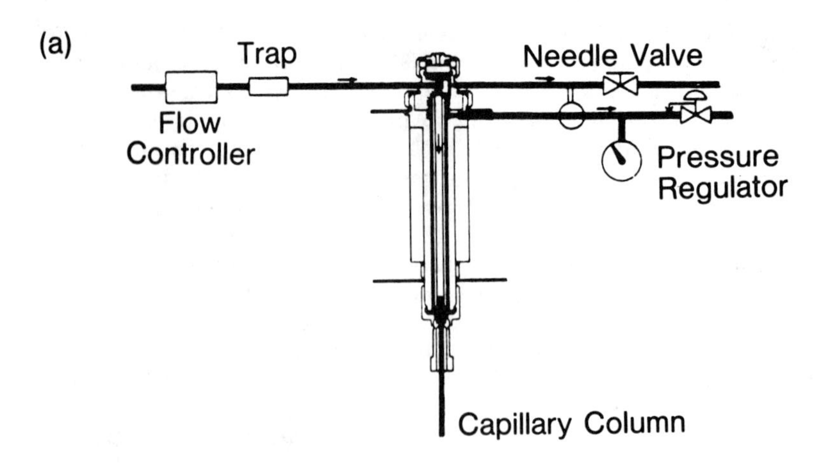

(b)

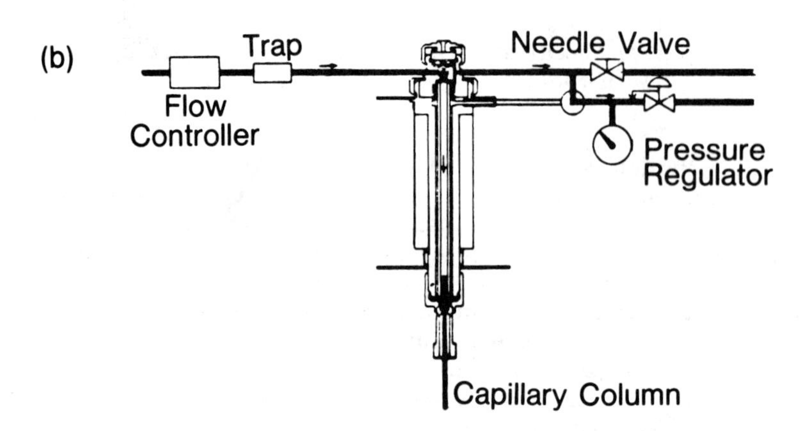

(c)

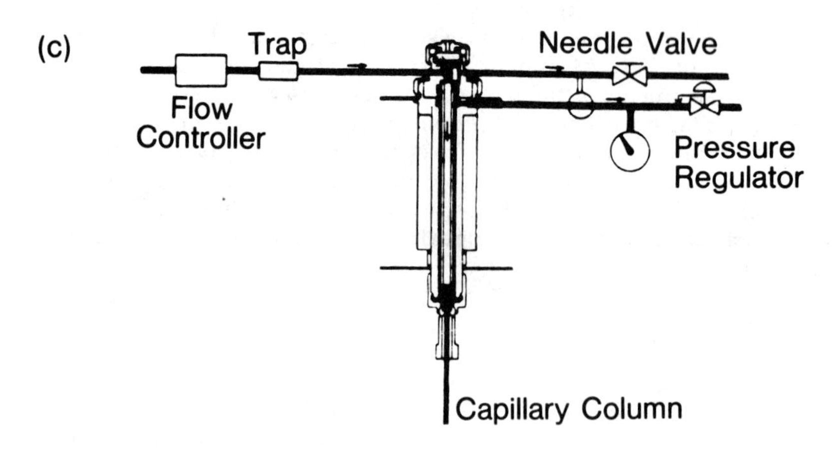

Figure 10. Basic splitless injector and pneumatic configuration: (a) prior to injection, (b) during injection and sample transfer, and (c) after injection and sample transfer.

a calibrated sampling loop. Figure 11 illustrates gas injection by gas sampling loop/valve. If these loop/valve systems are constructed from metal, then significant adsorptive losses to their inner walls can occur for certain types of gases. Such losses are most prevalent for low volatility compounds; polar compounds containing sulfur, nitrogen, and/or oxygen; and when the analytes are present at trace ppb/ppt (V/V) concentrations [68,69]. Sampling loop/valve systems constructed from Teflon are often advantageous in such situations; however, "memory" effects can cause problems if the analytes are somewhat permeable to Teflon. External heating should be used in either system in order to minimize losses due to condensation if the analytes have boiling points greater than approximately 40°C. It is also good practice to check the calibrated volume of any commercial sample loop or loop/valve systems [70]. This precaution is especially important if a commercially calibrated sample loop is purchased separately and then attached to an existing gas sample valve.

If feasible, the sample loop should be held at the same pressure as the column to minimize pressure perturbations when the sampling valve is switched. Solenoid-driven actuators for automated switching of the sample valves are also convenient [71].

An alternative to valve-based systems for precision gas sampling is pneumatic switching devices as initially described by Deans [72]. At least one such pneumatic gas sampling system is commercially available [73]. Pneumatic switching is based on the balance of flows by in-line flow restrictors and additional makeup gas. Such pneumatic switchers have three major advantages: (a) there are no in-line mechanical parts to cause potential contamination or leakage, (b) unswept void volumes are eliminated since the flow switching occurs inside of capillary tubing which is swept with carrier gas, and (c) all in-line parts can be made of glass or fused silica to provide a system as inert as possible.

Multidimensional Sample Introduction

Applications requiring high-resolution separation of certain components, or sample types, within highly complex samples have spurred the development of multidimensional gas chromatography (MDGC). In MDGC, two or more serially coupled columns of different selectivities, and sometimes different capacities, are employed. The desirable variations in selectivity can be due to different stationary-phase polarities, different phase ratios, and/or different column temperatures [6,14,22]. MDGC systems using packed-packed, packed-OTC, and OTC-OTC have been described in recent reviews [74-77]. Multidimensional chromatography is also possible by an on-line combination of either conventional HPLC, microbore HPLC, or SFC with capillary GC [22,78,79].

Whereas a GC with multidimensional capabilities can be operated in several modes, this discussion will only be concerned with MDGC for heart cutting and sample injection. In these specific MDGC modes, the initial sample is introduced into the first column and then only certain effluents from this precolumn are transferred to the second column. Figure 12 shows a schematic drawing of an MDGC system, and Figure 13 illustrates the heart-cutting application of MDGC to the separation of 2,3,7,8-

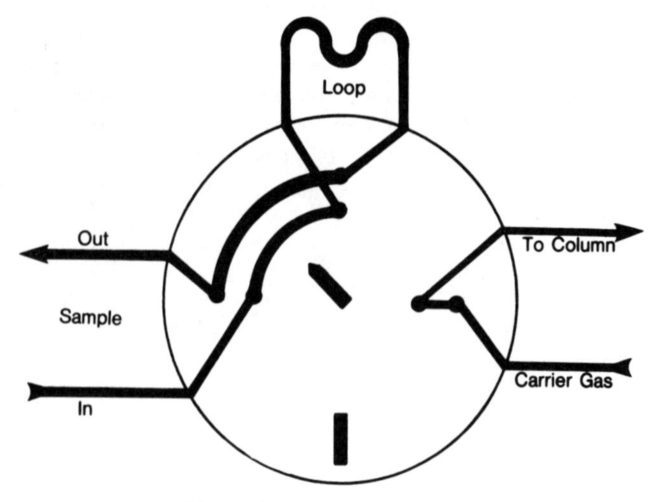

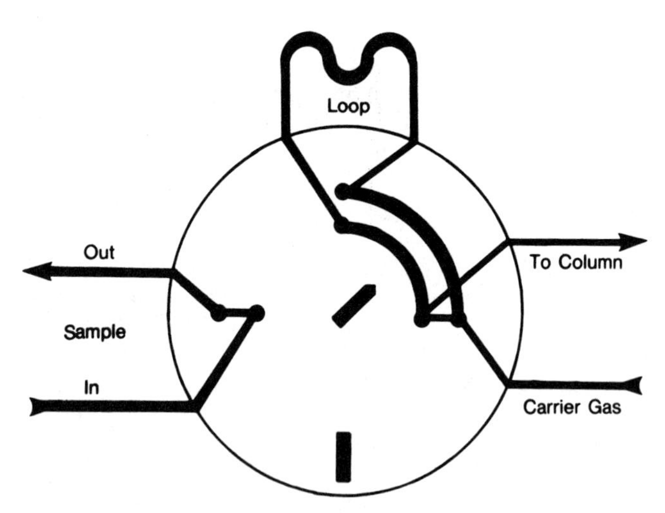

Figure 11. A combined gas sampling loop and valve system for injecting known volumes of gas samples.

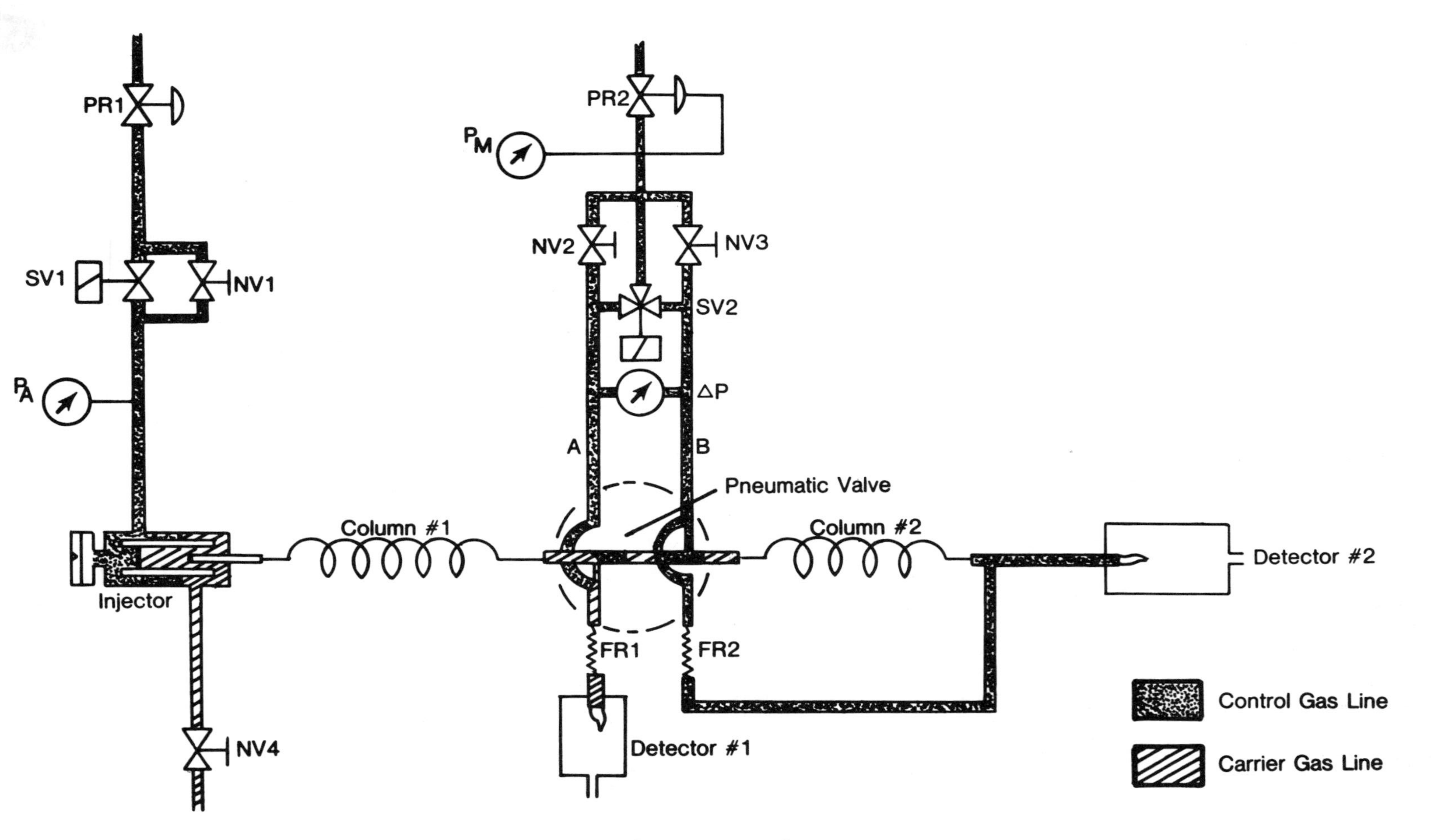

Figure 12. Diagram of a multidimensional GC system with a pneumatic flow switch [77].

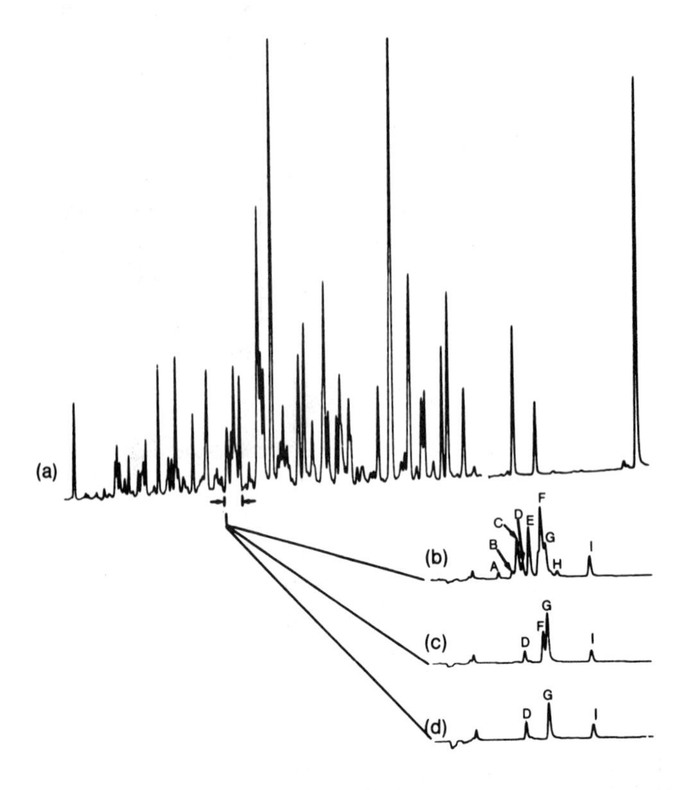

Figure 13. MDGC separation of 2,3,7,8-TCDD from a mixture of PCDD and PCDF [76,77]. Chromatogram (a) is PCDD/PCDF mixture on a 30 m × 0.75 mm Silar 10C (d_f = 0.18 μm) column. Chromatograms (b)-(d) are heart cuts of 60 sec, 20 sec, and 10 sec, respectively, from the Silar 10C column into an in-series 25 m × 0.32 mm OV-1 (d_f = 0.30 μm) column. Peak G is the 2,3,7,8-TCDD.

tetrachlorodibenzodioxin from a complex mixture of polychlorinated dibenzodioxins and dibenzofurans [76,77].

Although the various operational modes of MDGC can be performed by using either mechanical or pneumatic switching valves, the pneumatic Deans-type [72] valves are advantageous for high-resolution GC because of their reduced void volume, their relative inertness, and their lack of internal moving parts. However, the ability to temperature program one of the two ovens may be severely restricted if special design provisions are not made [22,76]. Less demanding applications of MDGC using packed and/or wide-bore WCOT columns can be adequately performed with more conventional, multifunctional mechanical valves [80]. An intermediate cryogenic trap must be used to refocus the analytes during the column transfer procedure if the volume of column effluent requires a lengthy transfer time, for instance, from a packed column to a narrow-bore WCOT column.

This cold trap must be capable of rapid heating, both to ensure a slug injection of volatilized analytes and to retain the inherent efficiency of the second column.

The required sophistication of MDGC instrumentation varies as a function of the application. Simple two-dimensional gas chromatograms for heart cutting and precolumn enrichment can be obtained on most modern single-oven GCs, which have been converted by adding several auxiliary components [81]. A more convenient approach for such multidimensional modifications is via the purchase of an MDGC conversion unit [82,83]. These commercial conversion systems cost in the range of $4000 to $6500. Gas chromatographic systems are also available that have been designed for MDGC. Most of these MDGC systems have two ovens for independent control of the column temperatures, two separate detectors, pneumatic switching valves that are controlled by the GC's microcomputer, and intermediate cold traps that use either liquid CO_2 (≈-60°C) or N_2 (≈-140°C).

Cryogenic and Adsorbent Trap-Based Injection Systems

Cryogenic Traps

Cold traps are used for one of the following functions: (a) precolumn enrichment of gaseous analytes at trace concentrations [69], (b) to focus injection or transfer volumes that would otherwise contribute excessive band broadening to the analyte peaks [26,74,84-89], and (c) to establish a precise injection point or time for the measurement of retention indices [8,23]. The cryogenic fluids most commonly used for such purposes are liquid nitrogen (-196°C), liquid oxygen (-183°C), liquid argon (-186°C), and liquid carbon dioxide (-78.5°C). The choice of the cryogenic fluid, and hence the temperature of the cold trap, should take into consideration the fact that cryogenic trapping methods have collection thresholds determined by the temperature coefficients of the analyte vapor pressures as expressed by the Clausius-Clapeyron equation [69]. In addition, analyte breakthrough losses due to microfog formation can be minimized if the cryogenic trap has a gradual temperature gradient [23]. When cryotraps are used in enrichment sampling of air, a temperature-gradient sampling tube can be constructed so that most of the sample water and CO_2 are spatially separated from the more volatile analytes [69]. Various chemical drying agents or Nafion drying tubes can also be used to remove water [69,90]. The trap design should also be capable of very rapid heating to desorb the trapped analytes into the OTC column as a narrow band [23,85]. Resistive heating of the actual trap, or of metal wire or tubing wrapped around the trap, by a high current during thermal desorption is usually preferred [23,85,91]. At least two commercial, cold-trap sample enrichment-desorption systems are apparently available for GC use [92,93], in addition to those noncommercial designs that are described in the literature.

Adsorbent Traps

The use of solid adsorbents for the preconcentration of different gases has continued to expand, as has the number of adsorbent materials and the commercial thermal desorption systems as GC accessories. Guidelines for selecting the proper adsorbent material for different analytical applications are presented in Ref. 69 and should be followed, rather than a priori discussions to use a particular adsorbent just because of its current popularity. The list of potential adsorbents includes graphitized carbon (e.g., the Carbopacks), activated charcoal, treated and untreated silica gel, various Amberlite XAD resins, various Chromosorbs, Ambersorbs, Tenax-GC, Thermosorb, Porapaks, alumina, polyurethane foam, Eu-Sorb, etc. [5,6,14,69].

A number of papers have described GC inlet interfaces for thermal desorption of volatiles from adsorbent-filled sampling tubes [94-100]. The actual thermal desorption chamber can be constructed from a modified, heated GC injection port as shown in Figure 14. However, a number of commercial devices designated specifically for automatic desorption of such traps are also available for approximate prices of $4000 to $7000 [101-108]. Desirable features for these automated thermal desorption units include (a) programmable, ballistic heating for rapid desorption, (b) heated transfer lines constructed from fused silica or nickel to minimize surface losses of certain analytes, (c) convenient assembly for insertion and removal of adsorbent cartridges, (d) intermediate cold trapping with subsequent thermal desorption for transferring a narrow injection volume to the head of the capillary GC column, (e) electronic flow controllers for each trap's purge gas, (f) compatibility with all standard sizes of adsorbent tubes, (g) low void volume switching valves, (h) microprocessor control with programmable entry of operational parameters, and (i) a standard RS-232 communication port for possible connection to the computer of the host GC system. Figure 15 shows the gas-flow paths for the Tekmar automatic desorber.

A new thermal desorption instrument based on microwave desorption has been described by Rektorik [109]. This microwave desorber has the advantages of very fast heating and direct transfer of the desorbed volatiles from the solid adsorbent trap to the capillary GC column without significant peak broadening effects. However, porous polymer adsorbents such as Tenax and the Chromosorbs do not absorb sufficient microwave energy for thermal desorption by this technique [109]. At present the microwave desorption method is limited to activated charcoal, graphitized carbon, and silica gel adsorbents.

Purge and Trap Analysis

Another major use of both adsorbent and cold traps is in the purge and trap analysis of volatile organic compounds at trace concentrations in aqueous samples. This purge and trap technique combines inert-gas-phase stripping of volatile organics from liquid samples and their simultaneous enrichment in an adsorbent trap [110,111]. Figure 16 illustrates the basic apparatus and sampling sequence for a typical purge and trap GC sampler. After the volatile organics in the liquid sample are purged into the adsorbent trap, they are rapidly thermally desorbed and backflushed into the open tubular GC column. During the GC separation of the desorbed sample analytes, the adsorbent trap is vented and heated to remove any contaminants in preparation for another run sequence. Purge and trap equipment is available from several commercial sources [112-116]. Desirable instrumental features include (a) choice of solid adsorbents, (b) automated purge,

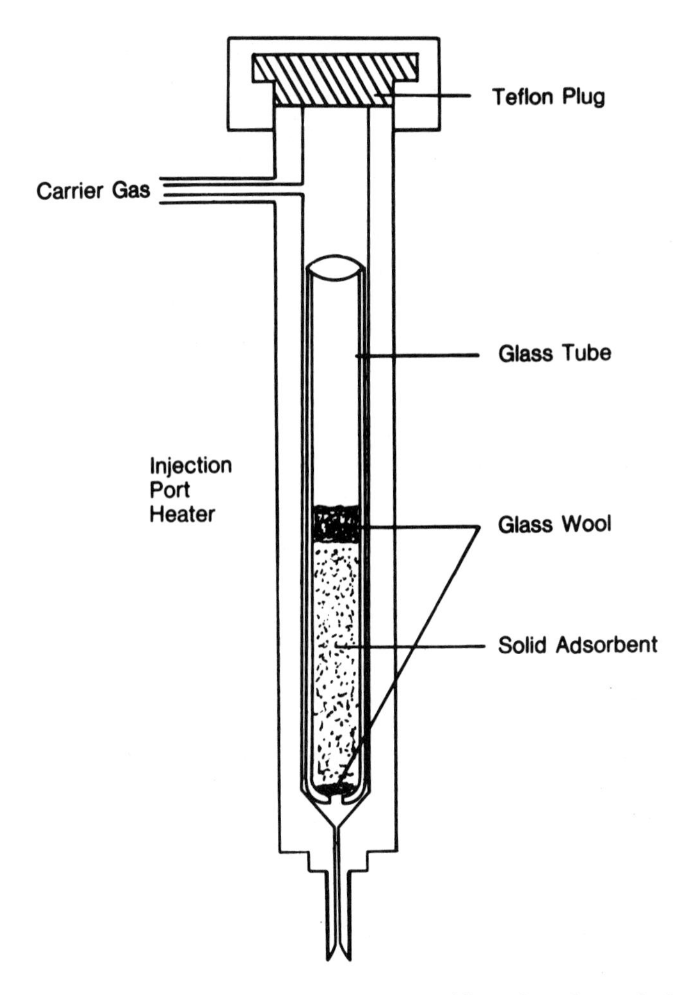

Figure 14. GC injection port modified for thermal desorption of solid adsorbent tubes.

desorb, and cleaning cycle, (c) ballistic heating of the adsorbent trap, (d) on-line removal of water by either a condenser or a Nafion drying tube [117], (e) an external coolant flow to cool the trap after thermal desorption for faster cycle times, (f) an intermediate cryogenic trap to refocus the analytes thermally desorbed from the adsorbent tube, (g) a low-void-volume, heated FSOT interface to the GC column, and (h) programmable set points, times, temperatures, and gas flows.

Purge and trap methods have two major advantages. First, they have low detectability since they function as a preconcentration technique. Second, the technique is easily automated for high sample throughput. However, it should be remembered that this gas stripping technique is a nonequilibrium method, and lower-molecular-weight organics are generally

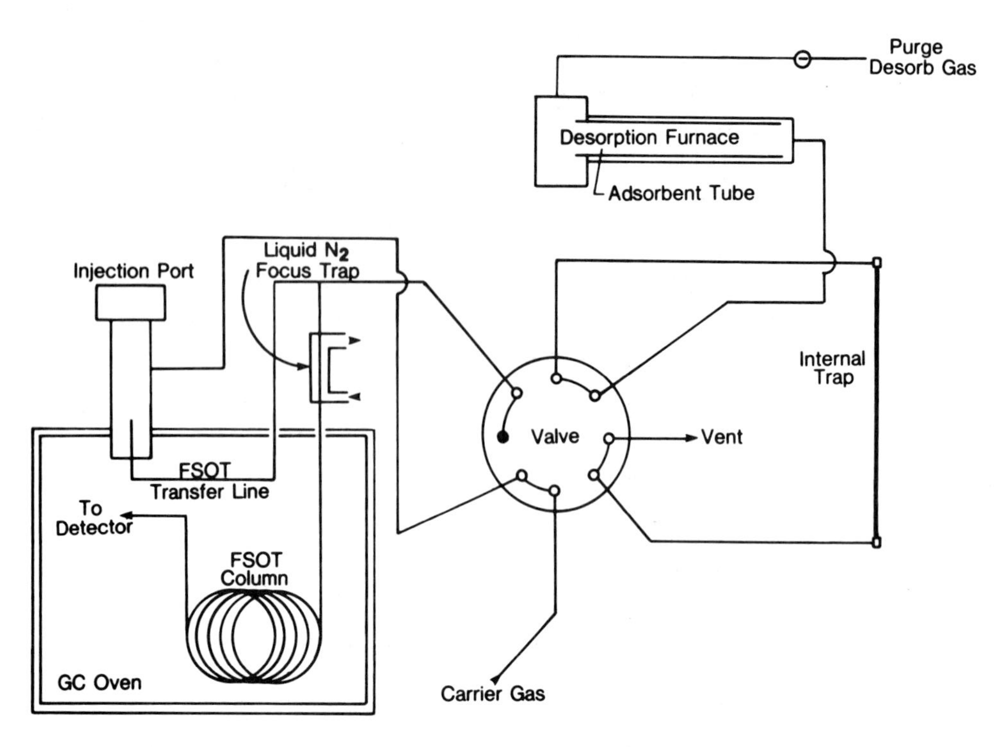

Figure 15. Gas flow diagram of the Tekmar model 5000 automatic desorber [102]. Volatiles are thermally desorbed from the adsorbent tube into an internal cryogenic or packed trap at high purge flow rates. Internal trap contents are then desorbed at lower flow rates into the cryofocusing trap, which is then flash heated for GC injection.

purged more rapidly than those of higher molecular weight. Careful standardization under controlled stripping and trapping conditions is essential for reliable quantification. The use of several representative internal standards is a good practice when feasible.

Headspace Samplers

Headspace GC analysis has become a widely used technique for the determination of volatile compounds in samples that are difficult to analyze by conventional GC methods, such as when the sample matrix is solid, when the sample is a liquid that would require extensive sample cleanup prior to analysis, and when the vapors above the sample are of direct interest [118]. To perform a headspace analysis, a measured amount of the sample is placed in a glass vial and sealed with a septum and a crimped cap. The sample vial is placed in a carousel where it is thermostatted to equilibrate a known volume of inert gas with a known amount of the sample. The internal pressure in the closed headspace vial increases because of

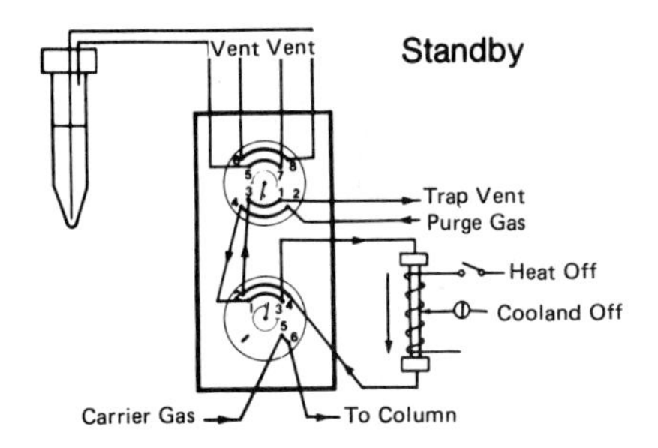

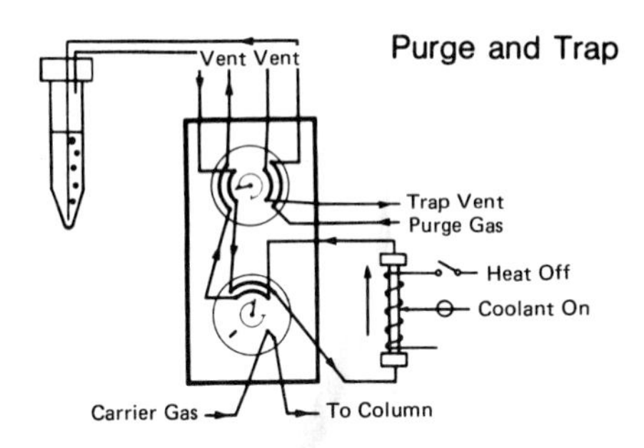

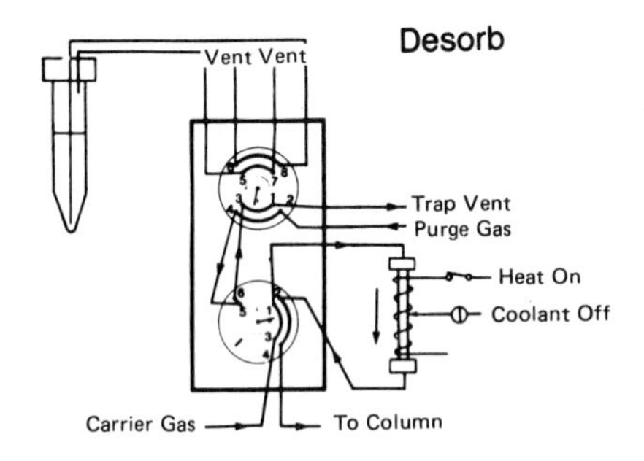

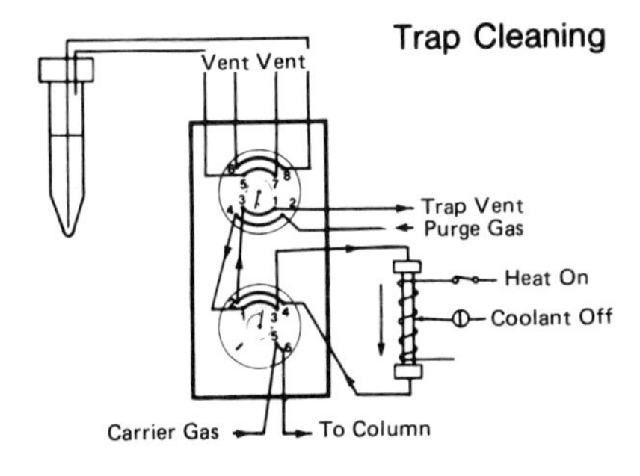

Figure 16. Sequential operations of a typical purge and trap GC sampler.

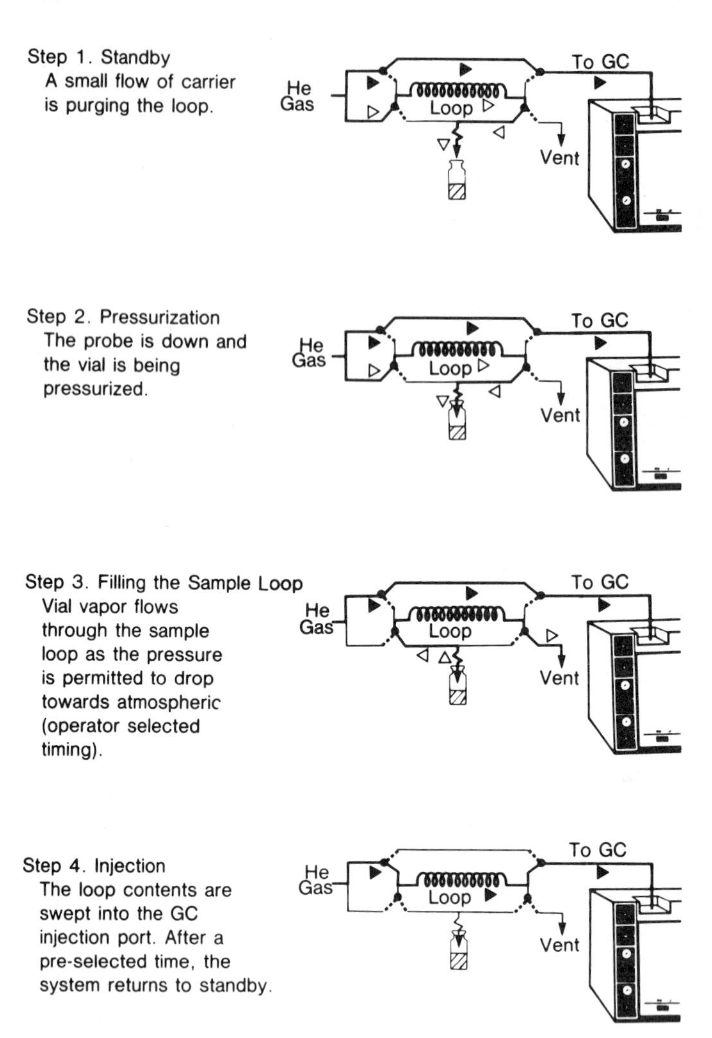
Step 1. Standby
A small flow of carrier is purging the loop.
He Gas
Loop
To GC
Vent
Step 2. Pressurization
The probe is down and the vial is being pressurized.
He Gas
Loop
To GC
Vent
Step 3. Filling the Sample Loop
Vial vapor flows through the sample loop as the pressure is permitted to drop towards atmospheric (operator selected timing).
He Gas
Loop
To GC
Vent
Step 4. Injection
The loop contents are swept into the GC injection port. After a pre-selected time, the system returns to standby.
He Gas
Loop
To GC
Vent

Figure 17. Operational sequence for an automated GC headspace analyzer.

the partial vapor pressures of all the volatile components within the sample. Although the headspace gas in the vial can be sampled with a gas-tight syringe and subsequently injected into the GC, most headspace analyses are performed with automated systems. The automated procedure is diagrammed in Figure 17. Automatic headspace samplers developed as GC accessories are available from several of the GC manufacturers [119-122]. The practical aspects of headspace sampling into capillary columns have been discussed with emphasis on its influence on the injected bandwidth in the column inlet [123].

Whereas headspace analysis is a relatively simple operational procedure, reliable calibration of the method can be a potential problem. That is, the concentrations of a particular volatile analyte in the liquid and gas phases are related to each other by a corresponding partition coefficient, which is matrix dependent. If the sample matrix cannot be exactly matched for the preparation of calibration standards, then the method of standard additions should be used [124].

For samples where matrix matching or the method of standard additions is inappropriate, the method of multiple headspace extraction may be a simple answer. The multiple-headspace extraction procedure is based on a stepwise sampling of the headspace from a single sealed sample [123-125]. The analyte concentration in the headspace becomes smaller after each equilibrated extraction, and continuous headspace extractions would eventually remove all of the analyte from the sealed sample. The sum of all the GC peak areas ΣA_n obtained for that analyte during these multiple extractions could be used to quantify the amount of analyte initially present in the sample. This total peak area can also be calculated from the sum of a geometric progression [126]. Fortunately, such exhaustive procedures are not required since it is usually possible to estimate accurately the total peak area ΣA_n from only two determinations for each sample via

$$\Sigma A_n = A_1^2/(A_1 - A_2) \qquad (6)$$

where A_1 and A_2 are the peak areas obtained for the analyte during the first and second headspace analyses, respectively. Thus, before purchasing a headspace sampler, one should carefully examine the utility of multiple headspace extraction in the desired analysis; if appropriate, then only consider those models capable of operating in this powerful mode.

C. Injection Systems for Solid Samples

Headspace Analysis of Volatiles in Solid Samples

The headspace techniques discussed in the preceding section can be used for the analysis of volatile compounds in solid samples [123]. For example, the multiple-headspace extraction technique was originally developed for the quantitative analysis of volatiles such as vinyl chloride monomer and water in a PVC resin [127].

Pyrolysis Sampling Systems

Pyrolysis GC is achieved by attaching a thermal pyrolyzer at, or before, the GC injection port. Its major use is in the identification of parent materials by their characteristic volatile fingerprints formed by thermal decomposition of the sample [6,22,128-131]. The samples analyzed by pyrolysis GC are usually quite insoluble and nonvolatile, such as soil, rocks, polymers, papers, inks, coal, drugs, textiles, microorganisms, or biological fluids. Pyrolysis fingerprints can be obtained on very small samples, an important factor in forensic applications [128-130], with only minimal sample preparation. The problem-solving capabilities of pyrolysis GC profits greatly from the combination of high-resolution GC with pattern recognition [132].

There are two different types of pyrolyzers used in GC: continuous-mode and pulse-mode pyrolyzers [6,22,128]. In the continuous-mode design, the sample is placed in a boat or cup that is subsequently heated in a

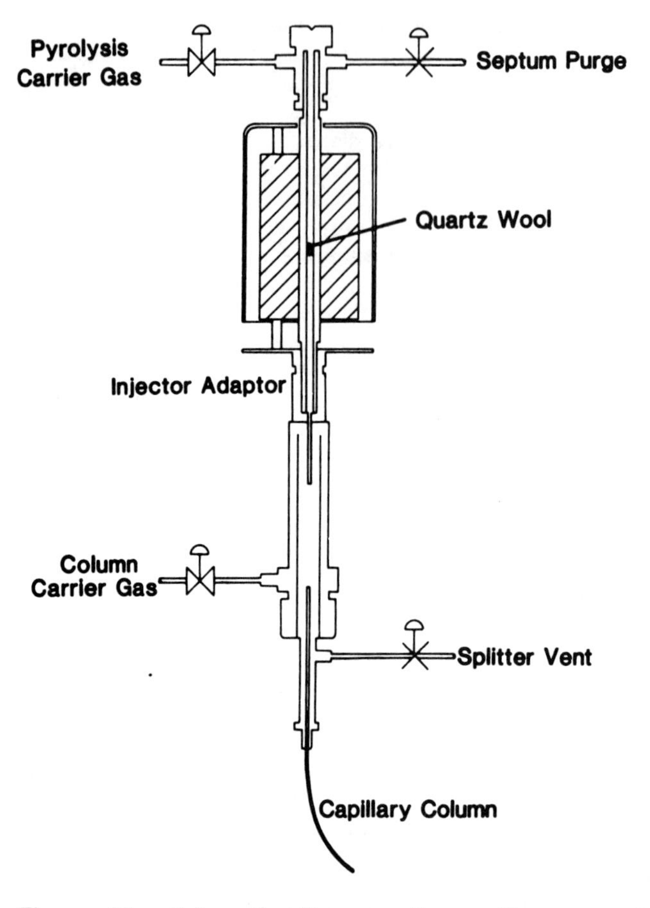

Figure 18. Schematic diagram of a continuous-mode pyrolysis GC system.

furnace. The sample is rapidly pyrolyzed and the volatile products are split and then swept onto the head of the FSOT column. Figure 18 shows a schematic diagram of a continuous-mode pyrolyzer.

Pulse-mode pyrolyzers are based on applying the sample either to resistively heated filaments or to Curie-point materials that are inductively heated in a radiofrequency field [6,22]. Schematic diagrams of these two types of pulse-mode pyrolyzers are shown in Figure 19.

Commercial pyrolysis equipment for GC is available from at least three manufacturers [133-135]. Factors to consider in purchasing pyrolysis GC equipment are (a) relative inertness of internal surfaces and transfer lines, (b) total void volume up to the head of the FSOT column, (c) compatibility with particular types of samples, (d) an intermediate cryogenic or adsorbent trap to refocus sample gases for narrow-band injection, and (e) the production of reproducible pyrograms. Major factors contributing to irreproducibility in pyrolysis GC are varying sample size, pyrolysis temperatures, pyrolysis atmospheres, surface contamination, and gas flows that influence residence times in the pyrolysis area. Small sample sizes (i.e., 10-20 μg) usually yield the best reproducibility, especially if splitting can then be eliminated in favor of complete sample enrichment in the intermediate sample trap [128-130].

The choice of pyrolyzing temperature is important and must be determined for each new sample type. If the temperature is too low, then little decomposition occurs. If the temperature is too high, then the decomposition products are mainly small molecules with little qualitative detail. The desired pyrolytic fingerprint region is somewhere in between these two temperature extremes.

Other Solid Injection Systems

Several other types of injectors besides pyrolyzers have been described for solid samples [5,136]. In general, these devices usually contain a plunger mechanism to place the solid sample within the heated area of a

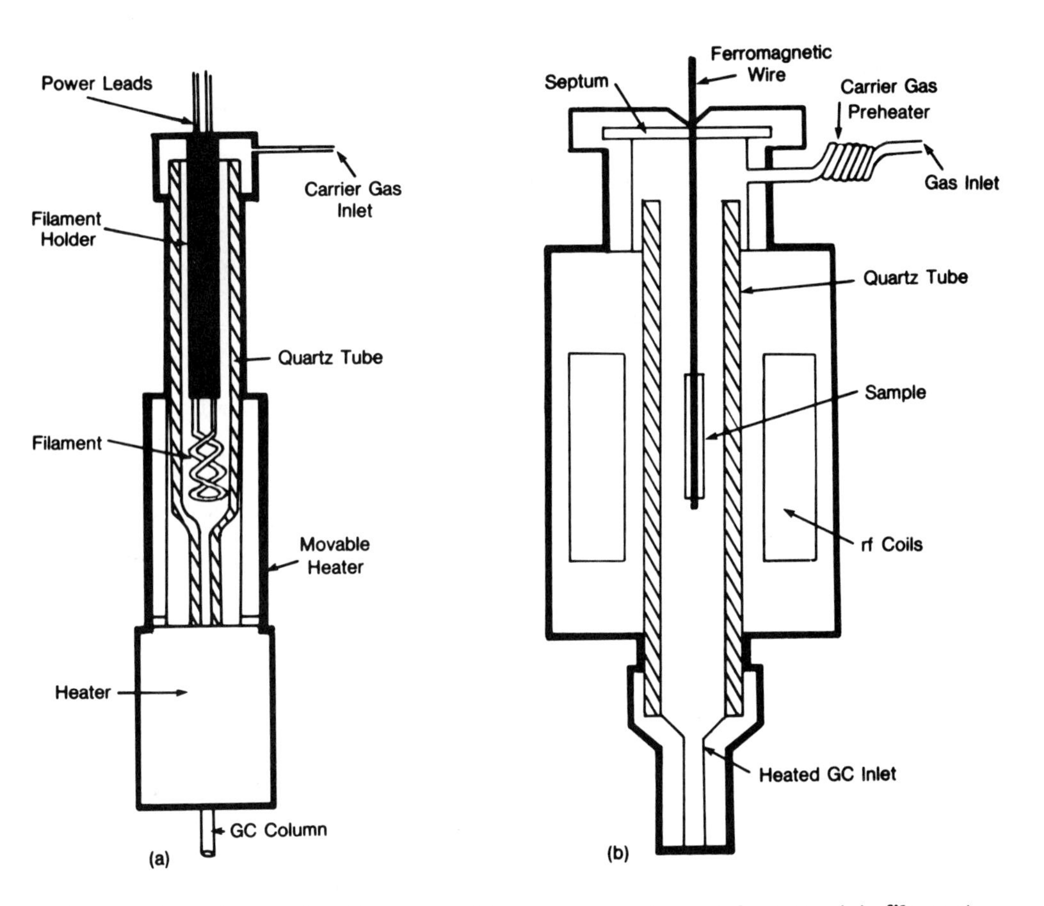

Figure 19. Schematic diagrams of two pulse-mode pyrolyzers: (a) filament or ribbon type and (b) Curie point.

conventional injection port. An all-glass solid sampling system that uses a quartz needle has been developed for use with open tubular columns [137]. In other designs the solid sample is initially sealed in either a glass tube or metal capsule. After insertion into the heated injection port, the sample is exposed by either mechanically breaking the glass tube, piercing the metal capsule, or melting the Wood's metal (m.p. = 61°C) or indium (m.p. = 155°C) capsule. However, such techniques do not provide instantaneous injection of the volatile components onto the column, and chromatographic efficiency will severely deteriorate unless intermediate trapping is used to focus the gaseous sample components prior to actual introduction into the FSOT column.

Although these latter solid-sample injection techniques have the advantages of solvent-free injection and no sample fractionation in syringe needles, they are generally limited to qualitative determinations. Hence, whenever possible a more attractive approach is to dissolve the solid sample in a suitable solvent and treat it as a liquid sample. A suitable solvent, or solvent mixture, must completely dissolve the solid sample or quantitatively extract the analytes of interest from the solid sample. In addition, the solvent components should not react or coelute with the analyte compounds.

D. Automatic Sample Injections

Pneumatically operated, automatic sample injectors are available for liquid, gas, and solid samples. Column D of Table 1 lists the commercial suppliers of automatic samplers. The major advantages of automatic samplers are (a) greater sample throughput, since the GC system can now operate unattended for periods up to 24 h per day, (b) an enhanced injection repeatability of ±0.5% compared to the typical ±5% repeatability of manual liquid injections, and (c) decreased GC conditioning effects since the system is exposed to sample injections on a more regular time basis.

Factors that should be considered when purchasing a particular automatic liquid sampler include:

Its mounting compatibility with your GC.
Its flexibility for use in different injection mdoes, such as split, splitless, or on-column.
Maximum number of sample vials in the carousel.
The programmable range of replicate injections per sample.
Adjustable injection speeds.
Adjustable injection volumes.
Adjustable injection dwell times, that is, the time the needle spends in the GC injection port.
Whether the sample syringe is loaded by pumping or flushing.
Bubble-free filling of the syringe
Sample volume actually required per injection, that is, flushing plus injection volumes.
Degree of cross-contamination between samples.
Simplicity of the procedure for aligning the syringe needle with the injection port.
Ease of syringe and/or needle replacement.

Selectable operation as either a repetitive autosampler or a single-sample injector.
Cost of the sampler instrumentation and consumable parts such as vials, caps, syringes, trays, etc.
Operation in either a stand-alone control mode or remote control via the GC itself or the GC's computer system.

Automatic sample injectors that appear to be suitable include the Hewlett-Packard 7673A, the Carlo Erba AS-550 or AS-V570, the Perkin-Elmer AS-8300, the Varian 8000, and the Synatech Precision Sampling GC-311.

V. COLUMN OVENS

A. General Requirements for Optimum Performance

Precise control of GC oven temperature is necessary because of its strong influence on analyte retention times, peak resolution, peak shape, and carrier gas flow rates. For example, a 1°C change in oven temperature may cause the retention times to vary by approximately 3 to 5 percent. The relationship between carrier gas flow rate and column temperature is given by

$$\frac{F_{Tf}}{F_{Ti}} = \left(\frac{T_i}{T_f}\right)^{1.7} \tag{7}$$

where F_{Ti} is the carrier gas flow rate at the column outlet at the initial column temperature (T_i) and F_{Tf} is the carrier gas flow rate at the column outlet at the final column temperature (T_f). Hence, variations in column temperature affect the flow rate of carrier gas through the column, and thereby affect resolution because of its van Deemter dependence on flow rate.

Conventional packed columns have relatively large thermal masses and therefore do not respond rapidly to temperature variations. In comparison, FSOT columns have very small thermal masses and closely follow small, rapid changes in oven temperature. Because of this latter fact, very precise and stable GC ovens must be used for reproducible qualitative and quantitative determinations with modern FSOT columns. Oven stability at any set temperature within the operating range should be at least within ±0.1°C, and ±0.01°C temperature stability is required to obtain retention time reproducibility of 99.9%.

Oven stability is determined by two variables: thermal noise, which is largely attributable to the inherent cycling caused by the oven temperature control electronics, and thermal drift, which is primarily caused by the controller's susceptibility to external fluctuations in the ambient temperature. As shown in Figure 20, thermal noise is the short-term, periodic fluctuations in temperature that are superimposed upon the longer-term "drift" component. Both thermal noise and drift can produce retention-time reproducibility problems because of phase differences between successive injection points. For instance, peak retention times will be different for two separate injections when their respective column temperatures correspond to the maximum and minimum amplitudes of thermal noise and drift.

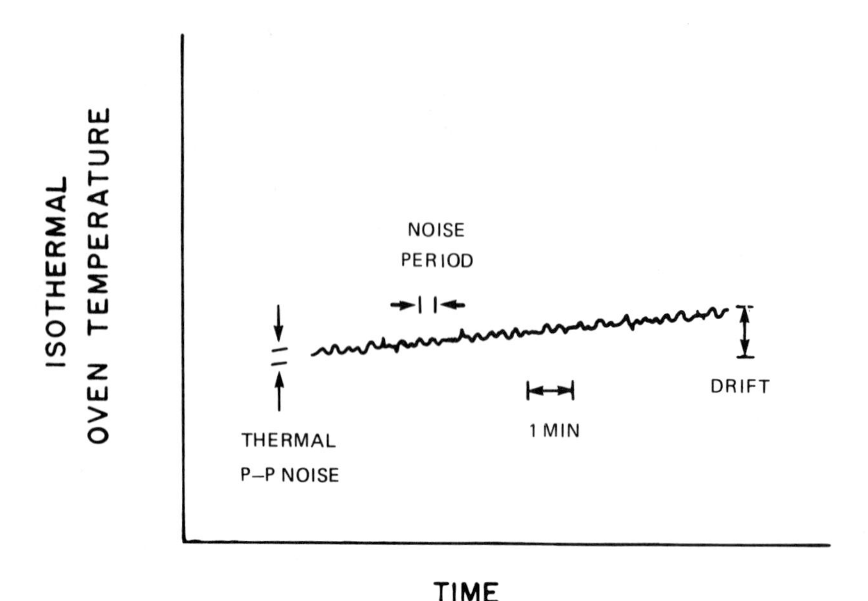

Figure 20. Graphical representation of thermal noise and thermal drift in GC ovens.

The specification term "ambient rejection" or "ambient temperature coefficient" defined by

$$\text{Ambient rejection} = \frac{\text{ambient temperature change}}{\text{oven temperature change}} \tag{8}$$

is often used to quantify drift. A good-quality column oven should display less than 0.02°C change for a 1°C change in ambient temperature [99]. The specification for thermal noise is usually determined by measuring the peak-to-peak band width for a time interval of 1 min. A good-quality column oven should display a thermal peak-to-peak noise that is not greater than 0.1°C for temperatures in the 50-150°C range, and no more than 0.3°C for temperatures in the 150-300°C range. Excessive thermal noise will contribute to retention time irreproducibility, in addition to causing significant baseline noise, which in turn will result in poorer detection limits.

Besides adequate temperature stability, it is also important to select a GC oven with a high degree of internal temperature uniformity. Most of today's high-quality GC instruments use a microprocessor-controlled, proportional integral temperature control system in combination with an efficient circulation fan to minimize temperature cycling and gradients within the volume of the oven occupied by the column [22,136,138,139]. If properly designed, the oven temperature should not be noticeably affected by the independent temperature control systems for the sample injection ports and the detectors. The worst-case situation, that is, when the injection ports and detector block are set at maximum temperatures of approximately 350°C, should be one of the manufacturer's specifications. Direct radiative heating of the FSOT column by the oven heating element or internal surfaces should also be avoided to minimize temperature gradients along the column and potential peak splitting problems. To ensure good peak shapes with FSOT columns, the average column temperature measured from one end of the column to the opposite end should be less than ±0.05°C at 50°C and less than ±0.15°C at 150°C.

Potential buyers of high-performance GC instrumentation should read Ref. 140 for a more detailed description of oven evaluation procedures. Manufacturer's oven specifications are often of limited comparative value because there is no standard set of test conditions. Thus, it may be necessary to perform one's own comparative testing program in the laboratory to obtain the performance data of interest. Again, Ref. 140 should be consulted for specific testing guidelines.

Figure 21 shows the design of a modern GC oven that is capable of excellent retention-time reproducibility with FSOT columns. Such state-of-

Figure 21. Column oven of a HP-5890A GC. (Reproduced with permission from Hewlett-Packard.)

the-art GC hardware and high-performance fused-silica WCOT columns have advanced qualitative analysis via GC because of this reproducibility. For example, it is now possible to identify many unknown compounds through the use of the Sadtler Capillary GC Standard Retention Index Library [141]. This particular library contains retention index data for approximately 2000 compounds on four different WCOT columns. Two types of retention indices are included in this Sadtler data base: isothermal, which correspond to the standard Kovats retention indices [142], and the temperature programming indices.

Commercial GCs have column ovens of different size and geometry. The best choice will obviously depend on the application. Accessibility of the oven for column installation/removal is an important factor because of its influence on the temperature of the chromatographer. However, ovens with "usable" volumes larger than about 1300 cubic inches should be avoided because of their larger internal temperature gradients and their larger thermal masses that affect maximum heating and cooling rates. Most manufacturers list the "usable" volume rather than the entire oven volume, since as much as half of the oven may be filled with fans, heaters, etc. and is consequently unusable for installation of columns and other accessories.

B. Temperature Ranges and Programming

The GC oven should allow operation within the range from about 30°C to approximately 400°C with temperature-setting increments of 1°C over the entire operating range. The minimum operating temperature without the use of a subambient accessory may be an important specification. Because of heat dissipation from heated zones external to the oven, a noncryogenic GC oven cannot operate with normal stability at ambient room temperature. Thus, the lowest oven temperature for ±0.1°C short-term stability is usually 5-10°C above ambient. For the analysis of high-molecular-weight compounds with aluminum-clad FSOT columns [2,3], the oven should have a maximum temperature of about 500°C. Subambient capability to operate at minimum temperatures in the -50 to -100°C range should also be available via optional cryogenic accessories. Liquid CO_2 may be used efficiently down to -50°C, but liquid N_2 should be the coolant of choice for temperatures in the -50 to -100°C range. The cryogenic oven configuration should not spray the coolant droplets directly onto the FSOT column. Instead, the coolant should be premixed with air in a shielded fan chamber and then circulated uniformly at high velocity within the oven to minimize thermal column gradients during cryogenic operation. Most commercial subambient accessories automatically control the amount of cryogen that enters the oven by periodic opening and closing of a solenoid valve that is attached to the cryogenic source. Further instrumental considerations for cryogenic GC have been described in a recent review paper [143].

The ability to temperature program the oven is essential in modern GC. The most common type of program is a linear ramp with isothermal hold times of selectable duration before and after the temperature ramp. Multi-step linear programs are also available on most microprocessor-controlled GCs. This latter capability is useful when using FSOT columns with on-

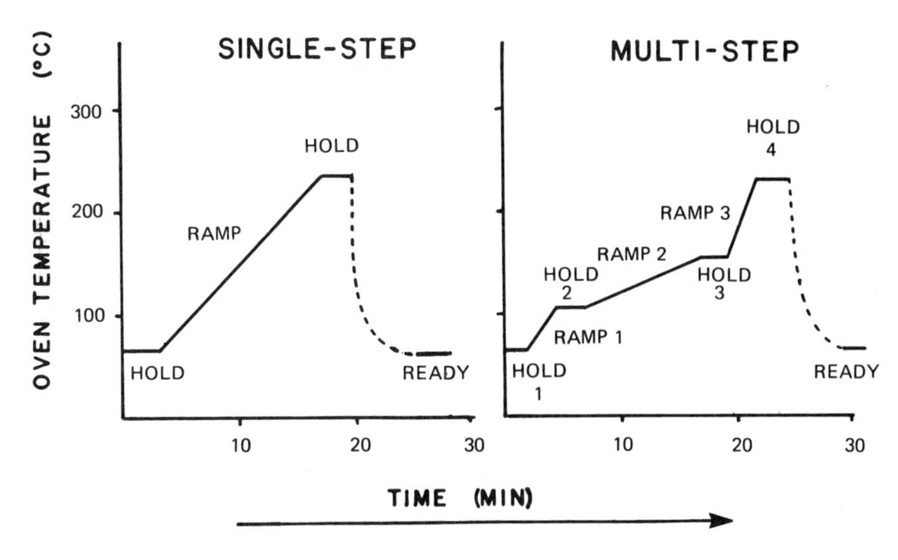

Figure 22. Two common types of linear temperature programs, single-step and multi-step.

column injections, since a low initial column temperature is required during injection, followed by a short rapid temperature ramp to initiate the chromatographic process, and then a slower temperature ramp during the rest of the chromatographic run time. Figure 22 illustrates both single-step and multiple-step linear temperature programming operations. Ramp rates from 1°C/min to 30°C/min, selectable in 0.1°C increments within this range, are desirable for general-purpose applications. The optimum value of the temperature ramp rate is a necessary compromise between resolution and analysis speed. To avoid significant losses in resolution, the maximum heating rate (r) should not usually exceed the value calculated from the following simple relationship [144]:

$$r = 12°C/t_M \tag{9}$$

where t_M is the retention time of an unretained component. However, certain multicomponent separations by FSOT columns do require faster heating rates than determined by equation (9) [136,145].

Both rapid heat-up and rapid cool-down of the column oven temperature are important for fast overall analysis and maximum sample throughput. Whereas the heating rate is determined by the desired resolution and the nature of the injected sample, the cooling rate at the end of the actual chromatographic run is limited by the thermal mass of the oven [22,140]. A GC whose oven is opened automatically and cooled with ambient air will reduce the time required for cool-down and equilibration at the desired starting temperature. A typical cool-down from 250°C to 50°C and re-equilibration at 50°C should require no more than 6-7 min. The more demanding cool-down from 400°C to an equilibrated 30°C should require no more than 25-30 min. Use of a subambient accessory can be used to significantly reduce these cool-down times.

Some kind of indication that the oven has reequilibrated at the end of the cool-down cycle is also a desirable feature. Certain instruments have a "ready" indicator connected to an internal delay circuit of 2-3 min to ensure that the oven has thermally stabilized before signaling that it is ready for another sample injection. Significant retention time variations can otherwise occur if injections are made at various times before the GC oven has reached thermal stability at the initial programming temperature [140].

VI. DETECTORS

A. Basic Performance Requirements

The most basic requirement of an "ideal" chromatographic detector is that its sensitive electrical response profile (i.e., the time integral of the transduced output signal) should be identical to the analyte concentration profiles at the end of the column (i.e., the detector input). In theory, then, the detector should not affect the number of theoretical plates, the analyte retention times, or the gas flow rates and gas flow patterns. Unfortunately, these theoretical requirements cannot be exactly satisfied with practical detectors. Therefore, nonideal, yet usable detectors must have response-time constants and effective volumes that are compatible with the particular GC conditions, such as column type, flow rate, etc.

The tremendous gains in column technology have placed increasing demands on detectors to achieve matching levels of performance. For example, (a) small-bore FSOT columns, as shown in Table 3, have relatively small sample capacities that in turn translate to the need for improved detector sensitivity and detectability, (b) detector cell volumes and response-time constants must be decreased concomitantly with column internal diameters to prevent significant contributions to extracolumn peak-broadening processes, and (c) the practical use of capillary columns with their high, yet finite, resolution for the quantitative analysis of exceedingly complex multicomponent samples requires detectors that are more and more selective to certain classes of compounds. In general, the development of GC detectors designed specifically for optimal compatibility with small-bore open tubular columns has received insufficient attention. Even when recognized in the past, this incompatibility problem has often been circumvented experimentally rather than conceptually solved. An example of this experimental approach is the use of detector makeup gas to decrease the effective detector volume (V_e) by the relationship

$$V_e = \frac{VF}{F + F_a} \tag{9}$$

where V is the effective detector volume without the added makeup gas, and F and F_a are the flow rates of the column effluent and added makeup gas, respectively. Whereas the makeup gas can decrease the effective detector volume, its use also dilutes the analyte concentration within the detector volume and thereby degrades detector sensitivity and detectability.

This detector compatibility issue, however, has finally influenced the design of several new commercial GC detectors. Section VI.C focuses on this new breed of detector designs rather than those designs that are largely unchanged from their conception during the packed column era. Because numerous review papers [146-150] and books [5,6,22,151-153] already describe in detail the general performance characteristics, proposed response mechanisms, and general designs for prominent GC detectors, no attempt has been made here to duplicate such information. However, basic detector specification terms are briefly defined in Section VI.C since an understanding of this detector nomenclature is essential in valid detector comparisons and purchasing decisions.

B. General Response Classification Schemes

Two common concepts are often used to classify detectors: as either concentration sensitive or mass-flow-rate sensitive, or universal (i.e., nonselective) and selective. The attempt to classify detectors as mass-flow or concentration sensitive seems meaningful only when the detectors are operated in a continuous monitoring mode and not when they are operated as chromatographic detectors [154]. As explained by Sevcik and Lips [155], the detector input is the varying concentration profile of the components as they elute from the GC column. During this elution time the concentration changes along with the mass-flow rate and the total mass of the analyte. Thus, concentration, mass flow, and total mass are obviously interrelated when a detector's response is used to profile the time-varying column output [155]. This realization has caused the author to question this particular detector classification dogma and its associated uses in basic detector terms such as sensitivity and detectability [154].

The general classification of detectors based on their response selectivity is, on the other hand, both valid and useful. In this scheme detectors fall into one of three response categories: universal, selective, or specific.

Universal

A detector that will respond to the presence of any and all substances besides the carrier gas is termed universal or nonselective. The only common GC detector that has a truly universal response is the thermal conductivity detector (TCD). The TCD measures changes in the thermal conductivity of the carrier gas, perturbed by any separated analytes eluting in sufficient concentration to cause a detectable variation in the temperature, and hence the resistance, of the filament sensor. Although the TCD has an universal response behavior, different compounds will have different relative molar responses (RMR) [5,6,148,151,153]. Thus, the universal TCD response is dependent on both the molecular weights of the eluted sample components and the carrier gas. In general, the relative molar TCD responses will decrease with increasing molecular weights of the eluted components [151]. For accurate quantitative analysis it is still necessary to calibrate the universal TCD for each analyte of interest.

Other commercial detectors with either universal or quasi-universal response capabilities are the mass spectrometer (MS) when operated in the total ion detection mode [5,6,22,148,156], the helium ionization detector (HeID) because of the high [19.8 eV] ionization potential of He [148,151,157], the photoionization detector (PID) with the 11.7-eV UV lamp [148,150-153, 158], the discharge ionization detector (DID) [197], and the ultrasonic detector (USD) [148,159,160].

Selective

This category of detectors includes those whose greatest response is to a certain class of compounds, for example, only those compounds containing a particular element or chemical combination of elements. Most common GC detectors fall into this selective designation. Examples include the flame ionization detector (FID) with a selective response to hydrocarbons; the electron-capture detector (ECD) for strong electron-absorbing compounds such as the multihalogenated organics; the flame photometric detector (FPD) for the selective detection of compounds containing either sulfur or phosphorus; the thermionic ionization detector (TID) for compounds containing nitrogen and/or phosphorus; the thermal energy analyzer (TEA) for the highly selective detection of *N*-nitroso compounds [161,163]; the redox chemiluminescence detector (RCD) for oxygenated hydrocarbons, olefins, aromatic hydrocarbons, amines, mercaptans, sulfides, hydrogen, ammonia, hydrogen peroxide, organic phosphates and phosphites, and sulfur dioxide [152,164,165]; and the electrolytic conductivity detector (ElCD) for halogen-, sulfur-, and nitrogen-containing compounds [148,151,152,166].

Two other selective detector systems of somewhat greater flexibility, complexity, and expense than those in the preceding list are (a) the mass spectrometer when operated in either the single or multiple ion-detection (MID) mode [148,156] and (b) the Fourier transform infrared (FT-IR) spectrometer [167]. Less expensive, bench-top GC/MS units [168,169] such as the Hewlett-Packard 5970 mass-selective detector (MSD) [170] and the Finnigan ion trap detector (ITD) [171] will undoubtedly continue to impact the GC detector field [172]. A similar bench-top GC/FT-IR system, the Hewlett-Packard 5965 IRD infrared detector [173,174], has brought FT-IR detection from the research bench to the practical chromatography lab.

Benefits derived from the use of selective detectors include more reliable qualitative and quantitative data, lower detection limits because of enhanced signal-to-noise ratios, and potentially faster analyses on shorter columns where coelution of analyte and nonanalyte sample components may otherwise cause interference problems.

Specific

A more restrictive response class of selective detectors is the specific detector that responds only to a single element or compound. Few detectors can be classified in this specific category, but the author will dismiss his normal trepidation in this matter and suggest that linked systems such as GC/FT-IR/MS [175-177] and GC/MS-MS [178] can yield specific molecular detection. Specific elemental detection can be approached for certain elements via GC coupled with plasma emission or atomic absorption spectrometers [152,198].

The potential capabilities of atomic emission detectors (AED) are available in the new Hewlett-Packard 5921A, which is based on a microwave-induced plasma and a photodiode spectrometer [199,200]. This automated AED can determine picogram quantities of up to 15 elements including C, N, H, O, S, P, and Cl. Other recent detectors that may produce specific responses are the O-FID for oxygenated hydrocarbons [201], the sulfur chemiluminescence detector [202], and the nitrogen chemiluminescence detector [203].

C. Performance Specification Terms

Sensitivity

Detector sensitivity is an often misused term. For proper use, sensitivity (S) is defined as the change in the measured detector signal (s) resulting from a change in the concentration (c) of the eluted analyte i:

$$S = \frac{ds_i}{dc_i} \tag{10}$$

The sensitivity must remain constant for the detector's output to have the same proportional value to the detector's input. Dimensional units of sensitivity are dependent on the units used for the detector input and output, such as Coul/g, amp-L/g, or Hz-L/g [151,155]. The sensitivities of individual detectors will depend on their design and will be different for different analytes. This dependency of sensitivity on analyte type is sometimes also referred to as compound response factors [148].

To avoid confusion, the word "sensitivity" should not be used synonymously with detection limits. A detector with high sensitivity and a high noise level may have a higher detection limit than one with lower sensitivity and a lower noise level.

Detectability

As noted in the preceding paragraph, the detector's signal-to-noise ratio for a particular analyte peak determines the lowest detectable concentration (g/liter), mass-flow (g/sec) or mass (g). The most common GC definition of minimum detectability is the analyte concentration, mass-flow, or mass in the detector that yields a peak response that is equal to twice the peak-to-peak noise determined for a time period equivalent to the analyte's peak width [148,151,152]. It should be emphasized that this definition refers to the amount of analyte in the detector and not to the amount in the injected sample. Furthermore, the minimum detectability for a given detector is not constant, but will vary as a function of the experimental conditions. For example, peak heights (i.e., the "signal" in the signal-to-noise ratio) are dependent on retention time and column efficiency. Likewise, the magnitude of the measured detector noise is dependent on the actual measurement method and the measurement period [155].

Because of the preceding variables in establishing a standard chromatographic measurement procedure for the minimum detectability specification, the American Society for Testing and Materials (ASTM) E-19 Committee on Chromatography has recommended that continuous sample injection methods should be employed for the determination of such quantitative detector specifications [179]. Whereas the detector specifications determined by this continuous sampling method are useful for the purpose of relative detector performance, these specification values often cannot be used to accurately predict detector performances for analyte peaks eluted from chromatographic columns.

The lack of an accepted standard method for the determination of minimum detectability is further complicated by the absence of a universal definition of minimum detectability. The author prefers the following general definition proposed by Sevcik and Lips [155]:

$$MDL = K \frac{N_{rms}}{S} \tag{11}$$

where MDL is the minimum detection limit, N_{rms} is the detector noise expressed as the root mean square of the noise signal, S is the detector sensitivity for a specific analyte, and K is a constant whose magnitude is determined by the analyte peak width and the desired statistical confidence level. Rather than setting K = 2 for any chromatographic peak as in the common definition of minimum detectability, equation (11) accounts for different experimental conditions by adjusting the magnitude of K.

Knoll (180) has more recently described an alternative method for estimating minimum detectability, or limit of detection (LOD). The LOD is determined via the formula

$$C_{LOD} = K_{LOD} h_n C_s / h_s \tag{12}$$

where C_{LOD} is the LOD quantity; the reciprocal of C_s/h_s, i.e., h_s/C_s is the analyte peak height per unit amount of analyte; h_n is the largest noise fluctuation (either positive or negative) observed in the noise fluctuation interval; and K_{LOD} is a tabulated constant whose magnitude depends on the peak width multiple. This simple, yet sound, approach is attractive because it circumvents the practical problems that often accompany the measurement of the standard deviation of the detector noise [155,180].

Until all manufacturers agree on both the definition and the measurement method for detectability specifications, potential customers must be careful when comparing such detector specifications. Thus, it is important to know the basis for quoted minimum detectabilities to ensure fair performance comparisons between different commercial designs of the same type of detector. If minimum detectability is of prime concern, then a side-by-side laboratory comparison using samples with the analytes of interest is recommended.

Linear Dynamic Range

Reliable, efficient quantitative analysis of real samples usually requires detectors with linear dynamic ranges (LDRs) that are as large as possible. However, there is no accepted definition for linear dynamic range among detector manufacturers, which in turn makes it difficult to directly compare quoted values for this important specification.

The ASTM recommended definition for LDR is the range of analyte concentrations, mass flows, or masses over which the detector sensitivity is constant to within ±5% [179]. Thus, LDR is not synonymous with the related specification of dynamic range (DR). The dynamic range is that range of analyte concentrations, mass flows, or masses over which a change in concentration, mass flow, or mass will produce a detectable change in detector signal. The dynamic range specification value for a detector is usually greater than its linear dynamic range. The LDR is the term of practical significance to the analyst, not the DR.

LDR values vary considerably with the type of GC detector. Table 7 lists typical specification values, including LDRs, for 10 commercial types

Table 7. Typical Specification Parameters and Values for Commercial GC Detectors

Detector	Selectivity	Sensitivity	Detectability	LDR	Internal volume (or effective)	Destructive
TCD	"Universal	4-7 V·ml/mg	0.4-1 ng/ml	10^5-10^6	3.5-150 μl	No
HeID	"Universal"	0.01-0.1 C/g	Low ppb (V/V)	10^4	100-200 μl	No
USD	"Universal"	. . .	0.2 ng/sec	10^6	180 μl	No
FID	Hydrocarbons	0.01-0.03 C/g	1-10 pg C/sec	10^6-10^7	(10-20 μl)	Yes
PID	Various organics and inorganics	0.3 C/g	1 pg-1 ng	10^4-10^7	40-50 μl	No
ECD	Electrophores	100-500 Hz·ml/pg (lindane)	0.1 pg/sec (lindane)	10^3-10^4	0.3-4 ml	No
FPD(S/P)	10^3-10^6 g C/g S	2-20 A/(g S/sec)^2	10 pg S/sec	10^2-10^3 (S)	0.1-10 ml	Yes
	10^4-10^5 g C/g P	20-200 C/g P	0.5-1 pg P/sec	10^3-10^5 (P)	0.1-10 ml	
TID(N/P)	10^3-10^4 g C/g N	0.1-1 C/g N	0.4 pg N/sec	10^4-10^5	(10-50 μl)	Yes
	10^4-10^5 g C/g P	1-10 C/g P	0.2 pg P/sec	10^4		
ElCD(N/S/Cl)	10^5 g C/g N	. . .	10^{-12} g N/sec	10^4 (N)	. . .	Yes
	10^5 g C/g S	. . .	10^{-12} g S/sec	10^4 (S)	. . .	
	10^6 g C/g Cl	. . .	10^{-13} g Cl/sec	10^5-10^6 (Cl)	. . .	
RCD	Oxygenated HCs, NH_3, SO_2, CO, H_2O_2, H_2, H_2S, etc.	. . .	0.2 ng (2,6-dimethylphenol)	10^3	. . .	Yes

of detectors. Columns E to T in Table 1 show the commercial sources for various GC detectors in 1986-1987.

Detector Volumes and Time Constants

Unless properly designed and used, detectors can make serious contributions to the extracolumn peak broadening processes in open tubular column GC. Four factors that contribute to such detector effects on band broadening are the detector's sensing volume, its flow patterns, time-dependent response mechanism, and the electrical time constants in the electrometer and recording system. Two classic papers on these subjects are recommended reading for any chromatographer [181,182].

Peak fidelity (PF) is used to describe the ratio of the "input" peak height to the "output" peak height and is expressed by

$$PF = \left(\frac{\sigma^2}{\sigma^2 + \tau^2} \right)^{1/2} \tag{13}$$

where σ^2 is the true peak variance and τ^2 is the extra column contribution from the overall detector time constant corresponding to the detector mixing volume and the electronic time constant (i.e., $\tau^2 = \tau_V^2 + \tau_e^2$). The earliest-eluting, and therefore narrowest, peaks in a modern OTC chromatogram pose the most severe challenge to peak fidelity. A peak fidelity of 95 percent requires that $\sigma > 3\tau$. Figure 23 demonstrates the importance of a sufficiently small detector time constant on peak height, shape, and resolution. GC detectors used with high-performance open tubular columns should have time constants of 50 msec or less [19,60,155]. This 50 msec or less specification should be the *combined* time constant of the actual detector and the electrometer, not just the time constant of the electrometer.

The detector volume within which the actual transduction reaction occurs must also be considered for good peak fidelity. This volume is often called the effective detector volume (V_e) and is simply related to the residence time of analytes (t) within the detector under given flow rates (F) by

$$V_e = tF \tag{14}$$

The relationship between detector volume and peak fidelity is quite obvious when one remembers that the detector output is effectively its response integrated over a time interval corresponding to the effective detector volume divided by the flow rate [182]. If the chromatographic peak widths become comparable in magnitude to the detector residence time, peak fidelity is severely degraded.

The following relationship between residence time and peak width at half-height ($W_{1/2}$) is a useful guideline:

$$t > \frac{W_{1/2}}{10} \tag{15}$$

For example, if the minimum peak width is 0.5 sec, then the maximum detector residence time should be 50 msec. Equation (14) can then be used to calculate the compatible values of V_e and F. Whereas the use of

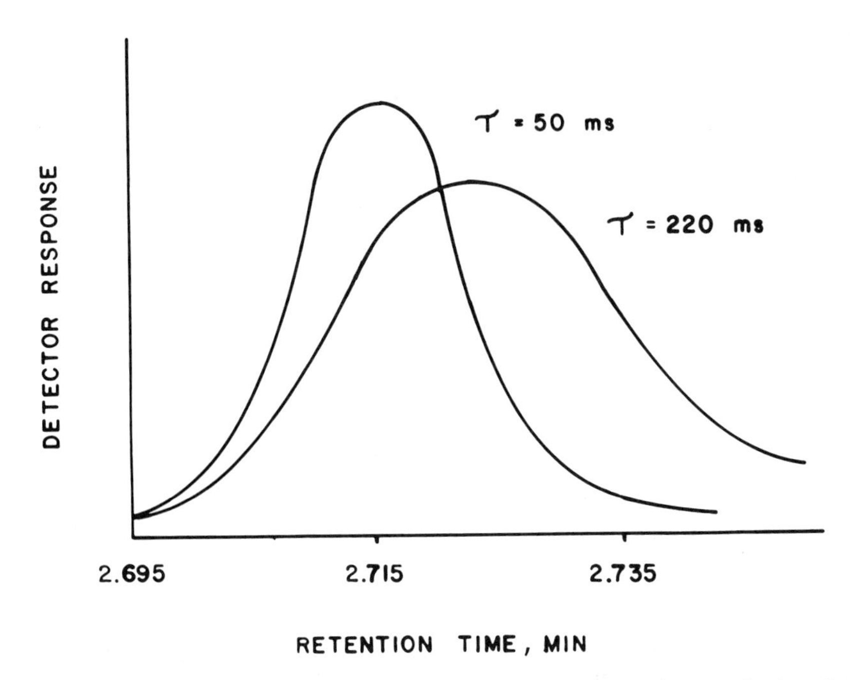

Figure 23. OTC peaks recorded on an expanded time scale to show the effects of 50 msec versus 220 msec time constant on peak height, shape, and resolution.

makeup gas can be used to increase F and thereby reduce the requirement for small detector volumes, there is a "price" associated with this approach. In this instance the "price" is higher detection limits, since the additional makeup gas dilutes the effective analyte concentration within the sensing volume of the detector [155]. Consequently, the wise buyer should obtain GC detectors with small effective volumes for two reasons, peak fidelity and minimum detectability.

In the case of detectors that employ a pulse-modulated method of operation (such as all constant-current ECDs and certain TCD designs), the modulation frequency may become a significant factor in the detector's time constant. For example, Simon and Wells [183] showed that a 5-μl volume TCD is limited to the detection of peaks with minimum widths of 1 sec when the TCD modulation frequency is 10 Hz. Thus, modulated TCDs and ECDs should be operated at relatively high pulse frequency, in addition to having small internal detector volumes, for high-fidelity peak detection in open tubular column GC.

VII. INTEGRATING RECORDERS AND DATA SYSTEMS

The number of commercial products available for the acquisition, processing, and recording of GC data has mushroomed during the past five years.

Columns U and V of Table 1 show that over 50 companies offer such chromatographic data equipment. Thus, choosing the optimum product for a particular application can be quite confusing. This section will categorize the products and then offer some general guidelines for the selection of a proper system. Although some top-of-the-line models of commercial GCs (e.g., the PE-8000 series and the HP-5880) contain various degrees of data handling and graphics capability, such sophisticated GCs will not be specifically discussed. Any attempt to describe such specific capabilities and features of particular GCs would be quickly outdated.

A. System Approaches

Current GC data system products can be divided into four general categories: (1) independent, stand-alone recording integrators such as the Hewlett-Packard 3396A, Spectra-Physics 4290/4270/4200, Perkin-Elmer LCI-100, and Shimadzu C-R5A; (2) chromatography workstations based on specific software/hardware additions to common personal computers (PC), such as Dynamic Solutions' Baseline, Nelson Analytical's PC Integrator, Hewlett-Packard's 5895A, and Perkin-Elmer's OMEGA; (3) several stand-alone reporting integrators networked to a personal computer, such as Spectra-Physics' ChromStation/2; and (4) a large number of dedicated analog-to-digital converters (ADC) networked to either a central microcomputer or minicomputer, such as Nelson Analytical's Turbochrom, Hewlett-Packard's 3350, VG's Multichrom, and Perkin-Elmer's LIMS 2600. In general, categories (1) and (2) are systems designed for dedication to a single GC, category (3) systems are designed for a small number (1 to 10) of on-line GCs, and category (4) systems are designed to operate with a relatively large number (10 to 75) of instruments.

The selection of an optimum system should consider several key items. First, which general type of system (i.e., categories 1-4 above) is best suited to the laboratory application, both now and in the near future? The major arguments for stand-alone integrators are the proven reliability of such devices, their user friendliness, the on-line level of interaction between the chromatographer and the GC, the increased capabilities of each new generation, and their relatively inexpensive prices [184]. Prices for these stand-alone integrators range from approximately $2500 to $3500. Significant disadvantages of stand-alone integrators include their limited abilities to archive the data; reprocess the data; incorporate the data in other general-purpose formats such as spread sheets, database managers, and word processors; and their usual lack of graphics for simultaneous comparison of several chromatograms, visual observation of baselines for peak quantification, rescaling for visual examination of very large or small peaks, etc. The readers should be aware that some of these integrator limitations may no longer be applicable to certain commercial units because of evolving technology. For instance, several stand-alone integrators can now be configured with either extra memory or external disks for additional chromatographic storage capability. Basic programmability is available on some models, and replotting/reintegration is offered as either standard or optional features.

The small yet independent PC-based systems in category (2) are an inexpensive option for greater data storage, calculation, recalculation, and archival and graphic capabilities. The additional price of these PC-based data systems compared to "stand-alone" integrators is approximately $2000 to $4000.

When it is desirable to network several GCs together within a laboratory for enhanced archiving and the generation of sophisticated report materials, then several stand-alone integrators connected to one personal computer may be an optimum solution. Major advantages of this approach include local plotting with complete annotation, local control of each GC's operation, local generalization of chromatographic reports, continuing operations even in the case of a CPU failure in the central PC, computer-based storage and graphics, and overall system flexibility and expandability [185]. The prices for such a networked PC/integrator system are approximately $10,000 for the PC and operating software plus about $2500 for each required integrator.

The larger data system approaches in category (4) are designed to work efficiently with a larger number of GCs where maximum system management, security, and CPU power are necessary. On a per-GC basis, the initial costs of a larger category (4) system are comparable to other approaches. For example, category (4) type systems intended for use with 20 or more GCs will cost approximately $3500 to $5000 per instrument. An issue of *Chromatography Forum* was dedicated to articles describing several of these larger chromatographic data systems and is highly recommended reading for potential buyers [186-189].

B. General Evaluation Data

Regardless of the particular data system selected, all employ digital processing of the analog chromatographic detector signals. Whereas such computerized signal processing has many advantages over similar analog approaches, it has become more difficult for the average chromatographer to understand and document the actual signal processing procedures. Answers to questions regarding the accuracy, repeatability, time constants of digital filters, overall response time constants, and operator ruggedness are of fundamental importance in the evaluation of chromatographic integrators, workstations, and data systems. Such information is not generally available from manufacturers. Further complications arise if the vendors refuse to give purchasers the algorithms that describe what the software does in these marvelous, yet invisible, data treatments [192]. As pointed out by Thompson and Dessy [193], it is imperative that users of digital signal processors have a basic understanding of how these devices condition the signal to avoid errors due to overfiltering, amplitude quantization, and aliasing.

A recent project has been initiated by Papas and Delaney [194] to address the heretofore lack of an objective quality assurance program for evaluation of the performances of chromatographic integrators and data systems. The initial report from this project describes the test results for four different integrators, although the integrators are not identified as to manufacturers or models. Significant performance differences between integrators occured in baseline construction with noisy, fused, skewed peaks, which in turn affected peak quantification. Other differences were noted for the degrees of warning provided to the operator when improper integrator parameters were selected. Hopefully, the results of further studies with additional chromatographic data systems will be available in the near future. The need for such quality assurance information has been vividly illustrated by reports that some stand-alone integrators may report different areas for the same peak from reprocessed data they originally generated. Likewise, some networked integrator/computer systems may report different areas for the same peaks when the same raw data are processed by both the integrator and the central computer software.

C. Specific Evaluation Criteria

Certain criteria apply to the evaluation of both large and small chromatographic data processing products. These "common" considerations are summarized in the following sections.

ADC Hardware

All modern GC integrators and data systems rely on an analog-to-digital values. Two general types of ADCs are used in this application, integrating (or low-level) converters and fast converters [195]. The more commonly used integrating ADCs are either the voltage-to-frequency (V/f) or the dual-slope designs. However, there seems to be no significant difference in performance between integrating and fast ADCs for chromatographic signal applications [186]. Typical specification values for ADCs used in chromatographic integrators and data systems are listed in Table 8.

Several of the ADC parameters in Table 8 deserve further discussion. For example, accurate representation of narrow OTC chromatographic peaks with widths less than 1 sec requires relatively fast sampling rates. From information theory, the criterion for adequate sampling states that the minimum Nyquist sampling frequency must be twice the highest frequency component in the signal, or twice the bandwidth of the signal waveform [195]. In general chromatographic terms, this means the ADC must be

Table 8. Typical ADC Specification Parameters and Values

Parameter	Value
Voltage range	-10 mV to 1 V
Noise	< 0.1 μV
Voltage resolution	0.1 to 1 μV
Sampling rate	0.1 to 60 Hz
Voltage drift	< 1 μV/°C
Linearity	0.1%
Dynamic range	> 10^6

capable of a sampling rate sufficient to obtain a minimum of 20 data points across the narrowest peak of interest in the chromatogram, that is, a minimum ADC sampling rate of 20 Hz for a peak width of 1 sec.

Those ADCs that allow full bipolar operation from -1 V to +1 V are necessary for the accurate integration of both negative and positive peaks in the same chromatogram. Most ADCs do not have true bipolar operation, which can be a problem when used with TCD, HeID, and FPD (S-mode) signals.

Another ADC consideration with category (3) and (4) data systems is how they are interfaced to the central micro- or minicomputer. Two types of interface protocols involve the current loop concept and the RS-232C serial port, where each has certain advantages and disadvantages [186].

Peak Detection Algorithms

Some peak detection algorithms use the first derivative of the signal with respect to time. As illustrated in Figure 24, the start and ending of a peak are indicated when the first derivative exceeds given threshold values. The threshold range is limited by the signal-to-noise ratio in the detector signal. This first-derivative approach is taxed to detect small peaks in the presence of baseline noise because small threshold settings result in

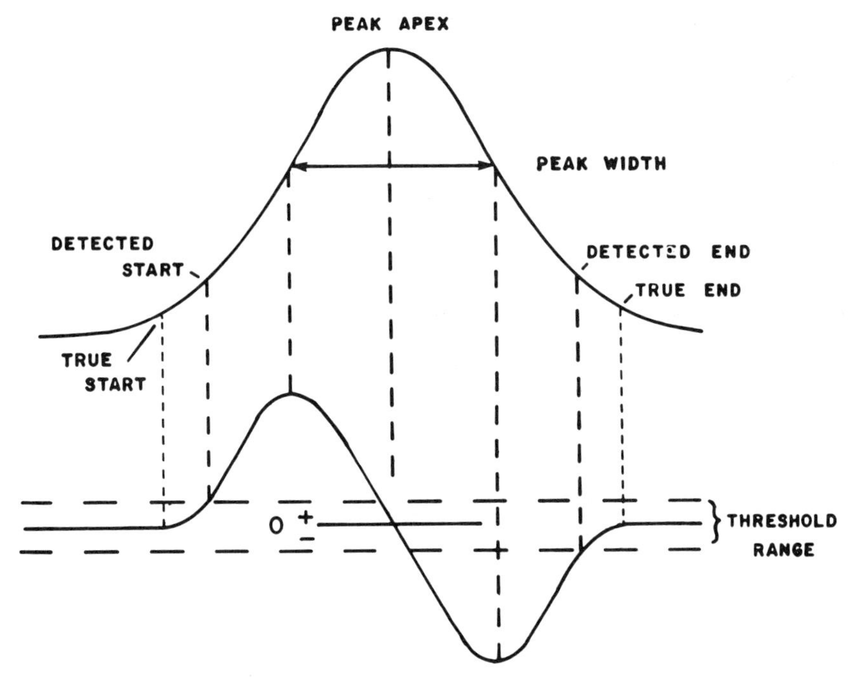

Figure 24. Basis for peak detection with a general first-derivative algorithm.

the detection of noise and drift as apparent analyte peaks. It is also impossible to determine accurately the true start and end points for the peaks via just the first-derivative algorithms. An additional second-derivative algorithm or curve-fitting routine is employed for this purpose. Other problems that tax the peak identification power of the first-derivative procedure are in the cases of unresolved peaks, badly tailing peaks, or drifting baselines.

Other peak detection algorithms beside the derivative method are used in many integrators and data systems. Most of these algorithms are proprietary, although it is usually possible to obtain a general idea of how each functions from the specific manufacturer. Some of these newer peak detection algorithms are reportedly difficult to fine-tune in trace analysis applications [186]. Thus, the performance characteristics of a particular data system's algorithm are best evaluated by using it to quantify typical samples analyzed with your lab's GCs.

Processing Software

One of the most important features in a data system is the ability to perform both peak height and area measurements accurately. Systems that use the maximum amplitude of a curve fitted to the data points rather than the actual maximum data point collected across the detected peak will generally yield more reliable peak-height and retention-time measurements. The measurement reliability of peak areas, on the other hand, is limited by the software's ability to locate the true start and end points of each peak. Potential problems often arise with merged peaks and/or when the baseline changes during peak elution. Desirable software packages should offer several choices of standard integrating procedures for merged peaks, including dropped perpendiculars, linear tangent skim, and exponential skim. In addition, the ability to let the user position a cursor to select integration start and stop points is very useful. This latter feature is most valuable when combined with a reintegration capability that allows the user to reprocess the raw chromatographic data until the desired quantification is achieved. Graphics that allow visual observation of drawn baselines, cursor positions, peak expansion/contraction, and replotting are a major advantage of modern computer-based data systems.

Most, if not all, of today's stand-alone integrators and data systems can perform at least four standard data calculations: (a) percent area, (b) area normalization, (c) internal standardization, and (d) external standardization [6,22,196]. Regardless of the quantification method, one or more GCs operating continuously over a year can produce a mountain of data. In such cases, programs that allow the user to efficiently store, sort, summarize, and output data tables and plots can be quite important. If no such standard software packages are available, then the system should at least allow the users to develop their own programs using BASIC.

For just such reasons, even some integrators now have BASIC programmability. However, most integrator BASIC implementations are interpreted, limited in capability, slow, and awkward to use. The BASIC used in the HP-3393A/3396A has largely overcome these obstacles [184].

Intangible Considerations

Before making a final purchasing decision one should consider other less tangible factors, such as (a) proven versus unproven system reliability, (b) local familiarity with the equipment (especially in the case of computer-based systems), (c) potential quality of customer service, product warranties, installation service, personnel training programs, etc, and (d) some indication of whether the manufacturer will still be in business in 5-10 years. As usual, it is worthwhile to remember *caveat emptor*!

Acknowledgments

Special acknowledgments are due to Dr. J. M. Hales at Battelle Pacific Northwest Laboratories and Dr. J. Shreeve at the University of Idaho for their support of my sabbatical leave, which made the preparation of this manuscript possible. I also gratefully acknowledge the patient and skilled assistance of my best colleague, Judy Farwell, who typed, proofread, corrected, retyped, and, above all, encouraged. Drafting for the figure drawings was meticulously performed by Jason Lingg.

REFERENCES

1. A. T. James and A. J. P. Martin, *Biochem. J.*, *50*:679 (1952).
2. S. R. Lipsky and M. L. Duffy, *HRC & CC*, *7*:376 (1986).
3. S. R. Lipsky and M. L. Duffy, *LC-GC*, *4*:898 (1986).
4. J. Drozd, *Chemical Derivatization in Gas Chromatography*, Elsevier, Amsterdam (1981).
5. J. A. Perry, *Introduction to Analytical Gas Chromatography*, Marcel Dekker, New York, pp. 178-222 (1981).
6. C. F. Poole and S. A. Schuette, *Contemporary Practice of Chromatography*, Elsevier Science Publishing, New York, pp. 29-143 (1984).
7. W. R. Supina, Packed Columns/Column Selection in Gas Chromatography, in *Modern Practice of Gas Chromatography*, 2nd ed. (R. L. Grob, Ed.), John Wiley & Sons, New York, pp. 117-158 (1985).
8. M. A. Kaiser, High-Resolution Gas Chromatography, in *Modern Practice of Gas Chromatography*, 2nd Ed. (R. L. Grob, Ed.), John Wiley & Sons, New York, pp. 159-185 (1985).
9. M. J. E. Golay, Theory and Practice of Gas Liquid Partition Chromatography with Coated Capillaries, in *Gas Chromatography (1957 Lansing Symposium)* (V. J. Coates, H. J. Noebels, and I. S. Fagerson, Eds.), Academic Press, New York, pp. 1-13 (1958).
10. M. J. E. Golay, Theory of Chromatography in Open and Coated Tubular Columns with Round and Rectangular Cross-Sections, in *Gas Chromatography (Edinburgh Symposium)* (R. P. W. Scott, Ed.), Butterworths, London, pp. 139-143 (1960).
11. L. S. Ettre, The Evolution of Open Tubular Columns, in *Applications of Glass Capillary Gas Chromatography* (W. G. Jennings, Ed.), Marcel Dekker, New York, pp. 2-3 (1981).
12. B. A. Jones, K. E. Markides, J. S. Bradshaw, and M. L. Less, *Chromatog. Forum*, *1*:38-44 (1986).

13. L. S. Ettre, *Introduction to Open Tubular Columns*, Perkin-Elmer Corp., Norwalk (1978).
14. W. J. Jennings, *Gas Chromatography with Glass Capillary Columns*, 2nd ed., Academic Press, New York (1980).
15. R. C. M. de Nijs and J. de Zeeuw, *J. Chromatogr.*, *279*:41 (1983).
16. J. de Zeeuw and R. C. M. de Nijs, Molsieve 5 Å PLOT Column: Separation of Permanent Gases with Capillary Columns, *Chrompack Topics*, *12*:1 (1985).
17. P. A. Larson and B. L. Ryder, "Series 530u Chromatography Columns: An Alternative to Packed Columns, " Hewlett-Packard Technical Paper No. 115, Avondale, Pennsylvania (1986).
18. How to Choose the Right Internal Column Diameter for Your Capillary GC Analysis, *The Supelco Reporter*, *V*:1 (1986).
19. Guidelines for Capillary Column Selection: Relationship between Internal Diameter and Column Performance, *All-Chrom Newsletter*, *25*:7 (1986).
20. "The Packed Column Alternative: MEGABORE," J & W Scientific, Inc., Catalog, Folsom, California (1986).
21. J. R. Berg, *LC-GC*, *5*:206 (1986).
22. M. L. Lee, F. J. Yang, and K. D. Bartle, *Open Tubular Column Gas Chromatography: Theory and Practice*, John Wiley & Sons, New York (1984).
23. R. R. Freeman, *High Resolution Gas Chromatography*, 2nd ed., Hewlett-Packard Co., Avondale, Pennsylvania (1981).
24. W. J. Jennings, *Comparisons of Fused Silica and Other Glass Columns in Gas Chromatography*, Huethig Publishing, Mamaroneck, New York (1981).
25. R. D. Dandaneau and E. H. Zerenner, *HRC & CC*, *2*:351 (1979).
26. S. O. Farwell, S. J. Gluck, W. L. Bamesberger, T. M. Schutte, and D. F. Adams, *Anal. Chem.* *51*:609 (1979).
27. C. J. Barinaga and S. O. Farwell, *HRC & CC*, *9*:388 (1986).
28. W. O. McReynolds, *J. Chromatogr. Sci.*, *8*:685 (1970).
29. M. S. Klee, Optimizing Separations in Gas Chromatography, in *Modern Practice of Gas Chromatography*, 2nd ed. (R. L. Grob, Ed.), John Wiley & Sons, New York, pp. 187-207 (1985).
30. G. Takeoka, H. M. Richard, M. Mchran, and W. Jennings, *HRC & CC*, *6*:145 (1983).
31. R. L. Dillard and R. C. Gearheart, "Chromatography Using a Mixed Phase 0.53 mm I.D. Capillary Column," Hewlett-Packard Application Note #228-46, Avondale, Pennsulvania (1986).
32. R. J. Laub and J. H. Purnell, *Anal. Chem.*, *48*:799 (1987).
33. R. J. Laub, *Physical Methods in Modern Chemical Analysis*, Vol. 3 (T. Kawana, Ed.), Academic Press, New York, pp. 249-341 (1983).
34. J. H. Purnell and P. S. Williams, *J. Chromatogr.*, *325*:1 (1985).
35. P. Sandra, F. David, M. Proot, G. Diricks, M. Verstappe, and M. Verzele, *HRC & CC*, *8*:782 (1985).
36. K. Grob, G. Grob, and K. Grob, Jr., *J. Chromatogr.*, *211*:243 (1981).
37. G. Schomburg, H. Husmann, S. Ruthe, and M. Herraiz, *Chromatographia*, *15*:599 (1982).
38. C. Madani, E. M. Chambaz, M. Rigaud, J. Durand, and P. Chebroux, *J. Chromatogr.*, *126*:161 (1976).

39. H. Frank, G. J. Nicholson, and E. Bayer, *J. Chromatogr. Sci.*, *15*: 174 (1977).
40. I. Benecke and G. Schombury, *HRC & CC*, *8*:191 (1985).
41. P. Sandra, *LC-GC*, *5*:236 (1987).
42. L. S. Ettre, *Open Tubular Columns: An Introduction*, Perkin-Elmer Corp., Norwalk, Connecticut (1973).
43. P. J. Schoenmakers and F. C. C. J. G. Verhoeven, *TrAC*, *6*:145 (1983).
44. F. Pacholec, H. T. Butler, and C. F. Poole, *Anal. Chem.*, *54*:1938 (1982).
45. F. Pacholec and C. F. Poole, *Chromatographia*, *17*:370 (1983).
46. C. F. Poole, H. T. Butler, M. E. Coddens, S. C. Dhanesar, and F. Pacholec, *J. Chromatogr.*, *289*:299 (1984).
47. J. B. Angell, S. C. Terry, and P. W. Barth, *Sci. Am.*, *44* (1983).
48. S. Saadat and S. C. Terry, *Am. Lab.*, *16*:90 (1984).
49. M. J. Mignano, P. R. Rony, D. Grenoble, and J. E. Purcell, *J. Chromatogr. Sci.*, *10*:637 (1972).
50. F. Poy, S. Visani, and F. Terros, *J. Chromatogr.*, *217*:81 (1981).
51. G. Schomburg, H. Husmann, F. Schulz, G. Teller, and M. Bender, *J. Chromatogr.*, *279*:259 (1983).
52. J. V. Hinshaw, *Instr. Res.*, pp. 14-24, March 1985.
53. Halfmil Capillary Column Catalog, Restek Corp., Port Matilda, Pennsylvania, 1987, pp. 7-10.
54. K. Grob and G. Grob, *HRC & CC*, *2*:109 (1979).
55. K. Grob, G. Karrer, and M. L. Reikkola, *J. Chromatogr.*, *334*:129 (1985).
56. High Resolution Chromatography Products, J & W Scientific Catalog, Folsom, California 1987/88, pp. 123-124.
57. W. Jennings, "Selecting the Injection Mode," in *Sample Introduction in Capillary Gas Chromatography* (P. Sandra, Ed.), Vol. 1, Huethig Publishing, Mamaroneck, New York, 1985, pp. 23-33.
58. HRGC Mega Series Catalog, Carlo Erba Strumentazione, Milan, Italy, 1986.
59. K. Grob and K. Grob, Jr., *J. Chromatogr.*, *151*:311 (1978).
60. F. J. Yang, D. Guidinger, R. Matthews, D. DeFord, R. Iwao, C. Ray, and G. Ogden, *Am. Lab.*, *16*:102 (1984).
61. G. Schomburg, "Temperature Programmed Sample Transfer in *Sample Introduction in Capillary Gas Chromatography* (P. Sandra, Ed.), Vol. 1, Huethig Publishing, Mamaroneck, New York, 1985, pp. 55-76.
62. F. Poy and L. Cobelli, Programmed Temperature Vaporizer (PTV) Injection, in *Sample Introduction in Capillary Gas Chromatography* (P. Sandra, Ed.), Vol. 1, Huethig Publishing, Mamaroneck, New York, 1985, pp. 77-97.
63. W. Vogt and K. Jacob, Programmed Temperature Injection System—Use with Large Sample Volumes, in *Sample Introduction in Capillary Gas Chromatography* (P. Sandra, Ed.), Vol 1, Huethig Publishing, Mamaroneck, New York, 1985, pp. 99-131.
64. D. H. Desty, A. Goldup, and B. A. F. Whymann, *J. Instr. Petroleum*, *45*:287 (1959).
65. K. Grob and G. Grob, *J. Chromatogr. Sci.*, *7*:584 (1969).

66. K. Grob and G. Grob, *J. Chromatogr. Sci.*, *7*:587 (1969).
67. K. Grob and K. Grob, Jr., *J. Chromatogr.*, *94*:53 (1974).
68. J. A. Jonsson, J. Vejrosta, and J. Novak, *J. Chromatogr.*, *236*:307 (1982).
69. D. F. Adams and S. O. Farwell, Sampling and Analysis, in *Air Pollution*, 3rd ed., Vol. VII (A. C. Stern, Ed.), Academic Press, New York, 1986, pp. 47-142.
70. J. E. Cuddeback, S. R. Birch, and W. R. Burg, *Anal. Chem.* *47*:355 (1975).
71. "Sampling Valves for Gas Chromatography," Technical Data Bulletin, Valco Instruments Co., Houston, Texas, 1980.
72. D. R. Deans, *Chromatographia*, *1*:18 (1968).
73. "UNIVAP: Precision Gas Sampling for Gas Chromatography," SGE, Austin, Texas, 1986.
74. D. R. Deans, *J. Chromatogr.*, *203*:19 (1981).
75. G. Schomburg, Multidimensional Gas Chromatography as Sampling Technique, in *Sample Introduction in Capillary Gas Chromatography* (P. Sandra, Ed.), Vol. 1, Huethig Publishing, Mamaroneck, New York, 1985, pp. 235-260.
76. G. Schomburg, *LC-GC*, *5*:104 (1987).
77. R. W. Slack and A. C. Helm, *Am. Lab.*, *18*:80 (1986).
78. T. V. Raglione, N. Sagliano, Jr., T. R. Floyd, and R. A. Hartwick, *LC-GC*, *4*:328 (1986).
79. H. J. Coates and C. D. Pfeiffer, *Chromatogr. Forum*, *1*:29 (1986).
80. Z. Naizhong and L. E. Green, *HRC & CC*, *9*:400 (1986).
81. R. J. Phillips, K. A. Knauss, and R. R. Freeman, *HRC & CC*, *5*:546 (1982).
82. SGE Pneumatic Column Switching System, Scientific Glass Engineering Inc., Austin, Texas.
83. Multiple Switching Intelligent Controller, Analytical Controls Inc., Delft, The Netherlands.
84. A. Zlatkis, H. A. Lichtenstein, and A. Tishbee, *Chromatographia*, *6*:6 (1973).
85. B. J. Hopkins and V. Pretorius, *J. Chromatogr.*, *158*:465 (1978).
86. A. Ducass, M. F. Gonnord, A. Arpino, and G. Guiochon, *J. Chromatogr.*, *148*:321 (1978).
87. J. Sevcik and T. H. Gerner, *HRC & CC*, *2*:436 (1979).
88. G. Schomburg, F. Weeke, F. Muller, and M. Oreans, *Chromatographia*, *16*:87 (1982).
89. E. L. Anderson, M. M. Thomason, H. T. Mayfield, and W. Bertsch, *HRC & CC*, *2*:335 (1979).
90. J. D. Pleil, K. D. Oliver, and W. A. McClenny, *J. Air Pollut. Control Assoc.*, *37*:244 (1978).
91. J. A. Settlage and W. G. Jennings, *HRC & CC*, *3*:146 (1980).
92. "AID Concentrator Model 380," Analytical Instrument Development, Inc., Avondale, Pennsylvania, 1982.
93. "Model 1000 Capillary Interface," Tekmar Co., Cincinnati, Ohio, 1984.
94. W. Bertsch, R. C. Chang, and A. Zlatkis, *J. Chromatogr. Sci.*, *12*:175 (1974).
95. E. D. Pellizzari, B. H. Carpenter, J. E. Bunch, and E. Sawicki, *Anal. Chem.*, *9*:556 (1975).

96. J. Russell, *Environ. Sci. Technol.*, *9*:1175 (1975).
97. W. V. Ligon, Jr. and R. L. Johnson, Jr., *Anal. Chem.*, *48*:481 (1976).
98. B. Versino, H. Knoppel, M. DeGroot, A. Peil, J. Poelman, H. Schauenburg, H. Vissers, and F. Geiss, *J. Chromatogr.*, *122*:373 (1976).
99. W. K. Fowler, C. H. Duffey, and H. C. Miller, *Anal. Chem.*, *51*:2333 (1979).
100. J. F. Pankow, L. M. Isabelle, and T. J. Kristensen, *Anal. Chem.*, *54*:1815 (1982).
101. "Envirochem, Inc. Model 850 Thermal Tube Desorber," Supelco, Inc., Bellefonte, Pennsylvania.
102. "Tekmar Model 5000 Automatic Desorber," Tekmar Co., Cincinnati, Ohio.
103. "Chrompack Thermo-Desorption Cold-Trap Injector," Crompack International B.V., Middelburg, The Netherlands.
104. "Nutech 320 Thermal Desorption System," Nutech Corp., Durham, North Carolina.
105. "CDS-330 Trapping Concentrator," Chemical Data Systems, Inc., Oxford, Pennsylvania.
106. "Century Programmed Thermal Desorber," Analabs, North Haven, Connecticut.
107. "Valco Automatic Trace Organics Concentrator," Valco Instruments Co., Houston, Texas.
108. J. H. Glover, *Chromatogr. Int.*, *15*:27 (1986).
109. J. Rektorik, Thermal Desorption of Solid Traps by Means of Microwave Energy, in *Sample Introduction in Capillary Gas Chromatography* (P. Sandra, Ed.), Vol. I, Huethig Publishing, Mamaroneck, New York, 1985, pp. 217-233.
110. T. A. Beller and J. J. Lichtenberg, *J. Am. Water Works Assoc.*, *66*:739 (1974).
111. "The Analysis of Trihalomethanes in Finished Waters by the Purge and Trap Method," Method 501.1, U.S. EPA, Cincinnati, Ohio, 1979.
112. "HP-7675A Purge and Trap Sampler," Hewlett-Packard, Avondale, Pennsylvania.
113. "OIC Model 4460 Purge and Trap Concentrator," O-I-Corporation, College Station, Texas.
114. "Tekmar Model 4000 Dynamic Headspace Concentrator," Tekmar Co., Cincinnati, Ohio.
115. "CDS Model 330-2 Water Sparging and Trapping System," Chemical Data Systems, Inc., Oxford, Pennsylvania.
116. "Chrompack Purge and Trap Injector," Chrompack International B.V., Middelburg, The Netherlands.
117. Th. Noij, A von Es, C. Cramers, J. Rijks, and R. Dooper, *HRC & CC*, *10*:60, 1987.
118. B. V. Ioffe and A. G. Vitenberg, *Head-Space Analysis & Related Methods in Gas Chromatography*, John Wiley & Sons New York (1984).
119. "H-P Model 19395A Headspace Sampler," Hewlett-Packard, Avondale, Pennsylvania.
120. "HS-100 Automatic Headspace Sampler," Perkin-Elmer, Norwalk, Connecticut.
121. "HSS-2A Headspace Sampler," Shimadzu Scientific Instruments, Inc., Columbia, Maryland.

122. "HS-250 Automatic Head-Space Analyzer," Carlo Erba Co., Rodano, Italy.
123. B. Kolb and P. Pospisil, Headspace Gas Chromatography (HSGC) with Capillary Columns, in *Sample Introduction in Capillary Gas Chromatography* (P. Sandra, Ed.), Vol. I, Huethig Publishing, Mamaroneck, New York, 1985, pp. 191-216.
124. L. S. Ettre, B. Kolb, and S. G. Hurt, *Am. Lab., 15*:76 (1983).
125. B. Kolb, *Chromatographia, 15*:587 (1982).
126. L. S. Ettre, E. Jones, and B. S. Todd, *Chromatogr. Newsletter, 12*:1 (1984).
127. B. Kolb, M. Auer, and P. Pospisil, *Appl. Chromatogr.*, No. 35E (1981), Perkin-Elmer.
128. S. A. Liebman, T. P. Wampler, and E. J. Levy, Developments in Pyrolysis Capillary GC, in *Sample Introduction in Capillary Gas Chromatography* (P. Sandra, Ed.), Vol. I, Huethig Publishing, Mamaroneck, New York, 1985, pp. 165-189.
129. E. J. Levy and T. P. Wampler, *J. Chem. Ed., 63*:A64 (1986).
130. D. Wright and P. Dawes, *Am. Lab., 18*:92 (1986).
131. T. P. Wampler and E. J. Levy, *LC-GC, 4*:1112 (1986).
132. M. A. Sharaf, D. L. Illman, and B. R. Kowalski, *Chemometrics*, John Wiley & Sons, New York, 1986, pp. 179-295.
133. "Pyrojector," SGE USA, Austin, Texas.
134. "Pyroprobe," Chemical Data Systems, Inc., Oxford, Pennsylvania.
135. S. Tsuge and T. Takeuchi, *Anal. Chem., 49*:348 (1978).
136. R. Schill and R. R. Freeman, Instrumentation, in *Modern Practice of Gas Chromatography*, 2nd ed. (R. L. Grob, Ed.), John Wiley & Sons, New York, pp. 293-357.
137. P. M. van der Berg and T. P. H. Cox, *Chromatographia, 5*:301 (1972).
138. G. Cox, *Chromatogr. Intl., 17*:18 (1986).
139. F. Rowland, *Am. Lab., 14*:110 (1982).
140. P. Welsh, "Evaluation of Commercial Gas Chromatograph Ovens," Hewlett-Packard Publication No. 43-5952-5769, Hewlett-Packard Co., Avondale, Pennsylvania (1977).
141. J. F. Sprouse and A. Varano, *Am. Lab., 16*:54 (1984).
142. E. Kovats, *Helv. Chim. Acta., 41*:1915 (1958).
143. T. A. Brettell and R. L. Grob, *Am. Lab., 17*:19 (1985).
144. J. C. Giddings, *J. Chem. Ed., 39*:569 (1962).
145. C. J. Barinaga and S. O. Farwell, *HRC & CC, 10*:538 (1987).
146. L. S. Ettre, *J. Chromatogr. Sci., 16*:396 (1978).
147. S. O. Farwell, D. R. Gage, and R. A. Kagel, *J. Chromatogr. Sci., 19*:358 (1981).
148. M. J. O'Brien, Detectors, in *Modern Practice of Gas Chromatography*, 2nd ed. (R. L. Grob, Ed.), John Wiley & Sons, New York, pp. 211-291.
149. F. J. Yang and S. P. Cram, *HRC & CC, 2*:487 (1979).
150. P. L. Patterson, *J. Chromatogr. Sci., 24*:466 (1986).
151. J. Sevcik, *Detectors in Gas Chromatography*, Elsevier, Amsterdam (1976).
152. M. Dressler, *Selective Gas Chromatographic Detectors*, Elsevier, Amsterdam (1986).

153. D. J. David, *Gas Chromatographic Detectors*, John Wiley & Sons, New York (1974).
154. S. O. Farwell and C. J. Barinaga, *J. Chromatogr. Sci.*, *24*:483 (1986).
155. J. Sevcik and J. E. Lips, *Chromatographia*, *12*:693 (1979).
156. G. M. Message, *Practical Aspects of Gas Chromatography/Mass Spectrometry* (John Wiley & Sons, New York (1984).
157. F. Andrawes and R. Ramsey, *J. Chromatogr. Sci.*, *24*:513 (1986).
158. J. N. Driscoll and M. Duffy, *Chromatography*, *2*:21 (1987).
159. H. W. Grice and D. J. David, *J. Chromatogr. Sci.*, *7*:239 (1969).
160. K. J. Skogerboe and E. S. Yeung, *Anal. Chem.*, *56*:2684 (1984).
161. D. H. Fine, F. Rufeh, D. Lieb, and D. P. Roundbehler, *Anal. Chem.*, *47*:1181 (1975).
162. D. H. Fine and D. P. Roundbehler, *J. Chromatogr.*, *109*:271 (1975).
163. D. H. Fine, D. Lieb, and F. Rufeh, *J. Chromatogr.*, *107*:351 (1975).
164. S. A. Nyarady, R. M. Barkley, and R. E. Sievers, *Anal. Chem.*, *57*:2074 (1985).
165. R. S. Hutte, R. E. Sievers, and J. W. Birks, *J. Chromatogr. Sci.*, *24*:499 (1986).
166. S. Gluck, *J. Chromatogr. Sci.*, *20*:103 (1982).
167. P. R. Griffiths, S. L. Pentoney, A. Giorgetti, and K. H. Shafer, *Anal. Chem.*, *58*:1349A (1986).
168. F. W. Karasek, *Res. Develop.*, *27*:42 (1976).
169. G. C. Stafford, P. E. Kelley, and D. C. Bradford, *Am. Lab.*, *15*:51 (1983).
170. "HP-5970B Mass Selective Detector," Hewlett-Packard Publication No. 23-5953-8091, Hewlett-Packard Co. (1984).
171. "Ion Trap Detector," Perkin-Elmer Publication No. L-952, Perkin-Elmer Corp. (1986).
172. S. A. Borman, *Anal. Chem.*, *59*:701A (1987).
173. "HP-5965A IRD...the FTIR detector dedicated to gas chromatography," Hewlett-Packard Publication No. 23-5954-0665, Hewlett-Packard Co. (1986).
174. "GC/FTIR - The HP 5965A Infrared Detector for capillary gas chromatography," *HP Source*, *5*:1 (1986).
175. C. L. Wilkins, *Anal. Chem.*, *59*:571A (1987).
176. J. R. Cooper, I. C.Bowater, and C. L. Wilkins, *Anal. Chem.*, *58*:2791 (1986).
177. E. S. Olson and J. W. Diehl, *Anal. Chem.*, *59*:443 (1987).
178. J. V. Johnson and R. A. Yost, *Anal. Chem.*, *57*:758A (1985).
179. "ASTM Standard Recommended Practices E-516-74, E-594-77, E-697-79, and E-840-81," ASTM, Philadelphia, Pennsylvania.
180. J. E. Knoll, *J. Chromatogr. Sci.*, *23*:122 (1985).
181. L. J. Schumauch, *Anal. Chem.*, *31*:225 (1959).
182. J. C. Sternberg, Extracolumn Contributions to Chromatographic Band Broadening, in *Advances in Chromatography*, Vol. 2 (J. C. Giddings and R. A. Keller, Eds.), Marcel Dekker, New York, pp. 205-270 (1966).
183. R. Simon and G. Wells, Use of a Small Volume Thermal Conductivity Detector for Capillary Gas Chromatography, in *Ultrahigh Resolution Chromatography* (S. Ahuja, Ed.), American Chemical Society, Washington, D.C., pp. 59-76 (1984).

184. M. F. Miscoski, *Chromatogr. Contents, 2*:49 (1987).
185. T. A. Rooney, *Chromatogr. Contents, 2*:43 (1987).
186. L. D. Rothman, *Chromatogr. Forum, 1*:13 (1986).
187. J. Tookey and R. Hoberg, *Chromatogr. Forum, 1*:23 (1986).
188. S. J. Hawkes and M. D. Hamilton, *Chromatogr. Forum, 1*:28 (1986).
189. W. Kipiniak, D. Karlan, P. Mansfield, S. Wheaton, P. Berthrong, and R. Voelkner, *Chromatogr. Forum, 1*:30 (1986).
190. G. Lawler, M. McConnell, E. Long, and P. Batchelder, *Chromatogr. Forum, 1*:37 (1986).
191. E. Warwick and A. Flamberg, *Chromatogr. Forum, 1*:44 (1986).
192. G. H. Morrison, *Anal. Chem., 55*:1 (1983).
193. M. R. Thompson and R. E. Dessy, *Anal. Chem., 56*:583 (1984).
194. A. N. Papas and M. F. Delaney, *Anal. Chem., 59*:55A (1987).
195. H. Malmstadt, C. Enke, and S. Crouch, *Electronics and Instrumentation for Scientists*, Benjamin/Cummings, 1981, Chapters 9 and 13.
196. F. J. Debbrecht, Qualitative and Quantitative Analysis by Gas Chromatography, in *Modern Practice of Gas Chromatography*, 2nd ed. (R. L. Grob, Ed.), John Wiley & Sons, New York, pp. 359-421 (1985).
197. U.S. Patent 4,789,783.
198. P. C. Uden, *Chromatogr. Forum, 1*:17 (1986).
199. J. J. Sullivan and B. D. Quimby, *HRC & CC, 12*:282 (1989).
200. R. L. Firor, *Am. Lab., 21*:40 (1989).
201. "O-FID Oxygenates Analyzer," Carlo Erba Strumentazione (1989).
202. R. L. Benner and D. H. Stedman, *Anal. Chem., 61*:1268 (1989).
203. A. L. Britten, *Res. Develop.*, Sept. 1989, pp. 77-80.

22

Instrumentation for High-Performance Liquid Chromatography

RAYMOND J. WEIGAND, DAVID BROWN, and DENNIS JENKE / Baxter Health Care Corporation, Round Lake, Illinois

I. INTRODUCTION

During the short period since its introduction in the late 1960s, high-performance liquid chromatography (HPLC) has evolved into a methodology that has made and continues to make significant contributions to pharmaceutical, biochemical, clinical, and environmental analysis. The key to the emergence of this methodology as a primary force in these areas of analytical chemistry is related to the development of appropriate column technologies and advancement of the state of the art in terms of chromatographic apparatus. Important developments in column technology include the identification and production of suitable support materials, reduction in particle size of the support materials, and preparation of chemically bonded phases (in which the nature of the separation medium is modified by bonding organic groups to the support surface). As column technology/efficiency improved, greater performance was required from the supporting apparatus. The apparatus needed to carry out modern HPLC is very different from the simple and unsophisticated equipment used to produce classical LC separations. While useful and sometimes involved separations can be carried out with modest equipment, a relatively sophisticated apparatus is needed for the separation of complex mixtures or for producing highly accurate and precise quantitative analyses.

In modern HPLC, mobile phase from a solvent reservoir is pumped (after filtration, degassing, and pressurization, if necessary) through the chromatographic column. The sample, containing a mixture of solutes, is injected at the top of the column and separated into components upon traveling through the column. As they elute from the column individual solutes are monitored by the detector, whose response either is recorded as an analog signal (e.g., peaks on a recorder) or is converted to a digital response (for input into an integrator). Thus, as shown in Figure 1, the main components of a high-performance liquid chromatograph are a high-pressure pump, a column/injector system, a detector, connecting tubing, and appropriate data collection devices. Additional components such as the solvent reservoir, in-line filters, pressure gauges, pulse dampeners, thermostats, and reaction coils may also be required for particular applications. We observe in passing that in order to obtain truly high-performance operation in a specific application the individual components of the analytical

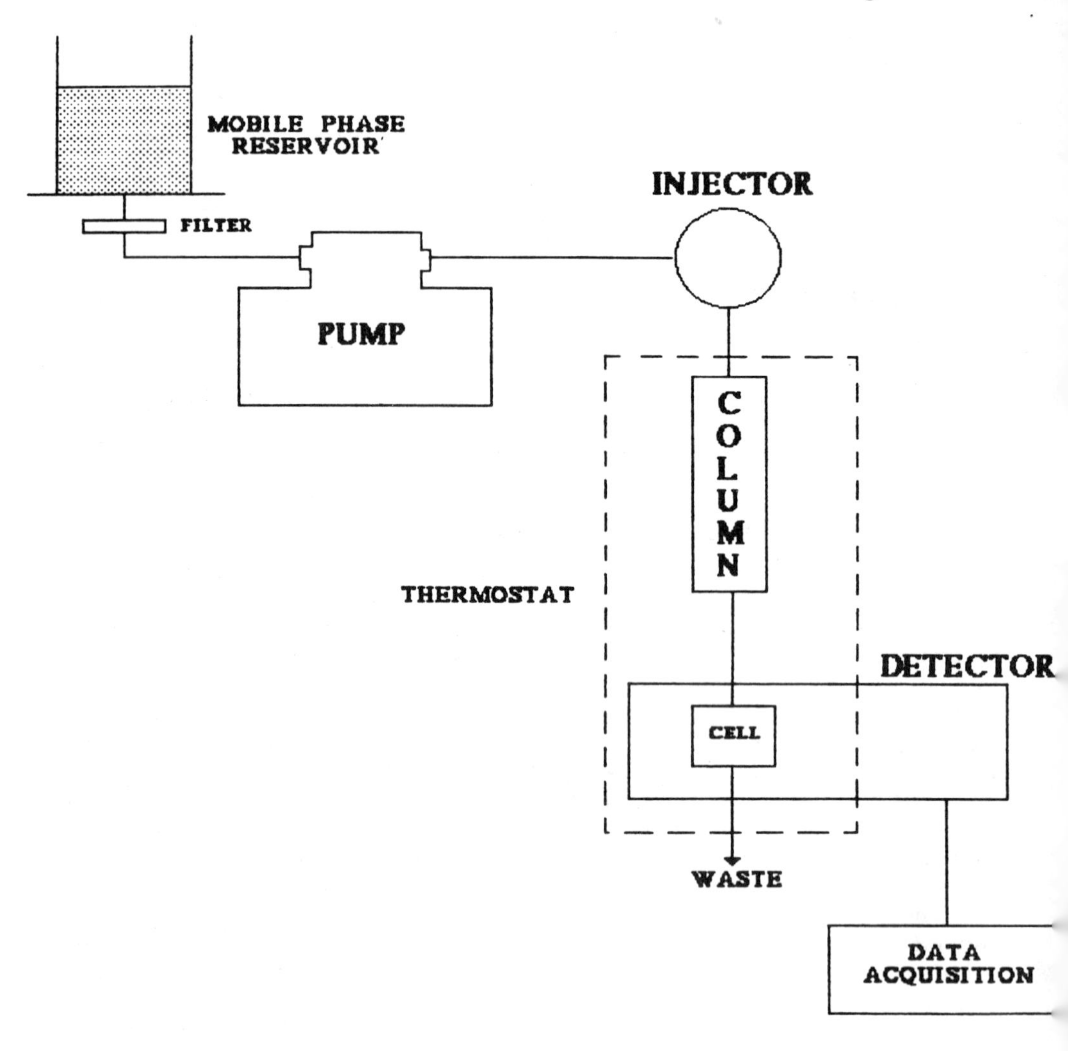

Figure 1. Main components of a high-performance liquid chromatography system.

system must be manufactured to precise specifications and the entire system must be constructed from compatible and effective components.

The purpose of this chapter is to provide the reader with an overview of HPLC apparatus and components that are in current use in analytical work. Emphasis is placed on design features and performance characteristics as opposed to a presentation of an in-depth derivation of theoretical concepts and an extensive compilation of device applications. While equipment that supports more or less routine applications is given the major emphasis, emerging technologies of potential significance are also described as appropriate. Although the chromatographic column is an important component of the system, the discussion in this text is limited to a description of the currently available technology and a commentary on the theoretical basis of column operation/performance. A detailed discussion of stationary phases and resin materials which are currently available can be found in any of the numerous texts devoted to liquid chromatography [1], while column descriptions and selectivity comparisons are common in the chemical literature (e.g., for octadecyl-bonded-phase columns [2]).

II. SOLVENT RESERVOIR

The purpose of the solvent reservoir is to contain and protect the mobile phase. It should be made of an inert material so as not to affect mobile phase composition. Both glass and stainless steel are used in reservoir construction; however, stainless steel should be avoided if halide ions are present in the mobile phase. Reservoirs are sometimes built into the pumping system, such as in syringe pumps, but more commonly they are found as free-standing devices of various size. The advantage to a free-standing reservoir is that the chromatographer can choose the size appropriate for the analysis. The reservoir should be large enough to hold sufficient mobile phase for the analysis at hand plus 20-25 percent more to account for equilibration of the column, test injections, etc. Adding mobile phase to the reservoir during the run should be avoided because the composition of the added mobile phase is apt to be slightly different, leading to retention-time changes. Also, pouring fresh mobile phase into the reservoir can introduce air into the solvent, leading to pump and detector problems.

The mobile phase that is put into the reservoir should be filtered through filters with pore size no greater than 0.5 μm and should be well degassed. Degassing is especially important for aqueous solution or aqueous/organic mixtures. Degassing is often achieved through vacuum ultrasonification (or ultrasonification alone), boiling, or helium sparging. Since transferring the filtered and degassed mobile phase from one vessel to another can introduce particulates or air into it, it is desirable to filter and degas the mobile phase in the same vessel that will be used as the reservoir. A common piece of laboratory glassware that serves this purpose well is an appropriately sized filter flask. It is thick-walled glass made especially with vacuum procedures in mind and can withstand boiling temperatures. As a safety precaution against vacuum implosions, it is good practice to place elastic mesh netting around the flask to reduce the hazard of flying glass. During the HPLC analysis the reservoir must be covered to prevent airborne particles from contaminating the mobile phase and to reduce or eliminate evaporation of volatile mobile-phase components.

Stirring of the mobile phase during the analysis is important to obtaining good chromatographic precision. Even miscible solvents will separate if their densities differ (a chloroform/hexane mixture is a good example), producing locally stronger or weaker pools of mobile phase within the reservoir. Stirring is generally accomplished by introducing a Teflon-coated magnetic stir bar into the reservoir and then placing the reservoir on a magnetic stir plate. Stirring should be at a speed that ensures thorough mixing without introducing air into the mobile phase. It is good practice to avoid setting the reservoir directly on the magnetic stir plate, as the heat generated by the stir plate can encourage evaporation of mobile-phase components. A cork ring placed between the reservoir and stir plate eliminates this problem.

A mobile-phase reservoir designed specifically for HPLC application is the Ultra-Ware system manufactured by Kontes. This system has features that make it ideal for both filtration/degassing and solvent delivery. The filtration/degassing apparatus as shown in Figure 2 can withdraw mobile phase directly from a solvent bottle, graduated cylinder, or other common piece of glassware for filtration. The filtered mobile phase passes into a uniquely shaped reservoir, which can also be used as the solvent delivery vessel. After filtration the stopcock on the solvent pickup adapter (Figure 2) can be turned to allow vacuum degassing (the reservoir may be placed in an ultrasonic bath, if desired). When degassing is complete, the filtration apparatus is removed from the reservoir and the solvent delivery cap is attached.

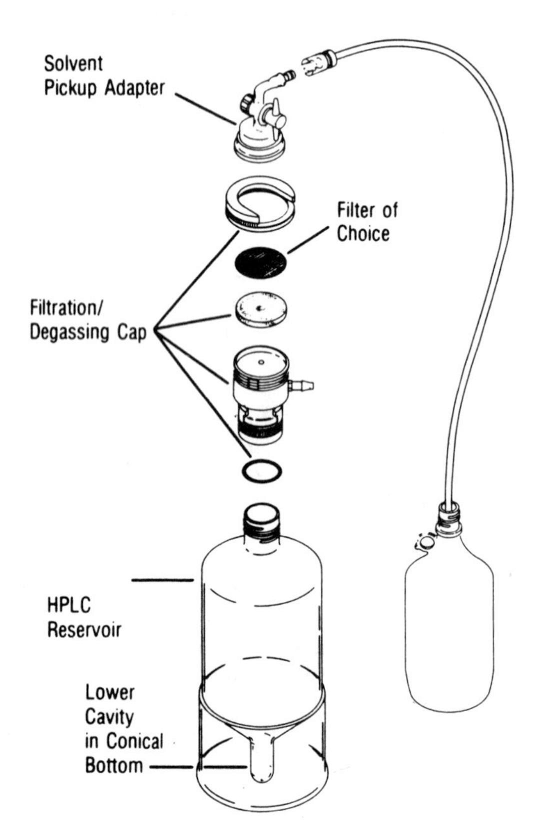

Figure 2. Kontes Ultraware solvent filtration/degassing system. (Courtesy of Kontes Life Sciences Products, Vineland, New Jersey.)

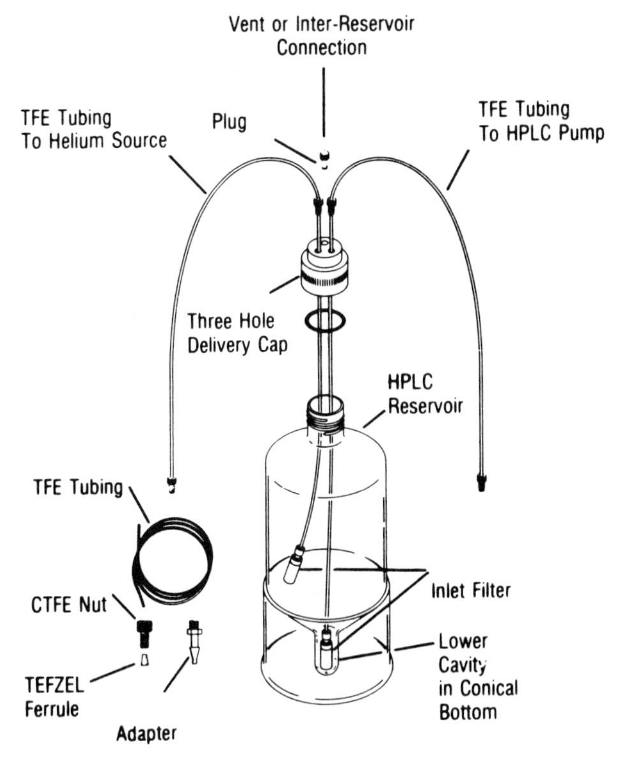

Figure 3. Kontes Ultraware solvent delivery system. (Courtesy of Kontes Life Sciences Products, Vineland, New Jersey.)

The apparatus used in solvent delivery is shown in Figure 3. The solvent reservoir filter is attached to the delivery cap and rests in the lower cavity of the conically shaped reservoir bottom. This unique shape ensures that all but the last traces of mobile phase will be delivered to the pump without tilting the reservoir, which is sometimes done with flat-bottomed vessels. A small, double-finned stirring bar can be placed immediately under the reservoir filter to provide adequate mixing of the mobile phase. Because of the conical shape of the reservoir bottom, the reservoir can rest directly on a magnetic stir plate without undue heating of the mobile phase. If desired, a helium sparge line can also be introduced through the delivery cap to provide a means of continuous mobile-phase degassing.

III. PUMPS

Pumping systems for HPLC come in a variety of designs. All, however, have a common purpose, which is to provide a continuous and reproducible supply of mobile phase to the chromatographic column. Requirements of a good HPLC pump include [1,3,4-6]:

1. Delivery of mobile phase through the chromatographic column at a constant rate necessary for accurate quantitation (this is especially important when peak area measurements are made).
2. Stability of flow with minimal pulsations.
3. Delivery of mobile phase over a wide range of flow rates.
4. Delivery of mobile phase against high column inlet pressure ($\geq$ 5000 psi).
5. Lack of reactivity of pump components with the mobile phase (pump components are typically constructed of stainless steel, Teflon, ruby and sapphire).
6. Ease of serviceability by the chromatographer.
7. Adaptability to gradient HPLC analyses.

HPLC pumping systems can be divided into two general categories: constant-pressure systems such as the gas displacement and pneumatic amplifier pumps, and constant-flow-rate systems, such as the reciprocating piston, piston diaphragm, and syringe pumps.

A. Constant-Pressure Pumps

Gas Displacement Pump

Constant-pressure pumps are not commonly used in present-day HPLC systems and are mentioned here mainly for historical reasons. The gas displacement pump [7,8], as shown in Figure 4, was the earliest and simplest of the HPLC pump designs. In its simplest form, this pump consists of a holding coil or reservoir containing the mobile phase to which a constant pressure is applied by means of a gas cylinder. In order to limit the amount of gas dissolved in the mobile phase, the gas/liquid interface is kept as small as possible. Dissolved gasses can cause problems by coming out of solution to form bubbles when the pressure is reduced. This pressure drop occurs as the mobile phase leaves the column, and often causes detector problems when the bubbles that form pass through the detector cell. The mobile-phase reservoirs on these pumps are usually of limited capacity because of the need to keep the gas/liquid interface small. Being of simple design, the gas displacement pump has low maintenance costs. A major advantage of all constant-pressure pumps is their ability to produce pulseless flow, which contributes to a stable chromatographic baseline. A limiting factor of the gas displacement pump is its maximum operating pressure, which is dependent on the gas cylinder pressure. This maximum pressure will decrease as the gas cylinder empties.

Pneumatic Amplifier Pump

In order to increase the maximum pressure of a gas-cylinder-operated pump, the gas pressure can be applied to a large-bore piston, which in turn pushes a smaller-bore piston. The smaller bore piston is in contact with the mobile phase and thus forces it through the column. The design of this pump, called the pneumatic amplifier pump, is shown in Figure 5.

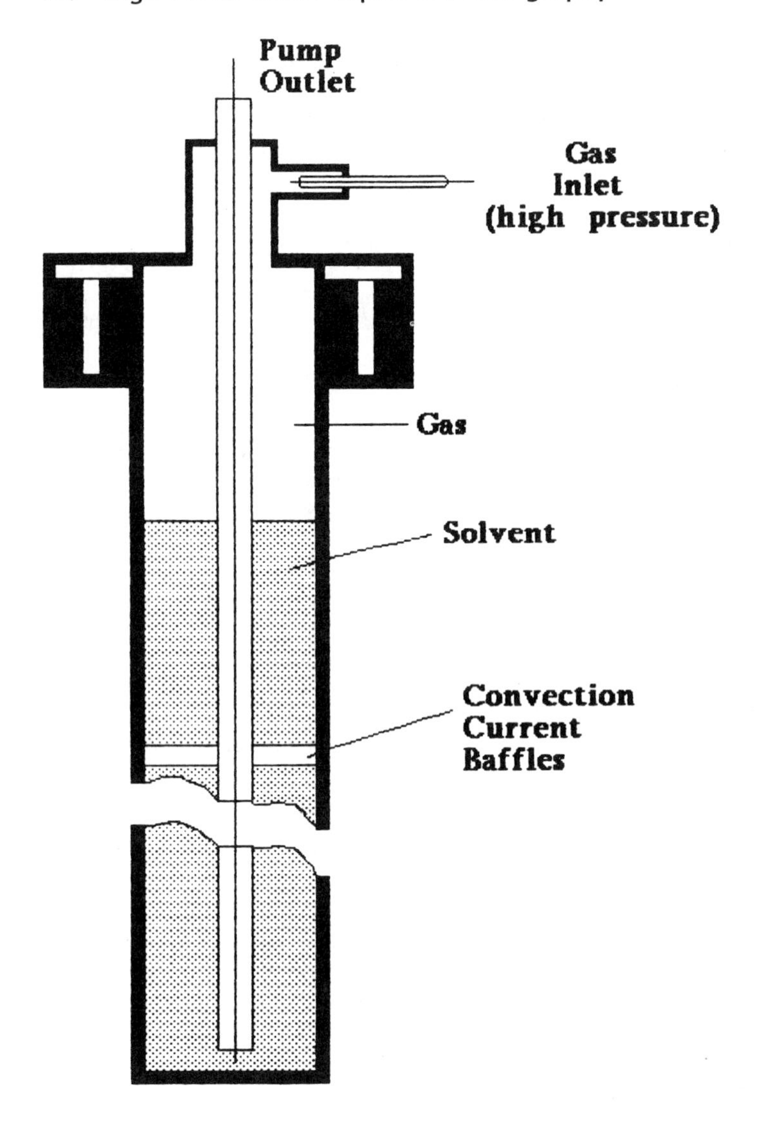

Figure 4. A typical gas displacement pump. (From Ref. 61, with permission.)

The pressure amplification over the gas displacement pump is equal to the ratio of the surface areas of the two pistons. The pneumatic amplifier pump is capable of rapidly refilling the reservoir when it is empty. This rapid refill stroke causes a momentary pressure drop and a short interruption of the flow. Normally this flow inconsistency does not cause chromatographic problems, and the advantages of rapid refill are great. With the introduction of this pump, chromatographers were given essentially pulseless flow and an unlimited supply of mobile phase. Some of the more technically

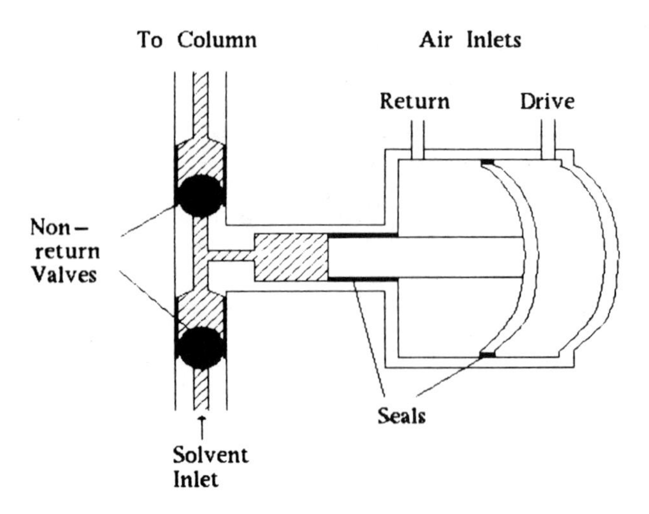

Figure 5. Pneumatic amplifier pump (Haskell). (From Ref. 61, with permission.)

advanced pneumatic amplifier pumps have flow controllers that monitor the flow rate and adjust the gas inlet pressure to produce a constant rate of flow. This device compensates for changing column backpressure and for changes in the viscosity of the mobile phase. A constant flow rate is essential to obtaining acceptable quantitative results.

Constant-pressure pumping systems will provide precise quantitation as long as the flow rate remains constant, that is, when the inlet pressure is regulated to compensate for changes in system backpressure or the backpressure of the HPLC system remains constant. If the flow rate changes, as will occur when column frits become partially blocked or the column temperature changes, constant-pressure pumping systems will lead to nonreproducible peak responses. This is especially true when peak area is the measured response [9,10]. For this reason, constant-flow-rate pumps, rather than constant-pressure pumps, are the choice of all contemporary chromatographers.

B. Constant-Flow-Rate Pumps

Syringe Pumps

Constant-flow-rate pumps are generally divided into three major groups: syringe pumps, diaphragm pumps, and reciprocating piston pumps. For applications that require high precision at low flow rates, such as with microbore columns, or applications that require high detector sensitivity where pump "noise" may be a problem, a syringe pump is the best choice. The design of a typical syringe pump [11] is shown in Figure 6. In a syringe pump the mobile-phase reservoir is contained completely within the pump itself. This is a necessary part of the pump design and leads to both the biggest advantage and disadvantage of this pump type. Because all of the mobile phase is contained within the pump, it can be compressed and pressurized to the level of the column inlet pressure at the beginning of the analysis. A piston, driven by a motor connected through worm gears, then forces the mobile phase through the column in a single pump stroke at the selected rate. There are no pulses due to pressure fluctuations and, hence, there is no need for pulse damping. Since all of the mobile phase is delivered in a single stroke of the piston, check valves, whose function is to keep mobile phase moving in the proper direction, are unnecessary. The advantages of pulseless flow, low maintenance and the simplicity of the syringe pump design are often overshadowed, however by the fact that the mobile-phase capacity of these pumps is generally limited to approximately 250-500 ml. The reservoir must then be manually refilled. This makes even moderately long analyses impractical with conventional size columns and flow rates. These pumps are, however, ideally suited to the very low flow rates and low-volume throughput associated with microbore columns, and it is here that they find their greatest use.

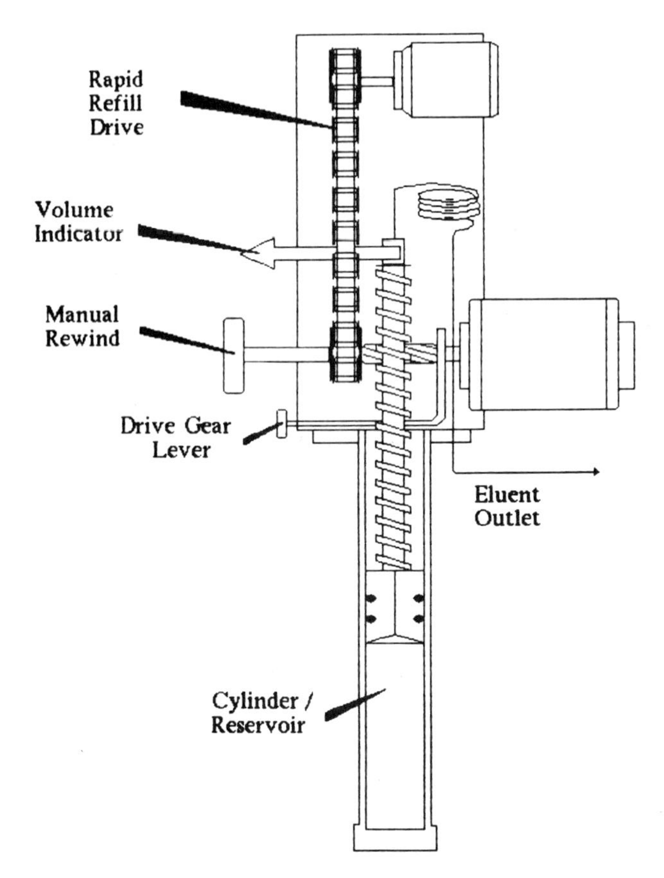

Figure 6. Schematic of a positive-displacement syringe pump. (Courtesy of Perkin-Elmer Corporation, Norwalk, Connecticut.)

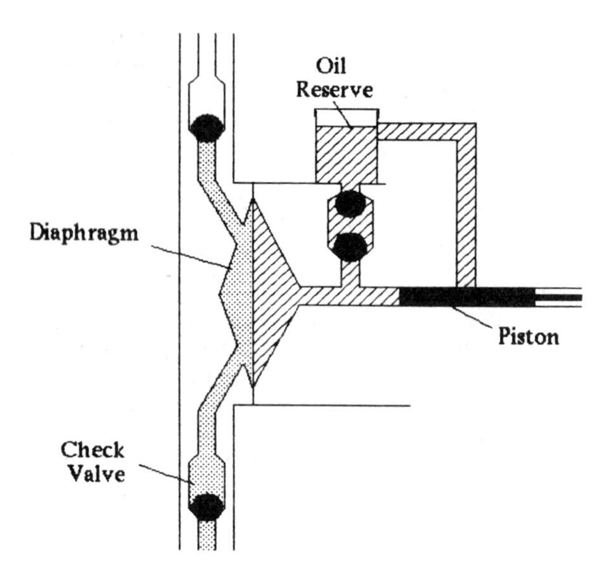

Figure 7. Schematic of a piston-diaphragm pump. (From Ref. 69, with permission.)

Piston-Diaphragm Pumps

Piston-diaphragm pumps have been used in HPLC analyses since the early 1970s [12]. The design of a typical piston-diaphragm pump is shown in Figure 7. In this type of pump, mobile phase is pulled into the pumping chamber through an inlet check valve by the suction of the retreating piston. The metal diaphragm is deflected toward the piston by the mobile phase entering the chamber. As the piston enters its delivery stroke, it traps hydraulic fluid between itself and the diaphragm. The continued forward movement of the piston pressurizes the hydraulic fluid, which forces the diaphragm forward, causing mobile phase to move out of the pump head. Check valves located at the inlet and outlet of the pumping chamber assure that mobile phase moves in the proper direction, that is, from the solvent reservoir and toward the column. The amount of deflection, and hence the amount of mobile phase being delivered with each pump stroke, is dependent on the extent of penetration of the piston into the pump head. The pump head itself is positioned by stepper motors to control the amount of solvent delivered and hence the rate of flow. As the pump head is moved farther onto the piston, the suction part of the refill lasts longer and a greater volume of mobile phase is pulled into the pumping chamber. Since the frequency of the piston stroke is constant, the flow rate increases. Flow-rate changes in these pumps are somewhat slow to take effect because of the time necessary to move the pump heads.

The frequency of the piston stroke in a piston diaphragm pump is relatively high (~300 strokes/min). This reduces the pulsations that occur as the flow rate drops each time the piston moves through its intake stroke. High-pressure seals normally used to prevent leakage of mobile phase along the piston are not used within the pump head. Although the piston fits snugly within the cylinder, hydraulic fluid will leak along the sides of the piston during the delivery stroke and return to the hydraulic fluid reservoir. The piston must cycle at high speed in order to minimize the pressure drop, and hence the reduction in flow rate, caused by the leaking oil.

The lack of seals does offer some advantages. Buffers, which normally cause piston seals to wear relatively quickly on other pump types, present no problems for piston diaphragm pumps. The lack of seals also reduces the cost and effort required to maintain these pumps. Less expensive steel pistons may be used, rather than the much higher cost sapphire plungers used to minimize seal wear in reciprocating pumps. Steel pistons are less subject to wear and breakage than sapphire, further reducing the maintenance of these pumps. Also, no specially shaped and expensive cams or gears are needed to compensate for the reduction of flow during the refill stroke because of the rapid piston stroke.

Piston-diaphragm pumps are considered low-maintenance pumps because of their lack of piston seals and because steel pistons are used in their construction. The most common problems associated with these pumps is air lodging in the check valves. This causes the pump to cavitate. Air, once lodged in the check valve, expands as the piston moves through its intake stroke, reducing the suction and preventing mobile phase from entering the pumping chamber. The air must be removed before normal pumping operation can continue. This can usually be accomplished by removing the stainless steel tubing from the outlet check valve and starting the pump, which bleeds the air from the check valve. It is sometimes necessary to attach a syringe to the pump outlet with proper chromatographic fittings to pull the air out of the check valves. Perhaps the most troublesome aspect of using piston-diaphragm pumps is that they are somewhat harder to troubleshoot and service than other pump types.

A modified version of the piston-diaphragm pump is manufactured by Hewlett-Packard. The model 79835A solvent delivery system [13], as shown in Figure 8, has become an important component of the HP1090 liquid chromatograph. This pump utilizes two pump types to deliver mobile phase to the chromatographic column. Three dual-syringe metering pumps are used to both measure and deliver an appropriate amount of mobile phase to a high-pressure piston-diaphragm pump. In order to prevent pressure spikes from damaging the syringe pumps, the mobile phase passes through a low-pressure compliance before entering the diaphragm pump. The low-pressure compliance allows mixing of the solvents from the three syringe pumps, besides acting as a damper for the pressure spikes of the diaphragm pump. There is also a pulse damping unit (Figure 9) between the high-pressure diaphragm pump and the injector. The pulse damper is a reservoir with two compartments, one of which is filled with water. The second compartment is the mobile-phase chamber. The two compartments are separated by a membrane. As the pressure rises with each delivery stroke of the piston, the water absorbs energy as it is compressed by the pressure of the mobile phase of the membrane. During the piston's intake stroke, this energy is released, mitigating the pressure and flow-rate drop that would otherwise be seen. Mobile-phase flow through the column during

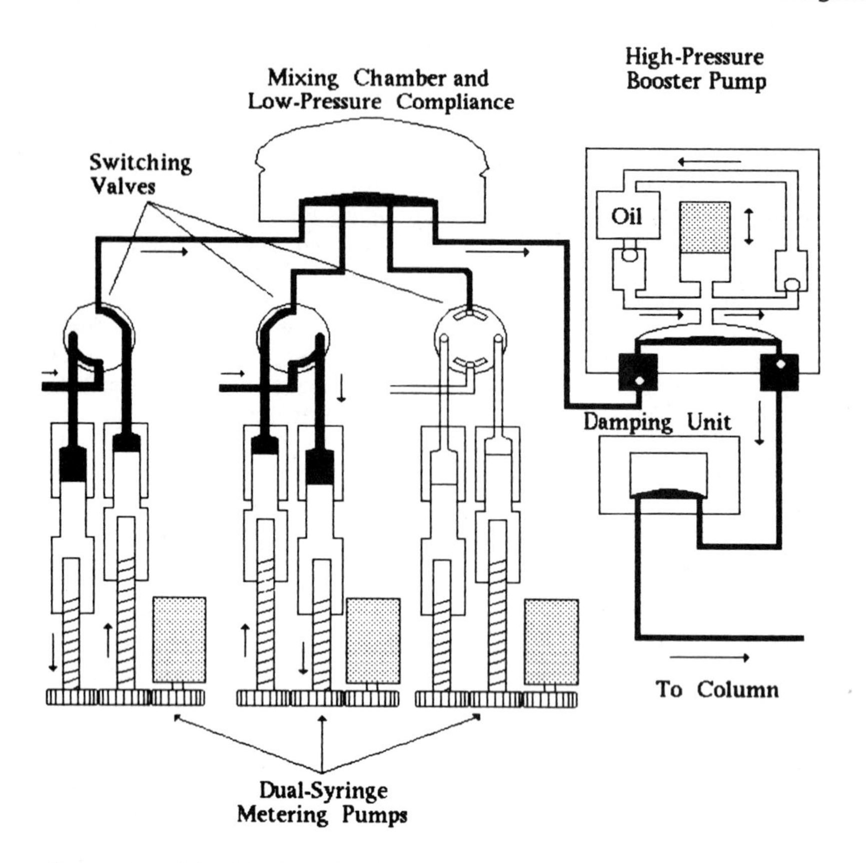

Figure 8. The Hewlett-Packard model 79835A solvent delivery system. (Courtesy of Hewlett-Packard, Avondale, Pennsylvania.)

the refill stroke of the piston is actually due to the action of the pulse damper. A detailed account of the workings of this pump can be found in Ref. 13. The main advantage of this pump type is that it combines highly accurate and precise syringe pumps for flow rate control with a low-maintenance piston-diaphragm pump for power generation.

Reciprocating Piston Pumps

Reciprocating piston pumps are by far the most common type of HPLC pump currently in use. These pumps derive their name from the fact that one or more pistons alternate between a filling and a pumping cycle. Almost all currently marketed reciprocating piston pumps use sapphire pistons. This material is used because it has a hard, extremely smooth surface, which creates a minimal amount of friction on the high-pressure seal [14]. These pumps differ from piston-diaphragm pumps in that with reciprocating piston pumps, the piston is in direct contact with the solvent. It is the piston itself, rather than hydraulic fluid, that provides the force

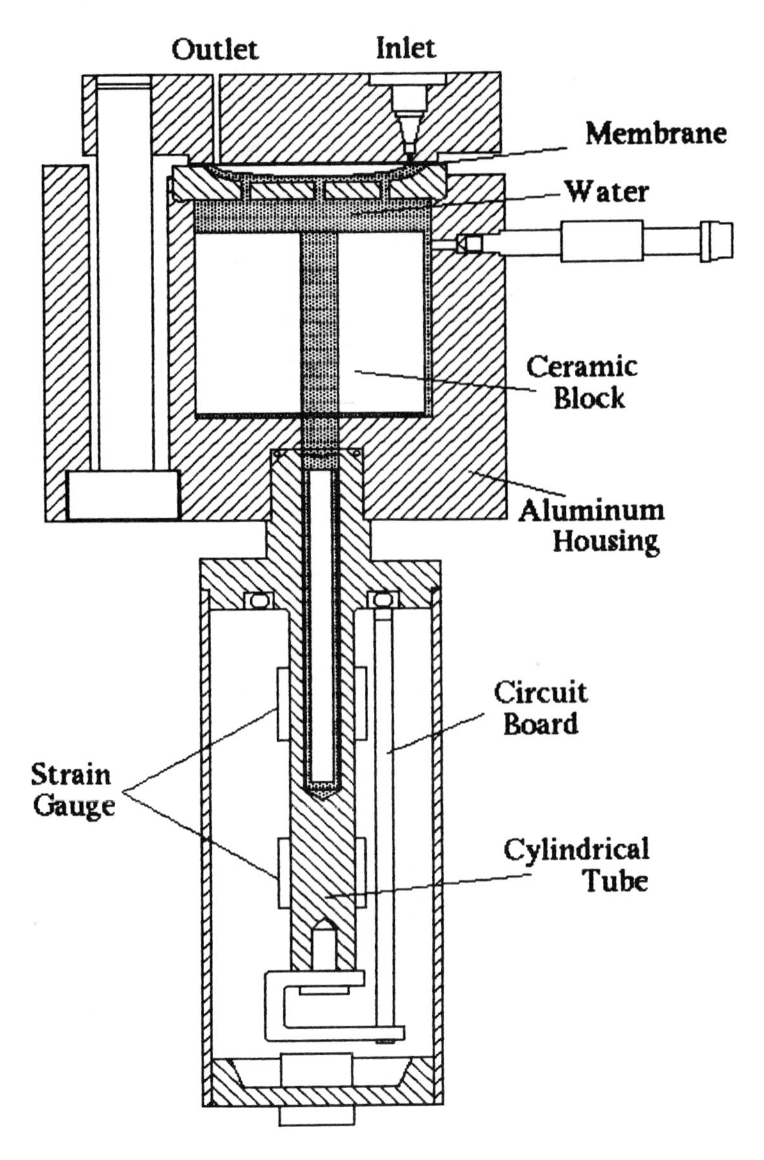

Figure 9. Schematic of the pulse-damping unit found on the model 79835A. (Courtesy of Hewlett-Packard, Avondale, Pennsylvania.)

to push mobile phase down the column. Flow-rate control can be accomplished by varying either the pump motor speed or the piston stroke length. Both methods are exploited in present day pumps. Some of the reasons that reciprocating piston pumps have become so popular are listed here [15]:

1. Ease of operation
2. Economical

3. Compatible with gradient elution
4. Amenable to high-pressure operation
5. Rapid solvent change (low internal volume)
6. Unlimited mobile-phase capacity
7. Good flow reproducibility
8. Limited maintenance

The biggest drawback to reciprocating piston pumps is the fact that they all produce flow pulsations because of the reciprocating action of the piston. This pulsing must be reduced or eliminated for the pump to be useful for present-day chromatographic applications.

Several different instrument designs exist within the reciprocating piston group. These designs include single-head, dual-head, and triple-head pumps, along with what is commonly known as the "piston-and-a-half" design. A rather recent introduction to the reciprocating piston design is the high-speed single-head pump, which produces essentially pulseless flow. The evolution of reciprocating piston pump design has been driven by the need to minimize and, if possible, eliminate pump pulsations.

Single-Piston Pumps

Early reciprocating piston pumps (Figure 10) were of rather simple design consisting of a motor connected to a circular cam. The cam, by means of a connecting rod, pushed a single piston back and forth through a pumping chamber. A high-pressure seal was used to prevent mobile phase from escaping from the back of the pumping chamber. Check valves at both the inlet and outlet to the pumping chamber kept mobile phase moving from the solvent reservoir toward the injector and column. Because the refill stroke of the piston took as long as the delivery stroke, mobile-phase flow would occur only during one half of the piston cycle. This led to pulsating flow along with erratic baselines and peak shapes.

Solvent compressibility was also a problem with early reciprocating pumps. As the piston entered its delivery stroke, the mobile phase in the pumping chamber would be compressed until the chamber pressure rose to the level of the column inlet pressure. At this point the outlet check valve would open and flow of mobile phase through the column would occur. During the time in which mobile phase was being compressed, there would be no flow. This lack of flow through the column thus extended from the start of the intake stroke to that part of the delivery stroke when the pumping chamber pressure reached that of the column inlet pressure.

In order to reduce the pulsations a pulse damper was placed between the outlet check valve and the injector. Pulse dampers store some of the energy provided by the delivery stroke of the piston and then use the stored energy to reduce the pressure drop and provide the impetus for continued flow during the piston's refill stroke. This is usually done by storing energy in one or more small reservoirs during the pumping stroke and then releasing this energy through a restrictor during the refill stroke. This can be accomplished with a coil of tubing that has been flattened in several places to provide the necessary resistance to flow required for the storage of energy. As the piston forces mobile phase through the

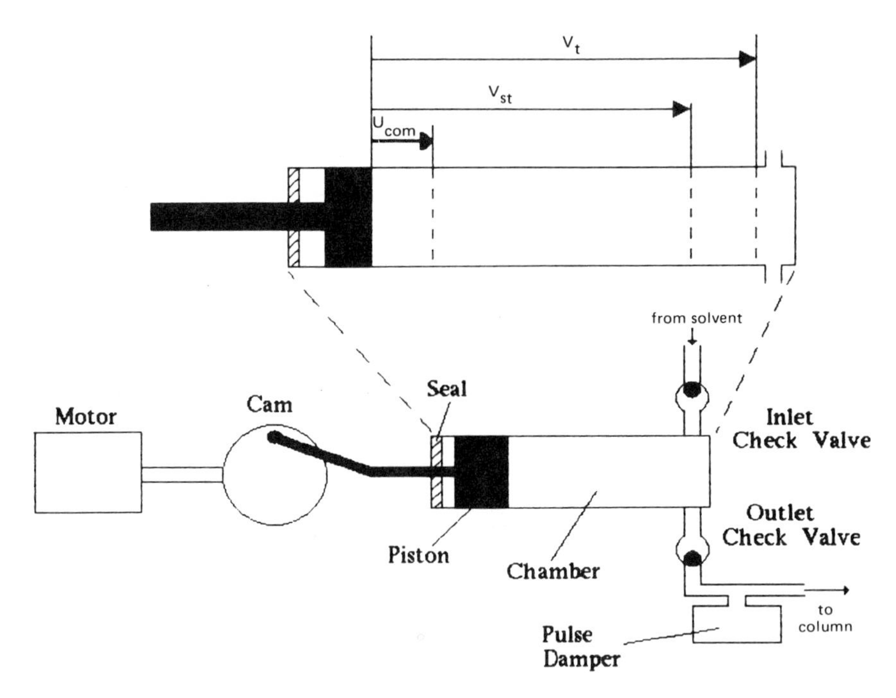

Figure 10. Schematic of a simple reciprocating pump showing an enlarged view of the pumping chamber. (From Ref. 1, with permission.)

column, the coil expands. The return of the coil to its original shape releases the energy necessary to damp the pulse. Because the time frame over which the refill stroke occurred was rather long, a great demand was placed on the pulse damper to smooth the flow pulsations. The flow versus time output of an early reciprocating pump is shown at A in Figure 11.

Later pump designs have all been concerned with producing a constant rate of flow by reducing pulsations as much as possible. Altex Scientific marketed an innovative single-piston reciprocating pump [15], as shown in Figure 12, in which the time frame for the refill and compression portion of the delivery stroke was greatly reduced. This allowed a more constant delivery of solvent to the column since, when measured on a time scale, most of the pumping cycle was devoted to the delivery stroke. The rapid refill of the pumping chamber was accomplished through two mechanisms. First, a specially shaped cam was used to allow the spring loaded piston to return quickly (0.2 sec) during the refill stroke. Second, the pump motor speed was increased dramatically to bring the cam to a position that would allow the delivery stroke to begin immediately after refilling the chamber.

To further reduce the pulsing, an electronic compressibility compensation circuit was added to this pump. This circuit measured the pressure near the end of each delivery stroke and sustained the increased motor

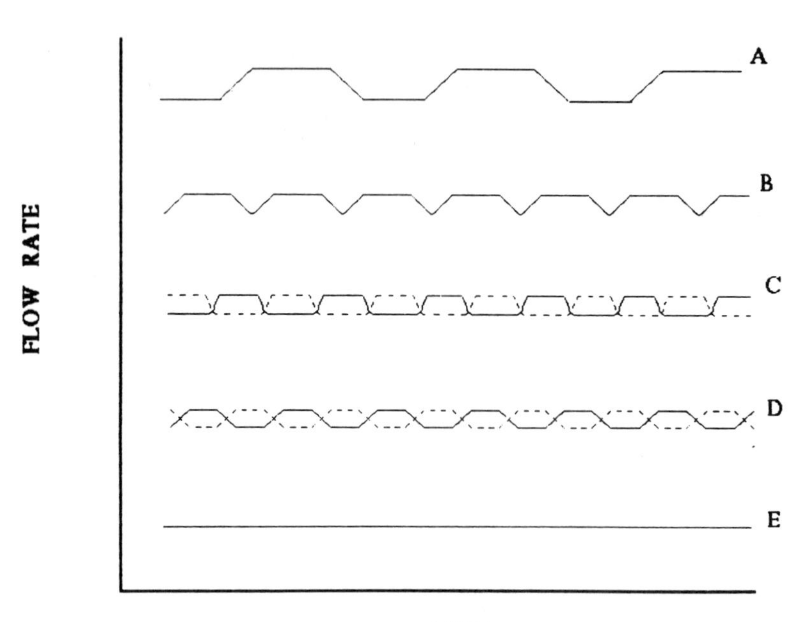

Figure 11. Pump flow output patterns from (A) a single-piston pump with slow filling cycle; (B) a single-piston pump with rapid filling cycle; (C) a dual-piston pump with filling cycles operated 180° out of phase; (D) a dual-piston pump utilizing noncircular gears (the filling time is slightly shorter than the time required for the pumping stroke); and (E) pulseless flow of a high-speed single-piston pump.

speed of refill through solvent compression until the chamber pressure was equal to that of the previous delivery stroke. The addition of rapid refill and a compressibility compensation circuit to the single piston design produced the flow versus time profile shown at B in Figure 11. Other advantages of the single-piston design are that it is relatively inexpensive (only a single pump head is required) and that it is mechanically simple.

The disadvantage to a rapid-refill pump is that it is subject to cavitation when mobile phases are used that have a high vapor pressure or that have not been thoroughly degassed. As the piston enters its refill stroke, a sudden decompression occurs in the pumping chamber, opening the inlet check valve and rapidly pulling mobile phase from the solvent reservoir. The sudden decompression can pull air or solvent vapor out of solution. Although these bubbles can cause compression problems in the pumping chamber itself, a more critical problem is that they sometimes lodge in the inlet check valve, causing it to malfunction. With the next intake stroke, the air in the check valve expands, diminishing the pressure drop and not allowing sufficient suction to properly draw mobile phase from the reservoir. The flow rate and pressure fall, sometimes to zero, and the analysis must be stopped until the air bubble is cleared. Manufacturers that market single-piston rapid-refill pumps include Beckman/Altex, Isco, Shimadzu, Perkin-Elmer, and Bio Rad.

Check valves are used on all reciprocating piston and piston diaphragm pumps. The function of check valves within the pump is to keep mobile phase flowing in the proper direction, that is, away from the solvent reservoir and toward the pump outlet. There are several competing check valve designs, but perhaps the best known and certainly the most common is the ball and seat design (Figure 13). As mobile phase is drawn into the pump head by the suction of the retreating piston, the ruby ball in the inlet check valve is pulled off of its seat (usually made of sapphire) and into the ball guide. The other end of the ball guide is either machined to have solvent channels running through it or has a few small clasps on it to prevent the ball from seating properly. This allows mobile phase to flow around the ruby ball and into the pumping chamber. On the delivery stroke of the piston, mobile phase is forced out of the pumping chamber and pushes the inlet check-valve ruby ball back into the ball seat, thus preventing mobile phase from returning to the solvent reservoir. Mobile phase is in this way forced through the outlet check valve (which operates in a similar fashion) and toward the pump outlet.

Some manufacturers feel it is more reliable to spring load the ruby balls in their check valves. This provides a more positive seating action for quicker response and increased flow accuracy and precision. Another

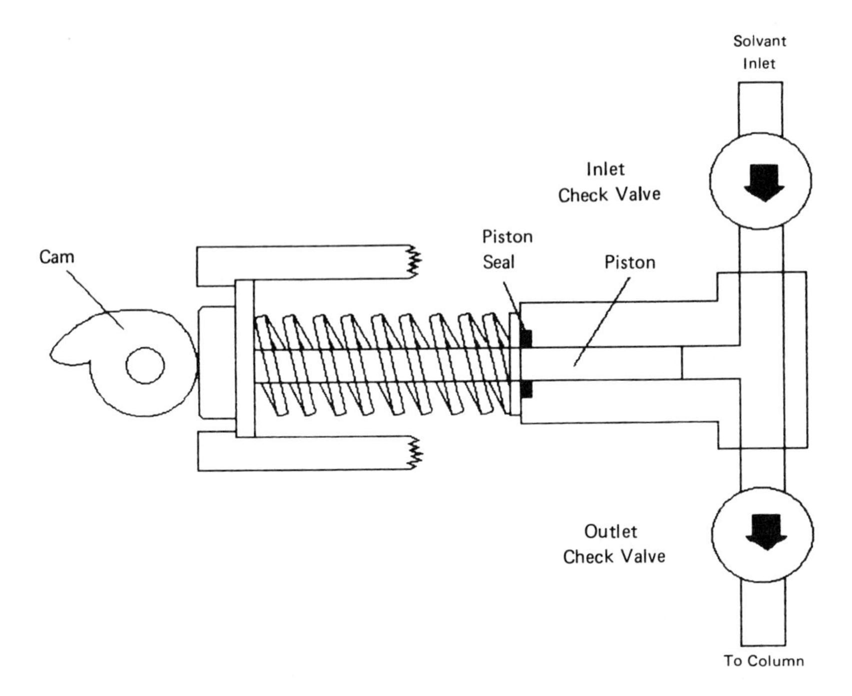

Figure 12. Schematic of a single-piston pump. (Drawing courtesy of the Perkin-Elmer Corporation, Norwalk, Connecticut.)

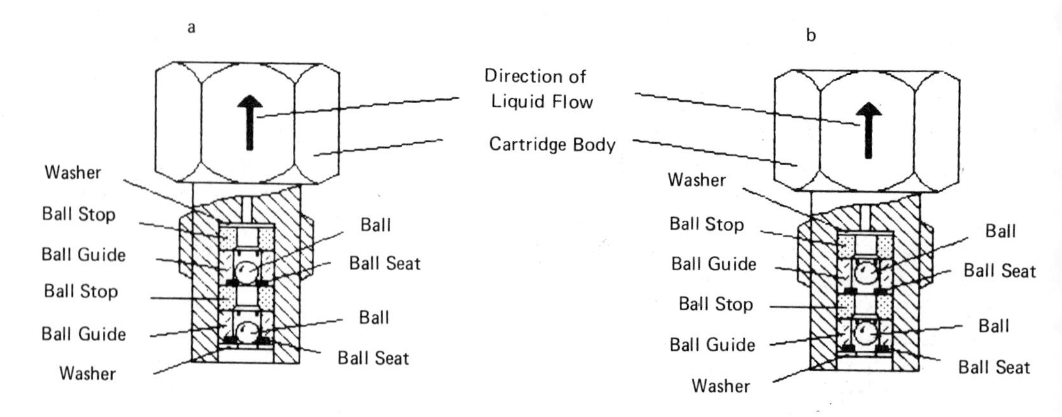

Figure 13. Double ball-and-seat style check valve (a) in the closed position, and (b) in the open position. (Courtesy of the Perkin-Elmer Corporation, Norwalk, Connecticut.)

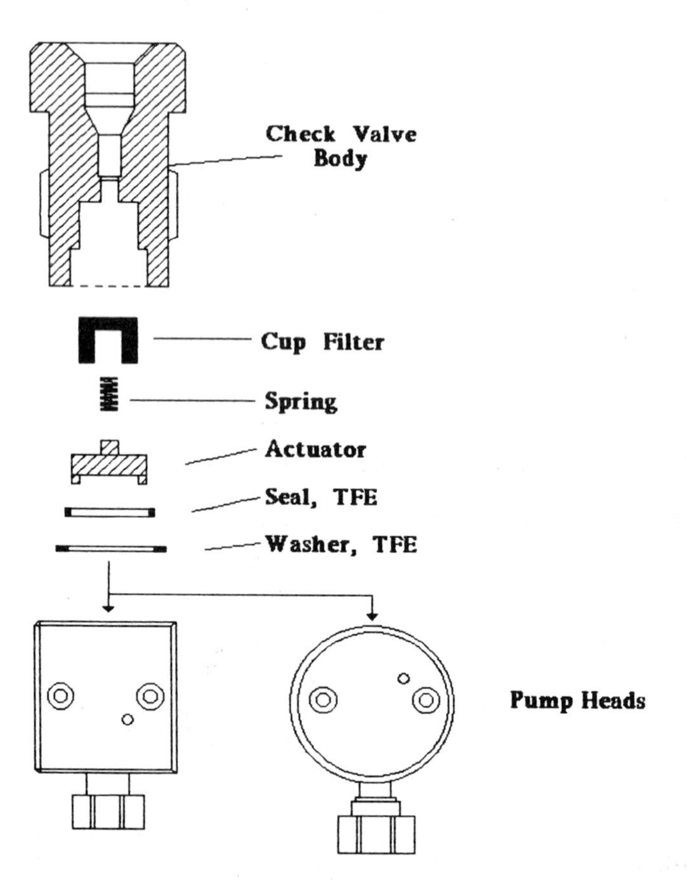

Figure 14. Actuator/spring/cup filter design used in outlet check valves on Waters Associates pumps. (Courtesy of Waters Chromatography, Division of Millipore, Milford, Massachusetts.)

way to improve check-valve reliability is to incorporate two sets of balls and seats in each check-valve body. This provides back-up protection in the event that one ball and seat fails.

A more recent check valve design used on the outlet side of pump heads manufactured by Waters Associates is the actuator/spring/cup filter design shown in Figure 14. In this style check valve the stainless steel actuator rests on a seal within the pump head. During the compression portion of the delivery stroke the pressure within the pumping chamber rises to the level of the column inlet pressure. At this point mobile phase pushes the actuator off of the seal and compresses the spring, allowing solvent to flow around the actuator, through the cup filter, and out of the check valve. As the piston enters its refill stroke, the pumping chamber pressure drops. The column inlet pressure then reseats the actuator on the seal, preventing the return of mobile phase. The advantage of this design is the elimination of ruby balls, which are subject to cracking or chipping leading to check-valve failure. This design is not used on the inlet side of the pump head because the spring-loaded stainless steel actuator requires too much force to open (~150 psi) to allow free flow of mobile phase from the reservoir into the pumping chamber.

Dual-Piston Pumps

Another approach to reducing pump pulsations is to add a second pump head positioned in parallel (as opposed to in series) to the first. This pump head contains a piston that is 180° out of phase from the first. As one piston is delivering mobile phase to the column, the other piston is pulling mobile phase from the reservoir. This eliminates the need for a rapid refill stroke. Without rapid refill, the pumps are less subject to cavitation. The addition of another piston and pump head assembly also reduces the wear that the piston seal experiences compared to that of a single-piston system.

The dual-piston design drastically reduces pump pulsations (C in Figure 11). A small pulse still occurs, however, at the "crossover point" when the pistons change direction. These pumps often rely on a pulse damper to sustain the flow at the crossover point.

A slight modification of the dual piston design led to the nearly complete elimination of pump pulsing. The modification involves driving the pistons with noncircular gears (Figure 15). The use of noncircular gears leads to a slightly faster intake than delivery stroke and shifts the piston strokes to be slightly more than 180° out of phase. This allows both pistons to deliver mobile phase to the column simultaneously for a short period of time. One piston accelerates into the delivery stroke as the other one decelerates. The combined output of the two pistons at this point is equal to that of either piston during the constant velocity portion of the cycle. The flow output of this pump type is shown at D in Figure 11. Waters Associates pump models M-6000, 510, and 590 are examples of this type.

As mentioned previously, solvent compressibility can be compensated for by measuring the chamber pressure near the end of the delivery stroke and then increasing the pump motor speed through refill and solvent compression. The speed increase persists into the next delivery stroke until

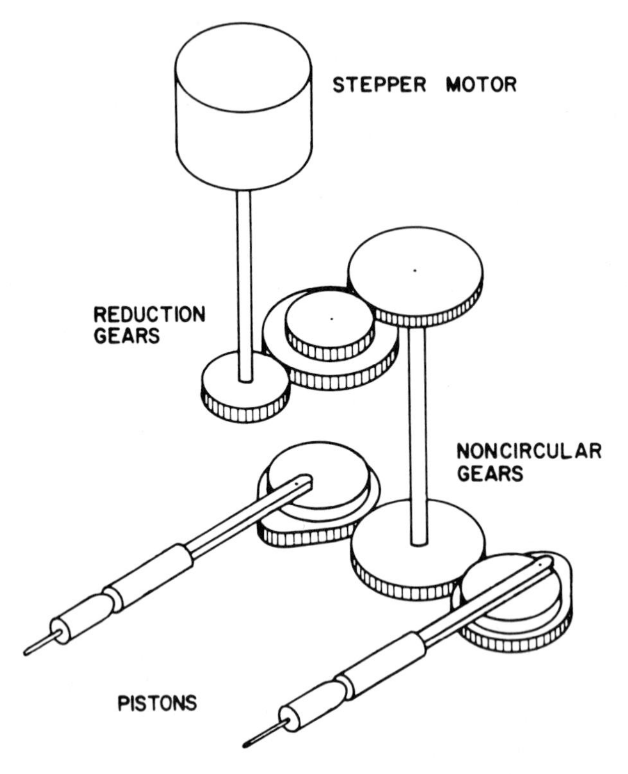

Figure 15. Mechanical design of a dual-piston pump (parallel pump head) manufactured by Waters Associates, showing the noncircular gear mechanism. (Courtesy of Waters Chromatography, Division of Millipore, Milford, Massachusetts.)

the pressure is returned to the value of the previous stroke. Some manufacturers of dual-piston pumps (E. M. Science/Hitachi), have retained the momentary speed increase method of correcting for solvent compressibility. The choice of other manufacturers (Waters Associates) has been to spread the speed increase over the whole piston stroke. In the Waters system, no pressure comparison is made. Instead of momentarily increasing the motor speed until the chamber pressure reaches that of the previous delivery stroke, a small increase in motor speed is sustained throughout the piston cycle. In this way, compression compensation becomes independent of pressure. The degree of compression compensation on Waters pumps can be adjusted through a potentiometer accessible from the front of the pump. Either method will increase the accuracy of flow delivery.

To produce accurate quantitative results, flow rate must remain constant through variations in the column inlet pressure. This is usually accomplished through electronic circuits that monitor the pressure fluctuations described by the pressure transducer and vary the power supplied to the motor as the backpressure of the system changes. For example, as the pressure within the HPLC system increases due to a column frit becoming blocked, the electronic circuitry senses a rise in the pressure and compensates by supplying more power to the motor to maintain a constant rate of flow. Supplying more power to the motor should not be confused with increasing the motor speed. In order to overcome increasing system backpressure, the pump needs more power to run at the current rate of flow, not an increase in the speed of the motor. The electronic circuitry in connection with currently available reciprocating pumps thus allows a constant delivery of mobile phase even under conditions in which the system inlet pressure changes or when a compressible mobile phase is used.

Triple-Piston Pumps

A few manufacturers have produced three-piston pumps (DuPont, Jasco). This design is shown in Figure 16. A three piston pump is designed to reduce pulsing to an absolute minimum. The pistons cycle 120° out of phase from one another and are driven by a uniquely designed cam. These pumps show minimal pulsing because at the crossover point of any two pistons, the third piston is in the middle of its delivery stroke. Thus no noticeable flow aberration occurs. This type of pump, however, is mechanically complex and can be somewhat difficult to troubleshoot.

Piston-and-a-Half Pumps

There is another pump design that has become increasingly popular over the past several years. It is often called the "piston-and-a-half" design

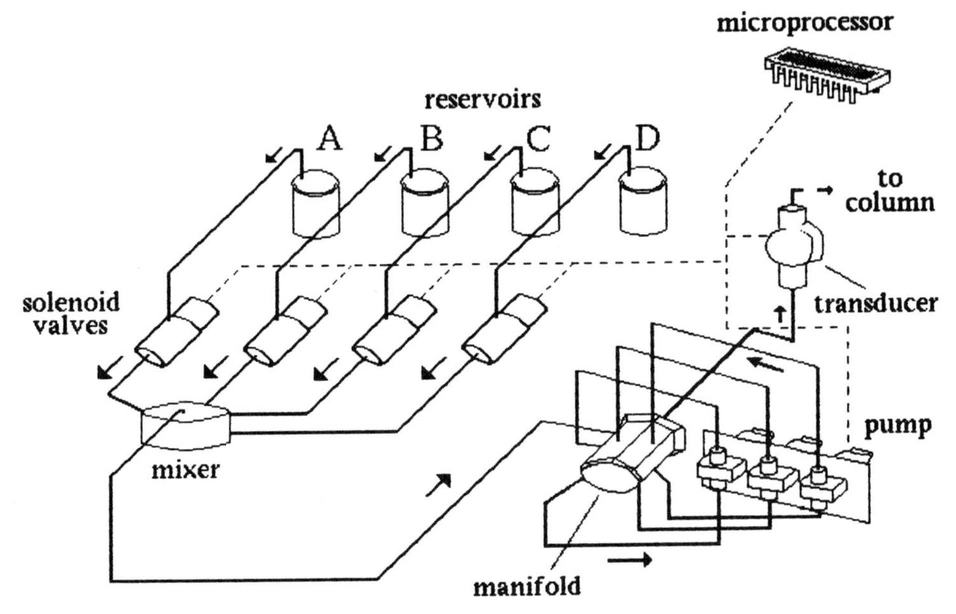

Figure 16. Mobile-phase flow path in the DuPont 8800 series triple-piston pump. (From Ref. 6, with permission.)

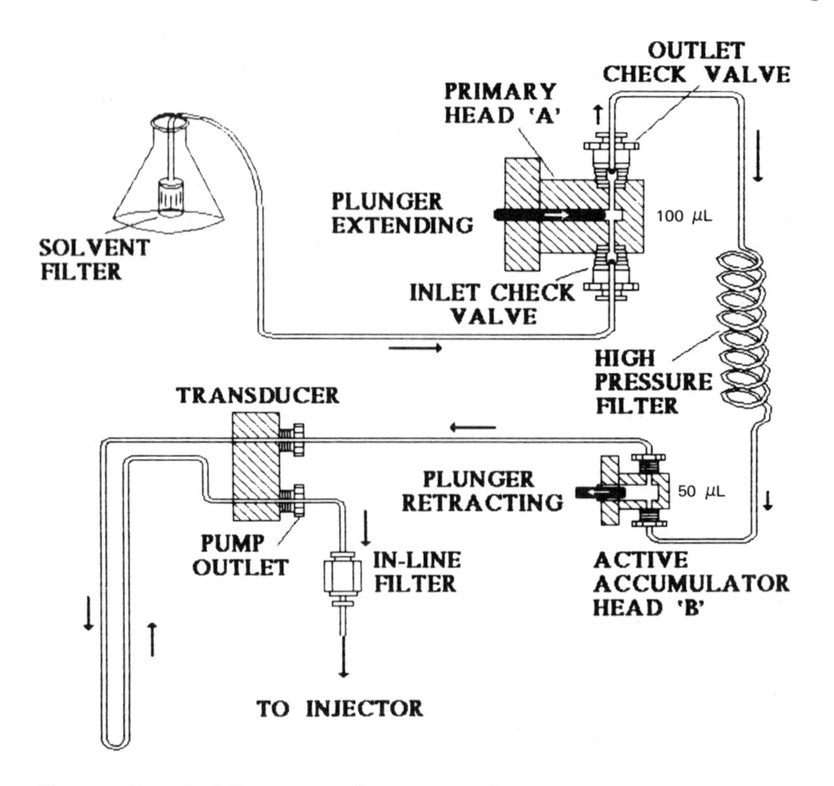

Figure 17. Mobile-phase flow path of the M-45 pump manufactured by Waters Associates. This model is representative of the "piston-and-a-half" pump design family. Note that the pump heads are in series and that they are of different volume displacements. (Courtesy of Waters Chromatography, Division of Millipore, Milford, Massachusetts.)

(Figure 17) because the dual pump heads that are employed are of different displacement volumes. The primary head holds a volume of either 80 μl (Kratos) or 100 μl (Waters, Hitachi, Pharmacia), while the secondary head holds half the volume of the primary head. The flow of mobile phase through this pump differs from that of previously mentioned dual piston pumps in that the pump heads in the "piston-and-a-half" design are connected in series rather than in parallel. As mobile phase is pushed out of the primary head by the action of the piston, one half of the output is taken up by the secondary head while the other half is delivered to the column. While the primary head is refilling, the mobile phase taken up by the secondary head is delivered to the column. Having pump heads in series provides one big advantage over other dual piston designs: only one pair of check valves (located on the primary head) is required in this design. Since the pump heads are in series, the outlet check valve on the primary head serves as the inlet check valve for the secondary head. No outlet check valve is needed on the secondary head because of the constant source of forward-directed pressure (from the action of the alternating pistons) at this point. A single set of check valves makes these pumps mechanically simple and easy to troubleshoot. The disadvantage of this design is that it delivers to the column only one-half the volume with each pump stroke that parallel dual-piston designs deliver. This means that the pistons need to operate twice as fast to deliver mobile phase at the same flow rate. This leads to more wear and, hence, more maintenance of piston seals and pistons than on parallel piston pumps. The "piston-and-a-half" design is perhaps best thought of as a design that combines the advantages of single- and dual-piston pumps: that is, the mechanical simplicity of a single-piston pump with the slow refill stroke of dual-piston pumps.

High-Speed Single-Piston Pump

The youngest member of the reciprocating piston family is a high-speed single-piston pump from Applied Chromatography Systems Ltd. This pump is of novel design (Figure 18) and overcomes the problems of pump pulsation and flow rate control in a unique manner. Unlike other reciprocating pumps, it uses a high-speed motor (1400 rpm) to drive a piston through a pumping chamber at a rate of 23 strokes/sec. Because of this high rate of piston movement, even flow-sensitive detectors are not capable of distinguishing these rapid flow-rate fluctuations. No pulses are seen (E in Figure 11), and hence pulse damping is unnecessary.

This high stroke rate of the piston has made a couple of changes necessary to the check valves and pressure seals. The check valves, both inlet and outlet, are spring loaded. Because the check valves open and close 23 times per second, the positive seating action that the spring provides is required for optimum performance. There are two high-pressure seals located within the pump's head. These seals are made of a specially designed glass-filled PTFE material, and are guaranteed for one year.

Flow-rate changes on this pump are made by changing the length of the piston stroke rather than by changing the pump motor speed. This is analogous to the way flow-rate changes are made on a typical piston-diaphragm pump, with one exception. On a piston-diaphragm pump, the pump heads themselves are moved by stepper motors. This is a time-consuming process, and flow-rate changes are relatively slow. On the ACS high-speed pump flow-rate changes occur by varying the length of the piston stroke, which can be done very quickly. As the stroke volume is increased, more solvent is pulled into the pumping chamber. Since the pump motor speed is constant, the flow rate increases by an amount proportional to the stroke volume increase.

On this pump, compressibility compensation is also handled by varying the piston stroke volume. The position of the piston in the pumping chamber is constantly monitored by an induction coil located on the motor drive shaft. The induction coil can determine when the piston begins and ends the delivery stroke. Part of the delivery stroke is devoted to compression of the mobile phase and high-pressure seals within the pumping chamber. Delivery of mobile phase to the column begins only after compression (pressurization) of the mobile phase to the level of the column inlet pressure. At this point, the outlet check valve opens and the flow of mobile phase begins. A strain gauge measures the pressure within the pumping

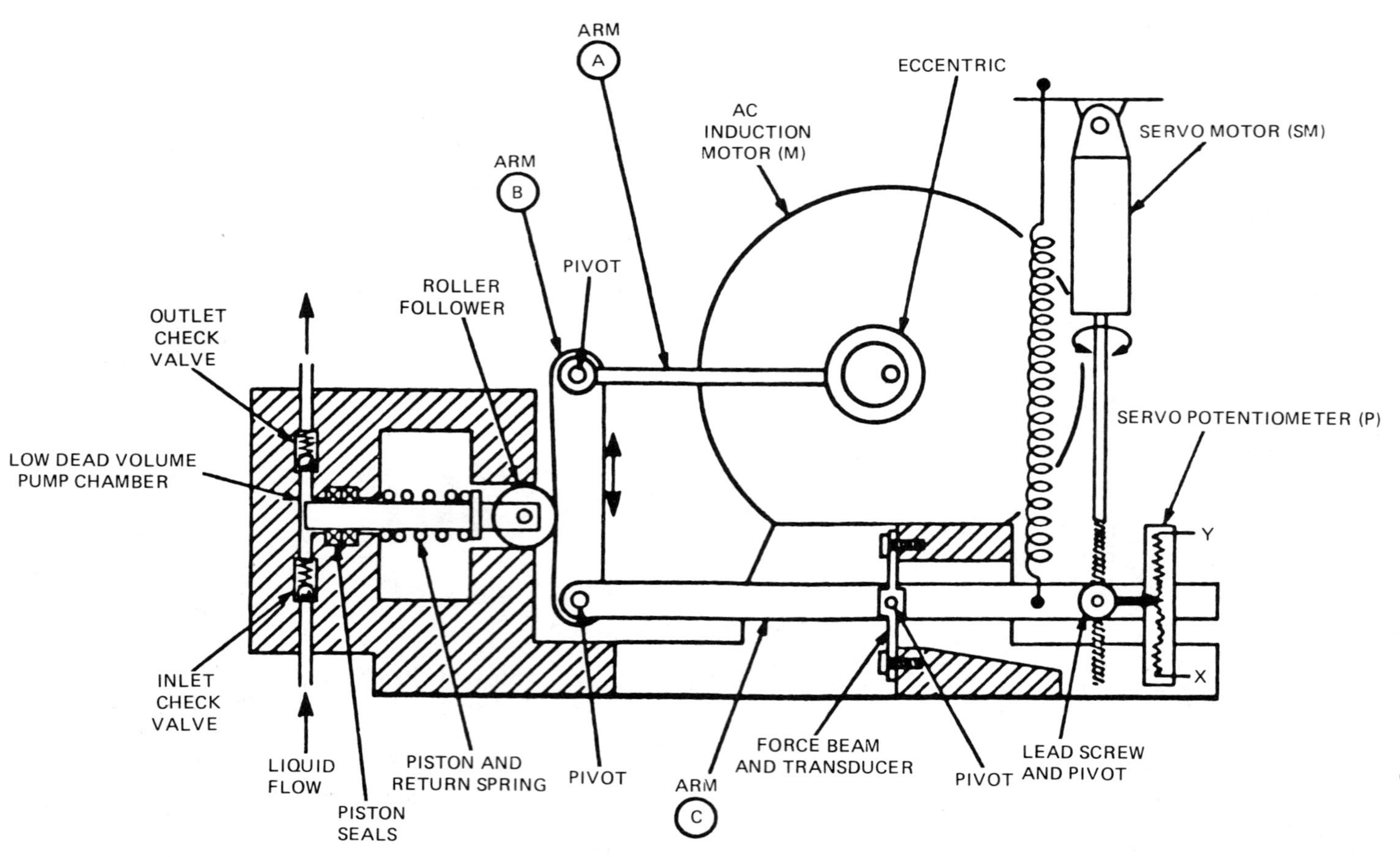

Figure 18. Mechanical design of the high-speed single-piston pump manufactured by Applied Chromatography Systems, Ltd. (Courtesy of Applied Chromatography Systems, Ltd.)

chamber. The pressure rises as compression occurs and then remains steady during solvent delivery. With a knowledge of the pumping-chamber geometry and the position of the piston when solvent delivery begins, the actual flow rate can be calculated. If this rate differs from the requested flow rate, the length of the piston stroke is changed to compensate. It is thus possible to keep the flow rate constant even with changes in solvent composition, column inlet pressure, or temperature.

A real advantage to a fixed-frequency pump of this type is that it is possible to drive two independent pumps from a single motor. These two pumps could have independent flow controllers and could be used in separate chromatographic systems. It is also possible to use both of these pumps in a single gradient system, producing an accurate, pulse-free, inexpensive gradient HPLC solvent delivery system. Since this system is free of pulse dampers (which normally add dead volume to a gradient system), it allows rapid response to subtle gradient profile changes. No mixing chamber is required in this system since the very low stroke volumes (1.4 μl at 2.0 ml/min) mix readily in the pump outlet tubing. This further reduces the system dead volume compared to conventional gradient systems, making this high-speed gradient system extremely responsive.

C. Gradient Elution Pumps

Sample mixtures often contain components that have a wide range of retention characteristics. An isocratic (constant mobile-phase composition) system is often not capable of separating these components in a reasonable time frame. For these types of separations a gradient system is necessary in which the mobile-phase composition changes with time. In almost all cases the system is set up to provide increasing solvent strength with time, so that components that elute late under isocratic conditions will now emerge from the column in a much shorter time frame while still providing separation of early eluting components. A gradient profile can be set up to provide maximum resolution for a given sample mixture. Gradients also have the added advantage of keeping the chromatographic bands narrow throughout the run which increases their detectability. Applications of gradient elution in HPLC include rapid method development for isocratic separation [16], and maximum resolution and detection of complex sample mixtures [17].

The types of gradient profiles that may be used are numerous. Step gradients (often used in trace enrichment analyses) produce sudden changes in mobile-phase composition periodically throughout the run. Gradient profiles come in a variety of shapes, but all are derived from three basic types: linear, concave, and convex (Figure 19). Combinations of these shapes are often used to produce the exact profile desired. Linear gradients are used when sample components have widely different retentions and are rather evenly distributed throughout the isocratic chromatogram. Concave and convex gradients are used to separate components that have similar capacity factors early in the chromatogram and late in the chromatogram, respectively.

Gradient HPLC systems come in two basic types: low-pressure and high-pressure gradient formers. In either system a gradient controller is used to control the shape of the gradient profile. The gradient controller

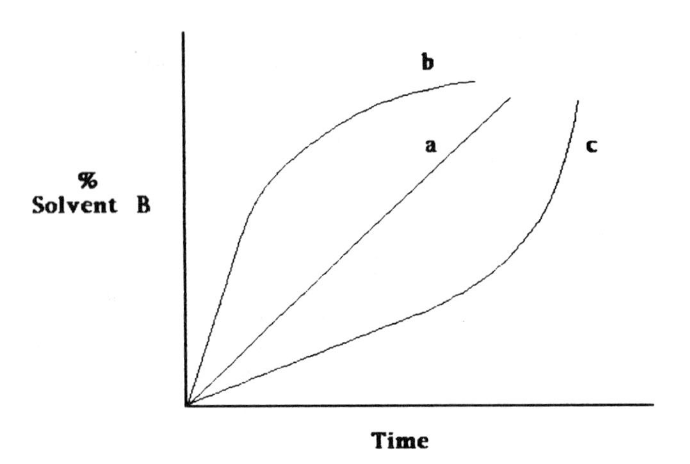

Figure 19. Three basic gradient profiles: (a) linear, (b) convex, and (c) concave.

is generally a separate unit, the specific purpose of which is to control gradient formation. Recently, however, Kratos Analytical has built a gradient controller into their microprocessor-controlled model 783 UV absorbance detector. This was a novel idea that reduced both the space requirements and the cost of a gradient HPLC system. An added advantage of the Kratos system is that it is capable of controlling pumps of many other manufacturers. In low-pressure gradient systems (Figure 20a) two or more solvents are mixed at ambient pressure prior to entering the HPLC pump. Mobile-phase composition is usually determined by microprocessor-controlled proportioning valves. In these systems, only one high-pressure pump is needed to deliver mobile phase to the column.

In high-pressure gradient systems (Figure 20b), the gradient is formed on the high-pressure side of the chromatographic system. One pump is needed for each solvent used to form the gradient, and the mobile-phase composition is controlled by varying the flow rate of the individual pumps. To obtain optimal results in either system, efficient mixing is required in as low a volume as possible. When operating normally, either design should be capable of producing precise and accurate gradient profiles. Desirable characteristics of either design include [18]:

1. Ability to generate complex gradient shapes
2. Ability to control flow rates and solvent compositions accurately and precisely
3. Ability to generate rapid mobile-phase compositional response
4. Capacity for automation

Many present-day liquid chromatographs with gradient capability have simple exponential and linear gradient profiles stored in the read-only memory (ROM) of the gradient controller. The microprocessor within the gradient controller also allows the generation and storage of complex user-defined profiles. This combines ease of use with the versatility that chromatographers demand. Accurate and precise flow rates and solvent compositions are requisite to obtaining meaningful peak responses in gradient elution chromatography. The accuracy and precision of flow rate control is dependent on the ability of the pump(s) to compensate for changes in mobile-phase compressibility and/or system backpressure.

In low-pressure gradient systems, flow-rate control is dependent on a single pump. The gradient is formed and mixed before entering the pump. Pump electronics therefore must compensate for changes in solvent compressibility as the solvent composition within the pump changes. Thermodynamic solvent volume changes due to mixing occur before entry of the mobile phase into the pump in a low-pressure gradient system and do not adversely affect the flow rate.

In high-pressure mixing systems the gradient is formed and mixed on the high-pressure side of the HPLC system. Each pump delivers a single

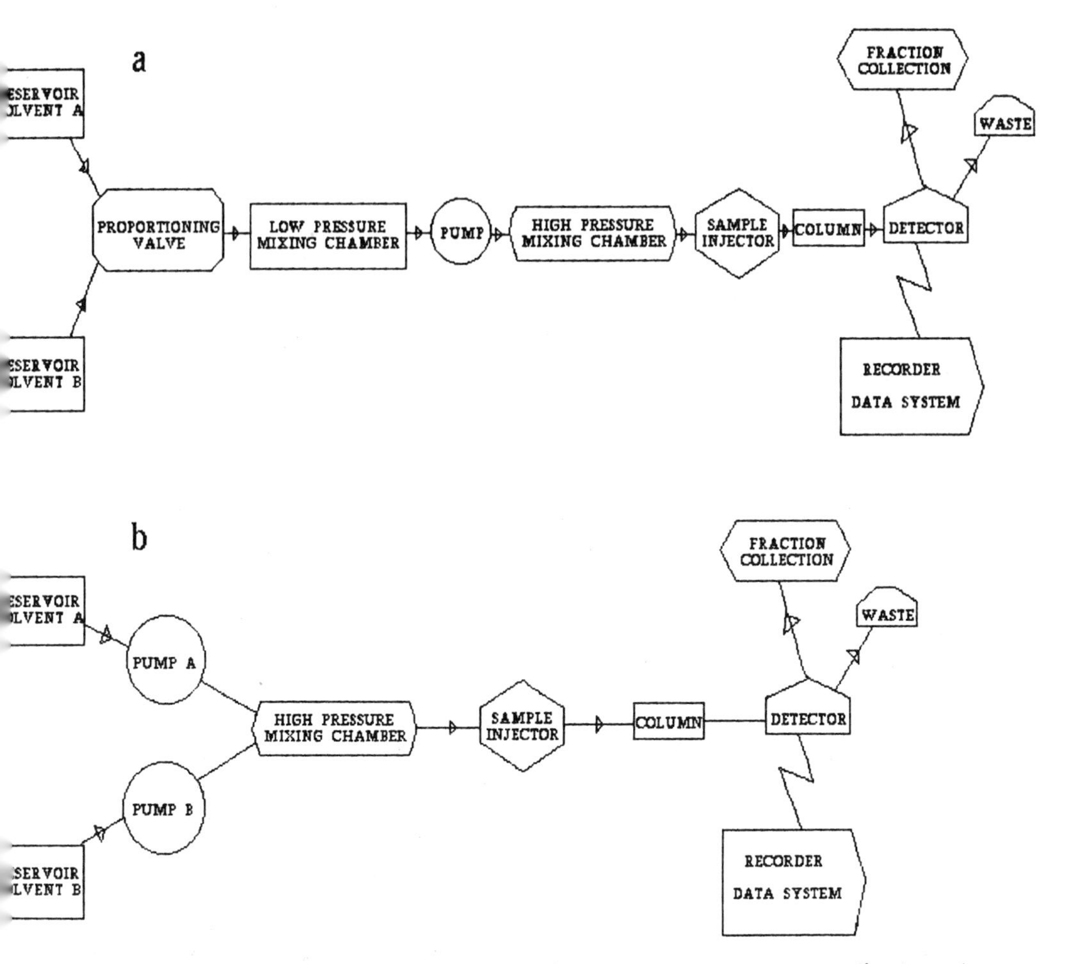

Figure 20. Schematic representation of (a) a low-pressure gradient system and (b) a high-pressure gradient system. (From Ref. 18, with permission.)

solvent (or at least a constant-composition solvent) throughout the gradient formation process. Therefore no solvent compressibility changes occur within the pump. Solvent volume changes that occur at mixing, however, do change the system backpressure. The pumps used in high-pressure gradient systems must be able to compensate for these pressure changes. In high-pressure gradient systems the flow rate is determined by the combined output of two pumps. Flow-rate control suffers if either pump malfunctions or is pushed to its specification limits. At a low percentage of either solvent one of the pumps will be operating at a low flow rate. For example, at a total flow of 1.0 ml/min and solvent ratio of 90:10, pump B will be operating at 0.1 ml/min. Most HPLC pumps are neither very accurate nor precise at this slow rate. Because of their simplicity and inherently better flow-rate control, single-pump low-pressure gradient systems are preferred by many chromatographers.

The accuracy of mobile-phase composition (when compared to the preset gradient profile) is determined by the amount of dead volume between the point at which the gradient is formed and the head of the column. Because gradients in low-pressure mixing systems are formed prior to the HPLC pump, the dead volume is usually significantly larger than in high-pressure mixing systems. The added dead volume not only delays the onset of the gradient but can also cause remixing of the formed gradient before entry onto the column. This can sometimes negate the improved accuracy that low-pressure proportioning valves have over two-pump gradient formation. It is essential in low-pressure gradient systems to use low dead volume pumps with a small but efficient mixing chamber. The volume of the pumping chamber is especially important in a low-pressure gradient system because the pump delivers discrete segments of the gradient in stepwise fashion. The volume of each segment is equal to the volume of the pumping chamber. The larger the volume of the pumping chamber, the less continuous and more step-like the gradient will be. Pumping chambers of 100 μl or less are preferred.

The precision or reproducibility of gradient formation is dependent on the precision of microprocessor-controlled proportioning valves in low-pressure mixing systems and on the precision of flow control of the two pumps in high-pressure mixing systems. Proportioning valves are generally more precise than is the flow control of two pumps, especially at the gradient extremes when one pump is delivering solvent at a very low rate. At these low rates of flow, friction around the high-pressure seal causes irregularity in the motion of the piston, leading to a lack of reproducibility of both flow rate and mobile-phase composition [14].

The decision as to which type of gradient system to purchase is often difficult to make. Low-pressure systems are preferred by many chromatographers because of their lower cost (only one pump is required) and because of the high precision and accuracy that the microprocessor-controlled proportioning valves provide. Also, the precision and accuracy of flow rate are dependent upon a single pump operating at easily controllable speeds. The major disadvantage to a low-pressure system is that to maintain an accurate gradient profile a low dead volume pump is required. High-pressure systems offer a reduced delay time between the programmed change and the actual compositional change on the column and do not have to account for constantly changing mobile-phase compressibility within the pump heads. The disadvantages to a high-pressure gradient system are its generally higher cost and its inability to deliver accurate and precise flow rates and compositional profiles at the gradient extremes.

IV. INJECTORS

Sample injection systems are the devices used to insert sample onto the stationary phase in column chromatographic systems. A sample injection system is comprised of an injection device and a sample storage and delivery unit. Integrated systems can be purchased or modular systems can be assembled from commercially available components. Areas of discussion will be general requirements, materials of construction, injection devices, and injection systems. The terminology used herein has been chosen to describe capabilities and does not necessarily correspond to that used by individual manufacturers.

There are three general types of performance requirements associated with the process of applying sample to a chromatographic system: chromatographic, instrumental, and mechanical. The chromatographic performance of a system addresses its ability to introduce sample without adversely affecting the separation efficiency of the column. Ideally, the sample is introduced as an infinitely narrow band or point at the top of the column. In practice, this ideal is difficult to achieve and manufacturers approach it through the minimization of the number and volume of dead zones (stagnant, unswept junctions) and the utilization of low-volume, cleanly swept ports. Instrumental performance addresses the more quantitative aspects of sample introduction. The sample must be applied reproducibly with no sample-to-sample carry-over or memory effect. In addition, it is helpful to have the capability of introducing variable volumes of known amount. Mechanical performance relates to the physical demands of the environment. The system must be chemically and physically inert to the mobile phase and sample. The sample introduction system must be capable of operation at pressures of 5000-6000 psi. In addition, it must be sufficiently rugged to endure thousands of injections with minimal operator intervention. While there are many performance demands for sample introduction systems, several types and models have been developed to meet these requirements in cost-effective ways.

A. Materials

Modern injection valves and injectors achieve high-pressure sealing at a mated interface between metal and a resilient polymer. The metal components are made from a stainless steel that is relatively inert and resistant to scoring. Rotors are made of polymers such as VESPEL, TEFZEL, or VALCON-H, all of which are nonporous, wear resistant, and inert to most chemicals under HPLC conditions. Some companies offer a choice between TEFZEL and VESPEL, since the latter is sensitive to attack by alkaline solutions. The tubing used for connections (internal and external) and sample loops is almost exclusively small-bore 316 stainless steel.

Recently, major manufacturers have introduced valves made from more inert metals to address the specialized needs of protein and ion chromatography. Known interactions with stainless steel include protein adsorption and ligand substitution from samples and dissolution/corrosion by mobile phases [19]. In response, injection valves are now available constructed from Hastelloy C (a nickel-containing alloy) and titanium. Connection tubing can be replaced with 0.010-in ID TEFZEL tubing for pressures of up to 3500 psi but not with inert metals due to the technical difficulties of extruding small-bore tubing [20]. All of these materials, while more expensive, are highly inert and readily available.

B. Sample Injection Devices

Sample injection devices can be classified into four main types: septum injectors, rotary valves, universal injectors, and syringeless injectors. The first three types permit the introduction of small, discrete volumes of sample directly into the chromatographic system. Syringeless injection is a hybrid of solid-phase extraction (SPE) sample preparation and liquid chromatography that incorporates the large-volume sample manipulation of SPE with the automation of LC.

Septum injection is the simplest method of sample introduction. As shown in Figure 21, a high-pressure syringe penetrates a self-sealing elastomer septum and deposits the sample in or on the top of the column bed. Since the mobile phase is flowing continuously, it sweeps the sample plug into the column bed with little dispersion. Septum injection is inexpensive, allows variable sample volume, and requires little excess sample. Septum injection can be chromatographically a very efficient sample introduction method but is sensitive to injection technique and needle position [21]. In addition, the septums and syringes have 1500 psi pressure limits, tend to fragment and leak with repeated use, and are not compatible with some HPLC mobile phases.

Rotary-valve injectors are now very popular because they are easy to use, versatile, and eliminate the problems of septum injectors. The rotary-valve injector is a two-position valve that holds sample in a fixed-volume loop. While there are differences in manufacture design, all operate basically as shown in Figure 22. The valve has two basic parts: the rotor and the stator. The stator is a highly polished steel body or plate with evenly spaced ports to accept connections from the column, pump, and injection loop. The rotor is precisely machined inert-polymer disk mated to the surface of the stator with grooves aligned to connect pairs of stator ports. As the valve moves between its two positions, the rotor connects alternate sets of stator ports. In this manner, the sample loop is either engaged or bypassed as mobile phase is delivered to the column. In operation, the loop is filled with sample at ambient pressure by syringe with the valve in the "bypassed" (load) position, and is then inserted into the mobile phase at the time of injection.

Injection loops come in discrete sizes and can be mounted either internal or external to the valve, depending on the injection volume desired. Internal loops span a range of 0.2-10 μl, while external loops deliver volumes of 6-1000 μl or more. In general, small loops are overfilled while large loops

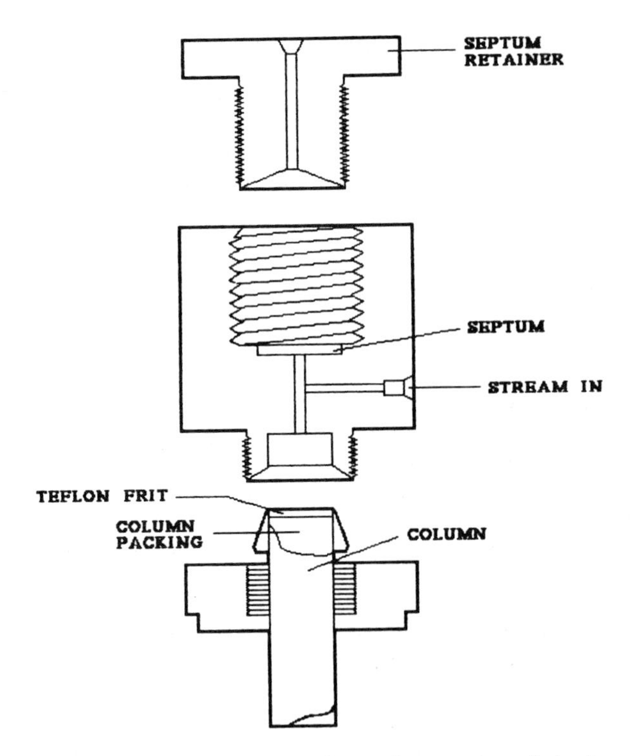

Figure 21. Exploded view of a septum injector.

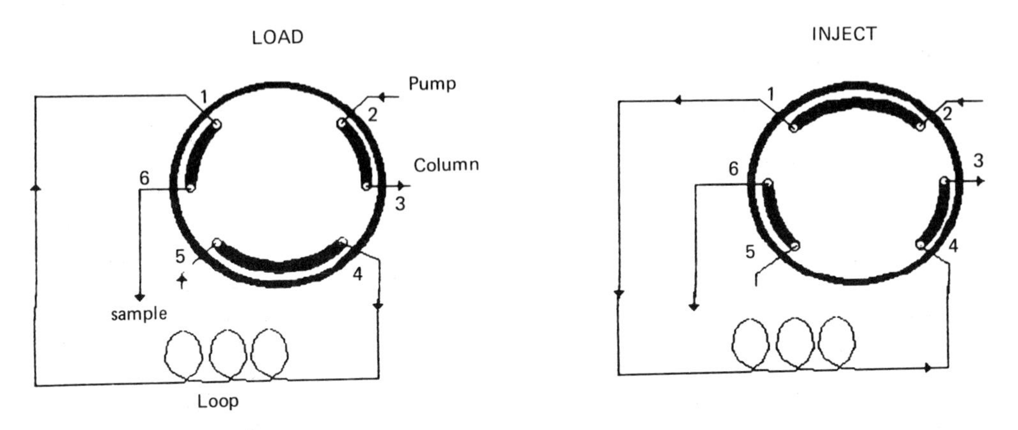

Figure 22. Schematic diagram of a six-port rotary valve.

can be overfilled or partially filled depending on the supply of sample. In an overfill mode (25-100 μl excess sample pushed through the loop) the injected volume is that of the loop size with excellent resultant precision (rsd < 0.5 percent) even at 0.2 μl. The precision of partial-fill injection depends on the volume injected with good performance (rsd < 1.0 percent) observed above the 10-15 μl range for manual injections.

Rotary valves can be actuated by manual, pneumatic, or electrical means. Since the flow of mobile phase to the column is interrupted during the transition from one valve position to the other, transition times can be an important characteristic. Typical transition times are 100-500 msec for manual and electrical actuation, with pneumatic actuation as fast as 20 msec [22]. At nominal flow rates of 1-2 ml/min, these transition times pose no problems but at the higher flow rates of fast LC (4-6 ml/min), slow transitions cause a pronounced pressure pulse that can lead to shorter column lifetime.

Versatility is an added dimension of the rotary valves since they are not limited to use as injection devices. Rotary valves come in six- and ten-port configurations and can be used in column switching (trace enrichment or sample cleanup), column selection or other high- or low-pressure switching applications [23]. Figure 23 shows a six-position valve used for column selection, and Figure 24 shows a ten-port valve that employs column selection and regeneration. With the proper control and interface equipment, these valves can be operated remotely with single or paired contact closures.

The universal injector combines the variable volume/minimal sample waste features of the septum with the durability and inertness of the rotary valve. The injector operates as shown in Figure 25. There are three valves within the sample loop and a restrictor coil parallel to the loop. Valves A and B are closed in the load position with a resultant mobile-phase diversion through the restrictor coil. Sample is introduced to the column end of the loop at ambient pressure, where it displaces residual mobile phase to waste through valve C. On initiation of injection, valve C first closes and then valves A and B open, allowing mobile phase to flush the contents of the loop onto the column. The mobile phase is split such that approximately 85 percent flushes the loop and 15 percent traverses the restrictor. In this way, there is no momentary mobile-phase pulse to the column when the valve is switched between its two positions. The precision of the universal injector is similar to that achieved in the partial fill of a rotary valve loop (rsd < 1.0 percent for 10 μl and above), but the band spreading is slightly worse due to the complex valve configuration.

Syringeless injection is an integrated system that employs a hybrid of solid-phase extraction and LC. The sample (any volume) is manually drawn through an adsorbant-filled cartridge with reduced pressure, as shown in Figure 26. The adsorbant is chosen such that the analytes concentrate on the adsorbant as the rest of the sample matrix passes to waste. The cartridge can then be washed with a weak solvent (to remove weakly adsorbed interferents) or loaded directly into the instrument. At the time of sample introduction, the cartridge is placed in a compression chamber and mobile phase is directed through the cartridge. The analytes are eluted from the adsorbant and carried into the analytical system for separation

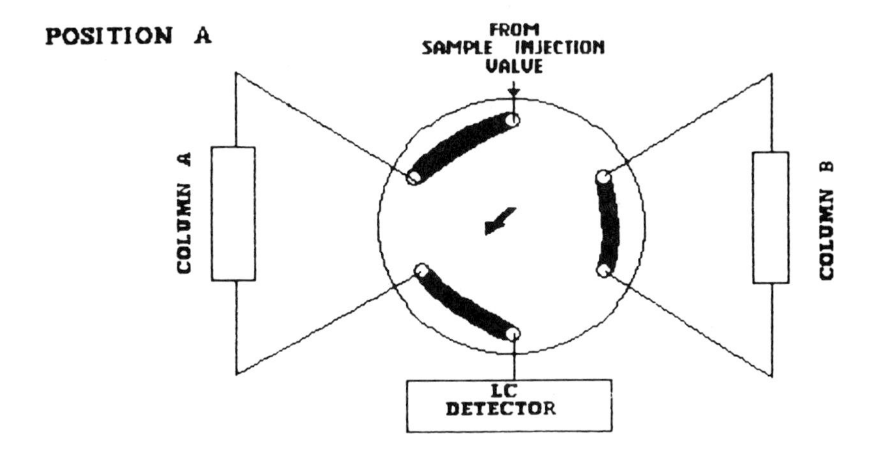

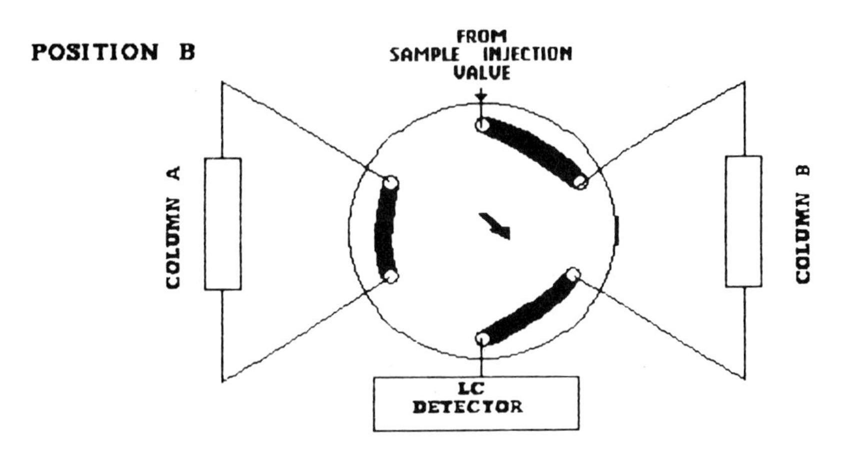

Figure 23. Two-column selection with a six-port rotary valve.

and detection. Ten cartridges are packaged in a plastic cassette and the instrument can hold 10 cassettes. In this way, 100 samples, once loaded onto cartridges, can be analyzed with no further operator intervention.

A similar preconcentration/sample cleanup system can be constructed for manual or automated operation with a short (1-3 cm) HPLC guard column and a 10-port rotary valve. The contents of the sample loop (sample volume independent) are pushed through the precolumn by a "loading" fluid with a HPLC pump, as shown in Figure 27. The loading fluid can be innocuous (purified water) or a mild wash solution to remove interferents in the sample matrix. The injection step inserts the precolumn into the analytical system, where the sample is backflushed with mobile phase onto the analytical column for separation and subsequent detection. During the separation,

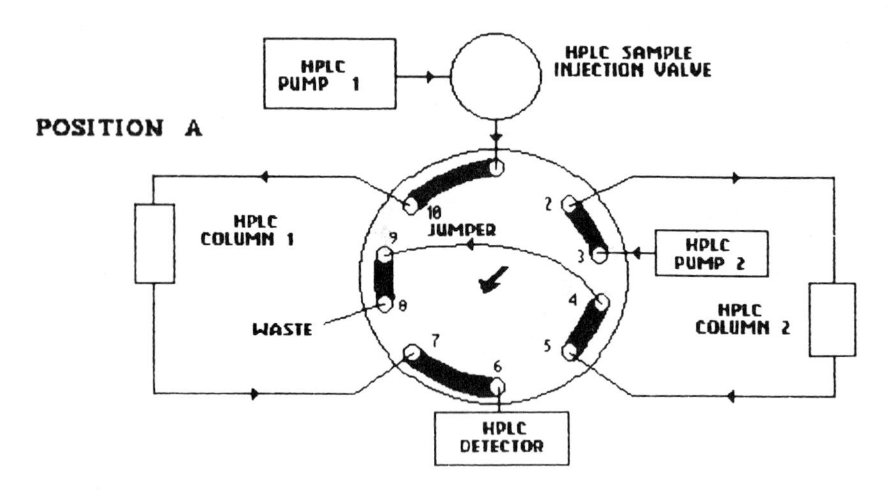

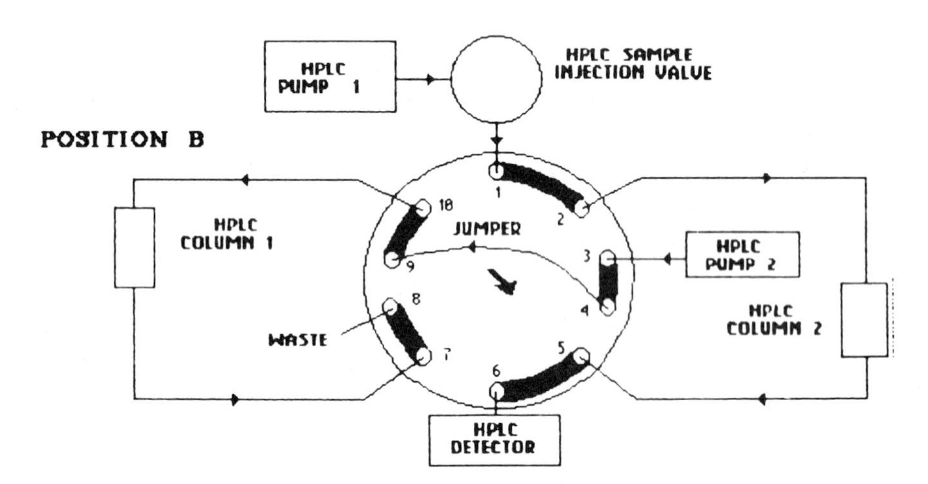

Figure 24. Two-column selection with a 10-port rotary valve. Pump 2 is configured to wash the off-line column.

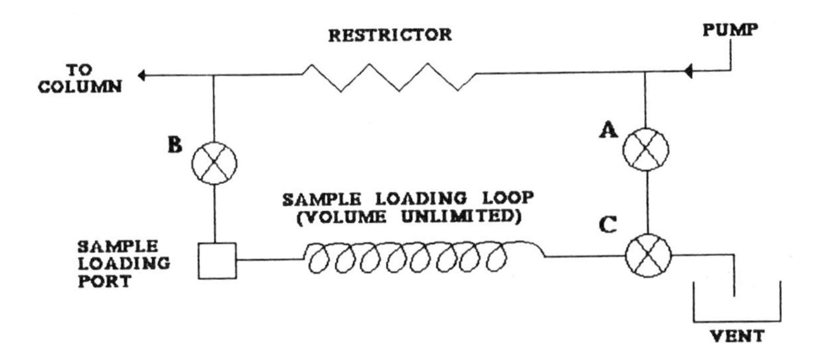

Figure 25. Schematic diagram of a universal injector.

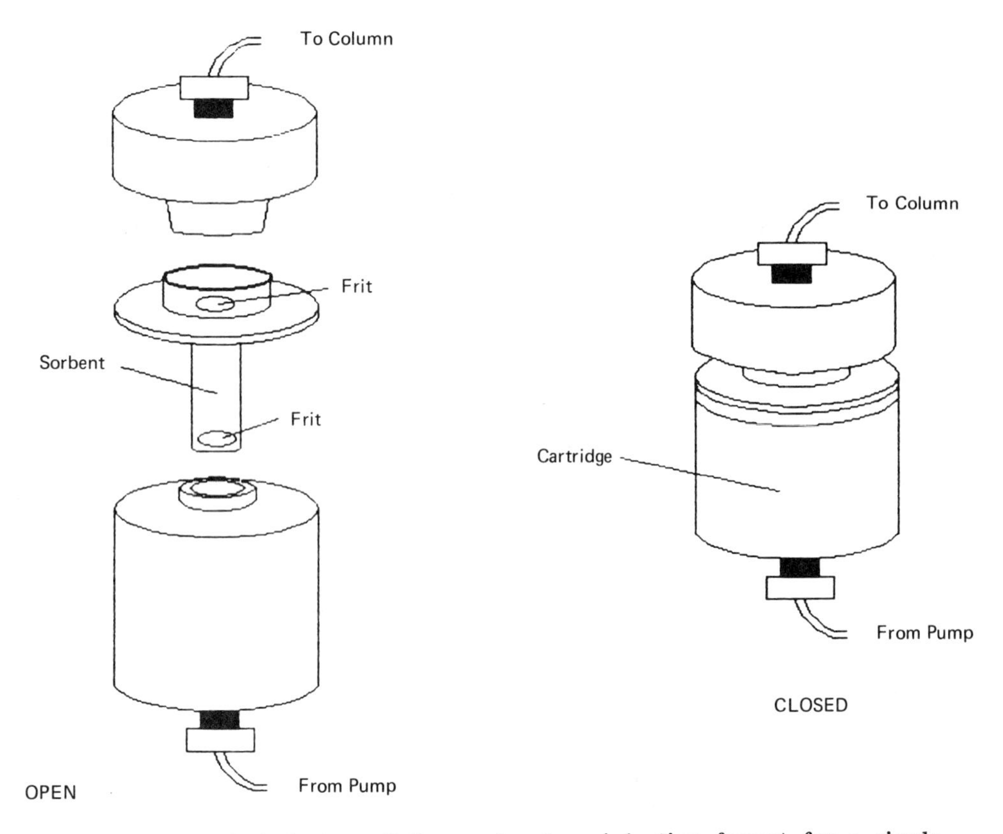

Figure 26. Exploded view of the syringeless injection format for a single-sample cartridge.

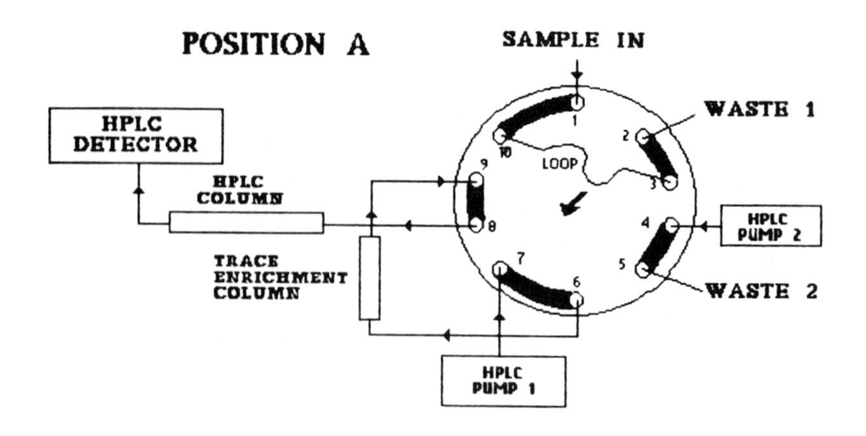

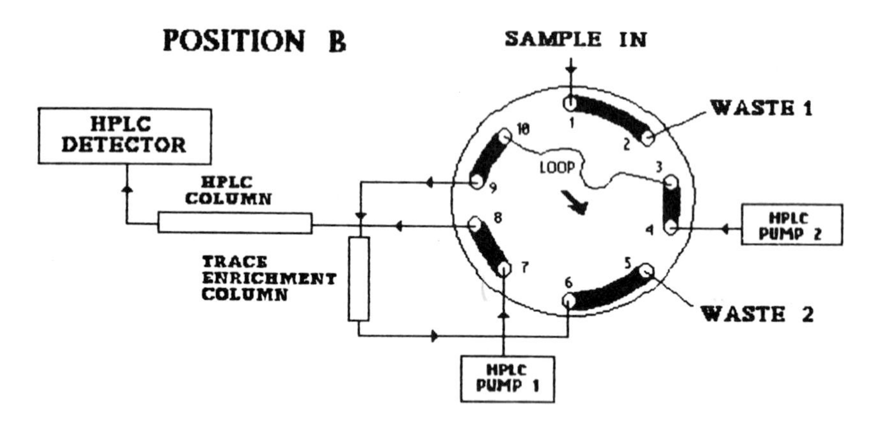

Figure 27. On-line preconcentration sample preparation with a 10-port rotary valve.

the sample loop can be refilled and the valve returned to the load position to allow the concurrent steps of sample 2 preparation and sample 1 analysis.

The integrated and modular forms of sample preparation through SPE both have several important advantages when working with dirty or trace-level samples. One can use a large volume of sample without the extended void response and reequilibration effects of direct injection. In addition, in some cases, a secondary selectivity may be accessible through the use of different stationary phases in pre- and analytical columns. Finally, one is afforded an opportunity to remove analytical interferents or corrosive sample matrix elements selectively through the judicious choice of solvents for washing and loading.

C. Sample Injection Systems

Sample injection systems, frequently called autosamplers, combine a sample injection device with a sample storage/supply system. While a system can be as simple as a siphon feed from a large-volume sample (repetitive or time-based measurements of the same sample), sample storage and supply systems usually take the form of many small, discrete storage vials arranged in a carousel or rack. The instrument mechanically moves the rack until the target vial reaches a designated spot where a needle penetrates the vial and delivers sample to the injector. After the injection is completed, the carousel is then indexed to the next sample position. Common features of sample introduction systems are multiple injections per vial, needle wash (to prevent cross-contamination of samples), and injection mark signal output.

Sample injection systems basically fall into two distinct types: dedicated autoinjectors and research-grade autosamplers. Autoinjectors are an economical way to make serial, unattended injections. They typically use fixed-loop rotary-valve injection in an overfill mode. Samples are stored at ambient temperature, and sample selection is dictated by the order of the sample vials in the carousel. Autoinjectors generally have minimum sample-volume requirements of 500 μl and from this can inject samples of 0.2-200 μl. For automated injection performance and precision, these instruments are unsurpassed.

Autosamplers, on the other hand, are capable of the same automated injection performance as the autoinjectors, with some marked advantages. Many autosamplers now offer optional sample storage under refrigerated, ambient, or even elevated temperature conditions. Sample selection can be random access, and sample retrieval almost exclusively uses syringe withdrawal (no sample waste) with introduction by either partial or filled-loop injection from either rotary valve or a universal injector. The injection volume is programmable and variable between vials and the sample volume overage requirements are low (typically 5-10 μl). In addition, recent advances can now incorporate a complex preinjection routine to move aliquots of sample or reagent between vials (with subsequent mixing) to allow for automated dilution, liquid-liquid extraction, or precolumn derivatization.

The sample storage/supply feature of the autosampler has, as the name implies, two functions: storage of sample prior to injection and delivery of the sample to the injection device. Storage conditions are important because the samples can sit for several hours (24 or more) prior to injection. Samples with volatile components can change in volume or composition, and labile samples can degrade. Most autosamplers can be equipped with a refrigerated storage compartment while autoinjectors cannot. One novel solution for autoinjectors is to change the ambient temperature of the entire instrument (not just the sample compartment) by operation from the inside of a compact refrigerator [24]. While not compatible with all types of instruments (those with optical vial sensors), this approach works well for some autoinjectors in our laboratories.

The sample delivery function of the autosampler can take one of three formats: pressurized vial, positive displacement, or syringe withdrawal. Pressurized vial and positive displacement formats are found in autoinjectors and use the overfilled loop injection format. In comparison, the syringe withdrawal format is found in autosamplers and used in conjunction with either partial or overfilled loop injection.

The pressurized vial format uses gas pressure to propel the sample from the vial to the injection loop. A hollow needle with two distinct channels handles both gas and sample simultaneously, as shown in Figure 28.

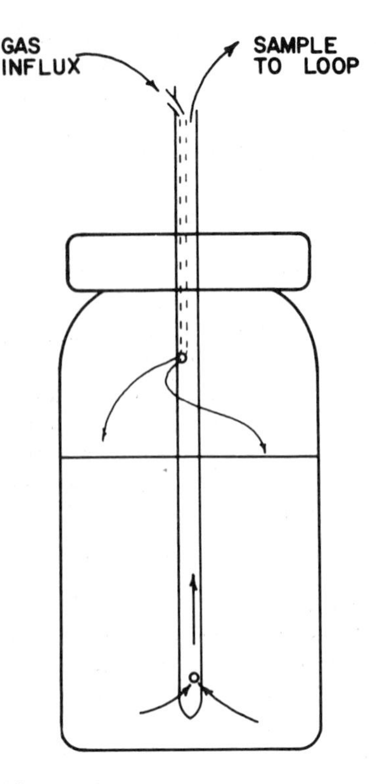

Figure 28. Pressurized vial sample delivery format.

The sample is stored in airtight septum-sealed vials. The needle assembly pierces the septum and the vial is pressurized with gas (either nitrogen or compressed air) from the top hole in the needle, for a preselected duration. This forces fluid out of the vial and into the loop through the bottom hole in the needle. After injection, the needle assembly is depressurized and removed from the sample vial. The pressurized vial format is simple and effective for samples of similar viscosity but samples of variable viscosity can cause problems. In addition, multiple penetrations of the septum can occasionally generate fragments that can clog the needle.

The positive displacement delivery format also uses vial pressure to force sample through the loop. Sample is stored in a vial that is tall and cylindrical in shape. A hollow plastic cap fits tightly in the top of the vial and seals against the sides. To deliver sample, as shown in Figure 29, a hollow plunger/needle assembly descends, pierces the cap and extends into the vial. As the plunger drives into the vial, the vial contents are forced (positively displaced) up through the needle and into the loop. The plunger then remains stationed in the vial between multiple injections to prevent delivery failure due to multiple holes in the vial cap.

The syringe withdrawal format employs a stepper motor driven microsyringe (typically 25-100 μl) to remove sample from a septum-covered vial. The microsyringe, located in an area remote to the sample compartment, is connected to the needle assembly with inert tubing. The needle descends into the sample vial and the syringe draws the preprogrammed volume of sample into the needle, which is then removed from the vial. At this point, there are two different ways to move the sample to the injection site or loop: sample segment transport or hardware transport.

Sample segment transport moves the sample through the internal tubing of the instrument with bubbles or segments of air. After the needle is removed from the sample vial, the syringe draws a bolus of air into the needle, as shown in Figure 30. This bolus moves the sample segment into the injector loop for delivery to the column. Since the bolus has the same volume as the internal volume of the needle and connection tubing, the sample is accurately positioned in the column end of the sample loop bracketed by tiny bubbles of air. After injection, the needle assembly and syringe are rinsed several times with wash solution to prevent sample cross contamination.

Hardware transport involves the physical movement of hardware to bring the injection site and sample-containing needle assembly together. There are two basic approaches to instrumental design for hardware

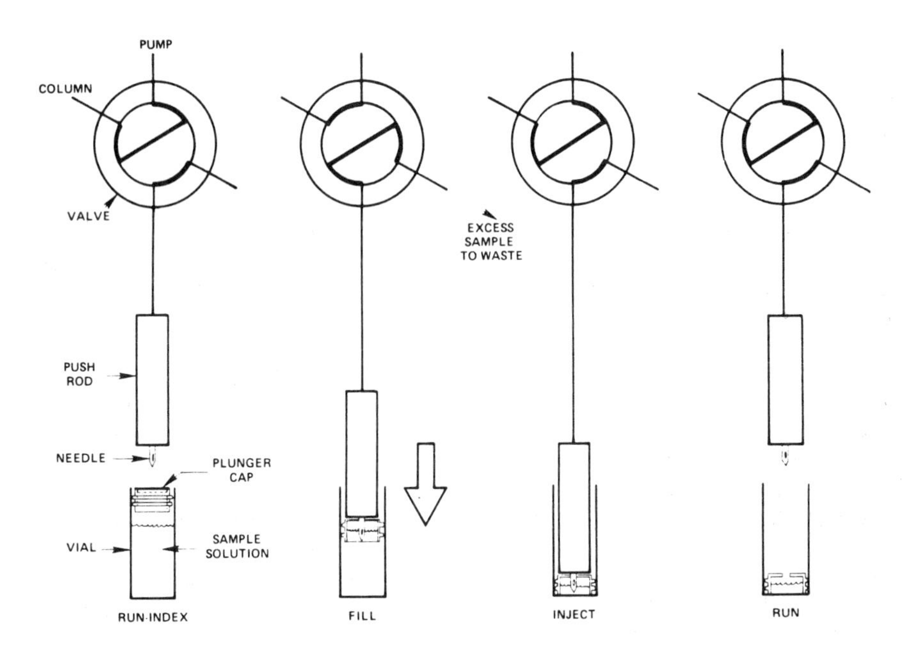

Figure 29. Positive-displacement sample delivery schematic. (Courtesy of Alcott Chromatography Company.)

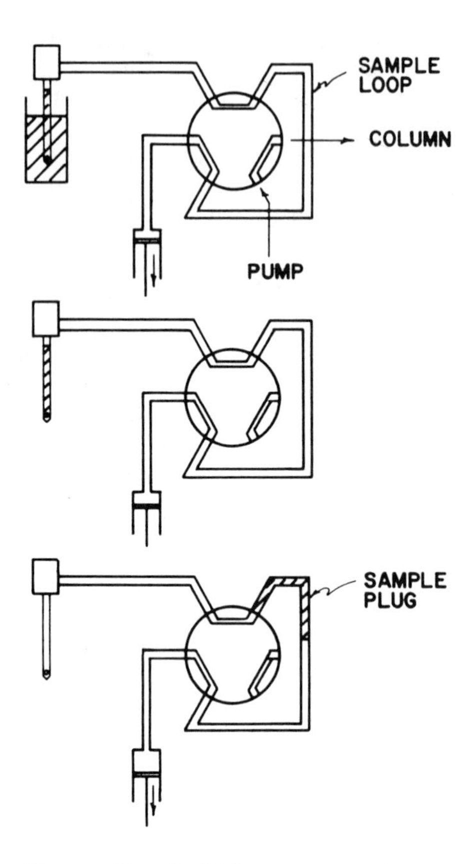

Figure 30. Bubble-segmentation sample transport in syringe-withdrawal sample delivery.

transport. One approach moves the needle assembly from the site of sample retrieval (from the vial) to the site of injection. The other approach, with an example shown in Figure 31, moves the site of sample retrieval to the site of injection. In this case, the injection needle forms a leak-tight high-pressure seal with an inert polymer seat. The needle assembly is lifted and the sample vial is mechanically moved to a position immediately above the seat. The needle assembly is then lowered into the sample vial and, once drawn into the needle, the sample remains stationary while the vial is moved and the needle is reseated. The rotary valve then turns, connecting the mobile phase with the needle assembly for delivery to the column. Postinjection flush is a continuous process since the mobile phase passes through the needle for all but the load cycle.

Syringe withdrawal, regardless of the sample transport method, is the most versatile of the sample delivery formats. The low overage requirements are sample conservative, allowing for the injection of small volumes of sample with little or no waste. The injection volume selection is controlled by stepper motors with high resultant accuracy (error $<$ 1.0 percent at

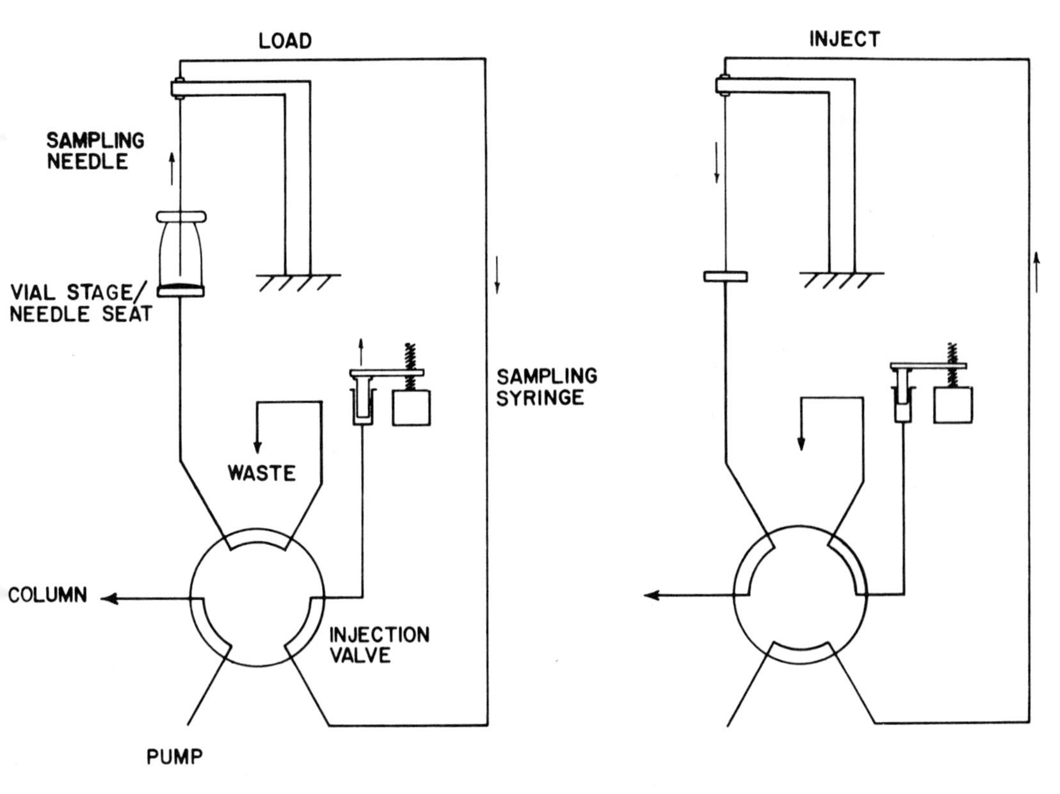

Figure 31. Schematic view of hardware-transport sample movement for syringe-withdrawal sample delivery.

10 μl) and precision (rsd < 1.0 percent for 10 μl). In addition, there is a programmable slow-fill cycle to compensate for viscous samples.

V. DETECTORS

In a very real sense, liquid chromatography can be considered a hyphenated technique, in that separation is only one component of the analytical process. The other components, quantitation and less frequently identification, are achieved by placing an appropriate detector after the chromatographic column. One can regard the separation process as being a method of sample preparation and cleanup before the analytical measurement is made at the detector. The purpose of the detector is to measure the analyte concentration in the mobile phase and to produce an electrical signal whose magnitude is proportional (preferably linear) to the solute concentration. Since separation by LC is a dynamic process, the detectors should work on-line and produce an instantaneous record of the column events. The detector must be able to monitor a separation without influencing the extent of resolution. The separation and detection processes must be considered concurrently in the development of an optimum analytical system. Poor chromatographic performance with excellent detection is as limited in application as is the situation in which outstanding separation is coupled to an insensitive detector.

While research aimed at the development of a truly universal detector for HPLC has been vigorously pursued since the inception of this technique in the early 1960s, it is fair to say that at this time the goal has not been achieved. The general problem encountered in HPLC detection revolves around the fact that quite often the physical properties of the solute and the mobile phase overlap. Detectors designed to overcome this problem can be classified into three broad categories:

1. Bulk property detectors that measure quantitative changes between the mobile phase and the solute within a given physical property
2. Detectors that sense a property specific to the solute against a negligible (or correctable) mobile-phase background
3. Transport detectors that remove the carrier phase and then measure some general solute property

A classification of currently available HPLC detectors based on these distinctions is shown in Figure 32; while the list is not exhaustive, in general most detectors suitable for routine application appear therein.

A. Performance Criteria for HPLC Detectors

Requirements for the "Ideal" Detector

Given the nature of the chromatographic process, there is a wide variety of properties that a detector must possess in order to be effective. A comprehensive list describing the requirements for an "ideal" detector would include the following constraints:

1. Universal response, applicable to all analytes
2. High sensitivity
3. Negligible noise
4. Large linear dynamic range
5. Response independent of variations in operational parameters (e.g., temperature, pressure, flow rate)
6. Response independent of mobile-phase composition
7. Low dead volume and response time
8. Nondestructive
9. Convenient and reliable to operate
10. Inexpensive (capital and operating costs)
11. Stable response over long periods of operation
12. Capable of providing qualitative information about the solute (e.g., identity)

At the present time, no HPLC detector is capable of meeting all (or even most) of these criteria. In a practical sense, requirements 1, 8, and 12 are usually considered to be ideals for which researchers continue to strive and do not usually dictate detector choice.

To be able to select a specific detector (type and/or vendor) for a particular laboratory environment or application, it is preferable to compare performance based on a uniform set of criteria. Such criteria include response (dynamic range), noise level, solute sensitivity, total dispersion, and sensitivity to operation conditions. Each of these will be considered in greater detail below.

Response

The chromatographer should be aware that detectors commonly have two specifications that deal with their ability to respond to the presence of a solute species. The dynamic range (D_r) is the analyte concentration range over which the detector will produce a signal whose magnitude is concentration dependent. At the low end of this range is the detection limit, while the high end is that point at which the detector becomes saturated. The dynamic range is usually expressed in terms of orders of magnitude of solute concentration and is thus dimensionless. The linear dynamic range (D_l) of the detector is that range of analyte concentration over which the detector response is linear. Typically this range is approximately one-tenth of the dynamic range and is expressed in similar units. The value of D_l can be determined by considering the equation

$$y = AC^r \qquad (1)$$

where y is the detector output, C is the analyte concentration, A is a constant, and r is the response index. A plot of the logarithm of response versus the logarithm of concentration has a slope of r; in the ideal case where the response is truly linear, r will have a value of 1. In practice, an r value of between 0.98 and 1.02 is considered to indicate a linearly responding detector.

Noise

Detector noise is here defined as any disturbance or change in detector output that is not related to the elution or presence of a solute. This characteristic of a detector is extremely important, since it impacts a variety of other operating characteristics including stability and sensitivity. Detector noise has somewhat arbitrarily been divided into three types based on the frequency of the disturbance (Figure 33). Short-term, or high-frequency, noise consists of signal fluctuations that have a frequency similar to or slightly lower than the analyte peak. Typically, a frequency of 10 cycles per minute is used to distinguish between these two types of noise. Of the two, long-term noise is of the most immediate concern since this type of noise limits the analyst's ability to even identify a chromatographic response while short-term noise limits the sensitivity and accuracy with which a response can be quantitated. Thus, short-term noise will not obscure an analytical response. Sources of short-term noise include the electronics associated with the detector/recorder (incorrect grounding, response set too fast, gain set too high, Schottky effect on electronic components) and pump pulsations and can usually be reduced

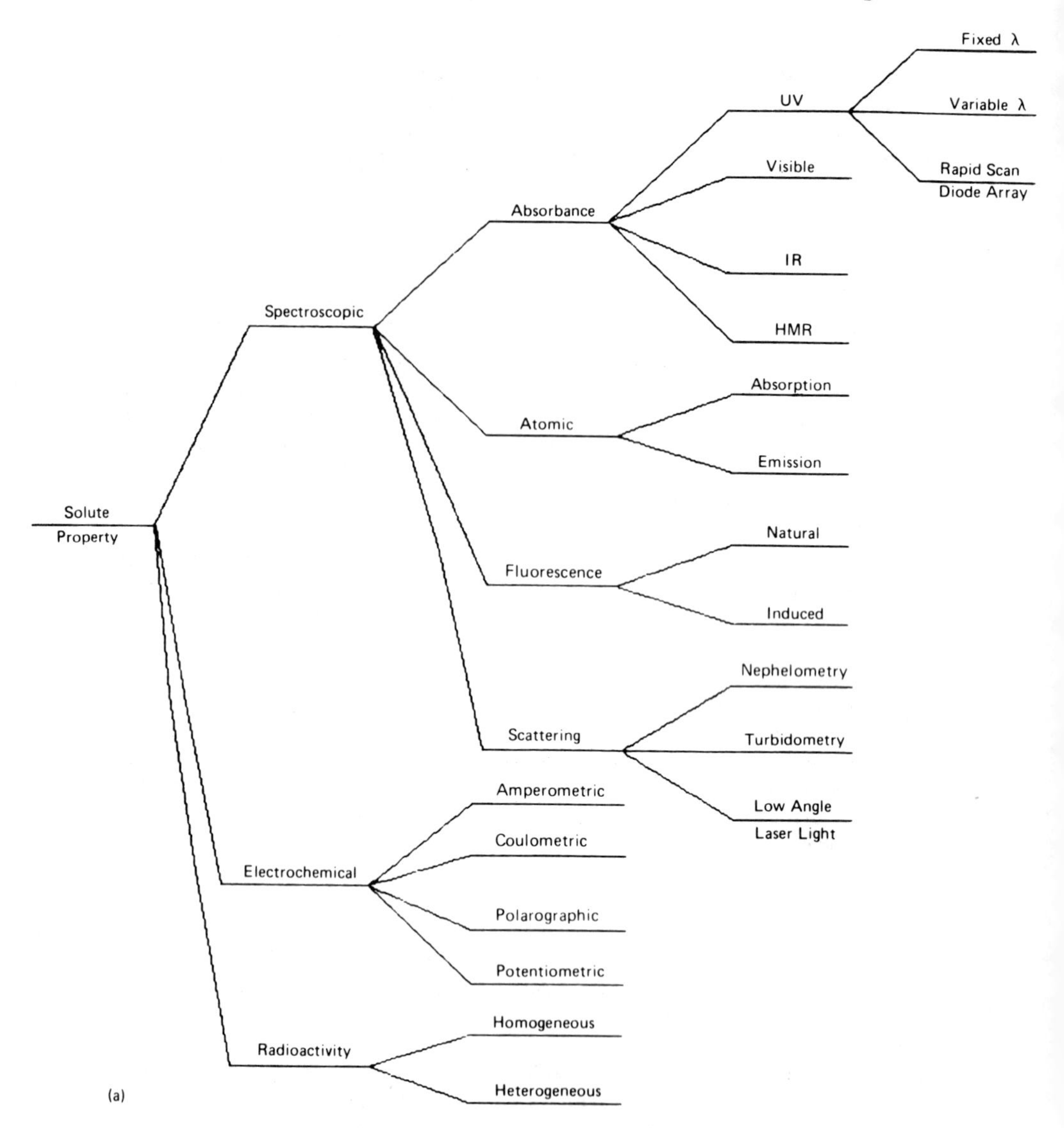

Figure 32. Classification of common HPLC detectors by (a) solute property

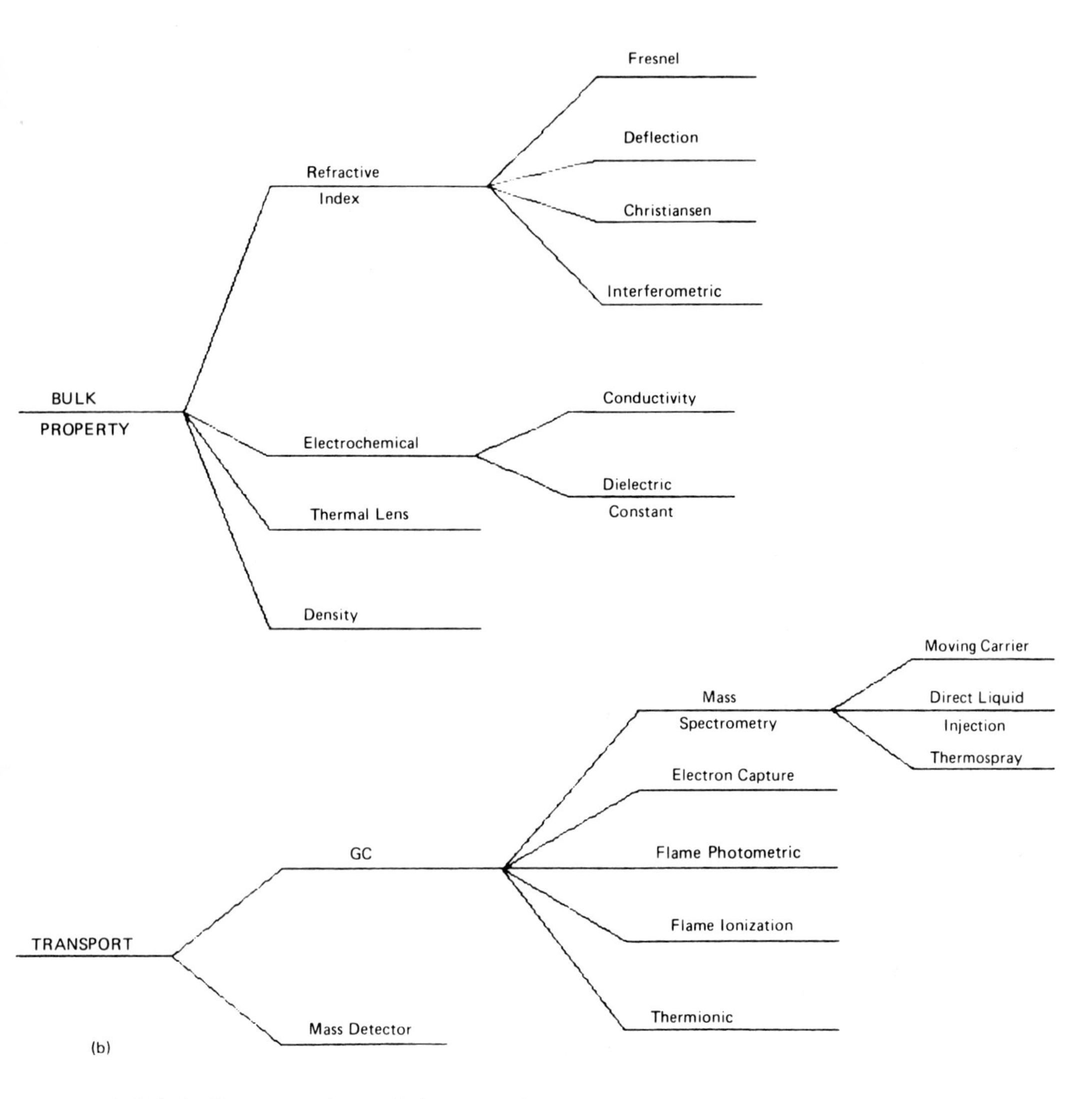

and (b) bulk property and transport.

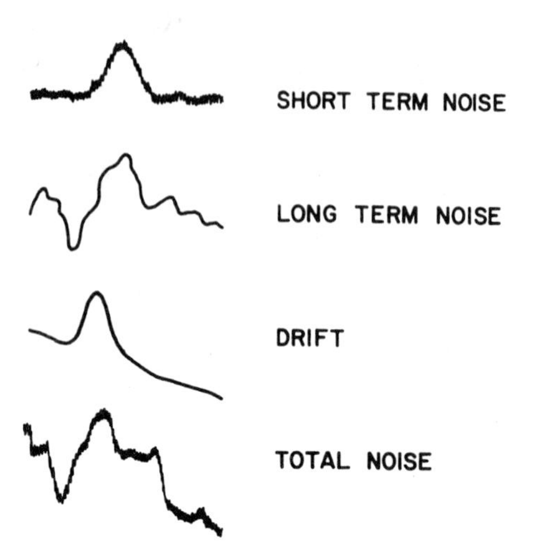

Figure 33. Various types of detector noise.

in magnitude by the use of appropriate noise filters. Sources of long-term noise include impurities in the mobile phase, column bleed, bubbles in the detector cell, and changes in factors affecting the sensing system of the detector (e.g., component instability, changes in temperature, pressure, etc.). Correction of long-term noise requires identification and elimination of the source, since any filter capable of eliminating the noise would remove the solute peak as well.

The third type of noise has a frequency much less than that of the eluted peak and is called drift. Drift is frequently recognized by a continuous increase or decrease in detector signal. While drift will not obscure a peak nor directly effect its integration, its presence is indicative of a poorly operating system and will require that the baseline be adjusted periodically. Drift can result from slowly changing detector performance (i.e., power level and lamp aging) but is more often the result of nonequilibrium conditions such as changes in ambient temperature or in mobile-phase composition.

All types of noise should be measured, under typical operating conditions, from strip-chart tracings of the detector output; noise should be expressed in terms of the variation in the physical property being measured by the detector per unit time of measurement. Clearly the unit of time used must increase inversely with the frequency of the noise.

Solute Sensitivity

Detector sensitivity may be defined as the minimum concentration of solute that can be discerned from the noise. Two expressions that are commonly used to define this quantity are limit of detection (LOD) and noise equivalent of concentration (C_n). Both are measured with respect to the level of the short-term noise; the limit of detection is the concentration of the solute that will produce a signal equivalent to twice the peak-to-peak noise, while C_n is that concentration of solute that produces a signal equivalent to the short-term noise. It is important to note that the sensitivity of a detector is not equal to the minimum mass of solute that can be detected. This latter property is dependent on chromatographic dispersion and is thus a property of a complete chromatographic system and not solely of the detector.

Total Dispersion

The best detection system from a viewpoint of providing highest resolution and sensitivity will be one that enables detection of an eluting component in the smallest possible volume, and that does not increase analyte dispersion by its configuration. Both these parameters (cell size and configuration) are important to consider in identifying an effective detector.

Sensitivity to Operating Parameters

It has been shown previously that a detector's sensitivity to external system variables (temperature and pressure) affects its long- and short-term noise level. The detector's sensitivity to these variables, expressed in terms of response change per unit change in operating variable, should be known prior to the identification of an appropriate detector. Of somewhat lesser concern is the nature of the detector's response to changing mobile-phase composition. This factor is primarily of importance when gradient elution separations are anticipated.

General Comments

It is clear from a review of the recent chemical literature that great care must be taken when comparisons concerning the operating characteristics of various detectors are made. Other than an ASTM procedure for verifying the performance for fixed-wavelength photometric detectors [25], there are no universally accepted definitions for performance criteria and quite often their units are not standardized. Instrument manufacturers have a poor reputation in terms of supplying truly useful performance data. This situation may be due not only to the obvious competition that exists in the instrument marketplace but also to the diversity of operating conditions used and physical properties measured in liquid chromatography. For example, UV sensitivity is clearly wavelength dependent while the sensitivity of a refractive index detector will change as a function of the refractive index of the solvent. While Scott [26] suggests that quinine sulfate in ethanol is an appropriate solute for use in comparing detector performance (it has a refractive index that is significantly different from most solvents, it is ionized and thus conductive in solution, it has a significant absorbance and fluorescence in the UV), no generally accepted means for comparing detector performance has been recognized. Potential purchasers of HPLC detectors are cautioned to understand their own requirements and the conditions under which a particular detector's specifications were obtained (and are thus valid) before comparisons between detectors are made in earnest.

B. Bulk Property Detectors

Bulk property detectors continuously monitor some property of the mobile phase and through the use of relatively simple electronics provides a voltage-time output whose magnitude is proportional to either the physical property being measured or the concentration of the solute. Any eluted solute will be detected by a bulk property detector provided that the magnitude of the physical property being monitored differs significantly between the solute and the mobile phase. It is this characteristic of the bulk property detector that is its greatest strength ("universal" applicability) and its most important weakness (sensitivity to changing or poor operating conditions). An additional drawback of this class of detectors is a somewhat limited sensitivity, which is due to the small differences in bulk properties of solutes and mobile phases commonly of interest in LC. Properties of the mobile phase most commonly monitored in commercially available bulk property detectors include, in order of popularity, refractive index, electrical conductivity, dielectric constant, and density.

Refractive Index Detectors

Most organic compounds of interest in HPLC (as either solutes or mobile phases) have a relatively small range of refractive indices (RI). This range can be exploited to provide a useful, if not overly sensitive, method of detection. Since refraction is a property shared by both the solute and the mobile phase, most RI detectors work on a differential principle in which the refraction of the mobile phase in a reference cell is compared to the refraction of the mobile phase plus the solute in the sample cell. There are four general methods of measuring refractive index which are applied in commercially available instruments. These include the Fresnel (reflectance) method, the interferometer method, the deflection (angle of deviation) method, and the Christiansen effect method, each of which will be described in somewhat greater detail below.

a. Fresnel (reflectance) method

The Fresnel refractive index detector responds to a change in the amount of light transmitted through a dielectric interface between the surface of a glass prism and the liquid being monitored; this change occurs when the refractive index of the liquid being monitored (the mobile phase) changes (due to the presence of the solute). A generalized optical schematic diagram of this type of detector is shown in Figure 34. Light from the tungsten source passes through an infrared filter (to prevent heating of the cell) and into a lens assembly that splits the light into two beams, which are focused respectively on both the sample and reference portions of the flow cell. The prism assembly emits the refracted light through a set of focusing lenses to a dual-element photocell, which is in turn connected to the appropriate amplifying electronics. The difference in the intensity of the light reflected in the two beams is related to the refractive index difference between the sample and reference cells by Fresnel's law. This type of design is the basis of detectors offered by both Perkin Elmer (Norwalk, Connecticut) and Laboratory Data Control (LDC) division of Milton Roy (Riviera Beach, Florida).

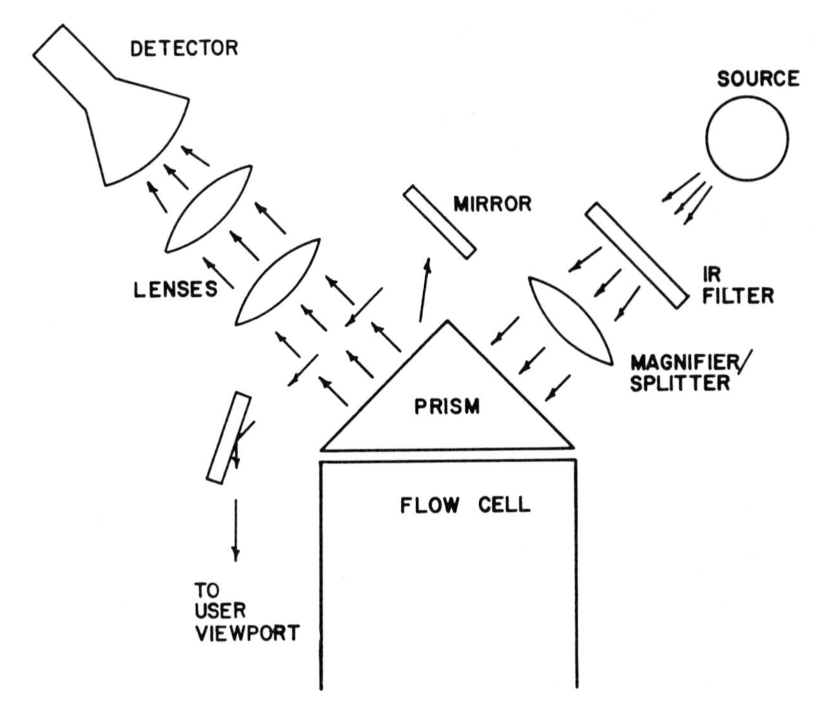

Figure 34. Schematic diagram of a differential refractometer.

b. Interferometer method

This type of detector measures the difference in the speed of light in a sample and reference cell by means of the interference of two light beams which have passed through the cells. As shown in Figure 35, light from the source (tungsten) passes through a beam splitter wherein it is divided into two orthogonally polarized beams of equal intensity that are made parallel to the optical axis of the detector by a lens placed in front of the flow cells. The beams pass through the respective cells and are focused by the second lens. The two beams are recombined by a second splitter and measured by a photomultiplier sensor. The detector signal is proportional to the amount of energy reaching the photomultiplier, which will be a maximum when the sample and reference beams are polarized in phase (i.e., the refractive index of both cells are the same). As the speed of light in one of the cells is changed, destructive interference occurs and the energy measured by the photocell decreases. Photomultiplier output will follow a sine curve if the refractive index in the two cells changes in a linear fashion as the result of constructive and destructive interference cycles. This type of detector is offered by the Bifok/Optilab division of Tecator AB (Sollentuna, Sweden).

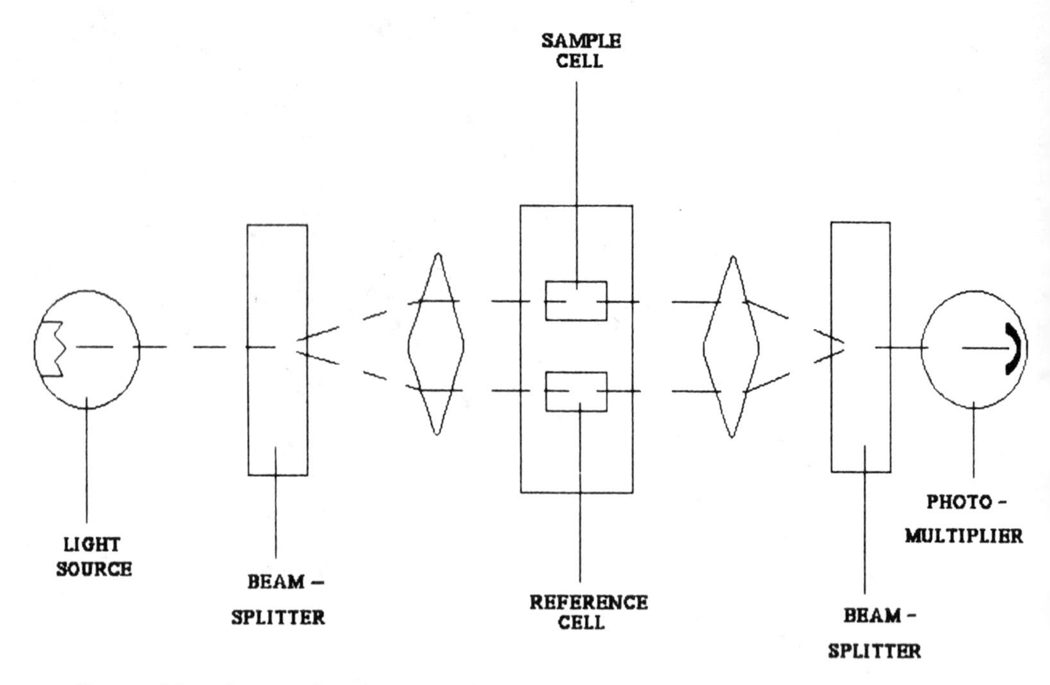

Figure 35. Schematic diagram of an interferometric refractometer.

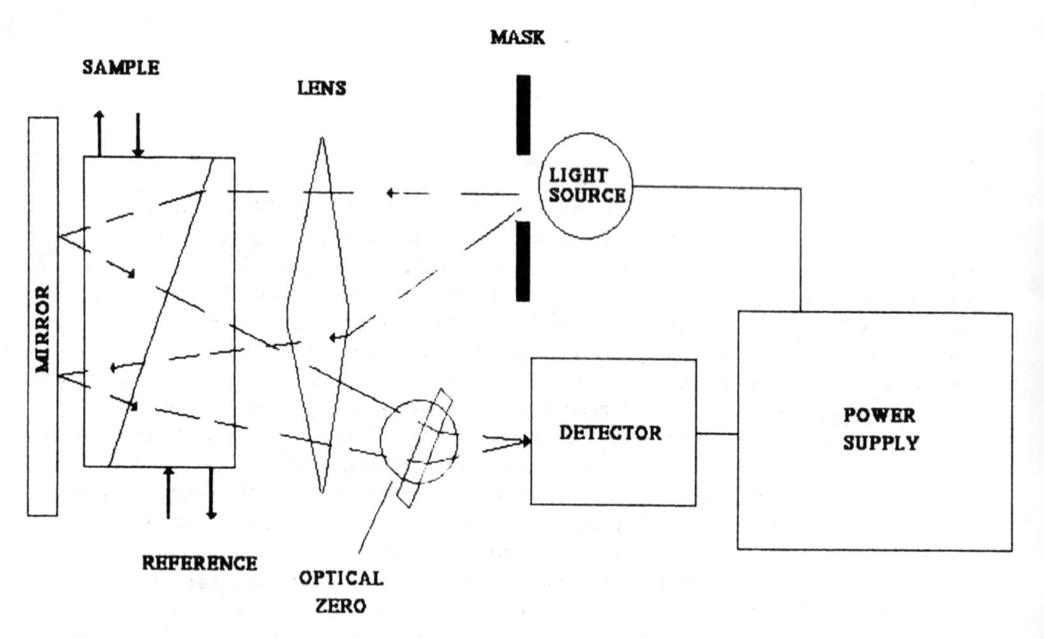

Figure 36. Schematic diagram of a typical deflection refractometer.

c. Deflection (angle of deviation) method

The deflection refractometer measures the bending of a light beam when it crosses a dielectric interface separating two media of differing refractive indices at a nonzero angle of incidence. In an HPLC detector this concept is embodied in a prismatic flow cell. A ray of light passing through the cell is refracted from its original path and focused onto a photocell. As the nature of the mobile phase changes (i.e., the refractive index changes due to the presence of a solute) the angle of refraction will change according to Snell's law and thus the amount of light falling onto the photocell will be changed and its output modified. An optical schematic of this type of detector is shown in Figure 36. The light from the source passes through an optical mask (which is used to confine the beam to a given region of the detector), a collimating lens, and the reference and sample cells. A mirror then reflects the beam back through the cells to the lens, which focuses the beam onto the photocell. The most common commercial form of this detector, manufactured by Waters Associates (Milford, Massachusetts), includes an optical zero control that consists of a rotatable glass plate, which provides for the lateral adjustment of the position of the light beam.

d. Christiansen effect method

This type of detector is based on the observation that light is transmitted directly through a mixed-bed cell containing both a liquid phase and solid particles only if the refractive index of the fluid is the same as that of the particles. As the refractive indices of the two phases diverge, multiple reflections and refractions of the light will occur at the fluid-solid interfaces with the net result that a portion of the light will be scattered away from the incident beam and thus that the intensity of the transmitted beam will be reduced. A detector based on this concept is manufactured commercially by the Gow-Mac Instrument Company (Bound Brook, New Jersey) and has a generalized optical diagram as shown in Figure 37. Light from an appropriate source is concentrated onto an aperture, made parallel by an achromatic lens, and split into two parallel beams by a double prism. The two beams pass through the sample and reference cells, respectively, and then go to separate photocells. It is noted that since the refractive indices of the solid particles and the fluid are matched over only a narrow range of wavelengths, the detectors are usually wavelength specific.

e. Relative performance

It is not the purpose of this section to discuss the relative performance of commercially available detectors in quantitative terms; a comprehensive (if somewhat dated) review of such information is provided by Munk [27]. Improvements made since that work are derived more from attention to mechanical detail and good thermal insulation than from fundamental changes in detector design. However, a few comments concerning each type of RI detector is warranted. The Christiansen detector is typically quite inexpensive and has low dispersion; however, its major disadvantages include a limited linear dynamic range (when compared to the other types) due to the complexity of the detection method and the necessity to change cell packing in order to deal with mobile phases of differing refractive

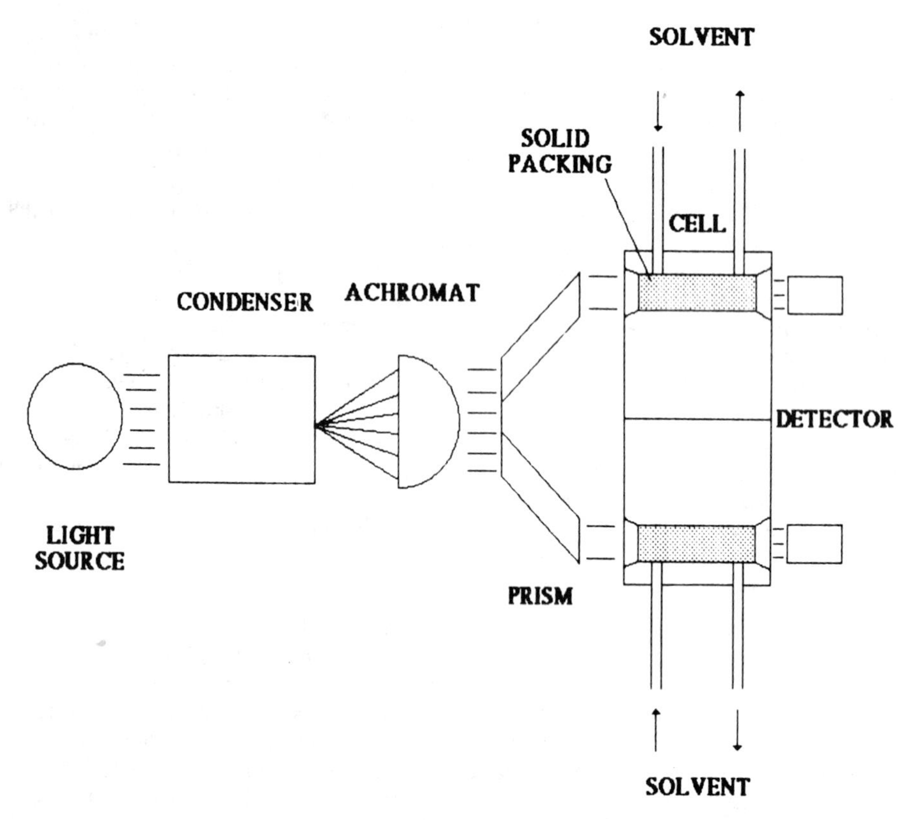

Figure 37. Schematic diagram of a Christiansen-effect refractometer.

index. Deflection detectors, in addition to having a wide linear range, typically are extremely rugged. In this design the entire range of useful refractive index can be measured using a single prism, small volume cells are readily produced, and the cells themselves are moderately insensitive to air bubbles or buildup of contaminants on the windows. Usually, however, the cell used in the deflection detector is larger than those used in the other methods and therefore this detector suffers from increased extracolumn dispersion. The Fresnel detector has the advantages of high sensitivity, compatibility with low flow rates, ease of cleaning, and availability of small-volume flow cells, but suffers from the disadvantage of being instrumentally more complex than the other designs. This complexity is reflected in the need to use two prisms to cover the full range of refractive index and the need to use a flowing reference cell to avoid pressure related changes in cell geometry. Finally, the main advantage of the interferometer design is increased sensitivity (nearly an order of magnitude better than the other designs).

It is important to comment on the effect of operating variables on the performance of RI detectors in general. While such detectors are relatively insensitive to pressure changes (a change in pressure of 1 atm typically produces a change of 10^{-6} RI units), temperature has a profound effect on refractive index. Typically, a change of 0.001°C causes a change of 10^{-6} RI units. In light of this dependence on temperature, commercial detectors normally are fitted with an internal heat sink, which often takes the form of a length of narrow-bore tubing placed in the inlet side of the detector cell. While a typical heat sink may provide temperature control to ±0.0001°C in a well-designed system, it does contribute 20-100 μl of dead volume to the chromatographic system.

Electrical Conductivity Detector

Acids, bases, and salts dissociate in aqueous media to form ionic species, which, in the presence of a voltage applied across two electrodes, will conduct an electric current. From Ohm's law it is observed that the electrical resistance between the two electrodes is equal to the applied voltage divided by the current. Thus as the ionic content of a mobile phase flowing past the two electrodes changes, so will the electrical resistance, and hence this property can be exploited in HPLC detection. Prior to 1975, conductivity detection of ionic solutes in HPLC mobile phases that were suitable for the separation of ions was extremely difficult because these mobile phases themselves were highly ionic and thus produced a high background response. It was not until the development of low-capacity ion-exchange resins and a means for reducing the background conductivity of the mobile phase that the development of the technique termed ion chromatography was made possible. Exploiting the first development, Gjerde and associates were able to produce effective ionic separations with mobile phases of low conductance (usually weak organic acids) [28]. In this approach, termed single-column or nonsuppressed ion chromatography, the background conductance of the mobile phase is sufficiently low that the separation could be monitored with more or less conventional conductivity detectors. Using the concept of mobile-phase suppression, Small and associates were able to produce chromatographically significant separations using more conventional eluents/resins [29]. In suppressed-conductivity detection, the mobile phase is chemically modified in such a way so as to reduce its ionic content without affecting the solute. To achieve this process, the eluent passes through a suppressor after it leaves the column but before it reaches the detector. Essentially, the suppressor acts as an ion-exchange medium for the counterions of the mobile phase. For example, as shown in Figure 38, a typical eluent used for the separation of anions might be the salt of a weak acid generalized by the notation Na_2P. In this situation, the P^{2-} and HP^- ions act as pusher species while the sodium ion contributes nothing to the separation process. In the suppressor, the mobile phase and a regenerant solution are separated by an ion-exchange barrier. For anion determinations, the regenerant is a strong acid (HA) and the barrier is permeable only to cations. Thus, as the mobile phase and regenerant flow countercurrent to one another, sodium in the mobile phase is replaced with hydrogen ions from the regenerant. The net effect of this process is to convert the mobile phase into a nonionic solution as the weak acid becomes completely protonated. Clearly the analyte must have a pK_a lower than the eluent weak acid so

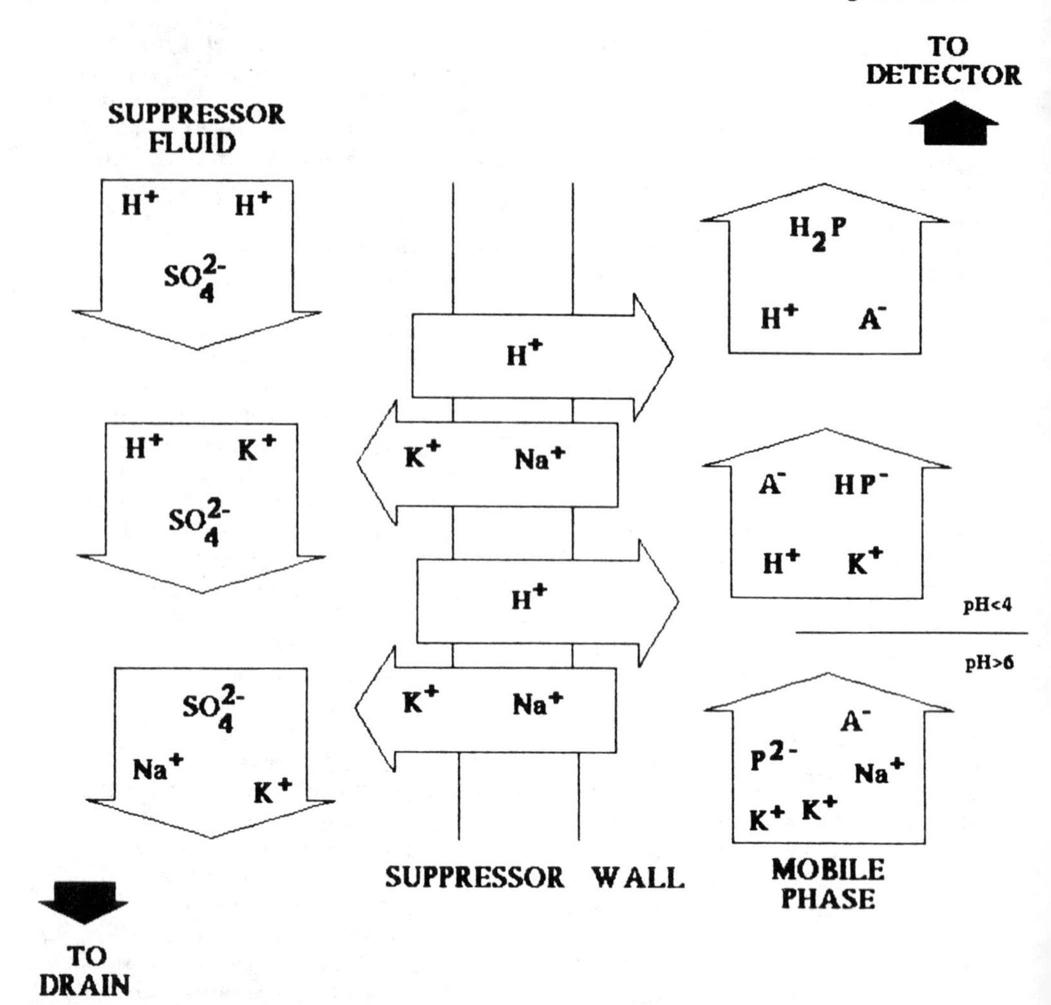

Figure 38. Operation of the suppressor in conductivity detection. K_2P is the weak-acid eluent, A^- is an analyte, and H_2SO_4 is the regenerant. The regenerant solution is shown to the left of the suppressor wall, the mobile phase to the right.

that it remains charged (and detectable) throughout the course of the suppression process.

The relative merits of the nonsuppressed versus suppressed detection systems remains a topic of considerable discussion among proponents of either system. Manufacturers of the nonsuppressed methodology identify the cost and relative complexity of the suppressor as major disadvantages. Additional drawbacks of this method include a slightly lower sensitivity in the analysis of alkali and alkaline earth metals, some minor dispersion that occurs due to the suppressor void volume, and a potentially smaller linear dynamic range. Strong points of the suppressed method include improved sensitivity in anion analysis, freedom from certain types of matrix interferences, use with gradient elution, and versatility. In a very real sense, these two methods should be considered as complementary approaches for the quantitation of ionic species by HPLC.

A block diagram of a simple form of the conductivity detector is shown in Figure 39a. In its simplest form, the conductivity flow cell consists of a small (5 μl or less) chamber fitted with two electrodes (Figure 39b). While such a configuration has the advantage of simplicity, a major disadvantage involves baseline instability. Because a single electrode acts as both the reference and detection electrode, the cell will produce an electrical current as the conductivity is being measured. Current in the flow cell results in an uneven distribution of ions in the solution and prevents the accurate measurement of the resistance between the electrodes. This problem can be circumvented using the multiple-electrode approach

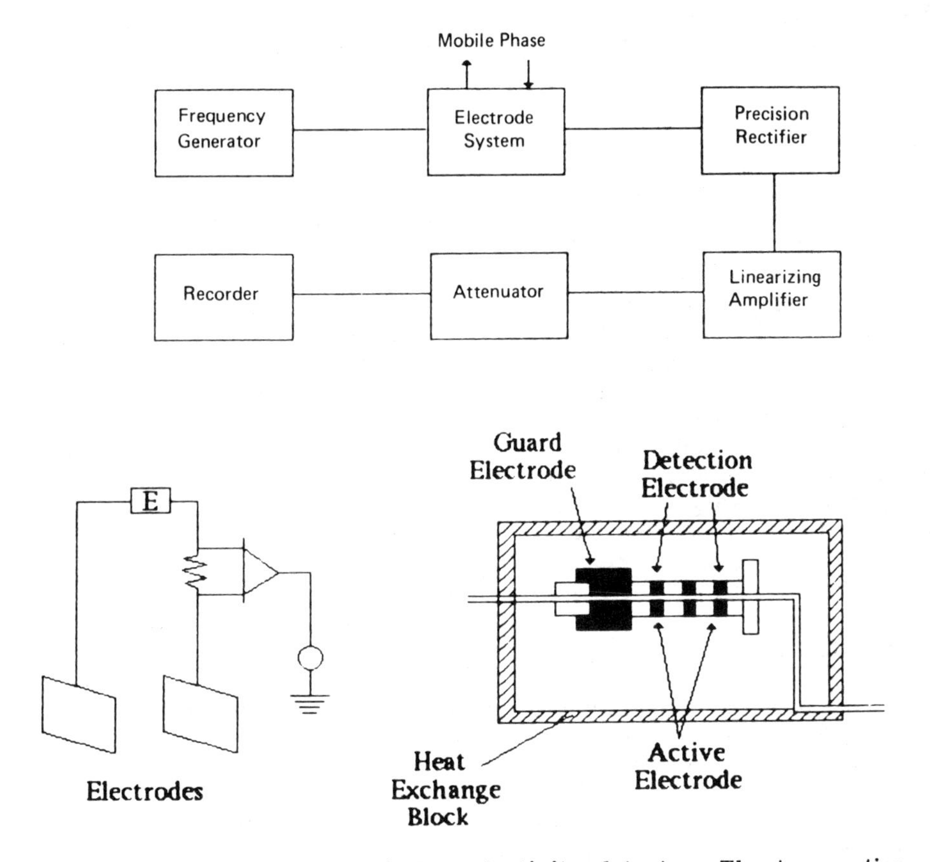

Figure 39. Details of a typical conductivity detector. The top portion shows a generalized block diagram applicable to all common detector designs, while the bottom portions show flow cells based on two- and five-electrode configurations, respectively.

shown in Figure 39c. In this case the flow cell contains two reference and detection electrodes and a guard electrode that serves as an electrical ground. This design prevents the current-related impedance problem. Regardless of cell design, the system operates in the following manner. An AC voltage is supplied across the electrode system in a resistance network that can take various forms (the most common of which is the Wheatstone bridge). In this form, the electrode pair takes the place of one resistance arm of the bridge while another arm can contain an adjustable resistance to permit zero control or can contain a reference cell to allow for operation in the differential mode. The resulting signal is passed to a precision rectifier, which produces an output proportional to the change in resistance of the cell. Since the concentration of ions in the cell is inversely proportional to the cell resistance, a linearizing amplifier is used to provide a concentration-dependent output.

When properly used, the conductivity detector is a reliable and sensitive means of monitoring HPLC column effluents for ionic solutes. A major drawback of the detector is its high temperature sensitivity; the temperature coefficient of conductivity is approximately 2 percent per degree Celsius [20]. It is clear that effective operation of such a detector requires careful control of temperature.

Dielectric Constant Detectors

When a conducting material is placed in an electric field, electrons are displaced from their equilibrium position and a separation of positive and negative charges (polarization) occurs. If a dielectric is placed between two plates of a capacitor, the capacitance is increased (due to the polarization) by a factor called the dielectric constant (ε). Since all materials possess a dielectric constant, this property can be used as the basis for a chromatographic detection method.

The most commonly employed detection circuit for the measurement of changes in dielectric constant involves the use of an AC bridge in which the detector cell is situated in one arm of the bridge. The flow cell takes the form of either a cylindrical or parallel plate condenser. The type of bridge used depends on the cell design and total capacitance. If the capacitance of the cell is large (greater than 100 pF) a Wein bridge is used; if the capacity is small (1-10 pF) a Schering bridge is a more appropriate choice (Figure 40). Both of these circuits can accommodate a reference cell in another arm of the bridge so that differential operation of the detector is possible.

A more recent method used for monitoring changes in dielectric constant is the heterodyne system. In this design, the cell is made part of an oscillator circuit. Changes in the dielectric constant will produce a change in the frequency of this circuit. A mixer is used to compare the variable frequency of the detector oscillator with the fixed frequency of a reference oscillator. The resultant difference signal thus will monitor variations in dielectric constant experienced by the mobile phase.

In many ways the RI and dielectric constant detector are nearly equivalent in performance. The sensitivity of the two is similar, and while the dielectric detector has a somewhat larger linear dynamic range, the range extends into relatively high solute concentrations that are of little practical use in analytical HPLC. Neither type of detector is suited for gradient

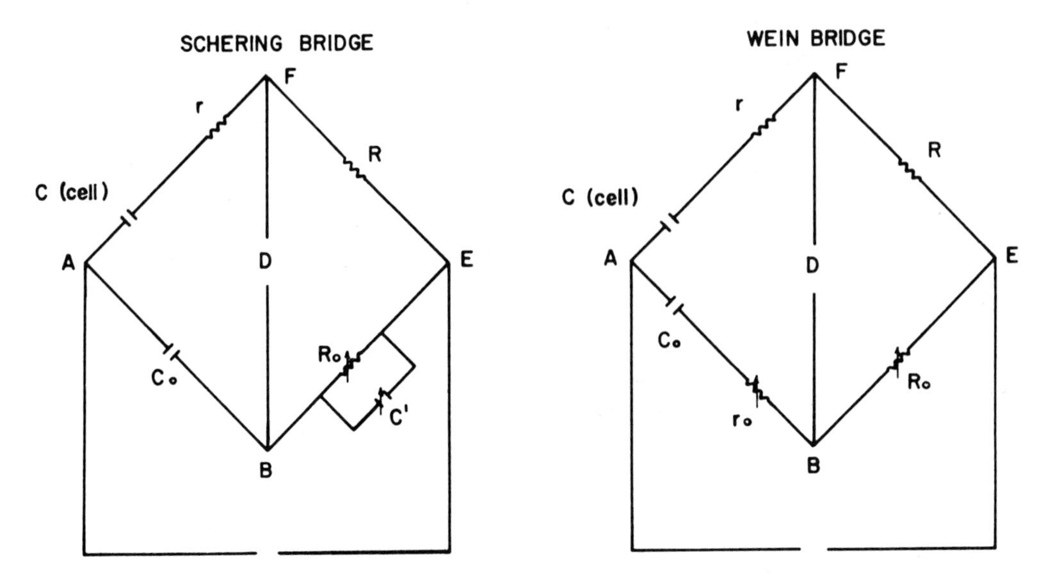

Figure 40. Typical bridge configurations used in a dielectric-constant detector.

use; however, the dielectric detector can be used in conjunction with flow programming. The response of the dielectric detector is more uniform among a group of compounds (e.g., hydrocarbons) than is that of the RI detector, and therefore use of response factors for quantitation with the dielectric detector is not necessary. Both detectors are temperature sensitive. A major disadvantage of the dielectric detector is its large cell size (greater than 20 μl); additionally, its use in reversed-phase applications may be limited due to the relationship between polarity and dielectric constant.

Other Bulk Property Detectors

There are a number of other bulk property detectors that have been described in the chemical literature but that are in limited routine use. The thermal lens detector is based on the effect that focused laser light will have on an absorbing material. As light is absorbed by the material, a localized increase in temperature occurs, a spatial variation in refractive index is produced, and a thermal "lens" is generated within the medium. A commercial detector based on monitoring changes in the viscosity of the mobile phase has been developed by Viscotek Corporation. Density changes in the mobile phase resulting from a dissolved solute can also be used as the basis for detection; however, the sensitivity of this method is extremely poor.

C. Solute Property Detectors

The ideal solute property detector measures some property that is exhibited solely by the solute. This class of detectors contains those that are the most widely used, the most sensitive, the most specific, and those with the largest linear dynamic range; however, these qualities are not found in a single detector. Solute property detectors are more versatile than bulk property detectors in that they can be used with gradient conditions; on the other hand, they are less versatile than the bulk property detector in the sense that they are compatible with a more limited number of mobile phases. Clearly the mobile phase cannot possess the property that the detector monitors. Important subclasses of this detector type to be discussed include spectroscopic, electrochemical and radioactivity monitors.

Spectroscopic Detectors

As used herein, a spectroscopic detector is one in which information (qualitative or quantitative) about the solute is transmitted as an optical signal in the ultraviolet, visible or infrared region of the electromagnetic spectrum. This optical signal is converted to an electrical signal by a radiation detector and is then transmitted to some read-out device. These detectors can be classified on the basis of their principle of detection and include light absorbance, emission, fluorescence or scattering.

a. Absorbance

Detectors based on the absorbance of light are governed by the Beer-Lambert law:

$$A = \log(P/P_0) = abC \tag{2}$$

where A is the absorbance, P_0 and P are the intensity (or power) of the light entering and leaving the cell, b is the optical path length, a is the molar absorptivity of the solute, and C is the solute concentration. This relationship is valid regardless of the wavelength of the radiation absorbed, but both A and a are wavelength dependent. Absorbance detectors are usually classified according to the type of radiation they measure.

A great many compounds absorb light in the *ultraviolet* (UV) range (180-350 nm). Not surprisingly, then, the UV detector has gained wide popularity in HPLC. Detector design is relatively straightforward and consists of a light source, a flow-through cell, and a photoelectric monitor. In this type of detector the design of the flow cell is particularly important, since turbulence, band dispersion, and cell design all affect the overall performance of the detector. Turbulence in the flow cell produces changes in the apparent refractive index and is manifested as a sensitivity to mobile-phase flow rate. The effect of cell design on performance is somewhat complicated in that sensitivity is increased as pathlength increases but dispersion increases with increasing volume. In practice, cell design is a compromise between these properties, with the H-cell, Z-cell, and tapered cell being the designs that have been most successfully commercially (Figure 41).

Several types of UV detectors are commercially available and can be classified into two general types: fixed wavelength and variable wavelength.

The *fixed-wavelength* detector operates with light of a single wavelength (in practice a narrow band of wavelengths) generated by a specific type of electric discharge lamp. Nearly monochromatic light is supplied by a

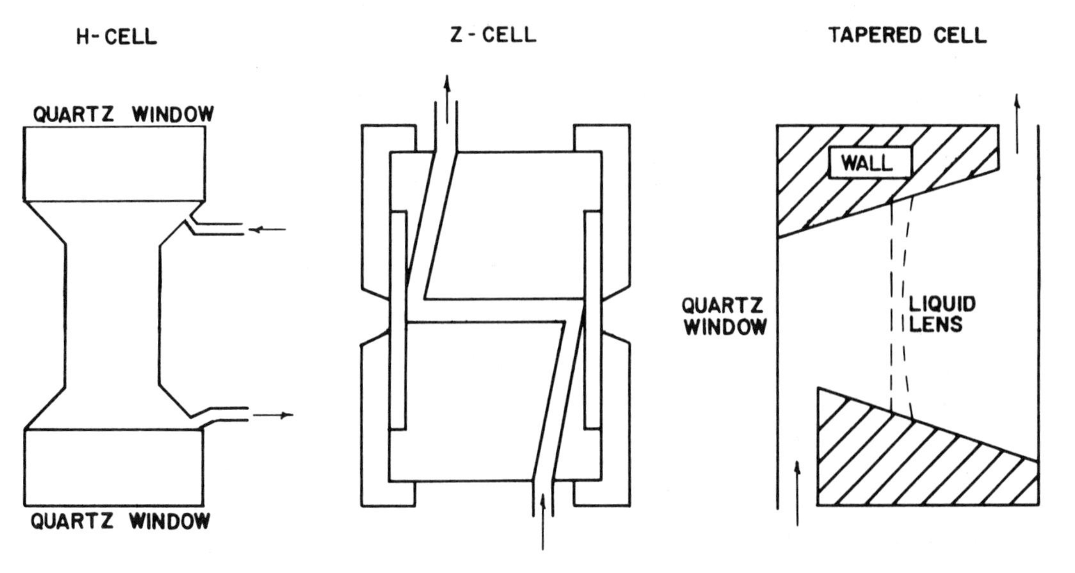

Figure 41. Typical flow-cell designs used in UV detectors.

source with distinct wavelength characteristics (e.g., Table 1), or with a continuum source (medium pressure mercury or deuterium lamps) used in conjunction with good-quality filters. Since many solutes exhibit some absorbance at a wavelength of 254 nm, the low-pressure mercury arc discharge lamp (which produces primarily this wavelength) is by far the most commonly employed source in fixed-wavelength detectors. The strong emission of both cadmium and zinc lamps at lower wavelength is useful for detecting species that have higher absorptivities at lower UV wavelengths and that are present in trace quantities. In cases where a longer wavelength is desired, it is possibie to use a phosphor to convert short-wavelength radiation (e.g., 254 nm) to a longer one (280 nm).

Fixed-wavelength detectors have several properties that are advantageous to HPLC, including

1. Low noise level (excellent sensitivity)
2. Low capital and operating cost
3. Ease of use and maintenance
4. Compatibility with gradient elution

A major drawback of the method is that it is difficult to optimize the detection of a multicomponent mixture unless all the solutes are very similar in character.

A logical progression in UV detector design was the development of *variable-wavelength* instruments in which the desired wavelength is obtained by operation of a diffraction grating, which isolates a distinct wavelength of light from a continuous source (usually a deuterium lamp). The major advantage of this detector is that specificity can be enhanced by proper choice of monitoring wavelength; a major sacrifice made when using this

Table 1. Summary of Typical Wavelengths for Sources Used in Fixed-Wavelength UV Detectors

Source	Emission line (nm)	Phosphor (nm)
Mercury	254	280
	313	300
	365	320
	405	340
	436	470
	546	510
	578	610
		660
Cadmium	229	
	326	
Zinc	214	
	308	
Magnesium	206	
Deuterium	190-350 (continuous)	

technique is a lower sensitivity than can be achieved with the fixed-wavelength design.

Variable-wavelength detectors are either single- or dual-beam spectrophotometers. In practice, single-beam instruments provide satisfactory performance provided that the mobile phase used gives better than about 70% transmission at the wavelength of interest. With lower transmission levels, or when gradient elution is used, the dual-beam detector offers some advantage in terms of being able to compensate for the background absorbance of the mobile phase.

There are two basic types of variable wavelength detectors, the dispersion detector and the photodiode array detector. In the former, the light is dispersed prior to entering the detector cell so that the sample sees virtually monochromatic radiation. The single detecting photocell used in this design also sees light of a single wavelength. In the photodiode array detector, dispersion of the light occurs after the detector cell; thus the sample sees all the wavelengths of light generated by the source. Light transmitted through the cell is dispersed over an array of photodiodes so that each will respond to light of a discrete wavelength.

The components of a typical dual-beam variable wavelength detector are shown in Figure 42. Light from the continuous source (deuterium or tungsten) is collimated onto the diffraction grating, where it is dispersed and focused onto the beam splitter. Here light is diverted to both the reference photocell (which compensates for any changes in energy level from the source) and the sample flow cell. The outputs from the sample and reference photocells are balanced against one another and the difference signal is amplified logarithmically so as to provide a detector output that is linear with respect to solute concentration. In order to eliminate thermal noise and flow sensitivity, the flow cell usually includes a heat exchanger.

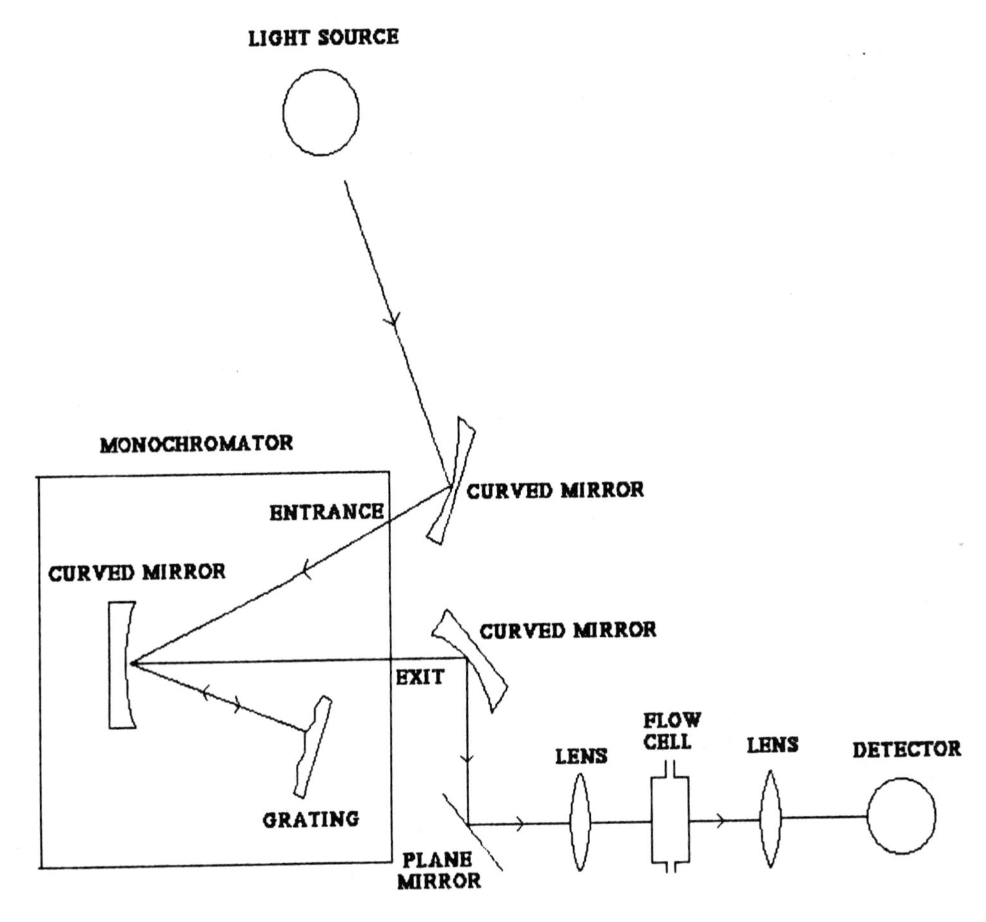

Figure 42. Optical diagram for a variable-wavelength UV detector.

The photodiode array detector is optically analogous to the fixed-wavelength detector up to and including the flow cell. After this point in the optical train, however, the light transmitted through the cell is directed to a holographic grating where it is dispersed to an array of photodiodes, each of which monitors a single wavelength. Consequently, light of all wavelengths generated by the source and passing through the cell is simultaneously monitored by the diode array. This detector thus represents the ultimate in terms of variable-wavelength detection and can be used for the identification of solutes by their UV spectrum. The detector is also quite useful in terms of confirming peak purity. By comparing UV spectral response over a wide range of analytical wavelengths at various points on the analytical peak, peak purity can be assessed by various means, including comparing overlays of absorption spectra [26], examining chromatograms obtained at different wavelengths for apparent shifts in retention time [31], and mathematically by a technique known as absorbance indexing [32].

Despite its many useful features, the diode array detector suffers from some drawbacks. From a practical standpoint, the detector is somewhat expensive and is typically less sensitive than more conventional variable-wavelength detectors. Operationally, the major drawback of the system lies in the fact that the simultaneous presence of multiple wavelengths of light in the sample cell increases the potential for a false positive response produced as the result of secondary fluorescence.

Detectors that operate in the *visible* wavelength range of 350-800 nm are commercially available for use either solely in this wavelength range or as an upgrade of a UV spectrometer (replacement of the conventional source with a tungsten lamp and use of a more appropriate detection photocell). The practical use of this wavelength range is somewhat limited since few analytes of interest exhibit a distinct color; however, the use of derivatization reactions to produce a colored product from a transparent solute has dramatically expanded the utility of this type of detector. While the ninhydrin reaction for the detection of amines and amino acids remains a classic example of this approach [33], numerous other applications have been reported in the recent chemical literature [34].

The ability of many organic compounds to absorb radiation in the *infrared* (IR, 4000 to 400 cm^{-1}) means that this property can be exploited for solute detection in HPLC. Utilization of this detection strategy suffers from two major drawbacks: limited sensitivity due to the relative weakness of IR absorptive processes, and high background noise produced since most chromatographically useful solvents absorb strongly in this region.

As recently as 10 years ago, these disadvantages were sufficiently severe that IR was considered unsuited for HPLC detection. The development of both size-exclusion chromatography, which allows for the separation of IR-active, high-molecular-weight compounds (e.g., polymers) with relatively IR-transparent mobile phases, and Fourier transform infrared techniques, which improve sensitivity, contribute ease of data manipulation, and increase the absolute magnitude of information obtained, has combined to make LC-IR a practical reality. While dispersive IR detectors (which in optical design are similar to variable-wavelength UV detectors) are commercially available, their use is somewhat limited. FT-IR monitoring of the chromatographic process has been accomplished by two means: direct detection via a flow cell, and detection after solvent removal. In the flow-cell approach, IR absorptions are usually measured in transmission cells, although attenuated total reflection (ATR) of a thin film produced over a reflective surface can be used. Clearly, when cells are used the cell material and path length are critical operating parameters.

Various means have been described to accomplish solvent removal. These include automated fraction collection/solvent evaporation, heated metal ribbons, light pipes, and membrane extraction. The use of the heated wire and light pipes essentially involves deposition of the LC effluent onto a device from which the solvent is evaporated. This suffers from poor precision due to uneven solute deposition on the surface. The first device, while effective, is a bit complex for routine applications. The use of segmented flow membrane extraction (similar in concept to suppression in ion chromatography is experimentally interesting but suffers from poor sensitivity and large band dispersion.

In general, then, IR is a suitable detection method if sensitivity and solvent constraints can be met. This detector provides the analyst with more solute-related information (e.g., functional-group identification) than do the other, more typical detectors discussed previously. The detector offers the advantages for versatility (it can be "universal" or specific), consistency of response (for a given class of compounds), and insensitivity to flow rate (pressure) changes.

b. Fluorescence

Fluorescence is the emission of radiation as the result of the transition of an electronically excited molecule back to its ground electronic state. Excitation is achieved by irradiation with light of the appropriate energy; relaxation involves a process that may include loss of excess vibrational energy (non-light-producing) and radiation emission. Radiation emitted by fluorescence possesses a longer wavelength than that originally absorbed. While the radiation may be emitted in any region of the electromagnetic spectrum, only that radiation produced by the absorption of visible or UV light is of immediate interest in HPLC detection. The difference in absorption and emission wavelengths is an important property with respect to the use of this effect in HPLC, since any excess light from the source can conveniently be removed by the optical system and the fluorescence can be measured against a near-zero background. This property translates into a sensitivity for the fluorescence detector that may be as much as three orders of magnitude higher than that obtained with direct UV detection.

Relatively few compounds exhibit a sufficiently high quantum yield of fluorescence to allow their detection directly; however, whether the solute fluoresces naturally or is made to fluoresce by some type of derivatization reaction (either pre- or postcolumn) is immaterial to this discussion. The reader is directed to the many excellent review articles available to pursue the details of the currently used derivatization reactions [35-37].

In many ways the fluorescence detector is optically and electronically similar to a variable-wavelength detector that is equipped with a second monochromator placed between the source and the sample cell. A typical design diagram of such a spectrometer is shown in Figure 43; the spectrometer consists of an excitation source, two wavelength selection devices, a sample compartment and a detection/output device. The excitation source can be split into two beams and the sample compartment can be constructed in such a way that differential measurements can be obtained. The radiation source most commonly used in HPLC detectors is the xenon arc, which has a relatively broad continuum extending into the UV region. The deuterium lamp has also been used as a continuous source. To achieve greater excitation at selected wavelengths, the mercury arc lamp can be employed. The use of lasers as excitation sources has received much attention in the recent chemical literature; its potential advantages include a high excitation energy, reduced signal-to-noise ratio, reduced level of stray light, and more accurate positioning and focusing of the beam (all of which improve sensitivity). Current disadvantages of this source include cost and a narrow operating wavelength range; both of these shortcomings will undoubtedly be improved in the face of steady advances being made in laser technology.

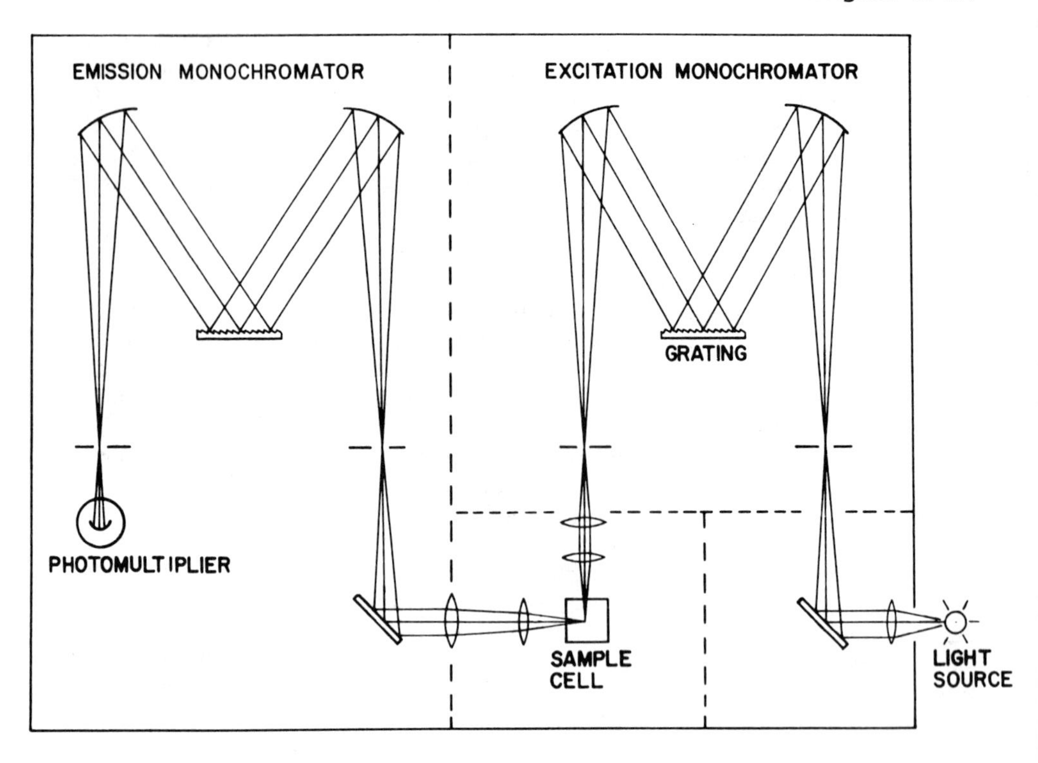

Figure 43. Optical diagram for a grating-based fluorescence spectrometer.

Since the fluorescence will emerge in a random fashion from an excited sample, the emitted light can be viewed in any direction; in most fluorometers the emitted light is detected at right angles to the excitation beam. This scenario not only results in the reduction of the background signal but also allows one to measure the absorption of the radiation by the solute independently of the fluorescence (i.e., simultaneous measurement of both properties). The wavelength selection devices (excitation and emission) can either be filters or grating monochromators. Filters frequently offer sufficient selectivity and have the advantages of simplicity, lower cost, and greater sensitivity. Conversely, the grating monochromators offer better resolution (and thus increased selectivity and greater versatility).

Flow cells designed for fluorometers tend to differ from those employed in a UV detector. While in absorbance measurements the optical path length plays a critical role in defining overall sensitivity, in fluorescence detection the most important property of the cell is total illumination volume. The most efficient fluorescence flow cells are those in which the illuminated volume is as close to the total cell volume as possible. Additional constraints on flow cell design for fluorometers include stray light (produced as a result of scattering at the liquid-cell interface) and washout characteristics (effecting carryover, dilution effects and dispersion). Flow cells used in fluorescence detection are usually constructed of narrow-bore quartz tubing and have a very small (less than 5 μl) volume.

The detection device used in the fluorometer is placed at the exit site of the emission wavelength selection system. A photomultiplier tube is usually preferred because of its high gain and relatively broad spectral sensitivity.

Despite the impressive advantages in sensitivity and selectivity that it has to offer, the fluorescence design does possess certain undesirable characteristics. First, the choice of solvents used in the separation process is very important when fluorescence detection is used. The magnitude of the fluorescence produced is dependent on solvent, buffer and other additives, pH, ionic strength, viscosity, temperature, and oxygen content, and is quite sensitive to certain deactivating or quenching species (e.g., highly polar solvents, many buffers, halide ions). In order to reduce the number of molecular collisions experienced by the solute (which tend to quench fluorescence), low operating temperatures and solvents with high viscosities are recommended. Second, it is clear that this detector is not as universal as those described previously. While the process of concentration quenching (in which the emitted radiation is re-absorbed by the solute or solvent) tends to affect response linearity at high solute concentrations, the linear dynamic range of the fluorometer is usually comparable to that of other popular detectors discussed previously. Finally, it is noted that the sensitivity of fluorescence to chemical environment may limit its utility in gradient applications.

c. *Others*

Spectroscopic detectors described to this point have been concerned with molecular interactions with electromagnetic radiation; atoms are also capable of absorbing and emitting radiant energy, and either property can be exploited in HPLC detection. The use of *atomic absorption* (AA) spectrophotometers in HPLC is well established and offers the advantages of selectivity and sensitivity for a variety of elements and compounds (primarily organometallics). Since commercially available AA instrumentation is used in this type of senario, its design and operating principles will not be discussed here. In a general sense, the measurement of atomic absorption requires that the analyte be converted to free atoms in the gas phase; this process can be accomplished either by a flame (into which the sample solution is aspirated via a nebulizer) or with an electrothermal atomizer (in which samples are placed in a conductive cavity, usually graphite, and heated by electrical conduction). The flame atomization process offers somewhat less sensitivity than does the electrothermal atomizer but has the major advantage of being compatible with on-line monitoring. Since electrothermal atomization produces free atoms in a series of heating/cooling steps, the HPLC effluent must be collected and stored prior to analysis. Additional problems associated with the electrothermal atomizer include matrix interferences, cuvette degradation, and a different response to various species of the same analyte.

Flame AA detection suffers primarily due to the disparity between the flow rate of the HPLC system (0.5-2 ml/min) and the sample uptake rate of most commercially available nebulizers (2-10 ml/min). Approaches designed to deal with this disparity, including nebulizer starvation, post-column mixing of the mobile phase with a makeup solution, and use of a flow injection process, have met with limited success. The most likely solution to this problem lies in the development of high-speed columns that can be operated at high mobile-phase flow rates. However, use of such columns with AA detection will require the development of low-dispersion interfaces between the column and AA detector.

Atomic emission (AE) can also be used as a means of monitoring HPLC eluates; however, production of this effect requires a higher excitation temperature than does AA. This higher temperature is usually achieved by using a plasma as the excitation source. In the inductively coupled plasma (ICP) source most commonly used in HPLC applications, the plasma is generated by inducing a magnetic field around the top of an assembly of coaxial silica tubes. A conducting gas (argon) flows through these tubes. The HPLC eluate is introduced into the plasma via a nebulizer. While the ICP detector is characterized by excellent sensitivity, selectivity, wide linear dynamic range, and (in most cases) multielement detection capabilities, it is compatible with only a limited number of organic mobile phases and is relatively expensive. Reasons for its limited compatibility with organic solvents include plasma ignition and stability problems, the potential for spectral interferences related to the emission/absorption properties of the solvent, and the fact that a number of HPLC solvents contain appreciable levels of metallic species (and thus produce a background signal).

Certain nuclei (e.g., ^{1}H, ^{19}F, ^{13}P, ^{15}N) have the ability to absorb radiation; the phenomenon of *nuclear magnetic resonance* (NMR) is a powerful technique for the identification and structural analysis of unknown organic compounds. While this methodology has been adapted to monitor HPLC eluates, the association of NMR and HPLC is characterized by four major problems. First, NMR signal intensity is inversely related to flow rate; while correct cell design limits the magnitude of this effect, it cannot be eliminated altogether. Second, attainment of high NMR resolution requires an extremely homogeneous magnetic field, which is usually achieved by spinning the sample tube at fairly high speeds. Since spinning has not been achieved with flow-through NMR cells, the NMR/HPLC couple suffers from reduced spectral resolution. Third, the choice of mobile-phase solvents is somewhat limited since, for example, if a proton spectrum is to be obtained the mobile phase cannot contain protons. Use of non-hydrogen-containing solvents impairs chromatographic efficiency, while use of deuterated solvents can be quite expensive. Finally, the sensitivity of the NMR spectrometer is quite limited; while sensitivity can be improved by using Fourier transform techniques, the cost of such instrumentation precludes its routine use as a HPLC detector. It is also important to note that most (if not all) HPLC/NMR interfaces and flow cells currently available contribute excessive dispersion to the chromatographic process. From this discussion it is clear that a major improvement in technology is necessary before HPLC/NMR becomes a viable analytical methodology.

In addition to molecular interactions with light, solutes can interact with light in a strictly physical sense: that is, they can cause the light to be scattered. *Light scattering* results from the interaction of visible light with a nonuniform medium. While solute/solvent systems are generally considered to be homogeneous, if the molecular mass of the solute is much larger than that of the solvent, scattering can occur. Use of this property for HPLC detection falls into two categories: nephelometry, in which the sample is irradiated and the amount of light scattered at some angle from the incident beam is measured, and turbidimetry, in which the sample is irradiated and the amount of transmitted light is measured. Both techniques suffer from poor sensitivity and high background and are thus not commonly employed in routine applications. However, one variation of the nephelometric approach, low-angle laser light scattering (LALLS), has proven to be quite useful as a detector in size-exclusion chromatography. This methodology, in which scattering measurements are made at angles less than 7 degrees from the incident path, has been particularly useful in detecting large molecules (e.g., plastics, polymers, and proteins). The combined SEC (size exclusion chromatography)/LALLS technique permits determination of solute molecular weight and intrinsic solute viscosity in addition to solute concentration.

Electrochemical Detectors

Electrochemical detectors are used for the quantitation of species that are electroactive—that is, those species that can be electrolytically oxidized or reduced at a working electrode. The output of the detector results from the electron flow caused by the electrochemical reaction that takes place at the surface of this electrode. These detectors can be classified according to the electrochemical property being measured and include potentiometric, voltammetric, and polarographic designs. Each of these configurations will be considered in greater detail below.

a. Potentiometry

In this design, an electrode is selected that contains an element that will enter into a chemical equilibrium with ions of the same element in solution. As a result of this interaction, a potential, whose magnitude is related to the activity of the ion in solution by the Nernst equation, is produced. The measurement of the potential developed by these ion-selective electrodes is called potentiometry. A pH, or hydrogen-sensing, electrode is the most commonly employed example of a potentiometric electrode.

The potentiometric design includes two electrodes: an indicator electrode capable of monitoring the activity of the species of interest, and a reference electrode that provides a known half-cell potential to which the indicator potential can be compared. Indicator electrodes in common use in HPLC detectors utilize a membrane to confine a reference solution (in the electrode) and to make electrolytic contact with the mobile phase. This membrane, which is typically either glass, a porous disk, or an ionic crystal, functions primarily via an ion exchange mechanism. With miniaturized electrodes, cell volumes can be made very small (5 μl) and therefore dispersion is greatly reduced.

The instrumentation required for potentiometric detection is simple and relatively inexpensive. The voltages produced are typically quite small and thus the measuring system must be very sensitive. The voltage produced by the detector must be measured in such a way that the amount of current drawn by the measuring circuits is minimized. In many cases, research-grade pH meters are appropriate for recording the potentials produced by potentiometric detectors.

b. Voltammetry

Voltammetric detectors are based on the principle that when a sufficiently high voltage is applied to an electrode, any electroactive material will undergo electrolysis and produce an electric current. This method gains its selectivity from the fact that specific organic functional groups will only be electrolyzed at specific applied potentials. Two different approaches to voltammetric detection can be used in the analysis of an electroactive species: coulometry or complete electrolysis of the solute, and amperometry or partial electrolysis of the solute. Amperometry is further broken down with respect to the type of electrode material; thus polarography is amperometry performed with a mercury electrode. As used in this chapter, amperometry will refer to the more limited class of applications that involves the use of a solid working electrode.

A conventional voltammetric detector consists of three electrodes: the working electrode (at which the electrochemical process occurs), a reference electrode (which compensates for changing mobile-phase properties), and an auxiliary electrode (which serves to carry the electric current). Generally the choice of material used for the reference and auxiliary electrodes is not critical with respect to system's performance but the type of material used for the working electrode has an important effect on performance. The dropping mercury electrode used in polarography is restricted to reductive processes due to its low oxidation potential. Primarily due to the large overpotential for the evolution of hydrogen, mercury electrodes can be used with potentials more negative than -2V. Another important advantage of mercury as a working electrode is that a renewed surface can constantly be supplied and thus surface contamination (and changing electrode response) is minimized.

The development of solid working electrodes is of primary importance in the routine application of voltammetric detection. These electrodes typically have a high anodic range, can be used for analytes that are easily reduced, and have low residual currents. Solid electrodes typically exhibit lower noise, higher sensitivity, and have a simpler cell design than their mercury counterparts but are more prone to surface contamination. Materials that have been used as working electrodes include carbon pastes, glassy carbon, and, less commonly, various noble metals (e.g., gold, platinum, silver).

Polarography has been extended to HPLC applications by rigorous design modifications, which are required in order to minimize cell volume. In most commercially available detectors based on this principle, the HPLC eluate impinges on a rapidly dispensed, specifically sized mercury drop whose drop time is 1 sec or less so as to maintain the chromatographic separation. A generalized design for such a system is shown in Figure 44.

While many designs have been devised for systems that use a solid working electrode, all are based on a three electrode design. The reference and auxiliary electrodes are placed downstream from the working electrode so as to prevent potential cross-contamination (e.g., leakage from the reference, and electrochemical products from the auxiliary). The reference is normally placed as close to the working electrode as possible so that the electrical resistance of the cell is minimized. The electrodes are fitted into the body of a nonconducting flow cell (Kel-F, PTFE) in either a thin-

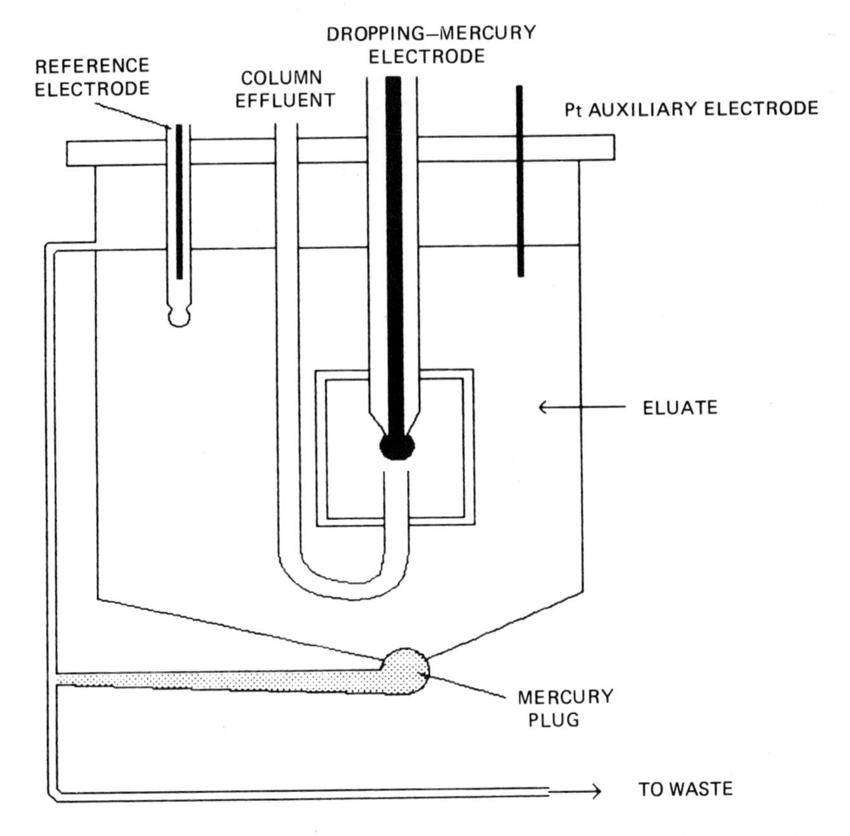

Figure 44. Design of a dropping-mercury-electrode flow cell for monitoring HPLC separations via polarography.

layer or wall-jet configuration (Figure 45). In both cases, the cell consists of two halves (separated by an insulator) into which a slot is cut so as to allow the mobile phase to wash over the electrodes. Spacer thickness defines the cell volume; up to a critical thickness of 50 μm, increased spacer thickness improves sensitivity by increasing the linear flow rate of the mobile phase.

The coulometric detector is similar in design to the amperometric detector except that the working electrode typically has a larger surface area so that complete electrolysis can occur. Coulometric detectors are not particularly popular because of their lower sensitivity (which is attributed to difficulty in achieving 100% reaction completion), increased background, larger cell volume, and increased potential for electrode surface contamination. Roe has recently used a "figure of merit" to compare amperometric and coulometric designs and concludes that the latter is capable of superior performance in routine applications [38].

The basic circuit used in an electrochemical detector is shown in Figure 46. The auxiliary electrode A is held at a fixed potential by an

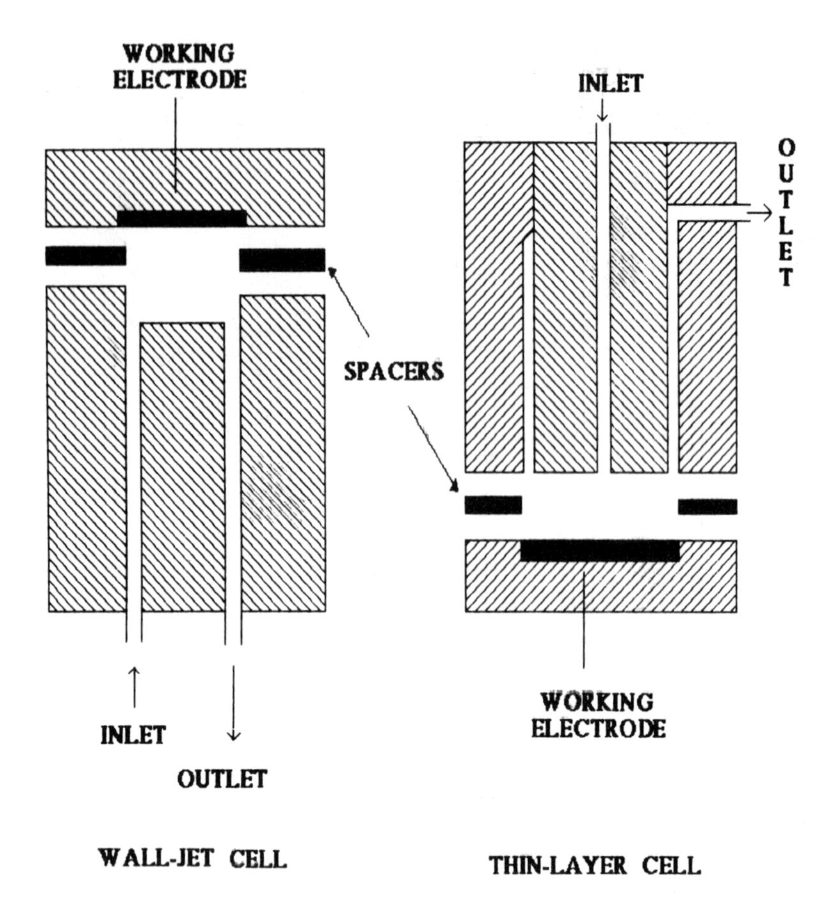

Figure 45. Cell designs used in amperometric detectors.

amplifier (B) and potentiometer (P). The potential of the mobile phase near the auxiliary electrode is sensed by the reference; when an electroactive solute interacts with the surface of the working electrode, an electron transfer process occurs. The current produced by such a process is converted to a voltage output (V) by an amplifier (C), after which it can be further attenuated and filtered if necessary.

In most applications of electrochemistry to HPLC detection, DC currents are used to electrolyze the analytes. Thus the current measured consists of two components: the faradaic current (I_f) that results from the electron transfer reaction occurring at the electrode surface, and the charging current (I_c), which is not related to the presence of the solute but results from the capacitive nature of the electrode surface. The charging current represents an undesirable contribution to both the background signal and the detection limit. Discrimination between these two currents can be obtained by the use of pulsed, AC, or rapid-scan square-wave voltammetry. The latter two methods have only a limited application, primarily due to design complexity and high capital cost. Pulsed, reversed pulsed, and differential pulse voltammetry (DPV) have all been successfully applied to HPLC detection and, in addition to improved sensitivity, are more stable, less sensitive to flow rate, and in some cases more selective than nonpulsed methods. These techniques are based on the observation that when a potential is applied to an electrode both I_f and I_c increase but when the potential is removed I_c decays at a much faster rate than does I_f. Pulsed and reverse pulse techniques are similar in operation (current is measured just before the applied voltage is removed) but differ in the polarity of the initial pulse. The reverse pulse approach, which utilizes a large negative potential, offers the advantage of producing a signal at a potential where dissolved oxygen is not reduced. In the DPV approach, the pulsed potential is applied as described previously but the current is measured twice (once before applying the pulse and again just before the end of the pulse). The output obtained thus represents the difference in current versus time.

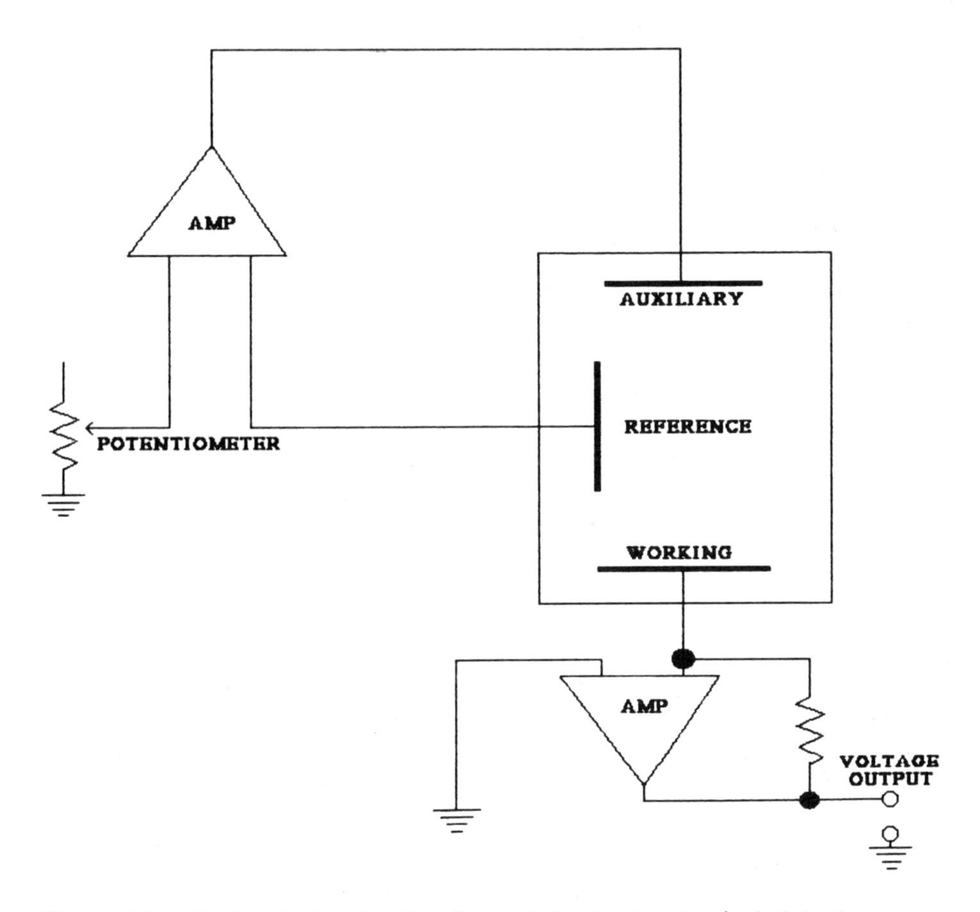

Figure 46. Basic electronic circuit used in electrochemical detectors.

In summary, then, electrochemical (EC) detectors offer the advantages of selectivity, reliability, relative simplicity, and sensitivity when compared with the other major classes of HPLC detectors. In addition to their lack of universality (solutes must be electroactive), the EC detectors suffer from some important operational drawbacks, which include.

1. The mobile phase has to be conducting (i.e., no normal phase).
2. The mobile phase and sample must be oxygen free.
3. The mobile phase should be free of reducible metals.
4. Response varies as reaction byproducts contaminate the working electrode.
5. Response is highly dependent on a consistent flow rate (i.e., a truly pulseless pump must be used), solution pH, ionic strength, temperature, and cell geometry.
6. There is limited applicability in situations requiring gradient elution.

Currently, EC detectors represent the fourth most popular design for routine applications. Current research interest in this field is high and one expects continuing improvements in terms of detector design and optimization.

Radioactivity

Radioactive tracers (e.g., compounds labeled with a radionuclide) play an important role in investigations concerning biological and metabolic processes and frequently can be readily separated by HPLC. Recent developments in detector design have resulted in the production of on-line radioactivity detectors for HPLC. These detectors are usually classified as being either homogeneous systems, in which the eluate is mixed with a liquid scintillation cocktail before passing through a flow cell located in a scintillation counter, or heterogeneous systems, in which the eluate passes through a flow cell packed with a solid scintillator. Despite obvious differences in design inherent in these two approaches, the analyzer used in both cases is essentially equivalent. When a radioactive compound disintegrates in the flow cell, the particle or ray produced will strike the scintillant and produce a flash of light, which is detected by the photomultiplier tube of the scintillation counter. A ratemeter is used to count the number of photons produced over a period of time. In order to discriminate between a true radionuclide distintegration and stray background radiation, most modern analyzers have two identical photomultipliers that "sandwich" the flow cell. These photomultipliers are connected to an electronic coincidence circuit, which transmits a signal to the ratemeter only when it receives a simultaneous signal from both photomultipliers.

Homogeneous detection systems require that a postcolumn reaction occur and are thus instrumentally more complicated than heterogeneous systems. Additionally, thc necessity that the mobile phase and scintillation cocktail both be miscible and produce minimum radioactivity quenching places some restrictions on the choice of mobile phase. 2,5-Diphenyloxazole (PPO) and 1,4-bis-2-(5-phenyloxazolyl)-benzene (POPOP) in dioxane are commonly recommended scintillants. Since sensitivity in a radioactivity detector is dependent on the total amount of activity in the flow cell, the cell volume, design, and flow rates are important factors that control sensitivity. Also, it is noted that since the homogeneous systems require postcolumn reactions, the sample becomes contaminated as a result of the detection process.

Heterogeneous systems, while they offer the advantages of instrumental simplicity, lower operating cost, and lack of sample contamination when compared to the homogeneous system, generally have a poorer counting efficiency and thus are less sensitive than their homogeneous analog. Cell design, particle size of the scintillator, and optical properties of the solute and solvent all affect sensitivity in these systems. Solid scintillators used in these designs include cerium-activated glass or yttrium silicate, europium-activated calcium fluoride, some plastics, and anthracene. Additionally, sodium iodide crystals can be used to monitor radiation produced by γ-emitters and have been incorporated in HPLC detection, particularly in applications involving the ^{125}I isotope.

D. Transport Detectors

GC Detectors

By convention, transport detectors involve two complementary processes: mobile-phase removal, followed by solute detection. The transport detector represents an attempt to adopt GC-type detection strategies to an HPLC separation; such a marriage is typically a difficult one to achieve since the GC detectors are not, for one reason or another, compatible with the large quantity of solvent associated with an HPLC separation. While the development of micro- and minicolumn technologies, which both represent a means by which high column efficiencies can be achieved at low flow rate, has improved HPLC/GC-detector compatibility somewhat, some type of solvent-removing interface is still commonly required. Gas chromatographic detectors that have been adapted for use in monitoring HPLC eluates include flame ionization detectors, electron capture detectors, flame photometric detectors, thermionic detectors, and mass spectrometers. The design, operation, and performance properties of these detectors are discussed in detail elsewhere; of primary importance to this chapter is a discussion concerning the nature of the mobile-phase-removing interface.

The most common type of interface involves entrainment (via dropping or spraying) of the eluate onto a moving carrier (wire, chain, belt, disk). The moving carrier passes through an evaporation oven wherein the more volatile solvent is removed from the carrier and the solute is left behind. The carrier, coated with solute, then passes to the detector, where the solute is volatilized by a means compatible with the detector being employed. Such a system, shown in Figure 47, generally suffers from mechanical instability, poor collection efficiency, excessive band spreading, contamination, and incomplete solute/solvent separation.

Direct volatilization is another approach that has been used as a means of interfacing HPLC with a GC detector. In this method, the HPLC eluate is atomized and volatilized via nebulization, with desolvation being accomplished primarily by heat. In applications of LC/MS, this type of interface has been termed a thermospray detector. When no attempt is made to

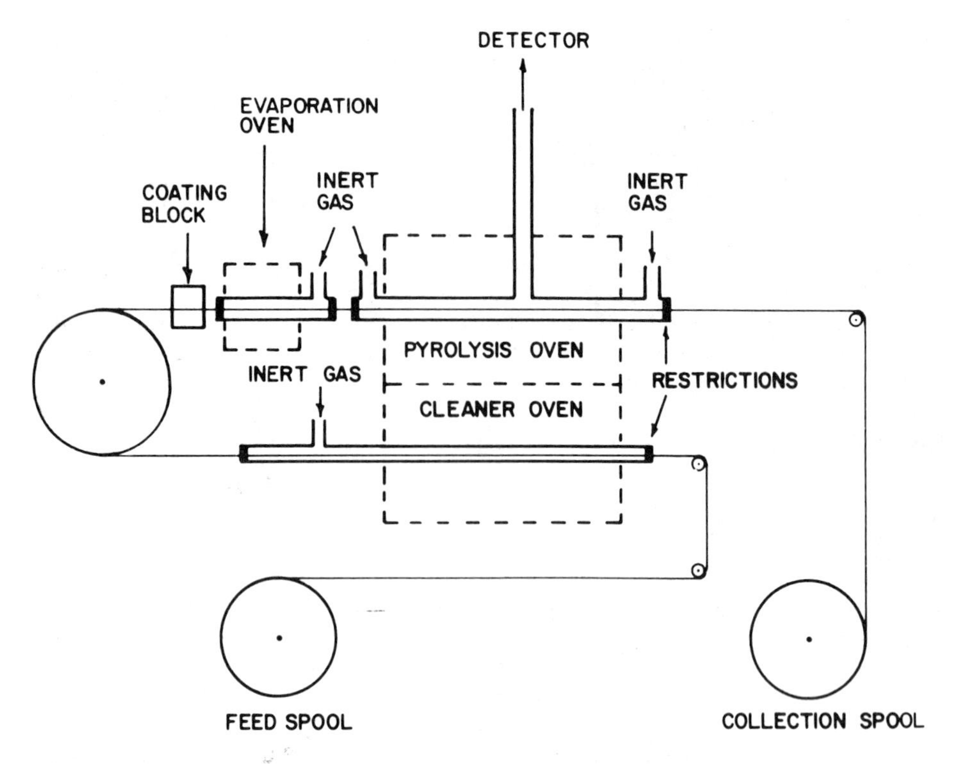

Figure 47. Schematic diagram of a wire-transport detector.

desolvate the nebulized eluate, the interface essentially becomes a direct liquid injection (DLI) system.

Coupling of a GC-type detector to the HPLC separation system offers the advantages of sensitivity, selectivity, and, in the case of the mass spectrometer, the ability to characterize analyte structure. However, numerous problems relating to the interface between separator and the detector (which include contamination, ruggedness, mobile phase compatibility, etc.) currently limit the routine application of this approach.

Mass Detector

The evaporative light scattering detector, or mass detector, works on the following principle (Figure 48). The HPLC eluate is nebulized and passes through an evaporation channel, resulting in evaporation of the solvent and the production of finely divided solute particles. These particles pass through a light beam; light scattered by the particles is detected at a fixed angle from the incident beam with an appropriate photomultiplier. Detector response is linearly related to analyte concentration as long as reflection and refraction are the primary scattering mechanisms. Above and below the linear operating range the nebulization and evaporation processes produce particles of varying sizes, which promote scattering by other mechanisms (e.g., Mie and Rayleigh scattering). Detector response is reported to be independent of the chemical structure of the solute, and thus this detector produces mass-dependent (as opposed to strictly concentration-dependent) output. The detector is universal in the sense that it responds to any nonvolatile solute and is reported to be roughly a factor of 10 more sensitive than most RI detectors. The major disadvantage of this detector is a relatively high operating cost, which is related to the large quantity of inert gas (N_2) used in the nebulization/evaporation process and the need to vent the exhaust gas away from the working environment.

E. Chemical Reaction Detection

In spite of the diversity in detection systems compatible with HPLC, in many applications there is still a need for greater sensitivity and/or

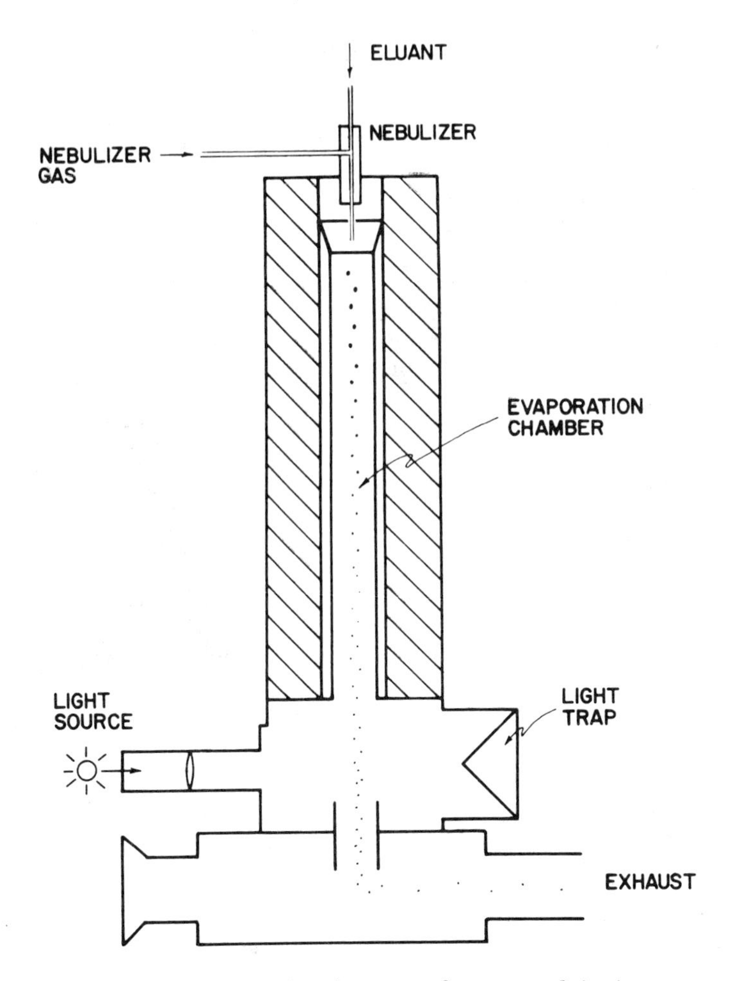

Figure 48. Schematic diagram of a mass detector.

selectivity. Two common types of supplement involve chemical reaction with analyte-selective reagents either prior to or following chromatographic separation but before detection. Precolumn derivatization utilizes a batch-type chemical reaction between the analyte(s) and detection-enhancing reagent during preinjection sample preparation. Postcolumn derivatization, or chemical reaction detection, utilizes an on-line chemical reaction subsequent to the separation through the continuous supply of reagent under appropriate reaction conditions. The two formats are in many ways both comparable and complementary, and the indications for respective use have been discussed elsewhere [39]. The focus of this text will be chemical reaction detection (CRD).

The practice of CRD has many attractive features. One interesting aspect is that the automated chemistry need only be reproducible; it need not result in a single, well-defined species or even a derivatized form of the analyte itself. Reaction byproducts, sometimes problematic in precolumn derivatization, are usually unimportant, and in some cases can even serve as the detectable species. While incomplete reactions and/or indirect detection can be somewhat unsettling at times, in practice they can be successfully implemented because the reaction stoichiometry is well defined and easily controlled in the clean postseparation environment. In addition, the chromatographic environment (e.g., mobile-phase pH) can be optimized for the separation and then changed postcolumn to enhance detectability.

The price of CRD is an increase in equipment complexity and some counterproductive peak broadening. Since the reagent is added continuously, there is reagent waste in unimportant regions of the chromatogram; additionally, the reagent can contribute to the background signal in important chromatographic regions. The reaction stoichiometry is governed by equipment that can change in performance over long periods of time. Finally, in some cases there may be an insurmountable incompatibility between the mobile phase and reagent.

The most common configuration of CRD, shown in Figure 49, employs a single reagent solution to effect the reaction. The reagent is added to and mixed with the column effluent by a metering pump and mixing unit. The reaction solution then passes through a delay or reaction system to allow sufficient time for the reaction to occur. The reaction products are then monitored by any one of several types of conventional LC detectors. Popular variants include the addition of supplemental detectors in pre- and postreaction positions and the addition of multiple solutions (buffers, reagents) to the column effluent to permit more exotic chemistries or improve reagent/analyte/mobile-phase compatibility. Recent reviews cover the chemistry in applications as diverse as thiols and amines [40], inorganic species [41], drugs in biological fluids [42], and carbohydrates, pesticides, phenols, and hydroperoxides [43,44].

There are several formats available with which one can "add" reagents. The system outlined above employs the direct addition of reagent by a pump so that this method can be called fluid addition (FA). Transmembrane addition (TM) also uses a reagent solution but passes or exchanges the reagent through a permeable ion-exchange membrane. In addition to the reagent solution formats, one can employ solid-phase reagents or catalysts

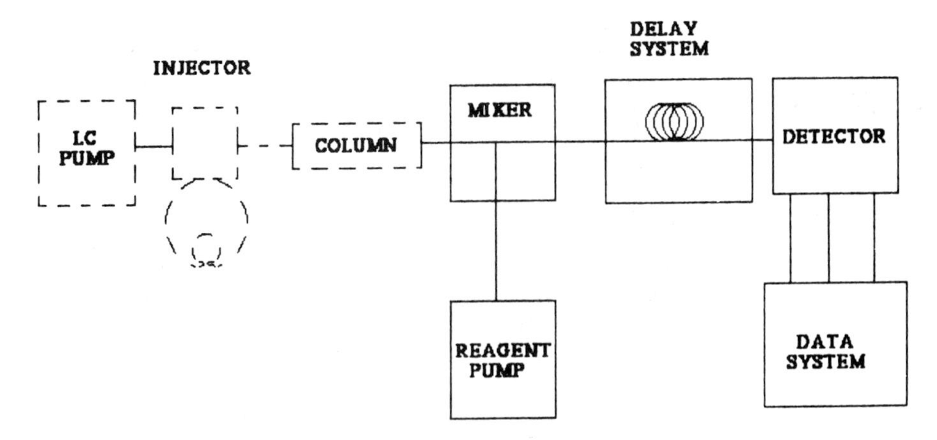

Figure 49. Typical instrumental schematic for chemical reaction detection.

as well as enzymes immobilized on a stationary support. These will be collectively referred to as the solid-phase procedure (SP). The last category uses an agent in the mobile phase to effect a reaction and will be called generated/doped format (G/D). In this case, the reagent is a mobile-phase component (present prior to separation) that is "activated" postcolumn either by heat, light, or electric current. A brief discussion of some of the salient features of each will follow.

Fluid Addition

Fluid addition is currently the most common way of implementing CRD. The system, described earlier, requires the use of a reagent pump, mixing system, and delay or reaction system. Each component is important and their requirements will be discussed.

a. Pumps

The general features of both the mobile phase and reagent pumps are described in Section III. In this particular application though, it is especially important that both pumps (mobile phase and reagent) deliver consistent, pulse-free flow. This is necessary to insure constant reaction conditions and a smooth, quiet background signal. A significant difference in absorbance between the mobile phase and reagent in conjunction with fluctuations or pulsations in either pump will lead to changes in postmixing reagent concentration and a background modulation at the frequency of the errant pump. A general observation is that the mobile-phase pumps are the lesser problem, and this is probably due to the fact that there is more effective pulse damping at the higher precolumn pressures (1000-2000 psi). In our experience, most LC pumps, including inexpensive single-piston pumps, can be very effective in reagent addition applications if properly pulse-damped. Either the "hockey puck" or Bordon tube pulse dampers are suitable when combined with an adequate source of backpressure.

For example, a 10-ft piece of 0.005 in ID stainless steel tubing will develop a 500-psi pressure drop with water at 1.0 ml/min. In some cases the pressure drop of the reaction coil alone may be enough to supply the pressure needed to suppress pulsation.

b. Mixers

Mixers are the devices used to combine and mix the column effluent with the reagent solution. The mixing effect in the CRD system has two sources: the site of fluid combination (the mixer), and the subsequent turbulence or dispersion introduced by the system (usually the delay system). The mix time will depend on the efficiency (from both sources), the relative volume ratio of the solutions, and the difference in their viscosities. Since the required delay for a given reaction will be the sum of the reaction time and the mix time, minimization of the mix time will reduce the delay-system requirements and help preserve chromatographic efficiency and maximize sensitivity. In this spirit, there are several types of devices available to meet this requirement, and these usually combine reagent addition with efficient mixing.

There are basically four types of mixers for use in CRD: "tee" or "Y," cyclone, vortex, and the micromixer. The tee mixer is the simplest and most readily available through any chromatographic supply company. While adequate for adding fluids together, this device does not provide efficient mixing under most circumstances. As shown by Poppe et al., the mixing effects for a simple tee can be quite poor [45]. Their experimental design employed a splitter tee downstream of the mixing tee. The flow was split and carefully adjusted so that one-half was directed to a UV detector and the other half to waste. Figure 50 shows the detector response of water and dilute phenol solutions pumped into the inlets of the mixing tee as it is rotated through different relative angular positions with respect to the splitter. There is a definite sinusoidal shape to the detector response as the splitter position "phase matches" the transverse concentration distribution within the cross section of the mixing tee exit tube. Unfortunately, the exact experimental conditions were not specified so the viscosity differences could not be determined. However, these data indicate spatial inhomogeneity in the "mixed" solution leaving a simple tee. One solution recommended by these workers was the inclusion of a small (10 cm × 1.0 mm ID) packed-bed tube filled with 110-μm beads to mix the combined solutions prior to delay-system entry [45].

There are modified versions of the tee that are more effective in mixing the two combined solutions. These are the Y and the vortex mixer using flow-induced turbulence to create a mixing effect. Modified tees and Y's are, to date, homemade and, as shown in Figure 51, use variants of intersection geometry or proximity to initiate turbulent flow at the site of reagent addition. Some of these devices have been well characterized and can be used at eluent:reagent ratios as high as 10:1 [46]. The reader is referred to general discussions [43,47] or the primary references [46,48] for more details.

The vortex mixer, which also uses flow-induced turbulence at the mixing site to create a mixing effect, is commercially available. As shown in Figure 52, the two inlet branches impinge on the outlet port from

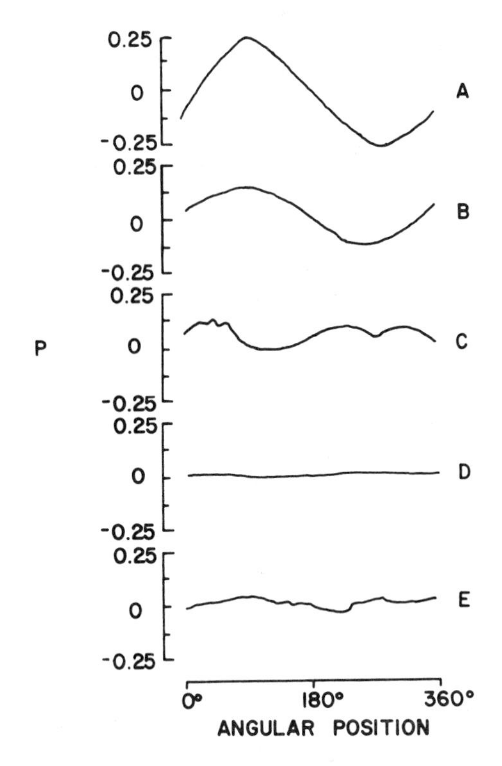

Figure 50. Dependence of the observed inhomogeneity (P) on the relative angular position of the mixing and splitting tees. Conditions: flow rates, 20 ml/h aqueous phenol solution (0.1 g/liter) and 20 ml/h water; 1:1 split ratio at outlet tee between detector and waste. (A) Open tubing, 10 cm × 0.5 mm ID. (B) Open tubing, 10 cm × 0.1 mm ID. (C) Packed tubing, 5 cm × 1.0 mm ID with 25-μm glass beads. (D) Packed tubing, 10 cm × 1.0 mm ID with 110-μm glass beads. (E) Packed tubing, 5 cm × 1.0 mm ID with 110-μm glass beads. The quantity P is defined by $P = (A_{obs} - A_{calc})/A_{calc}$ where A is the absorbance measurement, either observed, or calculated for perfect mixing. (From Ref. 45, with permission.)

tangential positions. As the two fluids enter, a vortex is formed that mixes the fluids. Unfortunately, few data are available to evaluate its efficiency.

The micromixer is a relatively new device that combines a simple tee with vigorous secondary mixing. The reagent and eluent streams are combined and directed through a set of waffle-shaped discs containing tiny "spin chambers." As shown in Figure 53, the mixture is subjected to a highly tortuous flow path involving several changes in spin direction and forced turbulence in three dimensions. While interesting in concept, this device is as yet uncharacterized in the literature, with no data available to evaluate its effectiveness.

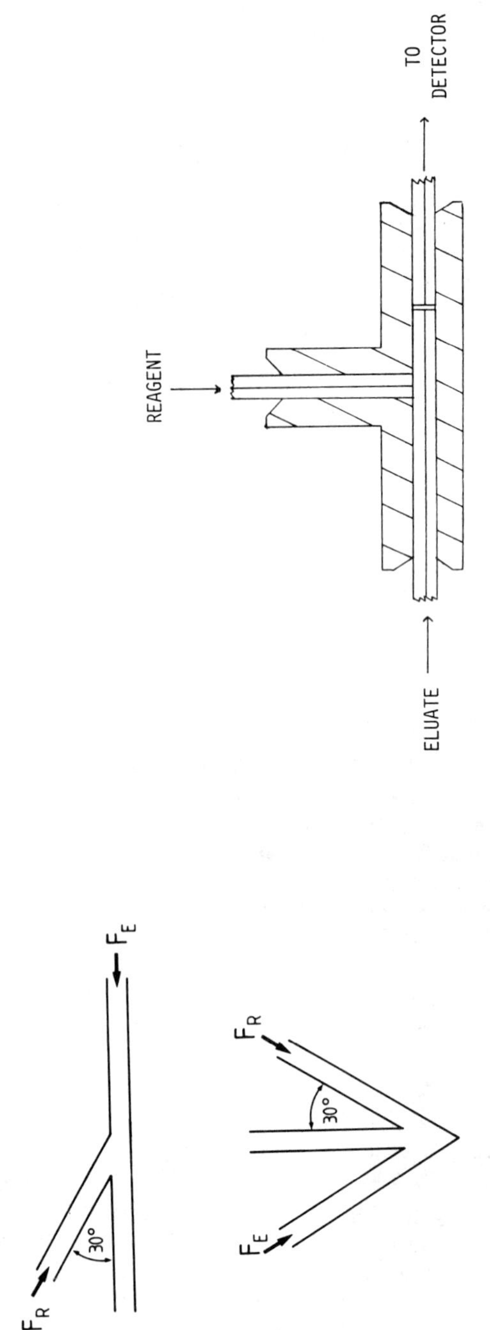

Figure 51. Diagrams of modified Y and tee mixers. (From Ref. 46 and 41, respectively, with permission.)

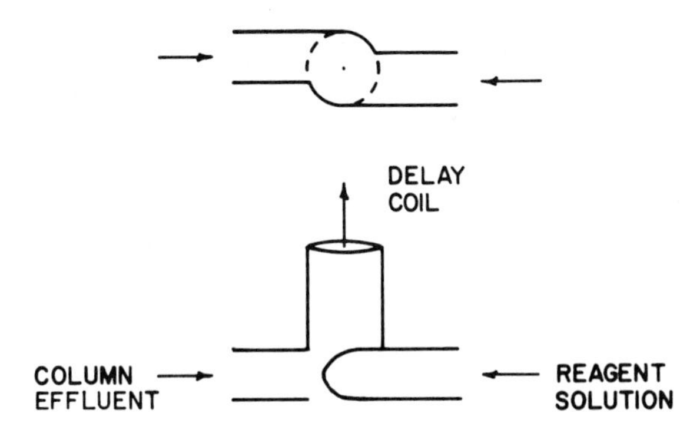

Figure 52. Top- and side-view drawings of a vortex mixer.

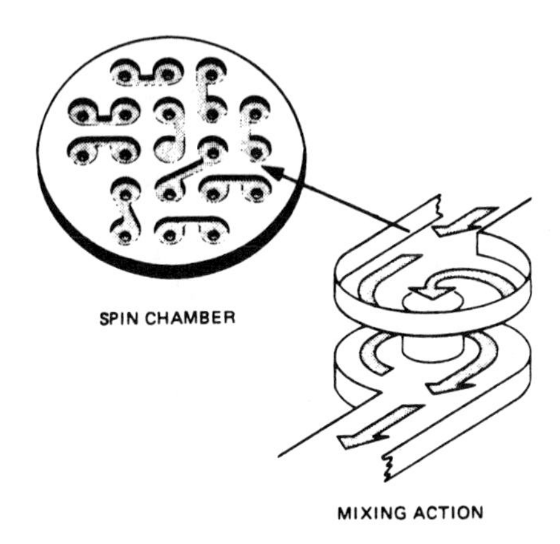

Figure 53. Schematic and flow diagrams of a MicroMixer. (Reproduced with the permission of The Lee Company, Westbrook, Connecticut.)

c. *Delay systems*

There are three basic types of delay systems in CRD: open tubular, packed bed, and fluid segmentation. The choice of a particular system is roughly determined by the required delay time, allowable dispersion, and complexity of the reagent addition protocol. Guidelines for the design of specific reactors can be found in more detailed discussions and references therein [43,47]. In general though, tubular systems are used for short durations (0.1-3 min), packed beds are used for medium durations (1-5 min), and fluid segmentation systems are used for long times (5-20 min) or complex reagent addition schemes.

Tubular delay systems are made from small-bore (0.01-0.02 in ID) stainless steel, Teflon, or PTFE tubing in lengths (and therefore volumes) selected for the required delay. In straight, open tubes, flowing fluids develop a parabolic velocity profile across the diameter of the tube. At the wall there is essentially zero velocity, with a velocity maximum at the center. This velocity gradient, called laminar flow, predominates as the dispersive force because diffusion in liquids is slow [43]. The net result, as shown in Figure 54, is that a plug of sample is longitudinally dispersed as it moves through the tube, and this appears as a badly broadened peak as the plug is cross-sectionally averaged by the detector [49]. One must introduce turbulence across the tube diameter to defeat laminar flow dispersive effects and preserve the detector sensitivity and chromatographic resolution developed by the column. Simple ways to do this are to coil or knit the delay tubing.

The coiled tube (CT) is the simplest method to use. By coiling the delay tubing, one introduces a centrifugal component that initiates a secondary flow perpendicular to the main direction of flow. As shown in Figure 55, this causes radial mixing and helps break up the laminar flow profile. One can determine an approximate coil radius based on the flow rate and tubing radius through the following:

$$H = 7ur^4/12R^2yD_M$$

where H is the height equivalent to a theoretical plate, u is the volumetric flow rate, r is the tube radius, R is the coil radius, y is an obstruction factor (y = 1 in open tubes) and D_M is the molecular diffusion coefficient.

The knit tube (KT) delay systems are commercially available from several sources and offer superior suppression of laminar flow effects. Results of systematic deformation studies of the cross-sectional shape and tubing orientation led to a simple but effective weave pattern [49,50]. The KT, shown in Figure 56, combines a series of half loops of small radius with a continual orientation change through three dimensions to create highly efficient secondary flow even at low flow rates. The advantages of a KT are shown in Figure 57 [47]. At a combined flow rate of 1.5 ml/min (22 cm/sec), the dispersion values in straight open tubes and tightly coiled

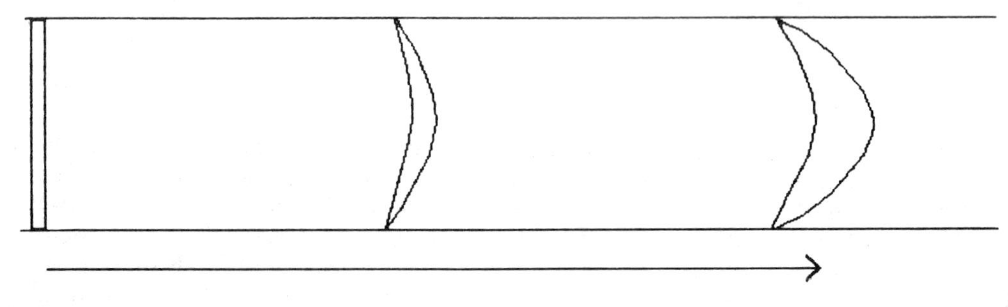

Figure 54. Laminar flow profiles of a sample plug during various stages of development. This profile is presented as a longitudinal slice of the tubing.

Figure 55. Secondary flow pattern in the cross section of a coiled tube. (From Ref. 43, with permission.)

Figure 56. Knit open tube. (From Ref. 47, with permission.)

tubes are roughly 20 and 5 times the dispersion in the KT, respectively. In addition, the good performance of the KT is independent of flow rate over the range most useful to columns of standard dimension.

Packed-bed (PB) delay systems are glass or stainless steel columns packed with an inert (usually glass) bead material. Their behavior is essentially that of a chromatographic column operating under nonretention ($k' = 0$) conditions, with dispersion being described in terms of packing geometry, particle size, tortuosity, fluid velocity, and diffusion coefficients [51]. As in chromatography, the least dispersion will be obtained with small (5-10 μm) particle size [43]. However, 15-20 μm particles are a popular size because they can be dry packed and develop small pressure drops. For example, a 30 × 0.4 cm column packed with 15-20 μm material will give a delay volume of 1.6 ml at a pressure drop of 150 psi [47]. Packed-bed delay systems offer low dispersion, but care must be used to avoid particulate

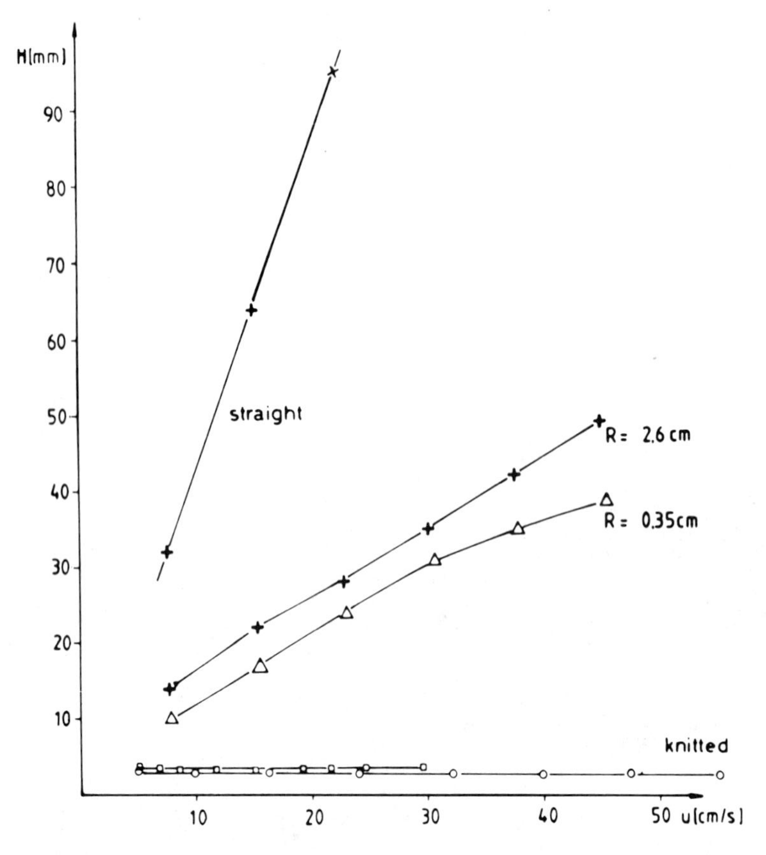

Figure 57. Peak dispersion in ideal, tightly coiled, and knit open tubes: H, theoretical plate height; R, coil radius. (From Ref. 47, with permission.)

or precipitate clogging or bead solubility in alkaline solutions. In addition, due to a lack of radial dispersion, complete mixing of the reagent and column effluent is essential prior to bed entry [45].

Fluid segmentation (FS), based on the Technicon Autoanalyzer technology, operates by the physical segmentation of the mobile-phase/reagent mixture by the periodic injection of fluid plugs. The segmentation fluid which can be a gas or an immiscible liquid, serves to isolate a given segment and prevent longitudinal dispersion. Band-broadening effects are introduced at mixing tees, debubblers (phase separators), and the segment-to-segment dleakage within the delay coil itself. Becausc the leakage is a minor component, long reaction times (5-20 min) are possible [51]. In addition, the segmentation approach allows the use of large-bore (1-2 mm) tubing and a peristaltic pump. In this way, more sophisticated chemistries (up to four reagents) are accessible with simple equipment and little loss of resolution.

A recent discussion of all three delay systems includes a comparison of their general applicability, pressure drop, and level of band broadening [47]. In general, conditions exist under which all three delay systems can produce equivalent behavior. In addition, like the introduction to CRD in this chapter, guidelines and conditions are given to aid in the selection of one type of system over another when they are not equivalent. The reader is referred to Ref. 47 for further details.

Transmembrane

Transmembrane CRD uses Dionex (Dionex, Sunnydale, California) suppressor technology to "add" reagent. As shown in Figure 58, the column effluent is passed through a hollow, semipermeable membrane or fiber that is immersed in reagent solution. The membranes operate on an ion exchange/exclusion principle so that reagent ions are added to the column effluent by means of transmembrane exchange with mobile-phase ions. Analyte ions are maintained within this fiber by ion exclusion effects. In addition to reagent addition, fiber suppressors have been used to remove selected interferents, so care must be exercised in the choice of membrane so that the analytes will not be lost too [52]. The membranes are compatible with strong acids and bases and moderate concentrations of low-molecular-weight alcohols. Reagent supply is continuous but now can be driven by pressurized reservoir, peristaltic pump, or even gravity feed. One interesting application uses TM to raise the postcolumn pH and degrade a β-lactamase inhibitor (clavulanic acid) to a more detectable form [53].

Solid-Phase Reactors

Solid phase is a relatively new aspect of CRD that uses a PB delay system containing a reactive (noninert) packing material. The bed can be consumptive

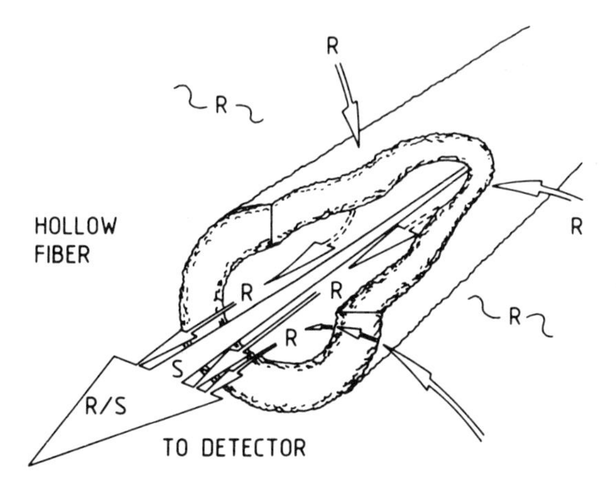

Figure 58. Schematic of transmembrane reagent addition. (From Ref. 41, with permission.)

(with periodic regeneration) or catalytic in nature. In either case, the equipment requirements are essentially those of the basic HPLC system. Solid-phase reactors can be chemically or enzymatically based, and discussion of their application is better served by recent reviews [54,55].

Generated or Doped Systems

Generated or doped systems employ an agent in the mobile phase that is "activated" postcolumn for reaction with the analyte. Doped systems require slow room-temperature reaction kinetics because the reagent is present as a mobile-phase component. The column effluent is then passed through a heated reactor to promote the reaction. One interesting example uses ninhydrin in the mobile phase followed by postcolumn heating to promote the detection of amines at the picomole level [56]. This approach could also be applied as a means of adding one of multiple reagents to simplify the reagent addition protocol of exotic chemistries.

Generated reagent systems convert an innocuous mobile-phase component to an active reagent through interaction with current or light. One scenario, shown in Figure 59, features bromide ion in the mobile phase with postcolumn oxidation to bromine. This type of approach has been applied to the determination of unsaturated fatty acids at the subnanomole level by electrochemically monitoring the change in background bromine levels [57].

F. Final Comments

Given the lack of a truly ideal HPLC detector, an analyst will frequently be faced with a choice among numerous viable detector candidates when attempting to construct an optimized HPLC system for a specific application. To aid in the selection process, the strengths and weaknesses of most detector designs have been discussed previously; characteristic performance

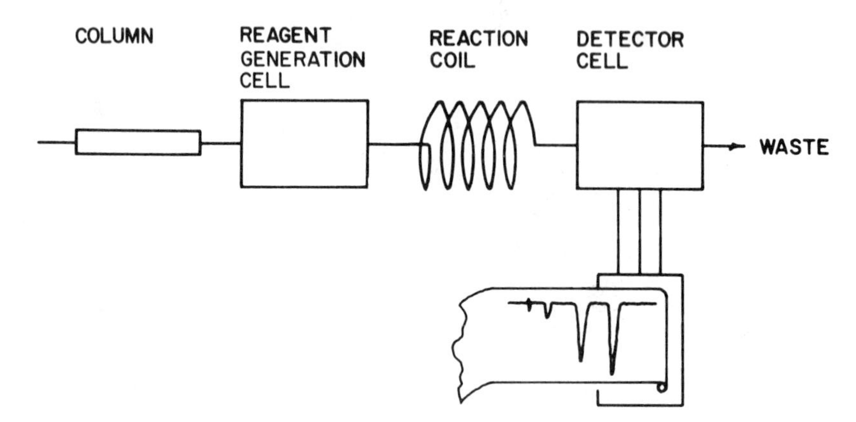

Figure 59. Schematic of a system for the generation of a derivatizing reagent.

Table 2. Characteristic Performance Features of Common HPLC Detectors

Detector	Range of application	Minimum detectable quantity[a]	Linear dynamic range	Inherent flow sensitivity	Temperature sensitivity	Gradient compatibility
RI	Universal	0.1-10 μg	10^3-10^4	No	10^{-4} units/°C	No
UV	Moderately selective	0.1-1 ng	10^4-10^5	No	Low	Yes
Fluorescence	Selective	1-10 pg	10^3-10^4	Yes	Low	Variable
Electrochemical	Selective	1-10 pg	10^6	Yes	1.5%/°C	Variable
Conductivity	Moderately selective	10 ng	10^4	Yes	2%/°C	Variable
LC/MS	Very selective	0.1-1 ng	10^3-10^4	Yes	—	No

[a]For commercially available equipment.

features of the most routinely used detector types are summarized in Table 2. While the present authors do not intend to make blanket generalizations concerning detector applicability, we feel that the usage or popularity ranking of commercially available detectors (variable-wavelength UV, fixed-wavelength UV, RI, electrochemical, and fluorescence, in order of decreasing popularity [37] accurately reflects the relative strengths and weaknesses of each design.

Given the scope of this chapter, each detector type was discussed with a conciseness of detail; to supplement the material contained herein, the reader is directed to one of the many excellent review texts that summarize the current state of the art in HPLC detector design, capabilities, and performance [26,27,33-36,58-61]. We have carefully avoided any recommendation or in-depth discussion concerning specific instruments offered by a particular manufacturer; such comments fall outside the scope of this text, and in any event firm generalizations are difficult to defend given the rapidly changing instrument marketplace. Given the popularity of HPLC as an analytical technique, the number of manufacturers offering HPLC detectors is quite large. Rather than attempt to list these manufacturers here (since the list is somewhat long and is updated quite frequently), the authors direct interested readers to equipment-related summaries that frequently appear in the chemical literature [5,62].

VI. CHROMATOGRAPHIC FITTINGS AND CONNECTING TUBING

Chromatographic fittings and connecting tubing serve as the link between the various chromatographic components that comprise the HPLC system. The present trend in HPLC is to use short, efficient chromatographic columns. These columns minimize chromatographic band broadening, thus allowing extracolumn band-broadening effects to present themselves more clearly. It is therefore becoming more and more important to use properly prepared chromatographic couplings. In order to maintain the system efficiency, it is important to keep extracolumn band-broadening effects to a minimum through use of the proper fittings (compression screws, ferrules, and unions) and connecting tubing.

Problems caused by misuse of connector hardware generally fall into three categories: leakage, contamination, and dead volume [63]. For purposes of clarity, an HPLC connector will be defined as the tubing and fittings that when put together allow mobile phase to flow from one component of an HPLC system to the next. Connectors are generally assembled as needed by the chromatographer. When assembled properly these links provide a leak-free environment and maintain the chromatographic efficiency that the HPLC column generates. If misused, or if a connector is used to join components other than those for which it was intended, problems of leakage or increased dead volume (a cause of poorly shaped chromatographic peaks) can occur.

In order to create a leak-free, chromatographically efficient connection, a compression screw is placed onto a piece of stainless steel tubing that has been cut to the proper length. A ferrule is then swaged in place.

There are several manufacturers of chromatographic fittings, including Parker-Hannifin, Rheodyne, Swagelok, Upchurch Scientific, Valco, and Waters Associates. Although there is a fair degree of interchangeability of fittings from different manufacturers, it is good practice to match the type of compression screw and ferrule to the port for which it is being made. For example, if a connector is being made to link a Waters Associates column to a Rheodyne injection valve, the connector should have a Waters compression screw and ferrule on one end and a Rheodyne compression screw and ferrule on the other end. The positioning of the ferrule on the tubing during swaging is the critical step in creating a leak-free connector without affecting the system dead volume. Although, in general, all female ports used in HPLC have standard 10/32 threads, the depth to which these ports are machined differs between manufacturers. If a ferrule is swaged onto tubing at the proper location for one port and the connector is then used in the female port of another manufacturer, a leak or increased dead volume is likely to be the result (Figure 60a,b). The length of tubing extending beyond the ferrule when properly swaged for different manufacturers' unions is shown in Figure 61. As shown, the distances for fittings from Swaglok, Parker-Hannifin, and Upchurch are the same, indicating that connectors made from these manufacturers' fittings are interchangeable. The distances for Waters, Valco, and Rheodyne are

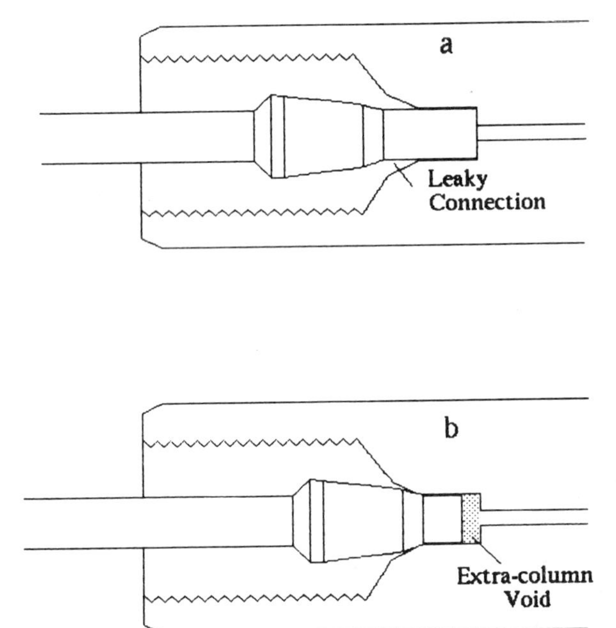

Figure 60. The result of using a connector in a port for which it was not intended. (a) The ferrule is swaged in place too far from the end of the tubing, leading to a leaky connection. (b) The ferrule is swaged in place too close to the end of the tubing, forming a small mixing chamber leading to decreased chromatographic efficiency. (Reproduced with the permission of Upchurch Scientific, Oak Harbor, Washington.)

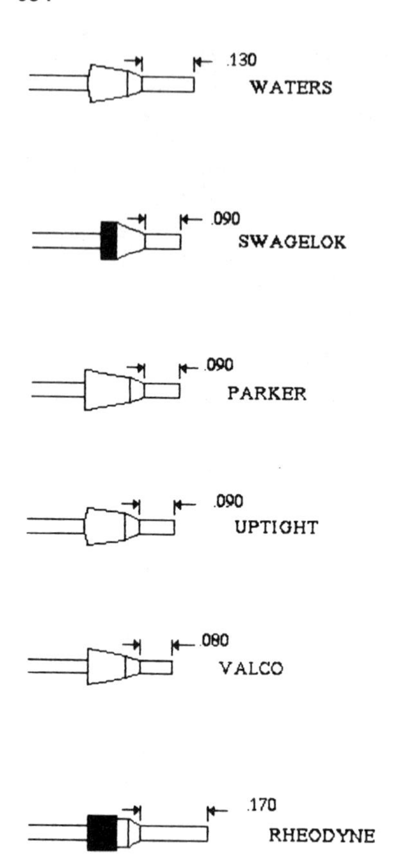

Figure 61. The length of tubing extending beyond the ferrule when properly swaged for different manufacturers' unions. (Reproduced with the permission of Upchurch Scientific, Oak Harbor, Washington.)

all different. Connectors made by swaging these manufacturers' ferrules in place in one of their own ports will not properly fit into other manufacturers' ports. The smart chromatographer will label both ends of each connector with the type of port for which that connector was made.

In the last few years, ferrules made from a synthetic polymer have come into use. Instead of irreversibly locking onto the chromatographic tubing, these ferrules compress around the tubing to form a leak-free connection during use and then return to their original shape when the connector is removed. This allows the polymeric ferrule to be moved along the tubing (or completely removed, if desired) to be reseated at the proper distance for the next port in which it will be used.

Carrying this idea one step further, Alltech Associates distributes a universal fitting made from a synthetic polymer in which the compression screw and ferrule are molded as one piece. This piece fits into a knurled thumb-wheel housing, which when hand tightened provides a leak-proof connection to 5000 psi. The position of the ferrule can be adjusted when tightening to provide a connection that will not add to the system dead volume. When disconnected, the universal fitting is easily removed from the tubing to be used again in a different port. These universal fittings have the added advantage of being compatible with both male and female column end fittings.

The stainless steel tubing used in HPLC connectors comes in several different inside diameters (ID) ranging from approximately 0.007 to 0.050 in. To minimize extracolumn band spreading, it is essential to use short lengths of narrow-bore tubing (0.007-0.020 in.) wherever sample diffusion will adversely affect the chromatography (i.e., between the sample injector and the detector cell). Wide-bore tubing (0.050 in.) is generally preferred prior to the injector and as the outlet tubing from the detector, to reduce the risk of blockages prematurely ending the chromatographic analysis.

In order to minimize system dead volume and eliminate nonswept areas within the chromatographic system, it is essential to use connecting tubing with ends that are cleanly cut perpendicular to the tubing length. On the tubing end there should be no burrs or nicks that might prevent the tubing from seating properly within the port. Tubing of various lengths whose ends have been properly cut can be purchased or prepared in the laboratory.

When cutting tubing in the laboratory, an electric tubing cutter is perhaps the best choice. It produces straight cuts and does not reduce the size of the tubing lumen or produce out-of-round tubing. Electric tubing cutters come with a deburring tool to remove any burrs that form during the cutting process.

C-Clamp cutters, which are actually miniature pipe cutters, are also available to the chromatographer. These mechanical cutters do not reduce the lumen size or produce out-of-round tubing. If electric or mechanical tubing cutters are not available, tubing can still be cut with common laboratory tools. A triangular file can be used to score the tubing at the appropriate length for the cut. Two pairs of long-nose pliers (one pair on either side of the score) can then be used to snap the tubing at the score. This procedure usually produces burrs, which must be removed with the triangular file. If not done carefully, this method of cutting tubing can close the lumen or produce out-of-round tubing, making it difficult or impossible to place the compression screw and ferrule on the tube.

VII. COLUMNS

A. Hardware

The column is that part of the chromatographic system wherein the actual separation process occurs. The column accommodates (i.e., serves as a physical container for) the two chromatographic phases (the stationary phase, which remains in the column, and the mobile phase, which is transported through the column). As such, the column consists of four basic components: the column body (which holds the stationary phase), the column fittings (which allow the column to be connected to the other components of the chromatographic system), porous plugs or frits (which are used to retain the stationary phase within the column) and the stationary phase. Considering the latter, in practical HPLC the stationary phase typically consists of solid particles (the resin), which are made of some material that either itself acts to produce the desired interaction with the solute or that has been chemically modified to facilitate the interaction process. Numerous stationary phases exist and are commercially available, and even a qualitative discussion of those in most common use is beyond the scope of this manuscript. The reader is directed to various texts that provide information related to available phase types [1,63a-66]. A few generalizations concerning the nature of stationary phases used are, however, appropriate. Chromatographic resins are classified either as pellicular (i.e., the individual resin particles consist of a thin porous outer shell of stationary phase that coats a solid core) or porous (i.e., a thin porous shell over a core of stationary phase). Pellicular materials are generally characterized by low column resistance factors (thereby decreasing the magnitude of the pressure drop across the column and increasing efficiency) and tend to minimize the contribution of intraparticle diffusion to the separation efficiency (again improving performance), and thus they were extensively used in early applications of HPLC. However, pellicular materials have a rather large minimum particle size, are difficult to generate reproducibly (the surface layer must be even to achieve high efficiencies), and are typically less rugged than their porous counterparts. This increased ruggedness and the ability to attain small particle size (and thereby increase efficiency) are the major reasons that most modern high-efficiency chromatographic phases are porous.

Porous phases are either irregularly or spherically shaped. Spherically shaped resins are currently more popular (by nearly a factor of 2) among modern chromatographers (66) and offer the advantages of slightly lower pressure drops across the column, a somewhat greater ruggedness (primarily a greater resistance to shear), and a more reproducible packing density. They are, however, more expensive than their irregularly shaped counterparts. Definitive comparative studies have demonstrated that when particle size is roughly equivalent, irregular and spherical resins have similar efficiencies.

Resin particle size, chemical nature of the resin, and the mode of packing all affect column efficiency to some extent. Simplistically, the smallest available particles will produce the highest efficiency. More realistically, as particle size decreases, the pressure drop across the column is increased (decreasing efficiency), ruggedness is decreased, and cost is increased. Most typical HPLC resins can be obtained in particle sizes ranging from 2 to 30 μm; however, the most commonly used particle sizes are 5-7 μm (54 percent of the respondents in a recent survey) and 10-15 μm (26 percent of the respondents) [67].

The chemical nature of the resin backbone also has some effect on its utility in HPLC. Resins are most commonly made from either inorganic materials (i.e., silica or alumina) or organic, polymeric materials. The use of polymer-based columns has increased in recent years in response to dramatic improvements in their efficiency and manufacturing reproducibility. The most widespread uses of polymeric resins are in reversed-phase chromatography, size-exclusion chromatography, and ion (exchange and exclusion) chromatography. These polymeric materials offer excellent ruggedness, a wide operational pH range, and a rather unique (and frequently less complicated) selectivity when compared to their silica-based counterparts.

Finally, the way in which a column is packed may influence its performance somewhat. In general, columns packed via a slurry method are more efficient than those that are dry packed. The slurry packing process tends to minimize the column's resistance factor (minimizing particle fracture during packing), results in a more uniform bed of stationary phase, and is more reproducible than the dry packing process.

Considering the other pieces of column hardware, column bodies are essentially pieces of large-bore (1-10 mm inner diameter for analytical applications) tubing whose sides are extremely smooth so as to prevent channeling in the packed resin from occurring at or near the column wall. The choice of an appropriate material for the column body depends on the operating pressure and the chemical reactivity of the mobile phase, sample matrix, and the solutes. Precision-bore type 316 stainless steel remains the most common body material by virtue of its high pressure tolerance (6000 psi or better), its moderate chemical resistance, and its low cost. However, in lower-pressure applications or where the integrity of the stainless steel is compromised by the nature of the mobile phase (i.e., highly acidic), a variety of alternate materials (e.g., glass, Teflon, polymer, titanium, fused silica) is available. Liquid chromatographic columns are almost universally straight, although coiling does not effect efficiency significantly if the ratio of the coil radius to the column radius is greater than 130 [68]. Analytical columns are commercially available in lengths ranging from 3 to 50 cm and internal diameters from less than 1 mm to somewhat greater than 8 mm. Despite the fairly wide range of possible combinations of diameter and length, a recent user survey shows that a column of 250 × 4.6 mm is by far the most popular among practicing chromatographers (46 percent of the respondents routinely choose this size) [67].

Column fittings must fulfill two requirements: they must not leak, and they must contribute minimal dead volume to the total chromatographic system. Strategies for producing leak-free connections and minimizing extracolumn volume have been discussed previously; the use of modern compression-type fittings (available on most commercial columns) produces essentially zero dead volume connections. Recently, cartridge systems have been introduced by vendors primarily to reduce column cost (specifically by eliminating fittings). In these systems a reusable holder tube houses a replaceable column cartridge (fitted on both ends with polymeric endcaps); the cartridge is sealed in the holder with screw-type compression fittings.

The column frit is a porous plug, placed at the end of the resin chamber, that serves as a transition between the resin and the column fitting and thus serves to "keep the resin in" and reduce voiding. The frit is typically a porous stainless steel plug (although Teflon is acceptable when the pressure is less than 4000 psi) that has a pore size just smaller than the particle size of the stationary phase.

B. Column Design

Column design is the process of defining and producing a column whose performance characteristics are optimal (or at least acceptable) for the desired application. Optimization criteria routinely used include column length, internal diameter, and characteristics of the stationary phase (particle size, type, preparation process, functionality). Of interest in some applications are solvent consumption, desired detection sensitivity, and system compatibility. For practical chromatography, design is simplified since the analyst desires the ability to analyze mixtures with adequate resolution in an acceptable time with a reasonable pressure drop across the column. Theoretical relationships between performance criteria and column properties have been established in the chemical literature (e.g., Ref. 68). For example, equations relating performance, column length, and particle size can be summarized as follows:

$$hv = (\text{\textasciicircum}P)\ d_p^2/@nND_m \tag{3}$$

$$hv = 2 + v^{1.33} + (4 \times 10^{-2})v^2 \tag{4}$$

$$L = Nh\ dp \tag{5}$$

$$t_R = N^2h^2@n(1 + k')/(\text{\textasciicircum}P) \tag{6}$$

where h is the reduced plate height, v is the reduced linear velocity, ^P is the pressure drop across the column, d_p is the particle diameter, @ is the column resistance factor, n is the mobile-phase viscosity, N is the number of theoretical plates, D_m is the diffusion coefficient of solute in the mobile phase, L is the column length, t_R is the retention time of the solute, and k' is the capacity factor. It is noted in passing that equation (4) is an approximation of the reduced performance curve for well-packed columns containing porous particles [68].

The utility of these expressions can be demonstrated by the following example. Consider a chromatographic system with the following characteristics:

$n = 10^{-2}$ g cm^{-1} sec^{-1} (water)
$D_m = 10^{-5}$ cm^2 sec^{-1} (typical of small molecules in water)
@ = 1000 (typical of columns packed with a porous resin)

The analyst desires a separation wherein N = 7000, k' = 5, the column pressure drop is 100 atm, and the resin particle size is 5 microns. The questions that arise are (1) what is the appropriate column length and (2) when will the solute elute? Both are answered as follows:

1. Appropriate values of ^P, d_p, @, n, N, and D_m are placed in equation (3) and hv is calculated.
2. Equation (4) is solved for v and then h.
3. Equation (5) is solved for column length.
4. Equation (6) is solved for t_R.

For the example given, the appropriate column length is 9.9 cm and the analyte retention time is 205 sec.

Column diameter affects performance primarily due to the lateral dispersion of solute across the bed of the resin. As the solute approaches the column wall, it reaches a zone characterized by relatively poor packing order (producing different flow velocities across the cross section of the column), with the result that band broadening occurs. Thus it is desirable that wall effects be minimized (i.e., that the solute never approaches the wall) and thus that the column diameter be optimized. One defines the radial peak width (w_r) arising from a point injection onto the column as

$$w_r^2 = 2.4 d_p z + 32 D_m/u + w_i^2 \tag{7}$$

where z is the solute migration distance, u is the linear velocity, and w_i is the initial band dispersion (typically 2.5 μm for a 10-μl injection).

In order for the column to exhibit the "infinite diameter effect" (and thus prevent the solute from reaching the walls), the column diameter (d_c) should be chosen so that

$$d_c > w_r \tag{8}$$

For most practical applications, this implies that an internal diameter of 4.6 mm is acceptable when a 10-μl injection is to be made onto a 25-cm-long column.

REFERENCES

1. L. R. Snyder and J. J. Kirkland, *Introduction to Modern Liquid Chromatography*, Wiley, New York, 1979.
2. M. D. Walters, *J. Assoc. Off. Anal. Chem.*, *70*, 465-469 (1987).
3. R. W. Yost, L. S. Ettre, and R. D. Conlon, *Practical Liquid Chromatography: An Introduction*, Perkin-Elmer, Norwalk, Connecticut, 1980.
4. J. W. Munson, *Drugs and Pharmaceutical Science*, Vol. 11 (Pharmaceutical Analysis Part B), Chapter 2, Marcel Dekker, New York, 1984.
5. H. M. McNair, *J. Chromatogr. Sci.*, *22*, 521 (1984).
6. W. S. Hancock, Ed., *CRC Handbook of HPLC for the Separation of Amino Acids, Peptides and Proteins*, Vol. 1, CRC Press, Boca Raton, Florida.
7. R. E. Jentoft and T. H. Gouw, *Anal. Chem.*, *38*(7), 949 (1966).
8. L. R. Snyder, *J. Chromatogr. Sci.*, *7*, 595 (1969).
9. S. R. Bakalyar and R. A. Henry, *J. Chromatogr.*, *126*, 327 (1976).
10. J. H. M. Vandenberg, C. B. M. Didden, and R. S. Deelder, *Chromatographia*, *17*, 4 (1983).
11. A. Y. Tehrani, *LC Magazine*, *1*(4), 230 (1983).
12. J. W. Dolan and L. V. Berry, *LC Magazine*, *2*(3), 210-212 (1984).
13. W. Geiger and H. Vollmer, *Hewlett-Packard J.*, *35*(4), 13 (1984).
14. R. W. Allington, private correspondence, August 1986.
15. R. Stevenson, R. Henry, H. Magnussen, and P. Mansfield, *Am. Lab.*, *10*(1) (1978).
16. P. J. Schoermakers, H. A. H. Billet, and L. DeGalan, *J. Chromatogr.*, *205*, 13 (1981).
17. J. J. Kirkland and J. L. Glajch, *Anal. Chem.*, *54*, 2593 (1982).
18. M. Savage, *Am. Lab.*, *11*(5), 49 (1979).
19. R. A. Mowery, *J. Chromatogr. Sci.*, *23*, 22 (1985).

20. *All-Chrom Newsletter* from Alltech-Applied Science, *25*(1), 2 (1986).
21. J. J. Kirkland, W. W. Yau, K. S. Stoklosa, and C. H. Dilks, *J. Chromatogr. Sci.*, *15*, 303 (1977).
22. M. C. Harvey and S. D. Stearns, *Anal. Chem.*, *56*, 837 (1984).
23. M. C. Harvey and S. D. Stearns, HPLC Sample Injection and Column Switching, in *Liquid Chromatography in Environmental Analysis* (J. F. Lawrence, Ed.), Humana Press, Chicago, 1983, p. 301.
24. R. C. Simpson and P. R. Brown, *Liquid Chromatogr.*, *3*(6), 537 (1985).
25. American National Standard ANSI/ASTM E685-79, in *Annual Book of ASTM Standards*, American Society for Testing and Materials, Philadelphia, 1979.
26. R. P. W. Scott, *Liquid Chromatography Detectors*, Elsevier, New York, 1986, p. 12.
27. M. N. Munk, *Chromatogr. Sci.*, *23*, 165-204 (1983).
28. D. T. Gjerde, J. S. Fritz, and G. Schmuckler, *J. Chromatogr.*, *186*, 509-519 (1979).
29. H. Small, T. S. Stevens, and W. C. Bauman, *Anal. Chem.*, *47*, 1801-1809 (1975).
30. C. A. Pohl and E. L. Johnson, *J. Chromatogr. Sci.*, *18*, 442-452 (1980).
31. E. S. Yeung and R. E. Synovec, *Anal. Chem.*, *58*, 1237A-1254A (1986).
32. A. F. Poile and R. D. Conlon, *J. Chromatogr.*, *204*, 149-152 (1981).
33. P. C. White and T. Catterick, *J. Chromatogr.*, *280*, 376-381 (1983).
34. P. C. White, *Analyst*, *109*, 677-697 (1984).
35. H. Lingeman, W. J. M. Underberg, A. Takadata, and A. Hulshoff, *J. Liquid Chromatogr.*, *8*, 789-874 (1985).
36. D. C. Shelly and I. M. Warner, *Chromatogr. Sci.*, *23*, 87-123 (1983).
37. H. G. Barth, W. E. Barber, C. H. Lochmuller, R. E. Majors, and F. E. Regnier, *Anal. Chem.*, *58*, 211R-250R (1986).
38. D. K. Roe, *Anal. Lett.*, *16*, 613-631 (1983).
39. R. W. Frei and U. A. T. Brinkman, *Trends Anal. Chem.*, *1*, 45 (1981).
40. K. Imai, T. Toyo'oka, and H. Miyano, *Analyst*, *109*, 1365 (1984).
41. R. M. Cassidy and B. D. Karcher, Post-Column Reaction Detection of Inorganic Species, in *Reaction Detection in Liquid Chromatography* (I. S. Krull, Ed.), Marcel Dekker, New York, 1986.
42. R. Weinberger, Drug Determination in Biological Fluids by Liquid Chromatography-Fluorescence, in *Therapeutic Drug Monitoring and Toxicology by Liquid Chromatography* (S. H. Y. Wong, Ed.), Marcel Dekker, New York, 1985.
43. R. W. Frei, Reaction Detectors in Liquid Chromatography, in *Chemical Derivatization in Analytical Chemistry* (R. W. Frei and J. F. Lawrence, Eds.), Plenum Press, New York, 1981.
44. T. D. Schlabach and R. Weinberger, Solution Chemistry for Post-Column Reactions, in *Reaction Detection in Liquid Chromatography* (I. S. Krull, Ed.), Marcel Dekker, New York, 1986.
45. J. F. K. Huber, K. M. Jonker, and H. Poppe, *Anal. Chem.*, *52*, 2 (1980).
46. R. W. Frei, L. Michel, and W. Santi, *J. Chromatogr.*, *142*, 261 (1977).
47. B. Lillig and H. Engelhardt, Fundamentals of Reaction Detection Systems, in *Reaction Detection in Liquid Chromatography* (I. S. Krull, Ed.), Marcel Dekker, New York, 1986.

48. R. M. Cassidy, S. Elchuk, and P. K. Dasgupta, *Anal. Chem.*, *59*, 85 (1987).
49. H. Engelhardt and U. D. Neue, *Chromatographia*, *15*(7), 403 (1982).
50. K. Hofmann and I. Halasz, *J. Chromatogr.*, *199*, 3 (1980).
51. R. W. Frei, *J. Chromatogr.*, *165*, 75 (1979).
52. D. Brown, R. Payton, and D. Jenke, *Anal. Chem.*, *57*, 2264 (1985).
53. J. Haginaka, J. Wakai, and H. Yasuda, *Anal. Chem.*, *59*, 324 (1987).
54. S. T. Colgan and I. S. Krull, Solid-Phase Reaction Detectors for HPLC in *Reaction Detection in Liquid Chromatography* (I. S. Krull, Ed.), Marcel Dekker, New York, 1986.
55. L. D. Bowers, Enzyme Reaction Detectors in HPLC, in *Reaction Detection in Liquid Chromatography* (I. S. Krull, Ed.), Marcel Dekker, New York, 1986.
56. J. N. LePage and E. M. Rocha, *Anal. Chem.*, *55*, 1360 (1983).
57. W. P. King and P. T. Kissinger, *Clin. Chem.*, *26*, 1484 (1980).
58. R. L. Stevenson, *Chromatogr. Sci.*, *23*, 23-86 (1983).
59. T. M. Vickrey and R. L. Stevenson, *Chromatogr. Sci.*, *23*, 322-353 (1983).
60. N. A. Parris, *Instrumental Liquid Chromatography*, Elsevier, New York, 1984.
61. C. F. Poole and S. A. Schuette, *Contemporary Practice of Chromatography*, Elsevier, New York, 1984.
62. H. M. McNair, *J. Chromatogr. Sci.*, *25*, 564-582 (1987).
63. J. Dolan and V. V. Berry, *LC Magazine*, *2*(1), 20 (1984).
63a. R. E. Majors, *LC-GC Magazine*, *6*, 382-388 (1988).
64. K. K. Unger and B. Anspach, *Trends Anal. Chem.*, *6*, 121-125 (1987).
65. L. J. Lorentz, High Performance Liquid Chromatography, in *Modern Methods of Pharmaceutical Analysis*, Vol. III (R. E. Schirmer, Ed.), CRC Press, Boca Raton, Florida, 1982.
66. R. E. Majors, *LC-GC Magazine*, *6*, 298-302 (1988).
67. H. Barth, E. Dallmeier, and B. L. Karger, *Anal. Chem.*, *44*, 1726-1732 (1972).
68. J. H. Knox, Theory of HPLC. Part I: With Special Reference to Column Design, in *Practical High Performance Liquid Chromatography* (C. F. Simpson, Ed.), Heydon and Son, New York, 1976.
69. H. H. Willard, L. L. Merritt, J. A. Dean, and F. A. Settle, *Instrumental Methods of Analysis*, 6th ed., Wadsworth, 496-500, 1981.

23

Supercritical Fluid Chromatography Instrumentation

THOMAS L. CHESTER / The Procter & Gamble Company, Cincinnati, Ohio

I. INTRODUCTION

Supercritical fluids have been known since the nineteenth century. However, they have only recently been used as chromatographic mobile phases [1]. At first, supercritical fluid chromatography (SFC) was done exclusively with packed columns. However, the trend in recent years, especially in the published literature, has been with the use of open-tubular, or capillary, columns. This chapter will describe instrumentation, techniques, and guidelines specifically for capillary-column SFC. However, much of the information presented is also applicable to packed-column SFC.

A. Supercritical Fluids

We usually think of matter existing in three states: solid, liquid, and gas. Imagine that we take a pure gas and put it in a vessel with variable volume. We adjust the temperature to just above the normal boiling point of this gaseous material, and we adjust the volume until the pressure is 1 atm. Now, if we isothermally compress the gas, we find the pressure increases steadily to a point. The pressure stops increasing when it reaches the vapor pressure of the corresponding liquid, and condensation of the gas begins. At that point liquid appears and the material separates into two distinct phases. Further reduction in the volume reduces the mass in the gas phase and increases the mass in the liquid phase. But no change in pressure is observed as long as two phases coexist in equilibrium at the fixed experimental temperature.

If the experiment is repeated at a slightly higher fixed temperature, we find that the pressure required to achieve condensation is higher than before. This trend continues, with increasing temperature giving an increased vapor pressure, until a temperature is reached where no amount of pressure can cause condensation to occur. This behavior is illustrated in the pressure-temperature diagram shown in Figure 1. The highest temperature where gas and liquid phases can coexist is the critical temperature. Above the critical temperature the gas is a "supercritical fluid." We can compress it until it becomes as dense as a liquid and begins to act like a liquid in its ability to dissolve and solvate other materials. Yet it never undergoes condensation and retains some characteristics of gas.

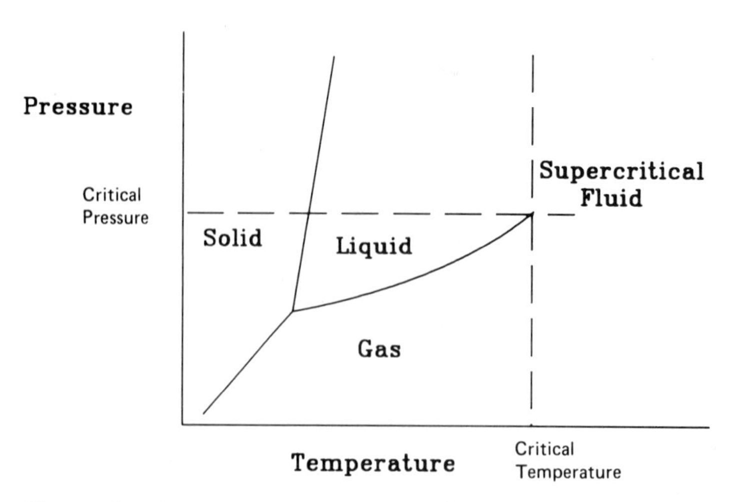

Figure 1. Pressure-temperature diagram for CO_2.

Another way to describe the formation of a supercritical fluid is shown in Figure 2. Here, 1 g of CO_2 is sealed in a vessel with a 2-cm^3 volume. At 0°C it exists in two phases. As the temperature is raised, the volumes and individual masses of the two phases change as required by the total mass and volume used in the experiment. In this example, the volume of liquid increases in going from 0°C to 20°C. But if the total mass of CO_2 were somewhat smaller, the volume of liquid would decrease with the same temperature change.

Also, as the temperature is raised, the densities of the two phases approach each other. When the temperature reaches the critical temperature the densities of the two phases become equal, the phase boundary disappears, and a single supercritical-fluid phase exists as the temperature is raised further. Cooling the vessel reverses the process and a phase separation occurs upon lowering the temperature through the critical point. Experiments like this are fairly easy to perform [2,3] and give an impressive visual display—density and refractive index change rapidly around the critical temperature, so small temperature differences within the vessel result in gravity-driven fluid motion that is easy to see.

Supercritical fluids have physical properties between those of liquids and regular gases. The more a supercritical fluid is compressed, the more liquid-like it becomes. In general, diffusion coefficients of solutes dissolved in supercritical fluids are between the values of the same solute dissolved in a liquid solvent or evaporated into a gaseous carrier. The density of a supercritical fluid is usually between that of liquid and gas, but depends on the temperature and pressure. Solvent strength of a supercritical fluid increases proportionally with density at a given temperature. However, increasing the density lowers the values of solute diffusion coefficients. Viscosities of supercritical fluids are near the values of gases.

B. Supercritical Fluid Chromatography

In addition to the physical properties just mentioned, many common supercritical fluids have chemical properties (or, perhaps more accurately, lack chemical properties) making them especially valuable as chromatographic mobile phases. A list of chromatographically useful supercritical-fluid mobile phases is given in Table 1, along with pertinent physical data for each.

Carbon dioxide has been the most popular supercritical-fluid mobile phase to date. It has a low critical temperature and is compatible with a large variety of gas chromatography (GC) and high-pressure liquid chromatography (HPLC) detectors. Its compatibility with the flame ionization detector (FID) is particularly noteworthy. CO_2 is also readily available in high purity, has a low price compared to other solvents, is nontoxic, and is easily and safely disposed of by venting.

Supercritical fluids are compressible. Since the solvent strength of a supercritical fluid at a given temperature depends on its density (which, in turn, depends on temperature and pressure), programmed elution in SFC is usually done with some form of pressure programming. Typically, the temperature is fixed and the pump delivering the mobile phase is programmed to raise the pressure as a function of time. Commercial instruments usually provide density programming as well. Here, the density desired throughout the chromatogram is specified, and the pump controller continuously computes and updates the required pressure. Temperature can also be programmed, either alone or in concert with a pressure or density program, and is useful in separating samples with both high- and low-volatility components. However, temperature programming should be used cautiously—the influences of (isothermal) pressure and density changes on solute retention are straightforward, but temperature changes can affect

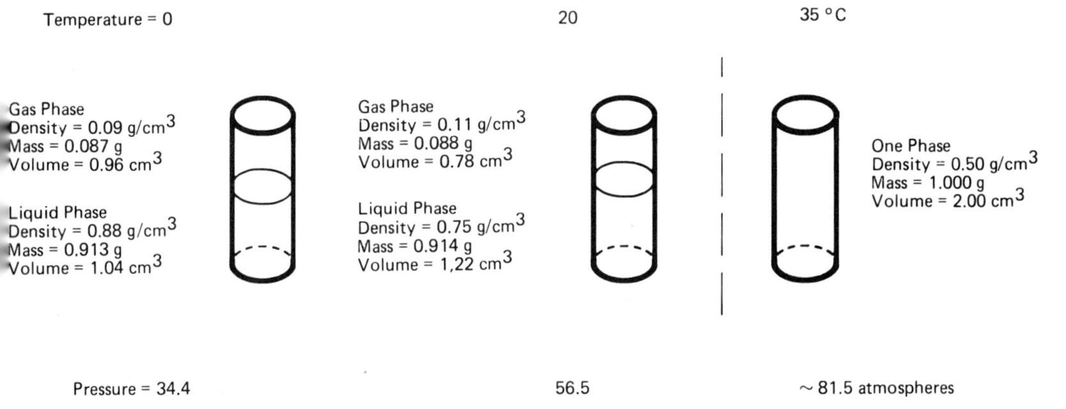

Figure 2. Formation of a supercritical fluid in a sealed container.

Table 1. SFC Mobile Phases and Physical Data[a]

	Critical temperature	Critical pressure	Dipole moment
Xe	16.6	58.	0
CO_2	31.0	72.9	0
N_2O	36.4	71.7	0.167
SF_6	45.5	37.1	0
NH_3	132.4	112.5	1.47
SO_2	157.8	77.7	1.63
H_2O	374.1	218.3	1.85
n-Pentane	196.6	33.3	0
Methanol	240.0	78.5	1.70

[a]*Handbook of Chemistry and Physics*, 65th ed., CRC Press, Inc., Boca Raton, FL, 1984-85, except for data on SF_6 which is taken from W. Braker and A. L. Mossman, *Matheson Gas Data Book*, 6th ed., Matheson Gas Products, Inc., Secaucus, NJ, 1980.

solute selectivity, making retention of individual solutes difficult to predict. Solvent composition programming (gradient elution) is also possible in SFC, and is commonly used with packed-column systems equipped with ultraviolet absorbance detectors. However, the effect of organic cosolvents on retention when CO_2 is the mobile phase is small in capillary SFC compared to packed-column SFC. Presumably, this is because of the tremendous reduction in relative surface area and superior deactivation of capillary columns compared to packed columns. The addition of organic solvents also severely limits the use of the FID, eliminating (or greatly restricting) a big advantage of SFC over HPLC. Thus, solvent programming is rarely used in capillary SFC.

C. Guidelines for Choosing SFC

The first practical capillary SFC instrument was described in 1981 [4]. Only a few groups were actively working in SFC through its first 5 years, and progress was relatively slow. Naturally, as with any new technique, many of the first applications were not particularly challenging, even by the standards of the day. Thus, a fair amount of doubt was raised about the long-term value of the technique based on those early applications. The prospects are considerably brighter today as an ever-increasing number of difficult applications, impossible by GC or HPLC, reach the literature.

The first guideline in choosing a separation technique is: "Use GC if it can solve the problem." GC is the fastest separation technique and, by far, the most efficient when analysis time is limited [5]. Even the most enthusiastic SFC proponent will agree that GC should be used every time it can do the job. The development of aluminum-clad columns and improved stationary phases allows column temperatures exceeding 400°C to be routinely used [6,7]. Thus, alkanes as heavy as C_{90} can now be separated by GC.

In practice, many solutes of interest are not as thermally stable as alkanes and may not survive such high temperatures during the course of the separation. More experience in GC at temperatures in the 350-450°C range is required to find out where the analysis limits exist for various solute classes. Thermally unstable analytes cannot be eluted by GC. They require the use of a solvating mobile phase and low separation temperatures. In addition, there is a practical analyte size limit in GC, even for thermally stable analytes, due to the requirement of a minimum analyte vapor pressure at the maximum column temperature. The absence of sufficient volatility requires the use of a solvating mobile phase for elution.

Compared to GC, SFC is not absolutely dependent on volatility, and thus has a much larger mass range—a 3000-dalton range is fairly easy to achieve with capillary columns and a CO_2 mobile phase. A 7000-dalton range has been demonstrated with a flame ionization detector (FID) in use. Prospects for an even higher mass range are good. Since highly elevated temperatures are not necessary (with the right choice of mobile phase), SFC is applicable to problems concerning thermally unstable solutes. And all of this is possible while retaining the use of most GC detectors. The cost for doing SFC instead of GC is 100 to 1000 times longer analysis times to achieve the same column efficiency on identical columns, or the use of much smaller column diameters with the concomitant loss in maximum sample size, or the sacrifice of separation efficiency in exchange for shortening the analysis time. In practice, a compromise solution is common: column diameters used (typically 50-100 μm) are smaller than is common in GC, and mobile-phase velocities well above optimum are used to reduce analysis time. Overall, efficiencies are lower than in typical capillary GC and analyses take about five times longer.

When an instrument-based separation is needed and GC is impossible, HPLC is usually the second choice. HPLC, compared to GC, is virtually limitless in mass range. (The actual limit is about 10^5-10^6 daltons.) In addition, HPLC is applicable to both nonionic and ionized solutes. (GC only works on nonionic solutes. SFC may work for some classes of ionized solutes, but the scope is unknown.) HPLC is usually limited to short (<1 m), packed columns, and thus is not capable of achieving the column efficiencies common in capillary GC and SFC. This is not as severe a limitation as it might seem because of the wide selectivity variation possible through mobile-phase composition options.

However, an even bigger limitation in HPLC is the absence of a versatile, universal detector analogous to the FID. A variety of sensitive but selective detectors exists. But these are of no good to a chromatographer unsure of the identity or properties of the species being determined. Enough knowledge of the problem must exist to guide HPLC detector selection. Often, only a little information is needed. For example, when trying to separate an off-color impurity from a product, spectroscopic detection is obvious, even if the chemical nature of the impurity is unknown. But an HPLC detector cannot be rationally selected when too little is known about the problem.

SFC is the only technique available with the combination of a low-temperature, programmable-strength mobile phase, capillary column efficiency, and universal detection for addressing problems within its mass range. Thus, the second guideline is: "If the sample is nonionic and you can't rationally choose a specific HPLC detector, then use SFC-FID if the mass range allows it."

When on-line mass or infrared spectrometric detection is needed for a particular application, other factors come into consideration. Supercritical-fluid chromatographs are relatively easy to interface to mass spectrometers (MS) [8-12] and infrared spectrometers [13-18]. It is easier to remove highly volatile supercritical fluids than liquids from a mass spectrometer source, so direct fluid injection works better with SFC than with HPLC. Thermospray ionization has not been reported in SFC-MS (at the time of this writing), but is widely used in HPLC-MS. Even so, it works best for polar analytes and is complemented nicely by the good performance of SFC-MS with low-polarity analytes. Infrared transmission is much better in selected supercritical fluids than in most liquids, thus significantly widening the applicability of on-line, postseparation, infrared detection.

Further guidelines are best made by individuals for their particular situations. SFC is *complementary* to GC and HPLC, and all three share some overlap, but each also has a unique area of capability. In many cases, the technique choice will be a toss-up between these separation techniques, or between chromatography and direct spectroscopic techniques for solving a particular problem. Only with experience will we learn what is best or most cost-effective for particular types of problems.

II. INSTRUMENTATION

A. General Features

The functional components of a typical capillary supercritical fluid chromatograph (with flame ionization detection) are shown in Figure 3. Commercial instruments often have fewer visible parts simply due to the packaging. However, every SFC instrument has these same basic components.

Details of the components appear in the next few sections. Basically, they work together like this: A programming device directs a pump to supply mobile phase at a selected, or programmed, pressure. Sample is introduced with a sampling valve. The injected sample may be split in some fashion to avoid overloading the column with injection solvent, or to expedite the process of clearing sample and solvent from the volume of the valve and precolumn fittings. Sample components are separated on a fused-silica column, similar to a GC column, placed in an oven. With low-pressure detectors, the column outlet is interfaced to the detector with a flow restrictor. Pressure is maintained over the entire system to within about a millimeter of the restrictor outlet, then dropped abruptly just ahead of the detector. With pressurized detectors, the column and

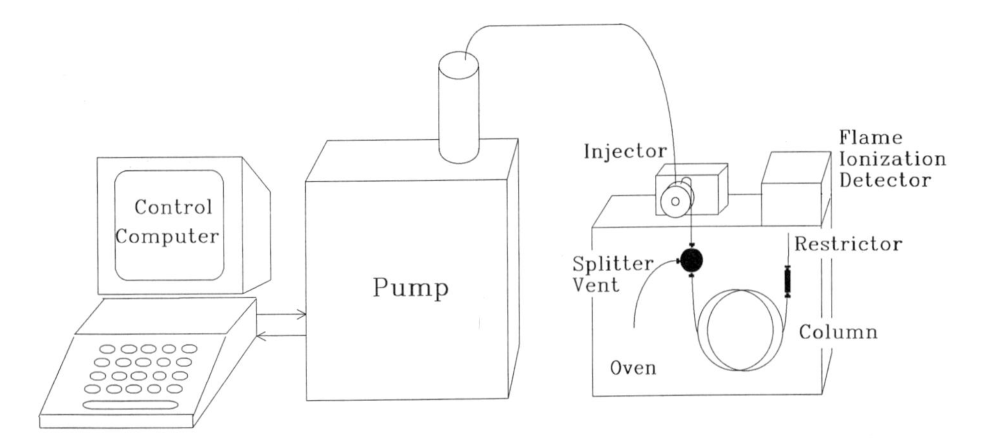

Figure 3. Schematic diagram of a capillary supercritical-fluid chromatograph with a flame ionization detector.

detector are directly connected, or detection is done "on-column." A restrictor or valve follows the detector to reduce the pressure and vent the mobile phase from the system. Some packed-column instruments work differently. The pump controls flow, and pressure is adjusted after the detector with a regulator. However, low-pressure detectors are most frequently used with a pressure-controlling pump.

With the usual arrangement, solvent strength is controlled by the regulation of oven temperature and pump pressure. Mobile-phase velocity is not directly controlled but depends on the column and restrictor dimensions, pressure, and mobile-phase viscosity. Velocity varies somewhat throughout a pressure-programmed chromatogram, and is adjusted by changing the restrictor dimensions.

Operating pressures are typically in the 70-400 atm range. The pressure drop (over a capillary column) required to give the desired mobile-phase flow velocity range is only around 1 atm. Thus, the selected system pressure is maintained practically unattenuated over the entire column.

B. Pump

Because of the small pressure drop in capillary columns, the mobile-phase flow must be free of pulses. Reciprocating pumps and diaphragm pumps, typical of modern HPLC instruments, have been used successfully with packed-column SFC in many cases [for example, 19-23]. High-pressure syringe pumps have been most successful in capillary SFC [24,25]. They may range in volume from about 10 ml to about 250 ml. A small-volume pump may be perfectly adequate in many applications, but requires refilling more frequently than larger pumps. In choosing a pump, you must be sure you have enough deliverable volume (also accounting for filling efficiency) to complete one chromatogram under conditions of maximum mobile-phase usage. Another alternative is a dual pump system with automatic refilling—one side pumps while the other fills. But, again, a single syringe pump of sufficient volume is the simplest option.

The pump must be able to deliver pressures accurately up to about 400 atm. It should be able to run at several milliliters per minute in order to rapidly change pressure when pressure steps are programmed. (Even though the pump volume changes rapidly at this rate, the mobile-phase flow rate is dictated by the column outlet pressure. The rapid volume change only compresses the mobile phase.)

Even though most supercritical-fluid mobile phases are liquid at room temperature, they are still fairly compressible. Pumps generally work better when refrigerated, since lowering the temperature reduces the mobile-phase compressibility. However, a well-designed pump will work acceptably at room temperature in nearly all practical circumstances.

Good pressure control is an absolute necessity in SFC. The pressure should be programmable in increments no larger than about 0.2 atm, and the pump should be able to control the pressure over its entire pressure range to within 0.2 atm of the set point. A block diagram of a simple pump and pressure control system is shown in Figure 4. The circuit introduced by Van Lenten and Rothman [24] has been widely used to convert a stepper motor-driven, flow-programmed pump to pressure-controlled operation. An improved version of that circuit is shown in Figure 5. The external signals required are simply a dc voltage from a pressure transducer (scaled to 10.0 mV/atm) and a dc reference voltage (also scaled to 10.0 mV/atm).

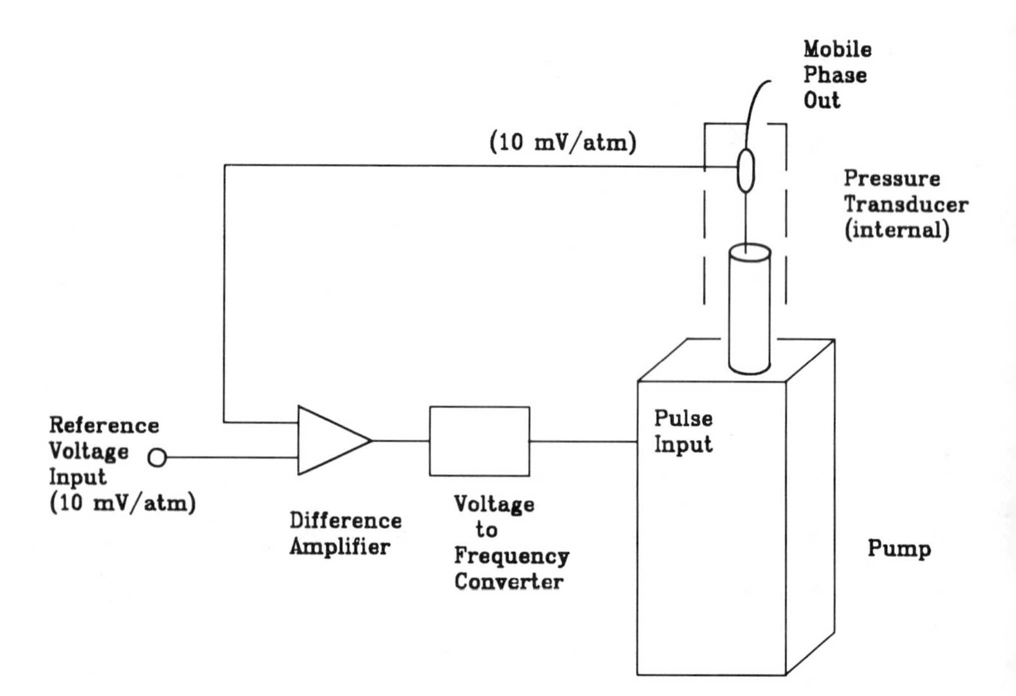

Figure 4. Block diagram of a pressure-controlled pump for SFC.

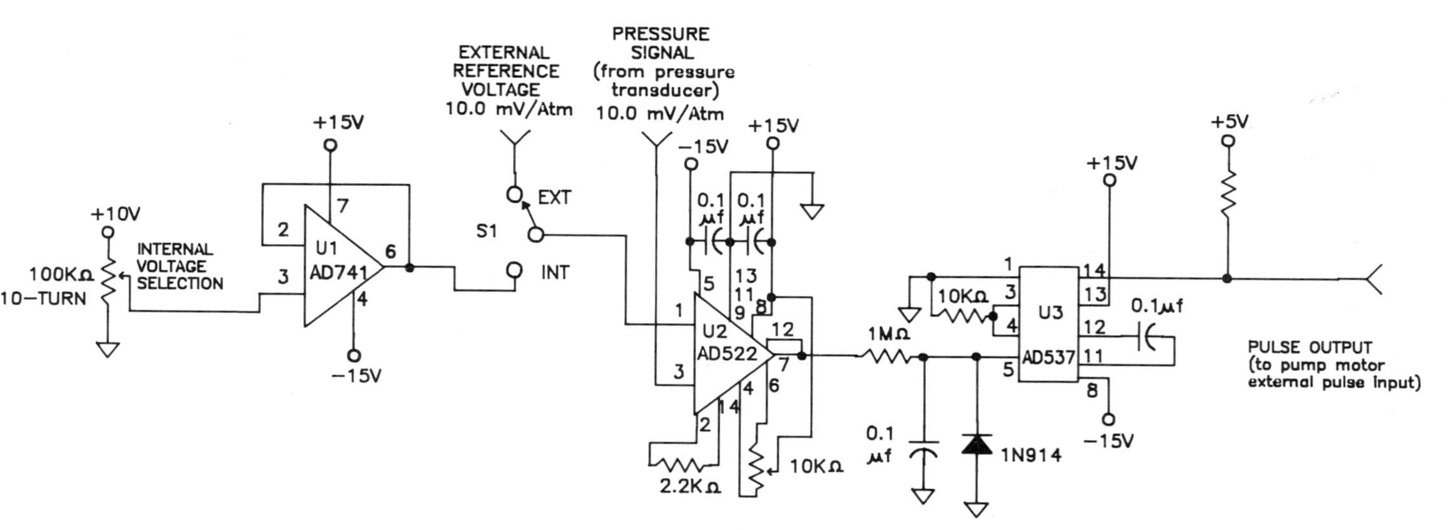

Figure 5. Schematic diagram of a simple analog electronic pump controller for pressure control. Switch S1 selects the reference voltage from an internal source (U1) or from an external source such as a computer-programmed digital-to-analog converter. Amplifier U2 compares the selected reference voltage with the pressure transducer output and generates an error signal. This is converted to a frequency by U3 and then applied to the pump external pulse input.

Pressure programming is simply accomplished by programming the reference voltage from an external source. The most versatile source is a computer-programmed digital-to-analog converter. Thus, relatively simple computer programming can produce the desired pressure program. Density programming (or any other desired pressure variation) is accomplished by computing the pressure required as a function of time throughout the duration of the chromatogram and sending the appropriate reference voltage to the pump controller. A digital-to-analog converter with 12 bits scaled from 0 to 4 V (corresponding to 0-400 atm) provides pressure resolution of 0.1 atm.

Alternatively, the analog control illustrated can be done digitally. An analog-to-digital converter can be used to read the pressure signal. The same computer used for programming the pressure could calculate the motor speed needed and send pulses directly to the stepper motor controller (or may use a programmable oscillator as a buffer). This approach seems to be preferred by the commercial suppliers, not only for increased sophistication, but for ease of changing or updating the control algorithms. One advantage to increased sophistication is that the pump motor may be controlled in both directions. This is not possible with the simple analog system, described earlier (Figure 5), where pressures can only be reduced with manual override of the pump motor. Dual direction control also allows for slightly faster pressure equilibration after a big pressure change, especially when the pump is not refrigerated.

Programming for the analog control system is very easy. Since the computer does so little, a BASIC interpreter gives more than enough computing speed. Commercial instruments usually offer stepwise-linear pressure and density programming, and may offer more sophisticated programming options. In practice, there is very little difference between linear pressure and density programs for CO_2 when the temperature is about 100°C or higher. So choosing between linear pressure and density programming becomes important only if many applications are of thermally unstable materials when temperatures must be kept low. Other programming schemes (such as simultaneous density-temperature programming) are beginning to be offered on commercial instruments.

C. Injector

Injection in SFC has aspects of GC and HPLC. In its simplest form, a high-pressure sampling valve is used to switch a small-volume sample into the mobile-phase stream. However, the volume available with the smallest commercial valves (0.06 μl) still causes some problems, especially when solvent-sensitive detectors are used. For example, using an FID with direct injection of 0.1 μl of solvent from a typical valve can produce a solvent peak over 15 min wide with normal mobile-phase velocities.

The room-temperature, flow-splitting injector [26] of Figure 6 is used to reduce the volume effectively injected to a few nanoliters to avoid the big solvent overload. The increased carrier velocity through the injection valve and fittings helps the solvent peak shape as well.

Rather than splitting the valve volume in flow, it is possible to directly connect the column to the valve and split the sample volume in time [27,38].

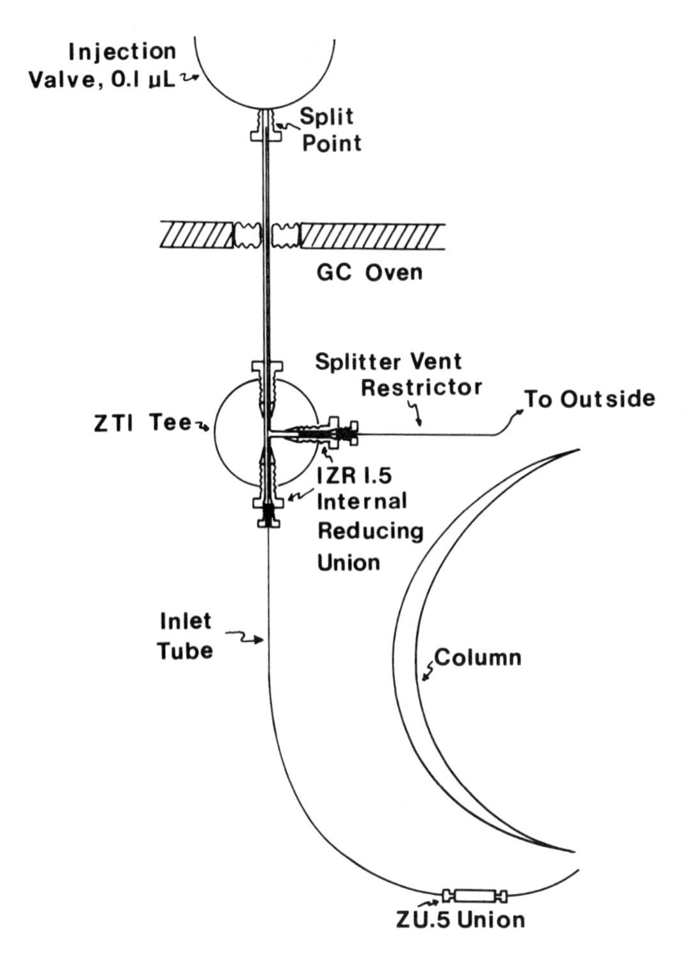

Figure 6. Schematic diagram of a flow-splitting injector for SFC. (Reprinted with permission from Ref. 36.)

That is, by rapidly switching the injection valve from the load to inject position and back, less than the full contents of the valve will be injected on the column. The remainder is flushed out the valve waste vent when the next sample is loaded. Although there is absolutely no benefit with this design over flow splitting in terms of the effective volume injected, there is a benefit in terms of accuracy. Flow-splitting injectors suffer from discrimination effects due to the differences in diffusion coefficients between the solutes, combined with viscosity differences between the injection solvent and the mobile phase. Split ratios can change between components as well as with changes in injection conditions when flow splitting is used. Discrimination should not occur with the time-splitting injector, since the split point is actually the outlet side of the sample channel in the valve rotor—virtually no mixing of injection solvent and mobile phase occurs before the entire injection process is over. Nonetheless, suspicion is warranted and freedom from discrimination should be verified in every case until sufficient experience with this injector is published.

Little is known about the effects of inlet flooding by the injection solvent in capillary SFC. However, only 50 nl of solvent would *completely fill* a 25-mm length of capillary column with a 50-μm inside diameter. Obviously, this plug of solvent will travel some additional length down the column before the liquid dissipates. Since the liquid injection solvent will be a much stronger solvent than the mobile phase (at least at the time of injection), solute could be carried well into the column by the liquid, and be spread over the entire liquid path. When this happens on the column there are no easy prospects for getting the sample components reformed into a narrow band. If this spreading were done over a deactivated, but otherwise uncoated, inlet tube [28], then refocusing on the head of the column would be trivial. Solute would begin moving on the inlet tube at much lower pressures than required to move solute along the column. Thus, when solute reached the head of the column, it would partition into the stationary phase and effectively stop there, temporarily, until the pressure was raised sufficiently to begin solute migration. Refocusing would occur automatically due to the change in phase ratio at the inlet tube-column interface if the injection pressure were low enough, the pressure program allowed for it, and the length of the inlet tube were correct. The entire operation is illustrated in Figure 7. The use of an inlet tube is not particularly important with low-effective-volume injectors giving minimal solvent flooding. But with the desire to use larger volumes to achieve lower detection limits, the effects of inlet flooding will not always be negligible.

The ultimate injection goal for capillary columns is true, splitless injection with narrow solvent peaks and conveniently large sample volumes. Work is in progress in several labs on new injection ideas for SFC, although none have been reported at the time of this writing.

D. Oven

At this point the SFC instrument starts looking more and more like a GC. In fact, nearly all of the capillary SFC instruments reported use GCs to provide the oven and detector. There are no special requirements for the oven in a basic SFC instrument that cannot be met by a rudimentary GC: isothermal temperature control to about ±0.1°C, and a temperature range of about 35-250°C (which is about as high as anyone has done SFC so far; this is not too important since virtually any commercial GC or SFC you can get will have a 400°C or higher upper limit for its oven). For most applications, oven temperature programmability is not necessary, although it is needed for simultaneous density-temperature programming, and may be needed for some of the splitless injection techniques under development.

One other requirement of the oven is its size. There are more things for the operator to do inside an SFC oven than in a GC—there are more fittings to make and take care of, and some of the injection hardware will be in the oven compartment taking up room. So it is important to have a large oven by GC standards.

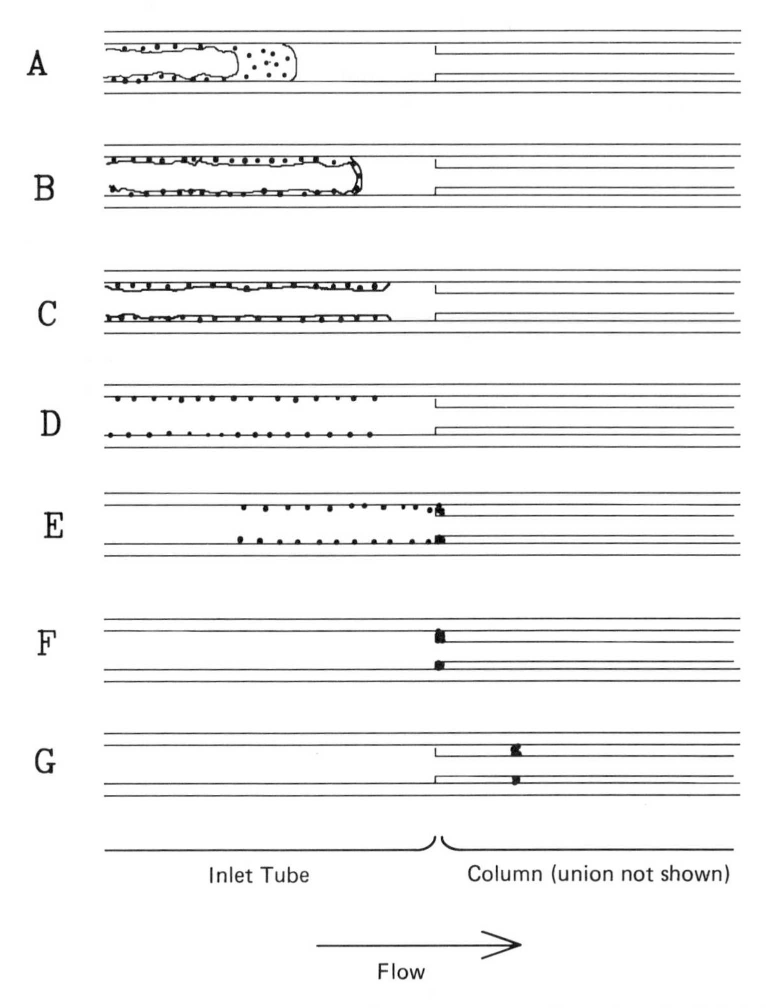

Figure 7. Representation of sample spreading due to inlet flooding, and the refocusing effect possible with the use of an uncoated inlet tube. In this figure, a phase separation between injection solvent and mobile phase is assumed. However, a similar effect will occur with mobile-phase-miscible solvents—sample spreading can occur due to the high solvent strength of the mixed mobile phase temporarily produced with injection, and lasting until the injection solvent is sufficiently diluted by mobile phase. (A) Solute is carried over a length of inlet tubing (or column) by the liquid injection solvent. (B,C) Sample spreading continues until injection solvent is depleted (or sufficiently diluted). (D) Solvent-free solute is left spread over what was the flooded zone. (E,F) If flooding occurs on an uncoated inlet tube, solute migration will begin at mobile-phase strength too low for significant migration on the column stationary phase. Solute reaching the stationary phase is refocused into a narrow band. (G) Solute migration along the column begins as the mobile-phase strength is raised further.

E. Column

All capillary SFC is done today using fused-silica columns. They are almost identical to modern GC columns, varying only in diameter and film thickness. The column and stationary phase must be inert and insoluble with respect to the mobile phase. Thus, the biggest single step in the development of capillary SFC was the development of nonextractable, "cross-linked" GC columns. In addition to cross-linking, many column manufacturers are also "bonding" the stationary phase to the inside surface of the capillary tube. This adds even more mechanical stability to the stationary phase and, presumably, prevents the stationary phase from shearing off the surface due to mobile-phase flow across any film irregularities.

The number of suitably stable stationary phases is relatively small, and nearly all are substituted silicones. Nonetheless, the available stationary phases encompass virtually the same polarity range available in GC. Table 2 lists some of the stationary phases available with commercial SFC columns.

Ideally, when optimum velocity is used, the diameter of a capillary column for SFC could be made very small (a few micrometers, depending on the length) before the detrimental effects of column pressure drop outweighed the advantages of analysis time and efficiency afforded by small diameters [29]. However, the disadvantage in sample capacity associated with very small diameters would be disastrous—with very-small-diameter columns, detection limits would be too high to be useful because the maximum sample size would be so small. So the trend today is to use columns somewhat smaller in diameter than GC columns, but still manageable in terms of the sample sizes and analysis times allowed. Inside diameters of 50-100 μm are popular and available commercially from several sources.

Optimum mobile-phase velocities are slower in SFC (around 0.1 cm/s for a 50-μm i.d. column) than in GC. This is due to the lower solute diffusion coefficients in supercritical fluids than in gases. However, the diffusion coefficients of the solutes in the stationary phase are about the same with both techniques. Since the mobile phase moves more slowly in SFC, peak broadening due to resistance to mass transfer in the stationary phase is not as significant as in GC. In other words, it is okay to use columns with a relatively thicker stationary phase film in SFC than is practical in GC [30]. The phase ratio, β, is related to the film thickness and is defined as the ratio of mobile phase to stationary phase volumes on the column. For the film thickness range used in chromatography, a simplified expression gives the phase ratio $\beta = r/2d_f$, where r is the column radius and d_f is the film thickness (expressed in the same units).

Being able to use columns with a lower phase ratio partially offsets the sample capacity disadvantage of having to use smaller-diameter columns in SFC than in GC. Film thicknesses available on commercial 50-μm-diameter columns range from 0.05 μm to 0.5 μm. With further column development, thicker-film columns may become available.

Films of up to 1 μm can be used with 50 μm i.d. SFC columns with negligible loss in efficiency [30]. But there is a disadvantage to be considered before going to such an extreme. SFC should not duplicate the molecular weight range of GC (except when applied to thermally unstable analytes). If the rewards of SFC are at higher molecular weight, then the column phase ratio has to be kept high: Peak retention is expressed

Table 2. Stationary Phases for Capillary SFC

100% Methyl polysiloxane
50% *n*-Octyl, 50% methyl polysiloxane
5% Phenyl, 95% methyl polysiloxane
50% Phenyl, 50% methyl polysiloxane
7% Cyanopropyl, 7% phenyl, 86% methyl polysiloxane
30% Biphenyl, 70% methyl polysiloxane
25% Cyanopropyl, 25% phenyl, 50% methyl polysiloxane
50% Cyanopropyl, 50% methyl polysiloxane
Polyethylene glycol (similar to Carbowax 20M, bonded)

in terms of capacity ratio, $k = (t_r - t_0)/t_0$, where t_r and t_0 are the elution times of the peak of interest and an unretained peak, respectively, under isothermal, isobaric conditions. The equilibrium distribution coefficient, K, is simply the ratio of solute concentration in the stationary and mobile phase. The value of K is fixed for a particular stationary- and mobile-phase selection at a specified temperature and pressure, and $K = \beta k$. Since K is constant for a given temperature and pressure, increasing the film thickness (decreasing β) increases the retention (k) of a given solute at those conditions. Solute retention also increases with molecular weight. So if we want to maximize the molecular weight range, then the thinnest film possible is called for. But thin films limit the maximum sample size.

This situation leads us to another compromise: sample size must be traded for mass range. In practice, 0.1-0.2-μm film thicknesses are a good starting point with 50-μm column inside diameters.

Capillary columns should be deactivated as completely as possible, during manufacture, to limit solute retention by adsorption to polar sites on the fused silica surface. This is especially important in thin-film columns where the silica surface is more accessible to the solutes. In addition, deactivation must be done with an appropriate agent to leave a surface wettable by the stationary phase to be coated [31-33]. Otherwise, a uniform stationary phase cannot be deposited. Since all of this is out of the hands of a commercial column user, additional detail will be spared—but know what you are buying.

Frequently, a new, commercial column will still be too active for some SFC applications, or a column, once good, may become active in use. Just as in GC, it is often possible to further deactivate columns, in place, by injecting silylating agents. Make sure the mobile phase will not react with the chosen reagent. (CO_2 is inert to silylating agents.) Set the oven temperature to about 150°C and the pressure to about 100 atm. Then simply make several injections of silylating agent in rapid succession.

Extensive conditioning is not required with most new, commercial SFC columns. After installation, a solvent injection should be made under typical injection conditions to make sure the velocity is in the right range. Then the pressure should be slowly programmed over the range desired for use. If necessary, a short hold can be made at the maximum pressure until the baseline stabilizes. Baseline slope in SFC is not necessarily indicative of column bleed. With CO_2, baseline slope may result from low-level impurities, or, with an FID or other flame-based detector in use, from flow- or pressure-related changes in the flame size and shape. Baseline slope is usually not a serious problem except at very low detector attenuation. Even then it can be circumvented by the use of pressure steps (and baseline readjustment) connecting several isobaric program segments.

F. Mobile Phase

Table 1 gave a list of mobile phases applicable to SFC. Despite the options this list represents, anyone getting started in SFC should begin with carbon dioxide (unless following published work that used another mobile phase). Besides being safe and cheap, most of the capillary SFC work published has been with CO_2. It connects the new user with the biggest body of previous experience.

If a good grade of CO_2 is used, no further cleanup is necessary as long as the connecting tubing is clean. When assembling a new system, the connecting tubing between the supply cylinder and the pump should be extensively flushed with liquid CO_2 prior to connection to the pump. You will need a supply cylinder equipped with a dip tube for this operation, although you may not need a dip tube for chromatography, especially with refrigerated pumps. For standard $\frac{1}{16}$-in. stainless steel tubing, it is usually sufficient to connect the inlet to the supply cylinder, clamp down the outlet securely with it pointing safely away (preferably into a fume hood), and (widely) open the supply valve. Only gas will be expelled at first, but liquid should start jetting through the outlet within several seconds. Flush for about another 10 sec, then close the supply valve. Allow any visible ice to melt or sublime, then wipe the outlet with a clean, white paper towel to check for any traces of oil or dirt. Continue flushing and checking, as necessary, until confident the tube is as clean as possible.

In purchasing a mobile phase, the important impurities to look for are oxygen, water, and hydrocarbons. Specifications for a commercial SFC-grade carbon dioxide are given in Table 3 and can be used as guidelines.

G. Detectors

There are only two basic kinds of detector for SFC: column-pressure and low-pressure. Column-pressure detectors include UV absorbance detectors, fluorescence detectors, and some infrared absorbance detectors. Low-pressure detectors include the flame ionization detector, thermionic detectors of various sorts, flame photometric detectors, and mass spectrometers. The column-pressure detectors listed all happen to be concentration detectors, that is, they respond to solute concentration, and peak area is a function of the mass within a peak and its time in the detector. Thus, slow flow rates produce larger peak areas than fast flow rates. The low-

Table 3. Typical Specifications for SFC-Grade CO_2

Ar	< 5 ppm (v/v)
H_2	< 5 ppm (v/v)
O_2	< 50 ppm (v/v)
CO	< 5 ppm (v/v)
CH_4	< 5 ppm (v/v)
H_2O	< 3 ppm (v/v)
Nonvolatile organic residue	< 0.1 ppm (w/w)
Particulates	< 1 ppm (w/w)

pressure detectors listed all happen to be mass detectors. Peak areas correspond only to the solute mass delivered to the detector, regardless of the flow rate.

With column-pressure detectors, mass transfer is trivial—any solute eluting from the column enters the detector. Decompression occurs after detection (through a valve, pressure regulator, or restrictor) where it is of no direct consequence to the chromatography. However, with low-pressure detectors, the situation is more complicated, as explained in the next section.

H. Restrictors for Low-Pressure Detectors

With low-pressure detectors the pressure must be lowered at some point between the column outlet and the detector. When this occurs, the mobile phase not only loses solvating power, but undergoes Joule-Thompson cooling. Both effects work in opposition to mass transfer and promote solute condensation. Condensation leads to noisy peaks as discrete solute particles enter the detector and each one produces a sudden response. This is generally known as "spiking," named for the appearance of the peaks. In extreme cases, the system can be plugged by condensate.

There are several steps that can be taken to prevent condensation and its consequences:

1. Postpone the decompression to the last possible opportunity.
2. Accomplish the decompression in the shortest possible time.
3. Accomplish the decompression over the smallest possible surface area.
4. Delocalize the cooling effect.
5. Add enough heat at the point of decompression to counter the cooling effect.

Once again, one of the important factors, delocalizing the cooling, is antagonistic with the first three points, and some sort of compromise is needed for best results.

There are four common pressure-dropping flow restrictors used for interfacing SFC columns with postcolumn detectors: straight-walled,

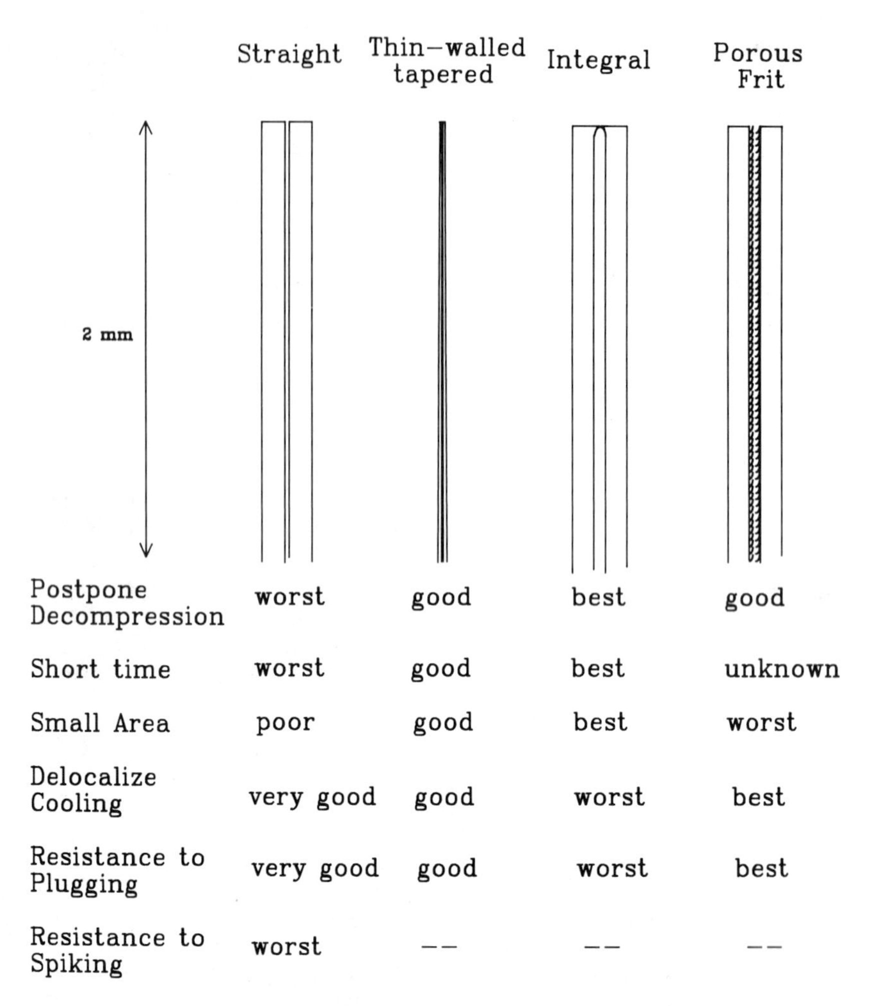

Figure 8. Restrictors for SFC.

narrow-bore capillaries [34]; thin-walled, tapered capillaries [35,36]; integral restrictors [37]; and porous-frit restrictors [38]. They are illustrated in Figure 8.

With all of these restrictors, decompression must occur as close to the point of detector signal generation as possible. For example, with flame-based detectors, a *packed-column* jet should be used. It has an inside diameter large enough for the restrictor to pass completely through the jet with the restrictor outlet positioned at the base of the flame. Most capillary jets have inside diameters too small to pass the restrictor. With them the last few centimeters of solute travel must be accomplished in the gas phase. This is not very efficient for solutes not volatile enough for GC.

The simplest (and earliest) restrictors were straight, narrow-bore capillary tubes with a diameter somewhat smaller than the column. A 15-cm length of 10-15 μm i.d. tubing is a reasonable starting place when using a 50-μm i.d. column. Small-diameter tubing varies considerably from the nominal inside diameters stated by the manufacturer, so a check with a calibrated light microscope is a good idea. Poiseuille's law is often used to estimate flow rates at a tube outlet: $f = \pi p r^4 / 8 \ell \eta$, where f is the flow rate, p is the pressure drop across the tube, r is the tube radius, ℓ is the tube length, and η is the viscosity of the fluid. Flow resistance varies with the fourth power of the radius, but only with the first power of the length, so a small difference in diameter between two restrictors may require a considerable difference in length to achieve the same flow resistance. The equation does not strictly apply to SFC restrictors because of the large changes in viscosity inherent with decompressing a supercritical fluid to atmospheric pressure (or vacuum). However, it can be used to estimate relative resistance of different restrictors.

Unfortunately, straight-walled capillary restrictors do not work very well in the molecular weight range where SFC is most useful. Most of the decompression occurs in about the last 1-2% of the length of straight restrictors, but the diameter is large compared to other restrictors, and the velocity is comparatively slow. Thus, decompression takes a relatively long time and surface area exposure is relatively large. Cooling is delocalized reasonably well, but the wall thicknesses are large, providing significant insulation from external heating. Preheating the mobile phase transports heat to the point of decompression and improves the performance of restrictors [35,39].

The tapered-capillary restrictor is a tremendous improvement over straight tubes. Decompression occurs in the last 1-2% of the *taper*, not of the entire tube. The taper length is typically 2-4 cm, but the inside diameter near the outlet is 2-4 μm, giving much faster outlet velocities for a particular column velocity. So, in terms of time required for decompression, the tapered restrictor is about 100-1000 times faster than straight tubes. The smaller diameter and shorter length (through the volume of decompression) also significantly reduce the surface area compared to straight tubes. Cooling is not delocalized as well as in straight tubes, but two other important advantages make up for it: First, the walls are very thin at the outlet—5-10 μm thicknesses are typical. This minimizes the insulation and aids heat replacement from the surrounding sources. Second, the inside diameter of the tube the tapered restrictor is made from can be larger than is practical in a straight tube. A 25-50 μm i.d. range is typical. This means that the residence time of the mobile phase is up to 10 times longer in the hot detector prior to decompression compared to a 15-μm-diameter straight tube. Thus, mobile-phase preheating is much more effective with a tapered restrictor.

Tapered restrictors can be made by hand-pulling fused silica tubes in a hot flame [35]. However, the best ones are made with automation. The first published report used a laboratory robot [36]. However, countless cheaper ways are only a matter of ingenuity.

Tapered restrictors must be microscopically sized and individually cut to give the desired outlet diameter. The cut must be perfectly square and should be done by a suitable method described in the next section. Restrictors should never be cut with scissors or a razor blade. These will simply crush the outlet and direct the outlet flow off axis.

The one potential disadvantage to tapered, fused-silica restrictors is that the polyimide coating is burned off the tapered section in preparation. Because of this, tapered restrictors have gained an unearned (and incorrect) reputation of being too fragile to use successfully. In actual fact, the polyimide coating on fused silica imparts little or no additional strength to the tube. Its main function is to protect the tube from being scratched, and it may prevent breakage across microfissures, should they occur in the silica. Any glass tube, once scratched, is easily snapped. Once installed, tapered restrictors have long lives—6 months is typical—always ended by plugging rather than breakage. The only trick to know in handling tapered restrictors is to avoid scratching them during installation. This is easily accomplished by inserting the restrictor inlet through the detector outlet, thus imparting any damage to the end that you can safely and easily shorten once it appears in the oven. In fact, this is the best approach to install any silica restrictor.

The integral restrictor shares many of the same attributes as the tapered restrictor. In fact, it is a tapered restrictor with a very short taper length. Thus, it is even better in terms of postponing and minimizing the time of decompression, and the surface area over which the decompression occurs is similarly reduced. Cooling takes place in a very small volume that is surrounded by thick walls. Thus, effective preheating of the mobile phase is required. Just as with the tapered restrictor, the inside diameter of the tube used for restrictor preparation need not be small. It is possible to prepare an integral restrictor directly on the end of the column in order to save making a fitting and to avoid the associated peak broadening. This advantage is usually outweighed by the desirability of leaving a working restrictor in place when it is necessary to change the column. The integral restrictor has the smallest outlet orifice and is thus the easiest to plug with foreign particles.

The porous-frit restrictor is completely different from the others. It is similar to the tapered restrictor in postponing decompression, but the velocities, and the time required for decompression, are not well understood. The surface area encountered by the decompressing sample is large, but the cooling is delocalized better than with any of the other restrictors. The large number of sample paths available imparts resistance to plugging.

No single one of these restrictors is best for every application. However, it is clear that a straight tube is worst and should not be considered for serious SFC with low-pressure detectors. The choice among the other three, or a move from one of them to another to circumvent a particular problem, depends on circumstances. Knowledge of solute properties (such as surface affinity, and any hints of the heats of vaporization or crystallization), combined with the information given in Figure 8, should be used to guide restrictor selection.

I. Making Connections

The smaller column diameters used in SFC, compared to GC, make quality connections all the more important. A small dead volume that would go unnoticed in GC, with a 300-μm column i.d. and a 3-ml/min flow rate, can be a serious problem on a 50-μm column with a flow rate of only 10 μl/min. To make matters worse, SFC connections must be leak-free over the full range of pressures to be encountered. Perfection is the goal in making SFC connections. Fortunately, good-quality, high-pressure fittings for capillary tubes are available from several commercial sources.

The most important step in making good connections is cutting the tube. Scoring and breaking is the best method to insure a square cut. However, the diamond-tipped pencils commonly sold by GC supply companies usually do not work reliably by SFC standards, especially on tapered-capillary restrictor tips. The best results are produced using a fiber-optic cleaver with a sapphire cutting element. (Harder, more expensive cutters are available, but they don't work any better.) Excellent results are also obtained using a freshly broken edge of a silicon wafer (from semiconductor manufacturing—rejects from the electronics industry work fine). Every cut should be inspected with a 20× magnifier and recut if any burr or raggedness can be detected.

Attention must be paid to the ferrules, especially when undersized tubing is to be used. For example, if a 150-μm o.d. tube is to be secured with a ferrule having a 400-μm hole through it, a great deal of compression will have to take place. This often results in destructive crushing of the ferrule and misalignment of the tube before the fitting can be tightened enough to prevent leaks. It is also difficult to completely remove a ferrule crushed under these conditions, and an expensive union may have to be discarded. Therefore, the ferrules should be matched closely to the tube whenever possible. When the right ferrule is not available, adjustments should be made. For example, Valco FS.4 ferrules (with a 400-μm hole) can be made to hold 150-μm tubes without destructive crushing by removing approximately 0.2 mm of material from the ferrule tip with a razor blade. This allows a small volume in the bottom of the fitting for the ferrule to flow into as it is compressed. Naturally, removing too much of the ferrule will produce an unswept dead volume, so care, skill, and experience must guide the chromatographer in these circumstances. There is also (chromatographic) danger in the use of "through" fittings, as shown in Figure 9. In these, the ends of the tubes to be joined are butted against each other. While this is usually a great way to make connections in GC, pressure programming and the frequent need to join tubes of different outside diameters make this a poor method in SFC if done improperly. Sleeves may be used to minimize dead volume. But any unswept or poorly swept dead volumes are compressed during pressure programming. Fractions of the peaks can be held up even to the end of the chromatogram. Decompression, in preparation for the next injection, will cause the trapped mobile phase to expand from the dead volumes and release any solutes they held, reinjecting various peaks of the old sample into the next run. Even though the total volume of a regular capillary union may be larger than in a "through" union, it is completely swept with proper assembly, and is the preferred connector for tubes of widely different outside diameter.

To avoid plugging problems in making a union connection, it is best to pass the nut (or compression screw) and ferrule over the tube, and then cut off a short length of the tube. This insures no snagged bits of ferrule are present on the orifice. The tube is then inserted into the

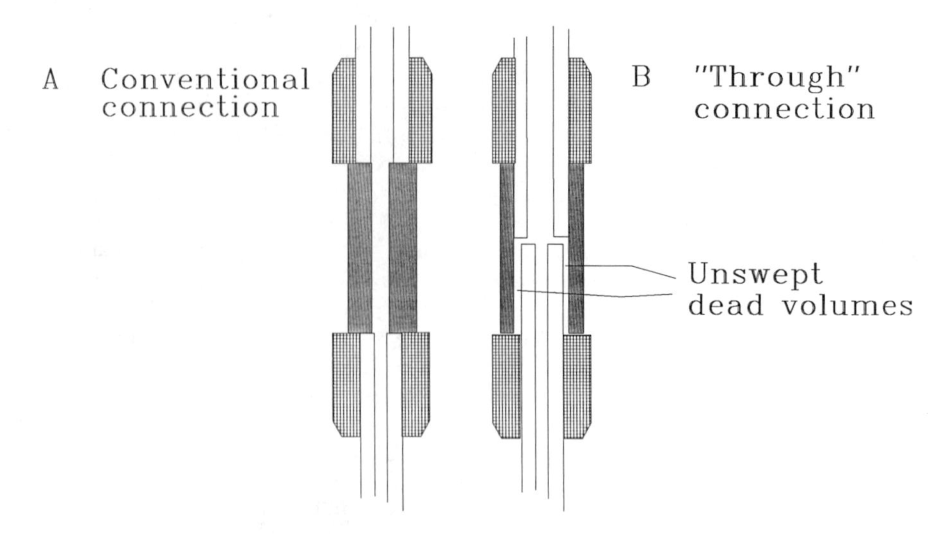

Figure 9. "Through" fittings may be worse than conventional fittings when coupling tubes of different outside diameter, or any time an unswept dead volume exists in the fitting.

union to the proper depth, and the assembly tightened in the usual manner. It is a good idea to make any new fittings one at a time, moving toward the detector, and to blow out the assembled parts of the system with mobile phase (if the instrument allows it) as each new connection is made. This extra care is a lot less trouble than changing a clogged restrictor.

Caution must be exercised whenever pressure is applied to a fused-silica tube, especially when the system is being worked on and the oven door is open. Column explosions are very rare but do occur, especially with old columns. Column explosions are not particularly dangerous because the volume of the column is so small. (This is not to be confused with oven explosions, which are possible in both GC and SFC when a flammable gas leaks into the column oven and ignites. Oven explosions are dangerous.) The rarity of column explosions can lead us into a false sense of security and relaxed safety practices. Nonetheless, precautions, including wearing safety glasses and tying down all unions to the column cage before pressurizing the system, must be routinely followed. Only very low pressure is required for the "blowing out" step during column installation. Liquid CO_2 vapor pressure is more than adequate. Excessive expulsion of gas or the use of very high pressure in this step can create microfissures in the capillary tubing. These may cause the tube to fail when pressurized.

J. Flow Behavior with a Pressure-Controlled Inlet

For people coming to SFC from an HPLC background, pressure control (rather than flow control) may be difficult to get accustomed to. For example, if a small leak develops in the connections between the pump and injector, there is often no significant consequence. The pump may run a little faster because of the increased flow demand caused by the leak, but often the chromatography is perfectly normal. If a small leak develops in the injector, at the column inlet, or in any other connections between the injector and the column inlet, retention times will still be normal since the pump will simply run faster than usual to maintain the set pressure. However, the leak represents a splitting path for the sample, and peak areas may be reduced significantly while nothing else appears to be wrong. A small leak at the column outlet will result in increased mobile-phase velocity and early-eluting peaks combined with smaller-than-normal peak areas. A partial plug anywhere between the pump and the column will always cause delayed peaks, since the flow velocity is reduced and a pressure drop at the plug point will lower the column pressure and increase peak retention. A partial plug at or beyond the column outlet will often cause delayed peaks. But since the plug lowers column velocity but maintains column pressure, a partial plug on the outlet would not delay the peaks as much as if it were on the inlet side of the column. Partial plugs in the inlet tube (if used), column, restrictor, or the intervening connectors will change the split ratio when splitting injectors are used, lowering the mass on column. Partial plugs (or complete plugs) may also occur in the vent of flow-splitting injectors, thus increasing the mass on column, and possibly broadening the peaks beyond usefulness.

III. OPERATION

A. Mobile-Phase Selection

There are two criteria for mobile-phase selection: polarity and critical temperature. Although there is a large number of potentially useful mobile phases, carbon dioxide should be the first choice in most instances, at least for initial experiments. In spite of the fact that it is nonpolar, CO_2 has a low critical temperature and a wide range of programmable solvent strength. It works well for analytes soluble in organic solvents, but not for those that are only water-soluble.

The biggest reason to reject CO_2 would be the potential for reaction with one or more of the solutes. Many basic solutes, such as amines, can be separated successfully with CO_2 but should be approached cautiously. It is a good idea to bubble gaseous CO_2 through a solution of a potentially reactive solute before risking an injection. Lack of a reaction under these conditions does not extrapolate to elevated temperature and pressure, but any sign of reaction is reason enough to try a different mobile phase. After the bubble test, it is best, if possible, to inject a small mass of a pure test compound as similar as possible to the solute(s) of interest. The goal is to minimize the risk of permanently altering the column or plugging the system at this point. If this single material does not elute as expected, then be extremely reluctant to inject any more of it, or any of the actual sample of interest, unless you are ready to deal with the possible consequences.

Nitrous oxide is a good substitute for CO_2, especially for separating bases. But it is not as desirable in general because of slightly increased background in the FID. Pentane has been used frequently, especially for the separation of polynuclear aromatic hydrocarbons with UV or fluorescence detection [26]. Sulfur hexafluoride has been used for hydrocarbon group separations [40]. Ammonia has been a curiosity for several years, and is being used with increasing frequency. When using ammonia, capillary column life is shortened, and only *n*-octyl-substituted polysiloxane stationary phases have shown resistance to attack [41]. Polyimide (Vespel) must be kept out of the flow since it dissolves readily in ammonia. Leaks are a nuisance, and the entire instrument must be kept in a hood. Finally, the advantage of low-temperature separations with SFC is progressively lost as increasingly polar mobile phases are considered—critical temperature, in general, increases with polarity.

When solute solubility in CO_2 is a limiting problem, one alternative is solute derivatization. This approach, when possible, sidesteps the solubility problem while allowing the benefits of CO_2 to be retained. This will be illustrated in Section IV.

B. Column Selection

When CO_2 or another nonpolar mobile phase is used, GC-derived McReynolds constants can be used as a guide in column selection. This approach is far from perfect, but does offer some idea of the affinity of a stationary phase for certain solute functional groups. Stationary-phase selection guidelines also parallel GC. For example, in separating alcohols and diols with the same carbon number, a polar column is called for. It is best to avoid the extremes of stationary-phase polarity if the solutes contain extremes of polarity within their functionalities. For example, polar lipids can be separated well using a methylsilicone (nonpolar) or a (90% cyanopropyl-)-methylsilicone (very polar) stationary phase. However, polar lipids have poor solubility in both of these stationary phases and tend to overload at low levels. An intermediate-polarity stationary phase, like poly(50%phenyl-methyl)siloxane, gives better results.

With no other clues to follow, a low to intermediate-polarity stationary phase is the first one to try. More applications have been successfully reported with lower-polarity stationary phases than with higher-polarity phases. And low-polarity columns tend to be better deactivated and coated more uniformly than high-polarity columns.

C. Temperature Selection

To do SFC, the temperature has to be above the critical temperature of the mobile phase in use. However, Schwartz [42] and Lauer et al. [43] have shown there is no thermodynamic discontinuity between supercritical-fluid chromatography and near-critical liquid chromatography. But with subcritical temperatures, it may be impossible to adequately attenuate the mobile phase solvation power by lowering the pressure—the solvent would boil and split into two phases. Although this has not been a problem experienced with packed-column SFC, much lower solvation strengths are required in capillary SFC because of the lower relative retention of the columns compared to packed columns. (The relative retention difference, in turn, is primarily responsible for the larger molecular weight range experienced with capillary columns than with packed columns.) The only benefit in doing near-critical liquid chromatography instead of SFC would be the slightly lower temperature. So, in general, and especially with capillary columns, the minimum column temperature should be a few degrees higher than the critical temperature. A good minimum for CO_2 is 35°C.

When thermally unstable solutes are to be separated, a low column temperature has to be selected. However, when the solute is stable, two benefits can be realized by elevating the temperature: diffusion coefficients improve, and, with low-pressure detectors, more heat is delivered to the point of decompression in the detector interface. Together, this means better column efficiency in less time with less potential for detection problems. In addition, the selectivity between solutes often changes as a function of temperature [44]. Thus, the separation may be improved substantially with a few experiments at different temperatures. (Any changes in solvent strength caused by a temperature change can be counteracted by a change in the system pressure or in the pressure program.)

High-temperature SFC (above 200°C) is largely unexplored. Density at maximum system pressure will drop with increasing temperature, but the loss in solvent strength is opposed by increased solute vapor pressure. At constant pressure, capacity ratios may increase or decrease with temperature, depending on the solute. Therefore, selectivity benefits may be found at higher temperatures.

When using CO_2 mobile phase, 100°C is a good starting point. If additional selectivity is needed, then two additional experiments should be run, at 50 and 150°C, to look for selectivity changes. Other experiments can follow, as required.

D. Mobile-Phase Velocity

In isobaric SFC, column efficiency is affected by velocity exactly as in GC. For capillary columns, the Golay equation (see Appendix) describes how column efficiency is affected by mobile-phase velocity. It is impractical to operate a capillary SFC column at optimum mobile-phase velocity in most situations because of the excessively long analysis times that would be required, especially without pressure programming. Therefore, velocities 10 to 30 times greater than optimum are common, and the resulting loss of column efficiency is tolerated.

In programmed elution, however, the peak widths are largely determined by the program rate for a given column. Resolution of peaks is affected by column length, mobile-phase velocity, and program rate. Faster program rates, faster velocities, and shorter columns all decrease resolution by lowering the mean capacity ratio and the number of theoretical plates. The effect has not yet been generally described for programmed, capillary SFC (although it has been described in the limit of rapid pressure programming [45]). However, it empirically resembles the behavior of resolution vs. capacity ratio in isobaric SFC. This effect is shown in Figure 10.

A velocity range of 1-3 cm/sec is common for 50-μm SFC columns. The actual velocity has little effect on the chromatogram if pressure

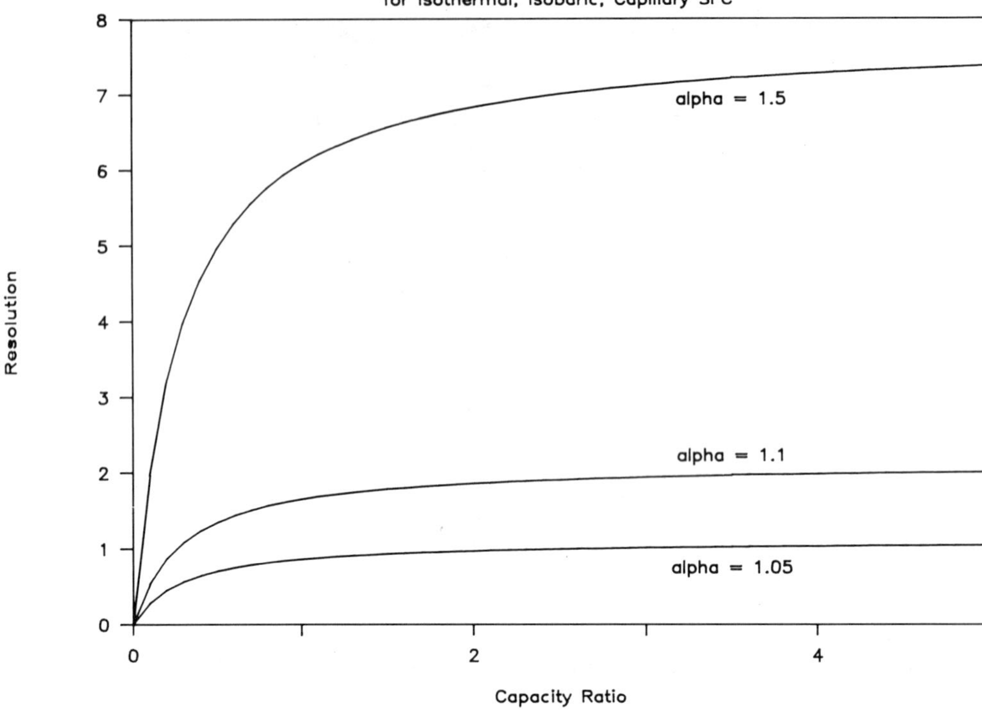

Figure 10. Theoretical resolution possible as a function of capacity ratio (k). Selectivity factors (α) are assumed independent of k. Other assumptions: solute diffusivity = 5×10^{-5} cm^2/sec, mobile phase velocity = 2 cm/sec, column diameter = 50 μm, column length = 10 m.

programs around 1-4 atm per minute are used: For a given program rate, any change in velocity affects the mean capacity ratio in a way counteracting the change induced in column efficiency, as long as all the parameters stay near the ranges stated. Similarly, going to a longer column will have virtually no effect on the chromatogram if column length is the only parameter changed. The improvement in column efficiency will be nearly canceled by a lowering of the mean capacity ratio unless the program rate is slowed by the same factor.

Snyder has explained the relationships between velocity, column length, and program rate in detail for HPLC [46]. The interested reader can apply Snyder's model to capillary SFC if the following distinctions of capillary SFC from HPLC are kept in mind: First, mobile-phase velocity increases with increasing pressure due to compression of the mobile phase and increased mass flow through restrictors. Second, solute diffusion coefficients decrease with increasing solvent strength (if derived by a pressure change). Third, the entire column experiences uniform solvent strength at any single point in time.

E. Programmed Elution

Programmed elution is used to cover a wide range of solutes with a single analysis or to save analysis time compared to isobaric operation. Usually, temperature is fixed and pressure is varied. The pressure may be programmed directly or indirectly by specifying the density. It is possible to simultaneously program the temperature with pressure or density.

Linear and stepwise-linear pressure and density programming are widely used. It is typical to inject at a low, fixed pressure (density) and hold the conditions for a few minutes after injection to allow the injection solvent to clear the column inlet. Remember, the injection solvent is strong and will tend to spread solute over the head of the column rather than depositing it in a small, focused band. So low injection pressure is usually called for to minimize the overall strength of the resulting mixed mobile phase. After a suitable wait, usually 1-5 min, the pressure can be increased either by beginning a pressure ramp or by stepping the pressure to the initial value of a ramp to follow.

For the first run of a new type of sample, a wide, linear pressure or density ramp is best. Ramp rates are ultimately chosen based on the tradeoff between resolution and analysis time described earlier. However, for the first few range-finding injections, a rapid ramp rate (of 5-10 atm/min) gives the best use of time. The starting and ending pressure (density) values and the ramp rate can be refined in subsequent trials to achieve adequate resolution without wasting time at the beginning or the end.

For the routine analysis of samples not containing a family of oligomers, it is likely that some parts of the chromatogram will contain peaks of interest while other parts are uninteresting. To save analysis time, the program should be rapidly advanced or stepped through most of the uninteresting pressure (density) range, but not advanced to what would be the elution pressure (density) of the next peak of interest at the maximum ramp rate giving the resolution required. The pressure has to be stopped and the proper ramp rate reestablished before significant migration of the next peak occurs. Otherwise, resolution will be lost.

More elaborate schemes for programmed elution are possible. Asymptotic density programming (in which the density asymptotically approaches a limiting value) can be used to elute oligomers with even spacing between all the peaks of the series [47]. Pressure or density programs can be combined with temperature programming. This is especially useful if both low-boiling and high-boiling compounds are present in the sample [48]. However, the combination of isothermal temperature with linear pressure or density ramps is versatile, easily understood, and applicable to most analysis situations.

F. Testing for Proper Operation

First, all connections between the injector and the detector must not leak at the maximum pressure to be experienced. Any leak represents a split in the sample path. Connections that leak at high pressure but not at low pressure will split the late-eluting peaks, but not the early ones, thus causing discrimination.

Solvent peaks (when seen by the detector) should rise squarely from the baseline on the ascending side. With typical split ratios (or injection volumes) and mobile-phase velocities they should be 2-10 min wide at the base for about a 10-min holdup time, measured at the rising edge. The trailing edge will not be as steep or square as the ascending side, but should not have excessive tailing. Figure 11 shows an example of a tailing solvent peak due to a poor connection. In Figure 12 good connections give much improved peak shapes.

Naturally, peaks should be narrow and symmetric. Peak splitting or uneven baselines on either side of a peak are indicative of connection

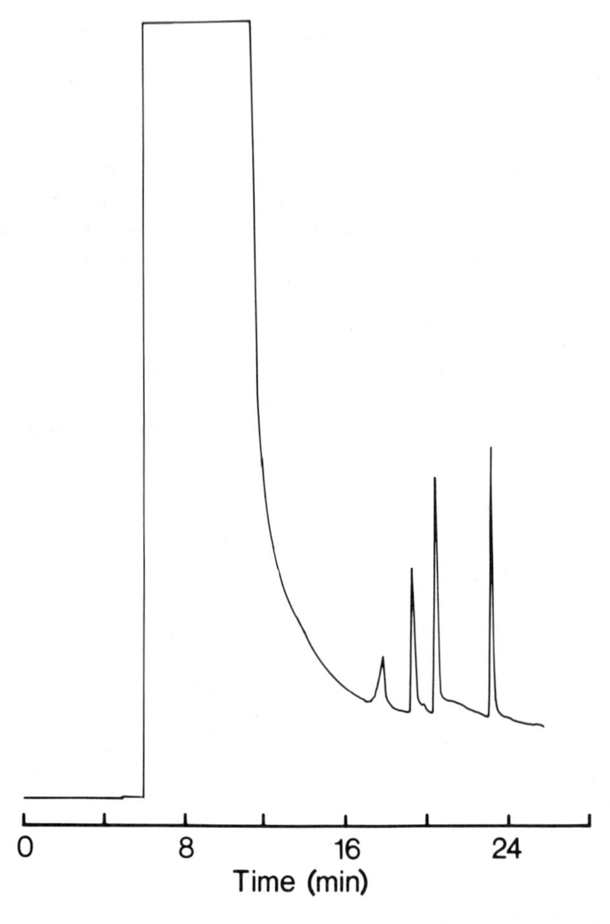

Figure 11. Effect of a poor connection on peak shape. An SFC test mixture containing C_{16} acid, phenanthrene, C_{22} alcohol, and tri-C_{12} amine was used. Conditions: 10 m × 50 μm i.d. × 0.1 μm film DB-17 column (J&W Scientific), CO_2 mobile phase at 100°C, FID at 310°C, injection at 70 atm with a 15-min hold, then a 4 atm/min linear pressure ramp. Compare with Figure 12.

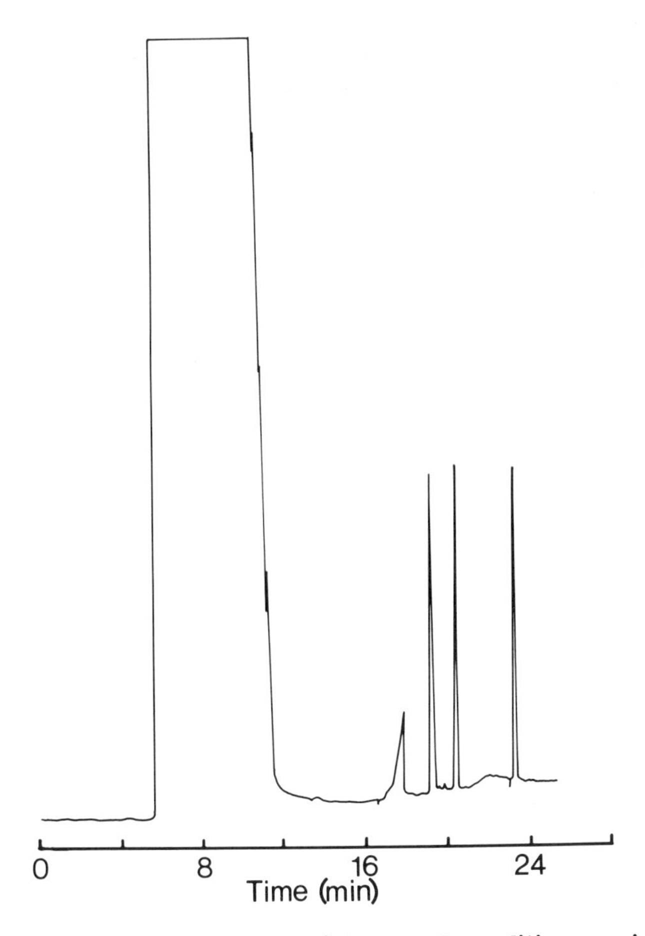

Figure 12. Same test mixture and conditions as in Figure 11, but with good connections.

dead volumes. Polarity test mixtures are valuable for examining inertness of the SFC system. However, commercial Grob test mixtures sold for testing GC columns are too volatile for properly testing an SFC instrument. Higher-molecular-weight analogs are more appropriate. A typical chromatogram of a simple test mixture is shown in Figure 12.

IV. EXAMPLES OF CAPABILITIES

At this still somewhat early stage in capillary SFC development, conventional techniques like GC and HPLC are easier to do than a technique in infancy and not supported by large instrument companies. So it is smart to consider why SFC might be of value before investing time and resources in it.

Here is the indisputable conclusion resulting from such consideration: SFC brings together combinations of solute molecular weight, temperature, and detectors that are not possible with GC and HPLC. Thus, SFC is uniquely able to solve certain classes of separation problems beyond the capabilities of these conventional techniques. The nature of these problems and the role of SFC in the analytical laboratory are covered in an earlier review [5]. The remainder of this chapter will simply demonstrate these capabilities.

A. Molecular Weight Range

The most stringent test of molecular weight range is with the use of a low-pressure detector. Even if the separation works, mass transfer problems

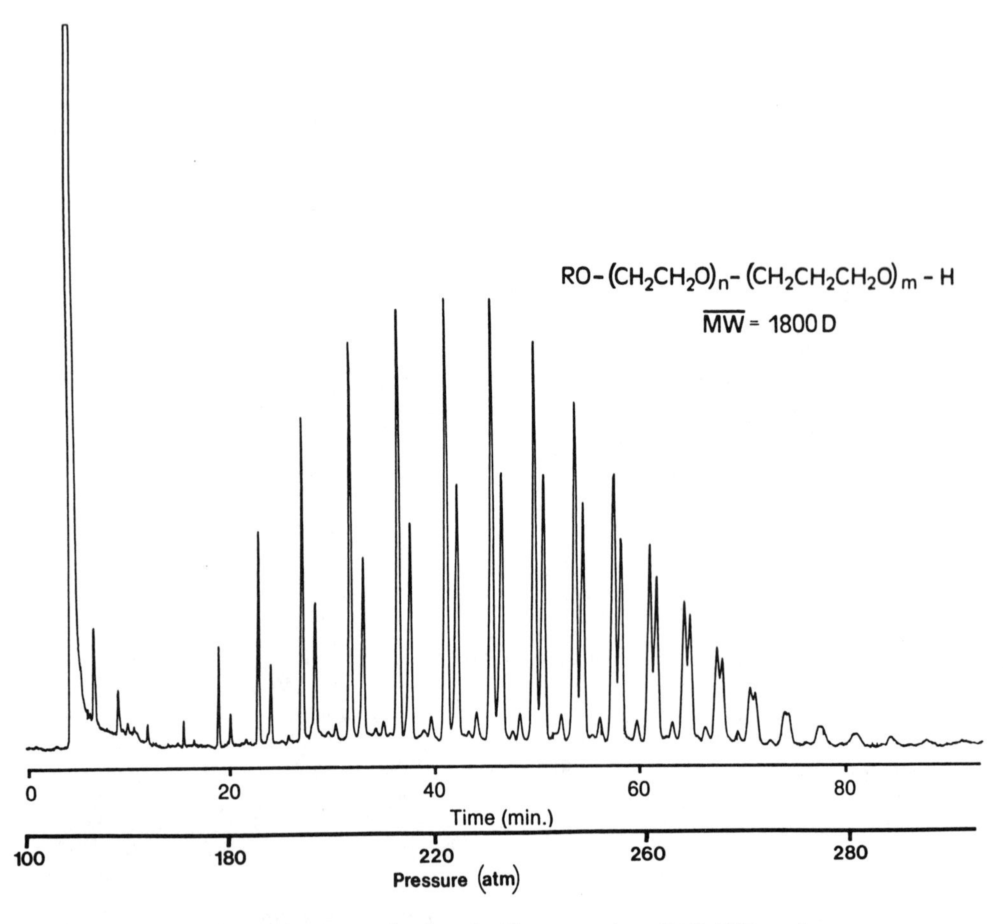

Figure 13. Chromatogram of glycol oligomers by SFC-FID using a porous-frit restrictor. Conditions: 10 m × 50 μm i.d. × 0.25 μm film SB-Methyl-100 column (Lee Scientific), CO_2 mobile phase at 160°C, FID at 375°C. (Chromatogram courtesy of Dr. B. E. Richter, Lee Scientific, Inc., and reproduced with permission.)

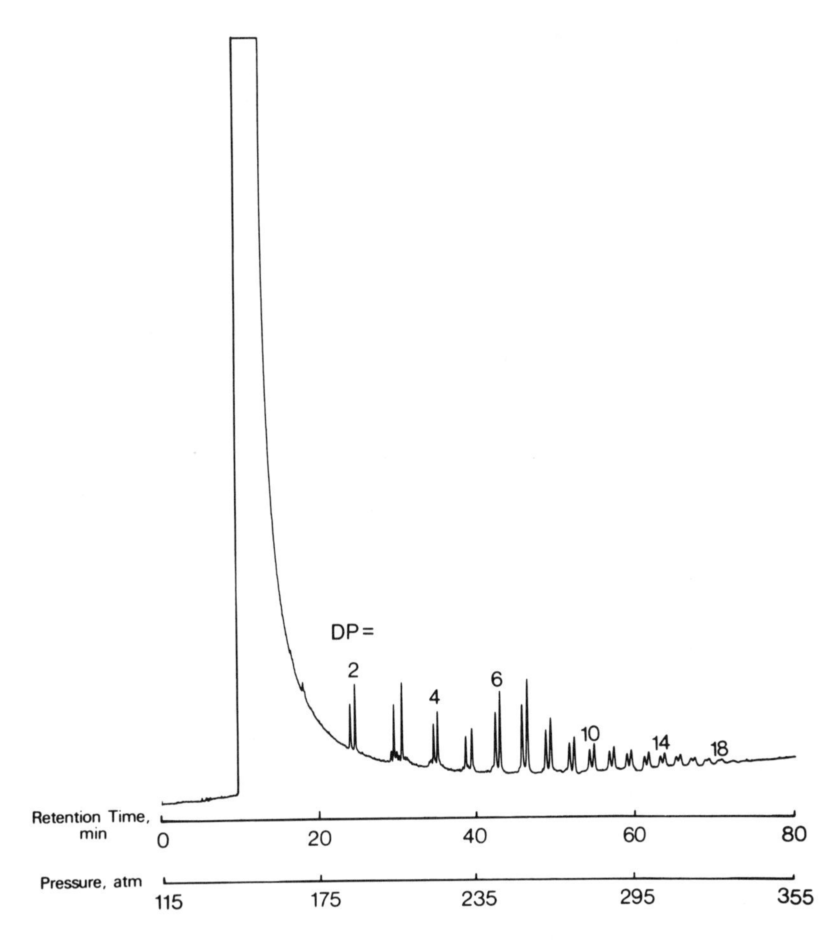

Figure 14. Chromatogram of silylated Maltrin 100 by capillary SFC-FID using a thin-walled tapered-capillary restrictor. The DP = 18 peak has a molecular weight of 6966. Conditions: 10 m × 50 μm × 0.2 μm film DB-1 column (J&W Scientific), CO_2 mobile phase at 89°C, FID at 400°C. (Reprinted from Ref. 50 with permission.)

through the detector interface could effectively limit the molecular weight range.

Figures 13 and 14 demonstrate the molecular weight range achieved by two different laboratories, on two different types of sample, with two different types of detector interface. Figure 13 is an SFC-FID chromatogram of a polyglycol material. The average molecular weight is 1800 but peaks of at least 2300 daltons are clearly visible. This was done using a frit restrictor. Polyglycols exceeding 5000 daltons have been analyzed by SFC-FID using a frit restrictor with no signs of spiking [49].

Figure 14 is a chromatogram of silylated corn syrup solids containing oligo- and polysaccharides [50]. The molecular weight of the DP = 18 derivative is 6966. This was also detected using an FID and a thin-walled tapered-capillary restrictor.

It is safe to say that SFC can easily cover a mass range well beyond all present and envisioned limits of GC while providing detection capability impossible today with HPLC.

B. Low Temperatures

With CO_2 mobile phase, SFC can be done just slightly above ambient temperature. This makes it possible to separate thermally unstable materials while

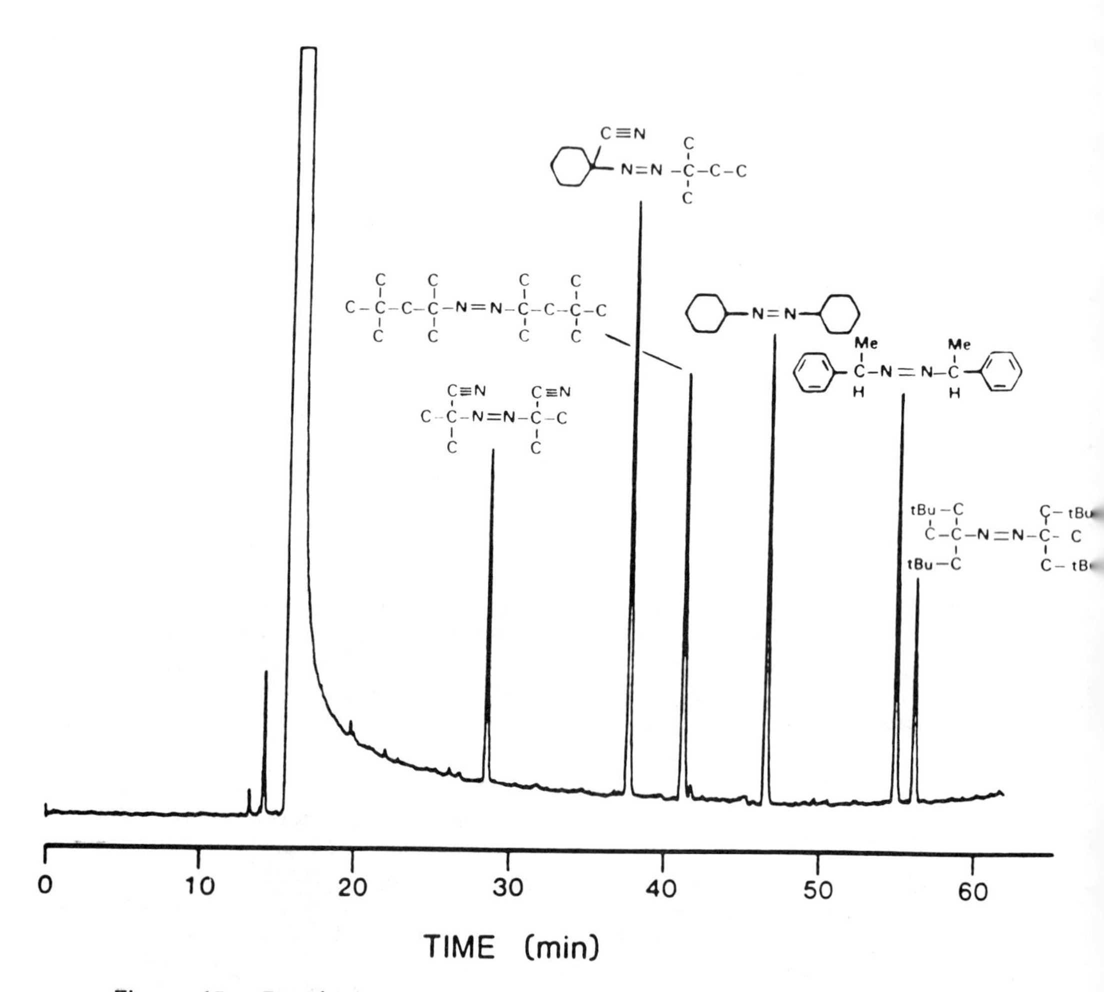

Figure 15. Density-programmed SFC-FID chromatogram of some azo compounds. Conditions: 34 m × 50 μm × 0.25 μm film SE-54 column, CO_2 mobile phase at 40°C, FID at ca. 350°C, straight capillary restrictor. (Reprinted from Ref. 34 with permission.)

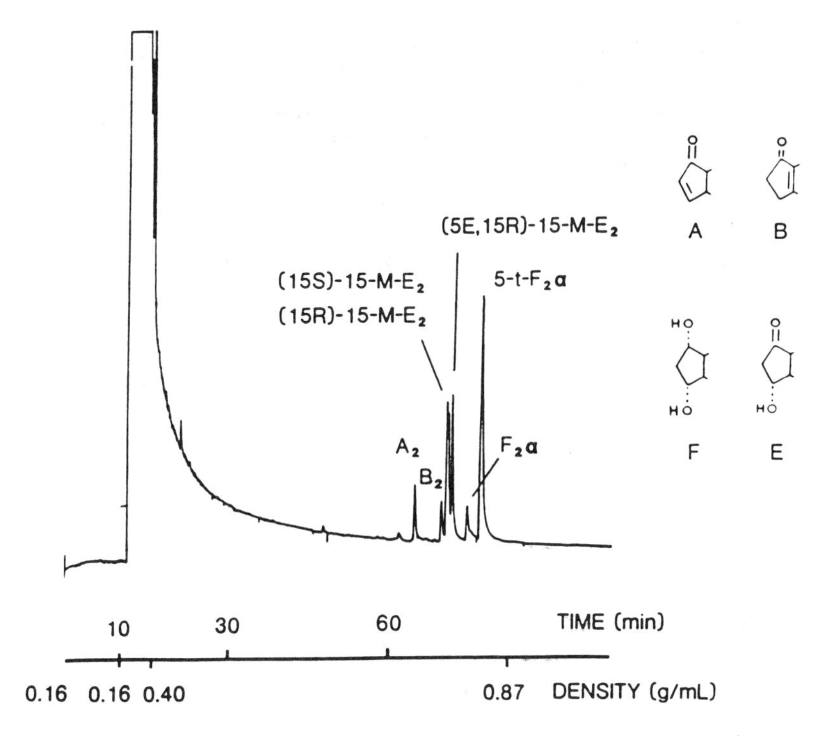

Figure 16. Density-programmed SFC-FID chromatogram of a prostaglandin mixture. Conditions: 12 m × 50 μm i.d. × 0.25 μm film column coated with 50% cyanopropyl methylpolysiloxane, CO_2 mobile phase at 62°C, frit restrictor. (Reprinted from Ref. 51 with permission.)

using a wide variety of detectors beyond the realms of HPLC. Figures 15 and 16 are separations of materials too labile for GC.

Figure 15 is a chromatogram of azo compounds separated with a column temperature of only 40°C [34]. Figure 16 is an SFC-FID separation of several prostaglandins [51]. These materials are thermally labile and often degrade during derivatization reactions aimed at increasing volatility for GC analysis or at improving detection in HPLC [51].

C. Solute Derivatization

Carbon dioxide works very well as SFC mobile phase, but does not provide enough solubility of polar solutes to be universally applicable. The use of more polar mobile phases is not only feasible, but often desirable. However, the advantages of CO_2 (especially inertness toward the stationary phases, low critical temperature, and wide detector compatibility) are hard to give up.

Instead of changing the mobile phase to solve a solubility problem, it is often possible to change the solutes by chemical derivatization. The large body of readily available literature on the subject, as applied to GC and HPLC, is directly applicable to SFC. The best example to date is the corn syrup solids chromatogram in Figure 14. Water is the only common solvent that readily dissolves the neat sample. Yet the sample reacts readily with silylating agents, producing a derivative soluble in a variety of solvents, including supercritical CO_2. Another example is shown in Figure 17. Here, a mixture of mono- and diethoxylated compounds was separated first without, and second with, silylation. Ethoxylates are usually very well behaved in SFC, and the poor results were a surprise. Even more surprising was the remarkable improvement with silylation.

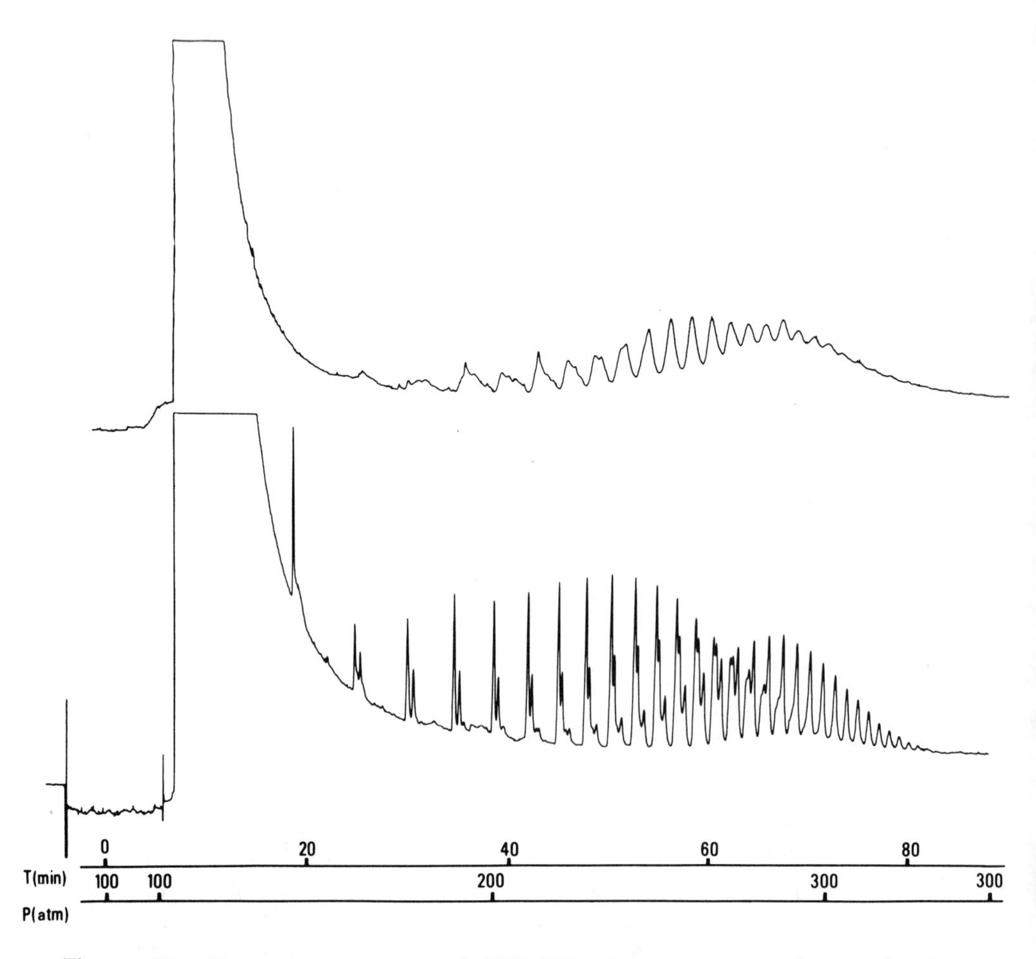

Figure 17. Pressure-programmed SFC-FID chromatograms of underivatized (upper) and silylated (lower) mono- and diethoxylate-chain surfactant mixture. Conditions: 10 m × 50 μm i.d. × 0.2 μm film DB-1 column (J&W Scientific), CO_2 mobile phase, thin-walled tapered-capillary restrictor at 400°C.

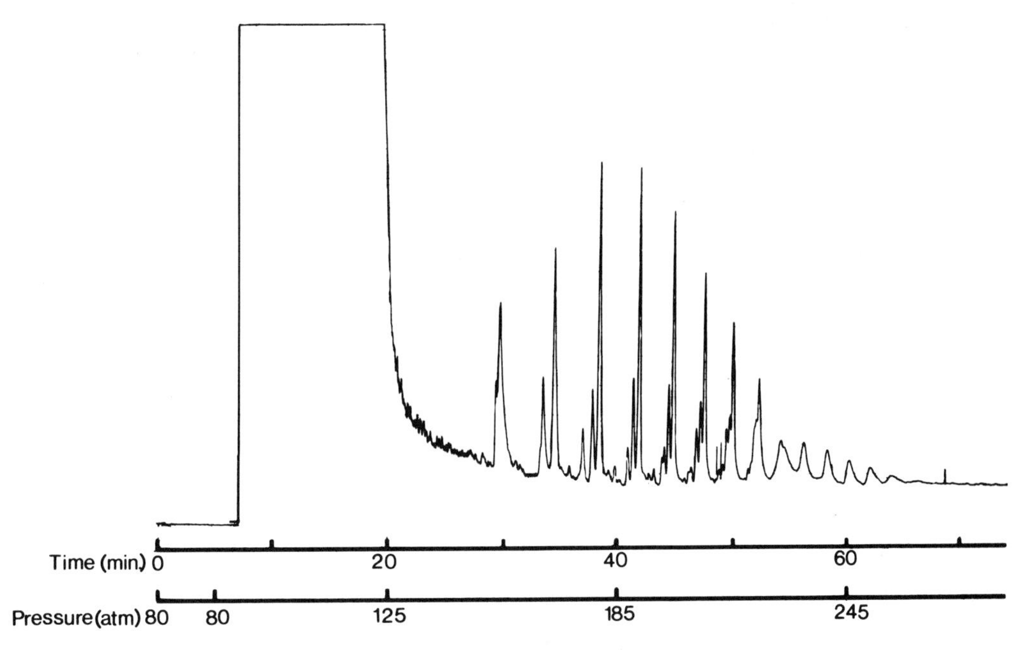

Figure 18. Pressure programmed SFC-FID chromatogram of a silylated, phosphate-terminated ethoxylate mixture. The sample was silylated with BSTFA/pyridine/dichloromethane, 9:2:2 by volume, for 1 h at 80°C. Chromatographic conditions: 14 m × 50 μm × 0.25 μm film SE-33 column (Lee Scientific), CO_2 mobile phase at 120°C, FID at 400°C, thin-walled tapered-capillary restrictor.

It is even possible to extend SFC to the analysis of ionic materials capable of being derivatized. For example, Figure 18 is a chromatogram of a silylated phosphate-terminated oligomer.

Silylation has been the most frequently used derivatization in SFC. However, any derivatization of low-molecular-weight compounds yielding products stable enough for GC will probably work just as well for higher-molecular-weight analogs and SFC.

D. Tolerance of Polar Functional Groups

In some instances, derivatization simply is not feasible. For example, a sample may contain a variety of components varying greatly in polarity, reactivity, and actual chromatographic need for derivatization. If complete separation of the components of an unknown mixture is the problem, rather than determination of just several known components, derivatization is usually not appropriate.

SFC is surprisingly tolerant of at least a few polar functional groups on a solute, and can often produce a separation of adequate quality for

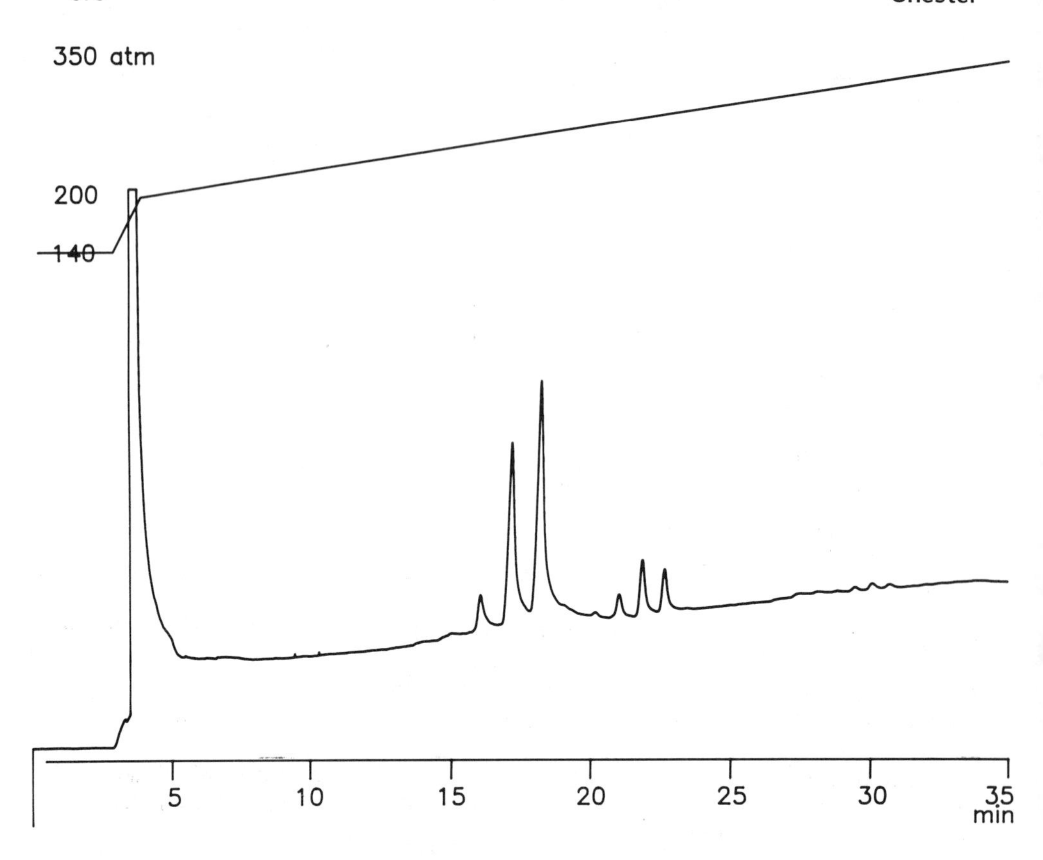

Figure 19. SFC-FID chromatogram of a polyfunctional surfactant (underivatized) containing primary and secondary amines, free hydroxyl groups, free fatty acids, amides, and esters.

the needs at hand without derivatization [52]. The quality of the peak shapes goes down as the number of polar groups goes up, but useful information can often be obtained in situations considered hopeless with conventional chromatography. For example, Figure 19 is the separation of a mixture of components containing free primary and secondary amines, hydroxyl groups, free fatty acids, fatty acid amides, and fatty acid esters. Aggressive derivatization to completely convert all of the free polar groups present was considered risky with respect to the amides and esters already there. SFC worked amazingly well without derivatization.

ACKNOWLEDGMENT

The help of my co-workers Grover Owens, David Pinkston, and Pete Rodriguez in reviewing this manuscript is gratefully acknowledged. In addition, much credit is due to the other past and present members of our SFC group: Claudia Smith, David Innis, Leisa Burkes, and Tom Delaney.

APPENDIX: THEORETICAL COLUMN EFFICIENCY AS A FUNCTION OF PHYSICAL PARAMETERS

Column efficiency is expressed by the number of theoretical plates, N, generated by the column under isothermal, isobaric conditions:

$$N = 5.54(t_r/t_{w,1/2})^2$$

where t_r and $t_{w,1/2}$ are the retention time and width at half height of a peak. In addition,

$$N = L/H$$

where L is the column length, and H is the height equivalent to a theoretical plate (both expressed in the same units). L is known for a given column, and H is given, in theory, by the Golay equation:

$$H = 2D_m/\nu + \frac{(1 + 6k + 11k^2)d_c^2\nu}{96(1 + k)^2 D_m}$$

where D_m is the diffusion coefficient of the solute in the mobile phase, ν is the mobile-phase velocity, k is the solute capacity ratio (= $[t_r - t_0]/t_0$, where t_0 is the time required to elute an unretained peak), and d_c is the diameter of the column. Stationary-phase effects are usually negligible in SFC, and the usual term was omitted. The best column efficiency corresponds to a minimum in H, and

$$H_{min} \approx 0.9d_c$$

for well-retained peaks. This minimum is a column property and is independent of the mobile phase used—gas, supercritical fluid, or liquid. However, the mobile-phase velocity corresponding to H_{min} is

$$\nu_{opt} = 4.2D_m/d_c$$

REFERENCES

1. E. Klesper, A. H. Corwin, and D. A. Turner, *J. Org. Chem.*, *27*:700-701 (1962).
2. E. F. Meyer and T. P. Meyer, *J. Chem. Ed.*, *63*:463-465 (1986).
3. V. Berry, *Am. Lab.*, *17*(10):33 (1985).
4. M. Novotny, S. R. Springston, P. A. Peadon, J. C. Fjeldsted, and M. L. Lee, *Anal. Chem.*, *53*:407A-414A (1981).
5. T. L. Chester, *J. Chromatogr. Sci.*, *24*:226-229 (1986).
6. S. R. Lipsky and M. L. Duffy, *J. High Resolution Chromatogr./Chromatogr. Commun.*, *9*:376-382 (1986).
7. S. R. Lipsky and M. L. Duffy, *J. High Resolution Chromatogr./Chromatogr. Commun.*, *9*:725-730 (1986).
8. B. W. Wright, H. T. Kalinoski, H. R. Udseth, and R. D. Smith, *J. High Resolution Chromatogr./Chromatogr. Commun.*, *9*:145-153 (1986).
9. B. W. Wright, H. R. Udseth, R. D. Smith, and R. N. Hazlett, *J. Chromatogr.*, *314*:253-262 (1984).
10. R. D. Smith, H. R. Udseth, and B. W. Wright, *J. Chromatogr. Sci.*, *23*:192-199 (1985).

11. A. J. Berry, D. E. Games, and J. R. Perkins, *J. Chromatogr.*, *363*: 147-158 (1986).
12. R. D. Smith and H. R. Udseth, *Anal. Chem.*, *59*:13-22 (1987).
13. P. R. Griffiths, S. L. Pentoney, Jr., A. Giorgetti, and K. H. Shafer, *Anal. Chem.*, *58*:1349A-1364A (1986).
14. K. H. Shafer, S. L. Pentoney, Jr., and P. R. Griffiths, *J. High Resolution Chromatogr./Chromatogr. Commun.*, *7*:707-709 (1984).
15. S. V. Olesik, S. B. French, and M. Novotny, *Chromatographia*, *18*: 489-495 (1984).
16. M. E. Hughes and J. L. Fasching, *J. Chromatogr. Sci.*, *24*:535-540 (1985).
17. P. Morin, M. Caude, H. Richard, and R. Rosset, *Chromatographia*, *21*:523-530 (1986).
18. K. H. Shafer, S. L. Pentoney, Jr., and P. R. Griffiths, *Anal. Chem.*, *58*:58-64 (1986).
19. D. R. Gere, *Science*, *222*:253-258 (1983).
20. Y. Hirata and F. Nakata, *Chromatographia*, *21*:627-630 (1986).
21. T. Takeuchi, D. Ishii, M. Saito, and K. Hibi, *J. Chromatogr.*, *295*: 323-331 (1984).
22. A. Wilsch and G. M. Schneider, *J. Chromatogr.*, *357*:239-252 (1986).
23. K. R. Jahn and B. W. Wenclawiak, *Anal. Chem.*, *59*:382-384 (1987).
24. F. J. Van Lenten and L. D. Rothman, *Anal. Chem.*, *48*:1430-1432 (1976).
25. J. C. Fjeldsted and M. L. Lee, *Anal. Chem.*, *56*:619A (1984).
26. P. A. Peadon, J. C. Fjeldsted, M. L. Lee, S. R. Springston, and M. Novotny, *Anal. Chem.*, *54*:1090-1093 (1982).
27. M. C. Harvey, S. D. Stearns, and J. P. Avarette, *LC*, *3*:434-440 (1985).
28. K. Grob, Jr., G. Karrer, and M.-L. Riekkola, *J. Chromatogr.*, *334*: 129-155 (1985).
29. S. M. Fields, R. C. Kong, J. C. Fjeldsted, M. L. Lee, and P. A. Peadon, *J. High Resolution Chromatogr./Chromatogr. Commun.*, *7*:312-318 (1984).
30. S. M. Fields, R. C. Kong, M. L. Lee, and P. A. Peadon, *J. High Resolution Chromatogr./Chromatogr. Commun.*, *7*:423-428 (1984).
31. C. L. Wooley, R. C. Kong, B. E. Richter, and M. L. Lee, *J. High Resolution Chromatogr./Chromatogr. Commun.*, *7*:329-332 (1984).
32. C. L. Wooley, K. E. Markides, and M. L. Lee, *J. Chromatogr.*, *367*: 23-34 (1986).
33. C. L. Wooley, K. E. Markides, M. L. Lee, and K. D. Bartle, *J. High Resolution Chromatogr./Chromatogr. Commun.*, *9*:506-514 (1986).
34. J. C. Fjeldsted, R. C. Kong, and M. L. Lee, *J. Chromatogr.*, *279*: 449-455 (1983).
35. T. L. Chester, *J. Chromatogr.*, *299*:424-431 (1984).
36. T. L. Chester, D. P. Innis, and G. D. Owens, *Anal. Chem.*, *57*:2243-2247 (1985).
37. E. J. Guthrie and H. E. Schwartz, *J. Chromatogr. Sci.*, *24*:236-241 (1986).
38. B. E. Richter, Pittsburgh Conference and Exposition, March 10-14, 1986, Paper No. 514, Atlantic City, NJ.
39. B. E. Richter, *J. High Resolution Chromatogr./Chromatogr. Commun.*, *8*:297-300 (1985).

40. H. E. Schwartz and R. G. Brownlee, *J. Chromatogr.*, *353*:77-93 (1986).
41. M. L. Lee and K. E. Markides, *J. High Resolution Chromatogr./ Chromatogr. Commun.*, *9*:652-656 (1986).
42. H. E. Schwartz, *LC-GC*, *5*:14-22 (1987).
43. H. H. Lauer, D. McManigill, and R. D. Board, *Anal. Chem.*, *55*:1370-1375 (1983).
44. T. L. Chester and D. P. Innis, *J. High Resolution Chromatogr./ Chromatogr. Commun.*, *8*:561-566 (1985).
45. R. D. Smith, E. G. Chapman, and B. W. Wright, *Anal. Chem.*, *57*:2829-2836 (1985).
46. L. R. Snyder, Gradient Elution, in *High Performance Liquid Chromatography, Advances and Perspective*, Csaba Horvath, Ed., Academic Press, New York (1980).
47. J. C. Fjeldsted, W. P. Jackson, P. A. Peadon, and M. L. Lee, *J. Chromatogr. Sci.*, *21*:222-225 (1983).
48. S. M. Fields and M. L. Lee, *J. Chromatogr.*, *349*:305-316 (1985).
49. B. E. Richter, private communication, 1987.
50. T. L. Chester and D. P. Innis, *J. High Resolution Chromatogr./ Chromatogr. Commun.*, *9*:209-212 (1985).
51. K. E. Markides, S. M. Fields and M. L. Lee, *J. Chromatogr. Sci.*, *24*:254-257 (1986).
52. T. L. Chester and D. P. Innis, *J. High Resolution Chromatogr./ Chromatogr. Commun.*, *9*:178-181 (1985).

V
MISCELLANEOUS METHODS

24

Mass Spectrometry

GALEN WOOD EWING* / Department of Chemistry, New Mexico Highlands University, Las Vegas, New Mexico

There are two general classes of instruments included under the heading of mass spectrometers. The first consists of instruments that are capable of separating gas-phase ions according to their masses, or most commonly according to the ratio of mass to charge, m/z. The second class of mass spectrometers detect the presence of ions of selected m/z ratio by a resonance technique, without actually separating them.

As produced in the mass spectrometer, the ions usually have unit charge, that is, a single electron is removed from or added to a neutral atom or molecular fragment, so that sorting by the m/z ratio is equivalent to sorting by mass. Sometimes, however, a singly charged ion can be converted by the same process to a doubly charged species. The m/z ratio is generally referred to as the *mass number*, and is often designated in terms of *mass units* (mu), sometimes called *daltons*.

I. DISPERSIVE (CLASS I) MASS SPECTROMETERS

Any Class I mass spectrometer must have the following components:

- A sample inlet system, which may be designed to handle gases, liquids, or solids, converting them all to gases.
- A means for ionizing the gas molecules.
- Provision for extracting the ions of a given sign from the plasma, and accelerating them in a given direction.
- A dispersing element to separate, either spacially or temporally, ions of differing m/z ratio.
- A detector to quantify the number of ions of a given mass number passing through the exit aperture.
- A high-vacuum system.
- A data-processing and control computer, with suitable read-out facilities.

It is evident that the most critical of these components is the dispersing element, as all the others will normally be designed around it. Therefore we will consider this segment of the instrument first.

*Retired

II. DISPERSING SYSTEMS

A. Magnetic Sectors

It is well known that charged particles moving through a static uniform magnetic field will follow circular trajectories with a radius r given by

$$r = mv/zB \tag{1}$$

where v is the velocity of the particle of mass m and charge z, and B is the magnetic field strength. This can conveniently be rearranged as

$$m/z = r(B/v) \tag{2}$$

In a mass spectrometer it is most convenient to maintain the radius constant, since it is determined by the geometry of the magnet. Thus in order to observe ions with a particular m/z ratio, one must change either the velocity of the ions or the magnetic field strength. The ionic velocity is dependent on the accelerating potential, hence can be varied readily. In most modern mass spectrometers the magnetic field is produced electrically, and so it also can be varied. (In some early instruments a permanent magnet was used, rendering B invariant; a recent example of a permanent magnet mass spectrometer is described in Ref. 1.)

It can be demonstrated either mathematically or geometrically that a divergent beam of ions of equal m/z can be brought to a focus by passage through a sector-shaped magnetic field, as shown in Figure 1a. Note that of the three trajectories sketched, the radius r is the same; hence the upper curve takes a slightly longer path through the field than the lower, which accounts for the focusing property. In principle, the sector can have any apex angle, of which 60° and 90° are the most common.

A difficulty with magnetic sectors is the edge effect. There is always a region through which the ion beam must pass at both the entrance to and the exit from the field region, where the field is not uniform, and this limits the sharpness of focus obtainable. One way to avoid this is to utilize a sector of 180° (Figure 1b) so that both the entrance and exit slits are immersed in the magnetic field. This gives a significant increase in sharpness, hence resolution, in the resulting mass spectra, but manipulation of the source and detector buried between the pole pieces of a large magnet is decidedly inconvenient.

A radial electrostatic field will also produce a circular trajectory for charged particles (Figure 2), but this achieves resolution in terms of energy rather than mass. This leads to the relation

$$m/z = Er/v^2 \tag{3}$$

where E is the magnitude of the electric field. An electric sector can be combined with a magnetic deflection sector to produce a spectrometer with much greater resolution than is attainable with the magnetic sector alone. This is because the ion beam taken directly from the ionization source is never completely homogeneous with respect to velocities, whereas the electric sector acts as an energy filter, permitting passage of ions with only a very narrow range of kinetic energies.

Instruments using combined electric and magnetic sectors are known as double-focus spectrometers. Two implementations have become standard,

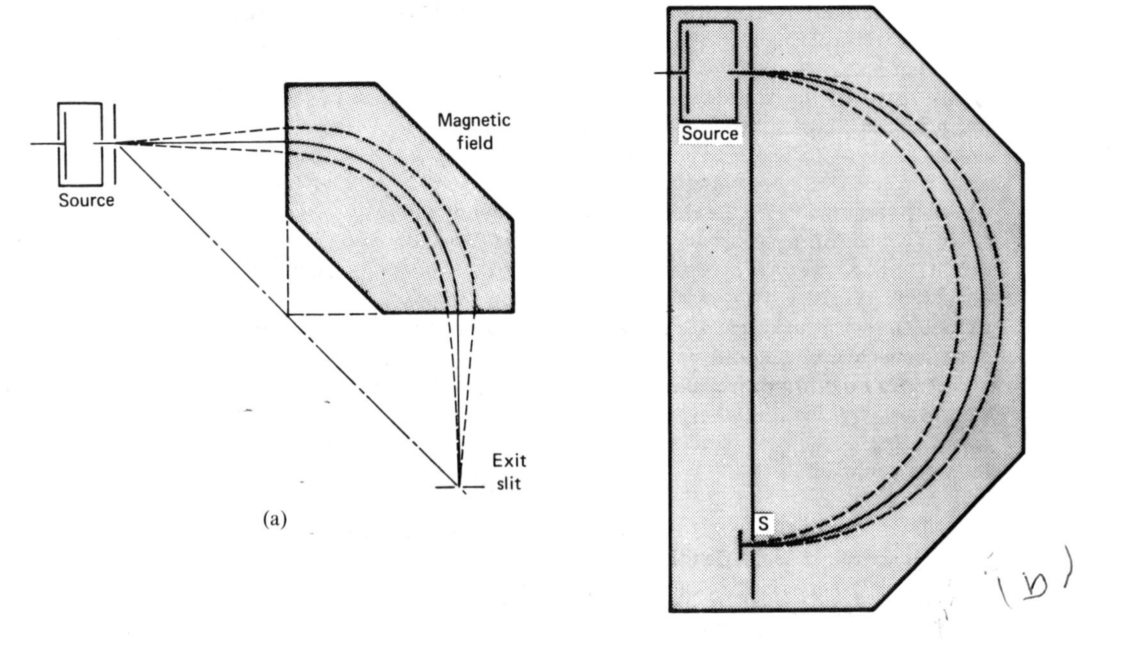

Figure 1. Geometry of a magnetic-sector mass analyzer: (a) a 90° sector (the source and exit slits and the apex of the sector must be on a straight line); (b) a 180° sector, with the source and both slits immersed in the magnetic field.

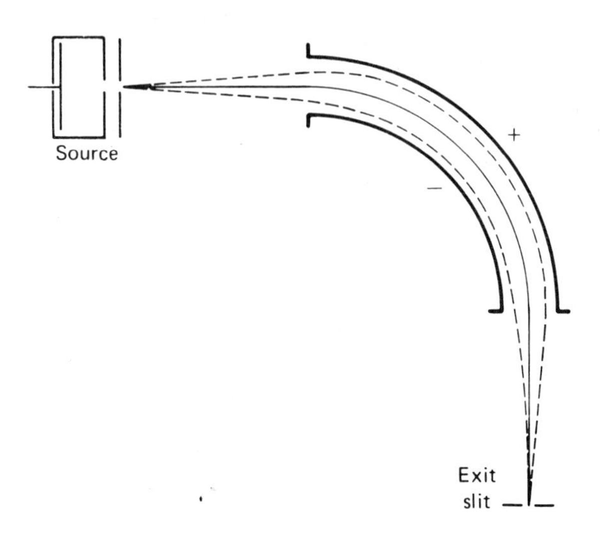

Figure 2. Geometry of an electric-sector energy discriminator.

though various modifications have been described. These are known as the Mattauch-Herzog (M-H) and the Nier-Johnson (N-J) designs. In both, the electrostatic sector precedes the magnetic. In the N-J model, the electric sector has an apex angle of 90°, arranged to bring the beam to a sharp intermediate focus before entering a 90° magnetic sector. An example is described in Ref. 2.

In the M-H spectrometer, the electric sector has an angle of 31°50', which produces a collimated (rather than convergent) beam. This beam enters a magnetic field at normal incidence without defocusing due to edge effects. Each separate species is brought to a sharp focus along a plane at a 135° angle with the first face of the magnet, which makes it possible to position a photographic plate in the focal plane, as has been done with some commercial models.

A drawback to the magnetic-sector instruments is the large size and weight of the required magnet. This can be reduced by the use of superconducting magnets, but this introduces the inconvenience and cost inherent in cryogenic cooling.

B. Quadrupole Mass Analyzers

The quadrupole mass analyzer consists of an array of four metal rods, which must be precisely straight and parallel, so arranged that the beam of ions is directed axially between them (Figure 3a). Ideally the rods should have a hyperbolic cross section (Figure 3b), but in practice less expensive cylindrical rods are nearly as satisfactory. Diametrically opposed rods are tied together electrically, and the two pairs are connected both to a source of direct potential and to a variable radiofrequency excitation. The DC voltages are of equal amplitude but opposite polarity with respect to ground, while the RF voltages are of equal amplitude and 180° out of

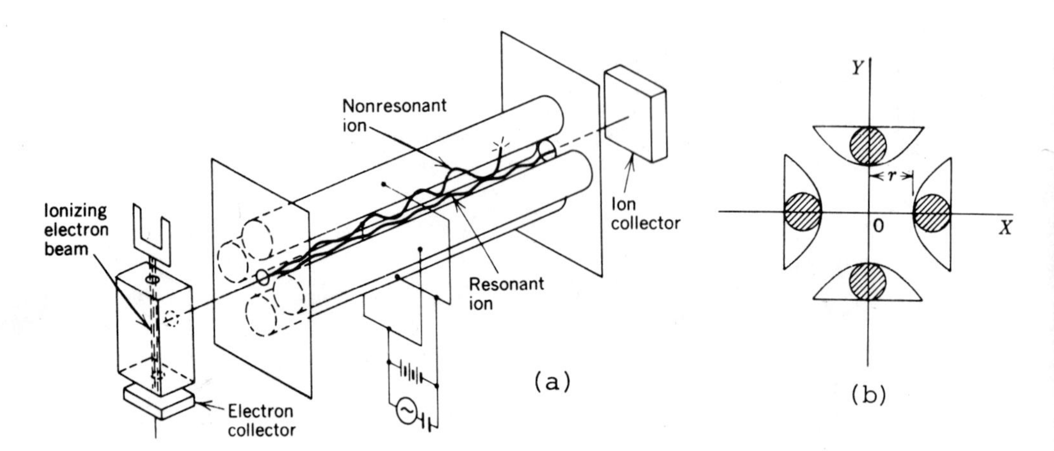

Figure 3. (a) Schematic diagram of a quadrupole mass spectrometer. (After Ref. 17.) (b) The geometrical arrangement of the rods in a quadrupole, showing the definition of r and the circular approximation to hyperbolic rods.

phase with each other. Typically the DC voltages can be anywhere between 0 and 200 V, and the RF between 0 and 1200 V, peak-to-peak. Maximum resolution is attained when the ratio V_{DC}/V_{RF} is slightly below a critical value of 0.168.

Neither the DC nor the RF field has any effect on the forward motion of the ions (along the z axis), but the ions are subject to a complicated lateral motion. The equation of motion of an ion in terms of the x coordinate becomes

$$d^2x/dt^2 + [(2z/mr^2)(V_{DC} + V_{RF} \cos \omega t)x] = 0 \quad (4)$$

where ω is the frequency of the RF field (in radians per second) and t is the time. The coordinate system is indicated in Figure 3b. Motion in the y-coordinate follows a relation identical to that of equation (4). It can be shown mathematically that there is a narrow range of frequencies for which the ionic trajectories are stable with respect to both the x and y coordinates. Outside this range, the trajectories will diverge, eventually striking one of the rods.

Time-of-Flight (TOF) Discriminators

It is possible to segment a stream of ions by means of an electrostatic shutter located adjacent to the ion source. Then if the ions are allowed to stream through an evacuated field-free drift tube, the ions of various mass-to-charge ratios will arrive at the exit aperture at different transit times given by

$$t_t = L(m/2Vze)^{1/2} \quad (5)$$

where L is the length of the drift tube, V is the accelerating voltage, and e is the electronic charge. Transit times of a few ions for representative instrumental parameters are (in microseconds) H^+, 1.58; N_2^+, 8.37; O_2^+, 8.94; Xe^+, 18.17.

A TOF mass spectrometer was first described by Wiley and McLaren [3] in 1955, but until recently this type has been eclipsed by rapid developments of other technologies. TOF instruments suffered from poor resolution and inefficient operation. These deficiencies have largely been overcome through the use of Fourier transform data treatment. Recent developments and references thereto can be found in Ref. 4.

III. ION SOURCES

A. Electron Impact Ionization (EI)

The traditional method, and still one of the most popular, for producing gas-phase ions in a mass spectrometer is by electron bombardment. A beam of electrons, emitted from a hot filament, is passed through a confined volume containing the sample gas at low pressure. Collision of an energetic electron with a sample molecule can result either in knocking out one or more electrons from the molecule to give a positive ion, or in adding itself to that molecule, giving a negative ion. It may also cause fragmentation of the molecule, the fragments in turn becoming ionized. In general, then, electron bombardment of a multiatom molecule results in the formation of an array of ionic species of both signs. Ions are extracted from the source volume by means of an electrostatic field, so that ions of only one sign are observable. Many mass spectrometers have provision for changing the polarity of the field, so that either positive or negative ions may be examined, but by far the largest amount of work is carried out with positive ions.

Figure 4 shows a typical EI mass spectrum, that of *n*-butane (C_4H_{10}, molecular weight = 58), both as determined with a magnetic-sector spectrometer, and as often represented in the form of a bar graph. Note that the molecular ion ($C_4H_{10}^+$) gives only a minor peak at $m/z = 58$, while the peak at 43, due to the fragment $C_3H_7^+$, resulting from deletion of a methyl group, is the largest, known as the *base* peak. Clearly, such fragmentation patterns can serve as highly reliable fingerprint information for identification of samples, and also can give information about the relative strengths of various interatomic bondings.

B. Chemical Ionization (CI)

In this technique the sample is diluted with a large excess (perhaps 10^4:1) of a "reagent" gas before being subjected to the electron beam. Because of the relative quantities, electrons are much more likely to collide with molecules of the reagent than of the analyte, so the primary ions formed are charged molecules of the reagent gas or of fragments thereof.

An often used reagent is methane. A series of reactions such as the following can take place:

$$CH_4 + e^- = CH_4^+ + 2e^-$$

$$CH_4 + e^- = CH_3^+ + H + 2e^-$$

$$CH_4^+ + CH_4 = CH_5^+ + CH_3$$

$$CH_3^+ + CH_4 = C_2H_5^+ + H_2$$

$$R{-}CH_3 + CH_5^+ = R{-}CH_4^+ + CH_4$$

In the last step, $R{-}CH_3$ represents a relatively heavy sample molecule. The product, $R{-}CH_4^+$, is seen to be larger than its parent (M) by one proton (H^+), and is often represented by the symbol $(M + H)^+$, sometimes also designated as "QM^+," for quasi-molecular" ion. The resulting mass spectrum, of course, includes peaks for the reagent (e.g., CH_4^+, CH_5^+, $C_2H_5^+$, etc.), as well as fragments formed from the analyte. Other commonly used reagent gases include *iso*-butane, ammonia, helium, and argon. The reagent ions, being smaller than the analyte, cause little spectral interference, and the major advantage of chemical ionization is the simplicity of the spectrum. Figure 5 shows two mass spectra of diketopiperazine, taken without and with the presence of methane. The second spectrum (CI) would make for easier identification of the molecule, either in the pure state or in mixture with other related compounds, but the first (EI) would give more information about the relative bond strengths within the molecule.

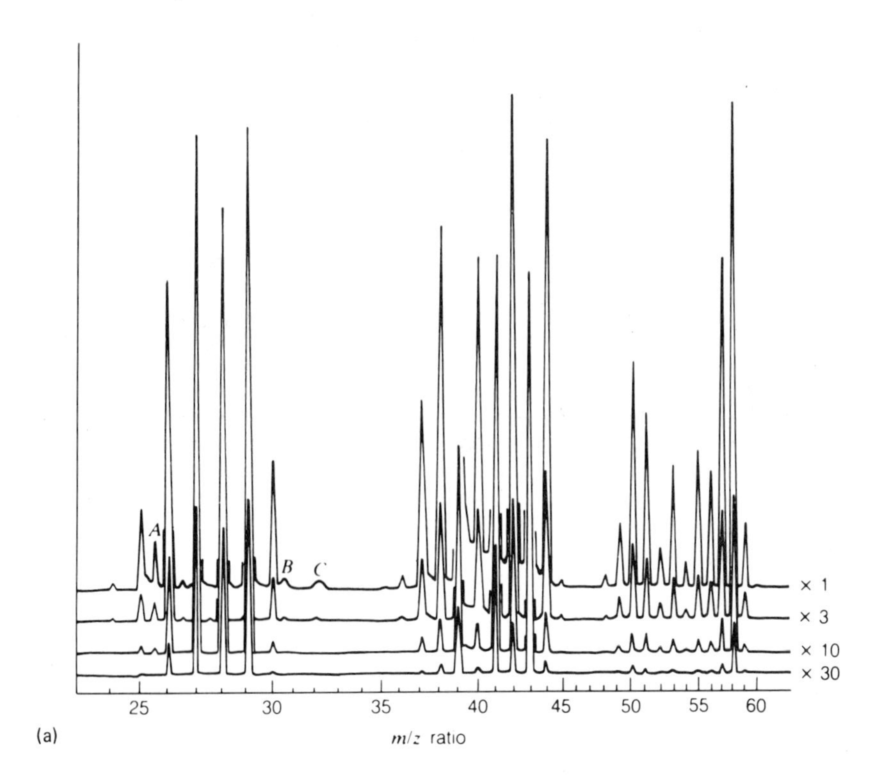

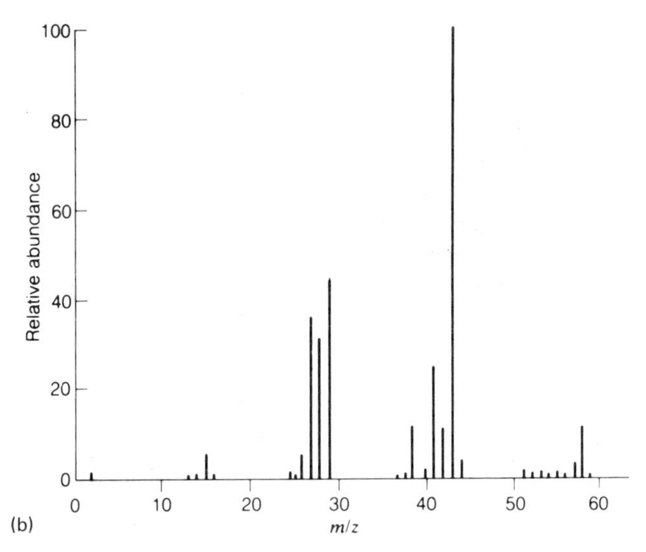

Figure 4. Mass fragmentation pattern of n-butane as recorded on a magnetic-sector mass spectrometer. (a) Four simultaneous tracings made by galvanometers of varying sensitivities (i.e., "× 1" is the most sensitive, the peaks of "× 3" must be multiplied by 3 to be consistent, and so on), thus providing a record with a dynamic range of about 30,000. The maximum marked A is due to a doubly charged ion of mass 51; those at B and C are due to metastable ions. (Courtesy of E. I. Du Pont de Nemours & Co.). (b) Bar-graph representation of the same spectrum.

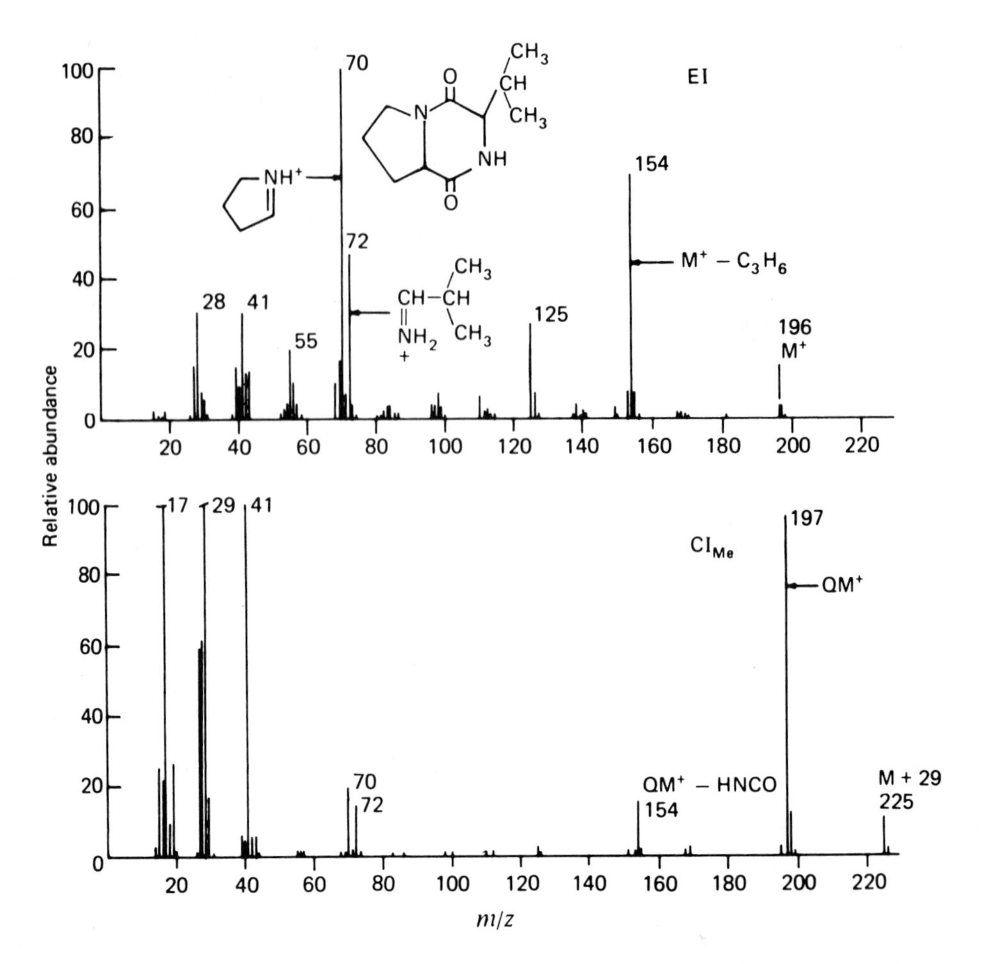

Figure 5. Mass spectra of diketopiperazine by EI ionization and by CI with methane as reagent. Note the great sensitivity of the $(M + H)^+$ peak, designated "QM^+." The peak at 225 represents the addition of $C_2H_5^+$, and serves to verify that the molecular ion must be 196. This is an example of a computer-reconstructed spectrum. (Courtesy of Finnigan Corporation.)

The ion source for chemical ionization is similar to that for simple electron bombardment. The demands on the pumping system, however, are more severe, usually requiring separate pumps for the ionization region and the analyzer proper, since it is necessary to maintain the reagent pressure at a level that is considerably higher than can be tolerated in the mass analyzer.

C. Desorption Ionization (DI)

The ion sources described above require a gas-phase sample. Solid samples can be introduced by means of a probe (described below), but must be evaporated or pyrolyzed prior to ionization. Most large molecules, however, do not have sufficient thermal stability to enter the vapor phase without decomposition. Clearly it would be desirable to have an ionization source that would operate directly on solid samples. Such a source is available through the technique known as desorption ionization.

In a DI source, a solid (or high-boiling liquid) sample is coated on a solid support, such as silver, often by adsorption. It is then subjected to bombardment by a beam of energetic particles. The energy of the projectile particles is utilized partly in removing the sample molecules from the surface and partly in ionizing them. Several versions of desorption are in use, varying mostly in the type of primary particle beam involved; several of them will be described briefly.

D. Secondary Ion Mass Spectrometry (SIMS)

The best established DI method is secondary ion mass spectroscopy. The excitation beam consists of a stream of ions with kinetic energy in the kilo-electron-volt range, produced in a specially designed ion gun. Commonly used ions include Ar^+, Cs^+, and O_2^+. If the sample and support are not electrically conducting, a positive electrostatic charge is likely to build up that will interfere with proper focusing. This can be eliminated by flooding the sample with low-energy electrons from an electron gun.

SIMS has the ability to detect all elements, including hydrogen. The formation of ions, which can be of either sign, can follow several mechanisms [5]: (1) a sample molecule, M, can attach itself, by a process known as cationization, to a charged atom of the substrate metal, such as Ag^+, to give an organometallic species $(M + Ag)^+$; (2) the molecule of M can ionize by a process of electron transfer to give the radical ions $M^{+\cdot}$ or $M^{-\cdot}$; (3) ions originating from inorganic salts can be ejected from the sample by direct transfer of momentum from the primary ions, giving ions such as $(M + Na)^+$. See Ref. 5 for further details.

E. Fast Atom Bombardment (FAB)

Fast atom bombardment resembles SIMS, but with the substitution of a beam of neutral inert-gas atoms rather than ions [6]. Argon is often used, but the more expensive xenon is more effective, presumably because its greater mass results in higher momentum. The atoms in the beam start out as ions from a gun such as that used in SIMS, but many of the ions are neutralized by electron capture. Those ions that remain are deflected out of the beam by a crossed electrostatic or magnetic field, and the uncharged atoms continue with their original momentum to impinge on the sample.

The sample is usually dispersed in a few drops of glycerol or similar low-volatility liquid. This has the advantage that as molecules are removed from the surface, they will be replaced by diffusion from within the body of the glycerol; thus the sample can be examined continuously over a longer period of time.

Both SIMS and FAB techniques are very successful in permitting mass spectroscopic examination of large, thermally labile, organic molecules, such as vitamins, drugs, etc. Both have found extensive use in secondary-ion microscopy [7], a subject too specialized to detail here.

F. Miscellaneous Forms of DI Sources

Desorption can be accomplished by several other methods not generally available in the marketplace. Thermal desorption has been found useful for inorganic solids, as has desorption by means of a laser beam [8]. One of the earliest DI methods was field desorption. In this technique the sample is placed on a carbon whisker, which is made one pole of an electric field. Because of its small size the whisker encounters an extremely large field gradient, contributing to ease of ionization of the sample.

IV. DETECTORS

The earliest mass spectrometers recorded the spectra directly on a photographic plate, giving an appearance reminiscent of similarly recorded optical spectra, thus giving rise to the name of the field. A few modern double-focus instruments have also used photographic recording, but for the most part they are obsolete.

A. Faraday Cup

The least expensive type of detector available is a cup-shaped metal plate that serves to collect ions and transmit the resulting current (a few microamperes) to an electronic measuring circuit or computer. This detector is called a Faraday cup, as it may be considered the direct descendant of the ice bucket with which Faraday carried out basic experiments in electrostatics. The cup shape lessens the likelihood of losing charge through emission of secondary electrons.

B. Electron Multipliers

Almost every mass spectrometer today is equipped with an electron multiplier, a device that bears the same relation to the Faraday cup that the familiar photomultiplier bears to a simple phototube. In its classical form, it is provided with a series of secondary electrodes, called dynodes, at successively higher positive potentials. As each dynode releases several electrons for each one it receives, the internal amplification may be as high as 10^6 or more. Many mass spectrometers make use of a modified form of electron multiplier called a channel multiplier. The external electronic amplification requirements are much less for either type of multiplier than for the Faraday cup detector.

V. SAMPLE-HANDLING SYSTEMS

A. Gas Expansion

One of the earliest methods of sample introduction is applicable to gases or to liquids with sufficiently high vapor pressure. The sample is allowed to expand into a heated evacuated vessel of 1 or 2 liters volume, then the resulting attenuated gas is admitted to the ionization chamber through a "leak," typically one or more pinpoint orifices in a gold foil sealed across the inlet tube. A dynamic equilibrium quickly establishes itself, with equal amounts of sample admitted through the leak and removed by the vacuum pumps, so that the pressure in the ionization source is held at the optimum value, 10^{-6} to 10^{-7} torr (higher, even up to atmospheric pressure, for CI).

B. Gas Chromatography

A much improved (and more expensive) method of admitting gases is through a dedicated gas chromatograph (GC), and several firms offer such integrated instruments. This combination, popularly known as GC/MS, provides an exceptionally germinal analytical tool, as it permits simultaneous observation of chromatographic retention times and mass-spectral identification.

The chief difficulty in this marriage of two major instrument types lies in the carrier gas (often helium) that is required in GC. If allowed to flow directly into the mass spectrometer, the excess helium would tend to interfere with the ionization process, and also would easily overwhelm the vacuum pumping system. The situation is not as critical with capillary GC columns, because the flow of helium is less. There are a number of devices available for removing the bulk of the helium, relying on its greater rate of diffusion compared to higher molecular weight sample components. It is advantageous to equip the mass spectrometer with a detector that is insensitive to helium and other low-molecular-weight gases.

Most mass spectrometers have provision for measuring the total ion current. This feature can be used as a universal GC detector. Indeed, it is a moot point whether the GC should be called a special type of inlet system for the MS, or the MS should be considered a glorified GC detector.

C. Liquid Chromatography (LC)

Interfacing the eluant from a liquid chromatograph to a mass spectrometer is more difficult than for GC. A number of devices have been proposed to eliminate the carrier liquid before the vaporization of sample constituents. One of the most successful, illustrated in Figure 6, includes an endless band of stainless steel or an inert polymer. A portion of the eluant (up to about 1 ml/min) is deposited on the moving belt (2) through an orifice (1). It is transported beneath a radiant heater (3), where the bulk of the solvent is evaporated, then through two successive chambers (4) where the solvent vapor is pumped off. At (5) the band is heated to desorb the solute molecules where they can diffuse into the ion chamber. The heater at (6) removes any residual material, thus preparing the belt to receive the next portion of eluate.

Another method of introducing LC eluate into the mass spectrometer, applicable with readily volatile liquids, is to bleed the column directly into the spectrometer through a capillary and pinhole orifice. In this case the solvent vapor serves as reagent in a chemical ionization process. This has been done successfully with such solvents as water, methanol, and acetonitrile.

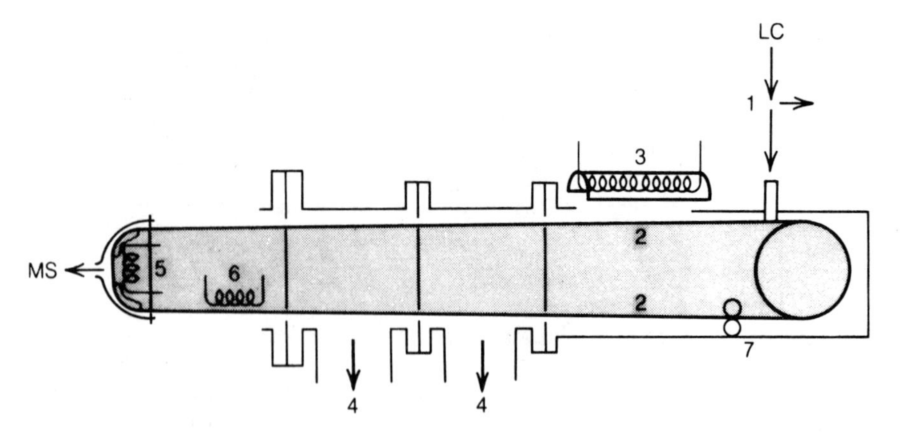

Figure 6. Schematic diagram (side view) of a moving-band GC/MS interface. (Courtesy of Finnigan Corporation.)

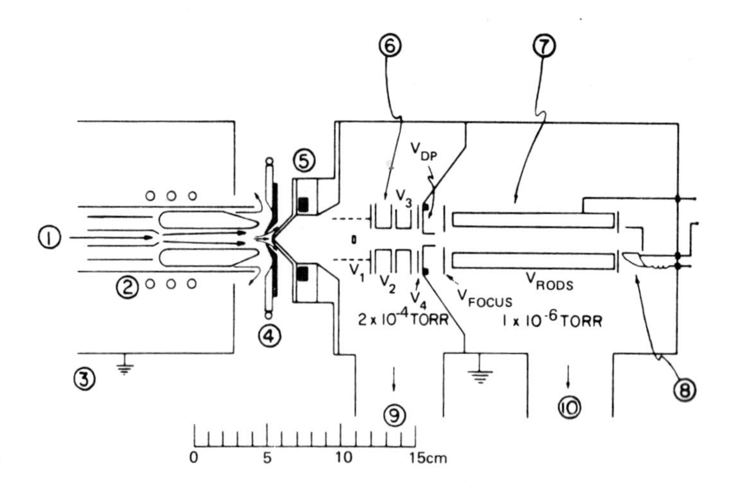

Figure 7. Schematic diagram of an ICP/MS interface. The nebulized sample is blown in at (1) through the ICP torch (2). Item (3) is a grounded protective shield; (4) is the skimmer that deflects most of the hot gases, allowing the sample to pass through its orifice and that in the sampler cone (5) into the quadrupole region (7). The various components at (6) are electrostatic lenses charged to appropriate voltages (V) to align the ion trajectories. The channel electron multiplier is (8); (9) and (10) are connections to the vacuum pumps. (After Ref. 18.)

D. Inductively Coupled Plasma (ICP)

The ICP is best known as a source of radiation for emission spectroscopy, but it is also useful as a source of metallic ions for examination in a mass spectrometer, a method that first came to the attention of the analytical community in 1986 [9,10]. The plasma torch is essentially the same as that used for atomic emission, with argon as the carrier. The torch is mounted horizontally, pointing toward the entrance to the mass spectrometer. The plasma encounters a water-cooled cone called a skimmer, with a small orifice at its tip, that deflects away most of the hot plasma (Figure 7). The gas that enters expands into an evacuated region (about 1 torr), and the central portion passes through the apex port of a second skimmer into the high vacuum of the spectrometer.

It has been shown [9] that the ICP is highly efficient in forming singly charged positive ions (M^+) from all metal atoms and some nonmetallic ones, and of course this is just what the mass spectrometer requires. Some background ions can be expected to appear in the mass spectrum. These may include ions formed from solvent (e.g., H_2O^+, ClO^+ from $HClO_4$, SO^+ from H_2SO_4) and from argon (Ar^+, ArH^+, ArO^+, among others). These may interfere with the determination of certain metal isotopes, but generally can be subtracted out by means of a blank run.

The ICP/MS technique is still in its infancy, but shows promise to be one of the best methods for general elemental analysis.

E. Solids Probe

Solids or high-boiling liquids can be injected directly into the ion source of a mass spectrometer by means of a heated probe. The sample is enclosed in a short section of glass capillary, closed at one end like a miniature test tube. This capillary is inserted into a cavity at the end of the probe. After insertion through a vacuum lock, heat is applied electrically, evaporating enough of the sample to provide a mass spectrum. Ionization can be either direct (EI) or by reaction (CI).

VI. THE ION-TRAP MASS SPECTROMETER

One difficulty with the traditional mass spectrometers described above is the short time slot during which an ion can be observed. This is particularly significant in chemical ionization, since at the desired low pressure the mean free path of a reagent ion between collisions with analyte species may be many meters, hence the required mean time may be measured in seconds, whereas it may take only a few microseconds for an ion to traverse the entire spectrometer.

This difficulty can be largely overcome by the device known as an ion trap. The ion trap makes use of combined radiofrequency and static electromagnetic fields in the configuration first developed for accelerating ions in the cyclotron. Hence the early spectrometers built on this principle were known as ion-cyclotron resonance (ICR) instruments.

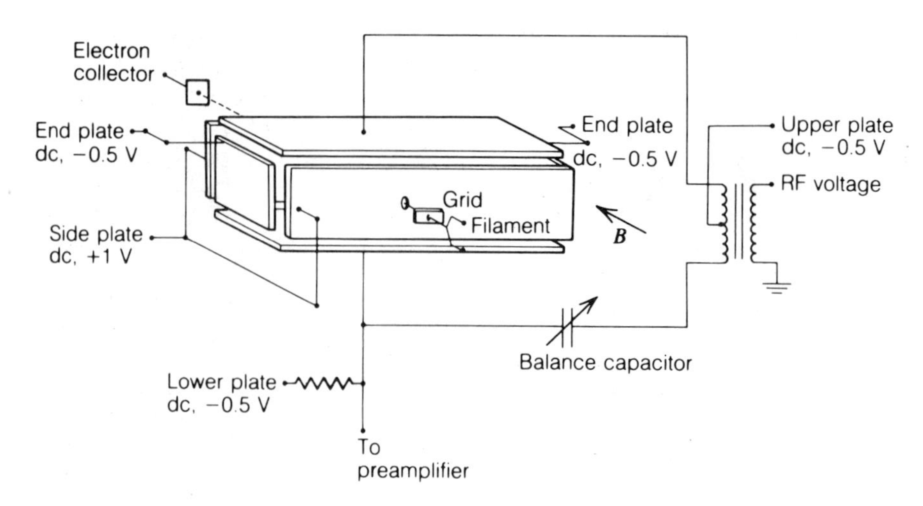

Figure 8. Ion trap of a Fourier-transform, ion-cyclotron resonance mass spectrometer. The total volume is about 10 cm^3. (After Ref. 19.)

Figure 8 shows the structure of an ICR spectrometer. It consists of a box-like chamber with an impressed static magnetic field, B. A more recent design (Finnigan-MAT) uses a cell with cylindrical symmetry. A beam of electrons, emitted from a filament, passes through the cell, ionizing sample molecules by impact. The ions as formed will be in random motion, hence constrained by the magnetic field to follow circular trajectories at a frequency, called the cyclotron frequency, given by

$$\omega_c = Bez/m \tag{6}$$

where B is the magnetic field strength. Thus the frequency is inversely related to the familiar m/z ratio. The kinetic energy of the ions can be increased by applying to the upper and lower plates of the cell a radio-frequency voltage at the exact same frequency, ω_c, causing the trajectory to enlarge spirally. Eventually the accelerated ions will strike one of the plates (in the older model), thus causing a small current that can be amplified and measured. In practice the applied voltage is swept through a range of frequencies, so that each ionic species will, in turn, be accelerated and measured.

In the newer model, the ions are not collected, but observed through the absorption of energy at the resonant frequency. The information so gained must be processed by an inverse Fourier transformation to convert it from the frequency domain to a form that can be read out in mass units. Hence ion-trap mass spectrometry is also referred to as FT/MS, for Fourier-transform mass spectrometry.

A. Metastable Ions

In any type of mass spectrometer there is always a chance that an ion will be altered, at some point during its flight from source to detector, in such a way as to change its mass. This may occur as the result of collision with a molecule of residual gas or with another fragment of the analyte. In these cases the changes to the spectrum are random, constituting a form of noise.

It may sometimes happen that a complex ion is inherently unstable (metastable), characterized by a half-life, so that there is a finite probability of its decomposing during transit through the spectrometer. One of the products of such an event will, in general, be an ion of the same charge type as the original, so that it will continue in flight, but with an altered trajectory. This will produce a perturbation in the recorded spectrum, usually consisting of an increased background over a short region. An example of this is visible in the spectrum of Figure 4a, at about 38 to 45 mass units, evident in the most sensitive trace (×1). The small peaks at approximately 30.8 and 32.3 (marked *B* and *C*) are also metastables, though with less spread. Metastable ions are more likely to be observed in double-focus spectrometers than in the simpler types, because of the longer dwell time of ions within the flight path.

VII. TANDEM MASS SPECTROMETRY (MS/MS)

Two mass analyzers can be operated in series to great advantage. The first analyzer selects ions of a particular mass number and sends them on to a second analyzer for identification. This is useful, for example, with isomeric compounds. As ionized by CI, they are indistinguishable, since only their molecular ions are produced in quantity. If these ions are subjected to further ionization by electron bombardment the mixture can be analyzed readily.

MS/MS can involve any types of mass analyzers, the most common being dual (or even triple) quadrupoles. These can be operated in various modes, to give a very powerful series of instrumental assemblies. For further information, the reader is referred to Refs. 11-13.

VIII. THE VACUUM SYSTEM

A. Pumps

The mass analyzer section of dispersive mass spectrometers must be maintained at a vacuum pressure not greater than about 10^{-7} torr to insure that the electrostatic lenses and other ion-optic components will perform adequately. For this purpose either oil diffusion pumps or ion pumps can be used. The throughput of ion pumps is superior to that of diffusion types, so they are usually selected when high precision and resolution are required. If diffusion pumps are used, it is important to avoid mercury or silicone oils as the working medium, as trace vapors of these materials will cause difficulties in the ion sources. Whatever pumping system is chosen, it is almost essential to have efficient cold traps, cooled either with liquid nitrogen or by a closed-cycle refrigeration device.

The throughput is particularly important with chemical ionization sources because of the excess reagent gas that must be expelled. Instruments designed especially for CI work generally employ a separate pump for the source region. The need for several stages of pumping in GC/MS and LC/MS have been mentioned previously.

B. Gauges

A variety of pressure sensors are suitable for use with a mass spectrometer. To monitor the rough vacuum produced by a fore-pump, a Pirani gauge is appropriate. For measuring the residual gas pressure in the analyzer itself, an ionization gauge is generally selected. For a complete discussion of vacuum gauges, the reader is referred to any text on vacuum techniques.

IX. SPECIFICATIONS OF MASS SPECTROMETERS

Mass spectrometers are described in terms of the mass range covered, the sensitivity, and the resolution. The first two of these are self-explanatory, but the concept of resolution requires special comment.

A. Resolution

According to the most widely accepted definition, the resolution is the ratio $M/\Delta M$, where ΔM is the difference in mass numbers that will give a valley of 10% above the base line between peaks of masses M and $M + \Delta M$ for two peaks of equal height. In many types of spectrometer the resolution becomes successively poorer as one goes to higher masses. In this case, the highest mass number that will just resolve two equal peaks is called the *unit resolution*, and can serve as a figure of merit. If this criterion is met for masses up to 600 and 601, for example, the spectrometer is said to have unit resolution of 600.

The unit resolution of a simple sector instrument may go as high as about 5000, whereas that of a double-focus type may reach 50,000 or even higher. This cannot be taken to mean that ions of 50,000 mass units can be vaporized intact—surely they cannot—but rather that ions of lower mass can be distinguished even if they are separated by much less than one mass unit.

A double-focus mass spectrometer can easily resolve peaks due to ions with the same nominal molecular weight but different elemental composition. Examples are:

N_2	28.0061	$C_5H_4N_4O$	136.0385
CO	27.9949	$C_6H_6N_3O$	136.0511
C_2H_4	28.0313	$C_7H_8N_2O$	136.0637
		$C_8H_{10}NO$	136.0762
		$C_9H_{12}O$	136.0889

B. Comparison of Spectrometer Types

There is considerable variation within any given type of spectrometer, but some general guidelines can be given. Table 1 lists the main kinds

Table 1. Typical Specifications of Mass Spectrometers

Spectrometer type	Max. mass number	Unit resolution
Magnetic sector	2000	5,000
Double focus (M-H)	2500	40,000
Double focus (N-J)	7200	70,000
Quadrupole	850	450
Ion-trap	650	650

of mass spectrometers with typical ranges for mass number and resolution. The sensitivity varies so much with operating characteristics, and the formula by which manufacturers describe it are so divergent, that it would avail little to include it in this table.

X. APPLICATIONS

Undoubtedly the most significant application for mass spectrometry is in the identification of chemical species, atomic or molecular. Nearly everything we know about stable isotopes has been acquired through mass spectra. In the realm of covalent compounds, especially those with appreciable vapor pressure, mass-spectral fragmentation patterns serve for fingerprint identification, though such identification may not be unique in the case of isomeric compounds with very similar structures. The utility of high-resolution mass spectra in distinguishing between isobaric compounds of different elemental composition has been demonstrated above.

The fragmentation patterns for the components of a mixture are additive; hence mixtures can be analyzed if spectra for the several components, run under the same conditions, are at hand. The calculation for n components involves a set of n simultaneous equations with n unknowns, easily solved with the help of a computer. An analysis of this kind can often be attacked more readily by GC/MS or LC/MS, where the components are separated prior to their quantitative determination.

A mass spectrometer can be used in tracer studies with reactants that have been enriched with stable isotopes of low natural abundance. The technique of isotope dilution, well known in connection with radioactive isotopes, is applicable here.

A. Physicochemical Applications

As has been noted earlier, a study of fragmentation patterns can give copious information about the relative strengths of interatomic bonds in covalent molecules. This can be investigated to advantage by the measurement of appearance potentials in EI spectrometry. The spectrometer is set to observe the mass number corresponding to a selected fragment, then the kinetic energy of the bombarding electrons is raised until that fragment just shows up; the energy in electron-volts is then a measure of the corresponding bond strength.

XI. DATA PROCESSING

Prior to the extensive use of dedicated computers, mass spectra were printed out on a rapid-response strip-chart oscillographic recorder (see Figure 4a for an example). Instruments in present manufacture almost universally use a computer to store the data from the detector. This is, of course, a much more flexible arrangement, permitting scaling, background subtraction, comparison with library files, and many other possible manipulations. Production of a bar-graphic representation of the spectrum, as in Figure 4b, is a common function of the computer.

There are a number of extensive collections of mass-spectral data available for computer-directed comparison with unknowns. The purchaser of a mass spectrometer should make certain that the instrument chosen is compatible with these library files.

Figure 9 shows a powerful method of three-dimensional plotting of GC/MS data. Each horizontal line is a mass spectrum taken on the portion of GC eluant corresponding to the elution time on the diagonal. This example was made with a privately assembled Fourier-transform ion-trap spectrometer together with a 0.5-mm-diameter open tubular GC column [14].

An example of modern data processing with the computer, as applied to ICP/MS, has been presented by Vaughan and Horlick [15]. They have utilized the graphic capabilities of the Apple Macintosh computer to create an interactive program for displaying elemental mass spectra and their probable interferences, taking the data from previously published information. The user is first presented with a periodic table on the computer screen. Clicking with the mouse on any element in the table will cause a mass spectrum of that element to appear with a notation as to the natural relative abundances of the isotopes. Another command will produce a list

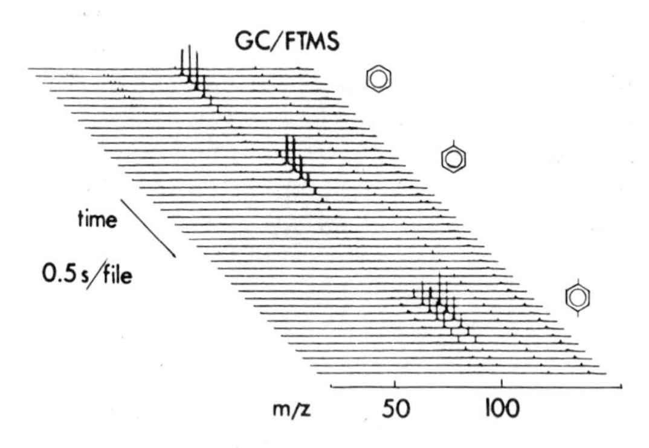

Figure 9. Series of GC-FT/MS spectra of equal amounts of benzene, toluene, and xylene. Twenty-five signal-averaged scans were taken for each file. (After Ref. 14.)

of interfering elements. Such a program could be expanded to include taking data automatically from an operating mass spectrometer and comparing it with the known spectra. One way in which this might be done is described (for ICP/AES rather than MS) by Karanassios and Horlick in a 20-page paper in the same issue of *Applied Spectroscopy* [16].

REFERENCES

1. J. R. Trow, *Rev. Sci. Instrum.*, *56*, 2163 (1985).
2. C. Brunnee, G. Jung, U. Markwardt, R. Pesch, and H. Wollnik, *Am. Lab.*, *18(11)*, 100 (1986).
3. W. C. Wiley and I. H. McLaren, *Rev. Sci. Instrum.*, *26*, 1150 (1955); W. C. Wiley, *Science*, *124*, 817 (1956).
4. W. N. Delgass and R. G. Cooks, *Science*, *235*, 545 (1987).
5. R. J. Day, S. E. Unger, and R. G. Cooks, *Anal. Chem.*, *52*, 557A (1980).
6. K. L. Rinehart, Jr., *Science*, *218*, 254 (1982).
7. M. T. Bernius and G. H. Morrison, *Rev. Sci. Instrum.*, *58*, 1789 (1987).
8. K. L. Busch and R. G. Cooks, *Science*, *218*, 247 (1982).
9. R. S. Houk, *Anal. Chem.*, *58*, 97A (1986).
10. H. E. Taylor, *Spectroscopy*, *1(11)*, 20 (November 1986).
11. R. G. Cooks and G. L. Glish, *Chem. Eng. News* (November 30, 1981, p. 40).
12. F. W. McLafferty, *Tandem Mass Spectrometry*, Wiley, New York, 1983.
13. K. L. Busch and G. C. DiDonato, *Am. Lab.*, *18(8)*, 17 (1986).
14. E. B. Ledford, Jr., R. L. White, S. Ghaderi, C. L. Wilkins, and M. L. Gross, *Anal. Chem.*, *52*, 2450 (1980).
15. M. A. Vaughan and G. Horlick, *Appl. Spectrosc.*, *41*, 523 (1987).
16. V. Karanassios and G. Horlick, *Appl. Spectrosc.*, *41*, 360 (1987).
17. D. Lichtman, *Res. Dev.*, *15(2)*, 52 (1964).
18. R. S. Houk, V. A. Fassel, G. D. Flesch, H. J. Svec, A. L. Gray, and C. E. Taylor, *Anal. Chem.*, *52*, 2283 (1980).
19. R. L. Hunter and R. T. McIver, Jr., *Anal. Chem.*, *51*, 699 (1979).

25

Thermoanalytical Instrumentation

DAVID DOLLIMORE / Department of Chemistry, University of Toledo, Toledo, Ohio

I. INTRODUCTION

A. Scope of Thermal Analysis

The name thermal analysis is applied to a variety of techniques in which the measurement of any property of a system is recorded as the system is programmed through a predetermined range of temperatures [1]. The plot of the physical property of the sample recorded as a function of temperature is said to be a thermal analysis curve. There is some confusion in the literature about this name, since it was initially applied to the specific technique in which the temperature of a sample was recorded against time as it was cooled down from a particular value. The use of the name in this way persists in physical chemistry textbooks, and in certain countries the name thermal analysis is reserved for this specific purpose. However, for this chapter, the term thermal analysis is taken in the wider sense, where it covers a group of techniques. There are further conditions that have to be satisfied in thermal analysis as it is usually practiced:

1. The physical property and the sample temperature should be measured continuously. It should be noted that this restriction makes the measurements of certain physical properties rather difficult. An example would be X-ray powder diffraction.
2. In practice both the property and the temperature should be recorded automatically.
3. The temperature of the sample should be altered at a predetermined rate. In many earlier textbooks it is stated that the sample should be cooled or heated at a uniform rate; however, the real basis of the use of these techniques is that it should be operated on a predetermined basis, as this allows various parameters to be followed—for example, one may compare directly an industrial process in which the temperature is raised and then held at a particular temperature and then raised again.

The purpose in making these measurements is to study the physical and chemical changes that occur in a sample or a system on heating. One therefore has to interpret a thermal analysis curve by relating the features of the property against temperature with possible chemical or physical events that have taken place in the system under observation. Possibly because of the difficulty in measuring some physical properties instantaneously at a particular temperature, the applications of certain techniques are somewhat limited. The most common property measured is that of mass loss, but calorimetry experiments predate this technique and give information concerning primarily the enthalpy changes that take place. The analysis and detection of evolved gas is also an important subject. Another group of studies comes under the heading of thermomechanical analysis. These deal with dimensional changes and with properties connected with the strength of materials when subjected to temperature changes. It should be noted that this group must by definition include the measurement of the density of samples subjected to a program temperature variation.

Those who practice these techniques of thermal analysis must be subjected to a discipline that needs definitions and conventions to provide an adequate description. Very often in commercial equipment these techniques are combined and this proves to be a very powerful combination in unraveling the events which occur when the systems are subjected to temperature changes. Generally thermal analysis techniques may be classified into three groups depending on the way in which the physical property is recorded:

1. The absolute value of the property itself can be measured, for example, the sample mass.
2. The differential method measures the difference between some property of the sample and that of a standard material, such as their temperature difference.
3. The rate at which the property is changing with temperature can be measured; this forms the basis of derivative measurements and very often may be interpreted on a kinetic basis.

The naming of the most commonly used techniques has become important because of the habit that scientists have of reporting their data not under the actual name but by using symbols derived from the name of the technique. It would not matter so much if authors took the trouble to define the technique they used or adopted systematic symbols. However, in the field of thermal analysis, organizations, national and international, have arisen where certain nomenclature abbreviations, definitions, and standards are recommended. These organizations have set up committees, which have formulated, in particular, a system of nomenclature that is adhered to in this text but is not always adhered to in journals and in certain fields of science.

B. Nomenclature and Definitions

The nomenclature to be recommended has been formulated by an international committee that is part of the International Confederation for Thermal Analysis (ICTA), and this committee has selected what it considers to be the most appropriate nomenclature definitions and conventions for thermal analysis. These recommendations have been widely circulated by Prof. G. Lombardi, past president of the Confederation, in the booklet "For Better Thermal Analysis." This booklet is regularly updated, and subsequent editions will be edited by others who will have the same authority as Prof. Lombardi [1].

The recommended names and abbreviations for the most commonly used techniques in thermal analysis are listed in Table 1. However such a table does not necessarily provide a short description of the techniques. Table 2 remedies this deficiency, giving a description and indicating methods of measurement. A careful examination of this table should enable the technique to be understood. Details of each technique are given later. As already noted, in principle any property may be plotted continuously as a function of imposed temperature, but in practice some properties that take time to measure might provide difficulties. An exhaustive review encompassing every possible method and measurement would obviously be impracticable. Furthermore, the use of these techniques extends well beyond what is conventionally referred to as "analytical chemistry." Many of the practitioners of the "art" would be upset at being described as analytical chemists or even chemists. The subject is really interdisciplinary. Nevertheless, the techniques are primarily used to describe the quality of a given product, and in this sense the data provided are analytical.

The definitions of these techniques from Tables 1 and 2 are as follows:

Thermogravimetry (TG) is one of the most widely used techniques of thermal analysis, in which the mass of a substance is measured as a function of temperature while it is subjected to a controlled temperature program. The record on the thermogravimetric or TG curve is the mass plotted against temperature (T) or time (t) if the variation of temperature with time can be indicated as well on the same graph. In decomposition reactions the reactant material degrades, often to be replaced by the solid product—an example of this is the decomposition of calcium carbonate to calcium oxide—and from a record of the mass of the system against the temperature the decomposition of the material can be easily followed. This is very often plotted in alternative ways such as the percentage mass loss or the fractional mass loss against the temperature or fractional decomposition (α) against the temperature.

Derivative thermogravimetry (DTG) is a technique derived by using the same instruments as for TG, but in DTG the first derivative of the TG curve with respect to either time or temperature is plotted. The DTG curve is plotted with the rate of mass loss on the ordinate plotted downward and temperature or time on the abscissa increasing from left to right. It should be noted that this can be done either as part of the instrumentation of a thermal balance, or by a computer analysis of the TG curve as a separate operation in the dedicated computer part of the equipment.

Isobaric mass change determinations refer to the equilibrium mass of a substance at a constant partial pressure of the volatile product(s) measured as a function of temperature while the substance is subjected to a controlled temperature program. The isobaric mass-change curve is plotted with mass as the ordinate decreasing downward and temperature on the abscissa increasing from left to right.

Evolved-gas detection (EGD) is a technique in which the evolution of gas from a substance is monitored as a function of temperature while the substance is subjected to a controlled temperature program. This can be performed quantatively with modern equipment, and then the term *evolved-gas analysis* (EGA) is applied and the amount of product and its identity are measured as a function of temperature while the substance

Table 1. Nomenclature in Thermal Analysis

Name	Abbreviation
1. General	
Thermal analysis	
2. Methods associated with mass change	
A. Static	
Isobaric mass change determination	
Isothermal mass change determination	
B. Dynamic	
Thermogravimetry	TG
Derivative thermogravimetry	DTG
C. Methods associated with evolved volatiles	
Evolved gas detection	EGD
Evolved gas analysis	EGA
3. Methods associated with temperature change	
Heating-curve determinations	
Heating-rate curves	
Inverse heating-rate curves	
Differential thermal analysis	DTA
Derivative differential thermal analysis	
4. Methods associated with enthalpy change (note that as classical calorimetry relies initially on noting a temperature change as the basis of the calculation of enthalpy, the distinction between categories 3 and 4 is often blurred)	
Differential scanning calorimetry	DSC
5. Methods associated with dimensional change or mechanical properties	
Thermodilatometry	
Derivative thermodilatometry	
Thermomechanical analysis	TMA
Dynamic thermomechanometry	

is subjected to a controlled temperature program. The method of analysis should always be clearly stated in these gas detection methods.

There are other techniques that relate to mass in thermal analysis. Thus *emanation thermal analysis* is a technique in which the release of radioactive emanation from a substance is measured as a function of temperature while the substance is subjected to a controlled temperature program.

A further set of techniques involves calorimetry. It has already been mentioned that historically and in some physical chemistry textbooks, thermal analysis applies to the determination of cooling or heating curves; these are techniques in which the temperature of a substance is measured as a function of the program temperature while the substance is subjected to a controlled temperature program. Usually these are cooling curves, and again these may be reported as the first derivative of the temperature curves with respect to time, that is, dT/dt. The function dT/dt should be plotted on the ordinate and T or t on the abscissa increasing from left to right. This technique has been used with regard to heating rate curves in which the temperature is increased, so it is important to say whether one is subjecting the sample to a cooling-curve program or a heating-curve determination.

Related to these methods are the two techniques of *differential thermal analysis* (DTA) and *differential scanning calorimetry* (DSC). Differential thermal analysis is a technique in which the temperature difference between a substance and a reference material is measured as a function of temperature while the substance and reference material are subjected to a controlled program. The record is a differential thermal analysis or DTA curve; the temperature difference ΔT should be plotted on the ordinate with endothermic processes downward, exothermic processes in the opposite direction, and temperature or time on the abscissa increasing from left to right. This equipment can very often be applied quantitatively, and here the area of the peaks can be made proportional to the quantity of the material decomposing or to the enthalpy of the process. However, another name, differential scanning calorimetry (DSC), has been applied to calorimeter experiments in which the background temperature of the calorimeter is raised through a programmed temperature regime being imposed on the system. Thus, differential scanning calorimetry may be defined as a technique in which the difference in energy inputs into a substance and a reference material is measured as a function of temperature while the substance and reference material are subjected to the same controlled temperature program.

Two modes of DSC are currently offered in instrumentation, namely, power-compensation differential scanning calorimetry (power-compensation DSC) and heat-flux differential scanning calorimetry (heat-flux DSC). The two methods need to be distinguished, as power-compensation DSC was for a long time a copyrighted term employed by one of the instrument manufacturers and as such the term DSC is often found in the literature applying to power-compensation equipment only [2].

The Nomenclature Committee has considered the distinction between quantitative DTA and heat-flux DSC; in its opinion a system with multiple sensors (e.g., a Calvet-type arrangement) or with a controlled heat leak (Boersma-type arrangement) would be heat-flux DSC, whereas systems

Table 2. Commonly Used Thermal Analysis Techniques

Technique	Property measured	Name of instrument	Measurement device	Comment
Thermogravimetry (TG)	Mass	Thermobalance	Automatically	Plot of mass (m) against temperature (T) under stated temperature program
Derivative thermogravimetry (DTG)	dm/dt where dm/dt represents the rate of change of mass (m) against time (t)	Thermobalance	Automatically recording balance	Note that in equipment controlled and directed by a computer work station the thermogravimetry trace can be generated from the plot of m against T
Differential thermal analysis (DTA)	$T_S - T_r \equiv \Delta T$ where T_S is the temperature of the sample, T_R is the temperature of the reference material, so ΔT is the difference between the temperature of sample and the reference material	DTA apparatus	Thermocouple or any temperature measurement device	Plot of ΔT against T under stated temperature program
Differential scanning calorimetry (DSC)	Heat flow dH/dt, the change of enthalpy (H) with time (t) subjected to a controlled temperature program	DSC calorimeter	Thermocouple or any temperature measurement device	In all types of DSC units the initial measurement in the recording of temperature of sample and reference material. The measurement of enthalpy change with time is a derived quantity based

				on calibration and utilization of these measurements. This blurs the distinction between DTA and DSC
Evolved gas detection or analysis (EGD or EGA)	Presence of evolved gases (EGD) depends on the amounts of individual gas species evolved	Name chosen is the actual unit of measurement, e.g., infrared gas detector or mass spectrometer	See column opposite	Plots of individual gas concentrations against temperature with sample subjected to controlled temperature program
Thermomechanical analysis (TMA) (e.g., dilatometry)	Deformation measured as either volume or length	Dilatometer	Various	Deformation plotted against temperature when sample is heated under a controlled temperature program. Sample may be heated under zero load (on sample) or sample may be subjected to a specified load.
Dynamic thermomechanometry or dynamic mechanical analysis (DMA)	Method uses sample subjected to an oscillatory load and measures the modulus and/or damping of the substance as a function of temperature	Various	See under method	Applied mainly to processed polymer pieces

without these or equivalent arrangements would be quantitative DTA. The conventional method of plotting DTA results is as already noted with endothermic peaks downward on the plot while the exothermic peaks are shown in an upward direction. However, because DSC is considered by some to measure thermodynamic quantities, the DSC plots are often found with the endothermic plots in an upward direction and exothermic plots in a downward direction. This is to conform with the IUPAC requirements on the presentation of thermodynamic parameters.

There are some unexpected nuances in the definitions of techniques grouped under the heading thermomechanical analysis (TMA). Thus *thermodilatometry* is defined as a technique in which the dimensions of a substance under negligible load are measured as a function of temperature while the substance is subjected to a controlled temperature program. The plot in the thermodilatometric curve is then the dimension plotted on the ordinate increasing upward, with temperature T or time t on the abscissa increasing from left to right. Related techniques are linear thermodilatometry and volume thermodilatometry, which are distinguished on the basis of the dimensions measured. These dimensions are actually indicated in the name of these two techniques. *Thermomechanical analysis* (TMA) is a technique in which the deformation of a substance under a nonoscillatory load is measured as a function of temperature while the substance is subjected to a controlled temperature program. It finds increasing use in polymer technology. The mode, as determined by the type of stress applied (compression, tension, flexure or torsion), should always be stated. *Dynamic thermomechanometry*, on the other hand, is a technique in which the dynamic modulus and/or damping of a substance under oscillatory load is measured as a function of temperature while the substance is subjected to a controlled temperature program. A related technique is *torsional braid analysis*, in which the sample is prepared by impregnating a glass braid or thread substrate with sample. The sample impregnated braid is then subjected to free torsional oscillations.

There are other techniques of thermal analysis that are not always in the instrument manufacturers' list of available equipment and most often have to be built for that particular purpose. One such technique is *thermosonimetry*, in which the sound emitted by a substance is measured as a function of temperature while the substance is subjected to a controlled temperature program. Another technique is *thermoacoustimetry*, in which the characteristics of the imposed acoustic waves are measured as a function of temperature after passing through a substance while the substance is subjected to a controlled temperature program. *Thermoptometry* is another unusual thermal analysis technique in which an optical characteristic of a substance is measured as a function of temperature while the substance is subjected to a controlled temperature program. *Thermophotometry*, *thermospectrometry*, *thermorefractometry*, and *thermoluminescence* involve measurement of total light, light of specific wavelengths, refractive index, and luminescence, respectively, as the system is subjected to a controlled temperature program. *Thermomicroscopy* refers to observations under a microscope while the sample is taken through a range of temperatures. Two other techniques may be mentioned, *thermoelectrometry*, in which an electrical characteristic of a substance is measured as a function of temperature, and *thermomagnetometry*, in which the magnetic susceptibility of a substance is also measured as a function of temperature.

C. Multiple Techniques

Obviously the combination of two or more techniques is better than one. It is easily seen that knowing the data on a TG curve (mass loss) can be vastly enhanced by knowing what the evolved species are and their extent (EGA). In some cases, such as the decomposition of limestones, it would seem to be obvious that the evolved gas is carbon dioxide, but in some cases the evolution of much smaller quantities of gaseous species such as water and carbon monoxide can give valuable indications as to possible mechanisms of decomposition.

The combination of thermal analysis techniques can be carried out in various ways. The most common approach is simply to obtain the two sets of data using different equipment and separate samples. This is probably the most economic way of approaching the problem but it has disadvantages. These can be listed as:

1. Samples may not be the same (always a problem in such cases as coals, limestones, cements, etc.).
2. The thermal environment and sample temperatures may not be the same in the two sets of equipment.
3. The ambient atmosphere and the flow rate may not be the same in both techniques and even when this is formally the same, atmospheric environment immediately above the sample may differ because of differences in the geometry of container and furnace around the sample.

Two more sophisticated approaches of "coupling" techniques are available. In the first, called "combined techniques," separate matched samples are used for each property measurement, subjected to a nearly identical thermal environment, that is, in the same furnace and subjected to the same temperature program. An example is a sample in a TG unit, while immediately below the sample crucible in the TG unit is a DTA cell containing sample and reference material. Simultaneous techniques, on the other hand, involve a single sample that is subjected to several "property" measurements while undergoing a programmed temperature change. Thus one can imagine the sample being simultaneously subjected to mass change determinations (TG), differential thermal analysis (DTA), and gas evolution measurements (EGA or EGD). In writing the names of simultaneous techniques these should be separated by the use of the word "and" when used in full and by a hyphen when the symbols are used, for example, simultaneous TG-DTA. Further distinctions are possible. Thus coupled simultaneous techniques covers the application of two or more techniques to the same sample when the two instruments involved are connected by an interface (e.g., simultaneous differential thermal analysis and mass spectrometry), and the term interface refers to a specific piece of equipment that enables the two instruments to be joined together. In coupled simultaneous techniques, as in discontinuous simultaneous techniques, the first technique to be mentioned is that in which the earlier measurement is made, for example, when a DTA instrument and a mass spectrometer are connected to an interface, DTA-MS is the correct form—not MS-DTA.

The term discontinuous simultaneous techniques covers the application of coupled techniques to the same sample when sampling for the second technique is discontinuous—for example, discontinuous simultaneous differential thermal analysis and gas chromatography, when discrete portions of evolved volatiles are collected from the sample situated in the instrument used for the first technique. A further "simultaneous" technique can easily be envisaged where two samples of the same specimen are located in the same furnace assembly but subjected simultaneously to two different techniques—for example, DTA and TG.

D. Reporting Results

In reporting results one must consider the recommendations of the standardization committee of ICTA [3]. This committee was set up because of the wide diversity of equipment and the need to correlate data. No single instrument is quite the same as any other. No single instrument represents the optimum design for all possible studies. The techniques are dynamic, very often involving kinetic factors, and produce data that are highly dependent on procedure. Before dealing with the actual reporting of data, a set of consistent terms relating to thermal analysis was drawn up. Certain arbitrary choices had to be made.

In DTA the *sample* is the actual material investigated, whether diluted or undiluted. The use of the word *sample* is common to all thermal analysis techniques. A reference material is a known substance, usually inactive thermally over the temperature range of interest in the DTA experiment. The *specimens* include both the sample and reference materials. The *sample holder* is the container or support for the sample, while the *reference holder* is the container or support for the reference material. These should be identical in size and shape in any single unit. *The specimen-holder assembly* is the complete assembly in which the specimens are housed. Sometimes the source of heating or cooling is part of the unit with the containers or supports for the sample and reference material, so this must then be considered as part of the specimen holder assembly. In certain units the specimen-holder assemblies are in the form of a relatively large mass of material in intimate contact with the specimens or specimen holders, and this is termed a *block*. The *differential thermocouples or ΔT thermocouples* are the thermocouples used to measure the temperature difference signal. There are units, however, where thermocouples are not used to generate this signal—in which case the thermosensing device should be named in referring to the ΔT signal.

In the case of TG equipment it should be noted that the balance assembly and furnace are referred to as a *thermobalance*. The sample in TG is rarely diluted, but it might be diluted in simultaneous TG-DTA. The sample holder in TG is then the container or support for the sample.

Both DTA and TG units employ a *temperature thermocouple* or *T thermocouple*, which measures the temperature. The position of this thermocouple with reference to the sample should be noted. It should be noted also if this thermocouple is used in addition to provide a central signal to enable temperature programming to take place. The *heating rate* is the rate of temperature increase (degrees per minute). If the equipment is undergoing cooling then it is termed the cooling rate. The heating rate is said to be constant when the temperature/time curve is linear. Not all equipment provides this information, so a check should be run every now and then to establish that "constant" heating rates really are constant.

The results from the DTA unit are produced as plots of ΔT versus T, termed the *differential thermal curve* or *DTA curve*. The plot of mass against T from the TG equipment is called the *thermogravimetric curve* or *TG curve*, and if differentiated it is called the *derivative thermogravimetric curve* or *DTG curve*. A complication is that the ordinate in DTA is labeled ΔT but the output from the ΔT thermocouple will normally be in volts. If the emf signal is represented by E it must be noted that while

$$\Delta T = bE$$

it is also true that $b = f(T)$, that is, b varies with temperature. A similar situation arises with other temperature sensor systems.

A set of definitions is recommended for DTA. All these definitions refer to a single peak shown in Figure 1. These definitions and their usage can be extended logically to multiple peak systems, showing shoulders or more than one maximum or minimum. The base lines on the DTA curve AE and CD represent regions where ΔT does not change significantly. An endothermic or exothermic peak is a portion of the DTA curve that has departed from and subsequently returned to the base line (EHC in Figure 1). In an *endothermic peak* or endotherm the temperature of the sample falls below that of the reference material, that is, ΔT is negative. In an *exothermic peak* or exotherm the temperature of the sample rises above that of the reference material, that is, ΔT is positive. The peak

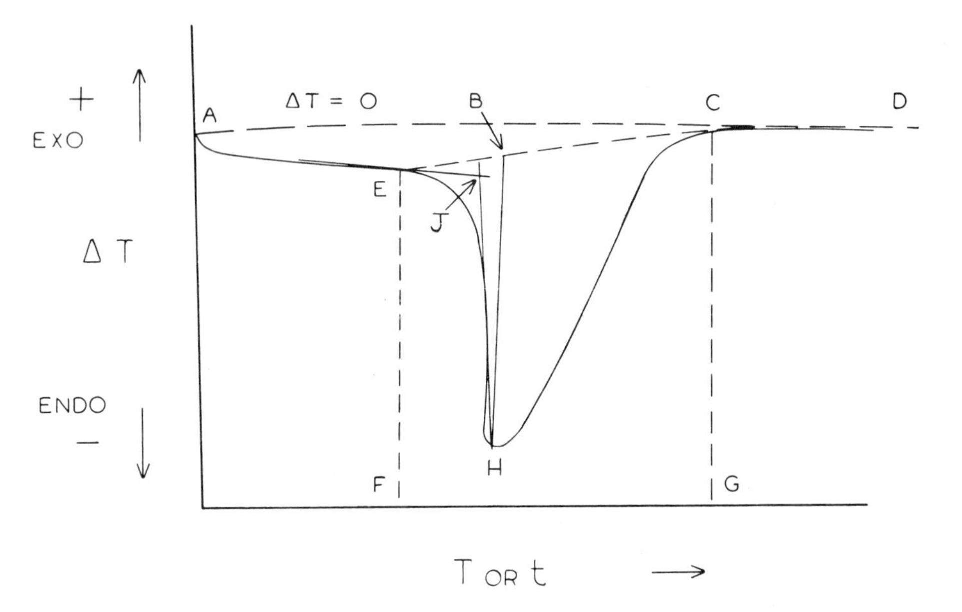

Figure 1. Schematic DTA peak demonstrating definitions mentioned in text.

width (EC in Figure 1) is defined as the time or temperature interval between the points of departure from and return to the base line. Peak height (HB in Figure 1) is the vertical distance between the interpolated base line and the peak tip (*H* in Figure 1). It is usually quoted in degrees Kelvin or Celsius. The peak area (EHCE in Figure 1) is the area enclosed between the peak and the interpolated base line. This area in a properly calibrated unit will be proportional to the enthalpy, and hence can be used to estimate amounts of material involved. Finally, with regard to DTA the extrapolated onset (B in Figure 1) is the point of intersection of the tangent (HJ in Figure 1) drawn at the point of greatest slope on the leading edge of the peak with the extrapolated base line (EJ in Figure 1).

With regard to TG all the definitions refer to a single-stage process as shown in Figure 2 but can be readily extended to a multistage process, which can simply be considered as a series of interconnected single-stage processes. Here a plateau (AB in Figure 2) is that part of the TG curve where the mass remains essentially constant. The *initial temperature*, T_i (*B* in Figure 2), is that temperature at which the cumulative mass change can be first detected on the thermobalance curve. Likewise, the *final temperature*, T_f (*C* in Figure 2), is that temperature at which the cumulative mass change reaches a maximum. The *reaction interval* is sometimes quoted, representing the difference between T_f and T_i.

The ICTA Committee on standardization has produced the following recommendations for reporting thermal analysis data in order to obtain the best presentation, both in terms of experimentation and reporting, so that information obtained and published is of maximum value. Unfortunately some editors of journals do not always adhere to either these recommendations, such as TG, DTA, etc., even though such recommendations carry the authority of IUPAC committees.

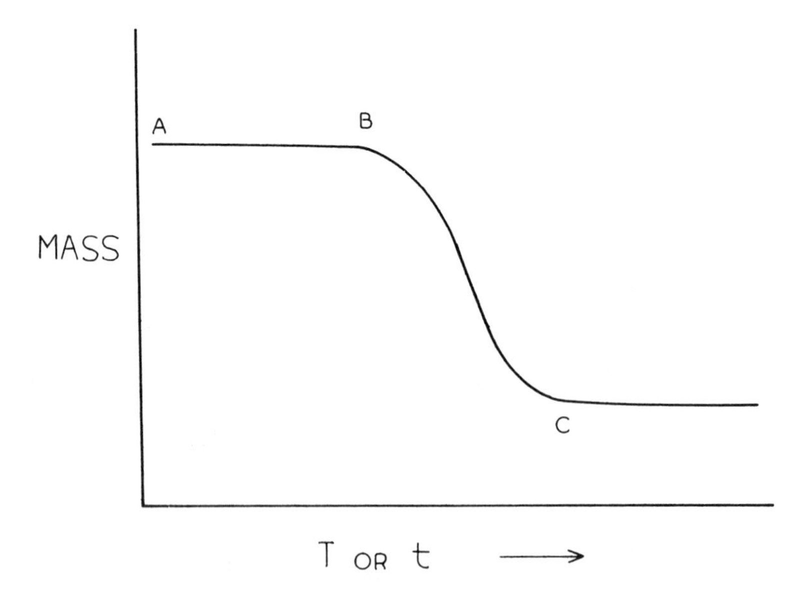

Figure 2. Schematic TG curve.

The recommendations on reporting data are as follows: To accompany each DTA, TG, EGA, EGD, or thermomechanical record, the following information should be provided.

1. Identification of all substances (sample, reference diluent) by a definitive name, an empirical formula, or equivalent compositional data.
2. A statement of the source of all substances, details of their histories, pretreatments, and chemical purities, so far as these are known.
3. Measurement of the average rate of linear temperature change over the range involving the phenomena of interest. Nonlinear temperature programming should be described in detail.
4. Identification of the sample atmosphere in terms of pressure, composition and purity; whether the atmosphere is static, self-generated, or dynamic through or over the sample. Where applicable, the ambient atmospheric pressure and humidity should be specified. If the pressure is other than atmospheric, full details of the method of control should be given. If the system involves a flowing gas supply, then the flow rate should be specified. Where purification of the flowing gas supply is practiced, the purification methods should be stipulated.
5. A statement of the dimensions, geometry, and materials of the sample holder.
6. A statement of the method of loading (quasi-static, dynamic) where applicable.
7. Identification of the abscissa scale in terms of time or of temperature at a specified location. Time should be plotted to increase from left to right.
8. A statement of the method used to identify intermediates or final products.
9. Faithful reproduction of all original records.
10. Identification of the apparatus used, by type and/or commercial name, together with details of the location of the temperature-measuring thermocouple. In reporting DTA or DSC data, additional details should be provided.
11. Wherever possible each thermal effect should be identified and supporting evidence stated.
12. Sample mass and dilution of the sample should be provided.
13. The geometry and type of thermocouples should be described.
14. The ordinate scale should indicate the deflection per degree Celsius at a specified temperature. The preferred plotting should indicate an upward deflection as a positive temperature differential and a downward deflection as a negative temperature differential with respect to the reference.

In reporting the TG data the following additional information should be provided:

11. A statement of the sample mass and mass scale for the ordinate. Mass loss should be plotted downward, but this follows automatically if the ordinate scale is simply that of mass of sample. Additional scales such as fraction decomposed, percent mass loss, or molecular composition are very useful and may be included as desired.
12. If DTG is employed then the method of obtaining the derivative should be indicated and the units of the ordinate noted.

If reporting EGA or EGD data then the following additional information is recommended:

11. A statement of the temperature environment of the sample during the reaction.
12. Identification of the ordinate scale in specific terms if at all possible. In general, an increasing concentration of evolved gas should be plotted upward. In the case of gas density detectors, the increasing gas density should be plotted upward.
13. The flow rate, total volume, construction, and temperature of the system between the sample and detector should be given, together with an estimate of the time delay within the system.
14. Location of the interface between the systems for sample heating and detecting or measuring evolved gases.
15. In the case of EGA when exact units are not used, the relationship between signal magnitude and concentration of species measured should be stated.

In reporting TMA data the following additional details should be given:

11. A clear statement of the temperature environment of the sample.
12. The type of deformation (tensile, torsional, bending, etc.) and the dimensions, geometry, and materials of the loading elements.
13. Identification of the ordinate scale in specific terms where possible. For static procedures, increasing expansion, elongation or extension, and torsional displacement should be plotted upward. Increased penetration or deformation in flexure should be plotted downward. For dynamic mechanical procedures, the relative modular and/or mechanical loss should be plotted upward.

E. Symbols

The abbreviations for each technique have already been introduced. The confusion that can arise however between the use of, for example, TG and T_g, the latter representing the glass transition temperature, has caused a number of investigators and instrument manufacturers to use TGA for TG and so avoid confusion. There are other aspects to the use of symbols, and a report of a subcommittee chaired by Dr. J. H. Sharp and presented to the nomenclature committee of ICTA is presented in substance here [4].

1. The international system of units (SI) should be used wherever possible.
2. The use of symbols with superscripts should be avoided if possible.
3. The use of double subscripts should be avoided if possible.
4. The symbol T should be used for temperature whether expressed degrees Celsius (°C) or in Kelvins (K). For temperature intervals the symbol K or °C can be used.
5. The symbol t should be used for time, whether expressed in seconds (s), minutes (min), or hours (h).
6. The heating rate can be expressed either as dT/dt, when a true derivative is intended, or as β in K min^{-1} or °C min^{-1}. The heating rate so expressed need not be constant and can be positive or negative.

The symbols m for mass and W for weight are recommended. The symbol α is recommended for the fraction reacted or changed.

The following rules are recommended for subscripts: Where the subscript relates to an object, it should be a capital letter:

m_S represents the mass of sample
T_R represents the temperature of the reference material

Where the subscript relates to a phenomenon occurring it should be in lower case:

T_g represents the glass transition temperature
T_c represents the temperature of crystallization
T_m represents the temperature of melting
T_σ represents the temperature of a solid-state transition

Where the subscript relates to a specific point in time or to a point on the curve, it should be in lower case or in figures:

T_i represents the initial temperature
T_f represents the final temperature
$t_{0.5}$ represents the time at which the fraction reacted is 0.5
$T_{0.3}$ represents the temperature at which the fraction reacted is 0.3
T_p represents the temperature of the peak
T_e represents the temperature of the extrapolated onset

F. Standards

Thermal analysis techniques are dynamic in nature and very flexible. The data are highly dependent on procedure. As a consequence, McAdie lists the prime requirements for standards as:

a. To provide a common basis for relating independently acquired data;
b. To provide the means for comparing and calibrating all available instrumentation, regardless of design;
c. To provide the means for relating thermoanalytical data to physical and chemical properties determined by conventional isothermal procedures.

McAdie goes on to say, "such standards must be applicable in whatever experimental design may be required for a particular purpose, otherwise the value of the results is isolated and the full potential of thermal analysis will not be realized. By including curves of standard materials obtained under conditions of the particular study, it will be possible for a reader to relate the subject matter to performance of his own instrumentation, to evaluate the quality of the published data and, hence, conclusions derived therefrom" [5]. The ICTA Committee on Standardization has organized the provision of reference materials, and several sets are available. The program leading to the provision of these standards involved interlaboratory testing of possible substances by over 100 laboratories in 13 countries using more than 35 different DTA or DSC instruments and types of thermobalances. The standards are provided by the U.S. National Bureau of Standards under the designation Certified Reference Materials. The range covered is -60°C to 940°C. These standards are primarily used to standardize DTA or DSC equipment. The first requirement is therefore a material obtainable in pure form showing a reproducible phase change. Since not all DTA/DSC units can cope with molten materials, the sets include eight inorganic materials exhibiting solid-phase transitions, two high-purity metals, and four organic compounds mainly of interest for their melting point. All materials are intended to be used in the heating mode, and used as temperature standards. Table 3 lists four sets of NBS-ICTA Certified Reference Materials available. The materials are intended for use in DTA units under normal operating conditions; consequently the temperature values quoted are generally higher than true equilibrium values given in the literature. A certificate is provided with each set defining the reference points on the DTA curve and giving the mean temperature values together with standard deviations. The certificates also include a discussion of the reasons for variation of the temperatures recorded in different instrument types. The substances used in the calibration sets can easily be obtained in a high state of purity. Table 4 lists details for inorganic materials and metals. However, for DSC in particular not only the temperature of the transition but the enthalpy change associated with the transition is helpful. Table 5 gives transition temperatures for selected organic compounds, and Table 6 gives enthalpy values for certain materials covering a range from 80 to 1065°C [6-12].

A further standard is available. It is a polymer standard of polystyrene with a highly reproducible glass-transition temperature at around 100°C. It is for use on DTA units, and the certificate provided gives the mean temperatures for the various initial points on the curve and the variations introduced by different specimen-holder assemblies.

Table 3. Four Sets of NBS-ICTA-Certified Reference Materials Available for Temperature Calibration of DTA Units

GM-757 -83°C-58°C	GM-758 125-435°C	GM-759 295-575°C	GM-760 570-940°C
1,2 Dichloroethane (melting)	Potassium nitrate	Potassium perchlorate	Quartz
Cyclohexane (transition, melting)	Indium	Silver sulfate	Potassium sulfate
Phenyl ether (melting)	Tin	Quartz	Potassium chromate
o-Terphenyl (melting)	Potassium perchlorate	Potassium sulfate	Barium carbonate
	Silver sulfate	Potassium chromate	Strontium carbonate

Table 4. Accepted Equilibrium Transition Temperatures and Extrapolated DTA Onset as Recommended by ICTA[a]

Element or compound	Transition type	Equilibrium transition temperature (°C)	Extrapolated onset temperature (ICTA standards set) (°C)
KNO_3	Solid-solid	127.7	128 ± 5
In	Solid-liquid	165.6	154 ± 6
Sn	Solid-liquid	231.9	230 ± 5
$KClO_4$	Solid-solid	299.5	199 ± 6
Ag_2SO_4	Solid-solid	424	242 ± 7
SiO_2	Solid-solid	573	571 ± 5
K_2SO_4	Solid-solid	583	582 ± 7
K_2CrO_4	Solid-solid	665	665 ± 7
$BaCO_3$	Solid-solid	810	808 ± 8
$SrCO_3$	Solid-solid	925	928 ± 7

[a]Refs. 3, 6, and 7.

Table 5. Transition Temperatures Shown by Selected Organic Compounds[a]

Compound	Transition	Equilibrium transition temperature (°C)	Extrapolated onset temperature (°C)
p-Nitrotoluene	Solid-liquid	51.5	51.5
Hexachloroethane	Solid-solid	71.4	—
Naphthalene	Solid-liquid	80.3	80.4
Hexamethyl benzene	Solid-solid	110.4	—
Benzoic acid	Solid-liquid	122.4	122.1
Adipic acid	Solid-liquid	151.4	151.0
Anisic acid	Solid-liquid	183.0	183.1
2-Chloroanthraquinone	Solid-liquid	209.1	209.4
Carbazole	Solid-liquid	245.3	245.2
Anthraquinone	Solid-liquid	284.6	283.9

[a]Taken from Ref. 105, p. 38; also Refs. 6, 7, and 8.

Table 6. Enthalpies of Fusion for Selected Materials[a]

Element or compound	Melting point (°C)	Enthalpy of fusion (J g^{-1})
Naphthalene	80.3	149
Benzoic acid	122.4	148
Indium	156.6	28.5
Tin	231.9	60.7
Lead	327.5	22.6
Zinc	419.5	113
Aluminum	660.2	396
Silver	960.8	105
Gold	1063.0	62.8

[a]From Refs. 6, 9-12. Richardson and Savill claim a more accurate value for indium of 29.2 J g^{-1} as cited in Ref. 12.

Standards for TG equipment are possible on the basis of specially formulated techniques. Four methods are available:

1. Use of materials with standard mass-loss points at stated temperatures
2. Use of materials with solid-liquid phase changes at given temperatures
3. Use of materials with magnetic transitions that can cause apparent mass loss related to specific temperatures, and
4. Independent calibration of furnace using either a separate calibrated temperature sensor or a material with a known phase transition

The first method receives few recommendations as TG curves are highly susceptible to experimental conditions. However, the decomposition of calcium oxalate monohydrate can be used to determine whether there are any changing or aging features associated with a particular unit. Calcium oxalate monohydrate goes through three separate decomposition stages:

$$CaC_2O_4 \cdot H_2O \rightarrow CaC_2O_4 + H_2O$$

$$CaC_2O_4 \rightarrow CaCO_3 + CO$$

$$CaCO_3 \rightarrow CaO + CO_2$$

Thus one can run the TG unit with this compound and then retest every month thereafter as a test of reproducibility and performance. This is really a test of both the balance and the furnace. However, it is an internal test and cannot really be used to compare one TG unit with another as regards temperature.

The use of materials with solid-liquid phase changes has been developed in such a way that on changing to a liquid the material is lost from the balance sample assembly and so there is a dramatic mass loss. McGhie and co-workers [13,14] have pioneered this method, termed fusible-link temperature calibration. The equipment is shown in Figure 3. In this a fusible metal standard is suspended from a coil of platinum wire. All this is held in position in the sample container of the DuPont balance. A hole is positioned in the bottom of the sample container so that when the fusible metal standard melts, it and the platinum coil are lost from the sample container, resulting in an immediate change of mass. A typical result using zinc is shown in Figure 4. Suitable materials and relevant data are shown in Table 7.

The use of ferromagnetic standards requires a small modification to the thermobalance, in that the ferromagnetic material is placed in the sample container and suspended within a magnetic field. At the material's Curie point temperature the magnetic mass diminishes and a mass loss is recorded by the thermobalance. Norem et al. [15,16] indicate the following criteria for the effective operation of this calibration technique:

1. The transition must be sharp.
2. The energy required should be small (to ensure its production under dynamic scanning conditions as a sharp transition).
3. The transition temperature should be unaffected by the chemical composition of the atmosphere and be independent of pressure.
4. The transition should be reversible so that the sample can be used repeatedly.
5. The transition should be unaffected by the presence of other standards so that a single experiment will determine several temperatures.
6. The transition should be observable when using milligram quantities.

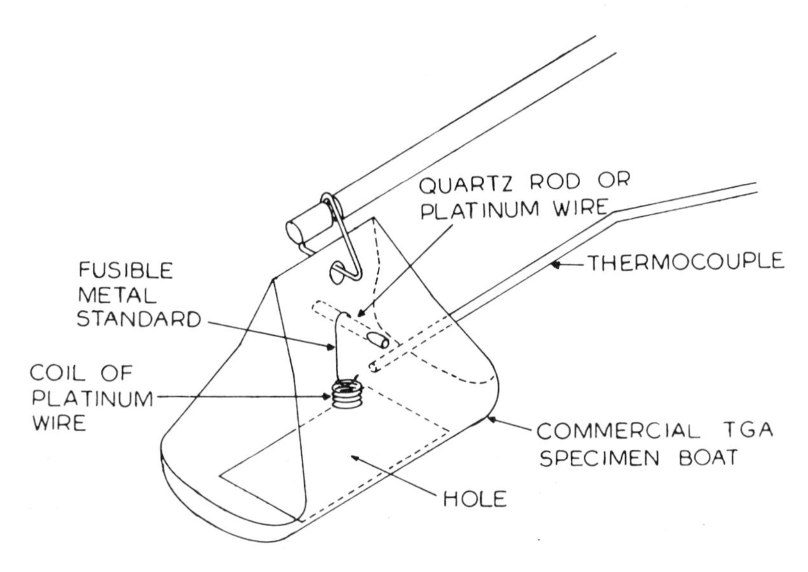

Figure 3. Fusible-link unit on Du Pont TG balance.

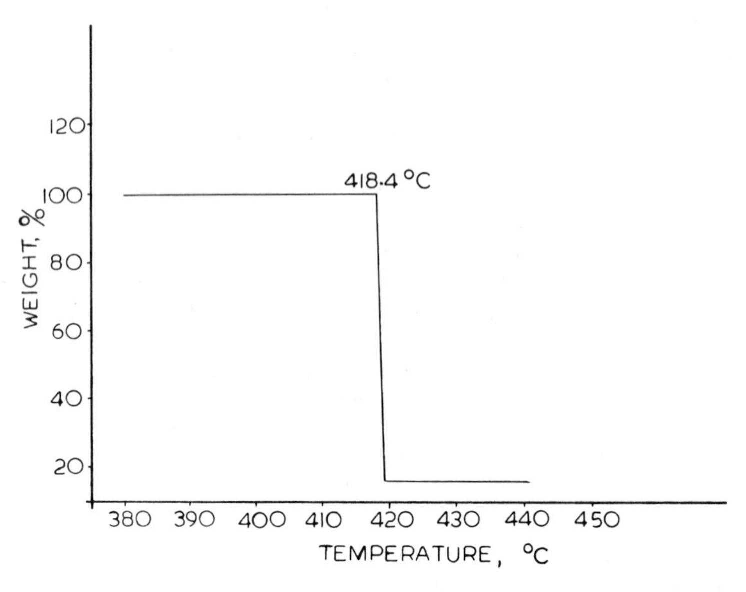

Figure 4. Fusible-link mass-drop curve for zinc.

Table 7. Fusible-Link Temperature Calibration Data[a]

Material	Observed corrected temperature (°C)	Literature temperature (°C)	Deviation from literature value (°C)
Indium	154.20	156.63	-2.43
Lead	331.05	327.50	3.55
Zinc	419.68	419.58	0.10
Aluminum	659.09	660.37	-1.25
Silver	960.25	961.93	-1.68
Gold	1065.67	1064.43	1.24

[a]From Ref. 13.

The data in Figure 5 show a typical run [16]. An ICTA Certified Magnetic Reference Materials Set GM761 was reported by Blaine and Fair [17] to give the results shown in Table 8. There is a dependence of the initial transition temperature on the rate of heating. This is small in the heating rate range from 5 to 20°C min^{-1} but appreciable for fast yeating rates [15,16,18,19]. The standards are intended for use in the heating mode, as they exhibit hysteresis on cooling.

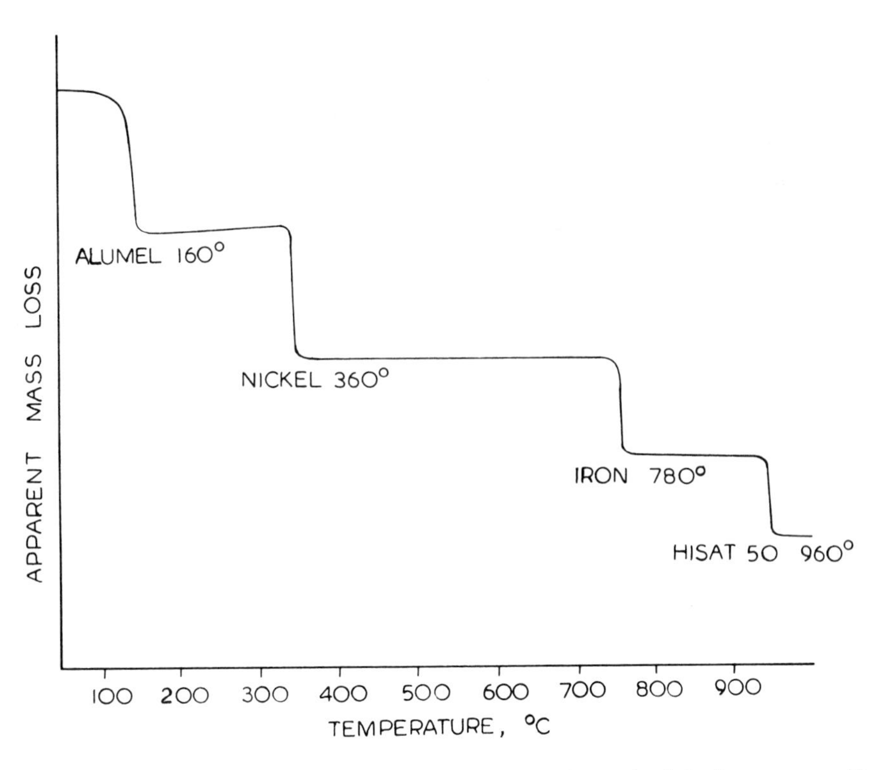

Figure 5. TG experiment using ferromagnetic materials in a magnetic field.

Substance	Observed temperature (°C)	Actual temperature (°C)
Alumel	160	163
Nickel	360	354
Iron	780	780
Hisat	960	1000

Table 8. Results Obtained by Blaine and Fair [17] for Magnetic Transition Temperatures of Magnetic Standards GM761

	Transition temp. (°C)		
Material	Experimental	Literature	Deviation (°C)
Permanorm 3	259.6 ± 3.7	266.4 ± 6.2	-6.8
Nickel	261.2 ± 1.3	354.4 ± 5.4	6.8
Mumetal	403.0 ± 2.5	385.9 ± 7.2	17.1
Permanorm 5	431.3 ± 1.6	459.3 ± 7.3	-28.0
Trafoperm	756.0 ± 1.9	754.3 ± 11.0	2.2

II. THERMOGRAVIMETRY

A. Development

As already noted, a thermobalance is an apparatus for weighing a sample continuously while it is being heated or cooled. There was a delay in the general production of commercial equipment until the middle 1950s. Reliable commercial equipment was not available until the introduction of the Chévenard thermobalance in 1943. Even this apparatus left much to be desired and, mainly due to the help of Duval [20], a more reliable modification was produced in 1947. Figure 6 shows schematically the requirements in contemporary thermogravimetric units. Older units did not possess computer

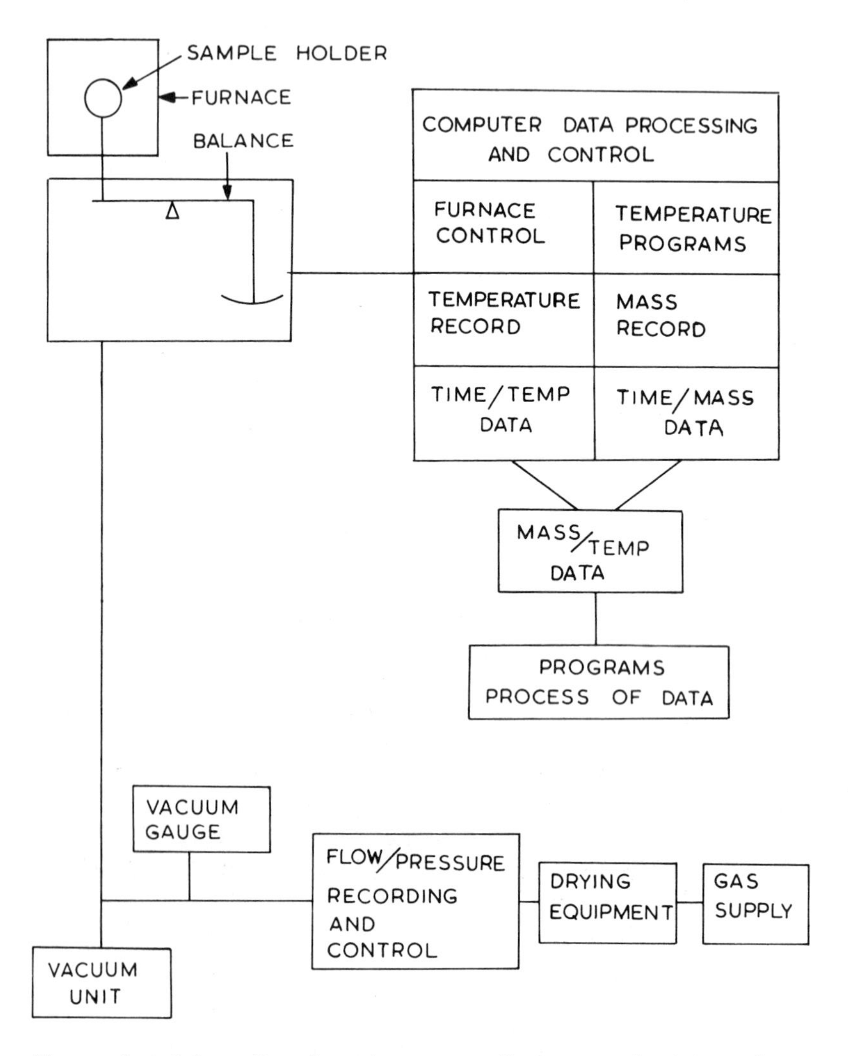

Figure 6. Schematic of contemporary thermogravimetric units.

Table 9. Applications of Thermogravimetry

Physical changes

- Sublimation
- Vaporization
- Absorption
- Adsorption
- Magnetic properties
  - Curie temperature
  - Magnetic susceptibility

Chemical changes

Solid → gas
- Thermal decomposition of many organic and polymeric substances
- Pyrolysis of coal, petroleum, and wood
- Thermal oxidative degradation of polymeric materials
- Carbon gasification with oxygen, steam, or carbon monoxide

$Solid_1$ → $solid_2$ + gas
or
$Solid_1$ + $solid_2$ → $solid_3$ + gas
- Thermal decomposition of many inorganic materials
- Roasting and calcining of minerals
- Determination of moisture, volatiles, and ash contents
- Dehydration studies
- Dehydroxylation studies
- Decomposition of explosives
- Development of analytical procedures
- Kinetic studies (also applicable to solid → gas)

$Solid_1$ + gas → $solid_2$
- Corrosion of metals

processing, control, and presentation capabilities and the data were displayed on conventional X-Y plotters. One feature of modern apparatus design is the modular approach, and this points the way to the probable form of future commercial equipment. It should be noted that one computer can collect, process, and control data from more than one thermal analysis unit. Applications of thermogravimetry are listed in Table 9. It can be seen that they are very extensive. The list is not exhaustive, and one can confidently expect many more examples to appear in the literature.

The early history of thermogravimetric balance design is not documented here—it is to be found in several publications [21,22].

B. Design Factors

The basic instrumental requirements for thermogravimetry are a precision balance, and a furnace capable of being programmed for a linear rise of temperature with time. If a computer is interfaced with the equipment then it should be dedicated to controlling the temperature program, data acquisition, data programming, and data presentation. Most instrument manufacturers present equipment that has access to this kind of dedicated computer. If not, then a suitable X-Y recorder is required.

These requirements can be incorporated into equipment in a variety of ways, and with a vast range of purchase price. Choice of equipment must be based on specific requirements but must take note of features of good thermobalance design and requirements of the operator. Points to be considered are:

1. Decide on the maximum working temperature to be required and then choose a model that can reach around 150°C in excess of this.
2. Check out the hot zone of the furnace. It should be uniform and of reasonable length. The word uniform is an elastic term, but probably for most purposes ±2°C would be satisfactory. The uniform hot zone should be of a size such that the crucible can be easily located within it, together with the measuring temperature sensors. If the balance is not a null deflection type and the crucible moves, then the uniform hot zone must be large enough to accommodate this movement.
3. The shape (or shapes) of the available crucibles must be considered. Some crucibles prevent the easy loss of volatiles. The rate of loss of volatile material from the crucible can be heavily dependent on the crucible geometry.
4. The winding of the furnace must be noninductive to avoid anomalous effects on mass with magnetic or conducting samples. This has been utilized as a method of calibration (see previous section).
5. The heating rate should be reproducible. A variety of heating programs should be available and the user must consider in advance the requirements imposed by the field of study. Temperature-jump programs, isothermal heating modes, holding a rising temperature experiment for required periods of time at a predetermined temperature, are all available. In kinetic determinations it is advantageous in theory to be able to impose parabolic and logarithmic heating rates on the samples under investigation. Finally, in considering the heating rate, the user should decide if a controlled cooling rate is required.
6. The sensitivity of the balance must be in accord with the mass of sample being used and the alteration in mass expected. There is considerable concern in some areas—coal testing, cement hydration, clay identification—that the choice of a suitable representative sample is jeopardized by the limited load capabilities of the balance.
7. The positioning of the balance with respect to the furnace is important in order to prevent radiation and convection currents from the furnace from affecting the recorded mass.
8. The recorded temperature should ideally be that of the sample. If possible this should also be the sensor that controls the temperature program.
9. Chemical attack by volatiles liberated in the decomposition process should be avoided. This is often most easily achieved by passing the carrier gas over the balance system, thus venting the products of decomposition away from the balance system.
10. Provision should be made for measuring the rate of flow of gas through the system and over the sample.

11. There should be in effect a continuous register of the change of mass of the sample, as a function both of time and of temperature.
12. There should be provision for the sample to be heated not only in a dynamic gas supply but also at pressures other than atmospheric and also in vacuum. Many instruments are claimed to meet this requirement but do so by continuous pumping. The user often requires that the unit be incorporated into an isolated high-vacuum system, maintaining this high vacuum under conditions of isolation.

Now the user will not need all these features to tackle any one program of research. These are features that must be considered. The choice of commercial equipment is limited and in some research studies the investigator might have to assemble the units required. This is possible, for the balances themselves are available, and so is equipment to operate them in vacuum or under pressures greater than 1 atm. Properly designed furnaces can be bought and also fairly easily constructed. It should also be remembered that access to other complementary techniques such as EGA might also be required.

As a guide to the correct selection of equipment, it is worthwhile considering in some detail the basic units of a thermobalance.

C. The Balance

The requirements for a good automatic, continuously recording balance are essentially those of any analytical balance plus the ability to continuously record the data. The specifications must include accuracy, sensitivity, reproducibility, and maximum load. The recording balance should have the capability of making an adequate range of mass adjustment, and it should show a high degree of mechanical and electronic stability. There should be rapid response to mass change, and the equipment should be mounted so that it is unaffected by vibration. The balance should also be unaffected by ambient temperature changes.

Various weighing systems can be discussed. Besides deflection and null-point balances, there are other methods available, notably those based on changes in a resonance frequency.

The principle of the deflection balances is simply that the deflection or movement can be calibrated to read in terms of mass. Several different systems of deflection balances are shown in Figure 7. These may be discussed separately.

The beam balance utilizes conversion of the beam deflection about the fulcrum into a suitable reproducible form. In early units a photographically recorded trace was used, but signals generated by suitable displacement-measuring transducers may be used or curves drawn electromechanically.

In the helical spring balance, Hook's law is utilized to convert an elongation or contraction of the spring into a record of mass change. The spring material is important; a quartz fiber is often employed, since this avoids anomalous results associated with temperature change and fatigue problems. The method has been used extensively in adsorption studies of gases and vapors on solids [23-25]. A balance of this kind was used by Loriers [26] for the study of metal oxidation. The Aminco Thermo-Grav

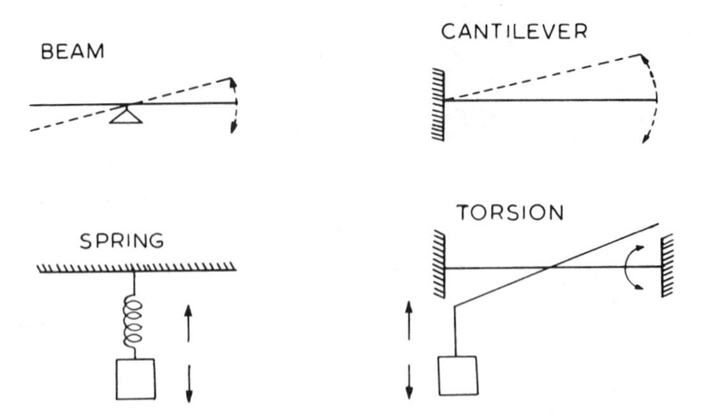

Figure 7. Different types of deflection balances.

unit employed a precision spring that could be selected to give the desired sensitivity and clamped magnetically or with a fluid dashpot to reduce oscillations [27]. The cantilevered beam has one end of the beam fixed, with the other end, on which the sample is placed, free to undergo deflections. The deflection can be translated into mass loss by methods utilized for beam type balances.

In the torsion-wire balance the beam is attached to a taut wire, which acts as the fulcrum. Deflections are proportional to changes in mass and the torsional characteristics of the wire.

In all these deflection-type balances various techniques may be used for the measurement and recording the deflection, which by calibration can be converted to mass change.

An optical lever arrangement can be used involving measuring the deflections by means of a light beam reflected from a mirror mounted on the balance beam. These deflections can be recorded photographically or measured electronically by means of a shutter attached to the balance beam, which intercepts a light beam impinging on a phototube. The light intensity is a measure of the beam deflection and is thus related to mass change. A further method of measuring deflection employs a linear variable differential transformer using an armature freely suspended from the balance beam into the coil of a differential transformer. Strain gauges can also be used to measure beam deflections.

The principle employed in the null-point balance is shown in Figure 8. This system is used almost exclusively in thermogravimetry. A sensor on the balance detects the deviation from the null position, and this must then operate a servomechanism to restore the deviation to the null position.

A review by Gordon and Campbell [28] summarizes the various methods that have been employed to detect the deviation of a balance beam, the methods of restoring the beam back to the null position, and the manner of recording the mass change. These are outlined in Table 10. Electronic laboratory balances are the subject of a review by Ewing [29].

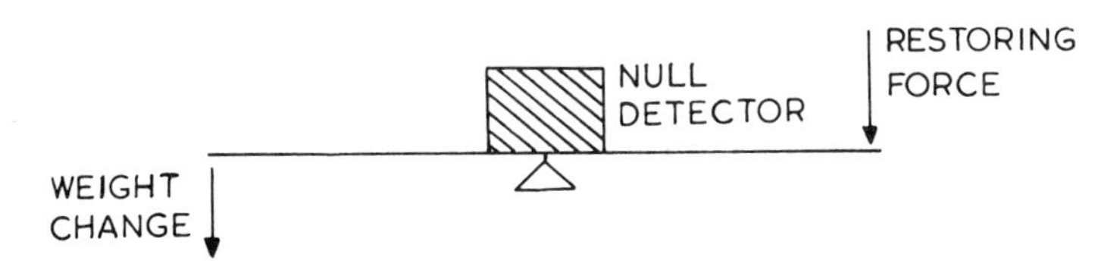

Figure 8. The null-point balance.

Table 10. Summary of Methods Employed to Detect and Restore Deviations in Null-Type Balances[a]

Detection of deviation

Optical
1. Light source: mirror, photographic paper
2. Light source: shutter, photocell

Electronic
1. Capacitance bridge
2. Mutual inductance: coil-plate, coil-coil
3. Differential transformer or variable permeance transducer
4. Strain gauge

Mechanical
Pen electromechanically linked to balance beam or coulometer

Restoration to null position

Mechanical
1. Addition or removal of weights, beam rider positioning
2. Incremental or continuous application of torsional or helical spring force
3. Incremental or continuous chain-weight operation
4. Incremental addition or withdrawal of a liquid (buoyancy)
5. Incremental increase or decrease of pressure (hydraulic)

Electromagnetic interaction
1. Coil-armature
2. Coil-magnet
3. Coil-coil

Electrochemical
1. Coulometric dissolution or deposition of metal at electrode suspended from balance beam

Method of recording

Mechanical
1. Pen linked to potentiometer slider
2. Pen linked to chain-restoring drum
3. Pen or electric arcing point on end of beam
4. Pens linked to servo-driven photoelectric beam-deflection follower

Photographic
1. Light source-mirror-photographic paper moving at a speed determined by temperature/time program

Electronic
1. Current generated in transducing circuit
2. Current passing through coil of an electromagnet

[a]From Ref. 28. Note that all signals can be collected by dedicated computer suitably interfaced.

D. Furnaces and Temperature Control

The furnace and control system should be capable of producing a wide range of temperature programs accurately. These should include isothermal runs, rising temperature, temperature jump, and parabolic and logarithmic temperature rise programs.

The position of the furnace with respect to the balance is important. Figure 9 indicates the possible positions for a furnace in a TG unit. The prime considerations are ease of operation, minimal convection effects, and rapid cooling.

A noninductively wound furnace should be standard and is usually provided in commercial equipment. Two common methods of making a furnace involve either (a) winding a suitable metallic wire or ribbon on a ceramic former and then covering the wound wire with a ceramic paste, or (b) coating a former with a thin layer of ceramic paste, winding the furnace wire or ribbon on top of this layer, and then covering the wire with a thicker layer of ceramic paste and withdrawing the former. The winding process is followed by careful baking, surrounding the baked furnace with insulating material, then incorporating the assembly in a suitable housing, and finally providing the necessary electrical connections.

The temperature required governs the choice of material. However, the atmosphere is an additional factor that must be considered. It is possible to construct furnaces going up to 2800°C, but about 1650°C represents an easily attainable maximum temperature using commercial equipment. Table 11 indicates maximum temperatures attainable for furnace resistance elements for use in thermobalance construction. From this table it can

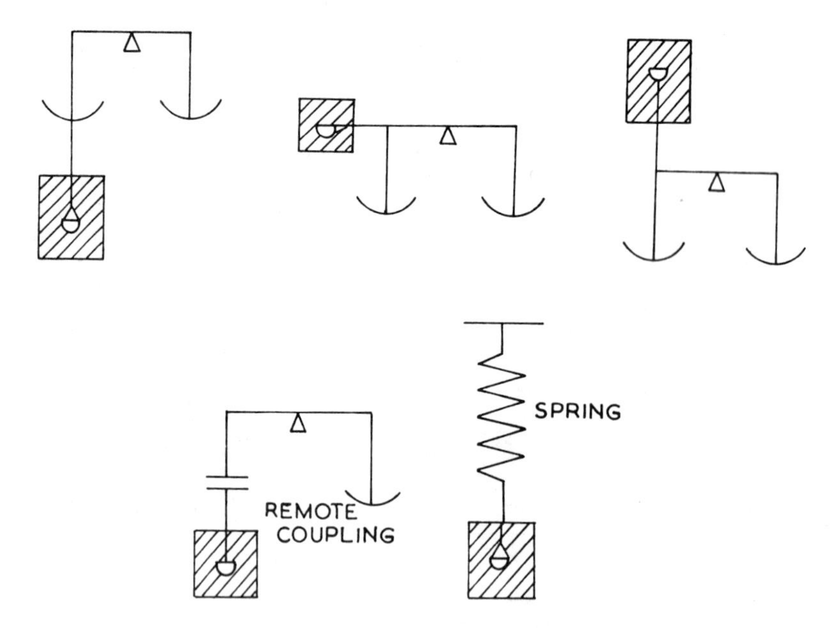

Figure 9. Position of furnace with respect to the balance.

Table 11. Maximum Temperature Limits in Furnace Construction for Thermobalance and Other Thermal Analysis Furnaces

Furnace element	Maximum temperature (°C)
Nichrome	1000
Chromel A	1100
Tantalum	1330
Kanthal or Nichrome	1350
Platinum	1400
Platinum-10% rhodium	1500
Platinum-20% rhodium	1500
Silicon carbide rods	1600
Kanthal Super	1600
Rhodium	1800

be seen that up to 1100°C, Kanthal or Nichrome wire or ribbon can be used. There are special winding techniques, such as "coiled-coil," to accommodate differential expansion problems as the temperature is altered. Various platinum alloys, especially with high rhodium content, allow temperatures to be reached as high as 1750°C. Higher temperatures are rarely required in thermogravimetry, but specialty furnaces employing tungsten, molybdenum, or graphite can be used for temperatures beyond 1750°C. These high temperatures require furnace operation in an inert atmosphere or even a reducing atmosphere. Not all furnaces used in thermobalances are wire-wound. Sestak [30] has, for example, constructed an infrared heater—its advantage being that it enables studying decompositions under vacuum conditions.

The ULVAC-RIKO TGD-3000-RH thermobalance also employs infrared heating. It has the advantage of very fast cooling (2-3 min). In this design an infrared image furnace is used where radiation from an infrared heater is concentrated on the sample by elliptical reflections to provide a hot zone 10 mm in diameter by 100 mm in length. The maximum temperature is about 1400°C on this unit, at a maximum heating rate of 999°C sec^{-1}!

It is not possible to construct one furnace suitable for all purposes. In a given commercial thermobalance assembly there may be as many as four furnaces used to cover the range (*a*) -150 to 500°C, (*b*) ambient to 1400°C, (*c*) ambient to 1600°C, and (*d*) 400 to 1800°C. The low temperatures are achieved by a combined heating-cooling unit, usually with the furnace working against cooling liquid nitrogen to allow accurate temperature programming. Furnaces have a definite lifetime, but without knowledge of the workload it is difficult to put this in terms of absolute figures.

The higher the temperature reached by the furnace and the greater its use with an oxidizing atmosphere, the shorter the lifetime expectancy.

Some units sold commercially have large, massive furnaces, whereas in other units the furnaces are miniature. Furnaces of low mass cool quickly but hold little heat, whereas the converse holds true for high-mass furnaces. In a low-mass furnace a linear temperature rise is relatively more difficult to control. A high-mass furnace may hold an isothermal temperature but takes a considerable time to achieve the required temperature. However, in everyday use, cooling requirements are paramount and the time required to reach a desired isothermal temperature must be as short as possible, and this usually means that the choice would be a system of low heat capacity. A uniform hot zone is required; this demands some skill in furnace winding, and is invariably based to some extent on trial and error. The high-mass furnaces should provide a larger, uniform, hot zone. This is important, and with smaller furnaces the positioning of the sample holder must be considered very carefully, especially if a deflection balance system is used, when movement of the sample holder might bring the sample out of the hot zone.

Temperature measurement is most commonly accomplished by use of thermocouples. Chromel/alumel thermocouples can be used up to 1100°C, while platinum metal alloys can be used up to 1750°C. Beyond this temperature, tungsten/rhenium thermocouples should be used. One must consider the possibility of the thermocouple reacting with the sample reactant or decomposition products. The highest signal output is achieved using a base-metal thermocouple with the additional advantage that the chromel/alumel thermocouples' response to temperature is approximately linear. The platinum/platinum alloy thermocouples have lower sensitivities and in the higher temperature ranges a nonlinear response.

Another temperature sensor that is employed is the resistance thermometer. The resistance of the furnace winding can be used if it has a high temperature coefficient. The Perkin-Elmer thermobalance has a small thermal mass furnace, consisting of an aluminum oxide furnace of 0.5 in diameter by 0.75 in length, wound with platinum wire. The wire is used in high-frequency cycles, first as a temperature resistance device in one half of each cycle, when the temperature is compared with the desired temperature program, followed by application of power in the second half of the cycle to null the difference between the actual and desired temperature. This eliminates thermal lag. Maximum temperature of the furnace is 1000°C. Optical and radiation pyrometers are rarely used to measure sample temperature [31].

It is important to position the temperature measurement device as close to the sample as possible. With the temperature sensor in other positions there can be a thermal lag between the furnace temperature and that of the sample. If the thermocouple is placed within the sample, the leads from the thermocouple can affect the sensitivity of the balance system. This was prevented by Dial and Knapp [32], who used inductive coupling. Schniztein et al. [33] solved the same problem by using the principle of a magnetic amplifier to sense at a distance the current caused by the potential generated by the thermocouple positioned in the sample.

The mechanical control of furnace temperature has never really been satisfactory. The most often used technique is to actuate temperature control from temperature sensor measurements of the furnace temperature.

In the next section the use of dedicated computers for temperature control is discussed.

E. Computer Work Stations to Record, Process, and Control the Acquisition of Data

Dedicated computers are increasingly being used both to control thermal analysis equipment and to record and process data. All the large instrument manufacturers supply such computer interfacing for thermogravimetric units. There are specialist computer firms that will interface any unit as required and allow the capability of processing thermal analysis data from commercially available programs or from programs derived by the investigators. These comments refer specifically not only to thermogravimetry but apply equally well to all other thermal analysis techniques. The matter is very much in the formulative stage at the moment, and to cite specific methods of computer usage is to invite criticism and make this review quickly out of date. One problem that the user should bear in mind is that the data acquired from the temperature sensors is not always linear and the computer acquisition system must accurately convert the data signal to a temperature. Most computer users will require a "change of scale" potential from their equipment and also a program to record percent mass change against temperature. They will also require the derivative (DTG) plot. These are minimum requirements, and most interfaced computers can process the data in many more ways, as required by the objectives of the particular experiment.

III. DIFFERENTIAL THERMAL ANALYSIS (DTA) AND DIFFERENTIAL SCANNING CALORIMETRY (DSC)

A. Basic Considerations

It may be recalled that differential thermal analysis is a technique in which the temperature difference between a substance and reference material is measured as a function of oven temperature, while the substance and the reference material are subjected to a controlled temperature program. The record is the DTA curve in which the temperature difference (ΔT) is plotted on the ordinate with endothermic reactions giving peaks in a downward direction and exothermic reactions in an upward direction, using an abscissa in terms of either T or t, increasing from left to right. If the technique is made quantitative, then logically it acts as a calorimeter, and lately the term differential scanning calorimetry is applied to such a unit. The term formerly was applied solely to a "power-compensated technique" in which the generated ΔT signal was compensated by an auxiliary heater and the energy required to neutralize the ΔT signal was measured [34]. It must be noted that equipment sold as DSC is, with one or two exceptions, simply quantitative DTA units, improved to make their calorimetric response more sensitive, with the calibration data stored in the dedicated computer station. The general signal is therefore a measure of enthalpy. There is now a conflict with the IUPAC Committee on Thermodynamics, which requires exothermic quantities (where the system loses energy) to be plotted in a downward direction and endothermic quantities (where the system gains energy) to be plotted in an upward direction. Actually this conflict is illusory—whenever measurements are reported in terms of ΔT then logically ΔT is negative for endothermic processes and positive for exothermic processes. When one plots the data in enthalpy units then logic demands that the directions be reversed.

Two points can now be made. The first is a confusion in reported data from DSC units with some instruments giving endothermic peaks in a downward direction, others upward, and yet a third group leaving it to the discretion of the user. The second point is that most users of DSC units do not use them quantitatively but as qualitative units when the advantage of using calibrated equipment is simply not utilized.

B. Design Factors

In classical DTA, a sample cell and a reference cell are subjected to the same temperature program. The names of the cells indicate their use. Thermocouples placed in each cell work against one another and record a ΔT signal against the temperature of the system. This is illustrated in Figure 10. It should be explained that DSC was originally the name given by Watson et al. [34] to a similar unit employing two individual heaters. Then in any heating program heat energy absorbed or evolved by the sample is compensated for by adding or removing an equivalent amount of energy in the heater located in the sample holder. Hence the power needed to maintain the sample holder at the same temperature as the reference holder provides an electrical signal opposite but equivalent to the varying thermal behavior of the sample. The schematic layout for this equipment is shown in Figure 11. This was developed commercially by the Perkin-Elmer Corporation. Because the contents of the two cells are maintained at an identical temperature by electronically controlling the rate at which heat is supplied to the sample and the reference materials, respectively, the ordinate of the DSC curve represents the rate of energy absorption by the test sample relative to that of the reference material; the rate depends on the heat capacity of the sample.

A third cell assembly has been recognized that, like the DSC unit of Watson, also gives a good calorimetric response. This is the "Boersma" cell [35], shown schematically in Figure 12. As already mentioned, the term "differential scanning calorimetry" was for a long time the property of the Perkin-Elmer Corporation, but nowadays the term is used by almost all instrument makers to describe DTA units that, like the Boersma unit, can be calibrated to give a good calorimetric response. As the calibration data are often incorporated into the dedicated work station, the user must pay attention both to the calorimetric design and the computer program. For qualitative use there is not much to choose between DTA and DSC. However, the available DTA units until recently covered a wider temperature range. For quantitative use the DTA units must be examined carefully

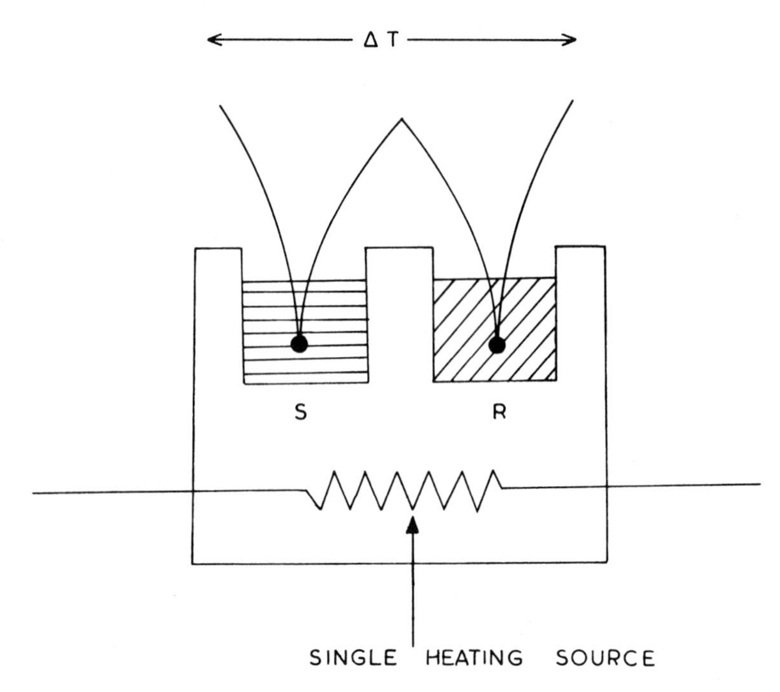

Figure 10. Classical DTA: S, sample cell; R, reference cell.

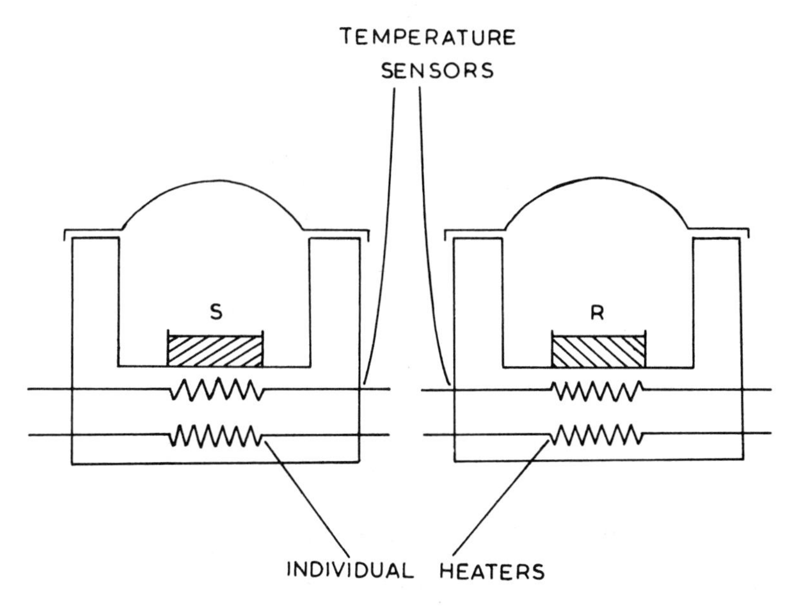

Figure 11. Differential scanning calorimeter: S, sample cell; R, reference cell.

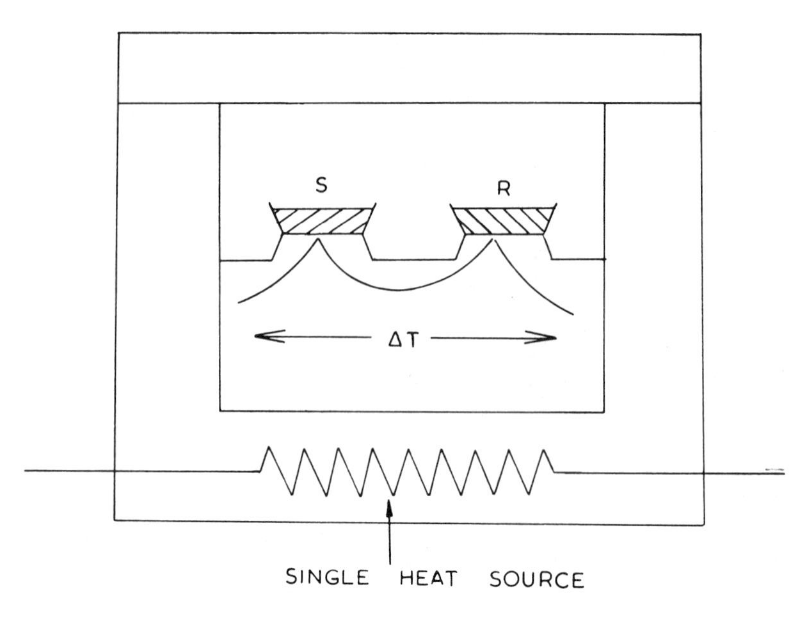

Figure 12. "Boersma" DTA.

for an accurate calorimetric response. The whole issue is made confusing by the change in the use of the name DSC.

A block diagram of a DTA unit is shown in Figure 13.

The sample holder measuring system comprises the thermocouples, sample and reference containers, and associated equipment. The furnace system varies in size from one instrument to another but must have a uniform temperature zone around the sample and reference cell.

A linear response with time was previously regarded as the major requirement in temperature programming. Now, however, the computer work station will include programs that allow a whole variety of optional temperature responses against time to be called on. These include stepped temperature jumps and, for example, logarithmic temperature time control. The range of temperatures available (although not in a single unit) is from around -200°C to around 2000°C. Generally equipment is commercially available as a single unit for the temperature ranges from -200°C to +500°C and for temperatures from ambient to around 1000°C. Units have also been introduced operating up to 2000°C. DSC units normally operate up to a maximum temperature of about 700°C. Beyond this temperature other heat flux conditions begin to operate, which complicates calibration, but this has not prevented units being developed up to about 1000°C. High-pressure systems have also been made available commercially.

This variety of operating conditions leads to a great range of cell designs. A selection of cell designs is given in Figure 14, all operating in the ambient to some high-temperature region [36]. They all give responses which are dictated by experimental conditions. The same is true for the low-temperature units and the high-pressure units.

C. Experimental Factors

DTA and DSC units are all responsive to changes in experimental conditions. These conditions include the packing of the sample, the container and thermocouple design, the ambient pressure, and the actual temperature program.

Thus disc-shaped thermocouples are utilized in some low-temperature DTA/DSC units. A major problem in operation is transporting heat uniformly away from the sample, which often precludes the use of ceramic blocks. The use of the disc-shaped thermocouples (usually either chromel/alumel

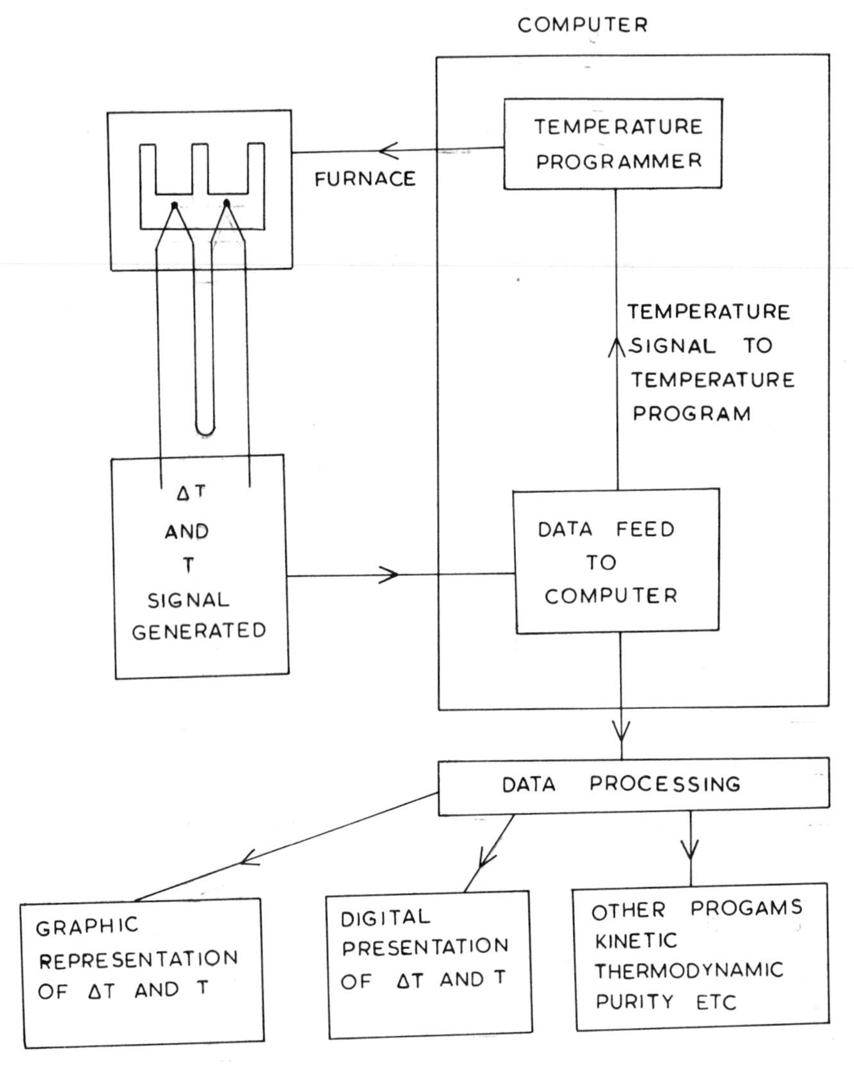

Figure 13. Schematic of contemporary DTA/DSC units.

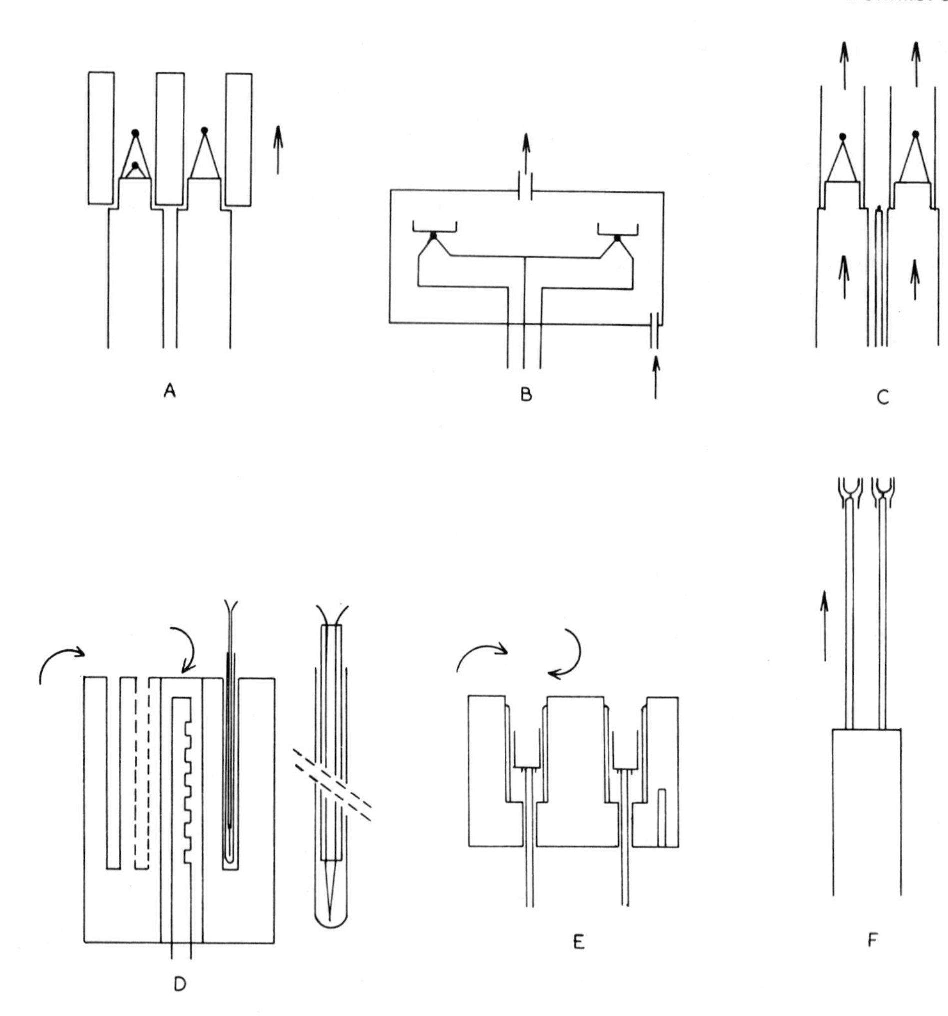

Figure 14. Typical DTA/DSC cell designs: (A) Netzsch standard cell (nickel block, wells approx. 8 mm diameter). (B) Stanton-Redcroft cell (aluminum pans, approx. 6 mm diameter). (C) Netzsch catalytic cell (ceramic sleeves, approx. 8 mm diameter). (D) Du Pont intermediate temperature cell (sample tube approx. 1 mm diameter). (E) Wilburn cell (nickel block, wells approx. 9 mm diameter). (F) Du Pont-high temperature cell with platinum liners in place (sample holder approx. 5 mm diameter). Arrows indicate normal direction of gas flow.

or iron/constantan) ensures optimum thermal contact, provided the sample container is similarly flat-bottomed. The early design by Jensen and Beevers [37] for low-temperature DTA/DSC units used a block of metal containing a liquid-nitrogen reservoir to transmit the ambient low temperature to the region around the sample and reference containers. The temperature program is then achieved by use of a furnace operating against the low ambient temperature achieved initially. In other designs the furnace assembly is initially cooled by a piped liquid-nitrogen pumping system. Thermal control is then achieved by keeping the nitrogen flow constant and programming the furnace heat input.

The use of a high pressure of gas above the sample in a DTA/DSC unit can alter the temperature at which phase changes or chemical reactions occur. Imagine a system;

$$A(s) \rightarrow A(g)$$

or

$$A(p) \rightarrow A(g)$$

Ignoring kinetic aspects of this kind of change (which can be important in liquid crystals and polymer systems), the Clausius-Clapeyron equation governs this kind of transition. This takes the form

$$\frac{dT}{dP} = \frac{T(V_1 - V_2)}{\Delta H}$$

where T is absolute temperature, P is pressure; $V_1 - V_2$ is the volume change in the material going from phase 1 to phase 2, and ΔH is the change in enthalpy associated with the phase transition.

Neglecting the volume of the solid with respect to that of the vapor at the phase transition allows the above relationship to be expressed as

$$\ln p = -\frac{\Delta H}{RT} + \text{constant}$$

The vapor-pressure thermometer is simply a manometer unit in which this temperature-pressure relationship can be explored for, say, an organic liquid-organic vapor system. The similar use of DTA experiments to provide data on such systems needs justification because much more experimentation is required, that is, a separate run at each pressure. The real justification for the use of high-pressure DTA units (or more properly controlled-pressure DTA systems) is their application to multiple-phase or complicated phase change systems.

The same is true in the application of the above thermodynamic relationships to chemical decompositions. At this stage the comment may be made that equipment for either DTA or DSC under pressure is available commercially [38]. Applications besides the instances cited above include standard tests on oils, lubricants, and explosives [39-42].

The next significant experimental feature of any DTA/DSC unit is the dependence of the results upon the rate of heating. Most investigations heed a careful consideration of this point. The effect on kaolin of lowering the heating rate [43] is demonstrated in Figure 15. The effect of the heating rate can be best studied in terms of the initial deviation temperature T_i, the magnitude of the ΔT signal at peak maximum (ΔT_p) (or minimum depending on endothermic or exothermic nature of change), and the actual temperature measured. There is an authoritative paper by Melling et al. on this subject [44]. In the case shown in Figure 15 it can be seen that the increase in heating rate increases T_i, ΔT_p, and T_f. The choice of

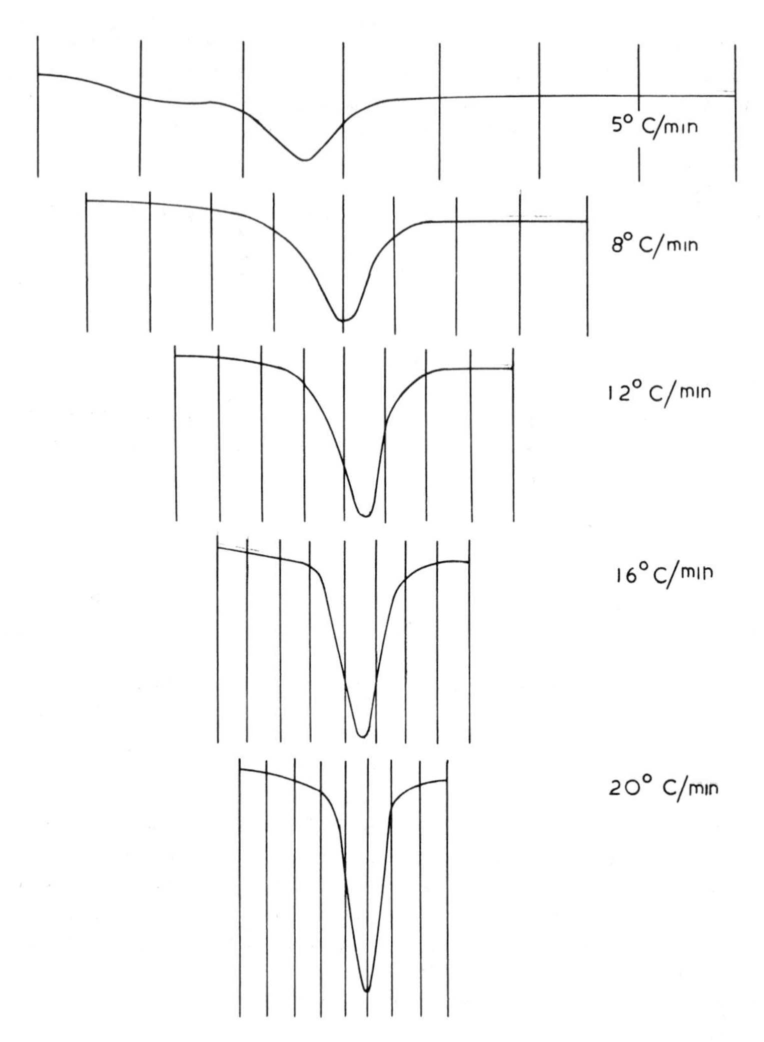

Figure 15. Effect of heating rate on DTA data for kaolin.

the temperature measurement is all-important in considering peak area. Melling et al. concluded that if ΔT (given by $T_s - T_r$, where T_s is the is the temperature of the sample and T_r the temperature of the reference material) is plotted against T_s then the peak area is proportional to the heating rate. If $T_s - T_r$ is measured against t then it is independent of the heating rate. Provided the peak area is defined by the latter method, then on a simple theoretical basis it must also be independent of the rate at which the reaction takes place and the specific heat of the sample, provided it remains constant. It does, however, depend on the heat of reaction per unit volume of the sample, on the thermal conductivity of the materials in the furnace and the sample, and on the heat transfer

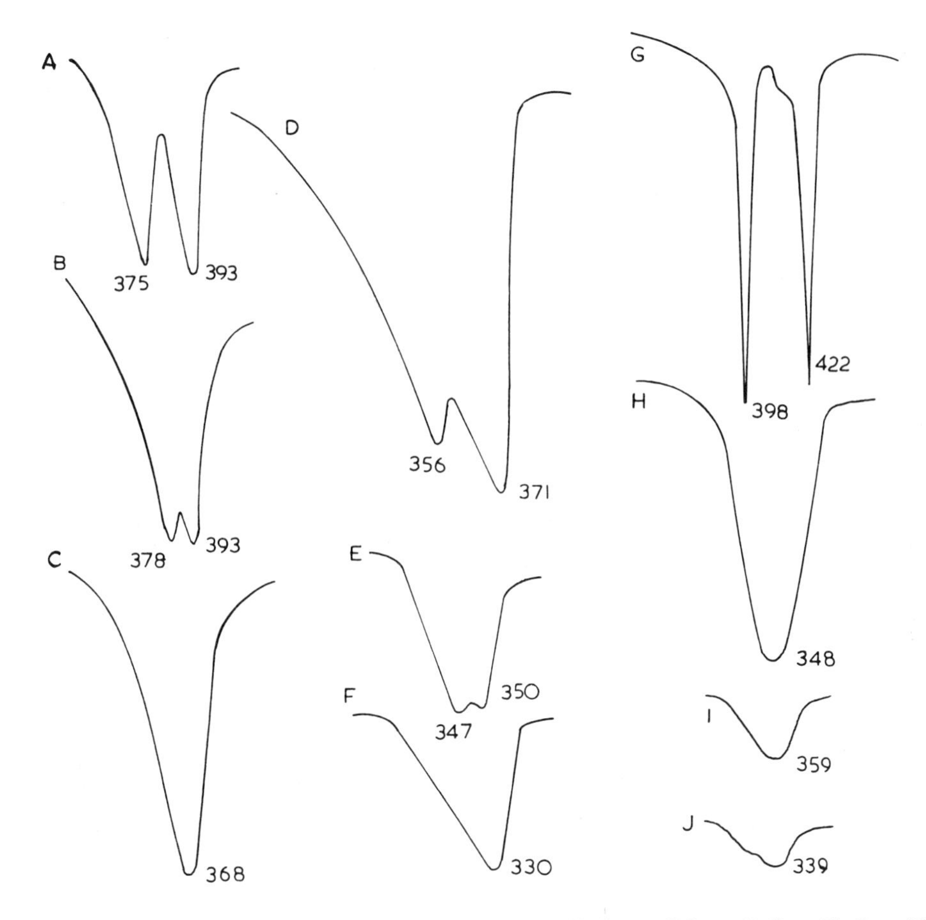

Figure 16. DTA curves for $UO_2 \cdot C_2O_4 \cdot 3H_2O \rightarrow UO_2 \cdot C_2O_4 \cdot H_2O + 2H_2O$ using various DTA cells (all results traced). (A) Netzsch standard cell, diluted sample, 300 ml min^{-1} nitrogen *over* sample. (B) Netzsch catalytic cell, diluted sample, 500 ml min^{-1} air *over* sample. (C) Netzsch catalytic cell, diluted sample, 5 ml min^{-1} air *through* sample. (D) Stanton-Redcroft cell, undiluted sample, 25 ml min^{-1} air *over* sample (E) Stanton-Redcroft cell, diluted sample, static air. (F) Stanton-Redcroft cell, diluted sample, 10 ml min^{-1} nitrogen. (G) Du Pont intermediate-temperature cell, undiluted sample, 400 ml min^{-1} nitrogen *over* sample. (H) Du Pont intermediate-temperature cell, undiluted sample, *in vacuo* (1.9 kN m^{-2}). (I) Du Pont high-temperature cell, with liner, diluted sample, 500 ml min^{-1} air *over* sample. (J) Du Pont high-temperature cell, without liner, diluted sample, 500 ml min^{-1} air *over* sample. Diluted samples: 20% w/w with α-Al_2O_3, peak temperatures in kelvins.

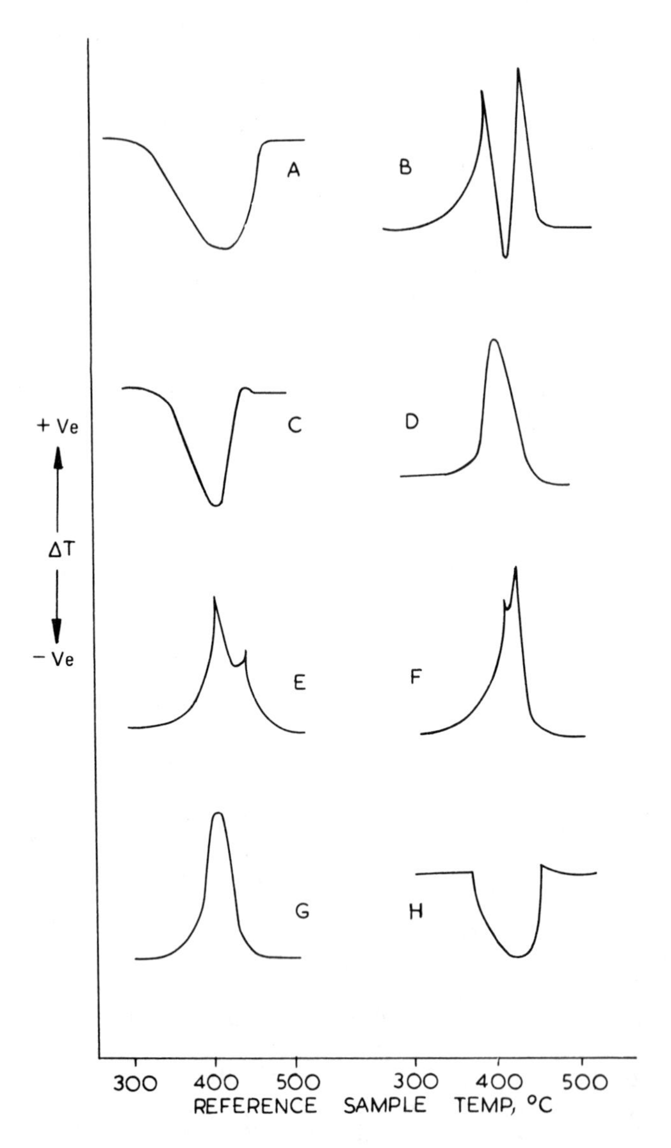

Figure 17. Differential thermal analysis of zinc oxalate dihydrate under various experimental conditions. (The dehydration peak is endothermic in all cases and this section of the curves has been omitted). (A) Sample weight 1.00 g, heating rate 5°C min^{-1}, static air. (B) Sample weight 95 mg, heating rate 10°C min^{-1}, static air. (C) Sample weight 82 mg, heating rate 10°C min^{-1}, nitrogen (1000 cm^3 min^{-1}). (D) Sample weight 82 mg, heating rate 10°C min^{-1}, oxygen (1000 cm^3/min). (E) Sample weight 97 mg, heating rate 5°C min^{-1}, static air. (F) Sample weight 80 mg, heating rate 5°C min^{-1}, static air. (G) Sample weight 66 mg, heating rate 5°C min^{-1}, static air. (H) Sample weight 206 mg, heating rate 5°C min^{-1}, static air.

coefficient between the holder and the furnace wall [45]. Furthermore, provided the ΔT signal is less than 10°C, then the addition of correction terms in refined treatments is not likely to seriously affect the conversion of peak area to reliable calorimetric data. It can be shown in addition that ΔT_p is dependent on heating rate.

The effect of the cell design upon results may be judged by inspection of Figure 16, which demonstrates the results for the decomposition of uranyl oxalate trihydrate using different DTA cells [36]. Figure 17 shows data obtained by Judd and Pope [46] for decomposition of hydrated zinc oxalate. It is at once apparent that the actual experimental conditions pertinent to the type of cell, heating rate, sample mass, and environmental gas should be considered in any DTA/DSC study that is proposed.

IV. EVOLVED GAS ANALYSIS

A. The Basic Units

Evolved gas analysis (EGA) refers to the quantitative analysis of gaseous materials, whereas evolved gas detection (EGD) refers to the detection of an evolved gas species without its actual measurement. Logically the terms bear no reference to thermal analysis and in fact the use of the methods described here has evolved merely with respect to the identification and analysis of gases evolved in any chemical process. Logically, too, the word detection could be held to refer to the identification of gaseous species evolved without quantitative analysis. However, in the following discussion the emphasis will be on evolved gas analysis of volatile product species from the thermal decomposition of materials subjected to programmed temperatures. It will be further seen that in this context EGA is a technique added to an already existing thermal analysis facility and is rarely used alone [47,48]. The detectors used for gas analysis include:

Mass spectrometer
Titration cell
Infrared cell
Various chemical detectors
Flame ionization detector
Thermal conductivity detector
Gas density detector
Pressure detectors

A precursor to the EGA unit is the appropriate thermal analysis system, including a pyrolysis unit or some other thermal "breakdown" device. Normally a carrier gas is used; this should be purified as required. Pyrolysis separator devices are often employed before the analysis stage. Properly, gas chromatography should be regarded as a separator and not a detector device, although often it is quoted as an analyzer or detector unit. The gas detector devices are often matched in pairs, one located in the inlet system and the other on the outlet. It is thus the difference signal that is measured. Figure 18 indicates the scheme outlined above. Certain gas detector devices, such as the mass spectrometer, and FTIR spectroscopy units are not used in the "paired" manner; the devices most often used in this manner are the flame-ionization and thermal conductivity detectors.

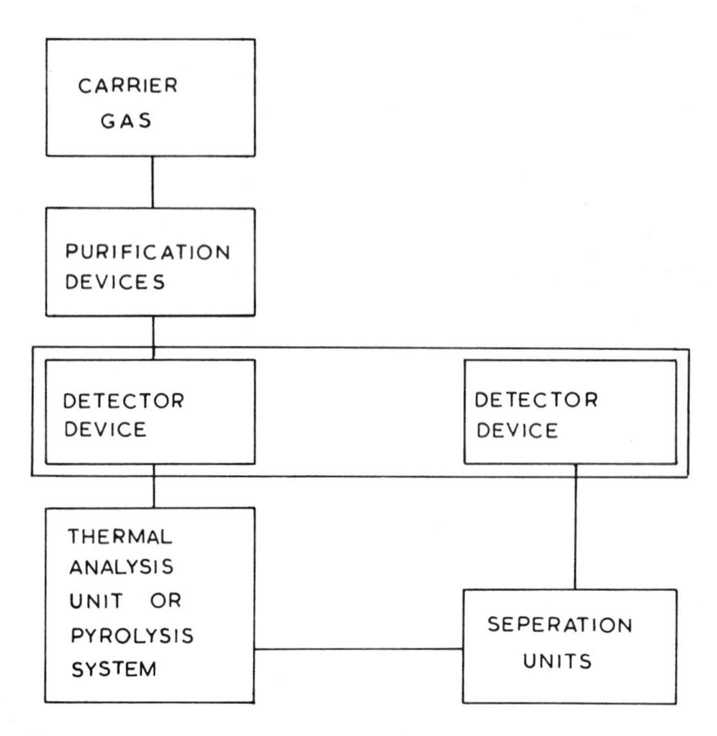

Figure 18. Component parts of EGA system.

B. Detector Devices

Detector units such as the flame ionization detector or thermal conductivity detector have found extensive use in chromatography and offer no problems in the context of EGA. They are almost always used in pairs—one on the inlet system, the other in the outlet stream [49,50]. Gas density detectors can be used in the same manner—their application in connection with thermal analysis has been described by Garn and Kessler [51].

In dynamic atmospheres the production of volatiles on decomposition causes a rise in pressure proportional to the rate of volatilization; in a closed system the increased pressure is proportional to the total amount of volatile produced. Various sensitive manometers can be used for the purpose of measuring the amount or rate of volatilization—the Prout and Tompkins experiments used in the study of the decomposition of potassium permanganate serve as a classic example [52].

The use of chemical detectors is usually important when the evolved species is difficult to detect by other methods. The Dupont Moisture Evolution analyzer utilized an electrochemical cell to determine water coulometrically. Another commercial unit made by Panametrics comprises a thin aluminum-alumina-aluminum sandwich and relies on alteration in the electrical impedance as water is adsorbed in the pores. Gallagher and Gyorgy [53] used this technique and reported excellent agreement with mass loss techniques.

The latest equipment available in the sphere of EGA is the FTIR spectroscopy technique. At least three commercial units are available with this capability associated with a thermal analysis unit—either a DSC/DTA or TG. This is still in the formulative stage and instrumentation is changing fast, so the latest commercial literature should be consulted and compared before making a choice of equipment.

A well-established infrared spectroscopy method is the use of an infrared cell—set to cover a particular absorption band. This is particularly useful for carbon monoxide, carbon dioxide, and water, which are difficult to analyze on a mass spectrometer. A review by Low [54] is especially useful covering nondispersive analyzers, dispersion spectrometers, band-pass filter type instruments, and interference spectrometers. One method of improving sensitivity is to convert the volatile product to another material that has a better response to infrared spectroscopy. A typical example is the use of calcium carbide to convert water to acetylene [55].

Mass spectroscopy enjoys a wide application in EGA. There is first the exciting prospect of placing the thermal analysis furnace directly in the mass spectrometer. This avoids interfacing problems, but the conditions of thermal decomposition under a programmed temperature regime are rather unique and results, although very basic, must be viewed cautiously.

A particularly useful design has been described by Gallagher [56] and applied by Kinsbrow et al. [57]. Another design employing a time-of-flight unit with a furnace incorporated within the unit is described by Price et al., who used the equipment to study kinetics of decomposition [58].

The equipment described by Lum uses Laser heating together with differential pumping to produce a molecular beam of volatile product [59]. The earliest report of the use of mass spectroscopy in this manner is probably the paper by Gohlke and Langer [60]. These instruments really combine the detector device and the thermal analysis equipment in a single unit. The more usual application of mass spectroscopy is as a detector device combined with either TG or DTA/DSC. Commercial units are available but there are many instances of equipment being interfaced in the laboratory. Zitomer [61] made early use of this technique, and a schematic of his equipment is shown in Figure 19. The important requirement is an efficient interface to reduce the pressure from atmospheric (or whatever pressure the TG or DTA/DSC unit is operating) to the low pressure of the mass spectrometer. This can be either a capillary tube [62] or a diaphragm. The aim should be to reduce the chance of secondary reactions and obtain the EGA data as soon after the event as possible. Commercial units are now available.

Dyszel [63] has used a combined TG atmospheric pressure chemical ionization mass spectrometry system. A Sciex TAGA 300 mass spectrometer was used. Initially ionization takes place with electron bombardment of nitrogen, and the N_2^+ ion so produced undergoes a charge transfer reaction with O_2 forming O_2^+. This then forms a cluster with moisture in the air to eventually produce a hydrated proton H_3O^+. The volatile products entering the ion source from the TG unit then react in an addition mechanism with the H_3O^+ and the ionized molecules pass to the quadrupole analyzer. Of course the evolved species can be trapped and transferred later to the mass spectrometer [64].

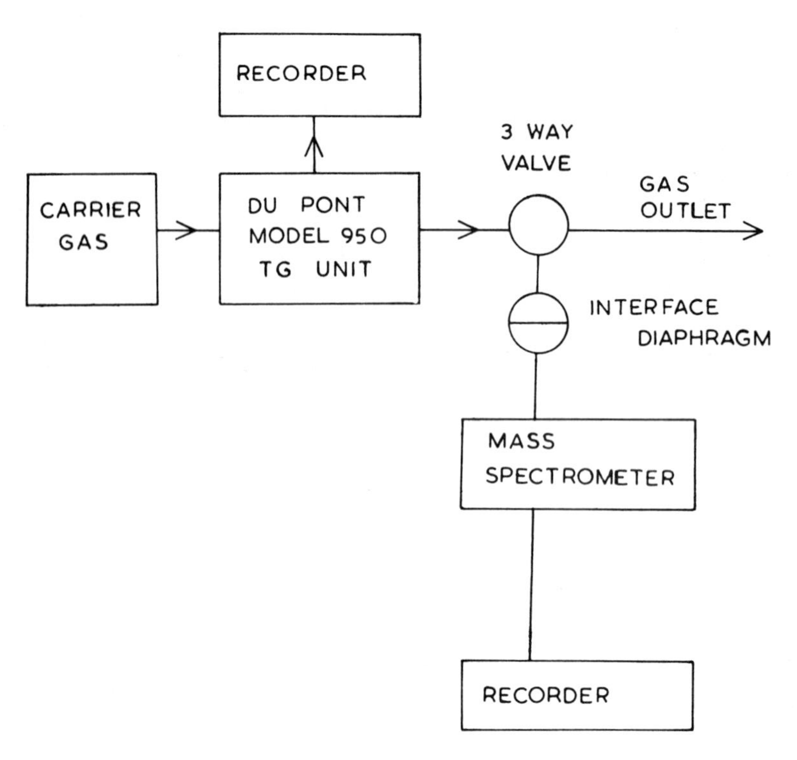

Figure 19. TG-MS apparatus described by Zitomer [61].

C. Precursor Considerations

Precursor units to the EGA are the TG units employed, or a pyrolysis chamber together with the carrier gas supply and a purification train. Pyrolysis techniques are now an essential part of organic chemistry characterization. A simple pyrolysis unit is described by Rogers et al. [65].

The equipment described by Rogers et al. and by Vassallo [66] were applied in studies on the pyrolysis of various polymeric materials. The aim of the Pyrochrom device is elemental analysis [67]. Automated pyrolysis equipment has been described by Nesbitt and Wendlandt [68]. A comprehensive guide on analytical pyrolysis has been published by Irwin [69].

Temperature-programmed reduction (TPR) is conveniently discussed in this section. In a typical application of this technique, 5% hydrogen in nitrogen is passed over a catalyst sample and the change in hydrogen content of the effluent gas is monitored as the temperature is continuously increased [70-74].

D. Separation Procedures

It is often found convenient to effect a separation or partial separation of volatile products from EGA before the analysis process. The most convenient tool for this is often gas chromatography. This separator is used with pyrolysis equipment or a thermal analysis unit. The following points may be made:

1. Time for a run leading to analysis is probably greater using a thermal analysis unit.
2. Production of secondary pyrolysis products is lower in the thermal analysis unit.
3. The loss of volatiles recorded on the GC can be matched with the process noted on the thermal analysis unit.
4. The temperature at which a particular volatile fraction or component is released can be recorded on the thermal analysis unit.

Equipment using TG-GC has been described by Chiu [75,76]. Special attention has been devoted to obtaining an efficient and reliable coupling

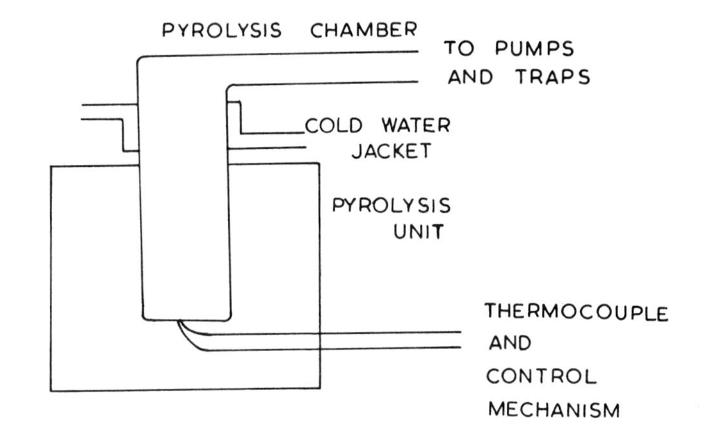

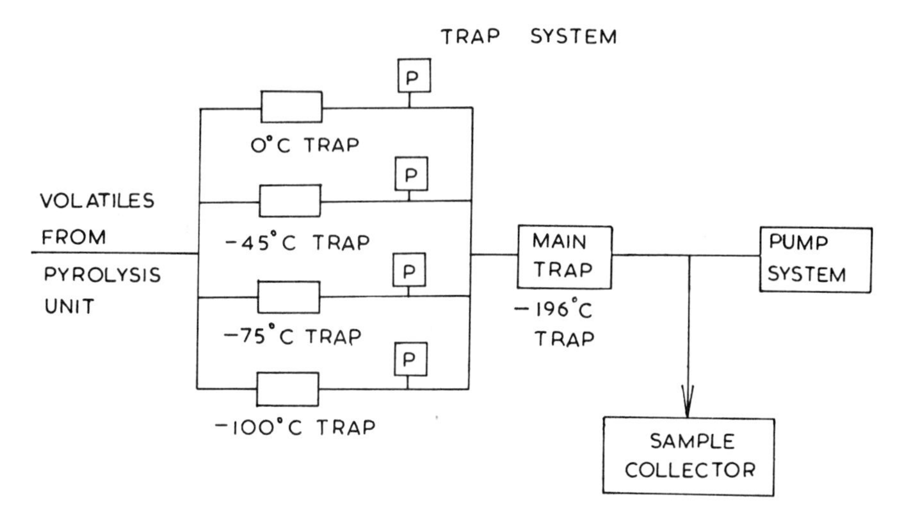

Figure 20. Thermal volatilization equipment.

system. Cukor and Persiani coupled a DuPont model 950 thermobalance to a Perkin-Elmer model 900 gas chromatograph [77]. Other descriptions of TG-GC have been given by Wiedemann [78] and by Uden et al. [79].

The equipment described by Uden coupled the thermal analyzer *and* a pyrolysis reaction system with on-line vapor-phase infrared spectroscopy, elemental and functional analysis, and mass spectroscopy. The partial separation of evolved gas using GC followed by analysis of separated fractions using MS has become common practice. Coupling of GC with DTA is also possible, but sampling time is greater than with DTA-MS [80].

A thermal analysis separator device devised by McNeill [81] has proved to be very useful. It is schematically portrayed in Figure 20. The thermal volatilization unit has a pyrolysis chamber containing the sample (usually a polymer), which is subjected to a predetermined temperature program while pumping with a vacuum pump unit. The effluent gas stream is passed through four isothermal traps (at 0, -45, -75, and -100°C) each with a pressure gauge before reaching the main trap at -196°C. This effectively allows the collection of fractions of different volatility. Each of the fractions can then be investigated separately. Any orthodox high-vacuum fractionation unit would also afford a similar separation.

V. THERMOMECHANICAL METHODS

A. Introduction

The term thermomechanical is somewhat misleading. It is defined as covering thermal analysis techniques that measure changes in volume, shape, length, and other properties relating to the physical shape of a solid substance subjected to a programmed temperature regime. However, logically the word covers other mechanical effects, such as the viscoelastic response of a sample, and the word dynamic here has come to mean oscillatory. The techniques most often found are (1) thermodilatometry (TDA) or dilatometry, (2) thermomechanical analysis (TMA), and (3) dynamic thermomechanometry, where the dimensional changes under zero load or under a minimal load are measured. Technically changes in density, including all the traditional devices such as a pycnometer, should be included under this heading. In TMA a nonoscillatory stress is applied to the solid sample and deformation under load is measured. DMA refers to techniques in which oscillatory stress is applied to the sample and the dynamic modulus or the mechanical damping of the sample is measured. There is probably as much application of these techniques isothermally as under rising temperature conditions. The techniques are increasingly used to characterize the behavior of "end products," and this is especially true of DMA.

B. Thermodilatometry and Thermomechanical Analysis

A typical TDA unit is shown in Figure 21. Instrumentation problems in TDA and TMA have been discussed by Paulik and Paulik [82], Gill [83], and Riesen and Sommerauer [84]. In the figure a linear variable transducer (LVDT) is shown as the instrument by which dimensional changes of the

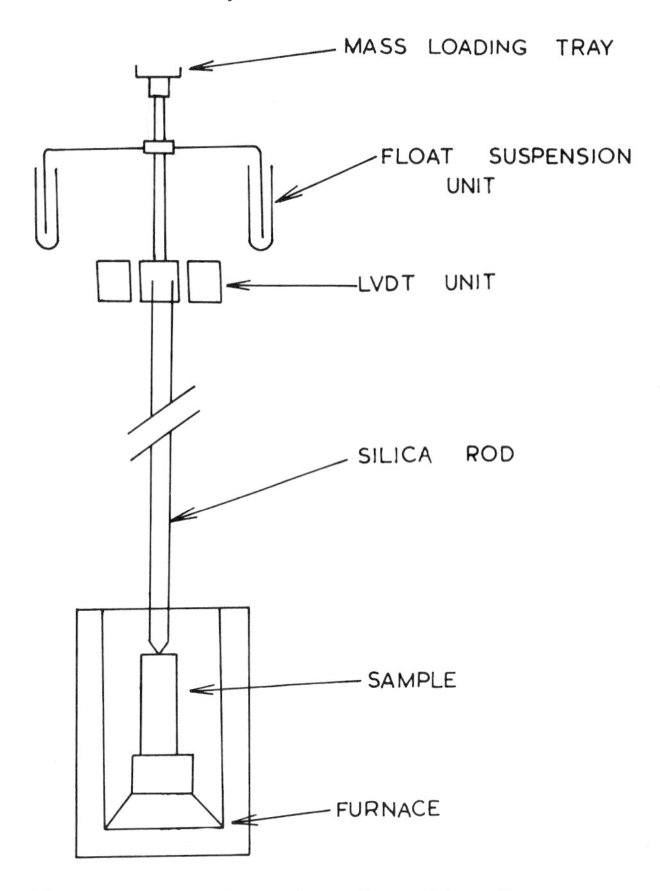

Figure 21. Basic design of a TDA unit.

sample may be measured. The movement is transferred into the LVDT unit via a rod usually of quartz or some material not subjected to an appreciable dimensional change on heat treatment. A mechanical guide is then necessary to position this rod and make sure it is placed correctly. The samples must be machined to convenient shapes, and the testing of powders is difficult but not impossible. The achievement of zero weight on the sample is also difficult to achieve but is lessened by some system such as the float system depicted. Provision for expansion against a defined load is generally provided. Equipment is available enabling measurements to be made over a temperature range from below ambient to around 1200°C, although not in the same unit.

C. Dynamic Mechanical Analysis

The word dynamic here refers to the fact that the instrument is used to study the viscoelastic response of a sample under oscillatory load. Commercial instruments use different techniques:

1. The sample is forced into oscillation or made to take on its natural frequency.
2. Stress can be applied in a variety of ways, for example, in flexure, tension, compression or torsion.
3. The sample has a load applied continuously to it and measurement is made of the modified oscillatory response.

The application of the instrument is to shaped samples, generally polymeric end products [85,86]. In experiments where the sample is in free vibration the dynamic modulus is related to the natural frequency (maximum amplitude).

In the Du Pont unit, Young's modulus is given by

$$E = \frac{(4\pi^2 f^2 J - K)}{2w[L/2 + D]^2} \left(\frac{L}{T}\right)^3$$

where f is the DMA frequency, J the moment of inertia of the sample arms (which are fixed to a rigid block via low-friction flexure pivots), K the spring constant of the pivots, D the clamping distance, w the sample width, T the sample thickness, and L the sample length. In the Polymer Laboratory unit a bar sample is used, clamped rigidly at both ends and subjected at its center point to a sinusoidol vibration using a drive shaft attached to the sample by a clamp. The stress on the sample is proportional to the current supplied to the vibrator, whereas the strain on the sample is proportional to the sample displacement. Special clamps allow the testing of soft materials (rubbers, adhesives, fats, etc). Films and fibers can also be tested while liquid polymers are tested on a support. Thus in cured epoxy systems the DMA technique can easily observe α, β and δ transitions, which are often beyond the capability of DSC techniques. Torsional braid analysis (TBA) is another technique employing oscillation imposed on the sample. It was introduced by Gillham [87,88]. In this technique the sample is prepared in solution form and impregnated onto glass braid or thread, followed by evaporation of the solvent. The sample is heated while subjected to free torsional oscillations. The relative rigidity parameter is noted (defined as $1/p^2$ where p is the period of oscillation) and used as a measure of the shear modulus. A measure of the logarithmic decrement, termed the mechanical damping index, $1/n$ (where n is the number of oscillations between two arbitrary but fixed boundary conditions in a series of waves), is also noted. It is used extensively in polymer systems to reveal glass transitions, melting, and the effect of chemical reaction.

VI. APPLICATION

There are other techniques of thermal analysis that have not been commercially utilized or extensively reported. These include those techniques coming under the heading of thermoelectrometry, covering the variation in the electrical characteristics of a substance when subjected to a controlled temperature program. Paulik and Paulik have written a review of this topic [89].

Another technique is the measurement of the optical characteristics of a substance measured as a function of temperature. This is called thermophotometry, although specific equipment receives individual names, such as thermophotometry, thermoseptometry, etc. These techniques find application especially in the study of solid-state surfaces. The subject has been extensively discussed by Wendlandt [90]. Optical microscopes can also have the sample subjected to temperature programming [91]. There are even thermal analysis techniques based on the sound emitted by the sample during heat treatment [92]. The applications of these minor thermal techniques and the major thermal analysis techniques rely basically on thermodynamics or on kinetics or a combination of both. In this review the emphasis is on techniques, but some examples of application may be cited.

A. Methods with a Thermodynamic Basis

The actual foundation of DTA/DSC is that of calorimetry—one of the main methods by which thermodynamic data is derived. In Section III.C the use of the Clausius-Clapeyron equation for phase transitions, liquid to vapor, and solid to vapor was mentioned. It should be noted that when the phase change is from a phase stable in a definite temperature range to another, stable in a different temperature range, the phase change is reversible, being endothermic in the heating mode but exothermic in the cooling mode. That is,

$$A(s) \xrightarrow[\text{heating}]{} A(l) \qquad \text{DTA/DSC, endothermic peak}$$

but

$$A(l) \xrightarrow[\text{cooling}]{} A(s) \qquad \text{DTA/DSC, exothermic peak}$$

The same is true for solid-solid transitions. Silica can be cited as an example. The transformations quoted in the Dawson and Wilburn review [93] were originally due to research by Fenner [94] and are with some slight changes still acceptable (Table 12). Experimentally, the changes from quartz to tridymite and tridymite to crystobalite are not observed with changing temperature programs using DTA. This is possibly because the complicated structural changes require time and a considerable kinetic factor is involved. The α and β transitions, on the other hand, involve minor structural alterations and can easily be noted from DTA/DSC experiments. The results of experiments show, however, that even though a system is reversible there is often a difference in the observed transition temperature depending on whether one is using a heating or a cooling mode temperature regime. There are numerous reports in the literature [93] that demonstrate this point and also show a variation from the temperatures recorded in Table 12.

If an amorphous (energy-rich) phase undergoes transition, then a stable crystal form may result. The same is true for crystallization in a glass. Essentially this reflects a change represented schematically as

Metastable form ⟶ Stable form

(amorphous phase glass, etc.) (crystalline)

Exothermic
Irreversible

As indicated, such transitions will be exothermic and irreversible in the thermodynamic sense. An example is the conversion of anatase to rutile [95], and the exothermic peak seen in the degradation of $CaHPO_4 \cdot 2H_2O$ at 530°C [96].

Thermal analysis techniques based on transitions of the kind discussed above find extensive application in organic compound studies, especially in investigating polymers. In the latter case there may also be extensive kinetic involvement, as these materials consist of extremely large macromolecules where rearrangement involved in a transition might be driven by energy transfer from one part of the molecule to another. Figure 22 shows for demonstrative purposes various phase changes and transitions that may occur in polymeric systems. It is doubtful if any one polymer would show all these effects in a single DTA trace. The first change (A) to notice is a second-order transition involving only a change in the baseline. The scale is exaggerated in Figure 22. For polystyrene and polyvinylchloride, if the ΔT value associated with this change is 0.1-0.2°C, then a typical first-order change would be 10-50 times as great as shown [97]. These second-order transitions are termed glass transitions. On the high temperature side the material has lost its rigidity and is plastic, whereas on the lower temperature side it is rigid. There is a degree of dependence of the T_g point in polymers on experimental conditions and the method of determination. Obviously TMA methods, DMA methods, and DTA/DSC can all be used in its evaluation. The subject has been reviewed by Kovacs [98].

The next peak represented in Figure 22 is an exothermic transition representing a recrystallization from the glassy state. It has been developed into a DSC method of determining the extent of crystallinity in a polymer material [99]. The melting peak of the polymer is represented by the endothermic peak C, but although reversible on slow cooling it may not be quantitatively so, as the comparative sizes of the endothermic peak area and the exothermic peak area in cooling will be affected by recrystallization. The degradation peak D may be exothermic in air, or endothermic in nitrogen. Cross-linking processes, for example in curing, will be exothermic [100]. The degradation studies or combustion studies in air can involve a whole range of thermal analysis techniques.

Degradation of inorganic materials likewise can involve a wide range of thermal analysis techniques. As already indicated for phase transitions, the relationship

$$\ln P = -\frac{\Delta H}{RT} + \text{constant}$$

can be demonstrated for, say, oxide dissociations, oxysalt decompositions, or dehydration processes. In the general reaction

$$A_{(s)} \rightarrow B_{(s)} + C_{(g)}$$

Table 12. Phase Transitions in Silica

Temperature (°C)	Transition[a]		
117	αT	$\rightleftharpoons$	β_1T
163	β_1T	$\rightleftharpoons$	β_2T
198-241	βC	$\longrightarrow$	αC
220-275	αC	$\longrightarrow$	βC
570	βQ	$\longrightarrow$	αQ
575	αQ	$\longrightarrow$	βQ
870 ± 10	Q	$\rightleftharpoons$	T
1470 ± 10	T	$\rightleftharpoons$	C
1625	βC	$\longrightarrow$	Silica glass

[a]Key: tridymite, T; cristobalite, C; quartz, Q; α polymorph stable at low temperature; β polymorph stable at high temperature.

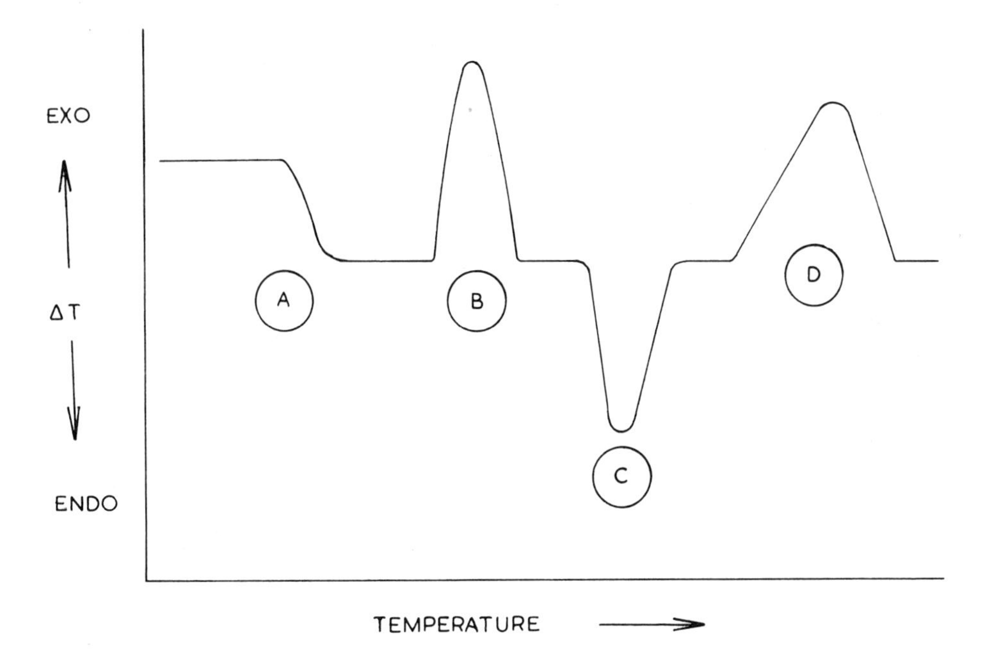

Figure 22. Hypothetical DTA trace for a polymer heated in air.

P is the equilibrium pressure of C above the degrading solid system, and T is the observed temperature of the dissociation. A detailed account of these dissociations has been given by Dollimore [101]. The application really only represents an advantage over more complicated systems when there are multiple dissociations or a complicated reaction sequence involved. Berg et al. [102,103] applied DTA methods to study complicated dehydration processes.

Finally, it should be noted that as the theory for DTA/DSC is formulated, the position of the "baseline" is determined by the heat capacity of the system [104]. This leads quite easily to a relatively simple method by which heat capacities may be determined at any temperature [105].

The introduction of added material to a sample causes an elevation of the boiling point or a depression of the freezing point, phenomena covered by classical methods, as described in any textbook on physical chemistry. An adaptation of this method to DSC techniques allows for the estimation of the purity of various materials. The subject has been covered in a book edited by Blaine and Schoff [106].

B. Methods with a Kinetic Basis

The kinetic parameters associated with the Arrhenius equation,

$$k = Ae^{-E/RT}$$

where k is the specific reaction rate constant and varies with temperature, A is the preexponential term, E is the energy of activation, R is the gas constant, and T is the absolute temperature, are classically determined by running a series of kinetic experiments at various temperatures. In solid-state reactions we then have:

$$\frac{d\alpha}{dt} = K(T)f(\alpha)$$

where α is the fraction decomposed at time t, $d\alpha/dt$ is the rate of reaction, $K(T)$ is the specific reaction rate at temperature T, and $f(\alpha)$ is some function of α describing the progress of the reaction. There is no reason why this classical method cannot be investigated by thermal methods, as almost all commercial thermal instruments can be operated under constant-temperature conditions. However, it is possible in theory to obtain the Arrhenius parameters in a rising-temperature experiment. If the temperature regime imposed in the system is represented by

$$T = T_0 + bt$$

where T is the temperature at time t, T_0 is the starting temperature, and b is the heating rate, then combination of the above equations gives

$$k(T) = \frac{(d\alpha/dT)b}{f(\alpha)}$$

and

$$\ln k = \ln A - \frac{E}{RT}$$

where k is defined as indicated. For some reason, instead of the traditional Arrhenius plot of ln k against $1/T$ this expression is given as

$$\ln\left[\frac{d\alpha/dT}{f(\alpha)}\right] = \ln\left(\frac{A}{B}\right) - \frac{E}{RT}$$

In this trivial change the plot is of $\ln [(d\alpha/dT)/f(\alpha)]$ against $1/T$. Either plot would lead to the evaluation of E and A. The differential signal $d\alpha/dt$ is, however, attended with difficulties, and most investigators would prefer an integral method. Yet the use of an integral method needs the evaluation of $\int e^{-E/RT}\,dT$ and this proves impossible analytically. Resort must be made to numerical methods or alternative similar expressions that can be analytically evaluated. The whole problem is reviewed in detail by Brown et al. [107]. The sample history is important in any solid-state kinetic problem, and there are articles dealing with the complications of thermal history. The article by Flynn [108] is particularly useful. There are many other applications in which the kinetics are employed more empirically or in such a way that they serve to characterize the system under study. Typical of these studies are those dealing with the stability (shelf life) or pharmaceuticals [109] and the proximate analysis of coal [110]. These examples are selective only, and there are many other applications to be found in the literature.

REFERENCES

1. G. Lombardi, *For Better Thermal Analysis*, 2nd ed., ICTA, p. 16 (1980).
2. E. S. Watson, M. J. O'Neill, J. Justin, and N. Brenner, *Anal. Chem.*, *36*:1233 (1964).
3. H. G. McAdie, *Therm. Anal., Proc. Int. Conf., 3rd, 1971*; p. 591 (1972).
4. See Ref. 1, p. 25.
5. See Ref. 1, p. 27.
6. B. Wunderlich and R. C. Bopp, *Thermochim. Acta*, *6*:335 (1974).
7. M. I. Pope and M. D. Iudd, *Introduction to Differential Thermal Analysis*, Heyden, London, p. 37 (1977).
8. A. Kambe, K. Horie, and T. Suzuki, *J. Thermal Anal.*, *4*:461 (1972).
9. M. J. O'Neill, *Anal. Chem.*, *36*:1233 (1964).
10. D. J. David, *Anal. Chem.*, *36*:2162 (1964).
11. E. E. Marti, *Thermochim. Acta*, *5*:173 (1972).
12. N. A. Lange, *Handbook of Chemistry*, 10th ed., McGraw-Hill, New York (1967).
13. A. R. McGhie, *Anal. Chem.*, *55*:987 (1983).
14. A. R. McGhie, J. Chiu, P. G. Fair, and R. L. Blaine, *Thermochim. Acta*, *67*:241 (1983).
15. S. D. Norem, M. J. O'Neill and A. P. Gray, *Proc. Third Toronto Symp. Thermal Analysis* (H. G. McAdie, ed.), Chemical Institute of Canada, Toronto, p. 221 (1969).
16. S. D. Norem, M. J. O'Neill, and A. P. Gray, *Thermochim. Acta*, *1*:29 (1970).
17. R. L. Blaine and P. G. Fair, *Thermochim. Acta*, *67*:233 (1983).

18. J. P. Elder, *Thermochim. Acta, 52*:235 (1982).
19. P. K. Gallagher, E. Coleman, and R. C. Sherwood, *Thermochim. Acta, 37*:291 (1980).
20. C. Duval, *Inorganic Thermogravimetric Analysis*, 1st ed., Elsevier, Amsterdam, p. 531 (1953).
21. C. J. Keattch and D. Dollimore, *An Introduction to Thermogravimetry*, 2nd ed., Heyden, London, p. 164 (1975).
22. R. C. Mackenzie, *Thermochim. Acta, 73*:307 (1984).
23. J. McBain and A. M. Baker, *J. Amer. Chem. Soc., 48*:690 (1926).
24. S. J. Gregg and K. S. W. Sing, *Adsorption, Surface Area and Porosity*, 2nd ed., Academic Press, New York, p. 303 (1982).
25. D. M. Young and A. D. Crowell, *Physical Adsorption of Gases*, Butterworths, London, p. 426 (1962).
26. J. Loriers, *Rev. Met., 49*:807 (1952).
27. T. Daniels, *Thermal Anal.*, Kogan Page, London, p. 49 (1973).
28. S. Gordon and C. Campbell, *Anal. Chem., 32*:271R (1960).
29. G. W. Ewing, *J. Chem. Educ., 53*:A251, A291 (1976).
30. J. Sestak, in *Thermal Analysis* (R. F. Schwenker and P. D. Garn, eds.), Academic Press, New York, Vol. 2, 1085 (1969).
31. D. R. Terry, *J. Sci. Instrum., 42*:507 (1965).
32. H. W. Dial and G. S. Knapp, *Rev. Sci. Inst., 40*:1086 (1969).
33. J. G. Schniztein, J. Brewer, and D. F. Fischer, *Rev. Sci. Instrum., 36*:591 (1965).
34. E. S. M. Watson, M. J. O'Neill, J. Justin, and N. Brenner, *Anal. Chem., 36*:1233 (1964).
35. S. L. Boersma, *J. Am. Ceram. Soc., 38*:281 (1955).
36. D. Dollimore, L. F. Jones, and T. Nicklin, *Thermochim. Acta, 11*:307 (1975).
37. A. T. Jensen and C. A. Beevers, *Trans. Faraday Soc., 34*:1478 (1938).
38. M. Kamphausen, *Rev. Sci. Instrum., 46*:668 (1975).
39. A. A. Krawetz and T. Tong, *Ind. Eng. Chem., Prod. Res. Dev., 5*:191 (1966).
40. F. Noel, *J. Inst. Petrol., 57*:354 (1971).
41. W. R. May and L. Bsharah, *Ind. Eng. Chem., Prod. Res. Dev., 10*:66 (1971).
42. J. N. Maycock, *Thermochim. Acta, 1*:389 (1970).
43. S. Speil, L. H. Berkelhamer, J. A. Pask and B. Davis, *U.S. Bureau of Mines Technical Papers, 664* (1945).
44. R. Melling, F. W. Wilburn, and R. M. McIntosh, *Anal. Chem., 41*:1275 (1969).
45. A. D. Cunningham and F. W. Wilburn, in *Differential Thermal Analysis*, Vol. I (R. C. MacKenzie, ed.), Academic Press, London, p. 31 (1970).
46. M. D. Judd and M. I. Pope, *J. Inorg. Nucl. Chem., 33*:365 (1971).
47. W. W. Wendlandt, *Handbook of Commercial Scientific Instruments*, Vol. 2, Marcel Dekker, New York (1974).
48. W. W. Wendlandt, *Thermal Analysis*, 3rd ed., Wiley, New York, Ch. 8, p. 461 (1986).
49. W. Lodding, *Gas Effluent Analysis*, Marcel Dekker, New York (1967).
50. H. G. Langer in *Treatise on Analytical Chemistry* (I. M. Kolthoff, P. J. Elving, and E. B. Sandell, eds.), Part I, Ch. 15, Wiley, New York (1980).

51. P. D. Garn and J. E. Kessler, *Anal. Chem.*, *33*:952 (1961).
52. E. G. Prout and F. C. Tomkins, *Trans. Faraday Soc.*, *42*:482 (1946).
53. P. K. Gallagher and E. M. Gyorgy, in *Thermal Analysis* (H. G. Wiedemann, ed.), Vol. 1, Birkhauser, Basel, p. 113 (1980).
54. M. J. D. Low, *In Gas Effluent Analysis* (W. Lodding, ed.), Marcel Dekker, New York, Ch. 6 (1967).
55. A. B. Kiss, *Acta Chim. Acad. Sci. Hung.*, *61*:207 (1969); *63*:243 (1970).
56. P. K. Gallagher, *Thermochim. Acta*, *26*:175 (1978).
57. E. Kinsbrow, P, K. Gallagher, and A. T. English, *Solid State Electron.*, *22*:517 (1979).
58. D. Price, D. Dollimore, N. J. Fatemi, and R. Whitehead, *Thermochim. Acta*, *42*:323 (1980).
59. R. M. Lum, *Thermochim. Acta*, *18*:73 (1977).
60. R. S. Gohlke and H. C. Langer, *Anal. Chim. Acta*, *36*:530 (1966).
61. F. Zitomer, *Anal. Chem.*, *40*:1091 (1968).
62. G. A. Kleinberg and D. L. Geiger, *Proc. 3rd ICTA Conf.* (H. G. Wiedemann, ed.), Vol. 1, Birkhauser, Basel, p. 325 (1971).
63. S. M. Dyszel, *Thermochim. Acta*, *61*:169 (1983).
64. J. Chiu and A. J. Beattie, *Thermochim. Acta*, *21*:263 (1977); *40*:251 (1980); *50*:49 (1981).
65. R. N. S. Rogers, S. K. Yasuda, and J. Zinn, *Anal. Chem.*, *32*:672 (1960).
66. O. A. Vassallo, *Anal. Chem.*, *33*:1823 (1961).
67. S. A. Liebman, D. H. Ahlstrom, T. G. Creighton, G. D. Pruder, and E. J. Levy, *Thermochim. Acta.*, *5*:403 (1973).
68. L. E. Nesbitt and W. W. Wendlandt, *Thermochim. Acta*, *10*:85 (1974).
69. W. J. Irwin, *Analytical Pyrolysis*, Marcel Dekker, New York, p. 578 (1982).
70. S. D. Robertson, B. D. McNicol, J. H. de Baas, S. C. Kloet, and J. W. Jenkins, *J. Catal.*, *37*:424 (1975).
71. S. Tsuchiya, Y. Amenomiya, and R. J. Cvetanovic, *J. Catal.*, *20*:1 (1971).
72. S. Tsuchiya and M. Nakamura, *J. Catal.*, *50*:1 (1977).
73. Y.-C. Chan and R. B. Anderson, *J. Catal.*, *50*:19 (1977).
74. C. A. Luengo, A. L. Cabrera, H. B. Mackay, and M. B. Maple, *J. Catal.*, *47*:1 (1977).
75. J. Chiu, *Anal. Chem.*, *40*:1516 (1968).
76. J. Chiu, *Thermochim. Acta*, *1*:231 (1970).
77. P. Cukor and C. Persiani, *Polymer Characterization by Thermal Methods of Analysis* (J. Chiu, ed.), Marcel Dekker, New York, p. 107 (1974).
78. H. G. Wiedemann, *Thermal Analysis* (R. F. Schwenker and P. D. Garn, eds.), Vol. 1, Academic Press, New York, p. 229 (1969).
79. P. C. Uden, D. E. Henderson, and R. J. Lloyd, *Proc. First European Symposium on Thermal Analysis* (D. Dollimore, ed.), Heyden, London, p. 29 (1976).
80. K. Yamada, S. Orra, and T. Haruki, *In Proc. 4th ICTA Conf.* (I. Buzas, ed.), Vol. 3, Heyden, London, p. 1029 (1975).
81. I. C. McNeill, *Eur. Polymer. J.*, *6*:373 (1970).
82. F. Paulik and J. Paulik, *J. Thermal Anal.*, *16*:399 (1979).
83. P. S. Gill, *Am. Lab.*, *16*:39 (1984).

84. R. Riesen and H. Sommerauer, *Am. Lab.*, *15*:30 (1983).
85. R. E. Wetton, *Anal. Proc.*, p. 416, Oct. 1981.
86. M. G. Lofthouse and P. Burroughs, *J. Thermal Anal.*, *13*:439 (1978).
87. J. K.Gillham, *Appl. Polymer Symp.*, *2*:45 (1966).
88. A. F. Lewis and J. K. Gillham, *J. Appl. Polymer Sci.*, *6*:422 (1962).
89. F. Paulik and J. Paulik, *Analyst*, *103*:417 (1978).
90. W. W. Wendlandt, in *Thermal Analysis*, 3rd ed., Wiley, New York, p. 559 (1986).
91. H. P. Vaughan, *Thermochim. Acta*, *1*:111 (1970).
92. K. Lonvik, *Thermochim. Acta*, *27*:27 (1978).
93. J. B. Dawson and F. W. Wilburn, *Differential Thermal Analysis*, Vol. 1 (R. C. MacKenzie, ed.), Academic Press, London, p. 977 (1970).
94. C. N. Fenner, *Am. J. Sci.*, *36*:331 (1913).
95. D. Dollimore, *Differential Thermal Analysis*, Vol. 1 (R. C. MacKenzie, ed.), Academic Press, London, p. 900 (1970).
96. J. G. Rabatin, R. H. Gale, and A. E. Newkirk, *J. Phys. Chem.*, *64*:491 (1960).
97. J. J. Keavney and E. C. Eberlin, *J. Appl. Polymer Sci.*, *3*:394 (1963).
98. A. J. Kovacs, *Fortschr. Hochpolym. Forsch*, *3*:394 (1963).
99. S. W. Shalaby, *Thermal Characterisation of Polymeric Materials* (E. A. Turi, ed.), Academic Press, New York, p. 235 (1981).
100. R. Bruce Prime, *Thermal Characterisation of Polymeric Materials* (E. A. Turi, ed.), Academic Press, New York, p. 435 (1981).
101. D. Dollimore, *Differential Thermal Analysis* (R. C. MacKenzie, ed.), Vol. 1, Academic Press, London, p. 396 (1970).
102. L. G. Berg and I. S. Rassonskaya, *Dokl. Akad. Nauk. SSSR*, *73*:113 (1950).
103. L. G. Berg, I. S. Rassonskaya, and E. V. Buris, *Izv. Sekt. Fiz-Khim. Analiza. Inst. Obstichei. Neorg. Khim.*, *27*:239 (1956).
104. M. J. Vold, *Anal. Chem.*, *21*:683 (1949).
105. M. I. Pope and M. D. Judd, *Differential Thermal Analysis*, Heyden, London, p. 39 (1977).
106. R. L. Blaine and C. K. Schoff, *Purity Methods by Thermal Methods*, ASTM Publication, Philadelphia, p. 151 (1984).
107. M. E. Brown, D. Dollimore and A. K. Galwey, *Reactions in the Solid State, Vol. 22 of Comprehensive Chemical Kinetics* (C. H. Bamford and C. F. H. Tipper, eds.), p. 99 (1980).
108. J. H. Flynn, *Thermal Analysis in Polymer Characterization* (E. A. Turi, ed.), Heyden, Philadelphia, p. 43 (1981).
109. A. Radecki and M. Wesolowski, *J. Thermal Anal.*, *17*:73 (1979).
110. R. J. Rosenvold, J. B. Dubow and K. Rajeshwar, *Thermochem. Acta*, *53*:321 (1982).

26

Automatic Titration

LARRY K. SVEUM / Department of Chemistry, New Mexico Highlands University, Las Vegas, New Mexico

I. INTRODUCTION

Titration techniques are quite old analytical methods. Historically, the methods are based on the idea of measuring the amount of an analytical reagent necessary to react quantitatively with the sample. In general, then, the necessary components of a titration instrument are the reagent delivery system, a means of measuring the quantity of reagent added, and the end-point detection system, so that the analyst knows when sufficient reagent has been added to react with the sample. The three components of a titration system will be treated individually, and subsequently the methods for combining the various components into commercial instruments will be discussed.

II. REAGENT DELIVERY SYSTEMS

The reagent may be either directly chemical in nature or indirectly chemical. If the reagent is itself a chemical, it is added to the sample as a standard solution by means of a buret, so that the amount of the reagent added is proportional to the amount of solution dispensed. If it is only indirectly chemical, the most common method of reagent delivery is by electrical generation, in which case the amount of reagent is related directly to the amount of electricity (coulombs) passed through the generator. Such a system is called coulometric titration. Thus, the two principal methods of reagent addition in titration are volumetric and coulometric.

A. Volumetric Reagent Addition

Traditionally, reagent is added to the sample by means of a buret, a calibrated piece of glassware in which the volume of solution delivered is read directly from the engraved graduations. At the end point of the reaction the amount of chemical in the sample is determined by multiplying the volume of solution added to that point by the concentration of the reagent and making appropriate adjustments for stoichiometry.

The concentration of the reagent must be known, as well as the amount of reagent added. The concentration of the reagent is most often determined by a standardization technique in which a known quantity of a pure sample is titrated with the reagent.

The primary innovation made by manufacturers in the delivery system of modern automatic titrators has been to develop methods to deliver reagent at a constant rate. Glass burets are inherently long narrow tubes, so that small quantities of fluid can be measured precisely. As a consequence the pressure on the exit orifice changes quite dramatically during the course of a titration due to the variation of the height of the fluid column within the buret. Noncommercial apparatus can be constructed with common laboratory glassware to approximate constant flow by minimizing the liquid pressure head variation. Several designs for such equipment exist, in general constructed as closed systems in which pressure-equalizing mechanisms operate. Once the apparatus is built, the flow rate must be calibrated. The flow rate is determined by the size of the exit orifice or capillary at the bottom of the delivery tube. Once the flow rate of the apparatus is calibrated, the elapsed time of the experiment is proportional to the volume of reagent delivered. The capillary tube is quite sensitive to the cleanliness of the solution. Small dust particles can cause the flow rate to vary considerably. This method of flow control is inexpensive to build and in general can be constructed from materials at hand in a laboratory. At the present time no commercial apparatus of similar construction is available.

Most commercial apparatus is somewhat more sophisticated and involves more complicated mechanics and electronics. By far the most common method for determining the quantity of reagent added is to use a device that delivers at a constant rate and to measure the elapsed time. This goal is accomplished by use of positive displacement pumps either of the syringe type or some other mechanical type such as a peristaltic or reciprocating pump. In general peristaltic or reciprocating pumps have a disadvantage in that they inherently produce pulsating flow. Most companies manufacturing systems for liquid transfer have opted for syringe-type positive displacement pumps in which the piston is driven by a screw and motor. Most systems presently use stepper motors, although synchronous motors may also be used. Stepper motors permit readily changing the flow rate by controlling the frequency of the driving pulses, something quite easy to do if the system is controlled by a computer.

Frequently in potentiometric titrations the reagent delivery system is controlled by an electronic circuit designed to anticipate the end point of the reaction based on the signal from the detector. A circuit that will calculate the second derivative of the sigmoidal potentiometric titration curve will accomplish this task. The maximum in the second derivative occurs slightly before the true end point, and if the buret valve actuator is energized at that moment the inherent mechanical delay will stop the titration at the proper point. These systems operate on the analog signal from the detector.

The newer systems tend to perform the operation digitally. Such systems slow the delivery rate as the end point is approached so that the likelihood of over-titrating is minimized. In general such systems also drive the X axis of a chart recorder in proportion to the amount of reagent delivered; that is, the same signal that drives the stepper motor of the pump also drives the stepper motor of the recorder.

B. Coulometric Reagent Addition (Coulometric Titrations)

Coulometric titrations are dependent on the generation of the chemical reagent *in situ* by electrochemical means. The chemical reagent then reacts with the constituent in the sample rapidly and quantitatively and is consumed and converted either back to the original nonreactive chemical species or to some other inert material. The amount of electricity that passes through the solution is proportional to the amount of chemical reagent generated, which in turn is proportional to the amount of sample present. The generation of the chemical reagent is always an oxidation or a reduction, and in general the reagent is either reduced or oxidized by reaction with the sample. However, a variety of electrochemically generated species will react by other mechanisms. For instance, either hydrogen ion or hydroxide ion can be electrochemically generated by the electrolysis of water. They in turn can react with the sample as either an acid or a base.

The generation of the chemical reagent requires a specified number of electrons to be liberated or consumed for each molecule. One mole of electrons is the equivalent of one farad, 96487 coulombs. One coulomb is defined as an ampere-second, so, in general, coulometric reagent generators operate as constant current sources and the amount of reagent generated is proportional to the time the generator operates. (See also Chapter 20.)

C. Gravimetric Reagent Addition

A very precise method of reagent measurement, inherently more precise than volumetric, is gravimetric addition. A reagent vessel and its delivery system are placed on a balance and the mass of the reagent added to reach an end point is measured. Since the reagent is in solution, it is necessary to know its density or its concentration per unit mass.

A gravimetric method could easily be automated if the balance used were either a recording or a digital type, able to transmit data to a computer. A typical recording device would then be an X-Y recorder in which the amount of titrant would be plotted along the X axis and the detector response along the Y axis. Currently, gravimetric titration systems are available from only one manufacturer, Luft Instruments.

III. DETECTORS

Knowledge as to when sufficient reagent has been added to the sample is required during the titration. As a consequence, some type of end-point detection is needed. A large variety of techniques is available, and often the titration method is named accordingly. Almost any type of transducer that will indicate a sharp change at the chemical equivalence point will suffice as a detector. Photometric detection requires a color change to occur at the end point, observable visually or by use of a spectrophotometer. A similar technique is based on light scattering: nephelometric detection, in which the intensity of the scattered light is measured, and turbidometric detection, in which the decrease in intensity of transmitted light is measured. Various electrical techniques are also used for end-point detection. The potential (which is proportional to the logarithm of the concentration of some ion in the solution), the conductance of the solution, and amperometric and biamperometric measurements are examples of electrical techniques used to detect titration end points. Thermometric detection is used in some commercial instruments. It measures the temperature change that occurs as heat is absorbed or emitted during the chemical reaction.

A. Photometric Detection

Acid-base indicators were the basis for some of the earliest titration techniques. Such indicators change color depending on the rapid change of hydrogen ion concentration at the equivalence point of an acid-base titration. The changing concentration of hydrogen ion causes the indicator to change from its acid form to its base form or vice versa. The two forms of the indicator have distinct colors. The natural detector for color changes is the analyst's eyes, assuming they can detect color normally and that the indicator colors are sufficiently different.

This detector does not lend itself to automation, however. The newer photometric titrators use detectors ranging from colorimeters to spectrophotometers and employ either flow-through cuvettes and a pump or use a fiber-optic system in which the detector unit contains a bifurcated fiber-optic cable. In the latter, one of the fiber bundles carries the light from the source to the sample while the other carries the unabsorbed light back to the photometric detector; the fiber-optic cable assembly is dipped into the sample and monitors the light absorbance during the titration. Most of these systems are equipped with some form of wavelength selection, ranging from color filters to monochromators. Brinkmann, Mettler, and Guided Wave are three manufacturers of this type of detector.

Photometric end points are useful beyond acid-base titrations. In fact, many types of analytical reactions suitable for titration have either natural color changes suitable for end-point detection or else suitable color-change indicators are available for them. For instance, titration analyses of most metal ions can be accomplished by use of EDTA or similar complexing agents, together with several indicator dyes that can be utilized for end-point detection. Indicators for acid-base, precipitation, complex formation, and redox titrations have been developed. Most are suitable for automated photometric titration, while others, especially adsorption indicators used for precipitation reactions, are only suitable for manual, visual titrations.

Light absorption is linear with concentration of the absorbing species (see Chapter 6). The titration curve obtained from a photometric detector depends on the species that absorbs light. For instance, if the titrant is the absorber in a particular titration, the detector will give no response until the end point of the titration when excess titrant is added (see Figure 1a). If the sample is the light absorber, the detector will indicate a decline in absorbance as the titration proceeds until the sample is completely reacted, beyond which point the detector will give no response (Figure 1b). If the reaction product is the light absorber, the detector will increase in absorbance in proportion to the amount of product that exists at any moment. Past the end of the titration the detector will give a flat response at a relatively high value of absorbance (Figure 1c).

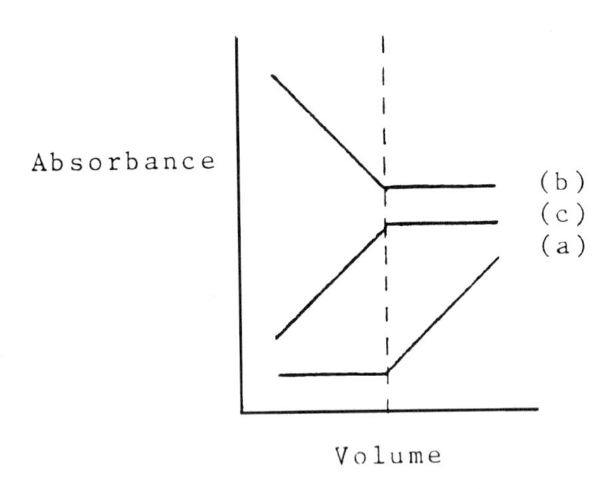

Figure 1. Typical photometric titration curves near the end point. (a) The reagent is the light absorber; it is only present and able to absorb after the end point is passed. (b) The sample is the light absorber, and is completely reacted and destroyed at the end point. (c) One of the reaction products is the light absorber; at the end point no more product can be created and the absorbance remains constant.

B. Nephelometric and Turbidimetric Detection

As mentioned earlier, both nephelometric and turbidimetric detection are special cases of photometric detection. The same equipment can be used for photometric titrations and for nephelometry or turbidimetry. Both nephelometric and turbidimetric detection depend on light scattering and, as such, are only useful for precipitation titrations. Because light scattering depends on the particle size of the precipitate, development of theoretical equations for the analysis requires more information than is usually available. However, from a practical point of view, if all of the experiments are performed in the same, reproducible manner, working curves can be developed that will give excellent results. It is difficult to compare experimental results obtained from different laboratories, however.

The two light-scattering detection methods differ from each other in a manner similar to the differences between fluorescence and absorption spectroscopy. Nephelometry corresponds most closely to fluorescence: the beam of light is directed into the sample and the detector is positioned at some angle with respect to it, usually at 90°, so that the amount of scattered radiation can be measured. Turbidimetry is similar to absorption spectroscopy, in that the detector is placed coaxially with the incident light beam and as the number of particles increases, the amount of transmitted light decreases because part of it is scattered by the particulate matter. From a theoretical point of view, the analysis of a nephelometric measurement is similar to that of fluorescence (see Chapter 8) in which the intensity of the scattered beam is proportional to the number of particles. The analysis of turbidimetric data is similar to that of absorbance (see Chapter 6), and an equation similar in form to Beer's law is applied.

At the present time no commercial equipment is available that is specifically designed for nephelometric or turbidimetric titrations. The use of fluorometers or absorption spectrometers along with an appropriate reagent delivery system can provide an acceptable instrument for nephelometric or turbidimetric analysis, however.

C. Potentiometric Detection

Historically, potentiometric detectors probably developed second to visual color-change detection. Their utilization was greatly enhanced with the development of the pH meter. The glass electrode was the first of several ion selective electrodes to be developed (see Chapter 16). Potentiometric measurements of concentration depend on a logarithmic function (the Nernst equation), and consequently the electrical potential is linear with the logarithm of the concentration of the ion of interest, the conventional *p* function of which pH is an example. The modern pH-meter is based on the glass electrode, which is sensitive to hydrogen ions and will conveniently cover the pH range of 1-12 with very little deviation from linearity. Quite a large number of ion-selective electrodes are presently available commercially. As a consequence, potentiometric detection of titration end points is currently one of the most popular and widely used techniques. Most of the commercial automatic titrators are based on potentiometric detection.

The titration curve is typically sigmoidal in shape due to the logarithmic dependence of the potential on concentration (see Figure 2a). The end point is at the steepest part of the curve. This phenomenon can cause some difficulty in end-point detection, especially if the change in potential is slow or gradual. The precision can be improved if the curve is differentiated with respect to volume by use of an electronic differentiator, acting either once (Figure 2b) or twice (Figure 2c), to generate curves in which the end point is easier to observe. The second derivative is also one of the methods used to anticipate the end of the reaction by closing the buret valve or stopping the drive motor when the second-derivative maximum is achieved.

D. Conductance Detectors

Titration detection based on the electrical conductance of ions in solution is a nonspecific detection method that has a wide range of application in aqueous solution but fairly limited application in nonaqueous solutions. In general, a highly polar solvent is required in order to dissolve and dissociate compounds into their ions, and most organic or nonaqueous solvents are incapable of causing dissociation. Electrolytic conductance is proportional to the total concentration of the ions in the solution (see Chapter 19).

Most ions have a similar equivalent conductance (i.e., normalized for ionic charge) at the same concentration, which is on the order of 60-70 S cm^2 mol^{-1}. The two notable exceptions are the proton, with a conductance of ~350 S cm^2 mol^{-1}, and hydroxide ion, ~200 S cm^2 mol^{-1}. A few other

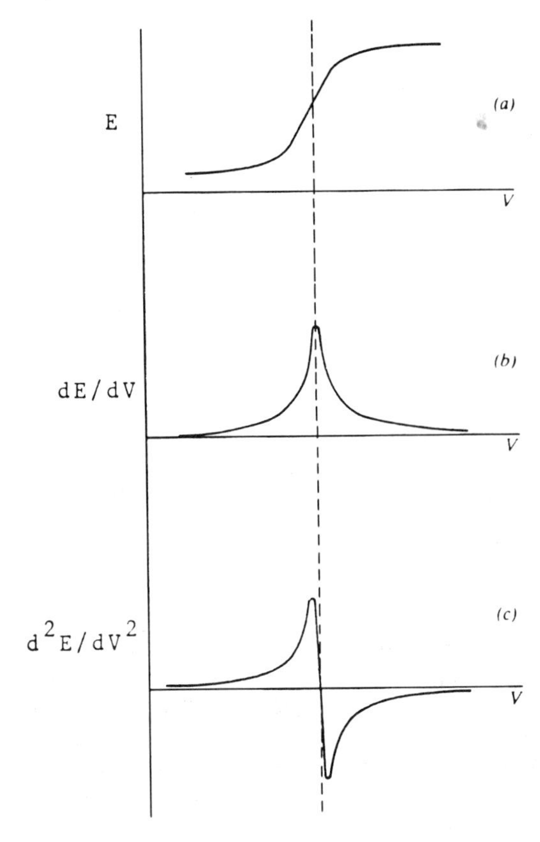

Figure 2. Typical potentiometric titration curves. (a) Potential (E) plotted against the volume (V) of titrant (sigmoidal curve). (b) The first derivative of potential with respect to volume, dE/dV, against volume. (c) The second derivative of potential with respect to volume, d^2E/dV^2, plotted against volume.

ions also have conductances that deviate significantly from the norm. In general those ions are quite bulky and tend to have low values of conductance. As a consequence of these facts, conductance detection is generally based on ion generation or consumption. The most common types of reactions to which it is applied are acid-base, precipitation, and complexation titrations; in general it cannot be used in redox reactions, since frequently the total number of ions remains constant during the titration.

The shape of the titration curve depends on the total concentration of the ionic species in solution, as well as their ionic conductance. The conductance of the solution is determined by the sum of the ionic conductances of each of the species present, multiplied by its concentration in the solution. Both factors must be considered in most reactions; however, in a strong acid-base titration, a good approximation of the titration curve can be obtained by simply calculating the concentration of the protons

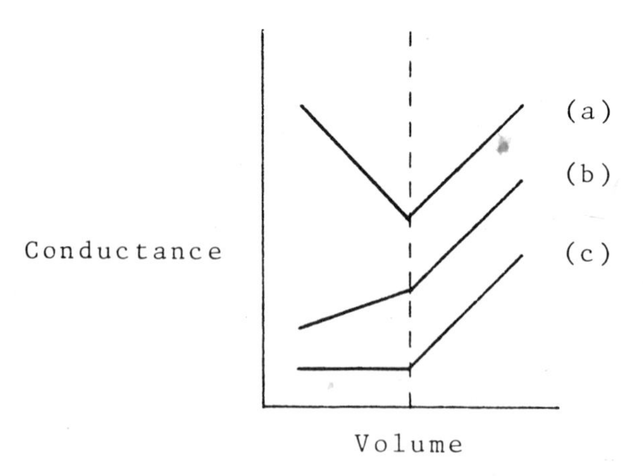

Figure 3. Typical conductometric titration curves in the vicinity of the end point. (a) Titration of a strong acid (or base) with a strong base (or acid). (b) Titration of a weak acid (or base) with a strong base (or acid). (c) A precipitation titration.

or hydroxide ions present in solution. In strong acid-base titrations, for instance, as the protons or hydroxide ions are consumed prior to the end point, they are replaced by ions having lower conductivity. The conductivity of the solution then declines linearly, neglecting dilution effects, until all of the protons or hydroxide ions are consumed. As excess reagent is added, the conductivity increases again, leading to a titration curve that is V-shaped, the apex of the V being the end point of the titration (Figure 3a).

The titration of a weak acid or base by a strong base or acid leads to a curve that increases slowly prior to the end point and steeply after the end point (Figure 3b). Curves for precipitation reactions will appear almost flat before the end point, inclining or declining slightly depending on the ionic conductance of the individual ions, and will increase sharply after the end point (Figure 3c).

Conductance detection is nonspecific and consequently provides a more general-purpose detector than potentiometric or colorimetric methods in which the electrode or the color change is dependent on specific species in the solution.

E. Amperometric Detection

Amperometric end-point detection in titration is based on electrochemical properties similar to those exploited in polarography (see Chapter 17). At least one chemical species must change concentration dramatically at the end point of an analytical reaction, and that species must be reducible either in its reactant form or in its product form. The detector consists of measuring the diffusion current for the reduction of the indicator species at a specified applied voltage. As the concentration of the indicator species changes, the diffusion current increases or decreases.

The apparatus requires two electrodes to apply the voltage and a circuit to measure the diffusion current. In general, one of the electrodes is either the dropping mercury electrode or a rotating platinum electrode. The dropping mercury electrode generates a new active surface on a regular basis so that diffusion of the electroactive species may achieve a steady-state level. The diffusion current is defined as the current that is due to steady-state diffusion of the ion to the electrode surface. The rotating platinum electrode accomplishes the same goal by minimizing the diffusion gradient through forced diffusion so that the actual concentration of the indicator species may be monitored on a continual basis. In order to obtain reproducible results, however, the speed of rotation must be carefully controlled, since the amount of diffusion must be the same in all experiments.

The titration curve that one obtains is similar in shape to those obtained with conductance detection and also depends on the type of reaction. The possible situations include a reducible reagent, in which the curve is horizontal until the reaction is over and then the current increases proportionally to the concentration of reagent (see Figure 4b), and a reducible sample, in which the curve decreases with the concentration of the sample until the end point is achieved at which point the curve becomes horizontal (Figure 4a). The third type involves both a reducible reagent and sample: the curve now appears to be a "V" in which the apex corresponds to the end point (Figure 4c).

Amperometric detection is quite general, can be used in most titrations involving a reducible species, and may include many redox, complexation, and precipitation reactions. The detection system can be made quite selective by choosing the proper voltage at which to measure the diffusion current.

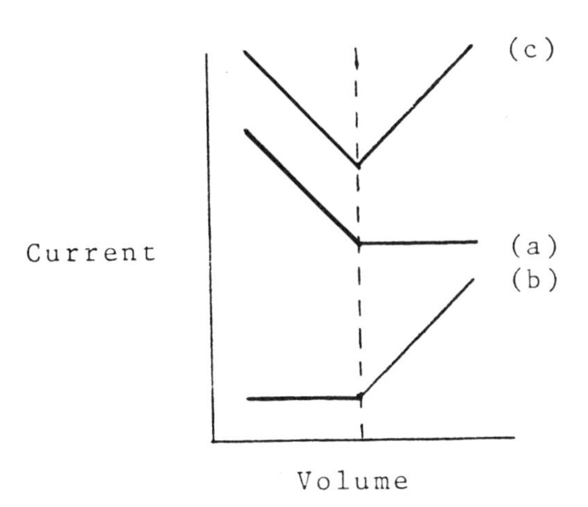

Figure 4. Typical amperometric titration curves. (a) Titration of a reducible sample; the current declines until the reaction is complete. (b) Titration with a reducible reagent; the current is constant (nearly zero) until after the end point, when it begins to increase. (c) Titration in which both sample and reagent are reducible.

F. Biamperometric Detection

Biamperometric detection is the final electrochemical detection method that will be discussed. It is similar to amperometric detection in that current is measured and its behavior both before and after the end point of the analytical reaction determines the end point.

The biamperometric method is used primarily with reversible redox pairs such as Fe^{3+}/Fe^{2+} and I_2/I^-, although some examples of nonreversible reactions can be found. For instance, H_2O_2 can be oxidized to give O_2 and reduced to give OH^-; also MnO_4^- is reduced to MnO_2 while its reduction pair Mn^{+2} is oxidized to MnO_2.

In the situation involved here, the potential required for oxidation is about equal to that for reduction and so the application of only a few millivolts to the electrodes will permit current to flow if both halves of the redox pair are present, and no current to flow if only one is present. The magnitude of the current will be proportional to the concentration ratio of the redox pair.

Once again, intersecting straight line segments identify the end point. Assuming that only one redox pair is involved in the reaction, the curve will decline and end with a horizontal section if the sample provides the reversible redox pair (Figure 5a), or it will start with a horizontal section and end with an increasing line if the reagent is the redox pair (Figure 5b). If both the sample and the reagent form redox pairs, a V-shaped curve will be formed as the end point is approached (Figure 5c). The end point in all cases occurs at the intersection of the two lines in which the current approaches zero because of the elimination of one species of each redox pair at the end point.

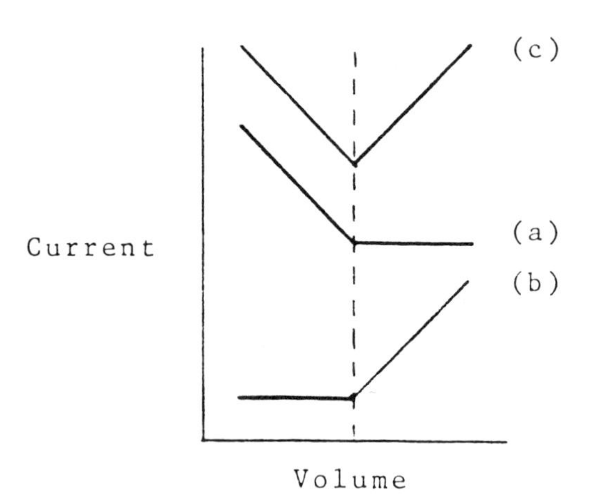

Figure 5. Typical biamperometric titration curves near the end point. (a) Titration of a reversibly reducible sample; the current declines until the reaction is complete. (b) Titration with a reversibly reducible reagent; the current remains low until the end point, then increases. (c) A titration in which both sample and reagent are reversibly reducible.

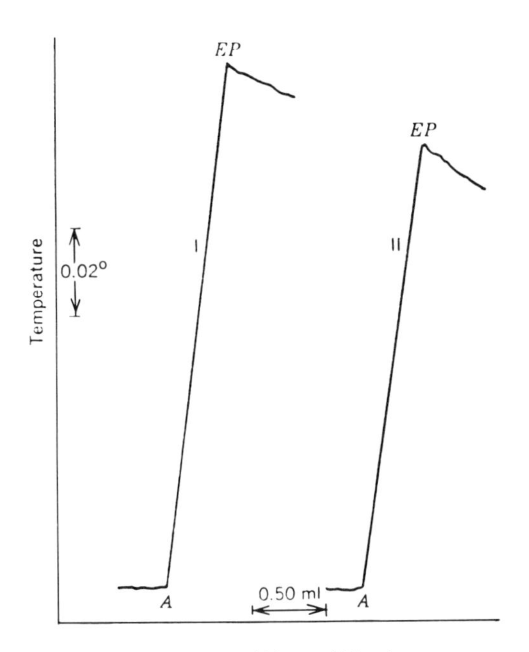

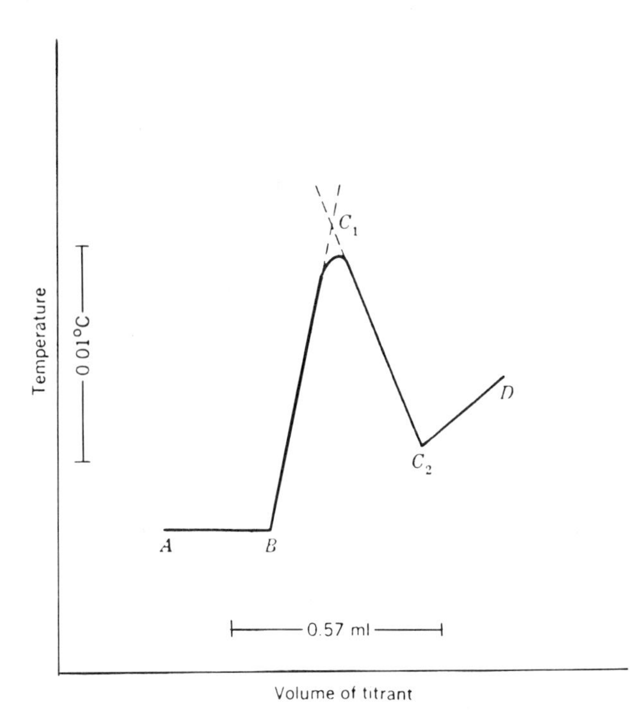

Figure 6. Thermometric titration curves. (a) Titration of 0.01 *M* aqueous solutions of (I) HCl and (II) H_3BO_3 with NaOH; the titrations start at A and reach end points at EP. [After J. Jordan, *Rec. Chem. Prog.*, *19*, 193 (1958).] (b) Titration of a mixture of 0.25 *mmol* each of Ca^{2+} and Mg^{2+} ions with EDTA. The titration was started at B, and showed a calcium end point extrapolated to C_1 and a magnesium end point at C_2. The region from C_2 to D represents the addition of excess reagent. [Redrawn from J. Jordan and T. G. Alleman, *Anal. Chem.*, *29*, 9 (1957).]

H. Thermometric Detection

All chemical reactions involve a heat of reaction, either heat evolution or absorption. If the reaction is carried out in an insulated vessel the heat of reaction will manifest itself as a temperature change that can be measured with a sensitive thermometer, thermocouple, or thermistor. Thermometric titration, then, amounts to monitoring the reaction progress by observing the temperature change that occurs during the process.

Usually the heat of reaction is quite significant, even if the Gibbs free energy is small, because of a compensating entropy change. Consequently, many substances that are impossible to titrate using other detection methods can be titrated using thermometric techniques. For instance, both strong and weak acids evolve roughly the same quantity of heat, and so acids such as boric acid that are too weak to titrate using any other detection method can be titrated thermometrically (Figure 6a). Also, a mixture of calcium and magnesium can be determined by titration with EDTA using thermometric detection even though analysis of the mixture is impossible by colorimetric methods. In fact, the formation of the EDTA complex of calcium is exothermic while that of magnesium is endothermic, so the titration curve of the mixture shows an increase in temperature during the titration of calcium followed by a decrease in temperature during the titration of the magnesium. The titration is consequently extraordinarily easy, looking like an inverted V with the calcium end point at the apex and the magnesium end point at a subsequent dip (Figure 6b).

IV. SYSTEMS

In general the purchaser of a titration system will have at least one of several possible goals in mind. It may be desirable to search out the individual components and then assemble a system in which several solvent delivery mechanisms or detectors may be available to select for titrations to which they are best suited. On the other hand, the pruchaser may wish to buy a completely assembled instrument in which the reagent delivery system is integrated with a particular detector. Many titrators exist, and generally companies manufacturing such equipment will provide large amounts of customer support, including tested methods. The third option available to the customer is even more specific. A variety of titrators have been designed to be totally dedicated to one particular analysis, such as water by the Karl Fischer method. In the following sections all three of these approaches will be discussed.

A. Component or Modular Titrators

Modular titrators tend to be the most versatile. In general the type of reagent delivery and the detector most suitable for a particular application and analysis can be purchased or built. Generally speaking, volumetric reagent delivery will satisfy most applications. Volumetric systems are the most widely available and are perhaps the least expensive.

However, coulometric delivery is more easily automated since it is electrical in nature and consequently more readily interfaced to computers and recorders. Coulometric titrators tend to be underutilized, although they have several distinct advantages. For instance, many reagents can be generated coulometrically, including redox reagents, acids and bases, and complexogens. As a consequence, coulometric titrators are much more versatile than normally thought. In addition, coulometric reagents do not require chemical standardization, a major advantage over volumetric methods. They have inherently good precision due to the high precision attainable through modern electronics.

As mentioned earlier, to my knowledge, there exists only one manufacturer of gravimetric reagent delivery systems. In general this is an obscure method of reagent delivery but one that certainly should be considered if extremely high precision is required.

Detectors can in general be classified into three types: photometric, electrochemical, and thermometric. The most commonly available devices that are easily adapted to titration end-point detection are the electrochemical detectors. Conductivity and pH meters are standard equipment in most chemical laboratories. Their adaptation to end-point detection simply requires inserting the electrode probes into the solution. As mentioned earlier, conductance is totally nonspecific, but potentiometric measurement of the end point can be as selective as the availability of specific ion electrodes. Amperometric and biamperometric detectors are very easy to construct, simply requiring a voltage source, a microammeter, and two inert electrodes such as platinum wires.

Photometric detectors usually require modifying colorimeters or spectrophotometers. In general the requirements are to purchase a flow-through cell and pump the solution through the cell as the titration proceeds. Once again, most chemical laboratories have such instruments. Flow-through cells are available from a variety of sources.

A major technological improvement in photometric detectors has been made within the last few years with the development of fiber optics. Several companies now market photometric detectors in which a bifurcated fiber optic carries light to and from the solution, using dip-type cells, allowing the remote detection of color intensity.

In principle, thermometric detectors are easily constructed from thermocouples or thermistors. The experimental setup consists of a reference and a measuring device, an amplifier, and a recorder or meter. The reference and the sample containers should be quite good thermal insulators, such as Dewar flasks or styrofoam coffee cups. Usually the titration is rapid enough that relatively poor insulators, such as nested beakers, can be substituted. Thermometric detectors are not commercially available as components for titration systems.

B. General-Purpose Systems

General-purpose automatic titrators are manufactured and distributed by many of the more well-known vendors of scientific equipment. The instruments are almost exclusively based on volumetric delivery of titrant and potentiometric detection. A general-purpose titrator in which the titrant is delivered coulometrically is available from Coulometrics, Inc. Several companies will provide detector modules so that end points may be detected by several techniques other than potentiometric.

While some of the titration systems are based around strip-chart recorders, most of the systems currently available commercially utilize dedicated microprocessors and computers (Fisher Scientific, Cosa Instruments and Radiometer-America), or they make use of personal computers [Sargent-Welch, Sanda, and ITI Seamark, who distribute the Kyoto Electronics (KEM) instrument]. Such computers may also be used for other purposes as well. The use of computers also allows method storage, with some instruments permitting the storage of up to 30 or more methods.

All manufacturers use piston pumps (operated by stepper motors), which can be refilled from a large reservoir of titrant, as the reagent delivery system. The precision of the titrant delivery is determined by the resolution of the stepper motor and the size of the buret. Typical burets range in size from 2.5 to 50 ml, having a precision of about 0.1% of the buret capacity. The manufacturers have taken several approaches to address the use of their instruments for back titrations and for use in several types of analyses on a routine basis. Some of them use modules containing the piston and requisite hoses but use a single pump to change titrant. Others require either cleaning the buret prior to addition of a second reagent or the purchase of additional pumps with their integral pistons and cylinder reservoirs for the delivery of several titrants. Both approaches have their advantages and disadvantages.

Potentiometric Titrators

The potentiometric detection systems are all based around pH or selective ion electrodes. Some of them use the first or second derivative as an end-point detection technique. Most contain either a strip-chart recorder or a printer to provide a hard copy of the titration curve. In addition, a digital printer will present the titration result in the units specified by the method. Each company provides a sample changer so that multiple samples can be analyzed unattended.

Multiple-Detector Titrators

Several manufacturers, most notably Mettler Instruments and Brinkmann Metrohm but also including Cosa Instrument Corp., Instrumentation Technology International, and Sanda, provide instruments that use a variety of detection techniques including potentiometric, photometric, luminescence, conductivity, and thermometric, biamperometric, and voltametric detectors.

The detectors tend to be modular in design so that the detector that is appropriate for a particular analysis can be chosen. All of these titration instruments also use piston pumps for titrant delivery. Again, a sample changer is available as an accessory for multiple unattended analyses. One company, Ionics, Inc., uses programmable read-only memory (PROM) devices to upgrade the equipment as needs change. Their PROMs can be programmed using a key-locked keyboard. Up to five different reagent additions can be programmed into a method. Each PROM can contain up to 99 methods. All of the instruments use either a strip-chart recorder or a digital printer or plotter to record the curves and the final results. Metrohm, Sanda, and COSA allow viewing the curve in real time on a computer screen.

Limited Systems

Tronac, Inc., provides an instrument that is limited to thermometric detection. For some titrations a thermometric detection technique is preferred because the other methods provide inadequate results. The Tronac instrument can also be used as a calorimeter. Their instrument uses a dedicated computer.

C. Specialized Systems

A number of automatic titrators that are designed to analyze a specific chemical species are available for either laboratory or quality control use. Among the most popular are those set up as Karl Fischer titrators for determination of water, generally as a minor constituent in a sample. Systems designed for the analysis of halogens or sulfur as well as calcium are also available.

Water Titrators

At least six manufacturers produce Karl Fischer titrators. About half of them are coulometric, while the remaining are volumetric titrators. They all require between 1 and 10 μg of water in the sample and can handle up to 10-250 mg of water, depending on the manufacturer. The concentration of water can range from a minimum of 1-10 ppm up to 100% in the sample. If water is a major constituent, only very small samples can be used. All of the automatic titrators are microprocessor controlled with digital display, and most have RS-232C interfaces for computer connection. Several will also record the sample size by interfacing with an electronic balance. The major concerns with Karl Fischer titrations involve the reagents. Traditionally pyridine, a noxious-smelling reagent, has been used. Recently, however, pyridine-free reagents have become available. These reagents use organic amines that have much less objectionable odors. A complexing base is necessary, however, because both I_2 and SO_2 must be strongly complexed so that their concentration remains constant with time. Amperometric detection is used almost exclusively for Karl Fischer titrations, although potentiometric and colorimetric detection can be used.

Automatic Karl Fischer titrators are available from Photovolt, Fisher Scientific, Mettler, Baird & Tatlock, Precision Scientific Group, and Quintel. Zymark has produced a system in which they provide a robot for automatic sample introduction into the Mettler reaction cell.

Halogen Titrators

Halogen titrators can be divided into two types: those used industrially and commercially for monitoring aqueous samples such as water from treatment plants or swimming pools, and those used for the analysis of halogens contained in organic materials. Two manufacturers, Fischer & Porter and Pennwalt, provide virtually identical instruments for the analysis of aqueous samples. Both systems deliver the titrant volumetrically and detect the end point amperometrically, and both are able to measure halogens to the 0.01 mg/liter (0.01 ppm) level. In both cases the free halogen dissolved in water is reduced to the halide ion by an appropriate reagent.

The other method, used for halogens in organic compounds, is provided by the Dohrmann Division of Xertex. Halogens in organic materials are usually covalently bound to the organic compounds. Two types of analyses are used. The first is total organic halogen (TOX), and the other is extractable organic halogen (EOX). The two methods differ primarily in the sampling techniques used. TOX involves the pyrolysis of the entire sample, converting it into simple gases that are adsorbed onto granulated activated carbon. EOX involves first extracting the sample with a reagent prior to pyrolysis so that only the extractable halogens are adsorbed onto the activated carbon. A third type of analysis, purgeable organic halogen (POX), involves purging the lower halogenated hydrocarbons from water by passing a carrier gas through the water sample and carrying the halogen compounds to the adsorber.

The activated carbon containing the halogen samples is then burned and the product gases are passed into a cell where silver ion is coulometrically generated to react with the halide ion. The silver ion concentration is monitored and maintained at a constant concentration. The amount of electricity (coulombs) necessary to achieve this reflects the amount of halide present in the sample.

The sensitivity of the TOX method is from 0 to 1000 μg/liter and uses samples in the 10-125 ml size range. It requires about 30 min for the adsorption and about 10 min more for the analysis. EOX is usually performed on solid or liquid samples of 2-30 μl containing extractable halogens in the 25-25,000 ng range with an analysis time of about 5 min. The interference rejection ratio in TOX is 50,000:1 for inorganic halide and residual chlorine and 5000:1 for chloramines, while no known interferences occur for EOX. The sample size for POX is about 10 ml and the method can handle a halogen range of 0-1000 μg/liter with a precision of about 2%. The sample time is about 10 min.

Sulfur Titrators

Sulfur titrators are also distinguished by two types: those oriented toward the sulfur analysis in general organic compounds or mixtures such as coal, petroleum, or other organic materials, and those designed for the analysis of specialized compounds such as those used to odorize natural gas.

Two companies, the Dohrmann Division of Xertex and Parr Instrument Co., provide general-purpose sulfur titrators. The methods used by them are decidedly different. The Dohrmann method generates SO_2 from the organic material by controlled oxidation. The resulting SO_2 is titrated by coulometrically generated I_2. The Parr method generates sulfate ion by complete bomb oxidation of the organic material. The resulting SO_4^{-2} is volumetrically titrated by lead perchlorate.

The Dohrmann method maintains a constant concentration of I_2 by regenerating it as it is consumed by the SO_2 passing into the solution. The method has a very large dynamic range of from 0.1 to 10,000 ppm in which the sample size is from 5 to 8 μl if the concentration of sulfur is greater than 2 ppm and from 30 to 40 μl if the concentration is less than 2 ppm. The instrument is microprocessor controlled and requires oxygen and nitrogen gases.

The Parr instrument is a volumetric titrator using lead perchlorate as titrant. The end point is detected potentiometrically with a lead-specific ion electrode. The method can analyze sulfur from 0.01% to 5% from a 1-g sample. The instrument can be interfaced to a balance as well as to an oxygen bomb calorimeter and, using the RS-232C serial interface, it can be connected to a printer or computer for data analysis. On the other hand, the pertinent data can be entered via the keyboard. After combustion, the sulfur analysis can be done in less than 5 min.

ITT Corporation provides a titrator for the analysis of odor-producing sulfur compounds that are added to natural gas. Their system uses coulometrically generated bromine to analyze directly a variety of sulfur-containing compounds including SO_2, H_2S, RSH, RSR, and RSSR. The gas containing the sulfur compound is passed through an electrolytic solution containing bromide. The amount of sulfur in the gas is proportional to the current necessary to maintain a constant concentration of bromine. The dynamic range of the instrument is from 1 to 1000 ppm as H_2S. It has a voltage output that can be attached to a strip-chart recorder for continuous monitoring of a gas stream.

Calcium Titrator

Precision Systems, Inc., manufactures a calcium titrator containing an automatic sampler in which predetermined volumes, ranging from 20 to 80 μl, can be aspirated into the sample chamber and then titrated with EGTA [ethyleneglycol bis(beta-aminoethylether)-*N*,*N*'-tetraacetic acid]. The detection technique involves a fluorometer. The indicator is a fluorescing complex of calcium-calcein. When the calcium is added to the calcein, the fluorescence is enhanced and is then quenched by the addition of the EGTA. The amount of EGTA needed to return the fluorescence to the level prior to the calcium addition is then directly related to the quantity of calcium in the sample. The method will measure concentrations from 0.5 to 70 mg/liter. Up to 30 samples can be titrated with each filling of the cuvette with reagent. Light modulation increases fluorometric stability, and narrow-band interference filters minimize the fluorescence interference from other biological fluids. This permits the instrument to be used both for aqueous solutions and clinical samples. Each analysis requires only 1 min, with the results reported as a digital readout on the instrument.

V. CONCLUSION

Automatic titration systems have been commercially available for several decades. The most recent innovations have involved the incorporation of microprocessors and computers into the system. As a consequence, automatic titration equipment and instrumentation is a mature field with incremental improvements rather than the revolutionary advances taking place in other fields. A mature field also has many manufacturers, so the competition among them is quite intense and the distinguishing characteristics are quite small. The primary decisions in choosing a titration system involve detection method and whether to build from modules or to buy a complete system. If the purpose of the system is quality control, the decision becomes whether to buy a dedicated system for use in analyzing a particular chemical species or general-purpose apparatus that may have applications beyond the specific analytical problem at hand.

Titration instruments from selected manufacturers have been discussed earlier. A list of manufacturers and distributors will not be included for several reasons: with a mature field such as this, it is very difficult to form a complete list of manufacturers, most manufacturers advertise widely in commonly read journals, and most manufacturers are listed in the *Thomas Register* as well as in the buyers' guides associated with such publications as *American Laboratory, Research and Development, Analytical Chemistry,* and *Science.*

REFERENCE

G. W. Ewing, *Instrumental Methods of Chemical Analysis*, 5th ed., McGraw-Hill, New York, 1985.

27
Continuous-Flow Analyzers

CHARLES J. PATTON* / Alpkem Corporation, Clackamas, Oregon

ADRIAN P. WADE / Department of Chemistry, University of British Columbia, Vancouver, British Columbia, Canada

"Continuous-flow analysis: just another sample down the tubes?"
A.P. Wade, 1986

I. PREFACE

Our aim in preparing this contribution is to indicate the diversity of real-world problems already solvable by continuous flow (CF) analyzers. We hope that you will be sufficiently encouraged to try some of the techniques presented and referred to here for yourself. The fact that this chapter is intentionally highly practical in nature should come as no surprise, since the vast majority of the papers published in this field have been "applications driven." That is to say, they have reported automation of tried and trusted manual wet chemical analyses using CF techniques. We include examples of actual methods and uses and have tried not to assume any prior knowledge of the principles or use of continuous-flow analyzers. Detailed theoretical treatments are kept to a minimum and, wherever appropriate, references for further reading and possible sources of materials are indicated. A guide to equipment suppliers and commercially available apparatus is also included. All companies we contacted have sent us material. We apologize to any firms we may have missed.

II. INTRODUCTION

Managers and senior scientists need strong reasons for any change to a working laboratory routine. They must be able to explain to senior management why the manual methods of analysis that the company has previously relied on are no longer "the best," and why more modern, automated alternatives such as continuous-flow methods are a better option. They must also be able to justify the expense of new equipment and staff training. Continuous-flow methodologies offer a number of significant

*Current affiliation: U.S. Geological Survey, Arvada, Colorado.

advantages. The level of automation by CF seen in clinical laboratories, as illustrated by the success of Technicon's various CF analyzers, is a sure testimony to this. Such automated methods of analysis are also gaining wider official acceptance, as shown by procedures now appearing in United States Geological Survey (USGS) and Environmental Protection Agency (EPA) manuals. There also now exist American Standard Methods (ASTM) and articles in the *Journal of Official Analytical Chemists* that describe CF procedures. Some larger industrial companies now maintain a laboratory for developing CF methods. Such methods are often (but not always) developed on research-grade equipment and may then be transferred to a standard, less expensive, commercially available apparatus or a "home-built" analyzer.

Continuous-flow methods enable wet chemical analyses to be carried out in an automated manner. The chemistries involved usually take place within flow systems made from narrow-bore tubing. A rudimentary CF system is shown in Figure 1 and has three basic components. These are (i) some means of sample introduction; (ii) some fluid transport mechanism to convey the sample to the detector, which may include chemical reaction(s); and (iii) a detector to indicate the concentration or other analytical value for the product(s), or for the disappearance of sample or reagent. This chapter will discuss, primarily, the two main classes of CF methods; these are air-segmented continuous-flow analysis (CFA) [C1] and flow injection analysis (FIA) [F1]. In addition to these, there exist both nonsegmented and segmented postcolumn reactors for liquid chromatography [F2,F3], respectively, which can also be roughly classified as FIA and CFA devices. The FIA classification is most appropriate in the few applications where the eluent from the column is sampled by injection into a separate flowing stream [F4]. Other nonsegmented mixing systems without injection have also been used to determine reaction stoichiometries [F5].

Continuous-flow analyzers can be considered as "conveyor belt" systems for wet chemical analysis. They share many of the throughput advantages of the "production line" approach to manufacturing. In industrial, clinical, and other applications, CF methods have replaced previous manual methods [F6,C2] and have also facilitated the automation of some sample pretreatment procedures, dilution and reagent solution mixing, that previously were carried out manually. Improved repeatability, lower reagent and/or sample consumption [F7], smaller sample sizes, and faster analyses have been reported [F1,F6,C2,C3]. Typical CF methods are capable of processing 60-200 samples per hour [F1,C4]. For simple systems, rates as high as 360 samples per hour are possible [C5]. The tubing used has an internal diameter (i.d.) of 2 mm or less for CFA (earliest systems were 3 mm) and 1 mm or less (typically 0.8 or 0.5 mm) for FIA. A few tens of microliters of sample may be all that is required per analysis, and thus considerable savings in costs may be possible. Where even the typical throughput rates are achieved, the time required for a sample to pass through a CF system is unlikely to be the rate-limiting step for most analytical methods; sampling and some sample pretreatment will almost certainly take longer. Sometimes, as a direct result of the highly repeatable timing of CF systems, the actual chemistries can be simplified [F6,F8].

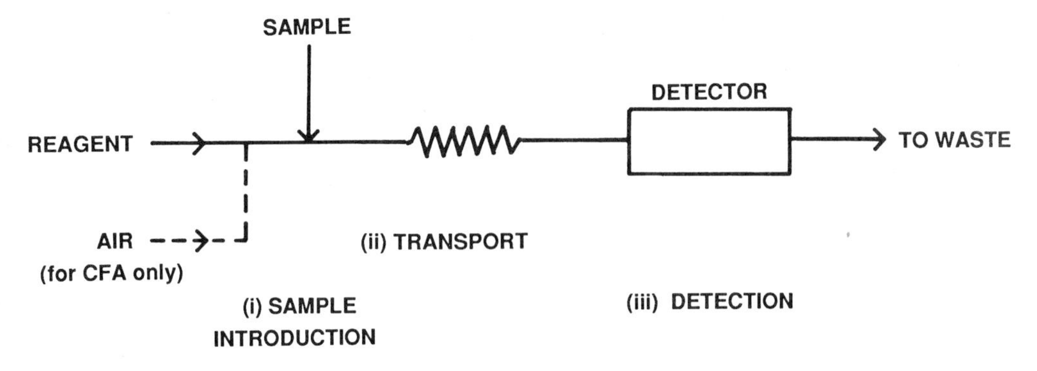

Figure 1. Schematic diagram of a continuous-flow analyzer.

Reproducible measurements can also be made before chemical reaction is complete [C6,C7]. The decreased number of solution manipulations that operators undertake minimizes labor costs and results in improved safety in the work place. Complex chemistries using toxic [F9], radioactive, or otherwise dangerous [F10] chemicals can often be carried out automatically in sealed systems. Since solution volumes are typically less than in manual analyses, disposal of chemical waste is less of a problem. Unstable reagents can be made *in situ*. Kinetic measurements are also possible [F11-F13,C2]. CF systems in process control applications facilitate on-line measurements at remote sites, 24 hours per day and without a full-time technician. Some continuous flow systems have unique capabilities that allow the design of dynamic experiments for which there are no manual analogs. All these points will be discussed in what follows.

A. Comparison of CF Methods with Manual Procedures

A simple manual colorimetric method involves addition of reagent(s) and buffer to a measured volume of analyte solution to form a colored species, which may then be observed and quantified in a spectrophotometer. Because this typically requires several manual steps, an analysis can take several minutes. Time spent cleaning the apparatus in preparation for the next experiment is an additional overhead. Complex manual methods may also include filtration and one or more solvent extraction stages. Continuous-flow methods have equivalents to each of the above operations, only many of them are automated.

On simple FIA systems a six-port, low-pressure valve is used for sample "injection." The operation of this is shown in Figure 2. When in the "fill" position, sample (analyte) solution is introduced into a fixed volume loop. This can be achieved manually using a syringe [F6] or in a more automated manner by a pump [F14]. Sufficient solution is introduced to completely displace the previous contents of the loop. This volume should be 2.5-5 times the volume of the loop. During this phase, the

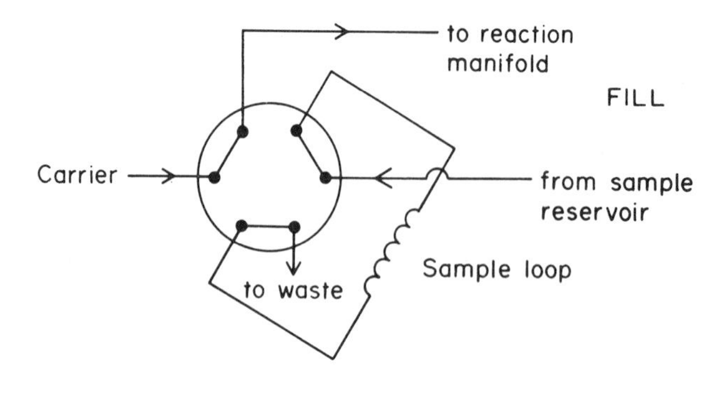

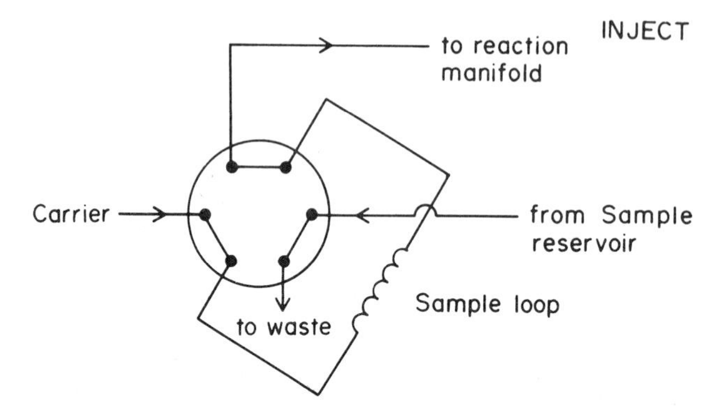

Figure 2. Internal workings of a six-port injection valve.

carrier stream passes directly on to the reaction manifold. Then, on switching the valve to the "inject" position, the carrier stream sweeps out the contents of the loop, transporting the sample downstream. The valve can then be returned to the fill position, ready for the next sample. Since the loop volume is fixed, no other measure of aliquot size is necessary. An alternative is to use two four-port valves as shown in Figure 3. An article reporting several ways of using a very versatile 12-port valve has recently been published [F15]. Valve operation may be automated using an actuator powered from compressed air or electrically. Sample introduction may be further automated by means of an "autosampler." This is a tray holding many small vials that contain samples, standards, or blanks, and that may be randomly accessed by an arm through which solution is sucked up.

On injection the sample is entrained in a narrow-bore tube and is propelled from its introduction point toward the detector by some carrier stream. Along the way, streams containing reagent solutions (and air, for CFA) merge and mix with the analyte and carrier in the reaction manifold or "analytical cartridge" (Figure 1). The amounts of the reagents "added" in this manner are determined solely by reservoir concentrations and relative flow rates. Passage of the sample through the manifold may include a separation step, such as solvent extraction or gas diffusion.

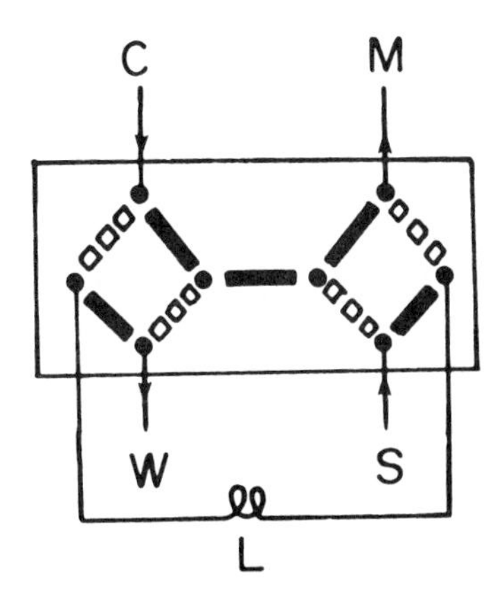

Figure 3. Internal workings of a dual four-port injection valve.

Detectors used in CF systems utilize flow-through cells (Figure 4). Since the flow is continuous, these cells (and indeed the whole apparatus) can be thought of as "self-cleaning." Some manufacturers make flow-through cells to fit existing manual spectrometers; thus conversion to a CF method need not entail the purchase of new detectors. Very inexpensive detectors that give more than adequate performance have been designed specifically for CF systems [F16-F18]. The signal observed by the detector is a transient, rather than steady state, and thus a chart recorder is the most common form of output device. Computerized data acquisition systems are becoming more common and more affordable. Eventually they will undoubtedly replace the chart recorder for many applications. An inexpensive digital readout has also been designed that may replace the chart recorder in some teaching laboratories and will allow FIA systems to be more widely used [F19].

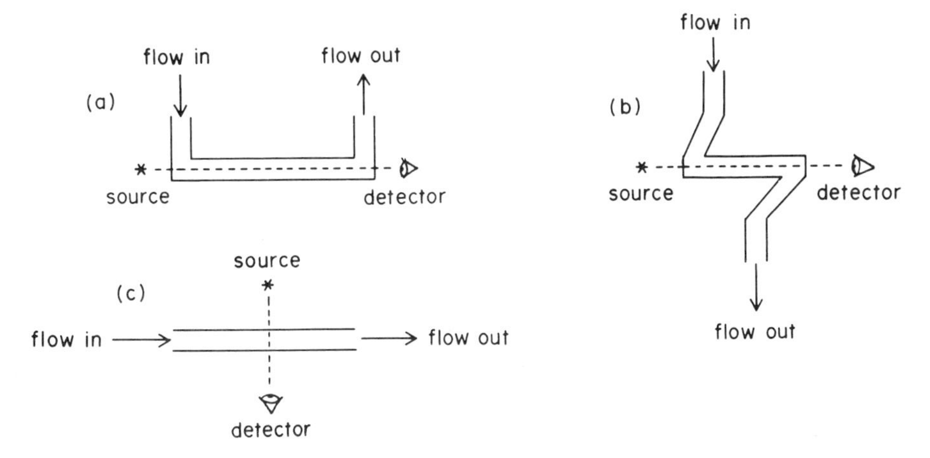

Figure 4. Various flow-cell designs for spectrophotometric measurements.

While improved sample handling and throughput capabilities of CF systems are important, they are seldom the only factors that must be considered. The repeatability of analyses may be equally or more important. Here also, CF methods score well. Since the internal geometry of any CF system is fixed, and time allowed for reaction is exactly determined by flow rate and the physical dimensions of the system, highly repeatable analyses can be obtained. In practice, replicate determinations with relative standard deviations of better than 1% are common. Methods with repeatabilities greater than 5% RSD are seldom published, although for some applications these would be quite adequate. Precision of analyses is typically as good as for manual methods, and may be better [F6,C8]. One would certainly expect them to be less prone to operator error. All in all, CF methods compare very favorably with manual wet chemical analyses.

B. Cost of Manual versus CF Methods

While the capital outlay for CF systems can be substantial, the payback can be rapid and is readily calculable as shown below. The various sources of expense should be considered for both the existing manual analysis and the continuous-flow alternative.

The total cost per analysis ($C_A = C_C + C_H + C_O + C_E + C_I$) may be calculated using the quantities defined below. The total chemicals cost per analysis, C_C equals the cost per liter of sample (including collection and pretreatment) multiplied by the volume of sample per analysis, plus the cost per liter of reagent(s), buffers, etc. (including time spent preparing solutions) multiplied by the volume per analysis. The human cost, C_H, is the cost per man hour including site overheads (C_M) multiplied by the time (in hours) per analysis (T_A) and by the fraction of time that the operator needs to be present during an analysis (F_T). The overhead, C_O, equals the C_M multiplied by the time (in hours) spent in cleaning and maintaining equipment (T_C) divided by the average number of samples processed per cleaning or maintenance session. The cost of expendables per analysis, C_E, should include sample cuvettes, pump tubing, etc. The initial cost, C_I, is the cost of analytical equipment divided by the anticipated number of samples to be processed during payback period for the equipment, plus the cost of replacing or servicing broken equipment (include glassware costs).

Since the volumes per analysis are likely to be smaller, C_C will usually be lower for CF. Preparation time for stock solutions will be about the same for both; however, for CF the mixing of reagent solutions will be automated. The term C_M may be lower for a CF method since, once the automated analysis is set up, the operator need not be as skilled an analyst. Also T_A will be lower and, for highly automated CF systems, F_T. Since a CF apparatus is essentially self-cleaning (it must still be rinsed through after operation) and the same vessel (i.e., the flow system) is used for each analysis, C_O should be lower for the CF system. The cost of a simple but durable "build-it-yourself" FIA system is not that much greater than that of the glassware for the manual analysis. Commercial CF analyzers are not inexpensive, but are a one-time expense. If the payback period is chosen sensibly, the CF method should appear financially advantageous in most instances. Since CF analyzers are rugged and essentially stay in one place, the likelihood of equipment breakage in routine use should be less for CF than for manual analyses.

C. Should One Purchase or Construct CF Instrumentation?

At the time of this writing, the authors have no affiliation to, or sponsorship from, any instrument company; this paragraph represents (we trust) our unbiased opinion. We consider it advisable to purchase commercially available instrumentation for routine analytical work wherever this is possible, but add the following caveat to this statement. CFA systems are naturally more complex than FIA systems, owing to the bubbles. Companies such as Technicon have been perfecting CFA systems for many years. It would be unreasonable to expect a laboratory analytical chemist without a great deal of experience to duplicate or exceed the specifications of a commercial CFA system within a reasonable time period. When the expense of man hours for developing instrumentation and any necessary operating software is taken into account, costs are likely to be comparable. However, for some simple single-channel FIA systems, the do-it-yourself approach may be quite viable and will save money. This reflects the fact that no company has so far produced a simple inexpensive FIA system. We hope that this will change.

Our recommendations are somewhat different for apparatus that is needed for research and methods development work. Development strategies for CF methods are discussed in a later section. It has been our experience that some commercial equipment is very good at doing what the manufacturer wants you to do with it, but can be totally ill-suited to the needs of a researcher. Essentially there are three paths around this problem: (i) You can pay an instrument company to develop your methods for you. They may do a better job than would be possible "in-house" since they will employ CF specialists and have research grade (i.e., more versatile and expensive) equipment available to them. They will then provide a developed analytical procedure to run on their hardware in your laboratory. (ii) You can choose commercially available equipment that either is versatile enough for research and development or can be modified for these purposes. This requires a degree of foresight, since the purchaser must anticipate all potential uses at the time of purchase. (iii) You can build your own research and development equipment, using some commercially available components. This is the approach that we, the authors, have taken. However, we do not recommend this to those who are not skilled in CF methods, instrumentation design, computer interfacing, and software, or to managers who do not have access to people with these skills.

D. Applications to Process Control/Continuous Monitoring

CF analyzers for process control require quite different characteristics from research instruments. They should be dedicated to a specific task and designed with this in mind. They should be robust, have the minimum number of moving parts, and require virtually no manual operations. The number of reagent and waste solutions, gas bottles, etc. that have to be changed or parts serviced should be minimal. Safety guidelines in force at the intended site of operation must be adhered to. This will require special design care and possibly certification. In extreme or remote environments, equipment may have to be battery powered or at least have a battery back-up facility to protect it from power outages. The long-term (days) reliability of unsupervised peristaltic pumps is open to question. Thus alternative means of solution propulsion should, where possible, be employed [F20]. One such is to use gas from cylinders to displace solutions from reservoirs and into the manifold. On exposed sites, potential problems due to temperature variation must be assessed. The skill level expected from a plant operator is less than that required of a skilled analytical chemist. Programs must be set up to train operators in routine analyzer maintenance. Since all these factors must be considered, it is little wonder that so few CF process control applications exist. Perusal of the Technicon Symposia proceedings [C38] between the years of 1965 and 1976 reveals over 70 articles describing on-line CFA monitors for applications that include pharmaceutical and metallurgical process control, *in vivo* analysis, and quality control monitoring of power plant boiler water, radioactive wastes, municipal, potable and wastewaters, and industrial effluents. Ruzicka estimated the number of FIA process monitors in use in 1986 at about 100 [F21]. Clearly we have a long way to go. However, when one recalls what can already be done, and that for FIA this progress has been made in 15 years, the outlook appears excellent.

Where process control by CF methods has been attempted, this has often been by the unsatisfactory route of taking a bench-top instrument, putting it into a box, and calling it a process instrument. CF process control analyzers have often been one-of-a-kind hand-crafted affairs. Ideally one would like to be able to purchase industrial standard equipment that has a proven safety and performance record. However, very few commercial systems have been designed from the ground up with process control in mind. In a recent review article in *Analytical Chemistry*, Callis et al. [F22] indicate that FIA is likely to play a much more prominent role in process analytical chemistry over the next few years and note that "process engineers will find the approach (FIA) particularly attractive because it so closely resembles the flow-stream concept used in automated chemical processing lines."

What is needed to design CF process control analyzers is what has been termed a "washing machine" philosophy. The domestic washing machine is a marvel: it is expected to withstand high stress and vibration, the effects of liquids and chemicals, and must work in a potentially hostile environment. In addition, it is operated by people without high technical ability and who have had minimal user training (only a brief instruction booklet). If it breaks down once in 3 years people complain; many machines still in use are 20 years old. If only analytical instrumentation were generally made to that standard—for reliable process control applications it has to be. The washing machine became that reliable partly because of the size of the market. Both CFA (Scientific Instruments and Technicon Industrial Systems) and FIA (Fiatron and Lachat) process monitors are available commercially. We are not in a position to judge the extent to which these systems meet the "washing machine" benchmark.

III. GAS-SEGMENTED CONTINUOUS-FLOW ANALYSIS (CFA)

The emphasis of this chapter is on applications, rather than history or background; thus this section (and the following section on FIA) is set in the order (A) principles, (B) an example system, (C) applications, (D) history, and finally (E) a brief discussion of the underlying theory. The reader is referred to supporting texts throughout.

A. Principles of CFA

The spatial arrangement of hardware required for a generalized bench-top CFA system is shown schematically in Figure 5. Further details are provided in the figure caption. The analytical cartridge (the portion of a CFA system between the pump and the detector) is represented by a mixing coil in this figure, but a wide variety of other components such as reagent addition tees, debubblers, dialyzers, gas diffusion cells, thermostated reaction coils, distillation heads, and phase separators are common in CFA applications. Furthermore, although a photometric detector is illustrated here, many other detector types have been successfully interfaced to CFA analytical cartridges.

In the discrete sampling mode, samples contained in disposable cups are loaded onto some type of automatic sampler with two generic mechanisms,

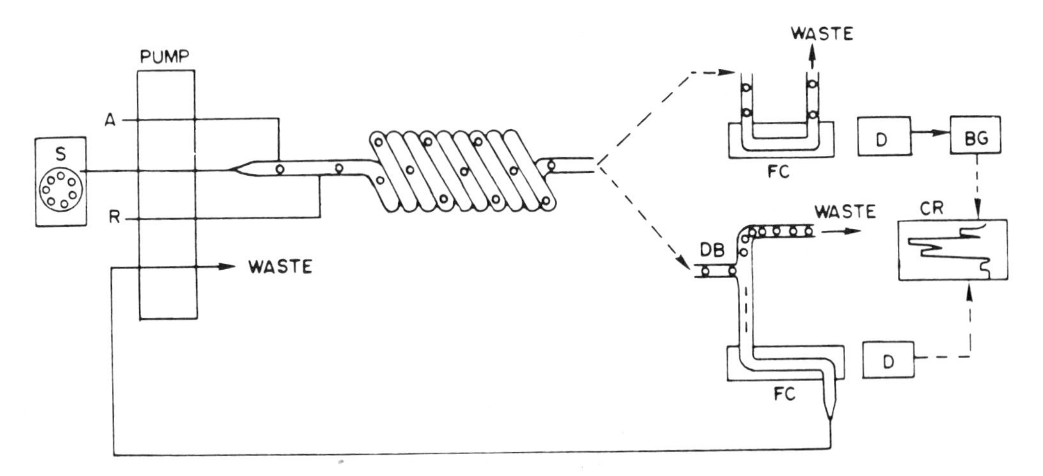

Figure 5. Analytical cartridge diagram for a generalized CFA instrument: S, sampler or sample; A, air; R, reagent; DB, debubbler; D, detector; FC, flow cell; BG, bubble-gate; CR, chart recorder.

one of which mvoes the sample withdrawal tube between the wash solution reservoir and the sample cups, while the other sequentially moves the cups positioned on it into the sampling position. The analyte-free wash solution should closely match the sample matrix. Other modes of sample introduction such as injection [C144] (usually associated with FIA) or continuous aspiration are also possible, and for some applications it may be expedient to reverse the role of sample cups and the wash reservoir, or even to omit the sampler altogether. The sample withdrawal tube connects to one of the peristaltic pump channels, itself a tube, which in turn connects directly to the analytical cartridge.

Samples and the wash solution are pumped into the analytical cartridge as discrete slugs that are separated by one or more air bubbles that form at the tip of the sample withdrawal tube each time it is exposed to the atmosphere. The introduction of several intersample air bubbles ("pecked sampling") at the beginning of each sample and wash cycle significantly reduces the interaction of slugs that would otherwise occur between the sampler and the analytical cartridge (see Figure 17). As each slug enters the analytical cartridge it is divided into many smaller, nominally identical subunits with bubbles of an immiscible fluid that is usually, but not necessarily, air. Alternatively, sample and wash slugs may be proportioned into an air-segmented diluent or reagent stream. In well-designed systems, the segmentizing gas is not added to the analytical cartridge continuously. Instead, it is admitted in short bursts, which are phased with pump pulsations. This practice minimizes proportioning errors that would otherwise occur [C2]. At this point a small portion of each slug is proportioned into many analytical stream segments, the actual number of which is simply the product of the sample or wash time and the segmentation frequency. Chemical reactions and separations are effected in appropriate sections of the analytical cartridge as the segmented stream flows in box-car fashion toward a recording detector.

Detection in CFA is complicated by the presence of gas bubbles, which are compressible, highly reflective, and electrically nonconductive. Because of these characteristics, their passage through most detectors is marked by artifacts that severely distort the analytical signal. Provided that the response time of a detector is sufficiently fast and the volume of its measurement cell is less than that of an individual liquid segment, it is often possible to remove such artifacts in real time by analog or digital signal-processing techniques known as *bubble-gating*, a term coined by Habig et al. [C9]. An obvious and frequently used alternative, of course, is simply to debubble the analytical stream just prior to detection (see Figure 5), but this expedient results in pooling of previously segregated liquid segments, which increases interaction and thus decreases the rate at which samples can be processed (see Figure 6a). Real-time analog or digital data reconstruction techniques, known as *curve regeneration* as first described by Walker [C10], can remove the effects of pooling that occur at flow cell debubblers and any other unsegmented zones of the system from recorded data. This permits sampling rates that approach theoretical maxima to be achieved with either debubbled or bubble-through detection (see Figure 6b). Specific details of bubble-gating and curve regeneration techniques, a discussion of their relative merits, and additional references can be found in Section VI.

Sample dispersion in CFA ultimately determines the rates of analysis and the accuracy and precision of results that can be achieved with a given system under a specific set of experimental conditions. For this reason CFA practitioners, theoreticians, and instrument manufacturers alike have expended much effort over the years to find its origins and quantitate its effects, which are referred to collectively and interchangeably as *interaction, carryover,* or *loss of wash*. One of the first ways, and still the easiest way, to measure the extent of carryover experimentally is with an *interaction test pattern* as first described by Thiers and Oglesby [C11] and shown in the rightmost group of peaks in each half of Figure 6a. Here two nominally identical low-concentration samples and one high-concentration sample are determined in the concentration sequence low (L_1), high (H), low (L_2). Percent interaction ($\%I$) is then defined as the fractional contribution made by H to the peak height recorded for L_2 as expressed by

$$\%I = 100(L_2 - L_1)/H$$

In more fundamental treatments, loss of wash is traced to two distinct processes, which can be separated and quantitated individually. The first, *longitudinal dispersion* [C12,C13], occurs in unsegmented zones of a system—that is, sample lines, flow-cell debubblers, flow cells. Here flow is essentially laminar and, as a result, recorded signals are deformed *exponentially* relative to the ideal square-wave output function that would be predicted in the absence of dispersion. The magnitude of this deformation, in units of time, can be expressed either as the *half wash time*, $W_{1/2}$, first described by Thiers et al. [C12], or as the analogous *exponential factor*, b ($b = 0.69W_{1/2}$), later derived by Walker et al. [C13]. For CFA systems that lack pecking samplers and bubble-through detectors, longitudinal dispersion is dominant, and once the value of b is extracted from the rise or fall curves of a particular instrument, its analytical performance at any chosen values of sample and wash time can be calculated from the (approximately) exponential relationship between absorbance and time [C14]. For example, peak heights within 90%, 95%, or 99% of steady state result when the sample time equals $2.3b$, $3.0b$, or $4.6b$, respectively; unless the wash time equals at least $0.7b$, $1.6b$, or $2.3b$, the second of two peaks observed for successive samples with concentration ratios of 2:1, 5:1, or 10:1, respectively, will appear as an ill-defined shoulder; and interaction of 7%, 1%, or 0.5% is predicted for cycle times (sample time plus wash time) of $2.65b$, $4.6b$, or $5.3b$, respectively. Note that all postprocess carryover correction schemes [C11,C15,C16] as well as curve regeneration algorithms [C17-C19] depend either explicitly or implicitly on a known value for b.

The second contributor to loss of wash is *axial* or *lag phase dispersion* [C20-C23], which arises from mass transfer between liquid segments via a stagnant wall film that wets the inner surfaces of the conduits through which they flow. Because gas-segmented streams depend on wetted surfaces for hydraulic stability [C2,C24], axial dispersion is fundamental to CFA as it is currently practiced. Its magnitude is expressed as the standard deviation (in units of time), σ_t, of the cumulative Gaussian peaks that are predicted in the absence of longitudinal dispersion (See Section III.E).

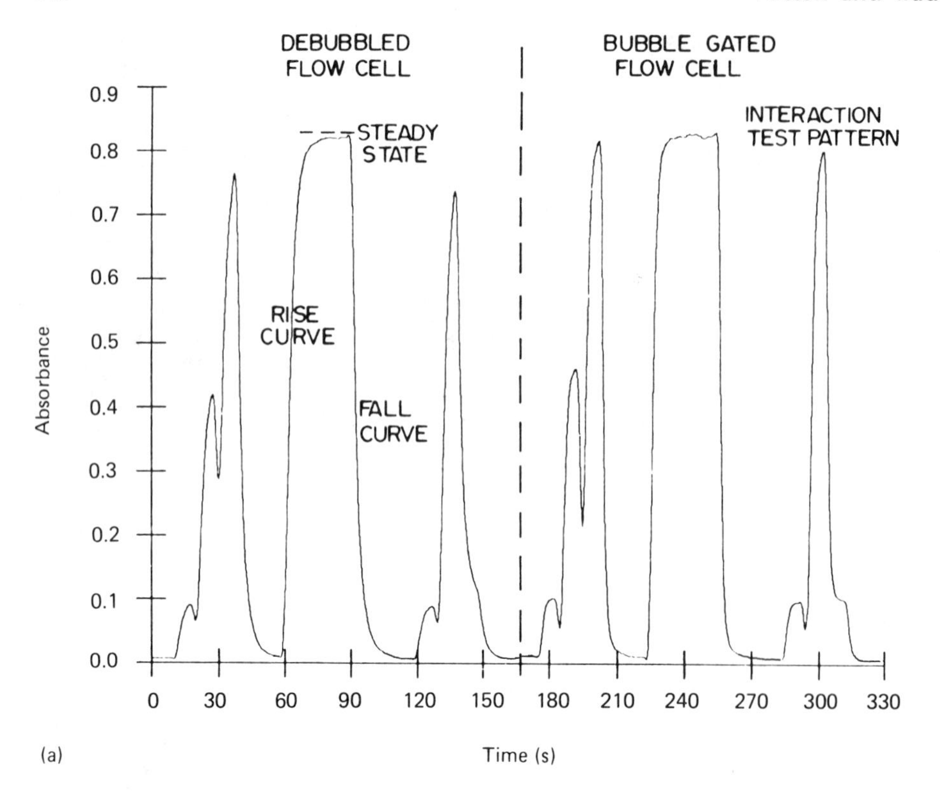
DEBUBBLED FLOW CELL
BUBBLE GATED FLOW CELL
STEADY STATE
INTERACTION TEST PATTERN
RISE CURVE
FALL CURVE
Absorbance
0.9
0.8
0.7
0.6
0.5
0.4
0.3
0.2
0.1
0.0
0
30
60
90
120
150
180
210
240
270
300
330
Time (s)

(a)

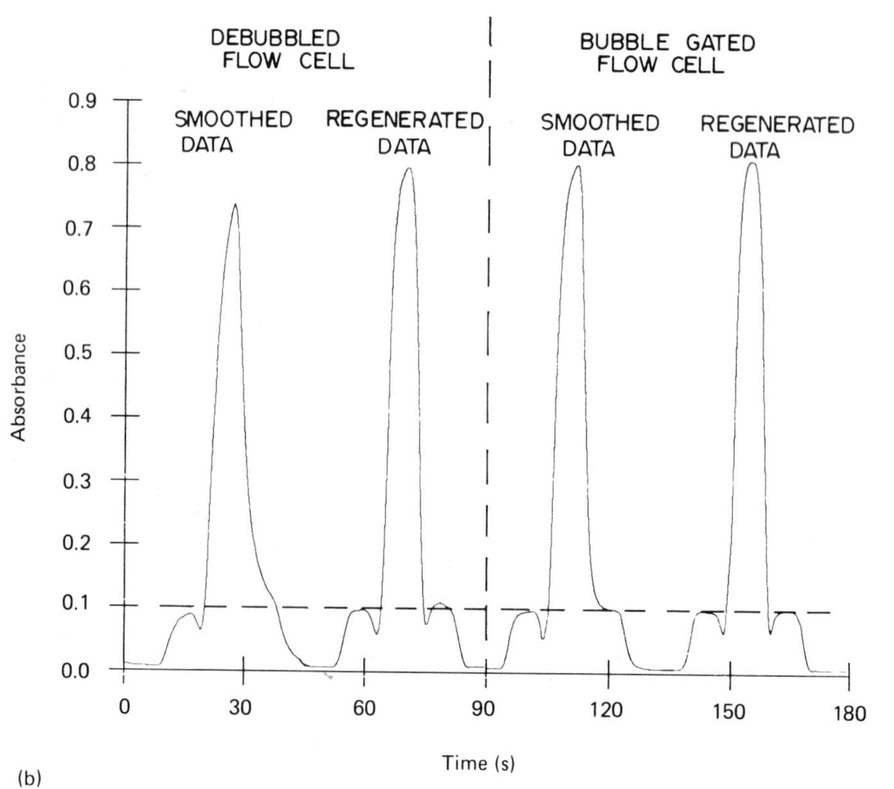
DEBUBBLED FLOW CELL
BUBBLE GATED FLOW CELL
SMOOTHED DATA
REGENERATED DATA
SMOOTHED DATA
REGENERATED DATA
Absorbance
0.9
0.8
0.7
0.6
0.5
0.4
0.3
0.2
0.1
0.0
0
30
60
90
120
150
180
Time (s)

(b)

Such peaks are never observed even in well-designed systems, however, unless the detector output is curve regenerated as it was for two of the interaction test patterns shown in Figure 6b. Only after the exponential deformation is removed from the data can σ_t be estimated, although it can be calculated a priori from experimental variables using the Snyder/Adler model [C25,C26]. When lag phase dispersion is dominant, cycle times of $4\sigma_t$ and $8\sigma_t$ are required to generate sharp peaks or peaks with flats [C4,C27]. Note that the experimentally determined value of σ_t for the curve regenerated peaks shown in Figure 6b is 1.3 sec. This suggests that a cycle time of about 10 sec should be sufficient to produce peaks with flats, as indeed is the case. The complete data sets from which Figures 6a and 6b were taken can be found elsewhere [C28].

B. An Example CFA System

Nitrate is one of the most commonly determined anions in water, seawater, wastewater, foodstuffs, and plant and soil extracts. In many automated procedures, cadmium metal is used to reduce nitrate to nitrite [C29], which is subsequently diazotized with sulfanilamide and then coupled with *N*-(1-naphthyl) ethylenediamine to form an azo dye [C30]. The dye formed has an absorbance maximum at 543 nm with a molar absorptivity of 4×10^4 liter mol^{-1} cm^{-1} [C31]. In the example presented here, an open tubular cadmium reactor as first described by Patton [C28] is used in conjunction with a commercially available third-generation CFA instrument to determine nitrate in water in the concentration range of 0.02-1.00 ppm. A diagram of the analytical cartridge is shown in Figure 7, and peaks from typical runs determined at rates of 120 h^{-1}, and 240 h^{-1} are shown in Figure 8a and 8b, respectively. The peaks shown from left to right represent two replicates of the 1.0 ppm calibrant, a blank, 0.1, 0.3, 0.5, 0.7, 0.9, and 1.0 ppm calibrants run in ascending and descending order, a blank, an interaction test pattern, and five replicates each of the 0.5 ppm nitrate calibrant, the 0.5 ppm nitrite calibrant, and an EPA check standard with a certified concentration of 0.18 ± 0.02 ppm. A blank precedes each of the three sets of replicates. Note that carryover in the run performed at a rate of 240 samples h^{-1} was minimized by digital curve regeneration as described in Section VI.E. The dwell time of this system is approximately 200 sec, and the flow rate of the analytical stream is approximately 0.8 ml min^{-1}. Analytical parameters and figures of merit for these data

---

Figure 6. Signals recorded when the *same* analytical stream passed first through a bubble-through flow cell and then through a flow cell equipped with a debubbler. Both flow cells had a path length of 1.0 cm and an internal volume of 2 μl. Samples were aqueous phenol red solutions, and reagent was 0.1 *M* borate buffer (pH = 10) containing surfactant. (a) Sample and wash times were 8 and 2 sec, respectively. The sample time for the central steady state peak was 30 sec. (b) Expanded-scale plots of smoothed and curve-regenerated interaction test patterns obtained from data recorded for the bubble-through and debubbled flow cells. See text for additional details.

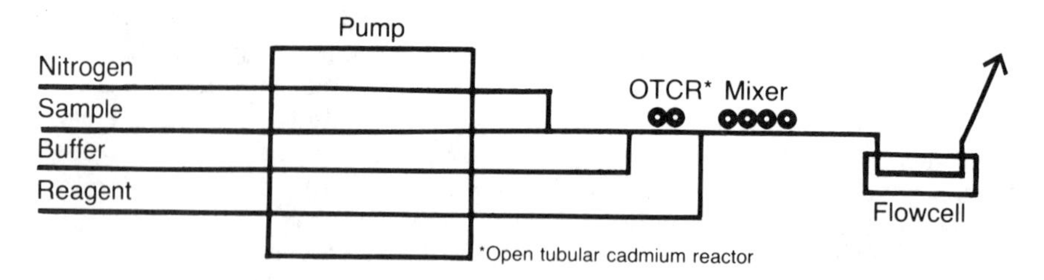

Figure 7. Analytical cartridge diagram for the determination of nitrate in water with the RFA-300 CFA system. (Courtesy of Alpkem Corporation.)

can be found in Table 1. This basic procedure, often with the addition of a dialyzer, has been adapted for the determination of nitrate in a wide range of sample matrices that include seawater, tobacco, cotton petiole, and soil extracts, pineapple juice, and pharmaceutical preparations. This example was chosen to illustrate that with modern hardware and data-processing techniques, rates of analysis much faster than those generally associated with CFA are readily achieved.

C. CFA: Key Literature and Applications for CFA

When FIA made its debut in 1975 [F24] CFA was already a mature and widely applied technology: the theoretical basis of its operation was completely understood; hardware required for its implementation was commercially available and, for the most part, perfected; and its bibliography contained almost 8000 references [C32]. Since 1975, however, literature pertaining to CFA has become more stagnant while that for FIA has proliferated. As of late 1985 FIA had nearly 1000 references to its credit [F21], with new ones appearing at a rate of one every other day [F35]. Readers whose interest in CF techniques is relatively new, therefore, may be unaware of the rich and extensive literature associated with CFA, much of which is still relevant to CFA and FIA practitioners alike.

The best entry point to the CFA technique is still William Furman's *Continuous Flow Analysis: Theory and Practice* [C33]. This volume includes almost 1000 references and presents a lucid and in depth review of the state-of-the-art in CFA up to mid 1973. Despite improvements in commercially available hardware that have occurred since that time, this book is still highly relevant. Also worth seeking out are Technicon's *AutoAnalyzer Bibliographies* and *Supplemental Bibliographies* [C34-C37], which cover the years 1957-1975. Technicon also sponsored a number of symposia between the years of 1963 and 1976, which served as a showcase for the application of CFA to the solution of problems in essentially all (clinical, pharmaceutical, food and beverage, agricultural, industrial, metallurgical, municipal, and forensic) areas of analytical chemistry. Papers presented were published privately by Technicon [C38], and they are still a valuable resource for anyone involved with developing CF procedures. Unfortunately, neither the bibliographies nor the symposia proceedings are still in print,

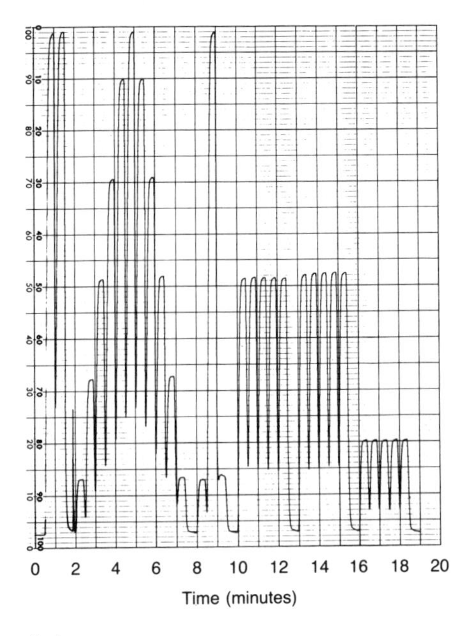

(a)

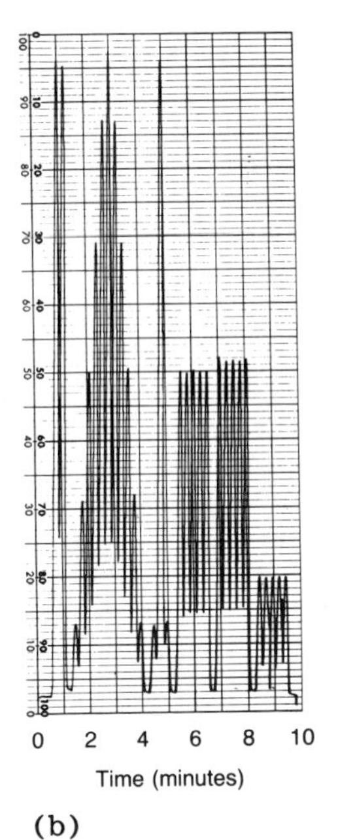

(b)

Figure 8. Peaks recorded for nitrate determinations in water using an RFA-300 CFA system and an RFAC data acquisition and processing system: (a) 120 samples h^{-1}; (b) 240 samples h^{-1}. (Courtesy of Alpkem Corporation.)

Table 1. Analytical Parameters for CFA Nitrate Determinations

Analytical rate (h^{-1})	120[a]	240[b]
Sample time (sec)	24	10
Wash time (sec)	6	5
Regeneration constant (sec)	—	1.75
Sample consumption[c] (μl)	90	38
Buffer consumption[c] (μl)	113	56.5
Reagent consumption[c] (μl)	113	56.5
Repeatability (%RSD)[d]	0.2	0.5
Percent interaction	0.6	0.5
Check calibrant[e] (ppm)	0.176 ± 0.000	0.175 ± 0.001
Correlation coefficient (r_{xy})	0.99998	0.99997

[a]Calibration function: $y = 0.016 + 4.923x - 0.154x^2$.
[b]Calibration function: $y = 0.016 + 4.602x - 0.149x^2$.
[c]Per test.
[d]Five replicates of the 0.5 ppm calibrant.
[e]EPA check standard with a certified concentration of 0.18 ± 0.02 ppm.

but they can usually be found in major university libraries throughout the United States.

CFA became, and to a large extent remains (on the basis of commercial success), the technique of choice for routine automated colorimetric analysis. The low dispersion and reproducible timing inherent in CFA facilitate the automation of manual procedures, even those that require long reaction times. The example method presented in the previous section is typical of 80-90% of all CFA applications. Readers should not assume, however, that CFA is limited to aqueous, single-phase colorimetric determinations. For example, CFA has long been used to automate the routine determination of sodium and potassium in human serum by flame photometry [C39]. Protein interferences are eliminated by dialyzing these elements into a recipient stream containing lithium, which serves as an internal standard. Calcium [C40] and lithium [C41,C42] can be determined similarly. CFA cold vapor mercury [C43,C44] and hydride generators [C45-C49] have also been developed for use with atomic absorption spectroscopy. Yamamato et al. [C48] classify their recent CF hydride generator as an FIA device, "combined with the gas-segmentation method [of Skeggs]." Thus, their system may be considered as a CFA device with sample injection [C144]. The same workers described a highly efficient tubular, porous Teflon gas-liquid separator [C49] for use with their hydride generator that reduces interferences from elemental metals and metal borides.

Many CFA applications that involve the separation of volatile or volatizable compounds from a liquid sample matrix (for preconcentration or to minimize interferences) have been devised. For example, various miniature flash distillation heads have been incorporated in the analytical cartridges of CFA systems used to determine ammonia [C50], aromatic amines [C51], fluoride [C52,C53], phenols [C54,C55], cyanides and metalocyanides [C45,C56], sulfate (after reduction to sulfide) [C57], alcohol [C58], and aldehydes and ketones [C59]. Also, so-called gas dialysis in which volatiles such as CO_2, NH_3, H_2S, HCN, and NO_x diffuse from a flowing sample stream across a porous or permselective membrane and enter a flowing recipient stream have been in use for some time. For example, in 1970, both porous Teflon [C60] and permselective silicone rubber membranes [C61,C62] were used for CO_2 determinations. The latter, however, were susceptible to interferences from some organic compounds such as salicylic acid [C61]. More recently, Fraticelli and Meyerhoff used this approach to increase the selectivity of ammonium-ion-selective electrodes [C63]. In a related application Martin and Meyerhoff [C64] demonstrated that a volatile analyte, such as NO_2, could be preconcentrated in a nonflowing recipient stream prior to detection with a nitrate-sensitive electrode. The apparatus described used CFA for sample pretreatment and FIA to sweep the recipient stream past the electrodes. This mode of separation and preconcentration seems particularly powerful and should be widely applicable. Pranitis and Meyerhoff [C65] have also reported using porous Teflon and porous polypropylene tubing filled with appropriate trapping solutions to concentrate trace-level gases from ambient air. FIA with potentiometric detection was used in this system, but for colorimetric detection as described in a similar application by Kawasaki et al. [C66] CFA could be used to advantage. Finally, Martin et al. [C67] describe the novel use of porous Teflon tubing to debubble analytical streams in CFA determinations.

CFA is particularly well suited for automating liquid-liquid extractions. Some of the earliest systems were described by Wallace [C68], Kuzel [C69], and Valentini [C70]. Air was omitted in these and other [C71] systems, and segmentation was provided by the organic solvent alone as is the case for current FIA extraction procedures. Note, however, that gas segments introduced by the air-bar technique regulate the volume of organic and aqueous segments, which improves precision of analytical results and can also facilitate phase separation by preventing the formation of emulsions [C70]. For additional references see the review articles by Burns [C3], Kuzel et al. [C7], and Kraak [C72]. Adler et al. [C73] provide a comprehensive discussion of solvent extraction in CFA that is particularly helpful with regard to strategies for the design of extraction coils and phase separators. Readers may also be interested in several more fundamental studies by Nord et al. [C74,C75] and Cantwell et al. [C145-C148].

Although Furman [C76] cites a number of CFA applications that use packed bed ion exchange columns to remove interference (e.g., Lazrus [C77] used BioRex-100 to continuously remove calcium and magnesium from samples determined for sulfate by the methylthymol blue procedure), no examples of preconcentration of analyte followed by rapid elution [F87] now used in a number of FIA applications could be found. This technique, along with the one that uses small ion exchange columns to achieve universal anion detection or to replace an anionic analyte with one that is more easily detected [C78], should be equally useful in CFA. Snyder [C27] cites two new techniques—coated tube separations [C79,C80], and evaporation to dryness followed by solvent exchange [C81]—that are particularly well suited to sample clean-up and preconcentration by CFA. Additional details can be found elsewhere [C3,C72].

Eichler [C82] described CFA gradient titration techniques in 1969 and showed how to generate linear and exponential gradients in stirred flasks (see Figure 15) or on a modified Technicon sampler. Examples are given for photometric and potentiometric end-point detection. Gradients have also been used to optimize reagent concentrations [C83-C90] and assess interferences [C91] during CFA methods development. Details can be found in Section VI.

Kinetic assays can be performed by CFA in a variety of ways. These include single point and single point with blank correction assays [C92], variable flow rate [C93,C94], and flow reversal [C95] procedures. With the advent of bubble-through flow cells and bubble-gated detectors [C9], the gas-segmented analytical stream could be passed through multiple detector stations with negligible increases in dispersion, and true multipoint kinetic assays became possible. In fact, Habig et al. [C9] demonstrated this capability in the first report of the bubble-gating technique to appear in the open literature. Technicon uses this multiple flow-cell technique for clinical enzyme determinations on its SMAC analyzer [C96].

CFA serves as a versatile adjunct to liquid chromatography (LC). As previously discussed, it facilitates the automation not only of precolumn sample clean-up, concentration, and derivatization [C3,C72], but also postcolumn derivatization, which can greatly enhance both the sensitivity and selectivity of LC detection. Further details can be found in a recent review article by Frei et al. [C97] and the references therein. Theoretically and experimentally derived design criteria for CFA postcolumn reactors (PCR) are provided by Snyder [C26] and Scholten et al. [C98], respectively. Readers with further interest in the capabilities of CFA post-column reactors are directed to a recent report by Ratanathanawongs and Crouch [C99]. These authors describe a state-of-the-art miniature CFA PCR with a 1-mm i.d. analytical cartridge and bubble-through flow cells that was used in conjunction with the determination of phenols. Major improvements in sensitivity and selectivity with negligible band broadening relative to UV detection at 254 nm are demonstrated.

D. History of CFA

Air-segmented continuous-flow analysis (CFA) was invented [C1] by Leonard Skeggs, and developed commercially by the Technicon Corporation. Skeggs's pioneering efforts provided novel solutions to problems posed by performing clinical analyses in a continuously flowing stream. He found, for example, that dialysis was particularly adaptable to CF, and he used it to remove proteins that interfered in many clinical assays. His most significant innovation, however, was unquestionably "the bubble." Intermixing of samples severely limited the performance of Skeggs's early systems. He soon discovered, however, that insertion of air bubbles both between samples and within the analytical stream reduced carryover dramatically. At this point "the bubble" became, as it remains today, the *sine qua non* of CFA. Readers interested in a more detailed account of Skeggs's invention of CFA and its ensuing commercial development at the Technicon Corporation are directed to a recent chronicle of these events by R. Stanley [C100], which spans the years from about 1950 to 1976.

Skeggs's first CFA patent [C101] issued in 1957, the same year that Technicon marketed a single-channel CFA system, the AutoAnalyzer (AAI), which was the first in a series of highly successful and profitable CFA instruments. Over the years Technicon continued to develop and improve hardware and technology associated with the basic CFA technique. Because these advances were protected by numerous U.S. and foreign patents, CFA developed and matured within the proprietary domain of a single company. This is not to say, however, that all technical and theoretical advances originated at Technicon. As the nearly 8000 papers concerning CFA that were published by 1975 [C32] attest, Technicon enjoyed a synergistic relationship with its customers, many of whom made significant contributions to the present state of the art.

While the basic components and general arrangement of a modern CFA instrument are little changed from the earliest systems, the hardware itself and the way in which data are acquired and processed have evolved through three distinct generations. First-generation instruments were general-purpose systems, which were readily adapted for nonclinical applications. The trend toward multichannel systems with rigid configurations that performed multiple diagnostic tests in parallel at a single work station is apparent, however, in Technicon's second-generation *simultaneous multichannel analyzer* (SMA 12/60) and third-generation *simultaneous multichannel analyzer with computer* (SMAC).

First-generation systems differed considerably in operation from present systems and were crude by comparison. The internal diameter of mixing coils, fittings, and flow cells was 3 mm; the dialyzer, which was the key to successful automated clinical assays at the time, was a cumbersome unit with a total groove length of 88 in; and flow rates in the range of 10-15 ml min^{-1} were common. Air, rather than the now customary analyte-free wash liquid, was aspirated between successive samples. Because these large intersample air slugs neither removed nor diluted sample residues that adhered to the walls of the tubing between the sampler and the analytical cartridge, sample interaction in excess of 10% was common [C11]. Placing cups of analyte-free liquid between each sample minimized this problem but halved the sampling frequency [C11]. Also, the movement of the sample crook, through which sample and air slugs were aspirated, was so sluggish that unless all cups were filled precisely to the same level, variations in the volume of sample aspirated caused the precision of analytical results to be poor [C11]. Furthermore, the compressibility of the intersample air segments as they passed under the pump rollers led to severe fluctuations in the volume and flow rate of the analytical stream, which in turn caused short-term variation in dialysis efficiency, hydraulic stability, and poor precision [C102]. Interestingly, these effects were masked to a large extent by pooling of previously segregated analytical stream segments that occurred in the debubbler and flowcell. The original debubbler was essentially a vented stand pipe connected to a gravity-fed, self-siphoning flow cell that filled and emptied in a manner analogous to a Sohxlet extractor. The large dead volumes of the debubbler and flow cell precluded steady-state operation, and sharp peaks without flats were the norm. By the early 1960s, however, design changes in the sampler, debubbler, and flow cell improved the wash of the AAI sufficiently to allow steady-state conditions (peaks with flat tops) to be approached within a 2- to 3-min sample cycle. The benefit of operating at or near steady state is that conditions of excessive carryover, insufficient sample, clogged sample line, etc., can be easily visualized for peaks with flats, but can often go unnoticed for sharp peaks [C27]. Furthermore, operation at steady state uncouples the precision of analytical results from the precision of sample and wash cycle timing [C103].

The first improvement on the original AAI sampler was the so-called double-crook modification [C102]. An auxiliary crook, mounted in opposition to the first, was connected to the pump tube that normally delivered air into the analytical cartridge. Nominally identical flow rates for the sample and air pump tubes are required. As the primary crook tilted out of a sample, the auxiliary crook dipped into an analyte free wash solution contained in a constant-level reservoir [C104]. Thus the roles of the sample and air delivery pump tubes were reversed during each sample and wash cycle. This resulted in a relatively constant delivery of liquid and air into the analytical cartridge, which in turn improved hydraulic stability. The problem of poor wash in the tubing between the sampler and the analytical cartridge remained. A sampler that improved both the wash characteristics and hydraulic stability was conceived by de Jong, who used a much more rapid mechanism to cycle the sample withdrawal tube between samples and a constantly renewed supply of analyte-free wash solution contained in a reservoir integral to the sampler. De Jong applied for a patent on this device in 1961. The patent, assigned to Technicon, issued in 1964 [C105] and its teachings were applied to the Sampler II (circa 1963) and all of Technicon's subsequent sampler designs. During this same time period Technicon phased out the original flow cell and introduced a more efficient debubbler and a lower-volume tubular flow cell [C106]. From this point on it became possible, and was deemed desirable, to wait until steady-state conditions were achieved at the flow cell before analytical data were recorded. As the wash of these hybrid first-generation systems improved, however, proportioning errors that were manifested as low-frequency noise superimposed on the flats of the steady-state peaks became more apparent. As Snyder has pointed out [C2], this problem was analogous to poor dispenser precision in discrete analyzers.

In 1964 Skeggs and Hochstrasser [C107] described a CF analyzer that performed eight clinical determinations in parallel at a rate of 20 samples per hour. Perhaps the most unique feature of this instrument was that only the steady-state portion of the detector signal for each channel was recorded sequentially on a single chart. The chart consisted of eight vertical columns, each scaled in appropriate units for the abnormally low, normal, and abnormally high range of each analyte in human serum. A single light source and photometric detector were mechanically multiplexed to eight flow cells mounted on a motor-driven stage. Suitable phasing coils were added to each analytical cartridge so that steady-state conditions were achieved in each flow cell sequentially at 20-sec intervals. Technicon described [C108] and briefly marketed a 12-channel version of this instrument in 1965 as the SMA 12/30.

Second-generation hardware first appeared in 1967 [C109] in the form of Technicon's SMA 12/60 clinical analyzer, which determined 12 serum analytes in parallel at a rate of 60 samples per hour (720 *determinations* per hour). Like the SMA 12/30, all 12 results were recorded in concentration units on a single, precalibrated chart. The improved performance of the SMA 12/60 was the result of a number of hardware refinements and innovations. Of major significance was Smythe's [C110] disclosure that uniform proportioning could be achieved with peristaltic pumps, provided that gas segments were added to the analytical stream in phase with pump pulsations that occur each time a roller leaves the platen surface. In the SMA 12/60, this was accomplished by passing the air-delivery pump tubes through a spring-loaded clamp that was linked mechanically to the pump drive mechanism [C2,C109]. This clamp opened briefly at the moment of roller lift-off to admit a uniform volume of gas into the analytical cartridges. Smythe applied for a patent on this device, now known as an *air bar*, in 1965. The patent [C110], assigned to Technicon, issued in 1967, and its teachings have been applied almost universally for CFA instruments that use peristaltic pumps. Other second-generation system refinements included reducing the internal diameter of analytical cartridge components to 2 mm, greatly miniaturizing the dialyzer, and reducing the flow rate of the analytical stream to an average value of about 2.5 ml min^{-1}. In addition, the reduced volume of the SMA flow cell and its integral debubbler improved the wash of the system significantly. Detector electronics were also greatly improved. Complete details can be found elsewhere [C109]. The SMA 12/60 was a major advancement beyond first-generation systems in terms of the quality of analytical results, the rates of analyses, and the economy of operation. In 1970, Technicon made second-generation technology available in modular, single-channel form as the now familiar AutoAnalyzer II (AAII). This instrument quickly became the industry standard for nonclinical automated wet-chemical analysis [C111]. Most approved automated wet-chemical methods in the procedural manuals of agencies such as the U.S. Environmental Protection Agency, the U.S. Geological Survey, and the Association of Official Analytical Chemists are based around the AAII. Only recently have moves been undertaken within these organizations to approve alternate continuous-flow hardware, methods, or techniques.

Third-generation hardware appeared in 1972 [C112,C113] as Technicon's 20-channel SMAC clinical analyzer in which the full potential of Skeggs's original CFA concept was realized. Pumps, dialyzers, and flow cells were further miniaturized and the inside diameter of the analytical cartridge was reduced to 1 mm. The average flow rate of the analytical stream was reduced to less than 1 ml min^{-1}. The segmentation frequency was increased from 0.5 Hz (30 bubbles min^{-1}) to 1.5 Hz (90 bubbles min^{-1}). In addition dispersion in previously unsegmented zones of CF systems, such as between the sampler and the analytical cartridge, and in flow-cell debubblers and flow cells, was reduced by using the pecked sampling technique and bubble-through flow cells [C113], respectively. Fiber optics and interference filters mounted on a rotating wheel were used to multiplex the 30+ flowcells to a single detector. A computer was used to remove the air-segment artifact from the detector signal, and to acquire data only from the plateau region of each peak. This eliminated the need to hydraulically phase the flow stream of each analytical channel for sequential attainment of steady state. The result was a low-dispersion system that permitted analysis rates on the order of 150 samples h^{-1} (3000 *determinations* per hour) and greatly reduced sample and reagent requirements. As Snyder has pointed out [C4], although theory [C25,C26] predicts the possibility of constructing even lower dispersion systems by reducing the inside diameter of tubing below 1 mm, decreasing flow rates, and increasing segmentation frequencies, the gains in speed and decreases in sample and reagent requirements would not offset the technical difficulties and costs associated with further miniaturization. Thus the third-generation hardware probably represents the culmination of hardware for the CFA technique.

Although plans for "do-it-yourself" third-generation hardware, most notably through the efforts of Neeley et al. [C114-C117] and later Patton and Crouch [C5,C28,C118], appeared in the open literature, a modular, single-channel, industrially oriented third-generation CFA system analogous to the AAII did not become commercially available until Alpkem marketed its RFA-300 analyzer in 1984. In 1986 Technicon introduced its own industrially oriented third-generation system, the TRAACS-800.

The conclusion of this section would be incomplete without brief mention of *zero dispersion* CFA. Because interaction limits sampling rates and therefore ultimately defines operational costs, the means to eliminate it might be seen as a sort of philosopher's stone with great potential for commercial exploitation and financial gain. The first hint about how this might be accomplished to appear in the open literature was provided by Chaney [C119] in 1967. He suggested that if the tubing in contact with aqueous samples and reagents were made from hydrophobic materials such as Teflon, the wall film, which is the sole medium for interaction in gas-segmented analytical streams, would be eliminated. Unfortunately, back pressure within a nonwetted tube increases in direct proportion to the number of air bubbles it contains and so attempts to use nonwetted conduits in "conventional" CFA systems of this era, which typically contained 6-12 bubbles per foot of tubing, were unsuccessful due to severe surging and bubble breakup [C120]. Chaney pointed out, however, that in the absence of a wall film few air bubbles were actually needed and in principle all those except the ones separating each reagent-treated sample slug from the next could be eliminated. Herein, Chaney suggested, might lie the key to operability of nonwetted systems.

In fact, 3 years prior to the appearance of Chaney's article, a patent application for a device based on such an approach had been filed by two inventors at Technicon. Ten years elapsed before this patent [C121], assigned to Technicon, finally issued. The inventors' preferred embodiment, which required neither intrasample gas bubbles (the hallmark of Skeggs's CFA concept) nor an analyte-free wash liquid, described an apparatus fabricated as much as possible from materials with nonwetting surfaces, such as Teflon. In operation, a short and a long slug of each sample and an intervening air bubble were aspirated by one channel of a peristaltic pump while an analogous pair of reagent slugs, also separated by an air bubble, was aspirated by another. With proper phasing, air combined with air and liquid combined with liquid at a downstream confluence point where matched pairs of sample and reagent slugs merged, and reacted as they flowed toward a photometric detector. Only the absorbance of the longer, following segment representing each sample was recorded for analytical purposes, while the shorter, leading segment served only to take up any residue deposited by the previous sample. The volume of the single analytical segment from each sample was sufficient to completely fill the flow cell so that the need to debubble the analytical stream prior to detection was eliminated and the integrity of each analytical segment was maintained. Thus sample interaction was eliminated and the output function of the device was a square wave. A commercial instrument based on this approach has not been marketed, and a recent report [C122] from a group of researchers at Technicon suggests that the lack of hydraulic stability inherent in gas-segmented streams flowing through nonwetted conduits renders such systems commercially unsatisfactory. Some readers may be interested in referring to two fairly new CF techniques, "rapid flow analysis" [C123] and "monosegmented flow injection analysis" [C124], which invite comparison.

In 1969, the same inventors were awarded two related patents [C125, C126], also assigned to Technicon, which disclosed how to eliminate sample interaction and still maintain hydraulic stability. The trick is to wet the inner surface of Teflon tubing with a liquid immiscible and inert to aqueous samples and reagents. A number of perfluorohydrocarbons meet these requirements and provide a wetted surface along which gas bubbles can glide with little friction. Technicon calls this technology "capsule chemistry," and has used it to eliminate interaction in the sample and reagent dispensing systems of its discrete random-access clinical analyzer, the RA-1000 [C127], and its newest random-access CF analyzer, the CHEM-1 [C122].

Compared to Technicon's third-generation CFA analyzer (SMAC), which in routine operation requires banks of multichannel peristaltic pumps, yards of tubing, and elaborate analytical cartridges fabricated to a large extent from glass, the ruggedness and simplicity inherent in Technicon's CHEM-1 "capsule chemistry" technology is impressive. Its analytical cartridge consists of a single Teflon tube along which are positioned eight photometric detectors and in which any of the 30+ resident tests can be performed in any order on any sample due to the fact that carryover is virtually eliminated. Test "capsules," which are surrounded by the perfluorohydrocarbon liquid and spaced apart by air bubbles, are formed at the rate of one every 5 sec and are drawn through the system by a single-channel peristaltic pump. Each "capsule" contains only 1 μl of sample and 7 μl each of two different reagents. Complete details of operation for a prototype system can be found elsewhere [C122].

Despite these impressive specifications, however, it is worth mentioning some inherent limitations of "capsule chemistry." For one thing, no opportunities exist for chemistry at interfaces or phase boundaries. Thus, applications that involve dialysis, gas diffusion, solvent extraction, immobilized enzyme or active metal packed bed or open tubular reactors, and electrochemical detectors, to name a few, are either not possible or extremely cumbersome relative to CFA. If the CHEM-1 can achieve its potential, it may well supplant CFA systems such as the SMAC for high-volume clinical assays. For most other applications, however, CFA should continue to be the method of choice into the foreseeable future.

E. Theory of CFA

Following conventions established by Thiers [C12], Walker [C13], and Snyder [C23], this discussion of dispersion in CFA begins by considering the square-wave output function that would be predicted in its absence. Transitions to or from some time-invariant signal level (*steady state*) proportional to the analyte concentration in samples or the wash solution would be instantaneous. In practice the peaks actually observed approach steady states (*plateaus*) more gradually along skewed *rise* and *fall curves* as shown in Figure 6a. Deconvolution of these deformed square-wave output functions reveals two components, one exponential, the other near-Gaussian (actually Poisson), which result from longitudinal and axial dispersion, respectively. Such data treatments permit the loss of wash for any CFA system to be characterized and quantitated in terms of two parameters, b and σ_t, both of which have dimensions of time. The exponential factor, b, is analogous to the RC time constant for a simple electronic filter consisting of a resistor and a capacitor. It is defined mathematically by the following equation: $y = A(e^{-t/b})$. Here y is the absorbance interval separating any point along a rise or fall curve from the steady state value it is approaching, t is the time, A is the absorbance interval between initial and final steady states, and b is simply the time required for the detector signal to change from any value of y to a value of $0.37y$ [C14].

Hrdina [C20] appears to have developed the first mathematical model of axial dispersion in CFA. His treatment begins with a series of uniform liquid segments designated by consecutive integers (k), separated by gas bubbles within a perfectly wetted tube. It is further assumed that initially the system is not flowing and that only the initial segment ($k = 0$) contains analyte molecules at concentration C_0, while the liquid segments that follow it ($k = 1,2,3, \ldots$) are analyte free. With these conditions established, the system is put in motion with uniform velocity. After an arbitrary period of flow, the concentration of analyte, C_k, in any segment k relative to the initial concentration is approximated by

$$C_k/C_0 = e^{-q}q^k/k!$$

Here q is a dimensionless parameter, which in simplest terms is the ratio of the volume of liquid that wets the walls of the *entire* flow system (V_f) to the volume of a single liquid segment (V_s). Note that the expression on the right-hand side of the above equation is the Poisson distribution. When analyte is contained in a series of liquid segments (the usual condition for CFA), the relative concentration distribution in any segment k is given by the summation of this equation, which for a large number of segments closely approximates the cumulative Gaussian distribution. It is therefore convenient to express the extent of axial dispersion as the standard deviation (σ_t) of the cumulative Gaussian curves (predicted or observed) for a given CFA system. The problem with this, and several other early models of axial dispersion in CFA that followed it (most notably those of Begg [C22], Thiers [C21], and Walker [C23]), is that they did not relate the extent of axial dispersion to major experimental variables such as residence time, flow rate, segmentation frequency, and inside diameter of tubing. A more detailed discussion of axial dispersion in CFA and some very useful figures are given elsewhere [C4,C26].

In 1976 Snyder and Adler [C25] experimentally verified Hrdina's model and presented a semiempirical derivation of it. The unique feature of their model was that V_f was expressed in terms of the length and inside diameter of the analytical cartridge. This allowed Hrdina's q (and therefore σ_t) to be related to major experimental variables. Snyder and Adler then extended this model to account for slow mixing between the stagnant film wetting the tubing walls and the liquid segments. Previous models [C20-C23] had assumed instantaneous mixing.

Later, Snyder [C26] recast the extended model into a form that made the effects of the major experimental variables on axial dispersion more obvious as given by

$$\sigma_t^2 = \left[\frac{538 d_t^{2/3} \; (F + 0.92 d_t^3 n)^{5/3} \; \eta^{7/3}}{\gamma^{2/3} \; F D_{w,25}} + \frac{1}{n}\right] \left[\frac{2.35(F + d_t^3 n)^{5/3} \; \eta^{2/3} \; t}{\gamma^{2/3} F d_t^{4/3}}\right]$$

Here σ_t is the standard deviation of a peak (in seconds), d_t is the internal diameter of the flow system (cm), F is the liquid flow rate (ml sec^{-1}), n is the segmentation frequency (Hz), t is the residence time of the sample in the flow system (sec), η is the viscosity of the liquid (poise), γ is the surface tension of the liquid (dyn cm^{-1}), and $D_{w,25}$ is an empirical diffusion coefficient (cm^2 sec^{-1}) that pertains only to diffusion in coiled tubes [C25]. The basic assumptions of the model follow:

1. Air segment volumes are the minimum required to totally occlude a tube of a given inside diameter ($7/24\pi d_t^3 = 0.92 d_t^3$).
2. The flow system is perfectly wetted.
3. Longitudinal mixing within the film is negligible.
4. The flow system is free from mixing effects (longitudinal dispersion).

Several other qualifying statements are in order. First, while temperature is not an explicit variable, values for all three minor variables (γ, η, and $D_{w,25}$) are temperature dependent. In general, however, these dependencies do not exert more than about ±15% variation in σ_t over the normal range of operating conditions—that is, 25-95°C. Second, it is necessary to add surfactants to the analytical stream to satisfy assumption 2, above, and to avoid changes in surface tension that would otherwise vary as a function of analyte concentration. Third, because σ_t varies as $t^{0.5}$, the residence time should be the minimum required for the chemical reactions involved to proceed to the required degree of completion. Therefore the only variables that can be controlled by the analyst are F, d_t, and n. The effects of these three variables are readily visualized by plotting σ_t as a function of n for different values of d_t and F, while t is held constant and representative values are chosen for the minor variables γ, η, and $D_{w,25}$. Many examples of such plots can be found in the literature [C2, C4,C26]. The most important result of Snyder's model is that for *any* value of F, optimum values of d_t and n exist where σ_t is minimum.

Furthermore, these optima are relatively flat, so that a variation in F or n by a factor of 4 affects σ_t by only about ±5% [C26]. Theoretically, in fact, σ_t can be decreased without limit by decreasing F, but d_t must be decreased and n must be increased simultaneously. Therefore practical constraints such as clogging or back pressure in narrow tubes, flow-cell volumes, and precision and delivery rates of pumps may limit further efforts to reduce dispersion in CFA systems [C26,C27].

IV. FLOW INJECTION ANALYSIS (FIA)

A. Principles of FIA

Since its inception in 1974 [F23,F24], FIA has become widely accepted as a convenient and very versatile form of continuous-flow analysis [F1,F21, F25-F35]. Over 1600 papers have now appeared in the literature. The vast majority of these deal with the development of FIA systems to replace conventional manual methods (e.g., colorimetry or titrations).

A simple FIA system consists of a single sample injector, a reaction manifold, and some form of detector. Two variations on this are shown in Figure 9. The injection valve is usually a six-port two-way valve (Figure 2), machined out of Teflon and stainless steel so as to be resistant to acids and corrosive organic solvents. The sample loop size is typically 20-250 μl [F5]. Less commonly, hydrodynamic injection is used [F36-F39]. Methods of sample introduction have been reviewed by Krug et al. [F40]. The reaction manifold is often of Teflon tubing of internal diameter 0.5 or 0.8 mm. Polyethylene, polypropylene, and nylon tubes are common,

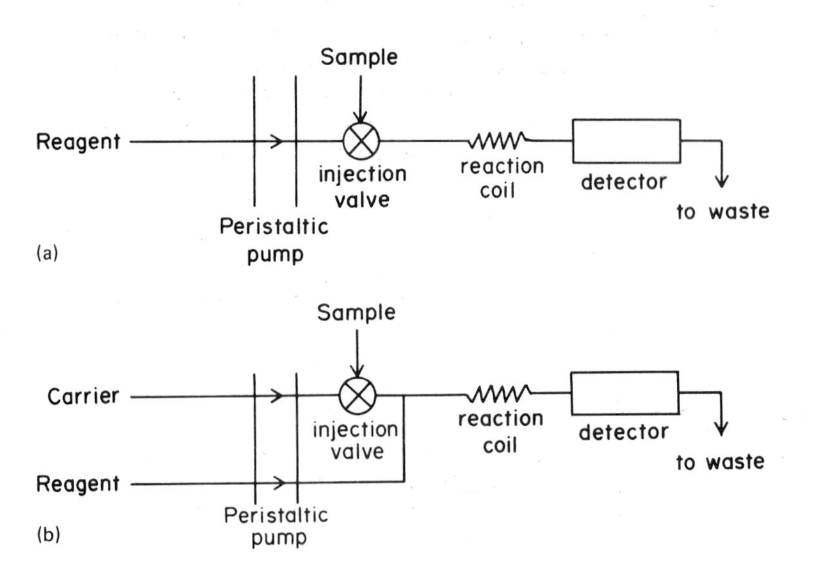

Figure 9. Simple FIA manifolds: (a) Injection of sample directly into reagent carrier stream; (b) merging of sample in nonreactive carrier with reagent stream.

but tend to become more easily discolored by organics. Some glass tubes are also used. While the typical length of an FIA manifold is less than 3 m, some systems using coils of up to 20 m long have been reported [F41]. Microconduit FIA reaction manifolds [F33,F39] have also been used and will increase in popularity due to their promise of improved repeatability and miniaturization. In these, the flow path is typically a multiple S-shaped groove cut out of the surface of a methylmethacrylate ("plexiglass" or "perspex") block. A plastic cover is placed over this to complete the enclosure.

The flowing stream passing through an FIA apparatus acts as a transport mechanism (carrier) for the samples and usually contains one or more reagents in an appropriate chemical buffer. On injection, the sample is a well defined "plug," but as it is carried downstream it mixes with the carrier and so allows reaction to take place. The sample "dispersion" is a consequence of the internal geometry of the tubing, the flow profile and molecular diffusion (Figure 10). Any product formed (or decrease in sample concentration) is then detected at some fixed point downstream. More complex FIA systems with multiple streams and several reagents are common. Analysis rates of more than 180 h^{-1} have been reported.

Ruzicka, one of the originators of FIA, has pointed out [F34] that the advent of FIA was responsible for the introduction of a new concept to analytical chemistry, that of "controlled dispersion," and that before then, complete mixing of sample and reagent was considered necessary for performing a chemical analysis. While the dispersion characteristics of FIA systems have been "controlled" and even modified (see below), they have yet to be fully exploited. A normal FIA peak is observed as a skewed Gaussian and represents a *gradient* of chemical concentration across its width. Almost all published FIA papers report procedures in which a single datum (usually peak height or peak area) is obtained from each injection, but this is not a fundamental limitation of the technique. That so few papers have got past this information barrier is probably due largely to the common practice of displaying FIA results as chart recorder output and measuring peak height by hand, at the maximum (where it is easiest!). Computerized FIA instrumentation and current research into better ways to exploit concentration gradients (such as the time-based studies of Stewart [F42], or the pH gradient work of Betteridge and co-workers [F43-F45]) may change this situation. Appropriately designed FIA systems can yield complex peak shapes that contain chemical information about more than one analyte species [F45]. System calibration using several readings from the tail of one sample peak has also been reported [F46].

B. An Example FIA System

Let us consider the FIA determination of isoprenaline, a drug used in the treatment of asthma [F6]. This reacts with potassium hexacyanoferrate(III) to produce a red product, *N*-isopropylnoradrenochrome, as is illustrated in Figure 11. The product then decomposes in a matter of hours to form an orange-brown solution of unknown composition. The previous manual procedure, as used by a major European pharmaceutical company, was somewhat complex, in that sample, pH 7.4 buffer, and

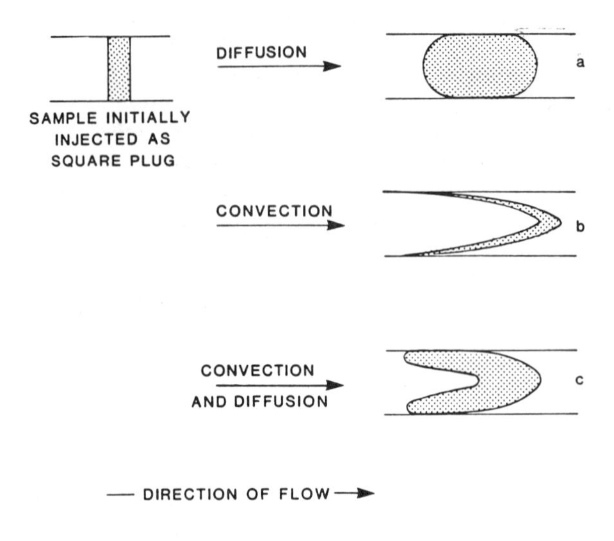

DISPERSION BY:
SAMPLE INITIALLY INJECTED AS SQUARE PLUG
DIFFUSION
a
CONVECTION
b
CONVECTION AND DIFFUSION
c
DIRECTION OF FLOW
a DISPERSION BY DIFFUSION ONLY
b DISPERSION BY CONVECTION ONLY
c DISPERSION BY DIFFUSION AND CONVECTION
ALL ABOVE FIGURES ARE 2D CROSS SECTIONS.
(ADAPTED FROM P. DAVID, Ph.D. THESIS, UNIVERSITY OF WALES, (1978))
PROCESSES OF DISPERSION OF A SQUARE PLUG OF FLUID INTRODUCED INTO A FLUID FLOWING UNDER LAMINAR CONDITIONS IN A STRAIGHT PIPE

Figure 10. Dispersion of a sample plug in a flow-injection analyzer. (a) Dispersion by diffusion only. (b) Dispersion by convection only. (c) Dispersion by diffusion and convection. (Adapted from Ref. F167, with permission.)

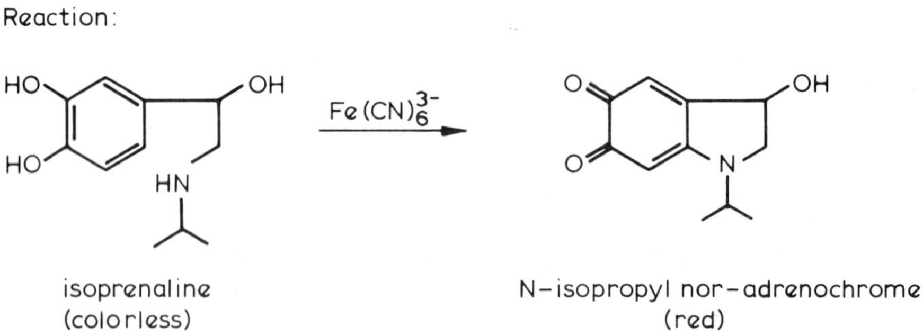

Reaction:
HO
HO
OH
HN
Fe(CN)$_6^{3-}$
O
O
OH
N
isoprenaline
(colorless)
N-isopropyl nor-adrenochrome
(red)

Figure 11. Reaction of isoprenaline with ferricyanide.

reagent solutions were mixed by hand. Then, 30 sec later, a stronger pH 4.0 buffer was added to stabilize the product. So many manual additions and the critical timing of those meant that the repeatability of the method was highly "operator dependent."

The FIA apparatus used to automate this analysis was as shown in Figure 9a. Reagent (ferricyanide) and buffer were premixed and formed the carrier stream. Isoprenaline sulfate solution, made up from capsules or from stocks of the pure compound, was injected into this. The red product could be seen forming in the reaction coil downstream of the injection valve, and was detected using an inexpensive photometer system [F17]. The reagent concentration, pH, flow rate, reaction coil length, and sample size were then optimized by both manual univariate and modified simplex optimizations [F6,F45]. Results from these were in good agreement as to the position and magnitude of the optimum. The timing repeatability inherent in the FIA method obviated the need for addition of the pH 4.0 buffer, and so simplified the method. Determinations were possible at the parts per million level, which was suitable for samples from individual capsules.

In a later study using an automated FIA system capable of self-optimization [F45], modified simplex optimization of this system was repeated, and an alternative response function that took into account both method sensitivity and analysis time was investigated.

C. FIA: Key Literature and Applications

The best sources of reference for publications on FIA are currently the FIA bibliography produced by Tecator [F47] and the recent review article by Ruzicka and Hansen [F21], and the second edition by the same authors [F1]. These really should be in the possession of anyone interested in developing or using FIA methods. At the time of this writing, Tecator distributes their FIA bibliography without charge. *Analytica Chimica Acta* publishes special editions reporting on papers presented at the regularly held conferences on flow analysis. One particular edition of this journal contains a 67-page article comprised of short papers by many of the most active researchers in FIA [F35]. The classic text book for FIA is Ruzicka and Hansen's *Flow Injection Analysis* [F1]. This covers in detail much of the theory that is not included here. Valcarcel and Luque de Castro have also published a textbook on FIA [F31] that should prove valuable. The bioanalytical potential of FIA has been assessed by Worsfold [F48]. Other review articles in the open literature are well worth reading [see Refs. F1,F21,F25-F35].

FIA, without any reaction, provides a means for transporting samples to a detector quickly and reproducibly. This can be used whenever measured properties are intrinsic to the sample. McLeod et al. used such a system for multielement serum determination [F7]. The detector in this case was an inductively coupled plasma optical emission spectrometer (ICP-OES). The method this replaced had involved continuous aspiration of serum, and matrix effects made sample pretreatment necessary. The FIA method used 20-μl samples and substantially simplified the analytical procedure. Matrix effects were found to be minimal and analyses could be carried out without sample pretreatment. Other workers have also combined FIA with ICP-OES [F49-F51] and atomic absorbtion spectroscopy [F52-F55].

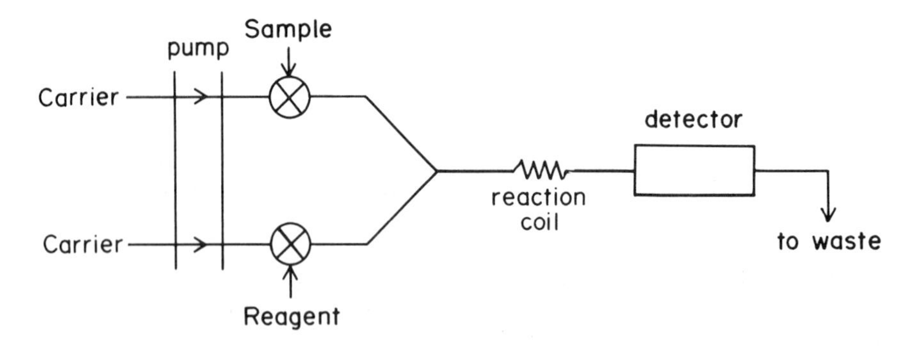

Figure 12. A merging-zones flow-injection analyzer.

Simple FIA systems can carry out sample dilution if a long length of tube (~5 m), a gradient chamber (internal volume ~1 ml) [F56,F57], or a diluent stream is incorporated into the manifold. In this way another manual sample pretreatment step can be avoided, and the dynamic range of concentrations capable of being analyzed can be extended by at least two or three orders of magnitude [F58]. Smaller and more dilute samples can also be obtained by "zone sampling" [F59], in which a small portion of a dispersed sample plug flowing in an FIA system is selected and injected into a second carrier stream.

It may be that much lower reagent consumption than sample consumption is desirable (e.g., cost, toxicity, inconvenience of replenishing reagent supply on a remote automatic FIA apparatus). In such circumstances the roles of "sample" and "reagent" can quite readily be reversed [F60]. Only sample throughput will suffer. If this is a problem then a merging zones approach should be used [F61-F69]. In this, plugs of sample and reagent are injected into two arms of a Y-shaped flow system (see Figure 12). These plugs then meet at a merging tee downstream and reaction takes place. Usually totally symmetric systems are used, in that the distances from the injectors to the tee, the flow rates in the two arms, and the plug sizes are the same, with simultaneous injection of sample and reagent. Mindegaard has shown the effect of making the system asymmetric by insertion of various coil lengths between one injector and the merging tee [F65]. Crowe et al. have studied other ways of introducing asymmetry, both with an experimental system and a random walk model [F69]. They have suggested that a reagent plug larger than the sample plug may be beneficial and that totally symmetric systems are not always optimal. This has also been recognized experimentally by other workers [F41]. It is known that the diffusion coefficient of a species in solution affects its dispersion in an FIA system. Indeed, FIA systems have been used to determine diffusion coefficients [F70]. In a merging zones system, sample and reagent with nonidentical diffusion coefficients will undergo slightly different dispersion and overlap at the tee may not be complete. However, this does not prevent the merging zones techniques from being a highly useful weapon in the FIA armory.

It is evident from the usual dimensions of FIA systems that most analytical methods adapted to FIA use fairly fast reactions. Typical colorimetric FIA methods use a sample size of 20-100 μl, a reaction manifold of 0.5-3.0 m tubing (0.5-0.8 mm i.d.), and a continuous carrier flow rate of 0.5-5.0 ml min^{-1}. These conditions allow sufficient mixing and time for a fast reaction to go almost to completion prior to product detection. The residence time of a sample in such a system is usually a few tens of seconds and is seldom more than 2 min. Longer times than this in a constant-flow-rate regime result in unnecessarily large dispersion and the magnitude of the analytical signal is consequently decreased. Some reactions, such as the chemical oxidation demand (COD) analysis of water [F71,F72], benefit from longer reaction times. When sensitivity is at a premium for slow reactions, it might be better to use "stopped flow," or go to a CFA method. In stopped-flow FIA the flow rate is programmed so as to (a) flush sample out of the injector and allow sufficient mixing for favorable reaction conditions to be established, (b) then stop the flow and wait a preset time for sufficient reaction to occur (reaction need not be complete), and (c) then restart the flow and so cause the product to pass into the detector and be measured. A variation of the stopped-flow approach is to stop the flow so that the product forms inside the colorimetric detector [F34]. This has enabled FIA to be used for kinetic analyses [F73]. In some instances (at present, where reaction rates vary by at least an order of magnitude) this could also be used for the determination of more than one sample component from one injection. Chemical kinetics has also been studied using multiple detectors [F12].

The sample throughput rate can be improved by using a long reaction manifold in which several samples are entrained concurrently [F74]. This is achieved by using a "stop, start, stop, start" flow program. During each stop a new sample is injected, and on each start, the foremost sample is expelled. The time interval between stops is carefully selected so as to allow no overlap of samples in the reaction manifold. An alternative to this is to have several "holding coils" in parallel [F75]. Reacting sample is allowed to enter one coil and this coil is then isolated from the flow. The next sample enters a second coil (etc.). In this manner perhaps as many as five coils may be used. Then the contents of the first coil are swept out and through the detector, where the product of perhaps 5 min of reaction is detected. Another sample may occupy the first coil in the same event. In this manner, residence times can be extended, but a high sample throughput rate is maintained.

pH gradients provide another handle on simultaneous multielement analysis [F43,F76,F77] and have been used for speciation studies [F76] and turbidimetry [F78]. Repeatable pH (and normally concentration) gradients can be obtained by injection of base into an acid stream (or vice versa), merging with a sample stream, and allowing a considerable degree of dispersion to occur. Reaction stoichiometry has also been determined by FIA and another nonsegmented continuous flow system [F5,F79].

Far more complex FIA systems than these are possible, and the number of potential combinations of tactics is vast. Systems involving multiple detectors [F12], multiple lines to allow several sequential chemical reactions [F80], parallel analyses each running the same chemistry [F81], straight and coiled reactors [F82], solvent extraction [F10,F83-F86], and in-line columns have been used for sample clean-up and preconcentration [F87]. Enzymatic reactors [F14,F88,F89], gas diffusion cells, and dialysis units [F89-F91] for FIA have also been reported. Single-bead string reactors (SBSRs) with enzymes chemically bound to them appear promising [F14,F88]. The enzyme reagent is essentially reusable and has a long lifetime.

Solvent extraction systems are also worthy of mention. A "generic" single-stage solvent extraction FIA system with postextraction reaction is shown in Figure 13. Betteridge et al. used one such as this for determination of uranyl ion in the presence of interfering metals and other species [F92,F93]. First, the aqueous sample (analyte plus interferents) is injected into an aqueous carrier stream. This is then merged with a stream containing a salting-out agent. This then merges with a stream of an immiscible organic solvent, which may contain a complexing agent for the analyte. This results in a segmented (aqueous-organic-aqueous-organic) stream. In this, the solvent extraction takes place. The salting-out agent pushes as much of the analyte as possible out of the aqueous and into the organic phase. After extraction, the aqueous and organic phases are separated, usually by a Teflon membrane separator [F94], although other designs have been reported [F95], and used [F93]. The aqueous solution, still carrying the interferents, is then passed to waste. Analyte in the organic stream may then be detected directly (usually spectrophotometrically) or, as shown here, after a further color-forming reaction has been allowed to occur. When one counts up the number of manual operations that have been automated by this type of system, with each analysis done in less than 2 min, one realizes that the methodology is quite impressive.

FIA has been coupled with other techniques for selectivity, separation, and simultaneous multielement analysis. Speciation studies using FIA have been reported [F96,F97]. Published "hyphenated techniques" include FIA-ICP [F7], FIA-AAS [F98,F99], FIA-GC [F100], FIA-FTIR [F101], and FIA-HPLC (both for precolumn derivatization [F95] and postcolumn reaction [F95,F102]). One worker is using FIA with an "individual droplet generator" (IDG) as a sample introduction system for laser-induced breakdown spectroscopy (LIBS) [F103]. Single-bead string reactors (SBSR) [F88,F104], packed beds, and flow reversals [F105,F106] have all been used to modify the dispersion characteristics of FIA systems. Coiled reactors are known to introduce secondary flow phenomena [F107,F108]. Some FIA methodologies allow several observations of the transient signal in various stages of dispersion or under altered chemical conditions [F12,F105,F109-F111]. These have been termed "multidetection" systems and have recently been reviewed [F105]. Automatic development of FIA methods by computer-controlled apparatus has been shown to be possible [F45,F112]. This will be discussed in more detail below.

Further examples of FIA systems, applications, and principles can be found in the literature [F1,F21,F25-F35,F47]. Lists of references to FIA methods for many areas and specific applications have been compiled and published [F21,F47].

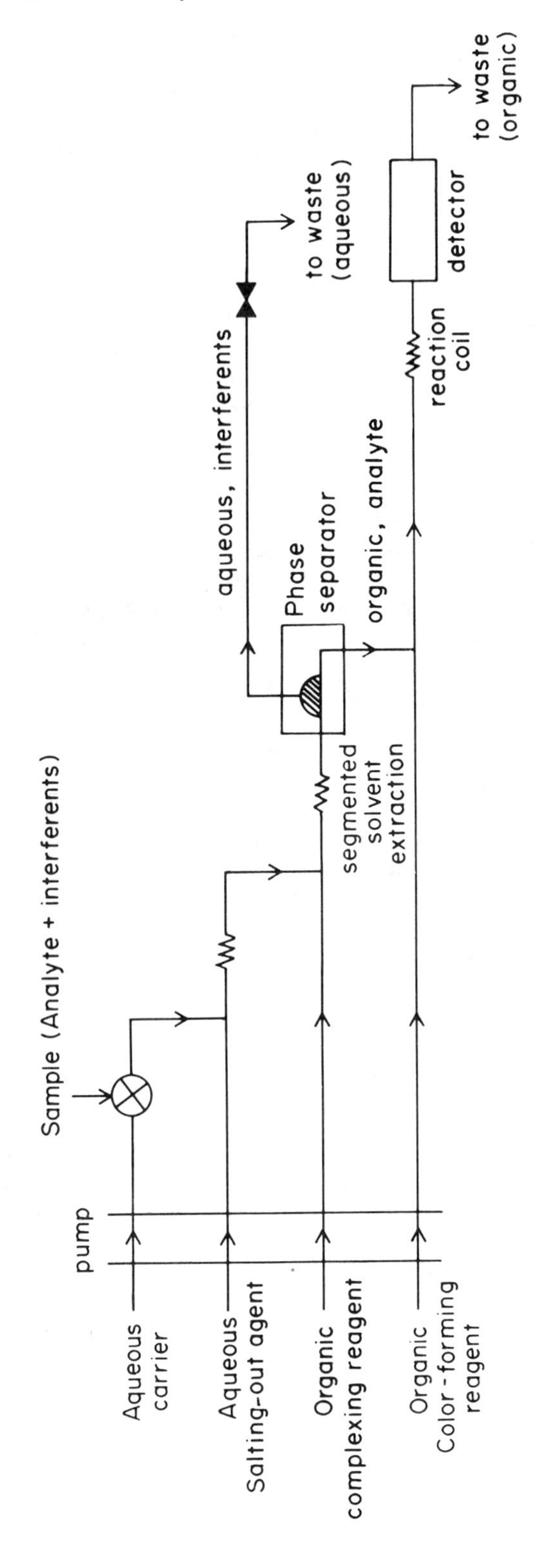

Figure 13. A flow-injection analyzer incorporating solvent extraction.

D. History of FIA

The history of FIA has recently been covered by Ruzicka [F21], and will be treated only briefly here. The principle of what we now know as flow injection analysis was first recognized by Ruzicka and Hansen [F23] and separately by Stewart et al. [F24] in 1974. Some commentators [F113] also note that "pioneering experiments" were published by Nagy et al. in 1970 [F114]. One would also not be too surprised if, on his way to inventing "the bubble" and CFA, Skeggs might have done some experiments that today might be called FIA. It was Ruzicka and Hansen who first suggested the name "flow injection analysis," and their names have become synonymous with developments in this area. Other workers have made significant contributions. An excellent review by Betteridge in 1978 [F25] served to focus the attention of the analytical community on the rapidly evolving potential of FIA. Karlberg pioneered FIA systems that incorporated solvent extraction [F84]. Mindegaard [F65] and Zagatto et al. [F61-F64] suggested the use of merging zones (discussed above). Betteridge and Fields used the controlled dispersion characteristics of FIA systems to produce pH gradients and then used these to demonstrate the possibility of multielement determinations [F43,F76]. Automated FIA systems have been published [F45,F105,F112,F115-F117]. Betteridge and co-workers also reported simplex optimization of FIA systems and the realization of automated methods development for FIA (including on-line simplex optimization) [F45,F105,F112,F117]. Nieman has been a major contributor in the use of chemiluminescence for FIA and immobilized reagents [F118]. Stewart has published time-based flow injection analyses in which the analytical measurement is not necessarily the peak maximum [F42].

Perhaps the main reason FIA is still less widely used than the authors feel it deserves is that no major company has championed FIA in the same way that Technicon has for CFA. It is a fact that at the time that CFA was conceived, the clinical automation market was large and yet untapped; CFA provided very elegant and profitable answers to these problems. FIA came along later, when CFA had already dominated the clinical market, and thus there has been less financial gain likely from its commercial development.

The FIA community has traditionally been a very open one—there are few secrets. This came about as a result of the different atmosphere engendered from the very beginning by its exponents. The lack of bubbles meant that simpler instrumentation was needed for FIA than for CFA. Thus many individuals have been prepared to build their own systems, often holding tubes in place with Lego plastic building blocks [F25], rather than buying commercial apparatus. This do-it-yourself approach is still much in evidence.

E. Theory of FIA

Flow injection analysis is a nonequilibrium technique—that is, the signal observed does not attain a steady state. In normal use there is an interaction of chemical reaction and physical dispersion. Once reaction is complete (or in the absence of reaction), dispersion causes a peak to decrease in height and become increasingly broad with time. The FIA manifold can be considered as a three-dimensional chemical reactor, within which there are an infinite number of smaller volumes that each have slightly different concentrations of sample and reagent, and that these concentrations change with time as a result of the flow profile, molecular diffusion, and chemical reaction. It is not surprising therefore that an exact mathematical solution for chemical reaction in even a simple single-channel system is far from trivial. For an ideal system, in addition to the flow rate and kinetic parameters, the effects of temperature, viscosity, sample size, and tubing internal diameter must be considered. Real systems are complicated by differences in viscosity and refractive index between sample and carrier, nonconstant flow rates of peristaltic pumps, and the effects of connector and detector geometries.

One explanation for molecular behavior in a simple single-channel FIA system (such as shown in Figure 9a) is given by the "random walk model" (RWM) proposed by Betteridge et al. [F119]. A random walk model for diffusion was used by Einstein to explain Brownian motion [F120]. Giddings [F121,F122] showed that such a model gave a satisfactory account of the chromatographic process.

One advantage of the RWM for FIA is that it deals with the fate of individual molecules. Rather than attempt to solve for the behavior of the complete sample plug using continuous mathematical functions, it uses discrete mathematics to simulate the behavior of a limited number (perhaps 1500) of sample molecules as they pass through the analyzer. This simplifies investigation of the effect of experimental factors such as sample size, reaction rate, and temperature, which other models [F104,F107,F123-127] have difficulty in dealing with. The model is conceptually easier than that based on a series of imaginary tanks [F124,F126].

The movement of a molecule in a laminar flow regime (such as is assumed to occur in some simple FIA systems) can be considered a combination of the effects of random walk (molecular diffusion) and laminar flow (convection). The RWM for FIA provides "snapshots" of the system after discrete time intervals. In each time interval, each molecule moves a distance along the tube which reflects the local flow velocity (nil at the walls, twice the mean velocity in the center) and then takes a random step, the length of which is determined by molecular diffusion. If after any time interval, local sample and reagent molecule "concentrations" determine that reaction is possible, a proportion of the molecules are allowed to react. This process is repeated many times. In the initial study [F119], the effects of 10 experimental variables were readily investigated. These were flow rate, tubing internal diameter, sample volume, reaction coil length, reagent concentration, temperature, viscosity, diffusion coefficient, reaction rate, and order of reaction. Its results were in good agreement with accepted theory and experimental evidence. Other predictions it made have been confirmed in subsequent experimental studies [F128,F129]. The model has been implemented as a BASIC computer program, which may be run on personal computers and produces animated sequences of snapshots. These serve to illustrate the dispersive processes, formation of product, and disappearance of sample in a highly graphical manner.

This random walk model has recently been extended to cover a merging zones FIA system [F69] and its predictions have been compared with results from a far from ideal experimental system. A good deal of correlation was found. This work has shown the importance of differences in diffusion coefficients [see F70] between reagent and sample plugs and an unnoticed pH gradient.

The difficulty of developing continuous models for FIA has not been considered insurmountable by some workers. Two papers of note are refs. F130 and F131. Other papers that discuss the theory of FIA and/or dispersion and reaction in flowing streams appear in the literature [F108,F113, F127,F131-F138]. These are also well worth reading.

V. COMPARISON OF FIA AND CFA

The challenge to CFA [C1] presented by FIA [F23,F24] has led to several lively exchanges in the literature [C128-C131] concerning the relative performance (rates of data production, sample and reagent consumption, and precision of analytical results) of the two techniques. Although theoretical comparison of CFA and FIA by Snyder [C26] predicts faster rates of analysis and lower sample and reagent requirements for CFA, exactly the opposite conclusion was reached in a recent review [C132] based on figures of merit for FIA and CFA determinations published in the literature. Recently, however, a direct experimental comparison of the two techniques was performed in which the same pump, flow cell, and detector were used [C5].

This study underscores the fundamental differences in the two techniques that arise not from the means of sample introduction, or differences in peak shapes, but from the presence or absence of gas segments in the analytical stream. In FIA the sample enters the reactor as a slug and begins to disperse into the carrier stream at the moment of injection. The sample loop is itself a mixing stage, as is the flow cell. Except in cases where samples are not required to undergo chemical reaction prior to detection, some dispersion of the sample slug is necessary. In single-line flow-injection systems (see Figure 9a), in fact, dispersion is the only mechanism by which the sample and reagent can interact. In dual-line flow-injection systems (see Figure 9b) where the sample and reagent are merged in a tee connector, mixing is improved but may still be incomplete. This can be a drawback to the FIA approach for spectrophotometric determinations, especially those that require more than about 30 sec to reach an appreciable degree of completion, because it is generally necessary to compromise between adequate mixing of the sample and reagent on the one hand, and excessive broadening of the sample slug on the other. There are several techniques, however, that enhance mixing and minimize dispersion in FIA which can make this compromise more favorable. These include inducing secondary flow by using tightly coiled or "wavy" tubes [C133] rather than loosely coiled open tubular reactors and by using single-bead string reactors (SBSR) [F104] rather than open tubes for mixers. Both these approaches are successful because they disrupt the parabolic velocity profile normally associated with laminar flow. Another effective way to increase residence time without excessive dispersion is simply to stop the pump after the sample and reagent have mixed sufficiently [C134].

Table 2. Comparison of Figures of Merit for Routine Colorimetric Chloride Determination by CFA and FIA

	CFA[a]			FIA[b]		
Analysis rate (h^{-1})	360	240	180	360	240	180
Sample time (sec)	5	10	15	—	—	—
Wash time (sec)	5	5	5	—	—	—
Sample volume (μl)	18	36	54	60	60	60
Reagent volume (μl)[c]	71	107	142	267	400	533
Precision (% RSD)[d]	0.43	0.25	0.34	1.72	0.44	0.67
Percent interaction	0.6	0.6	0.5	1-2	1	1
Slope	0.020	0.021	0.021	0.018	0.019	0.019
Y Intercept	0.032	0.034	0.033	0.036	0.040	0.037
r_{xy} (Correlation coefficient)	0.998	0.998	0.998	0.997	0.997	0.997

[a]Flow rates (ml min^{-1}): sample = 0.2, reagent = 0.4, air = 0.1; n = 2 Hz, t = 30 sec, d_t = 0.1 cm.
[b]Flow rates (ml min^{-1}): sample carrier = reagent = 1.6; 25 cm SBSR.
[c]CFA: The volume of reagent delivered during one sample and wash interval. FIA: The volume of reagent delivered during the time interval between sample injections.
[d]Calculated from 5 replicate determinations of the 15 ppm chloride standard.

During the "stop period," the sample zone may become more symmetrical due to radial diffusion, but longitudinal dispersion is essentially eliminated. After sufficient reaction time has elapsed, the pump is restarted and the reacted sample is propelled into the detector. This approach is effective, but operationally cumbersome, and it often decreases sample throughput.

The situation is quite different for CFA because the process of mixing the sample with reagent is not dependent on sample dispersion. In CFA gas bubbles are used to divide the sample slug into a number of nominally identical segments, each of which is proportioned with the same volume of reagent. Mixing within each segment is enhanced by the microcirculation pattern that develops naturally in a segmented, flowing stream, while mixing between segments is minimized [C25]. Furthermore, gas bubbles minimize longitudinal dispersion in other zones of a CFA system, such as between the sampler and the analytical cartridge (i.e., pecked sampling), and the detector (i.e., bubble-through flow cells). In addition, other factors being equal, dispersion decreases as the flow rate decreases which leads to conservation of samples and reagents.

Patton and Crouch showed [C5] that for equilibrium-based spectrophotometric determinations, CFA competes favorably with FIA even for relatively simple chemistries with fast reaction rates (see Table 2). It cannot be denied, however, that in the limit FIA is simpler, operationally, than CFA. Consider the case where a sample is treated with a single reagent and detected. The minimum requirements for a flow-injection system are a single-channel pump, an injection valve, and a recording detector. The same determination with CFA would require three pump channels (one each for sample, air, and reagent), and the detector signal would have to be gated to eliminate the air-bubble artifact; otherwise a fourth pump channel would be required to debubble the flow stream prior to detection. Also, for operation without an automatic sampler, introduction of samples with a valve in FIA is much less taxing than the manual, pecked sampling used for CFA, which requires precise timing. Thus for single-reagent determinations with relatively fast kinetics FIA may be preferred due to its operational simplicity. When multiple reagent additions, or reaction times that exceed about 30 sec, are required, however, the increased complexity of CFA may be offset by virtue of its high mixing efficiency, low dispersion, and conservation of samples and reagents. The point at which this tradeoff is reached is not clear-cut, and is strongly influenced by an analyst's ingenuity and predisposition to these two versatile and complimentary techniques.

VI. ANALYTICAL METHODS DEVELOPMENT FOR CF SYSTEMS

A. Optimization of CFA and FIA

When faced with any method development problem, including those for CF, the first step is usually to carry out a literature search. If a method has been published in a reputable journal, then the chances are it can be copied with little difficulty. Otherwise, use the experimental parameters cited for the manual procedure as a starting point for developing a CF procedure. Many times there are generic solutions for particular classes of analytical problems.

There are now several papers in the literature that report optimization of experimental conditions (flow rates, coil lengths, concentrations, etc.) of CF systems. These have indicated improvements in performance by as much as factors of 40 or more in relatively few (perhaps 30) experiments, and have transformed systems that initially gave unusable results into the required analytical engines. We certainly advocate the use of such methods, and discuss these in detail below. However, any optimization strategy can only find the best operating conditions for the apparatus and analytical methods it investigates.

Experimental Optimization

Experimental optimization techniques afford a means by which the performance of many chemical systems can be substantially improved, often by carrying out just a few experiments. Each CF system to be optimized will have some chosen merit factor (or response). Some simple examples of such responses are absorbance for colorimetry, current for polarography, and intensity of light emission for fluorometry. Of course, the magnitude of this response is likely to differ with the experimental conditions (pH, reagent concentrations, etc.) used. Optimization techniques attempt to find the set of experimental conditions that give the best response. Good optimization algorithms do this both rapidly and reliably by systematically changing settings for the experimental variables.

In this treatment, we consider optimization of a chemistry on a simple flow-injection analyzer (Figure 9a). Most of this text is equally applicable to CFA, and much of it is generic to quantitative analytical methodologies. Let us consider 10 questions that serve to guide us in effective experimental optimization.

1. What system response should be optimized?

This is the fundamental issue in experimental optimization. The most common factor to optimize is sensitivity. Therefore we require a readily measurable system response that relates directly or indirectly to sensitivity. For colorimetric analyses this is usually either product absorbance or decrease in sample or reagent absorbance, and for CF methods this is usually measured as a peak height at maximum, relative to the baseline. An alternative measurement sometimes used is peak area. A few FIA systems use peak width at half height as a response. Kinetic measurements (rates) are also possible.

Sensitivity alone, while usually adequate, may sometimes not be enough. Perhaps a combination of two or more system responses into some "response function" (or "objective function") would be better. This was certainly the case in the development of an FIA system for determination of uranyl ion [F93]. The system, which was similar to that shown in Figure 13, incorporated a solvent extraction step in which the organic component, containing the uranyl ion complexed with tri-*n*-butyl phosphate, was separated from the aqueous component containing possible interferents at high concentration. Conditions under which the flow rate into the separator exceeded 2 ml min^{-1}, while the most sensitive, brought with them a gross worsening in repeatability. Five to seven consecutive replicate results were used to characterize each set of experimental conditions tried. A better response to optimize for would have been $R = h/(1 + s_h)$, where h is the mean peak height for x replicates and s_h is their sample standard deviation. In this equation, any increase in the standard deviation of the peak heights results in a decrease in R, the response. The "1 + " in the denominator prevents problems when the standard deviation is zero (otherwise, R goes to infinity). The maximum value R can attain is the mean peak height and the function itself weights the results against conditions giving poor response repeatability. This function is specifically of use for CF systems incorporating components that can be unstable. To further weight the relative importance of repeatability and sensitivity, multiply the standard deviation by some constant.

If a straight-line calibration curve and good sensitivity are desired, an appropriate response function would be $R = h\ \mathrm{abs}(C)$, where h is the mean height and abs (C) is the absolute value of the correlation coefficient calculated for the plot of peak height versus sample concentration. Each experiment would entail running a range of sample concentrations (up to and including the top standard).

If one wishes to minimize an interference and maximize sensitivity simultaneously, then $R = h/[a + b(h - h')^2]$ may be used, where h is the peak height without interferent, h' is the height obtained with interferent (ideally $h = h'$, i.e., no interference effect), and a and b are constants that determine the relative importance of sensitivity versus minimization of interference, that is, the interference penalty we are prepared to pay for sensitivity.

Where sample throughput (as measured in samples per hour) is an important factor, $R = h/[a + b(t - t_{ideal})^2]$ may be used. In this, h is the peak height again, t is the time taken from injection to detection of the peak maximum, and t_{ideal} is the time we would like it to take (from the desired sample throughput rate). Alternatively, $R = h/(a + t)$ may suffice. These equations indicate that there will be a tradeoff between sensitivity and sample throughput. Other novel response functions have been derived empirically [F139-F141].

2. *What experimental variables are likely to affect this response?*

Parameters such as coil lengths, temperature, reagent concentrations, pH, sample size, and flow rates are all likely candidates. Do not include sample concentration as a system variable. Unless an unusual response function requires otherwise, use the top calibration standard as your sample throughout the optimization. Incorporation of variables that do not have any effect on the response will not alter the optimum values for those that do. However, whatever the method used, inclusion of redundant variables will extend the optimization process.

3. *Should parts of the flow system be redesigned to facilitate a better optimum?*

Yes. Modifications to the more fundamental nature of the flow system (which would not be considered a variable in a conventional experimental optimization) can result in greatly improved performance. Thus, once the apparatus is together and some results are being obtained, one may wish to carry out a limited preliminary investigation of the system design before attempting a rigorous experimental optimization of chemical conditions, flow rates, coil lengths, etc. In our experience, the gains to be made may equal those obtained by optimization. The optimization method can only make the best of the system that you give it.

4. *Which parameters can be varied on the existing apparatus?*

This analysis is important. It may indicate further aspects of the system that the user has not considered, and lead to a better system design prior to optimization [F93]. Factors such as tubing i.d. and system temperature may be difficult to vary.

5. *Which subset of these should be optimized?*

A typical subset might be reagent concentration(s), pH, flow rate(s), reaction manifold length(s), and possibly sample size. Choose those that can easily be changed. Ones that can be changed automatically are the best, since these significantly speed up the optimization process (see Section VI.F).

6. *What range of values for the variables chosen is allowed and of interest?*

Sensible limits should be chosen for each variable. A minimum flow rate of 0.2 ml min^{-1} and a reagent concentration of about half the expected sample concentration might be two suitable lower limits. The system temperature, if varied, should not be allowed to approach the boiling point of the solvent used. This can cause problems, particularly when using non-aqueous solvents such as ethanol, hexane, or ethanol/water mixtures, since vapor bubbles evolved can cause noise spikes at the detector and affect the dispersion characteristics of FIA systems. Note also that elevated temperatures may damage or degrade performance of detector electronics and flow cells and denature soluble or immobilized enzyme reagents. Other limits might be less easy to specify. Values of pH above some threshold may cause precipitation in the tubes [F8]. Obviously this should be avoided (unless one is detecting the analyte by nephelometry [F78]). Flow rates above a certain critical value may cause phase separators to malfunction.

7. *To what precision can the response and variables be measured?*

Note that precision "needs" and "wants" are not necessarily the same thing. A 1% repeatability might be normal practice, but 20% might be all that is *needed*, or even just a straight comparison—that is, is the concentration of this sample higher, lower or equivalent to that of some reference? There is little point in carrying out two experiments where, while the variable values are numerically different, that difference is less than the measurement precision of the system.

8. *Which optimization technique(s) should I use?*

It has been suggested that optimization methods can be divided into three classes—"the good, the bad and the ugly" [F139]. The good ones are those that work, improving conditions rapidly and reliably. The bad ones are unreliable, or take too many experiments to achieve a reliable optimum, or both. The ugly ones might work, but are mathematically so complex that nonmathematicians prefer to stay away from them and rely on the method they are used to instead. For these, cosmetic surgery is sometimes all that is required to see the beauty in the beast; some user-friendly computer programs achieve just this. In any continuous-flow method (whether CFA or FIA), in addition to the normal chemical variables such as pH, reagent concentration, etc., one must take into account the physical variables of flow rate, the lengths of tubing between points of injection and detection, the internal diameter and geometry of the system, and the sample size. Thus, even an experimentally simple single-channel CF system will have (at least) five separate variables [F6,F8], and in one solvent extraction system 12 were reported [F93]. In the conventional optimization procedure, called the univariate method, one variable at a time is altered while the rest are held constant. Univariate optimization is overly time consuming where more than two to three variables need to be considered and does not obtain the best set of experimental conditions reliably [see F6 and references therein]. This is particularly so when experimental variables interact, as they have been shown to do in CF systems [F61]. An alternative optimization strategy is to use a "modified simplex" [F6,F93,F142]. This is faster and is more reliable than the univariate method because it allows simultaneous variation of several factors and requires fewer experiments to achieve the same (or a better) improvement [F6]. The modified simplex approach has also been successfully applied to the automated self-optimization of separations for high-performance liquid chromatography (HPLC) [F143,F144]. While modified simplex optimization decreases the number of experiments required to reach the optimum, for a two-variable system, perhaps 25 experiments are still required [F139]. The number of experiments required by a simplex optimization increases with the number of variables to be optimized, but not as rapidly as with univariate optimization techniques [F140]. An analytical procedure using conditions on a flat portion of a response surface will be more stable and therefore will give more repeatable results than one on a sharp slope [F139]. It is pointless to do a univariate optimization in which 20 values of each variable are tried, when perhaps five might suffice. Similarly, if using a modified simplex algorithm, little will be gained from starting with too small a simplex.

9. When should the optimization be terminated?

The subject of "termination criteria" has been discussed in the literature [F93,F142]. The conventional wisdom of carrying on until a method converges onto an optimum may not work experimentally if the optimum lies on a plateau, since then the optimum will not be at a unique position, but will extend over a range of values. Making the many fine adjustments necessary to home in on an optimum to the last few tenths of a percent may not be justifiable to management, or necessary. Thus the authors suggest the following somewhat empirical guidelines. Carry out a modified simplex optimization using a predetermined number of experiments, perhaps 30-40. Then stop unless major improvements are still being found. At this point, you are likely to have obtained an improvement, and to have done it efficiently. If you need to be more sure that you have the global optimum, then restart another optimization using a larger simplex around the current best point. Alternatively, start a second optimization from a different set of initial conditions and see if the two converge to the same conditions. If you need to know the shape of the surface around the optimum (which is advisable), then carry out a limited factorial design.

10. What problems may be encountered?

Different optimization methods suffer from different problems. You should refer to a specialized optimization text for full discussion of these. However, generally one can assume that experimental noise is the *second* biggest problem. This can confuse search-type optimization algorithms such as modified simplex and can cause false conclusions to be made. The *biggest* problem occurs when an incorrect choice of instrumentation, flow system, or response function has been made in the first place.

Optimization of CF Systems by Modeling

It has been recognized that deeper theoretical studies would lead to the design of even more advanced continuous flow techniques, which would

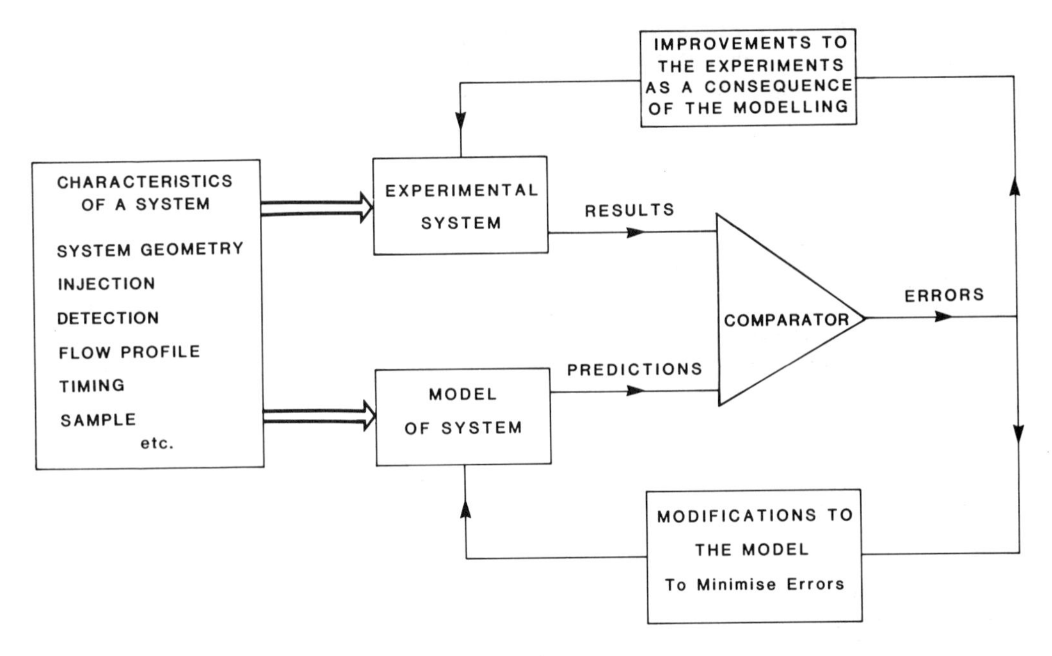

Figure 14. Modeling of a continuous-flow process.

allow chemical analyses to be performed in new ways [F34,C4,C21,C24]. These studies have inevitably relied to a great extent on computer models and simulations (Figure 14), and are not for those who shy from complex mathematics.

As discussed in a previous section, input of values for major experimental variables (d_t, n, F, t) into the Snyder-Adler dispersion model for CFA [C25,C26] yields estimates of σ_t (and therefore analyzer performance) that are in close agreement with values extracted from experimental data after the effects of longitudinal dispersion have been removed by curve regeneration [C4,C28]. This model is therefore a useful tool both for designing and evaluating new CFA systems. Models that also take into consideration the effects of chemical processes within the context of the fixed geometry and experimental variables of a particular analyzer have also been devised and used to optimize rather complex automated wet chemical determinations. Erni and Muller [F145], for example, combined relevant flow rates, concentration changes, mass and flow balance equations, etc. to model one CFA system. A few preliminary experiments were then undertaken to determine constants for their model. The mathematical technique of nonlinear programming was then used to optimize the model. This predicted the optimum conditions and that a 15-fold sensitivity improvement would be obtainable. Erni and Muller then did experiments at those conditions and actually obtained a 10-fold improvement.

Much insight into the complex interaction of reactive and dispersive processes within a simple FIA system has already been gained through a random walk model [F119]. Predictions from an extension of that model to a merging-zones FIA system led to a significantly improved experimental procedure for the determination of calcium [F69]. This and other models have been already detailed in the theory sections.

B. Optimization Specifics for CFA Systems

Automation of routine chemical determinations by CFA is often a straightforward procedure. Take, for example, a manual, colorimetric determination in which 1 ml of sample is mixed with 2 ml of reagent, diluted to a volume of 10 ml, and then allowed to react for at least 5 min before the absorbance of the resulting solution is measured photometrically. The same volume ratios of sample, reagent, and diluent could be achieved by using pump tubes with nominal delivery rates of 0.1 ml min^{-1}, 0.2 ml min^{-1}, and 0.7 ml min^{-1}, respectively. Mixing and delay coils with volumes sufficient to provide a 5-min delay between mixing and detection would be added to the analytical cartridge. The time between mixing and detection could be decreased significantly by using a thermostated reaction coil to elevate the temperature of the analytical stream, and thus increase the rate of the reaction. Alternatively, due to the reproducible timing inherent in an automated system, the absorbance could be measured before the reaction was complete. Compensation for decreased sensitivity could be achieved by increasing either the gain of the detector, or the path length of the flowcell. Manual procedures in which the analyte must be diluted or preconcentrated by more than a factor of 20 are more problematical, but can usually be accommodated.

For CFA determinations, dilution factors in the range of 15-150 are readily achieved by using either a dilution loop or a dialyzer. A dilution loop is an auxiliary analytical cartridge in which samples are proportioned into a segmented diluent stream, mixed, and returned to the inlet side of the pump. A small portion of each diluted sample is aspirated into the primary analytical cartridge(s), while the remainder flows to waste. The major disadvantage of the dilution loop technique is that the extent of longitudinal dispersion in the resampled stream as it makes a second pass through the pump may be large enough to significantly reduce the rate at which samples can be processed. In such cases, predilution by dialysis may be preferred. The extent of dilution by dialysis depends primarily on the groove length and surface-to-volume ratio of the dialyzer, the rate at which the donor and recipient streams flow through it, the type and thickness of the dialysis membrane, and the temperature of the donor and recipient streams. The main drawback to this mode of dilution is that temporal variations in membrane transport rates caused by temperature fluctuations or minor surface fouling may degrade precision of analytical results. Frequent calibration and/or strict temperature control can be used to compensate for these effects. Even higher dilution factors can be achieved by operating two (or more) dilution loops or dialyzers in series or in combination with each other. Coverly [C135] provides a complete and up-to-date discussion of predilution techniques for CFA. Solvent extraction, flash distillation, gas diffusion across microporous membranes, absorption of analyte or interferents on solid supports, and partial or total evaporation and resolvation can be implemented by the CFA technique

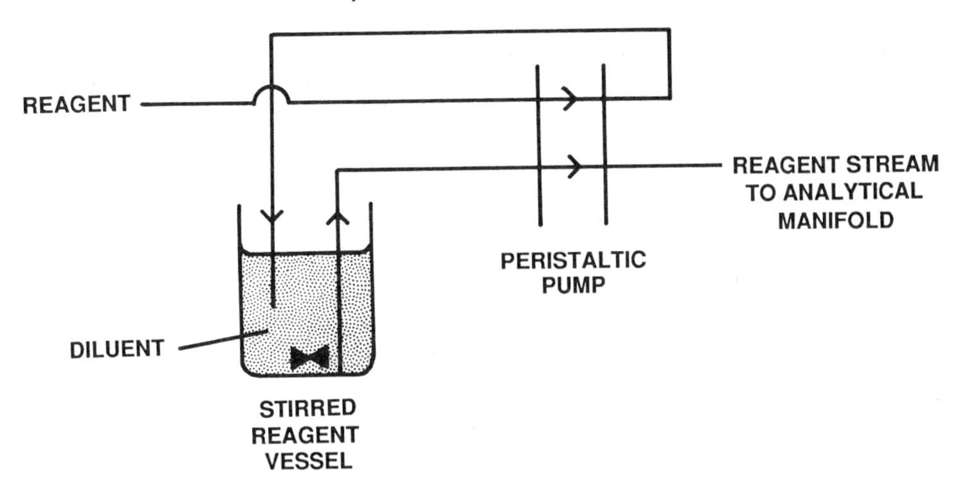

Figure 15. Typical *dilution flask* configuration for continuous variation of analyte, reagent, or interferent during development or optimization of CFA methods.

to concentrate or separate analytes from a wide variety of matrices as described in Section III.C. See Furman [C33] for additional references, as well as review articles by Burns [C3] and Kraak [C72].

Once the analyte concentration is brought into the required range, it is still necessary to optimize reagent concentrations and reaction times. Reagent concentrations can be optimized quickly by the exponential dilution flask technique described in detail by Furman [C83]. In this technique a reagent at some limiting concentration C_0 is pumped into a fixed volume ν of solvent at rate r contained in a magnetically stirred flask (see Figure 15). Liquid contained in the flask is pumped into the analytical cartridge at the same rate r, so that the volume of liquid in the stirred flask remains constant. The result is a concentration gradient that approaches the limit concentration C_0 exponentially. The concentration at any time, C_t, is calculated by the following equation:

$$C_t = C_0[1 - \exp(-rt/\nu)]$$

Generally, a midrange standard or sample is aspirated continuously while the reagent gradient experiment is in progress and the time at which a maximum signal is recorded, corrected for the lag between aspiration and detection, is used to calculate the optimum reagent concentration from the equation above. One word of caution is in order. Changes in the reagent background absorbance during such an experiment can lead to erroneous conclusions. It is therefore advisable to repeat such experiments with a sample blank to make sure changes in signal intensity are due to changes in the concentration of the product of the analytical reaction. After reagents are optimized, it is usually desirable to determine the optimal reaction time between reagent additions. This is most easily accomplished by the stopped-flow technique as described by Compton et al. [C136]. To implement this technique, a midrange standard is aspirated continuously, and after a steady-state detector response is achieved, the pump is turned off for a fixed time period. When the pump is turned off the volume of each mixing stage is essentially increased by an amount that corresponds to the product of the stop time and the flow rate of the analytical stream within the mixing stage. When the pump is turned on again, the effect of the added reaction time at each mixing stage, from last to first, appears sequentially on the chart recorder as positive or negative deviations from the original steady-state detector response. Data needed to generate a plot of signal level versus stop times for each mixing stage can be obtained by repeating the experiment for several different stop times. Results of such a treatment are then used to adjust the volume of each mixing stage to the optimum value. Using the exponential dilution flask and stopped-flow techniques it is usually possible to optimize analytical procedures in a relatively short period of time.

C. Calibration

Once experimental parameters are optimized, the next step usually is calibration. Calibration in CF requires the preparation of a set of solutions (calibrants) in which the concentration of analyte is known. The solutions are then determined in the normal way and the detector response (Y axis) is plotted as a function of nominal analyte concentration (X axis). Often such calibration functions are linear, but they need not be. It is good practice, in fact, to fit data to a second-order polynomial using a linear least-squares model. This often gives a better fit at higher concentrations, and in the event that curvature is negligible, the value of the quadratic parameter simply approaches zero. It is seldom necessary or prudent to fit data to higher than third-order polynomials, and as the order of the calibration function increases, so too must the number of calibrants. For second- or higher-order polynomials, use the Newton-Raphson method [G5-G8] to convert peak height (y) into concentration units (x) from calibration functions of the general form $y = a + bx + cx^2 + \cdots$. The common expedient of reversing the roles of the independent (x) and dependent (y) variables when higher-order least-squares calibration functions are determined, in order to dodge the added complexity (such as it is) of this calculation, can result in significant errors and should be avoided. If possible, the matrix of samples and calibrants should be closely matched: otherwise correction factors may need to be applied as described in the next section. Also, in the case of FIA, unnoticed pH or interferent gradients can cause nonlinearity, since they can affect the rate of reaction or compete with analyte for reagent [F69]. It may be possible to design a response function and reoptimize such that the effect of interferents is minimized while maintaining or enhancing method sensitivity [F141,F146].

Finally some mention of calibration frequency is in order. With older CF systems running at 20 samples h^{-1}, 5 calibrants reduce throughput by 25%. Running the same 5 calibrants on newer equipment at typical rates of 100 samples h^{-1} reduces throughput by only 5%. Thus calibrants can be run more often, improving the reliability of unknown determinations while maintaining or even reducing overhead costs. This is especially the case when data are acquired and processed by computer, since no technician time is required to calculate regression parameters or convert peak heights to concentration units. In light of the small price paid to run and process calibrants with modern CF instruments, it is advisable to establish baselines and recalibrate on an hourly basis. It is also good practice to include in each run several check standards that can be used to track long-term precision with control charts.

D. Evaluation of Interferences

In general, interferences in CF determinations can be grouped into two categories. Interferences of the first type occur when calibrants are prepared in a pristine matrix such as distilled water that lacks a substance or substances present in the samples that either enhances or suppresses the sensitivity of the analytical reaction. Such interferences are detected by spiking a sample with analyte and calculating the percent recovery using the equation

$$\%\ \text{Recovery} = \frac{[C_1(R - 1) + C_2]}{RC_3} \times 100$$

Here C_1 and C_2 are the assayed concentrations of the same sample before and after it is spiked, respectively, C_3 is the concentration of analyte in the solution used to spike the sample, and R is the volume of solution used to spike the sample divided by the total volume of the spiked sample. When preparing spikes it is usually best to add a relatively small volume of a very concentrated calibrant to a relatively large volume of sample. This minimizes dilution of the sample matrix. If all samples have the same matrix, as might be the case for seawater from the same general location, fruit juice from the same cannery, or paper pulp liquor from the same process stream, a constant factor can often be applied to all samples that accounts for such effects. When sample matrices are highly variable, it may be necessary to calculate concentrations by the method of standard additions.

Interferences of the second type occur when there is a mismatch between the physical properties of sample or calibrant matrices and that of the analyte-free sample carrier stream (FIA) or wash solutions (CFA). This can cause temporal variations in liquid junction potentials in the case of electrodes or refractive index in the case of photometric detectors that produce signals indistinguishable from those caused by the analyte. To check for such effects, omit a key reactant from one of the reagent solutions, and redetermine the samples. If peaks are observed, their heights should be subtracted from those previously measured under standard conditions. Then use the corrected values to calculate final results. As before, when the matrix of all samples is the same, a constant correction factor can be determined once and then applied to all samples. Otherwise correction factors must be determined individually for each sample as described above.

The exponential dilution flask techniques described in connection with optimization of reagent concentrations are also useful in determining the tolerance of particular assays to various substances. Details of this approach can be found elsewhere [C91,C137].

E. Data Acquisition and Processing

The output from both CFA and FIA appears as a transient waveform (peak) and is usually displayed on a chart recorder; alternatively, a microcomputer may sample the detector output, display this on a screen, and provide hard copy using a printer or plotter. The authors feel that, even where microcomputer data acquisition is available, it is useful to retain the chart recorder as both an immediate hard copy and as a "disaster indicator" to help ensure that the results seen actually represent what is occurring in the flow system. The authors have found that data acquisition rates of 2-5 Hz are quite adequate for most CF analyses, and thus the low-speed, inexpensive, 12-bit analog-to-digital convertors available as plug-in units for personal computers (PCs) are quite suitable. FIA titrations and other time-based CF procedures may require higher rates of data acquisition to achieve adequate precision.

Since for CFA within-segment variation is usually much less than between-segment variation, little is gained by high rates of data acquisition. Generally for systems with peristaltic pumps it is more effective to acquire data at the natural pump frequency—that is, at each roller liftoff or a multiple thereof—and then smooth the points either with a running average algorithm or, better, with a Savitzky-Golay polynomial least-squares digital filter [C138], which can also be used to calculate the first derivative of the data. The first derivative is useful not only for detecting the *rise*, *plateau*, and *fall* region of each peak, but also for minimizing carryover as described in the next paragraph.

In CFA, peak heights can be acquired most simply by the fixed time method. Data collection is usually synchronized with the fall time of the first peak. Falls for the peaks that follow are predicted to occur at intervals equal to the cycle time (sample time plus wash time). A single data point is then acquired a fixed number of seconds *before* each predicted fall time when the detector signal is expected to be at or near steady state. The problem with the fixed-time method is that it is susceptible to acquisition of spurious data. A more robust algorithm combines the fixed time approach with first derivative monitoring. Here *all* points with derivatives sufficiently close to zero (and therefore near steady state) are acquired and averaged. Any spikes superimposed on plateaus, however, are rejected. Having the derivative of the data available also allows its exponential component to be removed mathematically by the curve regeneration algorithm of Walker [C10] given by

$$A_r = A_t + b\,\frac{dA_t}{dt}$$

Here A_r is the absorbance predicted in the absence of longitudinal dispersion, A_t is the absorbance of the analytical stream at any time t, and b is a constant known as the *exponential factor* (see Sections III.A and III.B). The value for b can be determined, by empirical trial and error [C17], graphically from a plot of $\ln(A_{ss}-A_t)$ versus t [C13] where A_{ss} is the absorbance at steady state, or most efficiently by the method of Angelova and Holy [C16]. Digitally regenerated data can be output to a recorder through a digital-to-analog converter for display in real time. Curve regeneration generally allows sampling rates that approach the axial dispersion limit—that is, $8\sigma_t$ for peaks with flats. Carryover correction algorithms [C15,C16] that do not rely on point-by-point curve regeneration are just as effective, but since their results are not reflected in recorded peak profiles, interpretation of strip chart recordings becomes difficult (if not impossible) at high sampling rates. Readers who want to write their own software for CFA data acquisition and processing (including curve regeneration) will find articles by Macaulay et al. [C18] and Ames et al. [C139] excellent starting points. BASIC programs for FIA peak height and peak width determination can be found in appendices II and III, respectively, of Ruzicka and Hansen's *Flow Injection Analysis*, First Edition [F1].

Chemometric techniques are finding greater applicability in interpretation of CFA and FIA results. Massart has recently suggested that with the advent of "expert systems" in analytical chemistry, chemometrics should be redefined as "the chemical discipline that uses mathematical, statistical and other methods employing formal logic to (a) design and select optimal procedures and experiments; and (b) to provide maximum chemical information by analyzing chemical data" [G3]. Interest in chemometrics is growing, as is evidenced by the introduction of two specialist journals, *Chemometrics and Intelligent Laboratory Systems* and *Journal of Chemometrics*. As an example of what will be possible in the future, computerized curve deconvolution techniques and a training set of known results can be used to obtain the concentrations of several sample components in one injection [F45,F147]. One can also confidently expect expert-system-controlled continuous-flow analyzers, with the ability to self-calibrate, self-optimize, and run whole sequences of experiments unattended [F143,F148]. Such a system may interpret the observed "noise" characteristics and abnormalities in peak shape in much the way that a human expert would, and so would be able to detect and "flag" the occurrence of problems such as worn pump tubing, unwanted air bubbles, and unstable detector light source etc. (see "hints and tips" section below).

F. Automation and Automated Methods Development Systems (AMDS)

Automation of sample introduction is now commonplace. Automated report generation is becoming more common. Intelligent interpretation of those results and the scheduling of any confirmatory experiments these may suggest are probably the next steps. Total automation of CF analyzers relieves operators of tedious, repetitive tasks and the resulting boredom that this can bring. The results produced are uniform and usually more reliable since operator dependencies are no longer as major a factor. Moreover, automated instruments do not get tired (or exhausted) and are not prone to distractions, Monday mornings, Friday afternoons, or otherwise limited workdays. Against this, without adequate sensors and smart software, an automated instrument can be considered as severely handicapped when compared with a human operator; after all, it is blind, dumb, of low intelligence, has no common sense, and can take no responsibility for its actions. There is a growing need for the development of more intelligent instruments, in all areas of analytical chemistry. Those developed and under development in the authors' laboratories are a step in this direction and are termed automated methods development systems (AMDS).

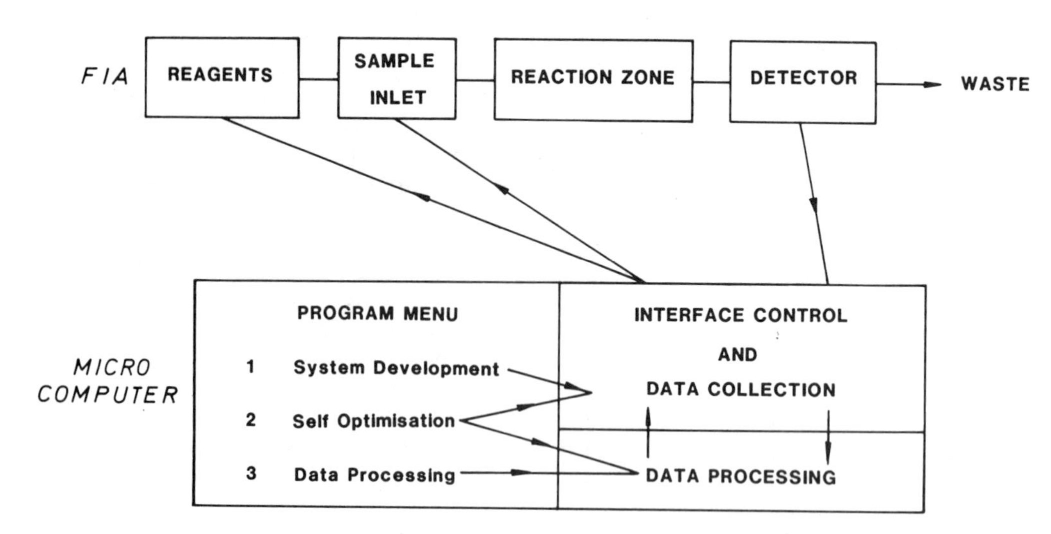

Figure 16. An automated methods development system for FIA.

Such a system for FIA, shown schematically in Figure 16, is a highly versatile continuous-flow analyzer on which chemistries can be investigated, automatically optimized, and then used for analysis of real samples [F45, F105,F117]. Such instruments are examples of "intelligent automation" in that they have a feedback loop that uses the results of the current and past experiments to influence (help determine) the next experiment(s) to be carried out [F168]. The control computer not only sets up the experimental conditions but also samples the detector(s) and interprets the readings it sees. Thus the instrument can be said to be able to "learn from experience" [F45,F143,F148]. The mode of operation of these instruments is programmable from software, which allows it to be put to many varied uses. Whole sequences of experiments can be preprogrammed and run automatically.

Several such instruments have been developed and recently published [F45,F105,F115-F117]. Those in our laboratories [F45,F105,F117,F169,F170] for FIA automatically mix several reagent streams, inject a specified size of sample, and, by means of flow reversals, can allow for continuous variation of reaction coil length. Individual and interacting concentration gradients can also be automatically generated, and the instrument may be used for various types of "step" and "stopped-flow" experiments. These have applicability to problems of multicomponent analysis by FIA [F45]. Also, mapping of chemical surfaces by means of FIA experimentation may be achieved using such a system [F169,F170].

For construction of future such systems we recommend that the controlling microcomputer(s) have at least the power of an IBM PC/AT and preferably have the capability for multitasking. Commercially available interface boards are quite adequate for this purpose and plug straight into empty slots in the PC. Some pumps are now made available with built-in RS232 serial communications, which allow direct computer control over a serial line. Digital input/output (I/O) is required for control signals to switch valves and for noticing that they have been switched. Also, digital inputs from liquid level detectors on reagent reservoirs that indicate levels are getting low *before* they run out are useful. Analog-to-digital conversion is required for reading simple detectors, and digital-to-analog conversion may be needed for setting detector offset and gain, and for output of results to a standard chart recorder. More modern (and expensive) detectors may have in-built microprocessor control and be able to receive commands and send data over a serial line or via an IEEE-488 interface. A good introductory text on computer-assisted experimentation has recently been published by Ratzlaff [G4].

Instrument control software has previously been written in FORTH [F115] and in BASIC [F45]. The newer compiled versions of BASIC (such as MicroSoft's QuickBASIC) have been found quite suitable in our laboratories. An alternative to writing ones own code in a high-level language is to use a higher-level language specifically designed for data collection and interpretation. Examples are the ASYST and ASYSTANT+ programs sold by MacMillan Software (866 Third Avenue, New York, NY). These provide data reduction and analysis routines, high resolution and color graphics, optional analog-to-digital conversion, and RS-232 and GPIB interfaces. The manufacturers' literature compares the capabilities of the various software options. As faster and cheaper implementations of languages more suited to artificial intelligence (e.g., LISP, Prolog) and powerful real-time expert system shells with the capability to talk to peripheral devices become available, we anticipate (and indeed are working toward) smarter automated decision-making software, and therefore smarter CF analyzers [F148].

G. Hints and Tips for Method Development and Fault Finding

As in any discipline that requires expertise, there are various tricks of the trade that, while seldom seen in publications or textbooks, can be of inestimable value to the user.

A CF system should be leak-tight. Fittings can be damaged by organic or otherwise corrosive solvents that a previous user "forgot to mention." This can leave them brittle and badly fitting. Worn fittings can leak, especially when there is a higher than normal pressure in the CF system. A roll of Teflon tape can be used to make worn fittings usable for a little longer. Wrap about 5 cm of the tape around the barrel of each fitting. Then screw it into the union. The tape will deform to make a good connection. Some fittings require that tube ends be flanged. Poor flanging on the end of tubes can also cause leaks. If the flange does not look even or circular, cut the end of the tube with a sharp knife and reflange it. Small-bore pump tubing may part company with the Teflon tube into which it is delivering solution. This will happen less often if barbed connectors rather than push fittings are used. Leaks can also be caused by too high a back-pressure. This may be due to precipitation in the tubes, or some other blockage (use your eyes).

A noisy baseline and poor experimental repeatability can also have several causes. Air bubbles in FIA systems will cause sharp off-scale spikes on the chart recorder. These can be introduced by poor injection technique, from a poor connection, or through failure to adequately degas solutions prior to use. In FIA systems, small air bubbles can sometimes become lodged within flow cells—running the pump at full speed for a few seconds will usually dislodge them. Worn pump tubing can cause an unstable baseline, as can precipitate in the tubes. Noise can also be electrical in nature. Cables carrying analog signals should be shielded. Even then, if someone is switching large currents in the laboratory next door, you may see noise spikes. All electrical connections should be reliable (soldered, screw terminals, or plug and socket—not two wires twisted together). If connections are good, check to see if (for absorbance measurements) insufficient light is reaching the cell. There may be a fault in the detector (is it switched on?); the light source may be getting old and variable. If this is so, then replace it.

Generally, you should try to make sure that reagents are freshly prepared, that sufficient mixing is taking place, dead volumes are minimized, all solutions are sufficiently degassed and filtered, reagents or sample do not attack the apparatus, and that expected temperature variations over a 24-h period do not cause unacceptable changes in the results.

One way of avoiding using specialist tubing for corrosive organic solvents that have low aqueous solubility is to use a displacement bottle. Here, water is pumped into a sealed chamber and displaces the organic solvent into the flow system. In one system [F93], the displacement vessel had an outer glass envelope through which circulating water (from a thermostated water bath) kept the organic solvent at a preset temperature.

It is advantageous to be able to specify and control the temperature of a CF system. Temperature will affect the rate of reaction and also the dispersion observed in FIA systems [F128,F129]. While adding to the complexity of the apparatus, heated reaction coils can decrease the analysis time and/or improve sensitivity for slow reactions. Typically, a 10°C rise in temperature may double the reaction rate, but remember that in FIA adequate mixing is equally important. Thermostatically controlled water baths may be used to hold solution vessels and the reaction manifold (analytical cartridge). In more complex systems [F150,F151] the flow system may be inside a larger-bore tube through which heated water is passed.

For chemical systems where generally trace-level concentrations are involved [see refs. G1, G2], preconcentration prior to analysis may be necessary. This may be achieved in a nonsegmented stream within the CF analyzer, by means of a chelate ion-exchange column or other preconcentration column. Alternatively, solvent extraction techniques can facilitate preconcentration factors of up to 20. These require phase separators [F94,F95], which may have to be machined "in-house." Finally, there exist special cells for gas diffusion and dialysis [F152]. These same techniques can be used for sample cleanup and the removal of likely interferents prior to determination of the analyte. Metallic reactors, such as the Jones Reductor, have been used for continuous-flow phosphorescence studies where, unless removed, any dissolved oxygen will quench the analytical signal. One problem here is that oxygen finds the walls of most tubing readily permeable, and thus special precautions still have to be taken [F153].

CFA Specifics

Sampling

Intersample air segments (IAS) greatly reduce interaction of sample and wash slugs as they flow from their containers into the analytical cartridge. This is demonstrated by the simple experiment [C28] illustrated in Figure 17b. By connecting a pump tube (nominal flow rate = 0.32 ml min^{-1}) directly to the flow cell (1 cm path length, 2 μl internal volume) or a recording photometer equipped with 540-nm interference filters, it was possible to observe the effect of IASs on the concentration profile of samples (an alkaline solution of phenol red in this case) that under normal conditions would enter the analytical cartridge. With reference to Figure 17c, curve 0 resulted when the IAS was eliminated by turning off the pump before the withdrawal tube was cycled between containers; curve 1 resulted from normal CFA sampling in which a single IAS separates each sample and wash slug; curve 3 resulted when two additional IASs were introduced at the leading edge of each sample and wash slug by the "pecked" sampling technique described in Section III.A. Pecked sampling can also be used to increase the efficiency of rinsing and filling the sample loops of injection valves, which should be connected as shown in Figure 17a for best results.

To further minimize longitudinal dispersion that occurs between the sampler and the analytical cartridge, keep the tubing connecting the sampler to the peristaltic pump as short as possible and make sure its internal diameter is no larger than that of the pump tube to which it is connected. Thin-walled polyethylene or Teflon tubing is a good choice for this application. The sudden appearance of malformed peaks, especially in third-generation systems, often results from IAS breakup or hang-up within this tubing. Replace it at once if this occurs. When multiple analytical cartridges are fed by the same sampler, a portion of each IAS must be aspirated into each one. Otherwise loss of wash will be excessive and precision will be poor. Commercially available, low-volume stream splitters, designed especially for this application, give the best results. It is worth mentioning that visually tracing the progress of intersample air bubbles as they flow toward the analytical cartridge is a good way to spot clogged transmission tubing or malfunctioning pump tubes. Briefly exposing reagent supply lines to the atmosphere allows the same approach to be used for trouble-shooting reagent transmission tubing and pump tubes.

Each time an IAS passes through the pump it is compressed and the flow rate of the solution ahead of it decreases briefly. As a result periodic proportioning errors occur at intervals equal to the sum of the sample and wash times, which may be manifested as "dips" or "notches" in the flats of recorded peaks. When peak heights are estimated by human operators these dips are little more than cosmetic annoyances, but they may seriously compromise the performance of computer-based peak height estimation algorithms. Such proportioning errors will not affect analytical results if they occur on the rising or falling edges of peaks, as in the case when the IAS for one sample enters the analytical cartridge at the same time that the IAS for the sample that follows it is compressed by a pump roller. Length of "sample" pump tubes can be adjusted on the down-stream (between the pump and the analytical cartridge) side to achieve this condition.

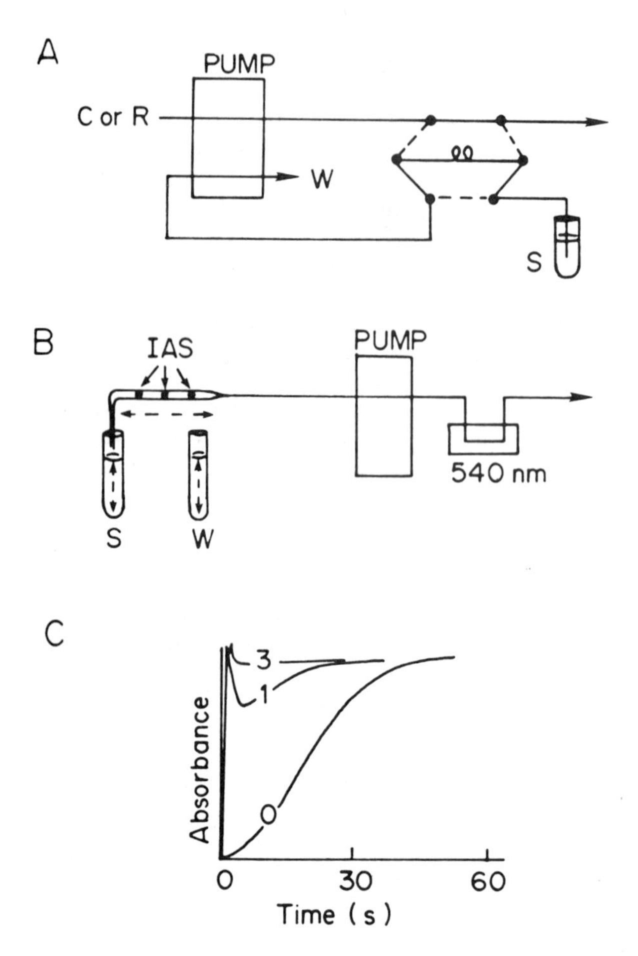

Figure 17. Sample introduction methods for CFA. (A) Recommended configuration to fill the loop of a sample injection device. (B) Sample aspiration with and without the introduction of intersample air segments (IAS). Here the pump tube is connected directly to the flow cell of a recording photometer to permit dye dispersion experiments described in the text. (C) Dye dispersion profiles showing the extent of sample zone broadening that occurs between the sampler and the analytical cartridge with and without the introduction of IAS.

The rationale of air phasing in CFA systems that use peristaltic pumps has already been discussed, but its importance cannot be overemphasized. Most peristaltic pumps designed for use with CFA systems are now equipped with some type of air bar. In the absence of such devices, however, uniform proportioning can still be achieved by using a "poor man's air bar," which was first described by Habig et al. [C9]. Two pump tubes are required, one of which (designated A) must have a delivery rate approximately

50% greater than that of the other (designated B). The outlet of tube A is connected to the inlet of tube B. The outlet of tube B is connected directly to the air inlet port of the analytical cartridge. Inside tube B pressurized air, delivered by tube A, is trapped between rollers. Each time a roller leaves the platen surface, this compressed gas rapidly expands into the analytical cartridge. This easily implemented technique is remarkably effective. Details of a more elaborate air bar with electro-optical phasing and solenoid valve gas flow regulation has been described elsewhere [C28]. As explained previously, the maximum allowable segmentation frequency (one per roller lift-off) for a particular pump is determined by the rate of rotation of its mechanism and the number of rollers it contains. Proportioning precision is maintained, however, when air segments are added at some whole-number fraction (1/2, 1/3, 1/4, ...) of this frequency. For some applications involving long path-length flow cells or low flow rates, the fact that the volume of each liquid segment can be doubled, tripled, or quadrupled simply by reducing the segmentation frequency by a factor of two, three, or four, is worth keeping in mind.

Pump pulsations can also adversely affect the performance of samplers and detectors. For example, even when sample and wash times are controlled to a high degree of precision, significant variation in the volume of sample aspirated into the system can occur if the movement of the sampler arm is not coordinated with pump pulsations. In addition, pump pulsations may also cause cyclic, low-frequency fluctuations in the outputs of many detector types that are impossible to remove by electronic *RC* filtering without severely distorting the analytical signal. In such cases, sampling the detector output intermittently and in phase with pump pulsations can decrease noise significantly. Opekar and Trojanek [C140] used this technique to good advantage for cathodic striping voltammetry experiments and provide details of the electronic circuitry necessary to implement it. This technique has much in common with bubble-gating, which is described below in more detail.

Thermostating

Thermostated reaction coils are frequently used in CFA determinations to accelerate the rate of chemical reactions. Failure to regulate the temperature to within ±0.5°C can result in cyclic changes in sensitivity, but this is seldom a problem with commercially obtained devices, which for the most part use proportional (as opposed to on/off) thermoregulators. It is good practice to cool the analytical stream to near ambient temperature before it enters the detector. This can be done by allowing it to flow through several mixing coils or some other type of heat exchanger such as a finned or water-jacketed length of tubing. It can also be beneficial to thermally insulate the tubing leading to the detector and the detector flow cell itself. This practice is particularly effective for reducing photometric detector noise when the analytical wavelength is above about 700 nm, particularly when using silicon photodiode detectors, which generally have maximum sensitivity in the range of 800-900 nm. It should also be mentioned that for kinetics-based determinations, thermal regulation for the entire system to within ± 0.1°C is required for good precision of analytical results. This degree of temperature regulation is seldom provided in commercial systems.

Debubbling and degassing

Outgassing of reagents or samples is seldom a problem in CFA determinations because the segmenting gas bubbles tend to scavenge any gas released from the analytical stream. When it is necessary to debubble the analytical stream—that is, applications that involve packed bed columns containing active metals or ion-exchange resins, or ungated detectors—glass debubbling tees in a variety of configurations are available commercially. Porous Teflon tubing (available through Anspec, Ann Arbor, MI) can also be used to debubble the analytical stream [C67] without recourse to debubbling tees or auxiliary pump tubes. Back pressures of approximately 1 psi are sufficient to force gas segments through the tube walls. [Porous polypropylene, which is available in sheet and tubular form (Enka America, Asheville, NC), may also be useful for such applications.] Porous plastics can become fouled and lose gas permeability. They are also permeable to many organic liquids. For these reasons they may not be suited to some applications.

Bubble-gate or regenerate?

Examples of bubble-gates that discriminate between air bubbles and liquid segments on the basis of conductivity [C9], software algorithms [C113], digital logic [C115,C118], and refractive index [C141] can be found in the open literature. Secondary bubble detectors based around commercially available opto-interrupter type switches have also been designed [C142]. Readers with an interest in constructing their own bubble-gates may find any of these approaches good starting points. For applications that require multiple detectors, such as automatic blanking or kinetics determinations, bubble-gating is the only way to go. If bubble-gating is being considered solely as a means to increase the sample throughput of existing second-generation CFA equipment, however, curve regeneration [C10,C139] often presents an equivalent and more easily implemented solution.

FIA Specifics

Wherever possible, if an FIA system is giving problems, initially test its operation by injection of an appropriate dye into a water carrier stream. In this way it is likely that equipment malfunction may be distinguished from problems with the chemistry. *Before* running the intended chemistry it is common to test out new FIA systems using a dye that absorbs at the same wavelength as the product.

One reason for excess tailing of peaks (and possibly a continuously rising baseline) is precipitation occurring in the tubes. In this case (and provided that the detector circuitry has sufficiently fast response) the trace is likely to appear noisy. If precipitation is occurring then usually the chemistry needs to be modified, perhaps by decreasing concentrations or a pH change. A rarer happening, likely only when a plastic flow cell is in use, might be the discoloration of the flow cell walls by precipitate or changing the pH. A rarer happening, likely only when a plastic flow cell plastic flow cells by flushing the system with dilute nitric acid, or filling it with the same and leaving it overnight.

Peristaltic pumps, particularly those with fewer than about 10 rollers, can cause substantial pulsation of the flow pattern. This may be observed by allowing a small air bubble into the tube and watching its progress through the system. Such pulsation can cause poor baseline and decreased repeatability in determinations. Generally, the more rollers a peristaltic pump has, the better its performance will be. Types with 10, 12, or 16 rollers give good results. The eight-roller or six-roller pumps are often acceptable. The cheap two-roller or three-roller "dispensing" peristaltic pumps are *not* recommended. One means of decreasing pulsation is to insert "depulse coil(s)" of perhaps 1.5 m length between the pump and the injection valve(s). The coils may be of the same material as the rest of the flow system, or even better, of the softer pump tubing, having a larger i.d. than the other tubing. These tubes act as a hydraulic buffer (damper) and, since they are inserted behind the sample injection valve(s), no contribution to dispersion is made. For some pumps, a greater pulsation effect appears to occur when the stream is pulled rather than pushed. If a suitable pump is not available, use gravity feed; it does not suffer from pulsation and can provide excellent results.

VII. HARDWARE

Many of the fittings and supplies used for CFA and FIA are common, and thus no distinction between them will be made in this section.

A. Pumps

The simplest form of propulsion for CF methods is a gravity feed. In this the carrier solution flows from a vessel placed at a level higher than the apparatus. Flow rate may be controlled by (i) the height of the reservoir relative to the manifold or (ii) a back-pressure-inducing flow constriction device at some point in the system—this may be, for example, a long coil of manifold tubing, a shorter length of narrower-bore tubing, a clamp on a piece of silicone tubing, or a needle valve.

Peristaltic pumps are by far the most commonly used means of propulsion in CF systems. As mentioned above, pumps with 10 or more rollers are to be preferred. Quality peristaltic pumps are made by Ismatec (Cole-Parmer, 7425 N. Oak Park Ave., Chicago, IL 60010; Phone 1-800-323-4340), Gilson International (Gilson Medical Electronics Inc., Box 27, 3000 W. Beltline Hwy., Middleton, WI 53562; Phone 608-836-1551), and Alitea USA, P. O. Box 26, Medina, WA 98038; Phone 206-522-8318.

The choice of peristaltic pump tubing to be used will depend on the solutions it is expected to pump. Common materials for pump tubing are polyvinyl chloride (PVC) and silicone rubber. However, these are not compatible with solvents such as chloroform or with strong mineral acids. In such cases "acidflex" tubing is much to be preferred. Pump tubings may be obtained from most vendors of FIA and CFA equipment, although their quality may vary greatly from supplier to supplier; *caveat emptor*.

Syringe pumps should find increasing use in CF analyzers. One exceptional system is Harvard Apparatus's pump no. 22 (Harvard Apparatus: An Ealing Company, 22 Pleasant Street, South Natick, MA 01760; Phone 617-655-7000). This unit is equipped with a modified RS-232 interface and easy-to-use front panel controls. The stepper motor drive is 1.5° per step, as opposed to the 7.5° systems used elsewhere, and thus pulsation is essentially minimal.

Pressurized vessels have been used to propel the streams [F20]. This is particularly useful in process control, due to the limited lifetime of pump tubing for peristaltic pumps and the level of operator attention that they need.

Reciprocating pumps, as used in liquid chromatography, have been used in FIA process control applications but generally are far from ideal. They are not used in CFA at all, as, in the presence of air bubbles, the degree of pulsation is unacceptable. This may be due to insufficient back pressure.

B. Valves and Injection Devices

The most common type of injection valve used is the six-port two-way valve, as manufactured by Rheodyne (Rheodyne Inc., P.O. Box 996, Cotati, CA 94928; Phone 707-664-9050). Valves for general stream switching, rather than sample injection, are also supplied. Operation of these has been discussed above and is shown in Figure 1. An alternative to this is to use two three-port solenoid valves. These switch an incoming stream one of two ways. Using two three-port valves for research equipment gives increased flexibility. Rheodyne also markets pneumatic actuators for their valves, and supplies electrical control for these. A totally electromechanical valve is made by P. S. Analytical Ltd. (Arthur House, Cray Avenue, Orpington, Kent, BR5 3TR, UK) and marketed in the United States through Questron Corporation (P.O. Box 2387, Princeton, NJ 08540-9978; Phone 609-587-6898). An alternative to actuated valves is the inexpensive timed solenoid valve proposed by Rothwell and Woolf [F149].

P.S. Analytical also makes an 80-position large-volume autosampler with in-built TTL logic interface as standard or an optional RS-232 interface. A septum-piercing probe and a magnetic stirrer are also available. The unit holds 80 30-ml sample bottles on a circular platter. Many other manufacturers also supply such equipment, although the layout and capabilities will vary considerably.

C. Manifolds/Reactors

Many suppliers of parts and accessories for liquid chromatography and clinical suppliers will sell tubes and fittings for manifolds for CFA and FIA. One company that offers a good selection, including Teflon tubing, couplings, end fittings, glass-to-Teflon connectors, plugs, and a flanging tool, is Supelco (Supelco Inc., Bellefonte, PA 16823-0048; Phone 800-247-6628). Omnifit (Omnifit Ltd., P.O. Box 56, 1187 Park Street, Atlantic Beach, New York, 11509; Phone 516-239-1655) provides an excellent selection of precision miniature fittings, junctions (tees, crosses, eight-way connectors, etc.), distribution valves, adapters, tubing, and an injection valve. Upchurch Scientific (P.O. Box 1529, Oak Harbor, WA 98277; Phone 800-426-0191) also sells fittings, junctions, unions, and tubing. Hamilton (Hamilton Company, P. O. Box 10030, Reno, Nevada, 89520; Phone 702-786-7077) manufactures a number of miniature inert valves and actuators and sells tubing and fittings. The choice of material for manifold tubing can be very important. In CFA systems, a sample line of polyethylene tubing is probably the best choice for stable flow, low back pressure, and low carryover. Highly acceptable mixing and delay coils for CFA can also be made from polyethylene tubing as described by Amadore [C143]. Note that it may be necessary to add more surfactant to reagent solutions when polyethylene coils are used. Polyethylene and Teflon tubes are commonly used for FIA systems. Porous Teflon [C67] (Anspec Co., Ann Arbor, MI) and polypropylene (Enka America, Asheville, NC) tubing may be used to remove unwanted gas bubbles from flowing streams.

D. Detectors and Detection Methods

Most manual detectors provide a single absorbance value or a single spectrum relative to some reference for what is usually a time-invariant signal. The detection systems used in CF methods monitor transient peaks and therefore there exist several other possible measurements. Spectrophotometric measurements are essentially limited to the UV/visible and/or near-IR wavelengths. In this laboratory simple single-wavelength photometric detectors [F16,F17] are used to measure *peak height* (i.e., the difference in absorbance between the product at its most concentrated part and the carrier stream) [F6,F8]. It is possible to read the absorbance (or absorbance difference) at some point other than the peak maximum. In the tail of the peak there is a continuous progression of concentration/absorbance measurements [F154]. One can also use a diode array spectrophotometer [F169-F170] to obtain an extra dimension of data, that is, the wavelength versus absorbance versus time profile of the peak. Peak area has also been used as the analytical measurement [F155], but is more difficult to calculate without computerized data processing. Peak area is the integration of signal minus background with time. Use of the differential of the peak has also been proposed as a means of improving the accuracy of results from FIA titrations [F156], and peak width is measured in some CF "titrations" [F157].

Types of detection for CF systems (other than UV/visible absorbance) abound; examples include FIA-FTIR [F101], ion-selective electrodes (ISEs) [F158], stationary mercury electrodes for voltammetry [F159], dropping mercury electrodes for polarography [F160], thermistors [F161], refractive index detectors [F16], and ChemFETS [F162]. Fluorescence [F163,F164] and chemiluminescence [F118,F152,F165] have also been used to great advantage. Most detectors designed for liquid chromatography will work for CF. However, since the CF systems we mention here all work at low pressure, high-pressure flow cells are unnecessary.

Vendors of flow-through electrode cells include Radiometer (Radiometer America, Inc., 811 Sharon Drive, Westlake, OH 44145; Phone 216-871-8900), Sensorex (11661 Seaboard Circle, Stanton, CA 90680; Phone 714-895-4344), and Lazar Research Laboratories, Inc. (920 N. Formosa Ave., Los Angeles, CA 90046).

VIII. VENDOR LIST AND PRODUCT DESCRIPTION

In this section we list manufacturers of laboratory CFA and FIA systems. The descriptions that follow are based largely on promotional material provided by each vendor.

A. Vendors of CFA Systems

Technicon Instruments Corporation

Technicon Corporation, 511 Benedict Avenue, Tarrytown, NY 10591; Phone 914-333-6142) has a full range of clinically oriented CFA systems and equipment. These include the third-generation SMAC and SMAC II analyzers, and the zero-dispersion, random-access CHEM I system. Industrially oriented systems now available through Bran & Lubbe Analyzing Technologies, Inc. (103 Fairview Park Drive, Elmsford, NY 10523-1500; Phone 914-524-8000) include second-generation AutoAnalyzer II and third-generation TRAACS-800. Both systems have provisions for automatic shutdown, and the TRAACS-800 is equipped with a random-access sampler and dilution loop that provides for automatic dilution and redetermination of samples that produce off-scale responses. Microprocessor-based data acquisition and processing systems that interface to an IBM-PC are available for both the AAII and TRAACS. These data systems provide for calibration plots, baseline drift correction, carryover correction, and full-page-size final results reports. An extensive library of developed methods is available to customers. In-house training, field advice, module exchange, telephone troubleshooting, and methods development services are available.

Scientific Instruments

Scientific Instruments (25 Broadway, Pleasantville, NY 10570; Phone 800-431-1956) offers its own line of second-generation CFA modules, including sampler, pump, and analytical cartridge. Spectrophotometric, potentiometric, and fluorometric detectors are available. They also sell replacement parts for modules for the Technicon AAII CFA system. A data acquisition and processing system is also offered, operating on an IBM-PC/COMPAQ PC, which provides for baseline correction, drift correction, carryover correction, check standards, and calibration. The interface provides for detector inputs but does not control the sampler.

Alpkem Corp.

Alpkem (P.O. Box 1260, Clackamas, OR 97015; Phone 503-657-3010) manufactures and sells a complete line of modular, third-generation, CFA instrumentation hardware and analytical cartridges under the trade name RFA-300. The RFA system is available in one-, two-, or three-channel configurations. Photometric (UV/VIS) and fluorometric detectors are available. An extensive list of developed methods is available to customers. A real-time, multitasking data acquisition and processing system, the RFA-PC SoftPac, implemented on an IBM PC (or compatible), is also available. The RFA-PC SoftPac controls two samplers independently and can acquire and process data from up to eight 0-5 V inputs with 12-bit resolution. This system is compatible not only with the RFA-300 product line but also with second-generation CFA hardware produced by other manufacturers. In addition to data acquisition, corrections for baseline drift and carryover are provided along with numerous calibration and QC plots, spikes, weight and dilutions calculations, etc. Also provided for is curve regeneration by the Walker algorithm. In-house training, field service, module exchange, telephone troubleshooting, and methods development services are available.

SKALAR Analytical

Skalar Analytical (Spinveld 2, 4815 HS, Breda, Netherlands; Phone 0-76-225477) manufactures and sells a complete line of second-generation CFA modules, hardware, and consumable supplies and has an extensive library of automated wet chemical methods. They also offer a variety of data acquisition and processing systems in configurations for 2-, 4-, 8-, and 16-channel applications. Seamark, Inc. (P. O. Box 1325, Buford, GA 30518; Phone 404-932-1300) is the distributor for Skalar products in the United States.

B. Vendors of FIA Systems

Lachat Instruments

Lachat (10500 N. Port Washington Road, Mequon, WI 53092; Phone 414-358-4200) offers a wide range of FIA systems and modules in its Quick Chem product line. Photometric, potentiometric, amperometric, and conductance detectors are available along with replacement parts, consumable items, and analytical cartridges. Training, field service, and analytical methods development services are available. A data acquisition and processing system implemented on an IBM PC-XT is also available that provides for calibration, baseline drift correction, and dilution and weight calculations. Over 50 chemical methods are available and custom development is an option. Lachat also markets an FIA process analyzer controlled by an IBM PC-XT and aimed at bioreactor applications.

Control Equipment Corporation (CEC)

Control Equipment Corporation (P.O. Box 2154, Lowell, MA 01851; Phone 617-459-0573) markets the MultiFlow line of FIA equipment, which features a pulseless gas-displacement pumping system and microprocessor control of operation parameters such as load and inject times, number of replicate injections per sample, and delay times. External control is possible through an RS-232 interface. Data acquisition is provided by a Spectra Physics 4200 integrator. Photometric, fluorometric, and amperometric detector modules are available, as are a variety of standard and custom-designed analytical cartridges.

Fiatron Systems

Fiatron (6651 N. Sidney Place, Milwaukee, WI 53209; Phone 414-351-6650) offers a line of FIA modules that include injection and switching valves, photometric, potentiometric, and amperometric detectors, and a variety of standard and custom analytical cartridges. An integral microprocessor in each detector module provides for data acquisition and processing. Raw data or calculated results can be displayed on a 16-digit fluorescent display or dumped to an optional printer via an I/O port. Fiatron also markets a FIA process monitor with gas-displacement pumps and microprocessor control.

Tecator offers the FIAstar flow-injection analyzer [F166]. In the United States this product line is distributed by the Alpkem Corporation (P.O. Box 1260, Clackamas, OR 97015; Phone 503-657-3010). Much published work has been done using this robust and effective unit. Tecator also distributes a bibliography of FIA applications and a number of information sheets detailing common analytical chemistries and their implementation on Tecator hardware.

In addition to the above companies, Hitachi produces FIA equipment in Japan and Microanal does so in Brazil.

REFERENCES

A. CFA References

C1. L. T. Skeggs, *Am. J. Clin. Pathol. 28*, 311-322 (1957).
C2. L. R. Snyder, J. Lavine, R. Stoy, and A. Conetta, *Anal. Chem. 48*, 942A-956A (1976).
C3. D. A. Burns, *Anal. Chem. 53*, 1403A-1418A (1981).
C4. L. R. Snyder, in *Advances in Automated Analysis: Technicon International Congress 1976*, Vol. I, Mediad Inc., Tarrytown, NY (1977), pp. 76-81.
C5. C. J. Patton and S. R. Crouch, *Anal. Chim. Acta 179*, 189-201 (1986).
C6. R. H. P. Reid and L. Wise, in *Automation in Analytical Chemistry, Technicon Symposia 1967*, Vol. II, Mediad Inc., White Plains, NY (1968), pp. 159-165.
C7. N. R. Kuzel, H. E. Roudebush, and C. E. Stevenson, *J. Pharm. Sci. 58*, 381-406 (1969).
C8. R. Robinson, T. J. Bonham, G. Poxom, and T. Kelleher, in *Advances in Automated Analysis: Technicon Symposia 1966*, Vol. II, Mediad, Inc., White Plains, NY (1967), pp. 211-214.
C9. R. L. Habig, B. W. Schlein, L. Walters, and R. E. Thiers, *Clin. Chem. 15*, 1045-1055 (1969).
C10. W. H. C. Walker, *Clin. Chim. Acta 32*, 305-306 (1971).
C11. R. E. Thiers and K. M. Oglesby, *Clin. Chem. 10*, 246-257 (1964).
C12. R. E. Thiers, R. R. Cole, and W. J. Kirsch, *Clin. Chem. 10*, 451-467 (1967).
C13. W. H. C. Walker, C. A. Pennock, and G. K. McGowan, *Clin. Chim. Acta 27*, 421-435 (1970).
C14. W. H. C. Walker, in *Continuous Flow Analysis: Theory and Practice* (W. B. Ferman, ed.), Marcel Dekker, New York (1976), pp. 217-218.
C15. R. E. Thiers, J. Meyn, and R. Wildermann, *Clin. Chem. 16*, 832-839 (1970).
C16. S. Angelova and H. Holy, *Anal. Chim. Acta 145*, 51-58 (1983).
C17. W. H. C. Walker, J. Townsend, and P. M. Keane, *Anal. Chim. Acta 36*, 119-125 (1972).
C18. M. Macauly, J. Mathers, C. Jacklyn, and C. Munro, *Clin. Biochem. 9*, 111-116 (1976).
C19. E. B. M. de Jong and A. H. Weyden, *Anal. Chim. Acta 114*, 311-317 (1980).

C20. J. Hrdina, in *Amino Acid Analysis: 6th Coloquium*, Monograph No. 3, Technicon Corp., Tarrytown, NY (1967).
C21. R. E. Thiers, A. H. Reed, and K. Delander, *Clin. Chem. 17*, 42-48 (1971).
C22. R. D. Begg, *Anal. Chem. 44*, 631-632 (1972).
C23. W. H. C. Walker and K. R. Andrew, *Anal. Chim. Acta 57*, 181-185 (1974).
C24. W. H. C. Walker, in *Advances in Automated Analysis: Technicon International Congress 1976*, Vol. I, Mediad Inc., Tarrytown, NY (1977), pp. 82-85.
C25. L. R. Snyder and H. J. Adler, *Anal. Chem. 48*, 1017-1027 (1976).
C26. L. R. Snyder, *J. Chromatogr. 125*, 287-306 (1976).
C27. L. R. Snyder, *Anal. Chim. Acta 114*, 3-18 (1980).
C28. C. J. Patton, Ph.D. Dissertation, Michigan State University, East Lansing (1982).
C29. F. Nydahl, *Talanta 23*, 349-357 (1976).
C30. M. B. Shinn, *Ind. Eng. Chem., Anal. Ed. 13*, 33-35 (1941).
C31. K. Bendschneider and R. Robinson, *J. Mar. Res. 11*, 87-96 (1952).
C32. M. K. Swartz, in *Continuous Flow Analysis: Theory and Practice* (W. B. Furman, ed.), Marcel Dekker, New York (1976), p. iii.
C33. W. B. Furman, *Continuous Flow Analysis: Theory and Practice*, Marcel Dekker, New York (1976).
C34. Technicon Instruments Corporation, *Technicon AutoAnalyzer Bibliography*, 1957/1967, New York (1968).
C35. Technicon Instruments Corporation, *Technicon Bibliography, 1967/1973*, New York (1974).
C36. Technicon Instruments Corporation, *Technicon Bibliography Supplement #1*, New York (1974).
C37. Technicon Instruments Corporation, *Technicon Bibliography Supplement #2*, New York (1975).
C38. Technicon Instruments Corporation, Technicon Symposia Proceedings 1963/1976, Mediad Inc., New York.
C39. J. Isreeli, M. Pelavin, and G. Kessler, *Ann. N.Y. Acad. Sci., 87*, 636-649 (1960).
C40. J. E. Steckel and R. L. Flannery, in *Automation in Analytical Chemistry: Technicon Symposia 1965*, Mediad Inc., New York (1966), pp. 116-122.
C41. D. B. Nevius and G. F. Lanchantin, *Clin. Chem. 11*, 633-640 (1965).
C42. R. H. Gaddy, in *Automation in Analytical Chemistry: Technicon Symposia 1965*, Mediad, Inc., New York (1966), pp. 215-219.
C43. B. W. Bailey and F. C. Lo, *Anal. Chem. 43*, 1525-1526 (1971).
C44. P. D. Goulden and B. K. Afghan, in *Advances in Automated Analysis, Technicon International Congress 1970*, Vol. II, Thurman Associates, Miami (1971), pp. 317-321.
C45. P. D. Goulden and P. Brooksbank, *Anal. Chem. 46*, 1431-1436 (1974).
C46. F. D. Pierce, T. C. Lamoreaux, H. R. Brown, and R. S. Fraser, *Appl. Spectrosc. 30*, 38-47 (1976).
C47. F. D. Pierce and R. Brown, *Anal. Chem. 48*, 693-695 (1976).
C48. M. Yamamoto, M. Yasuda, and Y. Yamamoto, *Anal. Chem. 57*, 1382-1385 (1985).

C49. M. Yamamoto, K. Takada, T. Kamamaru, M. Yasuda, and S. Yokoyama, *Anal. Chem. 59*, 2446-2448 (1987).

C50. J. R. Prall, in *Advances in Automated Analysis, Technicon International Congress 1970*, Vol. II, Thurman Associates, Miami (1971), pp. 395-398.

C51. R. A. Arend, J. E. LeKlem, and H. R. Brown, in *Advances in Automated Analysis, Technicon International Congress 1979*, Vol. II, Mediad, Inc., White Plains, NY (1970), pp. 195-199.

C52. R. H. Mandl, L. H. Weinstein, J. S. Jacobson, D. C. McCune, and A. E. Hitchcock, in *Automation in Analytical Chemistry, Technicon Symposia 1965*, Mediad, Inc., NY (1966), pp. 270-273.

C53. R. S. Manly, D. H. Foster, and D. P. Harrington, in *Automation in Analytical Chemistry, Technicon Symposia 1966*, Vol. I, Mediad, Inc., White Plains, NY (1967), pp. 250-253.

C54. H. O. Friestad, D. E. Ott, and F. A. Gunther, *Anal. Chem. 41*, 1750-1754 (1969).

C55. A. E. Goodwin and J. L. Marton, *Anal. Chim. Acta 152*, 295-299 (1983).

C56. R. K. Love and M. E. McCoy, in *Advances in Automated Analysis, Technicon International Congress 1969*, Vol. II, Mediad, Inc., White Plains, NY (1970), pp. 239-243.

C57. J. Keay, P. M. Menage, and G. A. Dean, *Analyst 97*, 897-902 (1972).

C58. R. Sawyer and E. J. Dixon, *Analyst 93*, 680-687 (1968).

C59. R. E. Duncombe and W. H. C. Shaw, in *Automation in Analytical Chemistry, Technicon Symposia 1966*, Vol. II, Mediad, Inc., White Plains, NY (1967), pp. 15-18.

C60. E. Seifler, D. Kambosos, and A. Chanas, in *Advances in Automated Analysis, Technicon International Congress 1970*, Vol. I, Futura Publishing Co., Mount Kisco, NY (1972), pp. 509-511.

C61. W. G. Evans and J. Bomstein, in *Advances in Automated Analysis, Technicon International Congress 1970*, Vol. II, Futura Publishing Co., Mount Kisco, NY (1972), pp. 233-239.

C62. M. A. Kenny and M. A. Cheng, *Clin. Chem. 18*, 352-354 (1972).

C63. Y. M. Fraticelli and M. E. Meyerhoff, *Anal. Chem. 53*, 992-997 (1981).

C64. G. B. Martin and M. E. Meyerhoff, *Anal. Chem. 59*, 2345-2350 (1987).

C65. D. M. Pranitis and M. E. Meyerhoff, *Anal. Chem. 59*, 2345-2350 (1987).

C66. E. H. Kawasaki, O. T. Leong, and T. M. Olcott, in *Automation in Analytical Chemistry, Technicon Symposia*, Vol. I, Mediad, Inc., White Plains, NY (1968), pp. 435-438.

C67. G. B. Martin, K. Cho, and M. E. Meyerhoff, *Anal. Chem. 56*, 2612-2613 (1984).

C68. V. Walace, *Anal. Biochem. 20*, 411-418 (1967).

C69. N. R. Kuzel, *J. Pharm. Sci. 57*, 852-855 (1968).

C70. L. Valentini, in *Advances in Automated Analysis, Technicon International Congress*, Vol. II, Mediad, Inc., White Plains, NY (1970), pp. 87-88.

C71. D. J. Blackmore, A. S. Curry, T. S. Hayes, and E. R. Rutter, *Clin. Chem. 17*, 896-902 (1971).

C72. J. C. Kraak, *Trends Anal. Chem. 2*, 183-187 (1983).

C73. H. Adler, D. A. Burns, and L. R. Snyder, in *Advances in Automated Analysis, Technicon International Congress*, Vol. IX, Mediad Inc., Tarrytown, NY (1973), pp. 81-85.

C74. L. Nord and B. Karlberg, *Anal. Chim. Acta 164*, 233-249 (1984).

C75. L. Nord, K. Backstrom, L. G. Danielsson, F. Ingman, and B. Karlberg, *Anal. Chim. Acta 194*, 221-233 (1987).

C76. W. B. Furman, in *Continuous Flow Analysis: Theory and Practice*, Marcel Dekker, New York (1976), pp. 149-150.

C77. A. L. Lazrus, K. C. Hill, and J. P. Lodge, in *Automation in Analytical Chemistry, Technicon Symposia 1965*, Mediad, Inc., New York, NY (1966), pp. 291-293.

C78. A. T. Faizullah and A. Townshend, *Anal. Chim. Acta 179*, 233-244 (1986).

C79. L. W. Snyder and J. W. Dolan, *J. Chromatogr. 185*, 43 (1979).

C80. J. W. Dolan and L. R. Snyder, *J. Chromatogr. 185*, 57 (1979).

C81. J. W. Dolan, S. Vanderwal, S. J. Bannister, and L. R. Snyder, *Clin. Chem. 16*, 871 (1980).

C82. D. L. Eichler, in *Advances in Automated Analysis, Technicon International Congress 1969*, Vol. II, Mediad, Inc., White Plains, NY (1970), pp. 51-59.

C83. W. B. Furman, in *Continuous Flow Analysis, Theory and Practice*, Marcel Dekker, New York (1976), pp. 158-161.

C84. C. W. Gehrke, J. H. Baumgartner, and J. P. Ussary, *J. Assoc. Off. Anal. Chem. 49*, 1213-1218 (1966).

C85. C. W. Gehrke, F. E. Kaiser, and J. P. Ussary, in *Automation in Analytical Chemistry, Technicon Symposia 1967*, Vol. I, Mediad, Inc., White Plains, NY (1968), pp. 239-251.

C86. C. W. Gehrke, F. E. Kaiser, and J. P. Ussary, *J. Assoc. Off. Anal. Chem. 51*, 200-211 (1968).

C87. C. W. Gehrke, J. S. Killingley, and L. L. Wall, *J. Assoc. Off. Anal. Chem. 55*, 467-480 (1972).

C88. C. W. Gehrke, J. P. Ussary, and J. H. Baumgartner, in *Automation in Analytical Chemistry, Technicon Symposia 1966*, Vol. I, Mediad, Inc., White Plains, NY (1967), pp. 171-176.

C89. J. P. Ussary and C. W. Gehrke, in *Advances in Automated Analysis, Technicon International Congress 1969*, Vol. II, Madiad, Inc., White Plains, NY (1970), pp. 89-94.

C90. A. R. Law, N. J. Nicolson, and R. L. Norton, *J. Assoc. Off. Anal. Chem. 54*, 764-768 (1971).

C91. J. E. Lindquist, *Anal. Chim. Acta 41*, 158-160 (1968).

C92. H. C. Pitot, N. Pries, M. Poirier, and A. Cutter, in *Automation in Analytical Chemistry, Technicon Symposia 1965*, Mediad, Inc., White Plains, NY (1966), pp. 555-558.

C93. S. Morgenstern, L. Chaparin, D. Vlastelica, and A. Kiederer, in *Advances in Automated Analysis, Technicon International Congress 1970*, Vol. I, Futura Publishing Co., Mount Kisco, NY (1972), pp. 85-90.

C94. M. Fleishner and M. K. Swartz, in *Advances in Automated Analysis, Technicon International Congress 1970*, Vol. I, Futura Publishing Co., Mount Kisco, NY (1972), pp. 92-93.

C95. D. A. Burns, U.S. Patent 3,876,374 (Apr. 8, 1975).
C96. S. Morgenstern, R. Rush, and D. Lehman, in *Advances in Automated Analysis, Technicon International Congress 1972*, Vol. I, Mediad, Inc., Tarrytown, NY (1973), pp. 15-19.
C97. R. W. Frei, H. Jansen, and V. A. Th. Brinkman, *Anal. Chem. 57*, 1529A-1539A (1985).
C98. A. H. M. Th. Scholten, V. A. Th. Brinkman, and R. W. Frei, *J. Chromatogr. 205*, 229-237 (1981).
C99. S. K. Ratanathanawongs and S. R. Crouch, *Anal. Chim. Acta 192*, 277-287 (1987).
C100. R. Stanley, *J. Auto. Chem. 4(6)*, 175-185 (1984).
C101. L. T. Skeggs, U.S. Patent 2,797,149 (June 25, 1957).
C102. W. B. Furman, *Continuous Flow Analysis: Theory and Practice*, Marcel Dekker, New York (1976), pp. 61-62.
C103. R. E. Thiers, in *Clinical Chemistry: Principles and Technics* (R. J. Henry, D. C. Cannon, and J. W. Winkelman, eds.), Harper & Row, Hagerstown, MD (1974), pp. 243-246.
C104. P. H. Scholes and C. Thulbourne, *Analyst 89*, 466-474 (1964).
C105. E. B. M. de Jong, U.S. Patent 3,134,263 (May 26, 1964).
C106. *Practical Automation for the Clinical Laboratory*, W. L. White, M. M. Erickson, and S. C. Stevens, eds., C. V. Mosby Company, St. Louis, MO (1968).
C107. L. T. Skeggs and H. Hochstrasser, *Clin. Chem. 10*, 918-936 (1964).
C108. E. C. Whitehead, in *Automation in Analytical Chemistry, Technicon Symposia 1965*, Mediad, Inc., White Plains, NY (1966), pp. 437-451.
C109. W. J. Smythe, M. H. Shamos, S. Morgenstern, and L. T. Skeggs, in *Automation in Analytical Chemistry: Technicon Symposia 1967*, Vol. I, Mediad, Inc., White Plains, NY (1968), pp. 105-113.
C110. W. J. Smythe, U.S. Patent 3,306,229 (Feb. 28, 1967).
C111. L. B. Lobring and R. L. Booth, in *Advances in Automated Analysis, Technicon International Symposia 1972*, Mediad, Inc., Tarrytown, ny 91973), pp. 7-10.
C112. J. Isreeli and W. J. Smythe, in *Advances in Automated Analysis: Technicon International Congress 1972*, Vol. I, Mediad, Inc., Tarrytown, NY (1973), pp. 13-18.
C113. H. Diebler and M. Pelavin, in *Advances in Automated Analysis: Technicon International Congress 1972*, Vol. I, Mediad, Inc., Tarrytown, NY (1973), pp. 19-25.
C114. W. E. Neeley and H. C.Sing, *Am. J. Clin.Chem. 61*, 840-809 (1974).
C115. W. E. Neeley, S. Wardlaw, and M. E. T. Swinnen, *Clin. Chem. 20(1)*, 78-80 (1974).
C116. W. E. Neeley, S. C. Wardlaw, and H. Sing, *Clin. Chem. 20(4)*, 424-427 (1974).
C117. W. E. Neeley, S. C. Wardlaw, T. Yates, W. G. Hollingsworth, and M. E. T. Swinnen, *Clin. Chem. 22(2)*, 227-231 (1976).
C118. C. J. Patton, M. Rabb, and S. R. Crouch, *Anal. Chem. 54*, 1113-1118 (1982).
C119. A. L. Chaney, in *Automation in Analytical Chemistry: Technicon Symposia 1967*, Vol. I, Mediad, Inc., White Plains, NY (1968), pp. 115-117.

C120. Personal communication, W. H. C.Walker (1986).
C121. W. J. Smythe and M. H. Shamos, U.S. Patent 3,804,593 (April 16, 1974).
C122. M. Cassaday, H. Diebler, R. Herron, M. Pelavin, D. Svenjak, and D. Vlastelica, *Clin. Chem. 31*, 1453-1456 (1985).
C123. P. W. Alexander and A. Thalib, *Anal. Chem. 55*, 497-501 (1983).
C124. C. Pasquini and W. A. De Oliveria, *Anal. Chem. 57*, 2575-2579 (1985).
C125. W. J. Smythe and M. H. Shamos, U.S. Patent 3,479,141 (Nov. 18, 1969).
C126. W. J. Smythe and M. H. Shamos, U.S. Patent 3,484,170 (Dec. 16, 1969).
C127. J. Smith, D. Svenmjak, J. Turrell, and D. Vlastelica, *Clin. Chem. 28*, 1867-1872 (1982).
C128. M. Margoshes, *Anal. Chem. 49*, 17-19, 1861-1862 (1977).
C129. J. Ruzicka, E. H. Hansen, M. Mosbaek, and F. J. Krug, *Anal. Chem. 49*, 1858-1861 (1977).
C130. M. Margoshes, *Anal. Chem. 54*, 678A-679A, 1106A (1982).
C131. C. B. Ranger, *Anal. Chem. 54*, 1106A (1982).
C132. B. Rocks and C. Riley, *Clin. Chem. 28*, 409-421 (1982).
C133. K. Hoffmann and I. Halasz, *J. Chromatogr. 199*, 3-22 (1980).
C134. J. Ruzicka and E. Hansen, *Flow Injection Analysis*, First Edition, Wiley, New York (1981), p. 21.
C135. S. C. Coverly, *J. Automatic Chem. 7*, 141-144 (1985).
C136. B. J. Compton, J. R. Webber, and W. C. Purdy, *Anal. Lett. 13(B10)*, 861-870 (1980).
C137. L. J. Kamphake and R. T. Williams, in *Advances in Automated Analysis, 1972 Technicon International Congress*, Vol. 8, Mediad, Inc., Tarrytown, NY (1973), pp. 43-49.
C138. A. Savitzky and M. J. E. Golay, *Anal. Chem. 36*, 1627-1639 (1964).
C139. H. S. Ames and R. W. Crawford, *Amer. Lab. 9*, 37-44 (1977).
C140. F. Opekar and A. Trojanek, *Anal. Chem. Acta 127*, 239-243 (1981).
C141. W. Vogt, S. L. Braun, S. Wilhelm, and H. Schwab, *Anal. Chem. 54*, 596-598 (1982).
C142. Personal communication, M. H. Rawlings, Ministry of the Environment, Water Quality Section, Roxdale, Ontario, Canada (1987).
C143. E. Amador, *Clin. Chem. 18*, 164 (1972).
C144. W. S. Gardner and J. M. Malczyk, *Anal. Chem. 55*, 1645-1647 (1983).
C145. L. Fossey and F. F. Cantwell, *Anal. Chem. 54*, 1693-1697 (1982).
C146. F. F. Cantwell and J. A. Sweileh, *Anal. Chem. 57*, 329-331 (1985).
C147. C. A. Lucy and F. F. Cantwell, *Anal. Chem. 61*, 101-107 (1989).
C148. C. A. Lucy and F. F. Cantwell, *Anal. Chem. 61*, 107-114 (1989).

B. FIA References

F1. J. Ruzicka and E. H. Hansen, *Flow Injection Analysis*, Wiley, New York, first edn. (1981); second edn. (1987).
F2. S. Al-Najafi, C. A. Wellington, A. P. Wade, T. J. Sly, and D. Betteridge, *Talanta*, in press.

F3. S. R. Crouch and S. K. Ratanathanawongs, *Anal. Chim. Acta 192*, 277-287 (1987).
F4. D. Betteridge, N. G. Coutney, T. J. Sly, and D. G. Porter, *Analyst 109*, 91-93 (1984).
F5. J. M. Calatayud, P. C. Falco, and M. C. P. Marti, *Analyst 111*, 1317-1320 (1986).
F6. D. Betteridge, T. J. Sly, A. P. Wade, and J. E. W. Tillman, *Anal. Chem. 55*, 1292-1299 (1983).
F7. C. W. McLeod, P. J. Worsfold, and A. G. Cox, *Analyst 109*, 327-332 (1984).
F8. G. C. M. Bourke, G. Stedman, and A. P. Wade, *Anal. Chim. Acta 153*, 277-280 (1983).
F9. J. A. Lown and D. C. Johnson, *Anal. Chim. Acta 116*, 41 (1980).
F10. T. P. Lynch, A. F. Taylor, and J. N. Wilson, *Analyst 108*, 470-475 (1983).
F11. D. Betteridge and B. Fields, *Frezenius Z. Anal. Chem. 314*, 386 (1983).
F12. D. J. Hooley and R. E. Dessy, *Anal. Chem. 55*, 313-320 (1983).
F13. V. V. S. Eswara Dutt and H. A. Mottola, *Anal. Chem. 47*, 357 (1975).
F14. C. L. M. Stults, A. P. Wade, and S. R. Crouch, *Anal. Chem. 59*, 2245-2247 (1987).
F15. B. C. Erickson, B. R. Kowalski, and J. Ruzicka, *Anal. Chem. 59*, 1246-1248 (1987).
F16. D. Betteridge, E. L. Dagless, B. Fields, and N. F. Graves, *Analyst 103*, 897-908 (1978).
F17. T. J. Sly, D. Betteridge, D. Wibberley, and D. G. Porter, *J. Automatic Chem. 4(4)*, 186 (1982).
F18. C. J. Patton and S. R. Crouch, *Anal. Chim. Acta 179*, 189-201 (1986).
F19. W. E. Bauer, A. P. Wade and S. R. Crouch, *Anal. Chem. 60*, 287-288 (1988).
F20. W. E. van der Linden, *Anal. Chim. Acta 179*, 91-101 (1986).
F21. J. Ruzicka and E. H. Hansen, *Anal. Chim. Acta 179*, 1-58 (1986).
F22. J. B. Callis, D. L. Illman, and B. R. Kowalski, *Anal. Chem. 59*, 624A-637A (1987).
F23. K. K. Stewart, G. R. Beecher, and P. E. Hare, *Fed. Proc. 33*, 1439 (1974).
F24. J. Ruzicka and E. H. Hansen, *Anal. Chim. Acta 78*, 145 (1975).
F25. D. Betteridge, *Anal. Chem. 50*, 832A-846A (1978).
F26. Proceedings of the International Conference on Flow Analysis, Amsterdam (1979), *Anal. Chim. Acta 114* (1980).
F27. C. B. Ranger, *Anal. Chem. 53*, 20A (1981).
F28. K. K. Stewart, *Talanta 28*, 789 (1981).
F29. Proceedings of the Second International Conference on Flow Analysis, Amsterdam (1982), *Anal. Chim. Acta 145* (1983).
F30. K. K. Stewart, *Anal. Chem. 55*, 931A-940A (1983).
F31. M. Valcarcel and M. D. Luque de Castro, *Flow Injection Analysis—Principles and Applications*, Wiley, New York (1987).
F32. Proceedings of the Third International Conference on Flow Analysis (1985), *Anal. Chim. Acta 179* (1986).

F33. J. Ruzicka, *Anal.Chem. 55*, 1040A-1053A (1983).
F34. J. Ruzicka and E. H. Hansen, *Anal. Chim. Acta 114*, 19-44 (1980).
F35. G. den Boef, R. C. Schothorst, J. Emneus, R. Appelqvist, G. Marko-Varga, L. Gorton, G. Johansson, Z. Fang, J. M. Harris, J. Janata, J. J. Harrow, B. Karlberg, W. E. van der Linden, J. N. Miller, V. K. Mahant, H. Thakrar, J. F. Tyson, H. A. Mottola, H. Muller, V. Muller, G. E. Pacey, C. Riley, J. Ruzicka, E. H. Hansen, K. K. Stewart, G. R. Beecher, J. T. Vanderslice, P. E. Hare, A. Townshend, M. Valcarcel, M. D. Luque de Castro, P. J. Worsfold, N. Yoza, N. Ishibashi, K. Ueno, F. J. Krug, E. A. G. Zagatto, and H. Bergamin F°, *Anal. Chim. Acta 180*, 1-67 (1986).
F36. C. Riley, L. H. Aslett, B. F. Rocks, R. A. Sherwood, J. D. M. Watson and J. Morgon, *Clin. Chem. 29*, 332 (1982).
F37. J. Ruzicka and E. H. Hansen, *Anal. Chim. Acta 145*, 1 (1983).
F38. P. W. Alexander and A. Thalib, *Anal. Chem. 55*, 497 (1983).
F39. J. Ruzicka and E. H. Hansen, *Anal. Chim. Acta 161*, 1 (1984).
F40. F. J. Krug, H. Bergamin F°, and E. A. G. Zagatto, *Anal. Chim. Acta 179*, 103-118 (1986).
F41. K. Uchida, D. Yoshizawa, M. Tomoda, and S. Saito, *Anal. Sci. (Jpn) 3*, 191-193 (1987).
F42. K. K. Stewart, *Anal. Chim. Acta 179*, 59-68 (1986).
F43. D. Betteridge, B. Fields, *Anal. Chem. 50*, 654 (1978).
F44. D. Betteridge and B. Fields, *Anal. Chim. Acta 132*, 139 (1981).
F45. D. Betteridge, T. J. Sly, A. P. Wade, and D. G. Porter, *Anal. Chem. 58*, 2258-2265 (1986).
F46. B. G. M. Vandeginste, private communication, 1984.
F47. "FIAstar Flow Injection Analysis Bibliography—Literature Reference List with Subject Index and References," Tecator Ltd., Cooper Road, Thornbury, Bristol, BS12 2UP, UK.
F48. P. J. Worsfold, *Anal. Chim. Acta 145*, 117-124 (1983).
F49. N. Tioh, Y. Israel, and R. M. Barnes, *Anal. Chim. Acta 184*, 205-212 (1986).
F50. Y. Israel and R. M. Barnes, *Anal. Chem. 56*, 1192-1194 (1984).
F51. E. A. G. Zagatto, A. O. Jacintho, F. J. Krug, B. F. Reis, R. E. Bruns and M. C. U. Araujo, *Anal. Chim. Acta 145*, 169-178 (1983).
F52. M. W. Brown and J. Ruzicka, *Analyst 109*, 1091-1094 (1984).
F53. J. Tyson, *Analyst 110*, 419-429 (1985) [Review Paper].
F54. J. Tyson, *Anal. Proc. 21*:377-378 (1984).
F55. J. F. Tyson, J. M. H. Appleton, and A. B. Idris, *Anal. Chim. Acta 145*, 159-168 (1983).
F56. H. L. Pardue and B. Fields, *Anal. Chim. Acta 124*, 39-63 (1981).
F57. H. L. Pardue and B. Fields, *Anal. Chim. Acta 124*, 65-79 (1981).
F58. K. K. Stewart and A. G. Rosenfeld, *Anal. Chem. 54*, 2368-2372 (1982).
F59. B. F. Reis, A. O. Jacintho, J. Mortatti, F. J. Krug, E. A. G. Zagatto, H. Bergamin F°, and L. C. R. Pessenda, *Anal. Chim. Acta 123*, 221-228 (1981).
F60. K. S. Johnson and R. L. Petty, *Anal. Chem. 54*, 1185-1187 (1982).
F61. E. A. G. Zagatto, F. J. Krug, H. Bergamin F°, S. S. Jorgensen, B. F. Reis, *Anal. Chim. Acta 101*, 17 (1978).
F62. E. A. G. Zagatto, F. J. Krug, H. Bergamin F°, S. S. Jorgensen, B. F. Reis, *Anal. Chim. Acta 104*, 279 (1978).

F63. E. A. G. Zagatto, F. J. Krug, H. Bergamin F°, S. S. Jorgensen, B. F. Reis, *Anal. Chim. Acta 107*, 309 (1979).
F64. E. A. G. Zagatto, F. J. Krug, H. Bergamin F°, S. S. Jorgensen, B. F. Reid, *Anal. Chim. Acta 104*, 279 (1981).
F65. J. Mindegaard, *Anal. Chim. Acta 104*, 185-189 (1979).
F66. J. Ruzicka and E. H. Hansen, *Anal. Chim. Acta 106*, 207 (1979).
F67. J. Ruzicka and E. H. Hansen, *ChemTech 9*, 756 (1979).
F68. C. S. Lim, J. N. Miller and J. W. Bridges, *Anal. Chim. Acta 114*, 183 (1980).
F69. C. D. Crowe, H. V. Levin, D. Betteridge, and A. P. Wade, *Anal. Chim. Acta 194*, 49-60 (1987).
F70. G. Gerhardt and R. N. Adams, *Anal. Chem. 54*, 2618 (1982).
F71. T. Korenaga and H. Ikatsu, *Anal. Chim. Acta 141*, 301-309 (1982).
F72. T. Korenaga and H. Ikatsu, *Anal. Chim. Acta 141*, 301 (1982).
F73. J. Ruzicka and E. H. Hansen, *Anal. Chim. Acta 106*, 207-224 (1979).
F74. J. N. Miller, *Anal. Proc. 21*, 372 (1984).
F75. B. F. Rocks, R. A. Sherwood, M. M. Hosseinmardi, and C. Riley, *Anal. Chim. Acta 179*, 225-231 (1986).
F76. B. Fields, Ph.D. Thesis, University of Wales (1981).
F77. S. Baban, *Anal. Proc. 17*, 535 (1980).
F78. S. Baban, D. Bettlestone, D. Betteridge, and P. Sweet, *Anal. Chim. Acta 114*, 319-323 (1980).
F79. A. Rios, M. D. Luque de Castro, and M. Valcarcel, *J. Chem. Educ. 63*, 552-553 (1986).
F80. T. A. H. M. Janse, P. F. A. vand er Wiel and G. Kateman, *Anal. Chim. Acta 155*, 89-102 (1983).
F81. J. Ruzicka and E. H. Hansen, *Flow Injection Analysis*, Wiley, New York (1981), pp. 65-67.
F82. C. C. Painton and H. A. Mottola, *Anal. Chem. 53*, 1713 (1981).
F83. B. Karlberg and S. Thelander, *Analyst 103*, 1154-1159 (1978).
F84. B. Karlberg and S. Thelander,*Anal. Chim. Acta 98*, 1-7 (1978).
F85. Y. E. Sahlestrom and B. Karlberg, *Anal. Chim. Acta 185*, 259-269 (1986).
F86. C. Silfwerbrand-Lindh, L. Nord and L. G. Danielsson, *Anal. Chim. Acta 160*, 11-19 (1984).
F87. S. Olsen, L. C. R. Pessenda, J. Ruzicka, and E. H. Hansen, *Analyst 108*, 905 (1983).
F88. C. L. M. Stults, A. P. Wade, and S. R. Crouch, *Anal. Chim. Acta 192*, 155-163 (1987).
F89. L. Gorton and L. Ogren, *Anal. Chim. Acta 130*, 45-53 (1981).
F90. E. H. Hansen and J. Ruzicka, *Anal. Chim. Acta 87*, 353 (1976).
F91. E. H. Hansen, J. Ruzicka, and B. Rietz, *Anal. Chim. Acta 89*, 241 (1977).
F92. S. Baban, Ph.D. Thesis, University of Wales (1981).
F93. D. Betteridge, A. G. Howard, and A. P. Wade, *Talanta 32(8B)*, 709-722 (1985).
F94. I. D. Cockshott, R. Payne, and P. B. Copsey, *Research Disclosure* 18721, November 1979.

F95. R. S. Rowles, Ph.D. Thesis, University of Wales (1984).
F96. T. P. Lynch, N. J. Kernoghan, and J. N. Wilson, *Analyst 109*, 839-842 (1984).
F97. T. P. Lynch, N. J. Kernoghan, and J. N. Wilson, *Analyst 109*, 843-846 (1984).
F98. J. Tyson, *Anal. Proc. 21*, 377 (1984).
F99. B. D. Mindel and B. Karlberg, *Lab Pract. 30(7)*, 719 (1981).
F100. L. Nord, S. Johansson, and H. Brotell, *Anal. Chim. Acta 175*, 281-287 (1985).
F101. D. K. Morgan, J. E. Katon, and N. D. Danielson, paper 293, presented at the FACSS XI Conference, Philadelphia (1984).
F102. D. Betteridge, N. G. Courtney, T. J. Sly, and D. G. Porter, *Analyst 109*, 91 (1984).
F103. S. R. Crouch and H. Archontaki, private communication, 1987.
F104. J. M. Reijn, W. van der Linden, and H. Poppe, *Anal. Chim. Acta 123*, 229-237 (1981).
F105. D. Betteridge, P. B. Oates, and A. P. Wade, *Anal. Chem. 59*, 1236-1238 (1987).
F106. A. P. Wade, S. R. Crouch, and D. Betteridge, paper 573, presented at the FACSS XIII Conference, St. Louis, MO, October 2 (1986).
F107. R. Tijssen, *Anal. Chim. Acta 114*, 71-89 (1980).
F108. J. H. M. van den Berg, R. S. Deelder, and H. G. M. Egberink, *Anal. Chim. Acta 114*, 91-104 (1980).
F109. V. V. S. Eswara Dutt, D. Scheeler, and H. A. Mottola, *Anal. Chim. Acta 94*, 289-296 (1977).
F110. M. D. Luque de Castro and M. Valcarcel, *Trends Anal. Chem. 5(3)*, 71-74 (1986).
F111. A. Rios, M. D. Luque de Castro, and M. Valcarcel, *Anal. Chem. 57*, 1803-1809 (1985).
F112. C. Z. Marczewski, D. Betteridge, R. M. Belchamber, M. P. Collins, T. Lilley, P. B. Oates, and A. P. Wade, paper XVIII-8, presented at Euroanalysis V, Krakow, Poland (1984).
F113. M. Gisin, C. Thommen and K. F. Mansfield, *Anal. Chim. Acta 179*, 149-167 (1986).
F114. G. Nagy, Z. Feher, and E. Pungor, *Anal. Chim. Acta 52*, 47 (1970).
F115. L. T. M. Prop, P. C. Thijssen, and L. G. G. van Dongen, *Talanta 32(3)*, 230-234 (1985).
F116. M. A. Koupparis, P. Anagnostopoulou and H. V. Malmstadt, *Talanta 32(5)*, 411-417 (1985).
F117. A. P. Wade, Ph.D. Thesis, University of Wales (1985).
F118. C. A. Koerner and T. A. Nieman, *Anal. Chem. 58*, 116-119 (1987).
F119. D. Betteridge, C. Z. Marczewski, and A. P. Wade, *Anal. Chim. Acta 165*, 227-236 (1984).
F120. A. Einstein, *Ann. Physik 17*, 549-560 (1905).
F121. J. C. Giddings, *J. Chem. Educ. 35*, 588 (1958).
F122. J. C. Giddings, *Dynamics of Chromatography, Part 1, Principles and Theory*, Marcel Dekker, New York (1965), pp. 26-35.
F123. G. Taylor, *Proc. Roy. Soc. (London) Ser. A 219*, 186 (1953).
F124. O. Levenspiel, *Chemical Reaction Engineering*, 2nd ed., Wiley, New York (1972).

F125. J. T. Vanderslice, K. K. Stewart, A. G. Rosenfeld, and D. J. Higgs, *Talanta 28*, 11 (1981).
F126. J. Ruzicka and E. H. Hansen, *Anal. Chim. Acta 99*, 37 (1978).
F127. M. J. E. Golay and J. G. Atwood, *J. Chromatogr. 186*, 353-370 (1979).
F128. C. L. M. Stults, A. P. Wade, and S. R. Crouch, *Anal. Chim. Acta 192*, 301-308 (1987).
F129. C. L. M. Stults, A. P. Wade, and S. R. Crouch, *J. Chem. Educ. 65*, 645-647 (1988).
F130. C. C. Painton and H. A. Mottola, *Anal. Chim. Acta 158*, 67-84 (1984).
F131. J. T. Vanderslice, A. G. Rosenfeld, and G. R. Beecher, *Anal. Chim. Acta 179*, 119-129 (1986).
F132. J. F. Tyson, *Anal. Chim. Acta 179*, 131-148 (1986).
F133. C. C. Painton and H. A. Mottola, *Anal. Chim. Acta 154*, 1-16 (1983).
F134. J. G. Atwood and M. J. E. Golay, *J. Chromatography 218*, 97-122 (1981).
F135. M. A. Gomez-Nieto, M. D. Luque de Castro, A. Martin, and M. Valcarcel, *Talanta 32*, 319-324 (1985).
F136. J. T. Vanderslice and G. R. Beecher, *Talanta 32*, 334-335 (1985).
F137. M. Valcarcel and M. D. Luque de Castro, *Talanta 32*, 339-340 (1985).
F138. J. M. Reijn, W. E. van der Linden, and H. Poppe, *Anal. Chim. Acta 114*, 105-118 (1980).
F139. A. P. Wade, *Anal. Proc. 20*, 523-527 (1983).
F140. A. P. Wade, *Anal. Proc. 20*, 108-111 (1983).
F141. R. M. Belchamber, D. Betteridge, A. J. Cruickshank, P. Davison, and A. P. Wade, *Spectrochim. Acta B. 41B(5)*, 503-506 (1986).
F142. D. Betteridge, A. G. Howard, and A. P. Wade, *Talanta 32(8B)*, 723-734 (1985).
F143. J. C. Berridge, *Techniques for the Automated Optimization of HPLC Separations*, Wiley, New York (1985).
F144. J. C. Berridge, *J. Chromatogr. 244*, 1-14 (1982).
F145. P. K. Erni and H. Muller, *Anal. Chim. Acta 103*, 189-199 (1978).
F146. T. P. Lynch, N. J. Kernoghan, and J. N. Wilson, *Proc. Extraction Metallurgy 1985*, Institute for Mining and Metallurgy, London, September 1985.
F147. D. Betteridge, R. M. Belchamber, T. P. Lynch, and D. M. Roberts, paper 445, presented at the FACSS XIII Conference, St. Louis, MO, September 1986.
F148. A. P. Wade, S. R. Crouch, and D. Betteridge, *Trends in Anal. Chem.* 7(10), 358-365 (1988).
F149. S. D. Rothwell and A. A. Woolf, *Talanta 32(5)*, 431-433 (1985).
F150. D. Betteridge, W. C. Cheng, E. L. Dagless, P. David, T. B. Goad, D. R. Deans, D. A. Newton, and T. B. Pierce, *Analyst 108*, 1-16 (1983).
F151. D. Betteridge, W. C. Cheng, E. L. Dagless, P. David, T. B. Goad, D. R. Deans, D. A. Newton, and T. B. Pierce, *Analyst 108*, 17-32 (1983).
F152. P. R. Kraus and S. R. Crouch, *Anal. Lett. 20(2)*, 183-200 (1987).
F153. P. R. Kraus, private communication, June 1987.

F154. J. Ruzicka and E. H. Hansen, *Flow Injection Analysis*, Wiley, New York (1981), pp. 63-65.
F155. W. R. Wolf and K. K. Stewart, *Anal. Chem. 51*, 1201 (1979).
F156. A. F. Fell, *UV-Spectrom. Group Bull. 7*, 5 (1979).
F157. J. G. Williams, M. Holmes, and D. G. Porter, *J. Automatic Chem. 4*, 176 (1982).
F158. A. J. Frend, G. J. Moody, J. D. R. Thomas, and B. J. Birch, *Analyst 108*, 1357-1364 (1983).
F159. J. Janata and J. Ruzicka, *Anal. Chim. Acta 139*, 105 (1982).
F160. S. J. Lyle and M. I. Saleh, *Talanta 28(4)*, 251 (1981).
F161. T. B. Goad, Ph.D. Thesis, University of Wales (1982).
F162. A. Hammerli, J. J. Janata, and H. M. Brown, *Anal. Chim. Acta 144*, 115 (1982).
F163. T. Imasaka, T. Harada, and N. Ishibashi, *Anal. Chim. Acta 129*, 195 (1981).
F164. K. Kina, K. Shiraishi, and N. Ishibashi, *Talanta 25*, 295 (1978).
F165. G. Rule and W. R. Seitz, *Clin. Chem. 25*, 1635 (1979).
F166. A. Shaw, Rapid automation of wet chemistry using a flow injection analysis, in *Laboratory News*, 21 June, 1985.
F167. P. David, Ph.D. Thesis, University of Wales (1978).
F168. A. P. Wade and S. R. Crouch, *Spectroscopy 3*(10), 24-31 (1988).
F169. P. D. Wentzell, A. P. Wade, P. N. Shiundu, R. M. Ree, M. Hatton, T. J. Sly, and D. Betteridge, *J. Auto. Chem.* (1990), in press. "Computer-controlled Apparatus for Automated Development of Continuous-Flor Methods."
F170. P. M. Shiundu, P. D. Wentzell, and A. P. Wade, *Talanta* (1990), in press. "Spectrophotometric Determination of Palladium with Sulfochlorophenolazo-rhodamine by Flow Injection Analysis."

C. General References

G1. J. C. Van Loun, *Selected Methods of Trace Metal Analysis: Biological and Environmental Samples*, Wiley, New York (1985).
G2. T. H. Risby (Ed.), *Ultratrace Metal Analyses in Biological Sciences and Environment*, Advances in Chemistry Series, Vol. 172, American Chemical Society, Washington, DC (1979).
G3. M. R. Detaevernier, Y. Michotte, L. Buydens, M. P. Derde, M. Desmet, L. Kaufman, G. Musch, J. Smeyers-Verbeke, A. Thielemans, L. Dryon, and D. L. Massart, *J. Pharm. Biomed. Anal. 4(3)*, 297-307 (1986).
G4. K. L. Ratzlaff, *Introduction to Computer-Assisted Optimization*, Wiley, New York (1987).
G5. N. R. Draper and H. Smith, *Applied Regression Analysis*, Wiley, New York (1966).
G6. L. M. Schwartz, *Anal. Chem. 48*, 2237-2289 (1976).
G7. L. M. Schwartz, *Anal. Chem. 49*, 2062-2068 (1977).
G8. L. M. Schwartz, *Anal. Chem. 51*, 723-727 (1979).

Abbreviations and Acronyms

AAS	Atomic absorption spectrometry, 139
AC	Alternating current
ADC	Analog to digital converter, 36
AES	Atomic emission spectrometry, 139
AES	Auger electron spectroscopy, 531
AFC	Automatic frequency control, 484
AMDS	Automated methods development system, 1027
APT	Attached proton test, 436
ARXPS	Angular resolved XPS, 535
ASCII	American standard code for information interchange, 21
ASTM	American Society for Testing and Materials, 2
ASV	Anodic stripping voltammetry, 620
ATR	Attenuated total reflectance, 263
BDS	Beam deflection spectroscopy, 337
BITBLT	Bit block transfer, 15
CAT	Constant analyzer transmission, 552
CCC	Controlled current coulometry, 665
CD	Circular dichroism, 367
CF	Continuous flow, 980
CFA	Continuous-flow analyzer, 980
CHA	Concentric hemispheric analyzer, 549
CHN	Carbon-hydrogen-nitrogen, 79
CI	Chemical ionization, 890
CID	Circular intensity differential, 376
CIDNP	Chemically induced dynamic nuclear polarization, 425
CL	Current loop, 70
CMA	Cylindrical mirror analyzer, 549
CME	Chemically modified electrode, 634
COSY	Correlated spectroscopy, 436
CPC	Controlled potential coulometry, 662
CPL	Circularly polarized luminescence, 369
CPU	Central processing unit, 6
CRD	Chemical reaction detection, 820

*This listing does not include terms that originated as acronyms but have attained status as separate words. Examples are LASER and FORTRAN. The pages noted correspond to first usage or to principal definition. Abbreviated names of companies (e.g., IBM) are not included here.

CRT	Cathode ray tube, 252
CSMA/CD	Carrier sense multiple access/collision detection, 22
CSV	Cathodic stripping voltammetry, 624
CT	Coiled tube, 826
CW	Continuous wave, 321
DAC	Digital to analog converter, 43
DBM	Double balanced mixer, 482
DC	Direct current
DCE	Data communication equipment, 18
DCP	Direct-current plasma, 116
DEP	Dynamic electron polarization, 515
DEPT	Distortionless enhancement by polarization transfer, 436
DI	Desorption ionization, 892
DLI	Direct liquid injection, 818
DMA	Direct memory access, 11
DME	Dropping mercury electrode, 603
DPP	Differential pulse polarography, 610
DPPH	1,1-Diphenyl-2-picrylhydrazyl, 500
DPV	Differential pulse voltammetry, 815
DR	Dynamic range, 728
DSC	Differential scanning calorimetry, 908
DTA	Differential thermal analysis, 908
DTE	Data terminal equipment, 18
DTG	Derivative thermogravimetry, 908
DTGS	Deuterated triglycine sulfate, 249
EAAS	Electrothermal atomic absorption spectrometry, 139
EBCDIC	Extended binary coded decimal interchange code, 24
EC	Electrochemical, 816
ECD	Electron capture detector, 726
EDL	Electrodeless discharge lamp, 176
EDTA	Ethylenediamine tetraacetic acid, 372
EGA	Evolved gas analysis, 908
EGD	Evolved gas detector, 908
EGTA	Ethyleneglycol-bis(aminoethylether)-N,N'-tetraacetic acid, 977
EI	Electron impact ionization, 889
EIA	Electrical Industries Association, 17
EICD	Electrolytic conductivity (GC) detector, 726
ELDOR	Electron double resonance, 514
ENDOR	Electron nuclear double resonance, 513
EOX	Extractable organic halogen, 976
EPA	Environmental Protection Agency, 980
EPR	Electron paramagnetic resonance, 357
ESCA	Electron spectroscopy for chemical analysis, 346
ESR	Electron spin resonance, 467
ETR	Electrothermal radiometer, 337
FA	Fluid addition, 820
FAAS	Flame atomic absorption spectrometer, 139

FAB Fast atom bombardment, 893
FAES Flame atomic emission spectrometer, 139
FAT Fixed analyzer transmission, 552
FDCD Fluorescence-detected circular dichroism, 376
FFT Fast Fourier transform, 127
FIA Flow injection analyzers, 980
FID Free induction decay, 395
FID Flame ionization detector, 726
FOCSY Fold-over corrected spectroscopy, 437
FPD Flame photometric detector, 726
FS Fluid segmentation, 828
FSOT Fused-silica open tubular column, 682
FT/IR Fourier-transform infrared, 238
FT/MS Fourier transform MS, 898
FWHM Full width at half maximum, 175

G/D Generated/doped format, 821
GC Gas chromatography, 673
GC/IR Gas chromatography combined with infrared, 235
GC/MS Gas chromatography combined with MS, 895
GCE Glassy carbon electrode, 607
GKS Graphic kernel set, 13
GPIB General-purpose interface bus, 17

HCL Hollow cathode lamp, 174
HDCOSY Homodecoupled correlated spectroscopy, 436
HeID Helium ionization detector, 725
HMDE Hanging mercury drop electrode, 609
HPIB Hewlett Packard interface bus, 21
HPLC High-performance liquid chromatography, 745
HV High voltage

IAS Intersample air segment, 1031
ICP Inductively-coupled plasma, 116
ICP/MS Inductively coupled plasma MS, 897
ICR Ion cyclotron resonance, 897
ICTA International Confederation for Thermal Analysis, 906
IDG Individual droplet generator, 1010
IEEE Institute of Electrical and Electronic Engineers, 17
IMFP Inelastic mean free path, 531
INADEQUATE Incredible natural abundance double quantum transfer experiment, 436
INEPT Insensitive nuclei enhancement by polarization transfer, 436
IR Infrared, 233
IRD Infrared (GC) detector, 726
ISE Ion-selective electrode, 593
ISO International Standards Organization, 16
ITD Ion trap (GC) detector, 726
IUPAC International Union of Pure and Applied Chemistry, 141

KT Knitted tube, 826

LALLS Low-angle laser light scattering, 811
LC Liquid chromatography, 745
LC/MS Liquid chromatography combined with MS, 895
LDR Linear dynamic range, 728
LEED Low-energy electron diffraction, 549
LIBS Laser-induced breakdown spectroscopy, 1010
LOD Limit of detection, 728
LVDT Linear variable displacement transducer, 950

M-H Mattauch-Herzog MS design, 888
MCT Mercury cadmium telluride, 250
MDGC Multidimensional gas chromatography, 705
MDL Minimum detection limit, 728
MFE Mercury film electrode, 630
MID Multiple ion detection, 726
MIP Microwave-induced plasma, 116
MOF-COSY Multiple quantum filter COSY, 437
MS Mass spectrometer, 885
MS/MS Dual or tandem MS instruments, 899
MSD Mass-selective (GC) detector, 726

N-J Nier-Johnson MS design, 888
NIR Near infrared, 213
NMR Nuclear magnetic resonance, 393
NOE Nuclear Overhauser effect, 437
NOESY NOE spectroscopy, 437
NPP Normal pulse polarography, 610
NQR Nuclear quadrupole resonance, 434

OAS Optoacoustic spectroscopy, 337
ODMR Optically detected magnetic resonance, 515
OIML International Organization of Legal Metrology, 53
OPD Path difference for an integral number of waves, 240
ORD Optical rotatory dispersion, 365
OSI Open systems interconnect, 16
OTC Open tubular column, 705

PAS Photoacoustic spectroscopy, 337
PB Packed bed, 827
PCR Postcolumn reactors, 996
PF Peak fidelity, 730
PID Position-integral-derivative (servo), 63
PID Photoionization detector, 726
PLOT Porous layer open tubular column, 682
PMT Photomultiplier tube, 185
POPOP 1,4-Bis-2-(5-phenyloxazolyl)benzene, 816
POX Purgeable organic halogen, 976
PPO 2,5-Diphenyloxazole, 816

PS	Pulse sequence, 428
PSA	Potentiometric stripping analysis, 620
PT	Photothermal technique, 337
PTR	Photothermal radiometer, 337
PTV	Programmed temperature vaporizer, 698
PVC	Polyvinyl chloride, 1035
PVDF	Polyvinylidene difluoride, 352
PZT	Lead circonate titanate, 352
QPC	Quadrature phase cycling, 429
RCD	Redox chemiluminescence detector, 726
RDE	Rotating disk electrode, 607
RF	Radiofrequency
RFA	Retarding field analyzer, 549
RG	Rosencwaig-Gersho model, 342
RI	Refractive index, 792
RMR	Relative molar response, 725
ROA	Raman optical activity, 374
RTP	Room-temperature phosphorescence, 306
RWM	Random-walk model, 1013
S/H	Sample and hold, 38
S/N	Signal to noise ratio, 214
SA	Successive approximation, 37
SAM	Scanning Auger microscopy, 548
SBSR	Single bead string reactor, 1010
SCE	Saturated calomel electrode, 581
SCOT	Support-coated open tubular column, 682
SECSY	Spin-echo correlated spectroscopy, 436
SEM	Scanning electron microscopy, 548
SERS	Surface-enhanced Raman spectroscopy, 318
SFC	Supercritical fluid chromatography, 843
SFC-IR	Supercritical fluid chromatography-infrared, 274
SFC/MS	Combined SFC and MS, 848
SHE	Standard hydrogen electrode, 580
SIMS	Secondary ion mass spectrometer, 893
SMDE	Static mercury drop electrode, 608
SNR	Signal to noise ratio, 214
SP	Solid-phase, 821
SPE	Solid-phase extraction, 774
ST-EPR	Stepped EPR, 511
SWP	Square-wave polarography, 610
SWV	Square-wave voltammetry, 610
TAC	Time-to-amplitude converter, 300
TCD	Thermal conductivity detector, 725
TCNQ	Tetracyanoquinone, 513
TCP/IP	Transmission control protocol/internet protocol, 16
TCSPC	Time-correlated single photon counting, 300

TEA	Thermal energy (GC) analyzer, 726
TFE	Thin-film electrode, 607
TG (or TGA)	Thermogravimetry, 908
TGA-IR	Thermogravimetric analysis-infrared, 235
TGS	Triglycine sulfate, 347
TID	Thermionic ionization detector, 726
TL	Total luminescence, 371
TLC	Thin-layer chromatography, 673
TM	Transmembrane addition, 820
TMA	Thermomechanical analysis, 908
TOF	Time-of-flight MS design, 889
TOX	Total organic halogen, 976
TPR	Temperature-programmed reduction, 948
TWT	Traveling wave tube, 483
UHV	Ultra-high vacuum, 538
UPR	Ultrasonic paramagnetic resonance, 515
USD	Ultrasonic (GC) detector, 726
USGS	U.S. Geological Survey, 980
USP	U.S. Pharmacopoeia, 2
UV	Ultraviolet, 213
VCD	Vibrational circular dichroism, 373
VIS	Visible spectral region, 213
WCOT	Wall-coated open tubular column, 682
WIG	Wax impregnated graphite electrode, 607
XPS	X-ray photoelectron spectroscopy, 346
YAG	Yttrium aluminum garnet, 385
ZFS	Zero field splitting, 517
ZPD	Zero path difference, 239

Index